W0245822

Hydrometallurgy '94

Papers presented at the international
symposium
'Hydrometallurgy '94'
organized by the Institution of Mining
and Metallurgy
and the Society of Chemical Industry,
and held in Cambridge, England,
from 11 to 15 July, 1994

SCI

Published for the Institution of Mining and Metallurgy
and the Society of Chemical Industry
by Chapman & Hall

SPRINGER-SCIENCE+BUSINESS MEDIA, B.V.

First edition 1994

Originally published by Chapman & Hall in 1994
Softcover reprint of the hardcover 1st edition 1994
ISBN 978-94-010-4532-2 ISBN 978-94-011-1214-7 (eBook)
DOI 10.1007/978-94-011-1214-7

A catalogue record for this book is available from the British Library

∞ Printed on permanent acid-free text paper, manufactured in accordance with ANSI/NISO Z39.48-1992 and ANSI/NISO Z39.48-1984 (Permanence of Paper).

Contents

Organizing committee

Dr A.J. Monhemius *(Chairman)*
Dr D.S. Flett
Dr N.M. Rice
Dr T.V. Arden
Dr R.F. Dalton
Professor D.J. Fray
D. Gosden
D. Naden
Dr C.V. Phillips
K.J. Severs

List of Sponsors

Generous support of the conference by the following organizations is acknowledged with gratitude:

Bechtel Limited
BHP Minerals International Inc.
Davy International
Outokumpu Metals & Resources International
Royston Lead Plc
RTZ Limited
Solvay Interox Ltd.
Techpro Mining & Metallurgy Limited
Wenmec Systems, Inc.
ZENECA Limited

Foreword

'Hydrometallurgy '94' is the fourth in the series of international conferences on hydrometallurgy that started in 1975 in Manchester. The preceding two conferences in the series, held in 1981 and 1987, respectively, were organized by the Solvent Extraction and Ion Exchange Group of the Society of Chemical Industry (SCI): this group also initiated the organization of 'Hydrometallurgy '94'. Following preliminary discussions about the scope of the conference by the SX–IX Group committee some two and a half years ago, however, it was soon concluded that to cover adequately the current field of hydrometallurgy and its associated disciplines it would be necessary to broaden the scope of the subject matter compared with the earlier conferences. Additionally, owing to an overlap of interests in the field of hydrometallurgy, it was felt appropriate to approach the Institution of Mining and Metallurgy (IMM) with a proposal for joint organization of the conference. IMM responded swiftly and positively to the suggestion and a committee that comprised members of both bodies was assembled to begin the task of organizing a meeting that would equal and, it was hoped, exceed the high standards that had been set by previous 'Hydrometallurgy' conferences.

Since the 'Earth Summit' in Rio the concept of sustainable development has been much in vogue. The associated ideas of cleaner technology, recycling and waste minimization have particular relevance to the extraction and processing of metals and other mineral products. The scientific principles of inorganic and physical chemistry on which are based most of the techniques and processes that are used in hydrometallurgy are precisely those which have to be employed to clean up the excesses of the past, to treat the effluents of today and to design the cleaner processes of the future. Thus, the separation of ionic species in solution by selective precipitation, ion exchange or solvent extraction—techniques that are very familiar to the hydrometallurgist—can be readily adapted to the treatment of industrial effluents and other waste waters containing toxic metals and other undesirable solutes. The thermodynamic principles that are used to measure and quantify the relative stabilities and instabilities of phases, solid, liquid and gaseous, and which, for the hydrometallurgist, are most visibly embodied in the ubiquitous

Eh–pH diagrams, are just those on which judgements have to be based about the environmental compatibility and stability of process wastes destined for long-term disposal.

Thus, it seemed to the Organizing Committee to be entirely appropriate to try to reflect the broad applicability of the principles and processes of the discipline by giving 'Hydrometallurgy '94' the subtitle, 'Environmentally Sustainable Technology'. The first circular and initial publicity, in which these ideas were put forward, seem to have struck a chord with workers in the field, as they resulted in well in excess of 150 abstracts being submitted for consideration. The Organizing Committee, though very gratified by this excellent response, was then faced with a dilemma: 'Hydrometallurgy' conferences have traditionally been run in single session so that all delegates could attend the whole conference. This format, however, severely restricts the number of papers that can be accommodated. The alternative—to go to parallel or multiple sessions to increase the number of papers presented—would fragment the audience and lose the intimacy and solidarity of interest that have characterized previous conferences in the series.

Eventually, a compromise was reached: the major part of the conference would retain the single-session format, but there would be some parallel oral sessions, plus a major poster paper session, in order to increase the number of papers that could be accepted. In spite of these changes, the Organizing Committee had the unenviable task of selecting for eventual presentation at the conference no more than half of the abstracts submitted. The results of this process are contained in this book, comprising 78 papers by authors from 30 countries, which we believe presents a comprehensive picture of the current state-of-the-art and future trends in the technology of hydrometallurgy and its rapidly expanding role in the field of environmental engineering.

For their help in bringing all this to fruition I am very grateful to my colleagues on the Organizing Committee here in London for their hard work, particularly in refereeing the papers. The members of the overseas advisory board have also made an important contribution to the event by soliciting support and providing publicity for the conference in their own countries or regions.

'Hydrometallurgy '94' is the first major conference for which the IMM and the SCI have collaborated in joint sponsorship. The division of responsibilities between the two Conference offices was clearly defined by agreement at the outset, IMM taking responsibility for the editing, refereeing and production of the proceedings, whereas SCI is dealing with the organization and management of the conference itself. To date, this arrangement has worked extremely smoothly and I wish to pay tribute to the dedicated efforts of the staff of both organizations. This experience augurs well for future collaboration between SCI and IMM, which have a number of areas of common interest.

Another major difference between 'Hydrometallurgy '94' and previous meetings in the series is the change of venue to Cambridge. We hope that, in choosing Churchill College, we have provided a setting of tranquillity that will enable delegates to obtain maximum benefits from the high quality of the papers that are being presented and from the company of their colleagues from all over the world. The picturesque town of Cambridge provides a wide choice of historic settings for the social events, which we hope will help to make 'Hydrometallurgy '94' a memorable event.

Finally, I wish to record two important votes of thanks from the Organizing Committee: first, to the sponsoring companies, listed elsewhere in this volume, whose generous donations have enabled us to provide first-class social events while still keeping the registration fees to reasonable levels; and, second, to all authors for their contributions to this volume and to the conference—your work has made ours worthwhile.

Dr. A. J. Monhemius
Chairman, Organizing Committee
London, April, 1994

Contributors*

L. M. Abrantes (409), Departamento de Química, Faculdade de Ciências de Lisboa, Lisbon, Portugal

Marcela Achimovičová (209), Institute of Geotechnics of the Slovak Academy of Sciences, Košice, Slovakia

Katerina Adam (291), Laboratory of Metallurgy, National Technical University of Athens, Athens, Greece

Alain Adjemian (3), Directorate General for Science, Research and Development, European Commission, Brussels, Belgium

S. Agatzini-Leonardou (193), Department of Mining and Metallurgical Engineering, National Technical University of Athens, Athens, Greece

M. Aguilar (725), Chemical Engineering Department, Universitat Politècnica de Catalunya, Barcelona, Spain

F. J. Alguacil (939), Centro Nacional de Investigaciones Metalúrgicas, Madrid, Spain

J. Alstad (701), Department of Chemistry, University of Oslo, Oslo, Norway

A. Anacleto (579), Department of Chemical Engineering, Instituto Superior Técnico, Technical University of Lisbon, Lisbon, Portugal

M. T. Anthony (13), St. Barbara Consultancy Services, Essex, England

Keith Atkinson (1011), Camborne School of Mines, University of Exeter, Redruth, Cornwall, England

A. A. Bagreev (517), Institute for Sorption and Problems of Endoecology, Academy of Sciences of Ukraine, Kiev, Ukraine

W. Bahl (869), Ruhr-Zink, Datteln, Germany

Peter Baláž (209), Institute of Geotechnics of the Slovak Academy of Sciences, Košice, Slovakia

A. Ballester (369), Departamento de Ciencia de los Materiales e Ingeniería Metalúrgica, Facultad de Ciencias Químicas, Universidad Complutense de Madrid, Madrid, Spain

K. Barbetti (253), A. J. Parker Cooperative Research Centre for Hydrometallurgy, Chemistry Centre (WA), Department of Minerals

*Initial page number(s) of authors' contributions given in brackets.

and Energy, Bentley, Western Australia
J. P. Barbosa (527), CETEM—Centre for Mineral Technology, Cidade Universitária, Rio de Janeiro, Brazil
O. Barbosa-Filho (425), Department of Materials Science and Metallurgy, Catholic University, Rio de Janeiro, Brazil
D. W. Barr (337), Advanced Technical Development, Bundoora, Victoria, Australia
Denise Bauer (675), Laboratoire de Chimie Analytique associé au CNRS, Ecole Supérieure de Physique et de Chimie Industrielles, Paris, France
E. Ben-Yoseph (683), Israel Chemicals, Ltd., IMI Institute for Research and Development, Haifa Bay, Israel
B. C. Blakey (159), Department of Chemical Engineering and Applied Chemistry, University of Toronto, Toronto, Canada
M. L. Blázquez (369), Departamento de Ciencia de los Materiales e Ingeniería Metalúrgica, Facultad de Ciencias Químicas, Universidad Complutense de Madrid, Madrid, Spain
D. V. Boger (971), Advanced Mineral Products Centre, Department of Chemical Engineering, University of Melbourne, Victoria, Australia
C. F. Bonney (313), Mineral Industry Research Organisation, Lichfield, England
Kjetil Børve (563), Norzink AS, Odda, Norway
B. S. Boyanov (859), Department of Chemistry, University of Plovdiv, Plovdiv, Bulgaria
I. Bustero (655), INASMET, San Sebastián, Spain
C. Byszewski (613), Graver Water, Union, New Jersey, U.S.A.
S. A. Cale (949), Knight Piésold and Partners, Ashford, Kent, England
C. Caravaca (939), Centro Nacional de Investigaciones Metalúrgicas, Madrid, Spain (presently, Department of Mineral Resources Engineering, Royal School of Mines, Imperial College of Science, Technology and Medicine, London, England)
J. M. R. de Carvalho (579), Department of Chemical Engineering, Instituto Superior Técnico, Technical University of Lisbon, Lisbon, Portugal
J. F. Castle (229), RTZ Consultants, Ltd., Bristol, England
A. M. Chekmarev (219), D. Mendeleev University of Chemical Technology of Russia, Moscow, Russia
Chen Jiayong (541), Institute of Chemical Metallurgy, Chinese Academy of Sciences, Beijing, China
T. T. Chen (125), CANMET, Ottawa, Canada
Y. Cheng (655), Department of Chemistry, University of Liverpool, Liverpool, England
André Chesné (635), Laboratoire de Chimie Nucléaire et Industrielle,

Ecole Centrale Paris, Châtenay-Malabry, France
S. V. Chizhevskaya (219), D. Mendeleev University of Chemical Technology of Russia, Moscow, Russia
Y. Choi (711), Delft University of Technology, Faculty of Mining and Petroleum Engineering, Department of Raw Materials Technology, Delft, The Netherlands
R. Cierpiszewski (675), Academy of Economics, Poznan, Poland
M. J. Collins (869), Sherritt, Inc., Fort Saskatchewan, Canada
J. L. Cortina (725), Chemical Engineering Department, Universitat Politècnica de Catalunya, Barcelona, Spain
Paz Cosmen (741), CIEMAT, Madrid, Spain
M. C. Costa (409), Departamento de Química, Faculdade de Ciências de Lisboa, Lisbon, Portugal
Gérard Cote (675), Laboratoire de Chimie Analytique associé au CNRS, Ecole Supérieure de Physique et de Chimie Industrielles, Paris, France
M. Cox (219), Division of Chemical Sciences, University of Hertfordshire, Hatfield, England
D. C. Cupertino (591), ZENECA Specialties, Blackley, Manchester, England
S. A. Curran (441), Dublin Institute of Technology, Dublin, Ireland
I. Dalrymple (1075), Environmental Division, E. A. Technology, Capenhurst, Chester, England
R. F. Dalton (601), ZENECA Specialties, Blackley, Manchester, England
R. P. Das (253), Regional Research Laboratory, Bhubaneswar, India
Francisco Delmas (1075), Materials Department, Instituto Nacional de Engenharia e Tecnologia Industrial, Lumiar, Lisbon, Portugal
J. S. J. van Deventer (483, 501), Department of Metallurgical Engineering, University of Stellenbosch, Stellenbosch, South Africa
D. Dimaki (193), Department of Mining and Metallurgical Engineering, National Technical University of Athens, Athens, Greece
O. N. Dimitropoulou (463), Laboratory of Metallurgy, National Technical University of Athens, Athens, Greece
R. I. Dimitrov (859), Department of Chemistry, University of Plovdiv, Plovdiv, Bulgaria
J. Doyle (1035), CRA—Advanced Technical Development, Bundoora, Victoria, Australia
A. W. L. Dudeney (325), Department of Mineral Resources Engineering, Royal School of Mines, Imperial College of Science, Technology and Medicine, London, England
Gérard Durand (635, 655), Laboratoire de Chimie Nucléaire et Industrielle, Ecole Centrale Paris, Châtenay-Malabry, France

J. E. Dutrizac (125), CANMET, Ottawa, Canada
Saskia Duyvesteyn (887), University of California, Berkeley, California, U.S.A.
W. P. C. Duyvesteyn (887), The Minerals Laboratory, BHP Minerals, Reno, Nevada, U.S.A.
D. R. East (961), Knight Piésold and Co., Denver, Colorado, U.S.A.
S. H. Eberle (767), Kernforschungszentrum Karlsruhe, Institute for Radiochemistry, Water Technology Division, Karlsruhe, Germany
S. Elinson (837), Institute of Physical Organic Chemistry of the Belarus Academy of Sciences, Minsk, Belarus
Blanca Escobar Miguel (385), Departmento de Ingeniería Química, University of Chile, Santiago, Chile
P. K. Everett (913), Intec Pty., Ltd., Chatswood, New South Wales, Australia
Fang Zhaoheng (541), Institute of Chemical Metallurgy, Chinese Academy of Sciences, Beijing, China
L. H. Filipek (961), Knight Piésold and Co., Denver, Colorado, U.S.A.
N. P. Finkelstein (683), Israel Chemicals, Ltd., IMI Institute for Research and Development, Haifa Bay, Israel
D. S. Flett (13), St. Barbara Consultancy Services, Essex, England
B. F. Foley (441), Dublin Institute of Technology, Dublin, Ireland
S. A. Foster (795), Arthur D. Little, Inc., Cambridge, Massachusetts, U.S.A.
Matthias Franzreb (767), Kernforschungszentrum Karlsruhe, Institute for Radiochemistry, Water Technology Division, Karlsruhe, Germany
G. Friedman (1059), Israel Chemicals, Ltd., IMI Institute for Research and Development, Haifa Bay, Israel
A. R. Gee (325), Formerly, Department of Mineral Resources Engineering, Royal School of Mines, Imperial College of Science, Technology and Medicine, London, England (now BHP Minerals, Mayfield, New South Wales, Australia)
D. Gilroy (655), E. A. Technology, Capenhurst, Chester, England
Inés Godoy Ríos (385), Departmento de Ingeniería Química, University of Chile, Santiago, Chile
C. Gómez (369), Departamento de Ciencia de los Materiales e Ingeniería Metalúrgica, Facultad de Ciencias Químicas, Universidad Complutense de Madrid, Madrid, Spain
F. González (369), Departamento de Ciencia de los Materiales e Ingeniería Metalúrgica, Facultad de Ciencias Químicas, Universidad Complutense de Madrid, Madrid, Spain
J. T. Gormley (777), Knight Piésold and Co., Denver, Colorado, U.S.A.
M. D. Green (971), Advanced Mineral Products Centre, Department of

Chemical Engineering, University of Melbourne, Victoria, Australia
R. J. Grolman (1087), E. P. & P., Chicoutimi, Quebec, Canada
N. J. de Guingand (971), Advanced Mineral Products Centre, Department of Chemical Engineering, University of Melbourne, Victoria, Australia
J. J. Gusek (777), Knight Piésold and Co., Denver, Colorado, U.S.A.
C. J. Haigh (1035), Charlestown, New South Wales, Australia
A. K. Haines (27), Minerals Technology, Gencor, Ltd., Johannesburg, South Africa
R. M. Hamilton (795), National Rivers Authority, Exeter, England
L. J. F. Harris (922), Process Systems and Safety Department, Babcock King-Wilkinson, Ltd. (formerly, Babcock Contractors, Ltd.), Crawley, England
G. P. Herz (613), Tokuyama Soda Company, Tokyo, Japan
J. B. Hiskey (43), Materials Science and Engineering Department, University of Arizona, Tucson, Arizona, U.S.A.
E. M. Ho (1105), A. J. Parker Cooperative Research Centre in Hydrometallurgy, Murdoch University, Perth, Western Australia
J. E. Hoffmann (69), Jan H. Reimers and Associates, Houston, Texas, U.S.A.
W. H. Höll (767, 983), Kernforschungszentrum Karlsruhe, Institute for Radiochemistry, Water Technology Division, Karlsruhe, Germany
G. C. Holywell (1087), Alcan International, Ltd., Kingston Research and Development Centre, Kingston, Ontario, Canada
J. S. Jackson (591), ASARCO, Inc., Salt Lake City, Utah, U.S.A.
M. Jaffari (613), Malek, Inc., San Diego, California, U.S.A.
A. Jakubiak (675), Institute of Chemical Technology and Engineering, Poznan Technical University, Poznan, Poland
M. A. Jordan (337), Camborne School of Mines, Faculty of Engineering, University of Exeter, Redruth, Cornwall, England
Diego Juan (1123), Department of Chemical Engineering, University of Murcia, Cartagena, Spain
K-P. Jüngst (767), Kernforschungszentrum Karlsruhe, Institute for Technical Physics, Karlsruhe, Germany
Roland Kammel (209), Institute of Metallurgy, Technical University, Berlin, Germany
P. F. Kavanagh (441), Dublin Institute of Technology, Dublin, Ireland
Ke Jia-Jun (807), Institute of Chemical Metallurgy, Academia Sinica, Beijing, China
F. M. Kimmerle (1087), Alcan International, Ltd., Arvida Research and Development Centre, Jonquière, Quebec, Canada
M. G. King (591), ASARCO, Inc., Salt Lake City, Utah, U.S.A.
Chris Kiranoudis (463), Laboratory of Process Analysis and Design,

National Technical University of Athens, Athens, Greece
O. M. Klimenko (219) , D. Mendeleev University of Chemical Technology of Russia, Moscow, Russia
L. Kogan (1059) , Israel Chemicals, Ltd., IMI Institute for Research and Development, Haifa Bay, Israel
Constantine Komnitsas (291, 351, 361) , Laboratory of Metallurgy, National Technical University of Athens, Athens, Greece
Antonios Kontopoulos (291, 463) , Laboratory of Metallurgy, National Technical University of Athens, Athens, Greece
Pertti Koukkari (139) , Kemira Oy, Helsinki, Finland
Theodoros Kritikos (463) , Laboratory of Process Analysis and Design, National Technical University of Athens, Athens, Greece
Mária Kušnierová (209) , Institute of Geotechnics of the Slovak Academy of Sciences, Košice, Slovakia
K. A. Kydros (547) , Department of Chemistry, Aristotle University, Thessaloniki, Greece
J. Kyle (1105) , A. J. Parker Cooperative Research Centre in Hydrometallurgy, Murdoch University, Perth, Western Australia
S. Lallenec (1105) , A. J. Parker Cooperative Research Centre in Hydrometallurgy, Murdoch University, Perth, Western Australia
A. A. Latre (1025) , Instituto de Beneficio de Minerales, Facultad de Ingeniería, Salta, Argentina
P. J. Leggo (815), Environmental Minerals (U.K.), Linton, Cambridge, England
Li Deqian (627) , Changchun Institute of Applied Chemistry, Changchun, China
H. Liao (159) , Department of Chemical Engineering and Applied Chemistry, University of Toronto, Toronto, Canada
H. T. Lieuw (711) , Delft University of Technology, Faculty of Mining and Petroleum Engineering, Department of Raw Materials Technology, Delft, The Netherlands
Houyuan Liu (887) , The Minerals Laboratory, BHP Minerals, Reno, Nevada, U.S.A.
L. Lorenzen (483) , Department of Metallurgical Engineering, University of Stellenbosch, Stellenbosch, South Africa
Ma Gengxiang (627) , Changchun Institute of Applied Chemistry, Changchun, China
M. Makwana (869) , Sherritt, Inc., Fort Saskatchewan, Canada
Dimitrios Marinos-Kouris (463), Laboratory of Process Analysis and Design, National Technical University of Athens, Athens, Greece
Zacharias Maroulis (463), Laboratory of Process Analysis and Design, National Technical University of Athens, Athens, Greece
S. Martínez (939), Centro Nacional de Investigaciones Metalúrgicas,

Madrid, Spain
I. M. Masters (869), Sherritt, Inc., Fort Saskatchewan, Canada
K. A. Matis (547), Department of Chemistry, Aristotle University, Thessaloniki, Greece
M. McClaren (993), Terra Gaia Environmental Group, Vancouver, British Columbia, Canada
C. F. McDonogh (825), Solvay Interox Research and Development, Widnes, England
R. O. McElroy (993), Fluor Daniel Wright, Vancouver, British Columbia, Canada
J. McLoughlin (441), Dublin Institute of Technology, Dublin, Ireland
C. McNamee (441), Dublin Institute of Technology, Dublin, Ireland
Meng Shulan (627), Changchun Institute of Applied Chemistry, Changchun, China
M. T. van Meersbergen (483), Department of Metallurgical Engineering, University of Stellenbosch, Stellenbosch, South Africa
Isabelle Michelet (755), Laboratoire de Chimie Nucléaire et Industrielle, Ecole Centrale Paris, Châtenay-Malabry, France
J. L. Mier (369), Departamento de Ciencia de los Materiales e Ingeniería Metalúrgica, Facultad de Ciencias Químicas, Universidad Complutense de Madrid, Madrid, Spain
N. Miralles (725), Chemical Engineering Department, Universitat Politècnica de Catalunya, Barcelona, Spain
P. B. Mitchell (1011), Camborne School of Mines, University of Exeter, Redruth, Cornwall, England
A. J. Monhemius (177, 425), Department of Mineral Resources Engineering, Royal School of Mines, Imperial College of Science, Technology and Medicine, London, England
Pedro Moya (395), Departmento de Ingeniería Química, University of Chile, Santiago, Chile
J. C. Mugica (655), INASMET, San Sebastián, Spain
D. M. Muir (1105), A. J. Parker Cooperative Research Centre in Hydrometallurgy, Murdoch University, Perth, Western Australia
K. Mullins (441), Minmet Plc, Dublin, Ireland
Bosko Nikov (1153), "Zletovo" Metallurgical and Chemical Company, Titov Veles, Former Yugoslav Republic of Macedonia
Carlos Nogueira (1075), Materials Department, Instituto Nacional de Engenharia e Tecnologia Industrial, Lumiar, Lisbon, Portugal
Terje Østvold (563), Institute of Inorganic Chemistry, NTH, University of Trondheim, Trondheim, Norway
P. Owens (441), Dublin Institute of Technology, Dublin, Ireland
E. Ozberk (869), Sherritt, Inc., Fort Saskatchewan, Canada
A. P. Paiva (409), Departamento de Química, Faculdade de Ciências

de Lisboa, Lisbon, Portugal
V. G. Papangelakis (159), Department of Chemical Engineering and Applied Chemistry, University of Toronto, Toronto, Canada
N. L. Papassiopi (291, 463), Laboratory of Metallurgy, National Technical University of Athens, Athens, Greece
Dominique Pareau (635), Laboratoire de Chimie Nucléaire et Industrielle, Ecole Centrale Paris, Châtenay-Malabry, France
J. Parkes (1075), Faraday Centre, Carlow, Ireland
Ioannis Paspaliaris (463), Laboratory of Metallurgy, National Technical University of Athens, Athens, Greece
Andrés Perales (1123), Department of Chemical Engineering, University of Murcia, Cartagena, Spain
F. W. Petersen (501), Cape Technikon, Cape Town, South Africa
C. V. Phillips (337), Camborne School of Mines, Faculty of Engineering, University of Exeter, Redruth, Cornwall, England
N. L. Piret (229), Stolberg Consult GmbH, Neuss, Germany
R. E. Pocovi (1025), Instituto de Beneficio de Minerales, Facultad de Ingeniería, Salta, Argentina
F. D. Pooley (291, 351, 361), University of Wales, Cardiff, Wales
N. A. Postlethwaite (795), Marcus Hodges Environmental, Ltd., Exeter, England
M. V. Povetkina (219), D. Mendeleev University of Chemical Technology of Russia, Moscow, Russia
R. Püllenberg (869), Ruhr-Zink, Datteln, Germany
P. M. Quan (601), ZENECA Specialties, Blackley, Manchester, England
Mohamed Rakib (755), Laboratoire de Chimie Nucléaire et Industrielle, Ecole Centrale Paris, Châtenay-Malabry, France
I. Raz (1059), Israel Chemicals, Ltd., IMI Institute for Research and Development, Haifa Bay, Israel
G. V. Reznik (517), Institute for Sorption and Problems of Endoecology, Academy of Sciences of Ukraine, Kiev, Ukraine
N. M. Rice (273), Department of Mining and Mineral Engineering, University of Leeds, Leeds, England
J. J. Robinson (253), A. J. Parker Cooperative Research Centre for Hydrometallurgy, Chemistry Centre (WA), Department of Minerals and Energy, Bentley, Western Australia
P. C. P. Rocha (527), CETEM—Centre for Mineral Technology, Cidade Universitária, Rio de Janeiro, Brazil
E. G. Roche (1035), Pasminco Research Centre, Boolaroo, New South Wales, Australia
J. F. Rodríguez (939), Estaños de Zamora S.A., Villaralbo, Zamora, Spain

M. L. Ruiz (741), CIEMAT, Madrid, Spain
Mohammed Samar (635), Laboratoire de Chimie Nucléaire et Industrielle, Ecole Centrale Paris, Châtenay-Malabry, France
Angel Sanhueza (395), Departmento de Ingeniería Química, University of Chile, Santiago, Chile
K. L. Sandvik (1049), Department of Geology and Mineral Engineering, NTH, University of Trondheim, Trondheim, Norway
A. M. Sastre (725), Chemical Engineering Department, Universitat Politècnica de Catalunya, Barcelona, Spain
J. W. Scheetz (777), Brewer Gold Company, Jefferson, South Carolina, U.S.A.
D. J. Schiffrin (655), Department of Chemistry, University of Liverpool, Liverpool, England
A. A. Shunkevich (837), Institute of Physical Organic Chemistry of the Belarus Academy of Sciences, Minsk, Belarus
F. A. Silva (655), Department of Chemistry, University of Porto, Porto, Portugal
O. A. Sinegribova (219), D. Mendeleev University of Chemical Technology of Russia, Moscow, Russia
Hannu Sippola (139), GEM Systems Oy, Espoo, Finland
O. A. Skaf (1025), Instituto de Beneficio de Minerales, Facultad de Ingeniería, Salta, Argentina
V. S. Soldatov (837), Institute of Physical Organic Chemistry of the Belarus Academy of Sciences, Minsk, Belarus
Song Wenzhong (627), Changchun Institute of Applied Chemistry, Changchun, China
Tomislav Stojadinovic (1153), "Zletovo" Metallurgical and Chemical Company, Titov Veles, Former Yugoslav Republic of Macedonia
Petar Stojanov (1153), "Zletovo" Metallurgical and Chemical Company, Titov Veles, Former Yugoslav Republic of Macedonia
V. V. Strelko (517), Institute for Sorption and Problems of Endoecology, Academy of Sciences of Ukraine, Kiev, Ukraine
Anna Sundquist (139), Kemira Oy, Helsinki, Finland
Trygve Sverreson (1049), NOAH Langøya A/S, Holmestrand, Norway
P. M. Swash (177), MIRO Arsenic Research Group, Department of Mineral Resources Engineering, Royal School of Mines, Imperial College of Science, Technology and Medicine, London, England
J. Szymanowski (675), Institute of Chemical Technology and Engineering, Poznan Technical University, Poznan, Poland
S. Tamburini (655), Istituto di Chimica e Tecnologia dei Radioelementi, Padua, Italy
Yu. A. Tarasenko (517), Institute for Sorption and Problems of Endoecology, Academy of Sciences of Ukraine, Kiev, Ukraine

P. A. Tasker (591) , ZENECA Specialties, Blackley, Manchester, England

N. Taylor (273), School of Chemistry, University of Leeds, Leeds, England

N. P. Tidy (291) , University of Wales, Cardiff, Wales

R. B. E. Trindade (527) , CETEM—Centre for Mineral Technology, Cidade Universitária, Rio de Janeiro, Brazil

S. D. Ukeles (683, 1059) , Israel Chemicals, Ltd., IMI Institute for Research and Development, Haifa Bay, Israel

G. Van Weert (711) , Delft University of Technology, Faculty of Mining and Petroleum Engineering, Department of Raw Materials Technology, Delft, The Netherlands

Tomás Vargas (395) , Departmento de Ingeniería Química, University of Chile, Santiago, Chile

S. Vigato (655) , Istituto di Chimica e Tecnologia dei Radioelementi, Padua, Italy

R. C. Villas Bôas (107), CETEM—Centre for Mineral Technology, Cidade Universitária, Rio de Janeiro, Brazil

Adriaan de Villiers (961), Knight Piésold and Co., Denver, Colorado, U.S.A.

Nikolaos Voros (463) , Laboratory of Process Analysis and Design, National Technical University of Athens, Athens, Greece

C. P. Waller (1011) , Camborne School of Mines, University of Exeter, Redruth, Cornwall, England

Wang Zhonghuai (627) , Changchun Institute of Applied Chemistry, Changchun, China

T. E. Warner (273), Department of Mining and Mineral Engineering, University of Leeds, Leeds, England

A. Warshawsky (725) , Organic Chemistry Department, Weizmann Institute of Science, Rehovot, Israel

Bradford Wesstrom (69), Phelps Dodge Refining Corporation, El Paso, Texas, U.S.A.

J. V. Wiertz (385, 395) , Departmento de Ingeniería Química, University of Chile, Santiago, Chile

T. R. Wildeman (961) , Colorado School of Mines, Golden, Colorado, U.S.A.

G. L. Yan (701) , North China University of Technology, Beijing, China (presently, Department of Chemistry, University of Oslo, Oslo, Norway)

A. I. Zouboulis (547) , Department of Chemistry, Aristotle University, Thessaloniki, Greece

Zhu Guocai (541) , Institute of Chemical Metallurgy, Chinese Academy of Sciences, Beijing, China

Hydrometallurgy and sustainable development

EU R&D activity and strategy for sustainable development in the mineral industries

Alain Adjemian
Directorate General for Science, Research and Development, European Commission, Brussels, Belgium

Abstract
After a brief history of European Commission research activities in the field of mineral processing, some EC shared-cost projects in this area are presented.

Beyond the general aim of strengthening the competitiveness of European industry, the specific objectives of EC research programmes are to support the effect of the European industry, to improve product quality and to promote sustainable development. Several successful projects have helped developing new flotation processes to increase yield and grades, new bio-hydrometallurgical processes to treat gold, complex sulphide, manganese ores and industrial minerals. Several others have investigated new clean routes to treat or recycle spent products, used materials, industrial solid and liquid wastes, residues, slags or scraps.

In the fourth Framework Programme, part of the budget will be allocated within the Industrial and Materials Technologies Programme to reinforce the trend towards clean production, sustainable mineral processing and extractive metallurgy, management of wastes and recycling of products and materials.
Keywords: European Commission R&D Activity

1 Introduction

European Raw Materials extractive and processing industries have been going through revolutionary change in recent decades and are still subject to a variety of fundamental pressures and constraints.

In the next 5 to 10 years these industries will have to comply with severe environmental regulations which will drive them towards a complete revision of their strategy. More than ever gas, liquid and solid effluents, muds, residues, tailings, slags and all other sorts of related wastes will be minimised, processed, controlled, valorised or recycled. Final deposits will decrease and will have to become more acceptable to the environment.

Previous European Community RTD funded actions in the field of Raw Materials and Recycling were already in line with promoting such trends. Recent strong recommendations for a sustainable development of industry have been included in the last EC Communication on "Growth, competitiveness, employment: the challenges and ways forward into the 21st century", December 1993 and will be underlined in the Workprogramme of the next EC R&D Industrial and Materials Technologies Programme (1994-1998).

2 The European Mineral Industries' Strength

The European Mineral Industries, with a 212 billion ECU production value and almost 2 million employees, represent a key European industrial sector (Table 1). It covers a wide variety of subsectors (Table 2) from extraction of metallic and non metallic ores to the production of metals, industrial minerals, construction materials, glass and ceramic goods.

Table 1. EC manufacturing industries

Manufacturing sector	Production (billion ECU)	Number of employees (million)
Food drink and tabacco	434	2.5
Chemical industry	295	1.8
Motor vehicles	262	1.8
Electrical engineering	258	2.8
Mechanical engineering	221	2.4
Mineral industries	212	1.7

Source: "Panorama of EC industry 1993" (EUROSTAT)

Table 2. Mineral industries

Manufacturing sub-sectors	Production (billion ECU)	Number of employees (thousand)
Extraction and preparation of metal ores	1	
Extraction of minerals other than metal ores	16	186
Production and preprocessing of iron and steel	64	398
Non ferrous metals	38	210
Manufacture of non metallic minerals	94	944
Total	212	1738

Source: "Panorama of EC industry 1993" (EUROSTAT)

EU is the world leading producer of non metallic minerals. Its production (90 billion ECU) is twice the size of that of the USA. About two-thirds is destined to the construction industry (concrete, cements, lime, plaster, ornamental stones). Glass and ceramic production feed other sectors like automotive, electronic, chemical industries.

EU has a leading importance in the production of ferrous and non ferrous metals. The production of the major non ferrous metals (aluminium, copper, zinc, lead) representing 24% of the world total, and the refining / fabricating capacity for precious metals is the largest in the world.

Moreover, due to the low level of metal mining production, the European metal industry has much experience in the processing of secondary materials (wastes, new scraps, spent products, etc.).

Despite differences throughout Europe common characteristics are clear. The expertise, the accumulated knowledge, the industrial competence, the historical leadership are still the main features of the European Minerals Industry. Behind these strong industrial assets we find a very highly skilled work force with significant national research facilities (Table 3) as well as strong capabilities in development with leading engineering companies and equipment producers. The Nordic States, not yet members of the European Union but already participating in EU R&D programmes, are strengthening this prominent position in all areas of the Minerals Industry.

To enhance this competitive position the European Union launched a succession of R&D programmes.

Table 3. National RTD centres with strong expertise in Mineral Processing[(1)] and Recycling

Institute of Geology and Mineral Exploration	GR
Bureau de Recherches Géologiques et Minières	FR
Centre de Recherche et de Valorisation des Minerais	FR
Warren Spring Laboratory (DTI)	UK
Instituto Nacional de Engenharia e Tecnologia Industrial	PT
Direcçao Geral de Investigaçao Cientifica e Tecnologica	PT
Centro Nacional de Investigaciones Metalurgicas (CSIC)	ES
Centro Tecnologico de Materials (INASMET)	ES
Consiglio Nazionale delle Ricerche - Istituto per il Trattamento dei Minerali	IT
Aachen University of Technology (RWTH)	DE
Institut de Recherche de la Sidérurgie	FR
Centro Sviluppo Materiali	IT
Centre de Recherche Métallurgique	BE
Verein Deutscher Eigenhuttenlente	DE

(1) Non exhaustive list

3 The EC Research activities in the field of Minerals extracting and processing

3.1 The EC R&D Programmes

EC R&D programmes have existed for the last 10 years. The Single European Act (1986) has initiated a European research and technological policy which has been consolidated in the Treaty on European Union in 1993 (Title XV, Article 130f-130p). Today EC funded R&D activities cover a broad range of areas the main emphasis being on Information Technologies, Industrial and Materials Technologies, Environment, Life Sciences and Technologies, Energy and Human Capital and Mobility.

For each area there is a specific R&D programme. All of them are part of a global RTD Framework Programme with a duration of four years.

3.2 Minerals and Metals processing in the EC R&D Programmes

The last three EC R&D Framework Programme (1984-1987, 1987-1991, 1990-1994) and the new Fourth Framework Programme (1994-1998) have all had a significant part of their budgets for selected R&D activities and projects in the mineral raw materials extracting processing and recycling sectors.

The objectives of these activities were centered around strengthening the competitivity of the European Raw Materials and Recycling industries through the development of innovative technologies and the modernisation of industrial tools. Impact on primary raw materials' supply, energy savings and environmental protection were the first three expected results. Technical areas for the Raw Materials and Recycling of Non Ferrous Metals programmes (1990-1002) and for the corresponding areas of the 1992-1994 Industrial and Materials Technologies Programme (Brite-EuRam) are listed in Tables 4 and 5.

Table 4. Areas of Research A and B for the Raw Materials and Recycling programme (1990-1992) (Second Framework Programme)

A. Primary raw materials
   1. Exploration
   2. Mining technology
   3. Mineral processing and extractive metallurgy

B. Recycling of non ferrous metals and strategic metals
   1. Characterisation and classification of secondary materials and physical separation and concentration
   2. Advanced pyrometallurgical processes
   3. Advanced hydrometallurgical processes
   4. Refining technologies and instrumentation on control of the processes

Table 5. Areas of Research[*] for the Industrial and Materials Technologies programme (1991-1994) (Third Framework Programme)

Area 1: Materials - Raw Materials

- 1.1. Raw Materials
- 1.1.1. Exploration Technology
- 1.1.2. Mining Technology
- 1.1.3. Mineral Processing
- 1.2. Recycling
- 1.2.1. Recycling and Recovery of Industrial Waste including Non Ferrous Metals
- 1.2.2. Recycling, recovery and reuse of advanced materials

[*] parts directly related to Raw Materials and Recycling

Calls for proposals were launched periodically (3 calls since 1990) and at each call several dozens of R&D projects were selected for funding in Exploration, Mining Technologies, Mineral Processing and Extractive Metallurgy, Recycling of Non Ferrous Metals and Processing of Industrial Wastes.

3.3 R&D project characteristics

Projects are launched for a duration of 2 to 4 years. For each project an average of three to six partners from different Member States, universities and industries, work together and share the results. In general 50% of project expenditure is reimbursed by the Commission to the partners. The average range for project total cost is between 0.4 MECU and 4 MECU. Commission scientific officers and contract unit staff monitor technical progress and contractual liabilities. Although know-how and results remain the project partners' priority, public domain synthesis reports are publicized and assessed. Regular technical workshops are organised by the Commission with the contractants to discuss the results achieved and promote synergy between projects.

4 The 1990-1992 EC Research Programmes on Raw Materials and Recycling of Non Ferrous Metals

The 1990-1992 European Community Research Programmes on Raw Materials and Recycling belong to the Second Framework Programme.

Many projects of the above-mentioned programmes are directly or indirectly aiming at sustainable development issues. If we consider only the projects just finished this year we already have quite positive results and success stories in this direction. New generic separation techniques, measurement systems, waste reducing and processing systems, reactor designs and biohydrometallurgy routes have been studied and developed. Some of these are listed below:

- A cryomagnetic separation system has been developed and patented to be used in the various stages of the hydrometallurgy of waste containing recoverable metals in particular for ultra fine and disseminated particles and liquid effluents.
- An eddy current separator has been successfully implemented on the reject line of a waste derived fuel plant processing domestic waste to extract aluminium efficiently and economically.
- High temperature solid electrolyte sensors have been developed for use in zinc and copper melts to measure impurities like arsenic and antimony. A new project will start soon to further develop this application.
- New operating conditions and rotary furnace design have improved the economy of the aluminium refining process reducing costs and pollutant fluxes.
- A first feasibility study has shown the potential benefit to vitrify goethite wastes orginating from zinc hydrometallurgical plants in order to produce glass and glass ceramic materials suitable for commercial exploitation. (A new larger project will soon build on these first results).
- An other clean solution for the electrolytic zinc plant iron residue combining hydro and pyrometallurgical techniques in order to meet environmental regulations has been developed. It transforms iron residue into unleachable compound which can be disposed of or even used in the cement industry.
- An hydrometallurgical route for the recovery of zinc from secondary materials containing zinc oxide and zinc sulfate has been successfully improved. The process is now technically feasible, environmentally clean, safe, economically attractive and ready for exploitation.
- Specifications for secondary lead used in the new sealed generation of lead acid batteries have been assessed in particular the effect of the content of impurities of recycled lead on the behaviour of the batteries.
- A new chloride hydrometallurgical process has been developed to treat lead flue dusts in a very simplified flowsheet characterised by a new procedure of leaching in pulp of $PbCl_2$ and lead electrolysis with impure electrolyte to recover lead, zinc and cadmium.
- An hydrometallurgical process able to eliminate one, two or all penalising elements (As, Sb, Te,...) contained in the anodic slimes from copper refining has given encouraging results for the recovery of copper, silver and selenium.
- Several routes have been developed to treat spent hydrodesulfurisation catalysts in order to recover Ni, Co, V, Mo on a continuous basis.
- Techniques for the reprocessing of spent zinc containing acid wastes from galvanising units have successfully been studied. A mobile pilote plant demonstration is now supported to test the retained flowsheet (evaporation and solvent extraction).
- A new approach for the selective concentration of valuable elements from concentrates and industrial wastes has consisted in the design and successful experimentation of a granulated pulsed bed electrochemical reactor.
- Two methods were developed for the separation of pyrite and arsenopyrite by flotation. A new collector has been designed and produced. It will allow to implement a simpler flowsheet with cheaper flotation operating conditions.

• Several biohydrometallurgical routes have been studied (high temperature bioleaching of complex mineral sulphides, removal of iron from industrial minerals, manganese recovery from ores, silica removal from bauxites) and are still ongoing (optimisation of a bacterial leaching process for the treatment of auriferous arsenical pyrites[*],...) to increase yields, concentrates grades and quality.

[*] A presentation on this activity will be given in this conference.

Catalogues describing the above-mentioned research activities (public domain synthesis reports, workshops proceedings) can be obtained from the Commission Services in DGXII-Brussels.

5 The "Industrial and Materials Technologies" Programmes

5.1 The Third Framework Research Programme (1990-1994)

Within the ongoing EC Third Framework Research Programme (1990-1994), raw materials and recycling are areas of the Industrial and Materials Technologies (Brite-EuRam) Programme.

Following the first Brite-EuRam II call for proposals (February 1992) new R&D projects concerning mineral processing and metal wastes recycling started end of 1992 or the beginning of 1993. Projects are in general three times larger in terms of financial support from the Commission compared to previous projects with a higher industrial participation and with an average EU funding per project of more than 1 MECU. Projects with a particular emphasis on environmentally clean minerals and waste processing are listed below.

• New process routes for the recovery of magnesite run of mines.

• New approach for valuable elements recovery from concentrates and wastes by selective chlorination through binary chlorides.

• Removal of iron from industrial minerals: mechanisms of dissolution and precipitation.

• Improvement and extension of the use of flotation columns flowsheet optimisation.

• Development of new or improved separation processes for the treatment of chromite ores indigeneous to the Community.

• New integrated flowsheets for separation and recovery of minor elements from sulphide ores.

• Optimisation of the rotary kiln operation in the production of Fe-Ni.

• Reclaiming of metals contained in spent catalysts used in petroleum refining petrochemistry and chemistry.

• Recycling of zinc and lead ferrous dusts from the electric arc furnace.

• Recovery of precious and base metals from industrial liquors using electrochemical ion exchange.

• Modelling of genetic biochemical cellular and microenvironmental parameters determining bacterial sorption and mineralisation processes for recuperation of heavy or precious metals.

- Lead recovery from lead oxide secondaries.
- The recovery and recycling of vanadium and nickel products from the combustion residues of orimulsion and other fuels.
- Magnesium metal recovery from asbestos and related waste materials.
- Innovative use of a proven technology for lead and zinc residues recycling.

After a second call for proposals (February 1993) a similar number of projects were selected in this area and started late 1993 and beginning of 1994. Some of them are listed below.

- Valorisation of Pb-Zn primary smelter slags.
- Recovery of industrial waste from surface treatment of aluminium to study its recyclability and its use in other industries.
- Elaboration de liants hydrauliques par valorisation de déchets provenant de l'industrie papetière.
- Recycling of jarosite waste in combination with granite scraps for producing glass and glass ceramic materials.
- Solid state sensors for on line control of high temperature metallurgical processes.
- Ultra fine grinding, delamination and characterisation of minerals.

5.2 The Fourth Framework Research Programme (1994-1998)

In the Fourth Framework Research Programme, the Industrial and Materials Technologies Programme, subtitled "A European Programme on Industrial Technologies for Sustainable Growth", includes a significant place for the Minerals and Metals Industries. R&D priorities for these industries are well addressed through all the areas of the programme and in particular in the following:

Area 1: Production Technologies for Future Industries
- 1.1: Incorporation of New Technologies into Production Systems
- 1.2: Development of Clean Production Technologies
- 1.3: Rational Management of Primary Raw Materials
- 1.4: Safety of Production Systems
- 1.5: Human Factors in Production Systems

Area 2: Technologies for Product Innovation
- 2.1: Materials Engineering
- 2.4: Technologies for Materials Recuperation at the end of Product Service-life

This new IMT programme will consider also research tasks linked with the Steel sector in order to progressively incorporate the actions covered under the ECSC treaty under the EC Framework Programme.

In our changing society we are moving towards a different development model in which more importance is attached to the quality of life and more rational use of human and natural resources.

The "Industrial and Materials Technologies" programme objectives will focus on the following three objectives:

- In the short term, priority should be assigned to research for the adaptation of existing technologies, or for the development of new industrial technologies, which provide competitive leverage, particularly in sectors where the level of technology is lower; small and medium-sized enterprises (SMEs) in these sectors account for a large proportion of European industry and provide the bulk of employment.
- In the medium term, research wil focus on industries which are already developing innovative technologies and strategies allowing better use of human resources while endeavouring to reduce the adverse environmental impact of production.
- In the long term, research will focus on new technologies for the production and design of products which allow new industries or markets to be created in a context of sustainable growth.

In the context of respect for human beings and the environment, and sustainable growth, the research will cover new process engineering methods, new manufacturing techniques, new inspection, diagnostic and quality maintenance and assurance systems. Special attention will be paid to business organisation, the incorporation of technologies addressing social concerns, the health and safety of workers (working conditions) and environmental concerns (clean technologies, rational use of resources) whilst taking into account their economic and industrial impact.

Budget allocated to this programme will be more than 1623 million ECU which is more than the double of what was allocated to the previous 1990-1994 IMT programme.

6 Conclusion

The previous, ongoing and future R&D programmes financed by the Commission allow several opportunities to strengthen the competitiveness of European Minerals and Metals Industries competitiveness.

For this purpose, since 1992, in each year an average of 10 European R&D projects have been started with the clear objective of developing new clean technologies for a sustainable minerals industry growth. This policy is presently reinforced with the new "Industrial and Materials Technologies" Programme under the Fourth Framework Research Programme (projects starting in 1995) which opens most of its areas to the minerals and metals extracting and processing and waste recycling R&D priorities.

The trend towards environmentally clean and safe productions, sustainable management of raw materials and wastes, recycling of products and materials is therefore promoted and supported by the European Commission as one of its key strategic areas for research and technology development.

Hydrometallurgy—an environmentally sustainable technology?

M. T. Anthony
D. S. Flett
St. Barbara Consultancy Services, Essex, England

Abstract
Hydrometallurgy has long been hailed as the environmentally friendly alternative to pyrometallurgy. There are, however, flaws in this argument, not least the difficulties, when hydrometallurgy is employed, of avoiding water pollution and the need to discharge solids which may be far from inert. A good example of this situation is that of jarosite, the main waste solid from the roast/leach/electrowin process commonly used for zinc recovery.

Pyrometallurgy, on the other hand, certainly has the potential to pollute the atmosphere, but the gaseous sulphur emissions can be positively dealt with (and in some circumstances can contribute to revenues) and the slag waste products can be of an inert nature.

Furthermore, apart from zinc, no major commercial plants for the hydrometallurgical treatment of primary copper or lead concentrates have yet been seriously considered, nor will be in the short-term future, despite claims for several new processes. Where hydrometallurgical operations are common, as mentioned previously for zinc, and also for aluminium and gold, environmental challenges still exist.

In this paper hydrometallurgy is examined from an environmental standpoint, in the light of Best Available Techniques (BAT) and Best Practical Environmental Options (BPEO). Conclusions with regard to the continuing development of hydrometallurgy for the extraction and purification of major metals are drawn.
Keywords: Environment, Hydrometallurgy, Metals, Processing, Residues, Sulphur, Tailings, Wastes.

1 Introduction

Hydrometallurgy has become an important part of the practice of extraction metallurgy. Over 200 specific hydrometallurgical processes and sub-processes are being applied, in addition to extensive application of generic hydrometallurgical processes, with a significant number of further techniques under development.

The technology now accounts for considerable percentages of the world's production of metals, over 90%, for instance in the case of gold, around 80% for zinc and 100% for alumina. Over 2000 books and papers have been written on aspects of hydrometallurgy in the last decade, and it finds increased application in secondary fields, as well as a growing employment in pollution abatement, while remaining the most popular area for research and development.

To illustrate the impact of hydrometallurgy it is only necessary to review for the major metals common hydrometallurgical practices which include:-

Gold oxidation - pressure leaching, flash chlorination, bioprocessing.

Gold recovery and purification - cyanide leaching, cyanide destruction and gold recovery.

PGM extraction - matte leaching.

PGM refining - PGM dissolution and solvent extraction.

Copper (oxide) ore and (off-grade) concentrate leaching.

Copper recovery - solvent extraction, electrowinning.

Zinc recovery - leaching, jarosite precipitation, solution purification, electrowinning.

Cobalt and nickel recovery - leaching, pressure leaching, solvent extraction, electrowinning

Alumina - alkaline leaching of bauxite and alumina precipitation.

Secondary treatment is provided extensively by hydrometallurgical processes for copper dusts, lead acid batteries, the production of zinc chemicals and silver residues and wastes.

Active research and development areas for precious metals include oxidation methods including the use of halides, nitrous and nitric acid and Caro's acid, bio-oxidation and pressure leaching using new lixiviants. Alternatives to the use of cyanide and cyanide regeneration processes as well as improved solvent extraction for PGM separations are also favoured subjects.

For base metals, advances are being made in in-situ, bacterial, chloride and catalysed leaching of copper, the use of chlorides and other novel lixiviants for lead and alternatives to jarosite for iron removal from zinc plus zinc solvent extraction.

It is thus clear that much has already been invested in hydrometallurgy and its ability to cope with both production requirements and environmental compliance demands. The question is whether confidence in the latter aspect can be justified in the context of future environmental legislation?

2 Environmental Process Considerations

Once ore has been mined, and thus excluding environmental problems which arise with this activity, the problems encountered may be summarised as:-

* Storage/disposal of tailings from concentrate production
* Treatment of liquors associated with tailings
* Recovery or disposal of sulphur contained in the ore/concentrate
* Recovery or disposal of heavy metal impurities contained in the ore/concentrate
* Storage or disposal of slags, residues or other waste materials generated during processing
* Recovery or disposal of waste liquors from processing
* Ultimate rehabilitation of the processing site

Hydrometallurgy is frequently referred to as "green" technology, but two points should be recognised which militate against this statement, and place hydrometallurgy in a more appropriate context.

Firstly, of the list above, each problem listed remains a consideration and does not disappear with the use of hydrometallurgy as opposed to any other technology.

In the second place the concept of net environmental efficiency must be considered. Put in simple terms, there may be a negative net global environmental impact if a high energy process is selected to produce an environmental gain at the metals extraction and refining stages if the environmental downside of that energy generation exceeds the benefits gained. This is of particular relevance to hydrometallurgy where it competes with highly energy efficient smelting processes.

Hydrometallurgy thus should not be seen as an overall panacea for curing the environmental problems of metals extraction and processing, and, in considering its application, its use must be judged on the individual merits of each application. Knowing what the problems are, and the principles to be applied to environmental compliance, together with some guideline emission levels, the status of hydrometallurgy can now be examined for each situation in turn.

2.1 Tailings

Clearly the production of concentrates where necessary is a mineral processing activity which is likely to be equally necessary for any extraction, purification and refining technology. In almost all circumstances it is cheaper to undertake the concentration step by physical separation techniques. The potential for hydrometallurgy is twofold. Firstly, and most commonly, it may have the ability to treat ores rather than concentrates which can improve the logistics of waste disposal as well as the economics of treatment. Typically this will occur with heap leaching for gold and copper, and has the advantage of leaving the vast majority of the waste material in one coherent mass for which a range of proven techniques exist for stabilisation and rehabilitation. Thus if heap leaching can be practised, the environmental requirement is for sealing and revegetation of the heaps, which is generally accepted as a low-impact, low-cost and practicable option.

In the second place, if in-situ leaching methods can be developed, hydrometallurgy has the potential to eliminate this requirement altogether, leaving a mine site with minimum environmental impact and no solids disposal problems. Where the correct ground structure and hydrogeological factors exist this technology is feasible, and has attracted much research focus. It is also already used with minerals such as potash, and the technology exists for all metals to create the necessary disruption of the orebody and access a suitable lixiviant. Unfortunately development trials conducted in the US and Australia have yet to show that in-situ leaching can be conducted in such a way as to guarantee acceptable recovery of the leach liquor, and therefore its widespread application must be considered remote at this time.

Sulphidic flotation tailings on the other hand present significant problems in relation to the generation of acidic, heavy metal-containing, liquors by bacterial oxidation of the contained sulphides. Iron oxidised to the ferric state also acts as a lixiviant thus solubilising even more heavy metals. This phenomenon, combined with acid mine drainage which is the product of similar oxidative leaching underground, is a strictly hydrometallurgical problem which can do untold environmental damage if not recognised and dealt with. The major factor in these problems is oxygen availability. If oxygen access to tailings dumps can be restricted then the problem can be contained. This can be done by subaqueous deposition of tailings or providing for flooding of tailings after mining and processing is finished. Dealing with acid mine drainage is less easy and treatment of the acid mine waters is necessary probably by liming and precipitation of the contained heavy metals. Alternatively such liquors may be treated by natural processes such as with sulphate reducing bacteria wherein heavy metals are precipitated as sulphides which can be recovered and returned to the process. Such a process is in commercial use at Budelco in Holland for treatment of contaminated groundwater.

Tests exist nowadays whereby the acid generation potential of tailings can be determined thus giving early warning of the magnitude of the problem and permitting cost effective strategies for prevention to be devised. Abandoned tailings are another matter and stabilisation can be a costly matter. These problems, it should be noted are a consequence of mining in the first place and not of hydrometallurgical activity per se. As noted above development of environmentally sound in-situ leaching technology could provide an environmentally acceptable hydrometallurgical alternative to mining and mineral processing of sulphide ores.

2.2 Sulphur

The major reason for hydrometallurgy being perceived as a green technology is the avoidance of sulphur emissions in gaseous form, which with its high visibility and track record of damage to the environment as acid rain, is a significant bonus. It should be remembered, however, both that sulphur must still be handled by a hydrometallurgy route, and that in certain situations, the generation of sulphur dioxide and its efficient conversion to sulphuric acid, actually provides extra income to a smelting operation.

Hydrometallurgy enjoys the advantages of largely recyclable use of the sulphide sulphur once oxidised and the ability to produce saleable labile sulphur, although in some instances the sulphur produced will still need treatment and disposal as part of the leach residue. To be marketable, elemental sulphur must be free of such impurities as As, Hg, Se, etc.

Sulphur recovery is relatively easy and is accomplished by flotation to produce a sulphur concentrate from which the sulphur can be separated by melting and filtration. Alternative methods exist, for example it can be recovered from the flotation concentrate by extraction with a solvent such as perchloroethylene and crystallised. Oxidation of sulphide sulphur to sulphate and its elimination as gypsum is unlikely to be accepted as BAT due to the large volumes of gypsum produced and adverse markets for such a by-product. Use of sulphate reducing bacteria to treat sulphate-containing waste liquors is possible however and, as noted above, is already used commercially for treatment of contaminated groundwaters.

2.3 Heavy metals

One of the consequences of metals smelting is that impurity metals with similar characteristics to those of the main metals behave in a similar fashion throughout the process, and can be collected fairly easily at the electrolysis stage as slimes. More volatile metals may be evolved during smelting, and these can be recovered, usually by scrubbing, at low cost. Gangue minerals report to slag.

Whereas in hydrometallurgy many impurities are brought into solution during leaching, their presence can greatly affect the electrolysis stage and thus intermediate purification is necessary. Hydrometallurgical techniques have a large capability to achieve this, but at the cost of large numbers of residue products. This is inevitable if an inert leach residue is to be produced which will pass appropriate leach tests for classification of such wastes.

The objectives of all purification processes therefore must be not only to achieve the degree of purification required but also to permit saleable products to be produced from the impurities such that further value is added to the process. At the very worst the impurities must be easily and cheaply converted into acceptable material for safe disposal.

The major impurities in hydrometallurgical circuits are iron and arsenic. In processing of bauxite to produce alumina the iron leaves the circuit as "red mud" which is impounded in vast tailings dams. This red mud contains alumina, silica, titanium and significant amounts of alkali salts. Thousands of tonnes per day are produced at a typical alumina plant using the Bayer process. Alkaline water from the dams can be recycled to the plant but no known use exists for the solid wastes which thus constitute a significant environmental problem.

In the zinc industry for the roast-leach-electrowin process and also for the newer pressure leach-electrowin process, iron removal is mostly achieved by jarosite precipitation. This is a very bulky precipitate and consequently contains significant quantities of entrained acid and zinc. The zinc industry is coming under strong environmental pressure to stop dumping jarosite and to clean up historical jarosite tailings dams. The problem is considered so severe that a "jarosite club" has been formed by several zinc producers to consider what they should do. Even after several years of endeavour no definite conclusions appear to have been reached. The alternative process wherein haematite is produced, although claimed to be the answer as it should produce (but doesn't) a saleable iron waste product, has now been abandoned by Ruhr Zink. A retreatment process is being operated by Espanola de Zinc at Cartagena in Spain wherein both historic and prompt jarosite is leached under controlled conditions to prevent significant iron dissolution and the zinc recovered by solvent extraction using DEHPA. The zinc strip liquor is used to leach calcine which is treated in the normal way to produce a tankhouse electrolyte for zinc electrowinning. The acceptability of the treated jarosite for ponding under Dutch or Canadian regulations is unknown but it would seem unlikely to be acceptable as no other company seems keen to adopt the Espanola approach. In fact Budelco are expected to smelt their jarosite and fume the slag to produce a final slag that will conform to Dutch regulations for building materials and thus be able to dispose of it. Thus zinc hydrometallurgy would seem to be under threat.

Interestingly although much research has been carried out and many processes proposed no hydrometallurgical processes for treatment of copper concentrates are in commercial use. Not only do they have to compete with highly energy efficient state-of-the-art smelting processes but, if zinc hydrometallurgy is in difficulties over iron disposal where the zinc:iron ratio is 1:3 to 1:5, how much more of a problem will iron disposal be in copper concentrate hydrometallurgy where for chalcopyrite concentrates the copper:ion ratio is 1:1.

2.4 Slags and residues

A feature of smelting is the production of waste largely as silicate slags with a good measure of stability and resistance to erosion and leaching. Hydrometallurgy can produce residues chemically to a degree to order, but the materials will be in particulate form and susceptible to being washed away. Their size will also make them more reactive should the conditions under which they are stored change.

2.5 Recovery and disposal of waste process liquors

Primary ore processing, metal extraction and the processes that are grouped under the generic title of metal finishing all generate wastes and effluents that need treatment prior to safe and legal disposal. For metalliferous effluents the usual approach until fairly recently has been to lime such effluents to produce a liquor suitable for discharge and a gypsum sludge containing precipitated heavy metals and to dispose of this sludge to landfill. Increasingly such landfilling practice is ceasing to be a cheap cost option and there can be little doubt that such practices will have to cease in the near future and thus cannot be considered BAT. The development of modern separation science and technology however offers clean options for effluent and waste processing and the opportunity to offset processing costs by recovery and recycling of valuable components and reagents back to the main process stream. These techniques can be used as add-on end-of-pipe techniques or applied in central processing facilities where effluents and wastes from industry within a convenient geographical area can be taken for treatment. Such plants are now beginning to appear particularly in North America and most use hydrometallurgical methods. The hydrometallurgical techniques used here are a) precipitation, b) ion exchange, c) biosorption, d) solvent extraction, e) membrane processes and f) electrolytic processes.

a) <u>Precipitation</u>

The treatment of metal-bearing effluents by precipitation processes followed by dumping of the precipitate follows well established principles and consists of a sequence of operations which are adapted to suit the effluent to be treated.

i) Separation of acidic and alkaline wastes. The latter may contain cyanides or other complexing agents and metals in the form of anionic complexes.
ii) Destruction of complexing agents using for example chlorine, hypochlorite or ozone.
iii) Reduction of anionic metal complexes by the use of reducing agents in acidic media.
iv) Combination of extracted acidic and alkaline streams.
v) Adjustment of pH using lime or water soluble alkali to precipitate the metal values as hydroxides or basic salts.
vi) Separation of precipitated solids by sedimentation, flotation or filtration.

This process may require additional steps to deal with particular metals such as cadmium which may require precipitation as the sulphide to effect the desired level of removal to meet effluent discharge standards. This process, though still widely practised cannot now be considered as Best Available Technology as in many instances it can only just meet discharge limits for treated effluent disposal and also produces a sludge that is unlikely to meet disposal criteria such as the EPA TCLP test.

b) <u>Ion exchange</u>

Ion exchange is technically an excellent method for effluent treatment. Its application to treatment of silver-bearing effluents from the photographic industry is a good example of such an application.

Use of mixed bed technology for effluent treatment can produce good quality water for discharge or recycle and by using appropriate regeneration techniques relatively strong strip liquors can be produced from which metals can quite readily be recovered. However it is a relatively expensive process due to the cost of the resins and the regeneration procedures and in the past has tended to be used only for high value metals in waste streams such as precious metals and PGM's. Fixed bed cartridge-based exchange systems are becoming quite popular. These are offered by specialist companies to whom the spent cartridges are returned and they elute and recover the metal values contained therein. Companies such as Eliminex in the UK and RMA Dornier in Germany and marketed in the UK by Water and Metal Recovery Ltd offer such services. Such a system allows for recycle of water within plants and should meet BAT requirements.

Some 15 years ago a new system called reciprocating flow ion exchange or Recoflo was developed and has been used for metals recovery and effluent treatment in the metal finishing industry. The system uses very short ion exchange columns filled with much smaller diameter ion exchange beads than other systems. This gives much faster exchange kinetics than conventional systems and much faster turnaround times. These systems are used for metal ion and acid recovery from a wide variety of process effluents. An example of the use of this system is the recovery of acids from pickle liquors. The Recoflo system using resins which absorb acids but reject salts affords a means of processing relatively concentrated acid solutions with little or no dilution. The first commercial application of this acid recovery process was in 1978. Since then over 100 such systems have been installed for acid recovery from sulphuric acid aluminium anodising, steel pickling with H_2SO_4 and HCl, and various aluminium, brass, copper and nickel etchants using HNO_3. It is significant that since mineral acids are cheap, justification for the purchase of these units is usually based on the reduction in neutralisation costs rather than acid savings. The installation also provides consistent process solution composition and therefore consistent process performance. Such a process clearly fits BAT principles.

c) Biosorption

The application of biological treatment of waste liquors to either remove trace quantities of toxic metals or to biodegrade non-metallics is not new. For example the activated sludge process is currently employed by most of the UK's regional water companies for domestic sewage treatment. However in the last two decades other biomasses have been studied for the specific purpose of recovery/removal of certain metallic/non-metallic species from either waste or process solution. It is claimed that there are technical and economic advantages in using relatively cheap biological materials, many of which occur naturally or are generated as by-products from other industrial processes, eg brewery yeasts, and these are now being employed to immobilise toxic species from industrial effluents. Other biomasses such as bacteria, filamentous fungi etc are also being considered as potential scavengers for pollutants.

The technical concerns for utilisation of natural products for metals recovery/removal include efficiency of removal, biosorbent metal loading capacity, reuse of the biosorbent, contamination of the recovery system leading to septic conditions and the physical and chemical stability of the biomass. In applying biosorption as a process in an industrial environment a choice has to be made as to the use of living or dead microorganisms. The use of dead biomass appears more attractive because toxicity of metals in solution will have no effect on the biomass, nutrients and other growth requirements are not needed and maintenance of the purity of the culture is not a concern.

Despite development of several "commercial" products eg AMT-Bioclaim, Biofix (name discontinued due to potential trade mark infringement) etc and the appearance of at least one company to exploit this technology (AMT Inc), now defunct, no significant use of such materials is made for base and precious metal-bearing effluent treatment at this time. The problem is as these materials behave basically like ion exchangers sorbing metal ions by ion exchange mechanisms and therefore regenerable by similar means, they compete head on with other processes including ion exchange which may well perform better without any significant cost advantage. Interest in removal of radionuclides by the use of biomass however continues.

The use of biohydrometallurgy for contaminated groundwater treatment however is, as mentioned above, practised commercially at the zinc refinery of Budelco BV in Holland. In this process, the plant for which was commissioned in early 1992, the groundwater is treated in a reactor with sulphate reducing bacteria to precipitate heavy metal sulphides. These are filtered off and returned to the main zinc refining process. Water soluble sulphides are reoxidised with aerobic bacteria to elemental sulphur in order to produce a final effluent that meets discharge limits criteria. The plant is said to be performing excellently and is a notable first for biohydrometallurgy in this field. Such a process also must be considered as meeting BAT criteria.

d) <u>Solvent extraction</u>

In early years it did seem as if solvent extraction would play a very important role in effluent and waste processing but few commercial plants are in operation today using this technology for such a purpose. This is probably due to the fact that solvent extraction cannot readily achieve the low heavy metal discharge values of ion exchange or indeed some precipitation processes which can meet statutory discharge limits, although a large number of processes have been proposed. Thus solvent extraction at best achieves raffinates containing ≥10 mg/l compared with ion exchange at 0.1-1.0 mg/l or better. The replacement of metal contamination with organic contamination is also highly undesirable. However applications do exist and include treatment of steel pickling liquors, treatment of rayon spinning wastes, treatment of zinc galvanising pickles, treatment of printed circuit board etchants and treatment of plating bath wastes. These processes however are clearly niche market applications.

e) <u>Membrane processes</u>

Membrane technology is well suited to waste water treatment and thus application of membrane processes to effluent treatment would appear to offer considerable potential. Membrane processes can be divided up into several classes or types viz:-
Microfiltration; Ultrafiltration; Dialysis (including Diffusion Dialysis and Electrodialysis); Liquid Membranes; Reverse Osmosis.

Microfiltration involves the removal of particulate materials from 10 microns to 0.1 microns, while ultrafiltration removes dissolved materials of high molecular weight. These processes will not be considered further here. Reverse osmosis (RO) will separate materials less than 0.001 microns and thus salts and water can be separated by this technology. RO can therefore be considered for so-called "zero discharge" systems on plating bath dragouts with recycle of the salts to the plating bath and the water to the rinse. It can also be used alongside other membrane technologies.

Emulsion liquid membranes and supported liquid membranes both appear to offer variations on solvent extraction technology for effluent treatment but neither technique has yet been commercialised. Ion exchange membranes on the other hand are used commercially for waste stream and effluent treatment. Typical areas for application of electrodialysis are the recovery of nickel from plating operations, copper from electroless copper plating, chromium from electroplating rinse waters and recovery of precious metals from plating rinse waters.

The recovery of acids and alkalis from waste waters produced in the metal finishing industry, particularly from pickling processes is also an area where membrane processes can find application, particularly diffusion dialysis and electrodialysis. While theoretically electrodialysis can achieve such an objective, in practice it is quite difficult to do this due to performance limitations with conventional monopolar membranes. Currently acid recovery from spent pickle liquors is practised commercially using diffusion dialysis. This process uses anion exchange membranes which are tailored to allow passage of hydrogen ions and anions but to prevent the passage of other cations. Acid recoveries in excess of 80% can be expected and, coupled with the very low running cost of such systems, diffusion dialysis is therefore a serious competitor for the ion exchange resin-based Recoflo process.

Bipolar membranes on the other hand can achieve ready recovery of both acids and alkalis. These are composite membranes consisting of a cation exchange region, an anion exchange region and the interface between them. On passage of an electric current in the presence of a salt solution water splitting takes place at the interface region thus generation acids and alkalis. This process has been commercialised for regeneration of spent stainless steel pickle liquors at Washington Steel, Pennsylvania where there is a plant for processing $6x10^6$ l/y of pickle liquor. The process uses recycled KOH solution to precipitate metal values from the pickle liquor. The clear filtrate containing KF and KNO_3 is processed in a bipolar membrane cell to generate a mixed HF/HNO_3 stream and a KOH stream for neutralisation. Bipolar membrane cells can also be used for the regeneration of sodium sulphate in a rayon plant, in flue gas desulphurisation and in battery acid recovery from the sodium sulphate produced in desulphurisation of lead acid battery paste. The use of this bipolar membrane technology for HCl and NaOH production for regeneration of full scale ion exchange units purifying boiler feed water at an electric power generation station is also feasible.

f) Direct electrolytic recovery

Direct electrolytic recovery of metals from waste streams has the following advantages a) reduction in volume of wastes to be dumped, b) no need for regeneration systems and c) recycling of recovered metal. The disadvantage is the need to treat mostly dilute solutions which cause concentration polarisation in conventional parallel plate cells thus requiring the use of cells of special design to achieve high removal of metal at acceptable current densities and current efficiencies. Several such special cells have been developed in recent years and direct electrochemical recovery of metals from waste streams is carried out in the photographic industry for silver recovery, in the gold and silver plating industry, in the plating industry in general and from spent etchants.

2.6 Rehabilitation

The major features for hydrometallurgy in site rehabilitation are the insurance of no ongoing impurity-containing discharges from heaps and residues as indicated above. The treatment of contaminated land however is an area for application of hydrometallurgical techniques, notably leaching. It should be noted that both electrochemistry and biohydrometallurgy can be applied to this problem with advantage depending on the nature of the contamination. Electrochemistry in particular can be used as an in-situ process and appears to have much promise in this important environmental application.

3 Conclusions

In conclusion, hydrometallurgy does offer some advantages in cleaner processing, but is not without some significant environmental drawbacks, specifically in the handling of the residues produced. Outside the areas where it is now practised there are also frequently economic reasons why its use, in comparison to smelting methods, is not favoured.

To meet ideal best practice criteria, a hydrometallurgical process should:-

* Utilise non-toxic reagents
* Take into solution from the ore or concentrate all metals to leave an inert leach residue
* Purify the value metal liquor stream to the requirements of the refining stages
* Remove all other metals into preferably saleable by-products, and failing that into inert and disposable residues
* Allow for total management of process streams and after-use site rehabilitation

Undertaking all of the above at a cost comparable with alternative technologies is the target for the next generation of hydrometallurgical processes.

The use of hydrometallurgical techniques in pollution abatement is seen to be widespread and essential ranging from wet gas scrubbing through a wide variety of effluent treatment processes to waste solids treatment. The development of modern separation science and technology clearly provides clean options for effluent and waste processing by hydrometallurgical means thus yielding the opportunity of offsetting processing costs by recovery of valuable components for in-process recycle or for sale. Increasingly strict environmental regulations and discharge limits require ever more sophisticated means of treatment with what was the cheapest cost option of neutralisation and precipitation of heavy metals in a gypsum sludge no longer acceptable under BAT. Thus if new, clean technology is not directly available, end-of-pipe solutions must be sought and here techniques such as ion exchange, solvent extraction, membrane processes and direct electrolytic processes are in the forefront of such applications.

This paper therefore illustrates that whereas hydrometallurgy does offer some advantages in the search for cleaner processes, it is not without its limitations and is certainly not a solution for all problems. The major barriers to be overcome are the development of non-toxic lixiviants and, in particular, providing routes for iron in major quantities, and other heavy metals in minor amounts, to be produced preferably as usable by-products, or at the very least as readily and safely disposable materials. As indicated pressures for this to happen already exists with the strong lobby against the storage of untreated jarosite residues providing a good example of such pressure.

Finally there is no case to be made for a confrontational situation of pyrometallurgy versus hydrometallurgy. Best practise is surely the appropriate combination of high temperature and low temperature technologies, taking into account the natural characteristics of the feed materials, that achieves the desired result at minimum environmental impact and cost.

Environmental impact of increasing production of gold from hydrothermal resources

A. K. Haines
Minerals Technology, Gencor, Ltd., Johannesburg, South Africa

Abstract
As the Witwatersrand Basin of South Africa loses its dominance in world gold production, and production from hydrothermal resources continues to increase so there is an upward trend in the exploitation of the so-called refractory zones associated with these deposits.

These ores are pyritic in nature, often containing appreciable quantities of arsenic as arsenopyrite. Much of the gold is occluded in the crystal structure of the sulphide minerals, particularly the arsenopyrite. As a consequence, a much higher level of processing is required when compared to conventional "oxidised" ores, with greater pressure on the environment.

Together with this trend in gold production, significant advancements have been made in the development of alternative recovery technologies. From the environmentally hazardous Edwards roasters through the more sophisticated fluosolids system, to second generation hydrometallurgical processes of pressure and biological oxidation.

The paper demonstrates how technological developments have not only been driven by the normal quest for lower costs and higher recoveries, but in this case by consideration for plant safety and the environment. The modern plant is considered "environmentally safe" with respect to current discharges. However, will they be just as safe in the long term?
Keywords: Refractory gold, environment, bacterial oxidation, pressure oxidation.

1 Introduction

Western world production of gold has been increasing rapidly over the past few years, and growth is expected to average some 4% per year over the 10 year period '86 to '96. During this period the Witwatersrand Basin's share of production has decreased from 49% in 1986 to 34% in 1993, and is expected to decline further to some 30% by 1996/97.

Production from resources outside the Witwatersrand Basin is predominantly from so-called hydrothermal deposits. These deposits are generally pyritic in nature often containing appreciable quantities of arsenic as arsenopyrite. More often than not the gold is locked into the crystal structure of the sulphide mineral and is difficult to extract by cyanidation. Thus the term refractory.

Currently most of the gold produced from hydrothermal deposits emanates from the surface oxidised zones and at this stage they do not give rise to processing difficulties. As these zones become depleted and the mines move deeper into the sulphide zones so we are seeing a rapid increase in production from the refractory sulphides. Figure 1 compares the growth in production from refractory ores with overall growth in the western worlds newly mined gold. This figure illustrates an average growth rate from refractory gold of some three times that of overall rate of increase in production.

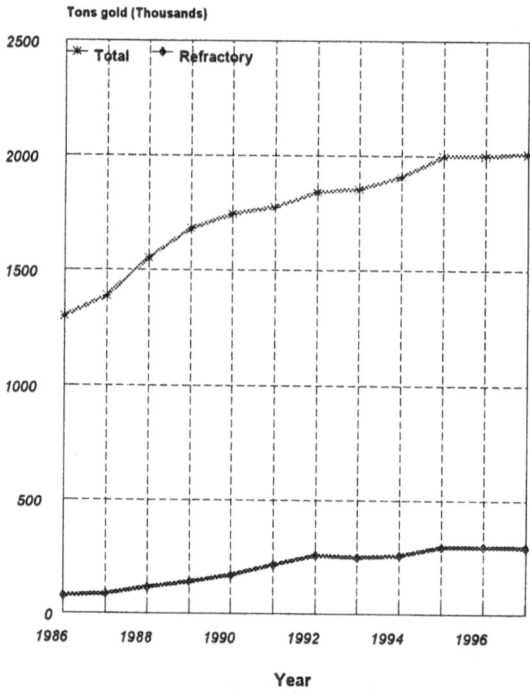

Fig. 1 Comparison of refractory versus total gold

2 Basic Mineralogy of Refractory Ores

The gold contained in the sulphide zones of a hydrothermal type deposit is generally associated with the minerals pyrite, arsenopyrite and pyrrhotite.

Figure 2 provides a good illustration of this association; the ore being that of the Sao Bento mine in Brazil. Although not always the case, the gold is occluded in the crystal structure of the sulphidic minerals with a tendency towards arsenopyrite.

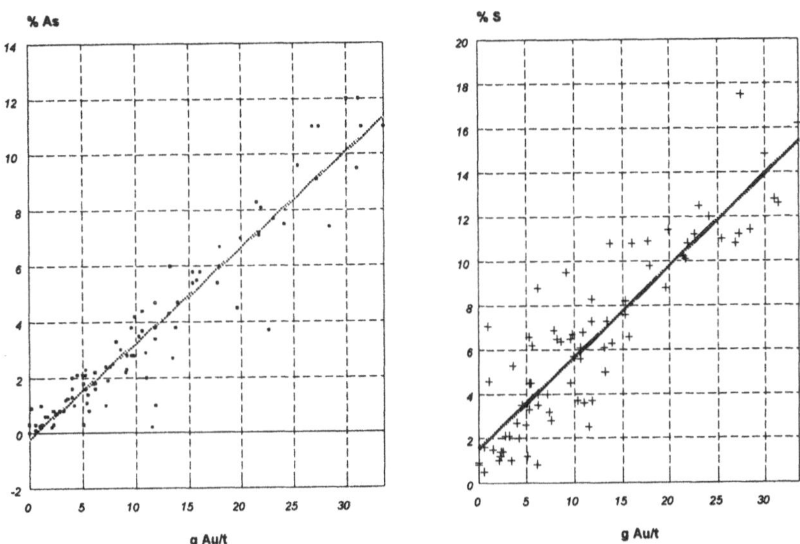

Fig. 2 Association of gold with arsenic and sulphur

Although it is dangerous to generalise another important phenomenon is that there appears to be a relationship between the size of the gold grains and its association with the individual minerals. Gold in pyrrhotite is of a large grain size and often simple grinding will liberate these grains. On the other end of the spectrum, gold associated with arsenopyrite is much finer and liberation more often than not requires oxidation of the arsenopyrite. Gold associated with the pyrite occupies an intermediate position and much of the gold associated with pyrite is liberated during initial grinding. Thus in general the higher the ratio of arsenic to gold the more difficult is the ore to process and the greater the level of oxidation required.

The actual size of the grain of gold occluded in arsenopyrite has been a subject of great debate. Early metallurgists tended to consider the gold as being in solid solution in the arsenopyrite [1].

This was due to the poor resolution of microscopes available at the time. However, with the advent of the Scanning Electron Microscope with a resolution of down to 0,1 to 0,2 µm, the general consensus became that the gold was not in solid solution but existed as discrete particles [2]. More recently, however, the thinking is back to the "solid solution theory" [3]. The development of ion probe micro-analysis techniques [3] has enabled the mineralogist to observe gold below the detection level of the SEM. Of vital importance in this work is the observation that the gold in solid solution or so-called invisible gold is strongly correlated with arsenic concentration which tends to verify the more practical findings that the higher the concentration of arsenic the more refractory the ore.

3 Recovery Techniques

Figure 3 is a generic flowsheet depicting the flour basic recovery steps, namely milling, flotation, oxidation, carbon-in-leach (CIL).

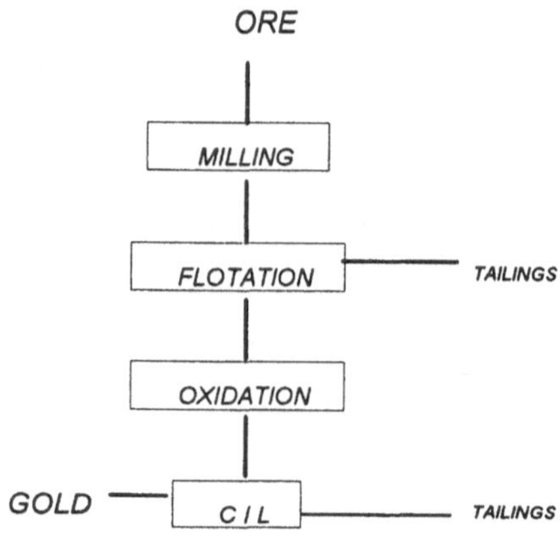

Fig. 3 Generic flowsheet of a refractory gold recovery plant

Generally it is the association of the gold with sulphide minerals which enables the use of flotation to upgrade the ore into a concentrate of much reduced volume. Recoveries of gold, although vary from ore to ore, are good. Table 1 shows numbers for operation with which Gencor has been associated.

Table 1. Flotation recovery of gold for various refractory gold plants

Fairview	90
Sao Bento	96
Wiluna	90
Ashanti	92
Harbour Lights	90

This is not always so, however, as is the case in areas of Nevada where oxidative treatment of the whole ore is required.

There are two basic means of oxidation thermal and chemical. Recovery of gold from the oxidised concentrate or ore is usually by way of conventional carbon-in-leach techniques.

3.1 **Thermal Oxidation**

Oxidative treatment of gold bearing pyrite and arsenopyrite concentrates, dates back to the early nineteen hundreds. These first generation processes involved roasting the concentrate in horizontal hearth coal fired roasters in which the concentrate was propelled through the roaster by way of slow moving horizontal rabble arms. Residence times were very long (14 hours) and initially roasting was performed in a single stage. However, as the industry progressed into the treatment of concentrates of higher and higher levels of arsenic so recoveries started to drop. This was considered due to the formation of ferric arsenate which tended to coat the discrete particles of gold contained in the concentrate, thus inhibiting cyanidation. Two stage roasting was introduced with the first stage carried out under conditions of limited oxidation which enabled the arsenic to be driven off as arsenic trioxide prior to oxidation.

The overall stoichiometry of the two stage thermal oxidation of arsenopyrite and pyrite is simply:

$$4FeS_2 + 11O_2 \rightarrow 2Fe_2O_3 + 8SO_2 \quad (1)$$

$$2FeAsS + 5O_2 \rightarrow Fe_2O_3 + As_2O_3 + SO_2 \quad (2)$$

The mechanism is however complex and involves a number of transformations.

Figure 4 which is reproduced from the classic paper by Swash and Ellis [2] shows very clearly the transformations that take place during the roast. In particular, it demonstrates the removal of the arsenic by the end of the reducing stage thus preventing the formation of ferric arsenate in the oxidation stage as mentioned previously.

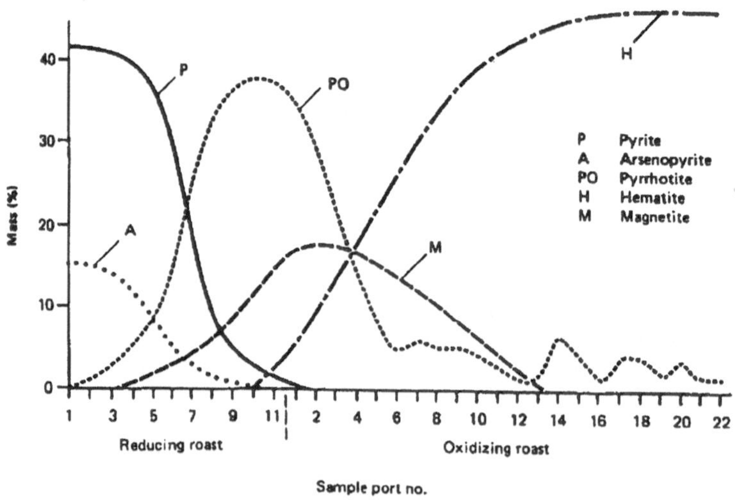

Fig 4. Mineralogical transformation in the Fairview roaster

The curves show almost complete conversion to hematite. In practice, however, some magnetite is allowed to remain; the so-called "chocolate roast". This is to enhance gold recoveries from the calcine.

The Edwards roaster is an extremely reliable machine and experience has demonstrated that high consistent recoveries of gold are achievable. However, the atmosphere within the roaster is difficult to control rendering it difficult to recover the arsenic and sulphur dioxide. Despite many attempts at using electrostatic precipitators and bag houses arsenic recovery from roaster flue gases has been a disaster, although a plant recently constructed at Ashanti would appear to be operating successfully.

Concern over the emission of arsenic led to the introduction of fluosolids roasting [4]. This system is well known for its close tolerance of temperature and reaction conditions and as a result recovery of arsenic has proved highly successful.

In principle, the fluosolid roasting system operates similarly to the Edwards roasters. Two stages are used where the concentrate comprises high levels of arsenopyrite. The arsenic is recovered as arsenic trioxide by way of a bag house and calcine carried over in the flue gas is usually recovered in hot electrostatic precipitators.

Generally the sulphur dioxide escapes into the atmosphere through a sufficiently high stack although it can of course be recovered as sulphuric acid should the economics dictate or the authorities require it.

Residence times are short and this characteristic has led to questions as to whether the system gives as good a gold recovery as the Edwards roaster particularly when treating high concentrations of arsenopyrite with concomitant high levels of invisible gold. Swash [2] in his analysis of the mechanism of roasting and its effect on gold recovery speaks of some form of "coalescence" of submicroscopic occluded gold taking place during the arsenic removal first stage and of these coalesced particles migrating to the surface of the pyrrhotite product of reaction.

In the fluosolids system with its good mass and heat transfer, oxidation is rapid particularly with the very fine (<38μm) particles. These particles tend to react very rapidly in the first stage and leave the reactor along with the off gases. It is conceivable that the reaction of these fine particles is too rapid for satisfactory coalescence to take place and that the coalescenced submicroscopic gold particles have not migrated to the surface where they would be exposed to subsequent cyanidation. Given that there is a tendency for the gold to be associated with the finer arsenopyrite, this could well explain the poorer recovery exhibited by the fluosolids system.

3.2 Chemical Oxidation

The need for a more environmentally acceptable process and a relatively poor recovery (92-94%) has led to the rapid development of chemical oxidation techniques as an alternative. Although there are a large number of approaches under development only two have achieved commercialisation; pressure oxidation and bacterial oxidation.

Pressure oxidation of pyritic ores was first successfully put into production at Homestake's McLaughlin mine in 1985. Since that time a further 6 have been commissioned and the technology can be considered well and truly proven. Of the seven plants only 2 treat concentrates [5,6] both of which use technology licensed from Sherritt.

The overall reaction stoichiometry is given as follows.

$$2FeS_2 + 7O_2 + 2H_2O \leftrightarrow 2FeSO_4 + 2H_2SO_4 \quad (3)$$

$$4FeAsS + 11O_2 + 2H_2O \leftrightarrow 4HAsO_2 + 4FeSO_4 \quad (4)$$

The iron and arsenic species are further oxidised to their higher oxidation states (III) and (V) respectively

Reaction takes place in an autoclave at a pressure of about 1600 Kpa and a temperature of 190°C using 95% pure oxygen as the oxidant. The products of reaction are separated in a counter current decantation washing system. The solution overflow containing soluble arsenic is neutralised to form basic ferric arsenate and calcium sulphate which is disposed of. The solid underflow is also neutralised and subjected to conventional cyanidation.

Pressure oxidation is no doubt a successful process. However, it is sophisticated and complex requiring a high level of skill to operate and maintain. Furthermore because of this inherent complexity capital costs are high. This is obvious when one looks at the conditions under which the plant has to operate; high temperature and pressure and relatively concentrated acid. On the other hand, the alternative bacterial oxidation process operates at basically atmospheric conditions of temperature and pressure with mild conditions of acidity and, although experience has shown that the process is not that simple that it can be ignored, a lower level of skill is required. The process is inherently suitable for remote locations. It is for this reason that a concerted effort is being directed into developing the bacterial process as an alternative to pressure oxidation. Obviously being less sophisticated the capital costs would appear lower.

The first commercial bacterial oxidation plant started up in October 1986. Since that time a further three have been bought into production and at the time of writing the huge Ashanti plant was being commissioned. All plants to date use Gencor's BIOX® technology.

The overall reaction stoichiometry is basically similar to pressure oxidation except that the operating conditions are vastly different. As previously mentioned, the bacterial process operates at near atmospheric conditions with a temperature of 40-45°c and a pH of 1,5. Oxygen is supplied at atmospheric pressure as air.

There have been over the past few years, a number of interesting developments in understanding the mechanism of bacterial attack on pyritic concentrates. It has been observed on a number of occasions that the extent of oxidation required to liberate an equivalent amount of gold is greater with pressure oxidation than bacterial oxidation particularly where the concentrate contains high levels of arsenic. Figure 5 is typical of the response of a concentrate which exhibits this phenomenon.

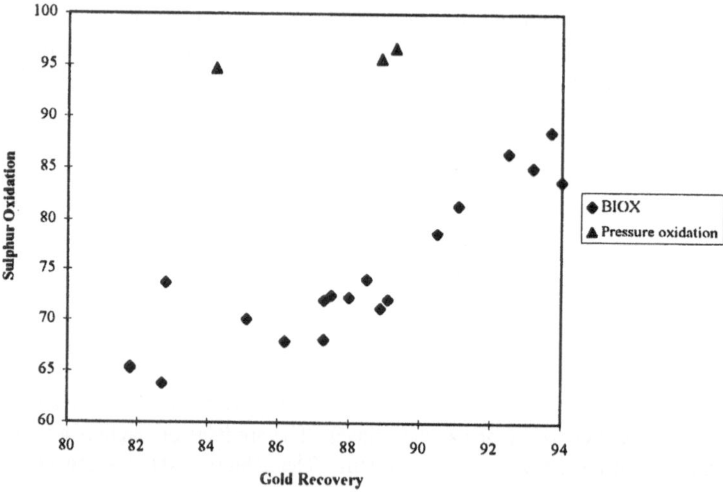

Fig 5 Effect of oxidation on gold recovery

Oxidation of pyrite and arsenopyrite under pressure tends predominantly to be indirect, by way of the ferrous/ferric couple. On the other hand bacterial oxidation comprises a dual mechanism both indirect via the ferrous/ferric couple and so called direct attack whereby the bacteria attach to the sulphide surface and facilitate electron transfer. It is probably this direct attack which leads to improved selectivity. This is because the bacteria tend to concentrate in regions of defect within the crystal structure of the sulphide. These defects occur along cracks, fissures and grain boundaries, but more particularly within zones of high arsenic content, and it is in these zones of course, that both the microscopic and sub-microscopic gold tends to concentrate. This phenomenon has been very clearly shown in the work by Claasens [7].

4 Environmental Impact

The rapid growth of the refractory gold industry particularly where arsenopyrite ores are involved is no doubt leading to increased environmental pressure. In fact, although through increased gold recovery the new generation of flowsheets can be justified economically, it is really the environmental problems that have been the driving force behind the development of alternatives to roasting.

The main problems associated with roasting are atmospheric pollution by the sulphur dioxide and arsenic trioxide. Generally sulphur dioxide has not been that large a problem and provided exhaust stacks are high enough such that ground level concentrations are below published TLV's the authorities have accepted direct discharge into the atmosphere. This situation is however changing and new roasting plants could well be required to include SO_2 recovery facilities.

Arsenic trioxide is a much greater problem, but is recoverable to acceptable levels by way of bag houses as has been demonstrated at Bogosu in Ghana and Campbell Red Lake in Canada [8]. However, for successful long term operation the roaster exit gas conditions have to be carefully controlled. As previously discussed, this is not always possible with Edwards Roasters and to date, operators of these plants have not been that successful in removing the arsenic in the same way as the two fluosolids roaster operators quoted above. Ashanti are admittedly using bag houses, but it is still somewhat early to judge the success of this operation.

Arsenic removal from roaster gases is by way of the trioxide, and the real problem is what to do with it. Whereas its production is on the increase, its consumption, because of its carcinogenic properties, is on the decrease. The extent of the problem can be guaged from the fact that world consumption is about 40 000 tons per annum whereas the potential production of just Sao Bento mine in Brazil and Ashanti is some 45 000 tons per year.

In the new generation of process, the problem has been shifted from air to water. No atmospheric pollutants are emitted, but both the arsenic and the sulphur dissolve on oxidation to produce a liquor phase containing arsenic (v) ferric and sulphate according to the general equations.

$$2FeAsS + 7O_2 + 2H_2O + H_2SO_4 \rightarrow Fe_2(SO_4)_3 + 2H_3AsO_4 \quad (5)$$

$$4FeS_2 + 15O_2 + 2H_2O \rightarrow 2Fe_2(SO_4)_3 + 2H_2SO_4 \quad (6)$$

The oxidised slurry passes through a counter-current decantation (CCD) circuit to separate and wash the acidic liquor from the solid product, containing the liberated metallic gold, prior to cyanidation. The soluble arsenic is subsequently removed from the diluted biological or pressure oxidation leach liquors, by precipitation as ferric arsenate, in a two stage neutralisation process employing limestone or lime to pH 4 to 5 and lime to pH 6 to 8. The overall chemistry of the neutralisation process is represented by the following equations:

Stage 1: Neutralisation to pH 4-5

$$Fe_2(SO_4)_3 + H_3AsO_4 + \tfrac{1}{2}H_2SO_4 + 3\tfrac{1}{2}CaCO_3 \rightarrow \underline{FeAsO_4} + \underline{Fe(OH)_3} + 3\tfrac{1}{2}\underline{CaSO_4} + 3\tfrac{1}{2}CO_2 + \tfrac{1}{2}H_2O) \quad (7)$$

or

$$Fe_2(SO_4)_3 + H_3AsO_4 + \tfrac{1}{2}H_2SO_4 + 3\tfrac{1}{2}Ca(OH)_2 \rightarrow \underline{FeAsO_4} + \underline{Fe(OH)_3} + 3\tfrac{1}{2}\underline{CaSO_4} + 4H_2O \quad (8)$$

Stage 2: Neutralisation to pH 6-8

$$H_2SO_4 + Ca(OH)_2 \rightarrow \underline{CaSO_4} + H_2O \quad (9)$$

Provided that conditions of precipitation are controlled to these levels the arsenic in solution is reduced to less than 0,1 ppm well below the limits set by the authorities throughout the world.

This method of co-precipitation of arsenic (v) with ferric hydroxide offers the most viable process for the removal of the arsenic from process waters. However there is still much controversy regarding optimum precipitation conditions, the structures and long term stability of these precipitates.

Robins [9,10] claims that arsenic, precipitated at pH values >2 and Fe/As mole ratios >1, only exists as hydrated ferric arsenate, $FeAsO_4 \times H_2O$, which is not suitable for long-term disposal as it slowly decomposes to form goethite and soluble arsenic at pH$\geq$ 2.2. Other investigators [11,12] have, however, found that the arsenic is present as an amorphous chemical compound of the type $FeAsO_4 \times Fe(OH)_3$, termed basic ferric arsenate, and furthermore is stable for extended time periods (<0.2ppm soluble arsenic after a test period of 3,7 years).

In our own work at Gencor Process Research we have also observed that the arsenic precipitates as basic ferric arsenate even at low Fe/As mole ratios ($\leq$ 2|1) [13].

In its assessment of the stability of the various residues produced by its plants, Gencor has adopted the USEPA Toxicity Characteristic Leaching Procedure (TCLP) as its standard test. This is due to its universal consistency with respect to experimental conditions and apparatus as well as its legal standing with the US Environmental Protection Agency. A modified procedure has also been developed which provides an indication of the effect of leach pH on the relative stability of the arsenic bearing residues. The TCLP has been designed to determine the mobility of contaminants in liquid solid and multiphase wastes [14]. The standard procedure comprises a buffered acetic acid leach under set conditions. The Gencor modified procedure utilises dilute sulphuric acid.

According to US EPA standards, a solid waste exhibits the characteristics of EPA toxicity with respect to arsenic, and thus cannot be land disposed, if the TCLP produces a liquid extract containing ≥ 5ppm arsenic [15].

Table 2 [13] provides results of stability testing of neutralisation residues from pilot plant testing of two of the BIOX® project that have gone into production.

Table 2: Results of the Testwork on BIOX® Neutralisation Pilot Plant Residues

Description	Final Operating Neutralisation pH	Storage Period (months)	Chemical Composition (%)			Species identified by IR Spectroscopy	Stability Testing	
			As(%)	Fe(%)	Fe/As mole ratio		Liquid Extract Description	As Value
		Fresh				$CaSO_4xH_2O$	TCLP	1,25
	10.8	-----				$Ca(OH)_2$	-----	-----
		1 week	4.8	15.6	4.4/1	$FeAsO_4xFe(OH)_3$	TCLP	0.26
		-----					-----	-----
Harbour Lights	(Tank B)	8 months					TCLP	<0.02
							pH3	0.07
							pH5	0.04
							pH7	0.07
							pH10	<0.02
	-5	3 months	5.8	15.0	3.4/1	$CaSO_4xH_2O$	TCLP	0.04
						$FeAsO_4xFe(OH)_3$	pH3	<0.02
							pH5	<0.02
	(Tank B)						pH7	0.05
							pH10	0.20
	-----	-----	-----	-----	-----	-----	-----	-----
	8.30	3 months	5.2	12.0	3/1	$CaSO_4xH_2O$	TCLP	0.13
						$FeAsO_4xFe(OH)_3$	pH3	0.13
Wiluna							pH5	0.06
	(Tank D)						pH7	0.14
							pH10	0.28
	-----	-----	-----	-----	-----	-----	-----	-----
	9.7	3 months	3.1	7.1	3/1	$CaSO_4xH_2O$	TCLP	0.24
						$Ca(OH)_2$	pH3	3.14
						$FeAsO_4xFe(OH)_3$	pH5	0.16
	(Tank D)					$Ca_3(AsO_4)_2$	pH7	0.09
							pH10	0.20

In Table 3 [13] results are given for TCLP tests on current neutralisation residues as well as material from the neutralisation residue disposal dam of Fairview mine.

Table 3: Results of the Testwork on Fairview BIOX® Neutralisation and CIP Residue Tailings

Description	Composition of Residues			Arsenic in TCLP Extracts (ppm)
	Fe (%)	As (%)	Fe/As Mole Ratio	
Neutralisation Residue:				
- Plant Tailings	13.9	5.4	3.4/1	2.62
- Slimes Dam	15.2	4.4	4.6/1	0.86
CIP Plant Tailings Residue	13.0	2.92	6.0/1	1.92

In all cases the arsenic values in the TCLP extracts are well below the limit of 5 ppm, although the plant values appear to be somewhat worse than those of the pilot plants.

The neutralisation residue is just one of the streams leaving the plant. Of equal importance is the CIP plant residue. In Table 3 TCLP results are also given for this stream at Fairview mine. Again the extract value is below the limit of 5 ppm.

Similarly acceptable figures have been observed at the Sao Bento mine which operates a pressure oxidation plant. These figures are shown separately in Table 4.

Table 4: TCLP test results for Sao Bento Mine

Sample Description	Operating Neutralisation pH	Storage Period after Sample Receipt	Fe/As-Mole Ratio in sample solid phase	As Conc in TCLP Extract (ppm)
Neutralisation thickener underflow	>10	20 months	29:1	0.02
	8	5 months	47:1	0.02
Blended CIL pulp and neutralisation slurry	8	2-3 weeks	10:1	0.02
	7	2-3 weeks	9:1	0.02
Samples taken from slimes dam in current discharge area	8	2-3 weeks	9:1	0.02
	7	2-3 weeks	10:1	0.02

In practice the neutralisation residue (thickener underflow) and the CIP tailings are blended prior to discharge to the dam.

Of vital importance, of course, is the actual concentration of arsenic in the dam return water. Table 5 below gives the average figures for the last six months for Fairview and Sao Bento.

Table 5: Arsenic concentrations in dam return water at Fairview and Sao Bento mine

MINE	PROCESS	AS IN DAM RETURN mg/l
Fairview	BIOX®	0.40
Sao Bento	Pressure Oxidation	0.31

The use of the method of the US EPA is somewhat controversial and it is accepted that the method does not necessarily provide a realistic indication of the long term stability of arsenic bearing residues. However, at this stage there is nothing else and it is consistent.

A number of modifications have been made to the procedure, but there is little standardisation amongst the various researchers. The Minerals Industry Research Organisation (MIRO) of the United Kingdom is investigating alternatives [16] but to date no suitable procedure has been identified. Consequently the EPA method continues to be used. The Gencor view is that despite the weaknesses of the method it remains acceptable given that actual slimes dam run off figures continue to be well below the legal limits in the countries in which the company operates.

5 Costs

Although the theme of the paper is environmental impact, it is considered worthwhile to include a short section on costs. Costs are obviously central to any technological development and generally increased environmental control means increased costs. Quoting costs is always difficult as the numbers are very site specific and generalisations can be misleading. Generic studies were undertaken in-house [17] two or three years ago. This study compared roasting, pressure oxidation and biological oxidation where, in order that one was comparing apples with apples, the roasting option included an acid plant, but also included revenue from the acid and the pressure oxidation plant included the oxygen plant. Table 6 sets out the results of this study in terms of rate of return. This approach is used because of the differing recovery figures and differing revenues resulting from by-product acid. A simple comparison of capital and operating costs is not really valid.

Table 6: Comparison of DCF rates of return

PLANT CAPACITY TPD	BIOX®	ROASTING	PRESSURE OX
150	1.00	0.59	0.53
300	1.15	0.86	0.75
600	1.23	1.08	0.93
1 200	1.27	1.22	1.35
2 400	1.29	1.30	1.50

These figures indicate that the biological process is the most viable at the smaller scale of operation. However, as concentrate production exceeds about 1500 tons per day, so pressure oxidation and the roasting become the preferred options.

The acid plant makes up some ... % of the capital cost of the roasting option. If one is allowed to discharge SO_2 directly into the atmosphere there is no doubt that roasting is the most economic option despite the loss of revenue from the acid. It is for this reason that Kanowa Belle in Australia is to install such a plant.

6 Conclusion

The exploitation of the refractory zones of hydrothermal type gold deposits is showing rapid growth worldwide. Treatment of these ores because of the association of the gold with pyrite and arsenopyrite is far more complex than for conventional ores and unfortunately the classic method of roasting is not exactly environmentally acceptable. However, concomitant with this growth has been the development of alternative treatment processes based on hydrometallurgy. These processes which involve chemical oxidation of the ore are not only economically superior to roasting, but overcome the environmental problems associated with the roasting process.

The monitoring and disposal of arsenic bearing effluents associated with this new generation of processes is a controversial subject. However, experience to date indicates that simple neutralisation with lime and precipitation of the arsenic as basic ferric arsenate is effective and the residues produced exhibit long term stability.

7 Acknowledgements

The contributions of my colleagues at Gencor are gratefully acknowledged.

8 References

1. Norwood, A.F.B. (1939) — Roasting and treatment of auriferous flotation concentrates. Proc. Aus.I.M.M. 116, 391-412.
2. Swash, P.M. and Ellis, P. (1986) — The roasting of arsenical gold ores: A mineralogical perspective. Gold 100: Proceedings of the International Conference on Gold, SAIMM, Johannesburg.
3. Chryssoulis, S.L and Cabric, L.J. (1990) — Significance of gold mineralogical balances in mineral processing. Trans. Inst. Min. Metall., 99, C1-C9.
4. Connel L. and Cross B. (1981) — Roasting process at the Giant Yellowknife mine. 20th Annual conference of Metallurgists. Canadian Institute for Mining and Metallurgy, Hamilton.
5. Da Silva, E.J. et al — Process selection, design commissioning and operation of the Sao Bento Mineracao refractory gold ore treatment complex. (Jounal title missing). World Gold '89, Reno, 322-332.
6. Frostiak, J. et al (1990) — The application of pressure oxidation at the Campbell Red Lake Mine. Randol Gold Forum. Scaw Valley.
7. Claasen, R. (1991) — The effect of mineralogy on the bacterial oxidation of refractory gold-bearing sulphides from a Barberton deposit. Colloquium on bacterial oxidation, SAIMM, Johannesburg.
8. Roberts, J.L. (1976) — Controlling roaster off-gases at Campbell Red Lake Mines. Canadian Mining Journal, 97, 54-56.
9. Robins, R.G. (1990) — The stability and solubility of ferric arsenate: An update. Proceedings EPD Congress, California, p93.
10. Robins, R.G. (1991) — Basic ferric arsenates - non existant Randol Gold Forum, Cairns, p197.
11. Krause, E. and Ettel, V.A. (1989) — Solubilities and stabilities of ferric arsenate compounds. Hydrometallurgy, 22, 311.
12. Papassioni, N., Stefanakis, M. and Kontopoulos, A. (1988) — Removal of arsenic from solutions by precipitation of ferric arsenates. Impurity Control and Disposal. Proceedings of the 15th Annual CIM Hydrometallurgy Meeting, Canada, 5-1.
13. Broadhurst, J.L. (1993) — The nature and stability of arsenic residues from the BIOX® process. International Conference and Workshop on Application of Biotechnology to the Mining Industry. Australian Mineral Foundation, Adelaide, 13.1-13.10.

14. US EPA, (1986) Toxicity characteristic leaching procedure. Appendix 1, Federal Register 51(216)

15. US EPA, (1990) Characteristics of EP toxicity. Paragraph 261.24, Federal Register 45(98).

16. Roberts, N.J. (1992) An evaluation of leach tests applied to metallurgical wastes. MIRO contract no. RC84.

17. Carter, A.J. (1991) Economic comparison of alternate methods for recovery of gold from refractory areas. Colloqium on Bacterial Oxidation, SAIMM, Johannesburg.

In-situ leaching recovery of copper: what's next?

J. B. Hiskey
Materials Science and Engineering Department, University of Arizona, Tucson, Arizona, U.S.A.

Abstract

The recovery of mineral and metal values by solution mining has been practiced for centuries. What appears at first glance to be a simple, empirical technique, is in actuality a very complicated process involving a large number of critical parameters that encompass several scientific and engineering disciplines. Hydrometallurgy, hydrology, geology and geochemistry, rock mechanics, chemistry, and environmental engineering and management are a few of the specialties utilized by modern operators.

Solution mining is conveniently divided into three main categories: heap leaching, dump leaching, and *in situ* leaching. *In situ* leaching (ISL) involves the application of a specific lixiviant to dissolve en masse minerals within the confines of a deposit or in very close proximity to its original geologic setting. Currently there is considerable interest in applying ISL technology to recover copper. Substantial research and engineering effort is being expended to recover copper from oxide ores. However, the greatest potential for copper ISL extraction remains with the deep seated deposits that would otherwise be left unmined by conventional methods.

This paper highlights some historical aspects of copper *in situ* leaching, and also reviews some commercial and experimental projects involving both oxide and sulfide deposits. In addition, large whole-core leaching experiments using a copper oxide ore will be described. This work has provided better understanding of the physical and chemical factors associated with the *in situ* leaching response.

Keywords: Copper, *in situ* leaching, solution mining

1 Introduction

Solution mining can be conveniently divided into three main categories: heap leaching, dump leaching, and *in situ* leaching. In general, solution mining is an extremely site specific endeavor. It is sensitive to such factors as deposit size and location, hydrologic conditions, rock type and ore mineralogy, and metal content. Hydrometallurgy when combined with solution mining techniques is acknowledged for its exceptional ability to extract metal values from dilute solid phases [1]. Numerous successful examples can be cited including: gold, copper and uranium. It is truly remarkable that gold can be extracted from oxidized ores assaying less than 1 g t^{-1} using cyanidation heap leaching. Equally impressive is the application of dump leaching to recovery copper from waste containing less that 0.2 % Cu. In fact, dump and heap leaching associated with modern solvent extraction and electrowinning currently account for 30 % of the total copper production in the United States [2]. Another example is the recovery of uranium from sandstone type deposits grading from less than 0.10 % U_3O_8 [3].

The historical fabric of copper solution mining is extremely rich: it is textured with many contributing threads. As noted in Table I, the history of copper solution mining can be traced back to ancient China.

The evolution of copper hydrometallurgy in China has been documented by Pu [4] and Lung [5] in considerable detail. Pharmacologists in the East Han Dynasty (25 - 220 A.D.) clearly understood the formation of gall (copper sulfate) springs and described in exquisite terms the precipitation of copper on iron from its solution. This early knowledge clearly contributed to the development of the gall-copper process practiced in China since 1086 A.D. The gall-process involved the collection of copper-bearing mine water and recovering the dissolved copper by means of cementation with iron. Near the end of the 13th century in the Yuan Dynasty it was observed [4]:

> *"Among all the prefectures and counties, those to the south in the lower reaches of the Yangtze River were the most prosperous in leached-copper production, where the total number of workshops exceeded 50".*

Another milestone in copper hydrometallurgy was the introduction of large-scale copper heap leaching at Rio Tinto, Spain as early as 1876 [6]. It is noteworthy that the methodology established at Rio Tinto is basically the same as that used today. Processing concepts such as solution management (sprinklers and leach/rest cycles) and heap construction techniques were introduced to maximize copper production.

A more recent major innovation that elevated the status of copper hydrometallurgy was the application of solvent extraction and electrowinning. High quality electron copper cathodes are presently competing for markets throughout the world. Solvent extraction technology nearly single-handedly allowed this to happen. Second generation copper extractants (salicylaldoximes and ketoximes) provided the impetus for efficient copper recovery from typical leach solutions at low pH. These reagents can be customized (with modifiers)

Table I Significant historical notes and events in copper solution mining

Period	Location	Comment
25-220 A.D.	China East Han Dynasty	Observations on natural oxidation and leaching of copper sulfide minerals and the formation of gall-springs. Precipitation of copper on iron.
1086	China Sung Dynasty	Gall-copper process
1254-1323	China Yuan Dynasty	Numerous copper leaching operations in lower Yangtze River region
1500	Hungary	Paracelsus the Great reported copper cementation in iron
1500	Lower Hartz	Basil Valentine in Currus Triumphaslis Antimonii discussed copper cementation at Rammelsberg
1556	Rio Tinto, Spain	Diego Delgado observed and reported copper cementation
1637	Peru	Alonzo Barba described the extraction of copper from mine water
1661	Rio Tinto, Span	Alvaro Alonso deGarfias granted a patent to recover copper from mine water
1876	Rio Tinto, Spain	Large-scale copper heap leaching
1888/1901	Butte, Montana	
1906	Bisbee, Arizona	Recovery of copper from underground mine water
1920	Cananea, Sonora, Mexico	*in situ* stope leaching
1930	Bingham Canyon Utah	Large scale dump leaching of open pit waste and recirculation of leach solutions.
1968	Miami, Arizona	Rancher's Bluebird mine first large-scale heap leaching and SX-EW of copper.

to achieve excellent Cu/Fe selectivities, fast kinetics, excellent pH behavior, and very good phase separation and crud generation characteristics [7-8]. The operating performance of copper electrowinning has also improved significantly over the last few years [9]. In fact, Magma Copper Co. secured COMEX and LME certification for their electrowon cathodes during 1993 [10].

As mentioned above, copper hydrometallurgy (i.e. leaching SX-EW) currently accounts for about one third of the total U.S. production of copper. Electrowon production was 510,296 t in 1992 and 541,182 t (preliminary estimate) in 1993 [11]. The growth in hydrometallurgical production has been the result of a number of factors such as rising costs associated with conventional mining, milling, smelting, and refining; tightening environmental restrictions; technological improvements, and declining ore grade. Ore grade depletion, a major problem associated with copper resources, is illustrated in Figure 1. This figure shows average copper yield (grade x recovery) for U.S. copper ores back to 1910. From 1940 to 1980, copper yield dropped 59 percent from 1.2 % to 0.49 % [12]. During the 1980's, copper yield increased slightly because of some high grading and improved mine planning. The situation with ore grade depletion is still critical, since metallurgical recoveries have increased noticeably over the years. With declining ore grade the opportunity for solution mining increases (i.e. hydrometallurgy is ideally suited to dilute mineral phases). This is especially true for large copper porphyry deposits which lack a sharp distinction between waste and ore. As such, copper oxide deposits along with sulfides contained in low- to medium-grade "halo" mineralization zones, deep-lying deposits, and low-grade waste dumps are all future targets for solution mining.

Hydrometallurgical extraction of copper by *in situ* techniques offers certain advantages over other methods and is being considered as a viable option for both oxide and sulfide deposits. The recognized advantages of *in situ* leaching over conventional mining and milling are listed in Table II. Disadvantages that could limit the application of this technology are also tabulated. The economic and operational advantages of *in situ* leaching can reduce the processing costs by as much as one-half that of conventional mining and processing [13]. Also of economic importance is that ore-reserves can be increased and augmented by this technology. For example, small low-grade or inaccessible ore-zones, gob and cave-fill areas, and dumps and tailing may become treatable by ISL techniques. However, the environmental advantages may ultimately be of more value. *In situ* solution mining does not have to cope with the disposal of enormous amounts of solid waste, the treatment of gaseous emissions,

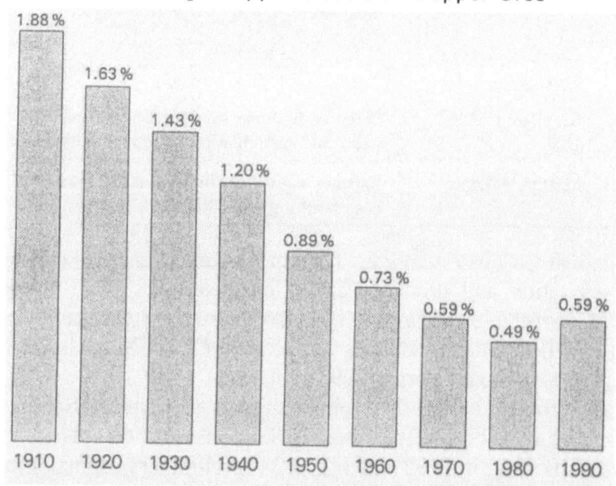

Figure 1. Average yield of copper ores mined in the United States during the 20th century. (Data for 1990 from 1991 USBM Minerals Yearbook)

or the stabilization and disposal of toxic dusts and sludges. Still and importantly, ground water quality must be protected as part of the operating strategy.

Table II Advantages and disadvantages of *in situ* leaching technology

Advantages	Disadvantages
1) combined capital and operating costs are normally lower than conventional mining and processing	1) physical and chemical constraints may limit metal recovery
2) energy and labor requirements are lower	2) laboratory testing and scale-up are difficult
3) start-up times faster	3) ground water and aquifer contamination
4) lessened environmental impact	
5) suitable for extracting metals from small, shallow deposits and especially suited to low-grade ores	
6) increase ore reserves	

2 *In situ* Leaching Principles

In situ leaching (ISL) is defined as chemical solution mining with a particular lixiviant to selectively dissolve metals of an ore deposit in or very near its original geologic position. As pointed out by Wadsworth [14], solution mining by its very nature requires interfacing various branches of science and technology. The multi-faceted character of *in situ* leaching is illustrated in Figure 2. Major areas contributing to, but not limited to, this technology are: hydrometallurgy, hydrology, chemistry, geology, environmental engineering, and process design and economics. Hydrometallurgy is critical to the understanding of leaching reactions and in developing the lixiviant chemistry. The hydrometallurgist is also essential to the process development related to metal recovery from solution. By far the greatest challenge facing the hydrometallurgist is accurate integration of chemical kinetic models, mass transfer models and hydrologic flow models to predict *in situ* leaching performance. Doing this successfully requires an understanding of the physical aspects of the deposit, mineralogy, reaction chemistry/kinetics, and hydrogeology.

2.1 Physical considerations

Figure 3 shows the principle *in situ* leaching approaches expressed in terms of their relative position to the water table. Extending these considerations, it is convenient to classify *in situ* leaching into basically three groups:

(i) no matrix modification
(ii) matrix modification (hydrofracturing, rubblizing, etc.)
(iii) low-grade remnants (abandoned pit walls, stopes and subsidence zones)

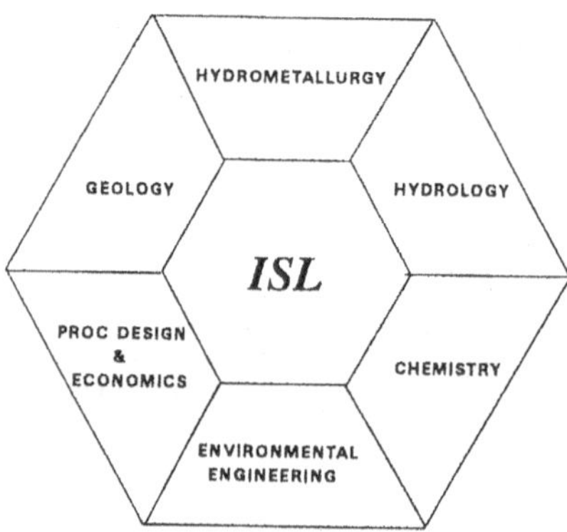

Figure 2. Technological interfaces in *in situ* leaching.

Under certain situations where the rock mass has sufficiently high permeability and porosity for adequate solution/mineral contact, there is no need for matrix modification. An excellent example is *in situ* leaching of uranium from permeable sandstone ore deposits. Uraniferous values are typically contained in ore zones confined between impermeable layers of clay. The mineralized zone may have porosities of 20 to 30% and permeabilities of 0.5 to 5.0 darcies [15]. Furthermore, it is a relatively simple task to derive hydrologic models for two-dimension flow in homogeneous isotropic deposits.

When required, permeability enhancement may be achieved using matrix modification methods. One scenario may involve drilling from the surface and performing short radius hydrofracturing. Occidental Minerals Corp. conducted hydraulic fracturing experiments at the Van Dyke copper oxide ore body near Miami, Arizona in 1976 [16]. These experiments were designed to create fluid flow paths and to improve the mineral/solution contact. The first experiment involved two wells 23 m apart and 305 m in depth. Both wells were hydrofracted. The second experiment utilized a 5-spot well pattern on 31 m centers. Hydrofracturing was initiated from each at a depth of 305 to 366 m [17].

Explosive fracturing from surface drill holes using a subsurface cavity to contain the explosive charge represents another method. Variations of these methods are: (1) drilling from existing underground entries and using either hydraulic fracturing or explosive fracturing to stimulate the deposit; and (2) drilling from new underground structures developed specifically for *in situ* leaching. One of the most ambitious efforts to rubblize a rock mass was that studied by the U. S. Atomic Energy Commission and Kennecott Copper Corp. during the mid 1960's [18]. The concept implicated the use of a large nuclear device to rubblize a deep-seated copper deposit. This approach would have surely created an enormous rubblized column; however,

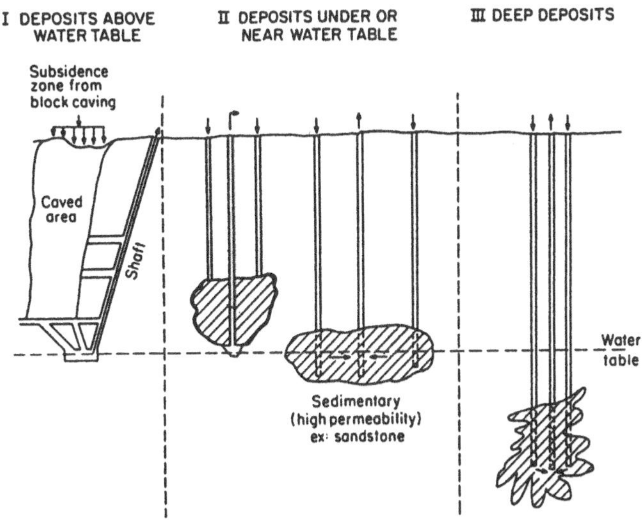

Figure 3. Conceptional *in situ* leaching systems in relation to water table level.

if used it would result in serious processing problems as pointed out by Malouf [19]. It was never pursued because of the following problems:

(i) *residual radioactive elements would be leached as well as copper*
(ii) *fractures in surrounding rock would be difficult to control*
(iii) *sequential and contiguous blasting and leaching would have serious limitations*

In situ leaching of remnant metal values associated with abandoned mine structures has been practiced for many years. In most instances there is considerable matrix modification caused by conventional mining activities. For example, block caving can produce a large subsidence zone with many fractures and considerable breakage. The caved area can be leached by applying solutions directly to the surface or injecting solutions at depth to contact specific mineralized zones. With old workings, solutions can be collected and recovered from the natural confines of an open pit mine, or from dams and sumps located in underground mines.

2.2 Mineralogical Factors

Porphyry copper deposits have been described in detailed terms by Titley and Hicks [20]. The term "porphyry copper" has several meanings, some based on ore genesis, some on mineralogical characteristics of alteration, and even others on economic and engineering criteria. Regardless, all porphyry copper deposits are characterized by disseminated low-grade copper mineralization. All are associated spatially with igneous rocks most commonly of the granodiorite and quartz monzonite composition.

2.2.1 Ore Minerals

Bateman [21] summarized the relative importance of copper minerals in a typical deposits. This information is listed in Table III.

Table III Copper minerals in zones of mineralization

Mineralized Zone	Mineral	Formula
Oxidized Zone (Secondary)	Native Copper	Cu
	Malachite	$Cu_2(OH)_2CO_3$
	Brochantite	$Cu_4(OH)_6SO_4$
	Antlerite	$Cu_3(OH)_4SO_4$
	Atacamite	$Cu_2(OH)_3Cl$
	Azurite	$Cu_3(OH)_2(CO_3)_2$
	Chrysocolla	$CuSiO_3 \cdot 2H_2O$
	Cyprite	Cu_2O
	Tenorite	CuO
Supergene Enrichment Zone (Secondary)	Chalcocite	Cu_2S
	Covellite	CuS
	Native Copper	Cu
Hypogene Zone (Primary)	Chalcopyrite	$CuFeS_2$
	Bornite	Cu_5FeS_4
	Enargite	Cu_3AsS_4
	Tetrahedrite	$Cu_{12}Sb_4S_{13}$
	Tennantite	$Cu_{12}As_4S_{13}$
	Covellite	CuS

Common copper minerals in the upper most portion (oxidized zone) of an ore body are classified as oxyanion minerals (i.e. hydroxides, carbonates, sulfates, chlorides, etc.). Naturally, these minerals are of special importance when *in situ* leaching is conducted in the oxidized zone of a deposit. Furthermore, similar phases may precipitate during the leaching of copper sulfides and sulfosalts, thus, limiting the concentration of copper in the lixiviant. Equilibrium constants are shown in Table IV for selected copper oxyanion minerals at 298 °K, taken from the WATEQ4F data base [22].

Table IV Reactions and equilibrium constants for selected copper oxyanion minerals

Species	Reaction	Log K
Malachite	$Cu_2(OH)_2CO_3+3H^+=2Cu^{2+}+2H_2O+HCO_3^-$	5.15
Azurite	$Cu_3(OH)_2(CO_3)_2+4H^+=3Cu^{2+}+2H_2O+2HCO_3^-$	3.75
Brochantite	$Cu_4(OH)_6SO_4+6H^+=4Cu^{2+}+6H_2O+SO_4^{2-}$	15.34
Atacanite	$Cu_2(OH)_3Cl+3H^+=2Cu^{2+}+3H_2O+Cl^-$	7.34
Tenorite	$CuO+2H^+=Cu^{2+}+H_2O$	7.62

The stability field of brochantite is presented in terms of pH versus log sulfate activity in Figure 4a. It is observed that cupric ion remains stable at pH values below approximately 4. The stability fields for azurite and malachite as a function of pH and log bicarbonate activity are shown in Figure 4b. Cupric ion shows stability upto about pH 5 for typical environments (e.g. P_{CO_2} = 30 Pa). As shown, precipitation of oxyanion salts of copper is unlikely under normal leaching conditions (i.e. low pH).

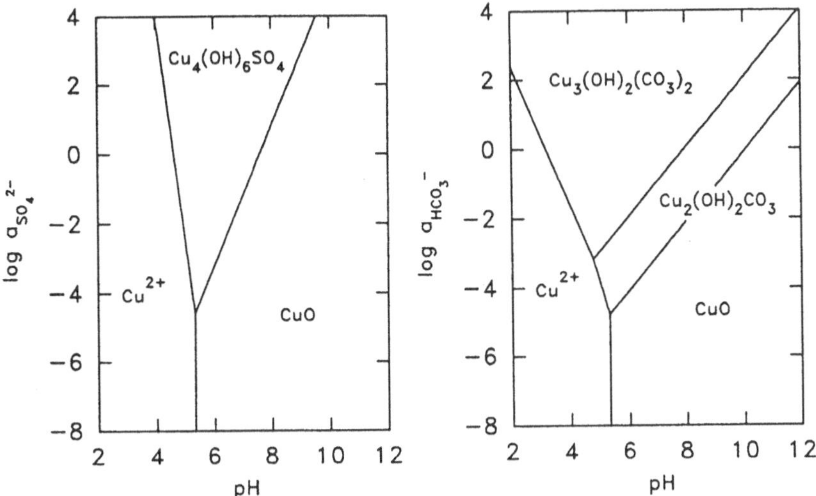

Figure 4a. Stability diagram of brochantite $a_{Cu^{2+}} = 10^{-3}$ M.

Figure 4b. Stability diagram for malachite and azurite at $a_{Cu^{2+}} = 10^{-3}$ M.

Supergene copper minerals result from the replacement of primary sulfides. Secondary enrichment reactions are shown for pyrite and chalcopyrite as follows:

Pyrite

$$14\ Cu^{2+} + 5FeS_2 + 12\ H_2O = 7Cu_2S + 5Fe^{2+} + 24H^+ + 3SO_4^{2-} \quad (1)$$

Chalcopyrite

$$11Cu^{2+} + 5CuFeS_2 + 8H_2O = 8\ Cu_2S + 5Fe^{2+} + 16H^+ + 2\ SO_4^{2-} \quad (2)$$

and

$$Cu^{2+} + CuFeS_2 = CuS + Fe^{2+} \quad (3)$$

Reactions (1) and (2) are redox sensitive, whereas reaction (3) ordinary metathesis [14]. The enrichment of $CuFeS_2$ to Cu_2S is represented by following cathodic reduction process

$$3\ Cu^{2+} + CuFeS_2 + 4e = 2\ Cu_2S + Fe^{2+} \qquad (4)$$

and the complementary anodic process

$$Cu_2S + 4H_2O = 2Cu^{2+} + 8H^+ + SO_4^{2-} + 10e \qquad (5)$$

The E_{redox} for reaction (2) is 0.33 V at $a_{Cu^{2+}} = a_{Fe^{2+}} = 10^{-2}$ M, $a_{SO_4^{2-}} = 10^{-1}$ M and pH 2. The overall secondary enrichment process occurs under reducing conditions immediately below the water table (i.e. $P_{O_2} = 10^{-48}$ Pa).

The dominant primary copper sulfides are chalcopyrite and bornite. However, the most abundant metal sulfide in the hypogene zone is pyrite. Pyrite is an important attendant mineral and its aqueous oxidation is critical to the overall success of copper sulfide leaching. In disseminated deposits, primary sulfide minerals appear as discrete grains throughout the host rock matrix and as fillings in micro-fractures and -seams. Chalcopyrite and bornite may display mutual boundaries or may exist as lattice-type exsolution textures.

2.2.2 Gangue Minerals

Gangue minerals are normally divided into three broad classes: primary, secondary and sulfide. Primary rock forming minerals encompass many complex silicates. Almost all secondary gangue minerals are the result of the alternation of primary rock forming minerals. Hydrothermal alteration, weathering, and metamorphism are some of the mechanism responsible for forming secondary minerals. In addition, some may form as products associated with ore mineralization. Sulfide gangue minerals may consist of pyrite, pyrrhotite, arsenopyrite, etc. Knowledge of chemical and physical parameters associated with gangue minerals is extremely important in developing *in situ* leaching. Of special importance are:

Chemical

(i) acid consuming/generating character
(ii) oxidant consuming character
(iii) ion exchange capacity
(iv) solid-liquid equilibria

Physical

(i) fracture type (both major and micro)
(ii) decrepitation behavior
(iii) ore mineral distribution

Clearly, the acid consuming nature of the host rock is of critical importance. In oxide copper deposits, acid is required from some external source to sustain the leaching reaction. If acid consumption is excessive, the economics of *in situ* leaching could be unfavorable. In the case of copper sulfides, hydrogen ion is necessary in maintaining a stable environment for ferric ion. As such, the sulfide gangue mineral, pyrite is important as a source of sulfuric acid.

3 Chemical considerations

Copper leaching takes place under conditions varying from powerfully oxidizing to moderately reducing and from strong acidic to strongly alkaline, depending on the kind of ore body to be leached, ore mineralogy, and the composition of the host rocks. Therefore, careful consideration of pH and oxidation potential is required to establish the conditions for solubilizing copper. The work of Garrels and Christ [23] and Peters [24] is noteworthy in the application of Eh-pH or Pourbaix diagrams in geochemical and hydrometallurgical systems. At present, there are a number of computer codes capable of calculating the chemical equilibria and constructing multicomponent Eh-pH diagrams [25-27]. Copper leaching can be divided into basically two chemical categories: oxide minerals and sulfide minerals.

3.1 Oxide minerals

Some oxidized copper minerals are water soluble. Simple dissolution of the mineral chalcanthite ($CuSO_4{\cdot}5H_2O$) is illustrated by the following reaction

$$CuSO_4{\cdot}5H_2O + H_2O_4 \rightleftarrows Cu^{2+} + SO_4^{2-} + 6H_2O \tag{6}$$

Hydrated copper sulfates are found in the oxidized near-surface zone as well defined crystals, stalactites, and crusts on mine surfaces. Except for native copper, the oxidized-zone minerals shown in Table III dissolve under acidic conditions; the rate of dissolution, however, varies. The following reactions represent the simple acid dissolution of malachite and tenorite.

$$Cu_2(OH)_2\ CO_3 + 4H^+ \rightleftarrows 2Cu^{2+} + CO_2 + 3H_2O \tag{7}$$

and

$$CuO + 2H^+ \rightleftarrows Cu^{2+} + H_2O \tag{8}$$

All oxidized-zone mineral, except chrysocolla and cuprite dissolve completely and at relatively fast rates with sulfuric-acid lixiviant. Chrysocolla leaching proceeds according to the following reaction

$$CuSiO_3{\cdot}2H_2O + 2H^+ \rightleftarrows Cu^{2+} + SiO_2{\cdot}nH_2O + (3\text{-}n)H_2O \tag{9}$$

This reaction is limited by the diffusion of H^+ and Cu^{2+} through a porous product layer of $SiO_2{\cdot}nH_2O$. Cuprite on the other hand, dissolves according to a disproportionation reaction under non-oxidizing conditions to yield half cupric ion and half metallic copper as follows:

$$Cu_2O + 2H^+ \rightleftarrows Cu^{2+} + Cu + H_2O \tag{10}$$

With cuprite, therefore, the leaching would be only about 50 % complete with simple acidic lixiviants. An oxidant would be required to dissolve the metallic copper produced by disproportionation.

Oxidized copper ores are especially suited to *in situ* leaching because of the fast kinetics associated with these minerals. As a result, processing strategies that involve leaching cycles ranging from a few weeks to several months are envisioned.

3.2 Sulfide minerals

Copper sulfide minerals are semi-conductors and dissolution generally occurs by electrochemical mechanism (Hiskey and Wadsworth) [28]. Table V provides resistivity data for selected copper sulfides.

Table V Resistivies of common copper sulfide minerals

Mineral	Formula	ρ(ohm-m)	Usual conductor type
Bornite	Cu_5FeS_4	1.6 to 6000 x 10^{-6}	p
Chalcocite	Cu_2S	80 to 100 x 10^{-6}	p
Chalcopyrite	$CuFeS_2$	150 to 9000 x 10^{-6}	n
Covellite	CuS	0.3 to 83 x 10^{-6}	metallic
Enargite	Cu_3AsS_4	0.2 to 40 x 10^{-3}	-
Tetrahedrite	$Cu_{12}Sb_4S_{13}$	0.3 to 30,000	-

Copper sulfides generally require the presence of an oxidant to achieve dissolution. One exception is the cyanide leaching of chalcocite (i.e. $Cu_2S + 4CN^- \rightarrow 2Cu(CN)_2^- + S^{2-}$). In acid solution, chalcocite reacts according to a two-stage mechanism, shown below with oxygen as the oxidant.

$$Cu_2S + 2H^+ + \tfrac{1}{2}O_2 = Cu^{2+} + CuS + H_2O \qquad \text{1st stage} \tag{11}$$

and

$$CuS + 2H^+ + \tfrac{1}{2}O_2 = Cu^{2+} + S^\circ + H_2O \qquad \text{2nd stage} \tag{12}$$

A detailed analysis of the oxidation of mechanism for Cu_2S to CuS indicates the sequential formation of a series of Cu_xS compounds Koch and McIntyre [29].

$$Cu_2S \rightarrow Cu_{1.93}S \rightarrow Cu_{1.83}S \rightarrow Cu_{1.67}S \rightarrow Cu_{1.4}S \rightarrow CuS \tag{13}$$

The second stage reaction involves the oxidation of CuS to cupric ion and elemental sulfur. The elemental sulfur produced is thermodynamically unstable; however, oxidation of sulfur to sulfate is limited by slow kinetics under ambient conditions. Elemental sulfur can coat the unreacted sulfide particle and retard the leaching reaction. Beside the complex mechanisms associated with sulfide minerals, leaching can be limited by the solubility of oxygen. At 298 K the solubility of oxygen in pure water in equilibrium with air is only 2.6×10^{-4} M ($\sim$8 ppm). As indicated by Hiskey and Bhappu [30], there is an associated decrease in oxygen solubility in the high ionic strength solution produced by the recycle of lixiviant. A 50% decrease in O_2 solubility results from ionic strengths equal to 1.5.

Ferric ion can exist, however, at activities much greater than that of dissolved oxygen and is kinetically more important in the oxidation of copper-sulfide minerals under normal conditions. Nonetheless, oxygen remains important in the oxidation of ferrous iron to ferric.

Finally, primary sulfide minerals, especially chalcopyrite ($CuFeS_2$), are of major economic importance. Chalcopyrite is the most refractory copper sulfide and reacts extremely slow under typical leaching conditions. The dissolution mechanism of chalcopyrite has been the subject of considerable study and intense debate [31-33]. It is beyond the scope of this general review to address the various sides and issues centered around the controversy.

It is, however, recognized that $CuFeS_2$ dissolves according to the following principle reactions:

$$CuFeS_2 + 16Fe^{3+} + 8H_2O = Cu^{2+} + 17Fe^{2+} + 2SO_4^{2-} + 16H^+ \quad (14)$$

$$CuFeS_2 + \tfrac{15}{4}O_2 + \tfrac{3}{2}H_2O = Cu^{2+} + \tfrac{1}{3}Fe_3(SO_4)_2(OH)_5 \cdot 2H_2O + \tfrac{1}{3}S + SO_4^{2-} \quad (15)$$

Reaction (14) is typical for dump leaching or *in situ* leaching under unsaturated conditions. While reaction (15) is reported by Braun et. al. [34] for leaching a saturated/flooded deposit by introducing oxygen under hydrostatic pressure.

Future *in situ* leaching targets containing primary mineralization may be deep-seated and positioned well below existing water tables. There are several issues worthy of discussion as relates to these candidate deposits. First, at high hydrostatic pressures, oxygen activities will exceed ferric ion activities. Second, the impact of extreme hydrostatic pressure and high oxygen activities on microbiological behavior.

The *in situ* leaching of deep seated deposits well below the water table may have a kinetic advantage because of the increased solubility of oxygen. At large hydrostatic pressures the leaching chemistry may change from ferric ion control to oxygen control. This is demonstrated in Figure 5 where the fraction (F) of chalcopyrite oxidation by dissolved oxygen is plotted as a function of hydrostatic head for various ferric ion concentrations. For pure oxygen, the system becomes oxygen dominant at between 30 and 70 meters depending on the ferric ion concentration. As shown, air can be just as competitive but at greater depths (F = 0.5 at 180 and 380 m for 1 and 2 g/l Fe^{3+}, respectively). Therefore, for deep seated deposits, well below the water table, air can be used effectively for the oxidation of copper sulfides.

The role and physiological involvement of microbes in the leaching of metal from sulfide deposits has been recognized for many years Colmer & Hinkle [35], Beck [36]. Microorganisms that function in this setting are classified as aerobic

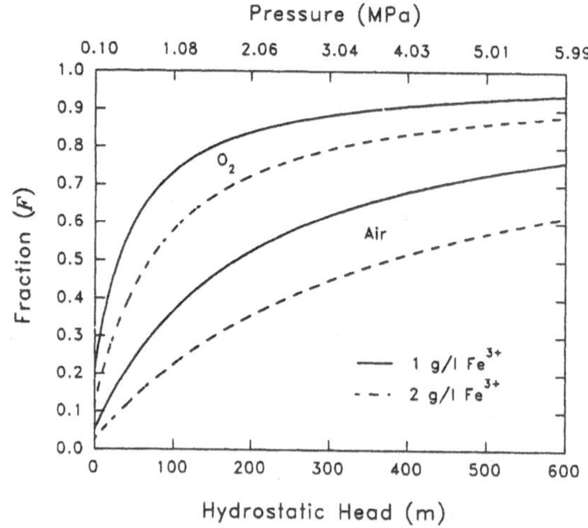

Figure 5. Fraction of $CuFeS_2$ oxide by dissolved oxygen as a function of hydrostatic head for air and pure oxygen at different ferric ion concentrations.

chemoanthotrophic. The organisms *Thiobacillus thiooxidans* and *Thiobacillus ferrooxidans* are capable of assisting the oxidation of ferrous ion and the reduced forms of inorganic sulfur. The exact mechanistic role that bacteria play in these reactions is still a matter of debate. Regardless, there is overwhelming evidence that they promote reactions which are important to the leaching of metal sulfides. *T. ferrooxidans* are virtually ubiquitous. They are especially indigenous whenever an acid environment is maintained in the presence of sulfide minerals. Bosecker, Torma, and Brierley [37] investigated the influence of hydrostatic pressure on the physiological activity of *T. ferrooxidans* during the leaching of chalcopyrite. Torma [38] had previously shown that these organisms were barotolerant when exposed to hydrostatic pressure of 15.2 MPa for 8h. However, deep *in situ* leaching of primary copper sulfides would necessitate the injection of oxygen at high pressures. In the later work, biological activity was determined by measuring the respiration rate at 690 kPa (air containing 1 % CO_2) as a function of time. The respiration rate of the unpressurized sample was 2.2 μl O_2 min^{-1} per gram of $CuFeS_2$ and at 690 kPa ranged from 1.7 to 2.0 μl min^{-1} g^{-1}. It was concluded that *T. ferrooxidans* are not greatly affected by overpressures of oxygen under these conditions. It must be recognized that deep *in situ* leaching will likely involve hydrostatic pressure well in excess of 690 kPa. The transition point in Figure 5 is approximately 1800 kPa for air and 1 g/l Fe^{3+}.

4 Copper *in situ* leaching - experience

To answer the question "Copper *In Situ* Leaching - What's Next?" one must briefly examine both past and present efforts to develop and commercialize this technology. Surely there could have been efforts to apply this technology that were performed anonymously. Notwithstanding, Ahlness and Pojar [39] compiled case histories of *in situ* copper leaching projects in the United States. A total of 10 commercial operations and 14 experimental/developmental projects was documented. Table VI list the commercial *in situ* mining projects that have operated since 1970.

Greenwood [40] reported that in Cananea, Mexico copper was recovered by leaching of mined out and caved underground stopes as early as 1920. About the same time (1922), the Ohio Copper Co. mine in Bingham Canyon commenced *in situ* leaching of old block caved areas [39]. This operation achieved a solution flow rate

Table VI Commercial *in situ* mining activities in the United States since 1970

Project	Dates	Principal Copper Minerals	Average Grade Percentage Copper	Preparation
Big Mike (NV) Ranchers	1973-74 1978-79	Cuprite, tenorite, and chalcopyrite	1.18	Blasted pit walls and bottom (terraced surface)
Old Reliable (AZ) Ranchers	1972-74 1979-81	Chalcocite, chalcopyrite, malachite, and chrysocolla	0.84	Blasted older underground workings (terraced surface)
Zonia (AZ) McAlester Fuel	1973-75	Chrysocolla	0.20	Blasted pit walls and bottom
Copper Queen Branch (AZ) Phelps Dodge	1975-	Chalcocite	0.29	None; Open pit underground workings
Inspiration (AZ) Inspiration Consolidated	1967-74	Azurite, malachite, and chrysocolla	0.50	None; Block-caved stopes
Miami (AZ) Magma	1942-	Chalcocite	NA	Surface prepared above block-caved areas
Lakeshore (AZ) Noranda	1983-	Chrysocolla, brochantite, and cuprite	0.45	None; Block-caved areas
San Manuel Magma	1987-	Chrysocolla	NA	None; Block-caved areas

of 273 to 318 $m^3\ h^{-1}$ (1200 to 1400 gpm) by November of 1923. Solution was applied at the surface via 46 m long wooden launders with regularly spaced 51 mm holes in the sides. The distribution launder was moved periodically in a regular pattern. Solution was collected in a underground tunnel approximately 90 m below the bottom of the caved zone. Pregnant leach solution (PLS) grades averaged about 2 gpl copper.

The Miami mine is another *in situ* leaching activity located in old mine workings. ISL copper recovery techniques have been applied there since 1942 [41]. The total copper production by *in situ* leaching is estimated to be 187,000 t. The operation is currently producing approximately 13.6 to 15.9 t d^{-1} of SX-EW copper.

Of the commercial operations listed in Table VI, three involved blasting to fragment the ore prior to leaching. These include: the Big Mike mine in Nevada and the Old Reliable and Zonia mines in Arizona. The blasts produced a rock size ranging from 20-36 cm. In each case, solutions were applied to the surface of the leaching area and allowed to percolate through the fractured ore. PLS was collected either from recovery wells or as in the case of the Old Reliable at the base of the rubblized ore body. Ward [42] indicated that the Old Reliable was designed to handle 227 m^3 h^{-1} (1000 gpm) of leach solution at an application rate of .015 m h^{-1} (0.006 gpm/ft^2). Design capacity was 9 t of copper per day. An estimated 20 % of the total copper was extracted from the Old Reliable deposit by *in situ* leaching. Even at this level of overall copper extraction the project was a financial success, however, the operation suffered from some technical shortcomings. The blast did not achieve the degree and uniformity of rubblization as planned. As a result, the deposit developed a perched water table that limited access of leach solution to much of the contained mineral value.

Several significant commercial copper *in situ* operations were developed after the report by Ahlness and Pojar [39], these include the operations at the San Manuel Division of Magma Copper Co. and the Lakeshore mine of Cyprus Mineral Co. *In situ* leaching is conducted at Lakeshore in the block caved stopes of the oxide ore body and in virgin ground using holes drilled from old underground workings to inject leaching solution Kline, Behnke and Musgrove [43]. An underground solution collection system was installed to accept a solution flow rate of (3000 gpm). It was reported that 590 t per month were produced by 58 employees. The technical problems associated with this operation were categorized as follows:

(i) *ore zone characterization*
(ii) *well completion*
(iii) *solution control*
(iv) *equipment failures*

Efforts to correct these shortcomings were successful and the operation continues to produce copper from solution applied to the caved areas via surface injection. The underground injection system was abandoned several years ago.

The *in situ* copper mining activity at Magma's San Manuel mine is depicted in Figure 6. This shows a well field where solution is injected into the deep oxide ore body from benches on the open pit mine [10]. A uni-well system with parallel holes is employed. This offers operational flexibility since wells can be used either in injection or production mode. At present there are several hundred wells handling a total flow of 568 m^3 h^{-1}. Copper production from this operation is about 910 t per

Figure 6. *In situ* leaching well field and located on a bench of the San Manuel open pit mine.

month. An accelerated *in situ* leaching program is proposed as the open pit mining operation is concluded. Additional wells will increase the capacity to 1820 $m^3\ h^{-1}$ and yield a target of 45,350 t of cathode copper per year.

Experimental *in situ* leaching has been attempted at a number of different sites. Probably the most important of these are the Kennecott Safford project and the ASARCO/Freeport Santa Cruz field experiment. The Safford *in situ* leaching test program was conducted by Kennecott starting in the mid-1970's. In fact, it was originally studied as a site for nuclear rubblization. This approach was dropped and a conventional ISL program initiated. Many aspects of solution mining were investigated during the course of this investigation, with extensive effort devoted to leaching chemistry and lixiviant development. A number of relevant patents have been assigned to cover this work. Hsueh et al. [44] described a method and apparatus for producing a two-phase ammoniacal leach solution containing oxygen bubbles which is forced under high pressure through an injection hole penetrating a deep-seated deposit. Huff and Huska [45] characterized a procedure to oxidize ferrous iron to ferric in the pregnant lixiviant during a gas-lift operation in the wellbore. Hsueh and co-workers [46] discussed the use of surfactants for enhancing the formation of fine bubbles and to minimize coalescence through the ore body. This procedure allowed the passage of a two-phase lixiviant through the natural hairline fracture openings. Without the surfactant gas blockage was a problem. Several other patents were issued covering the injection and recirculation of the two-phase lixiviant [47-48].

Likely the most advanced and extensive test program to date is the field experiment currently being conducted at a site near Casa Grande, Arizona. The Santa Cruz ore body is owned jointly by ASARCO Inc. and Freeport-McMoRan Gold Co. It is a moderately deep copper oxide deposit. The mineralized zone contains approximately 97 million t of ore averaging 0.7 % acid soluble copper. Chrysocolla ($CuO \cdot SiO_2 \cdot 2H_2O$) and atacamite ($Cu_2(OH)_3Cl$) are the predominant copper oxide minerals hosted in granodiorite porphyry and quartz monzonite, respectively. The bottom of the deposit lies at an average depth of 670 m, with an average thickness of about 100 m. True *in situ* leaching approach is being considered for this deposit [49]. The U.S. Bureau of Mines has performed considerable laboratory work in characterizing the physical, mineralogical, and chemical features of this ore body [50-51]. Field research is well advanced and includes a single 5-spot pattern. Hydrologic measurements and tracer tests have been completed. At the present time, the project is waiting approval (aquifer protection permit) to start injecting lixiviant for an actual leaching test.

Another test program of importance is that conducted at the Mammoth Mine of Gunpowder Copper Limited in Queensland, Australia [52]. The Gunpowder project consisted of leaching both broken ore (below an old open pit) and fractured ore in an test stope. The major copper minerals were chalcocite and chalcopyrite, with minor amounts of bornite, covellite and digenite. Chalcocite dominated the upper portions of the ore body. Solution was applied to the surface of the broken ore and allowed to percolate to a collection point in the old underground workings. In the test stope, solution was applied via sprays to the top of the ore in the stope. PLS was collected, sampled and allowed to commingle with PLS from the broken ore section. The status of this project is unknown at this time.

5 Large whole-core experiments

In order to minimize the costs associated with field testing and scale-up, it is critical to conduct properly designed laboratory experiments to evaluate *in situ* leaching parameters. The *in situ* leaching of broken ore under unsaturated conditions is similar to dump and heap leaching and can be evaluated conveniently by column leaching techniques. However, when a lixiviant flows through the fractures and connected pore spaces of an undisturbed deposit (true *in situ* leaching) experimental simulation becomes more difficult. Pressure controls the flow of lixiviant and lixiviant concentration controls mineral dissolution. These are the primary design parameters for an experimental system. In addition, the experimental design should simulate hydrostatic pressures for down hole conditions. A secondary design parameter is the flow orientation used in the simulator. One system utilizes axial flow conditions. The other experimental design is based on radial flow configuration.

Two groups have been active in conducting experiments to evaluate *in situ* leaching parameters and to gain phenomenological data. The Bureau of Mines (Twin-Cities) and the University of Arizona Copper Research Center have developed laboratory research programs using large whole core samples. Much of the USBM work is directed towards the Santa Cruz project of ASARCO and Freeport. Paulson et al. [53] has described in detail the experimental design and test results of axial flow experiments using ~0.1 m diameter core. Others have discussed the geochemical effects of long term recycle of lixiviant and secondary salt precipitation on the hydrology of the system [54]. Hiskey and Oner [55-56] have conducted long term experiments using Santa Cruz ore and other oxide ore samples under radial flow conditions. This work has yielded considerable data on copper extraction from fracture-controlled copper oxide ores. In addition, it has demonstrated the application of polymers to control fluid flow [57] and utilization of CT Scans to visualize copper distribution and leaching path formation [58].

It is necessary to maintain uniform solution flow into the ore formation for successful *in situ* leaching. The natural rock porosity and permeability are frequently low for many deposits. Furthermore, porosity and permeability are sensitive to:

(i) *ore mineral dissolution*
(ii) *gangue mineral dissolution*
(iii) *liberation, transport and compaction of fine particles*
(iv) *clay swelling*
(v) *precipitation of secondary mineral phases.*

The leaching of large whole-core of copper oxide with 5 g/l H_2SO_4 ad 138 kPa (ΔP = 13.8 kPa from injection to production side of sample) resulted in severe channelling after about 100 d and only 10 % copper recovery. Application of a macromolecular polymer to achieve closure of short-circuiting channels and flow-profile control was demonstrated to be very effective. The polymer solutions can be injected a multiple number of times. Overall copper recovery can be significantly increased by application of polymers to control solution flow. In these test, two treatments increased copper recovery from 10 % to about 25 %. Polymer injection prior to leaching could prove helpful in controlling the flow of solutions and excursions from the ore zone. In addition, they could be used during restoration programs to prevent leakage of contaminant ions into adjacent aquifers.

Copper distribution and leach path visualization by CT scan was carried on large core samples. X-ray computed tomography (CT) scan techniques provide a nondestructive procedure to analysis fracture distribution and to locate copper mineralization. The scanning experiments were performed at the University of Arizona, Medical School, Radiology department. A Siemens Somaton DRH model third generation full-body scanner was used. CT scanning tests were applied to the Santa Cruz core sample shortly after encapsulation. A series of 22 scans were performed at 4 mm intervals. Figure 7a shows a scan taken at 21 mm from the top of the core sample. This represents the pre-leach condition of the core. In general, the lighter areas represent the host matrix minerals which have a density <3 g cm^{-3}. Mineralized fractures are labeled A and B. The second CT scan depicted in Figure 7b illustrates the leaching paths/channels created for the same position shown in Figure 7a. Comparison of pre- and post-leaching CT results provides a clear picture of the copper leaching and the channeling phenomenon. Major leaching paths were

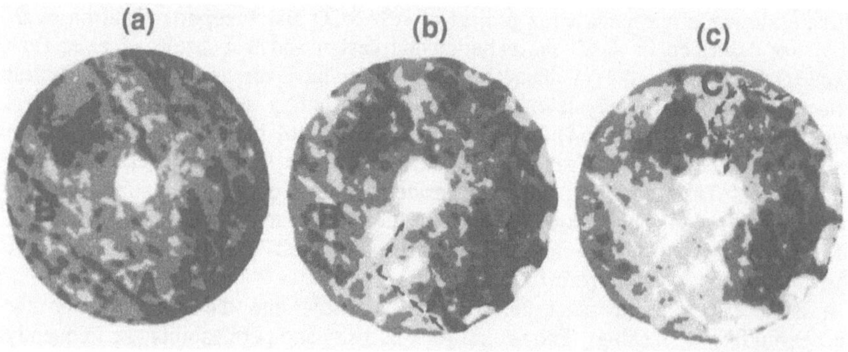

Figure 7. CT Scans of large whole core at various stages. (a) Pre-leach, (b) first stage first leaching stage pre-polymer (c) post-leach

along fractures A & B. Scan 7b shows some copper dissolution from the matrix. However, it also reveals considerable remnant non-leached areas containing copper. Figure 7c is a scan of the same location after polymer treatment and further leaching. It is apparent that leaching continued in close proximity to the original fractures. Furthermore, this CT scan indicates dissolution of copper from the highly mineralized periphery of the core and reveals formation of an additional short-circuiting path (c). This channel was closed by a subsequent polymer application. Individual scans can be reconstructed by the computer into 3-dimensional images and manipulated to yield completed visualization of pre- and post-leach features.

6 What's Next

Clearly *in situ* solution mining has the potential of recovering copper from resources that would otherwise not be mined. Copper has been recovered for many years by *in situ* leaching of mining remnants. As reserves of ore, treatable by conventional methods, become exhausted, opportunities for *in situ* leaching methods become available. For example, the accelerated *in situ* leaching program at San Manuel is anticipating the depletion of open-pit reserves. Similar potential exists elsewhere in the world. It is only a matter of time before solution mining (i.e. heap, dump, *in situ* leaching) becomes a part of long-term resource planning.

However, *in situ* copper leaching should not be regarded as a process of "last resort". As copper ore grades decline, the distinction between ore and waste becomes less obvious and the effectiveness of hydrometallurgy becomes more pronounced. In these instances, *in situ* leaching should be considered as the method of choice. The efforts of ASARCO and Freeport to develop the Santa Cruz oxide copper deposit are noteworthy. They have approached the development of this resource very carefully and systematically. Technical, environmental and community issues relating to this project have been throughly addressed. State approval to start an actual field test should be granted shortly. If this trial is successful, the next step would be securing an operting permit and to start actual *in situ* development of the resource. There are other oxide targest around the world especially in Chile, where the knowledge obtained from Santa Cruz could apply.

The greatest potential and most challenging goal of *in situ* leach mining is the treatment of deep-seated copper sulfide deposits. For example, the Safford deposit in Arizona is estimated to contain 2 billion t of ore averaging 0.41 % Cu. The efforts of Kennecott at Safford during the 1970's are commendable. However, much of the data and know-how obtained during this project never reached an outside technical audience. It is unfortunate that we must start a new learning curve. There are three main technical questions, aside from environmental compliance, affecting the feasibility of *in situ* mining deposits like Safford;

(i) *Can copper sulfides be leached at an economic rate?*
(ii) *Can economic amounts of copper sulfide be contacted by lixiviant?*
(iii) *Can adequate flow of lixiviant through the deposit be maintained?*

These are basically hydrometallurgical and hydrogeological factors. In addition, there are other important considerations: the copper ion-exchange capacity of the rock; long-term solution chemistry; and matrix rock decrepitation.

The future application of *in situ* leaching in the copper industry rests primarily on one factor: environmental acceptability. Protection and conservation of groundwater resources is a highly emotional issue and is regulated by a myriad of laws and governmental agencies. As such, the permitting structure is frequently very complex and time consuming. Furthermore, it is very costly and normally requires considerable legal help. To secure environmental approval and to obtain an operating permit, pre-operational information, operational controls and contingencies, and closure and post-closure strategies must be developed [59]. In the balance, the environmental benefits derived from this technology must be considered. It is interesting to note that the principles of *in situ* leaching can be applied to the clean-up and restoration of contaminated sites. Fletcher [60] reported on the application of ISL technology to solve a difficult pollution problem at Hughes Aircraft Corporation, Tucson, Arizona. Groundwater contaminated with trichloroethylene and chromium, has been remediated using 21 extraction and 14 recharge wells. A similar process is envisioned to remove heavy metals from soils contaminated by plating wastes at the same site.

In summary, *in situ* leach recovery of copper is still in its infancy. However, as pointed out by Agarwal and Burrows [13], "the eventual payoff is so great that the technology ***will*** be developed."

7 References

1. Wadsworth, M.E. (1994) Microprocesses in Hydrometallurgy. Lecture presented at the 1994 SME Annual Meeting, Albuquerque, NM, February 16, 1994.
2. Edelstein, D. (1994) **Mineral Industry Survey Copper in October 1993** U.S. Bureau of Mines, Washington, DC, January 1, 1994.
3. Bailey, R.V. (1980) Comments on Shirley Basin uranium deposits and on the age of all Wyoming uranium deposits in Eocene host rocks in **Proceedings of the Third Annual Uranium Seminar**, SME-AIME, New York, NY, 41-59.
4. Pu, Y. D. (1982) The history and present status of practice and research work on solution mining in China in **Interfacing Technologies in Solution Mining,** eds. (W.J. Schlitt and J.B. Hiskey) SME, Littleton, CO, 13-20.
5. Lung, T.N. (1986) The history of copper cementation on iron-The world's first hydrometallurgical process from medieval China. **Hydrometallurgy**, 17, 113-129.
6. Taylor, J.H. and Whelan, P.F. (1942) The leaching of cuperous pyrites and the precipitation of copper at Rio Tinto, Spain, **Inst. Min. Metall. Bulletin** 457, 1-36.
7. Kordosky, G.A. et al. (1987) A state-of-the-art discussion on the solvent extraction reagents used for the recovery of çopper from dilute sulfuric acid leach solution. **Separation Science and Technology**, 22(2&3), 215-232.
8. Atmore, M. G., Severs, K.J. and Voyzey, R.B.G. (1984) Past, present and future of solvent extraction of copper. **Proceedings International Conference Mineral Processing and Extractive Metallurgy**, Kunming, Yunnan Province, PRC, Oct. 1984, 261-273.
9. Jenkins, J.G. and Eamon, M.A. (1990) Plant practices and innovations at Magma Copper Company's San Manuel SX-EW plant in **Proceedings of the International Symposium on Electrochemical Plant Practice,** eds. (P.L. Claessens and G.B. Harris) Pergamon Press, New York, 41-72.
10. Jenkins, J.G. (1994) Private Communications, March 1994.
11. Edelstein, D. (1994) Private Communications, March 1994.
12. Sousa, L.J. (1981) The U.S. copper industry; problems, issues, and outlook, in **U.S. Bureau of Mines Mineral Issues Series,** October, 86 p.
13. Agarwal, J.C. and Burrows, J.C. (1980) Economic impact of in-situ solution mining of copper. **Presentation at the National and International Conference on Mineral Resources**, London, May 29, 1980.
14. Wadsworth, M.E. (1977) Interfacing technologies in solution mining. **Mining Engineering**, 29(12), 30-33.
15. Sherborne, J.E. et al. (1980) Uranium deposits of the Sweetwater Mine area, Great Divide Basin Wyoming in **Proceedings of the Third Annual Uranium Seminar**, SME-AIME, New York, NY, 27-37.
16. Anon. (1977) Oxymin details plans for *in situ* leach project in Arizona **Mining Engineering**, 29 (11), 13-16.

17. Huff, R.V., Axen, S.G. and Baughman, D. (1989) Case History: Van Dyke ISL Copper Project in ***In situ* Recovery of Minerals**, eds. (K.R. Coyne and J.B. Hiskey) Engineering Foundation, New York, 1989, pp. 165-176.
18. Hardwick, W.R. (1967) Fracturing a deposit with nuclear explosives and recovering copper by the *in situ* leaching method. **U.S.B.M. RI 6996,** 48 p.
19. Malouf, E.E. (1989) Case History of Old Reliable, Zonia, Big Mike Rubblization Projects for Leaching Copper in ***In situ* Recovery of Minerals**, eds. (K.R. Coyne and J.B. Hiskey) Engineering Foundation, New York, 1989, pp. 339-347.
20. Titley, S.R. and Hicks, C.L. (1966) **Geology of The Porphyry Copper Deposits - Southwestern North America**. The University of Arizona Press, Tucson, AZ.
21. Bateman, A.M. (1951) **The Formation of Mineral Deposits** John Wiley and Sons, New York, 371 p.
22. Ball, J.W. and Nordstrom, D.K. (1991) **User's Manual for WATEQ4F, With Revised Thermodynamic Data Base and Test Cases for Calculating Speciation of Major, Trace, and Redox Elements in Natural Waters** U.S.G.S. Open-File Report 91-183, 189 p.
23. Garrels, R.M. and Christ, C.L. (1965) **Solutions, Minerals, and Equilibria.** Harper and Row, New York, 450 p.
24. Peters, E. (1972) **Theory of Leaching: First Tutorial Symposium on Hydrometallurgy,** Golden, CO, 1972, 49 p.
25. Osseo-Asare, K. and Brown, T.H. (1979) A numerical method for computing hydrometallurgical activity diagrams. **Hydrometallurgy**, 4, 217-232.
26. Froning, H.H., Shaley, M.E. and Verink, E.D. Jr. (1976) An improved method for calculation of potential pH diagrams of metal-ion-water systems by computer. **Corrosion Sci.,** 16, 371-377.
27. Turnbull, A.G. and Wadsley, M.W. (1992) **CSRIO-MONASH Thermochemistry system Version 1.0.**
28. Hiskey, J.B. and Wadsworth, M.E. (1981) Electrochemical processes in the leaching of metal sulfides and oxides in **Process and Fundamental Considerations of Selected Hydrometallurgical Systems,** ed. (M.C. Kuhn) SME, New York, 304-325.
29. Koch D.F.A. and McIntyre, R.J. (1976) The application of reflectance spectroscopy to a study of the anodic oxidation of cuprous sulfide. **Journal of Electroanalytical Chemistry,** 71, 285-296.
30. Hiskey, J.B. and Bhappu, R.B. (1987) Role of oxygen in dump leaching in **Proceedings International Symposium on the Impact of Oxygen on the Productivity of Non-Ferrous Metallurgical Processes** CIM 26th Annual Conference of Metallurgists, Winnipeg, Manitoba, August 23-26, 1987.
31. Munoz, P.b., Miller, J.D. and Wadsworth, M.E. (1979) Reaction mechanism for the acid ferric sulfate leaching of chalcopyrite. **Met. Trans. B,** 10B, 149-158.
32. Dutrizac, J.E. (1981) The dissolution of chalcopyrite in ferric sulfate and ferric chloride media. **Met. Trans. B,** 12B, 371-378.
33. Majima, H. et al. (1985) The leaching of chalcopyrite in ferric chloride and ferric sulfate solutions. **Can. Metall. Q.,** 24 (4), 283-291.

34. Braun, R.L., Lewis, A.E., and Wadsworth, M.E. (1974) In-place leaching of primary sulfide ores: Laboratory Leaching Data and Kinetics Model. **Met. Trans.**, 5, 1717-1727.
35. Colmer, A.R. and Hinkle, M.E. (1947) The role of microorganisms in acid mine drainage - a preliminary report **Science**, 106, 253-256.
36. Beck, J.V. (1967) The role of bacteria in copper mining operations **Biotechnol. Bioeng.**, 9, 487-497.
37. Bosecker, K., Torma, A.E. and Brierley, J.A. (1979) Microbiological leaching of chalcopyrite concentrate and the influence of hydrostatic pressure on the activity of thiobacillus ferrooxidans. **European J. Appl. Microbiol. Biotechnol.**, 7, 85-90.
38. Torma, A.E. (1975) **Canadian Patent** 960463.
39. Ahlness, J.K. and Pojar, M.G. (1983) *In situ* copper leaching in the United States; case histories of operations. **U.S.B.M. IC 8961**, 37 p.
40. Greenwood, C.C. (1926) Underground leaching at Cananea **Engineering and Mining Journal**, 121 (13), 518-521.
41. Fletcher, J.B. (1971) In-place leaching - Miami Mine, Miami, Arizona **Presentation at AIME Centennial Annual Meeting SME-AIME**, New York, March 1971, AIME Preprint 71-AS-40.
42. Ward, M.H. (1973) Engineering for *in situ* leaching. **Mining Congress Journal**, January 1973, 21- 27.
43. Kline, J.T., Behnke, K. and Musgrove, P.M. (1985) Evaluation of underground copper leaching at the Lakeshore Mine **Presentation at the SME-AIME Fall Meeting**, Albuquerque, NM, October 16-18, 1985, SME Preprint 85-367.
44. Hsueh, L. et al. (1978) In-situ mining method and apparatus. **U.S. Patent 4,116,488**, September 26, 1978.
45. Huff, R.V. and Huska, P.A. (1975) Wellbore oxidation of lixiviants. **U.S. Patent 3,894,700**, July 15, 1975.
46. Hsueh, L. et al. (1977) In-situ mining of copper and nickel. **U.S. Patent 4,045,084**, August 30, 1977.
47. Huff, R.V. and Moynihan, D.J. (1978) Lixiviant recirculator for *in situ* mining. **U.S. Patent 4,079,998**, March 21, 1978.
48. Huff. R.V. and Davidson, D.H. (1978) Method for *in situ* mine fields. **U.S. Patent 4,125,289**. November 14, 1978.
49. Ahlness, J.K. and Millenacker, D.J. (1989) *In situ* copper mining field research project. **U.S.B.M. IC 9216 *In situ* Leach Mining**. Proceedings: Bureau of Mines Technology Transfer Seminars, Phoenix, AZ, April 4, 1989 and Salt Lake City, April 6, 1989, 4-6.
50. Dahl, L.J. (1989) Methods for determining the geologic structure of an ore body as it relates to *in situ* mining. **U.S.B.M. IC 9216 *In situ* Leach Mining**. Proceedings: Bureau of Mines Technology Transfer Seminars, Phoenix, AZ, April 4, 1989 and Salt Lake City, April 6, 1989, 37-48.

51. Paulson, S.E. and Kuhlman, H.L. (1989) Laboratory core-leaching and petrologic studies to evaluate oxide copper ores for *in situ* mining. **U.S.B.M. IC 9216 *In situ* Leach Mining.** Proceedings: Bureau of Mines Technology Transfer Seminars, Phoenix, AZ, April 4, 1989 and Salt Lake City, April 6, 1989, 18-36.
52. Butler, J.E., Ackland, M.C. and Robinson, P.C. (1982) Development of *in situ* leaching by Gunpowder Copper Limited, Queensland, Australia in **Interfacing Technologies in Solution Mining,** eds. (W.J. Schlitt and J.B. Hiskey) SME, Littleton, CO, 251-259.
53. Paulson, S.E., Dahl, L.J. and Kuhlman, H.L. (1987) *In situ* mining geologic characterization studies: Experimental design, apparatus, and preliminary results. **Presentation at SME Annual Meeting,** Denver, CO, February 24-27, 1987, Preprint 87-139.
54. Early, D. (1990) The effects of rock mineralogy, chemistry, and texture on *in situ* leaching of oxide copper ores. **Presentation at SME Annual Meeting,** Salt Lake City, UT, February 26 - March 1, 1990.
55. Oner, G. and Hiskey, J.B. (1992) Computer assisted *in situ* leaching experiments for copper oxide ores in **APCOM '92, Computer Applications in the Mineral Industries,** ed. (Y.C. Kim), SME, Littleton, CO, 73-84.
56. Oner, G. and Hiskey, J.B. (1992) Laboratory investigation of *in situ* leaching of copper oxide ores in **Proceeding 4th International Mineral Processing Symposium,** ed. (G. Ozbayoglu), Antalya, Turkey, Vol. 2, 700-709.
57. Hiskey, J.B. and Oner, G. (1993) Application of polymers to control fluid flow in *in situ* leaching in **Hydrometallurgy - Fundamentals, Technology and Innovation,** eds. (J.B. Hiskey and G.W. Warren), SME, Littleton, CO, 1073-1088.
58. Hiskey, J.B. and Oner, G. (1994) Copper distribution and leaching path visualization by CT scan. In preparation.
59. Weeks, R.E. and Millenucker, D.J. (1988) Environmental permitting considerations for true *in situ* copper mining in Arizona. **SME, Trans.,** 284, 1809-1813.
60. Fletcher, A.W. (1992) Industrial waste management at Hughes Aircraft Corporation, Tucson, Arizona, U.S.A. **IMM Bulletin Minerals Industry International**, No. 1008, 22-23.

Hydrometallurgical processing of refinery slimes at Phelps Dodge: theory to practice

J. E. Hoffmann
Jan H. Reimers and Associates, Houston, Texas, U.S.A.
Bradford Wesstrom
Phelps Dodge Refining Corporation, El Paso, Texas, U.S.A.

This presentation will first briefly review the major problems associated with the pyrometallurgical processing of copper refinery slimes. These include potentially environmentally abusive smelting emissions, substantial precious metals recycling in slags, lack of impurities bleed capabilities, and large in-process inventories of precious metals.

The process chemistry of a new hydrometallurgical process for treatment of sulfation roasted copper refinery slimes will be described. The process steps to be discussed include:

1) Silver extraction by metathesis leaching
2) Hydrolytic purification of silver bearing leach liquor
3) Recovery of high purity silver (>99.99%) by electrowinning
4) Lixiviant regeneration
5) Extraction of gold and platinum group metals by wet chlorination
6) Solvent extraction, purification, and reduction for recovery of >99.99% pure gold
7) Quantitative recovery of platinum group metals by tellurium collection
8) Effluent description and abatement techniques

The reduction to practice of the above technology is demonstrated by a description of the Phelps Dodge Precious Metals Refinery which employs this technology for the processing of their sulfation-roasted copper refinery slimes. Slides of the actual process equipment and operations are shown along with a discussion of the various unit operations and processes. A final process comparison sets forth the advantages of hydrometallurgical processing over conventional pyrometallurgical techniques.

Evolution of Slimes Processing

Slimes are the insoluble particulates which are obtained when copper is electrorefined. Some of these impurities report in the slimes in the same form in which they were present in the anode; most undergo some chemical change during the electrorefining process. Because slimes frequently contain significant precious metals values, in particular silver, gold, and palladium, technology has evolved for their processing. The initial step in virtually any slimes processing flowsheet is removal of their copper content which is usually the most abundant metallic constituent. For many years, copper removal was accomplished by oxidative roasting followed by leaching. This practice was gradually replaced by atmospheric leaching using air as an oxidant and refinery electrolyte, often augmented with sulfuric acid, as the lixiviant. The decopperized slimes were then smelted and fire refined into a impure silver bullion containing gold along with varying quantities of copper.

Because copper is particularly difficult to remove by fire refining and atmospheric decopperizing techniques usually left 2-5% copper in the slimes, pressure leaching techniques were developed. Pressure leaching not only readily reduced the copper content of the slimes to less than 1%, it also removed most of the tellurium in slimes, a metal nearly as difficult to remove by fire refining as copper. The removal of copper and tellurium greatly reduced the difficulty of smelting/fire refining slimes to Dore bullion but still left the selenium content of the slimes intact. The presence of significant selenium in slimes compels an additional smelting step as well as the recovery of selenium both from flue gases and slags. The capture of the selenium during smelting is a formidable task which can create costly workplace and environmental problems.

Over the years, a variety of processes have been developed for the removal of selenium from slimes. These include the acid and alkaline variants of the soda ash process, oxidative roasting technologies and sulfation roasting technologies. Of these, perhaps the most commonly applied are sulfation roasting technologies. The oxidant, sulfuric acid, is cheap; selenium levels of less than 0.5% in the deselenized slimes are readily attained; operating temperatures are low, circa 600°C; and emissions can be readily contained. Unfortunately, having developed technology for removal of copper, tellurium and selenium, the product from these treatments is still routinely smelted.

Slimes Smelting Technology

Process Description

Slimes from which copper, tellurium, and selenium have been removed are generally treated in two smelting steps. The first is a smelting step performed under neutral to slightly reducing conditions usually with silica and other fluxes to produce a slag containing lead, antimony, arsenic and other base metals and a foul bullion containing the copper, silver, gold and other precious metals. The smelting operation may require anywhere from 12 hours to several days depending upon the smelting technology and equipment employed.

The second step is essentially a fire refining of the foul bullion generated by smelting to produce an alloy of silver, gold and platinum group metals. This fire refining step removes much of the remaining copper and tellurium. In the past, fluxing was accomplished using soda ash followed by niter with the niter providing the requisite oxidant for removal of copper, tellurium and any remaining selenium. Because of environmental restrictions on the emission of NOX, pure oxygen has been substituted for niter at many slimes smelting facilities; however, the use of oxygen tends to increase the amount of silver and gold reporting in the refining slags.

Process Problems

Classical slimes smelting/refining practice as briefly outlined above, incurs a number of process problems. In the absence of alternate hydrometallurgical technologies, slimes processors deal with these problems with varying degrees of success; however, in all cases the costs associated with amelioration of them were and still are substantial. Some of these are discussed below.

1) Abatement of Smelting/Refining Flue Gases The smelting and refining of slimes in a fuel fired furnace, whether a classical Dore furnace or the more modern rotary furnace or top blown rotary converter, generates immense volumes of flue gases which require treatment for the capture of noxious gases and toxic metals. The noxious gases include sulfur dioxide, sulfur trioxide, and nitric oxides. The toxic metals usually include antimony, arsenic, lead, selenium and tellurium. With the advent of increasingly stringent regulations concerning emission of these metals to the atmosphere, the cost of equipment to capture and immobilize these toxic metals has increased greatly. Metallic selenium present in the furnace flue gases, particularly in those cases in which slimes which have not been deselenized, compels the use of elaborate scrubbing systems with their attendant reagent and energy consumptions, followed by electrostatic precipitators.

2) <u>Workplace emissions</u> Dust, gases, and fumes in the workplace must also be controlled, frequently requiring elaborate ventilation systems, the use of respirators, and routine monitoring of the toxic metals levels in the blood of those exposed to the operation. Because of the high levels of lead and silver in most slimes, it should be stressed that the hazard is not only associated with the smelting operation but with any aspect of charge preparation which may generate dust.

3) <u>Precious Metals and In-process Inventories</u> Smelting/refining cycles often last for several days, creating substantial in-process inventories of gold, silver and other precious metals. A greater in-process inventory problem is created by the precious metals which report in the various slags generated during the smelting and refining cycles. These may contain from 10 to as much as 30% of the precious metals content of the refinery slimes. The recycle of these slimes back to the smelter results in their remaining in the copper smelting/refining circuit for 30 to 90 additional days, resulting, in effect, in the loss of substantial interest charges on the metal ultimately recovered.

4) <u>Recycle of Impurity Metals</u> Because the precious metals content of the smelting and refining slags compels their return to the smelter (unless advantageous terms can be obtained for their sale, a very unlikely event) the return of the smelting and refining slags recycles many of the most undesirable impurities, such as antimony, arsenic and bismuth to the smelter. This ultimately results in higher impurity levels in the anodes sent to the copper refinery creating additional problems in electrolyte purity maintenance and cathode quality.

5) <u>Hexavalent Selenium Control</u> The pyrometallurgical processing of refinery slimes creates a unique effluent treatment problem because the soda ash slagging of smelted slimes results in the formation of hexavalent selenium. When these soda ash slags are leached for the recovery of selenium and tellurium, the hexavalent selenium passes into solution, ultimately reporting in the final process effluent from selenium and tellurium recovery. Hexavalent selenium cannot be removed from solution by conventional hydrolytic processing e.g. coprecipitation with ferric iron salts. As a result, costly chemical processing techniques are required to reduce the selenium to the quadrivalent form before it can be removed from solution. The reduction and removal of hexavalent selenium from solution is frequently the most costly process effluent treatment step in the processing of slimes.

Dore Bullion Refining Technology

Dore bullion from smelting and refining slimes usually contains 0.5-2% copper, 1-3% gold and the balance largely silver. Nearly universal practice for the refining of this bullion involves the use of electrolytic practices. These are discussed below.

Silver Refining

Silver refining is accomplished by the electrolysis of bullion in a weakly acidic solution of copper and silver nitrate. Two variants of the electrorefining technology are practiced: the Moebius technology which employs vertically disposed plane parallel silver anodes and stainless steel or titanium cathodes with mechanical scrapers for removal of the loosely adherent dendritic silver deposit; and Thum Balbach technology which employs horizontally disposed anodes and cathodes. Both processes are in common use today. Although capable of producing good quality silver, say 99.95%, they share common disadvantages.

1) Electrolyte Purification Frequent purification of the electrorefining electrolyte is required. This involves precipitation of the silver content of the electrolyte, either as silver chloride or elemental silver and disposal of the associated copper nitrate or nitric acid. This is a costly and time consuming operation with a significant burden.

2) In-process Silver Inventory The establishment of a silver electrorefining operation requires that an inventory of elemental silver in the form of impure bullion and pure silver cathode be maintained. This represents a permanent in-process inventory of silver metal that will not be recovered until the plant is dismantled.

3) Gold Holdup in Silver Usually far more costly than the silver inventory is the associated gold. Since conventional bullion refining practice frees the gold only after the silver is refined, the gold is also held in inventory until the silver is recovered. Frequently, the value of the gold in inventory in the silver refining process exceeds that of the silver.

Gold Refining

Gold and platinum group metals are recovered from the insoluble slimes remaining after silver refining, sometimes called parting plant mud or silver refinery slimes. This is accomplished by first separating the bulk of the silver from the gold either by raffination, the process of boiling the slimes in concentrated sulfuric acid to dissolve the silver, or by treating the slimes by chlorination chemistry to produce a gold bearing solution from which the gold is recovered in an impure form by chemical reduction. The gold product is cast to anode and electrorefined in a solution of chlorauric acid and hydrochloric acid, the so-called Wohlwill Process. This process technology although capable of producing 99.99% pure gold, has two major disadvantages.

1) The in-process inventory of gold is very costly. The electrolyte which must always be available for the electrorefining frequently contains in excess of 100 g/L gold. In addition to the solution, the gold anodes and cathodes increase the cost of the in-process inventory considerably.

2) Platinum metals present in the gold anodes dissolve and pass into solution where they gradually increase in concentration until they begin to contaminate the gold cathodes. At that point the electrolyte must be processed for platinum metals recovery. The recovery of the platinum metals must await their accumulation in the electrolyte to acceptable levels for recovery, creating still another in-process inventory of precious metals.

Targeted Characteristics of New Slimes Processing Technology

Based on the above discussion of present slimes processing technology, the important desirable characteristics of new slimes processing are defined below.

Elimination of Slimes Smelting and Fire Refining The new process technology should eliminate the use of smelting and refining. This will greatly reduce the environmental burden of the process as well as eliminating many of the workplace hazards.

Reduction of Precious Metals Recycle Irrespective of the process chosen, the first pass recovery of precious metals should exceed at least 95% of their aggregate value in the slimes.

Recovery of Base Metals For environmental reasons, accountability and immobilization of base metals extracted from the slimes should be essentially 100%. In particular, the selenium and tellurium extracted from the slimes should be essentially 100% recoverable at sufficient purity to meet commercial requirements.

Reduction of In-process Precious Metals Inventory The process chosen should minimize the in-process inventory of precious metals as much as possible. This requirement militates strongly against the use of electrorefining processes which usually require a substantial inventory of both pure and impure metal in the form of cathodic deposit and anodes.

Initial Separation of Silver and Gold The ideal process will separate silver and gold in the first process step, freeing both elements for immediate processing. This consideration virtually eliminates the use of electrorefining processes; however, electrowinning processes may still be acceptable provided they do not create a large inventory of precious metals.

Optional Impurity Metals Recycle Irrespective of the process chosen for extraction of the precious metals, their final level in the slimes treatment residue will require recycle of the final residue to the smelter. For instance, if the gold content of the slimes is 1% and first pass gold extraction is an excellent 99.5 %, assuming a 50% weight loss on processing, the gold content of the residue will be 0.01%, or 3.21 troy ounces of gold will be present in every ton of residue. This gold concentration will compel return of the residue to the smelter. However, if in addition to the return of precious metal values, all impurity values are also returned to the smelter it may create an intolerable impurity burden. For this reason, the process chosen should provide the opportunity to separate undesirable metal values from the slimes processing residue prior to its return to the smelter.

Characterization of Feed to Phelps Dodge Treatment Process

Phelps Dodge processes their copper refinery slimes by first pressure decopperizing/detellurizing then deselenizing the copper free slimes by sulfation roasting with concentrated sulfuric acid. Essentially the process consists of mixing dried decopperized slimes with sulfuric acid and roasting in a two hearth Herreshoff-type electrically heated roaster. As the slimes and sulfuric acid traverse the roaster hearths, elemental selenium and selenium present as intermetallic selenides are converted to volatile selenium dioxide, which exits with the roaster gases to a scrubbing circuit. The respective metal sulfates remain in the roaster calcine. Typical reactions are shown below.

1) $Ag_2Se + 2H_2SO_4 = Ag_2SO_4 + SeO_2 + 2H_2O$

2) $Se + 2H_2SO_4 = SeO_2 + 2H_2O + 2SO_2$

3) $CuSe + 4H_2SO_4 = CuSO_4 + SeO_2 + 3SO_2 + 4H_2O$

Arsenic and antimony which are present in the slimes in the pentavalent state and trivalent state respectively appear to remain substantially unchanged during sulfation roasting. Lead which is present in the decopperized slimes as lead sulfate appears unchanged. Elemental tellurium in converted to tellurium dioxide.

4) $Te + 2H_2SO_4 = TeO_2 + 2SO_2 + 2H_2O$

Gold does not react with sulfuric acid under the sulfation roasting conditions. Although both platinum and palladium both may react with boiling sulfuric acid, it appears that platinum and palladium sulfate are unstable at the terminal roasting temperature and decompose to their elemental form.

Idealized Process Flowsheet

Based on the preceding discussion of the desirable characteristics of a new hydrometallurgical technology and the compounds present in the sulfated slimes calcine, a block flow diagram can be defined. This is shown below.

Discussion of Block Flow Diagram: Idealized Hydrometallurgical Flowsheet

Step 1) The first step in the idealized hydrometallurgical process is a leach to separate the silver from the gold. This will allow the simultaneous recovery and purification of silver and gold.

Step 2) The leach residue is releached to solubilize the gold and platinum metals content of the leach residue.
Step 3) The silver bearing leach liquor is purified.
Step 4) Silver is recovered from the purified leach liquor.
Step 5) Gold is simultaneously recovered and concentrated from the gold-bearing leach filtrate by solvent extraction.
Step 6) The gold-bearing organic is purified and treated for gold recovery.
Step 7) Platinum group metals are recovered from the raffinate from gold extraction.

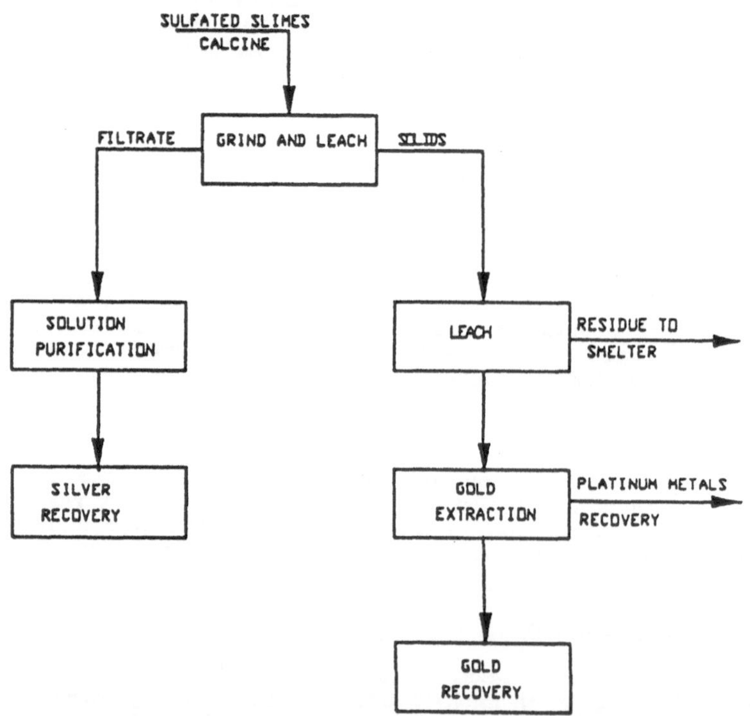

Because silver is separated from gold and platinum metals in the first step, the purification and recovery of gold, silver and platinum metals, steps 3-7, can take place simultaneously. Unlike present technology, gold and platinum metals recovery does not wait upon silver extraction, recovery and refining.

Chemistry Meeting Idealized Process Requirements

Silver Circuit

Leaching Silver is present in the sulfated slimes as silver sulfate. As a candidate for extraction of silver by leaching it is limited by its solubility which is shown below as a function of temperature.

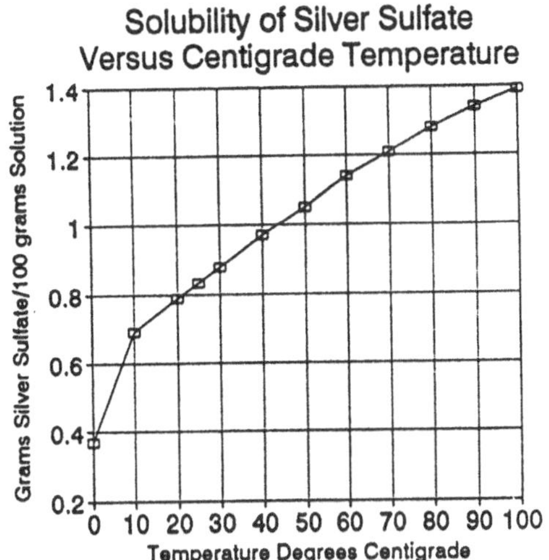

It is apparent from the data above that only dilute solutions of silver sulfate could be generated, with the attendant requirement of impractically large process reactors. A far more attractive salt for leaching is silver nitrate whose solubility curve is shown below.

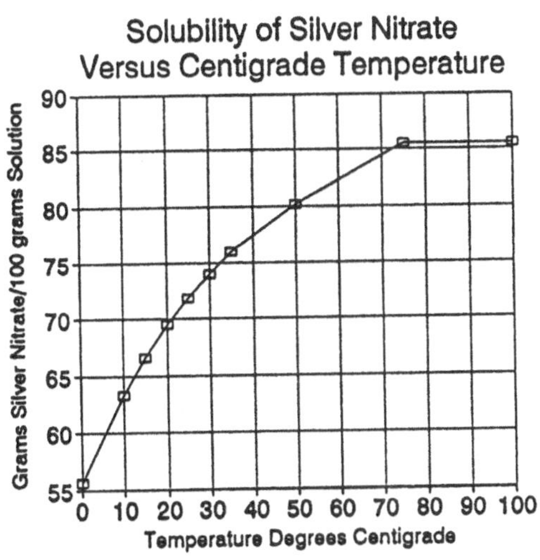

In examining the chemistry of the system, it became apparent that the chemistry which can convert silver sulfate to silver nitrate would make possible the extraction of silver from sulfated slimes calcine at high silver concentrations. The reaction identified for this conversion of silver sulfate to silver nitrate was the metathesis reaction between silver sulfate and calcium nitrate. This is shown below.

$$Ag_2SO_4 + Ca(NO_3)_2 = 2AgNO_3 + CaSO_4$$

Because calcium sulfate is much less soluble than silver sulfate, the sulfate ion is preferentially precipitated from solution resulting in the formation of the highly soluble silver nitrate.

Purification The silver nitrate bearing solution will contain other impurity elements present in the sulfated slimes, notably: selenium, tellurium, antimony, arsenic, bismuth, lead, and copper. Although the solubility of most of these elements is limited by the pH of the calcium nitrate lixiviant, they are sufficiently soluble that if they reported with the silver when it is recovered from solution, the silver product would be unacceptably high in impurities. To purity the silver bearing solution, advantage is taken of the difference in pH at which silver and the various impurities hydrolyze. These pH's were estimated from the Pourbaix diagrams for the impurity elements. All impurities except nickel and lead hydrolyze at least two to three pH units lower than the pH at which silver would begin to hydrolyze. To further decrease the concentration of the impurity metals in solution, ferric ion, either as ferric nitrate or ferric sulfate is added to solution. Ferric ion forms highly insoluble compounds of most of the impurities cited above. Additionally, the ferric hydroxide generated during hydrolytic purification acts as a collector of the highly insoluble compounds, effectively reducing their concentration in solution to innocuous levels.

Silver Winning The purified solution of silver nitrate and calcium nitrate is electrolyzed to recover elemental silver. The reactions are shown below.

Anode $H_2O = 1/2O_2 + 2H^+ + 2e$

Cathode $2AgNO_3 + 2e = 2Ag + 2NO_3^-$

Overall Reaction $2AgNO_3 + H_2O = 2Ag + 2HNO_3 + 1/2O_2$

Lixiviant Regeneration Inspection of the overall reaction shown above, indicates that the electrowinning of silver results in the generation of an equivalent amount of nitric acid. If the acid concentration were allowed to increase unchecked, it would reach a concentration at which silver would dissolve as rapidly as it was electrowon and current efficiency would fall to zero. The nitric acid concentration can be controlled by the periodic addition of calcium hydroxide to the electrolyte. This results in the generation of an amount of calcium nitrate exactly equivalent to that created by the metathesis reaction shown above. Thus a complete recycle of lixiviant is attained via the reaction:

$2HNO_3 + Ca(OH)_2 = Ca(NO_3)_2 + 2H_2O$

Gold

Gold Solubilization A powerful oxidizing agent combined with a suitable complexing agent is necessary for the solubilizing of gold and the platinum group metals. The use of chlorine and aqueous hydrochloric acid is commonly employed for this purpose. Chlorine requirements are very modest because only the gold, platinum, and palladium are present in the elemental form in the silver leach residue. The equations are shown below.

$2Au + 3Cl_2 + 2HCl = 2HAuCl_4$

$Pt + 2Cl_2 + 2HCl = H_2PtCl_6$

$Pd + Cl_2 + 2HCl = H_2PdCl_4$

Gold Concentration and Purification Because the gold concentration in slimes, is usually 0.1-1% by weight, the gold concentration in solution is limited by the solids to liquids ratio practically attainable. Thus, at the very high solids to liquids loading of 1000 grams of silver leach residue per liter of solution and a gold concentration of 1% in the residue, the maximum gold concentration attainable is only 10 g/L. To achieve practical gold concentrations for processing and to reject the preponderance of impurity elements in solution, gold is concentrated by solvent extraction of the filtrate from the wet chlorination operation described above. An organic reagent with a distribution coefficient O/A for gold greater than 1000 is desirable. This ensures greater than 99.9% gold recovery from the organic phase with two stages of countercurrent extraction. The solvent chosen to fulfil these conditions was dibutyl carbitol, which not only extracted gold extremely efficiently, but also completely rejected platinum, palladium and rhodium.

Gold Winning The large distribution coefficient which favors excellent extraction of gold into the organic phase, make back extraction of the gold into an aqueous phase impractical. For this reason, gold is reduced directly from the purified organic phase by reduction with oxalic acid.

$2HAuCl_4 + 3(HOOC)_2 = 2Au + 6CO_2 + 8HCl$

The finely divided gold metal is recovered by filtration, washed free of impurities and melted to ingot.

Platinum Metals

Collection All platinum, palladium, and rhodium dissolved during wet chlorination of the silver leach residue report in the raffinate from solvent extraction. The concentration of these metals in solution is quite small, rarely exceeding 100 ppm and they are quantitatively recovered from solution by treating the raffinate from solvent extraction with a reductant which will reduce tellurium from solution. (It should be noted that detellurizing as accomplished during pressure decopperizing does not remove all tellurium from the slimes. Usually 0.2-1% tellurium is present in decopperized slimes. This tellurium ultimately reports in the wet chlorination filtrate.) The reduction of tellurium causes the quantitative co-precipitation of gold, platinum, palladium, and rhodium from solution. This very powerful collection technique ensures that no precious metals escape from the process.

The small quantity of platinum group metals collected does not warrant processing to recover their metal values; instead the platinum metals bearing tellurium precipitate is recycled to the pressure decopperizing step. Here, the tellurium is dissolved and removed from the circuit but the platinum group metals are recycled to the wet chlorination where they are again dissolved; however their concentration has now doubled. This process is repeated until the level of platinum metals in solution warrants their recovery using conventional precipitation chemistry.

Impurity Rejection

The idealized hydrometallurgical process should provide a method of removing impurity metals such as: antimony, arsenic, bismuth, selenium, and tellurium from the circuit. Because all of these metals, with the exception of antimony, are readily soluble in dilute, say 1-2 molar, hydrochloric acid, they will be present in the detellurized raffinate. Process options for their isolation and immobilization include: precipitation as sulfides, removal by fractional hydrolysis, and concentration by distillation of hydrochloric acid. The optimum technology will be determined by tail liquor composition and process objectives. In any case, the opportunity for the isolation and immobilization of impurity metals is readily available in the proposed hydrometallurgical scheme.

Experimental Studies

Bench Scale Experiments Silver Circuit

Silver Extraction Initial process evaluation was performed on a bench scale using sulfated slimes calcine samples of 0.5 to 1.0 kilograms. Samples of various lots of calcine were ground to approximately 80 mesh and leached in a solution of 204 g/L calcium nitrate solution at a temperature of 80-90 °C. The results are shown below in Table II.

Table I Calcium Nitrate Leaching Studies

Lot No.	Time	Silver Content (g)					
	(min)	Filtrate	Repulp	Residue	Total	% Extn.*	Head %Ag (calc.)
40	40	229.4	9.0	22.9	261.3	91.2	23.8
240	70	230.9	1.7	20.2	252.8	92.0	23.0
242	45	260.7	1.6	21.5	283.8	92.4	25.8
244	60	281.0	1.4	19.0	301.4	93.7	27.4
252	52	295.7	10.2	19.0	324.9	94.2	29.6
252	65	304.6	10.5	18.8	333.9	94.4	30.4

* % Extn. includes silver recovered by reslurrying the cake.

These results confirmed the chemistry of the proposed silver leaching process and indicated that silver extractions were attainable which were much greater than those obtained by smelting of the calcine residue.

Solution Purification When superior extraction of silver was confirmed, testwork was undertaken to determine the optimum pH and iron addition, as ferric nitrate, necessary to ensure to ensure a pure solution for silver recovery. Samples of silver bearing leach liquor were treated with varying quantities of ferric nitrate and then adjusted to various pH values by addition of a slurry of calcium hydroxide. The treated solution samples were allowed to stand for about one hour, then they were filtered. The results are shown below.

Table II Assay of Filtrates from Hydrolysis Study

Sample No.	1	2	3	4	5	6	7
g/L Fe^{3+} added	0.0	0.0	0.34	0.68	1.01	1.69	3.38
final pH	0.8	3.4	3.1	3.4	3.4	3.4	3.5
Se (ppm)	37.0	1.6	1.6	1.7	1.6	1.6	1.6
Te (ppm)	416	7.4	2.5	11.3	11.9	5.7	1.1
As (ppm)	1450	16.1	7.4	18.7	14.3	2.6	1.2
Sb (ppm)	20.0	1.6	1.5	1.9	1.3	1.5	1.3
Bi (ppm)	45.0	0.5	0.7	0.8	0.7	3.3	0.7
Al (ppm)	1180	710	946	783	747	663	434
Cu (ppm)	1070	651	983	852	786	756	759
Fe (ppm)	57.0	4.0	5.7	15.2	29.9	15.9	10.4

The table above indicates that purification of the silver bearing leach liquor from arsenic, antimony, selenium, tellurium, and bismuth is readily achieved. Copper, aluminum, and lead reached limiting values of 760, 430 ppm and 2-3g/L respectively, quantities which are readily tolerated in the electrolyte since none them will electrodeposit with silver.

Silver Recovery Silver recovery was effected by the electrolysis of the purified silver-bearing leach liquor. A cathode current density of 380 amperes per square meter was employed and the anode current was slightly higher. The anode was expanded titanium mesh coated with platinum group metals, the cathode was stainless steel.

Silver purity The electrolytic silver deposit was scanned spectrographically for the following elements: Pb, Fe, Mn, Sb, As, Se, Te, Bi, Cu, Cr, Mg, Ca, Sn, Zn, Co, Pd, Pt. All were below 10 ppm in the silver product.

Bench Scale Experiments Gold Circuit

Gold Extraction Initial wet chlorination studies were performed on silver leach residue at a solid to liquids ratio of circa 900 g/L. The wet chlorinations were performed in 3 molar hydrochloric acid, at a temperature of 90°C. Chlorination time was 1-2 hours during which the slurry was sparged with chlorine gas. At the conclusion of wet chlorination, the slurry was filtered hot and washed with 1.5 molar hydrochloric acid. Gold extractions were consistently 97-98%, palladium 96%, and platinum 88%. It should be stressed that the level of platinum in the chlorination residue was so low (1-2 ppm) that the precision of the analytical method is suspect. Subsequent studies performed at a lower temperature of 60°C yielded substantially the same precious metals extraction.

Gold Concentration and Purification Gold was concentrated to 35-40 g/L and separated from the preponderance of soluble metal impurities by solvent extraction with dibutyl carbitol. The experimentally determined distribution coefficient O/A for gold was approximately 1000. The use of two stage countercurrent extraction was adopted. This provided an overall gold recovery during extraction of 99.9%. The impurity ions coextracted with gold were removed from the gold-loaded organic by three stage scrubbing of the organic phase with an equal volume of 1.5 molar hydrochloric acid. In all cases, three stage scrubbing reduced the level of selenium, tellurium, antimony, arsenic, bismuth, copper, lead, and silver to below less than 1ppm. Platinum and palladium were not coextracted to any measurable extent.

Gold Reduction and Recovery The use of oxalic acid was evaluated for the reduction of gold directly from dibutyl carbitol. An aqueous solution of oxalic acid was contacted with the loaded scrubbed organic and the two phases heated to approximately 85°C. Two to four hours were required to complete the reduction of the gold present in the organic. In all cases, the reaction was difficult to drive to completion. The gold product was washed free of dibutyl carbitol with ethanol, dried and assayed for metallic impurities. The results are shown below

Table III Analysis of Gold Generated by Oxalic Acid Reduction

Sn	5-10 ppm	Pb	< 2 ppm
Cu	1-5 ppm	Se	< 1 ppm
Mg	1 ppm	Bi	< 1 ppm
Cr	.1 ppm	Ni	< 1 ppm
Ag	5 ppm	Ba	< 1 ppm
Ca	.1 ppm	Te	< 1 ppm
Fe	5 ppm	Mn	< 1 ppm
Pt	<2 ppm	Sb	< 1 ppm
Co	<1 ppm		

Based upon impurities detected, the gold was 99.995% pure, superior in purity to most commercially produced gold.

Platinum Group Metals Collection

A variety of reducing agents were evaluated for the reduction of tellurium present in the gold extraction raffinate. These included, stannous chloride, sulfur dioxide, and hydrazine. In all cases gold, platinum, palladium, and silver were quantitatively precipitated from solution. Complete precipitation of the tellurium present in solution, usually in the range of 0.5 to 2 g/L, was not necessary to completely precipitate the precious metals. The use of tellurium precipitation was incorporated into the process as a vital part of the process chemistry to ensure complete recovery of precious metals.

Effluent Treatment

Figures 5 and 6 below show the solubility of the major impurity metals as a function of pH. These pH studies were only carried to pH 5.6. Subsequent studies showed that a final level of less than 1 ppm was readily achieved for these impurity metals at a final pH 8.3. The presence of circa 4 g/L of ferric iron in detellurized raffinate solution greatly assists in achieving a level of less than 1 ppm of these impurities metals. Examination of the curves below suggest that at least a partial separation can be made between the elements shown.

Conclusions Derived from Experimental Studies

1) First pass silver, gold, and platinum metals extraction were greatly superior to those obtainable by pyrometallurgical techniques.

2) The silver extraction and recovery circuit was operable as a closed circuit with complete recycle of lixiviant between electrowinning and sulfated slimes calcine leaching. The optimum concentration of silver in the pregnant purified leach liquor feed to silver electrowinning was 100 g/L. The return from electrowinning contained 50 g/L Ag, a net recovery of 50 grams per liter of solution processed.

3) Solvent extraction worked efficiently to concentrate and purify the gold content of the wet chlorination liquor generated from silver leach residue.

4) The gold bearing organic was readily purified by scrubbing with dilute acid and its gold content readily recovery by reduction directly from the organic phase. The gold produced was superior in quality to most commercially produced.

5) The use of tellurium collection techniques recovered all platinum, palladium and gold present in the raffinate from gold extraction.

6) The process technology showed sufficient promise to warrant construction and operation of a pilot plant to develop extended operating experience and design data.

Pilot Plant Studies of the Slimes Treatment Process

Information to be Developed

The pilot plant studies were performed to develop engineering data and to test all unit process and unit operations a sufficient number of times

Figure (5)

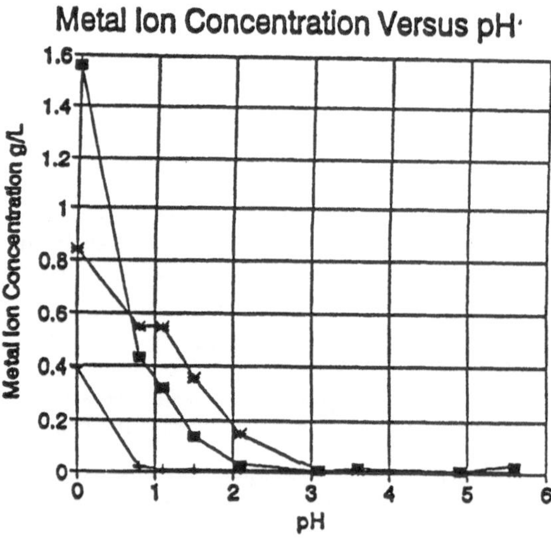

Figure (6)

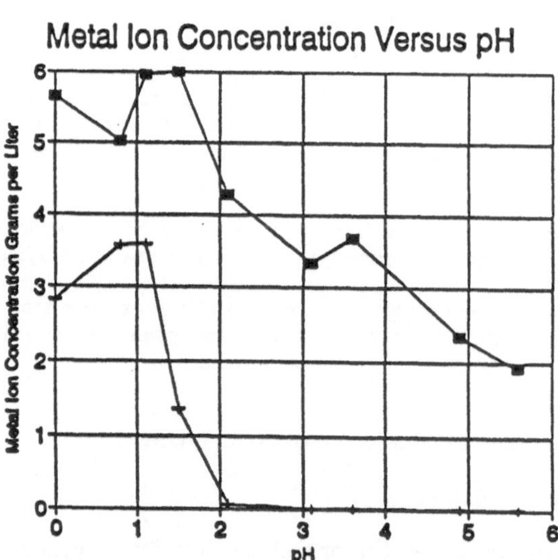

to confirm the robustness of the process chemistry and evaluate the rate and effect of impurity buildup in the system.

<u>Silver Circuit</u> Specific information to be developed included the following:

* The ability of the process to produce an electrowon silver product of consistent morphology and purity at high current density. Problems associated with the recovery and washing of the silver product.

* The durability of the DSA anodes used in electrowinning.

* The wash volumes required in the leach circuit necessary to ensure satisfactory silver extraction and the ability of the circuit to control the water balance. Optimum leaching and filtercake washing strategies.

* The magnitude of the loss of nitrates from the silver leach circuit and the impurity buildup in the recycled electrolyte.

* The robustness of the process chemistry, i.e. its sensitivity to changes in temperature, pH, impurity content, solids concentration etc.

* Safety and environmental problems associated with the process.

* The suitability of various materials of construction used in the pilot plant.

<u>Gold Circuit</u> Specific information to be developed included the following:

* The minimum chlorination time required for acceptable gold extraction.

* The optimum wet chlorination filtercake washing practice.

* The highest solids to liquids concentration at which consistently high gold extraction could be achieved.

* The lowest temperature at which the process could be operated successfully.

* The cause of the sluggish reaction kinetics when reducing gold from DBC with oxalic acid.

* Causes of crud formation during gold solvent extraction and how to deal with them efficiently.

* The durability of the solvent extraction reagent, that is how much solvent degradation occurred per extraction and reduction pass.

* The efficacy of steam distillation for solvent recovery.

* Safety and environmental problems associated with the process.

* The suitability of various materials of construction used in the pilot plant.

Because of space limitations, the silver and gold recovery circuits were evaluated sequentially. The silver recovery circuit was piloted first. Each silver leach residue was numbered then stored in water proof bags. This material constituted the feed to the gold recovery circuit. The pilot plant was approximately one twentieth the size of the anticipated full scale operation. The list of major equipment items for the gold and silver recovery circuit is set forth below to provide an indication of the scale of the pilot operations.

Equipment List Silver Pilot Plant Circuit

Item	Description	Capacity	Matl.of Const.
T-1	Calcine Slurry Storage Tank	500 gals	SS-316
T-2	Leach Cake Repulp Tank	300 gals	SS-316
T-3	Repulp Filtrate Storage Tank	1000 gals	Polypropylene
T-4	Calcine leach Reactor	300 gals	SS-316
T-5	Lime Slurry Storage Tank	55 gals	Polypropylene
T-6	Pregnant Leach Liquor Tank	1000 gals	Polypropylene
T-7	Hydrolysis Reactor	300 gals	Polypropylene
T-8	Purified Leach Liquor Storage	600 gals	Polypropylene
T-9	Acidity Adjustment Reactor	200 gals	Polypropylene
T-10	Spent electrolyte surge Tank	200 gals	Polypropylene
T-11	Spent electrolyte Storage Tank	1100 gals	Polypropylene
DP-1	Slurry Transfer Pump	10 gpm	Hypalon
DP-2	Slurry Transfer Pump	10gpm	Hypalon
DP-3	Diaphragm Pump	10 gpm	Hypalon & SS
DP-4	Transfer pump (portable)	5 gpm	Plastic
DP-5	Hydrolysis slurry filter pump	10 gpm	Hypalon & SS
CP-1	Transfer pump	25 gpm	SS-316
CP-2	Transfer pump	25 gpm	SS-316
CP-3	Metering pump	0-2 gpm	SS-316
CP-4	Electrolyte Recirculation Pump	10-20gpm	SS-316
CP-5	Transfer pump	5 gpm	SS-316
FP-1	Filter Press,washing	6ft3	Polypropylene
EC-1	Deposition Tank	160 gals	Polypropylene
	(SS-316 Cathodes, DSA/PM Anodes)		

Equipment List Pilot Plant Gold Circuit

R-2	Wet Chlorination Reactor	150 gals	Titanium
T-3	Chlorination Liquor Storage	300 gals	FRP
T-4	Polished Chlorination Storage	300 gals	FRP
T-6	Palladium Recovery	250 gals	FRP

T9/12	SX Reactor	165 gals	FRP/glass
D-1	Organic Containment Dam	685 gals	Polypropylene
T-10	First Stage SX Storage	30 gals	Polypropylene
T-11	First Stage Raffinate Storage	100 gals	Polypropylene
T-13	Stripped DBC Storage Tank	30 gals	Polypropylene
T-14	Final Raffinate Storage Tank	100 gals	Polypropylene
T-17	Gold Reduction Reaction	72 L	Pyrex
CP-1	Cl Liquor Transfer Pump	25 gpm	Plastic/Ceramic
CP-2	Cl Liquor Polishing Pump	30 gpm	Viton/Polypropylene
CP-3	Cl Liquor Transfer Pump	25 gpm	Plastic/Ceramic
CP-4	Organic Transfer Pump	0-3.5 gpm	Tygon
CP8/9	SX Raffinate Transfer Pump	5 gpm	Viton
CP-10	Neutralization Filter Pump	30 gpm	Hypalon
DP-10	Cl Slurry Filter Pump	20 gpm	Hypalon

Safety Review Prior to beginning pilot activities, a plant level and a corporate level safety review were required. Activities involved included: 1) identifying all possible chemical combinations which could inadvertently occur and providing for their prevention; 2) identifying the contents of all vessels including appropriate warnings concerning reactivity; 3) providing operators with sampling devices which were worn during operation of the pilot plant and periodically checked for exposure; 4) pressure testing of all equipment prior to plant startup; 5) a review of all electrical equipment and connections by certified inspectors; 6) training of all operators in the use of Scott Air Packs (portable self-contained breathing apparatus); 7) all personnel involved with the pilot plant operation were provided with MSDS sheets for all chemicals to be used along with lectures and participated in discussions concerning hazards associated with exposure to the chemicals; 8) extensive discussions were held among technical personnel and operating personnel to identify any hazards associated with the operation of pilot plant equipment. All tanks and vessels which contained volatile liquid or toxic liquids were lidded and vented, as required, to an external soda ash scrubber. Finally, detailed operating manuals were prepared for both the silver and gold recovery circuit. These manuals included all explicit operating instructions, all analytical procedures used on-site, and descriptions of all operating equipment and MSDS sheets for all chemicals used in the various unit processes.

Silver Recovery Pilot Plant

Studies Performed A total of 35 batches of sulfated calcine were processed through the silver recovery pilot plant producing 3395 lbs of electrowon silver Based on a average head of 25% silver, overall silver extraction was 92.6%. Typical assays of the electrowon silver produced are shown in Table below.

Table IV Analyses of Electrowon Silver Crystal

Impurities in ppm

Run	Pb	As	Se	Te	Bi	Cu	Ni	Fe	Sb
13	3	.9	<1	<1	<1	<1	<1	<1	3
14	3		<1	<1	<1	<1	<1	<1	
15	5				<1	1	<1	5	
16	6				<1	1	2	2	
17	9		<1	<1	1	<1	5	3	
18	25	<1	<1	<1	<1	<1	<1	2	6
20	8	<1	<1	<1	<1	1	<1	3	
21	7	<1	<1	<1	<1	1	<1	7	
22	1	<1	<1	<1		2	4	<1	
23	5	<1	<1	<1	4	2	<1		

The assays of the silver shown above, confirm the ability of the process to produce exceptionally pure silver.

Regarding specific information to be developed as set forth above:

DSA Anode Life - no deterioration of loss of the active precious metals coating could be detected. (At the time of preparation of this paper there is still no discernible erosion of the precious metals coatings.)

Optimum Filtercake Washing Strategies - The leach was divided into two countercurrent stages. Purification took place in the first leaching stage in which the purified feed to electrowinning was generated; the second leaching stage involved repulping of the leach residue to maximize soluble silver recovery. Optimum filtercake wash volume was determined to be 4 filtercake volumes. Washing to be performed countercurrently.

Nitrate Consumption - no measurable loss of nitrates were detected during operation of the pilot plant.

Impurity Buildup - The final assay of the electrowinning electrolyte after 35 recycles showed the following impurity concentrations: sodium 4.8 g/L, zirconium 0.2 g/L, magnesium 0.9 g/L, aluminum 0.05 g/L, potassium 1.5 g/L, zinc not detectable, strontium 0.014 g/L. Based on these impurity buildups, the bleed stream required for impurity control is less than 0.5% of the circulation volume daily.

Robustness of the Process Chemistry - was found to be excellent. This was dramatically demonstrated when the system overflowed over the weekend due to a ruptured deionized water line. The pilot plant concrete curbed area was completely filled with dilute process liquor. The operators simply siphoned up the solution, swept up all solids, including odds and bits of trash and dust which inevitably accumulate, and returned both to the leach reactor. The solutions were reconcentrated by boiling off excess water. The next batch of silver was as pure as the one which preceded it, with no additional process difficulties.

Temperature and pH control - control of pH was easily achieved; additionally, sufficient latitude in operating pH was available to simplify pH control. The pilot plant studies showed that the purified solution benefited from being allowed to stand and cool overnight. This allowed particulates which precipitated during cooling to settle out of the electrolyte, assuring a limpid feed to the electrowinning operation.

Safety - no injuries occurred during operation of the silver recovery pilot plant, nor did any unforseen reactions occur. Monitoring of the operators before, during and upon completion of the pilot plant campaign showed no measurable exposure to toxic substances associated with the process.

Materials of construction - chosen for the pilot operated without significant corrosion or wear. In particular, stainless steel 316 was completely unaffected by the service. The use of magnetic drive pumps was not recommended for the full scale facility because of failures in the pilot plant operation. This was not due to faulty or poorly designed pumps but rather the fact that the solution, by definition, is saturated in silver sulfate and calcium sulfate. Temperatures changes occurring with the intermittent use of these pumps in the pilot plant resulted in coating the magnetic drive rotors with a crystalline salt layer until they became bound.

<u>Gold Recovery Pilot Plant</u>

A total of 17 studies were performed. Each study consumed 2 batches of silver leach residue generated in the preceding evaluation of the silver recovery circuit. The average gold extraction obtained was 97.5%. The purity of the gold product obtained is tabulated below in Table V.

Table V Analyses of Gold Produced (ppm)

	Run 1	Run 2	Run 4	Run 5[1]	Run 6[1]	Run 9
As	3	<1	<1	2	<1	2
Te	3	<1	<1	<1	<1	<1
Bi	<1	<1	2	<1	<1	<1
Pt	<1	<1	<1	<1	<1	<1
Ag	<1	<1	<1	9	<1	3
Pd	<1	<1	<1	<1	<1	<1
Ni	<1	<1	1	<1	<1	<1
Ca	<1	<1	<1	<1	<1	<1
As	<1	<1	<1	1	3	4
Fe	<1	<1	<1	31	58	3
Si	<1	<1	<1	<1	<1	<1
Se	3	<1	<1	<1	<1	<1
Sn	<1	9	3	3	2	19
Mn	<1	<1		1	2	4
Sb	<1	<1	<1	4	<1	4
Mg	<1	<1	<1	<1	<1	<1
Pb	2	<1	<1	5	<1	2

[1] Assay after melting. Iron contamination occurred during melting.

The gold purity by difference is shown below in Table VI

Table VI Run	Gold Purity by Difference % Au
1	99.9976
2	99.9985
4	99.9981
5	99.9978
6	99.9952
9	99.9951

Regarding specific information to be developed as set forth above:

Minimum Chlorination Time - varied between three and four hours as determined by periodic sampling of the chlorination slurry and fire assay of the filtered and washed solids.

Optimum Filtercake Washing Practice - required first washing the filtercake with 1.5 molar hydrochloric acid, followed by three water washes. Each wash should be equal to the filtercake volume. Washing to be performed countercurrently.

Solids to Liquids Concentrations - as high as 1000 g/L of silver leach residue could be tolerated; however, the chlorination efficiency decreases at the higher solids concentration because of the increase in the superficial viscosity of the slurry. Because of the excellent O/A distribution coefficient of the reagent used for solvent extraction of the soluble gold, attaining high concentrations of gold in the chlorination liquor was less important that achieving maximum extraction of gold from the silver leach residue.

Wet Chlorination Temperatures - were evaluated from room temperature to 95°C. Room temperature was finally chosen for chlorination for several reasons: 1) lower temperatures made possible the use of polymeric materials of construction for the chlorination reactor, as opposed to the use of titanium or glass lined steel; 2) the use of cooling equipment was eliminated; 3) the solubilization of large concentrations of calcium sulfate which reprecipitated on cooling, requiring another solid/liquids separation was eliminated.

Crud Formation - was sufficiently great in some solvent extraction operations to necessitate the use of filtration. The constituents of the crud included lead, antimony, and silver salts, as well as silicic acid. The decision to incorporate a filter into the solvent extraction for filtration of the loaded organic was prompted by the results of the pilot plant studies.

Sluggish Reaction Kinetics - were determined to be due to the buildup of hydrochloric acid due to the stoichiometry of the reaction with oxalic acid.

$$2HAuCl_4 + 3(COOH)_2 = 2Au + 8HCl + 6CO_2$$

Substitution of sodium oxalate reduced the amount of acid generated allowing the reaction to proceed smoothly and rapidly to completion.

$$2HAuCl_4 + 3(COONa)_2 = 2Au + 2HCl + 6CO_2 + 6NaCl$$

Stability of Solvent Extraction Reagent - was tested by recovery of DBC from the raffinate by steam distillation and its repetitive recycling to the solvent extraction and gold reduction circuit. The recycled reagent was tested by measuring its density and gold extraction capacity and comparing this with fresh reagent. No significant degradation of the reagent was detectable based on this test.
Solvent Recovery - by steam distillation was evaluated. This process appeared to recover at least 95% of the soluble DBC (based on an assumed solubility of DBC in the raffinate). Steam distillation was chosen as the solvent recovery method for the commercial facility.

Safety and Environmental Problems - were minimal in the wet chlorination circuit provided that the reactors were lidded and the wet chlorination reactor properly vented to an alkaline scrubber. Blowing of the wet chlorination filtercake required that the air exiting the filter pass to the scrubber before being emitted to the atmosphere. Because of the comparatively low acidity at which the wet chlorination is performed, virtually no hydrochloric acid was lost to the scrubber. The choice of room temperature chlorination eliminated the hazard of operating with hot solutions. No problems were encountered with the DBC gold extraction solvent regarding flammability or toxicity. Monitoring of the operators before, during and upon completion of the pilot plant campaign showed no measurable exposure to toxic substances associated with the process.

Materials of Construction - were generally satisfactory, particularly the use of titanium for the chlorination reactor and the use of polypropylene filter chambers on the filterpress; however, based on the pilot plant studies the decision was made to use only polyolefins and fluoropolymers for containing and pumping DBC. This organic caused the swelling and softening of all rubber materials both natural and synthetic. Additionally, it appeared to soften the thermosetting plastics used in the various fiberglass/plastic laminate vessels.

<u>Instrumentation and Controls</u> The quantities measured and used for control during the pilot operation were temperature, amperage, pH and, oxidation/ reduction potential (ORP). Standard off-the-shelf instrumentation was sufficient in pilot plant service with the exception of ORP and pH measurement. In particular, the high silver concentration, elevated temperature, and high solids loading in the silver recovery circuit ultimately compelled the in-house development of a pH electrode capable of accurately measuring pH in the silver circuits. No ORP electrode was found which would withstand the aggressive chemical environment of the wet chlorination circuit near the completion of the unit process. The problem was circumvented by removing the probe near the completion of the reaction and periodically inserting it as required, for the minimum time needed to obtain a reading.

Final Process Flowsheet Based on the Pilot Plant Studies

Silver Extraction and Electrolytic Recovery

Based on the pilot plant campaigns the final form of the process flowsheet for the sulfated slimes treatment process was constructed. This is shown as the block flow diagram: Slimes Sulfation Roasted Calcine Treatment Process below. The reader is referred to the block flow diagram to assist in following the process description .

Calcine Grinding The calcine leach is performed in a countercurrent leach circuit. The silver bearing sulfation roasted slimes calcine, essentially free of selenium, is milled with a small fraction of the spent electrolyte returning from the silver electrowinning circuit. (The balance of the silver-depleted electrolyte passes to the second stage of the countercurrent leach.) The slurry of milled calcine advances to the leach/purify reactor.

Leach/Purify The calcine slurry is leached at or near a temperature of 90-100^{0}C for an hour in the repulp liquor generated in the second stage of leaching. The preferred pH during leaching is between pH 1-2. At the completion of the leaching period, an amount of iron salts sufficient to ensure the collection of all soluble impurities is added to the leach slurry and the pH elevated to precipitated the iron salts. The addition of the iron salts along with elevation of the pH results in the precipitation of heavy and base metal impurities present in the solution.

Filtration and Washing The solution is filtered hot and allowed to cool before advancing to silver recovery.

Solids Repulp The solids, filtercake, pass to the second stage of countercurrent leaching where they are repulped in the spent electrolyte returning from the silver electrowinning operation. The nominal silver concentrations in the circuit are: 100 g/L in the purified electrolyte passing to silver recovery, 50 g/L in the return spent electrolyte, and 65 g/L in the repulp liquor advancing to the first leach (leach/purify).

Silver Recovery by Electrowinning The purified electrolyte generated in the calcine leach/purify operation is the feed to the silver electrowinning operation. The electrolyte is allowed to cool and settle overnight prior to introduction into the electrowinning circuit. This results in small amounts of calcium sulfate precipitating as gypsum and reporting in the cooling tank. Any solids which may have escaped the filters also settle out, ensuring a limpid electrolyte entering the electrowinning tanks. Silver electrowinning is operated as follows:

1) Electrolyte at exit concentration of 50 g/L silver is continuously circulated between the pH adjustment reactor and the electrowinning cell.

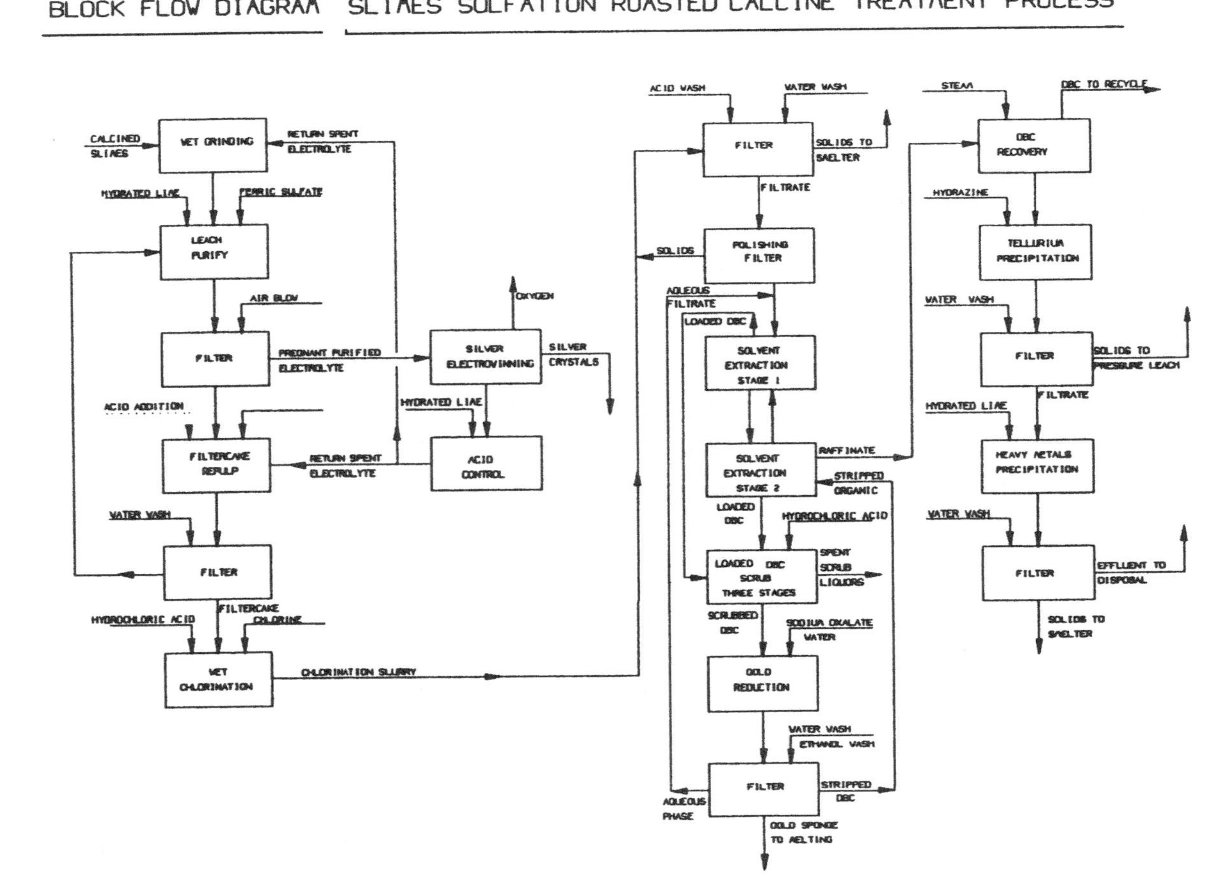
BLOCK FLOW DIAGRAM SLIMES SULFATION ROASTED CALCINE TREATMENT PROCESS
CALCINED SLIMES
WET GRINDING
RETURN SPENT ELECTROLYTE
HYDRATED LIME
FERRIC SULFATE
LEACH PURIFY
AIR BLOW
FILTER
PREGNANT PURIFIED ELECTROLYTE
OXYGEN
SILVER ELECTROWINNING
SILVER CRYSTALS
HYDRATED LIME
ACID CONTROL
ACID ADDITION
FILTERCAKE REPULP
RETURN SPENT ELECTROLYTE
WATER WASH
FILTER
HYDROCHLORIC ACID
FILTERCAKE
CHLORINE
WET CHLORINATION
CHLORINATION SLURRY
ACID WASH
WATER WASH
FILTER
SOLIDS TO SMELTER
FILTRATE
POLISHING FILTER
SOLIDS
AQUEOUS FILTRATE
LOADED DBC
SOLVENT EXTRACTION STAGE 1
SOLVENT EXTRACTION STAGE 2
RAFFINATE
STRIPPED ORGANIC
LOADED DBC
HYDROCHLORIC ACID
LOADED DBC SCRUB THREE STAGES
SPENT SCRUB LIQUORS
SCRUBBED DBC
SODIUM OXALATE
WATER
GOLD REDUCTION
WATER WASH
ETHANOL WASH
FILTER
STRIPPED DBC
AQUEOUS PHASE
GOLD SPONGE TO MELTING
STEAM
DBC TO RECYCLE
DBC RECOVERY
HYDRAZINE
TELLURIUM PRECIPITATION
WATER WASH
FILTER
SOLIDS TO PRESSURE LEACH
FILTRATE
HYDRATED LIME
HEAVY METALS PRECIPITATION
WATER WASH
FILTER
EFFLUENT TO DISPOSAL
SOLIDS TO SMELTER

2) Pregnant purified electrolyte containing 100 g/L silver is injected into the circulating spent electrolyte ahead of the electrowinning cell.

<u>Acid Control/Lixiviant Regeneration</u> The nitric acid generated by electrowinning is neutralized in the external pH adjustment reactor by addition of hydrated lime. Before returning to the deposition tank, the neutralized electrolyte is filtered to remove any insolubles that may have been present in the lime used for pH control. The circuit pH is generally maintained at a low enough value to ensure that impurities present in the lime will not hydrolyze and contaminate the electrowon silver and high enough that redissolution of electrodeposited silver is insignificant. Since, as discussed previously, the amount of lime added is directly proportional to the amount of silver electrowon and the amount of silver in solution is directly proportional to the amount of calcium consumed in the metathesis silver leaching reaction, the amount of lime added in the pH adjustment tank will exactly replenish that consumed in leaching, maintaining a constant calcium concentration in the leach/electrowin circuit.

<u>Gold Recovery Circuit</u>

<u>Chlorination of Calcine Leach Residue</u> The final leach residue from silver extraction is chlorinated for recovery of its precious metals. Because the sulfation roasting treatment oxidized virtually all metal values in the slimes other than gold, oxidant requirements are very modest. Chlorine gas is the oxidant of choice because it is both an inexpensive and highly aggressive oxidant.

<u>Filtration and Washing</u> The chlorination slurry is filtered at the final reaction temperature. This ensures that no chlorination liquor is entrained in crystallized salts which may form on cooling. Washing of the filtercake is first performed with dilute hydrochloric acid to avoid hydrolysis of soluble chloride compounds present in the chlorination liquor. Subsequent washes are performed with water. Water washes are performed counter currently. The chlorination residue is advanced to dechlorination.

<u>Cooling and Settling</u> The chlorination liquor is cooled and allowed to settle prior to a polishing filtration. This cooling and settling is provided to allow for any post precipitation of solids which would otherwise escape polishing filtration and become potential crud formers during gold extraction.

<u>Polishing Filtration</u> The solids intercepted by polishing are recycled back to the chlorination reaction where they ultimately leave the process in the chlorination residue. The quantity of material is so small that cake washing is not warranted. It should be stressed however,that the quantity of solids precipitated are dependent on the temperature at which the chlorination slurry is filtered.

<u>Solvent Extraction of Gold</u> Solvent extraction of gold is accomplished by two stage countercurrent extraction with dibutyl carbitol (diethylene

glycol dibutyl ether). Two stages of solvent extraction ensure recovery of 99.9% of the aqueous gold content into the organic phase. The ratio of organic to aqueous during extraction is defined by the gold concentration of the chlorination liquor and the desired concentration of gold in the organic phase. Usually a gold concentration of 35-50 g/L is targeted for the loaded DBC.

Phase separation is accomplished by gravity and usually requires approximately one half hour. The loaded dibutyl carbitol (DBC) from the first stage of extraction is advanced to scrubbing and the first stage raffinate is advanced to the second stage of extraction where it is contacted with the stripped DBC returning from extraction. The final gold content of the second, final, raffinate should not exceed 1-5 ppm gold. (Theoretical values would be less than 1 ppm gold; however, some loaded DBC is usually entrained in the aqueous phase.)

With a properly polished filtrate, crud generation may be negligible; however, provision should probably be made for the inclusion of a small filter press for removal of crud. If a crud product is obtained, it is be advanced to the DBC recovery step where the organic phase will be removed and the residual salts will pass into solution.

<u>DBC Recovery</u> DBC is recovered by steam distillation. The steam required for steam distillation may be generated by evaporating approximately 10% of the chlorination liquor by boiling. The vapor is condensed, the condensate forming two layers; the upper layer being DBC, the lower layer being water saturated with DBC. The recovered DBC is returned to the solvent extraction circuit. The DBC saturated water is used in the make up of the dilute hydrochloric acid which is used to scrub the loaded DBC.

<u>Scrubbing of Loaded DBC</u> Removal of the impurity metals coextracted with the gold is accomplished by staged countercurrent scrubbing with 1.5 molar hydrochloric acid. The actual number of stages are determined by the impurity levels found in the DBC. An aqueous to organic ratio of unity is sufficient for the countercurrent scrubbing operation.

<u>Gold Reduction</u> Gold is reduced directly from the organic phase by contacting the organic phase with an aqueous solution of the reducing agent, sodium oxalate. The reaction temperature is maintained between 70-80°C.The rate of reaction is controlled by the rate of reductant addition to produce a crystalline gold product which is readily filtered and washed.

<u>Filtration</u> The gold is filtered, water washed, then a final wash with ethanol to remove the final traces of DBC. Because of the value of the product, a Nutsch box filter is employed for solids/liquids separation. Both the organic and aqueous phases are filtered through the Nutsch filter and separation of the aqueous and organic accomplished on the filtrate. The gold-free DBC is returned to the solvent extraction circuit. The reduction liquors which contain sodium chloride, hydrochloric acid and small amounts of oxalic acid and gold are co-mingled with the DBC scrub liquors and returned to the solvent extraction circuit via the polished chlorination liquor.

Tellurium Precipitation Tellurium is recovered from the gold-free raffinate by reduction to elemental form using either hydrazine or sulfur dioxide. An effective comparison between these two reagents is required to determine the best choice. The comparison must include the cost of the reagents, the additional freight costs associated with the greater weight of sulfur dioxide which must be shipped, the ease of feeding a liquid reductant compared with introducing a gas, the amount of sulfuric acid which must ultimately be neutralized when sulfur dioxide is used, and the disposal of the gypsum product generated by neutralization.

Heavy Metals Precipitation Metals present in the detellurized raffinate will include: bismuth, antimony, arsenic, lead, copper, nickel, tin, and iron as well as varying concentrations of aluminum, and alkaline earth metals. These metals, other than the alkaline earth metals, are removed from solution by hydrolysis. Ferric sulfate is added to assist in the precipitation of the antimony, arsenic, and any traces of selenium and tellurium which may be present. Neutralization is accomplished using hydrated lime and vigorous agitation. Traces of lead or silver remaining in solution after heavy metal precipitation, are precipitated as sulfides.

Filtration The slurry is filtered and the filtercake washed only if the presence of soluble chlorides in the filtercake is a bar to disposal of the filtercake. Because of the calcium sulfate dihydrate (gypsum) precipitated during neutralization, the slurry is readily filterable. The final calcium chloride solution, after concentration by evaporation, may be sufficiently pure to be used for well drilling solution. It is important to remember that all effluents generated by the wet chlorination process should have a targeted use or destination. In this regard it should be noted that since no precious metals are present in the neutralization cake there is no compelling reason to recycle it back to the smelter. In returning the neutralization cake back to the smelter, the true cost of recycling impurity metals such as bismuth and antimony should be included in the economic evaluation.

The Production Facility

Conceptual Engineering

The success of the pilot plant campaigns justified the costs associated with the conceptual engineering of a commercial processing facility. The conceptual engineering was performed by defining a production rate based on the quantity of sulfated slimes generated yearly and the number of operating shifts yearly. The time for each unit process was broken into the individual steps required for the process and the estimated time required for completion. Based on material and energy balances, the time required for the various unit processes, the number which could be performed concurrently and those which required sequential operation, a preliminary estimate of equipment size was prepared. The in-process inventory of the sulfated slimes processing facility was determined based upon the minimum amount of process solutions available to produce the daily production of refined gold and silver. Transfer and filtration pumps were sized to minimize solution/slurry transfer time. In order to avoid solids remaining in the various processing reactors from batch to batch and to ensure high efficiency washing of filtercakes, filterpresses were sized to accommodate each reactor charge in an integral number of batches. Because of the deleterious effect of chloride ions on the extraction of silver as silver nitrate, the chloride and nitrate circuits were designed to prevent inadvertent transfer of solutions between them. Materials of construction were chosen based upon the results of the pilot plant studies. Heat transfer requirements were based on energy balances rather than the results of the pilot plant studies. This procedure was employed because of the difficulty associated with accurate measurement of steam consumption and accurate assessment of inlet cooling water temperatures. The accuracy and utility of this technique was ultimately confirmed during the startup of the commercial facility.

The equipment list and equipment layout, main floor and mezzanine, of the precious metals processing facility is shown below.

Major Equipment Within the Precious Metals Department

Equipment Designation	Volume	Type Construction	Chemical Exposure
Silver Production Area			
4 E. W. Tanks	1722 gal		Calcium/Silver nitrate
TS 7	3000 gal	Polypropylene	Nitric Acid (5%)
TS 4	1000 gal	Stainless Steel	Calcium/Silver nitrate
TS 3	5000 gal	FRP	" "
TS 5	5000 gal	FRP	" "
TS 2	5000 gal	FRP	" "
TS 1	5000 gal	FRP	" "
R 1 (Reactor)	5300 gal	Stainless Steel	Nitric Acid
R 2 (Reactor)	5300 gal	Stainless Steel	" "
Gold Production Area			
R 1 (Reactor)	1200 gal	FRP	Hydrochloric Acid/Cl_2
R 2 (Reactor)	1200 gal	FRP	" "
TG 10	1400 gal	FRP	" "
1st SX (North)	1400 gal	Polypropylene	Hydrochloric Acid
2nd SX (North)	1400 gal	Polypropylene	" "
Gold Pfaudler	200 gal	Glassed Steel	Dibutyl Carbitol
R 3 (Pfaud-SG)	300 gal	Glassed Steel	Hydrochloric Acid/Chlorine
Water Treatment			
TG 14	1124 gal	FRP	Hydrochloric Acid
TG 16	1124 gal	FRP	Hydrochloric Acid
Pfaudler/Evaprtr.	1000 gal	Glassed Steel	Hydrochloric Acid/N_2H_4
TG 17	650 gal	FRP/Polypropylene	"
TG 18	2036 gal	FRP/Polypropylene	" "
TG 19	2036 gal	FRP/Polypropylene	" "
TG 15	750 gal	FRP	" "

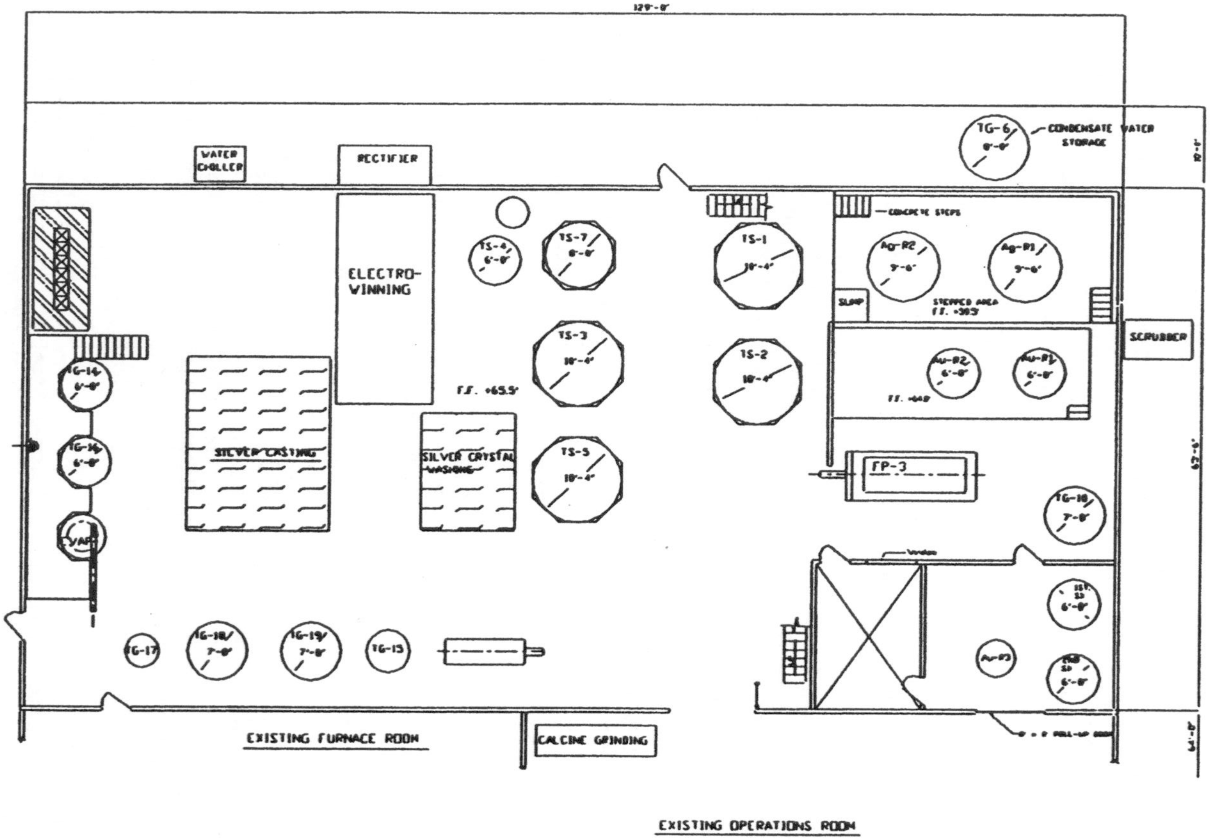

129'-0"

TG-6 8'-0" — CONDENSATE WATER STORAGE

WATER CHILLER

RECTIFIER

ELECTRO-WINNING

TS-4 6'-0"

TS-7 8'-0"

TS-1 10'-4"

TS-2 10'-4"

TS-3 10'-4"

TS-5 10'-4"

CONCRETE STEPS

Ag-R2 9'-6"

Ag-R1 9'-6"

SUMP

STEPPED AREA F.F. +38.5'

SCRUBBER

Au-R2 6'-0"

Au-R1 6'-0"

F.F. +44.0'

F.F. +65.5'

TG-14 6'-0"

TG-16 6'-0"

SILVER CASTING

SILVER CRYSTAL WASHING

FP-3

TG-10 7'-0"

TG-17

TG-18 7'-0"

TG-19 7'-0"

TG-15

EXISTING FURNACE ROOM

CALCINE GRINDING

EXISTING OPERATIONS ROOM

GOLD & SILVER RECOVERY PLANT--- LAY-OUT PLAN

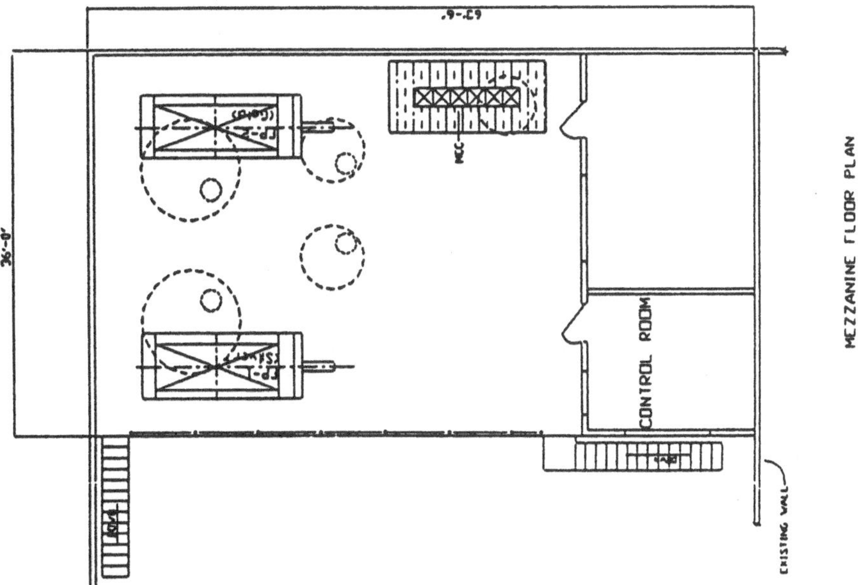

Capital Cost Estimate

The capital cost estimate for the processing facility was developed in-house as a collaborative effort with the Phelps Dodge Engineering Department. Equipment prices were solicited by bids, the structurals and building design were provided by a local architectural firm. Installed cost for various equipment items was done by factored estimates. Capital costs associated with workplace environment, safety, and emission control were developed at the local level and subject to corporate review.

Process Costs Estimate

Manpower requirements were developed from manning and scheduling information developed, part from the pilot plant operation and part from other operations at the Phelps Dodge Refinery. Reagent costs were developed from the material balances which had been calculated. Electrical costs were estimated based on the number and size of motors, rectifiers, and induction furnaces included in the plant design. Steam and cooling water requirements were developed from an energy balance.

Plant Operation

Plant Startup

Plant operations began when the silver recovery circuit had been partially completed. Because construction and installation required for the gold circuit had barely begun, plans called for the operation of the silver recovery circuit at 30-50% of its design capacity. The gold bearing silver leach residue was returned to the pyrometallurgical process. Although this process doubled the gold concentration of the Dore bullion, the gold concentration was still sufficiently low to allow parting in a conventional silver electrorefining operation. Initial results confirmed the pilot plant results both as regards the silver extraction obtained and the purity of the silver product obtained.

Startup Problems Problems encountered on startup of the silver recovery circuit included:

1) Plugging of the piping which conveyed the slurry of ground calcine from the grinding operation to the leaching circuit. Decreasing the solids to liquids ratio during the grinding operation relieved this problem; however the rate of grinding was reduced.

2) Excessive lime additions in the purification circuit due to the addition of dry lime. The slurry tank had not yet been installed for the addition of a dilute lime slurry.

3) Wide pH swings in the acid neutralization reactor in the silver electrowinning circuit. This was due to an excessively high delivery rate of lime slurry to the neutralization tank. Decreasing the pumping rate plus decreasing the percentage of hydrated lime in the slurry eliminated the wide pH swings.

4) Incomplete removal of silver bearing leach solution from the repulped leach residue. The countercurrent washing circuit had not been completed and filtercake wash volumes had been reduced below the design value to reduce the burden of water evaporation in the silver leach reactors. The problem was compounded by the fact that the device installed for grinding the calcine left a coarse partially ground phase present in the slurry which reduced washing efficiency in the filterpress.

5) Trace NOx emissions from the electrowinning cells and lower-than-design value current efficiency. The cause was identified as insufficient electrolyte flow to the electrowinning cells. The discharge weir was, in effect, lowered to allow a faster circulation rate to the cell without causing the cell to overflow.

Shutdown of the Smelter Facilities Because the new precious metals refinery was contiguous with the old pyrometallurgical facilities the ubiquitous dust associated with the smelting operation infiltrated the air in the new refinery, compelling all personnel to wear respirators. This included the contractor personnel involved with the installation of the gold refinery. This greatly hampered communication during the construction, resulting in falling behind schedule in the completion of the gold facility. Additionally, the fallout from the dust was a potential contaminant in the silver electrowinning circuit. The decision was therefore made to permanently terminate all smelting operations. The silver circuit continued to operate processing all sulfated slimes calcine and the silver leach residue was stored until installation of the gold recovery circuit was completed. This decision resulted in a backlog of approximately six weeks of silver leach residue. This backlog resulted in pressure to startup the gold circuit at full capacity immediately upon completion of its installation. Fortunately, the wet chlorination circuit was able to operate at full capacity immediately; however, problems in the solvent extraction circuit, particularly crud formation, hampered operations both in the SX circuit and the gold reduction. Raffinate exiting the gold recovery could not be treated for tellurium precipitation because this equipment had not been installed. As a result, the only option available was to evaporate the raffinate to reduce its volume and store the concentrated liquor until equipment was available for treating it. Viewed in retrospect, delaying the startup of the gold circuit until the downstream processing equipment had been installed would probably have been a better choice.

Full Onstream Operation

At the outset it should be stated that the precious metals refinery met expectations as regarded both precious metals extraction and precious metals purity. However to achieve those goals on a consistent day to day basis has required diligent attention to all aspects of the plant operation. Some of the operating difficulties encountered are discussed below.

Silver Leaching and Solution Purification

Silver Hydrolysis - During the purification of the pregnant silver leach liquor, ferric sulfate is first added followed by a hydrated lime slurry to elevate the pH and precipitate impurities. This hydrolytic purification /process developed for treating the pregnant leach has never failed to yield a highly pure feed for silver electrowinning. However, if the lime slurry is not rapidly mixed into the bulk of the solution, zones of local high pH will occur within the bulk of the solution, resulting in the hydrolysis of silver nitrate to silver hydroxide. At the terminal pH of the hydrolytic purification, the redissolution of the silver hydroxide is extremely slow. This results in silver hydroxide reporting in the filtercake sent to the repulping operation. To dissolve the silver hydroxide, additional nitric acid is added which ultimately results in the calcium nitrate concentration of the silver bearing solution gradually increasing, creating significant problems removing all nitrates from the filtercake prior to its processing by wet chlorination. Generation of silver hydroxide has been greatly reduced by: 1) programming the rate of addition of hydrated lime as a function of the pH in the reactor; 2) using as dilute a slurry of hydrated lime as possible consistent with maintaining the water balance in the system; 3) ensuring that the point of introduction of lime slurry into the reactor is at that location at which the most vigorous agitation occurs.

Repulp Filtercake Washing - In order to achieve the design silver extraction, virtually all soluble silver must be removed from the repulp filtercake. Additionally, and equally important, inefficient washing will result in nitrates remaining in the final repulp filtercake. These nitrates create profound problems in the DBC recovery circuit and the tellurium precipitation operation. Two causes of poor washing efficiency have been identified. The first is the presence of a coarse fraction of solids in the calcine leach residue. These solids settle to the bottom of the filtercake during filtration and result in short circuiting of the wash solution preferentially through the bottom of the filtercake. The presence of the coarse solids in the calcine is due to the use of a comminution device which unlike the pilot plant rod mill is a dispersion device rather than a true grinding device. Present plans call for the replacement of the present dispersing device with a ball mill or rod mill similar to that employed in the pilot operation.

The second cause of decreased washing efficiency is the use of a center feed press. This results in areas of the filterpress where little or no cake washing occurs. This is best addressed by replacing the present filterchambers with side ported feed chambers.

Silver Electrowinning - From time to time, the lead content of the electrodeposited silver increases to unacceptably high level, lead in the silver product of greater than 50 ppm. Lead is deposited on the anode as lead dioxide. The anodes are bagged in order to contain the lead dioxide; however, occasionally a bag is torn and lead reports in the silver deposit which collects at the bottom of the electrowinning cell. On rare occasions, lead dioxide crystals have actually grown through the anode bag and reported in the silver deposit. Attention to the condition of the anode bags is part of normal cell maintenance; additionally, the anode edges are covered with edge masking to reduce the likelihood of torn bags. Methods of precipitating the lead from the purified electrolyte have been considered; however, experience to date indicates that lead contamination can be controlled by proper anode bag maintenance.

Electrode contacts continue to require constant maintenance. The electrodes are considerably larger than those employed in a Mobius Cell and require substantially more current. Additionally, they weigh much less, reducing contact pressure. Methods of mechanically clamping them in place are presently being investigated.

Obtaining and maintaining a constant and uniform pH in the electrowinning cells has required changes in the hydraulic characteristics of the cell to allow a higher circulation rate through them.

Discharge of electrowon silver from the cell is accomplished by opening a hopper on the bottom of the cell and flushing the silver crystals into a filterbox. This requires that the cell be emptied and refilled. This is both time consuming and labor intensive. Additionally, since the cells are operated in series, all are taken off line simultaneously, resulting in lost production time. Other methods of silver discharge from the cell are presently being considered.

Wet Chlorination

Chlorine Efficiency - Chlorine efficiency has been less than anticipated based on the pilot plant studies. From the point of reagent cost, this is not a significant problem since so little chlorine is required. However, poor chlorine efficiency increases the time required for completion of chlorination. The problem is being addressed by relocating the chlorine sparge so as to direct the chlorine gas directly into the area of most intense mixing and changing the configuration of the chlorine sparge nozzle to increase the velocity of the chlorine gas at the point of injection into the slurry.

Handling Solutions Saturated With Chlorine - Because the chlorination operation is done at ambient temperature, the slurry contains a substantial quantity of dissolved chlorine. Because best gold extraction is obtained from the wet chlorination process when the solids/liquids separation is performed while the solution is still saturated with chlorine, the filtrate is also saturated with chlorine. This requires that all vessels which contain the chlorination be properly vented to the scrubber. Of more concern, is the fact that blowing the filtercake releases a large volume of chlorine in such a short time that it may overwhelm the scrubber capacity. This problem was solved by the expedient of performing the first filtercake wash prior to the cake blow. This has very little effect on the filtercake washing efficiency and resolved the potential chlorine emission problem.

Water Balance in the Wet Chlorination Circuit - The wet chlorination process, as envisioned, required four stages of countercurrent filtercake washing to ensure maximum gold extraction. Other process priorities have delayed the completion of the installation of the countercurrent washing circuit. As a result, the amount of solution fed to the solvent extraction circuit is greater than the design value creating problems with reactor capacity and solution storage. This condition is presently being resolved by completion of the countercurrent washing circuit.

Gold Solvent Extraction

Crud Formation - Solids which accumulated at the organic interface during solvent extraction, commonly called crud, have been the chief problem in the gold solvent extraction circuit. Crud problems increase as the ORP of the chlorination liquor decreases. At present, crud is controlled by filtration of the organic phase, the gold-loaded DBC, through a small filter press. The crud is washed as free of solvent as possible because of the high value of the gold content of the organic phase. Although filtration of the crud from the organic phase has proven successful, studies continue at identifying the mechanism of crud formation in order to eliminate it entirely.

Handling DBC - DBC is hostile to most polymeric substances exposed to it. In our experience, this includes Viton and Hypalon. Thermosetting vinyl ester resins are softened by it. This has limited materials of construction to polyolefins and fluoropolymers. This has created problems in the selection of pumps for filtration and transfer of DBC.

Gold Reduction

Assays of the gold powder produced and the final melted gold product made it apparent that contamination during melting of the gold was the cause of impurities in the final gold product. Contamination was most readily controlled if melting was performed in an oxidizing atmosphere using a silicon carbide crucible.

Silicon Contamination - Silicon contamination was apparent from time to time in the final melted gold product, despite the fact that none was detectable in the loaded DBC from which the gold was reduced. The cause of contamination was traced to the presence of small amounts of DBC adhering to the wet gold powder product from reduction. During induction melting of the gold powder in a silicon carbide crucible, the DBC pyrolyzed forming carbon. At the very high temperature on the inside surface of the crucible, this carbon reacted with the silica coating on the silicon carbide crucible forming small quantities of silicon which immediately alloyed with the molten gold. Silicon contamination was eliminated by the simple expedient of heating the gold powder in an oxidizing atmosphere to a temperature of 300-400 °C to eliminate all organic residue prior to melting. This procedure conveyed the additional benefit of densifying the gold product, making the subsequent melting operation considerably simpler.

Effluent Treatment

Hydrazine Consumption - Hydrazine was chosen for the detellurizing/precious metals collection process because of the ease of introduction, the absence of an odor in the final filtrate, and the fact that product of the reaction was nitrogen leaving no reduction byproducts requiring disposal. However, when nitrates were present in the chlorination liquor the consumption of hydrazine increased tremendously. The resolution of this problem, as mentioned earlier, was effected in the silver leach residue filtercake washing operation.

Hexavalent Selenium - Occasionally hexavalent selenium is detected in the gold free raffinate. The quantity of hexavalent selenium rarely exceeds 100 ppm. Two techniques have been employed for reducing hexavalent selenium to the quadrivalent state (the form in which it can be readily precipitated from solution). The first is simply concentrating the hydrochloric acid concentration in the raffinate to six molar by evaporation. This causes the spontaneous reduction of the hexavalent selenium to the quadrivalent form in which it is readily removed from solution. The condensate from evaporation can by recycled to the wet chlorination process. A second, simpler procedure is to perform the tellurium reduction; then, heat and agitate the slurry of tellurium particulates until the hexavalent selenium is reduced. Both methods have been employed successfully for selenium reduction; however, it would be preferable to avoid the formation of hexavalent selenium entirely.

Results to Date and Process Comparison

Furnacing production during the year of 1989, which was typical for any year when Dore was produced, will be used as a basis for comparison.

The silver and gold contained in the slag recycled to the Hidalgo smelter during the base year was 21% and 13% respectively of that contained in the copper refinery slimes. The total silver and gold contained in the final residue filter cake in the new hydrometallurgical operation is 9% and 2-3% respectively of the total silver and gold production.

The new hydrometallurgical operation does not require fluxing agents as did the old furnacing operation. This has resulted in a 44% reduction in reverts returned to the smelter.

Silver and gold inventories have been reduced 72% and 80% respectively. This was realized due to the decreased silver and gold in reverts and in process inventory.

As a result of the new hydrometallurgical operation, an expenditure of $1.5 million mandated for ventillation and offgas scrubbing for the old furnacing operation was avoided.

Production (troy ounces) since startup of the new hydrometallurgical plant is as follows:

	Silver	Gold	Palladium	Platinum
Startup Year	1,400,000	41,000		
First Year	2,400,000	78,000	1300	200
Second Year	3,100,000	86,000	2,000	200

The old furnacing operation produced Dore with the following assay:

Ag	Au	Pt	Pd	Cu	Te	Bi	Pb	O	Se	Fe
ppt	ppt	ppt	ppt	ppm	ppm	ppm	ppm	ppm	ppm	ppm
978	20.5	.12	1.0	395	40	44	20	187	10	2

Typical assays of the new hydrometallurgical operation for silver and gold are 99.95% and 99.995% pure respectively.

Silver Typical Assay (ppm)

Cu	Ni	Fe	Zn	Cd	Pb	Sb	Te	Se	Bi	Sn
1	<1	<1	<1	<1	45	1	<1	1	<1	<1

Gold Typical Assay (ppm)

As	Te	Bi	Pt	Cu	Fe	Si	Se
<1	<1	<1	<1	2	1	1	<1

The old furnacing operation used two Venturi scrubbers, one electrostatic precipitator, two cyclones, two quench towers, and two packed towers to scrub 40-50 tons of sulfur dioxide and 180-190 tons particulates per year. The new hydrometallurgical operation uses one lime scrubber to handle 1-2 tons of chlorine per year.

Materials production and the environment

R. C. Villas Bôas
CETEM—Centre for Mineral Technology, Cidade Universitária, Rio de Janeiro, Brazil

Abstract
Materials play a fundamental role in developing a nation and in maintaining or increasing its share in the world's economy. However, any material to be produced has in its transformation cycle, at least one extraction, processing, fabrication and manufacturing step in which releases of substances, gases, liquids, or solids, occur to the environment.
This paper addresses some environmental problems associated with the extraction and processing of some non-fuel, non-ferrous commodities that are of major interest to the hydrometallurgist attempting to design environmentally sound processes.

1 Introduction

The production and utilization of materials in general, and therefore, those of ores and metals, obey, within a given framework of industrial development, the economic cycles that are in effect during a given time period. These cycles have been well-discussed in the literature[1-4] and may reflect a world, a local, or a geopolitical trend.

As the selection of a given set of materials depends upon the predominant cycle in the industrialized countries, this determines, to a greater or lesser extent, the consumption pattern of a given commodity, inducing the market to adapt itself to such a new reality.

In materials-based industries, two general strategies arise: the search for materials that suit an available technology; and the development of technology for an available material. Recycled materials, the use of which in industry varies from economy to economy, need, as a general rule, lower capital and energy expenditures and more manpower than primary materials. Also, their processing involves lower pollution control costs than primary processes. Such recycling is more intense as the sophistication of the economy increases, since viable quantities of recycled materials must be available in order to reutilize them.

However, as important as they are in the world's economy, the production of materials promotes changes in the environment: materials require energy for processing, land for installation of process plant and disposal sites to receive tailings or waste, and the processes give off gases and dusts, and require water and earth moving. In fact, since earliest times, such environmental impacts have been recognized and some actions taken, here and there, to minimize them or at least to contain them within acceptable limits. Such acceptability, of course, changes from time to time, as social pressures increase, forcing legislative decisions that promote technological alternatives which, in turn, reflect on the economy.

Regarding the environment, two major questions are receiving worldwide attention: what are the effects linked to the production, disposal and use of materials, and what is their availability in the foreseeable future?[5]. This paper tries to focus on the first of these issues, reviewing some of the facts and figures of the environmental impacts caused by a limited range of commodities, within the non-fuel, non-ferrous industries.

2 Average metal recoveries and production steps

For any material to be produced, there are corresponding **steps** in which **discards** are also produced. These **discards** may be of two broad categories: **losses** and **effluents**. **Losses** are those discards readly identified with the main material produced, i.e., parts of that material that are left behind throughout the production steps. **Effluents** are the discards coming from these same steps and that are inherent to the applied technology within each production step, but not necessarily identified with the main material.

2.1 Average metal recoveries

In order to systemise the analysis of the environmental impacts of the discards, an attempt is going to be made to quantify such average metal losses. No universal claims are made about the reasonings used, but they may help to point out the importance of the facts and impacts.

It is well recognized that ore recoveries, from mining to final metal product, vary from country to country and from economy to economy, as a function of technology, skill, regulatory laws, finantial capability, etc.; so also do the environmenal impacts caused by the production of primary and secondary metals. Therefore, **recoveries** and **losses** vary substantially from metal to metal and even for the same metal from country to country, even when apparently similar technologies are used, due to the so-called "particularities" of the mining world: the cut-off grade and the compromise between grade versus recovery, making each orebody unique in its physical and economic characteristics.

Other things being equal, the lower the grade or the poorer the quality of the ore, the higher will be the cost of recovery of the valuable products. To the extent that there is a choice of the grade of the ore to be mined, there is also a choice of the total tonnage and of the total product recovered; the lower the permissible grade, the higher the tonnage. Therefore, the fixing of the cut-off grade in deposits of irregular grade distribution may require several computations of alternative tonnages and grades on the basis of different assumptions as to mineable limits.

Equally as important as grade is the workability of the ore, which is measured by the cost of physical removal of the rock. Other factors, such as accessibility from mine openings, thickness and regularity of the ore zone, hardness and toughness of the ore and the presence of interfering structures, such as faults, weak ground, etc., must all be evaluated when the decision on which ore should be mined is made.

Variations in the grade and in the workability of an ore body, may go side by side, or they may partly compensate each other. Ores of many different grades and many different costs, but sufficiently similar in other qualities to be amenable to the same treatment process, may be mined or blended to allow profitable recovery of what otherwise would be marginal ore.

Complete removal of all the available ore from a mine, or complete extraction from the mined ore is never achieved. Cost per unit recovered rises almost continuously and usually with increasing steepness as attempts are made to increase the proportion extracted. In the short run, with a given recovery plant, the percent extraction of metal will depend, to some extent, on the grade of the ore itself. The mining method usually limits the recovery of the ore in the mine[6] [7], so too does the processing technology utilized. For **gold leaching**, for example, the recovery figures are shown in Table 1.

Table 1. Gold leaching recoveries [8].

OPERATION	PARTICLE SIZE	METALLURGICAL RECOVERY	COSTS
Agitation	< 0,1mm	90 - 95% > 20h	↑IN ↑OP
Vat	< 10mm	70 - 80% 3 to 4h	↑IN
Heap	> 10mm	40 - 60% 3 to 4w	↓IN ↓OP

IN = investment costs
OP = operational costs
h = hour
w = week

Let us now look at some selected mineral commodities, with regard to recoveries and grades, as shown in Table 2. It can readily be seen that the problems associated with earth moving and tailings disposal are quite severe, since from the grade of the ore up to the production of a saleable concentrate, the mass of the produced concentrate, related to that of the ore total (MC) and the mass recovery itself (MR) are, of course, far from the sustainable target of total utilization, for the reasons mentioned above.

A very striking example of recovery, grade, mass recovery, earth moving and generated by-products is the production of phosphate fertilizers from volcanic rock, which, besides the usual earth-moving and disposal problems associated with the production of the concentrate, generates five times the mass of gypsum compared to the concentrate, P_2O_5 , when the concentrate is reacted with sulphuric acid to produce the fertilizer.

Table 2. Selected mineral commodity recoveries and grades

ORE	RECOVERY*	GRADE	COMPANY
Nb_2O_5 (3,0%) piroclore	MC = 3,3% Ore MR = 66%	60% Nb_2O_5 Conc.	CBMM [9]
TiO_2 (1,5%) ilmenite	MC = 2,2% Ore MR = 81%	55% TiO_2 Conc.	RIB [9]
Cr_2O_3 (17%) chromite	MC = 28% Ore MR = 65%	37% - 46% Cr_2O_3 Conc.	FERBASA [9]
WO_3 (0,5%) scheelite	MC = 0,49% Ore MR = 74%	75% WO_3 Conc.	TUNGSTÊNIO [9]
Sn (1,3%) cassiterite	MC = 1,9% Ore MR = 69,1%	48% Sn Conc.	RENISON [10]
Ta_2O_2 (0,16%) tantalite	MC = 0,22 Ore MR = 70%	49% Ta_2O_5 Conc.	BERNIC [10]

*MC stands for the mass of produced concentrate, as related to that of the ore total, in percent, and MR is the mass recovery, i.e., the recovered amount of the valuable commodity as related to the original amount in the ore.

2.2 The production steps

Let us describe such production steps and their discards, identifying four steps, namely **extraction, processing, fabrication** and **manufacturing**, as follows:

- **extraction step**, i.e., the mining and beneficiation of the ore to produce a commercial concentrate. The **losses** are dependent upon the mining method (open pit, cut-and-fill, room-and-pillar, etc.) and beneficiation techniques (gravity separation, flotation, etc.); the **effluents** generated are gases (CO_x and NO_x) from machinery and equipment, process waters and contaminated ground waters, particulate material, and earth moving disposals and rearrangements;
- **processing step**, i.e., the extractive metallurgy or chemical operations which convert a concentrate into a metal; the **losses** are depend upon the chosen technologies and skills involved (pyro-, hydro- and/or electro-metallurgical), and the **effluents** generated are gases (CO_X, NO_X, SO_X), liquids (heavy metal contamined waters) and solids (sediments, and heavy metal dusts);
- **fabrication step**, i.e., those operations devoted to the production of rods, bars, sheets, etc.; the losses are scrap materials resulting from those operations, denominated "home scrap" [11] endlessly recirculated, without any net loss of metal; the **effluents** are waste waters and gases;
- **manufacturing step**, i.e., the application of mechanical operations for the shaping of metals by machining, stamping and forging, other than those of the fabrication step; the **losses** are metal waste resulting from such mechanical treatments that do not produce the final product, being denominated "new scrap" or "prompt scrap" [11-14], for which recycling is well organized and efficient[12, 13]; the **effluents** are water vapors and gases.

The **production steps**, as indicated, are all illustrated in Figure 1. Such a flowchart, or Sankey diagram, helps to provide solutions related to the **discards** involved in each production step. A similar chart might be attempted in terms of overall mass flows (MC's), if defined for each production step, resulting in more realistic figures, since then earthmoving would be included.

The average metal recovery figures from ore to metal used in Figure 1, involving the **extraction** and **processing** steps are those of HASIALIS [15] and those for the **manufacturing** step are from MAR [14]. It is acknowledged that this last figure is historical for the U.S.A., where it was obtained in 1954; however, in other parts of the world, such a figure might still be reasonably valid. As for HASIALIS' data, these are average figures, and large departures from them, for a given case, may occur.

A detailed examination of each of the aforementioned **four** production steps will, hopefully, produce a clearer picture. For this, some explanations are needed in order to be able to follow Figure 1 and the remaining diagrams in the text:

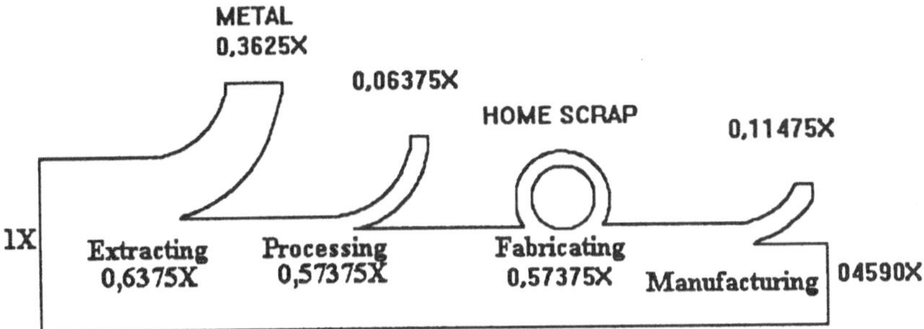

Figure 1. The production steps

X = the metal content of the *"in situ"* ore
L_E = is the **loss** of metal resulting in the extraction step, and is equivalent to 0,3625 X.
P_E = is the **product** in metal originated from the extraction step, and is equivalent to 0,6375 X.
L_P = is the **loss** in metal resulting from processing, and is equivalent to 0,0637^5 X.
P_P = is the metal in the **product** resulting from processing and is equivalent to 0,5737^5 X.
L_F = is the **loss** of metal resulting from fabrication and is equivalent to O X (endlessly recirculated).
P_F = is the metal in the **product** resulting from fabrication, and is equivalent to P_P.
L_M = is the **loss** of metal resulting from manufacturing, and is equivalent to 0,1147^5 X.
P_M = is the metal in the **product** resulting from manufacturing.
E_i = is the **effluent**, generated in each stage.

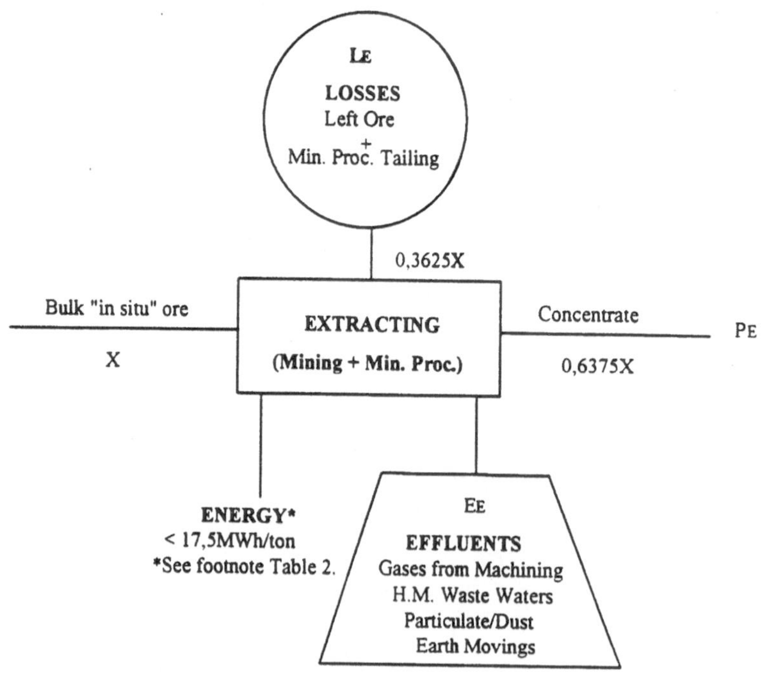

Figure 2. Input/output of the extraction step.

2.3 Identifiable environmental impacts and prospects in the extraction step
Consider Figure 2:

A. **Energy:**

Energy consumption - there is room for technical improvement. Figures in kWh (thermal), per tonne of primary metal, Al (10,175); Cu (17,420); Zn (1,240) [16].

B. **Losses:**

B.1 Unrecovered ore - function of cut-off and mining method - there is room for technical improvement.

B.2 Mineral processing tailings - room for gains pending improvements in the next step (processing), since commercial grade concentrates are inputs to a given processing technology.

C. **Effluents:**

C.1 Mining - earth moving impacts associated with land reclamation - room for improvement based upon compromises between legislation (function of social pressures) and costs of reclaiming. Physical disturbances are permanent; dust.

C.2 Mining - gases from machinery and equipment (as well as noises and vibrations) - there is room for technical improvement.

C.3 Mining - disruption of water regimes - little room for improvement with current mining methods.

C.4 Mineral processing - process waters and dust, still room for technical improvement.

C.5 Mineral processing - tailing disposals, solids, and control of acid generation.

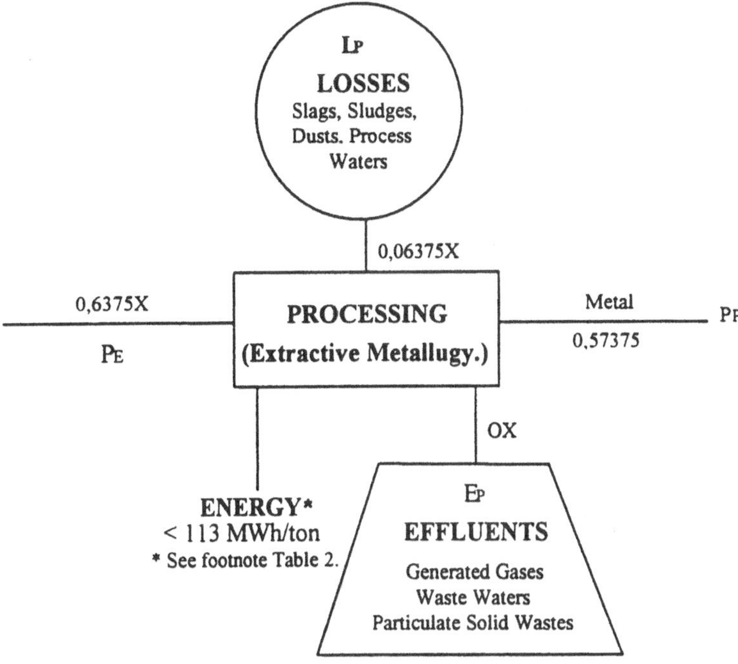

Figure 3. Input/outputof the processing step.

2.4 Identifiable environmental impacts and prospects in the processing step

Consider Figure 3:

A. Energy:

Energy consumption - there is room for improvement. Figures in kWh per tonne (thermal); Al (35,384); Cu (26,520); Zn (17,560); Mg (103,000) [16]. Other figures are reported for Al and Mg if hydro-based power is available (much lower figures).

B. Losses:

There is room for improvements, especially for processes devoted to the recovery of metal from slags, sludges and dusts from existing technologies, or new technologies designed to decrease the number of operations or equipment stages (e.g., continuous converting for copper and a still-awaited solution to the red mud problem in aluminium production).

C. Effluents:

Generated process gases (CO_x, NO_x, SO_x); waste waters after eventual removal of metal(s) from process waters; particulates throughout thc processing stages and solid

wastes other than slags, sludges, etc... (for the aluminium industry, for example, spent potlinings, drosses, electrodes, etc...) - still room for technical improvements.

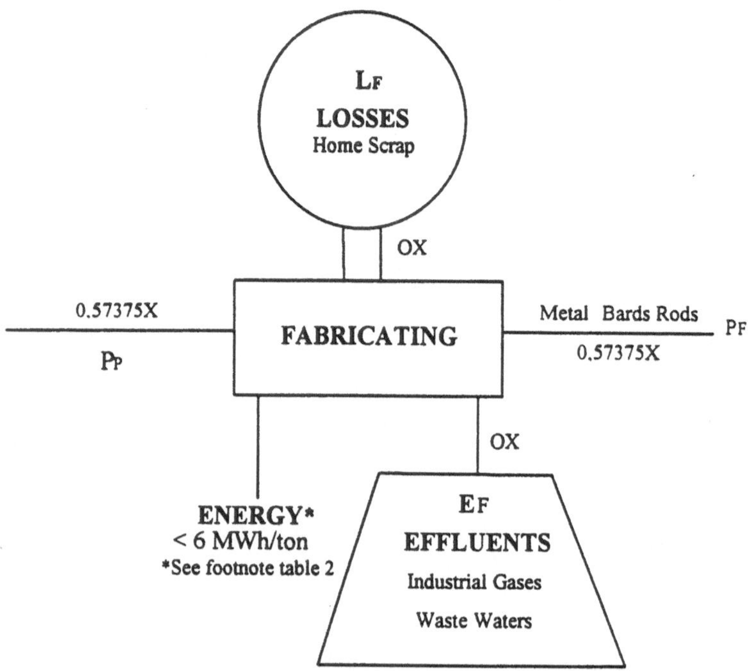

Figure 4. Input/outputof the fabrication step.

2.5 Identifiable environmental impacts and prospects in the fabricating step
Consider Figure 4:

A. Energy:

Energy consumption - room for some improvements. Figures in kWh/tonne (thermal); Al (4,937); Cu (5,970); Zn (1,492) [16].

B. Losses:

Generation of home scrap, no net losses. However, room to reduce such generation as fabrication operations and equipment become more efficient.

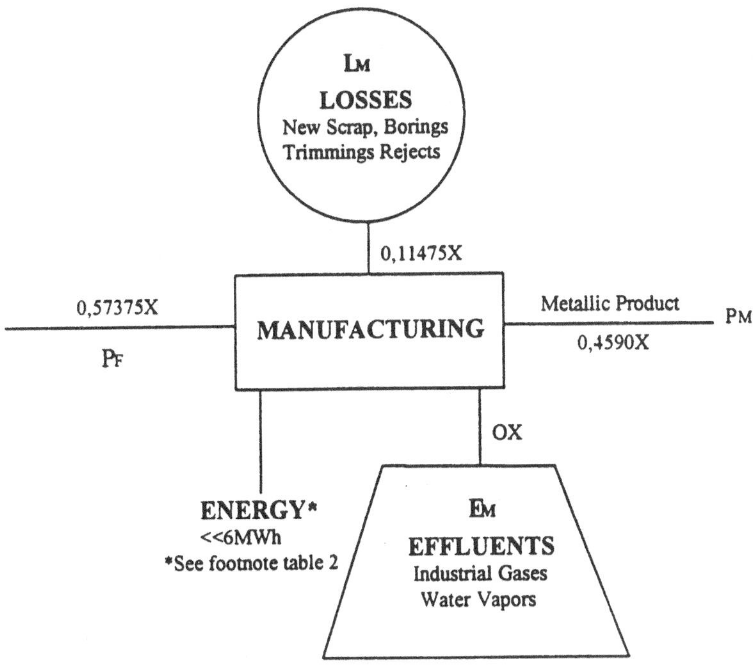

Figure 5. Input/output of the manufacturing step.

2.6 Identifiable environmental impacts and prospects in the manufacturing step

Consider Figure 5:

A. Energy:

Consumption is quite variable depending on the particular metallic product, through forging, stamping and machining. Much less consumption than any of the other previous production stages.

B. Losses:

The so-called new scrap that usually goes to secondary production.

C. Effluents:

Gases and water vapors.

3 Role of the hydrometallurgist

For an account of the role of hydrometallurgy in achieving sustainable development, the interested reader is referred to CONARD [17], where acidic mine drainage, metal removal from waste streams, arsenic management, reduction of gaseous pollutants and energy conservation, cyanide destruction, waste processing and product recycling are matters discussed through selected examples of hydrometallurgical technologies as applied to a better environment.

For those interested in research and novel techniques in hydrometallurgy and the aqueous processing of materials and industrial minerals, the recent review of DOYLE and DUYVESTEYN [18] is recommended, as well as that of NICOL [19] on electrometallugy. If the interest is in the energy requirements for the manufacture of some non-ferrous metals and for processing ingots and finished products, in addition to the previously mentioned references, the reader is also referred to HANCOCK [20] and WHITTER & HOSKINS [21].

The points to be raised in this section, however, are those of a general nature that may guide the hydrometallurgist towards a better understanding of the overall effect that a given process has upon the environment, thus hopefully enhancing his/her chances of designing environmentally sound processes.

Let us raise some major points in each of the **input/output** of the production steps, namely **energy, losses** and **effluents**.

3.1 Energy

Table 3 lists the energy consumed in each production step.

Table 3. Energy utilized in each production step

PRODUCTION STEPS	ENERGY (MWh [thermal]/ton)*
Extraction	< 17.5
Processing	< 113.0
Fabrication	< 6.0
Manufacturing	<< 6.0

* Figures are maxima, not averages, for a selected class of metals (Al, Cu, Zn, Mg, Ti)

The role of the hydrometallurgist is to seek processes that minimize energy consumption; his/her tasks are, thus, primarily devoted to the **processing step**, followed by the **extraction step** and, then, **fabrication** and **manufacturing**. Indeed, the efficiencies of the processing operations have been compared by CHAPMAN & ROBERTS [13], also in other papers dealing with the subjects of environment, metal production and energy, such as YOSHIKI-GRAVELSINS, et al.[16], and FORREST & SZEKELY [22].

For the purposes of this paper, the overall energy efficiencies in the processing step, i.e. the energy take up by the whole step and not just the direct one, as compared to the thermodynamic Gibbs Free Energy, G, for that same processing step, are of interest, since they give a good indication about where to search for process improvements. Table 4 lists some selected metals and their overall efficiencies [13],[16].

Table 4. Overall energy efficiences of the **processing step** for selected metals.

PRIMARY METALS	OVERALL ENERGY EFFICIENCIES* (%)
Al	13
Cu	1.4
Zn	5.5
Mg	6.1
Ti	4.1

* Energy take up by the whole step, as related to the Gibbs Free Energy.

Of great concern to the hydrometallurgist is the **source** of energy used, i.e., hydro or coal based, due to the greenhouse effect. Such a concern was extensively discussed by FORREST & SZEKELY [22].

3.2 Losses

Table 5 lists the average metal losses to the environment, in each production step. Here, the hydrometallurgist has to focus his/her attention on the **extraction step**, first, and to a lesser degree, on the **manufacturing step.**

Table 5. Metal losses to the environment per production step.

PRODUCTION STEP	AVERAGE METAL LOSSES*
Extracting	0,3625 X
Processing	0,06375 X
Fabrication	0
Manufacturing	0,11475 X

* Average metal loses as referred in the text.

It is worth pointing out, however, that these average figures may be misleading. For each particular metal/substance that the hydrometallurgist is studying, he/she has to refer to the actual values that are particular to the mining method, metal, process, skill, country, etc., as discussed above. Nevertheless, mining and minerals processing techniques are, in general, responsible for the greatest losses. In-situ mining techniques, which usually refer to the injection of a leach solution through boreholes into the ore, should be considered wherever possible [17].

The losses in the **manufacturing step** usually go to secondary recovery and aside from the strategic and economic aspect to the enterprise itself, as discussed by

CHAPMANN and ROBERTS [13] through the GER (gross energy requirement) concept, no major role for the hydrometallurgist can be foreseen, since such efficiencies are generally linked to the mechanical, electronic, or physical metallurgical aspects of the issue.

In the **processing step**, several improvements have been and continue to be made through process optimization and process improvements [16-18].

3.3 Effluents

Regarding effluents, discards to the environment are several, considering the liquid, the gaseous and the solid states, giving the hydrometallugist an extraordinary opportunity and offering several challenges.

Table 6 gives a list of problems that require solutions for each of the production steps, comparing in relative terms the impacts on land, water and air; the comparisons are made referring to acceptable environmental standards in the OECD countries, but they may vary considerably from country to country and from metal to metal.

Table 6. Comparisons between the impacts of the effluents for each production step.

PRODUCTION STEP		IMPACT	
	LAND	WATER	AIR
Extraction	S	S	M
Processing	M→S	M→S	S
Fabrication	L	L	S
Manufacturing	L	L	L

L ≡ low impact
M ≡ moderate impact
S ≡ severe impact

For the identification of the specific problems that face each particular metal industry, the reader is referred, for instance, to references [23] and [24].

Thus the role of the hydrometallurgist in developing environmentally sound processes has to be focused particularly on the **extraction step** (i.e., land disturbance, soil erosion, mine run-off water, water regimes, dust and tailing disposal, revegetation, etc...) and the **processing step** (i.e., acid generation, heavy metal effluents, disposal of solids, gas generation). For the specific techniques (biosorption, liquid-liquid exchange, electrowinning of dilute solutions, membranes, etc.) see references [17],[18] and [19]. Table 7 lists some environmental impacts associated with selected mineral industries.

Table 7. Major environmental impacts for selected mineral industries.

METAL	IMPACT
Al	Red mud slurry; HF; CO_2; tar pitch volatiles; spent pot linings; cyanide
Cu	SO_2; metal fumes; heavy metal effluents
Zn	Iron oxide; SO_2; Cd; heavy metal effluents
Mg	CHCs; dioxin
Ti	$FeCl_3$; volatile chlorides; CO_2
Ni	Metal carbonyl; heavy metal leachate;severe dusts and particulate emissions
P_2O_5	Gypsum, water consumption and disposal; radiation (whenever present)

4 Minerals that benefit the Environment

So far, discussion has been focused on the effects on the environment due to the production of minerals. It is worth remembing, however, that minerals can be viewed not just as villains, but also as assets to the environment. Tightening environmental legislation is forcing the regulations governing waste water treatment and disposals to be stricter; bentonite, lime, soda ash, magnesium hydroxide and zeolites are reported as environmental aids in the literature [25] and hence open a vast field of investigation to the hydrometallugist or mineral technologist.

5 Conclusions

It is hoped that this discussion of the **production steps** always present in the production of materials, namely, **extraction, processing, fabrication**, and **manufacturing**,which incorporate the inputs/outputs for each of these steps, namely, the **materials, energy, losses** and **effluents**, will help the hydrometallurgist to choose the relevant areas of his/her research interests in order to aid in the design of environmentally sound processes to promote sustainability.

No universal claims, regarding the average or maximum figures presented throughout this text, are made; instead the figures are for illustrative purposes only, in an attempt to develop some guidelines that the hydrometallurgist can use in his/her efforts to achieve sustainability.

6 Acknowledgements

The author wishes to thank the International Development Research Centre (IDRC), of Canada and the Conselho Nacional de Desenvolvimento Científico e Tecnológico (CNPq), of Brazil, for their financial support of the project from which this paper developed. The author, of course, has the sole responsability for the ideas herein expressed.

7 References

1. Malenbaum, W. (1978) **World Demand for Raw Materials in 1985 and 2000**; in University of Phyladelphia Publication Series, U.S.A.
2. Tilton, J.E.(1986) Atrophy in Metal Demand; **Materials and Society, 10,** nº 3.
3. Waddell, L.M. and Labys, W.C. (1988) Transmaterialization: Technology and Materials Demand Cycles; **Materials and Society, 12,** nº 1 .
4. Villas Bôas, R.C. (1987) Strategic Ores: Worldwide and Brazilian Prospectives; Second Southern Hemisphere Meeting on Minerals Technology, **Proceedings,** Rio de Janeiro.
5. Anon, (1993) Materials and Environment, where do we stand, **Minerals Today, our Materials World**: A Special Edition, U.S.B.M., April, 1993
6. Villas Bôas, R.C. (1976) Aluminium: Why Search for New Production Routes? **Proceedings of the IV National Meeting on Minerals Processing,** São José dos Campos, Brasil.
7. Corry, A. V. & Kiessling, O.E. (1938) Grade of Ore, Works Progress Administration, National Research Project, **Mineral Technology and Output per Man Studies,** USBM, Report E-6, August, p. 114.
8. Bahr, A. and Priesemann, Th. (1988) The Concentration of Gold Ores, **Workshop Rare and Precious Metals,** Castelo Ivano,Universitá di Trento, Italy.
9. Benvindo da Luz, A. et al. (1990); **Manual de Usinas de Beneficiamento,** publicação avulsa, CETEM, Rio de Janeiro.
10. Ottley, D.J. (1979) Technical Economic and other Factors in the Gravity Concentration of Tin, Tungsten and Tantalum Ores, **Minerals Sci. Engng.,** **11**(2), 99-121.
11. Beever, M.B. (1982) Materials, Technology Change and Productivity, **Materials & Society, 6,** nº 4.
12. Beever, M.B. (1976) The Recycling of Metals: I - Ferrous Metals; II - Non-Ferrous Metals, **Conservation & Recycling, 1.**
13. Chapman, P.F. and Roberts, F. (1983) **Metal Resources and Energy,** Boston, MA: Butterworth.
14. Mar, J.W. (1981) **Testimony at Hearings of the Subcommitee on Science, Technology and Space of the Comittee on Commerce, Science and Transportation of the Senate,** Washington. D.C., U.S.G.P.O.
15. Hasialis, M.D. (1975) Improvements in Minerals Recovery, National Materials Policy. **Proceedings,** National Academy of Science, Washington, D.C.

16. Yoshiki-Gravelsins, K.S. et al.(1993) Metals Production, Energy and the Environment, Part I: Energy Consumption, **JOM**, pp. 15-20, May.
17. Conard, B.R. (1992) The Role of Hydrometallurgy in Achieving Sustainable Devolpment, **Hydrometallurgy, 30,** 1-28.
18. Doyle, F.M. & Duyvesteyn, S. (1993). Aqueous Processing of Minerals, Metals, and Materials, 1993 Review of Extraction Processing, **JOM**, pp. 46-54, April.
19. Nicol, M.J. (1993) Progress in Electrometallurgy Research and Applications, 1993 Review of Extractive & Processing, **JOM**, pp. 55-58, April.
20. Hancock, G.F. (1984) Energy Requirements for Manufacture of some Non-Ferrous Metals. **Metal Technology, 11**, July, 290-299.
21. Whitter, W. and Hoskins, C. (1984) Energy Required to Process Ingots Semis, and Finished Products, **Metals Technology, 11**, July, 300-307.
22. Forrest, D. & Szekely, J. (1991) Global Warming and the Primary Metals Industry, **JOM**, 23-30, December.
23. UNIDO. (1987) **Pollution Problems and Solutions in the Non-Ferrous Metals Industry**, First Consultation on the Non-Ferrous Metals Industry, ID/WG. 470/3, Budapest-Hungary.
24. UNEP. (1993) **Environmental Management of Nickel Production: A Technical Guide**. Paris, (Technical Report, 15).
25. Harries-Rees, K. (1993) Minerals in Waste and Effluent Treatment, **Industrial Minerals**, 29-39, May.

Fundamentals

Reaction of galena in ferric sulphate–sulphuric acid media

J. E. Dutrizac
T. T. Chen
CANMET, Ottawa, Canada

Abstract
The reaction of galena (PbS) in $Fe(SO_4)_{1.5}$-H_2SO_4 media was investigated over the temperature range 55-95°C and for various $Fe(SO_4)_{1.5}$, H_2SO_4, $FeSO_4$ and $MgSO_4$ concentrations. Because of the low solubility of the $PbSO_4$ reaction product, the rates were monitored using the ferrous ion co-generated during leaching. Relatively slow, non-linear kinetics were consistently observed; in most instances, the parabolic rate law was closely obeyed. The non-linear kinetics were attributed to the formation of a tenacious layer of $PbSO_4$ and S° on the surface of the galena. The reaction of galena increased rapidly with increasing temperature, and the apparent activation energy is 61.2 kJ/mol. The rate increases as the 0.5 power of the ferric ion concentration, but is nearly independent of the concentration of the $FeSO_4$ reaction product. The rate is insensitive to H_2SO_4 concentrations <0.1 M, but increases at higher acid levels. The presence of neutral sulphates, such as $MgSO_4$, decreases the leaching rate to a modest extent. The overall galena reaction rate seems to be controlled by both the chemical reaction at the PbS surface and the progressive blockage of the surface by the $PbSO_4$ and S° reaction products.
Keywords: Elemental sulphur, ferric sulphate, galena, leaching, lead sulphide, parabolic kinetics, reaction products, sulphuric acid.

1 Introduction

In contrast to the extensive volume of documentation on the dissolution of galena in ferric chloride solutions [1], comparatively little is known about the reaction of galena in ferric sulphate-sulphuric acid media. Qualitative observations on the reaction of the galena contained in zinc concentrates during O_2-$Fe(SO_4)_{1.5}$-H_2SO_4 leaching at ~150°C suggest that the galena reacts rapidly to form $PbSO_4$ and elemental sulphur [2]. Despite its reported ease of reaction during pressure leaching, however, some galena is invariably present in the pressure leach residues, and the galena usually is rimmed by $PbSO_4$ [3]. Broken Hill Associated Smelters has developed an oxygen pressure leach process for PbS-Cu_2S mattes that is carried out in a sulphate-chloride medium. Recent work on this system, however, suggests that the PbS is actually attacked by the copper sulphate reaction product to generate $PbSO_4$ and CuS without any elemental sulphur formation [4]. Various efforts have also been made to use $Fe(SO_4)_{1.5}$-H_2SO_4 or O_2-$Fe(SO_4)_{1.5}$-H_2SO_4 media to leach both conventional lead concentrates and lead-rich pyritic bulk concentrates. The intent was to convert the galena to $PbSO_4$ which would be subsequently solubilized for lead recovery. Vizsolyi et al. [5] examined the reaction of unsized -45 μm (-325 mesh) lead concentrates in oxygenated H_2SO_4 media. The kinetics were significantly enhanced in the presence of dissolved ferric sulphate. In fact, the reaction was postulated to occur via a ferric ion intermediary; i.e., by ferric sulphate leaching. The electrochemical behaviour of galena in sulphuric acid media has also been investigated [6]; it was noted that the surface of the galena was readily passivated by the $PbSO_4$ or basic lead sulphate reaction products.

As noted above, the reaction of galena with ferric sulphate-sulphuric acid media is of considerable importance to lead hydrometallurgy. Relatively little work has been reported on this system, however, and this is at least partly a consequence of the low solubility of the $PbSO_4$ reaction product that makes the experimental work difficult. Accordingly, the present study was undertaken to help elucidate the kinetics of reaction of galena in $Fe(SO_4)_{1.5}$-H_2SO_4 media by following the ferrous ion reaction product co-generated during leaching and by using mineralogical methods to help explain the leaching results.

2 Experimental

2.1 Galena

Natural galena crystals from Galena, Kansas were cleaved into ~1 cm cubes, and "clean" cubes were selected. The galena cubes were crushed dry and the sulphide was then wet screened using acetone. The various size fractions were air-dried and, were then kept overnight under a dynamic vacuum to eliminate all traces of the acetone. Microscopic study of the galena did not detect any impurity phases. The measured bulk composition of the galena was 86.03% Pb and 13.63% S; this is essentially the composition of PbS (86.6% Pb and 13.4% S).

2.2 Leaching procedures

Because of the low solubility of $PbSO_4$ in sulphate solutions, the course of the galena leaching reaction was monitored by following the amount of co-generated ferrous ion. A 1- or 2-g sample of sized galena was agitated at 200 min^{-1} in 2.7 L of $Fe(SO_4)_{1.5}$-H_2SO_4 solution heated in a thermostated water bath. Samples were taken periodically, and were titrated for Fe^{2+} using ceric sulphate. In some experiments the solids were releached in 4 M NaCl at 60-70°C to ascertain the amount of $PbSO_4$ generated. The brine-leached solids were then air-dried prior to treatment with CS_2 in a Soxhlet reactor to recover the elemental sulphur. All parts of the leaching apparatus were rinsed with CS_2 to recover any sulphur adhering to the vessel walls, etc.

2.3 Mineralogical studies

Supporting mineralogical studies were carried out on both massive and sized galena which had been leached for various times and under different conditions. Both the as-leached surfaces and cleaved cross sections were examined using the scanning electron microscope (SEM) equipped with an energy dispersive X-ray analyzer (EDX). X-ray diffraction analyses were used to confirm the phases present. Details of the mineralogical techniques have been presented previously [7].

3 Results and discussion

3.1 The leaching reaction and reaction products

Figure 1 presents a typical reaction curve obtained when 2 g of 74-104 μm (-150 + 200 mesh) galena was leached at 95°C in 0.3 M $Fe(SO_4)_{1.5}$-0.3 M H_2SO_4 media. The results are reported as the percent galena reacted versus time. The percent galena reacted was calculated from the quantities of Fe^{2+} generated and the observation that each mole of PbS reacted produced approximately 2 moles of Fe^{2+} (see below). The use of sized galena powders results in ~25% reaction after 24 h at 95°C. The comparatively low rate of reaction in ferric sulphate-sulphuric acid media explains the common presence of unreacted galena in the residues from the leaching of zinc concentrates in O_2-$Fe(SO_4)_{1.5}$-H_2SO_4 media [3]. When the data of Figure 1 were replotted on a shrinking core basis (i.e., as $1-(1-\alpha)^{1/3}$ versus time where α is the fraction of galena reacted), non-linear curves were realized. The conclusion is that "linear" kinetics do not prevail during the leaching of galena in ferric sulphate media. When the data were replotted on a parabolic basis, (i.e., as $1- \frac{2}{3}\alpha-(1-\alpha)^{2/3}$ versus time) a good linear fit was obtained as can also be seen from Figure 1. The conclusion is that the reaction of galena in $Fe(SO_4)_{1.5}$-H_2SO_4 media obeys the parabolic rate law. For monosize powders, the relevant rate equation is:

$$1- \frac{2}{3}\alpha - (1- \alpha)^{2/3} = kt \qquad (1)$$

X-ray diffraction analyses of reacted galena powders indicated major amounts of $PbSO_4$, and sometimes, minor quantities of orthorhombic sulphur. Depending on the degree of reaction, various amounts of residual galena also were detected. Very rarely, goethite (α FeO.OH) was detected, and this phase may have formed during the filtration and washing of the leach residues. Surprising is the fact that lead jarosite was not detected.

To determine the reaction stoichiometry, 45-74 μm galena was leached for different times at 95°C and the various reaction products were determined. Table 1 summarizes the molar ratios of Pb^{2+}/SO_4, Pb^{2+}/S^o and Fe^{2+}/Pb^{2+} realized in these experiments.

Table 1. Molar ratios of the products formed during the reaction of 45-74 μm galena in 0.3 M $Fe(SO_4)_{1.5}$-0.3 M H_2SO_4 media at 95°C

Time (h)	Pb^{2+}/SO_4	$Pb^{2+}S^o$	Fe^{2+}/Pb^{2+}
3	0.95	1.05	2.73
3	0.94	-	2.10
4	1.02	0.80	2.53
6	1.00	1.07	2.01
17	1.03	1.27	2.06
17	1.00	1.22	2.07
28	1.01	1.25	2.38
30	1.19	1.33	2.85
40	1.19	1.33	2.85
40	1.05	0.83	2.70
Average	1.01	1.12	2.38

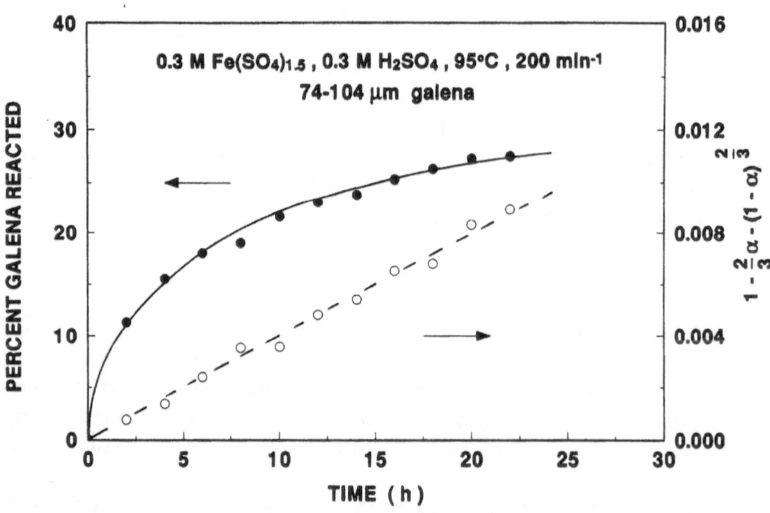

Fig. 1 Typical leaching curve realized when 74-104 μm galena was reacted in 0.3 M $Fe(SO_4)_{1.5}$-0.3 M H_2SO_4 media.

The data of Table 1 show that the Pb^{2+}/SO_4 molar ratio of the leached galena is 1.01; the conclusion is that the oxidized lead reports entirely as $PbSO_4$. This conclusion is supported by the X-ray diffraction analyses noted above. The average

Pb^{2+}/S° molar ratio is 1.12. This value indicates that elemental sulphur is the dominant sulphidic reaction product, and that the overall leaching reaction is closely given by:

$$PbS + 2\ Fe(SO_4)_{1.5} \rightarrow PbSO_4 + 2\ FeSO_4 + S^{\circ} \qquad (2)$$

That the average Pb^{2+}/S° ratio is >1.00, however, suggests that some sulphate may be produced during leaching.

$$PbS + 8\ Fe(SO_4)_{1.5} + 4\ H_2O \rightarrow PbSO_4 + 8\ FeSO_4 + 4\ H_2SO_4 \qquad (3)$$

The average molar ratio of Fe^{2+}/Pb^{2+} is 2.38, and this value suggests that $<10\%$ of the total amount of sulphide ion is oxidized to sulphate when galena is leached in ferric sulphate-sulphuric acid media. In total, the data of Table 1 indicate that Reaction 2 closely describes the leaching of galena in $Fe(SO_4)_{1.5}$-H_2SO_4 media. Preliminary tests showed that low concentrations of ferrous ion Fe^{2+} were not significantly oxidized at 95°C or 25°C in 0.3 M $Fe(SO_4)_{1.5}$-0.3 M H_2SO_4 solutions for times at least as long as 48 h. Hence, the co-generated ferrous ion reaction product is a useful indicator of the extent of galena reaction. Accordingly, a molar ratio of $Fe^{2+}/Pb^{2+} = 2.00$ was used to convert the measured ferrous ion concentrations to percent galena reacted in the remainder of the study. Although it is recognized that this is only an approximation, it provides a meaningful and consistent measure of the trends observed during galena dissolution.

3.2 Morphology of the reaction products

Figure 2 shows the surface of a galena particle after 30 min of reaction at 95°C in 0.3 M $Fe(SO_4)_{1.5}$-0.3 M H_2SO_4 media. Despite the short reaction time, the entire surface of the particle is covered with a thin layer of $PbSO_4 + S^{\circ}$. Figure 2 also illustrates a large irregular mass of S° which clearly has formed on both the euhedral $PbSO_4$ crystals and the $PbSO_4 + S^{\circ}$ layer covering the galena. Prolonging the leaching time results in the growth of the euhedral $PbSO_4$ crystals, the progressive thickening of the $PbSO_4 + S^{\circ}$ layer formed on the galena as well as the development of abundant globules of elemental sulphur which rest on the $PbSO_4 + S^{\circ}$ layer. In this regard, Figure 3 shows the fractured cross section of a galena particle reacted for 20 h at 95°C. The $PbSO_4 + S^{\circ}$ layer rests directly on the galena. Despite the 20 h reaction time, the product layer is $<10\ \mu m$ thick. Significantly, many spheroidal particles of elemental sulphur rest on the $PbSO_4 + S^{\circ}$ layer. The observed parabolic reaction kinetics could be a reflection of the diffusion barrier associated with the constantly thickening layer of $PbSO_4 + S^{\circ}$ or by the progressive blockage of the galena surface by tenaciously adhering grains of $PbSO_4$ and S°.

X-ray diffraction analyses of the various product layers were carried out. All the analyses indicated major $PbSO_4$, and this observation is in agreement with the SEM studies. No sulphur was detected in any of the products which were leached for <16 h. Orthorhombic sulphur began to be detected in the residues generated by leaching the galena for >16 h, and orthorhombic sulphur was usually detected in the samples when galena was leached for >20 h. The X-ray diffraction results for sulphur

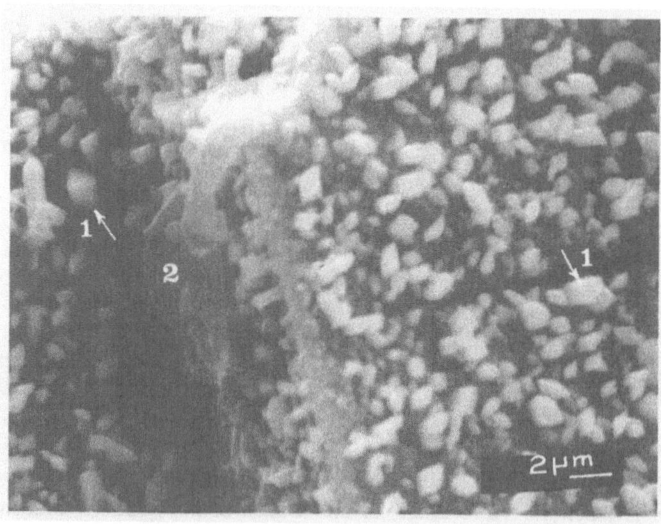

Fig. 2. Secondary electron micrograph illustrating a galena particle reacted for 30 min at 95°C in 0.3 M $Fe(SO_4)_{1.5}$-0.3 M H_2SO_4 solution. 1 - $PbSO_4$ crystals, 2 - S^o globule.

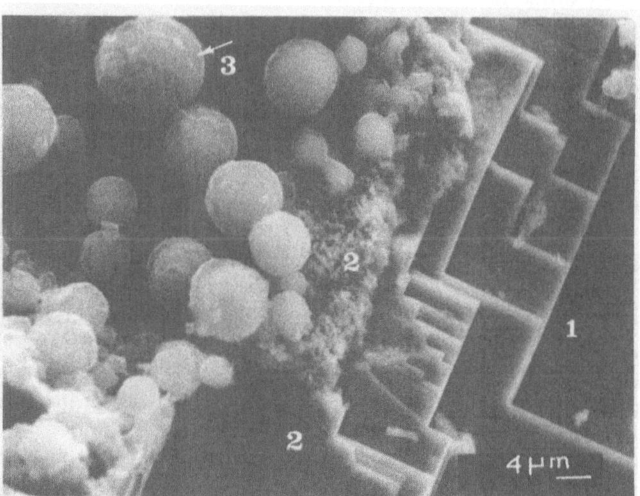

Fig. 3. Secondary electron micrograph showing the product layer formed on galena reacted for 20 h at 95°C in 0.3 M $Fe(SO_4)_{1.5}$-0.3 M H_2SO_4 solution (cleaved cross section). 1 - galena, 2 - $PbSO_4$ + S^o, 3 - S^o spheroids.

are not consistent with the abundance of sulphur detected in the samples by SEM-EDX study. The conclusion is that much of the sulphur, especially that made in the short time tests, is amorphous.

3.3 Effect of temperature

Figure 4 shows some of the leaching curves realized when 74-104 μm galena was leached at various temperatures in 0.3 M $Fe(SO_4)_{1.5}$-0.3 M H_2SO_4 media. The data were data plotted according to the parabolic relationship of Equation 1. Relatively good linear fits are obtained at all temperatures, and all the lines extrapolate through the origin or very near to the origin. The conclusion is that the parabolic rate law is closely obeyed over the entire temperature range studied. The extent of galena reaction is modest at all temperatures studied. For example, after 20 h of reaction, only ~7% of the galena was leached at 65°C and only ~25% was reacted at 95°C. It is also evident that the galena leaching reaction increases significantly with increasing temperature.

Parabolic rate constants were calculated from the slopes of the $1-\frac{2}{3}\alpha-(1-\alpha)^{2/3}$ versus time lines, and these rate constants are summarized on the Arrhenius plot shown in Figure 5. Although there is some scatter of the data, a linear trend clearly emerges. There is no discernable "break" in the Arrhenius curve, and the implication is that a single rate controlling process is operative over the entire temperature range studied. The apparent activation energy calculated from the data is 61.2 kJ/mol.

3.4 Effect of the ferric ion oxidant

Figure 6 shows some of the leaching curves, plotted as $1-\frac{2}{3}\alpha-(1-\alpha)^{2/3}$ versus time, realized when 1 g of 74-104 μm galena was leached at 90°C in 0.3 M H_2SO_4 solutions having various Fe^{3+} concentrations. The reaction generally obeys the parabolic rate law over the entire concentration range from 0.02 to 2.0 M Fe^{3+}. Furthermore, the rate increases moderately rapidly with increasing ferric ion concentration, but very fast rates are never observed. Parabolic rate constants were calculated from the slopes of the individual leaching curves. These rate constants were then plotted as a function of the ferric ion concentration, and the results are given in Figure 7. The rate increases systematically with increasing ferric ion concentration over the entire concentration range from 0.01 to 2.0 M Fe^{3+}. Regression analysis of the data yielded the following equations.

$$\log k = 0.752 + 0.467 \log[Fe^{3+}] \tag{4}$$

$$k \propto [Fe^{3+}]^{0.5} \tag{5}$$

Half-power ferric ion concentration dependencies are typical of many electrochemically controlled processes, although all such processes exhibit linear and not parabolic kinetics as is the situation in the current study.

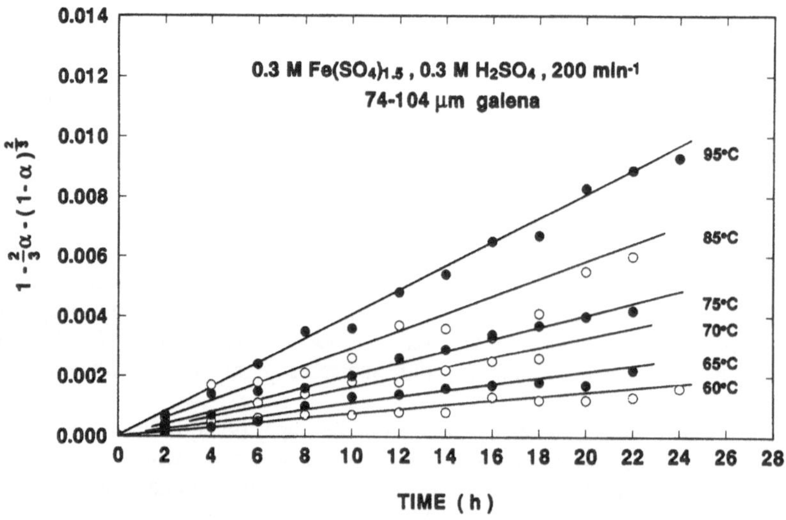

Fig. 4. Plots of the function $1-\frac{2}{3}\alpha-(1-\alpha)^{2/3}$ versus time for the reaction of 74-104 μm galena in ferric sulphate media at various temperatures.

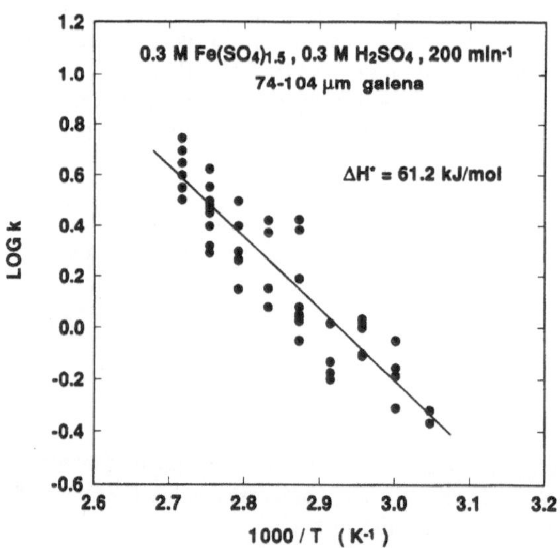

Fig. 5. Arrhenius curve for the reaction of 74-104 μm galena in ferric sulphate media.

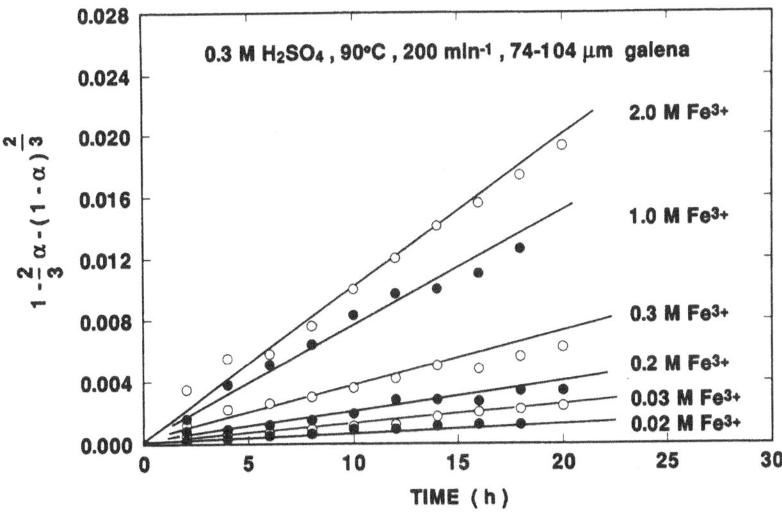

Fig. 6. Plots of $1-\frac{2}{3}\alpha-(1-\alpha)^{2/3}$ versus time for the leaching of 74-104 μm galena at 90°C in 0.3 M H_2SO_4 solutions having different ferric ion concentrations.

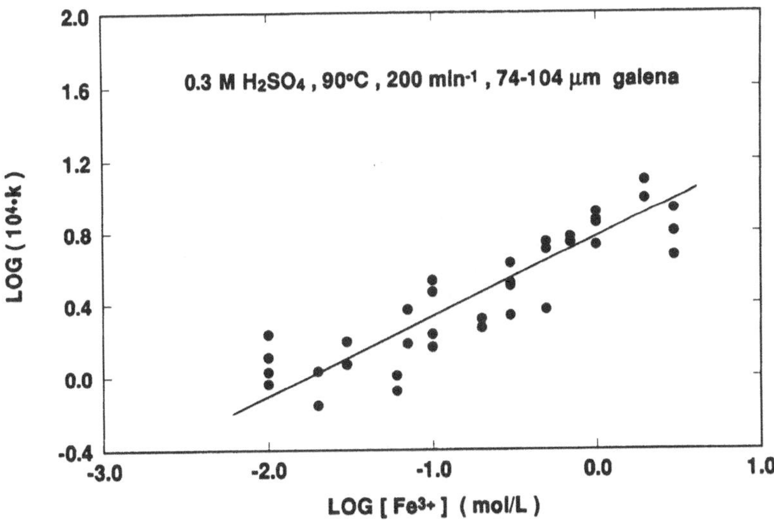

Fig. 7. Effect of the concentration of $Fe(SO_4)_{1.5}$ on the rate of galena reaction at 90°C in 0.3 M H_2SO_4 media.

3.5 Effect of the H_2SO_4 concentration

The effect of the H_2SO_4 concentration on the rate of reaction of 74-104 μm galena at 90°C in 0.3 M $Fe(SO_4)_{1.5}$ media was investigated for acid concentrations ranging from 0.01 to 4.0 M H_2SO_4. Figure 8 shows some of the parabolic leaching curves (i.e., plots of $1\text{-}\frac{2}{3}\alpha\text{-}(1\text{-}\alpha)^{2/3}$ versus time) realized in these experiments. In 0.3 M $Fe(SO_4)_{1.5}$ media, the rate of galena reaction is virtually insensitive to acid concentrations <0.1 M H_2SO_4; in fact, all the leaching curves obtained in dilute acid media are nearly superimposable. At acid concentrations >0.1 M H_2SO_4, however, the reaction rate increases and the parabolic rate law continues to be obeyed. For acid concentrations >0.6 M H_2SO_4, the parabolic rate law ceases to be followed, but the reaction rate continues to increase rapidly.

The acid concentration dependence suggests that the reaction of galena proceeds, at least in part, by the direct acid attack of the sulphide.

$$PbS + H_2SO_4 \rightarrow PbSO_4 + (H_2S)_{Dissolved} \tag{6}$$

It seems that the direct acid attack mechanism predominates at acid concentrations >0.6 H_2SO_4, where the parabolic rate relationship ceases to be obeyed. The $(H_2S)_{Dissolved}$ likely has a limited mobility in the leaching solution. That is, it could likely diffuse through the layer of $PbSO_4$ + S° formed on the surface of the galena before it was oxidized by ferric ions.

$$(H_2S)_{Dissolved} + 2\ Fe(SO_4)_{1.5} \rightarrow 2\ FeSO_4 + H_2SO_4 + S^o \tag{7}$$

The sulphur produced by Equation 7 would likely grow on pre-existing sulphur nuclei, and the result would be relatively few discrete particles of S° rather than an intimate mixture of $PbSO_4$ and S°. In this regard, Figure 3 shows a cleaved cross section of a galena particle leached for 20 h at 95°C in 0.3 M $Fe(SO_4)_{1.5}$-0.3 M H_2SO_4 media. That is, the sample was leached in a region where both ferric ion attack and direct acid dissolution likely occurred. The surface of the reacted galena is covered by a continuous layer of $PbSO_4$ + S°. Resting on the surface of the fine grained $PbSO_4$ + S° layer are large (to 10 μm) spheroids of pure elemental sulphur which are not mixed with $PbSO_4$. The size and morphology of the elemental sulphur indicate a formation mechanism involving a dissolved sulphur species. That is, H_2S forms and dissolves at the surface of the galena; the dissolved H_2S species diffuses through the thin $PbSO_4$ + S° reaction product layer. In the presence of ferric ions in the bulk solution, the $(H_2S)_{Dissolved}$ is oxidized to elemental sulphur. Such reactions are known to occur preferentially at active sites, and ideally, at a pre-existing sulphur surface. The consequence is sulphur spheroids or isolated sulphur globules of the type shown in Figures 2 and 3.

In some instances, the $(H_2S)_{Dissolved}$ oxidation process occurs sufficiently slowly that subhedral sulphur crystals form. Such examples are shown in Figure 9. The sulphur particles have crude crystallographic forms and rest on the $PbSO_4$ + S° reaction product layer. The presence of sulphur crystals with external crystallographic forms

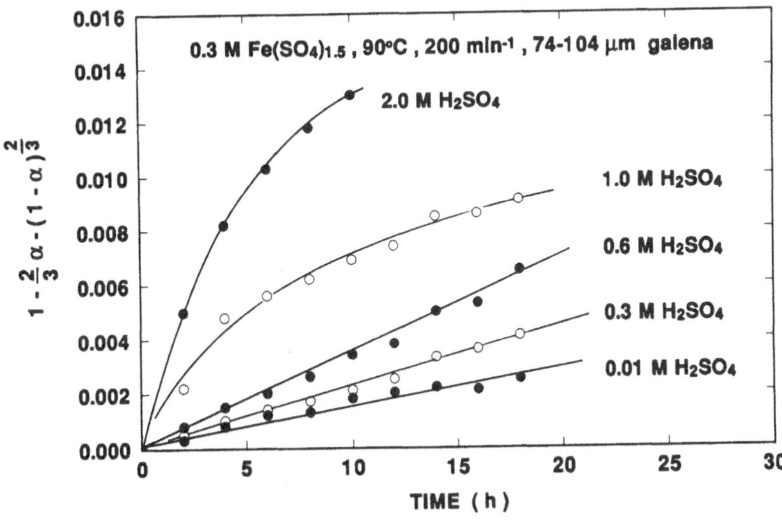

Fig. 8. Plots of $1-\frac{2}{3}\alpha-(1-\alpha)^{2/3}$ versus time for the leaching of 74-104 μm galena at 90°C in 0.3 M $Fe(SO_4)_{1.5}$ solutions containing various concentrations of H_2SO_4.

Fig. 9. Secondary electron micrograph showing the development of orthorhombic sulphur crystals on the surface of galena reacted for 22 h at 95°C in 0.3 M $Fe(SO_4)_{1.5}$-0.3 M H_2SO_4 solution. 1 - $PbSO_4$ + S^o, 2 - S^o crystals.

confirms that at least part of the elemental sulphur is produced by the ferric ion oxidation of dissolved H_2S which itself results from the direct acid attack of the galena.

3.6 The effect of $FeSO_4$ and $MgSO_4$

The reaction of galena in ferric sulphate media generates ferrous sulphate, and the effect of the $FeSO_4$ reaction product on the galena leaching rate was determined. Because it was impossible to analyze the small amounts of ferrous ion generated by the leaching reaction against the high concentrations of $FeSO_4$ initially added to the solution, these tests were monitored by measuring the quantity of $PbSO_4$ formed. The ferric sulphate leach residues were themselves leached in 4 M NaCl to solubilize the $PbSO_4$ reaction product. The concentration of Pb^{2+} was analyzed and the percentage of $PbSO_4$ formation was calculated.

Figure 10 shows the effect of the $FeSO_4$ concentration on the percent galena reacted after 20 h in 0.3 M $Fe(SO_4)_{1.5}$-0.3 M H_2SO_4 media. Although there is some scatter of the data, it seems that increasing $FeSO_4$ concentrations in the range 0.0 to 1.2 M $FeSO_4$ slightly accelerate the galena leaching rate. Certainly there is no suggestion that increasing $FeSO_4$ concentrations, and hence a reduced Fe^{3+}/Fe^{2+} ratio, decrease the rate of reaction.

To evaluate the effect of "neutral" sulphates, 1-g samples of 74-104 μm galena were reacted at 90°C in 0.3 M $Fe(SO_4)_{1.5}$-0.3 M H_2SO_4 media also containing 0.0 to 2.0 M $MgSO_4$. The course of the reaction was monitored by following the amount of co-generated $FeSO_4$, and parabolic kinetics (i.e., $1- \frac{2}{3}\alpha-(1-\alpha)^{2/3}$ versus time) were obeyed at all $MgSO_4$ concentrations studied. Parabolic rate constants were calculated from the slopes of the individual reaction lines, and these rate constants are plotted as a function of the $MgSO_4$ concentration in Figure 11. The presence of $MgSO_4$ reduces the galena reaction rate, and the rate decrease seems to vary in a linear manner with increasing $MgSO_4$ concentration.

4 Conclusions

The reaction of galena in $Fe(SO_4)_{1.5}$-H_2SO_4 media was investigated over the temperature range 55-95°C and for various $Fe(SO_4)_{1.5}$, H_2SO_4, $FeSO_4$ and $MgSO_4$ concentrations. The low solubility of the $PbSO_4$ reaction product dictated that the rates be monitored using the ferrous ion co-generated during leaching; also, mineralogical methods were employed to support the hydrometallurgical studies. The overall reaction of PbS with $Fe(SO_4)_{1.5}$-H_2SO_4 produced $PbSO_4$, $FeSO_4$ and S°. X-ray diffraction studies indicated that at least part of the sulphur is amorphous. Non-linear kinetics were always observed, and in most instances, the parabolic rate law was closely obeyed. SEM-EDX studies showed that the surface of the reacted galena was covered with an extensive compact layer of $PbSO_4$ + S°. Large globular masses and subhedral crystals of elemental sulphur are commonly detected on the surface of the $PbSO_4$ + S° layer. The implication is that the leaching mechanism involves, at least in part, the direct acid attack of the galena followed by the oxidation of the dissolved H_2S by ferric sulphate. This conclusion is supported by the strong dependence of the rate on acid

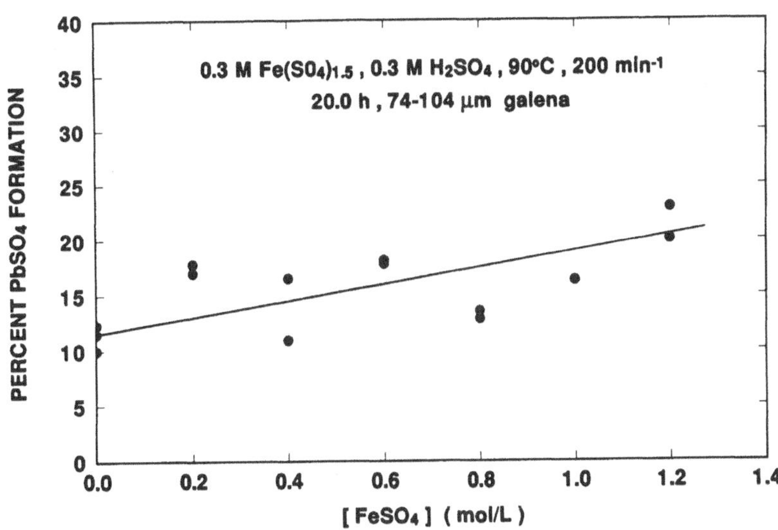

Fig. 10. Effect of the concentration of $FeSO_4$ on the rate of reaction of galena in 0.3 M $Fe(SO_4)_{1.5}$-0.3 M H_2SO_4 media.

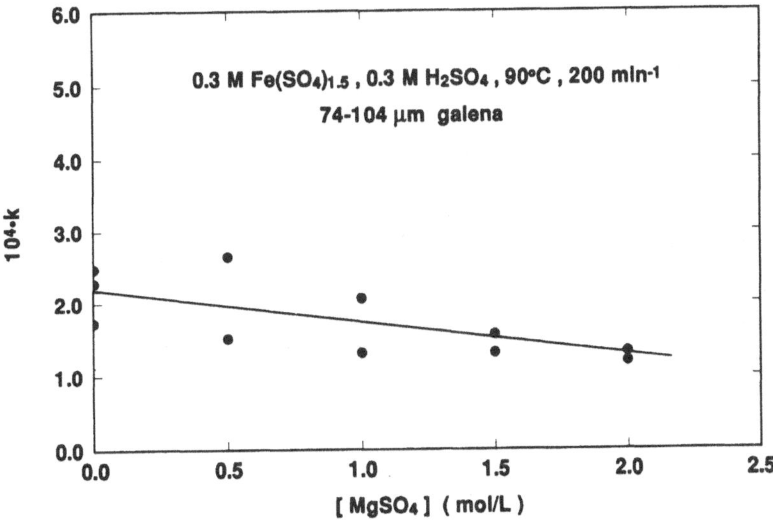

Fig. 11. Effect of the concentration of $MgSO_4$ on the rate of reaction of galena in 0.3 M $Fe(SO_4)_{1.5}$-0.3 M H_2SO_4 media.

concentrations >0.1 M H_2SO_4. The reaction rate increases significantly with increasing temperature; the apparent activation energy is 61.2 kJ/mol. The rate increases as the 0.5 power of the ferric ion concentration, but is nearly independent of the concentration of the $FeSO_4$ reaction product. The presence of neutral sulphates, such as $MgSO_4$, decreases the rate to a modest extent. The parametric reaction dependencies, coupled with the morphology of the reaction products, indicate a complex reaction mechanism involving both the ferric ion and direct acid attack of the galena as well as mass transport through the passivating product layer which progressively restricts the access of ferric ions and acid to the galena surface.

5 Acknowledgements

The authors would like to thank D.J. Hardy for his invaluable assistance with many aspects of the experimental work and P. Carrière for his X-ray diffraction analyses of the reaction products.

6 References

1. Dutrizac, J.E. (1992) The leaching of sulphide minerals in chloride media. **Hydrometallurgy** 29, 1-45.
2. Collins, M.J., Doyle, B.N., Ozberk, E. and Masters, I.M. (1990) The zinc pressure leaching process applications. in **Lead-Zinc '90**, eds. T.S. Mackey and R.D. Prengaman, The Minerals, Metals and Materials Society, Warrendale, PA, pp. 293-311.
3. Dutrizac J.E. and Chen, T.T. (1987) Mineralogical characterization of leach residues of a pyritic Zn-Pb-Cu-Ag concentrate. **Can. Metal. Quart.** 26, 189-205.
4. Sparrow, G.J., and Woods, R. (1991) The mechanism of leaching of copper-lead matte in oxygenated acidic chloride/sulphate solution. in **Hydrometallurgy and Electrometallurgy of Copper**, eds. W.C. Cooper, D.J. Kemp, G.E. Lagos and K.G. Tan, Pergamon Press, New York, pp. 45-60.
5. Vizsolyi, A., Veltman, H., and Forward, F.A. (1963) Aqueous oxidation of galena in acid media. **Trans. Metal. Soc. AIME** 227, 215-220.
6. Paul, R.L., Nicol, M.J., Diggle, J.M. and Saunders, A.P. (1978) The electrochemical behaviour of galena (lead sulphide). 1- Anodic dissolution. **Electrochim. Acta** 23, 625-633.
7. Chen, T.T. and Dutrizac, J.E. (1991) The use of mineralogical methods to elucidate the reactions occurring during the ferric ion leaching of sulphide minerals. in **Process Mineralogy XI**, eds. D.M. Hausen, W. Petruk, R.D. Hagni and A. Vassiliou, The Minerals, Metals and Materials Society, Warrendale, PA, pp. 117-132.

Multicomponent equilibrium calculations in process design: study of some acid digester reactors

Pertti Koukkari
Kemira Oy, Helsinki, Finland
Hannu Sippola
GEM Systems Oy, Espoo, Finland
Anna Sundquist
Kemira Oy, Helsinki, Finland

Abstract
The calculation of multicomponent thermodynamic equilibrium by Gibbs energy minimization has become an efficient tool in modern engineering practice. When calculating hydrometallurgical reactors a parametrized model for activities in concentrated aqueous solution is required. A comparative literature study was performed for several electrolyte models using standard deviations of the mean activity and osmotic coefficients with a number of electrolytes. The Pitzer model results with the smallest variation. Moreover its practical application area is wide due to the large number of published parameters.

With the activity coefficient model chosen the equilibrium calculation of a multicomponent aqueous reactor is based on known thermodynamic and activity data of the heterogeneous system. Omission of sime reactions from the calculation system may further be necessary because of slow reaction kinetics. In that case the minimization algorithm results with a close to equilibrium approximation.

The advantage of the thermodynamic algorithm is the possibility to follow both the chemical composition of all reactor phases and the energy balance of the reactor. Thus the calculation results can be verified by either composition or heat balance measurements. Simultaneously, quantified estimates of effluents and side-products can be made.

The well-known equilibrium routines Solgasmix and ChemSage were used for sulphate - sulphuric acid equilibria in concentrated solutions. A reactor calculation was performed for sulphuric acid digestion of ilmenite ore in TiO_2-pigment manufacturing and for series of reactors in the NPK-fertilizer process. The calculated maximum concentrations of such effluents as SO_2,H_2S and respectively NO_x NH_3 and fluorine were compared with Kemira Oy's process data.
Keywords: Thermodynamic Equilibrium, System Reaction, Kinetic Restrictions, Process Modelling, Titanium Dioxide, NPK-Process.

1 Introduction

Simulation of multicomponent and multiphase chemical reaction equilibria has become important means in the research of the process and materials technology. The computation of a multicomponent equilibrium as a rule is done by some numerical routine (Hillert 1981, Smith 1982). Most often Gibbs energy minimization is used (Eriksson 1975) and thence the equilibrium composition is reached for a state in constant temperature and pressure.

Since processes occurring at high temperatures are more likely to appear close to equilibrium, most of the calculation applications tend to concentrate in pyrometallurgy and other high temperature fields (Eriksson 1984, Magnusson 1980, Kaskiala 1989). As for lower temperature processes, in particular hydrometallurgy, the calculation of multicomponent equilibria has been applied to geological surveys or to a number of environmental systems (Dawson 1982). The thermodynamic theory of aqueous systems is also less well developed (Partanen 1989). Lack of tabulated thermochemical or activity data is often encountered when multicomponent and multiphase systems are studied. However, from the industrial viewpoint, the simulation of concentrated acid and base digestion processes would form an attractive range of study.

At the Institute of Materials Technology, Helsinki University of Technology, a number of well-known thermodynamic routines have been applied to calculate equilibria for concentrated aqueous solutions (Sippola 1992a, Hämäläinen 1992). As the research projects have been joint-ventures with the Finnish process industries some realistic approaches could be made, for instance, to simulate multicomponent digester reactors (Koukkari 1993a)

Among the reactors calculated were Kemira Oy's sulphuric acid digester of ilmenite ore in the so-called sulphate-route manufacturing of titanium dioxide pigment and the three reactors of Kemira Oy's NPK-fertilizer process. While the former digestion produces a solid 'sulphate cake' from concentrated acid and ore, the latter process operates with wide pH-range solutions from highly acidic to neutral.

The results of the calculations could be quantified in terms of, e.g., the reactor gas emission measurements. The calculated estimates of the offgas compositions could thus be verified and be further used as design parameters for necessary emission recovery investment.

Inherently, the phase compositions of the reactors were calculated by the thermodynamic approach. Their detailed conformity with the process experiments is a subject for more profound study. However, the results show that the physicochemical algorithms are appropriate to screen the process chemistry as part of more general simulation procedures.

2 Validity of the equilibrium-based reactor models

The thermodynamic equilibrium calculation results with the final composition of a reaction mixture, provided that the following conditions prevail:

1. Mass transfer does not limit the reactor processes (ideal mixing)
2. The reaction time is sufficient for reactions to reach equilibrium (thermodynamic control)

In a practical reactor system, these conditions are not often fully encountered. However, the close-to-equilibrium approximation is yet often legitimate. For instance, the chemical change proceeds close to equilibrium in several thermally stabilized batch reactors. Using a mechanistic approach, the complex chemistry of the actual processes often needs to be substantially simplified to reach good modelling results (Leppinen 1986). The computer-based multicomponent equilibrium method avoids the difficulty of the build-up of complicated reaction mechanisms. The equilibrium route becomes feasible when both the comprehensive thermodynamic substance data and sufficient process information are available. If such is the case, the equilibrium approach often provides a rewarding means of process approximation.

The standard method to reach a multicomponent equilibrium composition is the minimization of the system Gibbs energy. The calculation is performed for a change, where a number of known reactants from their well-defined initial states react to form a mixture of equilibrium products at given temperature and pressure. This change, for which all the calculation balances are written, may be called the 'system reaction'.

Most often the procedure is based on temperature-dependent heat capacities of the substances in the reaction mixture. Consequently the calculation produces also the heat balance of the process, or the respective 'system reaction'. Thence, essential criteria for the equilibrium models are the measured reactor temperatures, chemical substance concentrations, as well as calorimetric heat balances.

3 Kinetic restrictions

Thermodynamically, the chemical reactor processes are characteristic natural processes. The direction of the process is defined by the decreasing system Gibbs energy $dG^Z < 0$. The minimization principle thus is suitable in many cases to describe the reactor chemistry, though the true thermodynamic equilibrium would not be reached for all the system components. This implies, however, a mathematically well-defined system constraint applied to the calculation system. Kinetic constraints may be taken into account either when defining the composition of the calculation system or by developing separate computation algorithm to provide the constraints for the minimization procedure (Koukkari 1993b).

The composition of the model may be described with such constraints that the 'system reaction' does not proceed further than to a metastable state, known by process measurements. Strict thermodynamic rules would require the completeness of each chemical reaction. However, in practice one often encounters reactions, which would need presence of a catalyst to become completed. If there is no catalyst present in the system an approximate metastable state is formed. Such metastable state can be approximated by the modern minimization routines. This feature of the equilibrium routines, to the writers' experience, greatly enhances their applicability in solving practical reactor problems.

For example, the nitrogen oxides formed in an acid digestion reactor are slow to decompose to their respective equilibrium elements at low (< 100 °C) temperatures:

$$2NO(g) \;=\; N_2(g) + O_2(g) \qquad \Delta G^o(298) = -173 \text{ kJ/mol} \qquad \text{(I)}$$

$$NO_2(g) \;=\; 1/2N_2(g) + O_2(g) \qquad \Delta G^o(298) = -51 \text{ kJ/mol} \qquad \text{(II)}$$

Approximation of the formation of NO_x-species from nitric acid digesters may be reached by a multicomponent equilibrium program by omission of the relation between valence states of the nitrogen oxides and pure nitrogen in the system description. Yet, the rest of the system remaining untouched and a minimum value for the constrained multicomponent system Gibbs energy may be reached. The slow decomposition of the nitric oxides then becomes taken into account and the approximational calculation generally results with a concentration of NO_x-species comparable to the measured process results.

One of the most important tasks of the multicomponent equilibrium reactor modelling is to search the kinetic constraints of the practical process and to design the system description in such a way, that a practical approximation of the process conditions may be reached. The limits of such modelling procedure are then set by the accuracy of the multicomponent system approximation. This often results, however, with improved proximity to the experimental data when compared to a conventional kinetic [mechanistic] calculation.

4 Thermodynamic modelling of multicomponent aqueous systems

4.1 General principles for the calculation

In hydrometallurgical applications there are many unit processes where thermodynamic equilibria is desired and reached. In such systems three things are needed to calculate the final equilibria, i.e.,

Properties of incoming flows
Thermodynamic description of the system
Computer program for calculations.

Temperature, pressure and composition are required information for incoming flows. Thermodynamic description of the systems includes: temperature and pressure of the system, values of enthalpy ($\Delta H_{f,298}$) entropy (S_{298}) and heat capacity (C_p) for each species and an activity coefficient model to calculate activity coefficients in non-ideal phases.

Computer programs to calculate thermodynamic equilibria are based either on stoichiometric formulation where independent chemical reactions are given or on non-stoichiometric formulation where the Gibbs energy of the system is minimized (Smith 1982).

In Gibbs energy minimisation the Gibbs energy of the system

$$G_{sys} = \sum_{i,\alpha} n_i^{\alpha} \mu_i^{\alpha} \quad , \quad n_i^{\alpha} \geq 0 \ , \tag{1}$$

where summation goes over all phases (α) and species (i) and

G_{sys} = Gibbs energy of the system
n_i = amount of species i
μ_i = chemical potential of species i

is minimized with element balance as a constraint.

$$\sum_{i,\alpha} a_{ji} n_i^{\alpha} = b_j \quad j = 1..M \ , i = 1..N \ , \tag{2}$$

where

M = number of elements in the system
b_j = total amount of element j in the system
a_{ji} = number of atoms j in species i

In aqueous solutions chemical potential of ions is commonly expressed on molality scale:

$$\mu_i = \mu_i^o + RT \ln m_i + RT \ln \gamma_i \tag{3}$$

where

μ_i^o = chemical potential in standard state
m_i = molality (mol/kg H_2O)
γ_i = activity coefficient

Value of the chemical potential in the standard state can be calculated from the thermodynamic properties of the ion (table 1). The latter two terms in equation 3, the molality and the activity coefficient, are solved by the Gibbs energy minimizer.

Table 1. Sources for thermodynamic data

Wagman D.D. Et al. (eds.), The NBS Tables of Chemical Thermodynamic Properties (1982).
Cox J.D., Wagman D.D. and Medvedev V.A. (eds.), CODATA Key Values for Thermodynamics (1989)
Barner H.E. and Scheurman R.V., Handbook of Thermochemical Data for Compounds and Aqueous Species (1978)
Bard A.J., Parsons R. and Jordan J., Standard Potential in Aqueous Solution (1985)
Horvath A.L. Handbook of Aqueous Electrolyte Solutions (1985)

4.2 Activity coefficient models

The properties of aqueous solution differs greatly from ideal behaviour due to the electrostatic force between ions in solutions. One of the first presentations to describe this deviation from ideal behaviour was Debye-Huckel activity coefficient model introduced in 1929. Unfortunately Debye-Huckel model can be used only for very dilute solutions.

Nowadays there are several models based on Debye-Huckel theory, which describe at best the nonideal behaviour of the aqueous solution up to the solubility limit of the electrolyte.

These activity coefficient models have from one to four parameters per electrolyte and some of them have also parameters to describe interaction between two cations or two anions. A list of these models is represented in table 2.

Table 2. Activity coefficient models for aqueous solutions

Meissner equation (Meisner 1972)
Bromley equation (Bromley 1973)
Pitzer equation (Pitzer 1973)

Local composition models for aqueous solutions

NRTL (Cruz 1978, Ball 1985a)
NRTL (Chen 1982)
NRF-NRTL (Haghtalah 1988)
UNIQUAC (Sanders 1986)

Mean Spherical Approximation (MSA) (Ball 1985b)

The comparison of these models is illustrated in figures 1-3. As a conclusion we can estimate that the Pitzer model is still one of the best. Furthermore, it is superior to the other models in number of parameters published (Sippola 1992a).

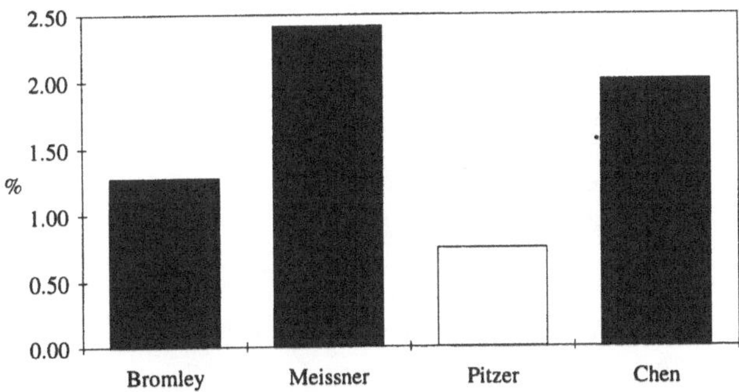

Figure. 1. Standard deviation of mean activity coefficients for 7 strong electrolytes (HCl, NaCl, KCl, NaOH, KOH, Na_2SO_4, $MgSO_4$). Data is taken from Zemaitis (1986).

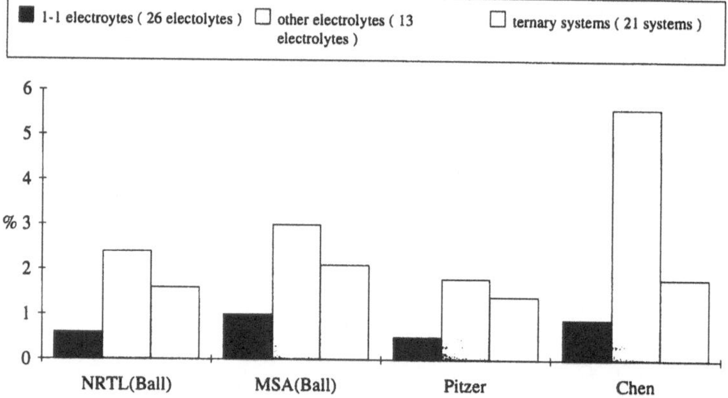

Figure 2. Standard deviation of osmotic coefficients. Data is taken from Ball (1973 and 1985). The value for MSA(Ball) for 'other binary electrolytes' is based only on ten binaries.

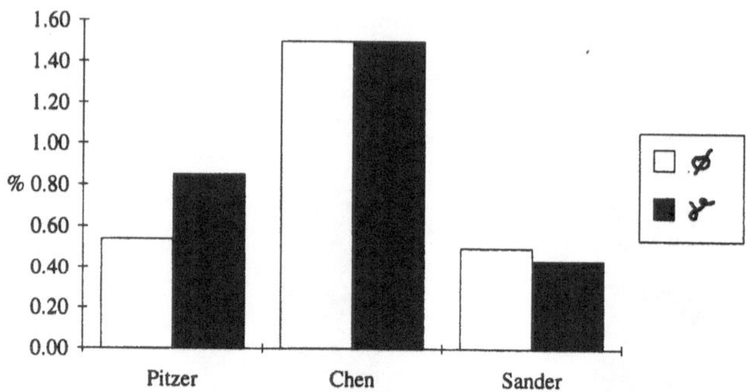

Figure 3. Standard deviation of osmotic and mean activity coefficients. Data is taken from Renotte (1989).

4.3 Computer Program

Calculating thermodynamic equilibrium in complex systems with Gibbs energy minimisation approach is more enhanced because there is no need to find the set of independent reactions. Also phase equilibria is considered without any extra effort.

In this work the thermodynamic equilibrium calculations were carried out with Solgasmix (Eriksson 1975) or with ChemSage (Eriksson 1990). ChemSage is based on the well-known Solgasmix, which has been introduced in many commercial thermodynamic data banks. The main difference from Solgasmix is that in ChemSage there is included a built-in activity coefficient model bank. The list of activity coefficient models and the application area of the models is summarized in table 3.

Table 3. Activity coefficients models in ChemSage

Model	Application area
Redlich - Kister - Muggianu Kaufman - Kohler Margules Kohler - Toop	empirical models
Sublattice formalism	solid alloys
Ionic sublattice	ionic alloys
Equilivalent sublattice	molten salts
Gaye-Frohberg	ionic oxidic mixtures
Blander-Pelton	ionic liquid mixtures
Wagner	metallic dilute solutions
Pitzer	aqueous solutions
Virial equation	gas phase

5 The ilmenite-sulphuric acid reaction

5.1 The sulphate process for titanium dioxide production

The first step in making white titanium dioxide (TiO_2) pigment via the classic sulphuric acid route is the acid digestion reaction. Ilmenite ($FeTiO_3$) ore is used as the principal raw material. The pretreated ore is reacted with concentrated sulphuric acid in large tanks:

$$FeTiO_3 + 2\,H_2SO_4 = FeSO_4 + TiOSO_4 + 2\,H_2O \qquad \text{(III)}$$

The reaction results with a dry, solid mass called the 'reaction cake'. Water is then added to dissolve the iron, titanium and other sulphates in the mass. Ferrous sulphate as well as the other impurity sulphates are removed by vacuum crystallization and the remaining liquor is then thermally hydrolyzed to yield hydrous titanium dioxide ($TiO_2{\cdot}nH_2O$) and sulphuric acid. The hydrous precipitate is filtered, dried and calcined to produce the final pure TiO_2.

The numerous minor metal impurities, which might degrade the whiteness of the end product, partly are removed with ferrous sulphate and partly stay with recycling acid.

The reactive core of the process is the acid tank, where the raw oxide/titanate mixture reacts with concentrated sulphuric acid (reaction III). The reactor is mixed with compressed air. The traditional raw material has been ilmenite ore with 42-46 % titania, but due to developing process technology and environmental reasons, enriched titania-slag with over 70 % TiO_2 is currently taking over. The main impurities, in excess to iron, are calcium, magnesium, manganese and to less extent chrome and vanadium.

Since the TiO_2-slag is produced by an iron removing smelting process, this raw-material also contains some metallic iron (0.4-0.8 % Fe). The reduced metal content is a concern for the pigment manufacturer, since it may, under un-controlled process conditions, be a source of gaseous effluents such as hydrogen, hydrogen sulphide and sulphur dioxide in the reaction.

In the present study the formation of these gases is screened by thermodynamic modelling. Special emphasis has was on the maximum concentrations for safety and environmental reasons.

5.2 Assumptions for the sulphate reactor calculation

The sulphate digester is a typical batch reactor and thus an interesting target for a thermodynamic multicomponent consideration. The thermodynamic description of this system is as follows:

1. The system is composed of three phases: the gas phase, liquid (acid) phase and the solid phase ('reaction cake')
2. For simplicity, both the gas phase and the acid liquid phase are assumed ideal.
3. The solid phase assumed to be a mixture of pure, condensed phases.

The most important metal cations are Ti(III)/Ti(IV), Fe(II)/Fe(III), Mg(II), Al(III) and the inert silica, SiO_2.

The second assumption with ideal liquid phase is not surprising when one takes into account the fact that the mass ratio of the raw materials is such that no liquid phase remains in the reactor at equilibrium.

In addition, the modelling of the reactor was based on the following assumptions:

4. The oxygen in the compressed air remains inert
5. The ferrous sulphate ($FeSO_4$) does not oxidize to ferric sulphate in the digester ($Fe_2(SO_4)_3$).
6. Ti(III) remains at its oxidation state

According to thermodynamics the oxygen in compressed air would oxidize all iron(II) and titanium(III) compounds, However, in the process conditions this is not observed due to the mass transfer resistance from gas into the solid cake.

The list of species included in the calculation are represented in table 4. All thermodynamic data was taken from HSC program (Roine 1988) except titanyl sulphates for which the thermodynamic data was estimated. Calculations were performed by Solgasmix routine (Eriksson 1975).

Table 4. Species considered in the slag reactor

Phase	Species
Gas :	$H_2O(g)$, $H_2SO_4(g)$ $H_2(g)$, $N_2(g)$, $O_2(g)$, $SO_2(g)$, $SO_3(g)$, $H_2S(g)$, $H_2S_2(g)$, $S(g)$, $S_4(g)$
Acid	H_2O, H_2SO_4
Solids	$Al_2(SO4)_3$, $Al_2(SO4)_3*6H_2O$, Al_2O_3, Fe, FeO, Fe_2O_3, Fe_3O_4, $FeO*TiO_2$, $FeSO_4$, $FeSO_4*H_2O$, $FeSO_4*7H_2O$, $Fe_2(SO_4)_3$, MgO, MgSO4, $MgSO4*H_2O$, $MgSO4*4H_2O$, $MgSO4*6H_2O$, $MgSO4*7H_2O$, SiO_2, TiO_2, Ti_2O_3, $TiOSO_4$, $TiOSO_4*H_2O$, $TiOSO_4*2H_2O$, FeS, Al_2S_3, MgS, SiS_2, FeS_2

5.3 Result of heat and chemical balance calculation

Using the known initial temperatures for the system reaction and experimentally based estimates for reactor heat balances, the calculated maximum temperatures of the exothermic reaction mixtures became 175 °C for ilmenite and 204°C for TiO_2-slag. These values compare to the measured process peak temperatures of 180 and 205 °C, respectively.

According to the model, the ilmenite reaction produces mainly hydrous sulphates. This result also could be confirmed by thermogravimetric analysis of the 'reaction cake'. The equilibrium model predicts the TiO_2-slag reaction, instead, to form anhydrous titanyl sulphate ($TiOSO_4$) which, however, becomes hydrated with decreasing temperature. It is known by experience, that the titanium compound produced in practice is a hydrous and amorphous sulphate, $TiOSO_4 \cdot nH_2O$ (Koukkari 1993a). The X-ray diffractometry applied to the slag reaction cake did not reveal any crystalline $TiOSO_4$.

5.4 Hydrogen formation in the slag reactor

The reaction system was repeatedly calculated with gradually increasing the sulphuric acid into the system so as to simulate local excess ratios of slag to acid. The reaction between the solids and sulphuric acid was assumed entirely controlled by thermodynamics, when the compound with the greatest affinity towards sulphuric acid is the one to react in the simulation. These compounds and the respective 'system reactions' can be identified by comparison the amounts of species before and after each sulphuric acid addition.

Under such strongly reducing conditions the metallic iron reacting with the acid produces hydrogen. Thus a set of calculational hydrogen concentrations in gas phase was achieved and they could then be compared with the very small $H_2(g)$-concentration of the equilibrium of complete mixing and to the maximum values, which have been detected from the reactor offgas.

According to the model, the system reaction with small acid increments would produce intermediate iron sulphide as follows:

$$5Fe + H_2SO_4 = FeS + 4FeO + H_2(g) \qquad \Delta G^o(110°C) = -428 \text{ kJ/mol} \qquad (IV)$$

The concentration of hydrogen in the reactor offgas would then be dependent on the rate of reaction (IV) and the further reactions of hydrogen. Since the main concern from the operational point of view is the theoretical maximum concentration, no assumptions of these further reactions were made. Assuming reaction (IV) to be much faster than the overall process reaction (reaction III), it could be estimated, that a maximum percentage of 2-4 vol-% of $H_2(g)$ in the gas phase is possible. Concentrations of this high range have only exceptionally been found in practice.

5.5 Formation of hydrogen sulphide and sulphur dioxide in slag reactor

The formation of hydrogen sulphide and sulphur dioxide were simulated by the same way as the generation of hydrogen. According to the calculation, hydrogen sulphide is not produced to great extent. However, it may be present as an intermediate product in several reactions. In figure (4), the partial pressure of $H_2S(g)$ is presented for a gas phase at 110 °C in terms of the reacted sulphuric acid. The maximum concentration of hydrogen sulphide in the reactor gas is 100 ppm-vol. According to process measurements the actual concentrations of $H_2S(g)$ have been 25 ppm-vol at their maxima (Koukkari 1993a).

Inferring by the calculation model, sulphur dioxide is produced with oxidation processes when an intermediate sulphide (FeS_2) reacts further with the sulphuric acid, giving ferrous sulphate and $SO_2(g)$. The SO_2-pressure estimated by calculation at 110°C is given in figure (5). The computed maximum concentration of sulphur dioxide is 15000 ppm-vol, The highest measured value in gas before the measuring probe failure due to the exhaust dust was 5000 ppm-vol (Koukkari 1993a).

Thus the thermodynamic calculation model enables the estimation of the hazardous effluent gas maxima from the reactor. The agreement of the simulation values and the process measurement is confined within the order of magnitude. More accurate results are possible, yet perhaps difficult to reach under varying process conditions. The model as such adds substantially to the process chemistry information and thus supports engineering design.

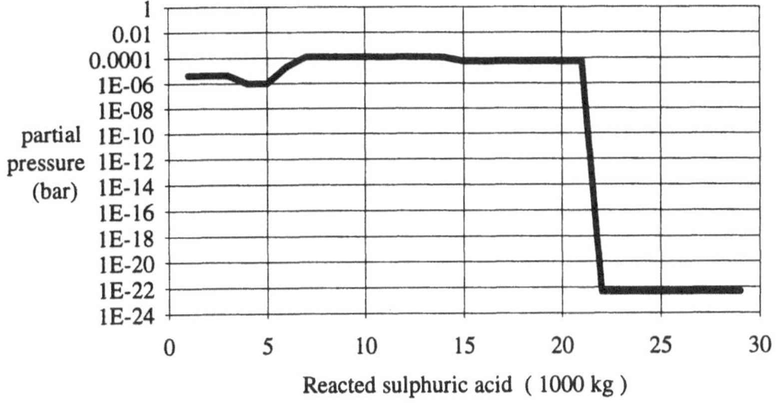

Figure 4. Calculated partial pressure for $H_2S(g)$ as a function of reacted sulphuric acid.

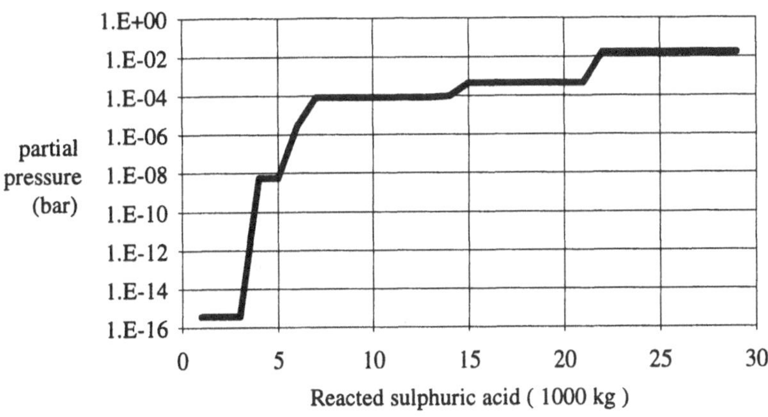

Figure 5. Calculated partial pressure for $SO_2(g)$ as a function of reacted sulphuric acid.

6 Treatment of acidic ferrous sulphate solutions

In the sulphate process acidic ferrous sulphate solutions is formed as a by-product in several units. These solutions are concentrated by vacuum evaporation. As a result some of the ferrous sulphate crystalizes out and the concentrated acidic ferrous sulphate solution is recycled.

The physicochemical behaviour of acidic ferrous sulphate solution was modelled by Harvie's (1980) modification of Pitzer equation in several temperatures from the experimental data. The calculation were performed by ChemSage, which was first modified for parameter fitting purposes (Sippola.1992a)

The solubility curve for iron sulphate in sulphuric acid at 25°C is represented in figure 6 and the partial pressure for water in the same system in figure 7 (Sippola 1992b). Thus Harvie's modification of Pitzer equation (HMW) is able to present the experimental solubility data of $FeSO_4$-H_2SO_4-H_2O system up to the acid concentration 10 mol/kg H_2O (50 w%). So figure 7 can be used to estimate the required vacuum for evaporation.

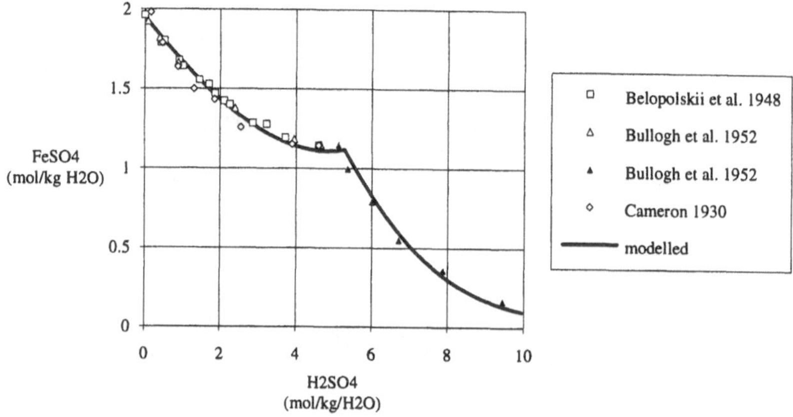

Figure 6. Solubility of ferrous sulphate in aqueous sulphuric acid at 25°C.

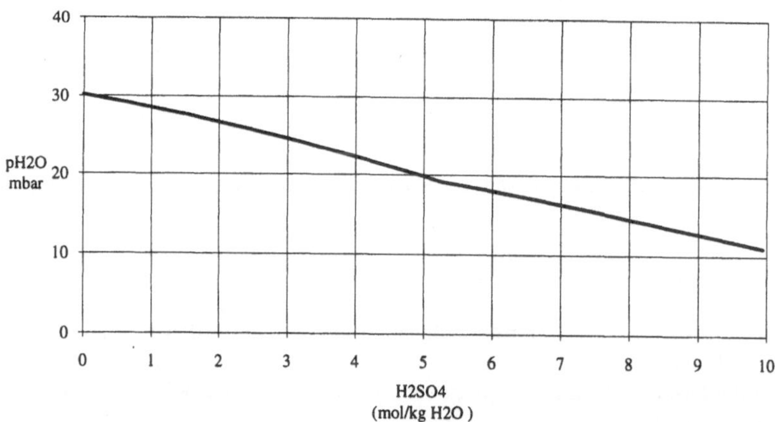

Figure 7. Partial pressure of water over saturated ferrous sulphate solution at 25°C.

7 The NPK-Simulator

The Gibbs energy minimization program may appear as the process chemistry subroutine in process simulation. Also within Kemira Oy, an NPK-process model consisting of three different Solgasmix-subroutines has been developed. Altogether 26 gas phase constituents and 70 other species were included in the thermodynamic description of the NPK-system. Such diversity is not without problems while the system is targeted to be processed in microcomputers. However, provided that the problem is convergent numerically in the Solgasmix procedure, the computing times do not become overwhelming.

The tentative flowsheet of the NPK-model is given in figure 8. The element balances in different phases of the NPK-reactors can be derived to a satisfactory verification with process measurement. The comparison of the process test results and calculation is given in table 5. A fair agreement between the calculation and experiment is seen, including the gaseous effluents such as nitrogen oxides and fluorine. The final component balance of the NPK-process could also be adequately reached by the Solgasmix-simulation. These results substantially add to the traditional main-component balance treatment of the multi-stage process (Swanström 1986). However, a detailed model for such complicated chemistry as in the NPK-reactors would be a subject of a separate, committed study.

Table 5. Comparison of the calculated results to measurements.

	Calculated	Measured
Reactor 1		
Slurry		
Solid content (%)	1	0.6-1.5
Aqueous Ca (%)	5	6
Gas		
N as NO_x (mol/s)	0.06	0.10-0.18
F (mol/s)	0.015	0.009
Reactor 2		
Slurry		
H_2O content (%)	23.0	not measured
Reactor 3		
Slurry		
H_2O content (%)	13.7	13.2
Gas		
N as NH_3 (mol/s)	7.15	6.15
N as NO_x (mol/s)	1.23	1.15
F (mol/s)	0.003	0.012
Fertilizer		
N as NH_3 content (%)	47	50

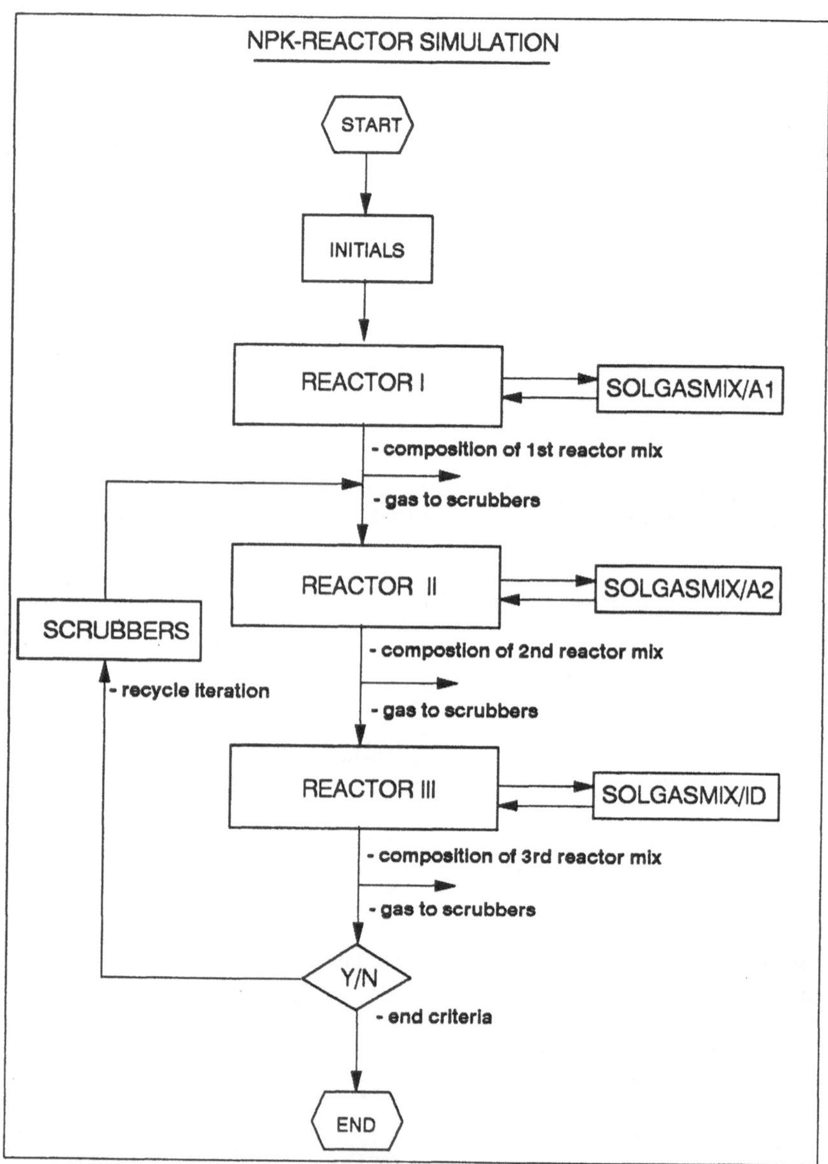

Figure 8. The NPK-Process simulation.

8 Discussion

The physico-chemical calculation provides a rewarding method to monitor chemical processes. The operator of such a routine constantly masters both the material and energy balances of the process. The additional advantage is the ability to follow the chemical composition of a multistage industrial system throughout the process. The calculation results but with the intermediate products appearing in different reactors, also with the composition of effluent gases and side streams.
The calculation is most practical in connection with test runs and process measurements. The benefits may be summarized as follows:

reduction of experimental work
discarding of 'impossible' experiment
screening of different process variations
quick balance checking in complicated systems
recognition of side streams and effluents and their elimination

The simulation results are used normally in the 'downstream' activities of process design, such as material balances and equipment design. The Solgasmix-type simulation can well be performed in conventional personal computers equipped with math co-processors. Thus the utilization of multicomponent simulation data in regular design spreadsheet software is straightforward.

The modelling of heterogeneous reactors by the Solgasmix-type routines is based on known thermodynamic and activity data. The practical computation also has to include a number of kinetic constraints, such as described above. High temperature pyrometallurgical processes are often straightforward to solve with the present programs and data-bases. Aqueous systems, instead, require more often preparatory work to design a satisfactory model in terms of substances and their activity data.

The thermodynamic multicomponent approach is a tool to align the theoretical chemistry with real industrially important problems. The methods, however, do not remain theoretical or 'concepts of the ideal world'. Instead, they enhance the understanding of the practical engineering options and thus may be used to substantially increase productivity of the process.

Acknowledgements:
M.Sc. Juho Jalava (Kemira,Pori)
M.Sc. Jussi Seikkula (Kemira,Pori)
M.Sc. Arie van der Meer (Kemira,Uusikaupunki)
Dr.Tech. Simo Liukkonen (Helsinki Univercity of Technology)
M. Sc. Ismo Laukkanen (Kemira, Helsinki)

9 References

Ball, F-X. Furst, W. and Renon, H. (1985a) An NRTL Model for Representation and Prediction of Deviation from Ideality in Electrolyte Solutions Compared to the Models of Chen(1982) and Pitzer(1973). **AIChE Journal**, 31, 392-399.

Ball, F-X. Planche, H. Furst, W. and Renon, H. (1985b) Representation of deviation From Ideality in Concentrated Aqueous Solutions of Electrolyte Solutions Using Mean Spherical Approximation Molecular Model,. **AIChE Journal**, 31, 1233-1240.

Bromley, L.A. Thermodynamic Properties of Strong Electrolytes in Aqueous Solutions, **AIChE Journal**, 19, 313.-320.

Chen, C-C Britt, H.I. Boston, J.F. and Evans, L.B. (1982) Local Composition Model for Excess Gibbs Energy of Electrolyte Systems, **AIChE Journal**, 28, 588-596

Cruz, J. and Renon, H. (1978) A New Thermodynamical Presentation of Binary Electrolyte Solutions Nonideality in the Whole Concentration Range of Concentrations, **AIChE Journal**, 24, 817-830.

Dawson, W.J. (1988), Hydrothermal Synthesis of advanced Ceramic Powders, **Ceramic Bulletin**, 67, 1673-1678.

Eriksson, G. (1975) Thermodynamic studies of high temperature equilibria XII, **Chem. Scripta**, 8, 100-103.

Eriksson, G. and Hack, K. (1984) Calculation of Phase Equilibrium in Multicomponent Alloy System Using A Specially Adapted Version of the Program Solgasmix, **Calphad,** 8, 15-24.

Eriksson, G. and Hack, K. (1990) , ChemSage - A Computer Program for the Calculation of Complex Chemical Equilibria, **Metallurgical Transactions B**, 21, 1013-1023.

Haglatah, A. and Vera, J.H. (1988) A Nonrandom Factor Model for the Excess Gibbs Energy of Electrolyte Solutions, **AIChE Journal**, 34, 803-815.

Harvie, C.E. Möller, N. and Weare, J.H. (1980) The prediction of Mineral Solubilities in Natural Waters': The Na-K-Mg-Ca-H-Cl-SO_4-OH-HCO_3-CO_3-CO_2-H_2O System to High Ionic Strengths at 25°C, **Geochimica et Cosmochimica Acta,** 44, 981-997.

Hillert, M. (1981) A discussion of methods of calculating phase diagrams, **Bull. Alloy Phase Diag.** 2, 265-268.

Hämäläinen, M. Rannikko, H. and Sippola, H. (1992) **Thermodynamic Modelling of Aqueous Systems**, Report TKK-V-C108 (in Finnish).

Kaskiala, M. Kemppinen, S. Niemelä, J. Taskinen, P. and Volotinen, H., (1989) Pöasma Chemistry and its applications in Chemical Processing, **Kemia-Kemi**, 16, 122-127 (in Finnish).

Koukkari, P. Sippola, H. Sundquist, A. (1993a) Calculation of Thermodynamic Equilibria for Process Design Part II, **Kemia-Kemi**, 20, 120-124 (in Finnish).

Koukkari, P. (1993b) A Physico-Chemical Method to Calculate Time-Dependent Reaction Mixtures. **Computers Chem. Engng.**, 17, 1157-1165.

Leppinen, J. (1986), **On the Interaction Between Thiol Collector Ions and Lead Sulfide Surface**, Thesis, University of Turku.

Magnusson, H. and Warnqvist, B. (1980) The NSP Project: An Alternative to the Conventional Recovery Furnace, **Pulp and Paper**, February, 54-56.

Meissner, H.P. and Tester, J.W. (1972) Activity Coefficient of Strong Electrolytes in Aqueos Solutions, **AIChE Journal**, 11, 128-133.

Partanen, J. (1989) Mean Activity Coefficient of Several Uni-univalent Electrolytes in Dilute Aqueous Solutions, **Acta Polytechnica Scandinavica**, 188.

Pitzer, K.S (1973) Thermodynamic of Electrolytes I. Theoretical Basis and General Equations, **Journal of Physical Chemistry**, 77, 268-277.

Renotte, J. Massilion, H. and Kalitventzeff, B. (1989) A new Model for the Simulation of Behaviour of Electrolyte Aqueous Solutions. Comparison with Three Well-Known Previous Models, **Computers & Chemical Engineering**, 13, 411-417.

Roine, A. (1988) **HSC-Software**, version 3.0, Outokumpu Research Centre, Pori, Finland.

Sanders, B. Fredenslund, A. and Rasmussen P. (1986) Calculation of Vapor-Liquid Equilibria in Mixed Solvent/Salt System Using Extended UNIQUAC Equation, **Chemical Engineering Science**, 41, 1171-1183.

Sippola, H. (1992a), **Solubility of Ferrous Sulphate in Sulphuric Acid**, Licentiate Thesis, Helsinki University of Technology (in Finnish).

Sippola, H. Koukkari, P. Rannikko, H. and Hämäläimem M. (1992) Thermodynamic Equilibrium Calculations to Process Design, **Kemia-Kemi**, 19, 734-737.

Smith, W.R and Missen, R.W. (1982), **Chemical Reaction Equilibrium Analysis**, John Wiley & Sons, New York, USA.

Swanström, S. (1986) The Kemira NPK Process, **Phosporus&Potassium**, 145, September, 34-36.

Zemaitis, J.F. Jr. Clark, D.M. Rafael, M. and Scrivner, N.C. (1986) **Handbook of Aqueous Electrolyte Thermodynamics**, AIChE, New York, USA.

Hematite solubility in sulphate process solutions

V. G. Papangelakis
B. C. Blakey
H. Liao
Department of Chemical Engineering and Applied Chemistry, University of Toronto, Toronto, Canada

Abstract
Iron rejection as hematite from hydrometallurgical process solutions appears to be advantageous for a number of reasons. These include its environmental stability, low residue volume, and potential use in the iron making, ceramic, and cement industries. For the design of a process that produces an iron oxide residue with controlled properties, suitable for disposal or secondary use, knowledge of hematite solubility is necessary.

In this paper, the solubility of hematite in sulphate solutions at elevated temperatures is investigated by thermodynamic modelling as well as by experimental measurements. The thermodynamic model, which is based on literature data, accounts for all known iron-sulphate-hydroxyl species present in the aqueous phase in equilibrium with ferric oxide. The model performs temperature extrapolations, incorporates recent thermodynamic data, includes foreign cations, and accounts for ionic strength effects. The experimental measurements are made using a titanium autoclave equipped with acid injection and sample withdrawal units. Reagent grade hematite is loaded into the autoclave, followed by injection of pre-measured quantities of sulphuric acid.

The results show that there is fairly good agreement between experimental measurements and theoretical predictions of hematite solubility within the temperature range of 170 - 200°C, and 30 - 100 g/L "free" sulphuric acid. The model explains the behaviour of hematite solubility change with temperature and with the addition of zinc sulphate in solution.
Keywords: Hematite, iron, solubility, speciation, sulphate, zinc.

1 Introduction

Iron precipitation as hematite from acidic sulphate solutions is encountered in several high temperature hydrometallurgical processes. Examples include the refractory gold pre-oxidation process (180 - 200°C) [1] and direct acid leaching of nickeliferous limonitic laterites (250°C) [2] where hematite precipitation takes place *in situ* through a rich-in-iron mineral dissolution-precipitation sequence. The most marked application, however, is found in the zinc industry where the so-called "hematite process" is used as a means of solution purification by iron rejection. This process currently competes against two other alternative iron removal processes; namely, the "jarosite" and the "goethite" processes [3].

Conceptually, the hematite process involves, first, an iron(III) to iron(II) reduction step, followed by neutralisation. The soluble ferrous solution is then heated to a temperature higher than 180°C, in an autoclave, where oxidation by oxygen gas and simultaneous hydrolysis to hematite takes place. Recently, there has been a renewed interest in the hematite process because of some distinct advantages in spite of the high cost associated with its operation. The merits of this process are include its ability to provide a more environment-friendly product due to the stability and compactness of hematite, and a marketable product as a reagent in the cement, pigment, and, potentially, in the steel making industries. The latter depends on the "cleanness" of hematite with respect to its zinc and sulphur contents. Until now, it has been impossible to produce hematite with impurity levels less than 100 ppm Zn (necessary level to feed blast furnaces [4]) and less than 0.6 wt. % S. So far, four plants around the world, of which two in the European Union, are employing the hematite process [4].

Whether hematite forms during a high temperature pressure leaching process or during a separate iron removal step, knowledge of its solubility under process conditions is highly desirable. This is particularly true for the "hematite process" where the nature and properties of the iron oxide product depend on the precipitation path which, in turn, depends on the (super) saturation level of Fe^{3+}, which is the driving force for precipitation [5].

One of the most comprehensive published experimental works on the solubility of hematite in zinc process solutions is that of Umetsu *et al.* [6]. In this work, solubilities of ferric oxide were determined in sulphuric acid solutions in addition to the solubility of hematite produced by the hydrolysis of an acidic ferric sulphate solution between 150 and 200°C in the presence and absence of zinc as $ZnSO_4$. Later, Tozawa and Sasaki [7] revisited the system in an attempt to explain the effect of coexisting metal sulphates on solubility on the basis of the buffering action of the bisulphate anion.

In the present paper, a prediction of hematite solubility is attempted using a thermodynamic modelling approach. The model performs a speciation analysis at the temperature of interest and calculates the solubility by accounting for all iron-sulphate complexes present in the solution. The model is validated by comparing its predictions with the data of Umetsu *et al.* and Tozawa and Sasaki as well as with experimental solubility measurements made by the authors at 200°C.

2 Experimental

The solubility tests were carried out in a 2 L PARR titanium autoclave. Agitation was provided by a magnetically driven, twin impeller stirrer. Temperature was controlled within ± 2°C. The controller was operated in the proportional-integral (PI) mode, manipulating both a heating mantle and a water cooling loop stream via a solenoid valve. A glass liner was used in the reaction chamber to avoid contamination of the reaction solution.

A typical experiment proceeded as follows. A mass of 50 g of reagent grade α-Fe_2O_3 and 950 g of water were placed in the reactor. This slurry was agitated under an impeller speed of 600 r.p.m. A total time of 35 minutes was required to achieve a stable temperature of 200°C.

Upon temperature stabilization, a known amount of concentrated sulphuric acid was injected into the autoclave under nitrogen pressure. The acid delivery device was then rinsed by further injecting 50 g of water. The amount of acid was calculated so that after injection, and the subsequent wash, the resulting solution would have a pre-determined concentration. Solution samples were withdrawn through a dip tube. The inlet of the dip tube was fitted with a Union Carbide 45 μm porous graphite filter to prevent solids from exiting with the aqueous samples. Sample streams were routed through a co-current heat exchanger, thus reducing their temperature to between 30 to 60°C at the outlet of the sampling system. Cooling was provided by flowing water on the shell side. Prior to obtaining an actual 30 mL sample, 15 mL of solution (approximately twice the internal volume of the sampling line) were withdrawn and discarded in order to wash the line out and obtain a sample representative of the solution in the autoclave.

Some solid material was found to pass through the filter and exit the autoclave when samples were taken. Thus, it was necessary to filter all samples using a 2.6 μm "Millipore" membrane filtration unit at the instant of sample acquisition. Flame atomic absorption (FAA) analysis was performed for determining iron concentrations.

3 Thermodynamic model formulation

3.1 Solubility

For the iron hydrolysis reaction

$$Fe^{3+} + {}^{3}/_{2}\, H_2O = {}^{1}/_{2}\, Fe_2O_3 + 3H^+ \qquad \text{(R-1)}$$

the equilibrium constant $K_{Fe_2O_3}$ is

$$K_{Fe_2O_3} = \frac{m^3_{H^+}}{m_{Fe^{3+}}} \qquad (1).$$

In the presence of complexing anions (ligands) like OH^- and SO_4^{2-}, ferric ion can undergo complexation reactions. For example,

$$Fe^{3+} + SO_4^{2-} = FeSO_4^+ \qquad \text{(R-2)}$$

with an equilibrium constant $\beta_{FeSO_4^+}$ given by:

$$\beta_{FeSO_4^+} = \frac{m_{FeSO_4^+}}{m_{Fe^{3+}} \cdot m_{SO_4^{2-}}} \qquad (2).$$

At equilibrium, equations (1) and (2) must be satisfied simultaneously. The solubility of iron, s, will then be given as the total concentration of all iron-bearing species in solution. That is,

$$s = m_{Fe^{3+}} + m_{FeSO_4^+} \qquad (3).$$

After substituting $m_{FeSO_4^+}$ from (2) we have

$$\begin{aligned} s &= m_{Fe^{3+}} + m_{Fe^{3+}} \cdot \beta_{FeSO_4^+} \cdot m_{SO_4^{2-}} \\ &= m_{Fe^{3+}} (1 + \beta_{FeSO_4^+} \cdot m_{SO_4^{2-}}) \end{aligned} \qquad (4).$$

In reality, more than one complex usually forms, particularly at high temperatures, and therefore the solubility expression has to be extended to include all existing complexes. This necessitates a speciation calculation.

3.2 Computational Method

The determination of species' concentrations in the present system was accomplished using the *equilibrium constant* approach [8]. In this method, all species containing the relevant metal ions and ligand molecules are divided into two groups: component and derived species. The group of components can be regarded as the chemical "building blocks" used to (algebraically) describe all species that the system contains in chemical equations. For the iron(III)-zinc-sulphate-water system the logical choices for component species are Fe^{3+}, Zn^{2+}, SO_4^{2-}, and H^+. Table 1 presents all species considered in this work along with their thermochemical data. The value of -4.6 kJ/mol (versus -16.9 kJ/mol [9]) was adopted for the $\Delta G°_{f,\,298}$ for Fe^{3+}, and every possible effort was made to ensure a consistent set of thermodata.

Next, a series of complexation reactions is set up for each "derived species", the concentration of which can be expressed as a function of the stoichiometric stability constant and the concentration of its components. Table 2 lists all equilibria that have been accounted in this work.

If the concentrations of all component species are known, then the concentration of any other species in the solution can be calculated. For example:

$$m_{OH^-} = \frac{1}{\beta_1 \cdot m_{H^+}}$$

Table 1. Species and thermodynamic data considered for this work (as cited in [9] and [10], unless otherwise indicated).

Species	$\Delta G^\circ_{f,\,298}$	$\Delta H^\circ_{f,\,298}$	S°_{298}	C°_p (J/K mol)		
	(kJ/mol)	(kJ/mol)	(J/K mol)	A	$B \times 10^3$	$C \times 10^5$
H^+	0	0	0	0	0	0
OH^-	-157.29	-230.12	-10.88	506.38	-1181.3	-246.02
H_2O	-237.18	-285.85	69.96	75.44		
SO_4^{2-}	-744.63	-909.18	20.08	874.6	-1759.7	-519.98
HSO_4^-	-756.01	-887.01	131.8	-547.29	1342.1	266.78
Fe^{3+}	-4.6	-48.5	-315.9	79.09	-219.35	15.4
$FeOH^{2+}$	-229.41	-290.8	-142.0	19.4	-21.22	29.4
$Fe(OH)_2^+$	-446.4	-543.8	-29.29			
$Fe(OH)_3^0$	-660.0	-795.73	75.4			
$Fe(OH)_4^-$	-830.0	-1050.4	25.5			
$Fe_2(OH)_2^{4+}$	-466.97	-611.38	-355.64	124.68	-619.0	14.0
$Fe_3(OH)_4^{5+}$ *	-926.6	-1225	-573.0			
$FeSO_4^+$	-772.8	-931.78	-129.7	37.7	317.0	-1.0
$Fe(SO_4)_2^-$	-1524.65	-1828.4	-43.07			
$Fe_2(SO_4)_3^0$	-2243.0	-2825.0	-571.53			
$FeHSO_4^{2+}$	-768.38	-894.29	-18.68			
$FeSO_4HSO_4^0$ **	-1514.4	-1757.7	159.86			
α-Fe_2O_3 †	-744.27	-825.4	87.40	98.282	77.82	-14.853
Zn^{2+}	-147.06	-153.97	-112.1	2.77	-39.14	34.6
$ZnOH^+$	-330.1	-357	5			
$Zn(OH)_2^0$	-522.73	-133.47	-133.5	-251.04		
$Zn(OH)_3^-$	-694.22	-844	107			
$Zn(OH)_4^{2-}$	-858.52	-1078	-218			
$ZnSO_4^0$	-904.9	-1047.3	5.02			

* Data calculated from [11].
** Data calculated from [12].
† Data calculated from [13].

Table 2. Equilibria of the Fe(III)-Zn-SO_4-H_2O system, complex stability constants, and Vasil'ev parameter b (calculated from [14]).
(Continued on next page)

Equilibrium	β or K	b
$H^+ + OH^- = H_2O$	$\beta_1 = \frac{1}{m_{H^+} \cdot m_{OH^-}}$	0.219
$H^+ + SO_4^{2-} = HSO_4^-$	$\beta_2 = \frac{m_{HSO_4^-}}{m_{H^+} \cdot m_{SO_4^{2-}}}$	-0.0478
$Fe^{3+} + OH^- = FeOH^{2+}$	$\beta_3 = \frac{m_{FeOH^{2+}}}{m_{Fe^{3+}} \cdot m_{OH^-}}$	0.277
$Fe^{3+} + 2OH^- = Fe(OH)_2^+$	$\beta_4 = \frac{m_{Fe(OH)_2^+}}{m_{Fe^{3+}} \cdot m_{OH^-}^2}$	0.652
$Fe^{3+} + 3OH^- = Fe(OH)_3^0$	$\beta_5 = \frac{m_{Fe(OH)_3^0}}{m_{Fe^{3+}} \cdot m_{OH^-}^3}$	1.049
$Fe^{3+} + 4OH^- = Fe(OH)_4^-$	$\beta_6 = \frac{m_{Fe(OH)_4^-}}{m_{Fe^{3+}} \cdot m_{OH^-}^4}$	1.249
$2Fe^{3+} + 2OH^- = Fe_2(OH)_2^{4+}$	$\beta_7 = \frac{m_{Fe_2(OH)_2^{4+}}}{m_{Fe^{3+}}^2 \cdot m_{OH^-}^2}$	0.492
$3Fe^{3+} + 4OH^- = Fe_3(OH)_4^{5+}$	$\beta_8 = \frac{m_{Fe_3(OH)_4^{5+}}}{m_{Fe^{3+}}^3 \cdot m_{OH^-}^4}$	0.905
$Fe^{3+} + SO_4^{2-} = FeSO_4^+$	$\beta_9 = \frac{m_{FeSO_4^+}}{m_{Fe^{3+}} \cdot m_{SO_4^{2-}}}$	0.216
$Fe^{3+} + 2SO_4^{2-} = Fe(SO_4)_2^-$	$\beta_{10} = \frac{m_{Fe(SO_4)_2^-}}{m_{Fe^{3+}} \cdot m_{SO_4^{2-}}^2}$	0.163
$2Fe^{3+} + 3SO_4^{2-} = Fe_2(SO_4)_3^0$	$\beta_{11} = \frac{m_{Fe_2(SO_4)_3^0}}{m_{Fe^{3+}}^2 \cdot m_{SO_4^{2-}}^3}$	*
$Fe^{3+} + H^+ + SO_4^{2-} = FeHSO_4^{2+}$	$\beta_{12} = \frac{m_{FeHSO_4^{2+}}}{m_{Fe^{3+}} \cdot m_{H^+} \cdot m_{SO_4^{2-}}}$	-0.247
$Fe^{3+} + H^+ + 2SO_4^{2-} = FeSO_4HSO_4^0$	$\beta_{13} = \frac{m_{FeSO_4HSO_4^0}}{m_{Fe^{3+}} \cdot m_{H^+} \cdot m_{SO_4^{2-}}^2}$	*
$Fe^{3+} + {}^3/_2\, H_2O = {}^1/_2\, Fe_2O_3 + 3H^+$	$K_{Fe_2O_3} = \frac{m_{H^+}^3}{m_{Fe^{3+}}}$	*

* Insufficient stability constant data available as functions of ionic strength. Equilibrium assumed not to vary with ionic strength.

Table 2. (Continued)

Equilibrium	β or K	b
$Zn^{2+} + OH^- = ZnOH^-$	$\beta_{14} = \frac{m_{ZnOH^+}}{m_{Zn^{2+}} \cdot m_{OH^-}}$	-0.087
$Zn^{2+} + 2OH^- = Zn(OH)_2^0$	$\beta_{15} = \frac{m_{Zn(OH)_2^0}}{m_{Zn^{2+}} \cdot m_{OH^-}^2}$	-0.163
$Zn^{2+} + 3OH^- = Zn(OH)_3^-$	$\beta_{16} = \frac{m_{Zn(OH)_3^-}}{m_{Zn^{2+}} \cdot m_{OH^-}^3}$	0.403
$Zn^{2+} + 4OH^- = Zn(OH)_4^{2-}$	$\beta_{17} = \frac{m_{Zn(OH)_4^{2-}}}{m_{Zn^{2+}} \cdot m_{OH^-}^4}$	1.113
$Zn^{2+} + SO_4^{2-} = ZnSO_4^0$	$\beta_{18} = \frac{m_{ZnSO_4^0}}{m_{Zn^{2+}} \cdot m_{SO_4^{2-}}}$	**

** The Davies' modified form of the Debye-Hückel equation was used to account for the ionic strength effect [10], *i.e.*: $\log\beta = \log\beta^\circ + A\Delta z^2\left(\frac{\sqrt{I}}{1+\sqrt{I}} - 0.2I\right)$.

$$m_{FeOH^{2+}} = \beta_3 \cdot m_{Fe^{3+}} \cdot \left(\frac{1}{\beta_1 \cdot m_{H^+}}\right)$$

$$\vdots$$

$$m_{ZnSO_4^0} = \beta_{18} \cdot m_{Zn^{2+}} \cdot m_{SO_4^{2-}}$$

There are four component concentrations, namely m_{H^+}, $m_{SO_4^{2-}}$, $m_{Fe^{3+}}$, and $m_{Zn^{2+}}$, hence, four equations are needed. These are:

$$[SO_4]_{total} = m_{SO_4^{2-}} + m_{HSO_4^-} + m_{FeSO_4^+} + 2m_{Fe(SO_4)_2^-} + 3m_{Fe_2(SO_4)_3^0} + m_{FeHSO_4^{2+}} + 2m_{FeSO_4HSO_4^0} + m_{ZnSO_4^0} \quad (5),$$

$$[Zn^{2+}]_{total} = m_{Zn^{2+}} + m_{ZnOH^+} + m_{Zn(OH)_2^0} + m_{Zn(OH)_3^-} + m_{Zn(OH)_4^{2-}} + m_{ZnSO_4^0} \quad (6),$$

$$0 = m_{H^+} + 3m_{Fe^{3+}} + 2m_{Fe(OH)^{2+}} + m_{Fe(OH)_2^+} + 4m_{Fe_2(OH)_2^{4+}} + 5m_{Fe_3(OH)_4^{5+}} + m_{FeSO_4^+} + 2m_{FeHSO_4^{2+}} + 2m_{Zn^{2+}} + m_{ZnOH^+} - (m_{OH^-} + 2m_{SO_4^{2-}} + m_{HSO_4^-} + m_{Fe(OH)_4^-} + m_{Fe(SO_4)_2^-} + m_{Zn(OH)_3^-} + 2m_{Zn(OH)_4^{2-}}) \quad (7),$$

and

$$K_{Fe_2O_3} = \frac{m_{H^+}^3}{m_{Fe^{3+}}} \quad (8).$$

Equations (5) and (6) are the sulphate and zinc ion mass balances, respectively. Equation (7) represents the charge balance of the solution and equation (8) is the constraint placed upon the hydrogen ion and the free (uncomplexed) ferric ion concentrations due to the solubility of hematite.

For most of the reactions considered here, the dependence of stability constants upon ionic strength has been determined by the Vasil'ev approach. Vasil'ev [15] proposed that the stability constant varies with ionic strength in the following way:

$$\log\beta_T = \log\beta_T^\circ + \frac{A\Delta z^2\sqrt{I}}{1+1.6\sqrt{I}} + bI \tag{9}$$

where β_T and β°_T are the concentration (molar)-based and thermodynamic stability constants at temperature T, respectively; A is the Debye-Hückel constant; Δz^2 is the algebraic sum of the squares of species' charges; I is the ionic strength on a molar basis; and b is a constant dependent on the given complexation equilibrium.

Because of the high temperature extrapolations made here, equation (9) is assumed to be valid for molal-based stability constants with the same b-value. This simplification was deemed necessary in view of the unknown densities of mixed electrolyte solutions at high temperatures. Hence, ionic strength was calculated on a molal basis. Table 2 also summarizes the Vasil'ev b-values obtained from the treatment of literature data by equation (9).

Extrapolations of the stability constants to elevated temperatures were made using Helgeson's method [16] for those species for which no heat capacity data were available (Table 1). Otherwise, the free energy change for a complexation reaction was calculated using the formula:

$$\Delta G_T^\circ = \Delta G_{298}^\circ + \int_{298}^{T} \Delta C_p^\circ dT - T\int_{298}^{T} \frac{\Delta C_p^\circ}{T} dT - (T-298)\Delta S_{298}^\circ \tag{10}$$

from which the stability constant at zero ionic strength was calculated via:

$$\beta_T^\circ = \exp\left(-\frac{\Delta G_T^\circ}{RT}\right) \tag{11}.$$

The Debye-Hückel constant, A, (on a molal basis) as a function of temperature in equation (9) was taken from Zemaitis [17]. Finally, the numerical solution of the system of equations (5) to (8) were performed using *Mathematica* [18].

4 Results and Discussion

The model predictions were compared with the authors' own experiments at 200°C, as well as with the findings of Umetsu *et al.* This comparison is shown in Figure 1.

Umetsu *et al.*, proposed the following equation that correlates the equilibrium iron concentration, in g/L, with that of "free" sulphuric acid, also in g/L, at room temperature:

$$\text{Log}[\text{Fe(III)}]_{\text{total}} = \text{a Log}[H_2SO_4]_{\text{free}} - \text{b} \qquad (12),$$

where a and b depend on temperature and on whether or not zinc (as zinc sulphate) is present. The parameters a and b of equation (12) are given in Table 3. These were obtained by least squares regression from Fe_2O_3 dissolution experiments (in the absence of zinc) and from $Fe_2(SO_4)_3$ hydrolysis experiments (in the presence of zinc) [6]. The "free" sulphuric acid of equation (12) was defined as the total sulphate ($[SO_4]_{total}$) less that which is stoichiometrically "bound" to ferric sulphate and zinc sulphate. That is:

$$[H_2SO_4]_{\text{free}} = [SO_4]_{\text{total}} - {}^{3}/_{2}\,[\text{Fe(III)}]_{\text{total}} - [\text{Zn}]_{\text{total}} \qquad (13)$$

[6] where $[\text{Fe(III)}]_{\text{total}}$ and $[\text{Zn}]_{\text{total}}$ are the total ferric and zinc concentrations, respectively.

In order to make comparisons between equilibrium solutions at different temperatures valid, the units of equation (12) were changed to g/kg of solvent using the following relationship:

$$[X]_m = \frac{[X]_M}{1000\rho - ([\text{Fe(III)}]_{\text{total}} + [\text{Zn}]_{\text{total}} + [SO_4]_{\text{total}})} \qquad (14)$$

where $[X]_m$ is either $[\text{Fe(III)}]_m$ or $[H_2SO_4]_m$ in g/kg, $[X]_M$ is the respective concentration in terms of g/L, and ρ is the density of the solution. At room temperature, the latter was taken to be equal to 1 g/mL for the zero Zn concentration case and 1.2 g/mL otherwise. These assumptions are in accordance with the experimental findings of Tozawa *et al.* [7].

Figure 1 shows that the experimental results of this work are in agreement with those of Umetsu *et al.* at 200°C. However, the thermodynamic model cannot accurately predict the experimental findings quantitatively, although it exhibits the correct general trends. It should be noted that numerous simulations with alternative thermodynamic data, alternative methods of temperature extrapolation (*e.g.*, Criss-Coble vs. Helgeson vs. heat capacity), molality-based description of equilibrium constants, and omission of the rather controversial $FeSO_4HSO_4^0$ complex failed to yield better and/or more consistent results.

This mixed (sulphate-bisulphate) complex has been identified by only one group of investigators [19] and has been used by McAndrew *et al.* [12] in the past. Its existence, however, was considered dubious in an earlier publication [10] and it was omitted from the speciation calculations up to 100°C. However, the thermodynamic prediction of

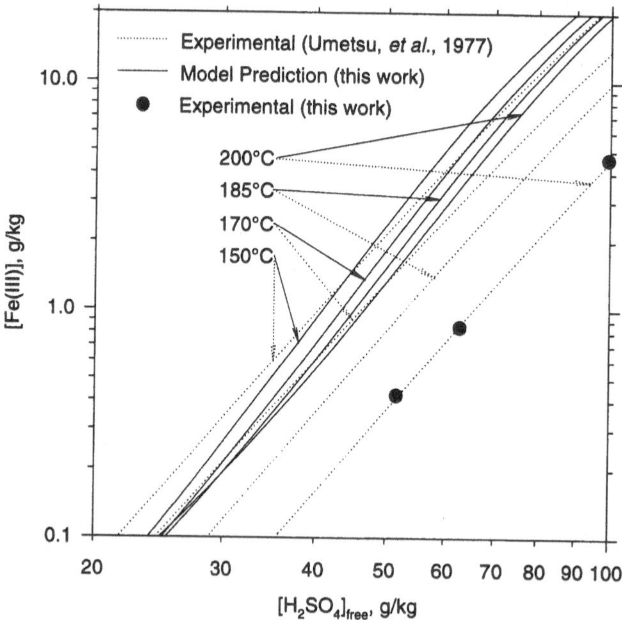

Fig. 1 Comparison between model prediction and experimental data from Umetsu *et al.* [6] and present work.

Table 3. Parameters of equation (12) [6].

[Zn] (g/L)	Temperature (°C)	a	b
0	150	3.8	6.05
0	170	3.7	6.10
0	185	3.9	6.65
0	200	3.9	7.00
68 - 101	185	6.03	11.02
68 - 101	200	5.34	10.09

hematite solubility within the temperature range 150 - 200°C fails dramatically* if this complex is unaccounted for. Hence, it may be postulated that $FeSO_4HSO_4^0$ is a true species under the conditions of temperature and acidity studied because of indirect experimental evidence in the form of solubility measurements. Furthermore, its existence, particularly in high temperature solutions, seems reasonable in view of the enhancement in complexation with an increase in temperature of an aqueous solution [20, 21]. This is supported by the lowering of the dielectric constant of water with increasing temperature, which results in a loss in the ability of water to stabilise highly charged species via solvation. Hence, reactions that lead to a reduction in total charge (like neutral complex formation, hydrolysis, *etc.*) are favoured [20, 21], and species that are not detectable at room temperature (the temperature at which complex identification by chemists is usually made) may very well form at an elevated temperature.

The same arguments lead to the speculation that mixed hydroxyl-sulphate complexes play an important role at high temperature systems. For example, complexes like $Fe(OH)SO_4^0$ have been suggested to exist at room temperature [22] but no reliable data are available to allow for inclusion into high temperature speciation calculations such as those in the present work. Hence, in the discussion that follows, the conclusions will have a qualitative, rather than a quantitative importance in view of the uncertainties involved.

In Figures 2a and 2b, a comparison is made between model predictions and experimental solubilities in the presence and absence of zinc as $ZnSO_4$. The simulations were performed with an average zinc concentration of 97 g/kg, representing the average range of zinc concentrations actually tested by Umetsu *et al.* (68 - 126 g/kg) [6]. Again, the model correctly predicts that the solubility decreases in the presence of zinc at 185°C (at $[H_2SO_4]_{free} < 60$g/kg) and 200°C (at $[H_2SO_4]_{free} < 80$ g/kg). Although it has been found that free acidities greater than about 60 g/kg yield $FeOHSO_4$ on hydrolysis rather than hematite [6, 7], the lines in Figures 2a and 2b have been extended to the full scale (*i.e.*, 100 g/kg "free" H_2SO_4) for consistency with Figure 1.

Figures 3a, b, c, and d depict the distribution of solution species at equilibrium with hematite versus free sulphuric acid at 150, 170, 185, and 200°C, and in the absence of any zinc in solution. It is interesting to observe that for the whole range of temperatures and acidities covered in Figure 3, $FeSO_4HSO_4^0$ and $FeHSO_4^{2+}$ are the two most dominant complexes. The former becomes more dominant with increasing acidity in all cases, while the latter is more favoured with increasing temperature. This behaviour is consistent with the observation that higher temperatures require higher "free" acidities to yield $FeOHSO_4$ on hydrolysis [7]. The $FeSO_4HSO_4^0$ species may act as a precursor for such a hydrolytic precipitation yielding $FeOHSO_4$ as the final product. On the other hand, iron(III)-hydroxyl complexes and the simple ferric sulphate and ferric bisulphate species are more dominant at low free sulphuric acid concentrations, which is the region of hematite precipitation.

---

* *i.e.*, solubility prediction is 2 to 3 orders of magnitude less and inverse solubility behavior is not observed.

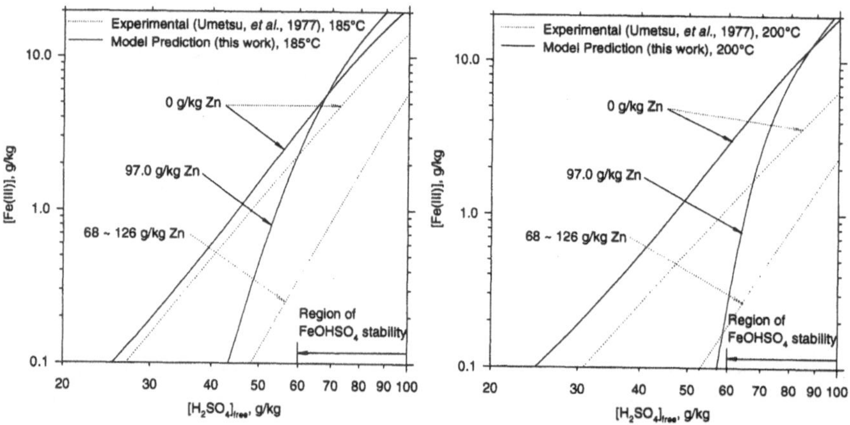

Fig. 2a The effect of zinc sulphate on hematite solubility at 185°C.

Fig. 2b The effect of zinc sulphate on hematite solubility at 200°C.

Figures 3a to 3d also depict the variation of the true acidity, which is plotted as -Log(m_{H^+}) versus "free" H_2SO_4 at equilibrium with hematite and in the absence of any zinc. At all four temperatures the relationship between -Log(m_{H^+}) and the logarithm of $[H_2SO_4]_{free}$ is linear, and the true acidity (m_{H^+}) does not appear to be affected by temperature significantly. This suggests that, at least for zinc-free solutions, the use of free sulphuric acid as an indicator of acidity is justified. Apparently, the explanation of why basic ferric sulphate starts forming at higher free acidities as the temperature increases [7] has to be attributed to the change in the relative abundance of the $FeSO_4HSO_4^0$ complex rather than acidity.

Figures 4a and 4b show the species distribution with 97 g/kg Zn in solution at equilibrium with hematite at 185 and 200°C, respectively. Again, the distribution is expressed as the percentage of total dissolved iron(III) or zinc. The domination of the $FeSO_4HSO_4^0$ complex in the aqueous phase is more profound in this case at both temperatures. Ferric hydroxyl complexes appear above the arbitrarily selected 0.01 % cut-off level at low "free" H_2SO_4 concentrations, and the abundance of the $FeHSO_4^{2+}$ complex is reduced markedly. On the other hand, the neutral $ZnSO_4^0$ complex is the dominant zinc complex, although appreciable amounts of uncomplexed Zn^{2+} coexist. The strongest effect of the zinc presence, however, is the lowering of true acidity (*i.e.*, m_{H^+}). For example, if one considers what happens at 200°C, an increase in "stoichiometric pH" from 0.24 to 1 at 60 g/kg "free" H_2SO_4 is observed. When Zn is present, this represents an almost eight-fold decrease in the acidity, which drives reaction (R-1) to the right, resulting in a decrease in hematite solubility, as shown in Figure 2b. Equivalent arguments can be employed to explain the decrease in hematite solubility at less than 60 g/kg "free" H_2SO_4 at 185°C, as shown in Figure 2a. Of

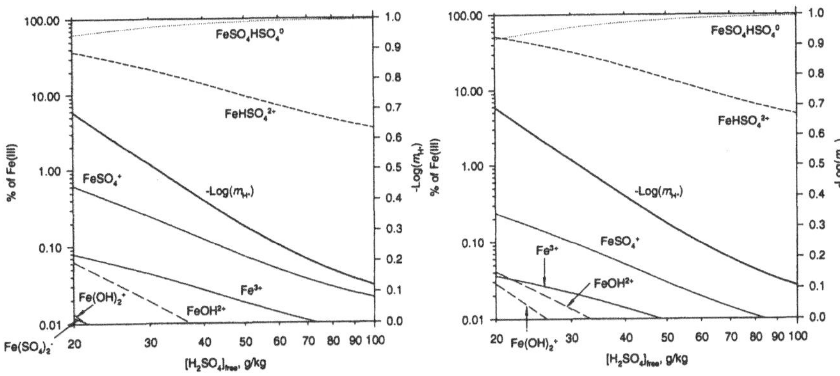

Fig. 3a Distribution of aqueous species at equilibrium with hematite at 150°C.

Fig. 3b Distribution of aqueous species at equilibrium with hematite at 170°C.

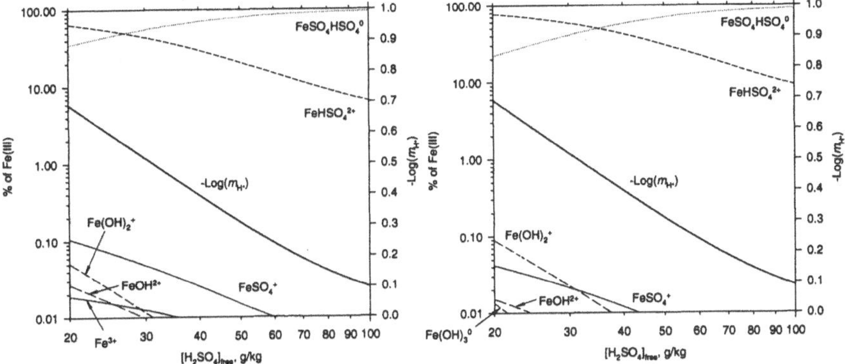

Fig. 3c Distribution of aqueous species at equilibrium with hematite at 185°C

Fig. 3d Distribution of aqueous species at equilibrium with hematite at 200°C

course, true acidity is not the sole determinant of solubility. The complexing ability of the ligand also plays an important roll in keeping ions in solution.

Hence, the decrease in hematite solubility is not proportional to the decrease in true acidity, as it is evident by comparing Figure 2b with Figures 3d and 4b at 200°C, for example. To better illustrate this statement, the absolute concentrations of the dissociation products of sulphuric acid should be plotted.

Figure 5a depicts the variation in HSO_4^-, H^+, and SO_4^{2-} concentration with "free" sulphuric acid at 185 and 200°C, at equilibrium with hematite. The concentrations of these species do not change dramatically with temperature, which implies that the

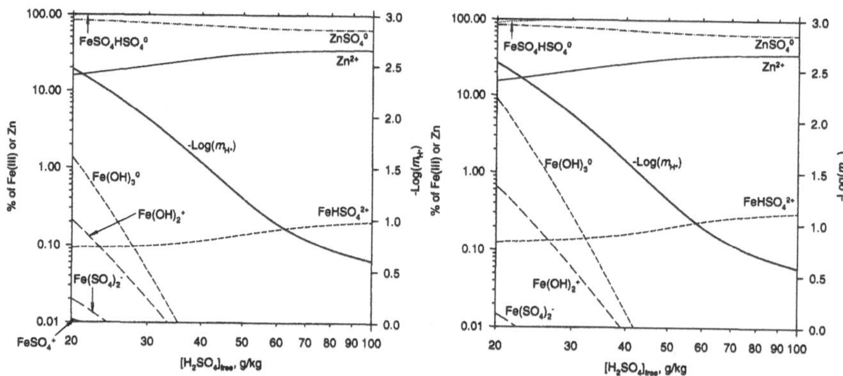

Fig. 4a Distribution of aqueous species at equilibrium with hematite in the presence of 97 g/kg Zn at 185°C.

Fig. 4b Distribution of aqueous species at equilibrium with hematite in the presence of 97 g/kg Zn at 200°C.

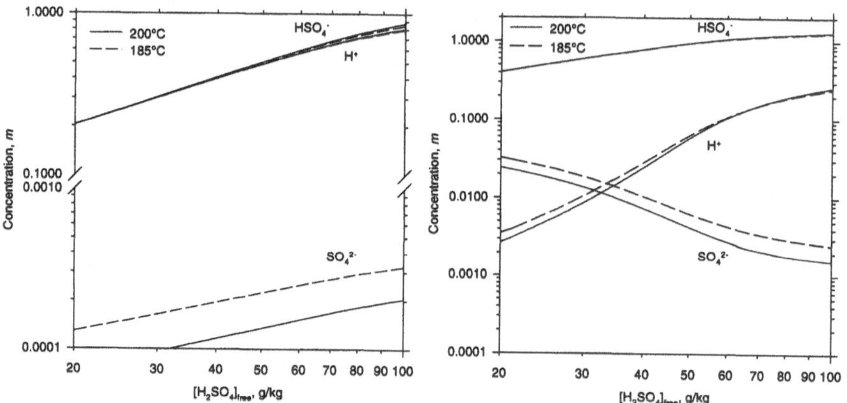

Fig. 5a Concentration of H_2SO_4 dissociation products at equilibrium with hematite.

Fig. 5b Concentration of H_2SO_4 dissociation products at equilibrium with hematite in the presence of 97 g/kg Zn.

solubility of hematite is mainly controlled by the way the equilibrium constant of reaction (R-1) (equation (1)) changes with temperature (*i.e.*, inverse solubility). In Figure 5b, however, with the presence of zinc there is a decrease in the acidity, an increase in SO_4^{2-} concentration, and an invariability in HSO_4^- concentration compared to Figure 5a. In essence, the decrease in acidity and the increase of SO_4^{2-} concentration drive the solubility of hematite towards two opposite directions by shifting reaction (R-

1) to the right, and by increasing complexation simultaneously, as shown by equation (4). It should also be made clear that the abscissae of Figures 5a and 5b do not correspond to the same total SO_4 in solution. As shown in both figures, and by equation (13), more total SO_4 exists in solution, when zinc is present, compared to when no zinc is present. For example, at around 70 g/kg "free" H_2SO_4 in Figure 2a, and at around 90 g/kg H_2SO_4 in Figure 2b, the predicted solubility of hematite is independent of the presence of zinc despite the fact that the true acidity is lower when zinc is present. This happens because more SO_4 exists in solution with zinc present, which compensates for the acidity effect on solubility.

The findings of this work appear to be at a disagreement with Tozawa and Sasaki who found that at 200°C, 60 g/L "free" H_2SO_4, and in the absence of Zn, and 100 g/L "free" H_2SO_4 in the presence of zinc, yield the same true acidities (*i.e.*, m_{H^+}) at the point where the precipitation of Fe_2O_3 gives way to that of $FeOHSO_4$. They based their calculations on estimating mean activity coefficients for the $ZnSO_4$ salt, and H_2SO_4. They were, thus, able to suggest that the onset of $FeOHSO_4$ stability occurs at the same true acidity due to the buffering action of HSO_4^-. Our findings, which are based on a speciation analysis, suggest that the addition of zinc has a marked negative influence on the acidity, and that the true transition from Fe_2O_3 to $FeOHSO_4$ stability has to be attributed to thermodynamic reasons (*e.g.*, $\Delta G^\circ_{f,\,FeOHSO_4}$). Unfortunately, there are no thermodynamic data for $FeOHSO_4$ in the literature to further substantiate this.

Finally, the so-called "buffering" action of HSO_4^- at high temperatures seems to act toward maintaining constant HSO_4^- ion concentration rather than maintaining constant acidity. The surplus of SO_4^{2-} ions when a foreign metal sulphate is present shifts the equilibrium of the reaction

$$HSO_4^- = H^+ + SO_4^{2-} \qquad \text{(R-3)}$$

to the left, consuming H^+ and lowering the effective (true) acidity of the system. The end result is extended stability of hematite in the presence of mixed metal sulphate solutions.

5 Conclusions

In this work, an attempt was made to predict hematite solubility in high temperature sulphate process solutions by a speciation approach. The findings of this work suggest that there is a lack of fundamental data on metal complexation for high temperature sulphate systems, and, as a result, this approach does not allow for an accurate quantification of the system. Hence, more research in identifying and assessing thermodynamic quantities for mixed sulphate and, quite probably, mixed hydroxyl-sulphate complexes at elevated temperatures is needed.

On the basis of the existing data, this work was able to qualitatively predict the trends observed in the solubility and stability of hematite over a temperature range of

150 - 200°C, and in the presence or absence of zinc sulphate. It was found that in pure iron(III)-sulphate solutions, the solubility is determined primarily by the change in the equilibrium constant of hematite hydrolysis with temperature. It was also found that for the same "free" sulphuric acid concentration in the presence of zinc (or presumably any other foreign metal sulphate) the solubility decreases (at low free sulphuric acid concentration) because the true acidity decreases. However, the solubility is also affected by the complexing ability of sulphate ion (particularly at high free sulphuric acid concentration) which has a positive effect. The end result depends on the relative strengths of these two opposing factors.

For acidities of less than 100 g/L "free" sulphuric acid, which is of interest to most hydrometallurgical applications, the solubility decreases in the presence of foreign metal cations, which provides an apparent "solubility enhancement" for hematite.

Acknowledgements

The Natural Sciences and Engineering Research Council of Canada (NSERC) is gratefully acknowledged for providing financial support for this research project. The authors also wish to thank Ms. M. Donque and Mr. J. Kambossos for their assistance with the data compilation and experimentation, respectively.

References

1. Demopolous, G.P. and Papangelakis, V.G. (1989) Recent Advances in Refractory Gold Processing. **CIM Bulletin**, 82, 931, 85-91.
2. Chou, E.C., Queneau, P.B., and Rickard, R.S. (1977) Sulphuric Acid Pressure Leaching of Nickeliferous Limonites. **Metallurgical Transactions B**, 8B , 547-554.
3. Dutrizac, J.E. (1980) The Physical Chemistry of Iron Precipitation in the Zinc Industry, in **Lead-Zinc-Tin '80**, ed. Cigan, J.M., Mackey, T.S., and O'Keefe, T.J., TMS-AIME, Warrendale, PA, USA, 1980, pp. 532-564.
4. Piret, N.L. and Melin, A.E. (1993) Impact of Environmental Issues on Iron Removal Process Evaluation in Electrolytic Zinc Production, in **Hydrometallurgy Fundamentals, Technology and Innovation**, ed. Hiskey, J.B. and Warren, G.W., TMS, Littleton CO, USA, 1993, pp. 499-520.
5. Dirksen, J.A. and Ring, T.A. (1991) Fundamentals of Crystallisation. Kinetic Effects on Particle Size Distributions and Morphology. **Chemical Engineering Science**, 46, 10, 2389-2427.
6. Umetsu, Y., Tozawa, K., and Sasaki, K. (1977) The Hydrolysis of Ferric Sulphate Solutions at Elevated Temperatures. **Transactions of the Metallurgical Society of the CIM**, 111-117.
7. Tozawa, K. and Sasaki, K. (1986) Effect of Coexisting Sulphates on Precipitation of Ferric Oxide from Ferric Sulphate Solutions at Elevated Temperatures, in **Iron**

Control in Hydrometallurgy, ed. Dutrizac, J.E. and Monhemius, A.J., Ellis Horwood, Chichester, UK, 1986, pp. 454-476.

8. I, T.-P. and Nancollas, H. (1972) EQUIL - A General Computational Method for the Calculation of Solution Equilibria. **Analytical Chemistry**, 44, 2, 1940-1950.
9. Papangelakis, V.G. and Demopolous, G.P. (1990) Acid Pressure Oxidation of Arsenopyrite: Part I, Reaction Chemistry. **Canadian Metallurgical Quarterly**, 29, 1, 1-12.
10. Filippou, D., Papangelakis, V.G., and Demopolous, G.P., (1994) Hydrogen Ion Activities and Species Distribution in Mixed Metal Sulphate Aqueous Systems. **American Institute of Chemical Engineers, Journal**, *in press*.
11. Smith, R.M. and Martell, A.E. (1976) **Critical Stability Constants**, vol. 4. Plenum, New York, NY, USA, p. 7.
12. McAndrew, R.T., Wang S.S., and Brown, W.R. (1975) Precipitation of Iron Compounds from Sulphuric Acid Leach Solutions. **CIM Bulletin**, 68, 753, 101-110.
13. Barner, H.E. and Scheuerman, R.V. (1987) **Handbook of Thermochemical Data for Compounds and Aqueous Species**. Wiley-Interscience, New York, NY, USA, p. 47.
14. Kotrly, S. and Sucha, L. (1985) **Handbook of Chemical Equilibria in Analytical Chemistry**. Wiley, New York, NY, USA.
15. Vasil'ev, V.P. (1962) Influence of Ionic Strength on the Instabiliy Constants of Complexes. **Russian Journal of Inorganic Chemistry**, 7, 8, 924-927.
16. Helgeson, H.C. (1967) Thermodynamics of Complex Dissociation in Aqueous Solution at Elevated Temperatures. **The Journal of Physical Chemistry**, 71, 10, 3121-3135.
17. Zemaitis, J.F., Clark, D.M., Ralaf, M., and Scrivner, N.C. (1986) **Handbook of Aqueous Electrolyte Thermodynamics**, AIChE, New York, NY, USA, pp. 84-97.
18. Wolfram Research Inc. (1992) ***Mathematica***, v.2.2, Wolfram Research Inc., Champaign, Illinois, USA.
19. Lister, M.W. and Rivington, D.E. (1955) Some Ferric Halide Complexes, and Ternary Complexes with Thiocyanate Ions. **Canadian Journal of Chemistry**, 33, 1603-1613.
20. Barner, H.E. and Kust, R.N. (1980), Application of Thermodynamics in Hydrometallurgy, in **Thermodynamics of Aqueous Systems with Industrial Applications**, ed. Newman, S.A., ACS Symposium Series, 133, Washington, DC, USA, 1980, pp. 625-641.
21. Turner, D.J. (1980), **Thermodynamics of Aqueous Systems with Industrial Applications**, ed. Newman, S.A., ACS Symposium Series, 133, pp. 653-679.
22. Lister, M.W. and Rivington, D.E. (1955) Some Measurements on the Iron(III)-Thiocyanate System in Aqueous Solution. **Canadian Journal of Chemistry**, 33, 1572-1590.

Hydrothermal precipitation from aqueous solutions containing iron(III), arsenate and sulphate

P. M. Swash
A. J. Monhemius
MIRO Arsenic Research Group, Department of Mineral Resources Engineering, Royal School of Mines, Imperial College of Science, Technology and Medicine, London, England

Abstract
A systematic programme of synthesis, characterisation and solubility testing of solids precipitated from solutions containing various ratios of Fe^{3+}, AsO_4^{3-} and SO_4^{2-} has enabled the relative stabilities of the various arsenical phases to be assessed empirically.

Over 90% of the arsenic can be precipitated from solutions of pH<1, Fe:As ≥1, at temperatures above 150°C; the solids formed are crystalline and usually consist of well-defined compounds or mixtures of compounds. While the crystallinities of the precipitates increase with temperature of synthesis, their stabilities do not always increase correspondingly due to the formation of different suites of compounds. The arsenical compounds formed include scorodite and two unknown compounds that have been designated Type-1 ($Fe_2(HAsO_4)_3.zH_2O$) and Type-2 ($Fe_4(AsO_4)_3(OH)_x(SO_4)_y$). Crystalline ferrihydrite can also be formed from solutions containing high concentrations of iron and this can readily incorporate substantial amounts of arsenic.

The relative order of stability (based on arsenic solubility) of the compounds synthesised at elevated temperatures is: Type-2 = Scorodite > Arsenical Ferrihydrite > Type-1. When a comparison is made of the stability of these high temperature crystalline solids with amorphous iron arsenates (Fe:As >3:1) precipitated at ambient temperature and at pH 5, the crystalline solids are marginally less stable. However, the crystalline compounds contain considerably more arsenic (~Fe:As 1:1) than the amorphous precipitates (Fe:As >3:1). These latter compounds consist mainly of amorphous ferrihydrite which on ageing can dehydrate and recrystallise and may release adsorbed arsenic due to reductions in surface area and active sites. Therefore the formation of stable crystalline compounds may be a more attractive disposal option for arsenic.
Keywords: Arsenic, ferrihydrite, hydrothermal precipitation, scorodite, solubility, stability, high temperature ferric arsenates.

1 Introduction

Previous work on the stability of iron-arsenate compounds [1-4] has been carried out on precipitates formed at ambient temperature as the solids produced under these conditions are analogous to those produced industrially during the neutralisation of arsenic-containing effluents. In this process, the addition of lime or limestone is used to raise the pH of the solution which causes arsenic to be precipitated together with iron as a gelatinous amorphous compound. Different Fe:As ratios in the effluent liquors form solid products that have different apparent stabilities. Previous workers have clearly demonstrated that the Fe:As molar ratio of the amorphous solids should exceed 3:1 for the product to be stable and suitable for disposal [5]. A possible alternative to lime neutralisation of arsenical solutions is high temperature hydrolysis. The work reported in this paper has modelled this process and examines the characteristics and stability of hydrothermally prepared solids in the Fe-AsO_4-SO_4 system at elevated temperatures (150-225°C) and low pH(<1). The ultimate aim of the current work is to identify and characterise such compounds and, with the understanding of the structure of these materials, to guide the processing of arsenical wastes so that practical and realistic solutions to the disposal of arsenic into the environment can be devised.

2 Methodology

In the initial part of this work, starting solutions of various compositions were made by mixing stock solutions of iron(III), sulphate and arsenate (0.5M $Fe(NO_3)_3.9H_2O$, 0.5M $Li_2SO_4.H_2O$ and 0.25M As_2O_5) in the appropriate proportions. In later work, stock solutions of $Fe_2(SO_4)_3.xH_2O$ and As_2O_5 were used in order to eliminate the potential interferences from lithium and nitrate ions and hence to produce solutions which were considered to be closer analogues to real hydrometallurgical liquors. The compositions of the solution mixtures are shown in Figure 1; they were used at their natural pHs, which were generally <1. The solutions were sealed in glass test tubes and heated in an autoclave at temperatures from 150-225°C for 24hrs. After cooling, the tubes were opened and the solids and mother liquors were separated and analysed; usually over 90% of the arsenic and iron were precipitated. The solids were air dried in a dessicator and characterised as described below.

The relative solubilities of the synthesised compounds were determined by a simple method called the MARG static bottle test (MIRO Arsenic Research Group). This involved contacting 0.1g of solid with 50mL of solution (L:S 500:1) at various pHs (3, 5, 7 and 11, with the pH adjusted using H_2SO_4 or NaOH) in plastic bottles which were sampled after predetermined times (1, 2, 4 and 7 days, and monthly intervals thereafter). At each sample time, a 5mL aliquot of solution was taken from each sample bottle and the pH of the solution adjusted and made up to the starting volume. The sample solutions were analysed for arsenic using standard atomic absorption for solutions greater than 3ppm As and the hydride generation method for those with less than 3ppm As. The solubility testing of a number of naturally occurring arsenate minerals was also carried out in a similar manner to acquire some comparative information between natural and synthetic arsenates. The EPA TCLP test (US Environmental Protection Agency, Toxicity Characterisation Leach Procedure), which is the most widely used test procedure, was also used on some of the solids. The test solution consists of a buffered acetic acid solution (glacial acetic acid and NaOH) at pH 4.93, L:S - 20:1, agitated for 20hrs at 30rpm (end-over-end bottle agitation). Due to the limited amounts of synthesised solids that were available, only 0.5g solids were used in the EPA TCLP tests.

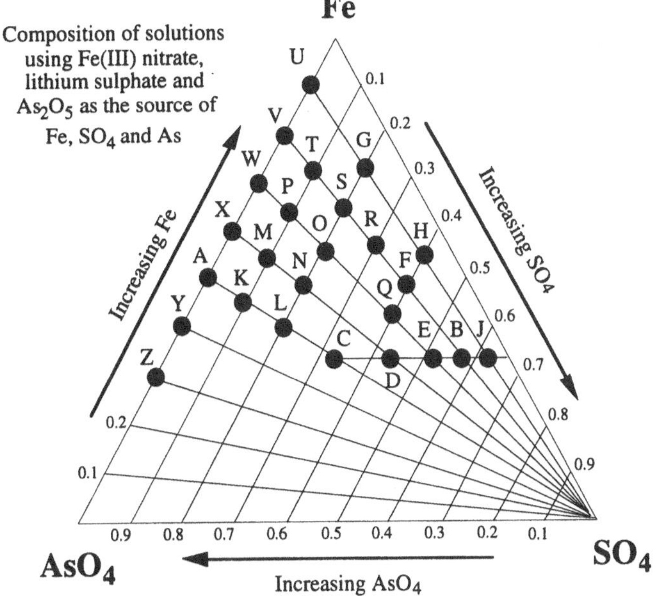

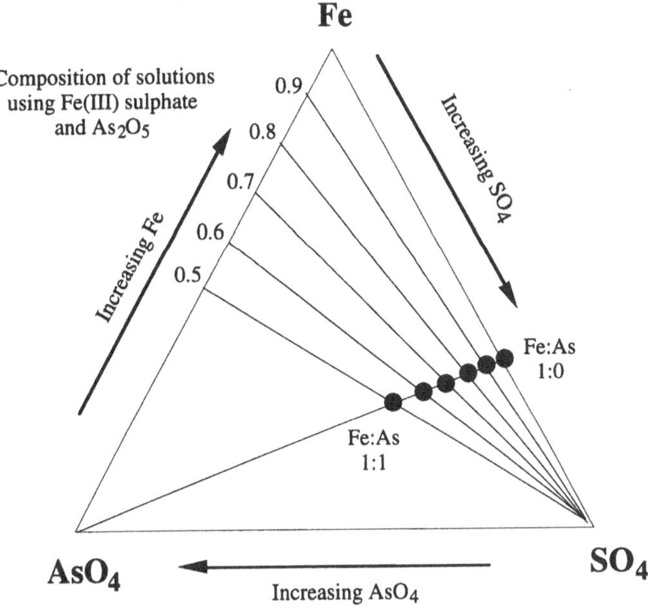

Figure 1. Composition of solutions used in the hydrothermal syntheses

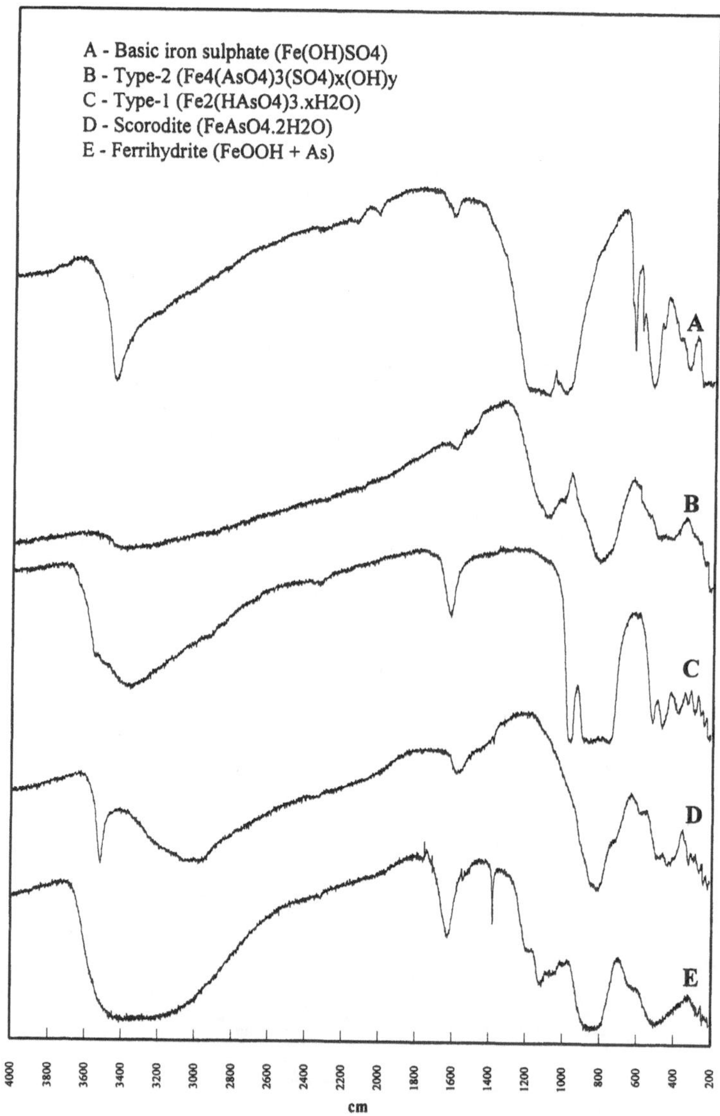

Figure 2. IR spectra of the distinct phases hydrothermally precipitated from natural pH solutions (<1)

Table 1. Chemical compositions of the arsenical compounds and the basic iron sulphate produced during synthesis

Compound	Fe (%)	AsO_4 (%)	SO_4[1] (%)	Average Fe:As (M)	H_2O[2] (%)	Size (µm)	Temperature (°C) of Decomposition[3]
Scorodite $FeAsO_4.2H_2O$	20.7 to 24.7	50.0 to 70.1	up to 4.0	1.02	11 to 16	up to 30	200 to 230 - hydration H_2O
Type-1 $Fe_2(HAsO_4)_3.zH_2O$ where z<4	21.4 to 23.0	56.3 to 75.6	<1.0	0.90	<5	up to 5	220 - hydration H_2O, 410 - OH
Type-2 $Fe_4(AsO_4)_3(OH)_x(SO_4)_y$ where x+2y = 3[4]	27.4 to 38.0	32.6 to 52.0	up to 13.0	1.30	<4	up to 50	490 - OH, 650 and 800 SO_4
Ferrihydrite FeOOH + As	<65	<20	usually <0.1	>3	<40	«0.1	<100 - OH
Basic iron sulphate $Fe(OH)SO_4$	31.0	-	57.0	-	nd	up to 20	560 - OH, 700 - SO_4

[1] dependent on solution composition. [2] Mass loss determined by DTA-TG , [3] water of hydration or constitutional water, [4] For charge balance, nd - not determined

Table 2. X-Ray Diffraction data for the unknown compounds identified during the synthesis programme

Type-1*		Type-2		Basic iron sulphate	
d-spacing	Intensity	d-spacing	Intensity	d-spacing	Intensity
8.77	42	4.89	5	4.79	13
6.42	63	3.38	100	3.58	9
5.08	12	3.25	64	3.27	5
4.02	23	2.64	19	3.22	100
3.83	14	2.35	6	2.29	2
3.67	22	2.32	20	2.23	9
3.39	13	2.06	15	2.05	21
3.27	11	1.89	9	2.00	2
3.23	24	1.76	2	1.84	2
3.16	100	1.69	8	1.78	2
3.09	43	1.63	11	1.64	4
2.97	69	1.62	28	1.60	14
2.96	48	1.47	9	1.59	13
2.91	63	1.45	3	1.56	5
2.87	17			1.44	13
2.84	13				
2.77	6				
2.71	16				
2.62	27				
2.59	18				
2.53	9				
2.41	11				
2.39	12				

* The XRD peaks were poorly defined in some of the hydrated Type-1 compounds

3 Characterisation

The solids were systematically characterised by a variety of techniques including X-ray powder diffraction (XRD), infra-red spectroscopy (IR), thermal analysis (DTA-TG), chemical analysis and scanning electron microscopy (SEM). Interpretation of the data was complicated by variable proportions of crystalline and amorphous components in the samples. A complete characterisation was only found possible through the integration of all the information from the different techniques.

The X-ray diffraction studies showed whether the samples were amorphous or crystalline, with known or unknown diffraction patterns, and provided semi-quantitative information on crystalline phases. SEM was used to identify the presence of amorphous or minor components (i.e. below the detection limits of XRD), as well as showing the crystal dimensions and morphology. Information on the molecular structure and chemical bonding characteristics of both crystalline and amorphous compounds was given by IR. Quantitative information on the compositions of the solids was obtained by chemical and thermal analysis, the latter allowing the amount of water of hydration and constitutional water to be determined. The mass loss measurements by thermogravimetry, in combination with the IR spectra of the heated samples, enabled the DTA-TG patterns to be interpreted with some degree of confidence.

During the study of the Fe-AsO_4-SO_4 system, two new arsenate compounds were identified. These compounds, which have been designated Type-1 and Type-2, have compositions approaching $Fe_2(HAsO_4)_3.zH_2O$ and $Fe_4(AsO_4)_3(OH)_x(SO_4)_y$ (where $z<4$ and $x+2y = 3$), respectively, and have distinct and well-defined XRD, DTA-TG and IR patterns. The precipitates formed during hydrothermal synthesis also contained other constituents including hematite, basic iron sulphate, scorodite and ferrihydrite-type phases. A study of over 100 precipitates formed from different solution compositions at various temperatures enabled the compounds to be fully characterised. Typical analytical composition ranges and thermal data for these phases are given in Table 1 and characteristic IR spectra are shown in Figure 2. The XRD data of the new compounds are given in Table 2. During the pressure oxidation of arsenopyrite at elevated temperatures, compounds identical to those of the Type-2 have also been identified [6, 7].

A major contribution to the interpretation of the IR characteristics of the synthesised solids was made by comparison of the IR spectra with natural arsenate minerals and with other synthetic arsenate compounds of known composition and atomic structure. IR examination of the samples following incremental heating and also deuteration (replacing water by D_2O) aided in interpretation of the spectra.

At the lower temperatures of synthesis, some of the more iron-rich solution compositions yielded solids that were partially amorphous in character, whereas at elevated temperatures, crystalline compounds predominated. By carrying out a series of syntheses at different temperatures, trends in the characteristics of the solids were observed and extrapolated into the low temperature amorphous region (i.e. moving from ordered to disordered) and this allowed the structures of the amorphous compounds to be better understood.

It was found that precipitation at temperatures above 100°C from iron-rich starting solutions often produced 6-line ferrihydrite (so-called because the structure of the compound shows 6 distinct peaks in its X-ray pattern), whereas at ambient temperature the ferrihydrite formed is the 2-line variety. The structure of ferrihydrite has not been fully established mainly because of its poor crystallinity, which is linked to the presence of vacant iron sites in the structure and to oxygen in the lattice being partially replaced by H_2O or OH. With time these compounds can transform into goethite, hematite, or a mixture of the two.

The dehydration of ferrihydrite reduces the surface area of the solid and the resulting material cannot as readily accommodate or adsorb arsenic. This results in the arsenic in the solid being more loosely bound and hence more soluble. However, when arsenic is adsorbed onto the amorphous 2-line ferrihydrite, its transformation to goethite is blocked or retarded [8], as has been found to happen when other anionic complexes e.g. phosphates, are adsorbed onto ferrihydrite [9].

Figure 3 shows the distribution of the various types of synthesised compounds relative to the composition of the starting solutions. The compounds that were precipitated from solutions using iron(III) nitrate, lithium sulphate and arsenic(V) solutions appeared to be mainly controlled by the Fe:As ratio of the starting solutions, while the sulphate content had little real influence. The Type-2 compound readily accommodated sulphate (up to 13%) into its lattice, whereas scorodite and Type-1 compounds contained only limited amounts (less than 4.0%). A change in compound distribution was apparent as the temperature of synthesis was increased.

When solution compositions were generated using iron(III) sulphate instead of iron nitrate and lithium sulphate (as a source of Fe and SO_4), there was a tendency to form a basic iron sulphate compound ($Fe(OH)SO_4$) at 175°C and above (Figure 4), whereas in the presence of lithium and nitrate, ferrihydrite preferentially formed under similar conditions. The basic iron sulphate compound was crystalline and unlikely to incorporate arsenic into its lattice and hence any arsenic present would most likely crystallise either as the Type-2 compound above 175°C, or as scorodite below 175°C.

4 Solubility Test Results

During initial solubility test trials, when $Ca(OH)_2$ was used in place of NaOH for the adjustment of the pH of the test solutions, major differences in the soluble arsenic levels were observed in the pH 7-11 solutions (see Figure 5). The reasons for these differences are thought to be due to the influence of calcium in solution which seems to lower the apparent solubility of arsenic. Reduced levels of soluble arsenic were also found in EPA TCLP tests when gypsum was deliberately added to test the influence of the presence of calcium. In the determination of the solubilities of arsenical compounds, the interaction of arsenic with calcium adds complications to the interpretation of the results. For this reason it is preferable to use NaOH to control the pH in solubility tests rather than lime and so avoid these interference effects.

The solubility test results showed that the relative solubilities of the individual solids measured after 7 days were comparable to those found after 180 days. Representative solubility data of the synthetic compounds are given in Figure 6, which illustrates the behaviour of scorodite, Type-1 and Type-2 compounds as a function of pH. These compounds were synthesised at various solution compositions and temperatures. For each compound type, a range of solubilities is apparent. This is most pronounced for the Type-1 compounds, where the solubility varies according to the hydration state of the solid, with the more hydrated solids being more unstable. It is evident that in all cases, arsenic solubility increases with pH, with relatively high solubilities occurring at pH 10.

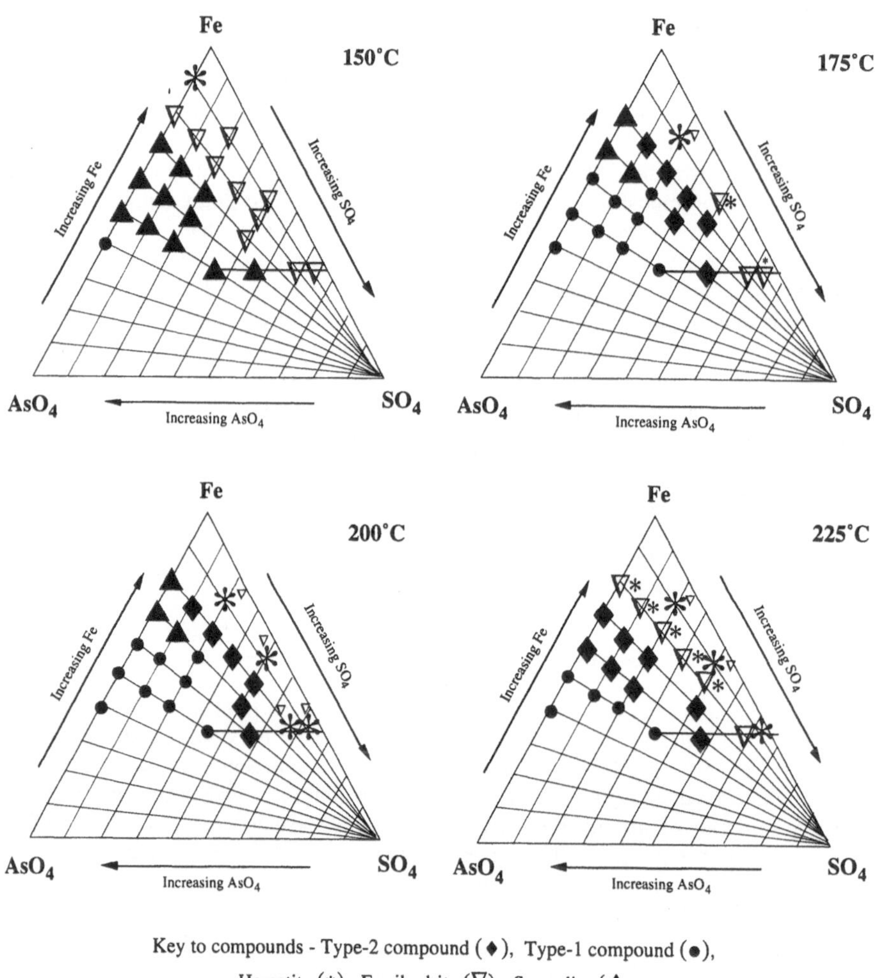

Figure 3. Distribution of compounds formed at 150-225°C with respect to the composition of the natural pH(<1) starting solutions

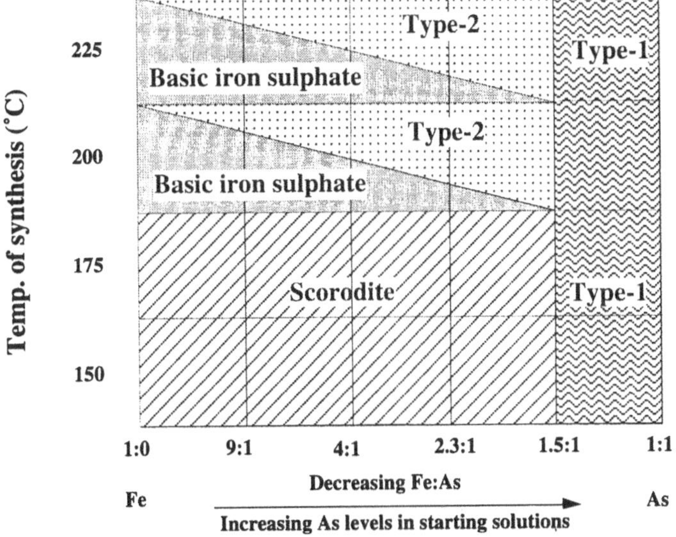

E.g. A solid synthesised from a 9:1 solution at 225°C yields ~90% basic iron sulphate and ~10% Type-2

Figure 4. Distribution of compounds formed from Fe-AsO_4-SO_4 solutions (using iron(III) sulphate) at different temperatures

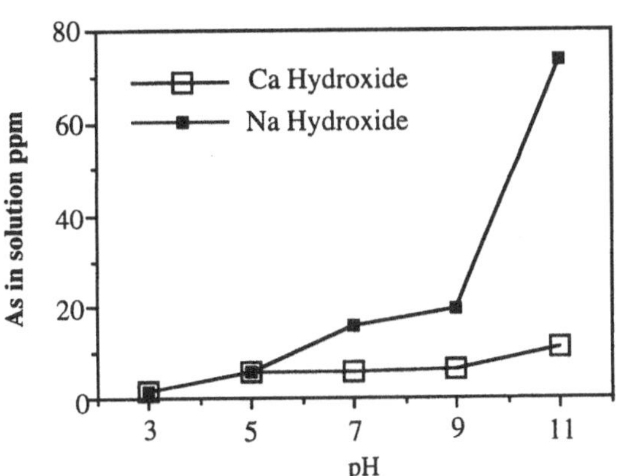

The tested solid was synthesised from a solution of Fe:As 2.3:1 at a temperature of 150°C and pH 5.

Figure 5. The influence of the use of calcium and sodium hydroxide for pH adjustment in the MARG static bottle test solutions.

Table 3. Solubility test results on finely milled (100% -38μm) natural mineral samples

Mineral Name	Ideal Formula	Solubility Test Data (As ppm)[1] pH 3	pH 5	pH 7	pH 9	pH 11
Bukovskyite	$Fe_2AsO_4SO_4OH\ .7H_2O$	0.19	0.10	0.07	0.51	84.30
Kankite	$FeAsO_4\ .3.5H_2O$	3.10	0.11	0.94	1.79	44.30
Scorodite[2]	$FeAsO_4\ .2H_2O$	4.60	0.10	0.07	0.08	2.05
Scorodite[3]	$FeAsO_4\ .2H_2O$	0.57	0.40	1.80	2.25	93.50
Scorodite[4]	$FeAsO_4.2H_2O$	0.12	0.04	0.62	1.85	61.00
Pitticite	Amorphous iron(III)-arsenate-hydrate	0.47	0.10	0.10	5.80	137.2
Yukonite	$Ca_6Fe_4(AsO_4)_2(OH)_6\ .5H_2O$	21.60	2.01	2.36	4.10	18.50
Arseniosiderite	$Ca_3Fe_4(AsO_4)_4(OH)_6\ .3H_2O$	2.16	0.40	0.62	0.71	2.17
Symplesite	$Fe_3(AsO_4)_2\ .8H_2O$	nd	nd	nd	2.42	50.50

[1] test solution compositions after 7 days, [2] crystalline scorodite, [3] porous scorodite, [4] scorodite-sericite schist, nd - not determined

Table 4. Solubility test results on typical crystalline synthetic compounds

Compound -Type	Temperature of Synthesis	Solubility test data (As ppm)* pH 3	pH 5	pH 7	pH 10
Scorodite	150°C	<0.5	<0.5	0.8	5.8
Type-2	175°C	<0.5	0.8	<0.5	7.3
Type-1	200°C	3.5	4.8	6.5	14.0

* test solution compositions after 7 days,

Table 5. EPA test results on Fe:As compounds synthesised using solutions of iron(III) sulphate and arsenic(V) at the natural pH(<1)

Temperature (°C)	EPA TCLP filtrate, As (ppm) Fe:As of synthesis solution used to produce solids* 1:0	9:1	4:1	2.3:1	1.5:1	1:1
225	nd	<0.10	1.04	0.44	4.70	11.9
200	nd	nd	1.58	0.34	4.90	16.7
175	nd	0.14	1.14	0.74	1.50	24.1
150	nd	nd	nd	0.80	1.06	30.6

nd - not determined, * the phases present in the solids can be found from Figure 4

Samples of natural minerals were tested by the MARG procedure and they showed; behaviour similar to the synthetic compounds (Table 3) with low solubilities at low pH and higher solubilities at elevated pH. Some of the crystalline synthetic compounds were found to have comparable solubilities (Table 4).

The solids formed during high temperature syntheses using solutions of iron(III) sulphate and As_2O_5 were solubility tested using the EPA TCLP method, the results of which are given in Table 5. In the EPA TCLP test the solution conditions are more

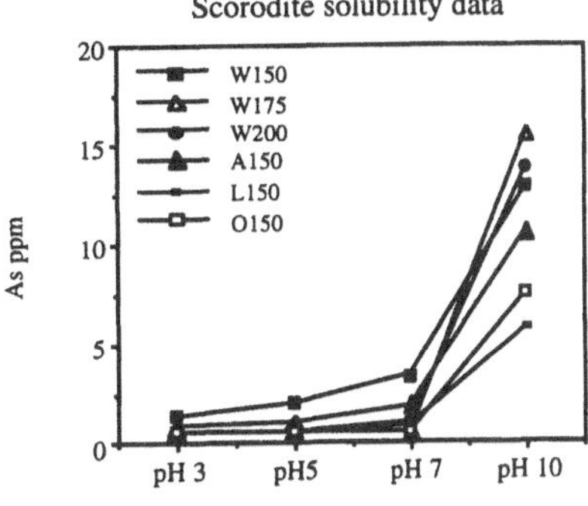

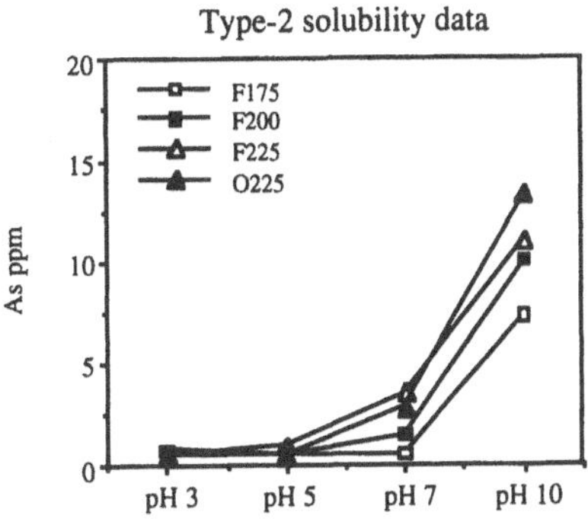

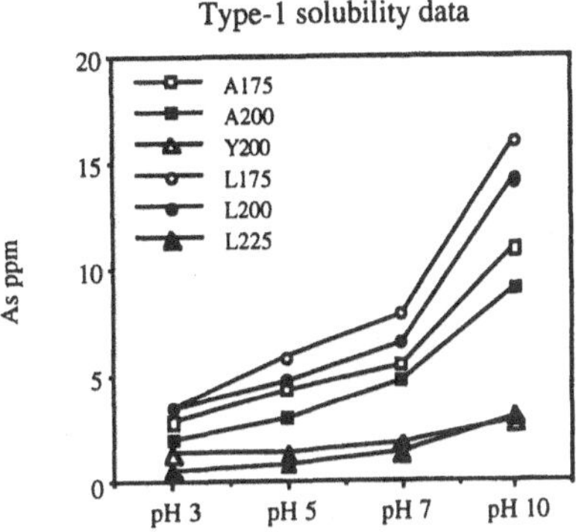

Note - Letters denote the composition of the starting solution (see Figure 1) and the numbers indicate the temperature of synthesis

Figure 6. Solubility plots of the distinct arsenate compounds

aggressive and a lower liquid to solid ratio is used compared with the MARG procedure. These differences result in higher apparent solubilities in the EPA tests, however, the trends are the same as those determined by the MARG tests. The scorodite and Type-2 compounds are of relatively low solubility (<5ppm) and are considered to be non toxic by the EPA TCLP method.

The EPA TCLP data for the high temperature crystalline compounds are compared with those for Fe:As solids precipitated at ambient temperatures and pH 5 in Table 6, the latter solids being analogous to those formed during the neutralisation of arsenical liquors in commercial operations. The high Fe:As ratio (>2.3:1), gelatinous, low temperature solids show lower solubilities than the high temperature crystalline solids precipitated from low pH(<1) solutions. However, low temperature solids precipitated from solutions with Fe:As <2.3:1 show much higher solubilities than the corresponding high temperature precipitates.

Table 6. Solubilities of high and low temperature precipitates measured by the EPA TCLP method

	Sample (conditions of synthesis)*	EPA TCLP filtrate (As ppm)
pH Natural (<1). High Temperature (>150°C)	Scorodite (Fe:As 2.3:1, 150°C)	<0.80
	Type-2 (Fe:As 2.3:1, 200°C)	<0.34
	Type-1 (Fe:As 1:1, 225°C)	11.9
pH 5 Low Temperature (20°C)	Ferrihydrite (Fe:As - 9:1)	0.4
	Ferrihydrite (Fe:As - 2.3:1)	1.2
	Ferrihydrite (Fe:As - 1.5:1)	50.2

* starting solutions of iron(III) sulphate and As(V)

5 Discussion

The combined structural and solubility results have shown that there are five main factors which appear to influence the stabilities of the arsenate compounds. These are:-

- type of arsenic containing species,
- crystallite/particle size,
- hydration of precipitated solids,
- presence of sulphate in the lattice, and
- Fe:As molar ratio.

The Type-2, scorodite-type, and the higher temperature, unhydrated Type-1 compounds, all showed reasonable stability under the test conditions used. The relative solubilities are strongly controlled by the way arsenic is chemically bound into the structure of the compounds (i.e. type of compound). The crystallinity factor cannot be easily quantified, yet this is important as it influences bond strengths and will dictate the rate at which solids attain equilibrium with a contacting solution. Those samples with a high surface area, such as the amorphous scorodite-type samples, will equilibrate at a more rapid rate than materials that are more crystalline in character. For the Type-1 compounds, it was found that the hydration state is an important factor which influences its solubility; the more hydrated the Type-1 compound, the more unstable the solid.

The synthesis work carried out at elevated temperatures suggests that a number of alternatives exist for the formation of low solubility arsenate phases; the most practical of which are Type-2 and crystalline scorodite compounds. The Type-2 compounds are essentially anhydrous, sulphate-containing, crystalline solids which can be produced from iron-rich solutions at temperatures greater than 175°C. Scorodite is also easily produced at lower temperatures (150-175°C) and shows solubilities similar to the Type-2 compound.

6 Processing Considerations

A fundamental understanding of the $Fe\text{-}AsO_4\text{-}SO_4\text{-}H_2O$ system at high and low temperatures is of practical importance to the hydrometallurgical processing of most arsenical solutions. This is particularly so for the processing of arsenical concentrates and dusts where the information may be used to help predict the types of compounds that can be formed if high temperature precipitation is considered a disposal option.

During bio-oxidation and pressure oxidation, the conversion of arsenical sulphides into Fe^{3+}, AsO_4^{3-} and SO_4^{2-} is common to both routes. During the bio-oxidation processing of auriferous arsenical concentrates, attempts are made through careful pH control to retain the arsenic in the liquid phase before separating the residue for subsequent cyanidation [10]. In contrast, during pressure leaching, a larger proportion of the arsenate can be precipitated during the high temperature oxidation, with the remainder retained in the liquid discharge. The precipitated iron arsenate compounds are expected to be moderately crystalline compounds whose composition and stability will be dependent on the temperature and the Fe:As:S ratio of the liquors. It can be predicted that solids comparable to Type-2 and the ferrihydrite/basic iron sulphate are the most likely compounds to be formed under process operating conditions (~200°C).

The liquor neutralisation steps of the two systems should yield solids of similar nature and composition if the solution characteristics are comparable (i.e. As(V)/As(III), Fe(III)/Fe(II), Fe:As etc.).

7 Conclusions

The synthesis, characterisation and stability test programme described in this paper shows that high temperature precipitation may be a practical route for treating liquors of high arsenic concentration. The work has enabled various compounds to be identified and their relative solubilities to be determined. It is concluded that coarsely crystalline scorodite and Type-2 compounds are the most suitable for the disposal of arsenic from solutions at high temperatures. While the high temperature solids are marginally more soluble in the short term than iron-rich arsenate compounds (Fe:As >3:1) formed at low temperatures by lime neutralisation, they appear to be sufficiently stable for disposal under the current regulations and have the advantage of containing higher proportions of arsenic than the low temperature precipitates.

8 Acknowledgements

The authors would like to thank the sponsors of the Mineral Industry Research Organisation (MIRO) RC59 Arsenate Stability Programme (American Barrick Resources Corp., Billiton Research B.V., Boliden A.B., Borax Consolidated Ltd, Codelco - Chile, Genmin Process Research, Lurgi GmbH, Minorco Services (UK) Ltd, Noranda Technology Centre, Outokumpu Research O.Y., Placer Dome Inc., MINTEK, RTZ Technical Services and Union Minière) who not only provided generous financial support to the study, but also gave practical direction to the work. Thanks also go to L. Burgess and M. Haddon of the Arsenic Research Group, Dr. R. Bowell of the British Natural History Museum and to Dr. G.B. Harris of MITEC for their enthusiasm and contributions to this study.

9 References

1. Robins, R.G. and Haung, J.C.Y. (1988) The Adsorption of Arsenate Ion by Ferric Hydroxide, in Reddy, R.G. et al., eds, *Arsenic Metallurgy: Fundamentals and Application,* Proceedings of an International Symposium. Phoenix, Arizona, TMS-AIME, 99-112.
2. Waychunas, G.A., Rea, B.A., Fuller, C.C. and Davis, J.A. (1993) Surface Chemistry of Ferrihydrite: Part 1. EXAFS Studies of the Geometry of Coprecipitated and Adsorbed Arsenates. *Geo. Cosmo. Acta.* **57**, 2251-2269.
3. Robins, R.G., Wong, P.L.M., Nishimura, T., Khoe, G.H. and Huang, J.G.N. (1991) Basic Ferric Arsenates - Non-existent, in *Cairns'91. Proceedings of the Randol Forum, Cairns,* Australia, 197-200.
4. Nishimura, T., Itoh, C.T. and Tozawa, K. (1988) Stabilities and Solubilities of Metal Arsenites and Arsenates in Water and the Effect of Sulphate and Carbonate Ions on their Solubility. op. cit., reference 1, 77-98.
5. Harris, G.B. and Krause, E. (1993) The Disposal of Arsenic from Metallurgical Processes: its Status Regarding Ferric Arsenate. in Reddy, R.G. and Weisenbach, R. eds, *The Paul E. Queneau International Symposium on Extractive Metallurgy of Nickel, Cobalt and Associated Metals*, Vol.I, Fundamental aspects. TMS, Warrendale, 1221-1237.
6. Ugarte, F.J.G. and Monhemius, A.J. (1992) Characterisation of High-Temperature Arsenic-Containing Residues from Hydrometallurgical Processes, *Hydrometallurgy*, **30**, 69-86.
7. Carageorgos, T. (1993) Acid Pressure Oxidation of Arsenopyrite, PhD Thesis, Imperial College, London.
8. Swash, P.M. and Monhemius, A.J. (1994) The Dehydration of Arsenical Ferrihydrite: Implications for its Long Term Stability, (in preparation).
9. Schwertmann, U. and Cornell, R.M. (1991) *Iron Oxides in the Laboratory: Preparation and Characterisation.* Weinham ; Basel, VCH.
10. van Aswegen, P.C., Godfrey, M.W., Miller, D. M. and Haines, A.K.(1991) Developments and Innovations in Bacterial Oxidation of Refractory Ores, *Minerals and Met. Proc.* **8**(4), 188-191.

Leaching

Heap leaching of poor nickel laterites by sulphuric acid at ambient temperature

S. Agatzini-Leonardou
D. Dimaki
Department of Mining and Metallurgical Engineering, National Technical University of Athens, Athens, Greece

Abstract
Column leach tests were conducted to evaluate the amenability of poor Greek nickel laterites to heap leaching with dilute sulphuric acid at ambient temperature. The ore tested was limonitic laterite with less than 1% nickel and it was subjected to a suitable pretreatment with water prior to it being loaded into columns, following findings of preliminary research work. The acid solution was recycled after suitable pH adjustment. Factorial designs were used to determine the effects and interactions of the following factors: sulphuric acid concentration, ore column height, ore grain size and ratio of the solution volume to the ore quantity. The responses which were investigated included nickel and cobalt recoveries, co-dissolution of iron, calcium, magnesium, aluminium, chromium and silica and also sulphuric acid consumption. Experimental error was also determined. The tests showed that nickel recovery of 86% could be achieved in 80 days.
Keywords: Cobalt recovery, heap leaching, laterites, nickel extraction, sulphuric acid leaching.

1 Introduction

Laterites comprise approximately 80% of the known nickel reserves and currently fulfil about 40% of the demand, mainly for stainless steel, alloys and the chemical industry. The dwindling reserves of sulphide ores, the other major source of nickel, demand that laterites play an increasingly important role in the industry [1-3].

Laterites are treated both pyrometallurgically and hydrometallurgically. Unfortunately, most of current industrial practice is energy intensive, e.g. matte smelting, smelting to ferronickel and the Caron's ammonia leaching, thus making uneconomical the treatment of low-grade ores. Sulphuric acid leaching at elevated pressure (Freeport Sulfur Company) is less energy intensive but it requires expensive equipment (autoclaves, flash tanks) and is highly corrosive, presenting severe engineering problems.

Although the grater part of the process development work done from 1970 to the present, was directed at improving on earlier developments, there were proposed some new approaches, such as the U.S. Bureau of Mines, Bethelehem Steel,

Universal Oil Products, AMAX and others [3][4]. Yet, none of them, not even those that were developed at a pilot-plant level, has been commercialized.

Greece has extensive low-grade limonitic laterite deposits which are currently being treated by a rotary kiln prereduction-electric furnace smelting-converter upgrading technique for the production of ferronickel. Because the rising costs of the fuel and electric energy and the declining nickel content, now being less than 1%, have rendered process uneconomic, the authors have undertaken research in order to examine the applicability of the heap leaching technique to low-grade laterites, for the first time worldwide, as far as the authors are aware, using dilute sulphuric acid.

Heap leaching was chosen because it involves low capital and operating costs and has a simple and continuous nature characteristics which are attractive for low-grade ores. Besides, it is technologically advanced as a result of its long and extensive application to low-grade copper, uranium and gold ores.

2 Experimental work

2.1 Materials

A low-grade limonitic type laterite from the "Litharakia" deposit on the island of Euboea, was used for this investigation. Chemical and size distribution analyses are given in Table 1. Ore microscopy, X-ray diffractometry and electron microprobe analysis performed on the ore showed that it is consisted of nickel chlorite, which was the main nickeliferous mineral, haematite, quartz, chromite, talc, illite and diaspore in small amounts. The ore was contaminated with small amounts of calcite during sampling because the boundary between the limonite deposit and the underlying limestone was, as usually, not well defined. The ore was characterized by a pissolitic texture, with pissolites composed of microcrystalline haematite aggregates set within a pelitomorhic red matrix. The matrix consisted primarily of quartz and chromite grains surrounded by chlorite and fine-dispersed haematite and secondarily of illite and diaspore.

2.2 Experimental procedure

In order to simulate heap leach conditions, a set of ten acrylic columns were used with 100 mm internal diameter and heights varying from 1 to 2.5 m. The columns contained about 8-22 kg of ore according to their heights. The ore was wetted with water to bring total moisture to about 10%, prior to loading into the columns. This pretreatment was arrived at after observations that leach solution percolation was poor to impossible when the ore was loaded dry, probably because the clayed constituents of laterite, being very absorbent, were swelled when leach solution started flowing through, thus closing the established porosity [5]. Wetting of the ore prior to loading formed a porosity which remained undisturbed during subsequent leaching. Besides, it caused agglomeration by the attachment of fines to the coarser particles, thus eliminating segregation of particles and percolation problems. A layer of at least 100 mm washed silica was placed at the bottom of each column to prevent the ore from plugging the pregnant solution outlet (Fig.1). A sulphuric acid leach solution was

Table 1. Chemical and size distribution analyses of the lateritic ore tested

Component	Percentage (% w/w)	Mesh Tyler		Percentage (% w/w)
Ni	0.73			
Co	0.047			
Fe_2O_3	18.11	+20	(+0.85 mm)	80.16
Cr_2O_3	2.19	-20 +40	(-0.85 mm +0.36 mm)	4.51
Al_2O_3	6.70	-40 +80	(-0.36 mm +0.18 mm)	4.77
MgO	2.30	-80	(-0.18 mm)	10.55
CaO	5.68			
SiO_2	45.34			
Ignition loss	13.00			

applied to the ore and the pregnant solution effluent was collected from the bottom of the column, each column having its own leach solution and pregnant solution effluent reservoirs. The leach solution was introduced onto the ore charge by a perforated distributor.

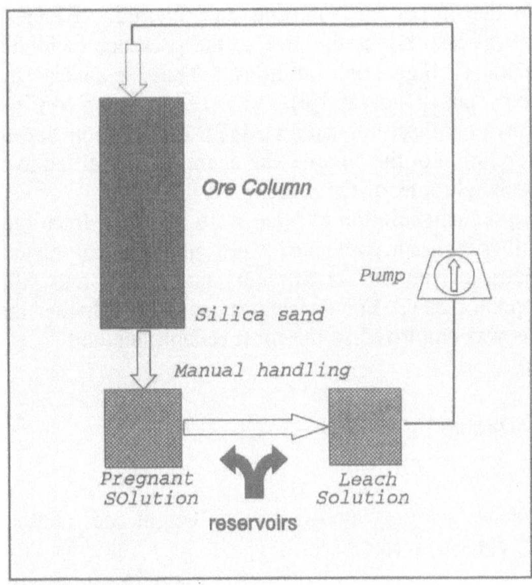

Fig. 1 Experimental Column

After the whole volume of the leach solution, employed for each column, had percolated through the column, thus completing one full leach cycle, it was reused according to the following techniques:

i) Leach solution was recirculated by a dosometric pump from its reservoir to the top of the column, without any pH adjustment. The solution performed so many leach cycles until nickel concentration in the pregnant solution remained constant. It was then replaced by fresh sulphuric acid solution of the same volume and the same initial normality for a new leach stage to start. The ore was thus leached till maximum possible nickel depletion (Technique I). Despite the disadvantages of this technique used (high water and acid consumptions, low concentrations of metals in resulting pregnant solution and long reaction times), it could give useful information on the abilities of the contained metal cations to dissolve out of the ore and their order of dissolution, a first insight into the leaching mechanism and also estimates of the acid consumption. Based on these data, Technique II was developed and applied in the subsequent stages of experimentation.

ii) Leach solution was recirculated by a dosometric pump from its reservoir to the top of the column, after pH adjustment to the initial predetermined pH value, thus performing a number of leach cycles, till maximum possible nickel extraction from the ore was achieved (Technique II). In order to adjust the acidity of the leach solutions at preset values and to calculate the sulphuric acid consumption, the determination of free acidity was necessary. This could not be done reliably by simply measuring the solution pH because of the high acidity and the relatively high ionic strength and density of the solutions encountered. Besides, titration with sodium hydroxide was also impossible due to the presence of hydrolysable cations (e.g. Fe^{3+}) in relatively high concentrations. Thus, a special titration method, developed by Rolia and Dutrizac [6], was used, based on the principle of hydrolysable cations complexation with a MgEDTA solution before titration with NaOH. The concentration of the complexing agent was modified to become suitable for the specific leach solutions of the present work.

Samples of the pregnant solution effluent were collected from the effluent reservoir at the end of each leach cycle and were analyzed for nickel, cobalt, iron, calcium, magnesium, chromium, aluminium, manganese and silicon by atomic absorption spectrophotometry. For concentrations of iron higher than 0.10 kg/m^3, volumetric analysis was employed as the most reliable method.

3 Results and discussion

3.1 Technique I

Six runs were conducted varying the ore column height and, consequently, the ore weight. Parameters chosen were:

1. Ore size: - 6.75 mm (- 3 mesh)
2. Leach solution flowrate: 5.2 m^3/s (4.5 l/day)
3. Sulphuric acid concentration in leach solution: 1 N
4. Leach solution volume/ore weight ratio (S/O): 0.8

The responses measured were the percent extraction of metal cations, the sulphuric acid consumption and the Fe/Ni ratio in the pregnant solution.

As Figures 2 and 3 show, nickel and cobalt recoveries obtained, under the above experimental conditions, could be as high as 80% and 65%, respectively. The nickel and cobalt leach rates were more or less constant and independent of the ore column height (ore weight) for a leaching period up to about 90 days. Thereafter, a dependence of the rates with the column heights, not obvious by mere visual inspection of the data in the graphs, was proven to exist by statistical analysis. This dependence could possibly be attributed to the different percentage of ore being in the zone, adjacent to the column wall, with unfavourable flow conditions, a fact usually occurring in small columns [7].

Figure 4 gives the percentage recoveries of all metal cations with time for the highest column. It can be seen that the highest extractions obtained were those for nickel, magnesium and cobalt. Their leach-rates curves were similar and close to each other, as all three of them occurred, mainly, in the same mineralogical phase, i.e. chlorite and mixed-layer chlorite-illite,as shown by detailed mineralogical and electron microprobe analyses of the ore and the leach residues. Co-dissolution values of aluminium and chromium, corresponding to equilibrium nickel recoveries, were around 38% and 9%, correspondingly. It should be stressed, at this point, that although their curves lie apart from each other and also apart from those of nickel, cobalt and magnesium, aluminium and chromium were shown by the mineralogical analyses to have been leached out of the same mineralogical phases (chlorite and mixed-layer chlorite-illite), as nickel, cobalt and magnesium. The reason, however, why their leach-rate curves lie apart from each other as well as apart from the nickel curve, is due to the fact that their percent extractions were calculated on the basis of their total contents in the ore and not on their contents in the leachable sheet-silicate minerals. It was found that chromium was distributed among the sheet-silicate minerals by 22%, the haematite by 16% and chromite by 62%. Aluminium was distributed among sheet-silicate minerals by 65% and the chromite by 35%.

Co-dissolution of iron, which was found to occur in haematite by 90%, in sheet-silicate minerals by 7% and in chromite 3%, was around 14%. The apparent selectivity of nickel to iron resulted in the ratios of Fe/Ni being approximately 2/1 in the leach liquor, for nickel recoveries up to 70%, compared to 17/1 in the ore. For nickel recoveries higher than 70%, the Fe/Ni increased due to iron leaching continuing with simultaneous nickel leaching retardation.

As was observed for nickel extraction, the effect of column height on the other cations percent extractions was found to be statistically significant and positive.

Figure 5 shows the total acid consumption (quantity of acid used up to form sulphate salts plus the free acid remaining in the leach liquor in the end) per kilogram of nickel dissolved as a function of the percentage nickel recovery. In the early stages of leaching, this consumption was high due to the dissolution of calcium in preference to nickel. As has already been mentioned above, calcite had contaminated the laterite ore during sampling. All dissolved calcium, however, was, immediately afterwards precipitated as gypsum in the vicinity of the calcite particles, due to fast kinetics. Thus, laterite particles were not found contaminated with gypsum and

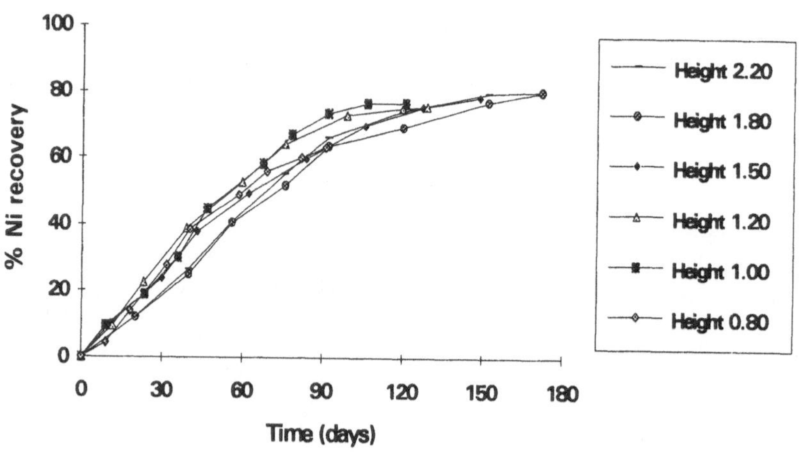

Fig. 2 Nickel recovery with time during laterite heap leaching

1N H2SO4, grain size - 6.75 mm, ratio S/O = 0.8, Technique I

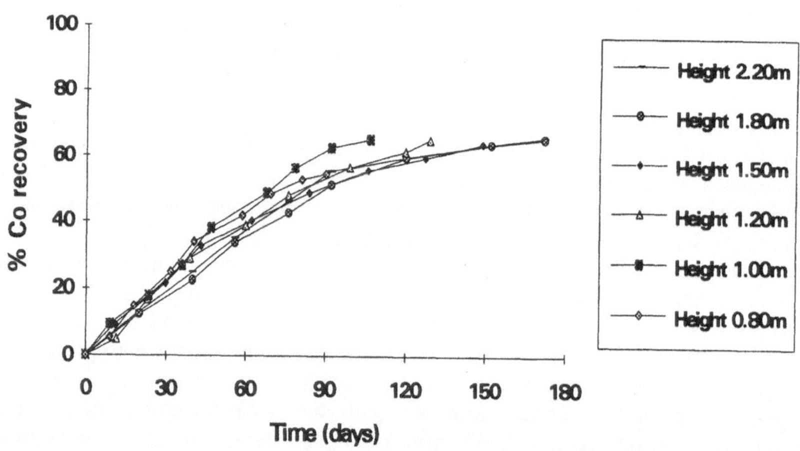

Fig 3. Cobalt recovery with time during laterite heap leaching

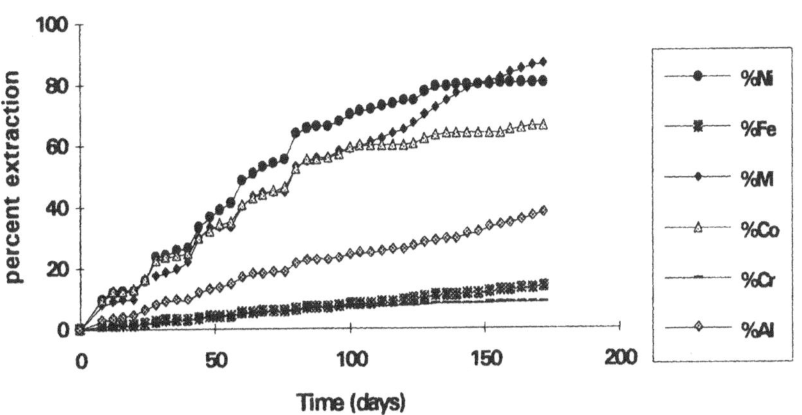

Fig. 4. Cations dissolution with time during laterite heap leaching

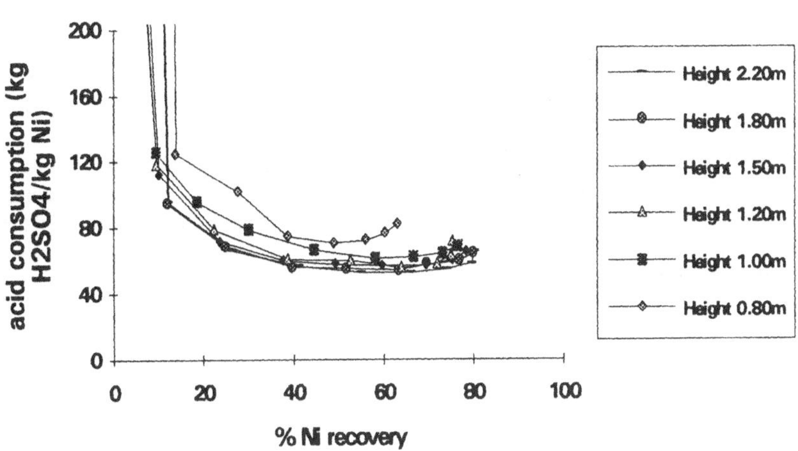

Fig.5. Sulphuric acid consumption as a function of nickel recovery during laterite heap leaching

permeability was not affected. The lowest consumption obtained was around 52 kg H_2SO_4/kg Ni in the highest column which corresponded to about 70% nickel recovery. Thereafter, in all columns, there was observed a tendency for increase which, compared to the experimental error of acid consumption (s = ± 1.56 with Φ = 8 degrees of freedom), was judged as statistically significant. This increase was attributed to a retardation in nickel leaching with a simultaneous progression of the other cations leaching. Figure 5 also shows a dependence of the acid consumption on the column height which exerted a negative effect (decrease in acid consumption with column height increase) as higher columns gave better nickel recoveries.

Having made an acquaintance with the particular leaching system, Technique II was developed and applied in order to improve leaching kinetics as well as sulphuric acid and water consumptions.

3.2 Technique II

Using Technique II, a number of runs were conducted to study the following factors:

1. Ore column height
2. Sulphuric acid concentration in leach solution
3. Leach solution volume/ore weight ratio
4. Ore size

The responses measured were the same as in Technique I. The experimental conditions of each run are shown in the relative figures.

Figure 6 gives nickel recoveries with time for ore column heights studied. As is easily observed, the rate of nickel extraction was almost the same for the different heights and the nickel recoveries obtained after 60-80 days of leaching, when equilibrium had been practically attained, were in the range 78-80%. The differences among those equilibrium values, however, were found statistically significant, although not highly (for nickel recovery s = ± 0.81 with Φ = 8 degrees of freedom).

The extraction of the other metal cations present in laterites, for the highest column height, is shown in Figure 7. Comparing this with Figure 4 of Technique I, it was concluded that the order of cation extraction did not change with the technique used, whereas the extraction rates almost doubled themselves with Technique II. Final extraction values were, however, similar, with the exception of iron, the extraction of which was obviously favoured under the stronger acid conditions, achieved by pH adjustment, used with Technique II. Again, as in Technique I, the effect of column height on the cation percent extractions was found statistically significant and positive.

Mere visual inspection of the graphical presentation of sulphuric acid consumption data (Fig. 8) showed no dependence of the response on the column height. It was, however, realised that right conclusions could only be drawn, by conducting a significant test in order to compare the differences observed among the consumption values for the different column heights with an estimate of the experimental error (for sulphuric acid consumption s = ± 1.56 with Φ = 8 degrees of freedom). Thus, a t-test, at a 5% significance level, applied to the minimum sulphuric acid consumption values obtained during leaching in the three columns (43.2, 46.6 and 49.2 kg

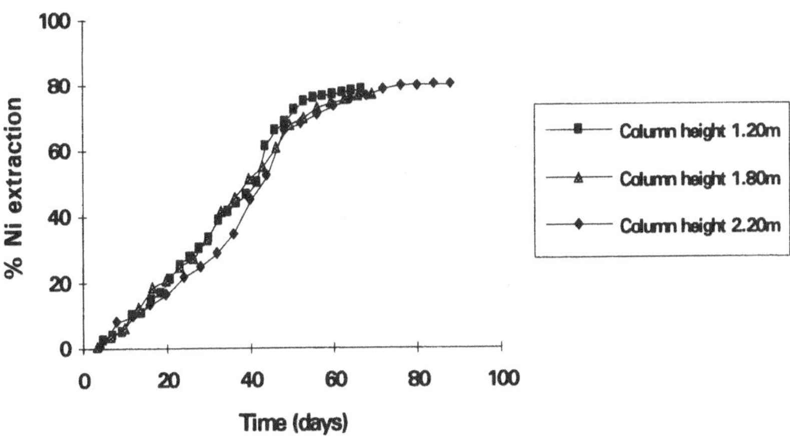

Fig. 6. Nickel recovery with time, at various column heights

1N H2SO4, grain size - 6.75 mm, ratio S/O = 0.8, Technique II

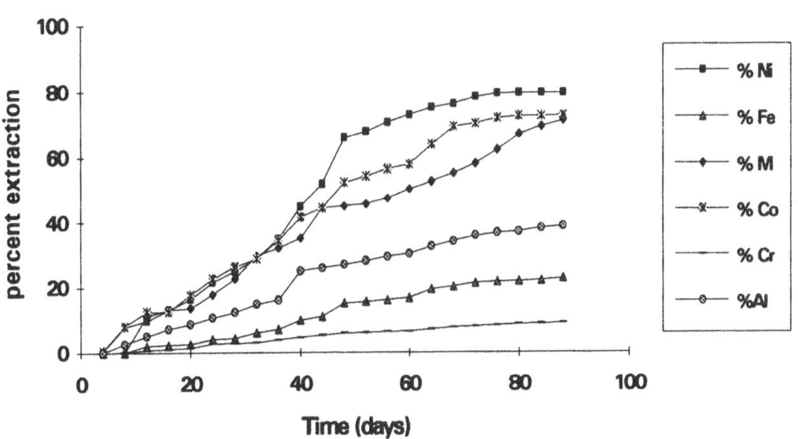

Fig. 7. Cations dissolution with time during laterite heap leaching

H_2SO_4/kg Ni corresponding to 1.20, 1.80 and 2.20 m column height) and revealed that a real difference in the sulphuric acid consumption existed between the lowest and the highest columns.

Figure 9 gives the dependence of nickel extraction on the sulphuric acid concentration in the leach solution. As is easily seen, 0.5 N concentration proved inadequate to extract nickel efficiently. Concentrations from 1 N to 3 N could, however, result in nickel extractions varying from 78% to 86%. The kinetics of nickel extraction was positively affected by sulphuric acid concentration, which was expected. Comparing the equilibrium nickel recoveries with the estimated experimental error (s for nickel recovery = $\pm$ 0.81 with Φ = 8 degrees of freedom) it was concluded that acid concentration had also a positive effect on equilibrium nickel recovery for concentrations up to 2 N. For higher concentrations, there was no effect as nickel extraction had already reached its maximum value, around 85%, for atmospheric pressure leaching. Acid concentration factor exerted a similar effect on the other metal cations leached out of the laterite ore. Sulphuric acid consumption (Fig. 10) was higher with high acid concentrations as more free acid remained in the leach liquor on completion of ore leaching.

Figure 11 shows nickel extraction with time at different values of the leach solution volume/ore weight ratio and sulphuric acid concentrations in the leach solution. It can easily be inferred that the above ratio had a positive effect on nickel extraction for all acid concentrations. This was explained by the fact that, during heap leaching, the continuous increase of cations concentrations in the recycled leach solution caused a reduction of the hydrogen ions activity, despite the relative constancy, at a predetermined value, of their concentration with an adverse effect on nickel recovery. The higher the leach solution volume/ore weight ratio, the lower the reduction in hydrogen ions activity and, consequently, the higher the nickel extraction. The effect seemed to be bigger at the lower value of acid concentration, (78% - 40% > 86% - 50.2%), indicating a negative interaction between the variables of leach solution volume/ore weight ratio and acid concentration. Similar effects of the ratio factor on the other cations co-dissolution, during leaching, was observed.

The effects of grain size on all cations extraction as well as on sulphuric acid consumption were also studied. Two representative samples of the laterite used were ground to -6.75 mm (- 3 mesh) and -30 mm and they were subsequently leached under the same conditions. The results of nickel and iron extraction with time, as well as the sulphuric acid consumption per unit mass of nickel are given in Figures 12, 13 and 14, respectively. These figures also present the data obtained from heap leaching of fractions <6.75 mm (- 3 mesh) and >6.75 mm (+ 3 mesh), screened out from the -30 mm material. The grain size, in the range studied, did not show any significant differences concerning either nickel extraction or sulphuric acid consumption. Apparently, there were no substantial differences in the reaction surface area available for leaching of the two ground samples. As far as iron extraction is concerned, it was clear that, both, extraction rate and final co-dissolution increased with finer material.

A quantitative analysis of the obtained pregnant solutions at the end of the leach process, at all experimental conditions used in this work, is given in Table 2, for

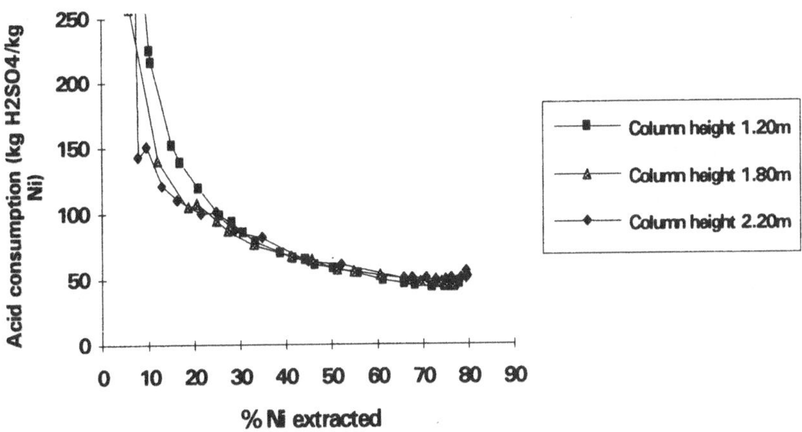

Fig. 8. Sulphuric acid consumption as a function of nickel recovery, at various column heights

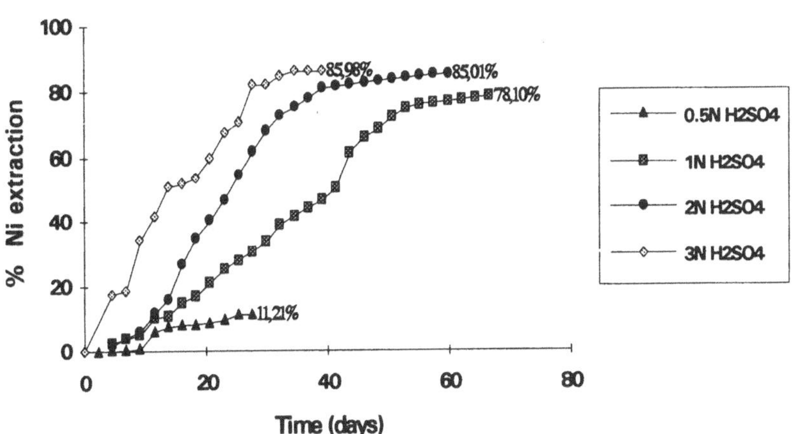

Fig. 9. Nickel recovery with time, at various acid concentrations

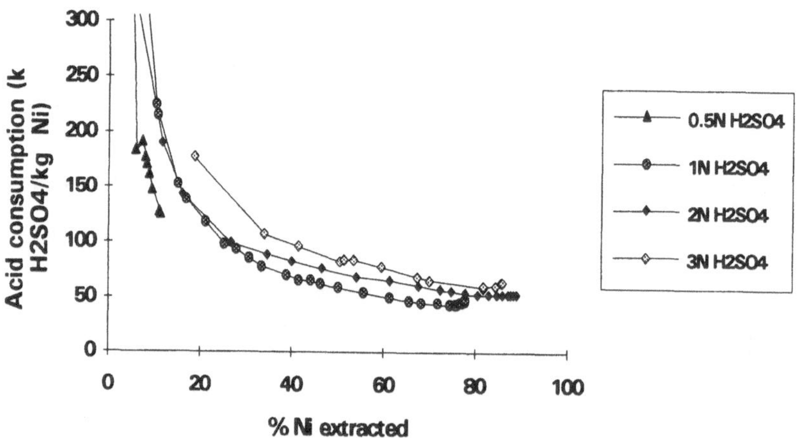

Fig. 10. Sulphuric acid consumption as a function of nickel recovery, at various acid concentrations

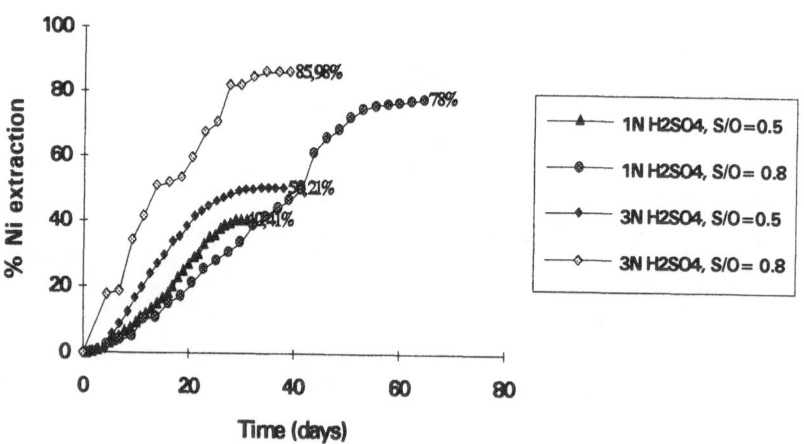

Fig. 11. Nickel recovery with time, at various combinations of acid concentration with solution volume/ore weight ratio

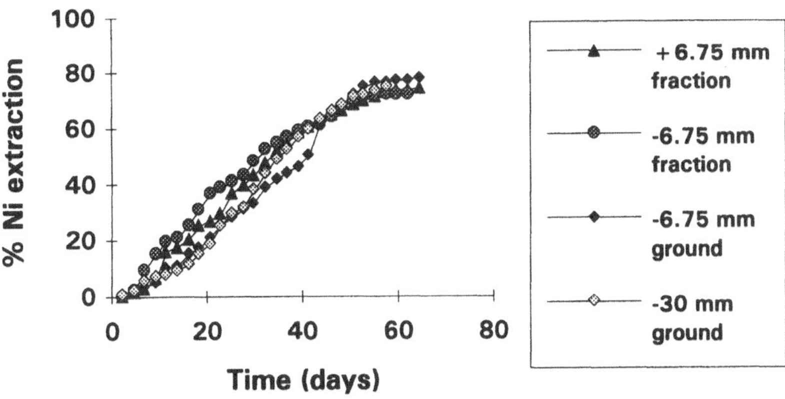

Fig. 12. Nickel recovery with time, at various grain sizes

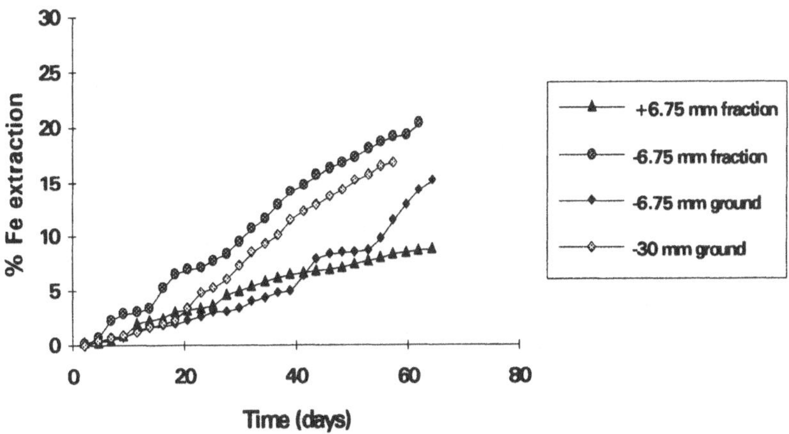

Fig. 13. Iron codissolution with time at various grain sizes

Table 2. Final cations concentrations in the pregnant solution using Technique II, at various experimental conditions

Experimental Conditions[1]	Ni (kg/m^3)	Co (kg/m^3)	Fe (kg/m^3)	Mg (kg/m^3)	Cr (kg/m^3)	Al (kg/m^3)
C.H.=1.20 m, G.S.=-6.75 mm, A.C.=1N, S/O= 0.8	7.125	0.429	25.830	12.900	2.141	13.630
C.H.=1.80 m, G.S.=-6.75 mm, A.C.=1N, S/O= 0.8	7.140	0.398	31.500	13.100	1.300	16.250
C.H.=2.20 m, G.S.=-6.75 mm, A.C.=1N, S/O= 0.8	7.110	0.416	35.200	12.900	1.661	16.710
C.H.=1.20 m, G.S.=-6.75 mm, A.C.=0.5N, S/O= 0.8	0.900	0.078	3.700	1.750	0.230	3.250
C.H.=1.20 m, G.S.=-6.75 mm, A.C.=2N, S/O= 0.8	7.447	0.430	35.000	13.680	2.400	11.400
C.H.=1.20 m, G.S.=-6.75 mm, A.C.=3N, S/O= 0.8	6.904	0.386	31.650	13.300	1.800	14.500
C.H.=1.20 m, G.S.=-6.75 mm, A.C.=1N, S/O= 0.5	5.910	0.370	28.100	11.800	1.110	15.875
C.H.=1.20 m, G.S.=-6.75 mm, A.C.=3N, S/O= 0.5	7.330	0.470	36.510	14.500	1.852	19.400
C.H.=1.20 m, G.S.=-30 mm, A.C.=1N, S/O= 0.8	6.010	0.330	23.800	11.100	1.430	10.100
C.H.=1.20 m, G.S.=-6.75 mm fraction, A.C.=1N, S/O= 0.8	6.300	0.368	30.500	13.500	1.880	11.150
C.H.=1.20 m, G.S.=+6.75 mm fraction, A.C.=1N, S/O= 0.8	6.740	0.477	26.400	13.680	1.735	9.550

[1] C.H.=Column Height, G.S.=grain size, A.C.=acid concentration, S/O=leach solution volume/ore weight ratio

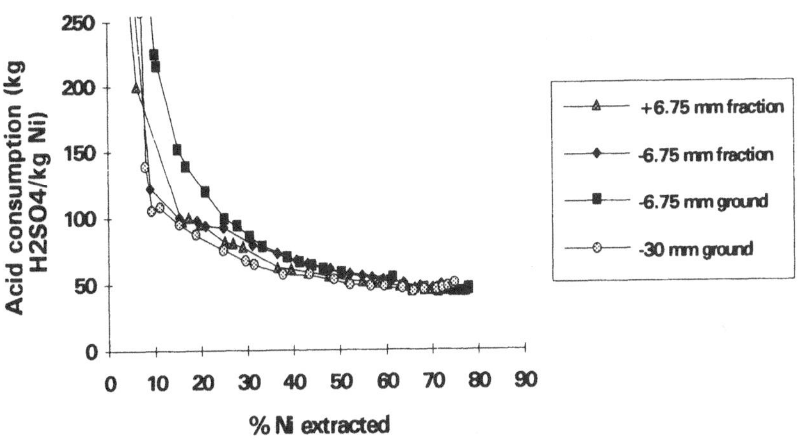

Fig. 14. Sulphuric acid consumption as a function of nickel recovery, at various grain sizes

the sake of completeness.

4 Conclusions

The present work has clearly shown that heap leaching, simulated by column leaching, can be applied to poor Greek laterites in order to extract nickel and cobalt efficiently. Maximum nickel recoveries achieved were 86% and 70%, respectively, in 40 days, by 3N sulphuric acid solution with iron co-dissolution around 20%.

Because nickel and cobalt in the Greek laterites are entirely contained in chlorites, associated with magnesium, iron, aluminium and chromium, contamination of the pregnant solution with those cations is unavoidable. Heap leaching, however, seems to be more selective for nickel and cobalt over iron than any other atmospheric pressure leaching technique. The ratio Fe/Ni resulted in the pregnant solution could be as low as 2/1, at certain conditions, compared to 17/1 in the ore. In no case, however, was that value higher than 5/1. This selectivity was attributed to faster kinetics of dissolution of chlorites compared to that of haematite, the main iron-bearing mineral in the Greek laterites, at the mild conditions of heap leaching at ambient temperature.

All factors studied, namely, ore column height, sulphuric acid concentration and leach solution volume/ore weight ratio had positive effects on both nickel extraction and acid consumption, in the range of values studied. The ore particle size did not seem to affect either nickel extraction or acid consumption for values < 30 mm.

A first understanding of the leach system has thus been achieved. However, further experimentation is presently conducted, using statistical techniques (response surface methodology) in order to fully optimize the heap leach process, not only with respect to nickel and cobalt recoveries bat also with respect to all those responses, such as sulphuric acid consumption, leach time, purity of pregnant solution etc, on which a simple, economic and competitive process will be based.

5 Acknowledgements

Many thanks are expressed to the General Mining and metallurgical S.A. LARCO for the laterite samples provided.

6 References

1. Buchanan, D.L. (1982) Nickel: a commodity review. **Occasional papers of the Institution of Mining and Metallurgy**.
2. Roodra, H.J. and Hermans, J.M.A. (1981) Energy constraints in the extraction of nickel from oxide ores (II). **Erzmetal**, 34(2), 186-190.
3. Simons, C.S. (1988) The production of nickel:extractive metallurgy - Past, present and future. Proceedings of a symposium on the **extractive metallurgy of nickel and cobalt**, 117th TMS Annual Meeting, Phoenix Arizona, January 25-28, 1988.
4. Kontopoulos, A. and Komnitsas, K. (1991) Extraction of nickel and cobalt from Greek laterites by sulphuric acid pressure leaching.In **International seminar on lateritic ore acid leaching technology**, Cuba, Nov. 1991.
5. Agatzini S. and Dimaki, D. (1991) Recovery of nickel and cobalt from low - grade nickel oxide ores by heap leaching with diluted sulphuric acid at room temperature. **Greek Patent 910100234**, 31 May 1991.
6. Dutrizac, J.E. and Rolia, E. (1984) The determination of free acid in zinc processing solutions. **Canadian Metallurgical Quarterly**, 23(2), 159-167.
7. Schlitt, W.J. and Nicolai, L.F. (1987) Nonvertical solution flow and its application in heap and dump leaching. **Minerals and Metallurgical Processing**, Feb. 1987, 1-7.

Mechano-chemical treatment of tetrahedrite as a new non-polluting method of metals recovery

Peter Baláž
Institute of Geotechnics of the Slovak Academy of Sciences, Košice, Slovakia
Roland Kammel
Institute of Metallurgy, Technical University, Berlin, Germany
Mária Kušnierová
Institute of Geotechnics of the Slovak Academy of Sciences, Košice, Slovakia
Marcela Achimovičová
Institute of Geotechnics of the Slovak Academy of Sciences, Košice, Slovakia

Laboratory investigations were performed with tetrahedrite concentrate from Mária-Rožňava mine in Slovakia having the chemical composition of 21% Cu, 11% Sb, 28% S, 1% As, 2% Zn, 0.2% Ag and 1% Hg. X-ray analysis reveal following proportions of dominating phases: 41% Tetrahedrite ($Cu_{12}Sb_4S_{13}$), 12% Chalcopyrite ($CuFeS_2$) and 25% Pyrite (FeS_2). Chemical leaching tests with Na_2S-solutions indicate that only half of the antimony in the concentrate can be recovered, whereas mechano-chemical leaching in a stirred ball mill under similar leaching conditions a nearly complete and selective antimony extraction is achieved. According to X-ray analysis the increased extraction by mechano-chemical leaching might be related beside the grinding action to the transformation of tetrahedrite into covelite (CuS). Depending on the degree of disorder in tetrahedrite the chemical and bacterial leaching of the solid residues after mechano-chemical treatment with $CS(NH_2)_2/Fe_2(SO_4)_3$ solution silver recoveries from 50-90% and with Thiobacillus ferrooxidans-medium copper recovery was 74-84%.
Keywords: mechano-chemical leaching, bacterial leaching, tetrahedrite, natrium sulphide, thiourea, antimony, mercury, silver

1 Introduction

Tetrahedrite is a sulphosalt of complex and variable composition corresponding to general formula M_1, $M_2(M_3)_4S_{13}$. The occurence of individual elements is varied and dependent on deposit. In Slovak tetrahedrites it is M_1, M_2 = = Cu,Hg,Fe,Zn,Ag,Cd,Co,Cr,Mo,Ni,Pb,S and M_3 = As,Bi,Sb. The great variability of various elements is caused by isomorphous substitution and is due to incomplete occupation of the lattice by atoms of some elements[1].

At present the metallurgical processing of tetrahedrite concentrates in Slovakia is focused on the recovery of mercury by thermal volatilisation [2-3] while the calcined residue with the valuable metal contents including copper of 18-20% are dumped and not further treated. This type of pyrometallurgical concentrate treatment is exposed to increasing pressure by environmental legislation.

During recent years there has been a great stimulus to the development of new metallurgical processing methods[4]. Among them direct hydrometallurgical techniques have been developed for the comercial recovery of non-ferrous metals from sulphide concentrates. In comparison with pyrometallurgical techniques the hydrometallurgical way offers larger flexibility, and more importantly, can process low-grade ores with recovery of all metals present.

The leaching step has a key position in the entire flowsheet of hydrometallurgical processing of sulphidic ores. Leaching can be effected by selecting the appropriate leaching agent and/or conveniently pretreating the solids. The last two decades have witnessed the intensive research on the effect of mechano-chemical pre-treatment on the leaching step. Mechanochemistry, as "science on acceleration and initiation of reactions in gases, liquids and solids by the effect of plastic energy"[5] has now great influence in various applications. Numerous applications in the field of sulphidic ore processing and hydrometallurgical treatment were cited in monographs[6-10].

Hydrometallurgical treatment of tetrahedrite is possible in oxidative[11-12] acid or in alkaline solution[13-14]. By acid oxidative leaching, e.g. in Fe^{3+} - medium, copper and iron enter into solution. Alkaline leaching in sodium sulphide medium dissolves selectively antimony, whereas copper and iron remain insoluble. Alkaline leaching with hot sodium sulphide solution was applied at the EQUITY[15] Silver Mines and the SUNSHINE Mining Company[16]. Leaching is performed at 110°C over a period of approximately 12 hours. At EQUITY antimony is removed from the leach liquor in autoclaves while electrowinning has been applied at SUNSHINE.

The structure and chemical composition of tetrahedrite and its refractory character require the application of concentrated leaching agents for efficient hydrometallurgical extraction of the valuable metal content. Also high temperatures and pressures as well, as complicated separation steps for individual metal winning are required. To avoid these unfavourable leaching conditions an alternative method comprising mechano-chemical leaching of antimony from tetrahedrite concentrate as well as subsequent selective leaching of copper, mercury and silver from solid residues has been tested.

2 Experimental

2.1 Materials

The investigations were carried out with a tetrahedrite concentrate (Mária-Rožňava deposit, Slovakia) of the following chemical composition: 21.0% Cu, 10.8% Sb, 28.3% S, 21.5% Fe, 2.15% Zn, 1.06% As, 1.05% Hg and 0.21% Ag. X-ray diffraction analysis showed the following proportions of dominant phases: 41% tetrahedrite ($Cu_{12}Sb_4S_{13}$), 12% chalcopyrite ($CuFeS_2$) and 25% pyrite (FeS_2).

2.2 Mechano-chemical leaching

The mechano-chemical leaching has been tested in an attritor mill "Molinex", type PE 075 (Netzsch, Germany) under the following conditions: 500 ml grinding volume, 2000 g grinding media of iron balls of 2 mm diameter, 56% of the volume of the mill filled with grinding media, 20 g concentrate, 200 ml of 100-200 gl^{-1} Na_2S solution at temperature 333-368 K with 60 minutes grinding time.

2.3 Chemical leaching

The solid residues filtered off and dried after mechano-chemical leaching were treated in a second step under the following conditions:

1) Extraction of mercury: glass batch reactor, 200 ml of leaching solution (10 gl^{-1} Na_2S + 5% NaOH), temperature 293-363 K, 1 g solid residue, stirrer revolution 300 min^{-1}.

2) Extraction of silver: glass batch reactor, 400 ml of leaching solution (10 gl^{-1} thiourea + 5 gl^{-1} $Fe_2(SO_4)_3$ + H_2SO_4 (pH=1.0), temperature 293 K, 1 g solid residue, stirrer revolution 300 minutes.

2.4 Bacterial leaching

The bacteria of the Thiobacillus ferrooxidans strain were used in bacterial leaching. This strain was isolated from mine water present in the Hodruša (Slovakia) sulphidic copper ore deposit. It was, to a certain extent, physiologically adapted to the substrate. The conditions of leaching were: solid/liquid ratio 2 gram/100 ml, 9KA salt leaching reagent at pH 1.5, concentration of bacterial cells 9.89×10^7 ml^{-1}, temperature 29 °C.

2.5 Physico-chemical characteristics

The X-ray phase analysis of the investigated samples was accomplished by using a diffractometer DRON 2.0 equiped with a goniometer Gur-5 (Techsnabexport, Russia) under the following conditions: Radiation $FeK\alpha$, 25 kV, 10 mA, time constant 2 s, limit of measurement 10^3 impulses s^{-1}, rate of counting tube 2° min^{-1}, rate of goniometer movement 1° min^{-1}, paper drive 1 cm min^{-1}.

The metal content in the leach liquours were determined by atomic absorption spectroscopy (Perkin Elmer,

Model 3030 and Spectr Varian, Model AA-30).

3 Results and discussion

3.1 Mechano-chemical leaching of antimony from tetrahedrite concentrate

The mechano-chemical leaching tests with tetrahedrite concentrate samples in the attritor were based on variations of temperature and sodium sulphide concentration. The results summarized in Table 1 reveal 52-100% recoveries of antimony (depending on experimental conditions) after sixty minutes of leaching. Antimony enters into solution selectively whereas the other valuable metals remain practically insoluble in the residue. Temperature 353 K and concentration 100-150 gl^{-1} Na_2S were found to be satisfactory for total recovery of antimony.

Table 1 Results of mechano-chemical leaching of tetrahedrite concentrate

Sample	Variable		Metal in leach (%)				
	Temperature (K)	C_{Na_2S} (gL^{-1})	Sb	Cu	Fe	Hg	Ag
1	333	100	52.09	0.1	0.1	0.1	0.60
2	333	150	83.90	"	"	0.16	0.46
3	333	200	82.64	"	"	0.33	0.45
4	353	100	99.64	"	"	0.62	0.40
5	353	150	99.78	"	"	0.11	0.47
6	353	200	92.76	"	"	0.11	0.38
7	368	100	84.31	"	"	0.12	0.45
8	368	150	95.02	"	"	0.1	0.49
9	368	200	92.96	"	"	0.14	0.74

On chemical leaching carried out under similar conditions the recovery amounts only to 40% of antimony and the selectivity of its extraction is lower with respect to mechano-chemical leaching (Fig. 1).

The investigation of leach residues after mechano-chemical treatment by X-ray diffractometry showed that the diffraction lines of tetrahedrite gradually disappear (Fig. 2). Moreover, the presence of the new phase, covelite CuS (ASTM-6-464), was observed in samples with high degree of decomposition. Because of the absence of tetrahedrite lines (T) on diffractograms with covelite (C) appearance preferential transformation of tetrahedrite → covelite and soluble antimony complexes is probable. But when comparing the most intensive diffraction lines the ratio of the present copper bearing mineral chalcopyrite

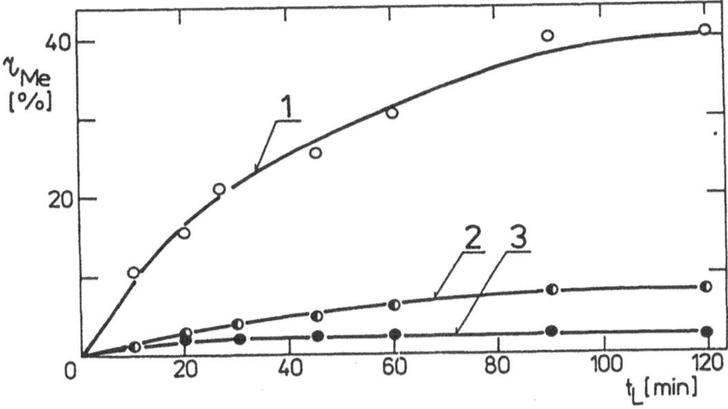

Fig. 1 Influence of chemical leaching time t_L on the recovery of metals η_{Me} from "as received" tetrahedrite concentrate. Temperature 369 K, symbols: 1 - Sb, 2 - Hg, 3 - Ag

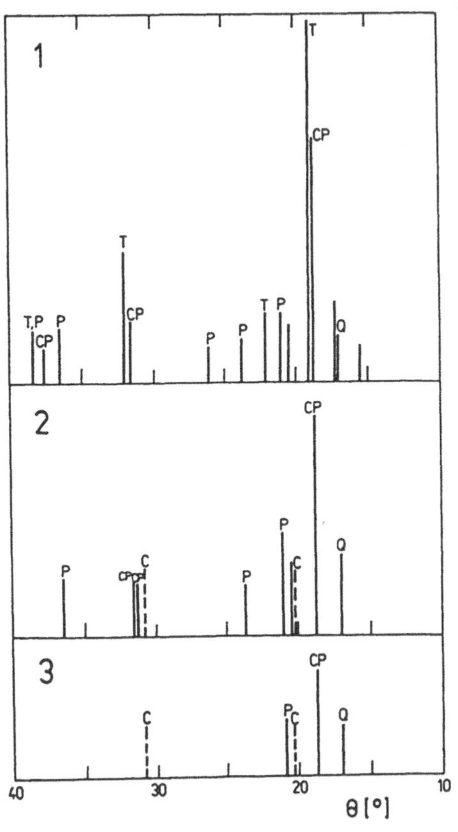

Fig. 2
Line diffractograms of tetrahedrite concentrate: 1-"as received" sample, 2-after mechano-chemical leaching (see sample 5 in Table 1), 3-after mechano chemical leaching (see sample 9 in Table 1). Symbols: T-$Cu_{12}Sb_4S_{13}$, P-FeS_2, CP-$CuFeS_2$, Q-SiO_2 C-CuS

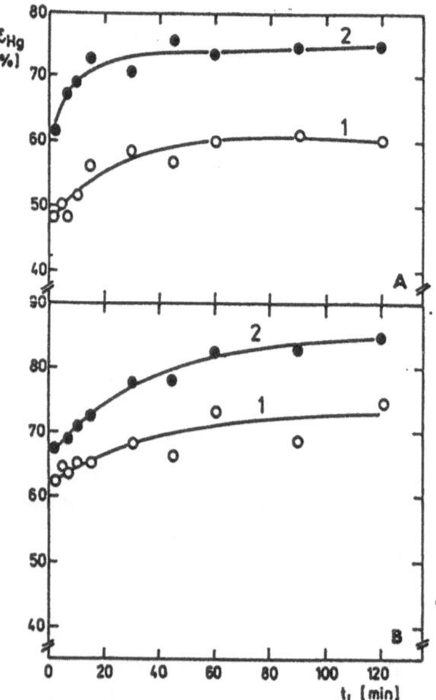

Fig. 3
Influence of leaching time t_L on the recovery of mercury ε^L_{Hg} from mechano-chemically treated tetrahedrite concentrate. Temperature: A-293 K, B-363 K, symbols: 1-sample 1 in Table 1, 2-sample 5 in Table 1

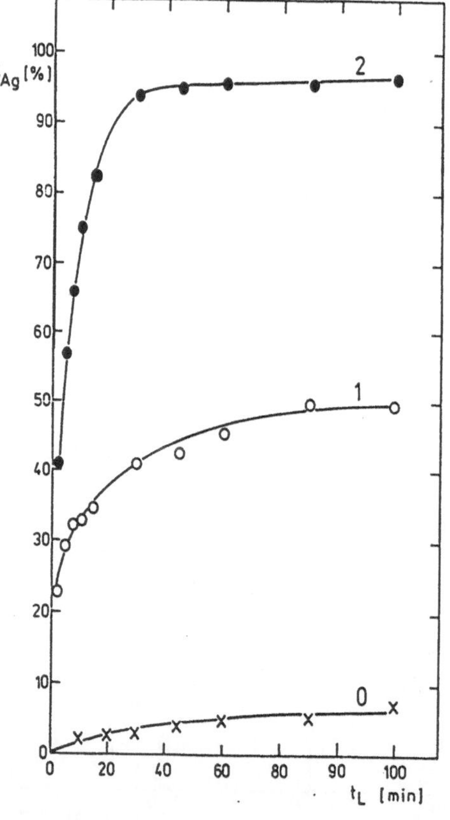

Fig. 4
Influence of leaching time t_L on the recovery of silver ε_{Ag} from mechano-chemically treated tetrahedrite concentrate. Temperature: 293 K, symbols: 0 - "as received" sample, 1 - sample 1 in Table 1, 2 - sample 5 in Table 1

(CP) and (P) (pyrite was selected as a standard), the following values of CP/P ratio can be obtained: 3.5, 2.2 and 1.7 for "as received" sample and samples after mechano-chemical treatment indicated in Table 1 as No. 5 and No. 9, respectively. It follows that during mechano-chemical leaching of tetrahedrite concentrate also changes in chalcopyrite occur. Because of the absence of copper in the leach solution (Table 1) these changes have probably involved a solid state transformation.

3.2 Chemical leaching of solid residues

3.2.1 Extraction of mercury

For the winning of mercury the solid residues after mechanochemical leaching of antimony were dried, analysed and chemically treated in alkaline solution of Na_2S. This reagent reacts with mercury sulphide to give soluble complex salts[17]. Experiments were performed at 293 K and 363 K with selected samples characterised by the minimum and maximum recovery of antimony in mechano-chemical leaching (samples 1 and 5 in Table 1). The results achieved by leaching the solid residues are presented in Fig. 3. The recovery of mercury ε_{Hg} varies depending on the degree of tetrahedrite disordering and applied temperature. From the time dependency it can be seen that after 1 h leaching recoveries ranging from 60-85% can be achieved. Our previous investigations[18] of the chemical leaching of tetrahedrite concentrate (without mechano-chemical pretreatment) in the same alkaline medium have shown mercury recoveries only of about 8% after 120 minutes leaching at 363 K.

3.2.2 Extraction of silver

Tetrahedrite belongs to silver-bearing ore types where silver is bound in different forms[19]. Silver in tetrahedrite of the Slovak provenience used in these experiments is part of the tetrahedrite structure where silver substitutes for copper cations[20].

Thiourea $CS(NH_2)_2$ as an attractive leaching alternative to direct cyanidation leads to silver soluble complexes. The efficiency of the silver leaching processes depends on ferric sulphate/thiourea ratio in the leach solution[21-23].

As starting materials for the leaching of silver in thiourea and ferric sulphate the solid residues after mechano-chemical leaching of tetrahedrite concentrate have been considered. In Fig. 4 extraction of silver into solution ε_{Ag} is plotted against leaching time. The recoveries obtained for "as received" sample (curve 0) are identical with the results published by Štofko and Štofková[20] for tetrahedrite of the same provenience. It can be seen that the application of thiourea for silver extraction without tetrahedrite pretreatment is not effective and the recoveries amount to less than 10% even after 120 minutes of leaching.

The recoveries from solid residues after mechano-chemical leaching are substantially improved. The extraction values depend on degree the disorder of tetrahedrite and can vary from 50 to 96% of silver. A comparison of minimum and maximum recoveries of antimony and silver for samples 1 and 5 (see Table 1) show, that similar values can be obtained for silver: 96.52/50.3 = 1.93 and for antimony: 99.78/52.09 = 1.92, respectively. From these values it can be elucidated that the amount of antimony and silver released from tetrahedrite depends on disorder of its structure and the sensitivity of leaching to this disordering is for both metals probably equal.

3.3 Bacterial leaching of solid residues (extraction of copper)

For the winning of copper the solid residues left after mechanochemical leaching were subjected to bacterial leaching. The results are presented in Table 2.

Table 2 Results of bacterial leaching of copper from mechano-chemically treated tetrahedrite concentrate

Sample	Copper in leach (%) Time of bacterial leaching (days)								
	1	3	7	9	11	14	18	21	24
1	41.60	48.84	52.60	54.33	60.40	63.87	67.92	68.78	73.70
2	37.35	46.47	67.94	70.59	75.00	75.00	77.06	77.35	78.80
3	42.86	51.65	57.42	60.99	67.31	70.33	76.65	79.02	84.30
4	37.87	43.73	52.27	59.73	68.80	70.93	74.13	82.38	80.00
5	22.80	32.90	49.74	61.91	70.02	80.31	82.64	82.10	81.87
6	35.39	41.02	45.31	47.45	50.40	53.08	58.18	71.15	75.34
7	47.72	51.58	73.68	77.54	81.40	82.10	82.10	81.08	81.05
8	40.15	43.63	57.14	66.41	76.45	81.08	81.47	79.00	81.85
9	20.73	23.62	47.24	72.70	75.33	79.53	80.84	80.15	79.53

The application of Thiobacillus ferrooxidans over 21 days led to extraction of 74-84% of copper from mechano-chemically leached samples of tetrahedrite concentrate. The presence of easily filtrabilitable jarosite $KFe_3(SO_4)_3(OH)_6$ (ASTM 22-827) was ascertained in the products of bacterial leaching by X-ray diffractometry.

4 Conclusion

The mechano-chemical leaching of antimony from tetrahedrite concentrate in an attritor and the chemical and bacterial leaching of the resulting solid residues have been studied. The results indicate recoveries of 52-100% Sb, 60-85% Hg, 50-96% Ag and 74-84% Cu can be achieved using alkaline solutions of Na_2S for leaching of Sb and Hg; $CS(NH_2)_2$ for leaching of Ag and Thiobacillus ferrooxidans for leaching of Cu. The recoveries obtained for metals are sensitive to disordering of the tetrahedrite structure as a result of combined effect of mechanical and chemical treatment. The hydrometallurgical processes under study might lead to the development of a potential and ecological technique which allows selective extraction of most of the metal values from tetrahedrite concentrates.

5 References

1. Kostov, I., Minceva-Stefanova, I. Sulphide Minerals, Mir, Moscow 1984 (in Russian)
2. Imriš, I., Komorová, L., Sehnálek, F. Complex tetrahedrite concentrates from Slovakia. In: Proc. Complex sulphide ores, Inst. Min. Metall., Roma 1980, p.63-70
3. Imriš, I., Komorová, L. Trends in pyrometallurgical processing tetrahedrite concentrates. Trans. Techn. Univ. Košice, No. 2, 1992, p. 57-69
4. Sohn, H.Y. The Coming-of-Age of Process Engineering in Extractive Metallurgy. Metall. Transactions B, Vol. 22B, 1991, p. 737-754
5. Butjagin, P.Y. The kinetics and essence of mechanochemical reactions. Uspechi chimiji, Vol. 40, 1971, p. 1935-1959 (in Russian)
6. Molčanov, V.I., Jusupov, T.S. Physical and Chemical Properties of Fine Ground Minerals, Nedra, Moscow 1981 (in Russian)
7. Avvakumov, E.G. Mechanical Methods of Chemical Processes Activation, Nauka, Novosibirsk 1986 (in Russian)
8. Molčanov, V.I., Selezneva, O.G., Žirnov, E.N. Activation of Minerals at Grinding, Nedra, Moscow 1988 (in Russian)
9. Kulebakin, V.G. Application of Mechanochemistry in Hydrometallurgy, Nauka, Novosibirsk 1988 (in Russian)
10. Tkáčová, K. Mechanical Activation of Minerals, Elsevier, Amsterdam 1989
11. Dutrizac, J.E., Morrison, R.M. The leaching of some arsenide and antimonide minerals in ferric chloride media. In: Proc. Hydrometallurgical Process Fundamentals (Ed. R.G. Bautista), Plenum Press, New York 1984 p. 77-112
12. Havlík, T., Škrobian, M., Dudáš, D. Study of acid oxidative leaching of tetrahedrite. Hutnícke listy, Vol. XLVI, 1991, p. 76-80 (in Slovak)

13. Melnikov, S.M. Antimony, Metalurgia, Moscow 1977 (in Russian)
14. Pawlek, F. Metallhüttenkunde, Walter de Gruyter, Berlin 1983
15. Dayton, S. Equity Silver on line with leach plant. Eng. Min. J., January 1982, p. 7-10
16. Holmes, W.C. How electrolytic Antimony is made at Sunshine plant. Eng. Min. J., vol. 45, 1944, p. 5-8
17. Melnikov, S.M. Mercury, Metalluria, Moscow 1951 (in Russian)
18. Baláž, P., Kammel, F., Kušnierová, M., Šepelák, V., Tkáčová, K. Extraction of Antimony and Copper from Mechanochemically Treated Tetrahedrite. To be published in Proc. Ist. International Conference on Mechanochemistry (Ed. K. Tkáčová et al.), Cambridge Interscience Publishing, Cambridge 1993
19. Gasparini, C. Silver-bearing minerals effect the recovery of primary silver. In: Precious Metals 1983 (D.A. Reese, Ed.), Pergamon Press, Toronto 1983, p. 197-203
20. Štofko, M., Štofková, M. Leaching capacity of Au and Ag in acid thiourea solutions. Trans. Tech. Univ. Košice No. 2, 1992, p. 127-131
21. Hiskey, J.B. Thiourea Leaching of Gold and Silver. Technology Update and Additional Applications. Miner. Metall. Process., November, 1984, p. 173-179
22. Wysloužil, D.M., Salter, R.S. Silver Leaching Fundamentals. In: Proc. "Lead-Zinc '90" (T.S. Mackey, R.D. Prengaman, Eds.), Min. Metal. Mater. Soc. 1990, p. 87-107
23. Acma, E., Bor, F.Y., Wuth, W. Gümüsköy - pasa cevherinden tiyoüre ile gümüs licinin kinetigi. Doga-Tr. J. Engin. Environm. Sci., Vol. 16, 1992, p. 95-99

Acknowledgement

The authors (P.B., M.K., M.A.) are grateful to the Slovak grant agency for science for partial supporting of this work.

Non-traditional methods of treating high-silicon ores containing rare elements

S. V. Chizhevskaya
D. Mendeleev University of Chemical Technology of Russia, Moscow, Russia
A. M. Chekmarev
D. Mendeleev University of Chemical Technology of Russia, Moscow, Russia
O. M. Klimenko
D. Mendeleev University of Chemical Technology of Russia, Moscow, Russia
M. V. Povetkina
D. Mendeleev University of Chemical Technology of Russia, Moscow, Russia
O. A. Sinegribova
D. Mendeleev University of Chemical Technology of Russia, Moscow, Russia
M. Cox
Division of Chemical Sciences, University of Hertfordshire, Hatfield, England

Abstract
Several methods of reducing or eliminating the problem of silica in solutions following leaching are available including modification of the leaching conditions, addition of flocculants or coagulants to improve filtration of the silicic acid, and modification of subsequent solvent extraction process to operate under organic phase continuous conditions. These have limited success and often reduce the metal content of the leachate by adsorption on the silica precipitate.

A novel technique is described which involves a combination of extractive leaching with nitric acid/tri-n-butylphosphate (TBP) coupled with intense mechanical activation in a specially designed reactor. Using the mineral eudialyte (>50% SiO_2) as a model for high silicate minerals it has been shown that high recovery of the rare metal content is possible and the silica released remains as an easily filtered residue. This technique should be widely applicable to other minerals where the desired metals can be extracted by TBP.
Keywords: Extractive leaching, high silicon ores, TBP/nitric acid leachant, silicic acid,

1 Introduction

It is well known that the main difficulties in the recovery of minerals from ores are associated with the presence of silica as an impurity in the ore. Leaching with mineral acids dissolves the silica as silicic acid which readily undergoes polymerisation and condensation forming gels which impede filtration, reduces the extraction of the desired components and inhibits any subsequent solvent extraction or absorption processes [1,2]. Even small amounts of silicic acid and its polynuclear species cause the formation of interfacial films in solvent extraction systems [3]. In some cases when highly acidic aqueous solutions containing these polynuclear silicic acid species are contacted with organic phases under intensive mixing conditions extremely stable emulsions are formed which are stable for a long time (weeks, months). Such emulsions as well as the films or gels formed by silicic acid set obstacles for the normal operation of mass transfer equipment [3,4]. The reasons for these phenomena are varied but generally result from the formation by silicic acid and its polynuclear species of a structural mechanical barrier (SMB) at the interface which block the surface of emulsion drops preventing coalescence [5]. This phenomenon has been studied in detail for the systems containing silicic acid and zirconium nitrate [6 - 10] as well as silicic acid alone [5]. The presence of silica in stock solutions for ion exchange or carbon absorption also complicates process operation due to blocking of the pores in the media [1], requiring the preliminary treatment of such solutions to decrease silica concentration.

Traditional methods to reduce or eliminate these problems caused by the presence of silica in hydrometallurgical solutions are as follows:

1. control of polymerisation reactions during the acid leaching to produce a silica species with good filtration properties;

2. coagulation or flocculation of silicic acid by the introduction of special reagents, sometimes in conjunction with method 1;

3. carrying out the purification and extraction of the desired component under conditions which neutralise the influence of silicic acid.

Thus Owens [11,12] showed that a coarse, easily filtered silica powder could be obtained from the nitric acid leaching of a sintered zircon by treatment with alkali if the nitric acid concentration was maintained above 3 mol dm^{-3} and not more than 40% of the water was evaporated. Layner [13] describes a process for the purification of an aluminium sulphate solution containing silica whereby residual silica (>80% SiO_2) at a solid:liquid ratio of 2:1 was mixed with the impure solution. After 5 hours mixing the silica concentration of the solution decreased to 0.155 - 0.193 g dm^{-3} allowing the recovery from the sulphate solution of high grade alumina.

References to the use of flocculants or coagulants for the removal of silica indicate the early use of natural materials like glue, flour or gelatine [14]. These have been supplanted by synthetic substances such as polyglycols, where, in the extraction of copper by hydroxyoximes, 90% of the silica was eliminated from an initial concentration of 4 - 6 g dm^{-3} [15], and polyacrylamide [16]. The choice of suitable flocculants is very dependent upon the nature of the system and flowsheet requirements. Thus some flocculants which effectively reduce the silica content of a solution may cause emulsification in subsequent solvent extraction operations due to their high residual solution concentration. Some flocculants reduce the silica content in acidic solutions to <0.1 g dm^{-3} which is approximately equal to the threshold concentration for polymerisation of silica in acidic media and thus allows the normal procedures of solvent extraction and absorption to be carried out. However sometimes it is necessary to reduce the silica concentration still further to <0.01 g dm^{-3} which is not achievable using traditional flocculants or coagulants. However by contacting the solution with a 'solid' flocculant consisting of a solid carrier containing an organo-silicon compound, a process has been developed to reduce the silica content of acidic leach solutions. The dependence of the residual silica concentration on the nature and concentration of the organo-silicon compound on a macroporous silica gel is shown, figure 1.

It is possible to avoid the formation of stable emulsions in extraction without prior removal of silica by ensuring a water in oil emulsion. Thus in the extraction of uranium from a sulphate pulp monododecylphosphoric acid (0.1 mol dm^{-3}) at an organic :aqueous phase ratio 750:1 was used [17]. A lower phase ratio (O : A = 1.5 - 2 : 1) is enough to allow normal extraction of zirconium by TBP from nitrate solutions with a nitrate concentration of 5 - 6 mol dm^{-3} and silica concentration of 0.1 - 3 g dm^{-3} [18]. Under these conditions the organic phase was continuous and no stable emulsion was observed. Application of special coalescing methods in the mixer-settlers [17] or long residence times in the settler [3] also may be used to avoid the prior removal of silica for the extraction of 3-d transition metals.

Thus the known methods of recovery of rare and non-ferrous metals from raw material in the presence of silica include either, the reduction of the silica content by one method or another or, the use of long residence times or special coalescing equipment with the need for large volumes of extractant. All these methods are rather complicated, lengthy and expensive.

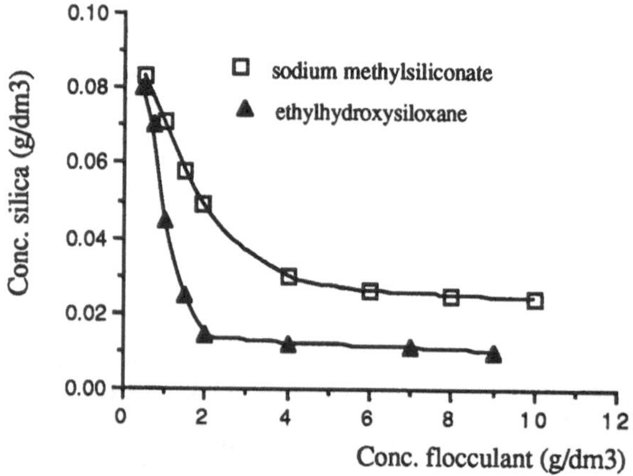

Figure 1. Dependency of silica concentration in solution on amount of added "solid" flocculant.
Initial silica concentration 0.1 g/dm3; nitric acid concentration 5 mol/dm3

Currently the search for new sources of raw materials for rare metals and the development of new economical and low-waste technologies are very important in Russia as elsewhere. With the depletion of the more easily worked ores, complex raw materials are becoming more important. Currently the main industrial sources of zirconium are zircon (60% ZrO_2) and baddeleyte (92 - 97% ZrO_2). Both of these are acid resistant minerals and require preliminary high temperature operations to facilitate the leaching operation. So the production of zirconium from these resources requires a multi-stage process and special preventive measures are required to minimise the detrimental effect of silica on the extraction and separation of zirconium and hafnium.

Russia has practically unlimited sources of another zirconium mineral, eudialyte, which, in contrast to zircon and baddeleyte, is easily dissolved in mineral acids. It is a complex zirconosilicate containing, in addition to zirconia (10 - 15% ZrO_2), rare earths, niobium, tantalum, and titanium. The mineral is found in rocks of the Lovozero alkali massif (Kolski peninsula) where luyavrytes include 5 - 10% eudialyte, 25 - 40% feldspar, and 25 - 40% nepheline. The average eudialyte content is about 9%, giving an average zirconia content in the rocks of 1.3%, ranging between 0.5 - 4.8%. Higher concentration of zirconia (5 - 9%) is typical for enriched rocks, eudialytites, (50 - 80% eudialyte). Eudialyte has the general formula: $Na_{12}Ca_6Fe_3Zr_3[Si_3O_9]_2[Si_9O_{24}(OH)_3]_2$ [19] with a layer structure similar to mica or zeolites consisting of twelve layers separated by oxygen atoms. Three mica-like fragments may be distinguished at the centre of which is a ZrO_6 octahedron. Above and below these are fragments which contain either triple, $[Si_3O_9]$, or ninefold, $[Si_9O_{27}]$, rings of silicon-oxygen tetrahedra. The $[Si_9O_{27}]$ ring structure was observed for the first time in this mineral. Partial isomorphous replacements are possible: Ca for Na, Mn, rare earths, K, Sr; Zr for Nb, Ta, Ti; Si for Fe, Al, etc.

The structural features of eudialyte cause the chemical composition to be complicated and varied and also explain its technological properties. Weak chemical bonds between the separate parts of the mineral lead to easy decomposition with mineral acids. But, because of the large proportion of silica (>50% SiO_2), acid leaching gives a pulp with a low concentration of the desired components which is very difficult to filter. Thus because sulphuric acid provides the best chance of coagulating the silicic acid formed during leaching, this acid has received most attention. A detailed review of the recovery and treatment of eudialyte has been published [20]. It should be noted that in general these methods are characterised by high cost of leaching agents, lengthy processing and low concentrations of the desired components in the leachates. This is because attempts have been made to dissolve the silica and provide easily filtered species which, as noted above, is difficult because of the nature of the mineral and its high silica content. The possibility of heap-leaching has been mentioned [21].

With this background it is obvious that if the potential of eudialyte is to be exploited a new approach to the leaching of the mineral is required. The results of such an investigation is presented in this paper.

2 Results and Experimental

2.1 Feed material:

The mineral used in this study was a eudialyte flotation concentrate with average particle size about 120 μm and percentage composition: 8.3% ZrO_2; 0.36% Y_2O_3; 0.49% Ce_2O_3; 0.57% Nb_2O_5; 0.056% Ta_2O_5; 4.3% Fe_2O_3; 5.4% Al_2O_3; 0.78% TiO_2; 1.9% MnO_2; 52% SiO_2. The concentration of silica was determined by spectro-photometry and the other elements by atomic emission spectrometry.

From the kinetics of extraction of this concentrate the following reaction scheme has been proposed:

1. ion exchange to form the unstable hydrogen form of the mineral;
2. destruction of the zirconium-silicon-oxygen skeleton with the dissolution of Zr^{4+} and release of silicon-oxygen fragments.

Destruction of the eudialyte lattice and depolymerisation of the silicon-oxygen fragments results in the formation of a large amount of silica which tends to form a polycondensed gel with a very long coagulation time. Thus treatment of concentrate with 11 mol dm^{-3} nitric acid at a solid:liquid phase ratio of 1:100 at 100C, coagulation is not complete even after 3 hours, even though the mineral was completely destroyed within 10 minutes.

2.2 Effect of mechanical treatment of the concentrate on leaching

Attempts to intensify the nitric acid leaching of the eudialyte concentrate by preliminary mechanical treatment showed that short-term (30 s) treatment in highly intensive activators increased by 1.5 times the extent of zirconium and rare earth extraction. The dependency of extraction on the mechanical activation time is shown in figure 2. After an initial increase in recovery, further activation causes a decrease in extraction but an enhancement of the filterability of the residual pulp was observed. The presence of a minimum on the curves obtained at very short activation times (curves 1,2 and 3) may be explained by the effect of silica which rapidly dissolves and then forms a sol which adsorbs the components of eudialyte. This is followed by the polycondensation of the silicic acid in solution to form a gel with Si-O-Si bonds and the elimination of water. The decrease of extraction following extensive mechanical treatment is probably caused by the reformation of the eudialyte structure. The results of X-ray investigation of the concentrate powder after 0.5 and 15 minutes activation are shown in figure 3. They confirm a change of crystal structure and show, in general, eudialyte becomes more amorphous, a state which apparently affects the tendency of the rare metals to leave the concentrate.

Thus a combination of mechanical treatment and leaching results in the decrease of amount of leaching agent required and increases the rate of filtrate of the resulting pulp ten-fold.

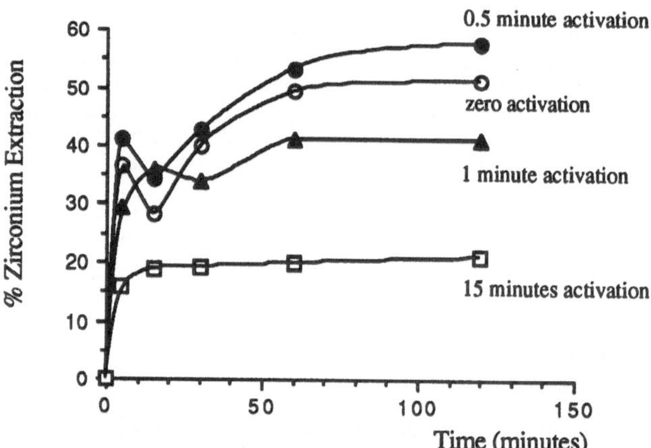

Figure 2. Dependency of time of mechanical activation (m/a) on the extent of zirconium extraction from eudialyte concentrate.
Solid:liquid ratio 1:8; nitric acid concn. 8 mol/dm3; temp. 100C.

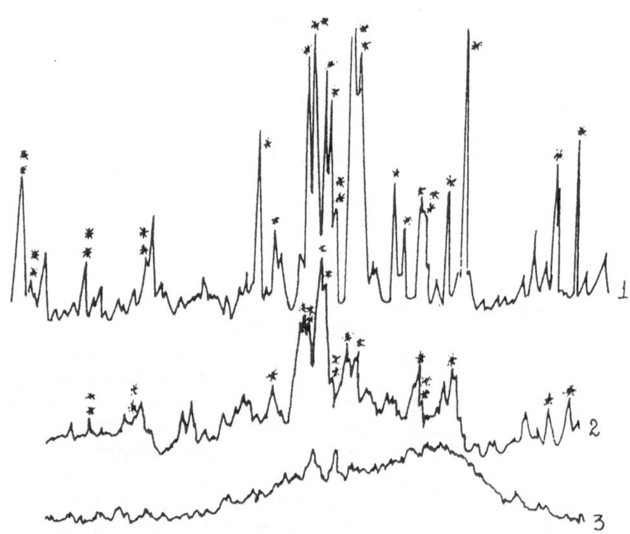

Figure 3. X-ray diffraction pattern of initial eudialyte concentrate (1), and after mechanical activation for 0.5 minutes (2) and 15 minutes (3).
(x = eudialyte, xx = nefelyne)

2.3 Effect of treatment with alkali

When treating raw material with a high concentration of silica it is sometimes useful to decrease the silica content by chemical treatment. The results of a study of the effect of alkali on the extraction of silica from the concentrate are given in table 1. These show that 32% was the maximum amount of silica which could be extracted into sodium hydroxide. This was believed to arise from other minerals in the concentrate and not from the eudialyte.

Table 1. Effect of sodium hydroxide treatment of eudialyte concentrate on silica extraction

[NaOH] (mol dm^{-3})	Time (hour)	Liquid:Solid Ratio	Temp. (C)	[SiO_2] (g dm^{-3})	%SiO_2 extracted
14.0	3	10	125	1.2	14.4
16.8	3	10	135	15.0	30.0
16.8	3	20	130	8.0	32.0
16.8	3	5	132	25.0	25.0
17.0	0.5	5	134	17.5	17.0
17.0	1	5	136	19.0	19.0
17.0	5	5	136	26.0	26.0

2.4 Solvent extraction of zirconium

In all the methods tried involving nitric acid extraction the maximum amount of zirconium recovered did not exceed 75%, apparently due to the large moisture content and absorption activity of silicic acid gel formed during leaching. To increase the extraction efficiency of the desired components different methods of solvent extraction or absorption from nitric acid were tried.

Using leachates prepared under standardised conditions it was found that solvent extraction by TBP from the cool pulp, to minimise acidic hydrolysis of the extractant, increased the extraction of zirconium to 95 - 97%, but most of the rare earth elements remained in the raffinate. Similar results were obtained using a TBP impregnated sorbent.

2.5 Development of a new method of extractive leaching

The results of the above experiments suggested a new technique for the extraction of eudialyte which could have advantages over the existing procedures, and may be applicable to other minerals with a high silica content.

The basis of the method is the ready dissolution of eudialyte in mineral acids and the ability to extract the desired components with TBP. Thus the leaching agent is 100% TBP saturated with nitric acid. The process is carried out in a specially designed extractor which subjects the concentrate simultaneously to a number of physical effects such as: intensive mixing, acoustic action, magnetostriction, and magnetic and electrical fields. The presence of a large number of hydrodynamic vortices causes the autogenous comminution of the concentrate and intensive mixing of the suspension increases mass transfer between the solid and liquid phases. Under optimal conditions both zirconium and the rare earths are almost completely extracted into the organic phase in under 3 minutes with subsequent phase separation taking only 2 minutes.

The dependence of extraction on nitric acid content of the TBP for some components, (figure 4), indicate that after 10 minutes leaching with 3.5 mol dm^{-3} nitric acid, complete recovery of cerium and yttrium is possible with 50% of zirconium being leached. Because

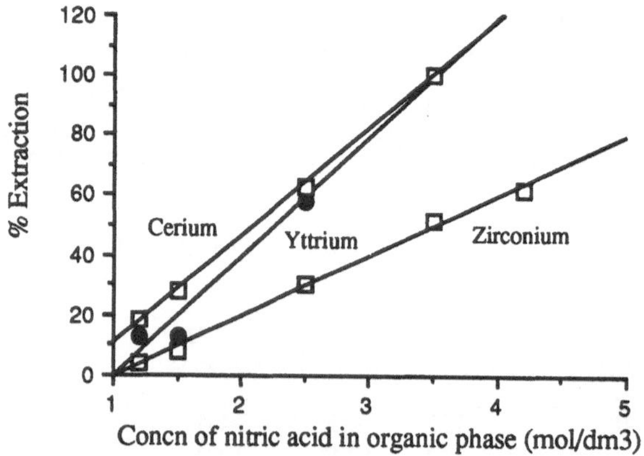

Figure 4. Effect of organic phase nitric acid concentration on extent of extraction of components of eudialyte concentrate. Liquid:Solid ratio 20:1; time 10 minutes.

of the absence of water the silica from eudialyte does not form silicic acid, but is released as large aggregates with a particle size about 0.3 μm.

Therefore after leaching the separation of the leachate from residue is instantaneous. Another advantage of the large particle size is that adsorption of the extractant is very low (1 - 2%).

Extractive leaching using nitric acid dissolved in TBP has also been used by Strelkov and Fillipov [22] for the extraction of uranium from uranium oxide. They found that extraction depended on the concentration of nitric acid and that oxidation of uranium to U(VI) by the acid aided the dissolution.

3 Conclusion

In the technique described above the presence of intensive mixing by mechanical agitation promotes the initial destruction of the crystal lattice and leads to the rapid dissolution/extraction of the desired components in the TBP/nitric acid. The inability of silica to dissolve and form complexes with the extractant provides a readily filterable product.

The technique should be applicable to other ores of similar composition, and reassuring preliminary results have been obtained on an ore from Western Australia, which has a silica content of about 60% [23].

4 References

1 Ritcey,G.M. (1986) 'Silica fouling in ion exchange, carbon-in-pulp and solvent extraction circuits' **Canad.Metall. Quart.** 25(1), 31

2 Sinegribova, O.A., Chizhevskaya, S.V., Bobyrenko, A.Yu. et al. (1988) 'Nature of stable dispersions appearing during solvent extraction of inorganic substances' **Proc ICHM'88**, China, 259

3 Solovkin, A.S. and Yagodin, G.A. (1970) 'Extraction chemistry of zirconium and hafnium' **Itogi Nauki. Neorg. Khim. VINITI**, Moscow, 46 (in Russian)

4 Nikolski, B.P., Nikipelov, B.V., Moshkov, M.M. et al. (1983) 'Conditions for formation and the nature of interfacial silica containing precipitates' **Atomnaya Energia** 55(5), 315 (in Russian)

5 Chizhevskaya, S.V., Sinegribova, O.A., Danilova, S.S., et al. (1977), 'Behaviour of silicic acid at the interface in the system nitric acid - silicic acid - decane' **Izvestia Vuzov, Chimia i Chim. Technologia,** V.XX (5), 694 (in Russian)

6 Sinegribova, O.A., Yagodin, G.A., Chizhevskaya, S.V., et al. (1980) ' The extraction behaviour of zirconium polynuclear compounds with certain tetravalent elements' **Proc ISEC'80**, Leige, 3, paper 80-162.

7 Sinegribova, O.A., Chizhevskaya, S.V. and Danilova, S.S., (1980) 'The interfacial layers in the system zirconium nitrate - nitric acid - decane' in: **Physico-chemical Phenomenon at the Interface,** MChTI after D I Mendeleev (Moscow), p 30 (in Russian)

8 Sinegribova, O.A., Chizhevskaya, S.V. and Danilova, S.S., (1980) 'Structural mechanical barrier at the interface in the system silicic acid - zirconium nitrate - nitric acid - decane' **ibid**, p46 (in Russian)

9 Sinegribova, O.A. and Chizhevskaya, S.V., (1992) 'Behaviour of some substances at the interface and time of aqueous and organic phase separation' **Radiochimia** (5), 76 (in Russian)

10 Sinegribova, O.A., Chizhevskaya, S.V. and Kotenko, A.A. (1989) 'Stable dispersions appearing in extraction systems containing zirconium and silicic acid' **Radiochimia** 31(6) 87 (in Russian)

11 Owens, W.H., (1967) **US Patent** 3351425

12 Owens, W.H., (1964) **US Patent** 976050 ; (1966) **US Patent** 3272590

13 Layner, Yu.A., '**Complex treatment aluminium - containing ores by acid techniques'** Moscow Nauka, p 127 (in Russian)

14 Choi, H.S., (1956) 'Preparation of pure zinconyl compounds from zircon caustic frit', **Canad. Mining and Metall. Bull.** (58) 193

15 Lland, I.M., (1980) 'Colloidal silica in the solvent extraction of copper with oximes' **Proc ISEC'80,** Liege, paper 80-74

16 Rutman, D.S., Toropov, Yu.S., Pliner, S.Yu., et al.(1985)'Fireproof materials from zirconium dioxide' **Moscow Metallurgia** 12. (in Russian)

17 Ritcey, G.M. and Ashbrook, A.W., (1979),'**Solvent Extraction Principles and Applications to Process Metallurgy**', part 2, Elsevier, 543

18 Sinegribova, O.A., Chizhevskaya, S.V. and Kotenko, A.A., (1989) **USSR Patent** 1701634

19 Golyshev, V.M., Simonov, V.I. and Belov, N.V., (1972) 'Crystallography of eudialyte' **Crystallographia** 17(6) 1119 (in Russian)

20 Chekmarev, A.M., Chizhevskaya, S.V., Masloboev, V.A. et al., (1989) 'Eudialyte as a potential industrial raw material for rare metals' in: '**Chemistry and Technology of Rare and Dispersed Elements**' edited A.A. Kopyrin, Leningrad, 65.

21 Hendrick, Y.B. and Templeton, D.A., (1989) 'Zirconium and Hafnium', **Minerals Yearbook,** USA 1176

22 Strelkov, L.A. and Fillipov, A.P., (1971) 'The investigation of oxidation and leaching processes for uranium compounds in nonaqueous media', **Radiochimia** 13(1) 58.

23 Chizhevskaya, S.V. and Cox, M., (1993) unpublished results

Scope and limitations for application of selectivity in oxidation potential-controlled leaching of metal sulphides

N. L. Piret
Stolberg Consult GmbH, Neuss, Germany
J. F. Castle
RTZ Consultants, Ltd., Bristol, England

ABSTRACT

Generally in the sulphate system, oxidative leaching of metal sulphides is not conducive to selectivity, due to the unselective oxidation of sulphide to elemental sulphur or to sulphate. However, in the chloride system, the ability to control the redox potential at a lower level on account of the stability of the cupric / cuprous ion species, permits the selective dissolution with respect to copper, of metals such as Pb, Ni, Zn, Fe by means of cupric chloride, according to the metathesis reaction, provided that the sulphide mineral is thermodynamically unstable under the conditions of leaching.

In the case of easily oxidisable sulphides, such as the components of copper matte or leadsulphide, under the oxidation potential conditions, at which cuprous sulphide is converted to cupric sulphide, according to the reaction:

$$Cu_2S + CuCl_2 \rightarrow CuS + 2\,CuCl,$$

selectivity of leaching of metals (Me) in respect to copper can be achieved, according to the simplified metathesis reaction

$$MeS + CuCl_2 \rightarrow MeCl_2 + CuS$$

Thereby, the cupric/cuprous chloride leach solution can be regenerated from the covellite formed by oxidation with Cl_2 or HCl and O_2.

In the case of a matte, the selective leaching and regeneration occur simultaneously, according to the overall reaction:

$$Cu_2S.MeS + Cl_2 \rightarrow 2CuS + MeCl_2$$

In the case of less easily oxidisable sulphide minerals, selectivity of leaching of metals with regard to copper can still be achieved by the metathesis reaction at elevated temperature. The oxidation of sulphide to sulphate is prevented by maintaining a high cuprous to cupric ratio in the chloride solution. An industrial application is the Falconbridge chlorine leach process for copper-nickel matte. Based on investigations, potential applications were identified in the field of complex or bulk metal sulphide concentrate treatment, copper-lead matte processing and also chemical converting.

The application of selectivitiy principle in sulphide leaching provides environmental benefits, since the sulphide itself is used for separation and the sulphur is brought in an environmental compatible form.

METALLURGICAL BACKGROUND

One important advantage of hydrometallurgical processing as applied to the winning of base metals, is the selectivity which can be achieved in the separation of base metals from each other and from iron. An adequate level of selectivity is often not achievable in pyrometallurgical processing without incurring metal losses or using multiple recirculations. Amongst other examples, this can be illustrated by the example of lead-copper smelters [1, 2], where there is a large recirculation of lead between the primary smelting and the matte converting step.

In copper-nickel smelters, to overcome the lack of selectivity in copper-nickel separation, methods other than the pyrometallurgical one are commercially applied. For instance, at Falconbridge Nikkelverk (FNV) copper-nickel matte was treated hydrometallurgically by the so-called matte leach process, in which selective dissolution of nickel was achieved by hydrochloric acid dissolution at high acid strength leaving copper as the sulphide in the residue [3].

Oxidative leaching also presents the possibility of achieving selectivity under certain circumstances, particularly between copper and the precious metals on the one hand and other non-ferrous metals and iron on the other. This has been clearly demonstrated by the successful commercial implementation of the chlorine leach process for copper-nickel matte treatment at FNV [4]. However, for kinetic reasons, in the practice of oxidative leaching of the common natural minerals present in sulphide concentrates, the oxidation conditions required are usually in excess of the ones at which selectivity could still be achieved. Therefore, in most of the hydrometallurgical treatment schemes which have been proposed and / or are commercially implemented, the base-metals and iron are transferred to solution unselectively, unless under the specific leaching conditions a particular metal is either precipitated out as a compound of limited solubility (e.g. such as lead in the BHAS process [5, 6] and iron in the low-acid version of the Sherritt Gordon zinc pressure leaching process [7, 8]), or associated with a stable mineral (e.g. such as iron in pyrite [9]). Besides the unleachable gangue the leach residue contains unleached sulphides, minerals, precipitated compounds, precious metals and, depending on the leaching method, elemental sulphur.

Thus, one general and important drawback of almost all of the hydrometallurgical treatment schemes for sulphide concentrates, is the necessity for separate residue treatment in order to recover precious metals or, at least, the gold [10]. Last but not least, the residual presence of leachable substances and unstable heavy metal compounds in the leach residue renders it environmentally more and more difficult to dispose of in a landfill.

These were also identified as problem areas in a recent publication [11] evaluating some of the better-known processes which had been proposed for the hydrometallurgical treatment of copper and complex copper-zinc-lead concentrates by means of oxidative leaching in the chloride system.

However, the application of selectivity for the separation of copper from other metals in sulphides, particularly by the method of oxidation potential controlled leaching, would present the advantage of eliminating the need to generate a residue from which firstly the precious metal must be recovered and which subsequently must be disposed of.

Instead, the generated copper - precious metal-sulphur product can be submitted to the copper smelting process. In the case that primary smelting had already been performed, the submission of a complex matte, such as copper-nickel matte or copper-lead matte, to a selective leach process, would result in an upgraded copper concentrate, which could be treated either by a pyrometallurgical or by a hydrometallurgical process [4].

This paper delineates the conditions under which selectivity in the oxidation potential controlled hydrometallurgical treatment of sulphides, both of concentrates and mattes, is or could be successfully applied.

Thereby, the term "selectivity" is restricted to the metal sulphide compounds. Whereas in the case of copper concentrates the selective leaching would only refer to iron, resulting only in an upgraded P.M.-bearing copper concentrate, the selectivity principle as applied to complex base-metal sulphide concentrates would result in the production of a copper concentrate suitable for smelting without encountering the drawbacks described above. Though recognising the advantages of bulk flotation rather than selective flotation of complex ores [12, 13], the majority of the proposed direct hydrometallurgical processes for the treatment of bulk concentrates do not apply selectivity in their schemes.

Limitations of Selective Dissolution

As outlined in [14], in each type of solution, four conditions have been identified under which a sulphide mineral might decompose:

⇒ oxidising versus reducing conditions,

⇒ acidic versus alkaline conditions.

With regard to the sulphur component, the decomposition products are:

⇒ under oxidising conditions: elemental sulphur or sulphate

⇒ under reducing conditions: metalsulphide, sulphide ion, hydrogen sulphide

The stability areas of sulphur in the S-H_2O system as a function of oxidation potential and acidity of solution at three different temperatures (25, 100, 150 °C) are represented by the well-known E_h-pH diagrammes of Figures 1A, 1B and 1C [15].

The reactions therein considered are as follows:

<u>Reactions:</u>

1) $HSO_4^- = H^+ + SO_4^{--}$
2) $H_2S_{(aq)} = H^+ + HS^-$
3) $HS^- = H^+ + S^{--}$
4) $HSO_4^- + 7H^+ + 6e = 4H_2O + S$
5) $SO_4^{--} + 8H^+ + 6e = 4H_2O + S$
6) $S + 2H^+ + 2e = H_2S$
7) $S + H^+ + 2e = HS^-$
8) $SO_4^{--} + 9H^+ + 8e = HS^- + 4H_2O$
9) $SO_4^{--} + 8H^+ + 8e = S^{--} + 4H_2O$
10) $SO_4^{--} + 10H^+ + 8e = H_2S_{(aq)} + 4H_2O$
11) $HSO_4^- + 9H^+ + 8e = H_2S_{(aq)} + 4H_2O$
12) $O_2 + 4H+ + 4e = 2H_2O$
13) $2H^+ + 2e = H_2$

FIGURE 1: System Sulphur - Water: E_h - pH Diagrams at 25, 100 and 150 °C [15]

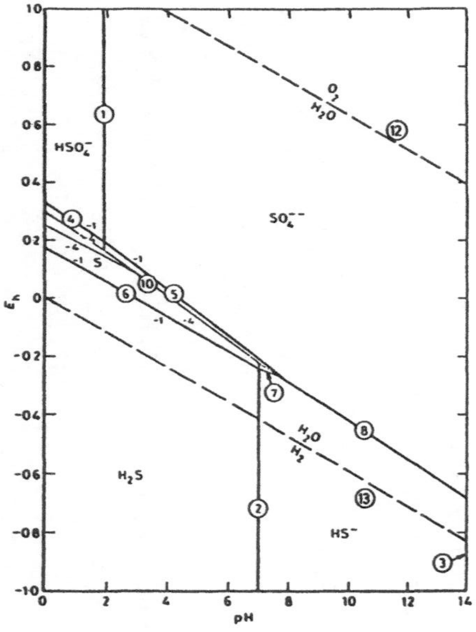

Fig. 1A: Temperature 298 K

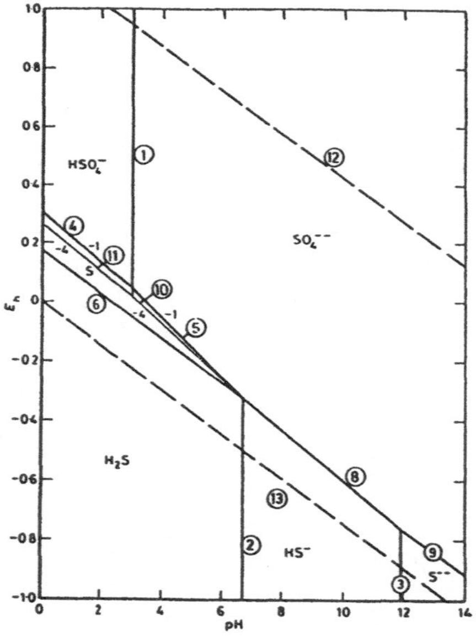

Fig. 1B: Temperature 373 K

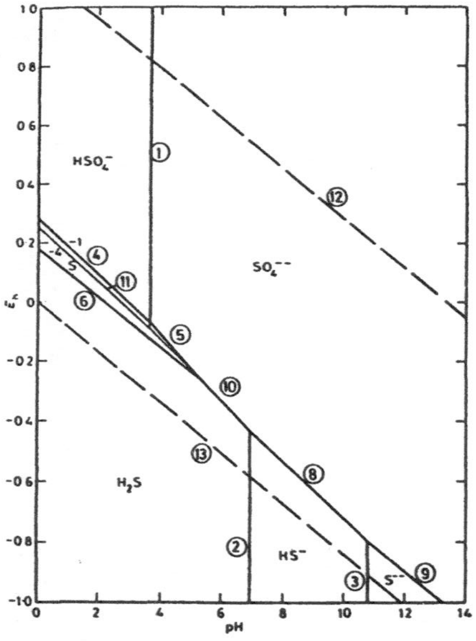

Fig. 1C: Temperature 423 K

<u>Conditions:</u>	activity of sulphur-bearing ions:	10^{-1} and 10^{-4} M
	pressure:	1 atm

It can be seen from these diagrams that:

1. above a given potential, which depends on the other conditions, only sulphate is stable.
2. elemental sulphur can only be formed in the acidic pH region.
3. the area of sulphate stability extends to lower potential at increasing pH.
4. the region of sulphate stability extends to lower potential at rising temperature.
5. the area of stability of elemental sulphur shrinks with rising temperature.

Accordingly, oxidative leaching processes at elevated temperatures, e.g. in excess of 150 °C [16, 17, 18] result in sulphate formation over the entire pH range. Oxidative leaching processes in alkaline media, e.g. ammoniacal leaching processes [19], result in oxidation to sulphate, even in the lower temperature range.

In oxidative leaching processes with sulphate formation, there are the important drawbacks that, ultimately, the sulphate has to be neutralised, usually by lime, which is unattractive from both the economic and ecological point of view.

Stability areas of sulphide minerals are presented in various publications in the form of E_h - pH diagrams [14, 15, 20]. In theory, selective leaching could be achieved if such leaching conditions (potential, pH, temperature) were selected under which one of the sulphide minerals is thermodynamically stable whereas the others are not.

However, in practice, the kinetics of decomposition of a particular mineral in a thermodynamically unstable region are often found to be so slow, so that, in order to obtain practical rates of dissolution, less selective conditions have to be selected, for the sake of faster rates. Conversely, selective leaching of a more reactive mineral can also be achieved under conditions of thermodynamic instability of the other mineral whose leaching rate is very low.

The factors which can influence the leaching kinetics are multiple, as was discussed in a comprehensive manner in [20], and thermodynamic factors are often less important in practice than these kinetic factors.

From the practical point of view it is a well-known fact that some distinct sulphide minerals are more easily-leachable than others. Several investigators have explained their findings in differences of leachability to be associated with the distances in crystal structure lattices [21]. Others have found the semiconducting properties to be decisive for the dissolution characteristics, whereby p-type semiconducting minerals are easily oxidised under formation of elemental sulphur, whereas n-type semiconductors are not [22]. Very often, however, the refractory nature of the sulphide mineral increases with the atomic ratio (S + As + Sb)/Me, whereby Me corresponds to Fe, Cu, Zn, Ni, Co, etc. (e.g. pyrite, arsenopyrite, etc.). The latter ratio, in fact, does not constitute an explanation of the behaviour, but is rather a reference for evaluation of the leachability characteristics. In practice, the following points are important with regard to achieving selectivity in oxidative leaching of sulphides:

1. Under the leaching conditions where the sulphide is converted to sulphate, in general no selectivity can be achieved [16, 17, 18] regardless of the type of mineral.

2. Under the leaching conditions where sulphide is converted to elemental sulphur, a certain degree of selective leaching can be achieved, particularly between minerals

with a lower ratio (S+As+Sb)/Me, which are dissolved and minerals with a higher ratio, which remain largely unaffected. An example is chalcopyrite or marmatite, (Zn, Fe)S, both having a ratio of 1, which can be selectively dissolved versus pyrite, which has a ratio of 2.

Under oxidative leaching conditions which result in the formation of S°, pyrite is known to be unreactive. Its behaviour in the ferric and the cupric chloride leaching system [e.g. 23] has been extensively discussed. However, minerals with lower ratio do leach well in such systems albeit at quite different leaching rates. This was shown for Cu-Ni concentrates [24], and for complex Cu-Zn-Pb concentrates [25, 26], and for a variety of sulphide minerals leached under the same conditions (see Fig. 2) [27].

In general, in oxidative leaching systems which result in the formation of elemental sulphur, there is usually no opportunity to achieve selectivity of leaching as far as the main base metals and iron are concerned, even though, on account of the different sulphur to metal ratio, selectivity versus some of the minerals (e.g. pyrite) is given. An exception is reported to be the oxidative leaching to S° of galena, (PbS), in the presence of chalcopyrite [28] or from complex concentrates at 95 °C at the oxidation potential of 300 mV [29].

It is, therefore, concluded that, in the practice of oxidative leaching of sulphides there is virtually no scope to achieve selectivity between the various base metals and iron, if thereby the sulphide is converted to sulphate or elemental sulphur. However, there is considerable scope for selectivity under the conditions in which the metal sulphide is maintained in the sulphide form as discussed hereafter.

Scope for Selective Dissolution

In order not to oxidise sulphide in a particular leaching system, in which the objective is to selectively leach one metal away from the other, the oxidation potential is not allowed to exceed a certain value which obviously depends on such variables as leach solution composition, temperature, pH, etc.

The value of the redox potential, which should not be exceeded, corresponds to the one at which the oxidation to elemental sulphur occurs. Thus, it is important that a leaching system is selected, in which the control of the redox potential can easily be performed at a low level. In oxidative leaching systems for sulphides, the two most common systems are:

- the system Fe^{3+}, Fe^{2+}, Cu^{2+}
- the system Cu^{2+}, Cu^{+}, Fe^{2+}

In the sulphate system the redox potential is controlled by the ratio Fe^{3+}/Fe^{2+} due to the low equilibrium concentration of the cuprous ion. In the chloride system the redox potential is either controlled by the Fe^{3+}/Fe^{2+} or by the Cu^{2+}/Cu^{+} ratio. Figure 3 shows the dependencies of the respective redox potentials on either of the ratios, in the chloride-free system and in the chloride containing system (the latter only in the case of Cu^{2+}/Cu^{+}).

FIGure 2: Oxidative Leaching of Sulphide Minerals by Means of $FeCl_3$ at Atmosph. Conditions (90 °C) [27]

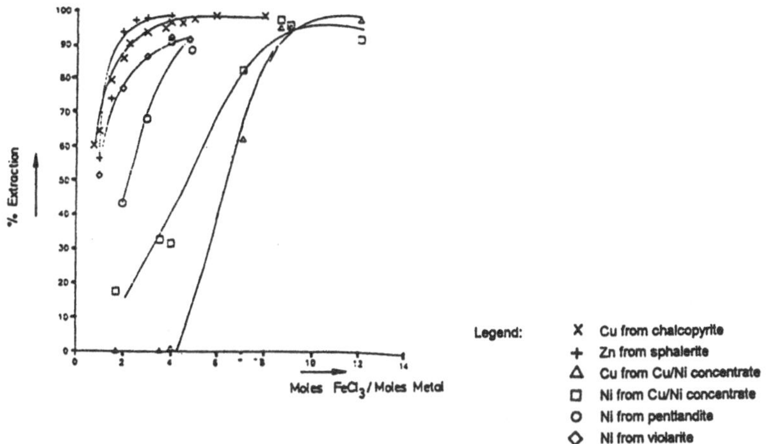

FIGURE 3: Dependency of Redox Potential from Fe^{3+}/Fe^{2+} and Cu^{2+}/Cu^{+} ratio (Temp.: 25 °C) [30]

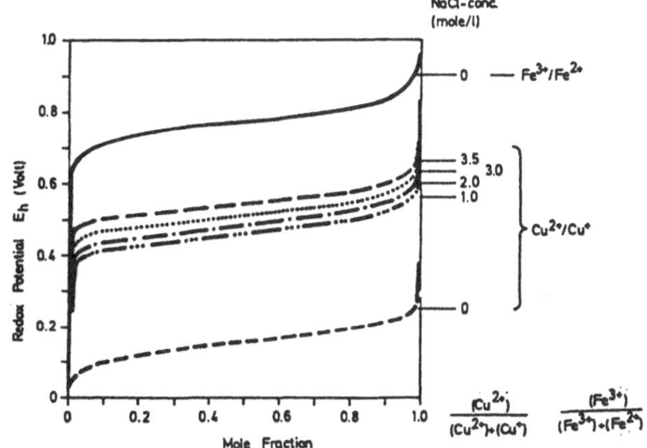

Legend: Data obtained from [30]:

NaCl-Concentration (mole/l)	0	1	2	3	3.5
Molefraction of Cu^{2+}	Redox Potential: E_h (mV)				
0.09	94	415	430	465	500
0.25	125	430	455	480	515
0.50	153	460	485	515	565
0.75	181	490	510	535	555
0.90	212	530	545	575	600

Thus, the chloride is preferable to the sulphate system for achieving selectivity in leaching because:

- the control of potential may be achieved more easily.
- and at a lower level, which is required for the purpose of selective leaching

In Figure 4 the standard oxidation potentials for the Cu^{2+}/Cu^{+} and Fe^{3+}/Fe^{2+} systems are presented (at 25°C) as a function of the total chloride concentration. Figure 5 represents the stability area of the chloro-complexes of Fe, Cu, Zn, Pb, Ag (at 25°C), at a given pH. It is interesting to note that at sufficiently high chloride concentrations the Cu^{2+}/Cu^{+} system can become more oxidising than the Fe^{3+}/Fe^{2+} system.

Whereas sulphide leaching applications in the chloride system have been extensively investigated [11, 33, 34, 35], despite the obvious scope for selectivity, there are only a few commercial applications of which the Falconbridge chlorine leach process [4, 36] is the most representative.

SELECTIVE LEACHING REACTION MECHANISMS

Simple Sulphide Systems (Mattes, Precipitates)

Overall Reaction Scheme:

It has been observed that under the conditions of oxidation potential at which the conversion of cuprous sulphide to cupric sulphide by means of cupric chloride takes place according to the reaction:

$$Cu_2S + CuCl_2 \rightarrow CuS + 2CuCl \quad \text{(1)},$$

a thermodynamically less stable mineral MeS can be leached selectively by means of the general and simplified exchange or metathesis reaction:

$$MeS + CuCl_2 \rightarrow MeCl_2 + CuS \quad \text{(2)}$$

Under these conditions of oxidation potential, there is clearly no formation of elemental sulphur, which, under certain circumstances, is known to inhibit the further reaction, or slow down the rate of reaction [37].

Whereas reaction (1) represents the oxidation of cuprous sulphide to cupric sulphide, reaction (2) proceeds due to the lower solubility of the cupric sulphide as compared to MeS.

Moreover, at the oxidation potential under consideration, the exchange reaction between the generated cuprous chloride and the less stable sulphide mineral, according to the exchange reaction:

$$MeS + 2CuCl \rightarrow MeCl_2 + Cu_2S \quad \text{(2 ')}.$$

The sum of reactions (1) and (2) represents the overall selective leaching scheme reaction (3):

$$Cu_2S\,.\,MeS + 2\,CuCl_2 \rightarrow 2\,CuS + MeCl_2 + 2CuCl \quad \text{(3)}$$

does not occur due to the fact that the potential, at which reaction (1) proceeds, is kept constant by maintaining an appropriate value for the ratio Cu^{2+}/Cu^{+}.

FIGURE 4: Standard Potentials in the System Cu^{2+}/Cu^{+} and Fe^{3+}/Fe^{2+} as a Function of the Total Chloride Concentration, [31]

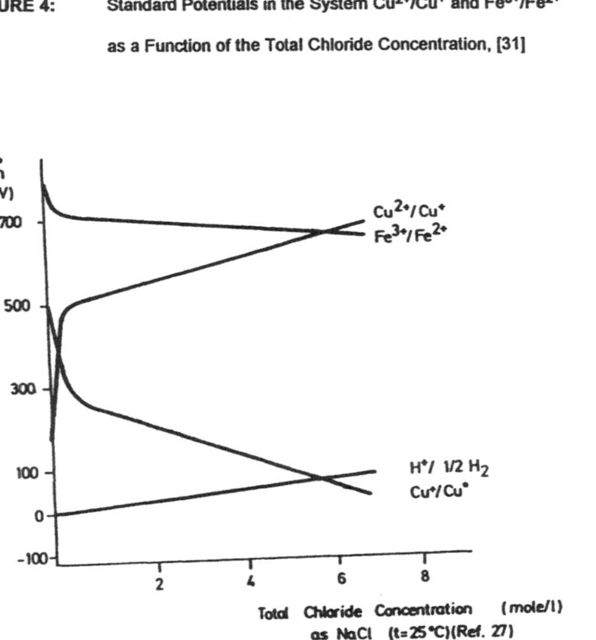

FIGURE 5: The Potential-Log (Cl^-) Diagram at 25 °C [32]

- Concentrations of Fe, Cu ions: 0.1 M
- Concentration of other elements: 0.001 M

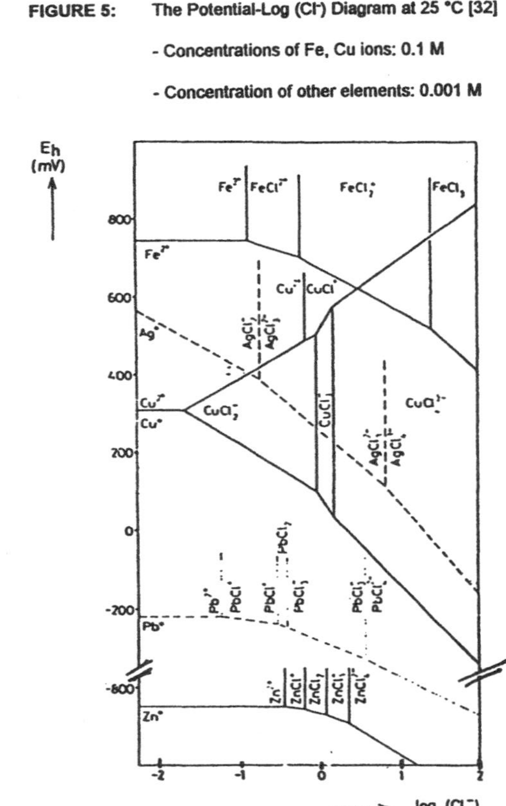

Conversion of Cuprous Sulphide to Covellite:

The preferential oxidation of a cuprous sulphide mineral species (chalcocite, Cu_2S, djurleite, $Cu_{1.96}S$ or digenite, $Cu_{1.8}S$) to covellite (CuS), has been studied extensively, particularly in the sulphate system [38, 39, 40, 41], but also in the chloride system [42].

In the sulphate system the reaction is generally represented by:

$$Cu_2S + H_2SO_4 + 1/2\ O_2 \rightarrow CuS + CuSO_4 + H_2O \qquad (4)$$

whereas in the chloride system the conversion is given by the reaction (1).

In the ammoniacal system the conversion of cuprous sulphide to covellite occurs under atmospheric conditions in a similar way as equation (4). This reaction represents the basis of the new process being implemented at La Escondida [43].

The complete conversion of a cuprous sulphide mineral to covellite is more easily achieved in the chloride system than in the sulphate system, as is shown from the parameters in the following Table 1:

TABLE 1: Conditions to Convert Cuprous Sulphide to Covellite [44]

System	Sulphate	Chloride
Reaction equation	(4)	(1)
Temperature (°C)	110-115	70
Potential E_h (mV)	600	575
O_2-partial pressure (atm)	10	atmospheric conditions
pH	1.5	1.5
Retention time (hr)	0.5	1.0

The Cu^{2+}/Cu^{+} ratio corresponding to a particular potential will obviously vary with the temperature, the total chloride concentration, the type and concentrations of the chloro-complex- forming metal ions.

The utilisation of exchange reactions as a process step in hydrometallurgical processing was first suggested in [45].

Exchange Reactions

It has been stated earlier that, to achieve good selectivity, the exchange reaction (or metathesis reaction) between the cupric ion and the metalsulphide should take place under the oxidation potential conditions at which cuprous sulphide is converted to covellite by means of cupric chloride, reaction (1).

The exchange reaction is characterised by the fact that it takes place at the surface of the mineral MeS. This means that the precipitated coppersulphide is gradually coating the original MeS particle which slows down the reaction because of the increasing diffusion path [46]. Hence, factors which will enhance the completeness of the exchange reaction are:

- fineness of the material
- exchange reaction temperature

In the case of copper matte, in which the copper is present as chalcocite (Cu_2S), the recrystallisation to covellite (CuS) appears to be essential to achieve separation between copper and the other metal (Me).

Two examples of actual commercial practice illustrate the above:

⇒ at FNV an autoclave step, operating at 145 °C, was incorporated in order to improve the selective leaching of nickel from Cu-Ni matte in the 100 % chloride system [36]

⇒ at INCO, in the 100 % sulphate system, the exchange reaction between coppersulphate and nickelsulphide in Cu-Ni matte was designed to take place in an autoclave at a temperature of 160 °C [47].

Natural Minerals

Common natural sulphide minerals, such as chalcopyrite and sphalerite, however, require considerably higher temperatures to render them conducive to the exchange process with the cupric ion. In order to achieve adequate exchange, temperatures at least above 150 °C, and generally in the range of 180 to 200 °C, are found to be required.

It has been shown that, at increasing temperature, the stability area of the cuprous ion in the sulphate system increases dramatically as shown in the E_h-pH diagram of Figure 6. This means that the cupric ion becomes oxidising and, at higher temperature, will oxidise sulphide to sulphate.

Reactions which have been described and were the basis of several proposed processes for copper concentrates [48, 49, 50], and zinc concentrates [51], and copper zinc concentrates [52] in the sulphate system are represented by the reaction equations (5) and (6), respectively. Thereby, the equation (6) represents the simplified version of equation (6').

$$6\ CuSO_4 + 3\ CuFeS_2 + 4\ H_2O \rightarrow Cu_9S_5 + 3\ FeSO_4 + 4\ H_2SO_4 \quad (5)$$

$$8\ CuSO_4 + 5\ ZnS + 4\ H_2O \rightarrow 4\ Cu_2S + 5\ ZnSO_4 + 4\ H_2SO_4 \quad (6)$$

$$9\ CuSO_4 + 6\ ZnS + 4\ H_2O \rightarrow Cu_9S_5 + 6\ ZnSO_4 + 4\ H_2SO_4 \quad (6')$$

The sulphide reaction product appears in all cases to be digenite (Cu_9S_5) under the above conditions. The exchange reactions at elevated temperature in the sulphate system have the considerable drawback of forming acid, which requires lime neutralisation and disposal of sulphate in the form of gypsum as the ultimate step.

In [53, 54] a process is described to up-grade copper concentrates based on the same principle. In [55], it has been proposed to carry out the exchange process, in the case of chalcopyrite concentrates, in the chloride system according to the reactions:

$$CuFeS_2 + 2CuCl \rightarrow FeCl_2 + CuS + Cu_2S \quad (7)$$

$$CuFeS_2 + CuCl_2 \rightarrow FeCl_2 + 2CuS \quad (8)$$

at a temperature of preferably 150 °C.

FIGURE 6: System Sulphur - Copper - Water: E_h-pH Diagrams at 25, 100 and 150 °C [15]

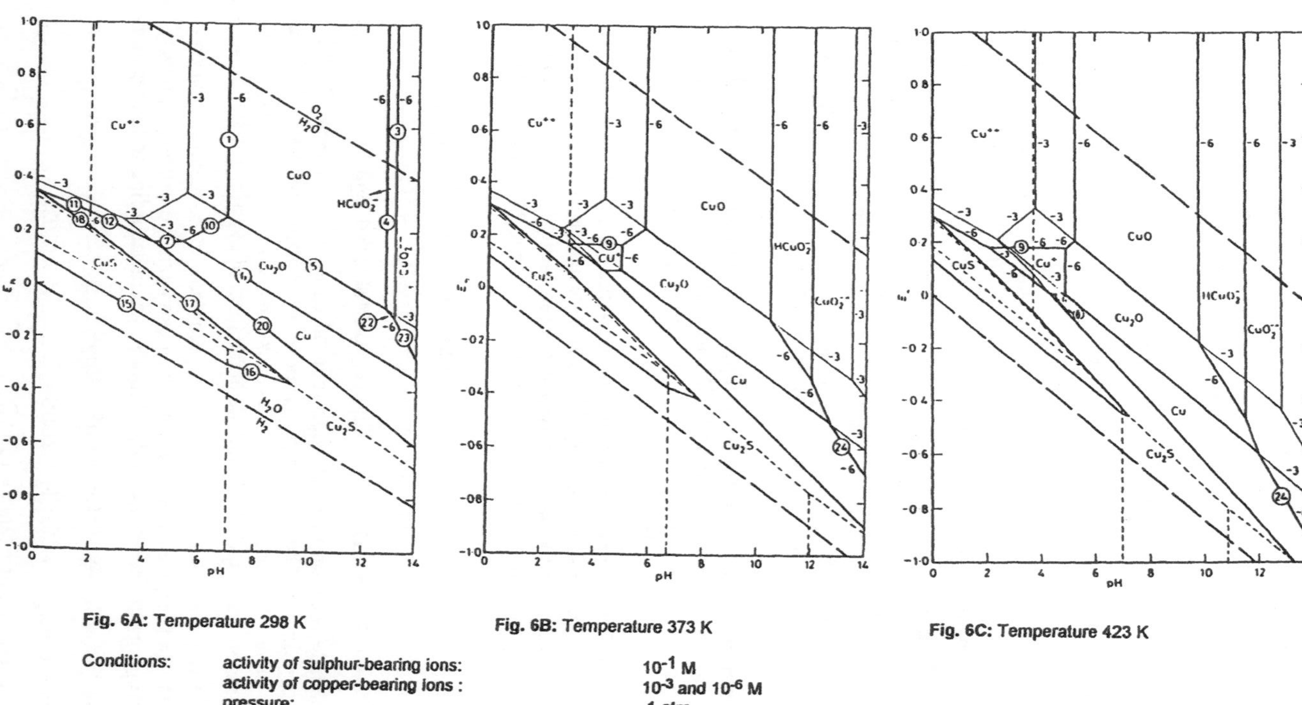

Fig. 6A: Temperature 298 K

Fig. 6B: Temperature 373 K

Fig. 6C: Temperature 423 K

Conditions:		
	activity of sulphur-bearing ions:	10^{-1} M
	activity of copper-bearing ions :	10^{-3} and 10^{-6} M
	pressure:	1 atm

In order to prevent the oxidation of sulphide to sulphate and the simultaneous generation of sulphuric acid, the ratio Cu^{2+}/Cu^{+} must thereby be selected low enough in accordance with the temperature. In [56], a predominant chloride system was selected to perform the exchange reactions at 180 °C in order to produce pure copper concentrates suitable for smelting from complex Cu-Zn-Pb concentrates. Thereby, the ratio of Cu^{2+}/Cu^{+} is so low that oxidation of sulphide to sulphate with the simultaneous generation of sulphuric acid is suppressed. The latter process has the objective of producing a pure copper concentrate containing the precious metals and the sulphur content of the original concentrate, in a closed, predominantly chloridic circuit from complex copper-zinc-lead concentrates.

The oxidation of sulphide to sulphate due to the oxidising power of cupric ion at elevated temperature can be suppressed in the presence of a reductant, such as Cu^0 and Fe^0 [57, 58] and SO_2 [59, 60]. However, the addition of a reductant increases the operating costs due to the cost of the reductant, the need to remove iron or to neutralise the sulphuric acid formed.

Concluding Remarks on the Selective Leaching Reaction Scheme

Summarising, it can be concluded that selective leaching can be achieved in the chloride system under the conditions presented in Table 1, at which cuprous sulphide is oxidised to covellite. This is the case with simple sulphide minerals such as galena, or the ones present in copper mattes:

- troilite (FeS) in Cu-Fe matte
- galena (PbS) in Cu-Pb matte
- haezlewoodite (Ni_3S_2) in Cu-Ni matte

As demonstrated in practice [36], high reaction temperature is beneficial to the selectivity, because of its effect on the kinetics of the exchange reactions.

With less reactive sulphide minerals, such as chalcopyrite, pentlandite, etc., a pretreatment step called activation is required in order to make them amenable to the method of selective leaching described here. In the seventies, activation of sulphide concentrates was investigated intensively, with the objective of improving leaching characteristics during the subsequent oxidative leaching. A comprehensive review of activation processes for chalcopyrite concentrates is presented in [9]. These processes were usually based on a thermal pretreatment in the presence of an element with high affinity for the less-stable part of the sulphur (such as H_2, Cu°, Fe°, CaO, etc). These will not be discussed here because they are beyond the scope of the present paper.

The separate thermal activation pretreatment can be overcome, if the exchange reactions are conducted in an autoclave at elevated temperature (180 - 200°C) in a predominant chloridic solution with a high ratio of cuprous to cupric ions.

A condition to achieve the selective leaching is thus the availability of copper chloride solution of suitable cupric to cuprous ratio.

In the case of simple sulphides, such as the ones associated with copper mattes, or minerals such as galena, the Cu^{2+}/Cu^{+} ratio may be relatively high since the exchange takes place with the cupric ion at moderate temperatures.

In the case of less reactive minerals at the elevated temperatures required to perform the exchange reaction, a very low Cu^{2+}/Cu^{+} ratio is required in order to prevent the oxidation of sulphide to sulphate and the simultaneous formation of sulphuric acid.

REGENERATION OF COPPER CHLORIDE SOLUTION

The regeneration of the copper chloride solution, which is required for the selective leaching can be achieved by relatively simple means, based on the treatment of the covellite or digenite concentrate, obtained from the selective leaching process.

For the sake of simplification digenite will be written as chalcocite (Cu_2S) in the chemical reaction equations.

It is well known that, above a certain oxidation potential, simple sulphides, i.e. covellite and digenite, are easily amenable to oxidative leaching by $FeCl_3$ or $CuCl_2$ with formation of elemental sulphur. Of all possible reactions the most important regeneration reaction in this scheme is reaction (9):

$$CuS + 2FeCl_3 \rightarrow CuCl_2 + 2FeCl_2 + S^{\circ} \quad \text{(9)}$$

or at lower oxidation potential with formation of covellite:

$$Cu_2S + CuCl_2 \rightarrow CuS + 2CuCl \quad \text{(1)}$$

The end composition of the regenerated copper chloride solution with regard to its Cu^{2+}/Cu^{+} ratio can be controlled as desired by adjusting the ratio of sulphides to the oxidants.

To maintain the desired level of oxidation potential, required for selective leaching, the Cu^{2+}/Cu^{+} ratio can be obtained by controlled addition of one of the more common oxidising agents: $FeCl_3$, Cl_2, $HCl{+}O_2$. These are normally being regenerated within the process scheme during the metal (Me) recovery or elimination step.

In the case of a matte, consisting of the two compounds MeS and Cu_2S in the stoichiometric combination $MeS.Cu_2S$, it is interesting to note that selective leaching and the solution regeneration may be accomplished simultaneously, as shown hereafter:

$$Cu_2S + CuCl_2 \rightarrow CuS + 2CuCl \quad \text{(1)}$$
$$2CuCl + Cl_2 \rightarrow 2CuCl_2 \quad \text{(10)}$$
$$MeS + CuCl_2 \rightarrow MeCl_2 + CuS \quad \text{(2)}$$

$$\text{Total: } Cu_2S.MeS + Cl_2 \rightarrow 2CuS + MeCl_2 \quad \text{(11)}$$

The Cl_2 is recovered from the metalchloride ($MeCl_2$) solution by electrolytic dissociation. This is illustrated later by means of two examples, one based on commercial practice (FNV, [4]), and one developed and reported in this work. Because of the simultaneous occurrence of the exchange reaction and the oxidation reaction by means of chlorine, this particular process has also been denominated "Chlorine Leaching" [61].

REDUCTION OF CUPRIC SOLUTION

Reduced copper chloride solutions are required in order to conduct the exchange reactions at the low cupric/cuprous ratio.

The reduction of a cupric chloride solution to the cuprous state becomes knowingly more difficult with lower Cu^{2+}/Cu^{+} ratio for two reasons:

⇒ the requirement of a stronger reductant,

⇒ the difficulty in preventing air reoxidation.

Disregarding the elements in the metallic state, which have the highest reducing capacity, but which are only known to be present in sulphur-deficient mattes, the reducing capacity of sulphides decreases with increasing sulphur to metal ratio.

In the case of digenite the reduction of the cupric chloride solution at low potential takes place according to reaction (1):

$$CuCl_2 + Cu_2S \rightarrow CuS + 2CuCl \quad (1)$$

Copper chloride solutions with a high copper concentration and a low Cu^{2+}/Cu^{+} ratio can essentially be obtained by application of one of the three above reactions, provided that:

⇒ the temperature of reduction is moderately high (in excess of 80 - 90°C),

⇒ an excess of reductant is used (e.g. 150 - 200 %),

⇒ and, last but not least, the reoxidation with air is effectively prevented.

PRECIPITATION OF COPPER FROM SOLUTION

It is possible that, after the selective leaching step, the metalchloride solution ($MeCl_2$) still contains a too high copper concentraction.

The precipitation of copper from that metal chloride solution can, in principle, be accomplished with the sulphide material itself, provided it is made available in stoichiometric excess for copper precipitation. The sulphide thereby enriched in copper is subsequently returned to the selective leaching step. The scheme would in that case consist of a two-stage counter current process.

In the case of sulphide concentrates elevated temperatures are required, e.g. 130 - 150 °C. The copper precipitation process was described by Dowa Mining [46].

The two-stage countercurrent process can be simplified significantly in the case of mattes. For instance in the case of Cu-Ni matte, the presence of a metallic phase [4] obviates the need to recirculate the material obtained, so a co-current two-stage process is used.

In the case of Cu-Pb matte, the copper concentration in the leach solution from the selective lead leach is sufficiently low so that no additional copper precipitation step is needed.

It should be noted that, while conducting the selective leach with complex Cu-Zn-Pb concentrates at elevated temperatures (180-200°C) in the chloride system, the leach solution contains less than 1 g/l copper [56].

COMMERCIAL APPLICATIONS OF SELECTIVE LEACHING AND OPPORTUNITIES

Copper Mattes

As extensively discussed before, selective leaching in the chloride system can be applied successfully on mattes:

⇒ Copper-nickel matte: as commercially applied at Falconbridge Nikkelverk, Norway,

⇒ Copper-lead matte: as outlined in the present work,

⇒ Copper-iron matte: as a replacement of pyrometallurgical converting; for this reason also called "chemical converting".

The flowsheet in simplified form for the selective leaching of nickel from Cu-Ni matte as practised at FNV, is presented in Figure 7. The metallurgical reactions are shown in Table 2. The data are selfexplanatory.

The case of the copper-lead matte is represented in Figure 8 and Table 3. The high selectivity obtained and the moderate leaching conditions required for the separation of lead and copper should provide enough incentive to future investigate this approach, despite the fact that lead must be recovered from $PbCl_2$-crystals. A comparison with the commercial BHAS - hydrometallurgical lead matte process in the mixed sulphate-chloride solution is given in Figure 9.

Complex Base Metal Sulphide Concentrates

In the case of complex concentrates, the present approach, involving autoclaving at 180 - 200 °C in the chloride system, is technologically much more complicated. The complete flowsheet in a closed, predominantly chloride circuit is presented in Figure 10, and explained with the help of the chemical reactions of Table 4 as well as the metallurgical balance of Table 5.

The metallurgical results show that good separation between copper, sulphur and the precious metals with respect to the other non-ferrous metals Zn, Pb and part of the iron is possible in only one step. The recovery of the metals from the leach solution can be considered as known technology and their recovery should not be limited to the approach showin in the flowsheet of Figure 10.

The digenite concentrate obtained from the selective leaching/exchange process is stagewise used for the copper chloride solution reduction and regeneration. The finally obtained metal-bearing copper concentrate contains the pyrite content of the original concentrate as well as elemental sulphur. This product is amenable to copper smelting after elemental sulphur removal or even to hydrometallurgical processing, should the latter prove to be more advantageous.

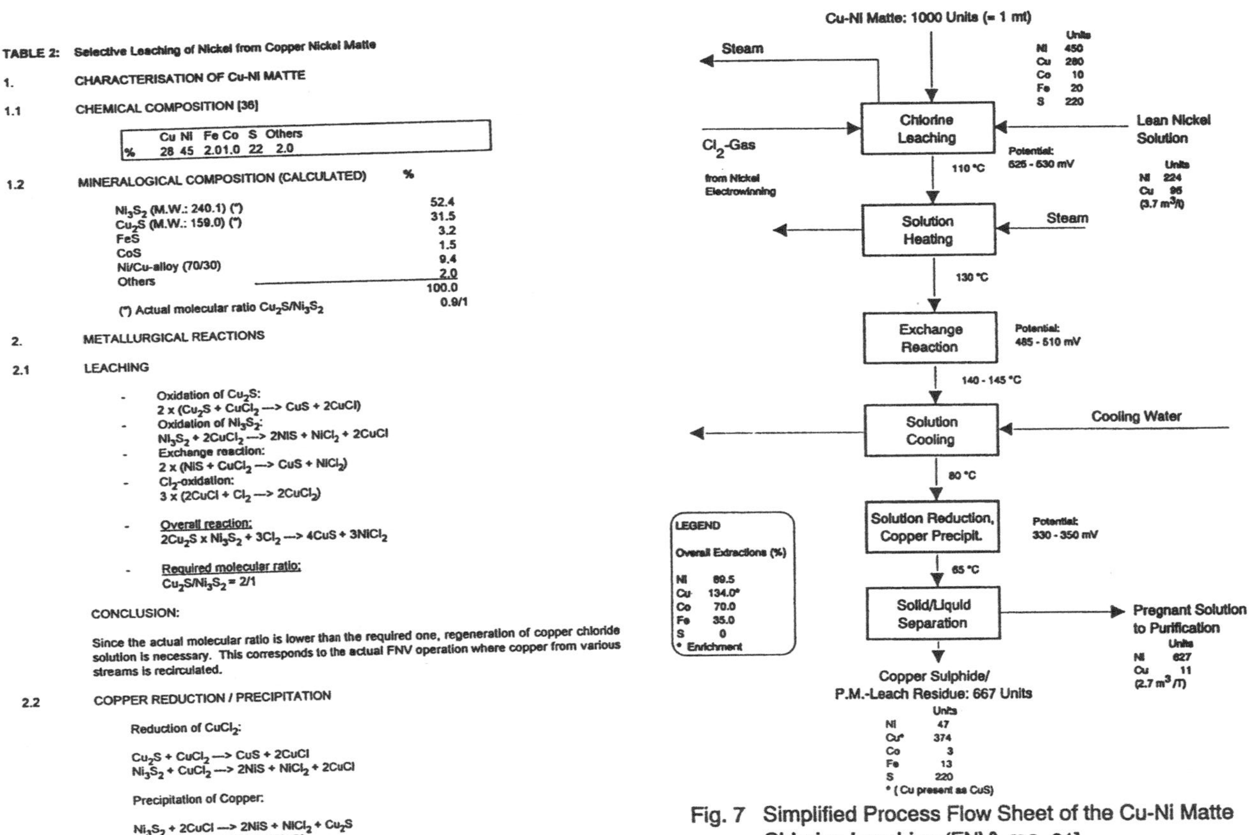

TABLE 2: Selective Leaching of Nickel from Copper Nickel Matte

1. CHARACTERISATION OF Cu-Ni MATTE

1.1 CHEMICAL COMPOSITION [36]

	Cu	Ni	Fe	Co	S	Others
%	28	45	2.0	1.0	22	2.0

1.2 MINERALOGICAL COMPOSITION (CALCULATED)

	%
Ni_3S_2 (M.W.: 240.1) (*)	52.4
Cu_2S (M.W.: 159.0) (*)	31.5
FeS	3.2
CoS	1.5
Ni/Cu-alloy (70/30)	9.4
Others	2.0
	100.0
(*) Actual molecular ratio Cu_2S/Ni_3S_2	0.9/1

2. METALLURGICAL REACTIONS

2.1 LEACHING

- Oxidation of Cu_2S:
  $2 \times (Cu_2S + CuCl_2 \longrightarrow CuS + 2CuCl)$
- Oxidation of Ni_3S_2:
  $Ni_3S_2 + 2CuCl_2 \longrightarrow 2NiS + NiCl_2 + 2CuCl$
- Exchange reaction:
  $2 \times (NiS + CuCl_2 \longrightarrow CuS + NiCl_2)$
- Cl_2-oxidation:
  $3 \times (2CuCl + Cl_2 \longrightarrow 2CuCl_2)$
- Overall reaction:
  $2Cu_2S \times Ni_3S_2 + 3Cl_2 \longrightarrow 4CuS + 3NiCl_2$
- Required molecular ratio:
  $Cu_2S/Ni_3S_2 = 2/1$

CONCLUSION:

Since the actual molecular ratio is lower than the required one, regeneration of copper chloride solution is necessary. This corresponds to the actual FNV operation where copper from various streams is recirculated.

2.2 COPPER REDUCTION / PRECIPITATION

Reduction of $CuCl_2$:

$Cu_2S + CuCl_2 \longrightarrow CuS + 2CuCl$
$Ni_3S_2 + CuCl_2 \longrightarrow 2NiS + NiCl_2 + 2CuCl$

Precipitation of Copper:

$Ni_3S_2 + 2CuCl \longrightarrow 2NiS + NiCl_2 + Cu_2S$
$Me^{\circ} + 2CuCl \longrightarrow 2Cu^{\circ} + MeCl_2$

Fig. 7 Simplified Process Flow Sheet of the Cu-Ni Matte Chlorine Leaching (FNV) [36, 61]

TABLE 3: Selective Leaching of Lead from Copper-Lead Matte [44]

1. CHARACTERISTICS OF COPPER-LEAD MATTE

1.1 CHEMICAL COMPOSITION

	Cu	Pb	S
%	40	40	17

1.2 MINERALOGICAL COMPOSITION (%)

- Chalcocite : 50
- Galena : 46
- Actual molecular ratio : $Cu_2S/PbS = 1.6/1$

2. METALLURGICAL REACTIONS:

- Oxidation of Cu_2S:

  $Cu_2S + CuCl_2 \longrightarrow CuS + 2CuCl$
- Exchange reaction:

  $PbS + CuCl_2 \longrightarrow PbCl_2 + CuS$
- Cl_2 - oxidation:

  $2CuCl + Cl_2 \longrightarrow 2CuCl_2$
- Overall reaction:

  $Cu_2S.PbS + Cl_2 \longrightarrow 2CuS + PbCl_2$
- Required molecular ratio:

  $Cu_2S/PbS = 1/1$

CONCLUSION:

The actual molecular ratio is larger than the required one, hence, there is no necessity to regenerate a copper solution.

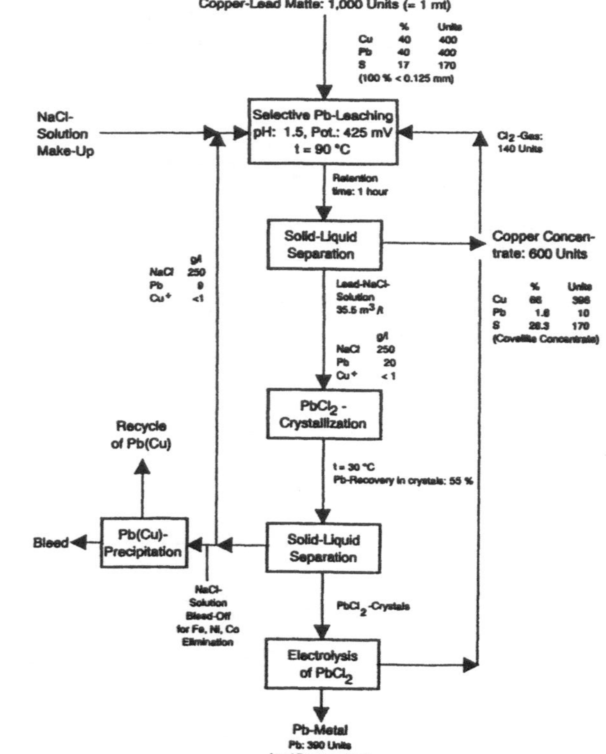

Fig. 8: Flowsheet for the Selective Leaching of Lead from Cu-Pb Matte [44]

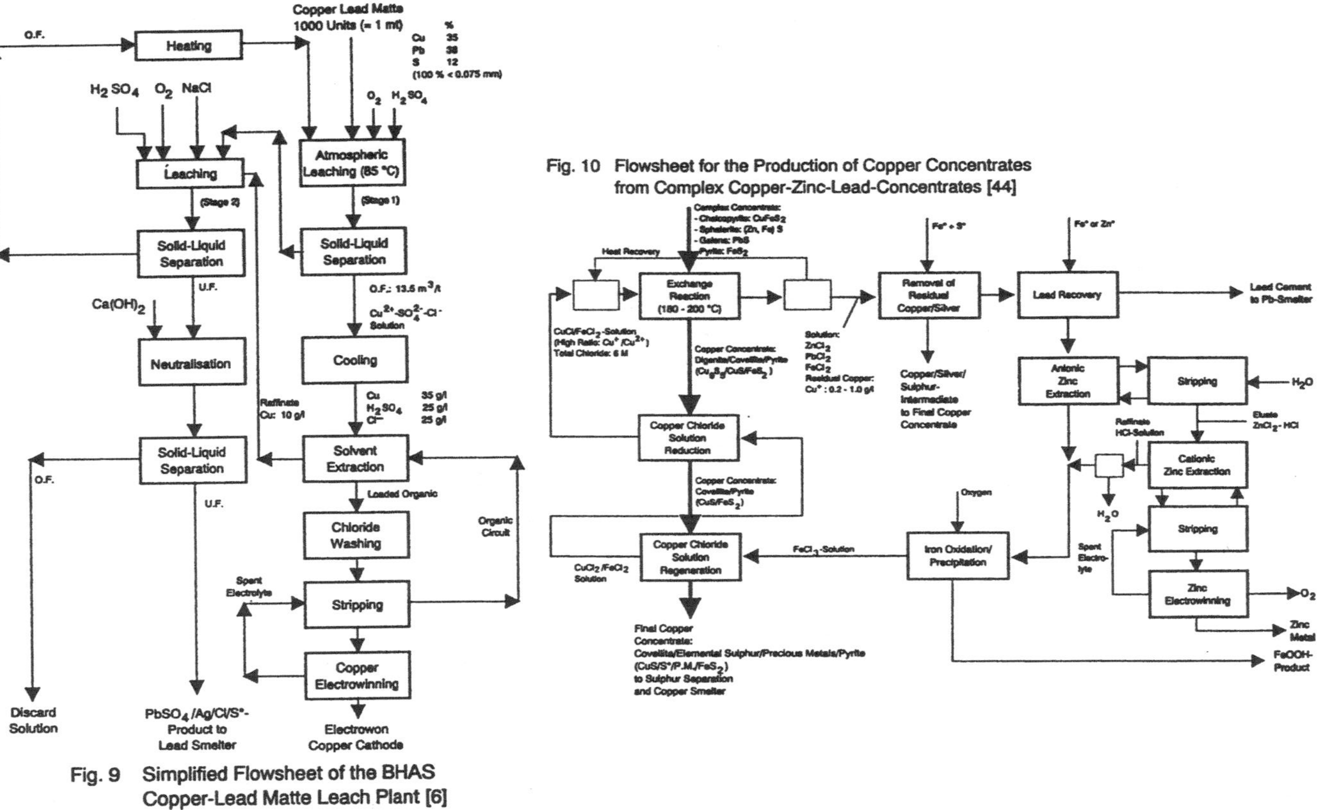

Fig. 9 Simplified Flowsheet of the BHAS Copper-Lead Matte Leach Plant [6]

Fig. 10 Flowsheet for the Production of Copper Concentrates from Complex Copper-Zinc-Lead-Concentrates [44]

TABLE 4: Production of Copper Concentrates from complex Copper-Zinc-Lead Concentrates : Reaction Schemes

1. EXCHANGE-REACTIONS: (180 - 200°C)

   $CuFeS_2 + 2CuCl \longrightarrow CuS + Cu_2S + 2FeCl_2$

   $(Zn, Fe)S + 2CuCl \longrightarrow (Zn, Fe)\,Cl_2 + Cu_2S$

   $PbS + 2CuCl \longrightarrow PbCl_2 + Cu_2S$

   $FeS_2 \longrightarrow FeS_2$

2. REMOVAL OF RESIDUAL COPPER (SILVER):

   $2CuCl + Fe° + S° \longrightarrow Cu_2S + FeCl_2$

3. LEAD RECOVERY:

   $PbCl_2 + Fe°(Zn°) \longrightarrow Pb° + Fe(Zn)\,Cl_2$

4. ZINC EXTRACTION:

   Anionic Extraction: $ZnCl_4^{2-} + X\text{-}(OH)_2 \longrightarrow X\text{-}(ZnCl_4^{2-}) + 2OH^-$

   Stripping: $X\text{-}(ZnCl_4^{2-}) + 2H_2O \longrightarrow X\text{-}(OH)_2 + ZnCl_2 + 2HCl$

   Cationic Extraction: $ZnCl_2 + 2H\text{-}C \longrightarrow Zn\text{-}C_2 + 2HCl$

   Stripping: $Zn\text{-}C_2 + H_2SO_4 \longrightarrow 2H\text{-}C + ZnSO_4$

   Anionic Extractant: $X\text{-}(OH)_2$

   Cationic Extractant: H-C

5. ZINC ELECTROWINNING:

   $ZnSO_4 + H_2O \longrightarrow Zn° + H_2SO_4 + 1/2\,O_2$

6. IRON OXIDATION/PRECIPITATION:

   $3FeCl_2 + 3/4\,O_2 + 1/2\,H_2O \longrightarrow FeOOH + 2FeCl_3$

7. COPPER CHLORIDE SOLUTION REGENERATION:

   $CuS + 2FeCl_3 \longrightarrow CuCl_2 + 2FeCl_2 + S°$

8. COPPER CHLORIDE SOLUTION REDUCTION:

   $CuCl_2 + Cu_2S \longrightarrow 2CuCl + CuS$

TABLE 5: Production of Copper Concentrates from Complex Copper-Zinc-Lead Concentrates: Metallurgical Balance [44]

1. COMPOSITION OF COMPLEX CONCENTRATE: 1,000 UNITS

1.1 CHEMICAL COMPOSITION

	Cu	Fe	Zn	Pb	S	Gangue	P.M./ppm
%	22.0	25.2	8.0	4.0	32.8	8.0	x

1.2 MINERALOGICAL COMPOSITION: %

	%
Chalcopyrite:	63.0
Sphalerite:	13.3
Galena:	4.6
Pyrite:	11.1
Gangue:	8.0
Total:	100.0

2. RECOVERIES: %

	%
Copper in Cu-concentrates	_100
Zinc as metal	90
Lead as intermediate	95
Precious metals in Cu-conc.	96 - 98
Iron in FeOOH-product*	90

* non-pyrite iron

3. PRODUCTS:

3.1 COPPER CONCENTRATE (incl. elemental sulphur): 700 units

3.1.1 CHEMICAL COMPOSITION:

	Cu	Fe	Zn	Pb	$S_{tot.}$	S°	Gangue	P.M./ppm
%	31.4	10.3	1.1	0.3	45.5	19.0	11.4	1.43 x

3.1.2 MINERALOGICAL COMPOSITION: %

	%	
Covellite:	42.5	
Elemental sulphur:	19.0	
Pyrite:	15.9	
Gangue:	11.4	
Chalcopyrite:	9.0	
Sphalerite:	1.9	unreacted
Galena:	0.3	
Total:	100.0	

3.2 ZINC METAL: 72 units

3.3 LEAD INTERMEDIATE: 42 units with 90 % Pb

3.4 FeOOH-PRODUCT: 300 units with 60 % Fe

ENVIRONMENTAL IMPLICATIONS OF THE SELECTIVE SULPHIDE LEACHING PROCESS

Environmentally the application of selectivity in sulphide leaching of mattes or complex sulphide concentrates would present benefit because of

- the utilisation of the sulphide itself as reagent for separation, obviating the need for introduction of other consumables,
- the possibility to bring out the sulphur as SO_2 in the case of smelting of the copper concentrate, or in the form of elemental sulphur in the case of hydrometallurgical processing,
- the oppurtunity to prevent the generation of solid residues if smelting of the final copper concentrate is adopted.

CONCLUSIONS

With this contribution it is hoped to have been able to draw attention to the potential and pitfalls of application of selectivity in the hydrometallurgical processing of complex sulphide materials such as mattes and bulk base metal sulphide concentrates.

The aim is the production of pure copper concentrates, enriched in copper, which are amenable to conventional copper smelting after a suitable elemental sulphur elimination step. Constraints are:

- that the conditions for selectivity need the process to be conducted in the predominant chloride system.
- to obtain selectivity by the metathesis reaction between copper and metal sulphide requires low oxidation potential to prevent formation of elemental sulphur or sulphate.
- Elevated temperatures are required to obtain acceptable kinetics.
- Much of the iron is rejected as a hydrometallurgical precipitate which may incur dumping problems.

The fact that the process:

- can be operated in a closed system with the main consumables oxygen and thermal energy,
- employs subsequent smelting of the copper concentrate resulting in high precious metals recovery and generation of a slag from the remaining iron/gangue that can be safely dumped,

may make it attractive for consideration in the treatment of particular high value feed materials with amenable mineralogy.

ACKNOWLEDGEMENT

The authors would like to sincerely express their thanks to the Management of DK-Recycling und Roheisen, Duisburg, previously Duisburger Kupferhütte, for the permission granted to disclose several aspects of the conclusions of the development work conducted there during the late 1970's.

REFERENCES

1 Leroy, J. L., Lenoir, P. J. and Escoyez, L. E., Lead Smelter Operation at N. V. Metallurgie Hoboken S.A., in: Extractive Metallurgy of Lead and Zinc, Cotterill, C. H. and Cigan, J. M., TMS-AIME, (1970), chap. 28, pp. 824-852

2 Franckaerts, A., Optimisation of the Lead-Sinter Plant and Blast-Furnace Operations at Metallurgie Hoboken-Overpelt, TMS-AIME Paper No. A86-53, (1986)

3 Thornhill, P. G., Wigstöl, E. and Van Weert, G., J.O.M., 23, (July, 1971), pp. 13-18

4 Stensholt, E. O., Zachariasen, H. and Lund, J. H., Falconbridge Chlorine Leach Process, in: Extraction Metallurgy'85, IMM, London, (Sept. 1985), Trans. IMM, C, V5, (1986), pp. 10-16

5 Lal, R. and Nicol, J. H., The BHAS Copper Leach Plant, TMS-AIME Paper No. A87-1, (1987)

6 Meadows, N. E. and Valenti, M., The BHAS Copper-Lead Matte Treatment Plant, Proc. of Non-ferrous Smelting Symposium, Port Pirie, S.A., (Sept. 1989), pp. 155-157

7 Au-Yeung, S. C. F. and Bolton, G. L., Iron Control in Processes developed at Sherritt Gordon Mines, Proc. 16th Annual Hydrometallurgical Meeting of CIM, in: Iron Control in Hydrometallurgy, 19, (1986), chap. 6, pp. 131-151

8 Berezowsky, R. M. G. S., Collins, M. J., Kerfoot, D. G. E. and Torres, N., The Commercial Status of Pressure Leaching Technology, J.O.M., (February, 1991), pp. 9-15

9 Dutrizac, J. E., The Leaching of Sulphide Minerals in Chloride Media, Hydrometallurgy, 29, (1992), pp. 1 - 45

10 Zunkel, A. D., Cuprex Metal Extraction Process (CMEP) Ready for Commercial Application, Engineering and Mining Journal, (December 1993), pp. 32 - ff.

11 Hoffmann, J. E., Winning Copper via Chloride Chemistry - An Elusive Technology, J.O.M., (August, 1991), pp. 48-49

12 Collins, D. N., et al., Role of Chloride Hydrometallurgy in Processing of Complex (Massive) Sulphide Ores, in: P. M. J. Gray, et al. (Editors), Sulphide Deposits - Their Origin and Processing, IMM, London, (1990), pp. 233 - 254

13 Craigen, W. J. S., et al., Evaluation of the CANMET Ferric Chloride Leach Process (FCL) for the Treatment of Complex Base-Metal Ores, in: P. M. J. Gray, et al. (Editors), Sulphide Deposits - Their Origin and Processing, IMM, London, (1990), pp. 255 - 269

14 Peters, E., Direct Leaching of Sulfides, Chemistry and Applications, Metallurgical Trans. B, Vol 7B, (Dec. 1976), pp. 505-517

15 Ferreira, R. C. H., High Temperature E-pH Diagrams for the System S-H_2O, Cu-S-H_2O and Fe-S-H_2O, in: Leaching and Reduction in Hydrometallurgy, Ed. Burkin, A. R., IMM, (1975), pp. 67-83

16 Enadimsa - Técnicas Reunidas S.A., Estudio Económico de un nuevo Procedimiento Industrial de Beneficio de Minerales Piríticos Complejos, Informe no. ITR/JM-4515/021/1978, Madrid, (Dec. 1978)

17 Druckard, W. J., Canterford, J. H., Dyson, N. F., et al., Oxygen Pressure Leaching of a Bulk Flotation Concentrate from a Complex Cu-Pb-Zn Sulphide Ore, Non-ferrous Smelting Symposium, Port Pirie, S.A., (September 1989), pp. 111 - 117

18 Dawson, P., Acid Pressure Oxidation of Sulfide Flotation Concentrates, TMS-AIME Paper No A86-8,(1986).

19 Kuhn, M. C., Arbiter, N., Kling, H., Anaconda's Arbiter Process for Copper, C.I.M. Bulletin, 67, (Febr. 1974), pp. 62 - 73.

20 Peters, E., The Physical Chemistry of Hydrometallurgy, in: International Symposium on Hydrometallurgy, TMS-AIME, Chicago, (1973), chap. 10, pp. 205-228

21 Gerlach, J., Pawlek, F., Rödel, R. et al, Der Einfluß des Gitteraufbaus von Metallverbindungen auf ihre Laugbarkeit, Erzmetall, Bd. 25, (1972), H.9, pp. 448-453

22 Daiger, K., Gerlach, J., Zur Kinetik der direkten Laugung Sulfidischer Erze, Erzmetall, Bd. 35, (1982), H.12, pp. 609-611

23 Holdich, R. G., Broadbent, C. P., Investigation of the Dissolution of Pyrite in Copper (II) Chloride Solutions, in: Extraction Metallurgy '85, IMM, London, (1985), pp. 645 - 658

24 Mukherjee, T. K., Hubli, R. C., Gupta, C. K., A Cupric Chloride-Oxygen Leach Process for a Nickel-Copper Sulphide Concentrate, Hydrometallurgy 15, (1985), pp. 25 - 32

25 Guy, S., Broadbent, C. P., Laugung eines komplexen Cu/Zn/Pb-Erzes mit Kupfer(II)-Chlorid, Aufbereitungstechnik, Nr. 9, (1983), pp. 539 - 547

26 Guy, S., Broadbent, C. P., Lawson, G. S., et al., Cupric Chloride Leaching of a Complex Copper/Zinc/Lead Ore, Hydrometallurgy, 10, (1983), pp. 243 - 255

27 Greig, J. A., Oxidative Chloride Leaching of Sulphide Concentrates, in: Separation Processes in Hydrometallurgy, Ed. Davis, G. A., (1987), pp. 35 - 48

28 Everett, P. K., The Dextec Lead Process, in: Hydrometallurgy - Research, Development and Plant Practice, Ed. Osseo-Asure, K., Miller, J. D., TMS-AIME, New York, (1982), pp. 165 - 176

29 Filmer, A. O., Briggs, G. G., Recovery of Lead from Mixed Sulphide Concentrates, MINTEK 50 Symposium, Johannesburg, (1984)

30 Bonan, M., Demarthe, J. M., Renon, H., et al, Chalcopyrite Leaching by $CuCl_2$ in Strong NaCl Solutions, Metallurgical Transactions B, Vol 12 B, (June 1981), pp. 269 - 274

31 Muir, D. M., Senanayaki, G., Principles and Applications of Strong Salt Solutions to Mineral Chemistry, in: Extraction Metallurgy '85, IMM, (1985).

32 Muir, D. M., Ritcey, G. M., Canterford, J. H., Recent Developments in Chloride Hydrometallurgy, in: Symposium on Extractive Metallurgy, Aus. IMM, (Nov. 1984), pp. 153 - 161

33 Peters, E., Applications of Chloride Hydrometallurgy to Treatment of Sulphide Minerals, in: Proc. on Chloride Hydrometallurgy, Benelux Metallurgie, Brussels (Sept. 1977), pp. 1-36

34 Canterford, J. H., Chloride Hydrometallurgy - Its Future Potential, Chemeca '83, The Eleventh Australian Conference on Chemical Engineering, Paper 2C, Brisbane, (Sept. 1983), pp. 73 - 82

35 Edmiston, K.J., An Update on Chloride Hydrometallurgical Processes for Sulphide Concentrates, SME-AIME, Paper No 84 - 114, (1984)

36 Stensholt, E. O., Zachariasen, H., Lund, J. H. and Thornhill, P. G., Recent Improvements in the Falconbridge Nickel Refinery, in: Proceedings of Symposium on Extractive Metallurgy of Nickel, Cobalt, TMS-AIME, Phoenix, Az, (Jan. 1988), pp. 403 - 412

37 Dutrizac, J. E., Chen, T. T., The Effect of Elemental Sulphur Reaction Product on the Leaching of Galena in Ferric Chloride Media, Metallurgical Transactions B, Vol 21 B, (Dec 1990), pp. 935 - 943

38 Dahms, J., Gerlach, J., Pawlek, F., Beitrag zur Drucklaugung von Kupfersulfiden, Erzmetall, Bd. 20, (1967), H.5, pp. 203-208

39 Kametani, H., Aoki, A., Potential -pH- Diagramme für das Spurstein / Digenit / Covellit -SO_4-H_2O Suspensionssystem bei 90°C, Erzmetall, Bd 29, (1976), H.9, pp. 394-402

40 Johnson, R. D., Miller, I. B., Meadows, N. E., Ricketts, N. J., Oxygen Treatment of Sulphidic Materials at Atmospheric Pressure in an Acid Chloride-Sulphate Lixiviant, Proc. of Non-Ferrous Smelting Symposium, Port Pirie, S.A., (Sept. 1989), pp. 163 - 166

41 Cheng, C. Y., Lawson, F., The Leaching of Synthetic Chalcocite and Covellite in Oxygenated Acidic Sulphate-Chloride Solutions, Proc. of Non-Ferrous Smelting Symposium, Port Pirie, S.A., (Sept. 1989), pp. 167 - 174

42 Clevenger, G. W., Pepple, G. W., US Pat. 4,384,890, (May 24, 1983), Cupric Chloride Leaching of Copper Sulphides

43 Duyvesteyn, W. C., et al., The Escondida Process for Copper Concentrates, in: Proc. of the Paul E. Queneau International Symposium, Extractive Metallurgy of Copper, Nickel and Cobalt, Denver, Co, 1993

44 Unpublished results of investigations by the author

45 McGauley, P. J., Roberts, E. S., US Pat. 2,568,963, (Sept. 25, 1951)

46 Yamada, M., (Dowa Mining), Jap. Pat. 49-123926, (Nov. 27, 1974), Process for the Recovery of Copper

47 O'Neill, C. E., Illis, A., Huggins, D. A., US Pat. 3,616,331, (Oct. 26, 1971), Recovery of Nickel and Copper from Sulfides

48 Johnson, R. K., Coltrinari, E. L., US Pat. 3,957,602, (May 18, 1976), Recovery of Copper from Chalcopyrite Utilizing Copper Sulfate Leach

49 McKay, D. R., Parker, E. G., US Pat. 4,024,218, (May 17, 1977), Process for Hydrometallurgical Upgrading

50 Swinkels, G. M., et al., The Sherritt Gordon - Cominco Copper Process - Part I: The Process, CIM Bulletin, (February 1978), pp. 105 - 121

51 Renken, H. C., Zegers, T. W., US Pat. 3,655,538, (Apr. 11, 1972), Process for Electrowinning Zinc from Sulfide Concentrates

52 Collier, D., et al, Comparative Economics of Sulphate-Based Hydrometallurgical Processes for the Treatment of Complex Sulphide Ores, Extraction Metallurgy '85, IMM, London, Sept. 1985, pp. 997 - 1014

53 Bartlett, R. W., et al., A Process for Enriching Chalcopyrite Concentrates, in: Metallurgical Reactor Design and Kinetics, Ed. Bautista, et al, TMS-AIME, 1986, pp. 227 - 246

54 Bartlett, R. W., Copper Super-Concentrates-Processing, Economics, and Smelting, EDP-Proceed-ings '92, TMS-Annual Meeting, San Diego, Ca., March 1992

55 Goens, D. N., Can. Pat. 1,065,615, (Jun. 11, 1979), Hydrometallurgical Purification Process

56 Piret, N. L., Höpper, M., Kudelka, H., US Pat. 4,260,588, (Apr. 7, 1981), Production of Sulphidic Copper Concentrates

57 Shirts, M. B., et al., Aqueous Reduction of Chalcopyrite Concentrate with Metals, US Bureau of Mines RI-7953, 1974

58 Hackl, R., et al., Reverse Leaching of Chalcopyrite, in: Proceedings of International Conference Copper '87, Viña del Mar, Chile, (1987), pp. 181 - 200

59 Sohn, H.-J., et al., Reduction of Chalcopyrite with SO_2 in the Presence of Cupric Ions, J.O.M., (Nov. 1980), pp. 18 - 22

60 Sequeira, C. A. C., Electrochemical Reductive Conversion of Chalcopyrite with SO_2, in: EMC '91: Non-Ferrous Metallurgy - Present and Future, Elsevier, 1991, pp. 219 - 228

61 Hougen, L. R., US Pat. 3,880,653 (Apr. 29, 1975), Chlorine Leach Process

Control of arsenic and selenium during nickel matte leaching

J. J. Robinson
A. J. Parker Cooperative Research Centre for Hydrometallurgy, Chemistry Centre (WA), Department of Minerals and Energy, Bentley, Western Australia
K. Barbetti
A. J. Parker Cooperative Research Centre for Hydrometallurgy, Chemistry Centre (WA), Department of Minerals and Energy, Bentley, Western Australia
R. P. Das
Regional Research Laboratory, Bhubaneswar, India

The ammoniacal leaching of Kambalda nickel mattes was investigated using a laboratory autoclave to study the process by which the impurities of arsenic and selenium were scavenged by the precipitating iron oxides. The permissible levels of these impurities in the leach liquors were governed by limits of 5 and 50 ppm, respectively, imposed on the final nickel metal product.

Initially, tests were conducted at 1 wt % pulp density and variations in the arsenic:iron ratio were simulated by, (i) adding ferrous sulphate to a leach of a high arsenic-containing matte, and (ii), making additions of soluble arsenic pentoxide to a matte of "normal" iron level. The results indicated that iron added as ferrous sulphate to the pulps scavenged arsenic similarly to iron originating from the matte.

At 11 wt % pulp density, an iron level in the matte of 4.7% was sufficient to scavenge 0.1% arsenic. In the same tests, the contained selenium was effectively reduced with slightly lower levels of iron. It was demonstrated that with increasing arsenic:iron ratios, the arsenic content of the residues also increased but the capacity of the residues for selenium decreased. It was noted that at the higher pulp density, arsenic was scavenged to a greater extent.

The mattes were composed mainly of heazlewoodite with smaller amounts of pentlandite, millerite and magnetite, while the residues consisted predominantly of ferrihydrite with about 15% by weight magnetite.

Electron microprobe analysis, thermal analysis and x-ray diffraction failed to reveal the presence of a ferric arsenate in the final residues. Instead the arsenic was found to be uniformly distributed throughout the ferrihydrite with little detected in the magnetite. It is suggested that arsenic and selenium were removed by adsorption on to ferrihydrite.

1 Introduction

Western Mining Corporation operates a Sherritt Gordon ammoniacal ammonium sulphate pressure leach - hydrogen reduction process at Kwinana in Western Australia [1, 2]. The matte is produced at the Kambalda smelter and freighted to the Kwinana refinery where the pressure leach oxidises the sulphides and complexes the nickel, copper and cobalt as soluble ammines. The iron is oxidised and is precipitated as hydrated iron oxides.

A number of impurities are present in the matte leached at Kwinana and of particular concern are arsenic and selenium. Market considerations - specifications and competitive product quality - dictate the permitted levels of these in the metallic product with concentration limits of 5 and 50 ppm, respectively. It is anticipated that arsenic concentrations in the matte will increase as a result of increasing levels of arsenic in the ores.

This investigation formed part of a project designed to provide a better understanding of the behaviour of increased levels of arsenic during leaching, thereby providing an opportunity to compensate for these changes and maximise impurity removal.

2 Review of leaching and impurity removal processes

Iron, although present as an impurity in many metallurgical processes, may be used for the control of many other impurities such as arsenic [3-8]. Waste water treatment has involved studies of adsorption of arsenic on amorphous iron hydroxide [9]. Pierce and Moore [10] and Robins et al. [11] also studied the removal of arsenic by ferric hydroxide. It was suggested that arsenic removal occurred through either precipitation as a basic ferric arsenate or arsenate adsorbed on ferric hydroxide; Robins [7] preferred the latter model.

Das [12] investigated the oxidation of Fe(II) in an ammonia - ammonium sulphate solution and studied the nature of the precipitated iron oxides. He concluded that arsenic reduced the rate of oxidation of the ferrous iron and stabilised the amorphous precipitate.

Sherritt Gordon [13] have recorded that both arsenic and antimony are adsorbed on the precipitating iron oxides during the ammoniacal ammonium sulphate leach of nickel matte, although it was stated that losses of nickel, cobalt and copper also occurred by this means. Osseo-Asare and Fuerstenau [14] referenced a number of authors who reported losses of nickel, copper and cobalt through adsorption on and coprecipitation with precipitating iron oxide in ammoniacal systems.

The iron oxides precipitated from ammoniacal leach systems have been variously described as hydrated iron hydroxide [13], hydrated iron oxide [15], ferric oxide [16], hematite [17] and hydrated ferric oxide [18]. A review of the literature indicated that little work has been published on the nature of the precipitates from ammoniacal leach systems. However, the iron oxide system has been studied in some detail by geochemists and, in particular, soil chemists. Schwertmann and Cornell [19] listed thirteen iron oxides, oxyhydroxides and hydroxides that were known at the time. Many of the early investigations on iron oxides referred to a particular phase as "amorphous ferric hydroxide". More recent work has confirmed that this phase should be more correctly known as ferrihydrite [20-24]. In natural systems ferrihydrite is regarded as an adsorbent with high reactivity and large surface area [25]. By using Extended X-ray Absorption Fine Structure (EXAFS) spectroscopy and kinetic measurements, Waychunas et al. [26] and Fuller et al. [27] determined that the loading of arsenate on ferrihydrite occurred by a chemical adsorption mechanism with no evidence of surface precipitation or solid solution formation. The tests were conducted over a pH range of 7 to 9.5 and it was shown that As(V) adsorption decreased over that range with increasing pH.

A number of environmental investigations have shown that selenite adsorption decreases with increasing pH in the range 4 - 9 and that selenate adsorption was low irrespective of pH [28, 29]. Waste water treatment to remove selenium by coagulation with ferric sulphate was not very effective but improved with increasing coagulant dose and decreasing pH [30]. In describing the Sherritt-Cominco copper process, Kawulka et al. [31] noted that selenium must be present in the Se^{4+} state to be effectively removed with the ferric oxide from the copper sulphate solution.

In ammoniacal leaching of nickel matte, oxidation of iron is believed to proceed through a ferrous ammine complex before precipitating as one or more iron oxides [32] Under the leaching conditions of high potential and pH, it is expected that arsenic and selenium, which most likely exist as arsenide and selenide in the matte, would oxidise progressively to arsenite (AsO_2^- or AsO_3^{3-}) and selenite ($HSeO_3^-$ or SeO_3^{2-}) and finally to arsenate (AsO_4^{3-}) and selenate (SeO_4^{2-}) [33]. Under the right conditions arsenic could be effectively scavenged by ferrihydrite although the high pH values of the ammoniacal system (>10) may not be favourable. There appear to be no available data on the ability of ferrihydrite to remove selenium although Hayes et al. [34] indicated that selenite was more strongly adsorbed on to α-FeOOH than selenate.

3 Experimental

The test program involved leaching four matte samples of differing composition at pulp densities of 1 wt % and 11 wt % in a 3.2 L Parr Pressure Reactor furnished with a PID Controller to regulate internal temperature. The pressure vessel, along with the internal components such as the cooling coil, thermowell and stirrer shaft and impellers, were constructed from titanium. Agitation was provided by a high torque magnetic drive stirrer. The autoclave was also fitted with gas inlet and sample collection ports.

3.1 Low pulp density testing

Initial tests on a high arsenic-containing matte, of approximately 1% As, were planned. The choice of a 1 wt % pulp density would provide an arsenic concentration in the pulp similar to that in practice and would minimise the concentration variations of the other matte components in the system. Supplementary additions of iron were made to the pulps in the form of iron (II) sulphate to adjust the As:Fe ratio.

The total solution volume was 1.7 L, and 17g of matte were added to produce a pulp density of 1 wt %. The required quantities of ammonium sulphate, ammonia, nickel matte and water (plus $FeSO_4.7H_2O$ or $As_2O_5.xH_2O$ where appropriate) were added to the autoclave which was then sealed and heating of the vessel was initiated. The autoclave was heated rapidly over the initial period to minimise the time taken to reach the reaction temperature. The contents of the vessel were stirred continuously at a rate of 500 ± 5 min^{-1}. Upon reaching the set temperature, a sample was withdrawn from the autoclave and immediately filtered. The filtrate was retained while the solid was washed firstly with about 25 mL of 10% v/v ammonia solution and then with 50 mL of distilled water before air drying. After sampling, oxygen was introduced into the autoclave at 140 ± 10 kPa and this level was maintained throughout the experiment by way of a regulator.

Further samples were collected after 0.25, 0.5, 1, 1.5 and 2 hours and were filtered. The solids were washed and dried as described above. Filtrates were also retained for analysis.

3.2 High pulp density testing

Pulp densities in the plant vary, but generally lie between 10 and 20 wt %. To assess the effectiveness of arsenic and selenium scavenging at a pulp density similar to that employed in the plant, several tests were performed at an arbitrarily chosen value of 11 wt %.

The calculation of the reagent additions required for the leaches were based on the composition of the matte such that the ammonium sulphate concentration at the end of the leach was 350 g/L and the desired molar ratio of "free ammonia" to the soluble metal (Ni, Cu and Co) concentrations was set at 4.5. In general, the total solution volume was 2 L and the mass of matte leached was 240 g. The leaching temperature was 90C and the partial pressure of oxygen was 140 kPa. Samples were collected at 0, 0.25, 0.5, 1, 2, 4 and 6 hours. The filtrate samples were immediately diluted with an equal volume of distilled water to prevent the ammine complexes from crystallising out of solution with decreasing temperature. The solids were washed with 50 mL of 10% v/v ammonia solution followed by about 75 mL of distilled water and then air dried.

3.3 Analysis of mattes and leach products

Samples of the mattes were dissolved in hydrochloric acid and hydrogen peroxide and the resultant solution quantified for iron concentration by flame atomic absorption spectroscopy (FAAS). Other matte samples were attacked with nitric and perchloric acids and the resultant solution analysed for arsenic and selenium concentrations by FAAS.

The analysis of the leach residues was accomplished by first leaching with a mixture of perchloric, nitric and hydrochloric acids and the resultant solutions were quantified for arsenic and selenium by FAAS. For nickel, the residues were dissolved in hydrochloric acid and quantified by FAAS. The same solution was used for the titrimetric determination of iron. The product leach solutions were analysed for nickel by FAAS after dilution or by X-ray fluorescence spectrometry. Iron, arsenic and selenium were determined by FAAS after their separation from the leach solution by coprecipitation with lanthanum hydroxide.

X-ray diffraction was used to characterise the solid products. Diffraction scans were run from 2° to 76° 2Θ with Co radiation using steps of 0.06° and 2 seconds counting per step, at 40 kV and 30 mA. A method developed by Parks [35] was used to quantitatively interpret the x-ray diffraction patterns.

The mattes and several leach residues were analysed by electron probe microanalysis (EPMA) and scanning electron microscopy. A leach residue produced from a high arsenic matte, at 1 wt % pulp density, was subjected to a Davis Tube magnetic separation and both fractions were analysed by EPMA. The non-magnetic fraction was also analysed by thermal analysis.

4 Results and Discussion

4.1 Chemical analysis of mattes

The four nickel matte samples used in this investigation consisted of the "normal" production matte (referred to as the KNR matte) and three special grade products. These four mattes and their chemical analyses are summarised in Table 1. For ease of comparison, the As:Fe ratios were included in this table.

Table 1. Chemical analyses of the matte samples.

Sample	Fe, %	As, %	As:Fe Ratio	Ni, %	Se, g/g	S, %	Co, %	Cu, %
KNR Matte	6.0	0.13	0.022:1	63.9	112	24.0	0.69	3.8
High As Matte	5.7	0.66	0.116:1	63.2	154	24.5	0.62	3.9
4.7% Fe Matte	4.7	0.10	0.021:1	66.3	90	24.9	0.84	3.5
3.3% Fe Matte	3.3	0.11	0.033:1	66.4	104	23.4	0.83	3.7

4.2 Concentration limits of arsenic and selenium in leach solutions

In order to determine whether the arsenic and selenium concentration limits imposed on the metallic nickel product were satisfied during these tests, concentration limits for arsenic and selenium in the leach solutions were calculated. The following assumptions were made: all of the nickel in the matte was solubilised; all of the impurity present in the solution after leaching reported to the nickel product and all of the nickel in solution was recovered as metal product. For the KNR matte and thc high As matte at 11 wt % pulp density, the concentration limits for arsenic and selenium of 0.38 mg/L and 3.8 mg/L, respectively, were determined. The values for both the 3.3% Fe and 4.7% Fe mattes were slightly higher at 0.4 and 4.0 mg/L. At 1 wt % pulp density, the limits were calculated at 0.03 mg/L (As) and 0.3 mg/L (Se).

4.3 Leaching at 1 wt % pulp density

To investigate the effect of arsenic:iron ratio on arsenic removal, a series of tests was undertaken in which additional iron was added as Fe(II) to a 1 wt % high arsenic matte leach. The overall iron concentration, measured in g/L of pulp, was varied from 0.6 (without any additional Fe(II)) to 13, while the corresponding As:Fe ratio ranged from 0.116:1 to 0.005:1. The soluble arsenic concentrations recorded with time are presented in Figure 1 and may be seen to decrease rapidly as the As:Fe ratio changed from 0.116:1 to 0.036:1. At a ratio of approximately 0.014:1, the soluble arsenic levels had been reduced to less than 0.03 mg/L.

The initial responses displayed by these plots may be explained by the iron content. The plots may be seen to fit three general types, i.e., (i) those plots which showed a fairly rapid increase in arsenic concentration with time (for iron concentrations of 0.6 g/L and 1.0 g/L), (ii) those which underwent an initial decrease but, with time, gradually increased in arsenic concentration (2.0, 3.0 and 4.0 g/L), and (iii), plots at high iron concentrations of 5.7, 7.5 and 13 g/L which revealed a continuing decrease in arsenic concentration with time.

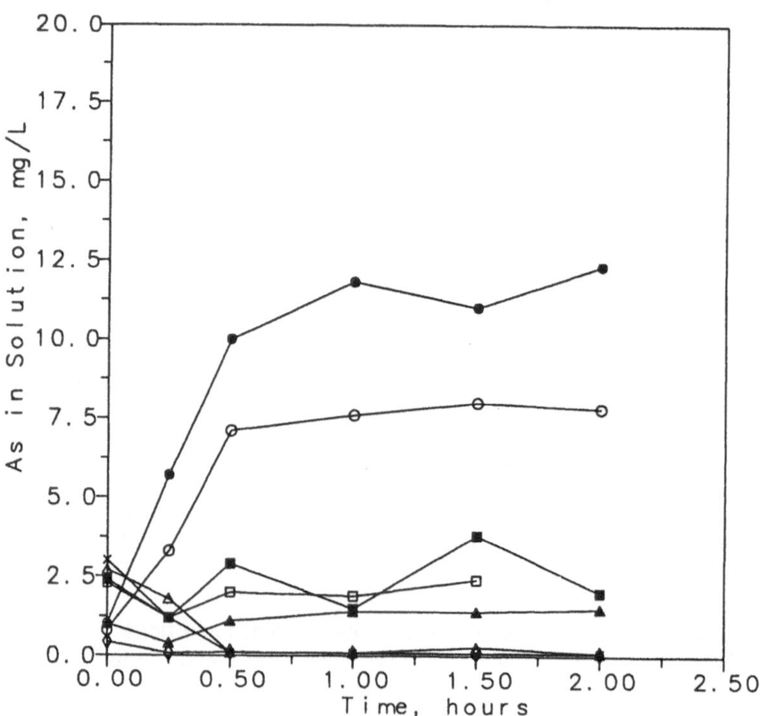

1 wt % pulp density, p_{O_2} = 140 kPa, Temp = 70°C,
$[NH_3]$ = 80 g/L, [Amsul] = 300 g/L

● As:Fe = 0.116:1
○ As:Fe = 0.067:1
■ As:Fe = 0.036:1
□ As:Fe = 0.026:1
▲ As:Fe = 0.016:1
Δ As:Fe = 0.014:1
x As:Fe = 0.009:1
◊ As:Fe = 0.005:1

FIGURE 1. Effect of As:Fe ratio (High arsenic matte with Fe(II) additions) on residual arsenic in solution at 1 wt % pulp density

During the initial stages of matte leaching, it is expected that arsenic dissolves and iron forms soluble Fe(II). Arsenic would not be scavenged until the iron had oxidised and had begun precipitating as oxide. With increasing concentrations of Fe(II) the extent of arsenic removal would be expected to increase but the rate would still be dependent on the rate of iron precipitation. Therefore for type (i) plots, there was insufficient iron present. The behaviour demonstrated by type (ii) suggested that some arsenic had been removed as the iron oxide precipitated but, with continued leaching, the level of soluble arsenic increased as a result of insufficient oxide precipitation. Type (iii) plot occurred because there was more than sufficient iron present and the iron oxide was precipitating at a rate greater than or equal to the rate of leaching of arsenic.

To compare the effect of different matte iron levels on arsenic removal, the four different mattes were leached under similar conditions and the change in soluble arsenic levels was measured with time. The results are plotted in Figure 2. Arsenic removal appeared to be directly related to the As:Fe ratio.

For comparison purposes, the arsenic concentrations in the final leach solutions from Figures 1 and 2 were expressed as a percentage of the total arsenic and were plotted against the corresponding As:Fe ratio in Figure 3. Up to an arsenic:iron ratio of approximately 0.015 essentially all of the arsenic was removed, with the soluble values less than the required limit of 0.03 mg/L. Above this ratio, arsenic concentrations remaining in the solution increased sharply. The correlation between the data suggested that iron added as Fe(II) performed an arsenic scavenging function similar to that of iron contained in the matte.

To establish the behaviour of increased arsenic levels in a normal matte, a series of leaches was carried out on the KNR matte with increasing quantities of arsenic added as a solution of arsenic pentoxide. The changes in the soluble arsenic concentration measured during these tests are recorded in Figure 4. The arsenic:iron ratio for the KNR matte was calculated at 0.022:1 and with increasing arsenic additions this ratio was progressively increased to 0.154:1. These ratios corresponded to equivalent arsenic concentrations in the matte of 0.13% and 0.93%, respectively.

It is evident that arsenic was scavenged rapidly and reached near constant levels after approximately 0.5 hours. Because the arsenic was introduced as a solution, the kinetics were different from that of arsenic originating from the matte and may be explained by the differences in the removal processes between the two different arsenic types. Arsenic pentoxide added in solution form is immediately available for adsorption by the precipitating iron oxides. However, arsenic in the matte must undergo oxidation during which iron precipitation may be occurring. Therefore the extent of arsenic adsorption may be limited to the final stages of iron oxide precipitation. This may explain the differences between the solution concentrations in Figure 4 and Figure 2.

4.4 Analysis of the matte samples and leach products by x-ray diffraction

The four matte samples were analysed by x-ray diffraction and were shown to be composed mostly of heazlewoodite with smaller amounts of pentlandite, millerite and magnetite. A small amount of a metallic phase, an alloy of nickel, iron and copper, was also identified. The KNR, High As and the 4.7% Fe mattes contained approximately 85 - 90% heazlewoodite while the 3.3% Fe matte was composed of approximately 95% heazlewoodite but with a lower proportion of pentlandite (about 3%).

The solid samples collected during leaching were analysed by x-ray diffraction. The results of a 1 wt % pulp density leach of a high arsenic matte, conducted at 90°C and at 80/300 g/L of ammonia/ammonium sulphate, are presented in Table 2.

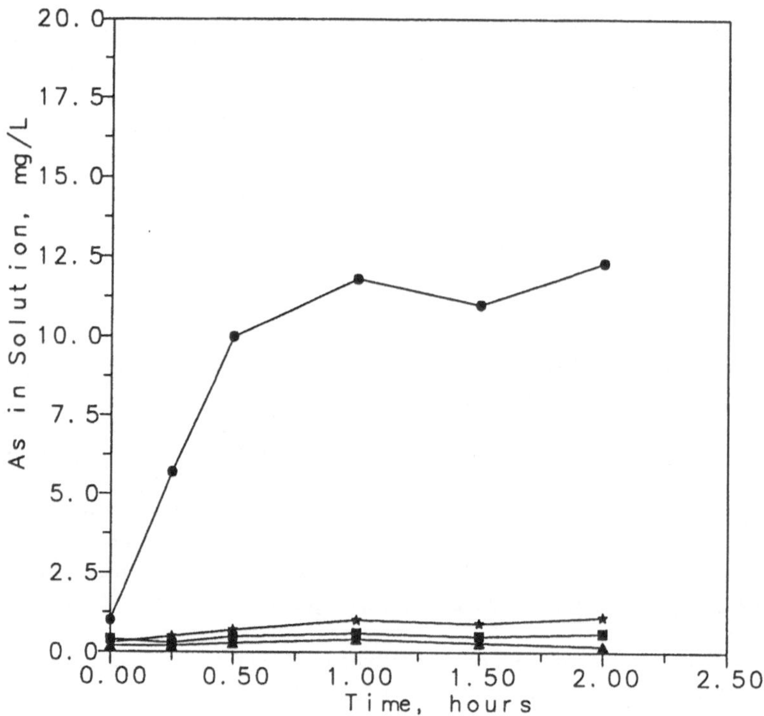

1 wt % pulp density, p_{O_2} = 140 kPa, Temp = 70°C,
$[NH_3]$ = 80 g/L, [Amsul] = 300 g/L

● High Arsenic Matte
■ KNR Matte
▲ 4.7% Fe Matte
★ 3.3% Fe Matte

FIGURE 2. Variation of arsenic concentration in solution with leaching time for different matte samples at 1 wt % pulp density

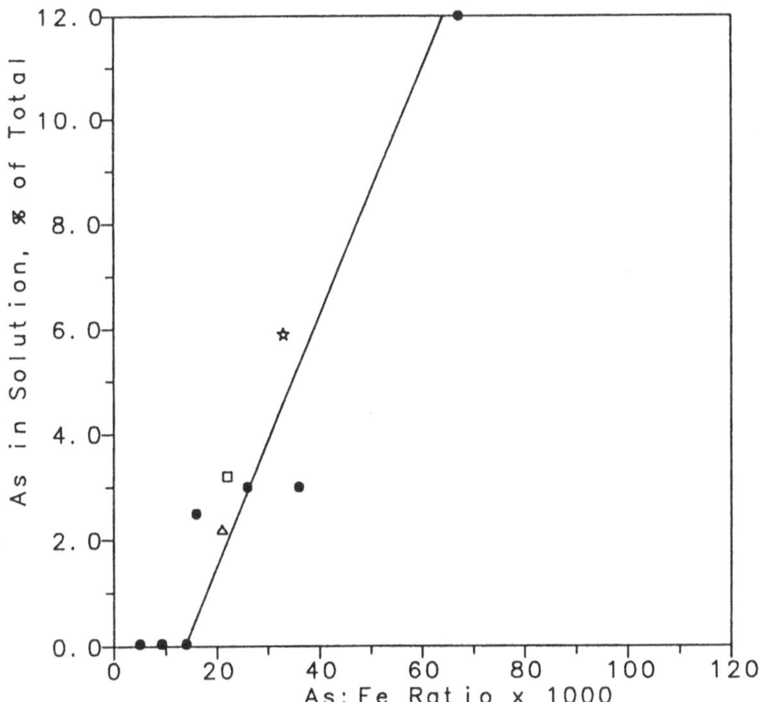

1 wt % pulp density, p_{O_2} = 140 kPa, Temp = 70°C,
$[NH_3]$ = 80 g/L, [Amsul] = 300 g/L

● High Arsenic Matte
□ KNR Matte
Δ 4.7% Fe Matte
☆ 3.3% Fe Matte

FIGURE 3. Final soluble arsenic concentration as a function of As:Fe ratio at 1 wt % pulp density

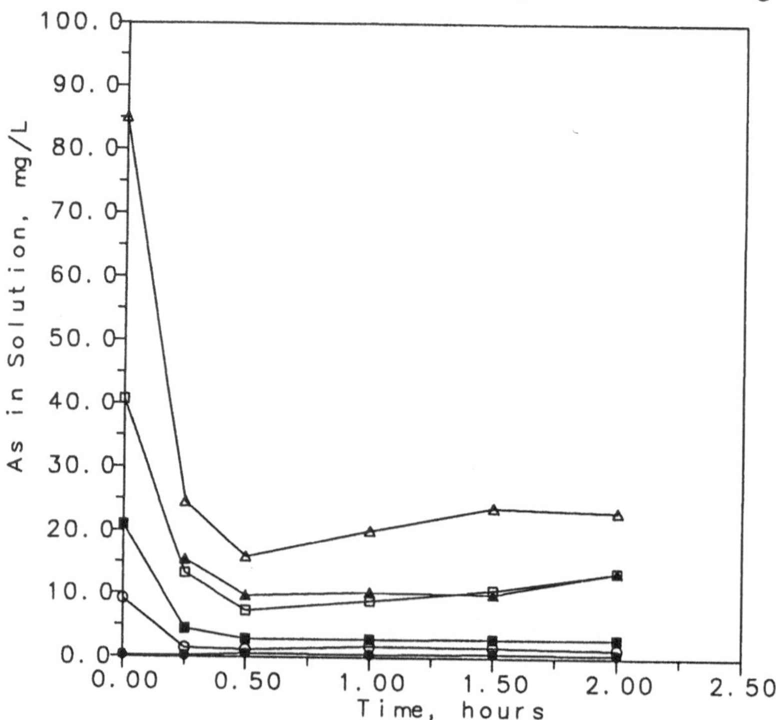

1 wt % pulp density, p_{O_2} = 140 kPa, Temp = 90°C,
$[NH_3]$ = 80 g/L, [Amsul] = 300 g/L

● As:Fe = 0.022:1
○ As:Fe = 0.044:1
■ As:Fe = 0.071:1
□ As:Fe = 0.104:1
▲ As:Fe = 0.122:1
Δ As:Fe = 0.154:1

FIGURE 4. Effect of As:Fe ratio (KNR matte with As(V) additions) on residual arsenic in solution at 1 wt % pulp density

The sulphide phases decreased in concentration with time except millerite which initially displayed a slight increase, presumably resulting from the decomposition of pentlandite. Thereafter the concentration fell rapidly to near zero. Pentlandite and millerite were no longer evident in the solid after approximately 0.5 hours but small percentages of heazlewoodite were present after 1.5 hours. Magnetite progressively increased from approximately 2%, the proportion contained in the original matte, to about 15%. The final residue consisted predominantly of ferrihydrite, the proportion of which was calculated at 85%. Both the "six-line" and "two-line" forms of ferrihydrite, as described by Schwertmann and Cornell [19], were identified. Residues also contained small proportions of other phases, including goethite, lepidocrocite and hematite.

Table 2. The change in the nature of the solid phases during leaching of the high arsenic matte.

TIME	Weight % of Mineral Phases in Leach Residues, by XRD (alloy phases excluded)					
(hrs)	Heazlewoodite	Pentlandite	Millerite	Magnetite	Ferrihydrite	Lepidocrocite
0	87	7	4	2	0	0
0.25	62	5	4.5	6	23	0
0.5	16	0	0	14	70	1
1.0	7	0	0	15	78	1
1.5	3	0	0	13	83	1
2.0	0	0	0	15	84	1

4.5 Electron probe microanalysis (EPMA) and thermal analysis of leach residues

A leach residue from a 1 wt % pulp density leach of a high arsenic matte was analysed by EPMA. It was shown that arsenic was uniformly distributed through the ferrihydrite at levels consistent with the bulk analysis but was not detected in the magetite particles.

A leach residue of the high arsenic matte was separated magnetically into two fractions, the weight ratio of which was calculated at approximately 1 (magnetic): 10 (non-magnetic). The magnetic fraction was shown to consist mostly of magnetite and the non-magnetic portion, ferrihydrite. The weight ratio of the two fractions corresponded roughly to the quantitative x-ray diffraction analysis of the leach residue.

The non-magnetic fraction was analysed by differential thermal analysis and thermogravimetric analysis. The results were similar to those reported in the literature for other ferrihydrites [19]. No endotherm at 550 - 600°C was observed where a decomposition of a ferric arsenate might have been expected. This suggests that arsenic, and possibly selenium, are removed by adsorption on to ferrihydrite, although further work is needed to confirm this.

4.6 Leaching at 11 wt % pulp density

Several tests were undertaken on the four different mattes at 90°C, an oxygen pressure of 140 kPa and for a duration of 6 hours. For comparative purposes the arsenic concentrations were plotted together in Figure 5 and the selenium values in Figure 6. Since it was noted that within the first 0.5 to 1 hour, sulphide oxidation occurred with the corresponding formation of iron oxide it was possible that the dissolution of arsenic

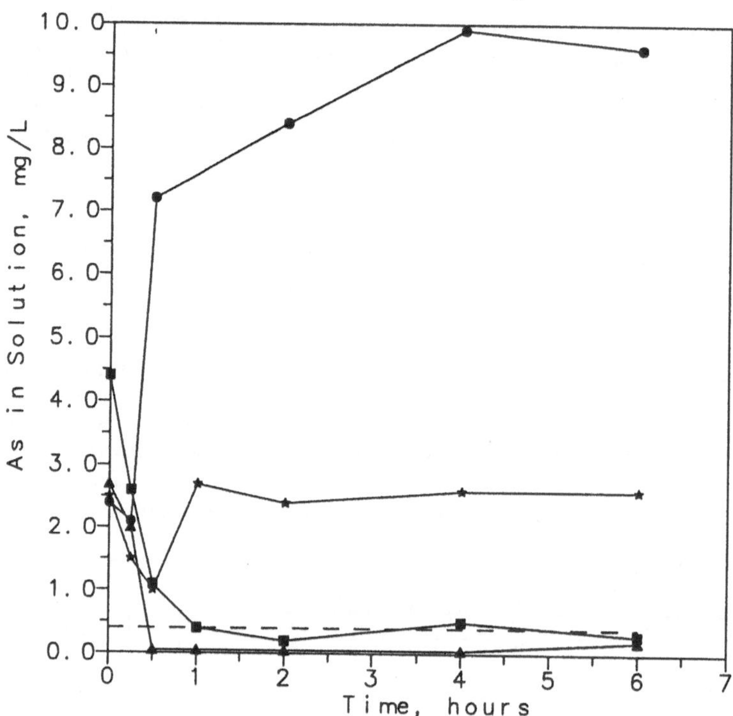

11 wt % pulp density, p_{O_2} = 140 kPa, Temp = 90°C,
$[NH_3]$ = 90 g/L, [Amsul] = 410 g/L

● High Arsenic Matte
■ KNR Matte
▲ 4.7% Fe Matte
★ 3.3% Fe Matte
- - - Limiting arsenic value

FIGURE 5. Variation of arsenic concentration in solution with leaching time for different matte samples at 11 wt % pulp density

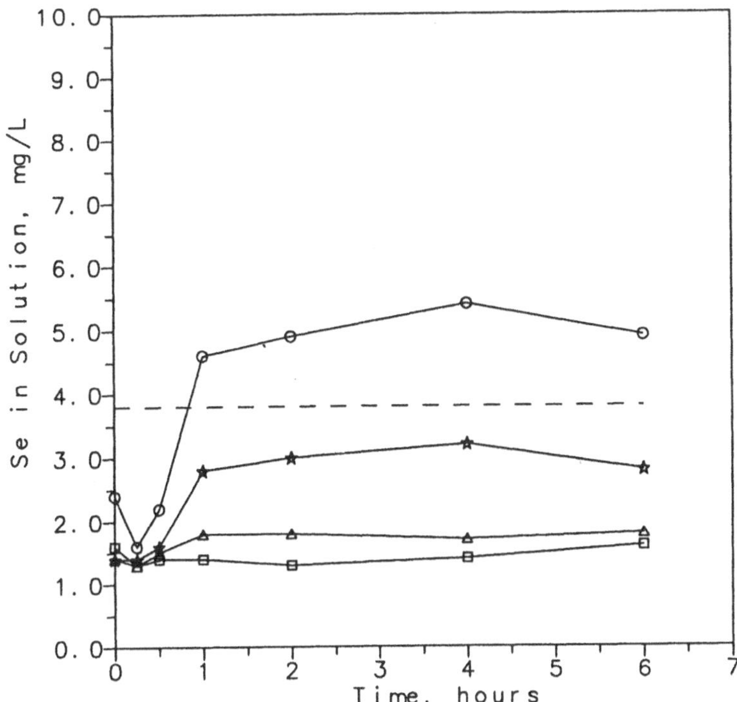

11 wt % pulp density, p_{O_2} = 140 kPa, Temp = 90°C,
$[NH_3]$ = 90 g/L, [Amsul] = 410 g/L

O High Arsenic Matte
□ KNR Matte
Δ 4.7% Fe Matte
☆ 3.3% Fe Matte
- - - Limiting selenium value

FIGURE 6. Variation of selenium concentration in solution with leaching time for different matte samples at 11 wt % pulp density

and selenium proceeded during this first period and scavenging by the iron oxides closely followed. This may explain some of the concentration changes noted in the early stages of the leach.

As expected, the high arsenic matte and 3.3% Fe matte produced the highest levels of soluble arsenic; the concentrations from the high arsenic matte were significantly higher. The residual arsenic concentrations formed from the 4.7% Fe matte and the KNR matte were similar and were less than or close to the desired limit (0.38 mg/L). The difference between the two plots may be explained by the lower level of arsenic in the 4.7% Fe matte (0.10% compared with 0.13%). The residual soluble arsenic values appeared to be related to the arsenic:iron ratio.

Selenium removal was most effective for the KNR matte with concentration levels well below the desired limit. In order of their residual selenium levels, the other mattes were ranked as follows: 4.7% Fe, 3.3% Fe and the high arsenic matte. Only the high arsenic matte resulted in selenium levels which exceeded the concentration limit of 3.8 mg/L. The higher selenium concentrations from the 3.3% Fe matte were predominantly a function of the low iron level while the poor performance of the high arsenic matte suggested that increased levels of arsenic had a detrimental effect on selenium removal.

To determine the effect of increasing arsenic levels on residual arsenic concentrations in solution and on the behaviour of selenium, several tests were performed on the KNR matte in which the arsenic levels were adjusted with additions of soluble arsenic pentoxide such that the arsenic:iron ratio was increased from 0.022:1 to 0.132:1. (These ratios were equivalent to arsenic concentrations in the matte of 0.13% (for no added arsenic) and 0.79%.) Figure 7 shows that soluble arsenic levels were reduced rapidly in the first hour of leaching. For the tests in which the arsenic:iron ratio was set at 0.022, 0.054 and 0.076:1 (corresponding to equivalent arsenic concentrations of 0.13, 0.32 and 0.45%), the residual soluble arsenic concentrations were close to the arsenic limit of 0.38 mg/L. With further increases in the arsenic:iron ratio to 0.105:1 and 0.132:1, the soluble arsenic concentrations increased to approximately 2 and 3.5 mg/L, respectively.

For the same As:Fe ratio, these results show lower levels of residual arsenic than those in which the arsenic concentration was varied as matte arsenic. The corresponding arsenic:iron ratio for the high arsenic matte was 0.116:1 and the residual soluble arsenic concentration after leaching for approximately 5 hours was markedly higher (see Figure 5). These differences were also observed in the 1 wt % pulp density tests.

The soluble selenium concentrations were also measured during the tests in which arsenic pentoxide was added to the leaches of the KNR matte. Figure 8 summarises these results. The results show that, for the arsenic:iron ratios investigated, the residual selenium concentrations were consistently lower than the desired limit. Generally, as the As:Fe ratio increased so did the soluble selenium levels.

The arsenic and selenium concentrations in the residues were plotted as a function of the arsenic:iron ratio (Figure 9). As the arsenic loading on the residue increased from 10 to 60 mg/g of residue, the capacity of the residue for selenium removal decreased from about 700 to 400 g/g.

The results show that the extent of arsenic removal was greater for the tests carried out at 11 wt % pulp density than for those at 1 wt %, even at similar As:Fe ratios. This suggests that the pulp density directly affected arsenic and possibly, selenium removal.

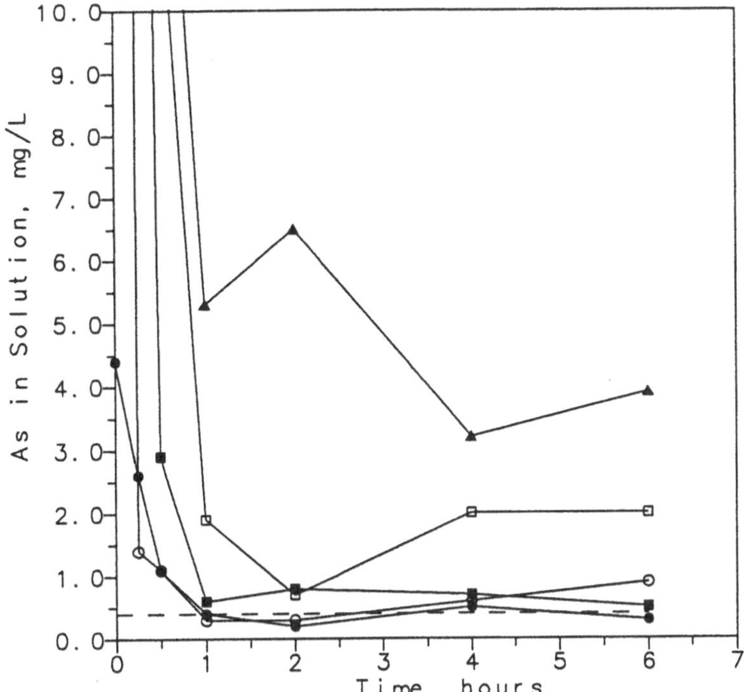

11 wt % pulp density, p_{O_2} = 140 kPa, Temp = 90°C,
$[NH_3]$ = 90 g/L, [Amsul] = 410 g/L

● As:Fe = 0.022:1
○ As:Fe = 0.054:1
■ As:Fe = 0.076:1
□ As:Fe = 0.105:1
▲ As:Fe = 0.132:1
- - - Limiting arsenic value

FIGURE 7. Effect of As:Fe ratio (KNR matte with As(V) additions) on residual arsenic in solution at 11 wt % pulp density

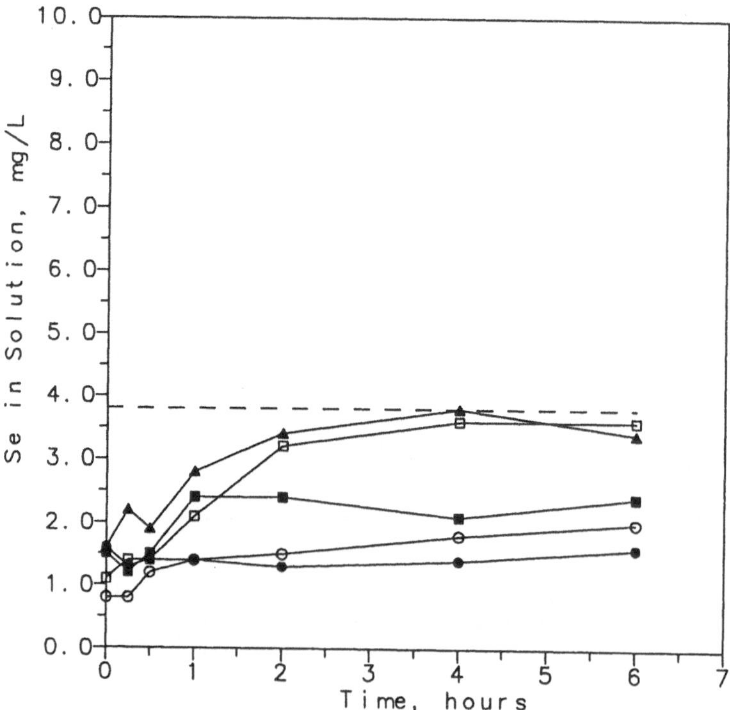

11 wt % pulp density, p_{O_2} = 140 kPa, Temp = 90°C,
$[NH_3]$ = 90 g/L, [Amsul] = 410 g/L

● As:Fe = 0.022:1
○ As:Fe = 0.054:1
■ As:Fe = 0.076:1
□ As:Fe = 0.105:1
▲ As:Fe = 0.132:1
- - - Limiting selenium value

FIGURE 8. Effect of As:Fe ratio (KNR matte with As(V) additions) on residual selenium in solution at 11 wt % pulp density

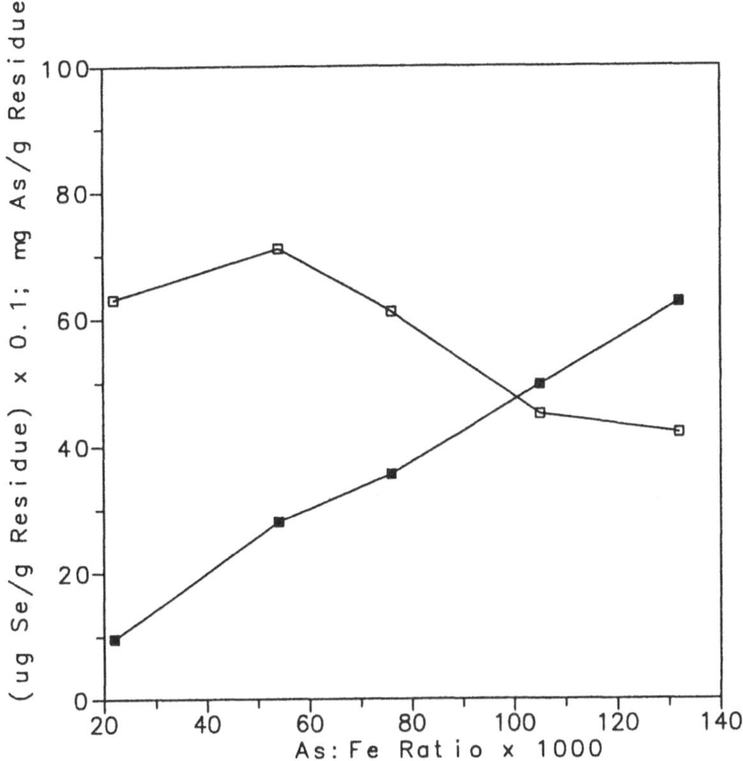

11 wt % pulp density, p_{O_2} = 140 kPa, Temp = 90°C, $[NH_3]$ = 90 g/L, [Amsul] = 410 g/L

■ Arsenic in residue
□ Selenium in residue

FIGURE 9. Arsenic and selenium concentrations in the residues as a function of As:Fe ratio for the KNR matte at 11 wt % pulp density

5 Conclusions

A limited number of tests were performed on the ammoniacal leaching of Kambalda nickel mattes in a laboratory autoclave at pulp densities of 1 and 11 wt %. At a pulp density of 1 wt %, the desired level of arsenic removal was achieved at a maximum arsenic:iron ratio of 0.015:1.

It was demonstrated in the 1 wt % pulp density tests that iron added as ferrous sulphate to the pulps simulated increased levels of matte iron.

At 11 wt % pulp density, matte arsenic levels of 0.1% were scavenged by an iron concentration of 4.7% or higher, resulting in a leach liquor containing less than the required level of 0.38 mg/L As. Within the same tests, iron levels in the matte as low as 3.3% reduced the concentration of selenium in the liquor below the required level of 3.8 mg/L.

Arsenic additions to the pulp as soluble arsenic pentoxide behaved somewhat differently from matte arsenic, with higher proportions removed. The influence of these different arsenic forms on the scavenging of selenium was also observed although the general effect of suppressing selenium removal was the same.

The residues generated at 11 wt % pulp density demonstrated that, as the arsenic loadings on the residues increased with increasing As:Fe ratios, the capacity of the residues for selenium decreased. Generally, a higher level of arsenic removal was achieved at the higher pulp density.

X-ray diffraction showed that the matte samples were predominantly composed of heazlewoodite with smaller amounts of pentlandite, millerite and magnetite. During leaching, the sulphide phases decreased in concentration with time, except millerite which displayed an initial increase. Thereafter its concentration fell rapidly to near zero.

The residues were composed mostly of ferrihydrite with about 15% magnetite. Small amounts of hematite and lepidocrocite were also identified in some residues.

No evidence of a ferric arsenate precipitate was found in the final residues. However, electron microprobe analysis showed that arsenic was uniformly distributed throughout the ferrihydrite with little detected in the magnetite.

6 Acknowledgements

One of the authors (RPD) is grateful to the A J Parker Cooperative Research Centre for Hydrometallurgy for awarding him a visiting fellowship. The assistance of Mr I Barrow and Mr D Herring (chemical analysis), Mr L Bastian (x-ray diffraction analysis), all of the Mineral Science Laboratory, Chemistry Centre (WA) and Dr T Parks, CSIRO Division of Mineral Products for the EPMA and for the quantitative analysis of the x-ray diffraction data, is gratefully acknowledged. The authors would also like to thank the management and staff of the Kwinana Nickel Refinery for their advice and assistance. This paper is published with the permission of Western Mining Corporation Ltd and the Director, Chemistry Centre (WA).

7 References

1. Boldt, J. R. and P. Queneau, 1967. *The Winning of Nickel*. Methuen and Co. Ltd. London.

2. Burkin, A. R., 1987. Ch.5: Hydrometallurgy of Nickel Sulphides in *Extractive Metallurgy of Nickel.* (Ed. A. R. Burkin). John Wiley and Sons, Chichester.

3. Comba, P., D. R. Dahnke and L. G. Twidwell, 1988. Removal of Arsenic from Process and Waste Water Solutions. *Proc. of Sym: Arsenic Metallurgy Fundamentals and Applications.* (Eds. R. G. Reddy, J. L. Hendrix and P. B. Queneau): pp 305-320, January 25-28, Arizona. TMS-AIME.

4. Dutrizac, J. E., 1990. Converting Jarosite Residues into Compact Hematite Products. *J. of Metals*, January: 36-39.

5. Gonzalez, V. L. E. and A. J. Monhemius, 1988. The Mineralogy of Arsenates Relating to Arsenic Impurity Control. *Proc. of Sym: Arsenic Metallurgy Fundamentals and Applications.* (Eds. R. G. Reddy, J. L. Hendrix and P. B. Queneau): pp 405-454, January 25-28, Arizona. TMS-AIME.

6. Papassiopi, N., M. Stefanakis and A. Kontopoulos, 1988. Removal of Arsenic from Solutions by Precipitation as Ferric Arsenates. *Proc. of Sym: Arsenic Metallurgy Fundamentals and Applications.* (Eds. R. G. Reddy, J. L. Hendrix and P. B. Queneau): pp 321-334, January 25-28, Arizona. TMS-AIME.

7. Robins, R. G., 1988. Arsenic Hydrometallurgy. *Proc. of Sym: Arsenic Metallurgy Fundamentals and Applications.* (Eds. R. G. Reddy, J. L. Hendrix and P. B. Queneau): pp 215-248, January 25-28, Arizona. TMS-AIME.

8. Rosehart, R. G., J. Y. Lee and G. W. Pattyson, 1972. Arsenic and its Removal from Gold Extraction Plant Effluents. Paper presented at *MMIJ - AIME Joint Meeting.* Tokyo, Japan. May.

9. Gulledge, J. H. and J. T.O'Connor, 1973. Removal of Arsenic (V) from Water by Adsorption on Aluminum and Ferric Hydroxides. *J. American Water Works Association*, August: 548-552.

10. Pierce, M. L. and C. B. Moore, 1980. Adsorption of Arsenite on Amorphous Iron Hydroxide from Dilute Aqueous Solution. *Environmental Science and Technology.* **14**(2), February: 214-216.

11. Robins, R. G., J. C. Y. Huang, T. Nishimura and G. H. Khoe, 1988. The Adsorption of Arsenate Ion by Ferric Hydroxide. *Proc. of Sym: Arsenic Metallurgy Fundamentals and Applications.* (Eds. R. G. Reddy, J. L. Hendrix and P. B. Queneau): pp 99-114, January 25-28, Arizona. TMS-AIME.

12. Das, R. P., 1993. *Precipitation of Iron(III) Oxides from Ammonia - Ammonia Sulphate Solution, and the Effect of Arsenic.* Report - A J Parker Cooperative Research Centre For Hydrometallurgy.

13. Au-Yeung, S. C. F. and G. L. Bolton, 1986. Iron Control in Processes Developed at Sherritt Gordon Mines, in *Iron Control in Hydrometallurgy.* (Eds. J. E. Dutrizac and A. J. Monhemius): pp 131-151, Ellis Horwood Ltd., Chichester.

14. Osseo-Asare, K. and D. W. Fuerstenau, 1979. Adsorption Phenomena in Hydrometallurgy. 1. The Uptake of Copper, Nickel and Cobalt by Oxide Adsorbents in Aqueous Ammoniacal Solutions. *Int. J. of Min. Proc.*, **6**: 85-104.

15. Veltman, H., 1977. Pressure Hydrometallurgy. Paper presented at *Symposium 77, The Indonesian Mining Industry, Its Present and Future*: Sponsored by the Ministry of Mines of the Republic of Indonesia, Jakarta, June.

16. Wishaw, B., 1989. *The Chemistry of the Sherritt-Gordon Nickel Extraction Process.* Report: Western Mining Corporation Ltd.

17. Kerfoot, D. G. E. and D. R. Weir, 1988. The Hydro and Electrometallurgy of Nickel and Cobalt. *Extractive Metallurgy of Nickel and Cobalt.* (Eds. G. P. Tyroler and C. A. Landolt): pp 241-267, TMS.

18. Naboichenko, S. S. and I. F. Khudyakov, 1968. Behaviour of Iron during Autoclave Ammonia Leaching of Nickel Matte. *Tsvetnye Metall.*, No. 3: 37-41 (English Translation).

19. Schwertmann, U. and R. M. Cornell, 1991. *Iron Oxides in the Laboratory*. VCH, Weinheim, (Fed. Rep. of Germany).

20. Schwertmann, U. and H. Thalmann, 1976. The Influence of [Fe(II)], [Si], and pH on the Formation of Lepidocrocite and Ferrihydrite during Oxidation of Aqueous $FeCl_2$ Solutions. *Clay Minerals*, **11**: 189-199.

21. Schwertmann, U. and R. M. Taylor, 1977. "Iron Oxides" in *Minerals in Soil Environments*. (Man. Ed. R. C. Dinauer): pp 144-180, Soil Science Society of America, Wisconsin, USA.

22. Lewis, D. G. and U. Schwertmann, 1980. The Effect of [OH] on the Goethite Produced from Ferrihydrite under Alkaline Conditions. *J. of Colloid and Interfacial Science*, **78**(2): 543-553.

23. Schwertmann, U. and E. Murad, 1983. Effect of pH on the Formation of Goethite and Hematite from Ferrihydrite. *Clays and Clay Minerals*, **31** (4): 277-284.

24. Johnston, J. H. and D. G. Lewis, 1983. A Detailed Study of the Transformation of Ferrihydrite to Hematite in an Aqueous Medium at 92°C. *Geochimica et Cosmochimica Acta*, **47**: 1823-1831.

25. Davis, J. A. and D. B. Kent, 1990. Surface Complexation Modeling in Aqueous Geochemistry. *Rev. Mineral.* **23** (Eds. M. F. Hochella and A. F. White). pp 177-260. Mineral. Soc. Amer.

26. Waychunas, G. A., B. A. Rea, C. C. Fuller and J. A. Davis, 1993. Surface Chemistry of Ferrihydrite: Part 1. EXAFS Studies of the Geometry of Coprecipitated and Adsorbed Arsenate. *Geochimica et Cosmochimica Acta*, **57**: 2251-2269.

27 Fuller, C. C., J. A. Davis and G. A. Waychunas, 1993. Surface chemistry of ferrihydrite: Part 2. Kinetics of arsenate adsorption and coprecipitation. *Geochimica et Cosmochimica Acta*, **57**: 2271-2282.

28. Bar-Yosef, B. and D. Meek, 1987. Selenium Sorption by Kaolinite and Montmorillonite. *Soil Sc.* **144**: 11-19.

29. Neal, R. H., G. Sposito, K. M. Holtzclaw and S. J. Traina, 1987. Selenite Adsorption on Alluvial Soils: II. Solution Composition Effects. *Soil Sc. Soc. Am. J.*, **51**, 1165-1169.

30. Patterson, J. W., 1980. *Wastewater Treatment Technology*. Ann Arbor Science, Michigan, USA.

31. Kawulka, P., C. R. Kirby and G. L. Bolton, 1987. The Sherritt-Cominco Copper Process. Part II: Pilot-Plant Operation. *Can. Min. Metall. Bull.*, February, 7(790): 122-130.

32. Monhemius, A. J., 1987. Ch. 3: Treatment of Laterite Ores of Nickel to Produce Ferronickel, Matte or Precipitated Sulphide in *Extractive Metallurgy of Nickel*. (Ed. A. R. Burkin). John Wiley and Sons, Chichester.

33. Pourbaix, M, 1963. *Atlas D'equilibres electrochimiques*. Gauthier-Villars and Co.

34. Hayes, K. F., A. L. Roe, G. E. Brown, K. O. Hodgson, J. O. Leckie and G. A. Parks, 1987. In Situ X-ray Absorption Study of Surface Complexes: Selenium Oxyanions on α-FeOOH. *Science*, **238**: 783-786.

35. Parks, T. C., 1994. Personal Communication. Division of Mineral Products, CSIRO. Wembley, W. A. 6014 Australia.

Electrochemical study of oxidative dissolution of synthetic violarite in aqueous media

T. E. Warner
Department of Mining and Mineral Engineering, University of Leeds, Leeds, England
N. M. Rice
Department of Mining and Mineral Engineering, University of Leeds, Leeds, England
N. Taylor
School of Chemistry, University of Leeds, Leeds, England

A study of the oxidative dissolution of the nickel-iron sulphide mineral violarite ($FeNi_2S_4$) by electrochemical techniques including potentiometry, linear sweep cyclic voltammetry, intermittent galvanostatic polarisation and chronoamperometry was made to clarify the mechanism by which violarite is leached in acidic iron (III) chloride solution. This forms a complementary study to that recently reported on the related mineral pentlandite ($Fe_{4.5}Ni_{4.5}S_8$). A mechanism for the oxidative dissolution of violarite is proposed. In acid solution, under potentiostatic conditions akin to iron (III) chloride leaching, violarite is oxidized to elemental sulphur. The formation of metastable amorphous sulphur as opposed to ortho-rhombic sulphur indicates that the system is substantially perturbed from equilibrium. The physical properties of the sulphur product layer cause an impediment to mass transport between the bulk aqueous solution and the mineral surface. However, the oxidation involves an intrinsically slow electron transfer for the S, Fe^{2+}, Ni^{2+} / $FeNi_2S_4$ couple with heterogeneous rate constant, $k_o = 10^{-9}$ m s^{-1}. Within the potential range relevant to iron(III) chloride leaching, this electron transfer is rate-determining for an appreciable part of the reaction. A comparison with similar studies on violarite is made.
Keywords: Chloride, Chronoamperometry, Cyclic voltammetry, Electrochemistry, Hydrochloric acid, Iron, Kinetics, Leaching, Nickel, Oxidation, Perchloric acid, Polarisation, Potential, Sulphide,
Sulphur, Violarite,

1 Introduction

The principal mineral associated with nickel sulphide ores is the nickel-iron sulphide pentlandite, with a nominal stoichiometry, $Fe_{4.5}Ni_{4.5}S_8$. Due to the recent economic climate, exploitation of new nickel sulphide deposits has focused on those associated with gold or platinum group metals (PGM). This created an impetus to develop industrial processes for the extraction and separation of both types of metal value, such as that operated at Impala Platinum Ltd., South Africa (Plasket and Romanchuk,1978). Likewise, the Sherritt Gordon Process (Forward, 1953; Forward and Mackiw, 1955; Warner,J.P., 1956) involves pressure hydrometallurgical technology for the separation of copper sulphide and the extraction of nickel and cobalt powders. Recently, there has been a growing interest in chloride hydrometallurgy. For example, use of copper(I) chloride as a lixiviant for chalcopyrite in the CLEAR process (Schweitzer and Livingston, 1982); the Falconbridge chlorine leach process for the treatment of nickel matte (Stensholt, Zachariason and Lund, 1986) and the Cymet process (Kruesi, Allen and Lake, 1973) where iron(III) chloride is used as a lixiviant for copper sulphide minerals. Furthermore, there is great interest in extending the use of these leaching agents to other chalcogenide systems, for example, anode slimes containing PGM (Luo, 1990) and in the treatment of low grade and complex sulphide ores. A suplphur layer can also retard the leaching of chalcopyrite in iron(III) sulphate medium unless measures such as attrition grinding to remove the surface product layer are carried out during leaching (Cobble, Jordan and Rice,D.A. 1993).

The use of acid iron(III) chloride as a lixiviant for pentlandite has been investigated by previous workers, notably Kelt (1975) and Tzamtzis (1976), using conventional chemical techniques. The results of their work indicated that the reaction rate is very slow, and that the nickel dissolution is incomplete. This was explained in terms of a mixed-kinetic regime involving mass transport control of the oxidant, iron(III) chloride, from the bulk aqueous solution to the pentlandite surface through a progressively thickening sulphur product layer. A complimentary study on synthetic pentlandite using electrochemical techniques, recently reported by us (Warner,T.E., Rice and Taylor, 1992), strongly suggested that within the potential range relevant to iron(III) chloride leaching (i.e. 0.4 to 0.7 V vs. SCE) and over the temperature range 293 to 343 K, and with a sufficient aqueous concentration of iron(III) chloride, the oxidation of pentlandite involves slow electron transfer for the S, Fe^{2+}, Ni^{2+} / $Fe_{4.5}Ni_{4.5}S_8$ couple itself and is rate determining.

The nickel-iron sulphide mineral, violarite, $(Fe,Ni)_3S_4$ frequently occurs as a supergene alteration product from the oxidation of pentlandite in certain massive nickel-bearing sulphide assemblages. Since these two minerals share an almost identical sulphide substructure, coupled with much evidence of an electrochemical mechanism for the supergene alteration process (Thornber, Allchurch and Nickel, 1981), together with the commercial importance of violarite, an electrochemical study of the oxidative dissolution of violarite in acid solution was undertaken. Thornber (1983) had previously performed a series of cyclic voltammetric and intermittent galvanostatic polarisation (IGP) experiments on naturally occurring violarite with the principal aim of investigating the mechanism of this supergene alteration process and therefore conducted these under conditions more akin to the relevant geological environment.

In the present work the mechanism for the dissolution of synthetic violarite under conditions relevant to envisaged hydrometallurgical processing has been explored using a series of electrochemical techniques. Since violarite is a good electronic conductor, much beneficial information concerning the dissolution mechanism can be revealed through various electrochemical techniques including, potentiometry, linear cyclic Voltammetry, IGP and chronoamperometry.

2 Experimental Techniques

2.1 Synthesis

Approximately 100 g of violarite with the nominal stoichiometry, $FeNi_2S_4$, was synthesised by a two stage dry *in vacuo* technique. This first involved the preparation of 100 g of monosulphide solid solution (MSS) with a net stoichiometry of $FeNi_2S_{3.25}$, by heating iron, nickel and sulphur powders together in the appropriate molar ratio in an evacuated double-bulbed fused "96%" silica glass vessel. The intermediate product MSS was ground and reacted with a further quantity of sulphur in another evacuated double-bulbed fused silica glass vessel at 568 K for several days. The product material was mildly sintered with a dull violet/grey colour, whilst the larger crystals showed a more metallic lustre. This product was ground, sealed in an evacuated silica tube and sintered at 568 K for 38 days to produce a compact rod of homogeneous violarite. Powder X-ray diffraction and EPMA revealed a distinct variation between two violarite compositions within the individual violarite grains: it had a core of Ni-deficient $Fe_{1.1}Ni_{1.9}S_4$ surrounded by a rim of Ni-rich, $Fe_{0.9}Ni_{2.1}S_4$. More explicit details concerning the synthesis are reported elsewhere (Warner,T.E., 1988; Warner, Rice and Taylor, 1992).

2.2 Electrochemical Techniques

The electrochemical experiments were done using a conventional three electrode cell configuration enclosed within a Metrohm vessel, with a thermostatic jacket, containing a working electrolyte volume of 0.05 dm^3. A Metrohm vessel lid provided ports for the working, platinum counter and saturated calomel reference electrodes, together with the thermometer and gas disperser unit. A full description of the fabrication of the working electrode, together with other aspects of the cell has been given previously (Warner, 1988; Warner, Rice and Taylor, 1992).

The experiments were performed using an EG&G PAR model 362 potentiostat /galvanostat. The potential control range was -9.999 to 9.999 V, with potential scan rates ranging from 0.0001 to 0.5 V s^{-1}. When operating in galvanostatic mode, cathodic or anodic currents could be applied in the range 1 μA to 1 A.

For IGP experiments an EG&G PAR model 175 Universal Programmer was used to trigger the galvanostat via an inlet modification to the model 362. Details of this are given elsewhere.(Warner, 1988; Warner, Rice and Taylor, 1992) A correction for a small offset in the *zero* current (which often occurred when in operation, i.e. when the "ON" button on the model 362 was pressed) was made by applying an appropriate potential from a Eurotherm millivolt source type 039, connected to the "EXT IN" on model 362. This was particularly important since the system investigated in this work often involved small exchange current densities.

Measurement of current and potential were made using a high input impedance Keithley model 171 digital multimeter and a Keithley model 616 digital electrometer, respectively. Current and potential profiles were recorded on a Brown Boveri multi-channel X-t chart recorder (Servogor model 460) and a Bryans X-Y recorder model 2600A4.

2.3 Analytical Techniques

A Vickers M55 microscope was used in the identification of the phases present in the metal sulphide samples. A Camscan Series 3 scanning electron microscope fitted with a Link Systems analytical 860 Series for energy dispersive X-ray analysis (EDXA) was used to study the surface morphology / composition of synthetic and electrolysed specimens. A Philips X-ray powder diffractometer and a Siemens Kristalloflex 2 powder camera were used for powder X-ray diffraction studies. A Joel JXA-50A electron probe microanalyser fitted with a Link 860-500 energy dispersive system was used for the compositional analysis of synthetic and electrolysed specimens.

3 Results and Discussion

3.1 Potentiometry

Measurements of the open-circuit potential of violarite $FeNi_2S_4$ in 1 M HCl solution over the temperature range 293 to 353 K gave poor reproducibility, with values drifting between 100 and 400 mV vs. SCE. This is to be expected since from thermodynamic considerations at ambient temperature it is impossible for hydrogen ions to oxidize this sulphur rich phase, whilst the alternative dissolution route with the conversion of all the sulphur to hydrogen sulphide would require reducing conditions (Warner 1988). Therefore, in a closed aqueous system of acid pH free from imposed oxidising or reducing conditions, violarite is considered to be metastable with respect to sulphur disproportionation:

$$4FeNi_2S_{4(c)} + 23H^+_{(aq)} + 4H_2O \rightarrow 8Ni^{2+}_{(aq)} + 4Fe^{2+}_{(aq)} + 15H_2S_{(aq)} + HSO_4^-{}_{(aq)} \quad (1)$$

hence the association of an ill-defined open-circuit potential.

When violarite is immersed in acid solutions with a defined oxidising potential (e.g. containing a FeC13/$FeCl_2$ couple), then reproducible initial open-circuit potentials are obtained. These can be crudely compared to the equilibrium potentials calculated from thermodynamic data at 298 K (Warner 1988) for the reaction:

$$FeNi_2S_{4(c)} = Fe^{2+}_{(aq)} + 2Ni^{2+}_{(aq)} + 4S_{(s)} + 6e^- \quad (2)$$

where

$$E(298\ K) = 0.303 + 0.0197 \log[Ni^{2+}] + 0.0099 \log[Fe^{2+}] \quad \text{vs. SHE.} \quad (3)$$

Taking $[Fe^{2+}] = 0.1$ M together with a conservative value for $[Ni^{2+}] = 10^{-3}$ M, and converting the reference from SHE to SCE gives E(298) = -8 mV vs SCE.

From Fig. 1 it can be seen that these open-circuit potentials closely correspond to those on an inert metal electrode such as platinum. In other words, the potential measured on the violarite is approximately equal to that of the solution, indicating that violarite is almost kinetically inert under these conditions.

The observation that electrode potentials of certain sulphide minerals, particularly those of polysulphides are very similar to the redox potential of the solution in which they are immersed was noted as long ago as 1914 by Wells (Wells, 1914). In the past, the mechanistic implication of these observations with regards to envisaged hydrometallurgical leaching may not always have been entirely appreciated.

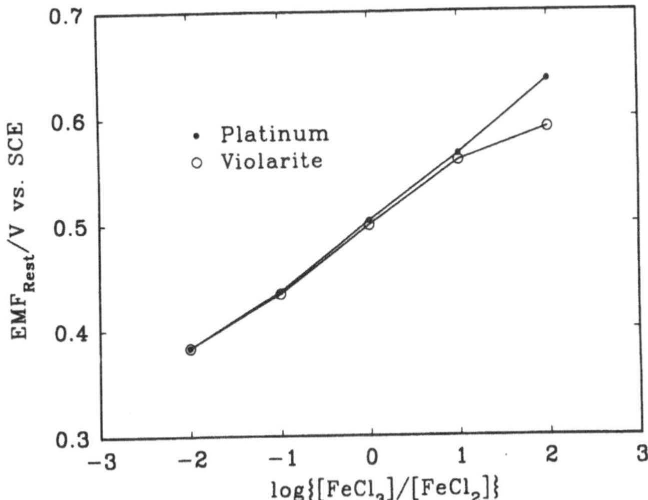

Fig. 1. Correlation of open-circuit potential in 1 M HCl solution with various $[FeCl_3]/[FeCl_2]$ ratios, 343K, 20 Hz RDE. 0.1 M total iron concentration.

Such observations can be viewed in terms of the electrochemical concept of a mixed-potential regime ((Warner, Rice and Taylor, 1992; Bard and Faulkner, 1980). From Fig. 1 the difference between the open-circuit potential on violarite and the redox potential of the solution for most $[FeCl_3]/[FeCl_2]$ ratios is very small. However, for the highly oxidising solution with $[FeCl_3]/[FeCl_2]$ ratio = 100, the disparity of 50 mV is due to the enhanced positive potential of the solution, such that the violarite experiences a greater over-potential leading to an increase in the anodic current. This manifests itself in an increased oxidative dissolution rate of the violarite. Nevertheless, the exchange current density is remarkably small even for this strongly oxidising solution. For a comparison, copper metal in the same solution (even at the colder temperature of 293 K) gives an initial mixed-potential E = - 0.23 V vs. SCE. This substantially lower value for the mixed-potential reflects the more facile electrochemical kinetics for the corroding $Cu°/Cu^{2+}$ couple, even after taking into consideration the greater "nobility" of $FeNi_2S_4$ compared with elemental copper (Warner, 1988)

These results indicate that the exchange-current density for the S,Fe^{2+},Ni^{2+} / $FeNi_2S_4$ couple is very small. In addition, since the concentration of metal atoms within the violarite crystal lattice is large (47 mol dm^{-3}), this implies that the heterogeneous rate constant for this couple is extremely small. Thus, it is reasonable to suggest that the oxidative dissolution of violarite in acid solution involves slow electron transfer over the temperature range studied (293 - 343 K). It is also interesting to note that this convenient (and relatively cheap) technique can yield important kinetic information. The results from more sophisticated electrochemical techniques described below tend to corroborate this point.

3.2 Linear sweep cyclic voltammetry

Fig. 2 shows a typical voltammogram for violarite in 1 M HCl solution at 343 K. The features are in many ways similar to those for the related mineral pentlandite [10]. Reasonable rates of oxidation are not obtained until relatively high potentials (0.8 V vs. SCE) which coincide with the formation of amorphous (plastic) sulphur on the violarite surface.

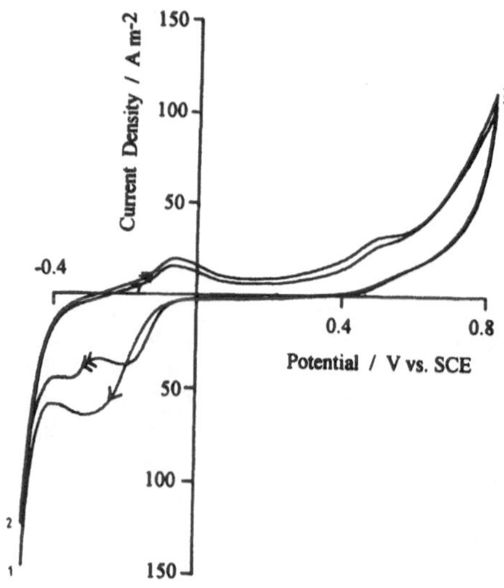

Fig. 2. Cyclic voltammograms for violarite electrode in 1 M HCl solution, 343 K, 5 mVs^{-1} sweep rate, 20 Hz RDE. Initial scan is cathodic.

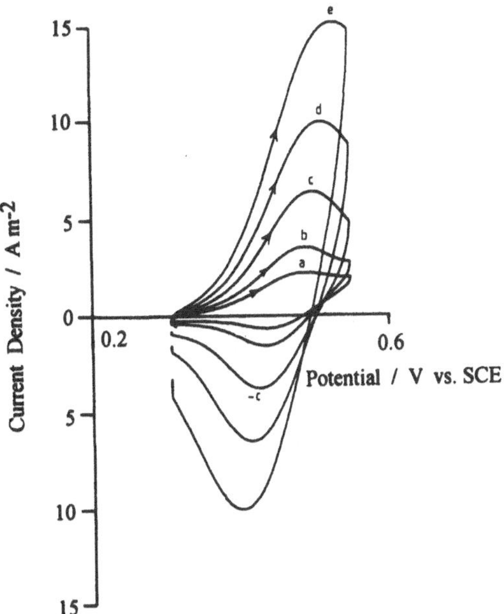

Fig. 3. Cyclic voltammograms for violarite electrode in 1 M HCl solution, 343 K, 20 Hz RDE. Initial scan anodic of E_{rest} (0.310 V) followed by return scan.
Repeated for various sweep rates (mVs^{-1}): (a) 1; (b) 2; (c) 5; (d) 10; (e) 20.

A few features arise in the cyclic voltammograms for violarite which are different from those for pentlandite. When the potential is initially scanned cathodic from the rest potential, a cathodic current is generated. However, potentials in the region 0.4 to -0.3 V vs. SCE are insufficiently negative for the evolution of hydrogen on the violarite surface. Furthermore, these currents cannot be attributed to the reduction of elemental sulphur since at this stage in the cycle none is presumedly present. Therefore, the reduction of violarite to hydrogen sulphide is implicated:

$$FeNi_2S_{4(c)} + 8H^+_{(aq)} + 2e^- \rightarrow Fe^{2+}_{(aq)} + 2Ni^{2+}_{(aq)} + 4H_2S_{(aq)}. \qquad (4)$$

The predominant cathodic current at potentials -0.4 V vs. SCE is due to the reduction of the acid solution resulting in hydrogen evolution. On the return (anodic) scan, the initial anodic current (with a peak at 0 V vs. SCE at high temperatures) is attributed to the oxidation of a metal rich surface, and compares favourably with the results of Thornber (1983).

An almost characteristic feature of violarite cyclic voltammograms is a "prewave" during the anodic excursion at approximately 0.5 V vs. SCE, with the appearance of associated cathodic currents on the reverse scan. This feature was studied by a series of cyclic voltammograms with a variety of scan rates to progressively greater anodic potentials. This revealed a clear association between the cathodic and the anodic "prewaves" (Warner, 1988). This "prewave" can be regarded as a superimposed anodic process in addition to the main anodic oxidation of violarite.

Results captured on a more sensitive current scale (see Fig. 3) show that the peak currents are proportional to the square root of the scan rate which is one of the diagnostic features of fast electron transfer (Bard and Faulkner, 1980).(see Fig.4) With reference to Fig. 3 the peak separation for the trace "(c)"
(i.e. $E_p(c) - E_p(-c) \approx 0.45$ V vs. SCE).

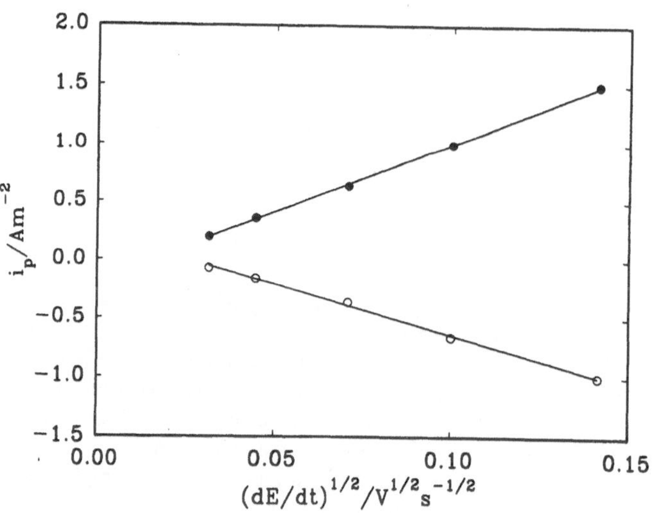

Fig. 4. A plot of the dependence of the peak current density on the square root of the potential sweep rate, for the results obtained from cyclic voltammetry on violarite (see Fig. 3).

At 323 K the peak potentials (E_p) are more independent of the scan rate than at 303 K, indicating increased reversible behaviour at this higher temperature. At 323 K the cathodic currents are of a smaller magnitude compared to anodic currents, suggesting that the oxidized species has a greater mobility to leave the surface of the violarite. However, a comparison between rotating disk and static electrode experiments showed that diffusion of the electroactive species in the electrolyte near the mineral surface was not rate determining (Warner, 1988). It is possible that the rate limiting currents are governed by the rate of formation of the electroactive species by a preceding reaction.

The above results strongly indicate the presence of an iron(III)/iron(II) couple. Since iron is absent from the bulk solution, the oxidation of an iron(II) species at the violarite surface as released by a preceding reaction (e.g. the main anodic oxidation process on violarite) is suggested here as the explanation for the superimposed "prewave" during the anodic polarisation of violarite. This helps to extend the earlier work by Thornber for which he ascribes this "prewave" to the oxidation of the "normal" violarite surface generating a sulphur rich layer (Thornber, 1983)

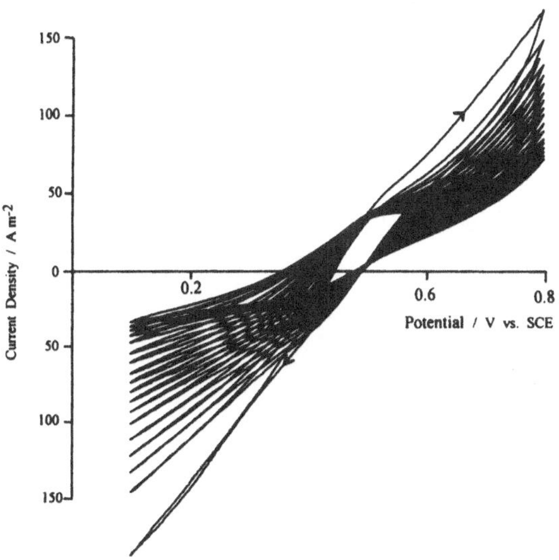

Fig. 5. Cyclic voltammograms for violarite electrode in 1 M HCl solution containing 0.05 M $FeCl_3$ and 0.05 M $FeCl_2$, 343 K, 5 mVs^{-1} sweep rate, 20 Hz RDE. Initial scan cathodic of E_{rest} (0.464 V) followed by anodic excursions and 18 repeated cycles.

Fig. 5 shows cyclic voltammograms for the $FeCl_3/FeCl_2$ couple on the violarite surface. These experiments are similar to those previously performed on pentlandite (Warner, Rice and Taylor, 1992). The effects on the observed current density during successive cathodic cycles were studied with progressively greater anodic excursions. These results indicate that the observed current densities decrease *pro rata* with the extent of anodic polarisation. The effects of an amorphous sulphur film on the mass transport of the Fe^{3+} and Fe^{2+} ions are implicated.

3.3 Intermittent Galvanostatic Polarisation (IGP)

Fig. 6. shows the result of a typical IGP experiment on violarite. Anodic potentials of 0.8 V vs. SCE are observed during the anodic polarisation cycle, which correspond to the major anodic peak in the cyclic voltammogram described above. The electrochemical process accompanying this potential is ascribed to the formation of elemental sulphur. The open-circuit potential decays with time to values akin to the initial rest-potential.

On reversal of the polarisation, it is inferred that the reduction of elemental sulphur and violarite occurs via the following two reactions:

$$S_{(s)} + 2H^{+}_{(aq)} + 2e^{-} \rightarrow H2S_{(aq)} \qquad (5)$$

and

$$FeNi_2S_{4\,(c)} + 8H^{+}_{(aq)} + 2e^{-} \rightarrow Fe^{2+}_{(aq)} + 2Ni^{2+}_{(aq)} + 4H_2S_{(aq)}. \qquad (6)$$

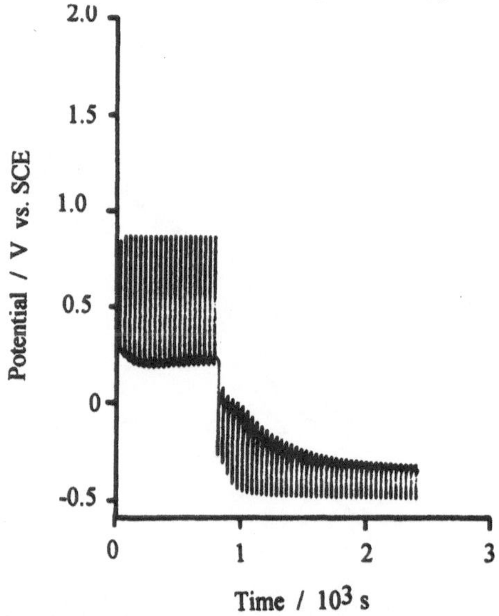

Fig.6. IGP trace for violarite in unstirred 1 M HCl solution, 343 K. Intermittent anodic polarisation for the first 900 s followed by intermittent cathodic polarisation. Current density 177 Am^{-2}.

At larger imposed cathodic current densities these reactions are accompanied by the reduction of the acid solution to hydrogen. During the intermittent spells in open-circuit, a transient potential arrest occurs at -0.3 V vs. SCE. It is possible that this corresponds to a mixed-potential regime between a "metal-rich" sulphide surface formed during the preceding Faradaic reaction (6) and the H^+/H_2 couple, i.e.

$$(Fe,Ni)_{1+x}S_{(c)} + 2(1+x)\,H^+_{(aq)} \rightarrow (1+x)\,(Fe,Ni)^{2+}_{(aq)} + H_2S_{(aq)} + x\,H_{2\,(g)} \qquad (7)$$

This interpretation compares favourably with that offered by Thornber (1983). However, in the absencee of additional surface analysis data, e.g. electron spectroscopy for chemical analysis (ESCA), this interaction must remain conjectural.

3.4 Chronoamperommetry

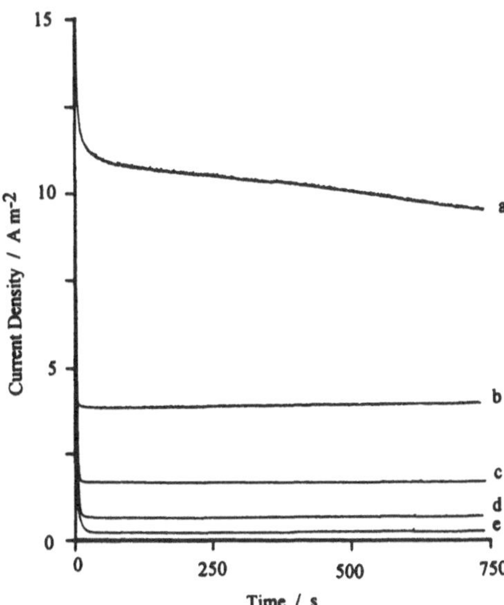

Fig. 7. Chronoamperograms for violarite electrode in 1 M $HClO_4$ solution, 353 K, 20 Hz RDE. Anodic polarisation for 720 s. Applied potentials / V vs. SCE: a = 0.9; b = 0.8; c = 0.7; d = 0.6; e = 0.5.

Fig.7 shows chronoamperograms for violarite in 1M $HClO_4$ solution at 353 K. An important feature of these results is that Cottrell type behaviour (Bard and Faulkner, 1980) is not observed. Steady state currents were rapidly obtained, indicating that diffusion processes are not rate limiting within the potential range 0.5 to 0.8 V vs. SCE. These data were used to construct a Tafel plot (see Fig. 8) in which the natural $\log_e$ of the observed current density (ln i) was plotted as a function of the overpotential (η):

$$\ln i = \ln i_o + \alpha_a nF\eta/RT \qquad (8)$$

where i_o is the exchange current density, α_a is the anodic charge transfer coefficient, n is the number of equivalents of charge per metal atom involved in the reaction (in this case $n = 2$), F is the Faraday constant and if the applied potential is E_{apl},

$$\eta = E_{apl} - E_{eq}$$

where the latter value was calculated using equation (11) below.

The following reaction (9) is assumed to represent the predominant process here:

$$FeNi_2S_{4(c)} \rightarrow Fe^{2+}{}_{(aq)} + 2Ni^{2+}{}_{(aq)} + 4S_{(s)} + 6e^- \qquad (9).$$

The possibility of the subsequent reaction (10):

$$Fe^{2+}{}_{(aq)} \rightarrow Fe^{3+}{}_{(aq)} + e^- \qquad (10)$$

is ignored in this treatment for the sake of simplicity. Unfortunately, the standard equilibrium potential, E°, for the Fe^{2+}, Ni^{2+}, S / $FeNi_2S_4$ couple is unknown at 353 K. Likewise, the activities of $Ni^{2+}{}_{(aq)}$ and $Fe^{2+}{}_{(aq)}$ are undefined in this working electrolyte. However, as a first order of approximation the equilibrium potential at 298 K was adopted from equation (11) (Warner, 1988):

$$E(298\ K) = 0.303 + 0.0197 \log[Ni^{2+}] + 0.0099 \log[Fe^{2+}] \text{ V vs. SHE,} \qquad (11)$$

such that taking arbitrary values of $[Fe^{2+}]$ and $[Ni^{2+}]$ both equal to 10^{-3} M, and converting the reference electrode from SHE to SCE gives an equilibrium potential E_{eq} = -28 mV vs SCE. This value was used to calculate the overpotentials plotted in Fig. 8.

The Tafel plot is linear in the potential range measured. The slope (= 9.2 natural decades V^{-1}) yields the anodic charge transfer coefficient, α_a = 0.14, and the intercept at = 0 (-6.2) yields the exchange current density $i_o = 2 \times 10^{-3}$ A m^{-2}. Taking the bulk metal atom concentration in violarite as 47×10^3 mol m^{-3} (Warner, 1988), gives an order of magnitude for the heterogeneous rate constant, k_o (Fe^{2+}, Ni^{2+}, S / $FeNi_2S_4$) = 10^{-9} ms^{-1},
which is well within in the domain of slow electron transfer.

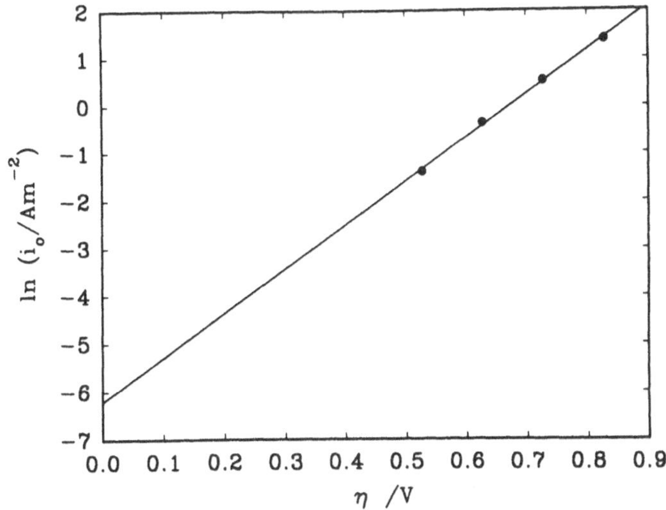

Fig 8. Tafel plot for the anodic oxidation of violarite electrode in 1 M $HClO_4$ solution, 353 K (see Fig. 7).

4 Conclusions

Violarite adopts a stray and undefined open-circuit potential in 1 M HCl aqueous solution. However, in the same solution with the addition of a defined oxidising potential e.g. the $FeCl_3/FeCl_2$ couple, violarite adopts an open-circuit potential very close to the redox potential of the solution in which it is immersed. This is also the case at the higher temperatures envisaged for hydrometallurgical leaching e.g. 343 K. Adopting the concept of a mixed-potential regime suggests that this corrosion couple involves an extremely low exchange current density for the reaction (9) viz.:

$$FeNi_2S_{4(c)} \rightarrow Fe^{2+}{}_{(aq)} + 2Ni^{2+}{}_{(aq)} + 4S(s) + 6e^-.$$

From cyclic voltammetric experiments on violarite in 1 M HCl solution it is evident that appreciable rates of dissolution are not achieved until potentials in excess of 0.8 V vs. SCE are applied. This presents a considerable overpotential for the above reaction which is in agreement with the results from the above potentiometric experiments. The "prewave" at approximately 0.5 V vs. SCE is interpretted as the oxidation of an iron(II) species on the violarite surface after release by a preceding reaction.

The physical properties of the amorphous sulphur product layer cause an impediment to mass transport for the oxidant species used in chemical leaching (e.g. $FeCl_3$) from the bulk aqueous solution to the violarite surface, and for the mass transport of the product aqueous metal ions from the mineral surface to the bulk aqueous solution. However, this mass transport process is not considered to be rate-determining during violarite dissolution under the conditions studied here.

Chronoamperometric data for violarite in 1 M $HClO_4$ solution yield a very small heterogeneous rate constant, $k_o = 10^{-9}$ m s^{-1}, which is consistent with an electron transfer controlled mechanism.

This study suggests that within the potential range relevant to iron(III) chloride leaching (i.e. 0.4 to 0.7 V vs. SCE) and over the temperature range 293 to 343 K, and with a sufficient aqueous concentration of iron(III) chloride, the oxidation of violarite involves slow electron transfer for the Fe^{2+}, Ni^{2+}, S / $FeNi_2S_4$ couple itself, and is rate determining. It is concluded that acid iron(III) chloride solution would be an inappropriate lixiviant for the dissolution of violarite. Chemical oxidants of higher potential would be required (e.g. persulphate). These might result in further oxidation with the formation of appreciable amounts of sulphate, and would therefore be undesirable from an environmental point of view. Further studies in sulphate and ammoniacal media are currently in progress.

5 Acknowledgements

The authors would like to thank the Science and Engineering Research Council of the United Kingdom for the provision of a Studentship in the form of a Research Quota Award. The authors also thank Sherritt Gordon Mines Ltd, Fort Saskatchewan, Alberta, Canada, for the donation of nickel powder; Kambalda Nickel Operations, Western Mining Co. Ltd., Australia, and INCO Ltd, Ontario, Canada, for the donation of natural specimens of nickel-iron sulphide minerals; and Dr. C.J.Stanley, of the British Museum (Natural History), London, for the provision of natural specimens of nickel-iron sulphide minerals on behalf of the British Museum.

6 References

Bard,A.J. and Faulkner,L.R. (1980) **Electrochemical Methods - Fundamentals and Applications,** Wiley, New York, NY.

Cobble,J.K., Jordan,C.E. and Rice,D.A. (1993) Hydrometallurgical production of copper from flotation concentrates. U.S.Bur. Mines Rept. Inv. RI-9472.

Forward,F.A. (1953) Ammonia pressure leach process for recovering nickel, copper and cobalt from Sherritt Gordon sulphide concentrate, **Trans. Can. Inst. Min. Metall.,** 56, 373-380.

Forward,F.A. and Mackiw,V.N. (1955) Chemistry of the ammonia pressure process for leaching nickel, copper and cobalt from Sherritt Gordon sulphide concentrates, **J. Metals,** 7 (3), 457-463.

Kelt,S.M. (1975) **The leaching of certain nickel sulphides.** Ph.D. Thesis, Univ. London.

Kruesi,P.R., Allen,E.S. and Lake,L.J. (1973) Cymet process - Hydrometallurgical conversion of base-metal sulphides to pure metals, **Can. Min. Metall. Bull.**, 66,(734), June, 81-87.

Luo Rong, (1990) **An electrochemical study of the oxidative dissolution of synthetic copper-silver-selenide minerals in aqueous media,** Ph.D. Thesis, Univ. Leeds.

Plasket,R.P. and Romanchuk,S. (1978) Recovery of nickel and copper from high grade matte at Impala Platinum by the Sherritt Process, **Hydrometallurgy** 3, 135-151.

Schweitzer,F.W. and Livingston,R.W. (1982) Duval's CLEAR hydrometallurgical process, in **Chloride Electrochemistry.** (ed. P.D.Parker), AIME, New York, pp. 221-227.

Stensholt,E.O., Zachariason,H. and Lund,J.H., (1986) Falconbridge chlorine leach process. **Trans. Inst. Min. Metall.**, 96, C10-C16.

Thornber,M.R., Allchurch,P.D. and Nickel,E.H., (1981) Variations in gossan geochemistry at the Perseverance nickel sulphide deposit, Western Australia: A descriptive and experimental study, Economic Geology, 76, 1764-1774.

Thornber,M.R. (1983) Mineralogical and electrochemical stability of the nickel-iron sulphides - pentlandite and violarite, **J. Appl. Electrochem.**, 13, 253-267.

Tzamtzis,G. (1976) **Leaching of some sulphide minerals of nickel under acidic oxidising conditions.** Ph.D. Thesis, Univ. London.

Warner,J.P. (1956) Nickel recovery at Fort Saskatchewan, **Ind. Chem.**, 32, 359-368.

Warner,T.E. **(1988) An electrochemical study of the oxidative dis-solution of synthetic nickel-iron sulphide minerals in aqueous media,** Ph.D. Thesis, Univ. Leeds.

Warner,T.E., Rice,N.M. and Taylor,N. (1992) An electrochemical study of the oxidative dissolution of synthetic pentlandite in aqueous media, **Hydrometallurgy,** 31, 55-90.

Wells,R.C. (1914) Electric activity in ore deposits, **Bulletin 548**, Dept. of the Interior, U.S. Geological Survey.

Bioprocessing

Stability of arsenical bacterial oxidation products

Katerina Adam
Laboratory of Metallurgy, National Technical University of Athens, Athens, Greece
Constantine Komnitsas
Laboratory of Metallurgy, National Technical University of Athens, Athens, Greece
N. L. Papassiopi
Laboratory of Metallurgy, National Technical University of Athens, Athens, Greece
Antonios Kontopoulos
Laboratory of Metallurgy, National Technical University of Athens, Athens, Greece
F. D. Pooley
University of Wales, Cardiff, Wales
N. P. Tidy
University of Wales, Cardiff, Wales

In the present study the stability of the arsenical solid residues emanating from the bacterial oxidation of auriferous arsenical pyrites was evaluated in relation to the operating parameters in the bioleaching circuit. The material examined as a case study was a refractory arsenopyrite concentrate assaying As: 21.0%, Fe: 26.0% and Au: 30.0g/t. The effect of pH-EMF and solution chemistry on precipitate formation during biooxidation and the subsequent stability of the arsenical solid residues was evaluated. The composition and morphology of oxidised precipitates was examined with acid treatment of the bioresidues, chemical and particle size analyses and scanning electron microscopy. Arsenic solubility of the solids under study was evaluated with the standard EPA leachability test. Based on the results of this study, an optimised set of biooxidation operating conditions was derived, including Fe/As ratios in the feed material and oxidised products and Fe^{3+} levels in solution, aiming at the production of stable arsenical bacterial oxidation products, suitable for environmentally safe disposal.
Keywords: Arsenic stability, bioresidues, arsenopyrite bioleaching.

1. Introduction

In recent years, one of the main developing areas of biohydrometallurgy is the extraction of gold from arsenical and other types of refractory sulphide ores and concentrates. Bacterial oxidation of sulphide ores is mainly based on the action of the aerobic, acidophilic, chemilithotrophic bacteria *Thiobacillus ferrooxidans (T.Fe)*. These bacteria, thriving in a temperature range between 30-40°C, are used to accelerate the rate of sulphides minerals oxidation by breaking down the sulphide lattice and thus liberating the encapsulated gold for subsequent recovery by conventional cyanide leaching.

Bacterial oxidation can offer advantages in cost, metallurgical performance and environmental considerations when compared with other oxidative pretreatment options such as roasting and aqueous pressure oxidation. The most attractive application of the process remains where selective oxidation can be explored, such as demonstrated with arsenopyrite-pyrite mixtures where gold is preferentially associated with arsenopyrite [12]. Even in cases where gold occurs in both arsenopyrite and pyrite, bioleaching still offers the possibility of selective oxidation of the gold bearing sites, given that these sites caused by slight variations in pyrite composition, lattice distortion, dislocations etc., have been shown to be preferential regions of bacterial corrosion [1,13,16].

The limited industrial application of this seemingly simple process for the treatment of arsenical pyrite concentrates was mainly attributed to the difficulties observed in effectively controlling the reaction mechanisms and the variability of the ore types [27]. Another issue of concern expected to also affect the commercialisation of this method, is the long term stability of the arsenical solid wastes resulting from the process and classified into two categories:

(a) arsenical sludge formed from the neutralisation of the acidic bioleachates
(b) cyanidation tailings after the cyanide leaching of the bioresidues

In a given circuit the relative amount of the above types of solid wastes depends on the sulphur content of the feed material and the selectivity of biooxidation. Should partial oxidation of the feed pyrites be performed, the volume of the neutralisation sludge is reduced with a parallel increase in the weight of the bioresidues and the respective cyanidation tailings.

Reported data from bioleaching applications on arsenical pyrites mainly refer to the stability of the neutralisation sludge; very limited information is available on the stability of the bioresidues and the respective cyanidation tailings. Regarding the neutralisation solids it is claimed that after treatment of the acidic liquor with lime, arsenic is fixed as stable ferric arsenate suitable for long term disposal [4,8]. Plant data for the first commercial application of bioleaching on a concentrate consisting of 70% pyrite and 30% arsenopyrite at the Fairview mines, showed that neutralisation of the acidic bioliquor with a Fe^{3+}/As [M] ratio in solution greater than 4 resulted in the precipitation of stable ferric arsenates [3].

These findings are in accordance with the results of comprehensive studies reported during the last decade on the stability of amorphous ferric arsenates. These compounds were produced by the neutralisation of purely chemical aqueous acidic systems at atmospheric conditions, with no emphasis paid in products received after the treatment of the bioleaching effluents. According to Robins, [23-25] the stoichiometric amorphous $FeAsO_4$ exhibits high solubility unacceptable for safe disposal. Following studies showed that when the Fe/As molar ratio in the aqueous solutions is higher than 3, the Fe-As precipitates produced were stable in a pH region ranging from 3-7 [6,7,9-11,21,26]. The stability of these precipitates was tested and confirmed with long duration tests, exceeding in some cases two years. Harris and Monette [6,7] showed that the stability region is expanded to the whole pH range tested, 4-10, with the coprecipitation of gypsum and base metals such as Cu, Zn and Cd.

Based on the results of these thorough investigation on the stability of ferric-arsenates precipitates from purely chemical systems it was concluded that in the bioleaching circuit prerequisite factor for the effective arsenic fixation is to maintain the ferric to arsenic ratio in solution to values greater than 3. Thus, it is claimed that in the case of selective biooxidation of arsenopyrite a ferric salt must be added to overcome the deficiency of dissolved ferric iron for the production of stable ferric arsenates [12].

It is noted that all the above studies refer to ferric-arsenate compounds, i.e. arsenic reporting in the oxidised pentavalent state. Arsenite compounds are known to be far more soluble than the respective arsenates [26].

However, evidence exists that during bacterial leaching of arsenical pyrites and depending on the composition of the feed material and the conditions prevailing in the oxidation circuit, an amount of the extracted arsenic reports in the trivalent form. In recently published data from a pilot-plant study of the Salsigne arsenical pyrite concentrate assaying As: 20% and Fe 36%, it is reported that both the bioresidues and the neutralisation sludge were unstable as far as arsenic was concerned [17,19]. The increased arsenic solubility was attributed to the combined presence of arsenite, 80% of total arsenic, and the low Fe/As molar ratio in solution, equal to 1.3. An oxidation step with hydrogen peroxide was incorporated in the biosolution and biopulp treatment to completely oxidise the remaining arsenite in the arsenate form combined with ferric iron addition to increase the Fe/As molar ratio to values higher than 3.0. This treatment resulted in the production of a stable neutralisation sludge. However, and despite this intense additional oxidation stage that added significant reagent costs to the process economics, the arsenic solubility of the bioresidue pulp remained in unacceptably high levels [17].

Similar conclusions were derived from a laboratory study for the treatment of another arsenic-rich refractory pyrite concentrate, the Olympias concentrate assaying As: 12% and Fe: 40% [2]. The neutralisation products exhibited arsenic solubilities above 5.0mg/L, despite the high Fe/As molar ratio of the bioliquor, ranging between 2.7-3.1. The presence of arsenite in the bioliquor, amounting to 20% of the total arsenic, was considered as the main factor responsible for the increased arsenic solubility; complete oxidation of arsenic to the arsenate form prior to neutralisation resulted in stable residues. The cyanidation tailings of the same process were also found to be highly unstable.

Finally, in another recent publication [5] the arsenic stability of the mixed neutralisation/cyanidation tailings from the pilot plant treatment of the Salmita ore assaying Fe: 2.9% and As: 0.8%, is reported. Oxidised iron/arsenic products had an average molar Fe/As ratio of 2.2, i.e. below the values recommended for optimum stability of amorphous ferric arsenate. At pH 10.5 the arsenic solubility of the mixed tailings was 6.0mg/L, and was gradually reduced to 0.2mg/L as the pH reached a value of 7.5. It is seen that with this low-arsenic feed the conditions prevailing in the bioleaching circuit allowed the production of a relatively stable final residue.

Given the limited number and often contradicting published information on the stability of arsenical solid bioresidues and the respective cyanidation tailings, the present works aims at examining arsenic solubility as related to the operating conditions in the bioleaching circuit.

The material examined as case study is a refractory arsenopyrite concentrate produced after differential flotation of a complex sulphide ore for the recovery of galena and sphalerite and a mixed pyrite-arsenopyrite concentrate. At the final stage of the process arsenopyrite, where refractory gold is quantitatively contained, is selectively separated from pyrite by flotation. The sulphide ore assays on the average Pb: 3.5%, Zn: 5.5%, As: 7.0%, Au: 10 g/t, Ag: 130.0 g/t, while the average grade of the arsenopyrite concentrate is As: 25% and Au: 30 g/t. This product is highly refractory in nature, since gold extraction in the as received material ranged between 10-30%.

Bacterial leaching study for the liberation of the contained gold was effected by the University of Wales, College Cardiff (UWCC). The bioresidues recovered from the leaching laboratory tests were examined by the National Technical University of Athens (NTUA) so as to define their stability as related to biooxidation operating conditions. Arsenic stability was also investigated in a limited number of cyanidation tailings. The nature of the oxidised products that precipitated from solution during bioleaching and report on the unreacted sulphides were also examined.

2. Materials and methods

2.1 Feed material

Arsenic stability studies were conducted on the residues obtained after the bioleaching of the refractory pyrites under study. The head assays and calculated modal analyses for the pyrite and arsenopyrite concentrates are given in Table 1. Both concentrates and particularly arsenopyrite contained an appreciable amount of calcium carbonate; the presence of sulphate sulphur $S(SO_4)$ indicates that small percentage of the feed sulphides was partially oxidised before the application of bioleaching. Particle size analysis of these concentrates, with Laser diffraction showed that the material under study was finely ground, d_{50} was 20 μm for pyrite and 12μm for arsenopyrite concentrate.

Table 1. Pyrite and arsenopyrite concentrate analyses

	Chemical, %								
	Fe	As	Cu	Pb	Zn	St	$S(SO_4)$	Ca	TOTAL
Pyrite conc.**(P)**	35.4	6.3	0.1	3.6	3.4	39.1	2.8	1.2	91.0
Arsenopyrite conc.**(A)**	26.3	21.2		1.6	3.8	22.6	2.6	3.4	80.4

	Modal, %					
	FeAsS	FeS_2	PbS	ZnS	$CaCO_3$	TOTAL
Pyrite conc.**(P)**	13.6	64.8	4.2	5.7	3.0	91.3
Arsenopyrite conc.**(A)**	46.5	21.6	1.7	5.7	8.5	83.6

2.2 Bioresidues

The bioresidues examined in the present study originated from leaching tests carried out in 1.7L air-stirred pachuca reactors at 35°C. Tests were conducted on pyrite, arsenopyrite concentrates and an 1:1 composite sample of both the above concentrates. The bacteria used was a Thiobacillus Ferrooxidans strain. The bioresidues studied are described as follows: (a) P 10% w/v: test on pyrite concentrate at 10% w/v solids, inoculated with bacterial ferric liquor, (b) A 2%, A 5% w/v and A 5% w/v inoculated: tests on arsenopyrite concentrate at 2% and 5% w/v solids inoculated with bacterial produced ferric liquor and bacteria culture respectively, (c) A 10% w/v releached: test on bioresidue from 10% A test, releached with fresh ferric liquor, (d) A + P 5% w/v and 10% w/v, inoculated: tests on mixed pyrite-arsenopyrite concentrate at 5% and 10% w/v solids inoculated with bacteria. Detailed description of the bioleaching tests procedures is given elsewhere [15].

The compositions of the bioresidues examined are given in Table 2.

Table 2. Bioresidues composition

	A 5%	A 5% inoc.	A 10% rel.	A+P 5%	A+P 10%	P 10%
			Assay, %			
Fe	28.5	15.6	12.9	21.6	28.5	25.9
As	18.8	5.0	5.0	1.0	2.0	1.0
Pb	2.2	2.7	2.7	7.6	3.8	4.6
Ca	3.0	3.6	5.9	0.7	2.2	1.3
St	27.0	23.4	21.5	28.3	36.2	34.3
$S(S^{2-})$	23.7	22.5	16.0	27.1	35.1	31.3
SO_4	10.9	2.6	16.5	3.7	3.3	9.0
TOTAL	*87.4*	*52.0*	*59.0*	*61.7*	*74.9*	*73.1*
Fe/As [M]	2.0	4.2	3.45	28.9	19.1	34.7
Weight loss(%)*	21.3	44.4	44.4	65.8	31.6	21.7
As extraction, % *	26.6	80.0	78.0	76.0	72.0	78.0
Fe extraction, %*	10.8	68.0	42.0	68.0	39.0	40.0

* achieved after biooxidation

2.3 Arsenic stability tests

To evaluate arsenic stability before and after biooxidation the pyrite/arsenopyrite concentrates, the examined bioresidues and a selected number of cyanidation tailings were subjected to the EPA (Environmental Protection Agency) toxicity test at pH 5.0 [30]. Before the EPA testing all samples were thoroughly washed, repulped and filtered. Acetic acid was used to buffer the pH; test duration was 24h.

2.4 Cyanidation tests

Given that the final solid wastes produced from the bioleaching circuit are the cyanidation tailings, cyanide leaching tests were effected on the bioresidues, A 5% and A+P 10%. The tests were carried out in magnetically stirred reactors at pulp densities 10%, retention time was 24h. Before the addition of NaCN the pH of the pulp was adjusted at 11.5 with lime. Initial NaCN concentration in solution was 4.0g/L. The cyanidation liquor was analysed for As and the solid tailings were further subjected to EPA test to evaluate their arsenic stability.

2.5 Oxidised precipitates

To determine the amount and nature of oxidised solids in the feed material as well as the bioresidues, the samples under study were subjected to an intense acid wash. Oxidised products formed during bioleaching are known to report in these soluble solids [23,24]. For the acid treatment 1 g of sample was immersed in a beaker containing 30 mL of 5N HCl. After 30 min of continuous stirring the solids were filtered and after micro filtration at 0.45 μm, the concentrations of Fe, As, As^{3+}, Pb, Zn, Ca and SO_4 were measured in the filtrates and the resulting soluble solids composition was calculated. The particle size distribution of the soluble solids was derived from the size analysis data of the as-received and the acid-washed bioresidues. Finally, the morphology and the composition of the oxidised precipitates was examined with Scanning Electron Microscopy (SEM), using high-magnification back scattered electron images combined with electron point microanalysis and X Ray Diffraction (XRD).

2.6 Analytical methods

Total iron Fe_t and arsenic As_t in solution were measured with AAS. The hydride generator technique was used for the low arsenic concentrations (detection limit 50 ppb). Ferrous ions Fe^{2+} in solution were analysed with the standard $K_2Cr_2O_7$ titration method and arsenite ions As^{3+} with a $KBrO_3$ titration method after the removal of Fe [28]. For solids analyses, S_t was determined in a LECO Sulphur Analyser, while $S(SO_4)$ was determined gravimetrically [14]; Fe, As, Ca, Pb, Zn assays in the solids were measured with AAS after acid-digestion in aqua regia.

3. Results and discussion

3.1 Arsenic stability, EPA tests

3.1.1 Feed & Bioresidues

Arsenic stability of the solids under study, feed material and bioresidues, was evaluated with the EPA test and the results are shown in Table 3.

Measurements of the initial pulp pH showed that the feed material exhibited alkaline values, possibly due to the partial dissolution of the carbonate minerals present. The oxidised bioresidues exhibited acidic pH and thus no acetic acid was added for pH regulation in the EPA tests, conducted in their natural pH.

Table 3. EPA* tests results

	P feed	A feed	A 2%	A5%	A5% inoc.	A 10% rel.	A+P 5%	A+P 10%	P10%
pH init. (1)	8.0	7.2	3.1	3.8	3.1	3.8	2.9	3.4	3.5
pH final (2)	5.1	5.1	3.5	3.6	3.6	4.1	3.5	3.4	4.1
As mg/L	3.0	4.0	17.0	36.0	17.0	25.0	5.0	5.0	1.0
(As^{3+})					(10.0)		(4.0)		

* EP Toxicity Test, US Federal Register, vol. 45, No. 98, May 19, 1980 [30]
(1) pH initial : before the addition of acetic acid, (2) pH final : after the completion of EPA test

EPA tests indicated that both pyrite and arsenopyrite concentrate are stable as far as arsenic is concerned. From the bioresidues examined only samples with increased Fe/As molar ratio in the feed material and the bioresidues, such as P 10%, A+P 5% and 10%, exhibited arsenic solubilities within the permissible levels set for safe environmental disposal, i.e. 5.0 mg/L. For the tests where arsenite (As^{3+}) level in the EPA filtrates was measured, it was noted that the majority of the soluble arsenic occurred in the trivalent state.

3.1.2 Cyanidation tailings

Given that the final residues emanating from the bioleaching of auriferous arsenical pyrites are the cyanidation tailings, two of the bioresidues under study, namely the unstable A 5% and the stable A+P 10% were cyanide leached and then subjected to the EPA test in order to evaluate the effect of this alkaline treatment on the corresponding arsenic stability. During cyanidation no arsenic solubilisation was observed, i.e. at pH 11.5 the arsenic levels measured in solution were below 0.5 mg/L, for both the stable and unstable bioresidues examined. Decreased arsenic solubility in the cyanidation stage is attributed to the formation of $Ca_3(AsO_4)_2$ that entraps the arsenic in the alkaline region. When these cyanidation tailings containing $Ca_3(AsO_4)_2$, $Ca(OH)_2$, $M(OH)_x$ and the unreacted sulphides are discharged to the environment, $Ca_3(AsO_4)_2$ in the contact with the atmospheric air will react with the CO_2 according to the following reaction:

$$Ca_3(AsO4)_2 + 3\ CO_2 + 4\ OH^- \rightarrow 3\ CaCO_3 + 2\ HAsO_4^{2-} + H_2O$$

Thus, a pH drop will be observed to an equilibrium value of pH 7.8 and the contained AsO_4 will be redissolved [18,23]. According to thermodynamic data the minimum solubility of $Ca_3(AsO_4)_2$ in contact with the atmosphere is observed at a pH of 7.8 and is estimated to more than 500mg/L [18, 23]. Therefore, this calcium arsenate compound is known to be soluble in the pH range of 5.0, where the EPA tests are effected.

EPA test results on the cyanidation tailings are shown in Table 4. No changes in the trend of stability were observed after the cyanidation treatment. On the contrary for the unstable bioresidue arsenic solubility was further increased. Again it was noted that the majority of the soluble arsenic reported in the arsenite state.

Table 4. EPA tests results on cyanidation tailings

	A 5%	A + P 10%
pH initial	6.4	6.2
pH final	4.3	5.1
As, mg/L	44.0	<1.0
(As^{3+})	(30.0)	

The relation between the arsenic stability of the bioresidues and the respective cyanidation tailings is illustrated in Figure 1. It is seen that the arsenic solubility of the cyanidation tailings is directly related to the stability of the respective bioresidues; for the unstable sample no improvement is observed after the alkaline treatment that precedes cyanidation. Thus, the operating conditions in the bioleaching circuit should be defined and effectively controlled in order that the final solid wastes produced from the process are suitable for safe long term disposal.

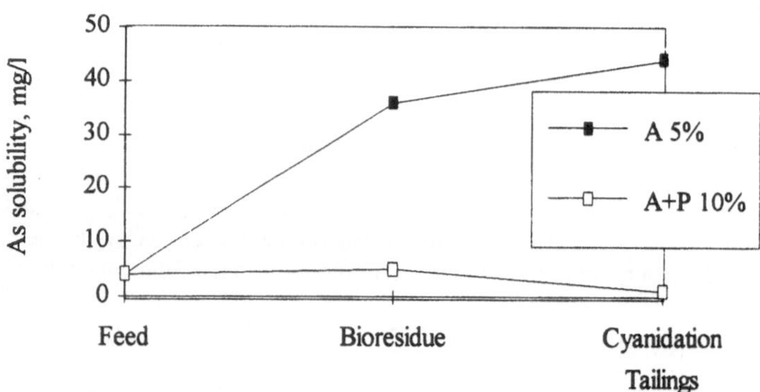

Fig. 1. Arsenic solubility of feed, bioresidues and the respective cyanidation tailings, for the bioleaching tests A 5% and A+P 10%.

3.1.3 Oxidised precipitates

The bioresidues produced from the treatment of the arsenopyrite concentrate were seen to be highly unstable as far as arsenic solubility is concerned. Given that the initial feed material exhibited low arsenic solubility it is concluded that the stability of the resulting bioresidues is solely dictated by the oxidised products formed during bioleaching. It is known that depending on the operating conditions of the bioleaching circuit, i.e. feed composition, pulp density, pH-EMF and solution chemistry, a percentage of the oxidised leached products, including iron, arsenic and sulphate sulphur, precipitates reporting on the unreacted sulphide solids. In order to examine the amount and nature of the oxidised products, the bioresidues under study as well as the feed material were subjected to acid treatment with 5N HCl. Based on acid wash liquor analyses the soluble solids composition was calculated; Fe, As, SO_4 percent content is given in Table 5 along with the modal analysis of the soluble solids.

Table 5. Soluble solids composition

ANALYSES	P feed	A feed	A 2%	A 5%	A 5% inoc.	A 10% rel.	A + P 5%	A + P 10%	P 10%
Soluble Solids (%)*	15.9	16.5	19.4	22.2	24.4	34.6	16.2	17.5	24.9
CHEMICAL, %**									
Fe	3.2	10.4	2.3	1.7	1.5	1.7	2.2	0.9	2.3
As (As^{3+})	0.3	3.4	1.9	1.2 (0.4)	1.1	1.0 (0.3)	0.5	0.4	0.1
SO4	11.9	17.7	35.5	47.6	60.4	43.3	36.7	43.9	34.2
Fe/As[M]	12.5	4.1	1.6	1.9	1.9	2.3	5.9	3.1	39.9
MODAL, %**									
$CaCO_3$	11.7	55.2							
PbS	21.6	8.5							
ZnS	2.7	3.7							
$CaSO_4.2H_2O$			54.8	72.2	64.2	71.1	18.9	52.3	28.4
$PbSO_4$			20.0	12.1	16.7	10.4	70.8	30.6	54.5
Fe-As oxidised products		36.3	25.2	15.7	19.1	18.5	10.4	17.1	17.1
TOTAL		100.0	100.0	100.0	100.0	100.0	100.0	100.0	100.0

* expressed as % of the bioresidue weight
**expressed as % of the soluble solids weight

The experimental data showed that the amount of soluble solids for both feed material and bioresidues was quite limited, ranging from 16-35% of the initial sample's weight.

The weight loss observed after acid washing of the feed material is attributed to the presence of acid consuming $CaCO_3$ as well as oxidised iron and arsenic compounds observed mainly in arsenopyrite and to a lesser extend to the pyrite concentrate. Thus, it is concluded that the feed material might be subjected to an acidic wash before the bioleaching stage.

Referring to the examined bioresidues it is observed that the soluble solids comprise mainly of sulphate compounds. For the arsenopyrite (A) bioresidues the majority of sulphates was associated with gypsum, $CaSO_4.2H_2O$, which accounted for more than 50% w/w of the soluble solids, whereas for the mixed (A+P) and pyrite (P) bioresidues significant amounts of $PbSO_4$ were also defined.

XRD analysis of the A 5% sample verified the presence of $CaSO_4.2H_2O$ and $PbSO_4$ on the bioresidues, subsequently reporting to the soluble solids; unreacted pyrite and inert SiO_2 were also identified.

The remaining quantity of the soluble solids was calculated to consist of oxidised iron-arsenic compounds which for all bioresidues examined accounted for less than 25% w/w of the soluble solids, corresponding to less than 7% of the initial weight of the bioresidue. Due to their limited quantity and their amorphous state these compounds were not identified with XRD.

The morphology of the soluble solids was examined with SEM, i.e. the bioresidues were observed before and after the acid wash. The secondary electron images for the feed material, the unstable A 5% and the stable A+P 10% bioresidues and the respective acid washed samples are presented in Figure 2, 3 &4. The composition of the examined particles was defined with elementary-point microanalysis. In both bioresidues, gypsum in the form of elongated rod shaped particles was identified as the main constituent of the soluble solids. Gypsum was not present in the feed material and disappeared after the acid-wash of the bioresidues. Due to their limited quantity and the coexistence of unreacted pyrites, iron-arsenic oxidised products were not clearly identified with SEM and no further attempt was made to describe them with a specific chemical formula. In Figure 3 the image of an oxidised arsenopyrite grain is presented in the acid-washed sample where preferred sites of biooxidation corrosion are revealed

To further examine the nature of the soluble solids the particle size analysis for the A 5% and A+P 10% bioresidues before and after the acid treatment was defined with Laser diffraction and the results are presented in Figures 5 and 6. From the experimental data it is deduced that for the A+P 10% residue, where the oxidation achieved amounts to 72% As and 38% Fe, the soluble solids are finer than the remaining unreacted sulphides as opposed to the coarser soluble solids observed for the A 5% sample, where oxidation proceeded to a limited extent, 27% As and 11% Fe.

a. The finely ground arsenopyrite concentrate A with a coarse arsenopyrite grain.

b. The pyrite P concentrate.

Fig. 2. Secondary electron images with SEM analyses of the arsenopyrite and pyrite concentrates

a. The A 5% bioresidue before the acid wash treatment. Gypsum is observed as the elongated rod particles .

b. An oxidised arsenopyrite grain in the acid washed A 5% bioresidue where the preferred sites of biooxidation corrosion are clearly seen.

Fig. 3. Secondary electron images with SEM analysis of the A 5% bioresidue before and after the acid treatment.

a. The A+P 5% bioresidue before the acid wash treatment.

b. The A+P 10% bioresidue after the acid treatment. The elongated grains of gypsum have disappeared.

Fig. 4. Secondary electron images with SEM analysis of the A+P 10% bioresidue before and after the acid treatment.

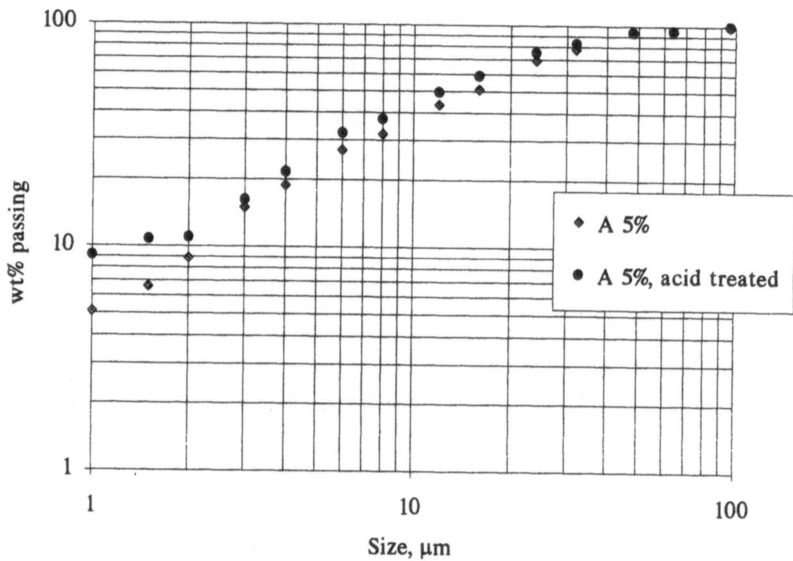

Fig. 5. Size distribution for the A 5% bioresidue before and after the acid wash

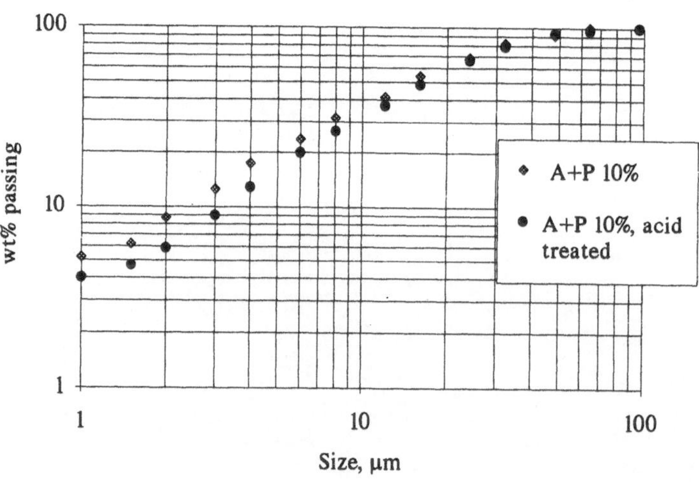

Fig. 6. Size distribution for the A+P 10% bioresidue before and after the acid wash

Fe/As molar ratios in the soluble solids, a critical factor for the subsequent arsenic stability of the solid residues, ranged from 1.6 for the unstable A 2% to 40.0 for the stable P 10% bioresidue. The relation between the Fe/As molar ratio in the soluble solids and the respective arsenic solubility of the bioresidues is illustrated in Figure 7.

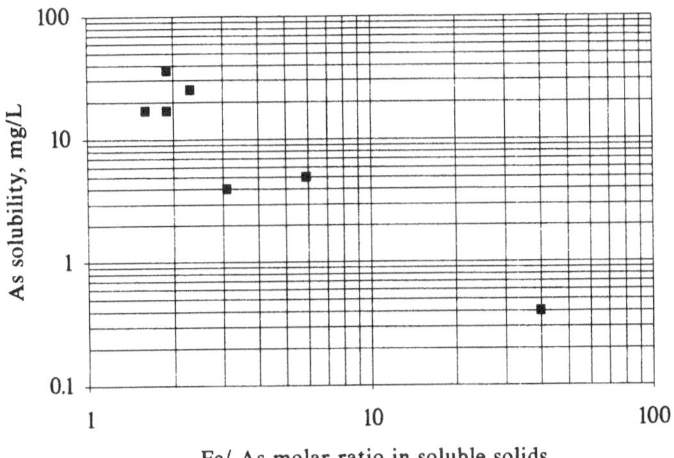

Fig. 7. Arsenic solubility as a function of Fe/As ratio in the soluble solids

A decrease in arsenic solubility is observed with increasing Fe/As ratios in the soluble solids. Stable bioresidues were obtained for Fe/As ratios greater than 3.0 in accordance with published data where it is stated that ferric iron excess is required for ferric-arsenic precipitates stable in a wide pH [10,11,21]. Thus, from the present study it is deduced that a factor significantly affecting the stability of the arsenical bioresidues is the Fe/As molar ratio in the oxidised soluble products.

Moreover, the presence of arsenic in the toxic trivalent state is another important parameter known to adversely affect the arsenic stability. For the samples examined, arsenite was detected in the unstable A 5% and 10% releach residues and accounted for 25-30% of the total arsenic content of the soluble solids as defined with the acid treatment. Arsenite levels close to the detection limit of the method employed (3 ppm) were also measured in the A 5% inoculated residue. However, and despite their limited contribution in the composition of the soluble solids, these arsenite products are another key factor dictating the stability of the bioresidues since arsenite accounted for the majority of soluble arsenic measured in the EPA leachabilty tests.

3.1.4 Arsenic stability as related to the bioleaching conditions

In order to elucidate the effect of the bioleaching operating conditions on the stability of the produced bioresidues, arsenic solubility data defined with the EPA test are presented in Table 6 as related to the major operating parameters including composition of feed material and solution chemistry.

Table 6. Bioresidues stability in relation to operating conditions in the biooxidation circuit

Bioxidation test			**Feed**	Final bioliquor			Bioresidue		
Test	As (%) extr.	Fe (%) extr.	**Fe/As [M] in feed**	pH	EMF mV vs SCE	Fe/As [M] in final biliquor	Fe/As [M] in bioresidue	Fe/As [M] in soluble solids	**As solubility mg/L**
A 2%	51.5	18.0	**1.7**	0.98	600	7.1	2.9	1.6	**17**
A 5%	26.6	11.0	**1.7**	1.16	445	5.1	2.0	1.9	**36**
A 5% inoc.	80.0	68.0	**1.7**	1.15	487	1.4	4.2	1.9	**17**
A 10% rel.	78.0	42.0	**1.7**	1.12	485	1.7	3.5	2.3	**25**
A+P 5%	76.0	68.0	**3.1**	0.83	500	2.7	28.9	5.9	**5**
A +P 10%	72.0	38.0	**3.1**	0.99	450	1.6	19.1	3.1	**4**
P 10%	78.0	40.0	**7.5**	0.75	639	6.1	34.6	39.9	**0.4**

For the system examined where bioleaching runs were conducted in very acidic pH ranging from 0.8-1.3, the secondary precipitation of dissolved iron and arsenic species was quite limited. Thus, no obvious correlation was found between the solution chemistry as recorded at the end of the run with EMF and Fe/As ratio measurements in the final bioliquor and the stability of the resulting bioresidues. Even in cases where ferric iron was externally introduced to the system with the initial addition of bacterially produced ferric liquor as with the A 2% and 5 % and A 10% rel. sample, and the EMF has reached values of 600mV as in the A 2% sample, the bioresidues obtained exhibited increased arsenic solubility.

This behaviour is attributed to the presence of arsenites in the solids formed either at the reaction site or precipitating from solution. These arsenite compounds despite their limited quantity were seen to adversely affect the arsenic stability. Based on the above, the performance of additional kinetic tests is deemed necessary in order to evaluate the evolution of solution chemistry as related to the stability of the bioresidues.

However, for the residues examined strong correlation was seen to prevail between the composition of the feed material, and more specifically the relative contribution of pyrite and arsenopyrite as expressed by the Fe/As [M] ratio in the feed solids, and the stability of the produced arsenical bioresidues illustrated in Figure 8.

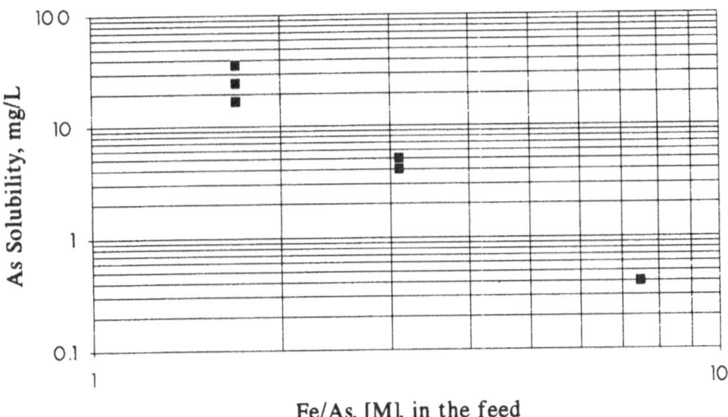

Fig. 8. Arsenic solubility as a function of Fe/As ratio in the feed material

As well documented in literature the presence of pyrite enhances the preferential bacterial oxidation of the less noble arsenopyrite, the latter acting as the anode whereas pyrite behaves as the cathode of the galvanic pair [8,17,20]. Reported data including mineralogical observations provide supportive evidence for intense preferential leaching of arsenopyrite in mixed grains with pyrite, demonstrating the result of the galvanic effect [1,17]. According to the findings of this study the presence of pyrite was seen not only to improve the selectivity and the kinetics of the arsenopyrite bioleaching but it also results in the formation of stable oxidation products with increased Fe/As molar ratios.

The stability of the oxidation products is dictated by the conditions prevailing during the oxidation of the arsenopyrite including feed composition, pulp density, bacterial activity, pH/EMF ferric iron levels in solution etc. These conditions should allow the complete oxidation of arsenic to the arsenate form before its deposition on the solid residue. Should unstable arsenical products be originally formed either on the sulphide surface or from the secondary precipitation of leached As^{3+}, no further treatment was seen to improve their stability even if strong oxidising conditions are subsequently established in the leaching circuit along with increased Fe/As ratios in solution. Reported data on the stability of the bioresidues emanating from the treatment of the Salsigne highly arsenical pyrite come to support this claim [17]. The unstable bioresidues despite an intense oxidative treatment with H_2O_2 and the addition of ferric iron to Fe/As 3, exhibited solubilities well in excess the environmental guidelines for safe long-term disposal.

Based on the above it is further concluded that in the cases of refractory arsenical pyrites where gold is preferentially associated with arsenopyrite, the attempt to effectively separate arsenopyrite from pyrite in a preceding beneficiation stage has to be carefully considered in terms of the process environmental performance and resulting costs should bacterial oxidation be the oxidative pretreatment method employed to liberate the contained gold from the sulphide lattice.

4. Conclusions

The stability of the bioresidues emanating from the bioleaching of arsenical pyrites is directly related to the occurrence of iron-arsenic oxidised products reporting on the surface of the unreacted sulphides.

This study verified that in order to obtain stable arsenical bioresidues the arsenic must be quantitatively oxidised in the pentavalent state. Incomplete oxidation of arsenic to arsenite subsequently reporting on the bioresidues was seen to result in increased arsenic solubility. Thus, careful control of the operating conditions in the bioleaching circuit are required. Should these unstable arsenite products be initially formed, no further treatment was seen to improve their stability even if strong oxidising conditions are subsequently established in the leaching circuit combined with the presence of increased Fe/As ratios in solution.

Furthermore, an excess of ferric iron is required to fix the arsenate ions in the oxidised solids. Stable bioresidues are obtained when the Fe/As ratio in the ferric arsenate precipitates is greater than 3.

For the system examined where the bioleaching treatment was performed in very acidic pH ranging from 0.8-1.3, the secondary precipitation of dissolved iron and arsenic species was very limited. Ferric-arsenic products remained in all cases under study below 7.0% of the bioresidue's weigh. Due to the limited secondary precipitation the solution chemistry as recorded at the end of the run was not seen to be directly related to the stability of the produced bioresidues.

In this study the main operating factor clearly effecting the stability of arsenical bioresidues was the Fe/As ratio in the feed. Arsenic solubility decreased by an order of magnitude from 20mg/L to 4mg/L and 0.4mg/L when the Fe/As ratio in the feed increased from 1.7 to 3.1 and 7.5 respectively. This observation indicates that the presence of pyrite in the feed material has a beneficial effect on the structure and stability of the oxidised products. Therefore, the attempt to produce an arsenopyrite rich concentrate, in order to reduce the total throughput in the bioleaching circuit, should be carefully examined in terms of the process environmental performance.

Given that the final residues emanating from the bioleaching of auriferous arsenical pyrites are the cyanidation tailings, a number of the bioresidues under study were subjected to cyanidation in order to evaluate the effect of this alkaline treatment on the corresponding arsenic stability. No change in the stability trend was observed after the cyanidation treatment. On the contrary for the initially unstable bioresidues the arsenic solubility was further increased.

No arsenic solubilisation was observed during cyanidation, i.e. at pH 11.5 the arsenic levels measured in solution were of the order of 0.5mg/L, for both the stable and unstable bioresidues examined. This observation is attributed to the formation of calcium arsenate, $Ca_3(AsO_4)_2$ that entraps the arsenic in the alkaline region. Calcium arsenate when in contact with the atmospheric air is decomposed to $CaCO_3$ releasing the contained arsenic.

Acknowledgements

The authors would like to acknowledge the financial support of the Commission of the European Communities, Contract No. EC MA2M- CT91-55. They would also like to thank Mr P. Ollivier and Dr D. Morin of BRGM for their kind co-operation during the execution of this study.

References

1. Adam, K., Prevosteau, J.M, Kontopoulos, A, Stefanakis, M, Errington, M , 1990: "Applications of Process Mineralogy on the Treatment of Olympias Pyrite Concentrate", Gold '90, SME-AIME, Littleton, Co , pp. 341-351
2. Adam,K. Taxiarhou, M, Stefanakis, M, 1991: " Application of Bioleaching for the treatment of the Olympias auriferous pyrites" Final progress report, GGET, PABE Programme, Contract No: 694-ERE-103, p.128, in Greek
3. Aswegen, P.C, Haines, A.K, Marais, H.J, 1988 " Design and operation of a commercial Bacterial Oxidation plant at Fairview", in PERTH GOLD 88, Proc of Randol International Ltd, Golden Co., pp. 144-147
4. Bruynesteyn, A, Hackl, R.P., Wright, F, 1986. : " The BIOTANKLEACH Process ", in GOLD 100, Proceedings of the International Conference of Gold, Ed C.E Fivaz, SAIMM, pp. 353-365.
5. Hackl, R.P., 1990 :"Stability of Arsenical Tailings from the Salmita Bioleach Pilot Project", Randol '90, p. 101-106.
6. Harris, G.B., and S. Monette, 1988: "The Stability of Arsenic-Bearing Residues", Arsenic Metallurgy-Fundamentals and Applications, Eds. R.G. Reddy, J.L. Hendrix and P.B. Queneau, The Metallurgical Society of AIME, 1988, p. 469-489.
7. Harris, G.B., and S. Monette, 1989: "The Disposal of Arsenical Solid Residues", Presented at Productivity and Technology in the Metallurgical Industries, TMS-AIME/GDMB Joint Symposium, Cologne, West Germany, Sept. 17-22, 1989, p. 545-559.
8. Karavaiko,G.I., L.K. Chuchatin, T.A.Pivovarova et al, 1986: "Microbiological Leaching of Metals from Arsenopyrite Containing Concentrates", Fundamental and Applied Biohydrometallurgy, eds R.W.Lawrence, R.M.R.Ebner, Elsevier Science Publishing, Amsterdam, pp. 115-126
9. Kontopoulos, A., N. Papassiopi and M. Stefanakis, 1988: "Arsenic Control in Hydrometallurgy by Precipitation as Ferric Arsenates", Proceeding of the 1st International Conference on Hydrometallurgy, ICHM '88, Eds. Zheng Yulian and Xu Jiazong, International Academy Publishers, 1988, p. 672-677.
10. Krause, E. and V.A. Ettel, 1985: "Ferric Arsenate Compounds: Are they Environmentally Safe? Solubilities of Basic Ferric Arsenates", Impurity Control and Disposal, Proceeding of the 15th Annual Hydrometallurgical Meeting of CIM,Vancouver, Canada, Aug. 18-22, 185, p. 5/1-5/20.

11. Krause, E. and V.A. Ettel, 1987: "Solubilities and Stabilities of Ferric Arsenates", Crystallisation and Precipitation, Eds. G.L. Strathdee, M.O. Klein and L.A. Melis, Pergamon Press, p. 195-210.
12. Lawrence, R. W and Marchant, P.B, 1988: " Biochemical Treatment in Arsenical gold ore processing", in Arsenic Metallurgy-Fundamentals and Applications, Eds.R.G. Reddy, J.L. Hendrix and P.B. Queneau, The Metallurgical Society of AIME, p. 199- 211.
13. Laser, M.J, Southwood, M.J, Southwood, A.J, 1986: " The release of refractory gold from sulphide minerals during bacterial leaching", in GOLD 100, Proceedings of the International Conference of Gold, ed C.E Fivaz, SAIMM, pp. 287- 297
14. Lenahan, W.C, Murray, R.L, Smith, 1986: "Assay and Analytical Practice in the South African Mining Industry", South African Instituet of Mining and Metallurgy, Monograph Series M6.
15. Ollivier P, Morin, D, Pooley, F.D, Tidy, N.P, Kontopoulos, A, Adam, K,1993: " Optimisation of a Bacterial Leaching process for the treatment of auriferous arsenical pyrites", EC Contract No MA2M- CT91- 0055, First progress report.
16. Morin, D., and Ollivier, P., 1989: "Pilot practice of continuous bioleaching of a gold refractory sulphide concentrate with a high As content", Biohydrometallurgy '89, Jackson, Wyoming, 1989
17. Morin, D., P. Ollivier, I. Toromanoff, A. Letestu, 1991: "Bioleaching of Gold ' Refractory Arsenopyrite Rich Sulphide Concentrate Laboratory and Pilot Case Studies", EPD Congress '91, D.R. Gaskell, The Minerals, Metals and Materials Society, 1991, p.p 577-589.
18. Nishimura, T., C.T. Itoh and K. Tozawa, 1988: "Stabilities and Solubilities of Metal Arsenites and Arsenates in Water and effect of Sulfate and Carbonate Ions on their Solubilities", Arsenic Metallurgy-Fundamentals and Applications, Eds.R.G. Reddy, J.L. Hendrix and P.B. Queneau, The Metallurgical Society of AIME, p. 77-98.
19. Ollivier, P, Morin, D, 1990: " Bioleaching of Sulfide concentrates and ores: Study of refractory gold and non-ferrous base metals ores", in SQUAW VALLEY 90, Proc of Randol International Ltd, Golden Co., pp. 93-99
20. Panin, V.V, Karavaiko, G.I., Polkin, S.I., 1985: "Mechanism and kinetics of bacterial oxidation of sulfide minerals", Biogeotechnology of Metals, eds. G.I. Karavaiko, S.N. Groudev, Moscow Center of International Projects, GKNT/UNEP, pp. 197-215.
21. Papassiopi, N., M. Stefanakis and A. Kontopoulos, 1988: "Removal of Arsenic from Solutions by Precipitation as Ferric Arsenates", in Arsenic Metallurgy-Fundamentals and Applications, Eds. R.G. Reddy, J.L. Hendrix and P.B. Queneau, The Metallurgical Society of AIME, 1988, p. 321-336.
22. Pinches, A., 1972: "The use of micro-organisms for the recovery of metals from mineral materials", PhD Thesis, University College Cardiff.
23. Robins, R.G., 1981: "The Solubility of Metal Arsenates", Metallurgical Transactions B, V12(B), March 1981, p. 103-109.

24. Robins,R.G., 1990: "The Stability and Solubility of Ferric Arsenate: An Update", EPD'90 Congress, TMS Annual Meeting, Anaheim, California, USA, Feb. 18-22, p. 93-104.
25. Robins,R.G., P.L. Wong, T. Nishimura, G.H. Khoe and J.C.Y. Huang, 1992: "Basic Ferric Arsenates - Non Existent", EPD'92 Congress, Eds. J.P. Hager, TMS, p.31-39.
26. Stefanakis, M, Kontopoulos, A, 1988: " Production of environmentally acceptable Arsenites- Arsenates" in Arsenic Metallurgy-Fundamentals and Applications, Eds. R.G. Reddy, J.L. Hendrix and P.B. Queneau, The Metallurgical Society of AIME, 1988, p. 287-304.
27. Suttill, K.R., 1989: "Bio-oxidation for Refractory Gold: Bio-oxidation Comes One Step Closer to Full-scale Commercial Operation" E&MJ, September 1989, pp. 31-32
28. Tan, L.K, Dutrizac, J, 1985: " Determinations of As III and As V in ferric chloride hydrochloric acid leaching media by Ion Chromatography", in Analytical Chemistry, 57, 1985, pp. 627-641
29. Toro, L et al, 1988: " The Influence of Precipitates on Chalcopyrite Bioleaching catalysed by Thiobacillus Ferrooxidans", in Chem. Biochm. Eng. Q 2(4), pp 253-258.
30. U.S Federal Register, 1980, Vol. 45, No. 98

Removal of iron from kaolin and quartz: dissolution with organic acids

C. F. Bonney
Mineral Industry Research Organisation, Lichfield, England

This research project has been co-funded under the European Community Raw Materials Programme (Contract No. MA2M-0014). The paper is based on work carried out by a consortium of research organisations and the principal researchers were:
A.Kontopoulos, National Technical University of Athens, Greece
G. Baudet, Bureau de Recherche Géologiques et Minières, Orléans, France
J.Berthelin, Centre de Pèdologie Biologique, Nancy, France
A.Marabini, Istituto per il Trattamento dei Minerali, Rome, Italy
W.Dudeney, Imperial College, London, UK

Abstract

Several minerals are of commercial interest due to their high level of purity. These include quartz sand which is used for glass manufacture and kaolin which is used as a coating material. However, these minerals contain traces of iron, usually as oxides, which cause discolouration and affect their commercial value. Mineral acids can be used to remove the iron but these are expensive, not always effective and produce effluents which require treatment for the removal of the iron. The objective of the work was to investigate the use of organic acids as a cheaper and more environmentally acceptable alternative.

The initial work involved the characterisation of the samples followed by leaching with both chemically and microbially derived organic acids. For the preparation of microbially produced organic acids, the micro-organism *Aspergillus niger* was used, with molasses as the carbon source. The chemically derived acids investigated were citric, oxalic and ascorbic.

The best results for both quartz and kaolin have been achieved with a mixture of citric and oxalic acids and at a ratio of 2:1 and temperature of 90°C, the iron removal was 37% and 40% respectively. This has reduced the residual iron contents to an acceptable level.

After leaching with organic acids the effluents contain ferrous iron, sulphate and residual oxalates/citrates. Testwork on a synthetic solution of these constituents has been carried out using an upflow anaerobic sludge blanket (UASB) reactor. Methanogenic bacteria in the sludge decompose the oxalates to CO_2 and CH_4, heterotrophic sulphate reducing bacteria cause the reduction of sulphate to sulphide and the sulphides of iron and other metals are precipitated in the presence of the HS^- anion.

Keywords: Iron, Kaolin, Leaching, Organic acids, Quartz

1 Introduction

This research programme (CEC Contract MA2M-0014) involved the use of organic acids which were either chemically or microbially derived, as a cheaper and more environmentally acceptable alternative and included the following activities:

- Characterisation of the quartz and kaolin samples;
- Preparation of microbially produced organic acids;
- Leaching testwork with chemically and microbially derived organic acids;
- Biological treatment of effluents.
- Development of Flowsheets

2 Characterisation of minerals

2.1 Chemical analyses

The minerals under investigation included two samples of kaolin and one sample of quartz, and Table 1 indicates the analyses of these materials:

Table 1 Analyses of iron-bearing materials under investigation

	Kaolin A	Kaolin B	Quartz
	Weight %		
SiO_2	46.80	46.10	99.40
Al_2O_3	36.00	37.00	0.27
Fe_2O_3	1.10	1.00	0.03
CaO	0.05	<0.10	<0.01
MgO	1.00	<0.20	<0.01
Na_2O	0.04	<0.20	0.03
K_2O	2.20	0.97	0.25
TiO_2	<0.05	0.06	0.03

2.2 Mineralogical analysis

Mineralogical investigations were carried out on the materials to determine the form and distribution of the iron and other major minerals.

2.2.1 Kaolin

The kaolin samples consist of kaolinite, muscovite and illite with minor traces of quartz. In the colloidal fractions (less than 0.125μm) the minerals are kaolinite (±60%), smectite and illite (±40%) with traces of chlorite and gœthite. Some of the iron is included in the lattice of the kaolinite while the majority is impregnated into the kaolin mycelium as iron hydroxide. It was also seen that there was an increase in the Fe_2O_3 content with decrease in particle size; e.g. -10+5 μm particles contained $\pm 1\%$ Fe_2O_3 whereas -0.125 μm particles contained $\pm 5\%$ Fe_2O_3.

2.2.2 Quartz

The quartz sample consisted mainly of α-quartz and feldspar, although a few particles of rutile, muscovite and biotite were also observed. The iron was not shown to be present as an oxide coating or as a staining around silica particles but was concentrated as small separate particles. The majority of the iron was present as Fe_2O_3 with less than 10% of it occurring as Fe_3O_4. A magnetic separation was carried out and the fractions were subsequently analysed by XRF. The results are shown in Table 2.

Table 2 XRF analysis of magnetically separated fractions

	Highly Magnetic Fraction	Magnetic Fraction	Non-Magnetic Fraction
		% Separated	
	0.14	2.68	97.18
		% Oxides	
SiO_2	68.32	98.11	99.00
Al_2O_3	9.48	0.98	0.78
Fe_2O_3	3.81	0.08	0.01
MgO	5.36	ND	ND
CaO	5.26	ND	ND
Na_2O	1.41	ND	ND
K_2O	0.33	0.26	0.18

ND = Not Detected (below 0.01)

This clearly shows that the highly magnetic fraction was high in Fe and analysis of this fraction by SEM-EDS showed that it consisted of 68% Quartz, 24% Muscovite, 6% Augite and 2% Magnetite.

It has also been shown that a refractory phase exists which contains iron that is insoluble. Further work to identify this phase is planned.

2.3 Particle size

2.3.1 Quartz

The quartz samples have a particle size distribution within a narrow size range with over 80% being in the range -0.35+0.18 mm.

2.3.2 Kaolin

The particle sizes of the kaolin samples are indicated in Table 3.

Table 3 Particle size distribution of kaolin samples

Particle Size - μm	Cumulative Undersize - % by mass	
	Kaolin A	Kaolin B
40	98.6	100.0
20	96.7	99.6
10	87.5	99.1
5	69.8	98.5
2	41.3	83.5
1	26.2	63.6
0.5	14.6	38.0
0.25	7.7	15.9
0.125	2.0	5.0

Analysis of the size fractions showed that for both samples the Fe_2O_3 content was highest in the fine fractions.

3 Preparation of microbially produced organic acids

The micro-organism *Aspergillus niger*, which is a fungus well known for its ability to ferment sugars for the production of organic acids, was used for the leaching experiments. Batch fermentation of *Aspergillus niger* has shown that sucrose as a carbon source will yield the highest production of citric acid. Testwork has focused on the use of molasses and desulphurised beet molasses has been identified as a most suitable low cost carbon source with an optimum concentration of 160 g/l being established, corresponding to an equivalent sucrose concentration of 90 g/l. The molasses solution was supplemented with 0.5 g/l $MgSO_4.7H_2O$ and 2 g/l $(NH_4)_2SO_4$ the latter being the best nitrogen source for the production of both citrate and oxalate.

Laboratory scale experiments have been carried out using this material and after 17 days fermentation a broth has been obtained containing 60 g/l citric acid and 11 g/l oxalic acid. This represents a conversion of sugars to acids of almost 80% which is one of the highest figures reported in the literature.

Scale-up of the process has been carried out in 20 litre fermentation vessels and Figure 1 shows the time profile for biomass and total sugars during a batch fermentation.

Figure 1 Batch Fermentation in a 20 litre reactor

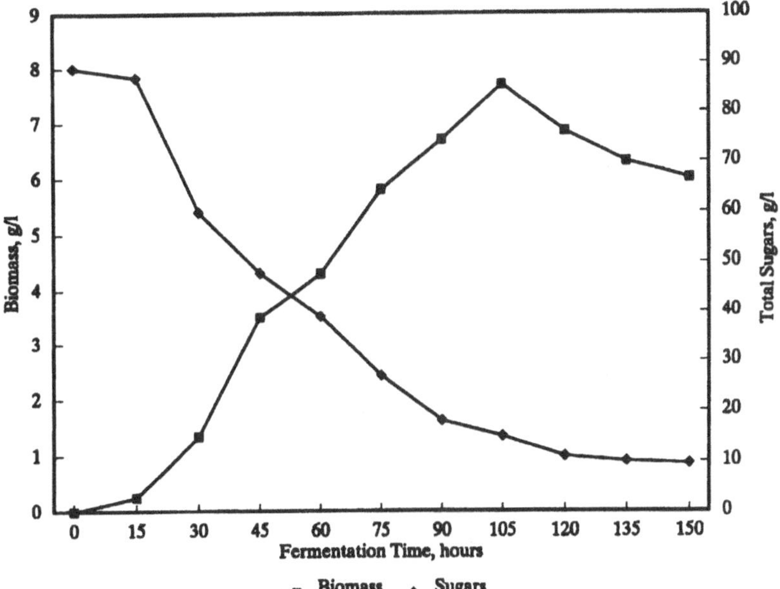

The data obtained from batch fermentation in the 20 litre reactor have been used to model the organic acid production. It has been shown that the production of both citric and oxalic acids follow the Leudeking-Piret model with a maturation time of 30 minutes and 60 minutes respectively.

To enhance the production of acids the effect of alcohol and oil additions were examined. The highest citric acid concentration resulted from the addition of 2% ethanol whereas the highest oxalic acid concentration was with the addition of 2 to 3% methanol. The addition of certain oils (e.g. olive oil, corn oil and soya oil) gave an increase in the citric acid production but a decrease in oxalic acid production.

The combined affect of both alcohol and oil was also examined. However, despite the positive effects of ethanol (increased oxalic acid production) and soya oil (increased citric acid production) the combination of these additives appeared to have no beneficial effect.

4 Leaching testwork

Leaching testwork has been carried out with chemically and microbiologically derived organic acids.

4.1 Tests with microbially produced organic acids

A limited amount of testwork has been carried out using the microbially produced acids due to the necessity to consider the reactions involving known concentrations of single and combined mixtures of organic acids.

A few tests have been carried out on quartz in 1 litre stirred spherical reactors. The organic solution was produced by fermentation with *Aspergillus niger* and contained 42 g/l citric acid, 15.5 g/l oxalic acid and was at a pH of 2.2.

The leaching which was carried out at a pulp density of 30% w/v, a temperature of 90°C and a residence time of 30 minutes, gave 35.7% Fe extraction.

4.2 Tests with chemically derived organic acids

The acids investigated were citric, oxalic and ascorbic both singly and in combination. In some of the experiments the acids were combined to simulate those which would be produced by fermentation with *Aspergillus niger*.

4.2.1 Tests on Quartz

Using the acids individually the testwork has shown, that at concentrations of 0.125 M to 0.5 M, iron removal can be effected with oxalic > ascorbic > sulphuric > citric.

Using a factorial design for the testwork and subsequently applying a linear regression it has been shown that the most significant factors which affect the extraction of iron are temperature and pulp density. The organic acid concentration did not appear to have a major influence on the iron removal.

In laboratory batch tests (1 litre spherical flasks) the best results for iron removal from quartz have been achieved with a mixture of citric and oxalic acids. The highest iron dissolution rates are at high temperature (90°C) and at a citric to oxalic ratio of 2:1, and the iron removal is more or less complete within 1 hour.

Continuous leaching tests have been carried out under ambient conditions in fixed bed columns (8 cm dia x 48 cm high). Using an oxalic acid strength of 0.5 M and percolation rate of 4.0 l/day, over 50% of the iron was removed after 5 days, as shown in Figure 2.

Due to these encouraging results a continuous bulk leach test was carried out. This involved the use of a drum (44 cm dia x 88 cm high) containing 170 Kg of quartz. 0.5 M oxalic acid was percolated through the bed at the rate of 9 l/day for 6 days after which the bed was washed with water for 6 days at 7 l/day. Analysis of the final product, which was confirmed by a mass balance, showed that 55.7% of the iron had been removed. This result indicates the feasibility of heap leaching quartz with oxalic acid.

4.2.2 Tests on Kaolin

The acids investigated were citric and oxalic and a series of tests have been carried out using a factorial design. A linear regression analysis of the results has shown that both temperature and acid concentration are significant variables. The highest iron extraction of 40.4% was obtained with an oxalic acid concentration of 40 g/l and at a temperature of 80°C.

Based on the factorial design testwork the optimum temperature and organic acid concentration to achieve 30 to 40% iron extraction is shown in Figure 3 on the previous page.

Figure 2 Column leaching of quartz with 0.5 M oxalic acid

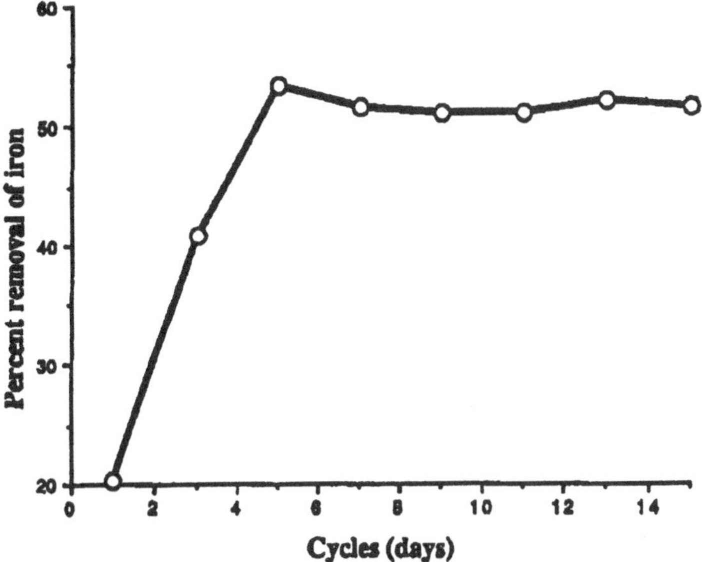

This has reduced the residual iron content of the kaolin to a level at which it is acceptable for use as a coating material.

To provide a reference point for the effectiveness of leaching with organic acids, bleaching tests were carried out. Bleaching of Kaolin A, using 10 Kg/t of sodium dithionite at pH 2.2 and a redox potential of -200 mV, gave a 27.7% removal of Fe_2O_3 and a reduction in the whiteness index from 14.2 to 4.0. Subsequent analysis of the results of these bleaching experiments showed that the extraction of iron takes place by a first order kinetic reaction. Diffuse reflectance spectroscopy has shown that the extractable iron probably occurs as a coating of iron hydroxide on the surface of the clay particles.

Figure 3 Optimum temperature and organic acid concentration for 30-40% Fe extraction

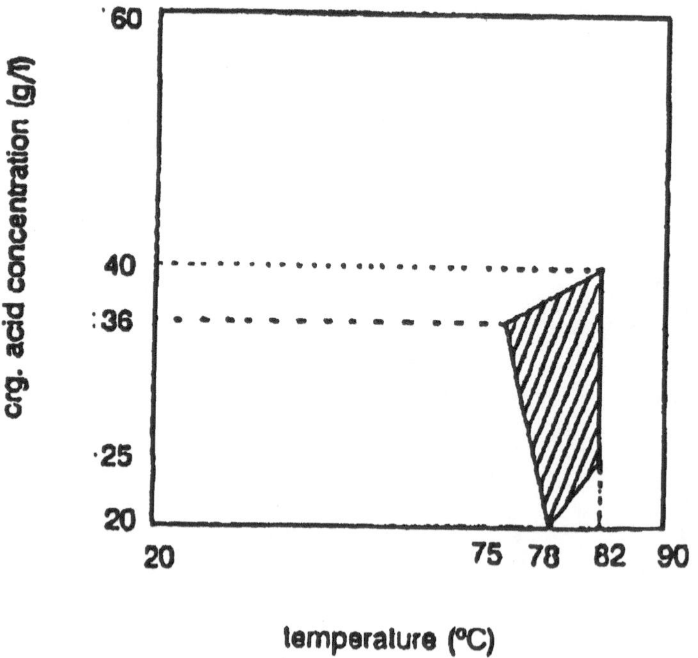

5 Biological treatment of effluents

After leaching with organic acids the effluents contain ferrous iron, sulphate and residual oxalates/citrates. A synthetic solution of these constituents has been made up and testwork has been carried out to investigate their removal using an upflow anaerobic sludge blanket (UASB) reactor. Four reactors were used and Figure 4 is a diagram of one of the Reactor columns.

5.1 Biodegradation of oxalates

Methanogenic bacteria in the sludge can decompose oxalates to CO_2 and CH_4 according to the following generalised reaction:

$$C_6H_aO_b + (n-a/4-b/2)H_2O = (n/2-a/8+b/4)CO_2 + (n/2+a/8-b/4)CH_4 \quad (1)$$

During the test period breakdown of oxalic acid and complexed oxalates occurred efficiently. For example after 65 days the effluent COD was near to zero.

Biodegradation of a mixed iron-bearing oxalic acid/sodium oxalate solution which was similar in composition to "real" bioleach liquors was examined. This showed that the UASB was efficient in reducing the concentrations of organics, sulphate and iron and after 21 days, removals of 66%, 60% and 90% respectively were recorded.

Figure 4 UASB column reactor

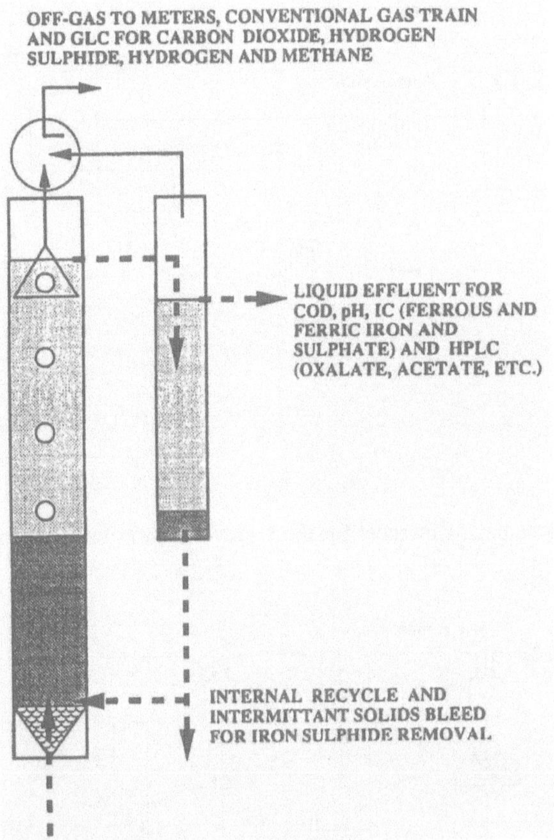

5.2 Biodreduction of Sulphate

Heterotrophic sulphate reducing bacteria cause the reduction of sulphate to sulphide according to the following equation:

$$2CH_3CH(OH)CO_2H + SO_4^{2-} = 2CH_3CO_2^- + 2CO_2 + HS^- + 2H_2O \qquad (2)$$

Figure 5 **Schematic process flowsheet for the removal of iron from kaolin**

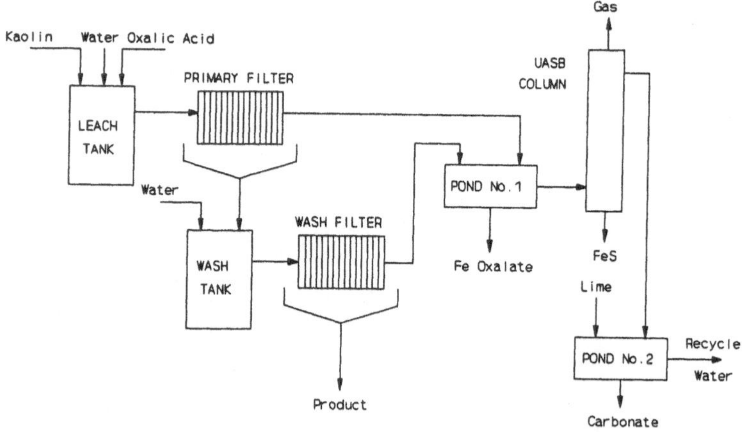

Figure 6 **Schematic flowsheet for the removal of iron from quartz**

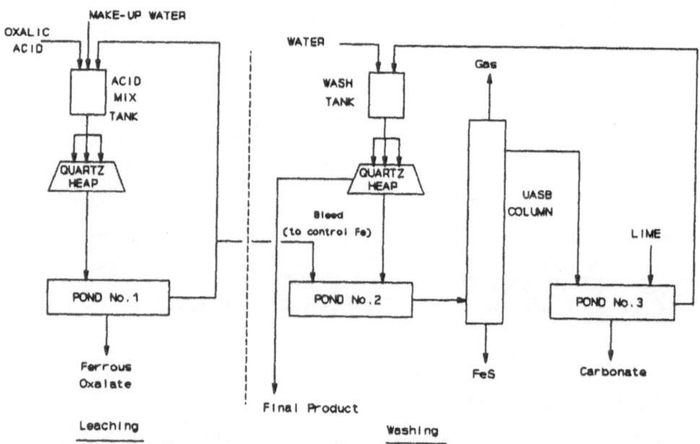

During the test period this sulphate bioreduction was erratic with sulphate removal of between 20% and 80%. Further testwork is required to provide an explanation for this.

5.3 Bioprecipitation of iron

The sulphides of iron and other metals are precipitated in the presence of the HS^- anion. Analysis of a fine-sized black suspension which had formed within the reactors was analysed and was shown to be partly composed of FeS. It appears that the process is very effective at removing iron.

5.4 Aerobic Treatment of UASB Effluents

Overflows from the UASB Columns were collected in an open tank which was inoculated with microorganisms contained in soil from around established rhubarb plants. This was selected as rhubarb plants generate oxalic acid in the leaves and is, therefore, likely to be associated with microorganisms able to metabolically degrade and utilise this acid. Healthy cultures of green algal cells (probably *chlorella)* soon became established. COD measurements of the overflow from the tank were compared with fresh UASB overflows and this showed a reduction from 900 mg/l to about 50 mg/l. Although this is a significant drop in COD, further aerobic treatment will be necessary to reduce it to the statutory level of 20 to 30 mg/l.

6 Development of flowsheets

Throughout the testwork programme attention was given to the production of data which could be used by the industrial partners for the development of process flowsheets. This will enable them to produce the design information necessary for carrying out a capital and operating cost estimate. The flow sheets under consideration are shown in Figures 5 and 6.

7 Acknowledgements

The author wishes to thank DGXII of the European Union, for permission to publish this paper. The copyright of this paper is retained by the European Union.

Bioreduction of gold on algae and with amino acids

A. R. Gee
Formerly, Department of Mineral Resources Engineering, Royal School of Mines, Imperial College of Science, Technology and Medicine, London, England (now BHP Minerals, Mayfield, New South Wales, Australia)
A. W. L. Dudeney
Department of Mineral Resources Engineering, Royal School of Mines, Imperial College of Science, Technology and Medicine, London, England

Abstract
A variety of instrumental techniques indicated that gold chloride reduction on *Chlorella vulgaris* cells, cell fractions and amino acids involved sulphur-containing amino acids. Diode array UV spectrophotometry provided in situ kinetic data on the rapid initial adsorption and reduction of gold as $AuCl_4^-$ to $AuCl_2^-$, which occurred during the first few minutes of reaction. Video-recorded light and electron microscopy gave respectively in situ and ex situ data on the slower reduction over several hours of $AuCl_2^-$ to metallic gold, which formed microscopic crystalline triangular and hexagonal platelets. A mechanism based on three main sequential steps: adsorption, reduction and disproportionation was proposed. A semi-continuous process in which 1-10 ppm gold chloride solution was pumped through a glass column loosely packed with *Chlorella vulgaris* or *Spirulina platensis* as dried, alginate immobilised, 1.5 mm pellets gave complementary data on the rate and mode of gold uptake on algae in bulk form. Typically the pellets retained over 99% of the gold as < 1 μm metallic crystals from passage of more than 200 bed volumes of the solution and apparently behaved as classic shrinking cores with regard to gold distribution.
Keywords: Bioreduction, mechanism, gold, amino acids, algae.

1 Introduction

Since Williams reported in 1918 that micro-fungi removed colloidal gold from solution [1], numerous papers have been published on the adsorption and reduction of gold on microorganisms and different forms of organic matter. Following the intensive studies in the last few years [2-9], the industrial potential of microorganisms, in particular algae and algal derivatives, has been recognised in recovering traces of gold from geothermal waters and process effluents.

Gold in the form of the tetrachloroaurate(III) anion ($AuCl_4^-$) is rapidly and efficiently adsorbed by algae in acidic media, particularly by species of *Chlorella* (commonly seen as a green stain on damp walls and greenhouse glass) and *Spirulina* (harvested, for instance, as a health food from alkaline lakes in Arizona). Darnall and co-workers [2-5] showed by a variety of techniques that adsorbed tetrachloroaurate(III) is reduced to gold(I) within a few minutes, probably by cell constituents containing nitrogen and sulphur, and to metallic gold over a period of days. Under certain conditions beautiful microscopic triangular and hexagonal laminae of crystalline gold were observed attached to the cells. The adsorption was endothermic (and therefore more favoured at elevated temperatures) and could be reversed by the action of the strong complexing agent thiourea, particularly in the presence of an oxidising agent such as iron(III) sulphate. Gee and Dudeney [7] reported complementary kinetic data for both reduction stages at ambient temperatures.

From these studies a basic understanding of the mechanism of gold bioaccumulation has been developed. However, a detailed model of the pathways of reaction has yet to be given. This provides one aim of the present paper.

Practical application was tested at room temperature in columns containing algae immobilised on silica [5], which indicated that non-living cell preparations could be employed over numerous cycles of gold adsorption and desorption. Analogous semi-continuous fluidised-bed column tests were also carried out on algal cells immobilised in beads of calcium alginate [7]. However, a problem remained with the lower biomass concentrations and rates of bioaccumulation achieved in these systems. For application at the higher temperatures experienced in geothermal waters, a proprietary stabilised adsorbant containing *C. vulgaris* was developed [6]. This material was stable over numerous cycles at 206°C while selectively removing gold and platinum at μg per litre levels from concentrated brines.

For possible use with refinery effluents, a modified semi-continuous system - based on a packed bed of alginate-immobilised *Spirulina* having an enhanced biomass content - has been developed. Significant intensification of ambient temperature adsorption is facilitated. A second aim of the present contribution is to describe the characteristics and modes of reaction of this system. The process may also be of interest to geologists concerned with the mode of deposition of proterozoic gold deposits, e.g., 'carbon leader' reefs,in the South African Witswatersrand.

2 Experimental procedures

2.1 Biomass preparation

Freeze-dried cultures of *C. vulgaris* and *S. platensis* (obtained from the Culture Centre for Algae and Protozoa, Cambridge) were typically revived and maintained in appropriate nutrient media [9] in 2 L flasks provided with light (from a number of 40 W bulbs) and air (via a sintered glass fritt and 0.2 μm filter) under sterile conditions. After 10 days the biomass was harvested by centrifuging (25 000/min), washing three times with water, and storing in a moist condition at 0° C until required. Prior to use, the material was redispersed in water and the pH adjusted as necessary. To determine the biomass content (normally 1 g/L) a small sample was evaporated to dryness and the residual mass determined. For larger scale work employing packed columns, *Chlorella* was prepared in 20 L flasks as above but under non-sterile conditions and *Spirulina* was obtained as a sun-dried filter cake product from Spirutec Inc., Arizona (and washed with water before use). The biomass was pelletised by stirring 100 g algae (dry mass basis) suspended in 1 L water for two hours at 75° C with 1 L 3% w/v sodium alginate solution; dripping the resulting mixture slowly into stirred 0.5 M calcium chloride solution to yield *ca* 3 mm diameter beads of biomass immobilised in calcium alginate; washing with water; drying on trays in an oven at 110 °C; and screening to obtain a narrow size distribution (normally - 1.5 + 1 mm) of the final black pseudo-spherical pellets (density 1.39 g/cm^3; biomass/alginate/water ratio 10:3:zero).

2.2 Measurement of biosorption and desorption isotherms

Different masses (5-200 mg) of biomass were equilibrated for 20 min with 10 mL aliquots of solutions of known gold concentrations (10-100 mg/L) at known pH. Equilibration was carried out in 20 mL stoppered tubes agitated on a wrist-action shaker and residual aqueous concentrations were determined by ICP analysis after removing the biomass by centrifuging for 15 min at 5000 /min. Desorption was carried out by adjusting the pH to 2 and reacting with thiourea, followed by metal analysis as before. To determine the form of gold adsorbed (see later), desorption was also carried out with 1 M sodium bromide solution.

2.3 Measurement of gold crystallisation

A Vickers optical microscope with oil immersion optics was employed with 400 or 1000x magnification in transmitted and reflected light. Gold precipitation and crystal growth were observed in a cavity microscope slide of working diameter 1.5 cm closed with a cover slip and typically containing 20 µL of a mixture of gold(III) chloride (10-1000 mg/L Au) and algal cells (gold/biomass ratio 0-1 g Au per g algae) at different initial values of pH (0-10). Recordings of the numbers and linear dimensions of the (laminar) gold crystals were made over periods up to 200 hr - either on a single 20 µL sample or on similar samples taken at set times from a separate bulk preparation - with an Olympus OM2 camera mounted on the microscope or a video-camera (similarly mounted), monitor and video-recorder, used in conjunction with a pre-calibrated eyepiece. Under the most favourable conditions (Au 1000 mg/L, gold loading 0.5 mMol per g biomass at pH <1) the amount of gold present after each period of time was calculated from a video-taped traverse of the 1.5 cm cavity diameter by (i) 'freezing' frames covering the traverse and measuring the linear dimensions of each crystal in the field of view and (ii) converting the linear dimensions to area dimensions, summing and converting to total mass formed on the basis that the crystals were flat-lying laminae of thickness about 1 µm. In this way, ten samples accumulated over 10 days were each measured in triplicate, thus amounting to still-frame examination of over 3 hr of video recording.

2.4 Gold adsorption and reduction on cellular fractions and amino acids

A Hewlet Packard 8451A diode array UV-visible spectrophotometer was used to measure the time dependence of the adsorbance of gold(III) [10] at 312 µm. Samples of cell wall fragments and intracellular fractions (obtained by ultrasonically rupturing *Chlorella* cells) and commercial amino acids (leucine, histidine, cysteine, cystine, serine and methionine) were mixed in a 1 cm spectrophotometer cell with gold(III) chloride (5-200 mg/L) at pH 3 and measured immediately against a suitable blank while being stirred magnetically. The biomass concentration was <200 mg/L to avoid interference with the spectral beam. The mole ratio of amino acid to gold was varied systematically from 1:10 to 10:1. Spectra were recorded at spaced time intervals over a period of 4 min to 24 hr, depending upon the rates of the particular adsorption/reduction reactions occurring.

2.5 Semi-continuous metal bioaccumulation

A vertical glass column 20 cm in height and 1.5 cm in diameter, loosely packed with rewashed *Spirulina* pellets at the required pH, was employed in open circuit at room temperature with a 1-10 mL/min gold chloride feed (nominal residence time 1.5-15 min). Samples of influents and effluents were taken for gold analysis at 20-30 min intervals.

3 Results and discussion

3.1 Adsorption, desorption and reduction fundamentals

Preliminary work confirmed that aqueous gold in weakly acidic chloride medium - largely in the form of the tetrachloroaurate(III) complex ion [11] - accumulated rapidly on the algal surfaces and thereafter slowly reduced to metallic gold. As in the earlier work, well-formed microscopic triangular and hexagonal platelets of crystalline gold were observed after several days adhering to the algal cells. Optimal adsorption occurred at about pH 3-5. At different pH values other ions, e.g., chloride in more acid medium and hydroxyl in alkali, retarded the process to a greater or lesser extent. Different cell fractions - 1-5 μm cell wall platelet fragments and intracellular colloidal/soluble material obtained by ultrasonic disruption of whole cells - reacted similarly with gold chloride. Thus, subject to the rate of dissolved ion penetration, adsorption and reduction occurred throughout the cell structure. Although true adsorption equilibrium was not achieved because of the various transport and reduction processes occurring, approximate adsorption isotherms (Fig. 1) confirmed that *C. vulgaris* was a rapid and highly efficient scavenger for gold at low initial concentrations with concentration factors well in excess of 1000. *S. platensis* was less efficient but nonetheless a good adorbent for gold.

Fig. 2 shows adsorption and desorption characteristics of the two organisms as a function of pH for a maximum loading of 0.05 mol Au per g biomass. The pH was initially 1.5 and increased in steps of one pH unit up to pH 8 with an equilibration time of one hour at each pH. The steps in pH were then reversed. Adsorption increased with pH and reached essentially 100% at pH 3 and 5 for *Chlorella* and *Spirulina*, respectively.

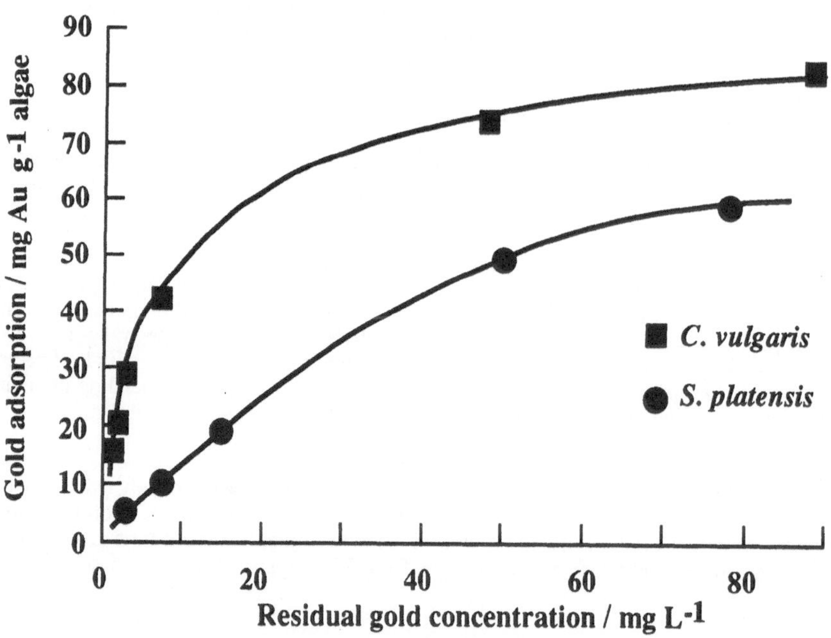

Fig. 1. Adsorption isotherms for tetrachloroaurate(III) on *C. vulgaris and S. platensis.*

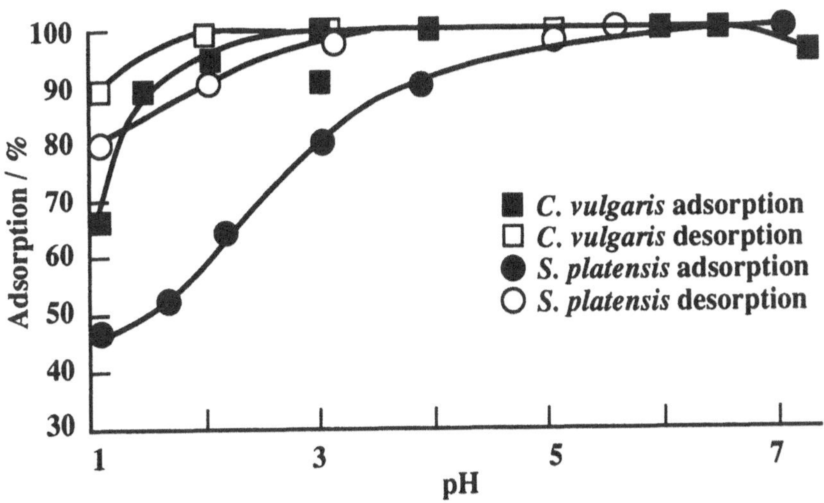

Fig. 2. Adsorption/desorption characteristics of tetrachloroaurate(III) on *C. vulgaris* and *S. platensis*.

Partial desorption resulted from the subsequent decrease in pH but more than 75% of the adsorbed gold was retained on the biomass, even at pH 1. Provided the mixtures were freshly prepared, some 95% of the gold was desorbed in 20 minutes by thiourea (a strong, non-oxidising, complexant for gold in oxidation state I [12]). If iron(III) ammonium sulphate was added as an oxidant, this desorption was successful even with mixtures aged overnight. These results were consistent with the two stage reduction of adsorbed gold outlined above, the relative stability of reduced gold on the biomass and the ability of ligands which form more stable complexes with gold than surface groups on the biomass to remove gold(I) into solution.

It is possible that amino acid residues, which comprise a substantial proportion of microbial constituents (protein, peptidoglycan, etc.), provide the main adsorptive and reductive potential. For *C. vulgaris* the following typical compositions (calculated as a percentage of the whole cells) have been reported [13-14]: (a) protein 51-58, carbohydrate 12-17, lipid 14-22 and nucleic acid 4-5 and (b) protein amino acids, alanine 4.1, arginine 3.0, aspartic acid 4.9, cysteine 0.5, glutamic acid 5.9, glycine 3.8, histidine 1.1, isoleucine 4.6, leucine 6.5, lysine 1.7, methionine 0.8, phenylalanine 9.1, proline 2.7, threonine 1.8, tryptophan 0.9, tyrosine 2.5 and valine 4.1. Protein amino acid residues are not easily extracted from the cell structure for direct testwork but laboratory reagent amino acids - which are known to reduce gold [12-15] - can be employed as model materials to investigate mechanistic aspects of the reactions occurring.

Table 1 compares present results of the amount of gold(III) reduced to gold(I) after set time intervals by representative members of the group.

Table 1. Moles of gold(III) reduced to gold(I) per mole after set time periods by six amino acids

Amino acid	10 minutes	1 hour	48 hours
Cysteine, $NH_2CH(CH_2SH)COOH$	0.7	1.5	2.9
Cystine, $(NH_2CH(CH_2S\text{-})COOH)_2$	0.06	0.35	2.4
Methionine, $NH_2CH(CH_2CH_2SCH_3)COOH$	0.65	1.0	1.0
Histidine, $NH_2CH(CH_2C_3N_2H_3)COOH$	0.01	0.03	0.1
Serine, $NH_2CH(CH_2OH)COOH$	0.0	0.0	0.1
Leucine, $NH_2CH(CH_2CH(CH_3)CH_3)COOH$	0.0	0.0	0.01

As can be seen the sulphur-containing acids were by far the most reactive, the groups -SH and $-SCH_3$ reacting within the first hour and groups S-S over about 2 days. The mole conversions fitted Eqns 1-3, where R represents the residue $NH_2CH(CH_2\text{-})COOH)$. These equations have been reported previously [12-13].

Cysteine $$2RSH + AuCl_4^- = RSSR + AuCl_2^- + 2H^+ + 2Cl^- \quad (1)$$

Cystine $$RSSR + 5AuCl_4^- + 6H_2O = 2RSO_3H + 5AuCl_2^- + 10H^+ + 10Cl^- \quad (2)$$

Methionine $$RSCH_3 + AuCl_4^- + H_2O= RSOCH_3 + AuCl_2^- + 2H^+ + 2Cl^- \quad (3)$$

Evidently, one mole of cysteine eventually reduced three moles gold(III) to gold(I). Methionine reacted largely in a 1:1 mole ratio, although the result at 48 hours had to be estimated because some reduction of gold(I) to metallic gold occurred.

Reduction to metal may be the result of disproportionation (Eqn 4), which is a well-known reaction of gold(I), or relatively slow direct reduction by the amino acid [13]. In the presence of excess of the reductant, amino acid - gold(I) complexes form [14-15], thus retarding reduction to metal. Conversely, when sufficient gold is present to utilise all of the reductant, disporportionation is more likely with the formation of metallic gold and gold(III) in a 2:1 ratio according to Eqn 4.

$$3AuCl_2^- = 2Au + AuCl_4^- + 2Cl^- \quad (4)$$

Analogous reactions are expected to occur at algal surfaces because (unlike amine and carboxylic acid groups) the -SH and $-SCH_3$ groups are not part of the backbone of proteins and should be little modified in reductive properties by inclusion in a polypeptide structure. Of course, the presence of a solid algal surface implies an initial adsorption process and precludes purely homogeneous reactions in the form represented in Eqns 1-3. Darnall, *et al.* [5] considered adsorption and reduction to gold(I) - accompanied by the expulsion of three chloride ions per mol of adsorbed gold - to occur almost as fast as the gold(III) complex was mixed with the algae. With the aid of polarographic titrimetry they showed that untreated algal surfaces contained free -SH groups while those treated with gold(III) did not, and, by means of X-ray adsorption spectrometry, that the adsorbed gold was also associated with nitrogen and/or oxygen on the algae.

In the present work, in situ UV-visible diode-array spectrophotometry was employed to gain more insight into the mechanistic steps governing the interaction between gold and *C. vulgaris* in acidic chloride medium. Under stirred conditions at low relative concentrations (1-20 mg/L or 0.5-10 x 10^{-5} M Au; <200 mg/L algae at pH 3) and overall gold/algae ratios (0.01-0.1 g or 0.05-0.5 mM Au per g algae), some 50% of the gold was adsorbed within 2 seconds. Up to this time, but not for much longer, a constant distribution ratio was observed between adsorbed and dissolved gold (Fig. 3) and the rate of adsorption was proportional to the initial gold concentration. Essentially complete adsorption required about four minutes. Plots of residual gold(III) concentration versus time up to four minutes and derived rate plots (not shown) indicated that, at sufficiently high initial concentration (6.6 x 10^{-5} M Au) to avoid the effects of solution transport control, the loss of gold from solution obeyed simple second order kinetics, while at lower concentration (0.8 x 10^{-5} M Au) first order transport kinetics predominated. Other experiments showed that, after a contact time of 20 minutes, roughly 15% of the gold could still be desorbed as gold(III) when concentrated chloride or bromide was added.

Thus, a relatively simple mechanism predominated initially but was soon complicated by gold depletion and other reactions. At sufficiently high concentration, however, a second chemically-controlled mechanism, presumably controlled by the concentration of gold and the numbers of reductive sites available on the bionass, became rate determining.

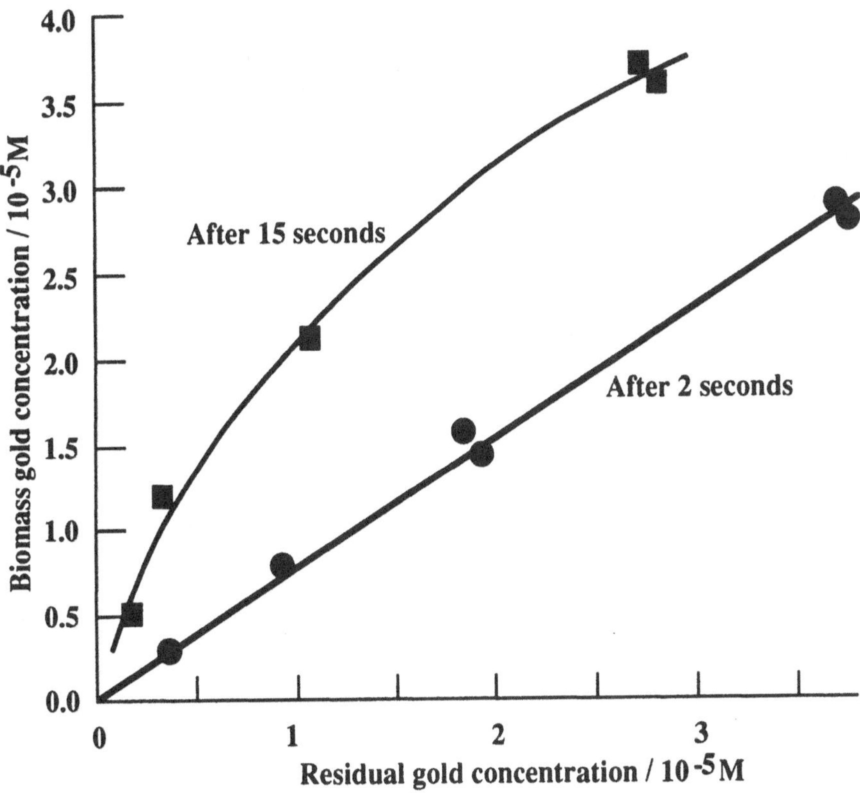

Fig. 3. Adsorption profiles for gold on *C. vulgaris* after contact times of 2 and 15 seconds.

These results, taken together with the slower reduction of gold(III) by amino acids observed in homogeneous solution according to Eqns 1-3, suggested that initial algal adsorption was significantly faster than, but overlapping with, reduction of gold(III) to gold(I). Under the conditions employed, some adsorbed gold remained in the +3 state even after pseudo-equilibrium had been established.

Gold crystallisation required much longer time periods. In order to study this process reliably, higher concentrations of gold (1000 mg/L Au at pH 0.5) and longer reaction times (up to 200 hours) were employed. Fig. 4 shows typical plots of the variation of the gold crystal 'area' (obtained by video microscopy) and residual gold concentration in solution versus time. As outlined in the experimental part, this 'area' gave a practical measure of the amount of gold present after set time intervals. The figure illustrates an initial lag phase of about 10 hours, followed by a pseudo-exponential (apparently autocatalytic) phase from 10-50 hours and a depletion phase after 50 hours as the gold concentration in solution approached zero. The lag phase was apparently due largely to slow initial formation of gold nuclei, but may be over-estimated in duration because of the difficulty of resolving first-formed, colloidal-sized, crystallites by light microscopy. The amount of gold adsorbed during the lag phase was 70-80 mg/L, somewhat less than for the higher pH values discussed above. The final uptake was about 100 mg Au /g algae.

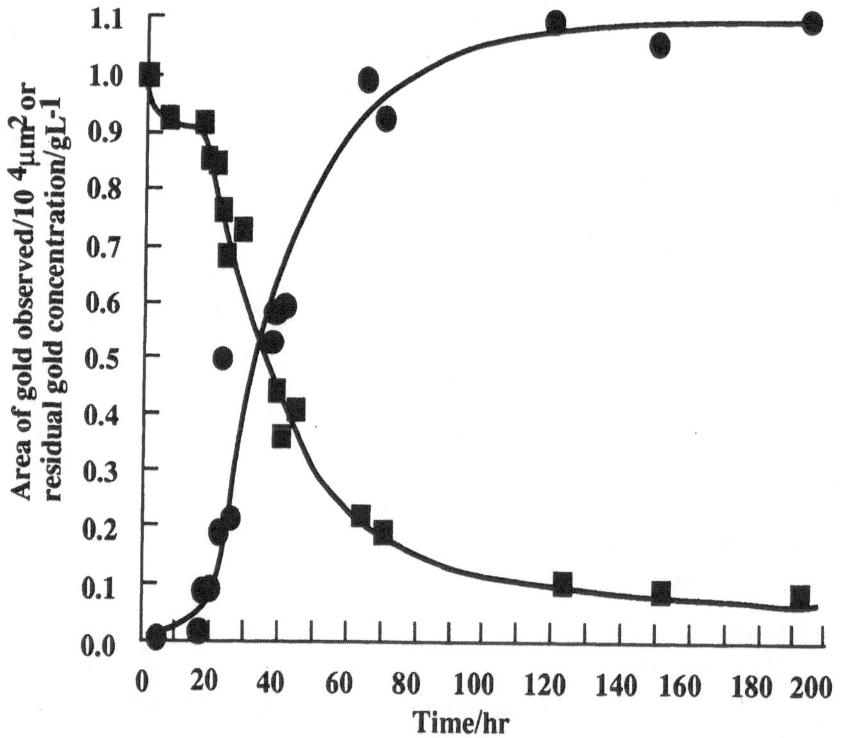

Fig. 4. Plots of gold crystal 'area' and residual gold concentration against time for *C. vulgaris*.

The bioaccumulation of gold on *Chlorella* cells is evidently a complex process. Complicating physical effects are bound to arise from the different possible depths of gold penetration into cell structure and from unquantifiable chemical effects which will result from the many different reactive groups likely to be present at various concentrations. It is nonetheless plausible from the available evidence to rationalise the bioaccumulation of gold(III) on *Chlorella* in terms of three main types of processes. These are represented (Fig. 5) by: (i) rapid reversible equilibration in seconds by anion exchange between tetrachloroaurate(III) and chloride ions on surface protonated amine residues (of which there are a large number) on proteins (ii) less rapid redox reaction over minutes at those (far fewer) adsorption sites near to reductive groups, leading to slower reversible gold(I) desorption/adsorption equilibration (iii) much slower disproportionation over hours of groups of three gold(I) species adsorbed in adjacent positions on the algal surface, followed by accelerated disproportionation of mobile gold(I) species at first-formed gold crystallites.

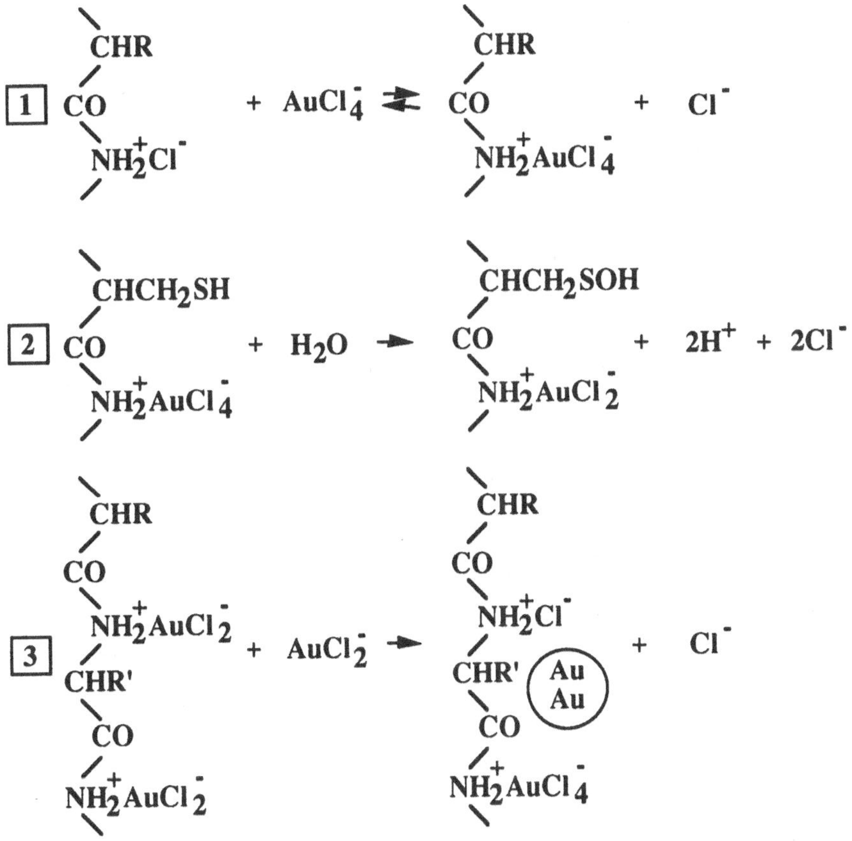

[1] $\diagdown CHR / CO \diagdown NH_2^+Cl^- /$ + $AuCl_4^-$ ⇄ $\diagdown CHR / CO \diagdown NH_2^+AuCl_4^- /$ + Cl^-

[2] $\diagdown CHCH_2SH / CO \diagdown NH_2^+AuCl_4^- /$ + H_2O → $\diagdown CHCH_2SOH / CO \diagdown NH_2^+AuCl_2^- /$ + $2H^+$ + $2Cl^-$

[3] $\diagdown CHR / CO \diagdown NH_2^+AuCl_2^- / CHR' \diagdown CO / NH_2^+AuCl_2^- \diagdown$ + $AuCl_2^-$ → $\diagdown CHR / CO \diagdown NH_2^+Cl^- / CHR'$ (Au Au) $\diagdown CO / NH_2^+AuCl_4^- \diagdown$ + Cl^-

Fig. 5. Mechanistic steps in the accumulation and reduction of gold on *C. vulgaris*.

In Fig. 5 water is shown as providing the oxygen required for (partial) oxidation of the group -SH to -SOH while gold(III) is being reduced to gold(I). As gold crystals of micrometer dimensions are eventually built up, desorption and solution transport of this gold(I) must occur. First formed gold nuclei act as preferential sites for gold(I) adsorption and further disproportionation, leading to autocatalytic kinetics. Overall three -SH groups result in the formation of two gold atoms.

Of course, such a mechanism cannot be proven and is not comprehensive. However, it provides a simple plausible model of the events occurring at the biomass surface.

3.2 Semi-continuous bioaccumulation

As mentioned earlier, a packed bed system employing dehydrated, approximately 1.5 mm, *Spirulina* or *Chlorella* pellets having enhanced (about 77%) biomass content was employed to provide greater gold adsorption efficiency than was achievable in a fluid bed sytem. Scanning electron microscopy of sections taken through the pellets indicated a highly porous structure containing networks of channels roughly 2-10 μm in diameter. Preliminary experiments showed that larger pellets gave marginally slower rates of gold uptake while smaller pellets were difficult to prepare. Higher biomass content also marginally increased uptake but the pellets were then mechanically unstable.

A series of tests employing 2-20 mg/L Au at pH 1-2 with 2-10 minute residence time gave adsorption values on *C. vulgaris* pellets initially > 99.8% Au, gradually diminishing with volume passed but generally > 92% after 100-250 bed volumes of solution had been fed to the column. For instance, a solution containing 10 mg/L Au at pH 1.8 having a residence time in the column of 10 minutes gave 99.2% adsorption (average loading 2.5 mg Au per g biomass) and an effluent concentration of 0.1 mg/L Au after passage of 120 bed volumes. No sharp 'breakthrough' of gold was observed, presumably because of continual replenishment of adsorption sites as gold on the biomass was reduced to gold(I) and metal. This replenishment was clearly indicated by running the column for five hours each day for three days while allowing it to stand undisturbed each night. The effluent gold concentration then increased from about 0.04 mg/L Au to 0.3-0.4 mg/L Au during the day and returned to the lowest value again after 'resting' overnight. After 2-3 days gold deposits were clearly visible in the lower part of the column: one analysis showed 9 mg Au per g algae near the inlet. Microscopic examination showed that fresh pellets changed in colour from black to brown during loading and then contained < 1 μm gold crystals. Cross sections through highly loaded pellets often showed complete gold penetration, whilst similar sections from lower loaded pellets indicated gold present only on the outer parts. Although not tested in detail, classic 'shrinking core' behaviour seemed to apply. From electron microprobe analysis and scanning electron microscopy, the outermost 1 μm or so of the pellets often had an altered (though porous) structure which was devoid of gold. This outer shell was probably the result of thermal degradation through 'baking' during the drying process. Metallic gold was readily recovered from the loaded pellets via incineration.

Separate tests showed that gold uptake was largely unaffected by the presence of the base metals copper, zinc and iron. Uptake was also little affected by sulphate but, as expected, was significantly reduced by strongly complexing anions like thiosulphate and cyanide. Pre-degradation of stable anionic complexes would be necessary in any process application.

4 Acknowledgement

A.R. Gee would like to acknowledge the award of a Stanley Elmore Fellowship from the Institution of Mining and Metallurgy which enabled this work to be carried out.

5 References

1. Williams, M., (1918) Absorption of gold from colloidal solutions by fungi. **Annals of Botany**, 32, 531-534.
2. Darnall, D.W., Green, B., Henzl, M.T., Hosea, J.M., McPherson, R.A., Sneddon, J. and Alexander, M.D., (1986) Selective recovery of gold and other metals from an algal biomass. **Environmental Science and Technology**, 20, 206-208.
3. Hosea, J.M., Green, B., McPherson, R.A., Henzl, M.T., Alexander, M.D., and Darnall, D.W., (1986) Accumulation of elemental gold on the alga *Chlorella vulgaris*. **Inorganica Chimica Acta**, 123, 161-165.
4. Green, B., Henzl, M.T., Hosea, J.M., Alexander, M.D. and Darnall, D.W., (1986) Interaction of gold(I) and gold(III) complexes with algal biomass. **Environmental Science and Technology**, 20, 627-632.
5. Darnall, D.W., Green, B. and Gardea-Torresdey, J. (1988) Gold binding to algae, in **Biohydrometallurgy** (eds. Norris, P.R. and Kelly, D.P.), Science and Technology Letters, London, 1988, pp. 487-498.
6. Darnall, D.W., McPherson, R.A.. and Gardea-Torresdey, J. (1989) Metal recovery from geothermal waters and groundwaters using immobilized algae, in **Biohydrometallurgy**, (eds. Salley, P., McCready, R.G.L. and Wichlacz, P), Canmet, Canada, 1989, pp. 341-347.
7. Gee, A.R. and Dudeney, A.W.L. (1988) Adsorption and crystallization of gold at biological surfaces, in **Biohydrometallurgy** (eds. Norris, P.R. and Kelly, D.P.), Science and Technology Letters, London, 1988, pp. 437-451.
8. Kuyucak, N. and Voleski, B. New algal biosorbent for a gold recovery process. (1988), in **Biohydrometallurgy** (eds. Norris, P.R. and Kelly, D.P.), Science and Technology Letters, London, 1988, pp. 453-463.
9. Gee, A.R. (1988) **Biosorption and biocrystallisation of gold**, PhD thesis, University of London, 1988, p. 33.
10. Lingane, J.J. (1962) Standard potentials of half reactions involving +1 and +3 gold in chloride medium. **Journal of Electroanalytical Chemistry**, 4, 332-342.
11. Puddephatt, R.J. (1978) **The Chemistry of Gold**, Elsevier, NY, 1978.
12. Natile, G. (1976) Chloroauric acid as an oxidant. Steriospecific oxidation of methionine to methionine sulfoxide. **Inorganic Chemistry**, 15, 246-248.
13. Shaw, C.F. (1980) Gold(III) oxidation of disulphides in aqueous solution. **Inorganic Chemistry**, 19, 3198-3201.
14. Isab, A.A. and Sadler, P.J. (1977) Reactions of gold(III) ions with ribonuclease A and methionine derivatives in aqueous solution. **Biochemica et Biophysica Acta**, 492, 322-330.
15. Sadler, P.J. (1982) The comparative evaluation of the physical and chemical properties of gold compounds. **Journal of Rheumatology**, Supplement No 8, 71-78.

Metal speciation in bioleach solutions during bacterial dissolution of a complex mineral sulphide concentrate

M. A. Jordan
Camborne School of Mines, Faculty of Engineering, University of Exeter, Redruth, Cornwall, England
C. V. Phillips
Camborne School of Mines, Faculty of Engineering, University of Exeter, Redruth, Cornwall, England
D. W. Barr
Advanced Technical Development, Bundoora, Victoria, Australia

The oxidation states of a number of ionic species were investigated following bacterial oxidation of a complex Cu/Zn sulphide flotation concentrate by six different acidophilic organisms. The major sulphide minerals comprising the concentrate were chalcopyrite, sphalerite, pyrite and arsenopyrite. The results of iron and sulphur speciation using the same concentrate described in this paper have already been published[1,2]. This paper concentrates on the speciation of copper only in bioleach solutions during mineral sulphide oxidation.

Copper speciation indicated the presence of Cu(I) in bioleach solutions from all of the organisms except *Metallosphaera sedula*. The appearance of Cu(I) during oxidation of the concentrate was unexpected since the pulp potential of the bioleach solution was in excess of 300mV and the Cu(I)/Cu(II) couple has an Eh of -153mV. The mechanism(s) behind the appearance of Cu(I) were not elucidated for the organisms investigated although reference to the literature demonstrates that certain acidophilic organisms are capable of reductive processes to utilise a different ion species as the terminal electron acceptor. The method for Cu(I) determination is discussed in light of the current experimental results.

Keywords: Bioleach, Metal Speciation, Complex Sulphide Concentrates, Bacterial Oxidation, Copper Speciation.

Introduction

The speciation of metal ions in either acidic or basic solutions is dependent not only on the stability of the species but also on a number of external factors such as pH and redox potential. Following the bacterial oxidation of complex sulphide minerals the release of ionic species such as Fe(II), Fe(III), Cu(II), Zn(II), is relatively common although the specific concentrations of each elements will be dependent upon the mineralogical characteristics of the concentrate. The presence of ionic species in two oxidation states, such as Cu(I)/Cu(II), may occur as the system approaches its equilibrium point[3] or possibly as the result of a reductive process[4-7].

The biological mechanisms resulting in the oxidation of sulphide concentrates (typically pyritic based), and subsequent release of metal ions into solution, can be summarised by the equations below[8,9];

Direct Oxidation - attachment of the different bacterial strains to the mineral surface is thought to be a pre-requisite condition to promote this route.

$$2FeS_2 + 7O_2 + 2H_2O \xrightarrow{\text{bacteria}} 2FeSO_4 + 2H_2SO_4 \quad (1)$$

$$4FeSO_4 + O_2 + 2H_2SO_4 \xrightarrow{\text{bacteria}} 2Fe_2(SO_4)_3 + 2H_2O \quad (2)$$

Indirect Oxidation - proceeds via the chemical action of Fe(III) on the mineral sulphide.

$$FeS_2 + Fe_2(SO_4)_3 \rightarrow 3FeSO_4 + 2S^\circ \quad (3)$$

$$2S + 3O_2 + 2H_2O \xrightarrow{\text{bacteria}} 2H_2SO_4 \quad (4)$$

As a result of the catalysis of the oxidation of sulphide concentrates/minerals by acidophiles, such as *Thiobacillus ferrooxidans* and *Sulfolobus*, metal ions such as iron, copper and arsenic are released in to solution. Broadhurst[10] noted that following oxidation of a pyrite/arsenopyrite gold-bearing concentrate, arsenic was present predominantly in its pentavalent oxidation state [As(V)]. This can be important since the toxicity of certain cations and anions, such as arsenic, is dependent upon oxidation state, with As(V) demonstrating reduced toxicity compared to As(III)[11]. Data is available regarding the bacterial dissolution of copper ores/concentrates although speciation data for both iron and copper within these systems[12] is often absent, hence comparison of work in the current study with the literature is limited.

Research by others has arguably been biased towards the application of strains of *Thiobacillus ferrooxidans* for the oxidation of sulphide ores/concentrates although interest in the application of moderate thermophiles for the oxidation of sulphide gold-bearing concentrates has been noted both on a laboratory scale[13] and commercially on pilot plant scale[8].

Application of alternative acidophilic bacterial strains in the catalysis of the oxidation of mineral sulphides has been investigated [14,15]. The mechanism utilised by these alternative acidophiles have not been studied as extensively as for *Thiobacillus ferrooxidans*. Literature is available on the possible oxidation mechanism utilised by a number of acidophilic thermophilic bacteria[1,16]. Our previous work[2] involving the thermophile *Sulfolobus* demonstrated that by due consideration of batch-leaching data resulting from the biooxidation of concentrates an indication of the possible route(s) of bacterial oxidation could be gained ie. degradation of an ore/concentrate by oxidation of iron and/or sulphur moieties.

Proliferation of bioleach plants using strains of moderate thermophiles or thermophiles is in the opinion of the authors dependent upon research extending the knowledge and understanding of these organisms in combination with continued pilot plant investigation of bioleach process within the minerals industry. This paper aims to highlight the importance of metal speciation in bioleach liquors in terms of both the possible mechanism utilised by different bacterial strains and the possible implications to downstream hydrometallurgical processes. A total of six bacterial strains were investigated using a complex copper/zinc sulphide flotation concentrate as the mineral substrate (energy source).

Materials and Methods

The cultures used were thermophiles *Sulfolobus*, strain BC (BC65, as described by Norris [17], *Metallosphaera sedula* (DSM 5348), *Acidianus brierleyi* (DSM 1651) and mesophiles *Thiobacillus ferrooxidans* DSM 583, ATCC 11820 and an uncharacterised culture from the acid mine drainage at the Wheal Jane mine, Cornwall.

All of the strains were grown in a nutrient salts medium containing (g/l): KCl (0.1), K_2HPO_4 (0.1), $MgSO_4$ (0.4) and $(NH_4)_2SO_4$ (0.4) in double glass-distilled water, adjusted to the correct pH with H_2SO_4. The mineral substrate (energy source) was a bulk flotation concentrate of a complex sulphide from Wheal Jane mine, Cornwall, (W.J.C), (21.2% Cu, 12.9% Zn, 34.3% Fe, 31.3% S, <250μm), at a pulp density of 5% w/v. The main sulphide minerals of chalcopyrite, sphalerite, pyrite and arsenopyrite were determined to be present following analysis of the substrate using a SIEMENS D5000 XRD and a JEOL 840 SEM and a LINK AN10,000 EDS analyser.

Batch growth of the *Sulfolobus* was in one litre volume, water-jacketed glass vessels, stirred at 200rpm and maintained at 68±1°C. *Metallosphaera sedula* and *Acidianus brierleyi* were batch grown in 500ml, water-jacketed pachuca reactors with internal draught tubes. Air containing 2%v/v CO_2 was passed into both vessels via an humidifier at 300ml/min. Batch growth of the mesophiles was carried out in 2 litre glass CSTR vessels maintained at 32±2°C.

Samples for total metal analysis were pre-treated with equal volumes of 5N HCl for 30 min (BS 1016 method, [18], centrifuged at 2000rpm and aliquots stored in 5% v/v HCl under airtight conditions at 4°C. Metal analysis was on a Pye Unicam SP9 atomic absorption spectrophotometer.

Solutions to be speciated for iron were placed in 0.9M H_2SO_4 (2ml), to maintain speciation, and allowed to stand for 30mins. Ferrous iron was determined by titration against standard $CeSO_4$ (5mM) using 1,10 phenathroline ferrous complex as the indicator.

Concentration of Cu(I) was determined using the bathocuproinedisulfonic disodium salt described by Blair and Deihl[19]. A stock copper solution was prepared by dissolving copper metal in HCl (11.0M). Standard solutions ranging from 0→10μg/ml Cu(I) concentration were prepared in double glass-distilled water, pH adjusted to 1.20 with dilute H_2SO_4, and used to produce a calibration curve from which subsequent Cu(I) concentrations of bioleach samples were determined. The Cu(I) standards were prepared by reducing an equivalent concentration of the stock copper solution by addition of 10ml of hydroxyammonium chloride solution (10%m/v). 11.0M HCl (1ml) was then added followed by 5ml of sodium citrate buffer (30%m/v) and 3ml of the bathocuproine reagent (0.2628g/100ml). Solution absorbance was then determined at a wavelength of 484nm using a Pye UNICAM SP6-550 UV/VIS spectrophotometer.

Determination of the concentration of Cu(I) in bioleach solutions was performed in an identical manner, replacing the stock copper solution with the bioleach sample. Total copper determination via the same method was carried on the bioleach samples by adding 10ml hydroxyammonium chloride solution (10%m/v) to the sample solution to reduce all copper present to the Cu(I).

Results

Iron and Sulphur Speciation

Concentrations of total iron and Fe(II) and sulphur and sulphate for the same sulphide substrate and range of organisms have already been published[1,2]. The preferential release of Fe(II) prior to copper and the corresponding balance of sulphur and sulphate have been discussed in terms of possible mechanisms of oxidation of the complex sulphide concentrate[1,2].

Copper Speciation

Prior to speciation of copper in bioleach solutions the method as described by Ayling[20] [for Cu(I) determination in Cl^- based solutions] was investigated for application in SO_4^{2-} systems. SO_4^{2-} interference on colour development of the Cu(I) complex was investigated having noted the results of Blair and Deihl[19]. Other ions (cations) investigated, both individually and in combination with each other, included Cu(II), Fe(II), Fe(III), Zn(II), As(III) and As(V). The ion concentrations used corresponded to the theoretical maximum concentrations of the ions present in the sample solutions as a result of the oxidation of the mineral concentrate. The stability of the bathocuproine reagent under laboratory conditions was also investigated.

The results of the investigations illustrated that at concentrations ≥0.36M SO_4^{2-} (in the final test solution) colour development of the reagent was completely inhibited. At sulphate concentrations of 0.18M → 0.36M colour development was impeded but some colour did develop after a 15 min period. Concentrations of SO_4^{2-} <0.15M were determined not to interfere with colour development. Since this concentration of SO_4^{2-} was in excess of that determined to be present in the bioleach solutions following dilution (maximum concentration was determined to be <10mM SO_4^{2-}/l) then interference of SO_4^{2-} was ruled out. The effect of biomass on colour development was not assessed but is discussed later. The effect of the different cations investigated, both individually and in combination, was also determined to be negligible. Stability of the bathocuproine reagent when stored under "normal" laboratory conditions was determined to be of the order of 24hours. Refrigeration of the sample at 4°C extended this period of stability to approximately 96hours.

It was initially proposed to use the Cu(I) method to determine both Cu(I) and total copper concentrations in the bioleach solutions and hence infer the concentration of

Cu(II), permitting the use of a single method for determination of copper speciation/concentration. Comparison of these results with AAS results would have then provided a series of "check" results. Figure 1 illustrates the effect of plotting total copper concentration as determined both by the VIS (colorimetric) method and by AAS following oxidation of the W.J.C in a CSTR vessel.

Figure 1 clearly demonstrates that the colorimetric determination of total copper was interfered with in the latter stages of copper release. This coincides with both an increase in copper concentration and an assumed increase in biomass concentration (not determined). The apparent difference in the two concentrations could not be eradicated by addition of an increased volume of hydroxyammonium chloride which, in the absence of biomass ie. organic components, had been conclusively proven to totally reduce Cu(II) to Cu(I). As a consequence total copper had to be determined using AAS whilst Cu(I) was determined using the VIS method. The speciation of copper during oxidation of the W.J.C was investigated in both CSTR vessels and airlift tubes for a variety of acidophiles, the results are summarised in table 1.

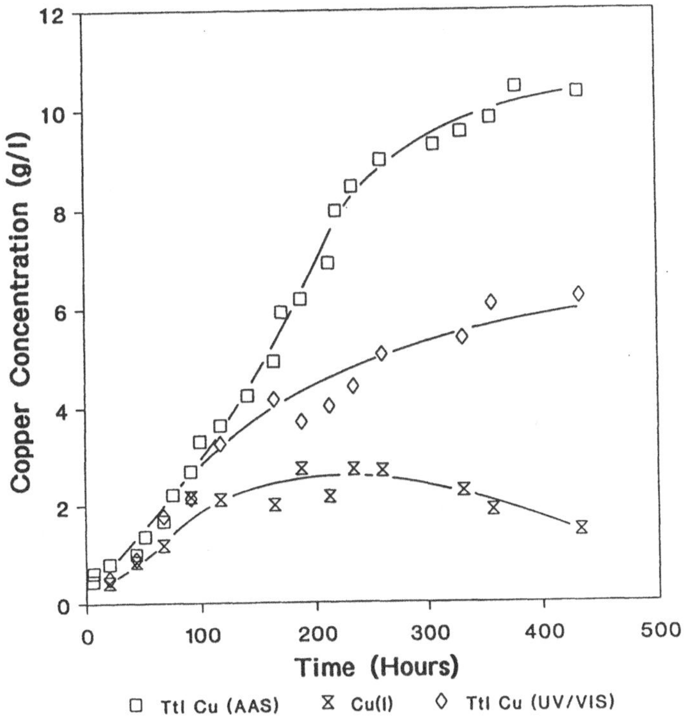

Figure 1: Comparison of total copper by colorimetric and AAS methods using a bioleach solution following mineral sulphide oxidation of the Cu/Zn concentrate by *Sulfolobus*.

Table 1: Summary of Copper Speciation Results for the W.J.C.

Strain Type	$Cu(I)_{MAX}$(g/l)	
	CSTR	Airlift
Thermophiles		
Sulfolobus	4.6	2.7
	6.0	-
Acidianus brierleyi (DSM 1651)	-	4.0
Metallosphaera sedula (DSM 5348)	-	0.3*
Mesophiles		
W.J. *T.ferrooxidans*	2.3	-
DSM 583	2.3	-
ATCC 11820	2.2	-

Note 1: Error during Cu(I) determinations were assessed to ±0.3gCu(I)/l.
Note 2: * Reading at the edge of the detection limit taking into account note 1.
Note 3: Supplementary addition of 8gFe(II)/l was made prior to inoculation of the CSTR vessel.

Following the preferential release of Fe(II) prior to copper noted previously[2] investigations of copper speciation during oxidation of the W.J.C by the thermophile *Sulfolobus* were completed simultaneously with iron speciation. Figure 2 illustrates copper speciation during batch-oxidation of the W.J.C by *Sulfolobus* in a CSTR vessel.

The presence of both Cu(I) and Cu(II) in solution during oxidation of the concentrate by *Sulfolobus* was observed, Cu(I) was also present on termination of the batch leach. Following the partial oxidation of the Fe(II) to Fe(III) in solution the concentration of Cu(I) decreased, the potential of the pulp correspondingly increased. The ratio of Cu(I) to total copper varied during the oxidation of the concentrate between 0.4-0.7 : 1.0. Following supplementary addition of Fe(II)[2], prior to inoculation of the vessel, the maximum concentration of Cu(I) determined to be present in solution was noted to increase by ≈1.4g/l. Similar experiments were carried out in airlift tubes (pachucas) with the W.J.C and the *Sulfolobus* culture. The peak concentration of Cu(I) in the was determined to be 2.7g/l a reduction of ≈40%. The ratio of Cu(I) to total copper was also decreased for the airlift control vessel varying between 0.1-0.4 : 1.0.

The differences in copper speciation seen during oxidation of the W.J.C. by *Sulfolobus* in both CSTR vessels and airlift tubes corresponded to a difference observed in the behaviour of the species of iron. The influence of Fe(III) concentration on Cu(I) concentration was apparent between the two reactor types during oxidation of the concentrate by the *Sulfolobus* culture. In the CSTR vessels (figure 2) where oxidation of Fe(II) was noted to be more pronounced, copper(I) concentration decreased between

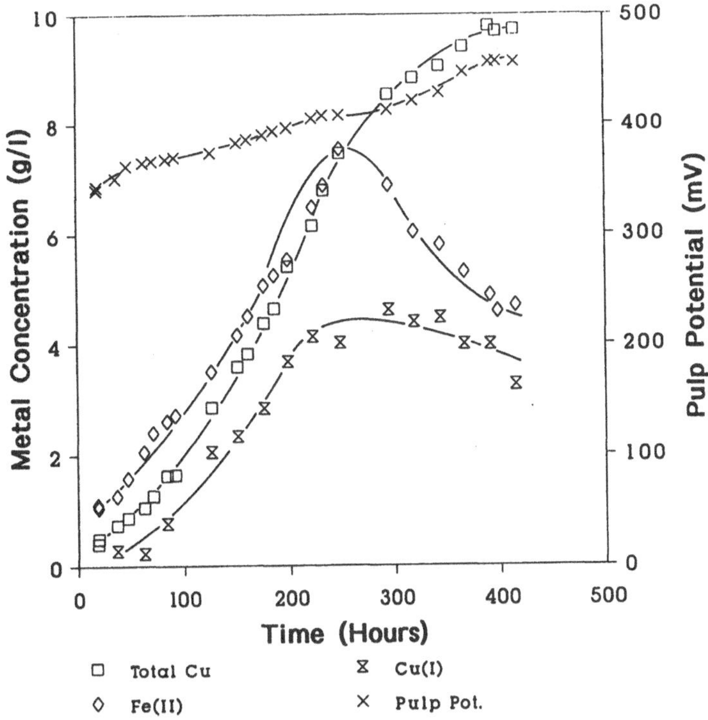

Figure 2: Copper speciation during oxidation of the Wheal Jane concentrate by *Sulfolobus* in a CSTR vessel at a 5%w/v pulp density and a temperature of 68±1°C.

17-25% (from its peak concentration of 4.6g/l) in 72hours. In comparison a decrease of only 3-12% was noted in airlift tube reactor (pachuca) experiments. In both reactor types following the onset of the oxidation of Fe(II) the concentration of Fe(III) continually increased as the result of the continued oxidation of the concentrate.

The changes in pulp potentials within the two reactors types were similar although oxidation of Fe(II) differed. In the CSTR vessels the potential increased gradually at 0.26mV/hr, rising from ≈340mV to ≈460mV (figure 2), the rate of increase in pulp potential rising to 0.83mV/hr during Fe(II) oxidation. Correspondingly the pulp potential during oxidation of the W.J.C in airlift tubes increased from ≈340mV to ≈460mV, even though oxidation of Fe(II) was limited in comparison to that of the equivalent CSTR vessels. The results suggest that the increase in Fe(III) concentration in CSTR vessels, in comparison to the Fe(II)/Fe(III) ratio, was not sufficiently significant enough to promote a change in pulp potential.

Similar experiments conducted in the airlift tube reactors with two other thermophiles, *Acidianus brierleyi* and *Metallosphaera sedula*, resulted in a noticeable difference in copper speciation. *Acidianus brierleyi* behaved in a similar fashion to *Sulfolobus*. Maximum concentration of Cu(I) in solution was determined to be 4.0g/l, an increase of 1.3g/l in comparison to the equivalent *Sulfolobus* batch airlift leach. Following the increase in Fe(III) concentration the Cu(I) concentration was again observed to decrease. The ratio of Cu(I) to total copper during oxidation of the W.J.C by *Acidianus brierleyi* was determined to peak at approximately 0.60 : 1.0. During this oxidation by *Metallosphaera sedula* the production of Cu(I) was negligible and on the edge of the analytical limit of detection. The results would suggest a significant difference in the physiology of the *Metallosphaera sedula* culture compared with the cultures of *Acidianus brierleyi* and *Sulfolobus*. Comparison of copper speciation following oxidation of the W.J.C in CSTR vessels by *Acidianus brierleyi* and *Metallosphaera sedula* could not be made. This was due to inability of the cultures to be grown under the conditions imposed in the CSTR vessels.

During oxidation of the same concentrate by *Sulfolobus* in a fed-batch CSTR (data not shown) the production of Cu(I) was determined to occur and in similar concentrations to that observed during batch experiments.

Copper speciation was also investigated during oxidation of the W.J.C in CSTR vessels by three different mesophiles. In all cases Cu(I) was detected in solution with the peak concentration of ≈2.3g/l varying only marginally between the different strain types. In comparison with the results presented for the thermophile *Sulfolobus* the maximum concentration of Cu(I) determined was reduced. More importantly on termination of the experiment the concentration of Cu(I) was <0.1g/l ie. below the limit of detection. This difference observed in copper speciation also corresponds to the difference in the behaviour of iron speciation[1,2]. Ferrous iron was initially seen to increase in concentration, although its maximum concentration and timing was noticeably different to that with the thermophiles. Following the onset of the oxidation of Fe(II), which resulted in the Fe(II) remaining in solution being <0.5g/l, Fe(III) dominated the bioleach solution with a corresponding stepped increase in pulp potential to in excess of 500mV. During oxidation of the mineral concentrate the pulp potential rose from ≈300mV to ≈530-570mV with a stepped increase in the region of 120mV coincident with the oxidation of Fe(II) (figure 3). The ratio of Cu(I) to total copper for strain type ATCC 11820 reached a maximum of 0.96 : 1.0 whilst the remaining mesophiles averaged ratios of 0.63 : 1.0 and 0.75 : 1.0 for the enrichment culture and strain type DSM 583 respectively. Figure 3 illustrates copper speciation for the mesophilic enrichment culture during the oxidation of the W.J.C.

In all batch CSTR where Cu(I) was determined, involving the acidophilic bacteria defined above, the concentration of Cu(I) was noted to fluctuate between sampling intervals, probably due to the relative instability of Cu(I).

Discussion

The speciation of ions in solutions resulting from the bacterial oxidation of a mineral sulphide flotation concentrate can be used to help indicate the mechanism/route of

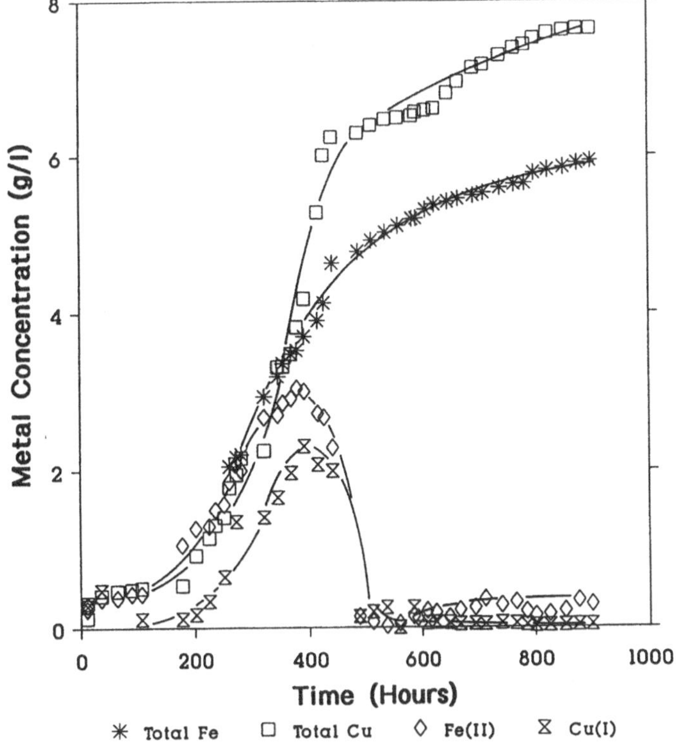

Figure 3: Copper speciation during oxidation of the Wheal Jane concentrate by the mesophilic enrichment culture ic CSTR vessels at a pulp density of 5%w/v and temperature of 32±2°C.

oxidation used by a particular bacterial strain[1]. The speciation of bioleach products may also be important in defining the impact on downstream processes, such as metal extraction from solution and the discharge of waste (effluent) solutions to tails.

Speciation of copper in bioleach solutions produced interesting results when assessing the experimental data available. The instability of Cu(I) is documented in the literature with the Cu(I)/Cu(II) couple having a potential of -153mV[3,21], although Ferreira and Burkin[22] do imply that at temperatures >100°C the stability of Cu(I) in SO_4^{2-} solutions increases [conversely Cu(I) becomes more unstable as temperature is decreased]. In terms of the speciation results presented questions remaining to be answered are why was Cu(I) found to be present and what does its presence indicate about the leaching mechanism utilised by the different bacterial strains during oxidation of the mineral concentrate?

The presence of Cu(I) was surprising since the pulp potentials of the batch bioleach experiments all exceeded 300mV, the Cu(I)/Cu(II) couple having a potential value of -153mV[3,21]. From the mesophilic batch leaches it can be inferred that oxidation of all Cu(I) will occur rapidly if the pulp potential exceeds 500mV. The presence of Cu(I) on termination of the batch experiments with the thermophiles *Sulfolobus* and *Acidianus brierleyi*, where pulp potential did not exceed 460mV, supports this theory. Interference from the cations and anions associated with the mineral concentrate did not effect the colour development from the complexing reagent did not occur. The influence of biomass/organics on the complexing reagent was not investigated. However, it is the opinion of the authors that interference of colour development due to their presence in the sample solutions was not apparent due to that behaviour of Cu(I). Should biomass/organics have been interfering with the complexing reagent one would expect the interference to both be continually present and to increase (possibly up to an upper limit) since the concentration of biomass in batch leaches of mineral concentrates increases following initial inoculation of the system. Interference of the Cu(I) method would appear to have been ruled out since during oxidation of the concentrate by the mesophiles Cu(I) was observed to oxidise and the release of Cu(I) during oxidation of the concentrate with *Metallosphaera sedula* was negligible ie a contrasting set of results. The inability of the method to reduce all copper present to Cu(I) following addition of hydroxyammonium chloride suggests that the presence of biomass/organics in the sample solutions, released from the bacterial cells during the oxidation process, prevented complete reduction of Cu(II) to Cu(I), perhaps by consumption of the reducing reagent.

A possible source of Cu(I) is the chalcopyrite mineral. Ferreira and Burkin[22] showed that the slow release of copper from chalcopyrite by *Thiobacillus ferrooxidans* can be attributed to the variation in the molecular formula $Cu^{2+}Fe^{2+}2S^{2-} \rightarrow Cu^{+}Fe^{3+}2S^{2-}$ for chalcopyrite, this alteration restricting the sulphur moiety as the only energy substrate. The release of Fe(II) from the concentrate prior to the main phase of the oxidation of pyrite was predominantly from chalcopyrite mineral, initial XRD and SEM analysis of the mineral concentrate indicating the absence of any iron-rich sphalerite and the molecular formula of chalcopyrite to be $CuFeS_2$, suggesting the appearance of Cu(I) to be the product of a reductive reaction. The presence of Cu(I) in solution at pulp potentials >300mV could only be explained if the rate of Cu(II) reduction exceeded the rate of Cu(I) oxidation. Re-oxidation of Cu(I) by iron-grown cells is acknowledged and involves the use of iron oxidases (a group of enzymes capable of oxidising iron) or the chemical oxidant Fe(III). Evidence of bacterial systems capable of performing reductive reactions is available in the literature[4-7,23]. Sugio et al.[23] have indicated that certain strains of *Thiobacillus ferrooxidans* can reduce Cu(II) to Cu(I) in the presence of S°. The optimum pH of reduction was 5.0 with the reduction proportional to the concentrations of Cu(II) and S° associated with the system. Since experimental conditions were different to that of Sugio et al.[23] such a mechanism for the mesophiles used in the current study cannot be inferred. Johnson and McGinnes[6] proved that certain acidophilic heterotrophic organisms are capable of Fe(III) reduction although no such system for copper reduction was noted. The presence of acidophilic heterotrophs in the cultures used in this study were not recorded and hence there role in Cu(II) reduction cannot be proven. It should be noted that in all figures illustrated Cu(I) concentration starts to decrease when total copper concentration reached 4-5g/l. The authors are aware of published work[24] that has shown the eradication of acidophilic heterotrophs using copper sulphate at concentrations of ≈5g/l from autotrophic cultures. This figure corresponds to the apparent start of the fall of the Cu(I) concentration in our batch experiments and may suggest the presence of acidophilic heterotrophs in certain of the cultures. The authors are not aware of any reports of the reduction of Cu(II) by acidophilic thermophiles such as *Sulfolobus* and *Acidianus brierleyi*, although Sehlin and Lindstrom[7] note the ability of *Sulfolobus acidocaldarius* to oxidise and reduce arsenic, and hence the explanation of the possible mechanism of mineral sulphide oxidation by certain thermophilic organisms noted in the current study and elsewhere[1] require further research.

It is apparent that confirmation of Cu(I) in bioleach solutions, presented in this paper, is required prior to explaination of its role in the oxidative cycle of chalcopyrite-based sulphide flotation concentrates. A definitive explanation for the presence of Cu(I) from the current study cannot be offered. If the source of Cu(I) cannot be the mineral substrate then this would favour the ability of certain bacteria to reduce Cu(II) this being supported by the evidence of organisms from the literature[4-7] capable of reductive processes. A question then remaining to be answered is why do the bacteria preferentially select a substrate to be reduced when alternative substrates exist? and could this infer a limiting condition in the system?

In terms of processing applications the detection of Cu(I) in solution, if confirmed, could possibly be of importance where electrowinning of copper from solution is subsequently applied. Power consumption in electrowinning would theoretically be halved for the Cu(I) species in comparison to the Cu(II), a cost saving benefit being derived. The behaviour of the Cu(I) following bioleaching would be a controlling factor with the distinct possibility that auto-oxidation of Cu(I) → Cu(II) would occur following removal of the metallic ions from the controlled environment present in the bioleach reactors. Auto-oxidation of Cu(I) prior to titration of Fe(II) against standard cerium sulphate (5mM) was determined to occur even though the solutions were highly acidic (sulphate). Such a reaction would eradicate any possible benefit from Cu(I) production in the bioleach process. Speciation of metals, like arsenic for instance is important in bioleach solutions when considering the discharge of materials to the tailings dam. Broadhurst[10] has indicated that following the oxidation of arsenopyrite-bearing concentrates arsenic is present predominantly as As(V). From a toxicity standpoint this is advantageous.

In summary by due consideration of metal speciation in bioleach solutions an indication of the possible mechanism of mineral sulphide oxidation may be gained. The implication of speciation data on downstream processes requires individual assessment for each process since the process variables may alter how a given species reacts to its environment.

Conclusions

1 Determination of metal speciation in bioleach solutions can give an indication of the possible mechanism(s) used during the bacterial oxidation of mineral sulphide concentrates.
2 The analytical method for the speciation of Cu(I) in SO_4^{2-} based bioleach solutions has been demonstrated.
3 Detection of Cu(I) at pulp potentials >300mV requires confirmation prior to further research on its possible implication on downstream hydrometallurgical processes.

Acknowledgements

The financial support of the Camborne School of Mines and the SERC, in the form of a quota award (MAJ), is greatfully acknowledged as is the assistance of Eric Crowther [Experimental Officer (Retd), CSM] in the development of the Cu(I) assay and Stephen McGinness for discussions on the microbial aspects of the work.

References

1 Jordan, M.A., Barr, D.W. and Phillips, C.V. (1993) Iron and Sulphur Speciation and Cell Surface Hydrophobicity During Bacterial Oxidation Of A Complex Copper Concentrate, **Minerals Engineering**, Vol. 6, Nos. 8-10, 1001-1011.
2 Barr, D.W., Jordan, M.A., Norris, P.R., and Phillips, C.V. (1992) An Investigation into Bacterial Cell, Ferrous Iron, pH and Eh Interactions During Thermophilic Leaching of Copper Concentrates, **Minerals Engineering**, Vol.3-5, 557-567.
3 Kust, R.N. (1979) Copper Compounds, **In:Kirk-Othmer Encyclopedia of Chemical Technology (3rd Edition)**, (Eds) Mark, H.F., Othmer, D.F., Overberger, C.G., and Seaborg, G.T., John Wiley & Sons (Chichester), Vol.7, 97.
4 Lovley, D.R. (1991) Dissimilatory Fe(III) and Mn(IV) reduction, **Microbiol. Rev.**, 55, 259-287.
5 Brock, T.D. and Gustafson, J. (1976) Ferric iron reduction by sulfur- and iron-oxidizing bacteria, **Appl. Environ. Microbiol.**, 32, 567-571.
6 Johnson, D.B. and McGinnes, S. (1991) Ferric iron reduction by acidophilic bacteria, **Appl. Environ. Micrbiol.**, 57, 207-211.
7 Sehlin, H.M. and Lindstrom, E.B. (1992) Oxidation and reduction of arsenic by *Sulfolobus acidocaldarius* strain BC, **FEMS Microbiology Letters**, 93, 87-92.
8 Pooley, F.D. (1987) Use Of Bacteria To Enhance The Recovery Of Gold From Refractory Ores, **In: Int. Symp. On Innovative Plant and Processes for Mineral Eng., MINPREP '87**, 1-14.
9 Lundgren, D.G. (1980) Ore Leaching by Bacteria, **Ann. Rev. Micriobiol.**, 34, 263-283.

10 Broadhurst, J. (March 1993), Neutralisation of arsenic bearing BIOX liquors, **Mining Environmental Management**, 4-5.

11 Silver, S., Budd, K., Leahy, K.M., Shaw, W.V., Hammond, D., Novick, R.P., Willsky, G.R., Malamy, M.N. and Rosenburg, H. (1981) **Journal of Bacteriology**, 146, 983.

12 Blancarte-Zurita, M.A., Branion, R.M.R., and Lawrence, R.W. (1987) Microbiological Leaching of Chalcopyrite Concentrates by *Thiobacillus ferrooxidans*. A Comparative Study of a Conventional and a Catalysed Process, **In:Biohydrometallurgy 1987** (Eds) P.R.Norris and D.P.Kelly, STL (1988), Proceedings of the International Symposium on Biohydrometallurgy (12-16 July 1987), 273-285, ISBN-0-946682-00-3.

13 Barrett, J., Ewart, D.K., Hughes, M.N., Nobar, A.M., and Poole, R.K. (1989) The Oxidation of Arsenic in Arsenopyrite: The Toxicity of As(III) to a Moderately Thermophilic Mixed Culture, **Biohydrometallurgy 1989**, Eds Salley, J., McCready, R.G.L., and Wichlacz, P.L., Proc. of the Int. Symp. at Jackson Hole, Wyoming, August 13-18, CANMET SP89-10, 49-57.

14 Pers.Comm., Andrew, C.J., (October 1993).

15 Le Roux N.W. and Wakerley, D.S. (1988) Leaching of Chalcopyrite ($CuFeS_2$) at 70°C using *Sulfolobus*, **In: Proceedings of Int. Symp. on Biohydrometallurgy**, (Eds) P.R.Norris and D.P.Kelly, STL (1988), Proceedings of the International Symposium on Biohydrometallurgy (12-16 July 1987), 305-318, ISBN-0-946682-00-3.

16 Larsson, L., Olsson, G., Holst, O. and Karlsson, H.T. (1993) Oxidation of pyrite by *Acidianus brierleyi*: Importance of close contact between the pyrite and the microorganism, **Biotechnology Letters**, Vol.15, No.1, 99-104.

17 Norris, P.R. and Parrott, L. (1986) High temperature, mineral concentrate dissolution with Sulfolobus, **Fundamental and Applied Hydrometallurgy**, Elsevier, Amsterdam, 355-365.

18 Atkins, A.S. (1978) Studies on the oxidation of sulphide minerals (pyrite) in the presence of bacteria, **Metallurgical Applications of Bacterial Leaching and Related Microbiological Phenomena**, (Editors) Murr, L.E. Torma, A.E. and Brierley, J.A., Academic Press, New York, 403-426.

19 Blair, D., and Diehl, H. (1961) Bathophenthrolinedisulphonic Acid and Bathocuproinedisulfonic Acid, Water Soluble Reagents for Iron and Copper, **Talanta**, Vol.7, Pergamon Press Ltd (N.Ireland), 163-174

20 Ayling, K. (1989) Chloride Leaching of Complex Sulphide Minerals and Recovery using the C.E.E.R Cell, **Ph.D Thesis, Camborne School of Mines**, Redruth, Cornwall, U.K.

21 Tuovinen, O.H. (1978) Metabolic Transitions in Cultures of Acidophilic *Thiobacilli*, **In: Metallurgical Applications of Bacterial Leaching and Related Microbiological Phenomena**, (Eds) L.E.Murr, A.E.Arpad and J.A.Brierleyi, Academic Press (London), 61-81.

22 Ferreira, R.C.H. and Burkin, A.R. (1975) Acid leaching of Chalcopyrite, **In:Leaching and Reduction in Hydrometallurgy**, Ed Burkin, A.R., IMM (London), 54-66.

23 Sugio, T., Tsujita, .Y, Inagaki, K., and Tano, T. (1990) Reduction of Cupric Ions with Elemental Sulphur by *Thiobacillus ferrooxidans*, **Applied and Environmental Microbiology**, Vol. 56, No.3 (June), 693-696.

24 Johnson, D.B. and Kelso, W.I. (1983) Detection of heterotrophic contaminants in cultures of *Thiobacillus ferrooxidans* and their elimination by subculturing in media containing copper sulfate, **J.Gen. Microbiol.**, 129, 2969-2972.

Bacterial oxidation of an auriferous arsenical concentrate

Constantine Komnitsas
Laboratory of Metallurgy, National Technical University of Athens, Athens, Greece
F. D. Pooley
University of Wales, Cardiff, Wales

Abstract
In this study the bacterial oxidation of an extremely refractory auriferous arsenical pyrite concentrate from Olympias, Greece, is examined. A solution exchange system comprised of two 5 L capacity air-stirred pachuca reactors was used in order to enhance oxidation kinetics and reduce retention times when operating at high pulp densities and finally achieve high percentages of oxidation for both sulphide phases. Inoculated with bacteria and bacterial ferric liquor tests were performed in order to determine the effect of the variation of the particle size of the concentrate on the percentage of oxidation for both mineral phases for a pulp density range varying between 5 and 30% w/v and to examine the importance of the bacterial ferric solution as leaching agent.
Keywords: Auriferous arsenical pyrite concentrate, Solution exchange system, Arsenopyrite oxidation, Gold recovery, Bacterial ferric leaching.

1 Introduction

Refractory gold ores and concentrates are characterised by low gold recoveries in traditional cyanidation and accompanied by high cyanide consumption (Jha 1987).

The main mineralogical factor influencing the refractoriness of gold in certain ores is the dissemination of fine grained or sub-microscopic gold within pyrite or arsenopyrite, which cannot be detected by optical microscopy or electron microprobe analysis. This mineralogy renders the gold inaccessible to cyanide leaching (Pooley 1987, Komnitsas and Pooley 1989). The bacterial oxidation of refractory ores and concentrates seems very attractive from an economical and environmental point of view compared to pressure oxidation or oxidation roasting. Despite this, a number of engineering problems, such as reactor design and process chemistry and control still remain to be resolved. As has been demonstrated in previous experimental studies (Komnitsas and Pooley 1990 and 1991) bacterial oxidation can be applied for the treatment of an extremely refractory auriferous arsenical pyrite concentrate, for a pulp density range 5-30% w/v and a concentrate particle size -63 to -32 μm. At 5 and 10% w/v pulp density gold recoveries achieved after 6-8 days of leaching reach 90%, whereas at 20 and 30% w/v pulp density gold recoveries are considerably lower and vary between 55 and 80% over a leach period of 2 weeks. The main reasons for decreased gold recoveries when operating at high pulp densities are the decrease in the number of free cells in suspension due to the increase of the reactive surface area present in solution and the inability of the cells to convert all ferrous iron produced to the ferric state, so that the lag period is extended to 3-5 days and suspension Eh values are prevented from reaching sufficiently high levels.

In order to overcome these problems a solution exchange system comprised of two air-stirred pachuca reactors was designed. The bacterial solution containing ultra fine concentrate particles was pumped from the first to the second reactor acting as source for the production of ferric ions, and fed back again to the first one, so that high reaction rates were attained in the system and therefore increased percentages of oxidation for both mineral phases over short retention times were achieved.

The objective of this paper is to examine the viability of biooxidation for the pre-treatment of an extremely refractory auriferous arsenical pyrite concentrate from Olympias, Greece, when operating at high pulp densities. Inoculated with bacteria and bacterial ferric liquor tests were performed, for a pulp density range varying between 5 and 30% w/v, in order to determine the effect of the variation of the particle size of the concentrate on the percentage of oxidation for both mineral phases and consequently on the degree of gold recovery.

Furthermore, the importance of bacterial ferric solution as a leaching agent enabling high percentages of oxidation for both mineral phases over short retention times had to be examined.

2 Exrerimental part

2.1 Material Used

The Olympias concentrate assays 27 ppm Au, 36 ppm Ag, 38.5% Fe and 13.1% As. Detailed chemical and wet screen analyses data has been reported in a previous work (Komnitsas and Pooley 1990). The main mineralogical components of this sample were pyrite (70%) and arsenopyrite (30%).

2.2 Bacterial Culture

The culture of iron oxidising *Thiobacillus Ferrooxidans* used in this experimental study have been continuously grown over 25 years in the Mineral Processing laboratory of the University of Wales College Cardiff, on a variety of metal sulphide concentrates and shown to be adaptive to growth in the presence of high concentrations of iron and arsenic.

2.3 Bacterial Ferric Solution

The bacterial ferric solution used in this experimental study was generated in 5 L air-stirred pachuca reactors with the use of pure pyrite concentrate, size -56+20 µm at 5% w/v pulp density. Leaching under these conditions produces a ferric liquor with a high bacteria concentration and capable of maintaining high oxidation rates for both pyrite and arsenopyrite over a long leaching period (Komnitsas and Pooley 1991 and 1994).

2.4 Leaching Technique

The solution exchange system used in this study was comprised of two 5 L capacity air-stirred pachucas, contained in a heated water jacket maintained at 35 ± 1 °C, into which air is diffused via a sparger arrangement at the base of each reactor at a rate of one litre per minute. The diffused air bubbles from the sparger pass up the inner column giving rise to aeration, circulation and suspension of the mineral particles.

Each reactor body and inner column are constructed from similar perspex tubing. The lower conical section of each reactor was threaded to accommodate the attachment of the sparger assembly.

The mineral concentrate was dispensed into the first reactor followed by 5 L of substrate-free 9K salts medium, acidified to pH 1.8 with a small amount of concentrated H_2SO_4. An inoculum density of $1x10^9$ cells per ml of suspension per 1% pulp density was employed when needed.

The second reactor contained either 5 L of 9K salts medium acidified to pH 1.7 with a limited amount of cells or 5 L of bacterial ferric solution of a known Fe^{3+} concentration.

A peristaltic pump was used to pump solution containing fine particles to the second reactor at a rate of 7 litre per day and the ferric solution produced in the second reactor was also pumped back to the first one at the same rate. The solution containing fine particles which was pumped to the second reactor was collected in a 1 cm diameter 30 cm long tube immersed in the first reactor, whereas the ferric solution pumped from the second reactor back to the first one was recovered from an outer cylinder as an overflow.

Correction for evaporation made prior to sampling by the addition of tap water acidified to pH 1.7.

2.5 Cyanidation Technique

All cyanidation tests were conducted in Erlenmeyer flasks. 50 g of mineral concentrate added to 100 ml of water. After the addition of the required amounts of calcium hydroxide and sodium cyanide, the flasks were agitated in an orbital incubator for 18 hours. The optimum pH range during cyanidation varied between 10.5-12.0.

2.6 Analysis

All elements in solution were measured using a PYE UNICAM model SP 9 Atomic Absorption Spectrophotometer (AAS). Digestion techniques were applied to the residues in order to perform mass balance determinations.

Suspension Eh was measured by a platinum/calomel electrode combination.

In order to examine the mineralogical changes occurred during leaching, the residues were examined by XRF and XRD techniques and also by optical microscopy.

Cells were recovered from the reactors using a MISTRAL 6L SME ultra-centrifuge and their concentration measured using a UNICAM SP 500 Series 2 ultraviolet and visible Spectrophotometer.

3 Results and discussion

Figure 1 illustrates the percentage weight distribution for As and Fe in the feed and in the leaching residue after a typical bacterial test at 20% w/v pulp density for the Olympias refractory arsenical gold concentrate, size -63 μm. The size fractions were obtained by wet cycloning 50 g of sample using a Warman Cyclosizer, Model SY 285.

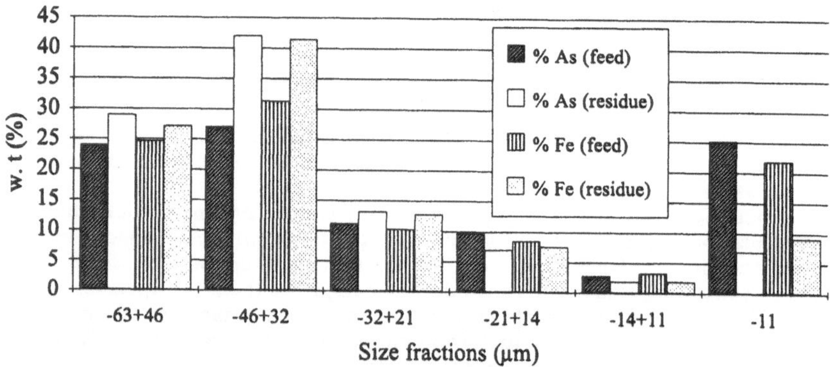

Fig. 1: As and Fe percentage weight distribution for various concentrate size fractions in feed and in leaching residues (pulp density 20% w/v, concentrate particle size -63 μm).

It can be seen that for both mineral phases the weight distribution of the finer fractions in the residue decreased dramatically, whereas it increased for the coarser ones. It is obvious that the finer particles are oxidised over a short leach period and therefore at the end of the test the coarser particles, although reduced in size, have an increased weight percentage.

Considering that in a concentrate of a typical particle size, such as -63 μm, a high percentage of fine particles exists, it is expected that oxidation in a single reactor will not reach high levels, due to the fact that the high surface area present in suspension causes a reduction in the number of free cells and therefore minimises their ability to oxidise ferrous iron to the ferric state. It is well known that in case where only a small proportion of the iron present in solution is in the ferric form, oxidation cannot progress to a great extent and therefore only small percentages of the mineral phases will be oxidised, especially when operating at high pulp densities.

The solution exchange reactor system was designed to solve some of these problems. Since solution containing only fine particles is pumped to the second reactor, oxidation of these particles takes place very rapidly and therefore the high Eh solution pumped back to the first reactor encourages oxidation of both mineral phases according to Equations (1) and (2)

For the oxidation of pyrite:

$$FeS_2 + Fe_2(SO_4)_3 = 3\ FeSO_4 + 2\ S \quad (1)$$

For the oxidation of arsenopyrite:

$$2\ FeAsS + 2\ Fe_2(SO_4)_3 + 3\ H_2O + 5/2\ O_2 = 2\ H_3AsO_4 + 6\ FeSO_4 + 2\ S \quad (2)$$

The ferrous iron which is produced from Equations (1) and (2) is simultaneously oxidised to the ferric state according to Equation (3)

$$4\ FeSO_4 + 2\ H_2SO_4 + O_2 \xrightarrow{\text{bacteria}} 2\ Fe_2(SO_4)_3 + 2\ H_2O \qquad (3)$$

Arsenopyrite dissolution starts immediately in the presence of bacteria and proceeds much faster than the dissolution of pyrite, indicating that arsenopyrite is the first sulphide phase to be oxidised. Eh values measured in such systems never exceeded 430 mV. Pyrite dissolution starts at higher Eh values, 460-500 mV, and is encouraged by the presence of high concentrations of ferric iron in suspension (Komnitsas and Pooley 1990).

Figure 2 illustrates the change of suspension Eh of the process in relation to leaching time, in a single reactor (inoculated with bacteria) and in each one of the solution exchange system (ferric test), during leaching of the Olympias concentrate at 20% w/v pulp density and -63 µm particle size.

It can be seen that suspension Eh values in the single reactor rise slowly and never exceed 470 mV enabling therefore oxidation mainly for arsenopyrite. In the latter stages of leaching they drop slowly indicating reduced bacterial activity, reduced Fe^{3+}:Fe^{2+} ratios in suspension and practically no further oxidation of the mineral phases.

In the solution exchange system the suspension Eh values rise quickly in the second reactor and subsequently in the first one, encouraged by the bacteria rich ferric solution pumped back and therefore high percentages of oxidation for both mineral phases can be achieved.

In order to examine the leaching behaviour of the Olympias concentrate in the solution exchange system two series of experiments at different concentrate particle sizes were conducted and the oxidation of each sulphide mineral studied. In the first series the second reactor contained only solution of substrate free 9K salts medium at pH 1.7, whereas in the second series it contained bacterial ferric solution with a concentration of 10 g/L Fe^{3+}. The pulp density maintained constant at 20% w/v and the particle size of the concentrate varied between -63 and -32 µm.

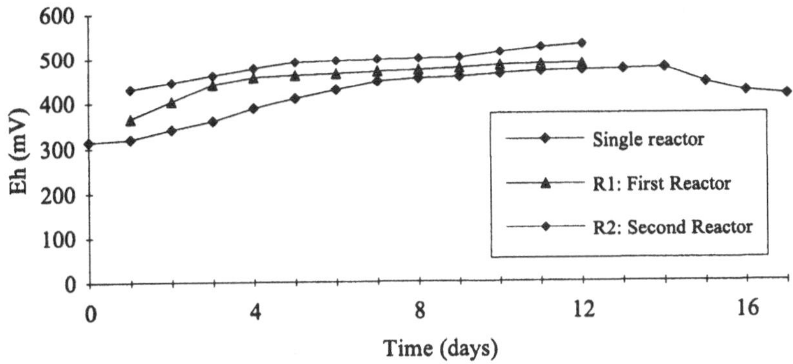

Fig. 2: Change of suspension Eh in relation to leaching time in a single reactor and in each one (R1/R2) of the solution exchange system (pulp density 20% w/v, concentrate particle size -63 µm).

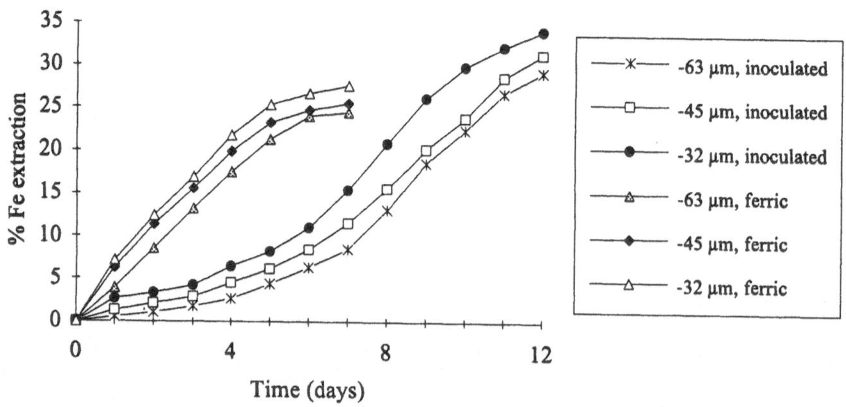

Fig. 3: Effect of concentrate particle size variation upon Fe extraction in relation to leaching time during inoculated with bacteria and bacterial ferric tests (pulp density 20% w/v).

Figures 3 and 4 illustrate the effect of the concentrate particle size variation upon the extraction of As and Fe respectively in relation to leaching time.

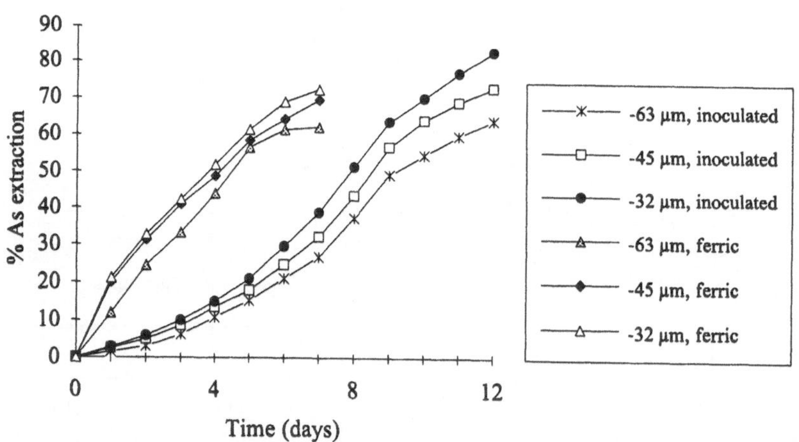

Fig. 4: Effect of concentrate particle size variation upon As extraction in relation to leaching time during inoculated and bacterial ferric tests (pulp density 20% w/v).

From Figures 3 and 4 the following conclusions can be drawn:
For inoculated with bacteria tests:

Both As and Fe extractions increase when the particle size of the concentrate is reduced and over a period of 12 days they reach 94% and 44% respectively when the concentrate particle size is -32 μm. These values represent 94% and 30% percentages of oxidation for arsenopyrite and pyrite respectively. For the above mentioned conditions the subsequent Au recovery was 89%, whereas the theoretical recovery according to equation described in a previous work (Komnitsas and Pooley 1990) should have been 91%. Gold recovery varied between 77 and 89% depending on the cyanide addition, which varied between 10 and 40 kg of sodium cyanide per tonne of concentrate. The actual cyanide consumption varied between 4.11 and 6.14 kg per tonne. Acid washing of the residues prior to cyanidation reduces the consumption of cyanide.
Over the same period of time, when the particle size of the concentrate was -45 μm, the percntages of oxidation for arsenopyrite and pyrite were 87% and 28% respectively and the subsequent Au recovery 86%. When the particle size of the concentrate was -63 μm, the percentages of oxidation for arsenopyrite and pyrite were 77 and 26% respectively and the subsequent Au recovery 76%.
Leaching tests conducted in the solution exchange system led to much higher percentages of oxidation for both sulphide minerals in comparison to the percentages achieved over a longer period of time, when leaching was performed in a single air-stirred pachuca reactor. For example, when the concentrate particle size was -63 μm, leaching in a single air-stirred pachuca reactor over a period of 17 days at 20% w/v pulp density led to 69 and 17% percentages of oxidation for arsenopyrite and pyrite respectively and to a subsequent 72% Au recovery, whereas when the particle size of the concentrate was -32 μm the percentages of oxidation for arsenopyrite and pyrite over a period of 2 weeks were 83 and 20% respectively and the subsequent Au recovery 81% (Komnitsas and Pooley 1991).

For initiated with bacterial ferric liquor tests:

When leaching took place in the solution exchange system and the second reactor contained ferric liquor with a concentration of 10 g/L Fe^{3+}, the oxidation of both mineral phases progresses much more rapidly. Therefore, when the concentrate particle size was -32 μm, leaching over a period of 7 days for the conditions mentioned previously, led to 72 and 14% percentages of oxidation for arsenopyrite and pyrite respectively, whereas when the concentrate particle size was -63 μm these percentages of oxidation drop to 62% and 11% respectively.
After the 7th day the oxidation for both mineral phases progresses slowly and dilution of the ferric liquor to about 10 g/L Fe^{3+} is required, in order to achieve higher percentages of oxidation for both mineral phases. When the particle size of the concentrate size was -32 μm the percentages of oxidation for arsenopyrite and pyrite over a leaching period of 10 days reached 97% and 49% respectively and the subsequent Au recovery was 95%.
From the experimental data presented, it is demonstrated that the maximum Fe and As concentrations the leaching liquor can tolerate are about 20 and 10 g/L respectively. When these concentrations exceed the above limits, practically no further oxidation occurs and therefore dilution of the liquor is needed.

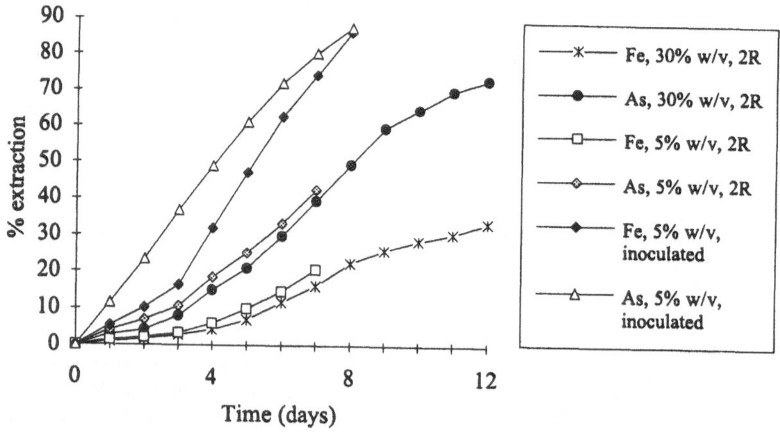

Fig. 5: Effect of pulp density variation upon Fe and As extraction in relation to leaching time (concentrate particle size -63 µm). 2R:solution exchange system

In order to examine the leaching behaviour of the concentrate at different pulp densities in the solution exchange system inoculated tests were conducted at both 5 and 30% w/v pulp density. The particle size of the concentrate maintained at -63 µm.

Figure 5 illustrates the effect of the pulp density variation upon the extraction of Fe and As respectively in relation to leaching time.

From Figure 5 the following conclusions can be drawn:

Leaching of the concentrate at 30% w/v pulp density progresses well and the percentages of oxidation for arsenopyrite and pyrite reached 73% and 34% respectively over a period of 12 days. The subsequent Au recovery was 74%. Leaching of the concentrate at 5% w/v pulp density over a period of 7 days led to percentages of oxidation as low as 37% for arsenopyrite and 6.5% for pyrite over a period of 7 days. These poor percentages of oxidation were mainly due to the inability of the diluted ferric liquor present in the system to further oxidise the mineral particles. In order to underline the effect of the ferric strength of the solution on the extraction of As and Fe, leaching data of the same concentrate under the same conditions in a single air-stirred pachuca reactor are presented in Figure 5. It can be seen that over a period of 8 days 87% of the arsenopyrite and 75% of the pyrite were oxidised, due to the quick "build up" of the ferric levels in the reactor and the increased kinetics of the system according to Equations (1) and (2).

4 Conclusions

The most important conclusions extracted from this experimental study for the bacterial oxidation of the arsenical gold sulphide concentrate from Olympias, Greece, using a solution exchange system are:

Increased percentages of oxidation for arsenopyrite and can be achieved for both inoculated with bacteria and bacterial ferric liquor tests in comparison to the percentages of oxidation achieved over a longer period of leaching in a single air-stirred pachuca reactor.
Reduction in the particle size of the concentrate has a positive effect on the percentages of oxidation for both mineral phases.
Inoculated with bacteria tests at 20% w/v pulp density over a period of 12 days lead to arsenopyrite and pyrite percentages of oxidation as high as 94% and 44% respectively, when the particle size of the concentrate is -32 μm. The subsequent Au recovery is 89%.
Bacterial ferric liquor tests using the solution exchange system achieve for both mineral phases, especially for arsenopyrite, high initial dissolution rates, but after a week dilution of the ferric solution is needed in order to maintain dissolution rates at high levels. 97% of arsenopyrite and 49% of pyrite are oxidised over a period of 10 days during leaching tests at 20% w/v pulp density, when the concentrate particle size is -32 μm. The subsequent Au recovery is 95%.
Oxidation of both mineral phases reaches high levels when leaching takes place at 30% w/v pulp density in the solution exchange system. On the other hand these levels become very low when pulp density is 5% w/v. In the latter case the diluted ferric liquor is unable to oxidise the mineral phases to a greater extent.
The solution exchange system offers significant advantages over single reactors, in terms of reduced retention periods and increased percentages of oxidation for both mineral phases studied, when leaching is conducted at higher than 20% w/v pulp densities.
The maximum concentrations a bacterial solution can tolerate, in order to maintain its oxidising strength, are about 20 g/L Fe and 10 g/L As. When these concentrations exceed the above limits practically no further oxidation takes place.

5 References

Jha, M.C. (1987) Refractoriness of certain gold ores to cyanidation: Possible causes and possible solutions, **Mineral Processing and Extractive Metallurgy Review**, 2, 331-352.

Komnitsas, C. and Pooley, F.D. (1989) Mineralogical characteristics and treatment of refractory gold ores, **Minerals Engineering**, 2(4), 449-457.

Komnitsas, C. and Pooley, F.D. (1990) Bacterial oxidation of an arsenical gold sulphide concentrate from Olympias, Greece, **Minerals Engineering**, 3(3/4), 295-306.

Komnitsas, C. and Pooley, F.D. (1991) Optimisation of the bacterial oxidation of an arsenical gold sulphide concentrate from Olympias, Greece, **Minerals Engineering**, 4(12), 1297-1303.

Komnitsas, C. and Pooley, F.D. (1994) Oxidation of arsenopyrite in bacterial ferric solutions, to be presented at the "**Hydrometallurgy '94**" Conference, Cambridge.

Pooley, F.D. (1987) Use of Bacteria to Enhance Recovery of Gold from Refractory Ores, in Minprep '87, **Int. Symp. on Innovative Plant and Processes for Mineral Eng**., 1-14.

Oxidation of arsenopyrite in bacterial ferric solutions

Constantine Komnitsas
Laboratory of Metallurgy, National Technical University of Athens, Athens, Greece
F. D. Pooley
University of Wales, Cardiff, Wales

Abstract
In this study the oxidation of arsenopyrite in bacterial ferric solutions is examined. Leaching tests were conducted in air-stirred pachuca reactors in order to define the major factors influencing the oxidation of arsenopyrite and to underline the importance of indirect leaching of the mineral by bacterially generated acid ferric sulphate solutions. The parameters studied were the particle size of the mineral concentrate and the initial ferric ion concentration in solution. The rate of oxidation of arsenopyrite during leaching was found to be almost directly proportional to the surface area of the concentrate. In addition, an initial ferric ion concentration of 10 g/l Fe^{3+} is considered mostly efficient for achieving high arsenopyrite oxidation degrees over a short leach period. The experimental data was used to predict the rate of arsenic release for the Olympias refractory arsenopyrite concentrate, assaying 13% As.
Keywords: Bacterial ferric leaching, Pachuca reactor leaching, Arsenopyrite oxidation, Reaction rate.

1 Introduction

In the bacterial oxidation of sulphide ores and concentrates, bacteria *Thiobacillus Ferrooxidans* are mainly used to enhance the rate of oxidation by breaking down the sulphide lattice. The efficiency of the bacterial attack is strongly dependent upon the mineralogical characteristics of both ores and concentrates and thus leaching conditions for different ore types may vary (Komnitsas and Pooley 1989).

The main reactions describing the oxidation of arsenopyrite in a bacterial system are:

$$FeAsS + 5/2\ O_2 + 3/2\ H_2O = H_3AsO_4 + FeSO_4 \quad (1)$$

$$2FeAsS + 2Fe_2(SO_4)_3 + 3H_2O + 5/2O_2 = 2H_3AsO_4 + 6FeSO_4 \quad (2)$$

$$2\ H_3AsO_4 + Fe_2(SO_4)_3 = 2\ FeAsO_4 + 3\ H_2SO_4 \quad (3)$$

The oxidation potential for arsenopyrite oxidation ranges between 390 and 430 mV. It is known that dissolution of arsenopyrite is initiated much more rapidly in the presence of bacteria and proceeds much faster than the dissolution of pyrite according to equations (1) and (2) (Komnitsas and Pooley 1989). It is therefore concluded that high concentrations of ferric ions in the bacterial system improve the kinetics of equation (2) and consequently the degree of oxidation of arsenopyrite. Furthermore, the steady increase of the ferric strength in solution during leaching, due to release of iron from the mineral, offers significant advantages over the use of chemical ferric solutions for oxidation of arsenopyrite. A disadvantage of bacterial leaching systems however, is the fact that reaction temperatures must be maintained in the range within which the micro-organisms will grow. In the case of *T. Ferrooxidans* this is approximately 35 °C.

There is considerable interest in the aqueous oxidation of arsenopyrite especially when the mineral occurs as a component in auriferous sulphide ores and concentrates and especially when gold is preferentially associated with it. In this case only partial oxidation of arsenopyrite is required to increase gold liberation and subsequent gold recovery.

The objective of this paper is the study of oxidation of arsenopyrite in bacterial ferric systems. The parameters examined were initial ferric ion concentration and particle size of the concentrate and their effect on the oxidation of arsenopyrite was studied. Leaching data has been analysed in order to predict the leaching behaviour of similar arsenopyrite concentrates.

2 Experimental part

2.1 Bacterial Ferric Solution

The bacterial ferric solution used in this experimental study was generated in 5L capacity air-stirred pachuca reactors, using pure pyrite concentrate, size -56+20 µm, at 5% w/v pulp density. The arsenic content of the bacterial ferric liquor never exceeded 0.5 g/L.

2.2 Leaching experiments

Leaching tests were conducted in 2 L capacity air-stirred pachuca reactors maintained at 35±1 °C (Komnitsas and Pooley 1990). Bacterial ferric solution at a given Fe^{3+} concentration was added into the reactor, followed by the dispension of the mineral concentrate. The approximate concentration of cells in the bacterial ferric solution was 1 x 10^9 cells per ml of suspension.

2.3 Material Used

The chemical and wet screen analysis of the arsenopyrite feed material used in this study are given in Table 1.

The copper and sulphur assays were 11.8% and 22.8% respectively. The main mineralogical components of the concentrate were arsenopyrite 69% and chalcopyrite 30%.

Table 1. Chemical and wet screen analysis of the arsenopyrite feed material

Size (μm)	Weight(%)	As(%)	Fe(%)
-106+63	03.36	41.35	31.77
-63+45	17.18	36.88	29.50
-45+32	14.79	34.67	27.77
-32+20	18.52	32.20	26.53
-20+10	45.14	27.69	25.24
-10	01.01	39.40	31.67
Total	**100.00**	**31.71**	**26.87**

2.4 Analysis
All elements in solution were measured using a PYE UNICAM model SP 9 Atomic Absorption Spectrophotometer (A.A.S).

3 Results and Discussion

3.1 Effect of the variation of the particle size of the concentrate
In order to examine the effect of the variation of the particle size of the concentrate on the degree of oxidation of arsenopyrite a series of experiments at different particle size fractions were performed and the oxidation of arsenopyrite studied. The size fractions examined were -106+63, -63+45, -45+32, -32+20, -20+10 and -10 μm. The bacterial ferric solution assayed 10 g/l Fe^{3+} and 0.20 g/L As and the pulp density was maintained at 2% w/v. Figure 1 illustrates the effect of the variation of the particle size of the concentrate upon the oxidation of arsenopyrite in relation to leaching time.
From the experimental data obtained the rate constants "k" were calculated using the following first order mass transfer equation

$$R = 1 - e^{-kt} \tag{4}$$

where
R : % arsenopyrite oxidised
t : leaching time (hr)
k : rate constant (hr $^{-1}$)

The resulting "k" values were plotted against 1/d (d: average size in microns for each size fraction) in Figure 2.

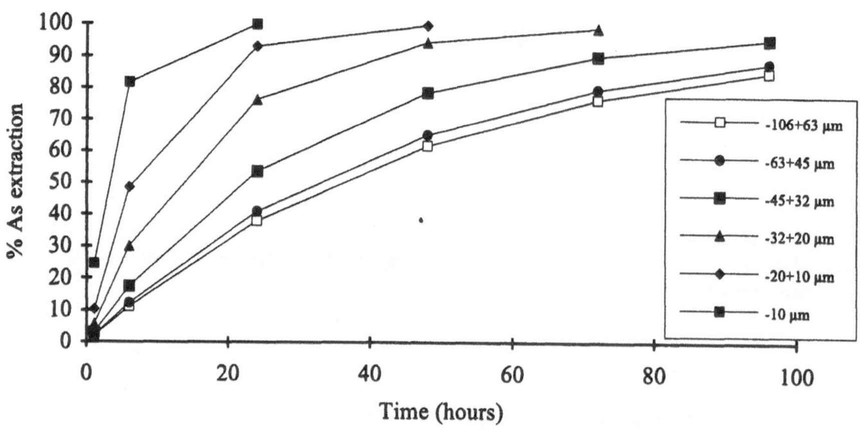

Fig.1: Effect of the variation of the concentrate particle size upon oxidation of arsenopyrite in relation to leaching time (pulp density 2% w/v, 10 g/L Fe^{3+})

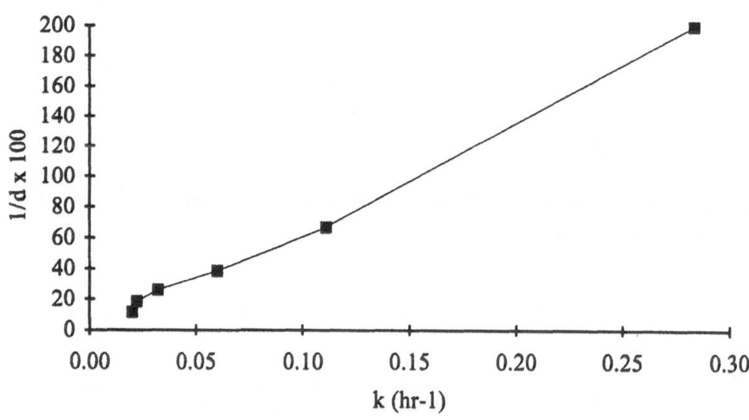

Fig. 2 : 1/d against "k" values (2% w/v pulp density, 10 g/L Fe^{3+})

From Figures 1 and 2 the main conclusions that can be drawn are:

Coarser size fractions, such as -106+63 and -63+45 μm, leach slowly and the arsenopyrite degree of oxidation does not exceed 78% over a leach period of 72 hours.
Finer size fractions, such as -45+32 and -32+20 μm, leach much faster, enabling over 80% oxidation of arsenopyrite within 48 hours.
Ultra fine size fractions with a great surface area, such as -20+10 and -10 μm, leach very rapidly, and within 24 hours the degree of oxidation exceeds 90%. It must be underlined however, that during leaching at higher pulp densities, a great percentage of such fractions present in pulp will reduce the overall rate.
Rate constants increase dramatically during leaching of finer fractions. "k" values are almost directly proportional to 1/d values and subsequently to the surface area.

The experimental data shows that in order to maintain high reaction rates during leaching of arsenopyrite the average particle size of the concentrate should not be greater than 63 μm. High percentages of ultra fine fractions however, especially -10 μm, will inevitably reduce the overall rate. The great surface area exposed in pulp will cause a reduction in the free cell population and subsequently minimise their ability to oxidise iron from ferrous to ferric state. Therefore, in plant conditions the milling circuit should be carefully controlled, in order to produce a concentrate with a normal particle size distribution.

3.2 Effect of ferric ion concentration

In order to examine the effect of the initial ferric ion concentration on the degree of oxidation of arsenopyrite, a series of experiments were performed using the above particle size fractions whereas the pulp density was maintained at 2% w/v. Initial ferric ion concentrations examined were 5, 10 and 20 g/L Fe^{3+}.
From the experimental data obtained the following conclusions can be extracted:

When the initial ferric ion concentration is 5 g/L an initial lag period of about 2 days is observed. After this period the rates of oxidation of arsenopyrite are similar to those observed when the initial ferric ion concentration is 10 g/L. Therefore the initial ferric ion concentration of 5 g/L is considered insufficient for fast arsenopyrite oxidation and a "build up" of the ferric level in the reactor vessel is required, before higher rates of arsenopyrite oxidation can be achieved.
When the initial ferric ion concentration is 20 g/L high initial arsenopyrite oxidation rates are observed, but after 48 hours the rates drop, indicating that the total Fe concentration existing in the reactor cannot be tolerated by the cells.
Therefore, the initial ferric ion concentration of 10 g/L is considered more efficient for achieving high arsenopyrite rates of oxidation. This initial concentration allows a ferric "build up" in the reactor, which allows cells to maintain their oxidising activity at high levels.

From the overall experimental analysis, it can be concluded that arsenic release is strongly dependent upon the surface area present in the reactor and the initial ferric ion concentration. However, when material with a very high surface area is present in the pulp, the efficiency of the ferric solution is expected to drop because the free cell population becomes considerably lower.

In order to determine whether the above experimental rate constants can be used to predict the As release rates for a different arsenopyrite concentrate, a number of experiments were carried out using a refractory arsenical pyrite concentrate from Olympias, Greece, assaying 13% As (30% FeAsS) (Komnitsas and Pooley 1994). The initial ferric iron concentration was 10 g/L Fe^{3+} and the pulp density 20% w/v. The obtained "k" values for concentrate particle sizes -63, -45 and -32 μm were 0.036, 0.045 and 0.066 hr^{-1} respectively. It can be seen from Fig. 2 that such values can be well predicted from the graph. The lower numerical values are expected due to the high pulp density and the coexistence of other minerals such as pyrite and chalcopyrite.

From the above analysis it can be seen that a good prediction of the As release can be made for arsenical concentrates given that the strength of the bacterial ferric solution and the operating pulp density of the mineral concentrate are known. There is no evidence that the As content of the concentrate plays an important role in oxidation rates of arsenopyrite. A little deviation from the predicted values though can be expected, when arsenopyrite coexists with other minerals and is not the main mineralogical component and when there exists in the reactor a high percentage of ultra fine material (-10 μm) which may cause a depletion in the free cell population.

From the data presented it can be seen that leaching of arsenopyrite in bacterial ferric systems offers the advantage of achieving high degrees of arsenopyrite oxidation over short leach periods.

Another advantage offered by the use of bacterial ferric solution, is its increased strength at the end of the run, which allows its re-use by dilution. The maximum operating As content in the bacterial ferric solution should not exceed 5 g/L. Leaching tests of the previously referred to pure arsenical concentrate with chemical ferric solution showed that the arsenic release is considerably lower over a much longer leach period. In order to maintain good arsenic release rates in chemical ferric solutions it is necessary to add other oxidising agents such as hydrogen peroxide or maganese dioxide to maintain the suspension potential within the optimum range for selected arsenopyrite oxidation.

4 Conclusions

The most important conclusions from this experimental study of the bacterial oxidation of an arsenopyrite concentrate are:

Rate constants for arsenopyrite oxidation are almost directly proportional to the surface area of the concentrate during leaching with an initial ferric ion concentration of 10 g/L Fe^{3+}.

Initial ferric ion concentration of 10 g/L Fe^{3+} is considered mostly efficient for achieving high arsenopyrite oxidation rates. This initial concentration allows a ferric "build up" in the reactor which allows cells to maintain their oxidising activity at high levels.

Arsenic release is strongly dependent upon the surface area exposed in the reactor and the initial ferric ion concentration. However, when material with a very high surface area is present in pulp, the efficiency of the ferric solution is expected to be reduced because the free cell population becomes considerably reduced.

The role of the bacteria in the system studied is to generate and maintain iron in a ferric state. Optimum leaching conditions are limited by the rate at which the micro-organisms are able to convert ferrous iron to the ferric state.
Extensive size reduction of the arsenopyrite concentrate will not necessarily enhance the predicted rate of leaching, especially when operating at pulp densities higher than 15% w/v, due to the fact that the high *surface area:volume* ratio existing in the reaction vessel will reduce the oxidising ability of the free cells and the oxidation potential of the ferric solution as well.
The use of a bacterial ferric solution of a considerable strength for the leaching of any arsenical concentrate offers significant advantages - such as increased reaction rates, selected arsenopyrite oxidation, short retention times, re-use capability - in comparison to leaching with chemical ferric solution or the use of a bacteria inoculum.

5 References

Komnitsas, C. and Pooley, F.D (1989) Mineralogical characteristics and treatment of refractory gold ores, **Minerals Engineering** , 2, 449-457.

Komnitsas, C. and Pooley, F.D (1990) Bacterial oxidation of a refractory gold sulphide concentrate from Olympias, Greece, **Minerals Engineering** , 3(3/4), 295-306.

Komnitsas, C. and Pooley, F.D (1994) Bacterial oxidation of an auriferous arsenical concentrate, to be presented at the **"Hydrometallurgy '94 "** Conference, Cambridge.

Effect of silver and bismuth on bioleaching of copper sulphide concentrates with thermophilic microorganisms

J. L. Mier
C. Gómez
A. Ballester
M. L. Blázquez
F. González
Departamento de Ciencia de los Materiales e Ingeniería Metalúrgica, Facultad de Ciencias Químicas, Universidad Complutense de Madrid, Madrid, Spain

Silver ion has been used successfully as catalytic agent in chalcopyrite leaching with mesophilic microorganisms to increase the copper dissolution rate. Although it could appear logical to add Ag^+ to accelerate the process at high temperatures, thermophilic bacteria are less resistant than mesophilic microorganims to silver and so its application is limited. Therefore other cations must be studied to increase dissolution efficiency at high temperatures. In this work, a comparative study is performed using silver and bismuth in the bioleaching of a copper concentrate with thermophilic (68°C) and mesophilic (35°C) bacteria. The experiments were carried out in shake flasks at 1% pulp density.

The quantity of silver added to both the thermophilic and mesophilic cultures was 1g Ag/kg concentrate. Silver was precipitated as Ag_2S, which was then dissolved by microorganism activity. The concentration of silver increased up to a maximum value of 18μg Ag/l causing bacterial inhibition. Then, there was a new precipitation of silver. It would prove silver dissolution is related to bacterial activity . Consequently, copper extraction was low (40% of mineral copper).

The thermophilic culture with 10g Bi/kg concentrate showed the highest copper dissolution rate, maximum extraction being attained in 48 hours (60% of mineral copper soulbilised). There was no effect when 1g Bi/kg concentrate was used. In both cases, soluble bismuth concentration increased, probably due to the dissolution of bismuth hydrolisis products.
Keywords: Bioleaching, Chalcopyrite, Thermophiles, Silver, Bismuth, Catalysis, Toxicity.

1 Introduction

Bioleaching is a very interesting possibility for the treatment of sulphides (ores and concentrates). However, it is a very slow process and is not therefore suitable for high grade ores.

Chalcopyrite bioleaching using thermophilic microorganisms leads to improved copper dissolution kinetics due the high temperatures used. Bacteria of the *Sulfolobus* genus have proved more effective than mesophilic microorganisms in the chemical attack of very resistant minerals, such as chalcopyrite and molybdenite.

One way of accelerating the dissolution process is to use metallic cations to increase the leaching rate, either chemically or biologically. Bjorling (1954) suggested That the dissolution rate of sulphides can be speeded up considerably by the addition of suitable catalytic ions to the solution. Such ions can have an important influence on the rate of the oxidation-reduction reactions, both homogeneous and heterogeneous, which take place during the leaching of sulphides; they also affect the nature of the reaction products.

It is necessary to emphasize the interactions between some potentially toxic metals and microorganisms and thus metal toxicity must be taken into account when considering the catalysis of the process. Until now, there have been few works on metal-thermophilic microorganism interactions in an attempt to find suitable chemical agents to catalyse bioleaching at high temperature.

Silver has been used successfully as a catalytic agent in the dissolution of chalcopyrite by mesophilic microorganisms (Ballester, 1987; Blancarte-Zurita, 1987; Ahonen and Tuovinen, 1990a; Ahonen and Tuovinen 1990b). Although it seems logical to use silver to accelerate copper dissolution in the presence of thermophilic bacteria, they show low tolerance to this metal (Norris, 1989), which is a limiting factor. Because of this, it is necessary to search for other cations, which are more effective than silver under such conditions. One such alternative is bismuth, which was used in previous bioleaching studies (Ballester et al., 1992) with interesting results on copper dissolution rates. However the way in which bismuth is involved in the mechanism of copper dissolution is still unknown.

In the present work, the effect of silver and bismuth are compared in chalcopyrite bioleaching with *Sulfolobus* to establish a possible mechanism for bismuth. The toxicity of silver on thermophilic cultures is emphasized by performing comparative tests at 35°C (mesophilic cultures) and 68°C (thermophilic cultures).

2 Materials and methods

2.1 Ore Samples

A chalcopyrite concentrate provided by Río Tinto Minera S.A. was used. Its chemical composition was (% wt): 19.17 Cu; 32.52 Fe; 28.89 S; 0.01 Pb; 0.04 Zn and 6.92 SiO_2. Analysis by X-ray diffraction showed that the most important mineral phases were chalcopyrite and pyrite. The particle size was 80% less than 33 μm.

2.2 Bacteria

A thermophilic culture of *Sulfolobus, strain BC,* isolated by Dr. Paul R. Norris (Warwick University U.K.) was used at 68°C. The mixed mesophilic culture consisted of *Thiobacillus ferrooxidans, Thiobacillus thiooxidans and Leptospi-*

rillum ferrooxidans and it was grown at 35°C.

The cultures were grown in orbital shakers on a chalcopyrite concentrate (1% w/v) and in the presence of a nutrient medium containing 0.4 g/l $(NH_4)_2SO_4$, 0.5 g/l $MgSO_4.7H_2O$ and 0.2 g/l K_2HPO_4. The pH was adjusted to 1.5. The salts used as sources of silver and bismuth were: Ag_2SO_4 and $Bi(NO_3)_3.5H_2O$.

2.3 Ferrous oxidation experiments

Ferrous oxidation experiments were carried out in a pneumatically stirred reactor as shown in Figure 1, with an air current (1% CO_2 v/v) supplied through the bottom. The reactor was placed in a silicone bath to maintain the test temperature (Figure 2).

Each reactor contained 95 ml of nutrient medium, 5 ml of bacterial inoculum (mesophilic or thermophilic bacteria), 50 mM Fe^{2+}, $1.5x10^{-3}$ M of potassium tetrathionate (as sulphur source) and/or the quantity of cation chosen to test. Ferrous ion concentration was analyzed by ceric sulphate titration. The percentage of ferrous ion oxidized was represented on a logarithmic scale to determinate clearly the maximum Fe^{2+} oxidation rate, which coincided with the maximum bacterial growth step.

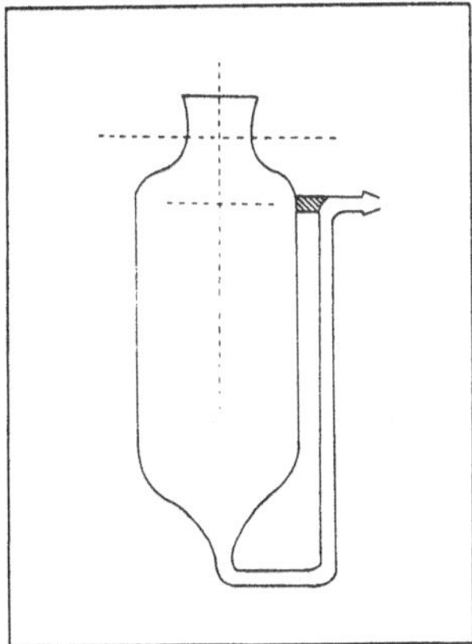

Fig.1 Pneumatically stirred reactor.

2.4 Leaching experiments

The experiments were carried out in orbital shakers at 100 rpm using 250 ml conical flasks. 90 ml of the nutrient medium with the ore (1% w/v) were introduced into the flasks. The chosen quantity of catalytic ion, if necessary, was added to the medium and the solution was shaken to achieve the maximun dissolution. Finally the solution was inoculated with 10 ml of mesophilic or themophilic culture and the experiment started.

For sampling, the suspended solids were allowed to settle for approximately 30 minutes and aliquots of the supernatant solution were filtered to remove the

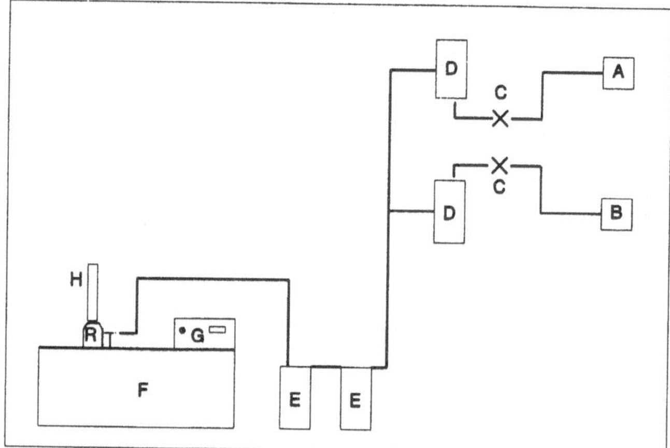

Fig.2. Schematic diagram of experimental equipment where the test of ferrous ion oxidation capacity by bacteria in absence of mineral were carried out. (A) Air compressor, (B) CO_2 cylinder, (C) regulating valves, (D) differential manometer, (E) mixing vessels, (F) silicone oil bath, (G) thermostat, (D) condenser and (R) pneumatically stirred reactor.

residual solids for analysis. Copper, iron, bismuth and silver by atomic absorption spectrophotometry. The graphite furnace technique was used in the case of silver and the flame technique with the three other catalysts. Ferrous iron concentration was measured by ceric sulphate titration and pH changes were determined to evaluate indirectly the bacterial activity.

2.5 Potential measurements in bismuth solutions

In order to study the influence of the bismuth concentration on the solution potential, different tests were carried out, over 20h and at 68°C, using abiotic solutions containing 0.1 g/l Fe^{3+} and the nutrient medium. Potential measurements were also performed in reactors with thermophilic cultures stored in a static incubator after five days from the inoculation and before maximum bacterial growth was reached.

3 Results and discussion

3.1 Silver

The presence of silver increased copper extraction in mesophilic cultures by up to 70%, whereas only 25% was obtained in the reference test (Figure 3A). In both cases, the iron concentration in solution was lower than that of the copper and decreased in the final stages due to the precipitation of potassium or ammonium jarosites (Figure 3B). The pH was higher in cultures with silver (Figure 3C), which also favoured jarosite precipitation in accordance with the following reactions:

$$K^+ + 3\ Fe^{3+} + 2\ SO_4^{2-} + 6\ H_2O \dashrightarrow KFe_3(SO_4)_2(OH)_6 + 6\ H^+ \quad (1)$$
$$NH_4^+ + 3\ Fe^{3+} + 2\ SO_4^{2-} + 6\ H_2O \dashrightarrow KFe_3(SO_4)_2(OH)_6 + 6\ H^+ \quad (2)$$

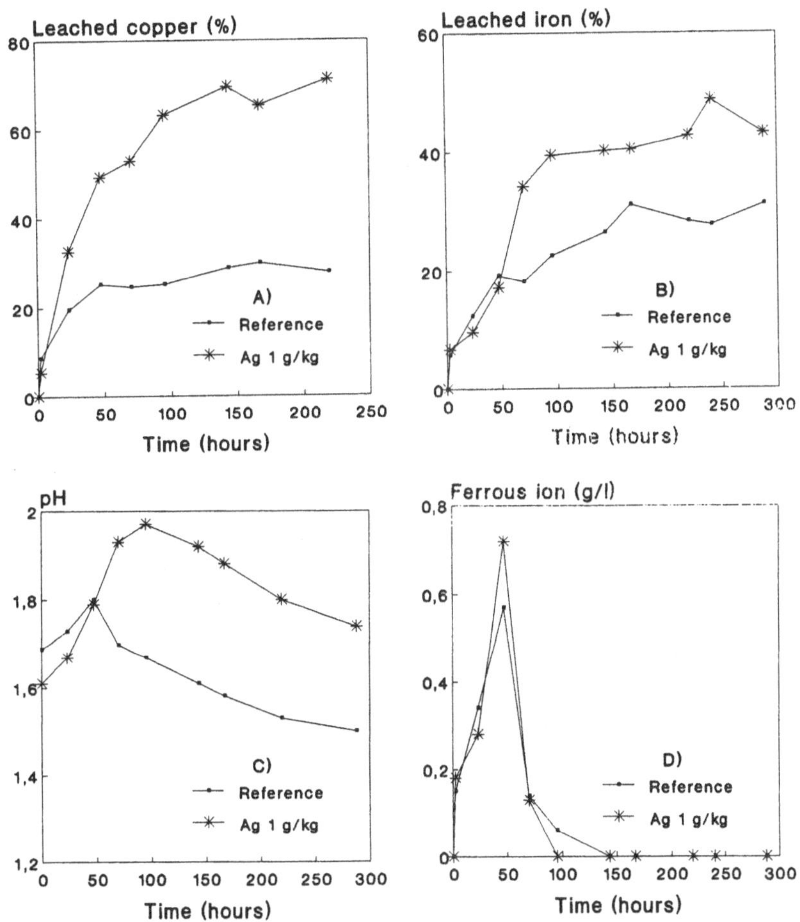

Fig.3. Evolution of copper (A), iron (B), ferrous concentration (D) and pH (C) during the leaching of a chalcopyrite concentrate by a culture of mesophilic bacteria in presence of silver (1g Ag/kg concentrate).

Figure 3D shows the highest concentration of ferrous iron obtained in the experiments using silver, probably as a consequence of the reaction:

$$CuFeS_2 + 4Ag^+ \text{------>} 2Ag_2S + Cu^{2+} + Fe^{2+} \quad (3)$$

in which silver sulphide is precipitated on the ore surface. After the maximum Fe^{2+} concentration was reached, the oxidation rate for this ion in both tets was similar, showing that silver accelerates the chemical dissolution without altering bacterial activity.

The capacity of mesophilic bacteria to oxidize Fe^{2+} in the absence of ore is represented in Figure 4. There was no significant silver toxic effect, even when 0.1 g Ag/l was added, although the slightly lower slope indicated some inhibition of the bacterial growth.

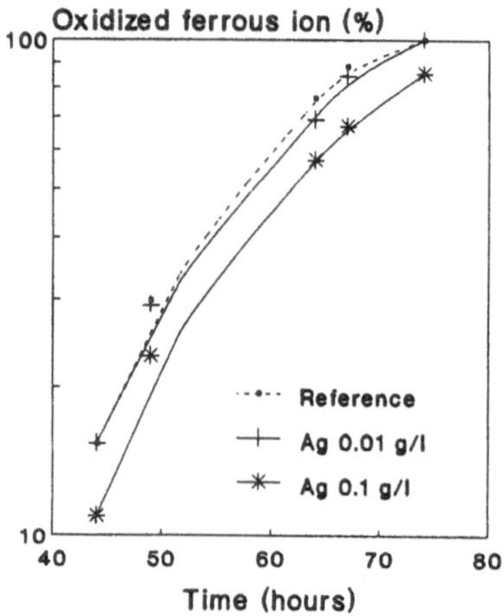

Fig.4. Oxidation of ferrous ion by mesophilic bacteria in the presence of silver and in the absence mineral.

Copper and iron dissolution was better in the thermophilic reference culture than in the corresponding culture with silver, which was the opposite of that seen at 35°C (Figures 5A and 5B). This is in accordance with the low tolerance of *Sulfolobus* to silver. The decrease in bacterial activity is revealed by both the ferrous concentration and the pH (Figures 5C and 5D), where the slow kinetics of oxidation of Fe^{2+} and sulphur can be seen. These data were confirmed by the bacterial oxidation test in the absence of mineral (Figure 6), where there was a slight inhibition of growth with only 0.01 g Ag/l and significant inhibition with 0.1 g Ag/l.

The mechanism of silver action is represented in Figure 7 (Miller et al., 1981; Miller and Portillo, 1981; Ballester, 1987; Ballester et al., 1988). When silver is introduced into the system, it reacts with the chalcopyrite surface precipitating silver sulphide (Ag_2S), which leads to a decrease in silver concentration at the beginning of the experiment. However, the oxidation of ferrous iron is catalyzed when bacterial activity begins. Ferric ion acts as oxidizing agent of Ag_2S and $CuFeS_2$, regenerating Fe^{2+} and increasing Ag^+ concentration. The Ag^+ again reacts with the chalcopyrite surface and the cycle starts again. When toxic levels of silver are reached, the bacterial activity ceases and the silver sulphide precipitates because Fe^{3+} is not being produced by the bacteria. The highest silver concentration reached during the experiments was 22 μM Ag (Figure 8), close to the inhibiton level of 8 μM Ag for *Sulfolobus solfataricus* reported by Grogan (1989).

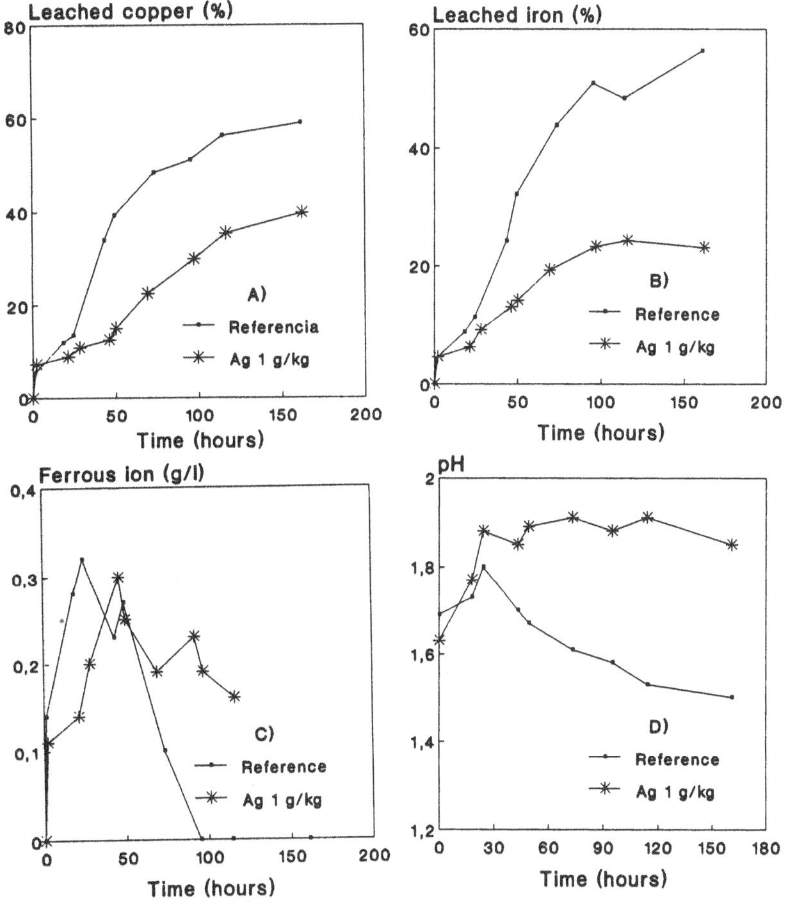

Fig.5. Evolution of copper (A), iron (B), ferrous (C) concentration and pH (D) during the leaching of a chalcopyrite concentrate by a *Sulfolobus* culture in the presence of silver (1g Ag/kg concentrate).

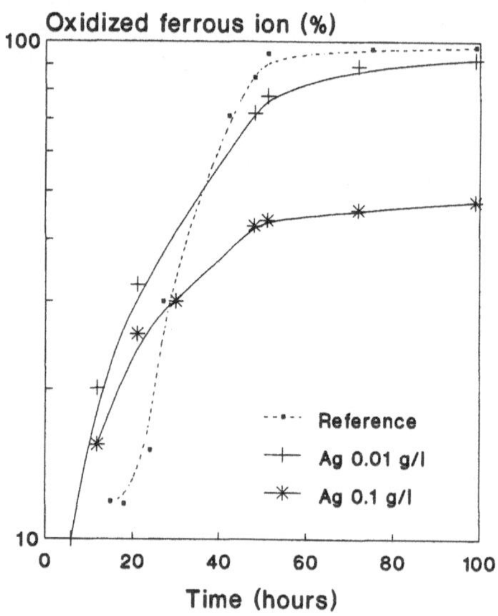

Fig.6. Oxidation of ferrous ion by *Sulfolobus* in the presence of silver and in absence of mineral.

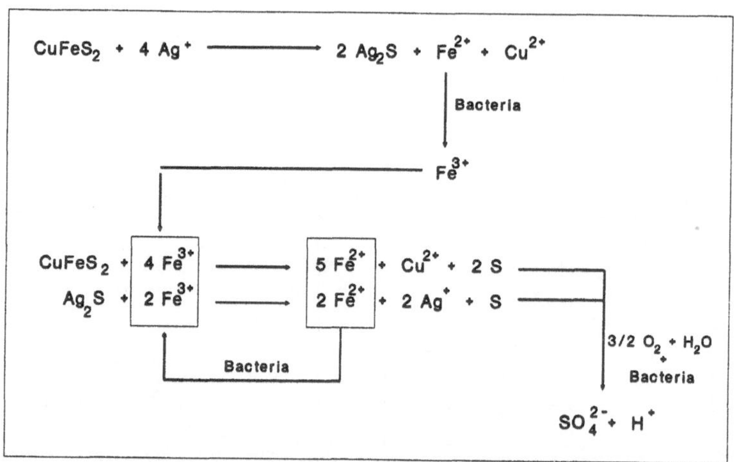

Fig.7. Diagram for mechanism of silver action on chalcopyrite bioleaching proposed by several authors (Miller et al., 1981; Miller and Portillo, 1981; Ballester, 1987; Ballester et al., 1988).

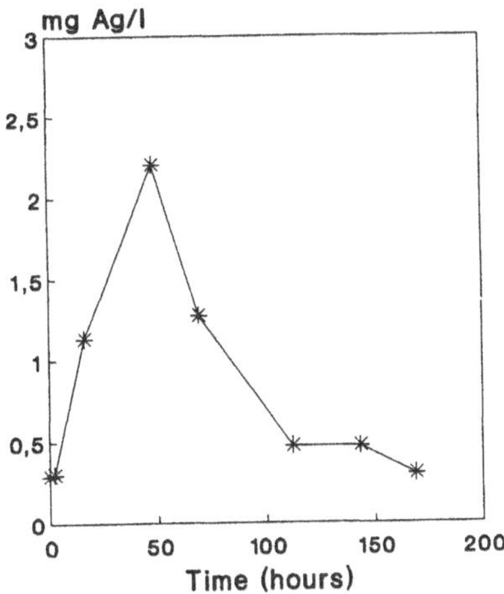

Fig.8. Evolution of silver concentration during the bio-leaching of a chalcopyite concentrate by a *Sulfolobus* culture in the presence of silver (1g Ag/kg concentrate).

3.2 Bismuth

Bioleaching tests were carried out with *Sulfolobus* cultures in the absence of and in the presence of either 1 g Bi/kg concentrate or 10 g Bi/kg concentrate. The test with the higher Bi concentration showed the best copper and iron efficiencies during the first days, whereas the reference test and that using 1 g Bi/kg cocentrate provided similar results (Figures 9A and 9B). However, no differences were apparent in the copper efficiencies when the tests were stopped. This might have been due to the formation of a layer of unknown composition on the chalcopyrite surface. This layer probably hindered the reaction between the chalcopyrite surface and the Fe^{3+} produced by the bacteria. The controlling step (electron or Fe^{3+} difussion through the layer) was not determined (Dutrizac et al., 1969; Ammou-Chokroum et al., 1977; Munoz et al., 1979).

In the bioleaching experiments on the chalcopyrite concentrate, the ferrous ion oxidation rate was faster in the presence of bismuth, which was possibly related with a higher degree of bacterial activity (Figure 9C). However, experiments concerning the ferrous ion oxidation capacity of *Sulfolobus* in the presence of bismuth showed no increase in the oxidation rate over the corresponding reference test (Figure 10). It appears, therefore, that the bismuth has no influence on bacterial growth, which contradicts the results obtained with chalcopyrite. In addition, thermophilic bacterial activity was slightly inhibited when 0.5 g Bi/l were added to the medium.

Ferric iron combines with phosphate, one of the nutrients of the medium, to produce complexing agents ($FeHPO_4^+$, $FeH_2PO_4^{2+}$, etc.) or precipitates ($FePO_4$) (Robins et al., 1991). Such compounds reduce the oxidation potential of the

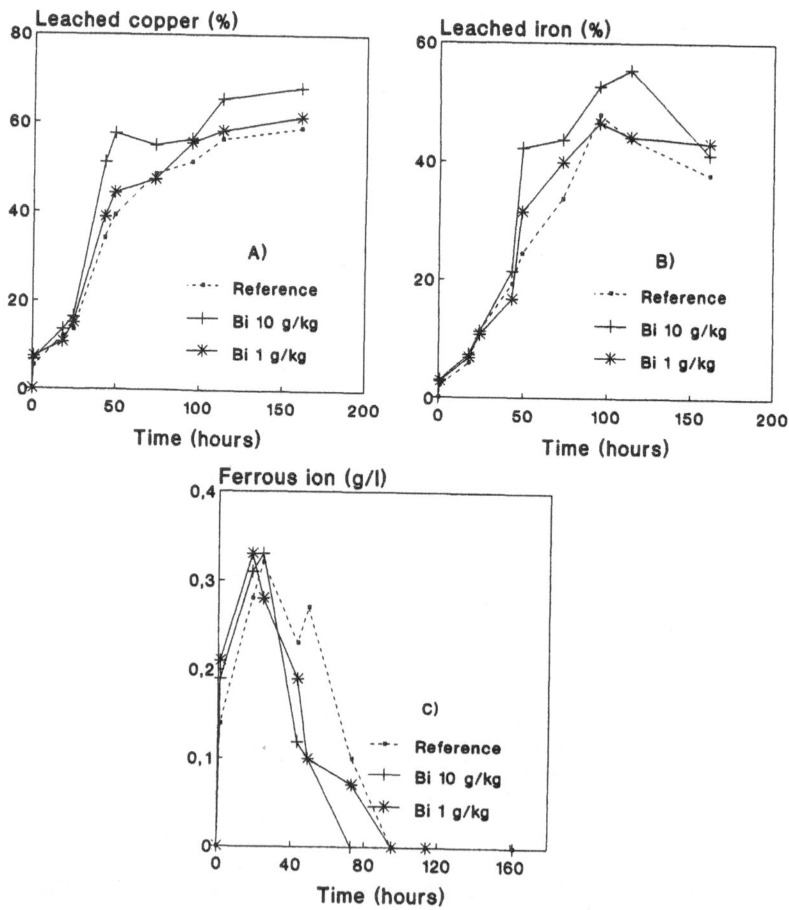

Fig.9. Evolution of copper (A), iron (B) and ferrous concentration (C) during the leaching of a chalcopyrite concentrate by *Sulfolobus* in presence of bismuth (1g and 10g Bi/kg concentrate).

Fe^{3+}/Fe^{2+} couple (Van Wazer, 1958). The presence of bismuth prevents the formation of iron-phosphate compounds because the $BiPO_4$ ($Ksp=1.3.10^{-23}$) is precipitated (Pascal, 1958). Thus, there is a higher concentration of Fe^{3+} and the Fe^{3+}/Fe^{2+} increases.

The influence of bismuth on an abiotic ferric solution (0.1 g Fe^{3+}/l) is shown in table 1. In the experiments with 0.1 g Bi/l, the solution potential was 54 mV higher than in the reference test, whereas this potential only increased 19 mV when 0.01 g/l Bi was added to the solution.

The same potential measurements were carried out in tests with thermophilic cultures just before maximum bacterial growth ocurred. In two different series of experiments (the second B was carried out one week later than the first A) in absence of and in presence of bismuth (1 g Bi/kg concentrate), the values were 18 mV and 15 mV higher than in the reference test (Table 2). This difference in potential was similar to that obtained working with the abiotic solutions.

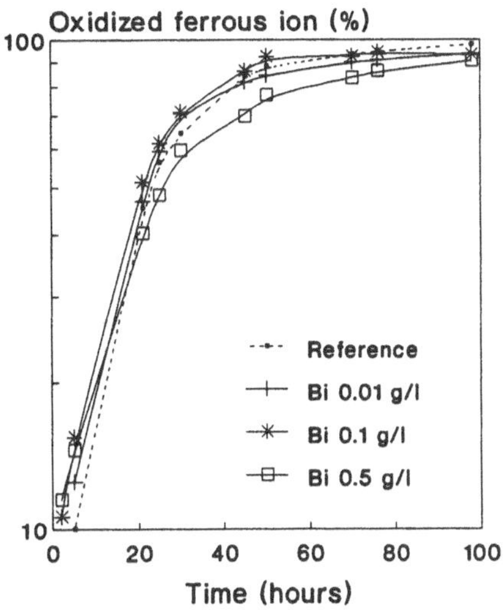

Fig. 10. Oxidation of ferrous ion by *Sulfolobus* in the presence of bismuth and withouth mineral.

Table 1. Bismuth influence on oxidizing potential (versus SHE) in abiotic solutions of Fe^{3+} (0.1 g/l) after shaking over 20 hours in nutrient medium at pH 1.5 and 68°C.

Bismuth added (g/l)	Potential (mV)
Without cation	660
0.01	679
0.1	714

Table 2. Potential measurements (versus SHE) in reactors stored in a static incubator after five days since the inoculation and before maximum bacterial growth. The test were carried out in two different series (A and B). The first A was carried out one week before the second B

Culture	Potential (mV)
Without cation (A)	445
Without cation (B)	428
With bismuth (A)	463
With bismuth (B)	443

Thus, the effect of bismuth seems to be more important in the first step of the experiments (in the lag phase). When bacteria start to grow the Fe^{3+} concentration increases until the quantity of free Fe^{3+} is considerably higher than the complexed or precipitated Fe^{3+}. The result is that the potential in the reference and in the bismuth tests reach maximum levels during the maximum growth step of the cultures.

Bacterial activity was not inhibited by the precipitation of bismuth as $BiPO_4$ which was detected by electron dispersion spectroscopy (EDS) microanalysis of residues. In addition, bismuth was dissolved during the bacterial growth (Figures 11A and 11B). The explanation for this behaviour could be the decrease in pH or the microbial interaction with bismuth phosphate (phosphate source).

A proposed mechanism for the action of bismuth on chalcopyrite bioleaching is shown in Figure 12. The first reaction:

$$CuFeS_2 + 4H^+ + O_2 \text{ --------->} Cu^{2+} + Fe^{2+} + 2H_2O + 2S^0 \quad (4)$$

would be the chemical leaching of the ore in the presence of an acid solution.

The second reaction:

$$CuFeS_2 + 4Fe^{3+} \text{ -------->} Cu^{2+} + 5Fe^{2+} + 2S^0 \quad (5)$$

would become predominant when bacterial growth takes place because the microorganisms are the main Fe^{3+} suppliers, metabolically oxidizing Fe^{2+}. The elemental sulphur formed in both reactions would be oxidized by bacteria leading to increased acid in the solution. As mentioned above, bismuth would tend to remove the phosphate, preventing the formation of ferric-phosphate compounds and, in this way, increasing the oxidation potential of the Fe^{3+}/Fe^{2+} couple.

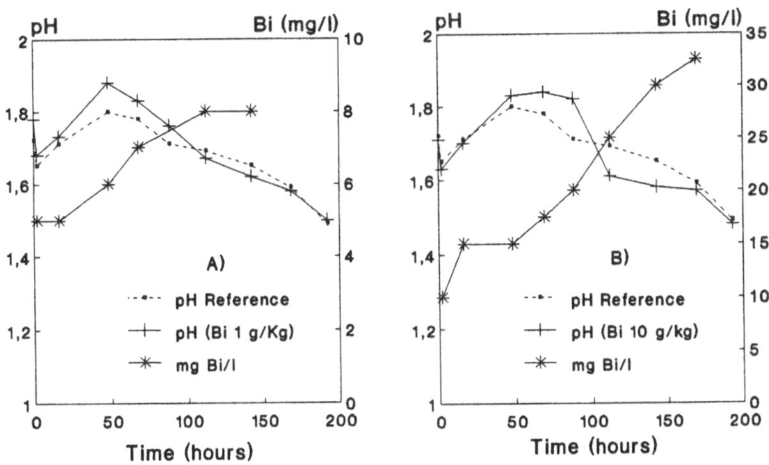

Fig.11. Evolution of bismuth concentration and pH during the bioleaching of a chalcopyrite concentrate by *Sulfolobus* in the presence of bismuth: (A) 1g Bi/kg concentrate, (B) 10g Bi/kg concentrate.

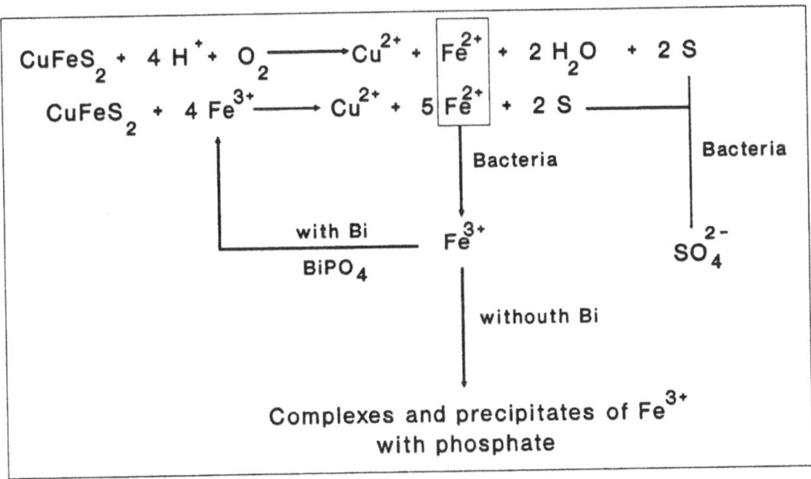

Fig.12.Proposed mechanism for bismuth action in chalcopyrite bioleaching.

4 Conclusions

The bioleaching of chalcopyrite by *Sulfolobus* was depressed by the presence of silver in the system. The dissolution of silver sulphide, once bacterial activity had started, increased Ag^+ up to toxic levels for the thermophilic bacteria. This behaviour was the opposite mesophilic bioleaching of chalcopyrite.

Bismuth increased the chalcopyrite bioleaching rate in thermophilic cultures because it prevented ferric iron from combining with phosphate to produce precipitates ($FePO_4$) or complexes ($FeHPO_4^+$, $FeH_2PO_4^{2+}$, etc.). Thus, the solution had a greater oxidizing capacity because there was more free Fe^{3+} available. The kinetics of copper dissolution was higher in experiments with bismuth at the beginning of the experiment before bacterial growth had started and when Fe^{3+} concentration was still low. The $BiPO_4$ was dissolved as time passed, which could be related to the bacterial activity.

5 Acknowledgements

The authors wish to express sincere gratitude to the Comisión Interministerial de Ciencia y Tecnología (Spain) and to the Commission of the European Communities for funding this research.

6 References

Ahonen, L. and Tuovinen, O.H. (1990a) Silver Catalysis of the Bacterial Leaching of Chalcopyrite-Containing Ore Material in Column Reactors. **Miner. Eng.**, Vol 3. 5, 437-445.

Ahonen, L. and Tuovinen, O.H. (1990b) Catalytic Effects of Silver in the Microbiological Leaching of Finely Ground Chalcopyrite-Containing Ore Materials in Shaked Flasks. **Hydrometallurgy.**, 24, 219-236.

Ammou-Chokroun, M. Cambazoglu, M. and Steinmetz, D. (1977) Oxydation Menagée de la Chalcopyrite en Solution Acide: Analyse Cinétique des Réac tions. II.Modèles Diffusionnels. **Bull. Soc. Fr. Mineral. Cristallogr.**, 100, 161-177.

Ballester, A. (1987) A Study of Silver-Catalysed Bioleaching of Chalcopyrite, in **Separation Processes in Hydrometallurgy** (ed G.A. Davies), Ellis Horwood Publishers, 99-110.

Ballester, A. Cooper, W.C. González, F. and Blázquez, M.L. (1988) Biolixiviación de Calcopirita: Mecanismo en Presencia de Iones Plata, in **XII Congreso Latinoamericano de Química** (eds R. Barriga and M. Figueroa), Sociedad Chilena de Química, Santiago (Chile), 7-12.

Ballester, A. González, F. Blázquez, M.L. Gómez, C. and Mier, J.L. (1992) The Use of Catalytic Ions in Bioleaching. **Hydrometallurgy.**, 29, 145-159.

Bjorling, G. (1954) Uber die langung von Sulfidmineralien unter Sauerstoffdruck. **Metallurgie.**, 8. 781-784.

Blancarte-Zurita, M. Branion, R.M.R. and Lawrence, R.W. (1987) Microbiological Leaching of Chalcopyrite by *Thiobacillus ferrooxidans*, A Comparative Study of a Conventional Process, **in Biohydrometallurgy 87** (eds P.R. Norris and D.P. Kelly), University of Warwick, Coventry (United Kingdom), 273-287.

Dutrizac, J.E. MacDonald, R.J.C. and Ingraham, T.R. (1969) The Kinetics of Dissolution of Synthetic Chalcopyrite in Aqueous Acid of Ferric Sulfate Solutions. **Trans. Metall. Soc. AIME,** 245, 955-959.

Grogan, D.W. (1989) Phenotypic Characterization of the Archaebacterial Genus *Sulfolobus*: Comparison of Five Wild-Type Strains. **J. Bacteriol.**, 171, no.12, 6710-6719.

Miller, J.D., McDonough, P.J. and Portillo, H.Q. (1981) Electrochemistry in Silver Catalyzed Ferric Sulfate Leaching of chalcopyrite, in **Process and Fundamental Considerations of Selected Hydrometallurgical Systems** (ed M.C. Kuhn), AIME, New York, 327-338.

Miller, J.D. and Portillo, H.Q. (1981) Silver Catalysis in Ferric Sulphate Leaching of Chalcopyrite, in **XIII Mineral Processing Congress Proceedings. Part A** (ed J. Lawskoski), Elsevier, Amsterdam, 851-894.

Munoz, P.B. Miller, J.D. and Wadsworth, M. E. (1979) Reaction Mechanism for the Acid Ferric Sulfate Leaching of Chalcopyrite. **Metall. Trans.**, 10B, 149-158.

Norris, P.R. (1989) Mineral-Oxidizing Bacteria: Metal-Organism Interactions, in **Metal-Microbe Interactions.** (eds R.K. Poole and G.M. Gadd), The Society for General Microbiology (IRL Press at Oxford University Press), Oxford United Kingdom), Chapter 7, 99-117.

Pascal, P. (1958) **Nouveau Traité De Chimie Minérale.** Masson et c[ie]. eds., Paris.

Robins, R.C. Twidwell, L.G. Dahnke, D.R. McGrath, S.F. and Khoe G.H. (1991) The Solubility of Metal Phosphates in **EDP Congress '91** (ed D.R. Gaskel), The Minerals, Metals & Materials Society, USA, 3-27.

Van Vazer J.R. (1958) **Phosphorus and its Compounds.** Interscience Publishers Inc, New York, 1958.

Dissolved iron equilibrium in bacterial leaching systems

J. V. Wiertz
Inés Godoy Ríos
Blanca Escobar Miguel
Departmento de Ingeniería Química, University of Chile, Santiago, Chile

Abstract
Bacterial leaching of metal sulphides is now recognized as an attractive industrial process and is successfully applied to recover copper from mixed and low grade sulphide ores. One of the main problems in the operation of bacterial leaching plants is the control of dissolved iron, generally present as a major impurity in all the treated sulphide ores.

The present study examines the behavior of the dissolved iron species in bacterial leaching systems and the different interactions observed between iron, bacteria and mineral solids. The rate of oxidation of ferrous iron by bacteria was studied for both free and attached bacteria, this showed a decrease of ferrous iron oxidation activity for bacteria attached to inorganic solids. On the other hand, the presence of attached bacteria on solid elemental sulphur and metal sulphides increases the reduction rate of ferric ions on the solid surface through oxido-reduction reactions.

In the presence of bacteria, the precipitation of oxidized ferric iron was increased and ferric ion precipitates even when the conditions in the bulk solution would permit its solubility. Moreover, the presence of solid particles increases iron precipitation. The precipitates were analyzed for iron and sulphate contents showing an iron/sulphate ratio very similar to that of jarosite. In the case of chemical precipitation of ferric ions by increasing pH, the sulphate content of the precipitate was lower, showing that in this case the precipitate was mainly ferric hydroxide.
Keywords: Bacterial leaching, ferric ion, iron precipitation, jarosite, sulphur oxidation.

1 Introduction

Bacterial leaching processes involve complex systems in which interact bacteria, mineral solids and dissolved species. Figure 1 shows the different mechanisms of leaching which involve bacteria and dissolved species, usually separated in direct and indirect effects.

The role of iron in bacterial leaching of metal sulphides has been studied for a long

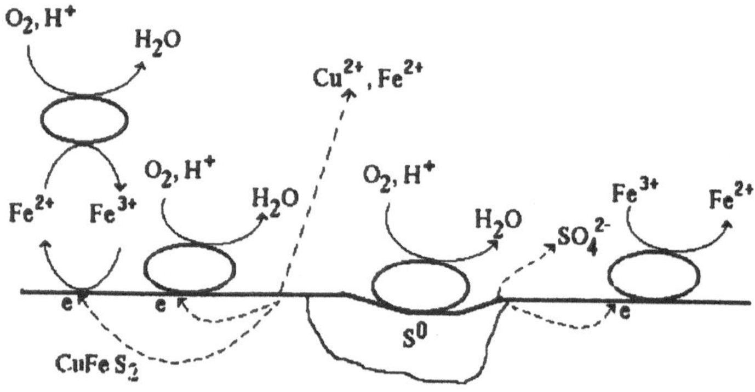

Fig. 1 Direct and indirect mechanisms of bacterial leaching.

time and seems to be essential in most associated processes. The indirect effect of bacteria on metal dissolution proceeds through the bacterial oxidation of dissolved ferrous iron and further sulphide oxidation by the produced ferric ion. The bacteria essentially continuously regenerate the oxidation agent. The direct effect of bacteria on metal dissolution involves the attachment of bacteria to the solid surface and the direct oxidation of sulphide. This process has been shown to be potential dependant and even in this case, the role of iron seems to be very important. The bacterial oxidation of ferrous ion to ferric ion generates a high oxido-reduction potential, condition under which the solid surface can be polarized [1]. Recent works have shown that the presence of ferric ion in the leaching solution accelerates bacterial oxidation of elemental sulphur [2].

The oxidation state of dissolved iron in leaching solutions is determined by the relative rate of the complex combination of oxido-reduction reactions involved in the process. When the oxidation rate of ferrous ions catalyzed by bacteria is lower than the reduction rate of ferric ions on the solid, the potential is low and the system can be considered as controlled by the bacteria. But when the ferrous oxidation rate is higher than ferric reduction rate, the potential is high and the system is controlled by the solid reaction. As both iron oxidation and reduction rates also depend on potential, the control of the system is quite complex and the Eh potential very difficult to predict.

The precipitation of iron is another problem of great importance in leaching processes. In most cases, the leaching solution is recirculated after passing through solvent extraction or copper precipitation, and it is essential to avoid a build-up of dissolved iron which would require a solution purge. Iron precipitation can also help to control solution pH and limit acid consumption. Iron acts as a buffer in leaching solutions and when the pH increases, iron precipitation generates protons which balance the consumption. On the other hand, the bacterial oxidation of ferrous ion to ferric is proton consuming and contributes to limit the pH decreases. Through these mechanisms, the presence of iron in bacterial leaching solutions contributes to regulate the free proton concentration.

The present work reviews the different interactions existing between dissolved iron, minerals and bacteria. Oxidation and reduction of iron was studied in different systems, using pure minerals and elemental sulphur. The influence of both bacteria and solid particles on iron precipitation was also investigated.

2 Material and methods

2.1 Bacteria

Thiobacillus ferrooxidans (ATCC 19859 strain) was used in this work. The bacteria was currently grown in MC medium with ferrous sulphate or elemental sulphur as energetic substrate.

2.2 Culture medium

MC medium was used in all experiments. The medium composition is the following:

$(NH_4)_2SO_4$	0.4 g/l
$K_2HPO_4.3H_2O$	0.056 g/l
$MgSO_4.7H_2O$	0.4 g/l

Initial pH was adjusted with sulphuric acid to 1.6 for ferrous sulphate cultures and to 2.3 for cultures with elemental sulphur.

2.3 Sulphur prills

Sulphur prills were prepared by chilling small droplets of melted sulphur in cold distilled water [3]. The melted sulphur was prepared by heating sublimated sulphur powder at 120°C. The almost spherical prills were then separated in different granulometric size fractions. The sulphur prills were sterilized before use with the liquid medium by heating twice at 100°C over 30 minutes.

2.4 Oxidation of ferrous iron

Bacterial oxidation rate of ferrous iron was determined by analyzing in time the remaining ferrous ion concentration using the o-phenanthroline colorimetric method [4].

3 Results and discussion

3.1 Rate of oxidation of ferrous iron by free and attached bacteria

The rate of oxidation of ferrous iron was determined for free bacteria in a suspension without solid particles and for bacteria previously contacted with mineral particles. 25 ml of bacterial suspension at an initial concentration of 4.8×10^7 bact/ml were contacted with 1 g of chalcopyrite (-100, +150 #). After 2 hours, the concentration of free bacteria in the solution stabilized at 1.50×10^7 bact/ml, the difference corresponding to

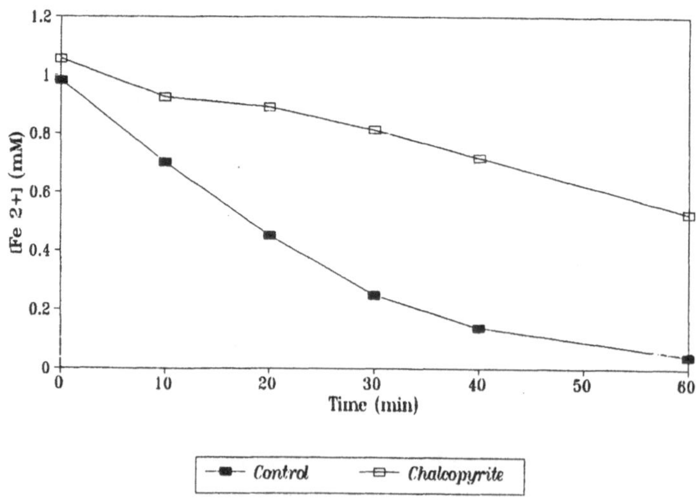

Fig. 2 Bacterial oxidation of ferrous iron in the presence of chalcopyrite particles.

the amount of attached bacteria. Then, ferrous sulphate was added to the solution in both free particle suspension and suspension previously contacted with solid particles at an initial concentration of 1 mM. The rate of ferrous ion oxidation was determined by analyzing the decrease of ferrous concentration.

Figure 2 shows the results of ferrous oxidation obtained for free bacteria and with some of the bacteria attached to chalcopyrite particles. In the presence of particles, 69 % of the bacteria attached to the solid while the other 31 % remained in the solution. The rate of ferrous iron oxidation was lower in the presence of particles. Supposing that the free bacteria is the same with or without particles, we can estimate the rate of ferrous oxidation by attached bacteria according to:

$$R = r_f \times N_f + r_a \times N_a$$

where R is the observed ferrous oxidation rate (in μmole Fe^{2+}/ml.min), r_f and r_a the specific oxidation rate (in μmole Fe^{2+}/bact.min) for free and attached bacteria, respectively, and N_f and N_a the corresponding bacterial concentration (bact/ml). N_f can be calculated from the results of the oxidation experiment without particles. This experiment shows that, under these conditions, the specific oxidation rate of attached bacteria, r_a was $4.1\ 10^{-11}$ μmol Fe^{2+}/bact.min, about 10 times lower than that of free bacteria ($r_f = 4.7\ 10^{-10}$ μmol Fe^{2+}/bact.min). These results confirm previous observations made by different authors with bacteria attached to different solids [5, 6]. At higher ferrous sulphate concentrations, the difference in oxidation rate between free and attached bacteria is lower (results not shown). It should be noted that the study of the kinetics of bacterial oxidation of ferrous ion reveals the possible existence of different oxidation systems which are activated at different ferrous concentrations [7, 8]. The present results suggested that the system involved at lower ferrous iron concentration is more affected when bacteria are attached to solid particles.

3.2 Bacterial oxidation of elemental sulphur in the presence of ferric iron
Bacteria were previously grown with sulphur prills (-8 +10#) as energetic substrate and maintained during 1 month in these conditions by transferring every week the prills colonized by attached bacteria into fresh sterile pH 2.3 medium containing a small amounts of ferrous sulphate (1 mM). The colonized prills were then used to study the effect of dissolved ferric iron on the rate of sulphur oxidation by bacteria. Similar samples of colonized prills were contacted with 100 ml fresh medium under different conditions. In the presence of 2.5 mM ferric sulphate $Fe_2(SO_4)_3$, the oxidation of sulphur to sulphate started immediately as observed by the proportional increase of sulphate concentration in the solution (Figure 3). The increase rate of proton concentration was constant during the whole experiment showing an almost linear kinetics of sulphur oxidation. When no iron was added to the leaching solution, the sulphur oxidation was much slower.

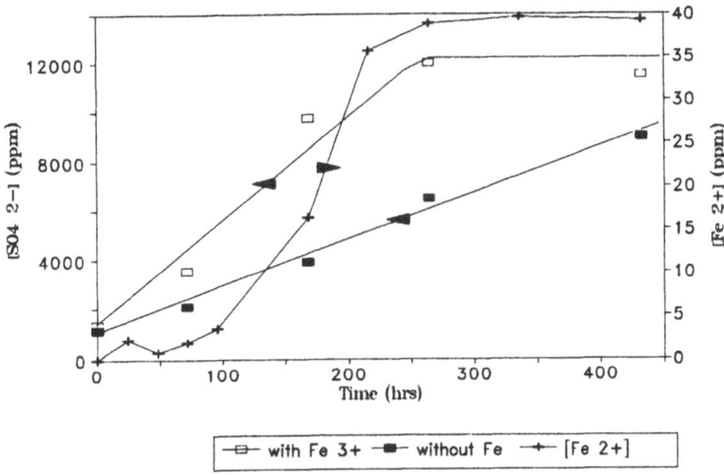

Fig. 3 Effect of ferric ions on bacterial sulphur oxidation.

In the first case, as the oxidation proceeded, an important decrease of pH was observed. At low pH value, the ferric ions were reduced to ferrous and a decrease of Eh was observed. The bacterial oxidation rate of sulphur decreased and when all of the dissolved iron was reduced to ferrous, no further oxidation of sulphur was detected. These results suggested that ferric ion could be involved as electron acceptor during bacterial oxidation of sulphur. Under normal conditions, the rate of reoxidation of ferrous ion by the bacteria is higher than the rate of reduction of ferric ion on sulphur oxidation and no change in the ferric concentration was detected. When the pH decreases, bacterial ferrous oxidation system is affected and inhibited so the rate of ferrous oxidation decreases and this ion then accumulates in the solution. As bacterial sulphur oxidation can also be observed without ferric ion in the solution, it should be noted that ferric ion is not the only possible electron acceptor for sulphur oxidation but that oxygen also could directly participate to the reaction. In the absence of bacteria, ferric ion does not oxidize elemental sulphur. Different authors suggested [9, 10] that ferric ion is directly involved in bacterial mechanisms of sulphur oxidation and that the associated enzymatic system involves a ferric ion sulphide oxidase. Sugio et al. [11] also showed that the presence of ferrous ion in the leaching solution could inhibit bacterial sulphur oxidation. We suggest that the oxidation of elemental sulphur could be regulated by the solution potential. When the potential is high, the presence of bacteria on sulphur surface catalyses the oxido-reduction reaction. In the case of metal sulphides, similar results were observed. Gaete et al. [12] observed that the presence of attached bacteria on pyrite surface enhances the reduction of ferric ion. Once again, under normal conditions, the bacteria reoxidize the ferrous ions produced and this effect is difficult to detect. At high ferric ion concentration (> 10 g/l Fe^{3+}), ferrous ion oxidation by bacteria is inhibited and then ferrous accumulation is observed in the leaching solution.

3.3 Iron precipitation in the presence of bacteria

Figure 4 shows the typical behavior of dissolved iron species during the batch bacterial oxidation of a ferrous sulphate solution. As the number of bacteria increases, the rate

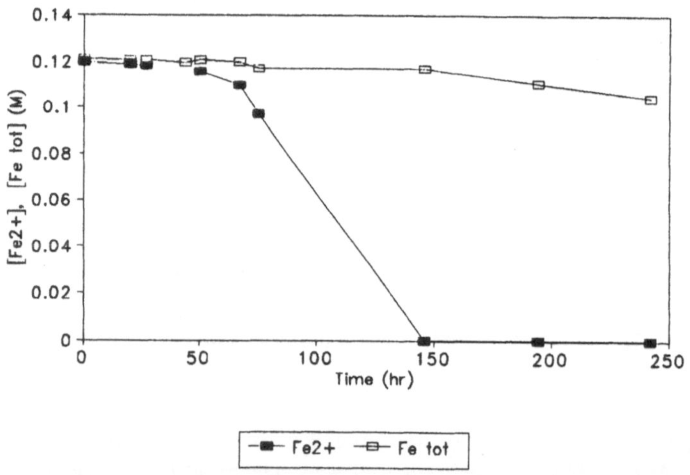

Fig. 4 Bacterial oxidation of ferrous iron and precipitation of ferric.

of oxidation also increases and the pH of the solution increases. The total iron concentration decreases as a result of ferric ion precipitation. Nevertheless, the final pH value is lower than the pH equilibrium value for a similar ferric sulphate concentration in a pure chemical system without bacteria. At higher initial bacteria concentration, iron precipitation is higher.
The oxidation reaction of ferrous sulphate to ferric consumes protons according to:

$$2\ FeSO_4 + 1/2\ O_2 + H_2SO_4 \dashrightarrow Fe_2(SO_4)_3 + H_2O$$

The increase of pH in the solution produces the precipitation of ferric ion in the form of jarosite or as ferric hydroxide. The amounts of iron precipitates is quite difficult to predict and does not only depend on the initial pH conditions and on the amounts of ferric ion produced. It also depends on the rate of oxidation related to the number of bacteria present in the solution and to the degree of agitation in the solution. As the oxidation reaction takes place at the interface between bacteria and solution, it produces a local condition very favorable to the precipitation. The bacterial surface could also act as a nucleation center for the precipitate.

3.4 Iron precipitation in the presence of bacteria and solid particles

Ferrous sulphate oxidation by bacteria was studied in the presence of solid particles of quartz. 5 g of quartz particles (-10 +14#) were added to a bacterial culture with an initial concentration of 0.1 M $FeSO_4$. The results of oxidation were compared with another experiment on the same conditions except for the addition of quartz. Figure

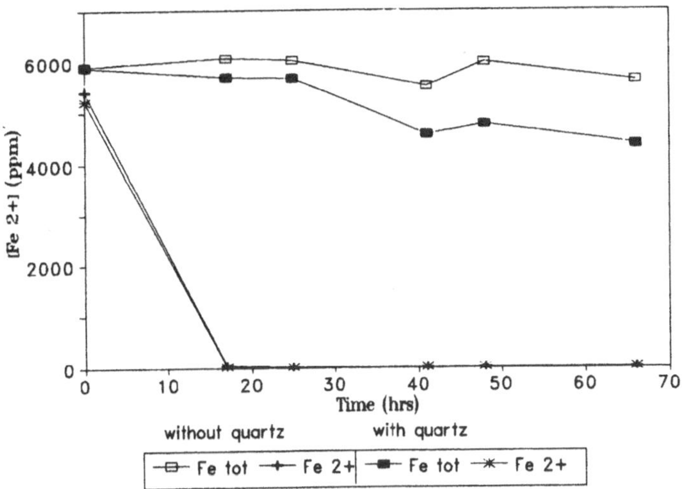

Fig. 5 Bacterial oxidaction of ferrous ion in the presence of quartz particles.

5 shows the results obtained on both experiments. In the presence of solid particles, iron precipitation was much higher and essentially proceeded after the complete oxidation of available ferrous ion.

The precipitates were separated from the solution washed on a filter and dried. There were then dissolved in a hydrochloric acid solution and analyzed for total iron and sulphate contents. The analyses of the precipitates formed in the presence of particles show sulphate and iron weight fractions of 0.31 and 0.40, respectively, values very similar to that of jarosite (0,34 for iron and 0,40 for sulphate). When precipitation of ferric ion is produced without particles, the sulphate weight fraction of the precipitate is much lower (0,20) showing that in this case, the precipitate is essentially composed by ferric hydroxide. The jarosite precipitated in the presence of bacteria and particles is quite stable and in the leaching solution, even when the pH decrease, it generally does not dissolve.

4 Conclusions

Bacterial leaching of metal sulphides involve complex multi-phases systems and cannot be considered as homogeneous. The dissolution of metal proceeds through oxido-reduction reactions in which dissolved iron plays a fundamental role. Ferric ion is an intermediate electron acceptor in both indirect and direct bacterial leaching and is continuously regenerated by bacterial oxidation of ferrous ion. Ferric ion precipitation is enhanced by the presence of bacteria and also by the presence of solid particles. In the last case, the precipitate formed is preferentially jarosite. The presence of ferric ion in the leaching solution has a buffer effect on solution pH and maintain it to low values. Oxidation and precipitation of iron by bacteria is also a possible way to control dissolved iron concentration in leaching solutions.

5 References

1. Vargas, T., Sanhuesa, A. and Escobar, B. (1992) Electrochemical studies of the bacterial leaching of pyrite with *Thiobacillus ferrooxidans*, in **EPD Congress 1992** (ed. J.P. Hager), TMS 1992, pp 273-284.
2. Wiertz, J.V. (1993) Ferrous and sulfur oxidation by *Thiobacillus ferrooxidans*, in **Biohydrometallurgical Technologies, vol II** (ed. A.E. Torma, M.L. Apel and C.L. Brierley), TMS 1993, pp 463-471.
3. Espejo, R.T. and Romero, P. (1987) Growth of *Thiobacillus ferrooxidans* on elemental sulfur, **Applied Environmental Microbiology**, 53, 1907-1912.
4. Muir, M.K. and Andersen, T.N. (1977) Determination of ferrous iron in copper-process metallurgical solutions by the o-phenanthroline colorimetric method, **Metallurgical Transactions**, 8 B, 517-518.
5. DiSpirito, A.A., Dungan, P.R. and Tuovinen, O.H. (1983) Inhibitory effects of particulate materials in growing culture of *Thiobacillus ferrooxidans*, **Biotechnology and Bioengineering**, 25, 1163-1168.

6. Kai, T., Takahashi, T., Shirakawa, Y. and Kawabata, Y. (1990) Decrease in iron oxidizing activity of *Thiobacillus ferrooxidans* adsorbed on activated carbon, **Biotechnology and Bioengieneering**, 36, 1105-1109.
7. Aguirre, R., Wiertz, J.V. and Badilla-Ohlbaum, R. (1991) An amperometric method for measuring iron oxidazing activity of *Thiobacillus ferrooxidans*, **Bioleaching: from molecular biology to industrial applications** (ed. R. Badilla-Ohlbaum, T. Vargas and L. Herrera), University of Chile, Santiago, Chile, pp 107-117.
8. Wiertz, J.V. (1992) **Lixiviation bactérienne de la chalcopyrite: role et importance du fer**, Thèse de Doctorat, Université de Liège, Belgium.
9. Sugio, T., Katagiri, T., Moriyama, M., Zhen, Y.L., Inagaki K. and Tano, T. (1988) Existence of a new type of sulfite oxidase which utilizes ferric ions as an electron acceptor in *Thiobacillus ferrooxidans*, **Applied Environmental Microbiology**, 54, 153-157.
10. White, Shute, E., Choate, D. and Blake, R.C. (1992) Existence of a hydrogen sulfide:ferric ion oxidoreductase in iron oxidizing bacteria, **Applied Environmental Microbiololy**, 58, 431-433.
11. Sugio, T., Hirose, T, Oto, A., Inagaki, K. and Tano, T. (1989) The regulation of sulfur utilization by ferrous ion in *Thiobacillus ferrooxidans*, in **Biohydrometallurgy 1989** (ed. J. Salley, R.G.L. McGready and P.L. Wichlacz), Canmet, Ottawa, Canada, pp 451-459.
12. Gaete, M., Flores, I., Campos, J., Lara, E., Razmilic, L., Guerrero, P., Wiertz, J. and Maturana, H. (1990) Comparación entre la lixiviación química por sulfato férrico y lixiviación bacteriana en el pretratamiento de un concentrado aurífero refractario, in **Iberomet-Conamet VI**, USACH, Santiago, Chile, pp 423-433.

Influence of substrate polarization on activity of *Thiobacillus ferrooxidans* in bioleaching

J. V. Wiertz
Pedro Moya
Angel Sanhueza
Tomás Vargas
Departmento de Ingeniería Química, University of Chile, Santiago, Chile

Abstract
In the bioleaching of sulphide minerals with *Thiobacillus ferrooxidans* an important fraction of the total bacterial population remains attached to the solid substrate. The attached microorganisms can both catalyze the oxidation of ferrous iron in solution or directly oxidize the substrate. In the present work we report experimental results on bioleaching of both sulphur and pyrite which show that the catalytic action of attached *Thiobacillus ferrooxidans* is influenced by electrochemical parameters in the system. Experiments in both cases were conducted with a pure strain of *Thiobacillus ferrooxidans* (ATCC 19859) in pH 2.0 nutrient medium at 30°C.

In biooxidation of sulphur prills in shake flasks it was observed that the rate sulphur oxidation by attached bacteria was dependant on the redox potential established in the leaching solution. Pyrite bioleaching experiments conducted under controlled electrochemical potential showed that the effect of the bacteria on pyrite dissolution was enhanced when the sulphide was anodically polarized.

A link between the influence of the electrochemical potential on the activity of *Thiobacillus ferrooxidans* and the biochemical mechanisms of electron transfer in the external membrane of this microorganism is proposed.
Keywords: Bacterial leaching, pyrite dissolution, sulphur oxidation, *Thiobacillus ferrooxidans*.

1 Introduction

Sulphide oxidation by *Thiobacillus ferrooxidans* is related to two types of mechanisms. The indirect mechanism proceeds by bacterial oxidation of dissolved ferrous iron which generates dissolved ferric iron, a powerful oxidizing agent able to dissolve most sulphides. Pyrite is chemically oxidized by ferric ion according to one of the following equations:

$$FeS_2 + 7Fe_2(SO_4)_3 + 8H_2O \longrightarrow 15FeSO_4 + 8H_2SO_4 \quad (1)$$

$$FeS_2 + Fe_2(SO_4)_3 \longrightarrow 3FeSO_4 + 2S \quad (2)$$

The bacteria then catalyze the oxidation of ferrous iron and regenerate the strongly oxidizing ferric ion:

$$4FeSO_4 + O_2 + 2H_2SO_4 \rightarrow 2Fe_2(SO_4)_3 + 2H_2O \quad (3)$$

The direct mechanisms require a direct contact of bacteria with the solid surface of mineral particles and are generally thought to proceed by transfer of electrons from the oxidized sulphide to oxygen through the bacteria. Direct oxidation of pyrite by *Thiobacillus ferrooxidans* is described by the equation:

$$4FeS_2 + 15O_2 + 2H_2O \rightarrow 2Fe_2(SO_4)_3 + 2H_2SO_4 \quad (4)$$

Attached bacteria can also directly oxidize residual sulphur according to:

$$2S + 3O_2 + 2H_2O \rightarrow 2H_2SO_4 \quad (5)$$

In the bioleaching of mineral sulphides or sulphur with *Thiobacillus ferrooxidans*, an important fraction of the total bacterial population remains attached to the solid substrate. These attached microorganisms can catalyze the sulphide dissolution by both indirect mechanism or direct oxidation of the sulphur moiety. For an efficient control of the dissolution bioleaching process, it is important to understand the different factors which govern the intensity and direction of the catalytic action of attached bacteria and to determine the relative contribution of direct and indirect mechanism. Panin et al. [1] suggested that the direct mechanism in pyrite is based on electrochemical interactions intensified by bacteria. There is also evidence of the direct utilization of pyrite sulphur moiety by *Thiobacillus ferrooxidans*. Silverman [2], reported that is probable that the two mechanisms of bacterial oxidation of pyrite operate concurrently.

The regulation of the iron and sulphur oxidation systems is also related to the understanding of the relative importance of both direct and indirect leaching by *Thiobacillus ferrooxidans* in pyrite dissolution. *Thiobacillus ferrooxidans* can oxidize iron and sulphur simultaneously [3-5]. However, other authors have observed independently that when both ferrous iron and sulphur are provided, *Thiobacillus ferrooxidans* preferentially oxidizes iron [6, 7].

Enzymatic mechanisms of dissolved ferrous iron biooxidation were extensively studied and are quite well understood. Different constituents of the associated electron transport chain were identificated and characterized [8]. In the case of enzymatic system associated to elemental sulphur oxidation, the existence and participation of a sulphite:ferric ion oxidoreductase (SFORase) were demonstrated [9, 10]. The authors also observed that the presence of ferrous ion in the solution inhibits bacterial sulphur oxidation [11]. Recently, Kuenen et al. [12], reviewed the enzymology aspects of bacterial sulphur oxidation and support that the SFORase is only of minor importance in the global mechanism and that the sulphur and iron oxidation mechanisms are not interlinked.

There is an increasing awareness of the need of understanding better the influence of the electrochemical aspects involved in bioleaching. Bioleaching of sulphides involves several phenomena of electrochemical nature. A mineral sulphide, a semiconductor, dissolves anodically while releasing electrons to an oxidant in solution, e.g. dissolved oxygen or ferric ion. The electrochemical nature of the metabolic processes of microorganisms has since long been recognized [13]. *Thiobacillus ferrooxidans* participates in the oxidation process, mediating in the transport of electrons from ferrous iron or reduced sulphur compounds to dissolved oxygen, so obtaining its energy for growth and cell maintenance [14].

Electrochemical measurements have been applied to study the bioleaching of several sulphide minerals [15-17]. The influence of galvanic interactions [18] and externally applied potentials [19-20] on the bioleaching process has been investigated. Different electrochemical aspects of pyrite oxidation by *Thiobacillus ferrooxidans* have been also studied. Recently, Vargas et al. [20, 21] observed that the pitting activity of bacteria attached to pyrite is activated by polarization induced by bacterial oxidation of ferrous iron in solution.

The present work reports experimental results of bioleaching of pyrite and sulphur. The study is focused on evaluating the influence of some electrochemical parameters, namely the solution redox potential and the substrate polarization, on the intensity and quality of the bacterial catalytic action. A particular aim is to discuss possible relations between these parameters and the biochemical mechanism of electron transfer in the external membrane of this microorganism.

2 Materials and methods

2.1 Bacteria

Thiobacillus ferrooxidans ATCC 19859 strain was used in this work. The bacteria was grown in a medium with the following composition:

$(NH_4)_2SO_4$	0.4 g/l
$K_2HPO_4.3H_2O$	0.056 g/l
$Mg_2SO_4.7H_2O$	0.4 g/l

Initial pH was adjusted with concentrated sulphuric acid.

2.2 Sulphur prills

Sulphur prills were prepared by chilling small droplets of molten sulphur in cold distilled water [22]. The melted sulphur was prepared by heating sublimated sulphur powder at 120°C. A fraction of -8 +10 # almost spherical prills were then selected for the experiments. Sulphur prills were sterilized with the liquid medium by heating twice at 100°C during 30 minutes.

2.3 Pyrite

High purity pyrite particles were manually separated from a copper ore sample from Los Bronces (Cia Disputada de Las Condes), and ground. The -115 +150 # fraction was used in the experiments. Pyrite samples were sterilized at high temperature under nitrogen. Unless otherwise is stated, the potentials reported in the text are with respect to the Ag/AgCl (sat.) reference electrode.

3 Results

3.1 Influence of redox potential on bacterial sulphur oxidation

Sulphur prills were initially colonized with bacteria and maintained during 1 month in these conditions by transferring every week the prills colonized by attached bacteria into fresh sterile pH 2.3 medium containing small amounts of ferrous sulphate (1 mM). Samples of 5 g of the colonized prills were then biooxidized at 30°C in shake flasks with 100 ml of three different solutions, each one presenting a different initial redox potential. Solution (a) was pure oxygenated basal medium, with 0.36 V initial redox potential; solution (b), oxygenated basal medium containing 5 mM $FeSO_4$, with 0.31 V initial redox potential; solution (c) oxygenated basal medium with 2.5 mM $Fe_2(SO_4)_3$ with 0.44 V initial redox potential. The biooxidation of sulphur was determined indirectly by monitoring the concentration of protons in solution, based on pH measurements. Redox potentials were monitored with a combined platinum electrode with Ag/AgCl reference.

Experimental results showing the changes in acid concentration and redox potential in solution are summarized in Figure 1.

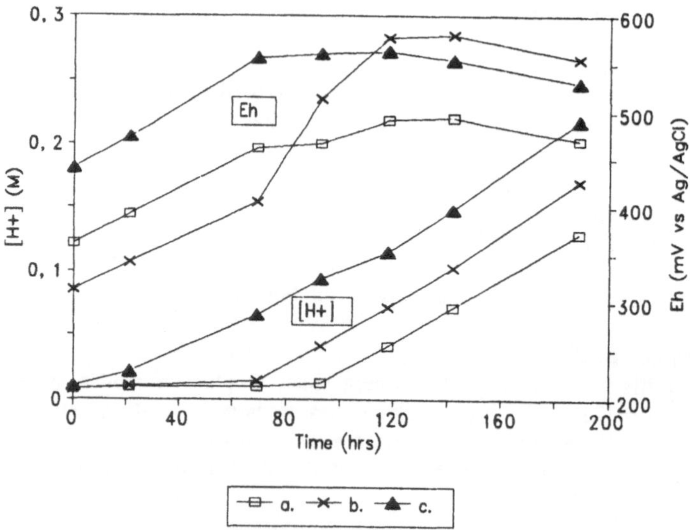

Figure 1: Sulphur oxidation by *Thiobacillus ferrooxidans*; a: without iron; b: with Fe^{+2}; c: with Fe^{+3}.

In the presence of ferric ion the oxidation of sulphur to sulphuric acid was initiated immediately after the start of the experiment (curve c in Figure 1). During this process the redox potential of the solution increased from 0.44 V to a maximum of 0.55 V. In the solution with ferrous ion, effective sulphur oxidation to sulphuric acid started after 60 hours of contacting the colonized prills with the solution (curve b in Figure 1). During these initial 60 hrs, however, the redox potential of the solution increased as a consequence of bacterial oxidation of ferrous ion. Sulphur oxidation was triggered after the redox potential reached a value of 0.4 V. When sulphur oxidation started, the redox potential continued to increase, indicating simultaneous oxidation of ferrous ion and sulphur, reaching a maximum value of 0.57 V.

In the case of pure basal medium, sulphur oxidation to sulphuric acid was observed only after 90 hours of contact (curve a in Figure 1). The redox potential of the solution also increased during this initial period presumably due to the oxidation of minor amounts of ferrous ion contained in the medium. Massive oxidation to sulphate started this time when the redox potential reached a value of 0.45 V. The redox potential continued its increase after triggering of sulphur oxidation, but reached a maximum value of 0.48 V, lower than the ones obtained when iron was present in solution.

In the three experiments, as the sulphur oxidation proceeded, the pH of the solution reached very low values (lower then pH 1.0). When these conditions were reached, there was an apparent inhibition of bacterial ferrous ion oxidation in the two experiments which were initiated with dissolved iron and the ferric ion reduced to ferrous, with a consequent decrease of solution Eh. Even when no iron was present in the solution, a same decrease of potential was observed. Then, the rate of sulphur oxidation decreased.

3.2 Influence of sulphide polarization on the bacterial leaching of pyrite.

Experiments with pyrite were conducted in a three-electrode cell, each one being in a separate compartment. The reference electrode was Ag/AgCl (sat.); the counter electrode was platinum foil; the working electrode was made of pyrite particles electrochemically activated through contact with a coil of platinum wire. The pyrite particles were suspended with a glass stirrer rotating at constant speed. Air was constantly bubbled in the pyrite compartment. The cell was immersed in a thermostatic water bath. The electrochemical potential of the pyrite particles was controlled with a EG8G Princenton Applied Research Model 363 Potentiostat/Galvanostat. All experiments were conducted at 30°C.

Leaching of the pyrite bed was conducted under three different conditions: (a) in abiotic iron-free oxygenated basal medium, with pyrite particles polarized to +0.6 V; (b) in inoculated ($4x10^7$ bact/ml) iron-free basal medium with pyrite in open circuit (+0.0 V); (c) in inoculated ($6x10^7$ bact/ml) iron-free basal medium, with pyrite particles polarized to + 0.6 V. In each case the dissolution of pyrite was evaluated from monitoring the evolution of sulphate and iron in solution.

Results of pyrite leaching under those different conditions are summarized in Figures 2. Analysis of these data shows that the rate of bacterial leaching of pyrite when the sulphide is anodically polarized (curve c) is larger than the rate of bioleaching

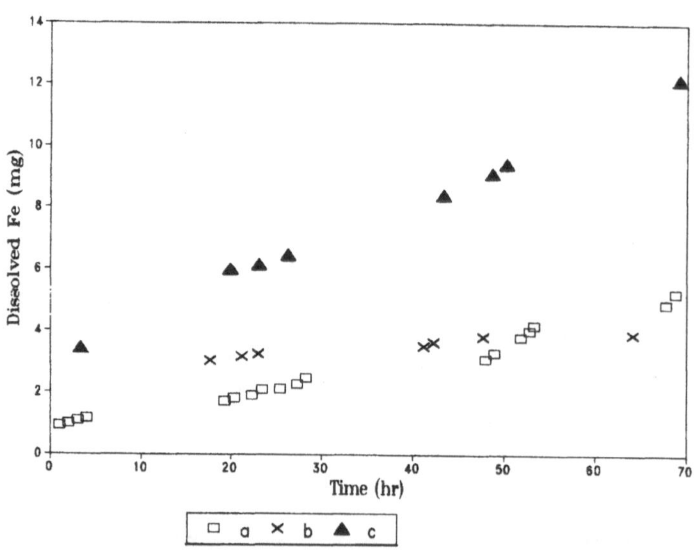

Figura 2: Iron dissolution from pyrite in MC medium; a: without bacteria, +0.6V; b: with bacteria, open circuit; c: with bacteria, +0.6V.

in open circuit conditions (curve b). On the other hand when pyrite is leached under anodic polarization but in abiotic conditions (curve a), the leaching rate is smaller than in the case when polarization is combined with bacterial inoculation. These results show that bacterial leaching is more efficient when the sulphide is anodically polarized. However this increased efficiency is not the result of the simple contribution of anodic dissolution, as shown by the fact that iron dissolution in this case is higher than in the polarized abiotic experiment (curve a). The results indicate that there is a synergistic influence between the presence of bacteria and the substrate polarization, which is presumably due to the modification of the bacterial activity, as will be discussed below. Influence of potential on sulphide bioleaching has been also reported by other authors [19].

4 Discussion

Experimental results reported above show that the behavior of *Thiobacillus ferrooxidans* in the bioleaching of two substrates, sulphur and pyrite, can be related to the evolution of electrochemical parameters. An attempt can be made to explain the dependence of

Thiobacillus ferrooxidans activity in both cases assorting to the mechanism of utilization of two substrates.

T. ferrooxidans is able to derive the energy required for growth and cell maintenance from the oxidation of reduced sulphur compounds, sulphur or ferrous ion, using O_2 as the oxidant. *T. ferrooxidans* grows optimally on Fe^{2+} at pH 2 with an electron-transport chain mediating between the two half-reactions:

$$2Fe^{2+} \dashrightarrow 2Fe^{3+} + 2e \quad (6)$$

$$2e + 1/2\, O_2 + 2H^+ \dashrightarrow H_2O \quad (7)$$

of the overall reaction:

$$2Fe^{2+} + 1/2\, O_2 + 2H^+ \dashrightarrow 2Fe^{3+} + H_2O \quad (8)$$

Ferrous ion is oxidized at the cell surface, or in the peripheral membranes of the cell [8], releasing electrons that are transported across the cell wall by biochemical paths [14, 15]. Hydrogen ions and oxygen diffuse through the cell wall in response to activity gradients, and within the cell the oxygen is reduced by the electrons released by ferrous ion. At 25°C, the electrode potential of reaction (7) is:

$$E = 1.23 + 0.0148 \log pO_2 - 0.0591\, pH - 0.0148 \log a_{H2O} \quad (9)$$

At pH 7 and an oxygen pressure of 10^{-2} atm, conditions inside the cell, equation (9) yields a potential of 0.787 volt [14, 23]. On the other hand, at 25°C the electrode potential for reaction (6) is:

$$E = 0.77 + 0.0591 \log (Fe^{3+}/Fe^{2+}) \quad (10)$$

When the (Fe^{3+}/Fe^{2+}) ratio is one, equation (10) yields 0.77 volt and the difference of 0.087 with reaction (7) potential is enough to move electrons from ferrous ions into the cell. However, as ferrous ion in solution undergoes oxidation, a point will be reached when the electrode potential for reaction [6] reaches 0.787 volt $(Fe^{3+}/Fe^{2+} = 29.65)$. In this situation, from a pure electrochemical perspective, the oxidation of ferrous ion by electron transport through the cell membrane should cease.

The reversible energy surplus is more favorable when bacteria oxidize sulphur or sulphides than in the case of ferrous ion oxidation. For instance for the oxidation of sulphur to sulphate:

$$S + 4\, H_2O \dashrightarrow SO_4^{2-} + 4H^+ + 6e \quad (11)$$

the reversible half cell potential for this reaction is:

$$E = 0.36 - 0.08\, pH + 0.01 \log SO_4^{2-} \quad (12)$$

For instance, for pH 2 and SO_4^{2-} concentration of 10^{-2} M, equation (12) yields E = 0.18 volt, which provides a potential difference of about 0.6 volt to drive electrons for oxygen reduction. However sulphur oxidation is a reaction much less reversible than ferrous ion oxidation and, in strictly electrochemical terms, much more positive potentials are necessary for this reaction to proceed. The bacteria, however, count with an enzyme system on the outer cell wall which mediates the sulphur oxidation, turning this reaction more reversible. According to Sugio et al. [9], *Thiobacillus ferrooxidans* oxidizes sulphur by reducing ferric ion bound to the external cell wall. These authors proposed that this enzymatic mechanism is in fact activated in the presence of high levels of ferric ion. Kuenen et al. [12], on the other hand, propose that sulphur oxidation by *T. ferrooxidans* occurs in an enzymatic system completely independent of the one used in ferrous ion oxidation.

The dependence of *T. ferrooxidans* activity on the electrochemical parameter observed in this work can be explained based on the bioenergetics phenomena involved in the metabolism of these microorganisms.

When bacterial oxidation of sulphur occurs in the presence of ferrous ion, the half cell potential for reaction [6] is very low, +0.31 volt vs Ag/AgCl (sat.) in experiment (b), and there is a favorable potential window of 0.26 volt for the microorganism to transfer electrons from ferrous ion in solution. When iron is mostly as ferrous ion sulphur is not oxidized, even considering the advantageous potential of this reaction. This, in terms of Sugio's enzymatic model [11], can be explained as a inhibition of sulphur oxidation by suppression of ferric ion from the outer cell wall.

When most of iron is present as ferric ion, it is obviously not possible for the microorganism to use directly ferrous ion as a source of electrons. In addition, a high Fe^{3+}/Fe^{2+} ratio contributes to keep a high concentration of ferric ions fixed to the external membrane and the oxidation of sulphur by this ion is encouraged, as observed by Sugio et al. [11]. In the case of sulphur oxidation in iron-free solutions, the activation of sulphur oxidation also requires a minimum redox potential. In this case, as the used bacterial inoculum came from several steps of growth in pure sulphur, the bacteria would require of the external potential to generate an adequate concentration of ferric ion in the external membrane.

The observations made upon the behavior of *T. ferrooxidans* during sulphur utilization in solutions with different redox potentials and the proposed mechanisms of bacterial activity regulation can be also applied to explain the influence of potential on the bioleaching of pyrite. In fact, bacteria attached to pyrite, apart from direct oxidation of reduced sulphur, are also in a position to oxidize both ferrous ion and sulphur released as subproduct during the decomposition of this sulphide. Evidence of the direct utilization of the sulphur moiety of pyrite by *Thiobacillus ferrooxidans* has been already observed [24].

Experiments showed that the bacterial dissolution rate of pyrite is faster when the solid is anodically polarized, that is when a high redox potential is established in the solution in the vicinity of the sulphide. The results also show that in this situation there is an increase of the catalytic activity of the bacteria. Evidence of the influence of polarization of pyrite on the catalytic activity of *T. ferrooxidans* has been also obtained from SEM observations on pyrite electrodes [21]. In those experiments it was observed that when a large population of bacteria attached to both open circuit and polarized pyrite, bacterial pitting of the pyrite surface was only triggered in the second case. On the other hand, simply application of +0.6 V in sterile conditions did not produce pitting of this sulphide. Considering the results of biooxidation of sulphur presented in this work, we can also assume that in the case of the pyrite, a high redox potential also enhances the ability of *Thiobacillus ferrooxidans* to oxidize sulphur. As sulphur as S^0 is an intermediate state in the oxidation of the reduced sulphur initially present in the lattice, one can postulate that the faster oxidation of elemental sulphur will contribute to the faster oxidation of the reduced sulphur. As the iron dissolution is coupled to this process, the global effect of an increase of the rate of sulphur oxidation by *Thiobacillus ferrooxidans* will be an increase in the rate of iron dissolution, as the one observed in our experiments.

5 Conclusions.

Thiobacillus ferrooxidans biological activity is closely interlinked to the electrochemical phenomena of oxidation of mineral sulphides, elemental sulphur and ferrous ions. The electronic and ionic transport phenomena in the outer and inner boundaries of the bacteria cell wall interferes and catalyses the oxidation reactions. A process of this nature is expected to be dependent on the electrochemical parameters established in the environment in which *Thiobacillus ferrooxidans* is working.

In this context, the experimental results presented in this work demonstrate in particular the importance of the level of the redox potential, and the polarization of the sulphide, on the activity of *Thiobacillus ferrooxidans*.

Bacterial oxidation of sulphur is activated when a high redox potential is established in the solution. When sulphur is in the presence of dissolved ferrous ion, ferrous ion will be first oxidized, and only when a high redox potential is reached, will sulphur oxidation start. This dependence of bacterial activity on electrochemical potential seems also to be relevant in the regulation of the catalytic action of bacteria in the oxidation of pyrite.

Acknowledgements
This work was supported by the University of Chile, FONDECYT Project 91-1219 and SAREC.

7 References

1. Panin, V.V., Karavainko, G.I. and Polkin, S.I. (1985), in **Biotechnology of Metals** (ed. G.I. Karavainko and S.N. Groudev), GKNT United Nations Environment Programme, Moscow, pp 115-125.
2. Silverman, M.P. (1967) Mechanisms of bacterial pyrite oxidation **Journal of Bacteriology**, 94-2, 1046-1051.

3. Landesman, J., Duncan D.W. and Walden C.C. (1966) Oxidation of inorganic sulfur compounds by washed cell suspension of *Thiobacillus ferrooxidans*, **Canadian Journal of Microbiology**, 12, 957-964.
4. Espejo, R.T., Escobar, B., Jedlicki, E., Uribe, P. and Badilla-Ohlbaum R. (1989) Oxidation of ferrous iron and elemental sulfur by *Thiobacillus ferrooxidans*, **Applied Environmental Microbiology**, 54, 1694-1699.
5. Wiertz, J.V. (1993) Ferrous and sulfur oxidation by *Thiobacillus ferrooxidans* in **Biohydrometallurgical Technologies, vol II** (ed. A.E. Torma, M.L. Apel and C.L. Brierley), TMS 1993, pp 463-471
6. Beck, J.V. (1960) A ferrous-iron-oxidizing bacterium. I. Isolation and some general physiological characteristics, **Journal of Bacteriology**, 79, 502-509.
7. Unz, R.F. and Lundgren, D.G. (1961) A comparative nutritional of three chemoautotrophic bacteria *Ferrobacillus ferrooxidans*, *Thiobacillus ferrooxidans* and *Thiobacillus thiooxidans*, **Soil Science**, 92, 302-313.
8. Blake II, R.C. and McGinness, S. (1993) Electron-transfer proteins of bacteria that respire on iron, in **Biohydrometallurgical Technologies, vol II** (ed. A.E. Torma, M.L. Apel and C.L. Brierley), TMS 1993, pp 615-628.
9. Sugio, T., Katagiri, T., Moriyama, M, Zhen, Y.L., Inagaki, K. and Tano, T. (1988) Existence of a new type of sulfite oxidase which utilizes ferric ions as an electron acceptor in *Thiobacillus ferrooxidans*, **Applied Environmental Microbiology**, 54, 153-157.
10. Sugio, T., White, K.J., Shute, E., Choate D. and Blake, R.C. (1992) Existence of a hydrogen sulfide:ferric ion oxidoreductase in iron-oxidizing bacteria, **Applied Environmental Microbiology**, 58, 431-433.
11. Sugio, T., Hirose, T., Oto, A., Inagaki K. and Tano, T. (1989) The regulation of sulfur utilization by ferrous ion in *Thiobacillus ferrooxidans*, in **Biohydrometallurgy 1989** (ed. J. Salley, R.G.L. McGready and P.L. Wichlacz), Canmet, Ottawa, Canada, pp 451-459.
12. Kuenen, J.G., Pronk, J.T., Hazeu, W., Meulenberg R. and Bos, P. (1993) A review of bioenergetics and enzymology of sulfur compound oxidation by acidophilic *Thiobacilli*, in **Biohydrometallurgical Technologies, vol II** (ed. A.E. Torma, M.L. Apel and C.L. Brierley), TMS 1993, pp 487-494.
13. Potter, M.C. (1911) **Proceedings Royal Society** (London), 84B, 260.
14. Ingledew, W.J. (1982) *Thiobacillus ferroxidans*: the bioenergetics of an acidophilic chemolithotroph, **Biochemica Biophysica Acta**, 683, 89-117.
15. Groudev, S.N. (1980) Complex bio-electrochemical oxidation system involved in the leaching of sulphide minerals by *Thiobacillus ferrooxidans*, in **Use of Microorganisms in Hydrometallurgy**, Pecs, Hungary, pp 193-200.
16. Chia, L.M., Choi, W.K., Guay R. and Torma, A.E. (1989) Electrochemical aspects of pyrite oxidation by *Thiobacillus ferrooxidans* during the leaching of a canadian uranium ore, in **Biohydrometallurgy 1989** (ed. J. Salley, R.G.L. McGready and P.L. Wichlacz) Canmet, Ottawa, Canada, pp 35-47.
17. Palencia, I., Wan, R.Y and Miller, J.D. (1991) The electrochemical behavior of a semiconducting natural pyrite in the presence of bacteria, **Metallurgical Transactions**, 22B, 765-774.

18. Metha, A.P. and Murr, L.E. (1983) **Hydrometallurgy**, 9, 235-256.
19. Natarajan, K.A. (1992) Electrobioleaching of base metal sulfides, **Metallurgical Transactions**, 23B, 5-11.
20. Vargas, T., Sanhueza, A. and Escobar, B. (1992) in **EDP Congress 92** (ed. J.P. Hager), TMS 1992, pp 273-284.
21. Vargas, T., Sanhueza, A. and Escobar, B. (1993) Studies of the electrochemical mechanism of bacterial catalysis in pyrite dissolvution, in **Biohydrometallurgical Technologies, vol I** (ed. A.E. Torma, J.E. Wey and V.L. Lakshmanan(, TMS 1993, pp 579-588.
22. Espejo, R.T. and Romero, P. (1987) Growth of *Thiobacillus ferrooxidans* on elemental sulfur, **Applied Environmental Microbiology**, 53, 1907-1912.
23. Peters, E. and Doyle, F. (1989) Leaching and decomposition of sulfide minerals, in **Challenges in Mineral Processing** (ed. K.V.S. Sastry), AIME-SME, pp 509-526.
24. Andrews, G. (1989) An examination of the kinetics of bacterial pyrite decomposition, in **Biotechnology in Minerals and Metal Processing** (ed. B.J. Scheiner, F.M Doyle and S.K. Kawatra) SME, pp 87-93.

Gold

Pretreatment by electrolytic or oxidative leaching for recovery of gold and silver from refractory sulphide concentrates

L. M. Abrantes
M. C. Costa
A. P. Paiva
Departamento de Química, Faculdade de Ciências de Lisboa, Lisbon, Portugal

Abstract
Anodic electroleaching or leaching using acid chloride media can be used to release the fine gold grains locked in a Portuguese arsenopyrite concentrate that exhibits refractory behaviour towards direct oxidative thiourea leaching. The efficiency of leaching or electroleaching processes as a pre-treatment, involving acidic chloride solutions, is evaluated in this work. The dissolution of silver during pre-treatment and its recovery from the liquor using solvent extraction with CYANEX 471X and triphenylphosphine has also been studied and an alternative process minimizing silver dissolution during pre-treatment has been investigated.
Keywords: arsenopyrite, cyclic voltammetry, electroleaching, gold, leaching, silver, solvent extraction.

1 Introduction

The development of hydrometallurgical methods for the recovery of metal values from refractory ores has received a growing interest from both researchers and the mining industry [1-6].

The economic significance of arsenopyrite is associated with its gold (and silver) contents [7,2] but arsenopyrite ores and concentrates are often unsuitable for direct leaching due to the great dissemination of minute size precious metal particles. Gold occurring as discrete grains between arsenopyrite or pyrite crystals and, as such, directly recoverable, ranges from 8 to 30% [8]. Since the major amount of gold is encapsulated, the sulphides must be subjected to pre-treatment in order to obtain satisfactory extraction [9,10]. The most successful hydrometallurgical process for this type of ore or concentrate is sulphuric acid pressure oxidation (O_2 pressure greater than 2200 kPa) which favours subsequent gold cyanidation [10]. Besides the cost of this pre-treatment and, as a result of environmental restrictions, there have been several investigations of noncyanide leaching processes and recently attention has been focused on the use of thiourea, thiosulfate, halogen and ammoniacal lixiviants [11,12]. Biological oxidative processes have also been proposed [13] but their application still seems rather limited for use as a general hydrometallurgical treatment.

As in many other hydrometallurgical operations the leaching or dissolution of metals and their compounds is an electrochemical process - it involves transfer of electrons or ions between the solid and the solution. The rate of such reactions is a function of the concentration of the reactants and of the electrochemical potential. The influence of the concentrations of oxidants and reductants on the rate of dissolution is well documented in published data [14,15], but relatively few studies [2,16] have been devoted to the effect of utilizing a higher potential than freely dissolving conditions, i.e. to evaluate whether an electrolytic leaching process can be used to enhance metal extraction.

The concept of utilizing electrochemical oxidation for metal recovery from refractory ores has been subject of our current research. Previous work on the electrooxidation of non-ferrous sulphide concentrates [17,18] has drawn attention to the possible application of this electrochemical method as a pre-treatment for auriferous arsenopyrite.

The process consists of electrolysis using a suitable leaching solution as the electrolyte and where the gold concentrate acts as a fluidized bed anode. The electrochemically induced reactions modify the complex matrix of the gold host mineral (usually pyrite and arsenopyrite) rendering the precious metal accessible to extraction with common, non-toxic leaching agents, e. g. thiourea solutions. The possibility of utilizing this method was first discussed and compared with other viable alternatives [19]. In addition, the effectiveness of the process for the recovery of gold from pyritic residues showed very promising results [20].

The arsenopyrite concentrate which is the focus of this paper is gold and silver bearing (about 185 and 470 $mgkg^{-1}$, respectively) and therefore an hydrometallurgical route to treat this raw material must be able to give reasonable extraction yields for both precious metals. In the present work an attempt has been made to discuss in parallel the gold and silver recoveries by a process consisting of leaching / electroleaching or pre-conditioning, followed by acidic thiourea leaching.

Taking into account that acid chloride systems show promise for concentrate pre-treatment (as regards gold extraction) and considering the present commercial availability of solvents suitable for silver extraction from these media, the results of an investigation on the solvent extraction of silver from that leach liquor are presented. The alternative route of leaving the major amount of silver in the residue to be recovered, along with gold, during subsequent thiourea leaching is also considered.

2 Experimental

2.1 Electrochemical Characterization

The typical chemical and mineralogical composition of the silver and gold-bearing concentrate used for the present investigation are given in Tables 1 and 2.

Table 1. Chemical composition of the concentrate

Element	Weight (%)	Element	(mg / kg)	Gangue	Weight (%)
Fe	27	Ag	470	SiO_2	8.8
As	16	Au	186	CaO	2.4
S	28			Al_2O_3	1.0
Pb	6				
Zn	4				
Cu	0.30				

Table 2. Mineralogical composition of the concentrate

Mineral	Weight (%)
Arsenopyrite	34.6
Pyrite	37.2
Galena	6.9
Sphalerite	6.0
Chalcopyrite	0.9

Compact disc electrodes were prepared from the concentrate (size range from 15 to 125 μm) by cold pressing (1.47×10^9 Pa) for 15 minutes using graphite powder (20%) as a binder.

Large arsenopyrite, pyrite and galena natural crystals were sawn into ~ 0.5 cm tubes for fabrication of electrodes. Only samples free from visible inclusions, cracks and voids were used.

Compact concentrate disc and mineral slabs were sealed with epoxy resin into "Teflon" holders and exposed surfaces were ground flat on 600 grit silicon carbide paper and polished on 1.0, 0.3 and 0.05 μm alumina.

After rinsing in distilled water the electrodes were installed in a three electrode glass cell where a large platinum sheet was used as the counter electrode and a saturated calomel electrode as the reference.

Electrolyte solutions were prepared from reagent grade chemicals with de-ionized double distilled water. Voltammetric measurements were carried out at room temperature and were performed with an EG&G PAR model 273A

Potentiostat / Galvanostat programmed with a Triudus PC Computer (electrochemical analysis model 270 software). Starting from the open circuit potential with a 50 mVs^{-1} scan rate, the potential domain from hydrogen to oxygen evolution was investigated.

2.2 Leaching / electroleaching procedures

All tests were performed using samples from a single batch of an arsenopyrite concentrate from the north of Portugal (Jales).

Chloride leaching experiments were carried out on the basis of 10% (w/v) of dried concentrate to leaching solutions prepared from a.r. reagents and double distilled water. Different oxidative leaching conditions were obtained at chloride ion concentrations (2.0 and 5.0 M), at varying pH and addition of a 3% hydrogen peroxide solution.

For the experiments in which the ore was subjected to the application of a potential, a two compartment pyrex cell was employed with a graphite rod anode (6.8 cm^2 total geometric area) and a stainless steel cathode (6 cm^2 geometric area). Stirring of the slurry in the anodic compartment with a teflon-covered magnetic bar provided the suspension of the solid particles. An Unilab (Blackburn) power supply was used and the potential (or current) of the anode maintained at the required value.

Thiourea leaching was performed on the residues from the previous treatment using the processing conditions summarized in Table 3.

Table 3. Processing conditions for thiourea leaching

Initial thiourea concentration	20 g / L
Initial acid (H_2SO_4) concentration	49 "
Initial oxidant ($Fe_2(SO_4)_3$) concentration	6 "
Pulp density	20% (w/v)
Temperature	40 ± 1°C
Stirring speed	> 1000 rpm
Reaction time	60 min *

* After this period, the ore concentrate slurry was filtered and subjected to a new thiourea leach for more 20 minutes.

Leaching / electroleaching efficiencies were determined by analysing the liquor for base metals and silver by atomic absorption spectroscopy (AAS) (Pye Unicam SP9 model associated to a Philips model 910 Computer) and for gold and arsenic by inductively coupled plasma (ICP) spectroscopy (ARL - Fison 3410). Matrix matched standards were prepared for the respective calibrations.

2.3 Solvent extraction experiments
Cyanex 471X (triisobutylphosphine sulphide - TIBPS - kindly supplied by the American Cyanamid Company) and triphenylphosphine - TPP (Merck, 98% purity), used as received were tested for extraction of silver from leaching solutions. The extractants were dissolved in 1,2-dichloroethane (4.5×10^{-2}M). Equal volumes of organic and aqueous phases were agitated in a thermostated reactor (25 ± 0.1°C). After an adequate time to reach the equilibrium (15 minutes) [21] the phases were separated and the aqueous removed for analysis. The organic phase metal content was deduced by mass balance.

The stripping of silver was achieved contacting the loaded organic phase with an appropriate solution of sodium thiosulphate [22], under constant temperature (25± 0.1°C) for 15 minutes.

Aqueous solutions were analysed using AAS.

3 Results and Discussion

This investigation was directed toward the recovery of silver and gold from concentrates containing primarily pyrite and arsenopyrite in an integrated flow sheet. As stated in the introduction electrooxidation in chloride media (2M) proved to be an adequate concentrate pre-treatment to enhance gold extraction. However, in order to avoid silver leaching limitation due to its low solubility in dilute chloride media the initial electrolyte selected for this work was HCl 1M + NaCl 4M.

3.1 Cyclic voltammetry
The oxidation (and reduction) reactions occurring at the concentrate / solution interface are complex due to the formation of a variety of soluble and solid products. An useful approach to obtain information on the electrochemical behaviour of the material under the leaching conditions chosen is the cyclic voltammetry technique. It was also used to characterize separately the behaviour of major mineral components in the ore concentrate in view of establishing the influence of the different mineralogical species on the electrochemistry of the system.

The voltammetric responses obtained at 50 mVs^{-1} for stationary electrodes in quiescent electrolyte are represented in figures 1 (concentrate electrode) and 2 (pure mineral electrodes). The scan was made in the anodic direction, starting from the displayed open circuit potential.

The highly oxidative nature of the electrolyte promotes some dissolution of pyrite and arsenopyrite mineral as well as of the concentrate, which was detected by the rapid increase in the open circuit potential.

This dissolution produces non-stoichiometric sulphides e.g. $Fe_{1-x}As_{1-y}S$ and $Fe_{1-x}S_{2-y}$ and iron (II) species, most probably chlorocomplexes, due to the high concentration of chloride ions in the electrolyte. Therefore, the broad anodic peaks detected at about +0.5 V (SCE) at the concentrate, (Figure 1), arsenopyrite and pyrite minerals surfaces, (Figure 2a and b) along with their corresponding reduction waves are accounted for Fe(II)/Fe(III) redox reactions. The similarity of behaviour for the arsenopyrite and pyrite concentrate electrodes is also found at higher anodic potentials where a considerable increase in the current is observed. In the case of pyrite this has been attributed to a transpassive oxidation with formation of S^0, S_x^{2-} and $S_xO_6^{2-}$ [6,16].

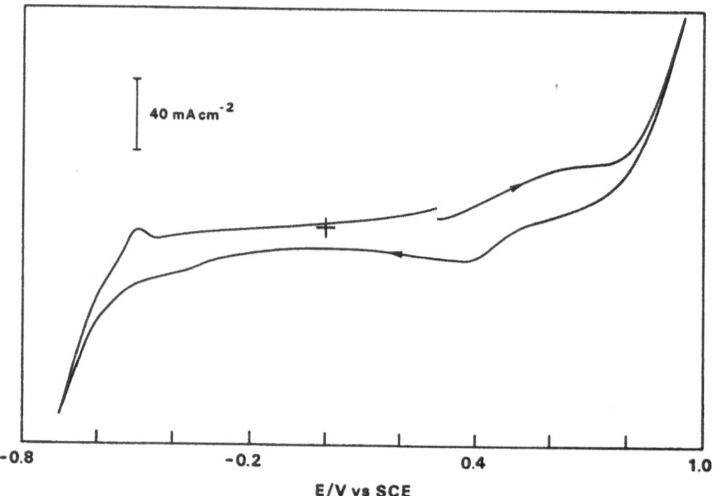

Fig. 1 Cyclic voltammogram of the concentrate in HCl 1M + NaCl 4M; v = 50 mVs^{-1}

Although the reduction of these products can only be observed in the cyclic voltammogram obtained for pyrite, at about -0.2 V (SCE) on the return cathodic peak, identical electrochemical reaction can be assumed to take place with arsenopyrite [16,23] and thus in the concentrate.

The anodic peak noticeable in the voltammetric response of the concentrate at about -0.5 V (SCE) is associated with the presence of galena in the raw material as can be easily concluded by comparing the PbS electrode current response - Figure 2c. In this case, the potential anodic scan started from -0.35 V (SCE) and the oxidation of the mineral is seen to take place at about +0.5 V (SCE); elemental sulphur rather than polysulphide or sulphoxy anions is considered to be generated during the oxidation [24].

Again the lead oxidation product is readily complexed by the chloride ions. Upon scan reversal and further in the cathodic potential domain S^0 and galena reduction take place with H_2S and Pb^0 formation. On the return positive direction sweep, the anodic current presents a slight shoulder before the main peak. It is reasonable to assume the dissolution of lead involving H_2S but mainly chloride ions from the electrolyte, as the associated oxidation reactions.

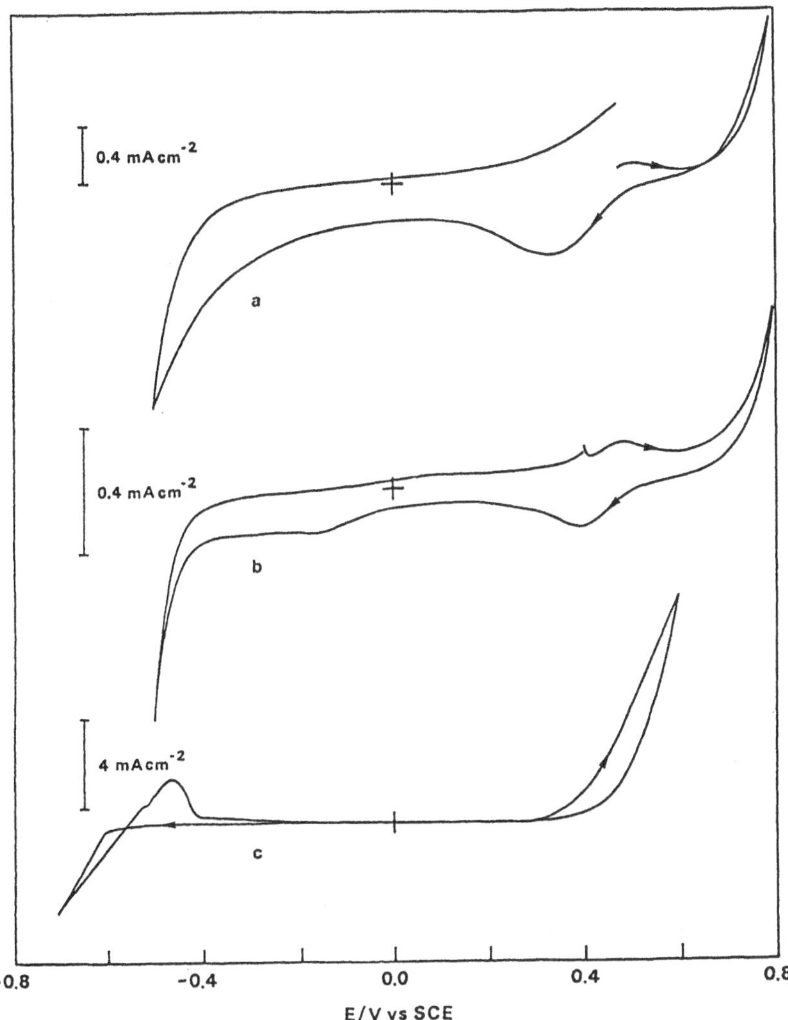

Fig. 2 Cyclic voltammograms of arsenopyrite (a), pyrite (b) and galena (c), pure minerals in HCl 1M + NaCl 4M; $v = 50\ mVs^{-1}$.

3.2 Silver extraction

The following data concerns the behaviour of the system with respect to silver extraction; different oxidative conditions, achieved by either the addition of hydrogen peroxide or by applying an anodic potential, are discussed independently of the eventual effect on the recovery of gold, which gave essentially the same response as in 2M chloride solution and will be discussed later.

Comparative leaching and electroleaching experiments carried out with the selected electrolyte (1M HCl + 4M NaCl) for 3 hours at room temperature provided solutions with the composition shown in Table 4.

Table 4. Dissolved elements after leaching / electroleaching of the concentrate (fluidized bed anode, E_a > 850 mV (SCE); HCl 1M + NaCl 4M electrolyte)

Element	Leaching (exp 1) (g / L)	Electroleaching (exp 2) (g / L)
Fe	0.73	0.93
Pb	4.10	8.60
Zn	0.12	0.17
Cu	9×10^{-3}	5×10^{-3}
Ag	10×10^{-3}	6×10^{-3}
As	0.32	0.17

Analysis of the table suggests that the electroleaching favours the dissolution of iron, lead and zinc over the leaching process whereas for the other metals an inverse tendency is seen. A plausible explanation is the occurrence of a passivation phenomenon which can be attributed to the amorphous S^0 formed by galena oxidation, in agreement with the electrochemical behaviour analysis presented in the previous section. In this case at least, a considerable amount of the silver contained in the concentrate is associated with the galena mineralogical species.

With the aim of achieving a better silver extraction during treatment of this type of concentrate the oxidative ability of the electrolyte was increased by the addition of hydrogen peroxide to the former solution and increase in temperature to 40 $\pm$ 2°C. The parallel electroleaching test employed 0.95 V (SCE) anodic potential.

Table 5 presents the metal content in the pregnant solutions obtained.

The same trends are observed by comparison of data in Tables 4 and 5 with the exception of iron dissolution which has been significantly enhanced, probably due to the higher anodic potential employed in the latter set of experiments.

Considering silver extraction, the use of a strong chemical oxidant media and increased temperature appear to be important factors. Table 6 compares the silver recovery efficiencies in the above leaching / electroleaching experiments.

Table 5. Dissolved elements after leaching / electroleaching of the concentrate (fluidized bed anode, E_a = 950 mV (SCE); HCl 1M + NaCl 4M + 3% H_2O_2 electrolyte)

Element	Leaching (exp 3) (g / L)	Electroleaching (exp 4) (g / L)
Fe	0.75	2.81
Pb	5.30	6.90
Zn	0.22	0.22
Cu	23×10^{-3}	29×10^{-3}
Ag	25×10^{-3}	22×10^{-3}
As	0.21	0.41

Table 6. Silver extraction by leaching / electroleaching

Experiment	Silver recovery (%)
1	22
2	13
3	59
4	51

In spite of the refractory character presented by the concentrate toward silver dissolution (it was also confirmed for all cases that the gold content in the electrolyte was below the limit of analysis detection, i.e. < 0.1 mg / L) a study of the application of solvent extraction for recovering silver free from impurities solution was carried out.

The choice of phosphorous extractants was based on previous research for silver recovery from chloride media [21,25,33]. Under the conditions described in the experimental section, the results obtained for silver extraction from liquor obtained in experiment 4 with TIBPS and TPP solvents are displayed in Table 7.

Table 7. Comparison of the efficiencies of silver extraction by TIBPS and TPP (feed solution from experiment 4)

Element	% of extraction	
	TIBPS	TPP
Ag	93.6	95.8
Fe	0.4	1.9
Pb	<0.1	<0.1
Cu	31.3	98.9

Both extractants exhibit good efficiency toward silver but TPP presents no selectivity. The relative high extraction of copper by both solvents is explained by the fact that all the copper in solution exists as Cu(I) species. It is known that the higher the concentration of chloride ion, the lower is the fraction of cupric species in the system [26]. $CuCl_2^-$, $CuCl_3^{2-}$ and $CuCl_4^{3-}$ compete with homologous silver chlorocomplexes for extraction with these ligands.

It is also worthwhile to point out that due to the high concentration of iron in the aqueous feed solution, considerable contamination of the solvent results, even with an extraction efficiency lower than 1% toward that element.

The impurity levels of copper and iron in the silver strip solution can be seen in Table 8, revealing the innefficiency of both extractants for purifying the leach solution. Further confirmation of this behaviour is given by Table 9 where extraction and stripping results observed in tests carried out with aqueous feed solution 3 and TIBPS are shown.

Table 8. Stripping efficiency from TIBPS and TPP loaded solutions (also shown are the metal contents in the strip liquor (mg / L))

Element	% of stripping	
	TIBPS	TPP
Ag	97.2 (20)	99.9 (21)
Fe	99.0 (16)	99.0 (54)
Pb	<0.1	<0.1
Cu	86.6 (9)	89.6 (28)

Table 9. Purification of leaching solution 3 by TIBPS (also shown are the metal concentrations in the strip liquor (mg / L))

Element	% of extraction	% of stripping
Ag	88.7	37.8 (8)
Fe	7.4	18.3 (10)
Pb	6.0	2.9 (9)
Cu	8.1	26.6 (0.5)

As a consequence of the lower iron content in aqueous feed solution 3, silver extraction by TIBPS decreases due to the absence of the synergistic effect of iron on silver extraction [21] which appears to apply also to the stripping process with the thiosulphate solution (38 vs. 97% silver recovery).

The unexpectedly poor performance of both extractants tested for removing impurities and producing an adequate silver strip liquor indicates that concentrate pre-treatment where most of the silver content remains in the residue, accessible for subsequent leaching (e.g. with thiourea) is a better route for recovery of this precious metal.

To evaluate this alternative, the residues of experiments 1 and 2 were subjected to acidic thiourea leaching. The metal dissolved was about 47 and 33 %, respectively. Taking into account that, depending on the ratio of thiourea and silver concentrations, several silver complexes can be formed in acidic thiourea solutions [27] some of which may also precipitate on the silver surface causing early passivation [28], these results are indeed susceptible to improvement and thus support the concept of using thiourea leaching.

Attempts to remove silver from thiourea solution by extraction with solvating extractants were unsuccessful. Cationic silver (I) thiourea complexes e.g. $Ag(NH_2CSNH_2)_n^+$ (n=1 to 4) and $Ag_2(NH_2CSNH_2)_m^{2+}$ (m=3 and 6) are stable [28] and other methods such as the use of resins or of granular activated carbon could be viable alternatives.

3.3 Gold extraction

The ore concentrate is refractory towards direct thiourea leaching. The recovery of gold was only 12.6% using the thiourea leaching procedure.

To proceed with the investigation of a pre-treatment to enhance the recovery of the valuable metal, a HCl 0.1M + NaCl 1.9 M electrolyte was selected. The use of higher chloride concentration solutions does not present any advantage as discussed in the previous section.

The effect of the pre-leaching treatment can be seen from the results presented in Tables 10 and 11. Gold recovery by thiourea leaching from the pre-treated concentrate has significantly increased.

Table 10. Dissolved elements after leaching of the concentrate in HCl 0.1M + NaCl 1.9M

Element	Concentration in the filtrate (g / L)
Fe	0.26
Pb	1.63
Zn	0.05
Ag	2×10^{-3}
Au	$<0.1 \times 10^{-3}$
As	0.18

A preliminary test on the advantage of an electroleaching pre-treatment (applied anodic potential E=1.15 V (SCE)) revealed that at least an improvement of 20% on gold recovery could be achieved, with no effect on the amount of silver dissolved in the presence of thiourea solution.

Table 11. Extraction by thiourea leaching of the pre-treated concentrate

Element	Total recovery (%)
Au	42
Ag	44

The efficiency of the electroleaching processes depends on the nature of the concentrate and on the correct choice of the experimental parameters, namely the applied potential / current [20, 29]. "Too high" an anodic potential can promote the formation of films with passivating properties, but the choice depends on the mineralogical composition. The presence of galena, for instance, can be determinant since its oxidation occurs at lower potentials than pyrite and arsenopyrite and gives rise to a passive layer of S^0 [24].

An electrooxidation carried out at i = 44A m^{-2} (E_a = 1.05 V vs. SCE) for two hours in the same electrolyte for another batch of concentrate from the same source gave over 80% gold recovery by thiourea leaching.

Considering the literature on the recovery of gold from roasted arsenopyrite by acidic thiourea leaching [30-32] and the results obtained with electroreacted concentrate, the electroleaching process must be considered a serious alternative pre-treatment for this kind of ore concentrates.

It must be pointed out that for the potential conditions used in this work, electrolytic in-situ generation of chlorine or oxychlorine species is unlikely and so an indirect chemical contribution to the reported effect can hardly be considered. Therefore, electrodissolution refers only to the ore participation as anode.

4 Conclusions

Electrochemically induced reactions can be used to enhance the response to hydrometallurgical treatment of gold bearing complex sulphide ores. By increasing the rate and improving the extent of dissolution of other constituents (e.g. Fe and As) the precious metal is released and can thereafter be readily leached in an appropriate media.

With respect to the extraction of silver, which often accompanies gold concentrates, any real improvement seems to be accomplished only by electrochemical pre-treatment. However, the best experimental conditions for gold extraction will also allow, at least, 50% silver recovery which compares favourably with direct chemical leaching, specially due to the difficulties in purifying the resulting electrolyte. Actually, the lack of selectivity showed by both extractants for the recovery of silver from chloride media compromises the use of solvent extraction at that stage of the process. As an alternative, the recovery of gold and silver from thiourea solutions may be considered.

5 References

1. Sanberg, R.G. and Huiatt, J.L. (1986) Ferric chloride, thiourea and brine leach recovery of Ag, Au and Pb from complex sulfides. **Journal of Metals**, June, 18-22.
2. Kirk, D.W., Gehring, R. and Graydon, W.F. (1987) Electrochemical leaching of a silver arsenopyrite ore. **Hydrometallurgy**, 17, 155-166.
3. Chandar, S. and Briceno, A. (1987) Kinetics of pyrite oxidation. **Minerals and Metallurgical Processing**, August, 171-176.
4. Dalton, R.F., Diaz, G., Price, R. and Zunkel, A.D. (1991) The Cuprex metal extraction process: recovering copper from sulfide ores. **Journal of the Minerals, Metals & Materials Society**, August, 51-56.
5. Dutrizac, J.E. (1992) The leaching of sulphide minerals in chloride media. **Hydrometallurgy**, 29, 1-45.
6. Zhou, X., Li, J., Bodily, D.M. and Wadsworth, M.E. (1993) Transpassive oxidation of pyrite. **Journal of the Electrochemical Society**, 140(7), 1927-1935.
7. Beattie, M.J. and Poling, G.W. (1987) A study of the surface oxidation of arsenopyrite using cyclic voltammetry. **International Journal of Mineral Processing**, 20, 87-108.
8. Collins, M.J., Berezowski, R.N. and Vardilt, W.D. (1993) The Lihir gold project: pressure oxidation process development, in **Milton Wadsworth International Symposium on Hydrometallurgy**, (eds.J. Hiskey and G. Warren) SME-AIME, Littleton, Co, 1993, pp. 611-627.
9. Sanchez, V.M. and Hiskey, J.B. (1991) Electrochemical behaviour of arsenopyrite in alkaline media. **Minerals and Metallurgical Processing**, February, 1-6.
10. Hanlen, Y and Guangxiang, X. (1993) A pre-treatment of pyritic gold concentrate with catalytic oxidation acid leach, in **Milton Wadsworth International Symposium on Hydrometallurgy**, (eds. J. Hiskey and G. Warren) SME-AIME, Littleton, Co, 1993, pp. 365-375.
11. Qi, P.H. and Hiskey, J.B. (1991) Dissolution Kinetics of gold in iodide solutions. **Hydrometallurgy**, 27(1), 47-62.
12. Han, K. N. and Meng, X. (1992) Extraction of gold/silver from refractory ores using ammoniacal solutions, in **Proceedings of Randol Gold Forum**, Randol International, 1992, pp. 213-218.
13. Lawrence, R.W. (1990) **Microbial Mineral Recovery**. McGraw-Hill Publishing Co., New York, USA.
14. Koslides, T. and Ciminelli, V.S. (1992) Pressure oxidation of arsenopyrite and pyrite in alkaline solutions. **Hydrometallurgy**, 30, 87-106.
15. Kumar, R., Das, S., Ray, R.K. and Biswas, A.K. (1993) Leaching of a pure and cobalt bearing goethites in sulphurous acid: kinetics and mechanisms. **Hydrometallurgy**, 32, 39-59.

16. Hamilton, I.C. and Woods, R. (1981) An investigation of surface oxidation of pyrite and pyrrotite by linear potential sweep voltammetry. **Journal of Electroanalytical Chemistry,** 118, 327-343.
17. Abrantes, L.M. and Araújo, L.V. (1993) An electrochemical study of the behaviour of chalcopyrite in acid and alkaline solutions. **Portugaliae Electrochimica Acta,** 11, 67-71.
18. Abrantes, L.M. and Araújo, L.V. (1993) Anodic dissolution of chalcopyrite concentrate in aqueous chloride solutions, in **Milton Wadsworth International Symposium on Hydrometallurgy,** (eds. J. Hiskey and G. Warren) SME-AIME, Littleton, Co, 1993, pp. 957-969.
19. Abrantes, L.M., Silva, L.M. and Costa, M.C. (1992) A hidrometalurgia na recuperação do ouro. **Boletim de Minas,** 29(4), 367-374.
20. Santos, L.S., Tavares, P.F., Costa, M.C. and Abrantes, L.M. (1993) Recuperação de ouro de resíduos piriticos, in **Proceedings of the Symposium on the Polymetallic Sulphides on the Iberian Pyrite Belt,** (eds Apimineral), Portugal, 1993, pp. 491-509.
21. Paiva, A.P. and Abrantes, L.M. (1993) Solvent extraction on silver recovery +from chloride leach solutions, in **Proceedings of the 4th Extraction and Processing Division Congress, 122nd TMS Annual Meeting,** (ed. J. P. Hager), 1993, pp.157-168.
22. American Cyanamid Company (1985) Cyanex 471X Extractant. **Technical Information on Cyanex 471X extractant.**
23. Gardner, J.R. and Woods, R. (1979) An electrochemical investigation of the natural flotability of chalcopyrite. **International Journal of Mineral Processing,** 6, 1-16.
24. Pretzker, M.D. and Yoon, R.H. (1988) A voltammetric study of galena immersed in acetate solution at pH 4.6. **Journal of Applied Electrochemistry,** 18, 323-332.
25. Paiva, A.P. and Abrantes, L.M. (1993) Molecular features in the solvent extraction of silver, in **Proceedings of the International Solvent Extraction Conference ISEC,** (eds. D. H. Logsdail and M. J. Slater) 3, 1993, pp. 1377-1382.
26. Kin, H.K., Wu, X.J. and Rao, P.D. (1991) The electrowinning of copper from a cupric chloride solution. **Journal of the Minerals, Metals & Materials Society,** August, 60-65.
27. Kuz'mina, N.N. and Sougina, O.A. (1963) Properties of the thiourea complexes of silver formed during amperometric titration. *Zhurnal Analiticheskoy Khimii,* 18, 323-328.
28. Huyhua, J.C., Zegarra, C.R. and Gundiler, I.H. (1988) A comparative study of oxidants on gold and silver dissolution in acidic thiourea solutions, in **Precious Metals '89,** (eds. M. C. Jha and S. D. Hill), 1988, pp. 287-303.
29. Abrantes, L.M. and Costa, M.C. (1994) Electrooxidation as a pretreatment for gold recovery. Abstract accepted by the **5th Extraction and Processing Division Congress, 123rd TMS Annual Meeting,** 1994, S. Francisco, USA.

30. Kirk, D.W., Gehring, R. and Graydon, W.F. (1987) Electrochemical leaching of a silver arsenopyrite ore. **Hydrometallurgy**, 17, 155-162.

31. Cabra, G. (1984) A kinetic study of the leaching of gold from pyrite concentrate using acidified thiourea , in **Precious Metals: Mining, Extraction and Processing**, (eds V. Kudryk, D. A. Corrigan and W. W. Liang) TMS-AIME, 1984, pp. 145-172.

32. Mossoulos, L., Potamianos, N. and Kontopoulos, A. (1984) Recovery of gold and silver from arseniferous pyrite cinders by acidic thiourea leaching, in **Precious Metals: Mining, Extraction and Processing**, (eds V. Kudryk, D. A. Corrigan and W. W. Liang) TMS-AIME, 1984, pp. 323-335.

33. Abe, Y., Flett, D.S. (1990) Solvent extraction of silver from chloride solutions by Cyanex 471X, in **Proceedings of the International Solvent Extraction Conference ISEC**, (ed.T. Sekine) 1990, pp. 1127 - 1132.

6 Acknowledgements

The authors wish to thank to E.D.M. for providing the arsenopyrite concentrate, to JNICT for the Ph. D. financial support to M.C.C. and A.P.P. and to Quimitécnica for the ICP analyses

Iodide–thiocyanate leaching system for gold

O. Barbosa-Filho
Department of Materials Science and Metallurgy, Catholic University, Rio de Janeiro, Brazil
A. J. Monhemius
Department of Mineral Resources Engineering, Royal School of Mines, Imperial College of Science, Technology and Medicine, London, England

Abstract

The leaching of gold by thiocyanate solutions is performed at pHs between 1 and 2, which allows the use of iron(III) as an oxidizing agent. The mechanism of dissolution of gold by iron(III)-thiocyanate solutions is directly linked with the autoreduction process, in which Fe^{III} is spontaneously reduced to Fe^{II} while oxidizing SCN^-. This oxidation proceeds through the formation of several intermediate species, particularly $(SCN)_2$ and $(SCN)_3^-$, which act both as oxidants and, upon reduction, as complexants for gold. The production of $(SCN)_2$ and $(SCN)_3^-$ must be continuous due to their fast decomposition by hydrolysis. The instability of these intermediate species towards hydrolysis is a major drawback of the iron(III)-thiocyanate leaching system, which can be overcome by additions of small amounts of I^- and/or I_2. Experiments revealed a synergistic effect of the thiocyanate-iodide mixture on the dissolution of gold. The dissolution rates in the mixed system were substantially higher than those obtained when either iron(III)-thiocyanate or iron(III)-iodide were used separately. The synergistic effect was attributed to the formation of relatively stable mixed iodine-thiocyanate species such as I_2SCN^- and $I(SCN)_2^-$, which participate in the mechanism of dissolution of gold.

Keywords: Gold leaching, thiocyanate, iodide, synergistic leaching, mechanism, mixed halide complexes.

1 Introduction

The effectiveness of thiocyanate for dissolving gold in the presence of a suitable oxidizing agent was first demonstrated in 1905 by White [1]. However, it was not until 1986 that research on the subject was resumed by Fleming [2]. A thermodynamic study of the thiocyanate system for leaching gold and silver ores was published by Barbosa-Filho and Monhemius in 1989 [3]; this was part of a more extensive investigation into this system, which was concluded in 1991 [4-6]. A further contribution on the thermodynamics of the iron(III)-thiocyanate system was published recently by Broadhurst and du Preez [7].

Thiocyanate is a technically viable and interesting lixiviant for gold. Leaching is performed between pH 1 and 2, which allows the use of iron(III) as an oxidizing agent. The mechanism of dissolution of gold by iron(III)-thiocyanate solutions is directly linked with the autoreduction process, in which Fe^{III} is spontaneously reduced to Fe^{II} while oxidizing SCN^-. This oxidation proceeds through the formation of several intermediate species, particularly $(SCN)_3^-$ and $(SCN)_2$, which act both as oxidants and, upon reduction, as complexants for gold. The formation of $(SCN)_3^-$ and $(SCN)_2$ must be continuous due to their relatively fast decomposition by hydrolysis into more stable oxidation products [4-6]. The instability of $(SCN)_3^-$ and $(SCN)_2$ towards hydrolysis is a major drawback of the iron(III)-thiocyanate leaching system. Experimental results with this system at 25°C yielded initial gold dissolution rates of the order of 10^{-10} mol/cm^{-2} s^{-1}. Rates as high as 10^{-9} or even 10^{-8} mol/cm^{-2} s^{-1}, depending on reagent concentrations, could be obtained by raising the temperature to around 85°C [4, 6], but this was seen as a limitation to the commercial use of thiocyanate as a gold lixiviant, as it would add costs to conventional agitated leaching.

Further research [4] was then oriented towards finding a method of increasing the gold dissolution rate at ambient temperature. Additions of small amounts of iodide or iodine were very successful in bringing about the desired results. The rates of gold leaching by iron(III)-thiocyanate solutions with the addition of small amounts of iodide/iodine at 25°C were comparable to those obtained at 85°C with the simple ferric thiocyanate solutions [4, 6]. The initial suggestion of adding iodine to the thiocyanate leaching system was made by Fleming [8].

2 Previous research on the iodide leaching of gold

Early applications of the iodide/iodine system were intended to recover gold from electronic or other scrap materials [9-12]. A process for the hydrometallurgical recovery of gold was patented by McGrew and Murphy [13], in which the lixiviant was a mixture of iodide ions and elemental iodine, the latter acting as an oxidizing agent. Gold was subsequently recovered on activated charcoal and the excess iodide formed during the process was reoxidized electrochemically to iodine in a special diaphragm cell. A study demonstrating the economic feasibility of the iodide/iodine system for in-situ leaching was published by Jacobson et al. [14] in 1987.

In a recent review of the dissolution chemistry of gold and silver in different lixiviants, Hiskey and Atluri [15] showed that both the aurous, AuI_2^-, and the auric, AuI_4^-, complexes can be formed with iodide. The aurous complex is the most likely to be found at redox potentials typical of iodine solutions, whose main oxidizing species are I_2 and I_3^-. Qi and Hiskey studied the electrochemistry and the kinetics and mechanism of the iodide leaching of gold [16-18]. These authors considered that the triiodide ion, I_3^-, was the oxidizing agent and found that the gold dissolution process was first order with respect to this species and half order with respect to I^-. A comparison of gold leaching between iodide and cyanide was also presented, in which a rate of about 2.6×10^{-9} mol/cm^2 s was obtained with 10^{-2} mol/dm^3 NaI and 5×10^{-3} mol/dm^3 I_2. This value is similar to those found in cyanidation experiments [17].

The iodide leaching of gold was also investigated recently by Davis and Tran [19]. Three different oxidizing agents were used: iodine, hypochlorite and hydrogen peroxide. Elemental iodine was found to be the best oxidant over the pH range from 2.7 to 11.5.

The optimum iodine-iodide mixture of 0.10 mol/dm^3 KI and 0.040 mol/dm^3 iodine could dissolve gold at a rate of 1.56×10^{-8} mol/cm^2 s from pH 2.7 to 10. They emphasized the importance of optimizing the iodine-iodide mole ratio for any given leaching conditions.

3 Some comparisons between the Au-I-H_2O and the Au-SCN-H_2O systems

Elemental iodine is almost insoluble in pure water, but its solubility increases in the presence of iodide ions. The dissolution process takes place through the formation of polyiodide ions such as I_3^-, I_4^{2-}, I_5^- and I_6^{2-}, besides aqueous I_2 [17]. Species higher than I_3^- are not relevant in the present case, as the iodide concentrations used in the leaching experiments are between 10^{-3} and 10^{-1} mol/dm^3. The relevant equilibrium is thus:

$$I_{2(aq)} + I^- = I_3^- \quad (1)$$

An advantage of the iodine system when compared with the thiocyanate system is that the oxidised forms I_2 and I_3^- are stable in water, which is not the case for the equivalent species, $(SCN)_2$ and $(SCN)_3^-$, which both readily hydrolyse and decompose to sulphate and hydrogen cyanide.

The (pseudo)halide character of thiocyanate lies between those of iodide and bromide. This is indicated by the standard potentials of several redox couples, listed in Table 1. Both bromide and iodide are lixiviants for gold and it is interesting to compare the Eh-pH diagrams published by Hiskey [15] for these two systems with that calculated for the gold-thiocyanate-water system [3, 4], as shown in Figure 1. The three diagrams are very similar, emphasizing the halide character of thiocyanate.

Table 1 Standard potentials for half-reactions of several bromide, iodide and thiocyanate redox couples in aqueous solution at 25°C.

Half-Reaction	E° (V)	Reference
$I_{2(aq)} + 2e^- = 2I^-_{(aq)}$	0.621	[21],[26]
$I_3^-{}_{(aq)} + 2e^- = 3I^-_{(aq)}$	0.536	[21],[26]
$(SCN)_{2(aq)} + 2e^- = 2SCN^-_{(aq)}$	0.77	[22],[23]
$(SCN)_3^-{}_{(aq)} + 2e^- = 3SCN^-_{(aq)}$	0.68	[24]
$Br_{2(aq)} + 2e^- = 2Br^-_{(aq)}$	1.0874	[21],[25]
$Br_3^-{}_{(aq)} + 2e^- = 3Br^-_{(aq)}$	1.0503	[21],[25]

It may be seen from Figure 1 that the potentials corresponding to the lines between elemental gold and the aurous complexes follow the same order as those for the redox couples in Table 1, i.e. $AuI_2^- < Au(SCN)_2^- < AuBr_2^-$. The reason for adding iodide to the iron(III)-thiocyanate lixiviant now becomes apparent. Table 1 and Figure 1 show that, even when the potentials are too low to permit the formation of $(SCN)_3^-$, iodide species such as I_2 and I_3^- are still stable in solution and, coincidentially, the complex

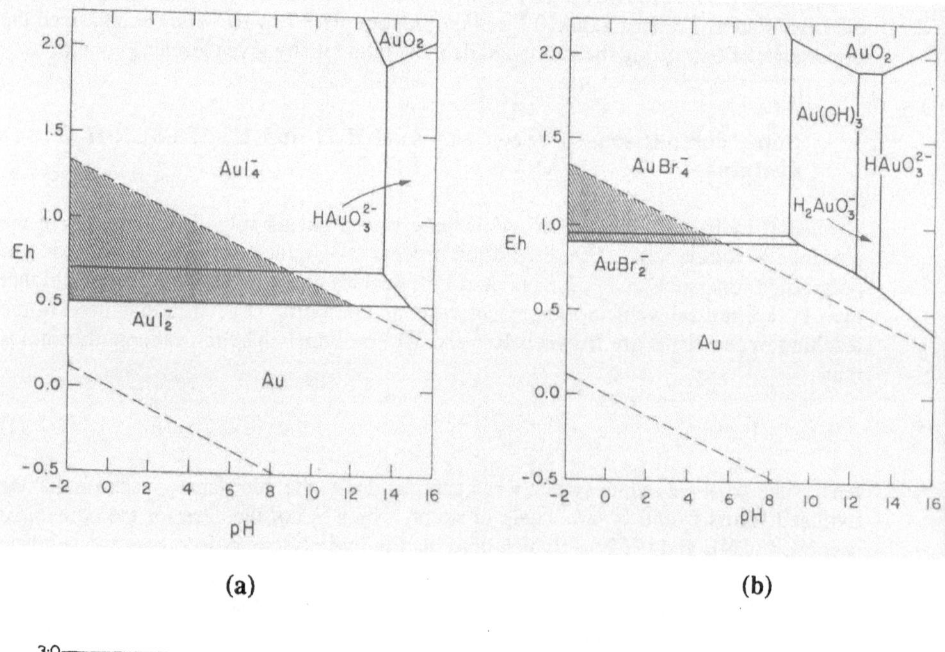

(a) (b)

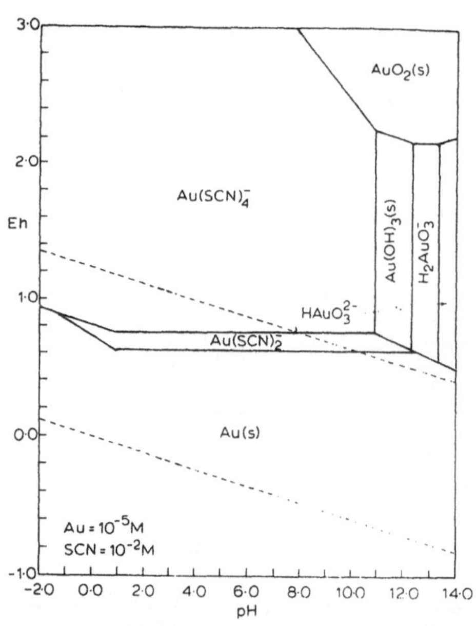

(c)

Figure 1 Eh-pH diagrams at 25°C for (a) $Au-I-H_2O$ system [15], (b) $Au-Br-H_2O$ system [15], (c) $Au-SCN-H_2O$ system [3,4]. In all cases the ionic activities are: {Au} = 10^{-5}M and {L-} = 10^{-2}M

AuI_2^- is also stable. Hence the presence of the iodide species in solution extends the potential range in which the dissolution of gold is thermodynamically feasible to values as low as *c.a.* 500 mV.

4 Mixed iodide-thiocyanate chemistry

The interaction between thiocyanate and iodine has received attention in the chemical literature [27-30]. As stated above, iodine is poorly soluble in water, but becomes readily soluble if iodide ions are present in solution, yielding equilibria such as that of reaction (1). Similarly, rather large quantities of iodine can be dissolved in aqueous potassium thiocyanate to give solutions which are yellow or orange in colour. These solutions are not very stable and undergo slow oxidation of SCN^- to HCN and SO_4^{2-}, over several hours. The overall process is irreversible and, in acidic media, corresponds to the reaction:

$$SCN^- + 3I_2 + 4H_2O = SO_4^{2-} + 6I^- + HCN + 7H^+ \qquad (2)$$

The rate of this reaction is not sufficiently high to account for the speed with which iodine dissolves in thiocyanate solutions. This observation led to the postulation of relatively stable intermediate species, particularly I_2SCN^- and ISCN [27].

Later work demonstrated that the complex I_2SCN^- is the predominant species [28, 30]. The following reversible equilibrium was found to be quickly established upon dissolution of iodine in acidic thiocyanate solutions:

$$I_{2(aq)} + SCN^- = I_2SCN^- \qquad (3)$$

A slightly different situation arose when iodide ions were added to acidic solutions containing thiocyanogen, $(SCN)_2$ [29]:

$$(SCN)_{2(aq)} + I^- = I(SCN)_2^- \qquad (4)$$

The species $I(SCN)_2^-$ has u.v. absorption characteristics which are very similar to I_2SCN^-, both showing absorption peaks at 303 nm [28-30]. These two species can be viewed as members of a substitution series whose extremes are I_3^- and $(SCN)_3^-$.

Both I_2SCN^- and $I(SCN)_2^-$ undergo irreversible decomposition, presumably to sulphate and other products of thiocyanate oxidation. The factors affecting the decomposition of both species are similar. Their rates of decomposition are greatly reduced in the temperature range 0-10°C and their stabilities also increase with ionic strength and H^+ concentration. Thus, with temperatures between 0 and 10°C, ionic strengths between 1 and 5 and H^+ concentrations from 1 to 2 mol/dm^3, the intensity of the absorption peak at 303 nm was observed to be practically constant (less than 0.1% decrease in 1 hour). Even at room temperature and with an acid concentration of 0.1

mol/dm^3, which are typical leaching conditions, the decrease in the intensity of the u.v. absorption peak of I_2SCN^- was less than 1% in 1 hour [28, 29]. These observations reveal a remarkable stability compared to that of $(SCN)_2$, whose hydrolysis is so fast that it is necessary to use organic solvents in order to observe it experimentally.

The foregoing discussion can be summed up in terms of the reactions given in Table 2 with their respective equilibrium constants. The value log K = 4.10 for reaction (4), which corresponds to an equilibrium constant of 1.25×10^4, indicates the propensity of I^- to stabilize $(SCN)_2$ through complexation.

Although I_2SCN^- and $I(SCN)_2^-$ are the main mixed iodide-thiocyanate species found in solution, they are not the only ones. Several other species may be present; ISCN is one of such species described in the literature [27-30].

Table 2 Equilibria involving iodine-thiocyanate species at 25°C in aqueous solution.

Reaction	Number	log K	Reference
$I_{2(aq)} + I^- = I_3^-$	(1)	2.86	[31],[34]
$I_{2(aq)} + SCN^- = I_2SCN^-$	(3)	1.93	[29]
$(SCN)_{2(aq)} + I^- = I(SCN)_2^-$	(4)	4.10	[28]
$(SCN)_{2(aq)} + SCN^- = (SCN)_3^-$	(5)	3.04	[4]

5 Experimental procedure

Gold dissolution experiments were performed using an Oxford Rotating Disc System Model MC5/88. The gold disc had a diameter of 1.4 cm and was made of 99.99% pure gold. The arrangement included a pH meter fitted with a glass electrode and an Eh meter fitted with a platinum electrode and an Ag/AgCl reference electrode. The glass reactor used had a capacity of 250 cm^3, was flushed with nitrogen and was immersed in a thermostatically controlled water bath. Iron(III) was the oxidizing agent used in all experiments, both with iodide and thiocyanate, which were added as the potassium salts. Gold in solution was analysed by atomic absorption spectrophotometry. Iron(II) was determined spectrophotometrically [32]. Iron(III) (in the stock solution), thiocyanate and iodide were determined by conventional volumetric techniques [20]. Details of the experimental arrangement and procedure are given elsewhere [4, 6].

6 Results and Discussion

The rates of dissolution of gold obtained in the runs discussed in this paper are listed in Table 3. In the absence of iodine, low dissolution rates of gold were obtained (see Table 3, expts. 1 and 2). Figures 2 and 3 illustrate these results and demonstrate, in particular, that the decrease in the dissolution rate with time is related to the autoreduction rate, which simultaneously decreases with time. The link between these observations is the role played by the intermediate thiocyanate species, particularly $(SCN)_3^-$, in the mechanism of dissolution of gold [4]. The autoreduction process itself is illustrated in Figure 3: the potential readings decrease (Fig. 3a) as Fe^{III} is reduced to Fe^{II} (Fig. 3b), with the corresponding oxidation of thiocyanate, resulting in the formation of the intermediate thiocyanate species.

Table 3 Gold dissolution rates observed with different concentrations of SCN^- and I^- (or iodine) and 0.055 mol/dm^3 Fe^{III} at 25°C and pH 1.5

Expt. No.	Concentrations (mol/dm^3) [SCN^-]	[I^-]	Initial Rate $J \times 10^9$ (mol/cm^2 s)	Rate after 1 Hour $J \times 10^9$ (mol/cm^2 s)	Average Rate $J \times 10^9$ (mol/cm^2 s)
1.	0.050	0	0.63	0.07	-
2.	0.10	0	0.78	0.08	-
3.	0	0.0022	0.32	0.23	0.28
4.	0	0.0055	0.89	0.72	0.89
5.	0	0.011	1.92	1.41	1.74
6.	0.050	0.0055	6.17	2.16	2.76
7.	0.10	0.0022	4.43	1.03	1.37
8.	0.10	0.0055	8.17	3.26	3.78
9.	0.10	0.011	10.7	5.42	6.27
Runs with addition of iodine solution (Normality units) instead of iodide ions:					
10.	0.10	0.001 N (*)	4.91	1.54	1.93
11.	0.10	0.005 N (*)	11.0	5.63	6.29

(*) 0.001 and 0.005 N of I_2 added as standard iodine solution correspond to 0.0022 and 0.011 mol/dm^3 of atomic I, respectively, which are the same concentrations used when I^- was added in some of the experiments above. (0.10 N iodine solution is made up by adding 12.7 g of solid I and 20 g of KI to one liter of water. This amounts to 0.22 mol/dm^3 of atomic I.).

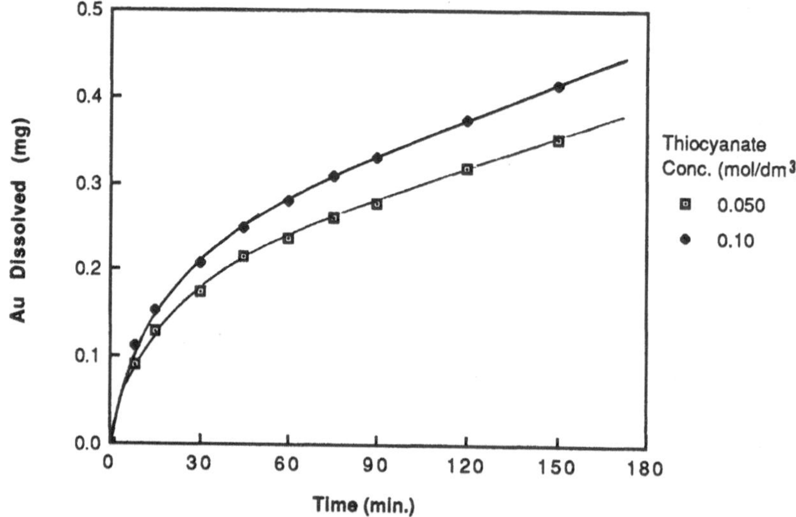

Figure 2 Gold leaching by iron(III)-thiocyanate solution. Initial concentrations: 0.055 mol/dm^3 Fe^{III}; pH 1.5; temperature, 25°C; disc rotation speed, 720 rpm.

The results of runs using iron(III)-iodide solutions (Table 3, expts. 3, 4 and 5) are shown in Figure 4. The corresponding potential readings and amount of reduction of Fe^{III} to Fe^{II} are given in Figure 5. The behaviour observed in these experiments using iodide was different from that seen with thiocyanate in two main respects: firstly, there was no marked change in the rate of gold dissolution with time; secondly, the rates obtained with iodide were much higher than those with thiocyanate, and compared well with those reported by Qi and Hiskey [17, 18]. The relative constancy of the gold dissolution rate with time within each run is due to the stability of the oxidised species, I_2 and I_3^- in solution.

The complete behaviour pattern of the iodide experiments emerges when Figure 4 is compared with Figure 5. It is interesting to note that, whereas there was a decrease in the potential readings in the ferric thiocyanate solutions due to the progress of autoreduction, the reverse occurred in the iodide system. The mixed potentials were low at zero time (shortly after the addition of Fe^{III} to the iodide solution) and then increased and tended to a plateau (see Figure 5a), due to the production of the species I_2 and I_3^-, generated by the oxidation of I^- by Fe^{III}. The corresponding Fe^{III} reduction is shown in Figure 5b. In contrast to $(SCN)_2$ and $(SCN)_3^-$, which are unstable and decompose through hydrolysis, I_2 and I_3^- are very stable and accumulate in solution as the iodide oxidation progresses.

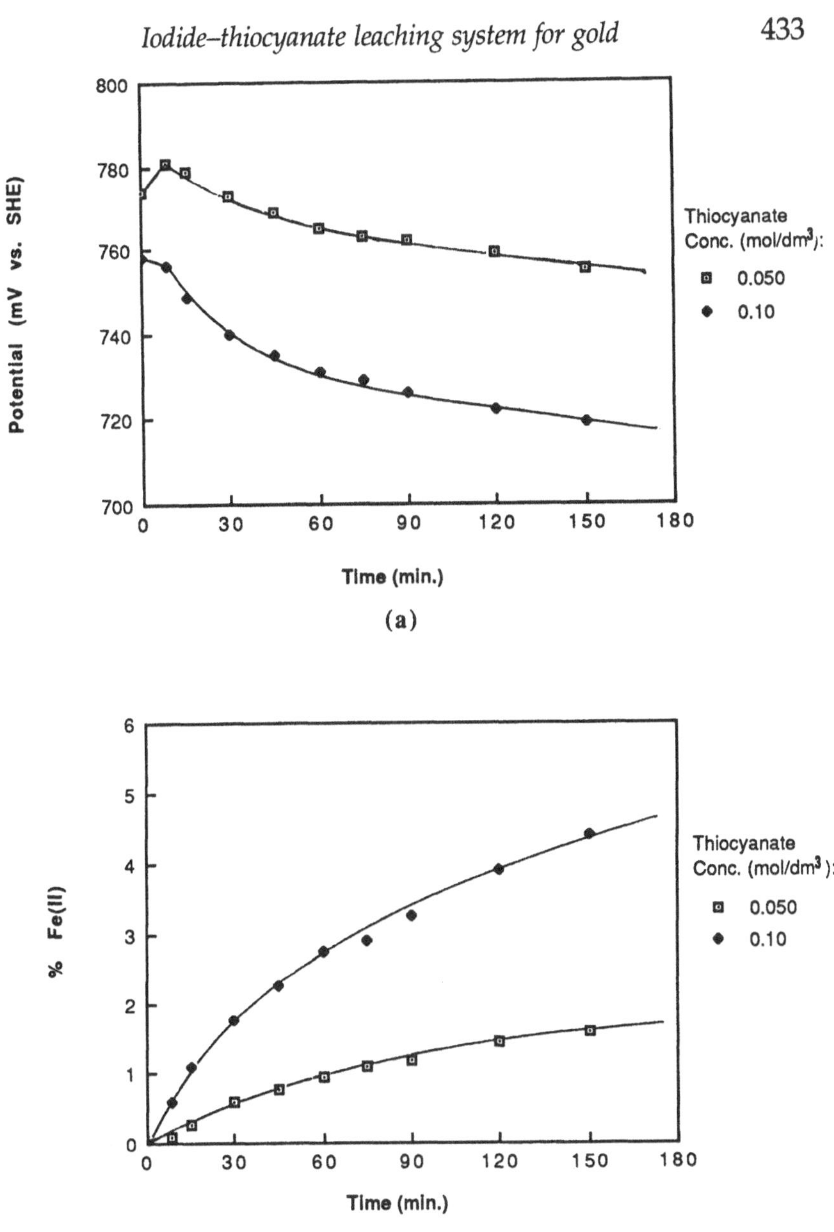

Figure 3 (**a**) Potential readings at the gold disc and (**b**) reduction of Fe^{III} to Fe^{II}, in the rotating disc runs with iron(III)-thiocyanate solutions shown in Figure 2.

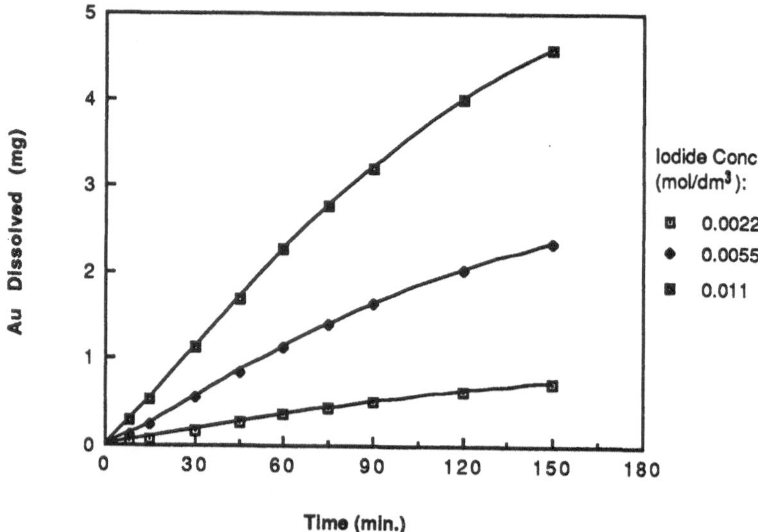

Figure 4 Effect of iodide concentration on the leaching of gold in presence of 0.055 mol/dm^3 (initial) Fe^{III} at 25°C. Rotation speed, 720 rpm.

The highest rates of dissolution of gold occur when both thiocyanate and iodide ions are used as a mixed leaching system. This is illustrated by Figures 6 and 7. These figures compare the separate behaviours of the iron(III)-thiocyanate and iron(III)-iodide system with that of the mixed iron(III)-thiocyanate-iodide system. The concentrations of thiocyanate and iodide were, respectively, 0.050 and 0.0055 mol/dm^3 (Table 3, expts 1, 4, and 6).

Figure 6 shows that a pronounced synergistic effect occurred when iodide ions were added to the iron(III)-thiocyanate system, the rates of dissolution of gold being much greater than those obtained when the two ligands were used separately. There was also, in the run with the mixed system, a considerably greater reduction of Fe^{III} to Fe^{II}, as shown in Figure 7b.

Both the extensive Fe^{III} reduction and the synergistic effect on the gold dissolution caused by the addition of iodide to the iron(III)-thiocyanate system can be explained in terms of reactions such as:

$$I^- + 2SCN^- = I(SCN)_2^- + 2e \quad (6)$$

$$2I^- + SCN^- = I_2SCN^- + 2e \quad (7)$$

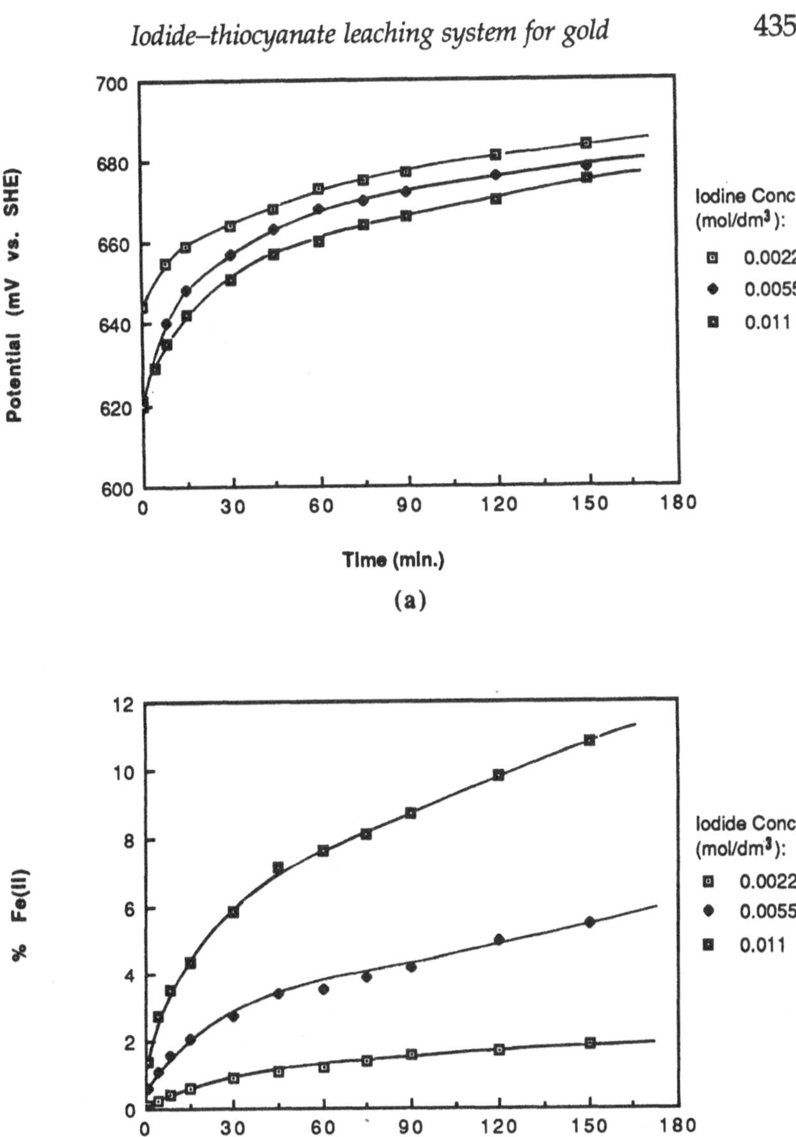

Figure 5 (**a**) Potential readings at the gold disc and (**b**) reduction of Fe^{III} to Fe^{II} in the rotating disc runs with iron(III)-iodide solutions shown in Figure 4.

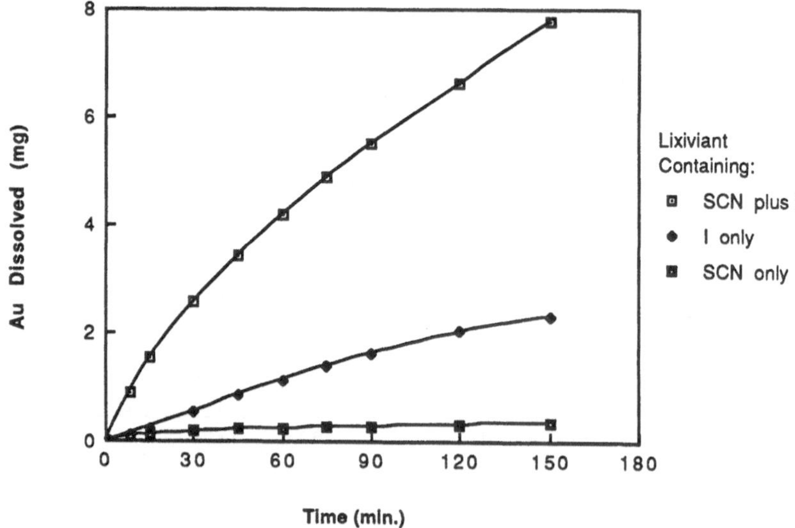

Figure 6 Effect of iodide addition on the leaching of gold by iron(III)-thiocyanate solutions at 25°C and pH 1.5. Lixiviant concentrations: 0.050 mol/dm^3 SCN^- and 0.0055 mol/dm^3 I^-. In all runs: 0.055 mol/dm^3 Fe^{III}, 720 rpm.

The experimental results suggest that these mixed iodide-thiocyanate species have important roles in the mechanism of dissolution of gold, similar to that of I_3^- and $(SCN)_3^-$ in the respective separate systems. The overall oxidizing ability of the mixed system is greatly enhanced by the accumulation of I_2SCN^- and $I(SCN)_2^-$, which are considerably more stable towards hydrolysis than the intermediate species of thiocyanate alone. Furthermore, as both SCN^- and I^- are ligands for gold, upon reduction these mixed intermediate species have not only oxidizing but also complexing ability.

Experiments with several iodide concentrations and 0.10 mol/dm^3 SCN^- are shown in Figure 8. The increase in the initial iodide concentration furthers the production of $I(SCN)_2^-$ and I_2SCN^- as shown by reactions (6) and (7) and redistributes the equilibria indicated by reactions (1), (3), (4) and (5), so that there is an increase in the overall concentration of X_3^- (where X = I and/or SCN) in the system. The active role of such species in the mechanism of gold dissolution is suggested by the considerable increase in the dissolution rates in this series of experiments. The highest initial dissolution rate, obtained with 0.10 mol/dm^3 SCN^- and 0.011 mol/dm^3 I^-, was 1.07×10^{-8} mol/cm^2 s, with an average rate of 5.42×10^{-9} mol/cm^2 s (see Table 3, expt. 9). This initial rate is similar to that reported for gold dissolution in thiourea (with 0.10 mol/dm^3 thiourea and 0.01 mol/dm^3 formamidine disulphide [33]). It is also close to the optimum rate of 1.56×10^{-8} mol/cm^2 s obtained by Davis and Tran [19] for gold leaching with iodine-

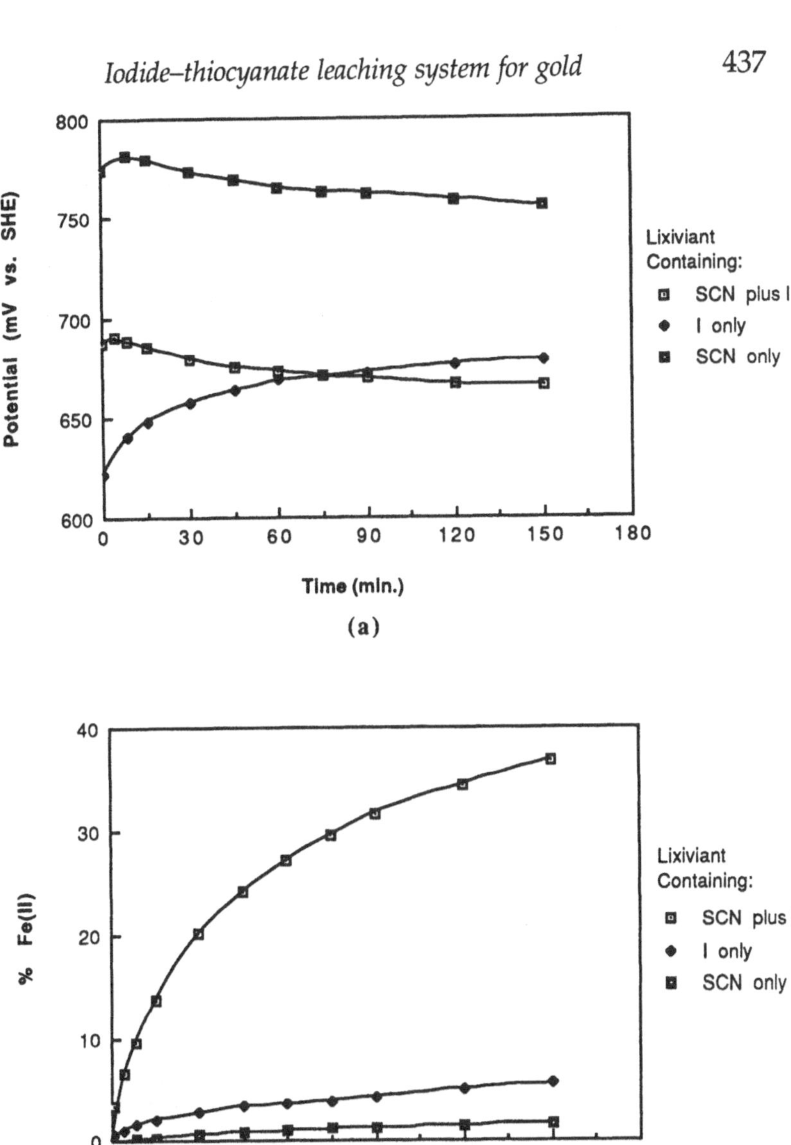

Figure 7 (**a**) Potential readings at the gold disc and (**b**) reduction of Fe^{III} to Fe^{II} in the runs with iron(III)-thiocyanate-iodide solutions shown in Figure 6.

iodide mixtures (0.10 mol/dm^3 KI and 0.040 mol/dm^3 iodine). However, the present lixiviant is more stable than thiourea and also contains much less I_2 than the latter solution, which means much less I_2 loss to the gas phase.

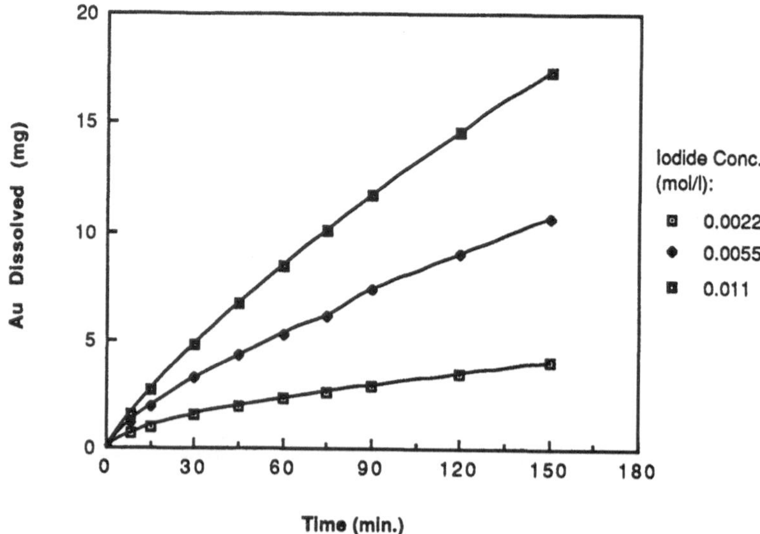

Figure 8 Effect of iodide concentration on the leaching of gold by iron(III)-thiocyanate solution at 25°C and pH 1.5. Lixiviant concentrations: 0.10 mol/dm^3 SCN^- and 0.055 mol/dm^3 Fe^{III} (initial). Disc rotation speed: 720 rpm.

7 Conclusions

Additions of small amounts of I^- and/or I_2 to iron(III)-thiocyanate solutions reveal a synergistic effect on the dissolution of gold. Thus, the dissolution rates obtained using the mixed iron(III)-iodide-thiocyanate system are substantially higher than those obtained when either iron(III)-thiocyanate or iron(III)-iodide solutions are used separately at the concentrations used in the mixture. The synergistic effect is ascribed to the participation of mixed iodine-thiocyanate species, particularly I_2SCN^- and $I(SCN)_2^-$, in the mechanism of dissolution of gold. Both I_3^- and the mixed iodine-thiocyanate species are more stable at lower potentials than $(SCN)_2$ and $(SCN)_3^-$, which extends the effective range of potential for efficient leaching. The mixed species show considerable stability towards hydrolysis and decomposition, in comparison with the corresponding thiocyanate species.

The dissolution rates of gold obtained by adding iodide/iodine to ferric thiocyanate solutions are as high as those obtained with similar concentrations of thiourea/formamidine disulphide and comparable with those obtained using similar total concentrations of iodide/iodine. However, the mixed thiocyanate-iodine system has an advantage over the iodide-iodine system in terms of permitting a considerable diminution of the I_2 concentration in solution, which minimizes losses of I_2 by vaporization. In comparison with thiourea, thiocyanate has the advantage of being more stable towards oxidation.

In the mixed system, thiocyanate acts as a "carrier" for halogenic species, which results in a considerable increase in the concentrations of X_3^- and X_2 (X = SCN and/or I) in solution. The overall halogenic character of these mixed species is not impaired (in respect to their reactivity towards gold) by the presence of thiocyanate in their structures, due the the pseudo-halide nature of thiocyanate. This produces a lixiviant with halogenic properties, which can be used in relatively high concentrations, but which is mainly composed of thiocyanate. In fact, the thiocyanate:iodide molar ratios used in the present work varied from about 10:1 to 20:1, with gold dissolution rates of the same magnitude as those obtained with similar overall concentrations of iodide/iodine.

8 References

1. White, H. A., The solubility of gold in thiosulphates and thiocyanates, *J. Chem. Metall. Min. Soc. S. Africa*, **6**, 109-111 (1905).
2. Fleming, C. A., A process for simultaneous recovery of gold and uranium from South African ores, in *"Gold 100 - Proceedings of the International Conference on Gold. Volume 2: Extractive Metallurgy of Gold"*, S. Afr. Inst. Min. Metall., Johannesburg, 1986, 301-319.
3. Barbosa-Filho, O. and Monhemius, A. J., Thermochemistry of thiocyanate systems for leaching gold and silver ores, in:*"Precious Metals'89"*, Jha, M. C. and Hill, S. D., eds., TMS-AIME, U.S.A., 1989, 307-357.
4. Barbosa-Filho, O., "Thiocyanate Leaching of Gold", PhD Thesis, Imperial College of Science, Technology and Medicine, London, U.K., 1991, 389 pp.
5. Barbosa-Filho, O. and Monhemius, A. J., Thiocyanate leaching of gold: Part 2: Redox processes in iron(III)-thiocyanate solution, *Trans. Inst. Min. Metall., Sect.C*, in press.
6. Barbosa-Filho, O. and Monhemius, A. J., Thiocyanate leaching of gold, Part 3: Rates and mechanism of gold dissolution, *Trans. Inst. Min. Metall., Sect.C*, in press.
7. Broadhurst, J. L. and du Preez, J. G. H., A thermodynamic study of the dissolution of gold in an acidic aqueous thiocyanate medium using iron(III) sulphate as an oxidant, *Hydrometallurgy*, **32**, 317-344 (1993).
8. Fleming, C. A., Private communication, 1988.
9. Homick, R. P., Gold reclamation process, U.S. Pat. 3,957,505 (1976).
10. Falanga et al., Recovery of gold and/or palladium from an iodide-iodine etching solution, U.S. Pat. 4,319,923 (1982).
11. Wilson, H. W., Noble metals solvation agents - hydroxyketones and iodine and iodide, U.S. Pat. 3,826,750 (1974).
12. Wilson, H. W., Process for separation and recovery of gold, U.S. Pat. 3,778,252 (1973).
13. McGrew, K. J. and Murphy, J. W., Iodine leach for the dissolution of gold, U.S. Pat. 4,557,759 (1985).

14. Jacobson, R., Murphy, J. and Whitman, D., The economics of small metallic in-situ mining operations, including uranium, gold, silver and copper, *"Proceedings of the Small Mines Conference"*, Royal School of Mines, London, 1987.
15. Hiskey, J. B. and Atluri, V. P., Dissolution chemistry of gold and silver in different lixiviants, *Min. Proc. and Extr. Met. Rev.*, **4**, 95-134 (1988).
16. Qi, P. H. and Hiskey, J. B., Dissolution kinetics of gold in iodide solutions. *Hydrometallurgy*, **27**, 47-62 (1991).
17. Qi, P. H. and Hiskey, J. B., Electrochemical behaviour of gold in iodide solutions. *Hydrometallurgy*, **32**, 161-179 (1993).
18. Hiskey, J. B. and Qi, P. H., Leaching behavior of gold in iodide solutions. In *World Gold '91 - Proceedings*. Parkville: Aust. Inst. Min. Met., 1991, 115-120.
19. Davis, A. and Tran, T., Gold dissolution in iodide electrolytes, *Hydrometallurgy*, **26**, 163-177 (1991).
20. Vogel, A. I., *"A Text-Book of Quantitative Inorganic Analysis"*, 3rd. edn., Longmans, London, 1961, 1216 pp.
21. Bard, A. J., Parsons, R. and Jordan, J., eds., *"Standard Potentials in Aqueous Solution"*, Marcel Dekker, Inc., New York and Basel, 1985, 834 pp.
22. Latimer, W. M., *"The Oxidation States of the Elements and their Potentials in Aqueous Solutions"*, 2nd ed., Prentice-Hall, Englewood Cliffs, 1952, 392pp.
23. Bjerrum, N. and Kirschner, A., Die Rhodanide des Goldes und das Freie Rhodan, *Kongelige Dansk Videnskabernes Selskab.*, **8** (1) 1-77 (1918); Thiocyanates of gold and free thiocyanogen, *Chemical Abstracts*, **13**, 1057-1060 (1919).
24. Itabashi, E., Spectroelectrochemical characterisation of iron(III)-thiocyanate complexes in acidic thiocyanate solutions at an optically transparent thin-layer electrode cell, *Inorg. Chem.*, **24**, 4024-4027 (1985).
25. Mussini, T. and Faita, G., Chap. I-2: Bromine, in: *"Encyclopedia of Electrochemistry of the Elements - Vol. I"*, Bard, A. J., ed., Marcel Dekker, Inc., New York, 1973.
26. Desideri, P. G., Lepri, L. and Heimler, D., Chap. I-3: Iodine and Astatine, in: *"Encyclopedia of Electrochemistry of the Elements - Vol. I"*, Bard, A. J., ed., Marcel Dekker, Inc., New York, 1973.
27. Griffith, R. O. and McKeown, A., Kinetics of the reaction between potassium thiocyanate and iodine in aqueous solution, *Trans. Faraday Soc.*, **31**, 868-875 (1935).
28. Lewis, C. and Skoog, D. A., Spectrophotometric study of a thiocyanate complex of iodine, *J. Am. Chem. Soc.*, **84**, 1101-1106 (1962).
29. Long, C. and Skoog, D. A., A thiocyanate complex of iodine(I), *Inorg. Chem.*, **5**, 206-210 (1966).
30. Országh, I., Bazsa, Gy. and Beck, M. T., Spectrophotometric study of the reversible iodine-thiocyanate interaction, *Inorg. Chim. Acta*, **6**, 271-274 (1972).
31. Sillén, L. G. and Martell, A. E., *"Stability Constants of Metal-Ion Complexes"*, *Special Publication No. 17*, The Chemical Society, London, 1964.
32. Fortune, W. B. and Mellon, M. G., Determination of iron with o-phenanthroline, *Ind. Eng. Chem., Anal. Ed.*, **10**, 60-64 (1938).
33. Groenewald, T., The dissolution of gold in acidic solutions of thiourea, *Hydrometallurgy*, **1**, 277-290 (1976).
34. Rice, N. M., Private communication, 1994.

Leaching and recovery of gold by use of acido-thioureation on copper-mine wastes: laboratory and pilot-plant tests and process modelling

P. F. Kavanagh
Dublin Institute of Technology, Dublin, Ireland
B. F. Foley
Dublin Institute of Technology, Dublin, Ireland
J. McLoughlin
Dublin Institute of Technology, Dublin, Ireland
P. Owens
Dublin Institute of Technology, Dublin, Ireland
S. A. Curran
Dublin Institute of Technology, Dublin, Ireland
C. McNamee
Dublin Institute of Technology, Dublin, Ireland
K. Mullins
Minmet Plc, Dublin, Ireland

Studies have been carried out in the Dublin Institute of Technology over the past few years on thiourea leaching and recovery of gold from mine wastes at the Avoca Copper Mines in Co. Wicklow. The work has been conducted in collaboration with Minmet plc who have operated a 5.0 te/d (solids) pilot installation on the site.

The paper addresses three aspects of this process;

a) Laboratory scale studies of leaching kinetics from both an empirical and mechanistic viewpoint. The kinetics and capacity of conventional and novel adsorbents for recovery of the complexion gold.
b) Computer simulation studies.
c) Operational data from the pilot plant.

The techniques employed to follow the fate of materials during the leaching adsorption processes include potentiometric titration and HPLC in relation to thiourea and its oxidation product, atomic absorption for gold and zeta potential and electronmicroscopy in relation to adsorbent surfaces. The software being employed is based on Aspen Plus Model Manager.

1 Introduction

1.1 Leaching

Leaching of finely disseminated gold from ores by the cyanidation process is by far the most important commercial process. However this process has a number of disadvantages, both operational and environmental. Refractory ores, ores containing base metals, sulphides, pyrite, arsenopyrite and carbonaceous sulphides have been uneconomical to treat directly because of poor extraction and high reagent consumption

[1, 2, 3, 4]. Environmental concerns over the large scale use of cyanide e.g. Croagh Patrick, Co. Mayo, Ireland, have heightened public awareness and increased opposition to planning permission applications for cyanide process plants. Recent fines levied on mining companies in the U.S. for cyanide leaks and alleged fish kills have highlighted the major concerns associated with cyanide usage [5]. A recent operation in British Columbia changed its initial cyanide based processing method to a gravity separation and flotation technique to circumvent environmental concerns and possible permit delays [6].

Extensive research has focussed on the search for an economically viable non-toxic alternative to cyanide. A number of possible agents have been considered over the past 100 years e.g. thiosulphate [7, 8], chlorine [7, 9], bromine [9, 10] and iodine [7]. However in more recent years attention has focussed on thiourea as the most likely alternative to cyanide.

The dissolution of gold in acidic thiourea solutions was first reported by Plaksin and Kozhukhowa [11] in 1941 and was again considered by the same researchers in 1960 [12]. Numerous investigations have since examined the use of thiourea, in the presence of a suitable oxidant, as a potential lixiviant for gold. Lodeishchikov [13] described a possible industrial process using acidic solutions of thiourea in the presence of ferric ions for the dissolution of gold in finely ground ore bodies. Despite high rates of extraction (92 to 98%) it was concluded that the process would be uneconomical compared to traditional cyanidation techniques except in specialised applications using refractory ores. Groenewald [14] carried out a comprehensive study of the effectiveness of various oxidants e.g. oxygen, hydrogen peroxide, ferric ion and formamidine disulphide. The ferric ion produced the fastest initial rate of gold dissolution but was considered impractical from the process standpoint (particularly for a recycling system) due to the build up of an iron - thiourea complex in the leach solution. Hydrogen peroxide was the favoured oxidant, however, a high rate of dissolution was counterbalanced by an extremely high consumption of thiourea. Chen [13] reported that the dissolution of gold in a suitable thiourea medium was over ten times faster than in cyanide solution. Also the dissolution of copper by thiourea is much less than that by cyanide making thiourea a more suitable agent for the extraction of gold from copper containing ores.

The main drawbacks in the use of thiourea are excessive reagent consumption due to chemical degradation of thiourea via formamidine disulphide and subsequent formation of elemental sulphur which in colloidal form coats the ore surface leading to passivation. Schulze [16] has proposed the addition of sulphur dioxide to the leachate to reduce thiourea consumption. The sulphur dioxide reduces formamidine disulphide back to thiourea during leaching preventing the generation of passivating sulphur. A deficiency of sulphur dioxide was advocated as a certain level of formamidine disulphide was seen as beneficial to the leaching efficiency.

1.2 Recovery

Removal of solubilised gold from pregnant leach liquors can be achieved in a number of ways. Cementation processes have been widely used in cyanide medium e.g. the

Merrill-Crowe process. Aluminium [17], lead [18] and iron [19] have all been used for gold cementation from thiourea solutions with varying degrees of success. However the use of activated carbon remains the most common method of gold recovery from solution. A number of studies have reported high levels of gold adsorption onto activated carbon from thiourea solutions [13, 20, 21]. However the adsorption of gold was also accompanied by a high loss of thiourea from the leach solution due also to adsorption onto the activated carbon. This could be a major disadvantage if the carbon-in-pulp (CIP) process is used in acidic thiourea solutions. Passivation of the carbon due to clogging of internal pores with colloidal sulphur has also been reported. Ion exchange resins offer another adsorption possibility as they are usually less susceptible to fouling by organic compounds compared with activated carbon. Cation exchange resins adsorb the gold thiourea complex very strongly but the lack of selectivity, poor abrasion characteristics and inefficient elution processes are major limitations of any resin based process.

1.3 Modelling

Several rate models have been applied to the leaching of low-grade ore and for the adsorption of gold onto activated carbon.

The purpose of this work is to critically examine the suitability of the models for application to the thiourea process, and their ability to simulate leaching/adsorption. A new homogeneous surface-diffusion model has been developed for the simulation of gold adsorption. The application of the flowsheet simulator, ASPEN PLUS, to the process has been investigated, bringing a link between bench scale and pilot plant.

The main objectives of this work were to investigate thiourea leaching and recovery of low gold levels from a copper mine waste, to optimise gold recovery, to minimise reagent consumption and finally to transfer the process from bench scale to a pilot plant.

2 Site Characterisation

The Avoca copper sulphide deposit is situated in County Wicklow on the east coast of Ireland. The sulphur bearing ores of the mining area strike in a NE/SW direction and are split by the Avoca river into east and west regions. The spoil heaps left from the mining activities on the eastern side were investigated in this project.

3 Dump Structure and Mineralogy

The Avoca copper mines have seen extensive workings over many periods since the Bronze Age. As a result of the mining activities a wide range of waste rock and tailings has been generated. Included in this are coarse waste rock, crushed and screened rock and hand cobbings. The gold bearing dumps investigated are generally fine-grained with coarse-to-fine grained rock fragments, and small amounts of timber are present which were originally used to support the heaps.

The composition of the dumps and their structure can vary quite considerably over a short distance. With the difference in parameters such as particle size, sulphide content and depositional history this can only be expected as the oxidation levels in the dumps will vary. As a result, a sulphide rich dump could have an outer core which is highly oxidised, an intermediate zone where oxidation is still occurring slowly and an inner unreacted core. Conversely the surface oxides can form a hard impervious layer which can inhibit oxidation of the dump. The intermittent mining of the area can be seen where highly oxidised dump material is overlain by unreacted sulphides.

The natural oxidation of the sulphides has caused elemental and mineralogical changes. In terms of elements the most noticeable loss occurs for iron, copper and zinc, whereas a concentration in lead, silver and gold occurs. Mineralogical changes are evident in an overall reduction in particle size, and an assemblage of silica, alkali leached chlorites, iron oxides and hydroxides, jarosite and plumbojarosite.

The size distribution of the gold in the oxidised dumps (50% < 7μm) suggests that much of the gold was originally present as fine exsolution blobs in pyrite and chalcopyrite which have been released by the dissolution of the host mineral.

4 Reserves

The gold grade in the heaps averages between 0.5 and 3.0g/tonne. Table 1 shows how the gold is distributed in the typical heap samples.

Table 1. Gold distribution in typical samples of the Avoca Mine Dumps

Size(μm)	Oxide Ore (%) wt	Au (mg/L)	Au (%) distribution	Sulphide Ore (%) wt	Au (mg/L)	Au (%) distribution
>6000	36.1	0.21	9.10	36.7	0.72	30.4
<6000>1000	26.4	0.33	10.6	20.0	1.23	28.1
<1000>300	11.1	0.30	4.0	10.0	0.66	7.50
<300>75	12.5	0.83	12.5	8.90	0.56	5.70
> 75	12.9	3.80	63.8	24.4	1.01	28.3
Head Grade		0.83			0.87	

The gold in the finer size fraction occurs as free gold, but in the coarser material appears to be locked in the sulphur lattices as grinding fails to liberate the gold. Of the 500,000 tonnes of material tested it has been shown that the likely mineral reserves are approximately 18,000 ozs. of gold.

There are also small amounts of silver present at grades averaging 30g/tonne.

5 Experimental

Equipment and Procedures

(i) Laboratory Tests

Thiourea, sulphuric acid, hydrochloric acid, nitric acid, ferric sulphate, aliquat 336 and sodium metabisulphite were all laboratory grade reagents and deionised distilled water was used. 2,6-dimethylheptan-4-one (DIBK) was technical grade. Commercial coconut shell activated carbon (1-1.4mm range) was used for adsorption studies.

Initial leach tests were performed in a 1 litre round bottomed reaction vessel with a wide flat neck which was connected to a 5-necked flat flange lid allowing the attachment of a stirrer (variable speed), thermometer, pH and redox electrodes and a reflux condenser. Subsequent tests were performed in stoppered glass conical flasks (250ml to 1L) and shaken on a flask shaker.

Typical solid/liquid ratio was 25% with pH varying between 1 and 2 and leaching times up to 6 hours. Thiourea and oxidant solution were mixed, stirred and added to the ore. Pulp samples were taken at varying time intervals e.g. 0.5hr, 1.0hr, 2.0hr, 4.0hr and 6.0hr. Changes in volume due to removal from addition to the leach solution were taken into account for all calculations.

Assays of the ore and filter cake for gold were performed by roasting, digestion in aqua regia, preconcentration with DIBK [22] and analysis by flame atomic absorption spectrometry.

The pregnant leach liquor and filter cake washings were analysed for gold also using the preconcentration step and AAS. All other metal analyses (Fe, Cu) were also performed by AAS.

Analysis of thiourea was performed by potentiometric titration with potassium iodate, using Pt/calomel electrodes connected to a mettler DL25 auto-titrator and Epson LX-800 printer.

Thiourea loss and formamidine disulphide concentration were calculated by the difference in thiourea concentration after regeneration by reduction of the formamidine disulphide. Regeneration was effected by contacting the solution with zinc powder. An alternative rapid method for simultaneous thiourea and formamidine disulphide analysis was also developed using HPLC.

Analysis of thiourea by titration with potassium iodate gives markedly elevated results in the presence of sulphur dioxide. This may be overcome by using correction factors or by sparging the solution with nitrogen gas for a few minutes [23]. The HPLC method is not affected by the presence of sulphur dioxide.

Initial thiourea level in the leach solution was generally at 12 g/kg of ore which is significantly below levels used in other similar studies (typically 30-100 g/kg have been used). As a once-through plant operation was envisaged a thiourea level capable of optimum extraction with minimum wastage was required. All solutions were made up in 0.01M sulphuric acid to minimise acid consumption (although higher acid concentration (0.1M) is reported to give slightly better leaching efficiencies [24]).

6. Results and Discussion

6.1 Leaching

(a) Influence of Ferric Ion

The importance of ferric ion, as an oxidising agent on gold extraction is shown in Fig 1. Without any added oxidant a leaching efficiency of over 60% was obtained over a given period. This was due to dissolution of oxidant from the ore. Addition of varying amounts of ferric ion up to a maximum of 27 mM gave a steady increase in total gold extracted over a 4 hour period up to a maximum extraction of ~83%. As clearly seen from Fig. 1 initial rate of dissolution is fast and falls significantly after the first hour. The decrease in rate is most likely caused by a decrease in thiourea concentration due to oxidation and complexation.

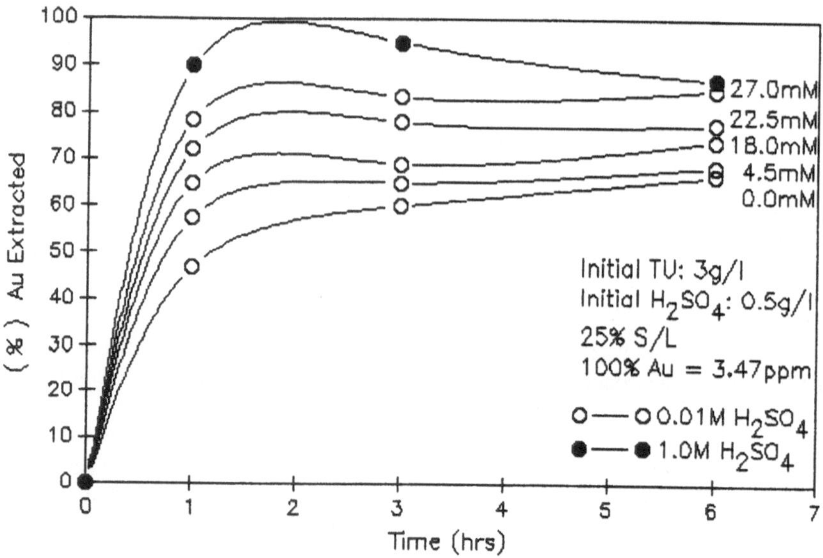

Fig. 1 (%)Au extracted vs Time(hr) at various levels of Fe^{3+} added (mM)

Furthermore, passivation of the ore surface can also be caused by adsorption of sulphur, a product of formamidine disulphide decomposition. Table 2 shows the consumption of thiourea with time at varying levels of ferric ion.

Table 2. The Average Consumption of Thiourea (g/kg of Ore) with Time at Various Levels of Ferric ion(mM)

Fe (III) (mM)	Ave. Thiourea Consumed (g/kg of Ore) 1 Hour	3 Hours	6 Hours
0.0	1.559	2.391	2.524
4.5	2.308	2.973	3.147
9.0	2.748	3.383	3.684
13.5	3.374	4.084	4.249
18.0	3.939	4.496	4.809
22.5	4.330	4.752	5.022
27.0	4.763	5.345	5.678

As also expected the ore is an acid consumer with a typical pH change of 1.8-2.5 over a 6 hr leach.

(b) Effect of Redox Potential

Optimum economic extraction has been shown to be dependent on redox potential. The redox potential of the leach system depends on the concentrations of both thiourea and formamidine disulphide which itself is claimed to be an active oxidant for gold [25]. However high redox potential also results in irreversible loss of thiourea due to oxidative degradation of formamidine disulphide. At high redox potentials passivation of the ore has also been reported [26] as well as an actual downturn in gold extraction possibly due to excessive thiourea consumption in the leach solution.

Schulze [16] proposed the use of sulphur dioxide as a specific reducing agent for formamidine disulphide to control the ratio of thiourea to formamidine disulphide in the leach system. This has the effect of reducing thiourea consumption and allowing the redox potential to be controlled. Addition of sulphur dioxide can be achieved by bubbling the gas into solution or by addition of sodium sulphite or sodium metabisulphite. All of these methods were attempted in this work. The most economical and effective method was found to be addition of sodium metabisulphite to the ore slurry followed by stirring for 30 minutes. After this conditioning period the required amounts of thiourea and acid were added to establish the necessary leaching conditions. Table 3 shows the effect of different initial sodium metabisulphite levels on gold extraction and redox potential over a four hour leach period with no added oxidant. In an attempt to improve the economics of the process the thiourea level in this series of tests was reduced to a 9 g/kg of ore.

Table 3. The effect of different initial sodium metabisulphite levels on gold extraction and redox potential over a four hour period

	0g/L $Na_2S_2O_5$		1.25g/L $Na_2S_2O_5$		1.88g/L $Na_2S_2O_5$		3.75g/L $Na_2S_2O_5$		8g/L $Na_2S_2O_5$	
Time (Hrs)	Redox Pot. (mV)	% Au	Redox Pot. (mV)	% Au	Redox Pot. (mV)	% Au	Redox Pot. (mV)	% Au	Redox Pot. (mV)	% Au
1	407	26	356	47	293	63	226	56	207	38
2	405	29	355	48	298	62	237	69	194	36
3	396	28	353	49	300	59	257	63	192	51
4	369	29	347	53	301	59	263	61	192	44

Under these conditions maximum gold extraction (~65%) occurs at a redox potential of approximately 250 to 300 mV. Fig. 2 illustrates average gold dissolved after 4 hours leaching at different redox potentials.

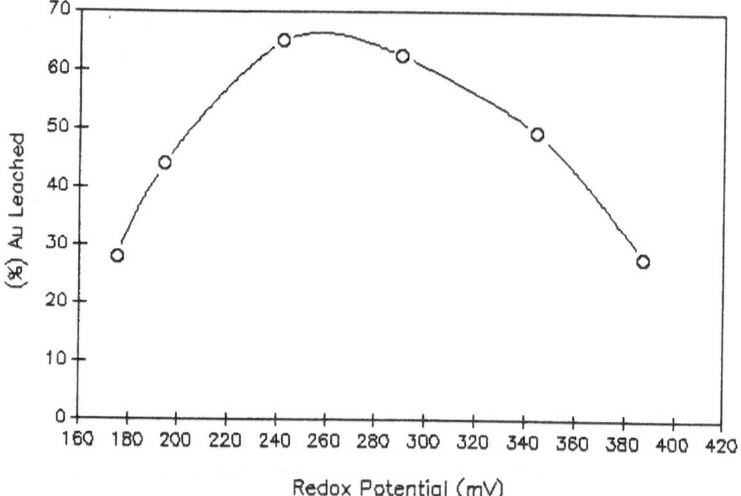

Fig. 2 % Gold Leached Versus Redox Potential (mV)

Schulze suggested that the formamidine disulphide to thiourea concentration ratio be maintained at ~0.5 to give a high gold dissolution rate. However with an initial thiourea concentration of 1g/l (0.013M) and assuming negligible irreversible loss of thiourea in the presence of sulphur dioxide a ratio of 0.5 corresponds to a redox potential of approximately 196 mV, clearly well below our optimum. Using the same assumption our optimum redox potential of 250 mV gives a formamidine disulphide to thiourea ratio of ~1.67 i.e., ~77% conversion of thiourea.

Further study is required on the irreversible loss of thiourea in the presence of sulphur dioxide and the consequent effect on the formamidine disulphide/thiourea ratio. The HPLC method developed for simultaneous analysis of thiourea and formamidine disulphide will be used for this study.

(c) Ore Pretreatment

Several workers [12,13, 14,15, 24, 27] have investigated the effects of subjecting a gold bearing ore to a preliminary treatment of washing before leaching. The primary washing reagents considered were water and varying concentrations of sulphuric acid.

The rationale for pretreatment is the removal of ore constituents which form complexes with thiourea hence reducing thiourea consumption in the subsequent leach. Some workers [24, 27] have found that pretreatment removes some of the native oxidants and has an adverse effect on gold leaching.

This work supports the latter findings and a significant decrease in gold extraction was observed (Table 4).

TABLE 4. % Gold Extracted after Various Pretreatments and Subsequent 60 min. leach

Prewash Type	Prewash Duration (Min.)	Gold in Leach (mg/L)	% Gold Leached
None	____	0.94	76.4
	5	0.10	8.13
Cold H_2O	20	0.12	9.76
	40	0.16	13.0
	5	0.10	8.13
Hot H_2O	20	0.10	8.13
(Reflux at 100°C)	40	0.13	10.6
	5	0.14	11.4
Cold Acid	20	0.15	12.2
(0.1M H_2SO_4)	40	0.19	15.4
	5	0.34	27.6
Hot Acid	20	0.37	30.1
(Reflux at 100°C)	40	0.41	33.3

(Note: 100 %Gold leached = 1.23mg/L)

There are two possible reasons for the drop in gold recovery, (i) the removal of base metal ions such as copper and iron and (ii) the loss of some leachable gold during the pretreatment step. Gold analysis of the ore and wash liquor showed that negligible amounts of gold are removed by the pretreatment step (Table 5).

TABLE 5 Gold Analysis of Filtrate after Pretreatment

Prewash Type	Prewash Duration (min.)	Gold in Filtrate (mg/L)
	5	< 0.01
Cold H_2O	20	< 0.01
	40	< 0.01
	5	< 0.01
Hot H_2O	20	< 0.01
(Reflux at 100°C)	40	< 0.01
	5	~ 0.01
Cold Acid	20	~ 0.01
(0.1 M H_2SO_4)	40	~ 0.01
	5	~ 0.02
Hot Acid	20	~ 0.02
(Reflux at 100°C)	40	~ 0.02

Of the major base metal components removed by pretreatment copper was found to have the greatest influence on gold extraction. All pre-wash conditions used in this work significantly reduced the amount of soluble copper. This was shown to have a direct effect on gold leaching. This finding is probably related to the reaction between the cupric ion and thiourea. This reaction, which is much faster than the corresponding ferric ion reaction [28] produces formamidine disulphide, an active oxidant for gold, and a soluble cuprous ion - thiourea complex. Obviously, cupric ion could also have an adverse effect on gold leaching if the level of cupric ion in solution was sufficient to reduce the thiourea level significantly.

Further evidence supporting the beneficial / adverse effect of cupric ion on gold leaching was obtained by pretreating the ore with a 60 minute water wash and doping the leach solution with varying levels of cupric ion. Fig. 3 clearly shows an optimum level of cupric ion at about 300 mg/L (200 mg/L added plus approximately 100 mg/L from ore) when using 12kg thiourea/te ore (all leach solutions had 1.5 g/L added ferric ion).

As shown, a deficiency or excess of cupric ion has a detrimental effect on gold extraction.

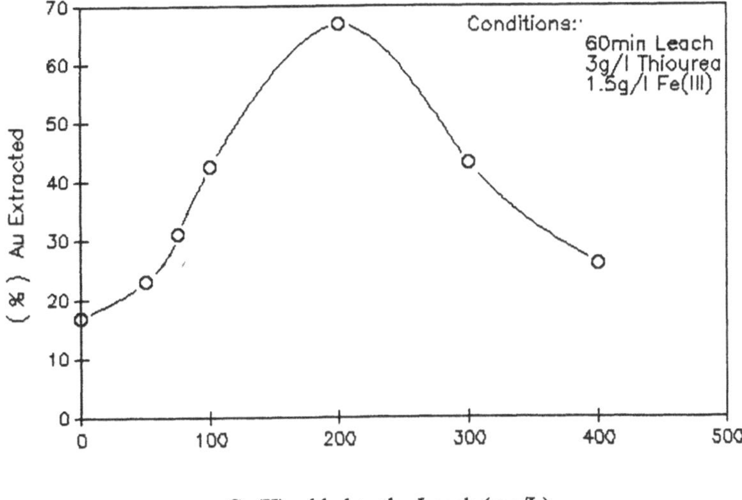

Cu(II) added to the Leach (mg/L)

Fig. 3 % Au Extracted Vs Copper added to Leach Liquor

(d) Adsorption Studies

The recovery of gold from thiourea solutions by adsorption onto activated carbon has been studied by a number of workers [29]. All report high levels of adsorption of gold from solution. Therefore the natural first choice for recovery of gold from solution appeared to be activated carbon. In this study three approaches were taken at laboratory. The first involved adsorption from an 'ideal' solution where the gold complex was generated from a rotating gold disc. The other experiments involved adsorption from leach filtrates and finally a Carbon-in-Pulp type process. The adsorption of thiourea and other possible co-adsorbates was also studied.

As expected very high gold loading onto activated carbon was achieved from the 'ideal' solution with values in excess of 70 g of gold / kg of carbon being recorded. However with the 'real' system loadings were drastically reduced. Adsorption from the leach filtrate gave a maximum of 0.62 g of gold / kg of carbon (Fig.4) after 15 hours. This result was not unexpected as the gold level in solution was very low (< 0.5 mg/L) compared to the level of thiourea and the other soluble cations. Not surprisingly a bench scale C.I.P. type process yielded even lower gold loadings than the leach filtrate tests probably due to fouling of the carbon with ore particles as well as co-adsorption problems. Maximum gold loading achieved with this process was 0.085 g of gold / kg of carbon. Analysis of variance (A.N.O.V.A.) was used to determine the most significant factors affecting the adsorption of gold from solution. The variables tested were Fe(III), Cu(II), thiourea and pH (time was assumed to have an obvious significance). Of these thiourea gave the most significant effect with significant interaction effects being seen. Loadings in excess of 30 g of thiourea / kg of carbon were readily achieved in a C.I.P. type process.

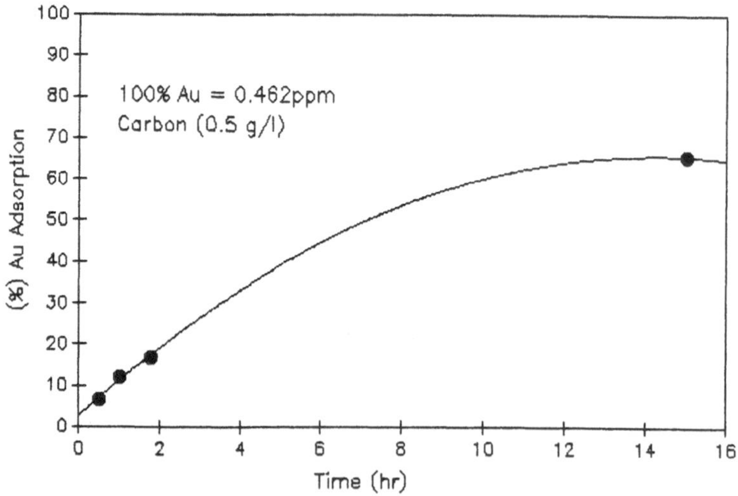

Fig. 4 (%) Au adsorbed from Leach Filtrate vs Time at a fixed level of Carbon (0.5 g/L) (100%) = 0.462 mg/L

Realistically the use of activated carbon did not appear to be a viable recovery process using the low grade ore coupled with thiourea as the lixiviant. Other recovery methods were tried and a cementation process was successfully developed on site which is now the subject of a patent.

7 The Process Operation

A purpose built pilot leaching plant was constructed in Avoca to test and optimise the technology of gold extraction from low grade and oxidised ores using thiourea. The plant was designed to operate at 100 tonnes/day of leach feed at 30% solids. It's operation can be described in two sections (as can be seen in Fig.5).
1) Washing / Classification - in this section the ore from the heaps was separated so that eventually the material going through the plant had a d_{50} cut of 150μm.
2) Leaching / Adsorption - here the gold was leached using the thiourea and subsequently recovered using activated carbon in a counter-current operation.

7.1 Washing / Classification

As was shown earlier over 80% of the gold is present in the -1000μm fraction of the oxidised ore dumps. Consequently it was necessary to remove the oversize material by a washing / classification process. With the exception of a shredder used to break up the coagulated ore no crushing was used to sort this material.

The ore collected from the spoil heaps around the site was added - via a front-end loader - to a 15cm grizzly/powerscreen unit with a shredder attached. The oversize material was stock piled and the shredded undersize ore was carried via a conveyor to a 7.5cm vibroscreen. Again the oversize was stockpiled and the undersize conveyed to the back of a washing barrel. The -7.5cm ore was thoroughly scrubbed in the washing barrel and the slurry was passed over a 1mm stainless steel screen. The +1mm material was rejected to a stockpile and the -1mm slurry, at a pulp density of 30% solids, passed through a hydrocyclone which gave a d_{50} cut of 150μm. The underflow from the hydrocyclone was sent to a tailings pond where it was retained for possible further processing. The overflow was fed to the leaching plant via a Delkor moving screen where any remaining +550μm material and wood chips were removed at the drive pulley using a belt wash spray, and were passed to the tailings pond via a gravity feed. The undersize material drained through the screen by gravity and formed the feedstock for the leaching section.

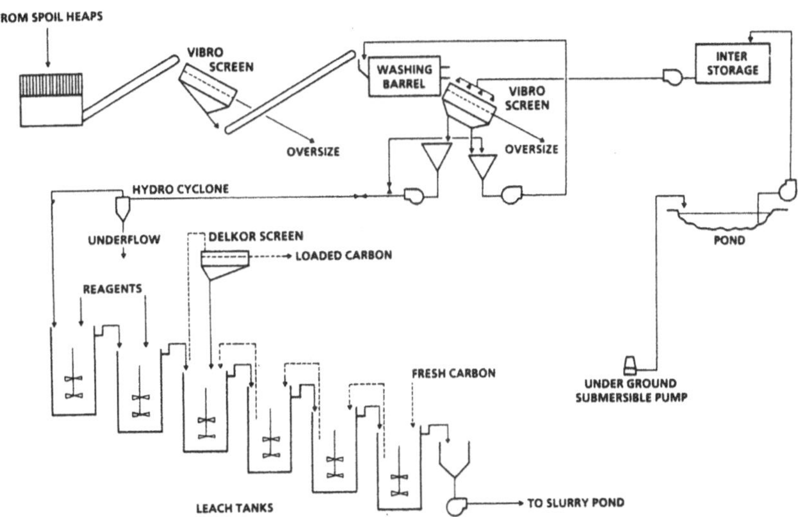

Fig. 5 Pilot Plant Flowsheet

7.2 Leaching

The leaching / adsorption section of the process was conducted in six 20 m³, rubber lined tanks in series. Mixing in each tank was achieved by twin 4-paddle agitators. The direction of mixing was reversible. The transfer of slurry from one tank to the next was done by air-lifting the slurry from the upstream tank through a labscreen into a launder feeding the receiving tank. At the end of the process slurry from tank 6 was air-lifted into a hopper and was pumped to a tailings pond.

The underflow from the Delkor screen flowed into tank 1. This tank was essentially used as a pretreatment tank where the reductant in the form of SO_2 (generated from sodium metabisulphite solution) was added to keep the redox level low enough to prevent rapid overoxidation of the thiourea when it was added, and sulphuric acid was added to bring the pH of the slurry to the required level. It should be noted that the source of water for the plant was minewater which when mixed with the ore gave the untreated slurry a pH of approximately 2.5, hence only small quantities of acid were added to the system to bring the pH to 2.0. The lag time in this tank, and the other 5 tanks, was between 30 and 45 mins., depending on the leaching time required and the pulp density, after which the slurry was air-lifted into the launder feeding tank 2. The thiourea was added in soluble form to this tank via the launder where primary leaching took place and, no other reagents were added. It was found that there was sufficient oxidant levels present in the ore to oxidise the thiourea in the system, so no oxidants like ferric sulphate or hydrogen peroxide were required. With thiourea being an acid consumer the pH of the slurry increases with leaching, thus the pH of the leached slurry going to tank 3 was readjusted with sulphuric acid to pH 2 0. After tank 3 no other reagents were added to the system to aid the leaching process. During leaching there were several variables monitored including pH, redox potential, pulp density, temperature, thiourea concentration, formamidine disulphide, gold in solution, solubilised gold adsorbed onto the ore, gold head grade and gold in the tailings.

7.3 Adsorption

The adsorption process used was a Carbon-In-Pulp / leach system. Activated carbon (> 1 < 3mm) was added to adsorb the leached gold from the slurry in a type of counter current operation. Initially 25 kgs of the carbon was added to each of tanks 3, 4, 5 and 6 after which any further additions to tanks 3, 4 and 5 were done by advancing the carbon, using air-lifts, from the next tank downstream and any fresh or reprocessed activated carbon added to the system was done so through tank 6. This transfer or addition of the activated carbon was normally done at the end of a days operation when the loaded carbon in tank 3 was removed from the system first. Here the loaded carbon in tank 3 was air-lifted onto the Delkor screen where it was washed on the cloth with water from spray bars. This normally took about one hour. The carbon was then recovered and stored for either washing and reprocessing or ashed to recover the precious metals adsorbed. Once the carbon was removed from tank 3 the air lifting from the other tanks was conducted.

8 Discussion

Before commenting on the results achieved from the pilot plant it should be noted that the plant in Avoca operated between September 1989 and March 1990. For the obvious reason of low temperature this was not the best time of the year to operate tests. With the temperature rarely rising above 15°C and no means of increasing the temperature of the system this had a detrimental effect on the leaching kinetics.

For years now efforts have been made to make thiourea a competitive alternative to cyanide for leaching gold from low grade mine ores like those in Avoca. The key to the success of these efforts is to reduce the consumption of thiourea to very low levels (i.e.,< 1kg Thiourea/tonne of ore). There are essentially two ways by which this can be done: add very low initial concentrations of thiourea (i.e., <1.0g/L solution) and limit reagent oxidation/consumption by the addition of a suitable reductant such as sodium metabisulphite ($Na_2S_2O_5$), or, use high concentrations of thiourea and recycle the solution at the end of the process. The latter would require the installation of a filtering system or some suitable flocculating system. In Avoca the former was chosen.

The early results of the plant trials paralleled laboratory tests with leaching figures of over 70%, however it proved difficult to improve on these results. With low amounts of thiourea present the oxidants in the ore, such as Cu^{2+} and Fe^{3+}, oxidised and complexed with the thiourea very quickly, hence the amount of thiourea remaining to leach the gold was insufficient. Efforts were made to rectify this by adding enough reductant to reduce the oxidation states of these metals, but, later laboratory tests (as were seen earlier) showed also that the starting Eh of the system (ie. after pretreatment) was too high. With the temperatures decreasing at the onset of winter the overall leaching kinetics were found to decrease. This led to increasing lag times in the tanks. By increasing the lag times more of the gold(I)thiourea complex was adsorbed onto the ore surface, and the oxidised form of the thiourea (formamidine disulphide) irreversibly decomposed and one of the products of decomposition included sulphur. Sulphur had a detrimental effect on the system because it coated the surface of the ore thus inhibiting leaching. As mentioned earlier the ore material proved to be very heterogeneous, and with the ratio of free to refractory gold varying, this meant that the extraction efficiency of the system was not always easy to determine.

It has been found in the laboratory that the adsorption of gold(I)thiourea complexes on to fresh activated carbon is rapid and high loadings of up to 2kg of Au/tonne can be achieved in relatively short times with low gold values in solution (0.2 - 0.4mg/L). In Australia, the New England antimony mine has successfully used activated carbon to recover gold from high grade antimony concentrates with short carbon contact times. In Avoca, however, the ore used is of a relatively low grade and consequently the contact time between the leached slurry and the carbon had to be increased considerably. This had a detrimental effect on the overall recovery of the gold from solution. The reason for this is that thiourea, thiourea-metal complexes and breakdown products of thiourea are readily adsorbed into the pores of the activated carbon thus blocking potential adsorption sites for gold(I)thiourea complexes. Other inhibiting factors included the organic material present in the ore. The maximum gold loadings achieved in Avoca were in the range 0.6 - 0.7 kg Au/tonne using up to 10 g/litre of activated carbon. Due to the inefficiency of the activated carbon there was considerable loss of solubilised gold from the back end of the plant. Attempts were made to improve the loadings by daily washes of the activated carbon at high temperatures, but this had little effect on the adsorption kinetics. Schulze [21] in 1988 claimed that the washing of the carbon to remove sulphur compounds and thus regeneration of the adsorption capacity is possible, but this has yet to be proved on a commercial scale. After a lot of effort at Avoca the use of activated carbon was abandoned.

Since then Minmet has tried to find a cheap alternative method of recovering the leached gold from thiourea solutions and has just recently patented a method that is economical. Compared to activated carbon, the kinetics are considerably faster when used with low grade ores.

9 Modelling

The modelling and simulation of Leaching / Adsorption processes has been of interest to the hydrometallurgy industry for a number of years and many mathematical models have been formulated to describe Copper Leaching from Sulphide ores [30,31] and Gold cyanide adsorption onto activated carbon [32]. The reason for such interest is due to the complexity of most Leach / Adsorption systems and a good model can be used to give an insight into the chemistry of the system, at the same time being able to predict the real time behaviour. Thus, the purpose of this work was to apply existing models to the Thiourea Leach / Adsorption onto activated carbon process and to formulate new models where necessary.

Separate models were used to define Leaching and Adsorption. There are two types of model:

1) Empirical - based on a first order rate model and
2) Mechanistic - a more complex approach to the problem involving a study of the transport mechanisms(diffusion processes) occurring on the ore particle or the carbon.

For Mechanistic leaching models, most involve variations on the diffusion equation for a spherical particle i.e. the ore particle. It is seen as a two way process whereby the reagents diffuse through the surface of the particle while the gold complex formed in the reaction diffuses out to solution. As the reaction proceeds the unreacted radius of the particle decreases giving rise to the Shrinking-Core theory [33]. There are two equations reported by Roach & Prosser [34] and these are used as a basis for the following equations which govern the process:

1) Initial Complete Reaction: Here all the grains of gold are immediately accessible and are all used up.

$$\ln(1-\phi) = \ln(6/\pi^2) - \frac{\pi^2 \, Deff \, t}{R^2} \qquad (1)$$

where

ϕ = proportion of reactive phase recovered in time t
R = radius of ore particle
Deff = effective diffusion coefficient inside the particle.

$$= \frac{Drb}{\tau} \left(\varepsilon + \frac{\rho\alpha}{\rho r} \right)$$

α = mass fraction of reactive mineral in the ore.
ρ = density of ore.
ε = porosity of the ore
ρ_r = density of gold.
τ = tortuosity factor.
Drb = bulk diffusion coefficient of gold.

2) Incomplete initial (shrinking-core model) Reaction:

$$\phi + 3/2(1-\phi)^{2/3} = k_1{}^2 t + 3/2 \quad (2)$$

where $k_1{}^2 = \dfrac{7M'CDeff}{\rho\alpha\, n'\, R^2}$

C = initial concentration of gold
M' = molecular weight of gold
n' = no of moles of reagent consumed per mole of Au.

These equations have been found to correlate quite accurately with the experimental concentration vs. time leaching curve for the ore (see Fig.6).

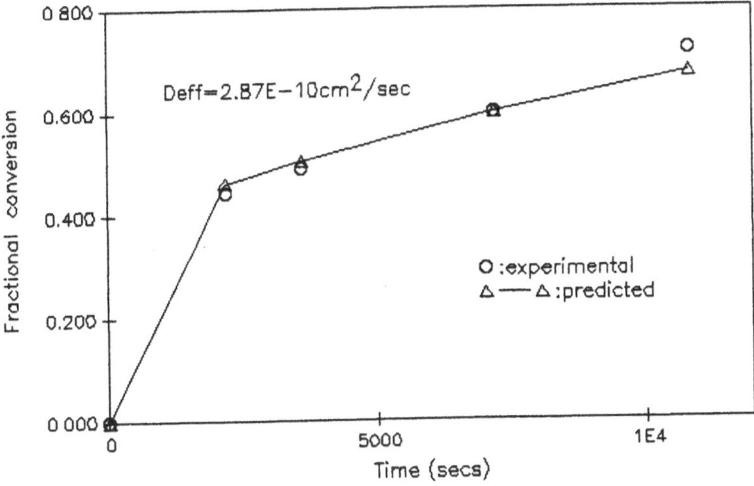

Fig. 6 Comparison of Experimental Leaching curve to curve predicted by Roach and Prosser's initial equation.

For the Adsorption model, various empirical model adsorption curves were compared to the experimental with varying degrees of success. Nicol & Fleming 's [35] model gave best results over a range of initial gold concentrations. The equation is given by:

$$q = k\, Co\, t^{n} \qquad (3)$$

where

q = mass of gold on the carbon
Co = initial conc of gold in solution
t = time
k,n = constants

Two mechanistic models for adsorption were proposed involving a film-transfer theory i.e. diffusion of gold from the bulk solution through a boundary of film surrounding the carbon particle - the Linear and Non-Linear film transfer models [32]. A third mechanistic model - the Homogeneous Surface Diffusion Model, which involved a spherical geometry approach to the diffusion equation was developed for the adsorption process. The two equations which govern the process are:

$$\varepsilon\, \delta c/\delta t + (1-\varepsilon-\alpha)\, \delta s/\delta t = 1/r^{2}\, \delta/\delta r\, [\, D\, r^{2}\, \delta c/\delta r\,] \qquad (4)$$

and

$$\delta s/\delta t = kc(s_{max}-s) - k_s/k_o \qquad (5)$$

where

ε = intraparticle void fraction
s_{max} = max adsorption
α = bulk liquid void fraction
s = gold complex conc in the solid phase
c = gold complex conc in the liquid phase.
k_o = Langmuir equilibrium constant.
k_s = Film coefficient.

These equations were solved using an orthogonal collocation technique and applying to a FORTRAN integration subroutine. Thus a plot of concentration vs. time was obtained for each initial gold concentration and these were compared with the experimental results. This gives very good correlation for a low initial gold concentration as can be seen in Fig. 7.

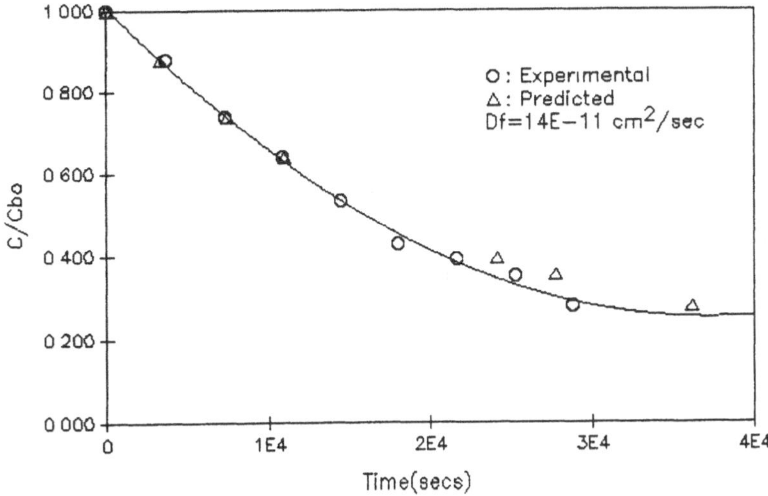

Fig. 7 Experimental vs Predicted Results for 9.967 mg/L initial gold concentration. Homogeneous Surface Diffusion Models.

Work was also done on the application of the process flowsheet simulator ASPEN PLUS [36] to the system. This was applied to the preliminary stages at the operation at Avoca, notably the screening and classification steps.

Overall, leaching and adsorption can be modelled by kinetic rate equations ranging from simple empirical models to the more complicated mechanistic models which accurately describe the actual processes occurring.

References

1. American Cyanamid Co. (1978) Chemistry of Cyanidation. **Mineral Dressing Notes,** No. 23, 7, .

2. Naggy, I., McKusic, P. and McCulloch, H.W. (June 1966) **RSA National Institute for Metallurgy,** Project C.28/62, Report No. 38, 4-11.

3. Henley, K.J. (1974) **Mineral Science and Engineering,** 7, No.4, 303-5.

4. Deschênes, G. (November 1986) **I.M. Bulletin,** 79 (895), 76-83.

5. **Mining Engineering,** (June 1992) Appendix I, 529.

6. Carter, R.A. (June 1992) **E & M J,** 29.

7. Von Michaelis, H. (June 1987) **E & M J,** 42-47.

8. Zipperian, J., Raghavan, S. and Wilson, J. (March 1986) **Proc. 115th A.I.M.E. Annual Meeting,** New Orleans.

9. Filmer, A.O. et al (Oct. 1984) **IMM Regional Conference: Proc. in Gold Mining, Metallurgy and Geology,** 1-8.

10. Kaloscal, G.I.Z. (1980) **Aust. Prov. Patent 30201/84.**

11. Plaksin, I. and Kozukhowa, M. (1941) **Doklady Akadomic Nark SSR,** 31, 671-4.

12. Plaksin, I. and Kozukhowa, M.A. (1960) **S. Nauchn. Tr. Inst. Isvetn. Metallov,** 33, 107-9.

13. Lodeishchikov, V., et al (1968) **Nauch. Tr., Inkutsk. Gros. Hauch. Insoled. Inst. Redk. Isvet. Metal.,** 19, 72-84.

14. Groenewald, T. (1976) **Hydrometallurgy, 1,** 277-90.

15. Chen, C.K., Lung, T.N. and Wan, C.C. (1980) **Hydrometallurgy,** 5, 207-12.

16. Schulze, R.G. (1984) **Journal of Metals,** 36, 62-5.

17. Van Lierde, A., Ollivier, P. and Lesoille, M. (1982) **Ind. Min., Les Tech., 1a,** 399-410.

18. Tataru, S. (1968) **Rev. Roam. Chim.,** 1043-9.

19. Kakovskii, I.A., Khmel Nilskaya, O.D., and Panchenko, A.F. (1982) **Fiz. Kh. Os. Pe. Mi. Syr.,** 148-55.

20. Gabra, G. (August 1984) Leaching of Gold from Pyrite and Chalcopyrite concentrates using Acidified Thiourea. **23rd Ann. C.I.M. Conf. of Met.,** Quebec.

21. Schulze, R.G., (January 1988) Gold Recovery in a Thiourea C.I.P. Process. **Randol Gold Forum 88,** Arizona.

22. Groenewald, T. and Jones, B.M (October,1971) **Analytical Chemistry,** 43(12).

23. Mullins, K. (1989) Precious Metal Extraction. **M.Sc. Thesis (N.U.I.).**

24. Deschênes, G. and Ghali, E. (1988) **Hydrometallurgy, 20,** 179-202.

25. Bilston, D.W., La-Brody, S.R., and Woodcock, J.T. (November 1984), **Aus. I.M.M., Symposium on "Extractive Metallurgy",** 51-60.

26. Groenewald, T. (1977) **J.S. Afr. I.M.M.,** 77 (II), 217-233.

27. Hendrix, J.L. and Pyper, R.A. (1981), **Extraction Metallurgy '81,** 57-75, (IMM- London).

28. Curran, S.A. (1991) Studies into Leaching of Gold from Avoca Ore. **M.Sc. Thesis (N.U.I.).**

29. Gallagher, N.P. et al (1990) **Hydrometallurgy,** 25, 305-316.

30. Madsen, B.W., Wadsworth, M.E. and Groves, R.D (1975). Application of a mixed kinetics model to the leaching of low grade copper sulfide ores. **Trans. Am. Inst. Min. Engrs.,** Vol.258, pp. 69-74.

31. Braun, R.L., Lewis, A.E. and Wadsworth, M.E. (1974) In-Place Leaching of Primary Sulfide Ores: Leaching Data and Kinetics Model. **Metall. Trans.,,** Vol. 5, pp. 1717-26.

32. Le Roux, J.D., Bryson, A.W. and Young, B.D. (1991) A Comparison of several Kinetic Models for the Adsorption of Gold Cyanide onto Activated Carbon. **Journal South African Institute of Mining and Metallurgy,,** Vol. 91, pp. 95-103.

33. Smith, J.M. (1970) **Chemical Engineering Kinetics**, 2nd ed., McGraw-Hill, pp.575-578.

34. Roach, G.I.D. and Prosser, A.P. (1978) Prediction of rates of chemical processes for treatment of low grade materials: theory and tests for mass-transfer controlled reactions. **Inst.Min.Metall.Trans.** pp. 129-138, no.87.

35. Nicol, M.J., Fleming, C.A. and Cromberg, G. (1984) The adsorption of gold cyanide onto activated carbon I. The kinetics of adsorption from pulps. **J.S. Afr. Inst. Mir. Metall.,** Vol 84, No. 2, pp50-54.

36. **Aspen Plus User Guide / Solids Manual** (1988) Release 8, Aspen Technology.

PRISMA—a hydrometallurgical process simulator: application in gold extraction from refractory pyrites

Antonios Kontopoulos
Laboratory of Metallurgy, National Technical University of Athens, Athens, Greece
Ioannis Paspaliaris
Laboratory of Metallurgy, National Technical University of Athens, Athens, Greece
N. L. Papassiopi
Laboratory of Metallurgy, National Technical University of Athens, Athens, Greece
O. N. Dimitropoulou
Laboratory of Metallurgy, National Technical University of Athens, Athens, Greece
Dimitrios Marinos-Kouris
Laboratory of Process Analysis and Design, National Technical University of Athens, Athens, Greece
Zacharias Maroulis
Laboratory of Process Analysis and Design, National Technical University of Athens, Athens, Greece
Theodoros Kritikos
Laboratory of Process Analysis and Design, National Technical University of Athens, Athens, Greece
Chris Kiranoudis
Laboratory of Process Analysis and Design, National Technical University of Athens, Athens, Greece
Nikolaos Voros
Laboratory of Process Analysis and Design, National Technical University of Athens, Athens, Greece

PRISMA is a software package for process design and simulation of hydrometallurgical processes. *PRISMA* features are summarised as follows: a) a user friendly interface for flowsheet construction and data introduction, b) flexibility in constructing and modifying the flowsheet, c) rigorous mathematical modelling of specific hydrometallurgical unit operations. *PRISMA* has been implemented in object-oriented language and runs on Apple Macintosh computers. Flowsheeting calculations are carried out using the standard sequential modular mode of computation. A salient feature of its integrated nature is that it supports the incremental development of a process design by interleaving modifications of the flowsheet structure with the simulations corresponding to them. Gold production from refractory arsenical pyrite concentrates is simulated as an application case. The main unit operations involved are: aqueous pressure oxidation, counter-current washing of the oxidised solids, neutralisation of the acid wash solution, cyanidation of the solid residues and gold recovery by CIP. Detailed simulation results concerning: a) gold liberation profiles and particle size distribution changes through the autoclave compartments, b) the influence of oxidised solids recycling on the energy balance of autoclave operation, c) the effect of CIP stages and initial carbon loading on the overall gold recovery are presented.

1. Introduction

Process flowsheeting and simulation has been focusing much research for a time span of more than three decades [1,2]. During this time, a large number of process simulation software packages have been produced, both in the commercial and academic field, i.e. ASPEN PLUS, SPEED UP etc., basically confronting the simulation of processes involved in the conventional chemical plants. Their differences lie mainly in the way that flowsheet equations are solved, and in the facilities they provide [3,4].

The need for detailed simulation of plants comprising units and streams that greatly differ in structure from those of a conventional chemical plant, leads to the construction of specific simulators which treat the particular processes in a more dedicated way, aiming at producing more accurate and robust results within an environment which is tailored to the needs of each flowsheet [5-9].

The use of conventional chemical process simulators in the case of hydrometallurgical processes is insufficient due to the particular problems that arise when solid particles and three-phase process streams are handled. It is desirable that a metallurgical flowsheeting package should be able to provide models describing how the particle size distribution of solid feed changes through the various unit operations of the metallurgical plant. During the last decade several general purpose software packages oriented to the mineral processing and metallurgical industry have appeared (METSIM [9], USIMPAC [10], etc.).

In the present paper the *PRISMA* Simulator, developed by the Laboratories of Metallurgy and Process Analysis and Design of the National Technical University of Athens is presented. *PRISMA (PRocess Integrated Simulator for Metallurgical Applications)* is a computer-based tool developed for simulation of hydrometallurgical processes. The program is a modular steady state simulator employing models for units and streams, which together comprise the flowsheet of a hydrometallurgical plant. *PRISMA* performs material and energy balances around individual unit operations or complex flowsheets with recycle streams.

2. Description of the *PRISMA* simulator

PRISMA is a simulator for hydrometallurgical applications. Pyrometallurgical applications are currently under development. The simulator runs on Apple Macintosh personal computers and has been developed in two versions. Version I was developed in Pascal, which is the programming language of Macintosh operating system. This environment has a remarkable advantage in speed and reliability when complicated flowsheet structures are simulated. Gold production from refractory pyrites, presented as application case in this paper, was developed in this version. In Version II, a high-level object-oriented language was used (Common LISP). This language offers a friendly framework and sophisticated programming tools for the fast development of new applications; furthermore, this environment will soon become portable and be able to run under Windows NT and X-Windows. Version II of *PRISMA* has been applied for the simulation of the following processes:

- Nickel extraction from laterites by sulphuric acid pressure leaching
- Leaching of yellow kaolin with oxalic acid
- Heap leaching of silica sand with oxalic acid
- The red mud thickening and washing circuit of the Bayer process. The entire Bayer process, including the digestion, precipitation and evaporation circuits are currently under development.

PRISMA offers some of the attractive features of a general purpose commercial simulator, but at the same time it is designed to satisfy the specific needs of particular processes. Each application is an integrated specific-purpose simulator. The available models of physical unit operations (i.e., comminution units, classifiers, thickeners, washers etc.), which are common in most hydrometallurgical plants, can be directly used in any application, but the chemical reactors are specifically designed for the particular processes. A special data base is also developed for each application, containing thermodynamic data and properties for the components involved in the specific process. The open architecture of *PRISMA* makes it highly configurable to the development of new applications.

Future development and enhancement of the *PRISMA* simulator involves:

- Implementation of a thermodynamic and physical properties database
- Generalisation of chemical processes (e.g. chemical reactors)
- Development of pyrometallurgical unit operations

The main features of the simulator and the steps followed to perform a simulation run are discussed in the following paragraphs.

2.1 Features of the *PRISMA* Simulator.

2.1.1 Integrated Graphical User Interface (GUI)

The graphics user interface was specially designed for *PRISMA*, using the Macintosh System 7 graphic environment. This friendly-user interface makes the use of *PRISMA* an extremely easy task. There is no need for the user to invest a lot of time to learn how to use the simulator. The interface consists of the selection menubar and the flowsheet worksheet as shown in figure 1. The *flowsheet drawing* is very easy. The user can select the unit operations from the menu, arrange the unit icons on the worksheet and create the appropriate stream connections between them. Each icon is connected with the mathematical model of the specific unit operation. The *data entry*, i.e. the description of the feed streams and the input of the operational parameters of each unit, are performed through dialogue boxes. The latter are activated clicking the mouse button above the unit icons. After the solution of the flowsheet, the user can read the *results of the simulation*, on any stream or unit of the flowsheet, through the dialogues which are again activated using the mouse. If desired, custom-made reports can be generated containing the information required by the user.

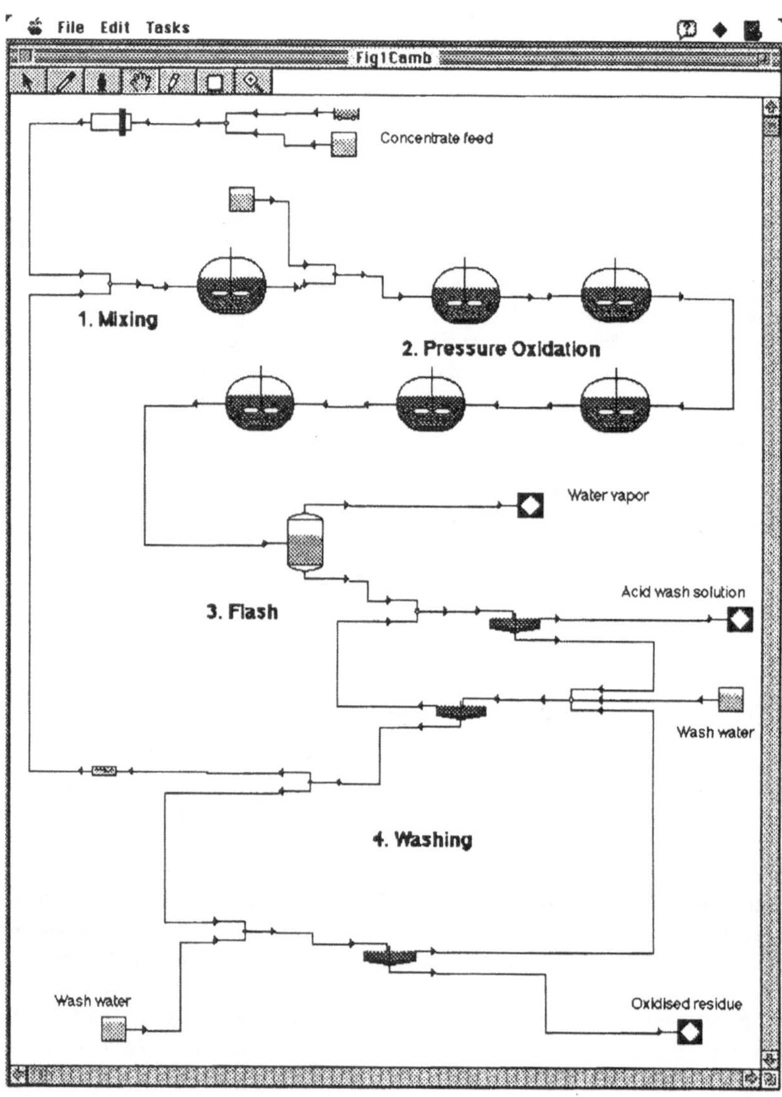

Fig. 1. Document window of the *PRISMA* simulator

2.1.2 Unit operations models

Unit operation models are the heart of any process simulator. Many of the requirements for simulating a metallurgical process are the same as for simulating other types of chemical manufacturing processes. New type of unit operation models, the coexistence of solid, liquid and gaseous phases, as well as the handling of solid particles are completely new requirements when dealing with specific hydrometallurgical processes. The unit operation models included in the *PRISMA* simulator up to now are listed in table 1.

Table 1. Unit Operation Models in the *PRISMA* Simulator

INPUT, OUTPUT UNITS
- Slurry feed and product units
- Carbon feed and product units
- Gas reject unit

PHYSICAL PROCESSES UNITS
- Stream mixers and splitters
- Comminution units (mills, crushers etc.)
- Classifiers (screens, hydrocyclones, filters)
- Thickeners
- Counter-current washers
- Flash tanks
- Heat exchangers

REACTORS
- CSTRs or PFR designed for specific processes
- Pressure oxidation autoclave
- Laterites pressure leaching autoclave
- Cyanidation reactor
- CIP-reactor
- Neutralisation tanks
- Desilication reactor

The models of physical processes are of general use. On the contrary, the reactors are specially designed for the specific chemical processes (the chemical reactions taking place in each unit and the associated thermodynamic data base and kinetic models are predetermined and cannot be modified by the user through the graphics interface). The models used may range from extremely simple to highly sophisticated ones, depending on the information available for the specific process. For instance, some chemical processes are described with a simple overall conversion factor. In this case the extent of reactions is specified by the user. When kinetic laws and equilibrium data are available, the extent is calculated by the model. The models for some crucial chemical operations included in the flowsheet of gold production will be presented in detail later in the article. Kinetic and equilibrium laws, as well as the influence of the initial particle size distribution of the feed material, are incorporated in these models.

2.1.3 Flowsheeting calculations

The *PRISMA* simulator uses the sequential modular technique to solve a flowsheet. Initial "guess" values must be entered for the recycled streams. The initial values given at the "tear" streams do not affect the final solution to which the flowsheet will converge. An iterative quasi-Newton algorithm is used in order to achieve fast and robust convergence.

2.2 The simulation procedure

The starting point in simulating a metallurgical process is the analysis and definition of the process flowsheet (figure 2). Once the problem is defined, the necessary information is introduced into the simulator. User input of the simulator consists of three parts:

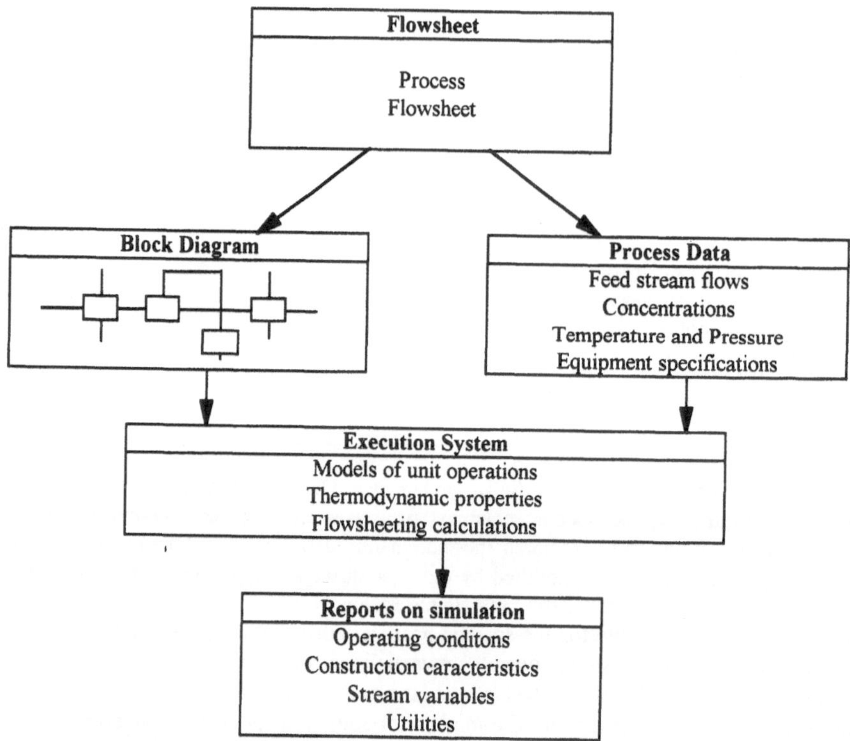

Fig. 2. Flow of information in a simulation run

a. the flowsheet description, i.e. the process units and the streams interconnecting them;
b. the design problem data, i.e. the feed stream data and the unit design variables specification;
c. the convergence algorithm specifications, i.e. the tear streams and the initial values of their properties.

As soon as all the appropriate information of the flowsheet is entered, the simulation run is carried out by the simulator execution system. The main components of the execution system are a) the unit operations models, b) the data base and the models to compute the thermodynamic properties of the components and c) the flowsheeting system to converge the integrated flowsheet with design constants and recycle streams.

Simulator output includes the composition and properties of all streams and the size and performance of the individual process units. Once a base case has been prepared, optimisation of the operating conditions of the plant can easily be achieved by simulating numerous alternative possibilities.

A complicated hydrometallurgical flowsheet, the aqueous pressure oxidation of refractory pyrite concentrates, has been selected as an example to show the capabilities of *PRISMA* simulator.

3. Application: gold extraction from refractory pyrite concentrates

The case of a gold extraction plant treating refractory arsenical pyrite concentrates with the aqueous pressure oxidation process has been developed on the *PRISMA* simulator. Pressure oxidation is used as an oxidative pretreatment step, aiming at liberating the refractory gold which is encapsulated in the sulphide lattice. Liberated gold is subsequently leached by cyanide and recovered by the Carbon in Pulp (CIP) method. This process has been developed by Sherritt Gordon, and described in detail in various publications [11-14].

3.1 Flowsheet description

3.1.1 Pressure oxidation circuit

A detailed description of the pressure oxidation circuit, as developed by Sherritt Gordon for various industrial applications, is given by Weir and Berezowsky [11]. The flowsheet of the circuit, as simulated on *PRISMA*, is presented in figure 1 through a document window of the simulator. The following unit operations are included:

1. *Mixing*. The concentrate is mixed with recycled oxidised solids and the resulting slurry is fed to the autoclave.
2. *Pressure oxidation in a 6-compartment autoclave*. The autoclave is simulated as a series of 6 continuous stirred tank reactors (CSTRs). The user can modify the autoclave configuration adding or deleting compartments in the flowsheet and specifying the size of each compartment.

3. *Flash tank.* The oxidised slurry is discharged from the autoclave through a flash tank and water vapour is released.
4. *Washing.* The slurry is then washed in a three stage counter-current decantation system. From the second wash thickener underflow a part of the oxidised solids is recycled to the autoclave feed tank and mixed with the fresh concentrate.

The *input streams* to the circuit are a) the concentrate feed and b) the wash water. The *output streams* are: a) the washed oxidised residue from the third thickener underflow, which is fed to the gold recovery circuit, b) the acid wash solution from the first thickener overflow which is directed to the neutralisation section and c) the water vapour from the flash tank.

3.1.2 Cyanidation- gold recovery circuit [11]

The washed oxidised residue is fed to the cyanidation reactor, where gold is leached from the solids with a sodium cyanide solution. The pulp is then directed to a cascade of six CIP reactors (figure 3). Carbon is fed to the last stage and transferred counter-currently, through the CIP cascade. The gold is transferred from the cyanide solution to the carbon. Output streams of the cyanidation-CIP circuit are a) the loaded carbon from the first CIP stage and b) the barren solution and solids from the last CIP stage.

3.1.3 Neutralisation Circuit [10-14]

The first thickener overflow containing the metals dissolved in the pressure oxidation is directed to the neutralisation circuit (figure 4). There, the solution is treated in the neutralisation tanks, first with limestone and then with lime, to precipitate ferric arsenate and metal hydroxides; the associated sulphate is removed as gypsum. The sludge is thickened and the underflow is mixed with the CIP tailing and rejected.

3.2 Components characterisation

Three types of components are present in the simulated flowsheet: solids, liquids and gaseous. A complete list is given in table 2. *Solid components* include feed material, oxidised residue, neutralisation solids and active carbon. *Liquid components* consist of water and soluble components, shown in the non-dissociated form in table 2. The concentration of corresponding ionic species is calculated in the reactors' models when equilibrium and precipitation reactions are involved. *Gas components* include oxygen, nitrogen and water vapour.

The *feed material* is described as a mixture of sulphides (arsenopyrite, pyrite and pyrrhotite described for simplicity as FeS), carbonates ($CaCO_3$ and $MgCO_3$) and insoluble components (SiO_2). These components are considered to represent, within a good approximation, the chemical and mineralogical composition of known arsenical gold concentrates (Red Lake, Sao Bento, Olympias etc. [14], for which the pressure oxidation process has been applied in an industrial or pilot-plant scale.

The route of gold from the sulphides, where it is initially disseminated, to the carbon, on which it is finally adsorbed, is described considering the following types of gold:

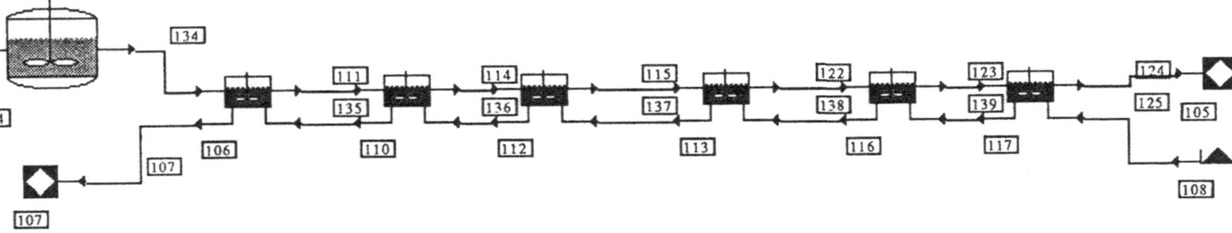

1 Fig. 3. Cyanidation and CIP circuit

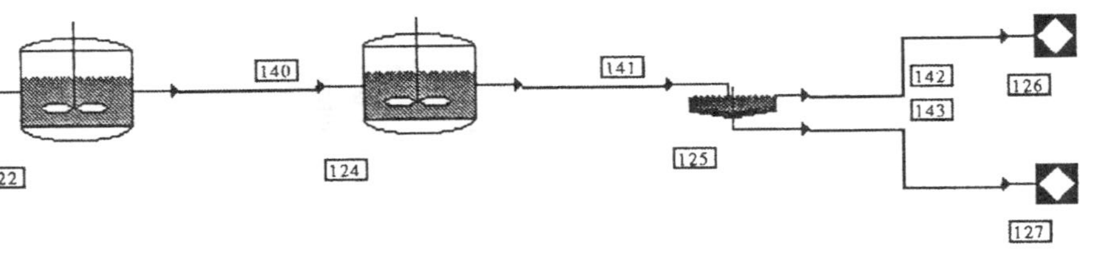

Fig. 4. Neutralisation circuit

Table 2. Components in the aqueous pressure oxidation process

SOLIDS
Feed material
FeS_2, $FeAsS$, FeS, $CaCO_3$, $MgCO_3$, SiO_2
Oxidised residue
$FeAsO_4$, $(H_3O)Fe_3(SO_4)_2(OH)_6$
Neutralisation solids
$Ca(OH)_2$, $CaSO_4.2H_2O$, $FeAsO_4$, $Fe(OH)_3$, $Fe(OH)_2$, $Mg(OH)_2$
Carbon
LIQUIDS
H_2O, H_2SO_4, H_3AsO_4, $FeSO_4$, $Fe_2(SO_4)_3$, $MgSO_4$, NaCN
GASES
O_2, N_2, H_2O

gold encapsulated in the sulphides (refractory gold), free gold, gold in cyanide solution and gold adsorbed on carbon.

The thermodynamic properties of the components have been taken from the HSC software package [15].

4. Mathematical models

To demonstrate the advanced process modelling that may be incorporated into the unit operations, the aqueous pressure oxidation of arsenical pyrites in autoclaves will be described.

4.1 Pressure Oxidation Autoclave

Pressure oxidation is a very complex hydrometallurgical process, involving three phases (gas, liquid and solid) and numerous physical and chemical processes taking place simultaneously between the three phases. The reactions and physical processes taken into account in the autoclave model are presented in table 3.

Physical processes include the mass transfer of oxygen [16-18] to the aqueous phase and vapour-liquid equilibrium between gaseous and aqueous phase

Oxidation reactions comprise the oxidative dissolution of sulphide minerals by the dissolved oxygen and the homogenous oxidation of ferrous to ferric ion in the aqueous phase [11-17]. The minerals dissolution reactions are heterogeneous, occurring on the solid - liquid interface. Thus, available surface area and subsequently particle size distribution of the feed is an important operating factor in pressure oxidation [16,17].

Precipitation reactions include the precipitation of ferric arsenate compounds and ferric hydrolysis products (iron oxides-hydroxides and jarosites) [13,14].

Table 3. Reactions in the pressure oxidation autoclave [11-17]

Physical Processes	
$O_2(g) = O_2(l)$	
$H_2O(l) = H_2O(g)$	
Oxidation Reactions	
Sulphides oxidation	*Gold liberation*
$FeS_2 + 3.5\ O_2 + H_2O = FeSO_4 + H_2SO_4$	$Au_{FeS_2} = Au_{free}$
$FeAsS + 3.25\ O_2 + 1.5\ H_2O = H_3AsO_4 + FeSO_4$	$Au_{FeAsS} = Au_{free}$
$FeS + 2\ O_2 = FeSO_4$	$Au_{FeS} = Au_{free}$
Ferrous oxidation	
$2\ FeSO_4 + 0.5\ O_2 + H_2SO_4 = Fe_2(SO_4)_3 + H_2O$	
Precipitation Reactions	
Ferric arsenate precipitation	
$Fe_2(SO_4)_3 + 2\ H_3AsO_4 = 2\ FeAsO_4 + 3\ H_2SO_4$	
Ferric hydrolysis	
$3\ Fe_2(SO_4)_3 + 14\ H_2O = 2\ (H_3O)Fe_3(SO_4)_2(OH)_6 + 5\ H_2SO_4$	

The development of the mathematical model was based on experimental studies [19,20] and on published modelling and experimental work on the Olympias pyrite concentrate [16,17] or similar pressure leaching systems [18]. Reaction kinetics and equilibrium are incorporated in the model as described in the following paragraphs.

4.1.1 Oxidation Reactions

The oxidation reactions control the rate and the thermal balance of the whole process. Four processes are considered to affect the kinetics, and the rate expressions incorporated in the model are shown in table 4. The following symbols are used in the equations: $C^*_{O_2}$ is the oxygen solubility in the aqueous phase related to oxygen partial pressure in the gaseous phase, P_{O_2}, through Henry's law, C_{O_2} is the oxygen concentration in the bulk of aqueous phase, k_L is the mass transfer coefficient and a is the gas-liquid interface; x is the conversion of the minerals, d_o the particle size and t the reaction time; $C_{Fe^{2+}}$ is the ferrous concentration in the aqueous solution; k^o are the chemical constants and E_a the activation energies.

The intrinsic kinetics of pyrite and arsenopyrite oxidation have been studied by Papangelakis and Demopoulos [16,17]. Conversions x are described using the shrinking core model (surface reaction control). Conversion is calculated for each size fraction and then integrated taking into account the initial particle size distribution (PSD). FeS dissolution is considered to be stoichiometric and not time dependent.

Table 4. Rate equations for the oxidation reactions [16,17]

Oxygen mass transfer:	$r_{O_2} = k_L a \cdot (C^*_{O_2} - C_{O_2})$	mole min^{-1}cm^{-3}
Pyrite conversion:	$x = 1 - \left(1 - k_1^o \exp\left(-\frac{E_{a1}}{RT}\right)\frac{C_{O_2} t}{d_o}\right)^3$	
Arsenopyrite conversion:	$x = 1 - \left(1 - k_2^o \exp\left(-\frac{E_{a2}}{RT}\right)\frac{C_{O_2} t}{d_o}\right)^3$	
Ferrous oxidation:	$r_{Fe^{2+}} = k^o_{Fe^{2+}} \exp\left(-\frac{E_a}{RT}\right) C_{O_2} C^2_{Fe^{2+}}$	mole min^{-1}cm^{-3}

Gold liberation rate is related to the dissolution rate of the auriferous sulphides. The initial distribution of gold amongst them can be specified by the user.

4.1.2 Precipitation Reactions

Although the precipitation reactions do not affect the kinetics of the pressure oxidation process and thus the autoclave design characteristics, they play a significant role in the overall plant performance, given that the precipitated oxidised solids are carrying the liberated gold.

The precipitation in the Fe-As-SO_4 system is an extremely complex process. A variety of ferric arsenate compounds, unidentified as yet [14,20] and ferric hydrolysis products are produced. The precipitation model developed for the *PRISMA* simulator is based on the following assumptions:

- The chemical composition of the solids can be described as a mixture of $FeAsO_4$ and H_3O-jarosite.
- Equilibrium conditions are attained and the extent of reactions can be calculated from the equilibrium concentrations of As and Fe in solution.

Analysing the available equilibrium and kinetic data from the experimental work previously conducted on the system [20-22], it was found that the As and Fe concentrations are correlated with the concentration of SO_4^{2-} ions in solution. The correlations are shown in figures 5 and 6 and the equations are given in table 5. (As) and (Fe) are the remaining concentrations in solution in g/l and (SO_4) is the total concentration of sulphate ions in g/l.

Table 5. Equilibrium equations for the precipitation reactions

$\log(As) = -3.011 + 1.66 \log(SO_4)$
$\log(Fe) = -1.478 + 1.42 \log(SO_4)$

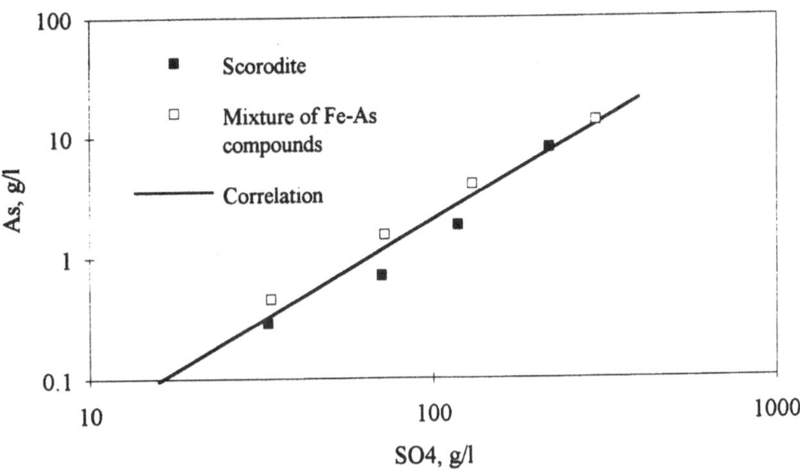

Fig. 5. Equilibrium concentration of arsenic at 190°C, in the Fe(III)-As(V)-SO_4-H_2O system [20].

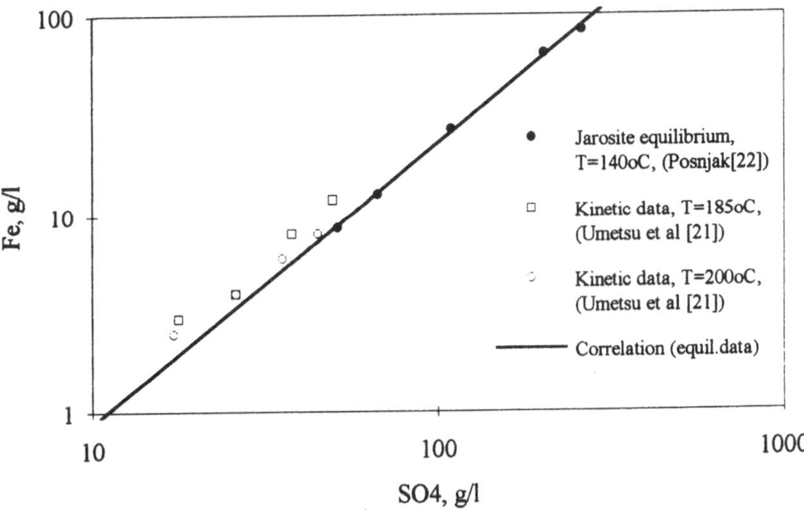

Fig. 6. Hydronium jarosite precipitation. Equilibrium data at 140 °C [22] compared with the metastable jarosite precipitation at 185°C and 200°C (one hour kinetic data, from [21]).

4.2 Cyanidation and CIP models

The kinetics of cyanidation is described using the Mintek expression [23]. In the equation given in table 6, r is the rate at which gold is leached from the solids into the solution, p is the concentration of the gold in the solids and p_o is the concentration after an infinite leaching time. This expression was developed for the case of direct cyanidation of ores and concentrates and p_o is related to the refractory portion of gold. The gold contained in the oxidised solids is assumed to be completely liberated and therefore p_o can be taken equal to zero.

Table 6. Rate and equilibrium equations in the cyanidation and CIP models [23]

Cyanidation

Leaching kinetics:

$$r = k_p(p - p_o)^2$$

CIP

Adsorption kinetics:

$$r = k_{ad}(x - x^*)$$

Adsorption equilibrium:

$$y = k_{eq}(x^*)^n$$

The CIP circuit is modelled as a series of continuous stirred tank reactors with a counter-current movement of pulp and carbon. The kinetics of adsorption is described with the classical model for film-diffusion mass transfer, while the Freundlich expression is used for the equilibrium [23]. In the equations given in table 6, r is the rate of adsorption per unit mass of carbon, x is the concentration of gold in solution and x* is the concentration in solution corresponding to the equilibrium with the carbon loading y. To have accurate predictions of the performance of the gold recovery circuit, the kinetic and equilibrium parameters must be estimated from plant operation data, taking into account that they strongly depend on solution chemistry and on the specific characteristics of the carbon used.

5. Simulation results

The simulation runs were performed for the treatment of Olympias gold concentrate, containing on the average 66% FeS_2 and 26% FeAsS and assaying 25g/t Au. 13 tonnes per hour of the concentrate are oxidised in a six compartment autoclave. The operating temperature and pressure are 190°C and 1800 kPa.

Percent sulphur oxidation through the autoclave compartments is presented in figure 7. Beginning with a 72% oxidation in the first compartment a 98.5% is achieved at the exit of the autoclave.

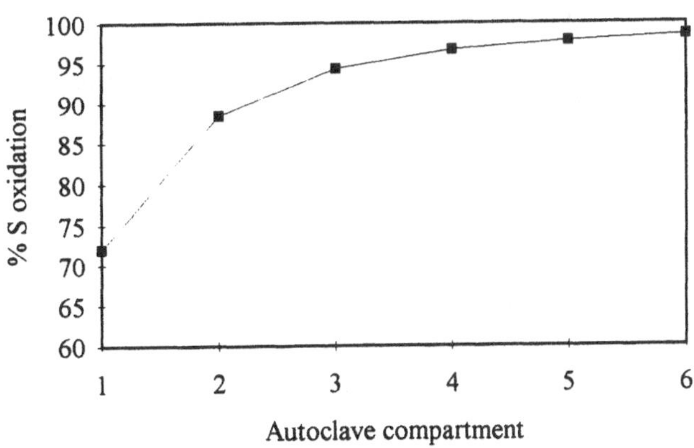

Fig. 7. Simulation results: percent sulphur oxidation in the autoclave compartments

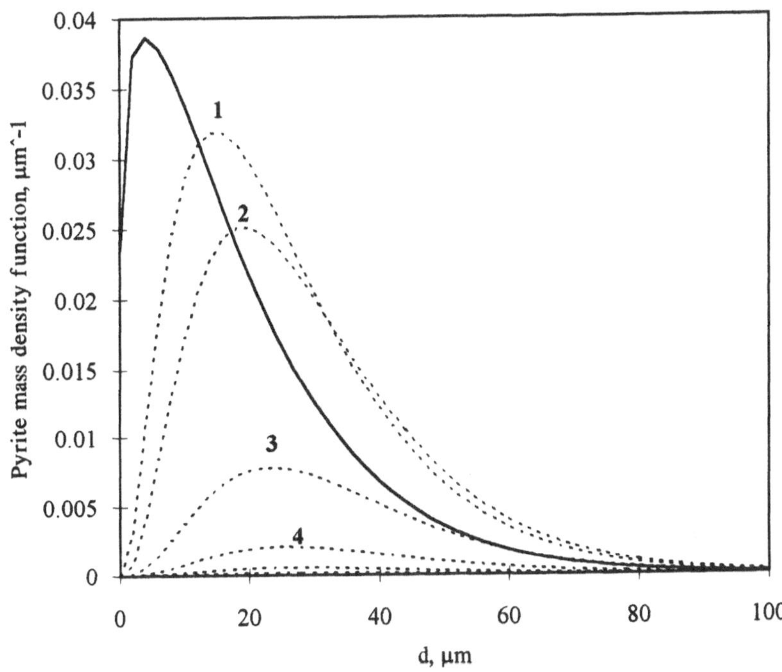

Fig. 8. Simulation results: mass density function of FeS_2 through the autoclave. Solid line: feed. Dashed lines: exit of each compartment.

The changes of mass density function through the autoclave is presented in figure 8. The solid line refers to the feed size distribution, while the dashed lines represent the output of each compartment. Small diameter particles are consumed in the first compartment and the size distribution shifts to greater diameters. Coarser particles, although they constitute a small percentage of the feed material, remain present even at the exit of the autoclave.

Recycling of oxidised solids from the washing circuit is incorporated in the pressure oxidation circuit in order to allow for the efficient dispersion of the sulphide minerals, eliminate elemental sulphur agglomeration during pressure oxidation and reduce the sulphide sulphur content of the feed down to the point where oxidation will be autothermal. The amount of recycled slurry is usually adjusted to maintain the operating temperature in the first compartment at the desired level. The sulphide oxidation reactions are highly exothermic and the excess heat produced must be removed. The oxidised slurry, cooled to about 30°C through the flash and washing operations, provides a heat sink for the exothermic reactions. The way that heat requirements in the first compartment vary with the recycle ratio of underflow oxidised slurry, is presented in figure 9. The curve has a linear form, so the required recycle ratio can be accurately calculated by means of only two simulation trials. 86.4% of the second thickener underflow slurry must be recycled in order to maintain the operation temperature to 190°C.

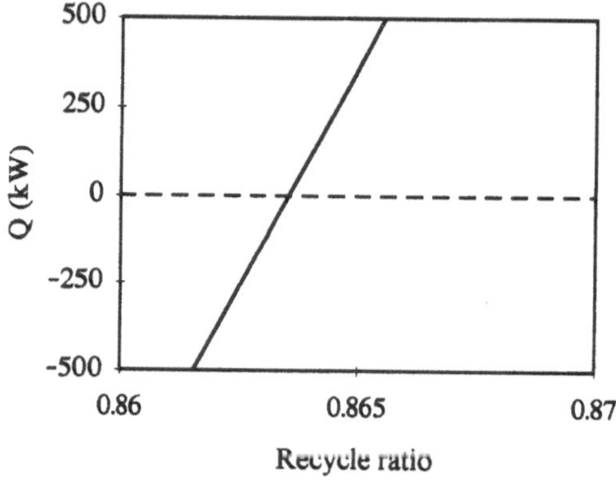

Fig. 9. Simulation results: variation of heat requirements in the first compartment of the autoclave with the recycle ratio of underflow slurry for steady-state operation at 190°C. Heat losses in the autoclave are neglected.

The amount of oxidised solids produced per tonne of concentrate is a function of the total liquid introduced to the autoclave per tonne of concentrate, as shown in figure 10. The total liquid flowrate includes the water introduced to the system with the concentrate and the recycled solids pulp, plus the cooling water, and it depends mainly on the energy balance requirements. Liquid flow rate will change should the temperature of any stream entering into the autoclave change (feed stream or recycled stream or cooling water). As seen in figure 10, the total amount of oxidised solids increases from 0.36 to 0.53 t/(t feed), when the liquid to concentrate ratio increases from 9.7 to 13.4 t/t. Gold concentration in oxidised solids drops respectively from 66 to 45 g/t. The decrease of gold concentration adversely affects gold recovery in the cyanidation circuit, as shown in figure 10, given that cyanide leaching kinetics strongly depends on the gold content of the oxidised solids.

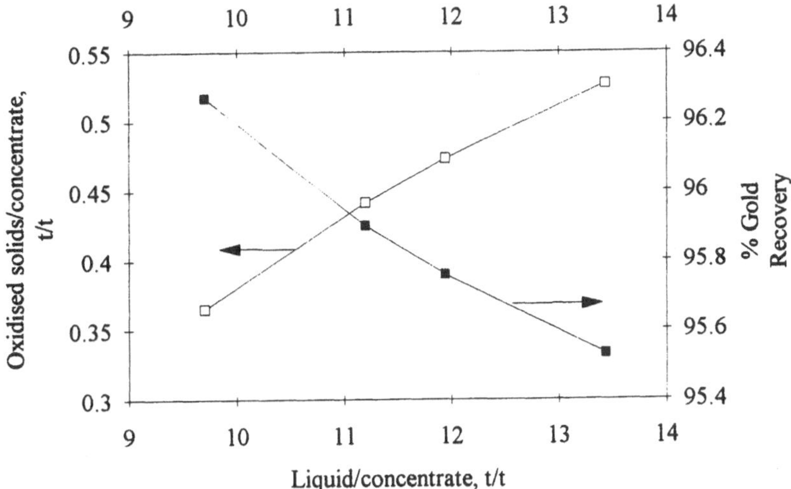

Fig. 10. The effect of total liquid to concentrate flowrates ratio on the amount of oxidised solids produced per tonne of concentrate. Effect on gold recovery during cyanidation

Examples of the simulation results of the CIP section are given in figures 11 and 12. In the case examined, 28 t/h of the leached pulp (25 wt%) are fed to a six-stage CIP circuit. Gold concentration in the feed solution amounts to 15 g/t. The carbon is fed to the last stage with an initial concentration of 25 g/t and is counter-currently transferred at an average rate of 26 kg/h. The profiles of gold in solution and carbon are shown in figure 9. Gold concentration in solution is seen to decline rapidly from the first to the fourth stage, while it remains almost constant in the last two stages. Five or even four stages, instead of six, would be sufficient in the case examined. Barren solution assays 0.002 g/t, which corresponds to 99.99% recovery in the CIP section. In this overdesigned circuit, the concentration of gold in the barren solution depends mainly on the initial concentration of gold in the carbon feed, as shown in figure 12.

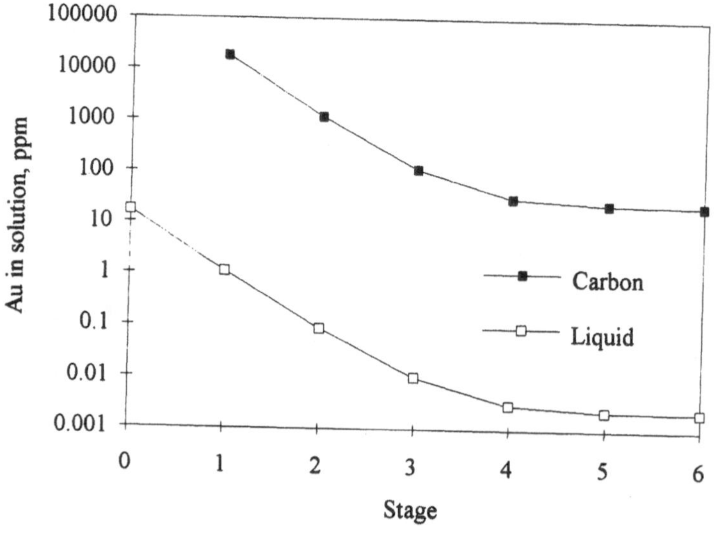

Fig. 11. Simulation results: gold profiles in solution and carbon in the CIP stages.

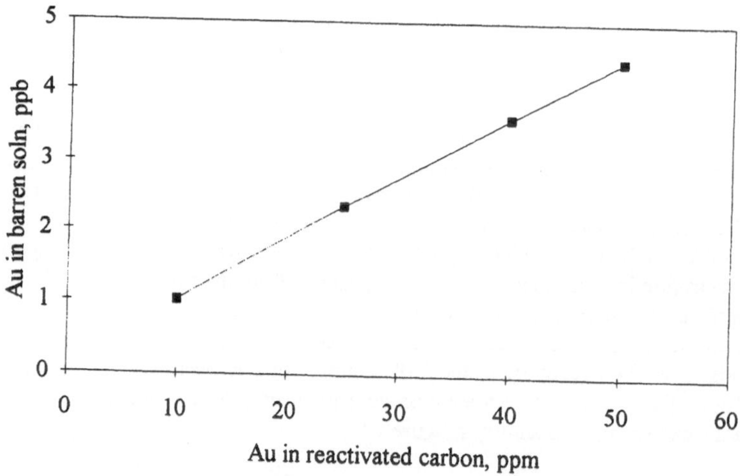

Fig. 12. Simulation results: the effect of gold concentration in the carbon fed to the CIP circuit on the concentration of barren solution.

6. Conclusions

PRISMA is a new simulation package for hydrometallurgical processes. It performs material and energy balances around individual unit operations or complex flowsheets with recycle streams. Its features are the friendly user interface, the flexible structure and the rigorous mathematical modelling of specific hydrometallurgical unit operations.

The simulation of gold extraction from refractory concentrates by the aqueous pressure oxidation process was presented as an application case. The modelling of some crucial unit operations of this process is described to demonstrate the advanced process modelling that may be incorporated into the unit operations. Detailed simulation results are presented to illustrate how *PRISMA* can be used to optimise the design and operation of the whole circuit.

Acknowledgements

The financial assistance of General Secretariat for Research and Technology of Greece, Aluminium de Grece SA, METBA SA and LARCO SA is gratefully acknowledged.

References

1. L.B. Evans, D.G. Steward, C.R. Spargue, "Computer aided chemical process design". *Chemical Engineering Progress.*, vol. 64, 1968, 39-46.
2. J.R. Flower and B.D Whitehead, "Computer aided design: a survey of flowsheeting programs". *The Chemical Engineer*, Part I, No272, Part II, No273, 1973.
3. J.L. Robertson, F.Park, "The ideal process simulator". *Chemical Engineering Progress*, vol. 85, 1989, 65-66
4. C.C. Pantelides, "SPEEDUP-Recent advances in process simulation", *Computers and Chemical Engineering*, vol. 12, 1988, 745-755.
5. M.J. Bush and P.L. Silveston, "Computer simulation of wastewater treatment plant" *Computers and Chemical Engineering*, vol. 2, 1978, 143-151.
6. D. Petrides, C.L. Cooney, L.B. Evans, R.P. Field and M. Snoswell. "Bioprocess simulation: an integrated approach to process development", *Computers and Chemical Engineering*, V13, 1989, 553-561.
7. K. Papafotiou, D. Assimakopoulos and D. Marinos-Kouris, "Synthesis of a a reverse-osmosis desalination plant. An object- oriented approach", *Transanctions of the Institution of Chemical Engineers*, V70, 1992, 304-312.
8. M. Spinos and D. Marinos-Kouris, "Integrated computer aided process design of waste water treatment plants on a PC system, *Water Science and Technology*, V25, 1992, 107-112.
9. J.T. Bartlett, "Process simulation and optimisation using METSIM", *Mineral Resources Management by Personal Computer*, Chapter 13, 105-116.
10. M.V. Durance, J.C. Guillaneau, J. Villeneuve, G. Fourniguet and S. Brochot, "Computer simulation of mineral and hydrometallurgical processes: USIM PAC 2,

a single software from design to optimisation", in *Modelling, Simulation and Control of Hydrometallurgical Processes*, Eds. V.G. Papangellakis and G.P. Demopoulos, CIM, Montreal, 1993, 109-121.

11. D.R. Weir and R.M.M.G.S. Berezowsky, "Aqueous pressure oxidation of refractory gold feedstocks", *International Symposium on Gold Metallurgy*, 26th Annual Conference of Metallurgist of the CIM, Winnipeg, Manitoba, August 1987.
12. A. Kontopoulos and M. Stefanakis, "Process selection for the Olympias refractory gold concentrate", *Precious Metals '89*, Eds. M.C. Jha and S.D.Hill, TMS, 1988, 179-209.
13. S.C.F. Au-Yeung and G.L. Bolton, " Iron control in processes developed at Sherritt Gordon Mines", *Iron Control in Hydrometallurgy*, Eds. J.E. Dutrizac and A.J. Monhemius, Ellis Horwood, Chichester, 1986, 131-151.
14. M.J. Collins, R.M. Berezowsky and D.R. Weir. "The behaviour and control of arsenic in the pressure oxidation of uranium and gold feedstocks", *Arsenic Metallurgy - Fundamentals and Applications*, Eds. R.G. Reddy, J.L. Hendrix and P.B.Queneau, TMS, Warrendale, Pennsylvania, 1988, 115-134.
15. HSC Chemistry for Windows, Ver.1.10, Outokumpu Research Oy, Pori, Finland, A. Roine, 1993.
16. V.G. Papangellakis, "Mathematical modelling of an exothermic pressure leaching process", Ph.D thesis, McGillUniversity, 1990.
17. V.G. Papangellakis and G.P. Demopoulos, "Reactor models for a series of continuous stirred tank reactors with a gas-liquid-solid leaching system. Part I-III", *Metallurgical Transanctions B*, Vol.23, 1992, 847-877.
18. E. Peters, "The mathematical modelling of leaching systems", *Journal of Metals*, vol..2, 1991, 20-26.
19. N. Papassiopi, M. Stefanakis, "Pressure oxidation of Olympias concentrate. Experimental study", Technical Report, METBA, Athens, 1991 (in Greek)
20. N. Papassiopi, M. Stefanakis, "Structure and stability of ferric arsenates", Research Project, METBA, Athens, 1991 (in Greek)
21. Y. Umetsu, K. Tozawa, K. Sasaki, "The hydrolysis of ferric sulphate solutions at elevated temperatures", *The Metallurgical Society of CIM*, Annual Volume, 1977. 111-117.
22. S.E. Posnjak and H.E. Merwin, "The system Fe_2O_3-SO_3-H_2O", *The Journal of the American Chemical Society*, vol.44, 1922, 1965-1995.
23. L.C. Woollacott, W. Stange and R.P. King, "Towards more effective simulation of CIP and CIL processes. 1. The modelling of adsorption and leaching", *Journal of the South African Institute of Mining and Metallurgy*, vol. 90/10, 1990, 275-282.

Interrelationship between lixiviants and galvanic interaction during dissolution of gold

L. Lorenzen
J. S. J. van Deventer
M. T. van Meersbergen
Department of Metallurgical Engineering, University of Stellenbosch, Stellenbosch, South Africa

ABSTRACT

Galvanic interactions between gold and associated substances are amongst the many factors that could influence the rate of gold dissolution in a cyanide, bromine or an acidic thiourea solution. Some of these galvanic interactions were investigated here by short-circuiting two-rotating disc electrodes either in one reactor or in two separate reactors linked by a salt bridge.

In cyanide solution it was established that the greatest decrease in the rate of gold dissolution was obtained when gold was in contact with copper, chalcopyrite and pyrite in the same reactor. On the other hand, in acidic thiourea solution, chalcopyrite and pyrite increased the rate of gold dissolution when both electrodes were in the same reactor. This was ascribed to sulphur species in solution that inhibited the degradation of thiourea to formamidine disulphide, one of the by-products of the dissolution of thiourea. Gold in contact with pyrite in bromine solution also enhances the gold dissolution rate.

Galena produced a very marked increase in the rate of dissolution of gold in cyanide solutions when both electrodes were in the same reactor. This could be ascribed to the action of Pb(II) ions on the surface of the gold. Galena causes the largest decrease in the rate of leaching when in contact with gold in bromine solution in the same reactor. This can be attributed mainly to the galvanic interaction between the two substances. However, galena had little or no affect on the rate of gold dissolution in an acidic thiourea medium.

As could be expected, the specific dissolution rate of gold on its own in acidic thiourea solution was higher than that in acidic bromine solution, which in turn was significantly higher than that in cyanide.

Keywords: Leaching; gold; cyanide; thiourea; bromine; galvanic interaction

1.0 INTRODUCTION

Over the last few years there has been an intense effort to identify lixiviants other than cyanide for gold leaching. Although traditional cyanidation remains the overwhelming choice for treating free milling gold ores, there are certain classes of ores and concentrates that are considered refractory. The inability of conventional cyanidation to treat these materials effectively, as well as the growing awareness of the detrimental effects of pollution on the environment, is still prompting the search for more powerful lixiviants.

During the leaching of gold in cyanide and thiourea solutions, gold is oxidized anodically to the aurous state. Subsequently, gold goes into solution either as the aurocyanide complex, $Au(CN)_2^-$ [1,2], or the aurothiourea complex, $Au[CS(NH_2)_2]_2^+$ [3-5], respectively. However, this is where the similarity between these electrochemical reactions ends. During the leaching of gold in bromine medium (N,N-halohydantoin Br_2(DMH)), gold is oxidized anodically to the auric state. Subsequently, gold goes into solution as the auribromide complex, $AuBr_4^-$, which has a stability constant of 10^{32} [6] in aqueous solutions.

Sufficient evidence exists for the cyanide medium to suggest that dissolved oxygen is reduced cathodically first to hydrogen peroxide and then to hydroxyl ion [7,8]. For the acidic thiourea system it is generally accepted that the oxidizing agents Fe(III), O_2 and H_2O_2 oxidize not only the gold, but also the thiourea to form formamidine disulphide [3-5], which may also take part in the oxidation [3]. Schulze [4] showed that when the formamidine was selectively reduced back to thiourea in order to maintain the oxidizing part of thiourea to about 50% of the initial concentration of thiourea, rapid dissolution of gold could be maintained. In a bromine medium, Br_2(DMH) is cathodically reduced to the very effective oxidizing agent HOBr (hypobromous acid). It is thus evident that all the reactions are electrochemical in nature, involving separate anodic and cathodic reactions irrespective of which of the three reagents is used.

Gold is closely associated with a variety of conductive minerals in the ore. These include amongst others pyrite, chalcopyrite, galena, pyrrhotite, arsenopyrite and sphalerite. Gold also often occurs in contact with oxide and gangue minerals such as haematite and magnetite. During milling, some of the gold may rub off, not only on these minerals, but also on highly conductive steels which are introduced during underground blasting. Various authors [9,10] have discussed the passivation of gold as a function of these minerals which are associated with gold. The method proposed by Lorenzen and Van Deventer [8] to investigate this phenomenon, differed from previous attempts in that both the gold and metal or mineral were in direct contact with each other and acted together as a working electrode, either in the same or separate reactors. The poor leaching characteristics of gold from various minerals are generally attributed to one or more of the following main factors:

(1) the presence of very fine particles of gold, which are occluded or trapped within the larger mineral particles, necessitating fine milling to liberate the gold,

(2) the formation of passivating films on the surface of the gold particles which isolate the gold from the lixiviant, and

(3) the gold ore is associated and in close contact with various conductive minerals causing galvanic contact between the gold and the associated mineral.

It is the objective of this paper to focus on the last point and to compare the leaching behaviour of gold in contact with various metals and minerals, and in solutions of cyanide, thiourea and bromine. The potential for using thiourea and bromine as replacements for cyanide in the leaching of gold was demonstrated by Chen et al [11] and Sergeant and Thanstrom [12], respectively. Their results have shown that the final levels of gold extraction were the same for cyanide, thiourea and bromine. The leaching kinetics, however, differ considerably. For the thiourea solution it was shown that the rate of gold dissolution is about 12 times as high as that for the corresponding cyanide solutions, while the leaching kinetics for gold extraction using bromine were about 4 times faster than those obtained for cyanide leaching.

2.0 EXPERIMENTAL

Rotating discs were used because this technique provided easily reproducible conditions and mass transfer from this geometry has been well established [10]. Pure gold, metal and mineral discs were used which avoided uncertainties regarding the effect of impurities and alloying elements. These discs were moulded into a teflon mould and the moulded disc was embedded in a teflon holder so that only its lower surface was exposed to the solution. The disc was then centrally mounted into a shaft at right angles to the plane of the disc. The surface of the teflon holder extended outwards, somewhat beyond the edges of the disc, and was flush with the disc surface.

The dissolution of a gold electrode in electrical contact with another electrode may be influenced by galvanic interaction, the species dissolved from the other electrode, or both effects. These dissolved species would either produce a passive film on the gold electrode, or could enhance the dissolution by another mechanism. If only the galvanic interaction was being investigated, the electrodes were placed in separate reactors as shown in Figure 1. If the combined effects of galvanic interaction and dissolved species were being investigated, the electrodes were palced in one reactor, as shown in Figure 2. All experiments involved a working electrode, the behavoiur of which was to be studied, a counter electrode (platinum) to complete the circuit and a reference electrode (calomel), brought into close contact with the working electrode by a Luggin capillary. With these electrodes connected to a voltammograph, current vs potential curves could be obtained for the working electrode in a particular test solution. The logging of data was accomplished by linking a microcomputer to the voltammograph via a data logging interface.

Before each experiment the working electrodes were prepared by polishing them with diamond paste and subsequently etching them in a 10 % ammonium persulphate solution. Finally the electrodes were conditioned for three minutes in a solutioncontaining either 10 % KCN, 10 % thiourea or 10 % Br_2(DMH), whichever was applicable. The physical and chemical conditions within the containers are summarised as follows:

Cyanide solution:

[KCN] = 0.2 g/l; pH = 10.3 - 10.5; [O_2] = 8.2 mg/l; Temperature = 20 °C; Rotational speed of electrodes = 100 rpm.

Thiourea solution:

[$Fe_2(SO_4)_3$] = 0.01 mol/l; [H_2SO_4] = 0.1 mol/l; [$CS(NH_2)_2$] = 0.1 mol/l; Temperature = 20 °C; Rotational speed of electrodes = 750 rpm.

Bromine solution:

[Br_2(DMH)] = 3.0 g/l; [NaBr] = 6.0 g/l, pH = 4-6; Temperature = 20 °C; Rotational speed of electrodes = 250 rpm.

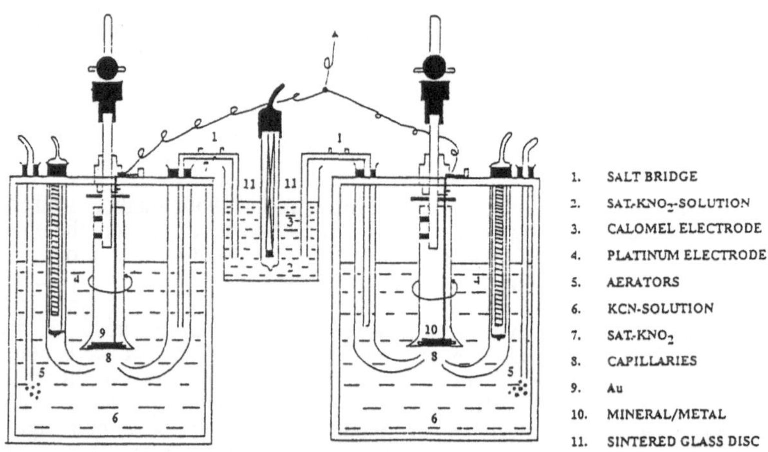

Figure 1 : Experimental equipment with the electrodes in separate reactors.

The mass of gold dissolved was monitored by taking samples at 30 minute intervals and analysing them for gold with an atomic absorption spectrophotometer. The results are presented in Figure 3.

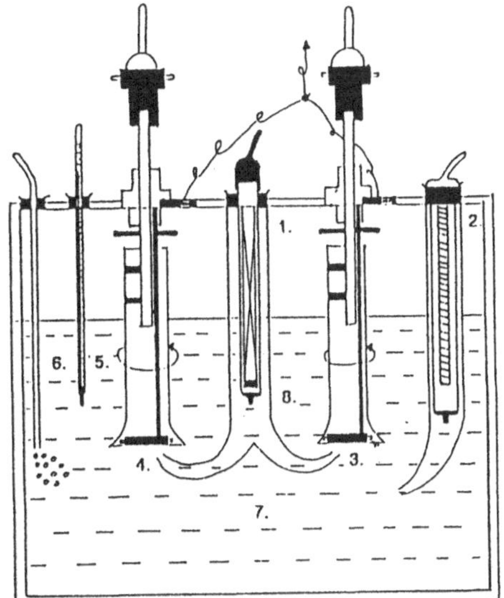

1. Standard Calomel Electrode
2. Platinum Electrode
3. Gold Disc
4. Mineral/metal disc
5. Thermometer
6. Aerator
7. KCN - solution
8. Luggin Capillary

Figure 2 : Experimental equipment with the electrodes in the same reactor.

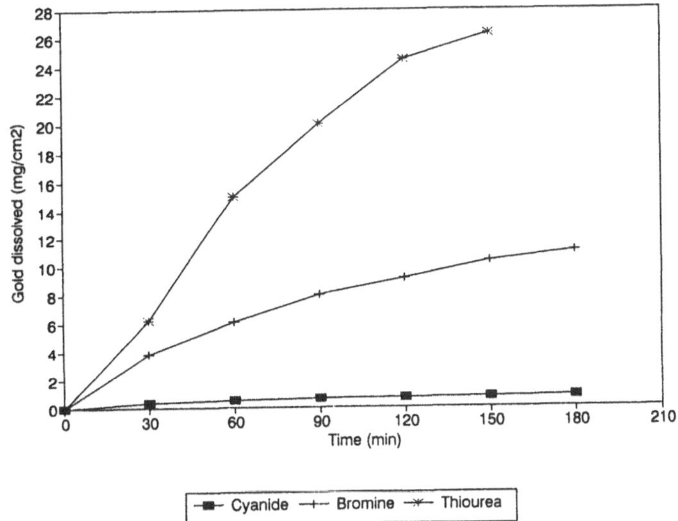

Figure 3 : Dissolution rate of gold (electrode area = 1.39 cm^2) in various lixiviants.

3.0 RESULTS AND DISCUSSION

From the analyses of the solution samples the average rate of gold dissolution, R, was calculated from the kinetic data as follows:

$$R = 1/(n\Delta t) \sum_{i=1}^{n} d_i/i \qquad (1)$$

where d_i = cumulative gold dissolved at time $i\Delta t$ in (mg Au/cm^2)

In this study n was 4 or 6 and Δt was taken at 30 min. By using this equation and the rate of gold dissolution alone as a reference, the entries in Table 1 could be calculated. A variety of minerals and metals were placed in galvanic contact with a gold disc in the same and in separate reactors, as described above. Table 1 presents the dissolution rates of gold in contact with such minerals, when both electrodes are in the same reactor and in separate reactors, respectively, as a percentage of the dissolution rate of gold on its own. The effect that each of the six metals and minerals in contact with the gold electrode exerted on the rate of gold dissolution, will be discussed below under separate headings.

Table 1 : A comparison of the relative dissolution rates of gold in contact with the various metals and minerals in cyanide, thiourea and bromine solutions.

Mineral/Metal in contact with gold	Cyanide		Bromine		Thiourea	
	One Cell	Two Cells	One Cell	Two Cells	One Cell	Two Cells
Gold	100	100	100	100	100	100
Copper	16	21	14	14	21	98
Mild Steel	59	71	10	15	5	83
Chalcopyrite	34	51	99	92	156	95
Pyrite	40	48	115	109	151	96
Galena	154	70	57	64	96	76
Haematite	91	91	102	97	80	95

3.1 Copper and mild steel in contact with gold

In all these experiments pure metallic copper and mild steel were placed in contact with gold respectively.

3.1.1 Cyanide solution

Gold in contact with a copper electrode in either one or two reactors produced a marked decrease in the rate of gold dissolution in Figure 4. The fact that film formation did not play a significant role when the two electrodes were in separate reactors, provides some evidence that galvanic interaction played the major role in retarding the dissolution rate. It is also clear from Figure 4 that the rate of dissolution decreased when gold was in contact with mild steel. Galvanic interaction decreased the rate of dissolution by 29 %. The additional decrease in the rate of gold dissolution when the electrodes were in the same reactor could be attributed to a greyish layer of iron cyanide which accumulated on the gold surface, hence forming a passivating layer. In the case of copper, the slightly lower rate of dissolution when both electrodes were in the same reactor points to a layer of a copper cyanide complex which possibly aggravated the passivation during the later stages.

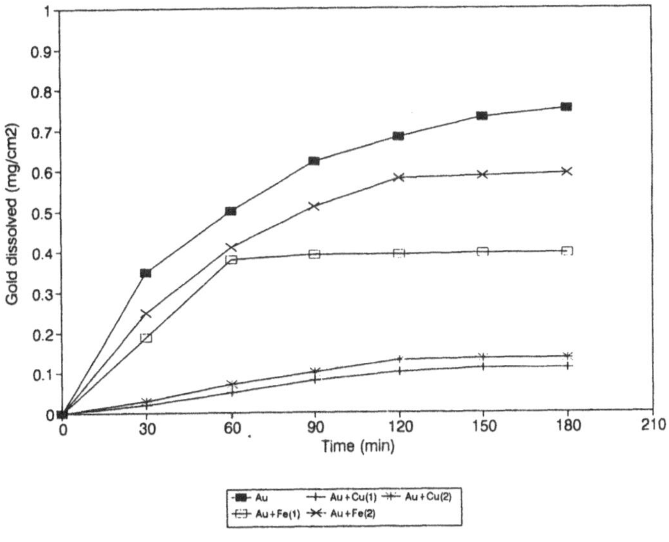

Figure 4 : Dissolution rate of gold in contact with copper and mild steel (electrode area = 3.8 cm^2) in cyanide solution.

3.1.2 Bromine solution

Figure 5 illustrates that when a gold disc was brought into contact with a copper disc (1 reactor), the dissolution rate of gold was distinctly reduced. When the two discs were placed in separate reactors, similar results were obtained. This provides enough evidence that the galvanic interaction between gold and metallic copper was the controlling factor in the reduction of the gold dissolution rate, and that the copper-bromide complexes which were probably formed on the surface of the gold only become a contributing factor to the decrease in the gold dissolution rate after a certain period of time. This was very similar to the results obtained with cyanide.

From Figure 5 it is also evident that Br_2(DMH) solutions react readily with metallic iron causing corrosion of the steel to a certain extent and that the rate of gold dissolution is markedly reduced. From the results it is also evident that galvanic interaction between mild steel and gold is the controlling factor in retarding the gold dissolution rate, and that the thin iron-bromide layers that may form only contribute marginally to the decrease in the gold dissolution rate.

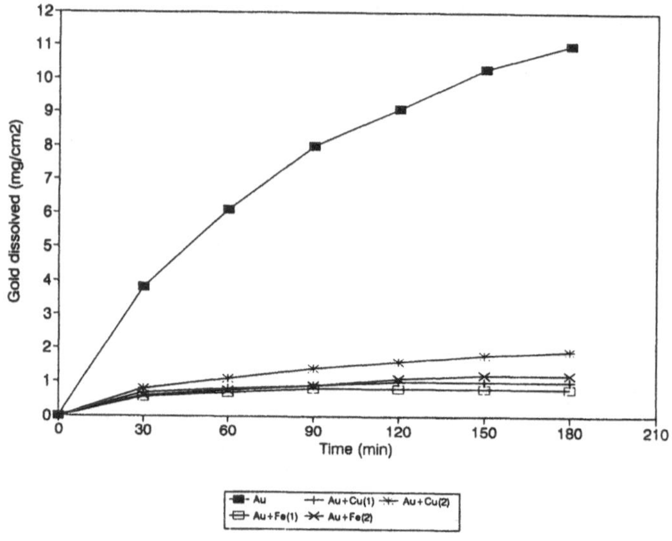

Figure 5 : Dissolution rate of gold in contact with copper and mild steel (electrode area = 3.8 cm^2) in bromine solution.

3.1.3 Thiourea solution

Whereas cyanide and bromine produced a large decrease in the rate of dissolution for the same as well as separate reactors, Figure 6 shows that the effect of copper in a thiourea solution was varied. If both electrodes were placed in the same reactor, gold was almost completely passivated. This was due to a brownish layer which covered the electrode. On the other hand, very little galvanic interaction was exibited when the two electrodes were in two different reactors. Similar results (see Figure 6) were obtained when mild steel was placed in contact with gold in thiourea solutions. Since the galvanic influence appeared to be less significant, the decrease could be attributed mainly to a light brown film (iron thiourea complex) that covered the gold electrode surface.

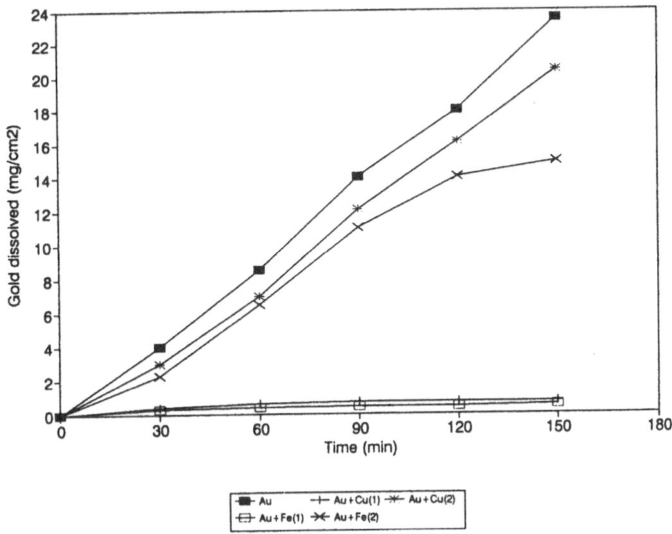

Figure 6 : Dissolution rate of gold in contact with copper and mild steel (electrode area = 3.8 cm^2) in thiourea.

3.2 Chalcopyrite and pyrite in contact with gold

The chalcopyrite (from the Orange Free State Gold Fields) used for the electrode contained Cu (as $CuFeS_2$) and Zn (as ZnS - separate phase) as the principal minerals, with traces of Mo, Mn, Ba, Co, Ni, Se, Rb, Sr and Pb also being present. The major impurity element in the sample of pyrite used was Cu, with trace elements Co, Zn, Se, Rb, Mo and Pb.

3.2.1 Cyanide solution

Figure 7 reveals that the galvanic interaction between gold and chalcopyrite or pyrite decreased the rate of gold dissolution significantly in cyanide solution. When the electrodes were in the same reactor, the rate was decreased even further owing to a layer of unidentified cyanide complexes which gradually caused an additional passivation of the gold electrode.

3.2.2 Bromine solution

As indicated in Figure 8, galvanic interaction decreased the rate of gold dissolution by 8% when chalcopyrite and gold were placed in separate reactors. This can be ascribed to the very high conductivity and low resistivity of the chalcopyrite used. On the other hand, when gold was in contact with chalcopyrite in the same vessel, an increase of 7% in the leaching rate was observed. The specific chalcopyrite was very soluble in the Br_2(DMH) medium, and it is possible that during leaching, the medium contained a significant amount of Zn^{2+} ions (Zn is the principal impurity) which enhanced the gold dissolution rate [6]. This possibly explains the higher leaching rate during the first hour or so. During this period sufficient chalcopyrite dissolved to form bromide complexes which probably formed a layer on the gold surface, which explains the sudden passivating trend after the first hour.

Pyrite is basically insoluble in bromine solutions and has a conductivity lower than that of many minerals associated with gold. From Figure 8 it is evident that an increase of 6% due to film formation and an increase of 9% due to galvanic interaction occurred. In both cases sulphur is formed during the reaction, which in aqueous medium will be oxidized to sulphate. In the first case the enhanced gold dissolution rate could possibly be ascribed to the high sulphur content, which, when oxidized to sulphate, probably formed a metal-sulphate layer on the gold surface which could lead to an increased conductivity. The passivation trend observed at 120 minutes is due to metal-bromide complexes which were probably formed on the surface of the gold and led to a decrease in gold dissolution rate.

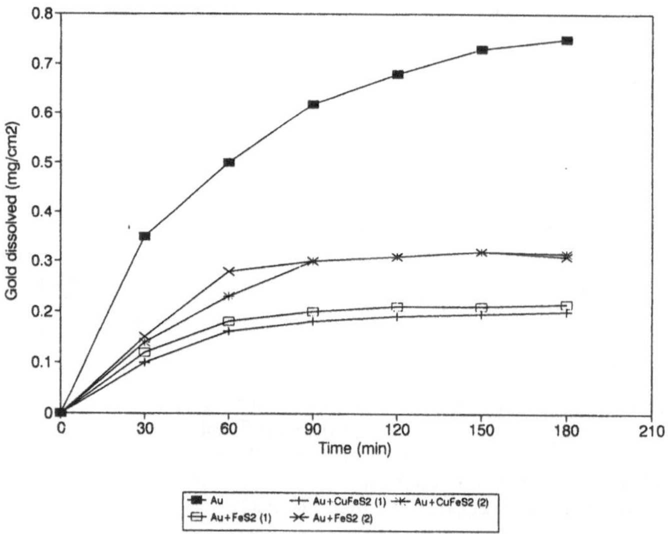

Figure 7 : Dissolution rate of gold in contact with pyrite (electrode area = 1.89 cm^2) and chalcopyrite (electrode area = 3.3 cm^2) in cyanide solution.

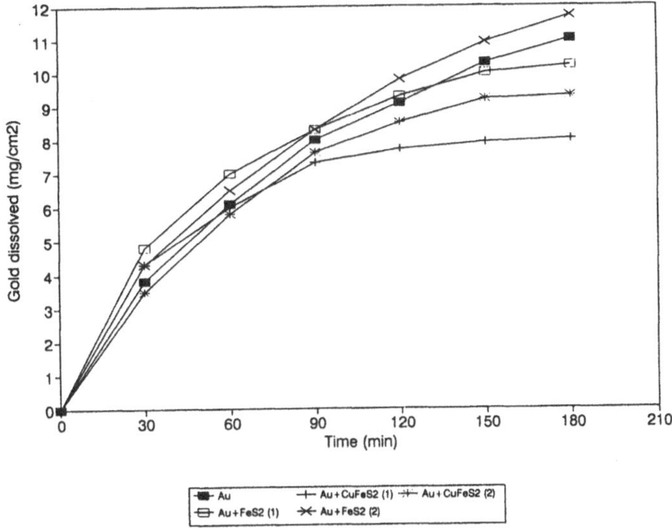

Figure 8 : Dissolution rate of gold in contact with pyrite (electrode area = 1.89 cm^2) and chalcopyrite (electrode area = 3.3 cm^2) in bromine solution.

3.2.3 Thiourea solution

When the electrodes were in separate reactors, the rate of gold dissolution in thiourea decreased only marginally for both pyrite and chalcopyrite (see Figure 9). This means that galvanic influences were more pronounced in cyanide than in thiourea. However, when both electrodes were in the same reactor, a large increase in the rate was produced, similar to those obtained in a bromine medium. This could be explained in terms of the regeneration of thiourea from formamidine disulphide by sulphur species dissolved from the sulphide mineral, and is in accordance with the results produced by Van Deventer *et al* [7].

3.3 Galena in contact with gold

The galena electrode contained Cu, Fe, Se and Zr as minor impurities with traces of Zn, Au and Ag.

3.3.1 Cyanide solution

As indicated in Figure 10, galvanic interaction decreased the rate of gold dissolution by 30% when the electrodes were in separate reactors. On the other hand, when the electrodes were in the same reactor the rate increased by 54%. This is ascribed to the lead ions in solution which increased the rate of gold dissolution [10]. This phenomnon could be attributed to an alteration of the surface character of the gold due to the alloying of Pb(II) ions with the displaced metals.

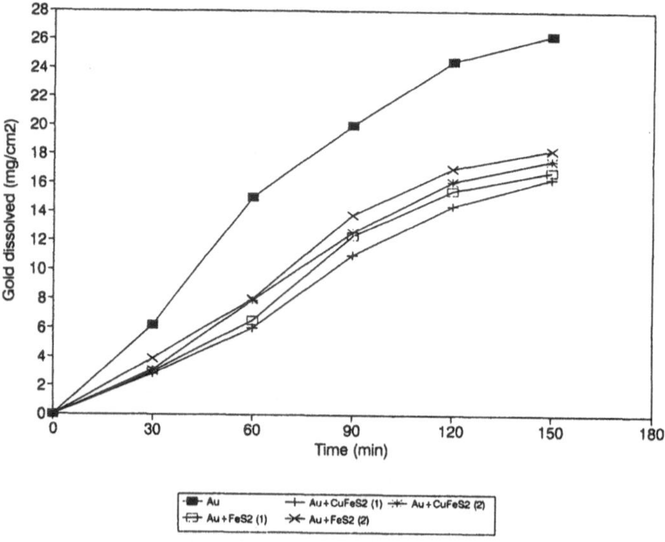

Figure 9 : Dissolution rate of gold in contact with pyrite (electrode area = 1.89 cm^2) and chalcopyrite (electrode area = 3.3 cm^2) in thiourea.

3.3.2 Bromine solution

Galena is only partially soluble in solutions containing Br_2(DMH) [6]. Furthermore, galena is very conductive with an extremely low resistivity, resulting in a reduced rate of gold dissolution (a total drop of 43% in dissolution rate of gold). From Figure 11 and Table 1 it can be seen that if the two discs were in separate reactors, but in electrical contact with each other, a drop of about 36% in the gold dissolution rate occurs. When the same experiment was carried out in the same reactor, an additional decrease in gold dissolution rate of 7% can be observed. This can be ascribed to the formation of lead-bromide complexes on the surface of the gold disc, which retarded the dissolution rate. The galvanic interaction between gold and galena was actually the controlling factor in retarding the gold dissolution rate. Obviously the Pb(II) ions did not exert the same enhancing effect in bromine solution as they did in cyanide medium.

3.3.3 Thiourea solution

A decrease of 24% in the dissolution rate was observed in Figure 12 when the two electrodes were in separate reactors, which indicates considerable galvanic interaction. The increased rate of dissolution when the electrodes were in one reactor, compared with the case of separate reactors, is not straightforward to explain. It was possible that sulphur species regenerated thiourea from formamidine disulphide, as was the case with chalcopyrite and pyrite in thiourea. This effect, however, was not as marked as in the case of chalcopyrite. A precipitate covered the galena elctrode during the later stages of dissolution, which probably decreased the concentration of sulphur species that could be used for regeneration. It could have been possible that the Pb(II) ions in solution acted in the same way as galena in cyanide.

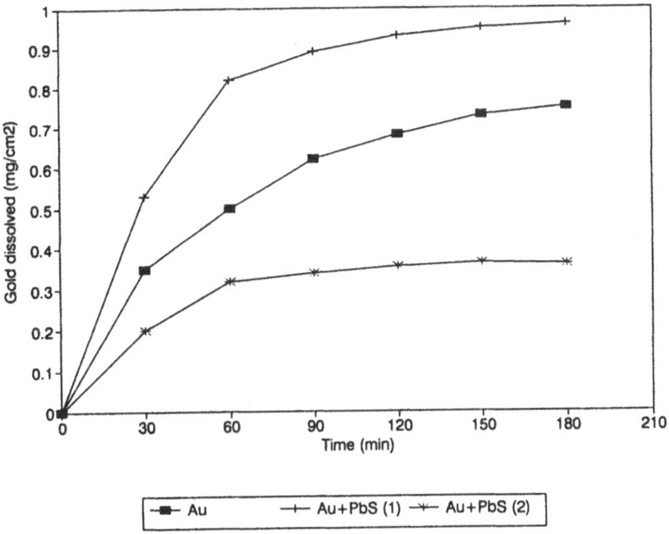

Figure 10 : Dissolution rate of gold in contact with galena (electrode area = 3.36 cm^2) in cyanide solution.

3.4 Hematite in contact with gold

Hematite is virtually insoluble in cyanide and does not really affect the gold dissolution rate when in contact with gold. From Table 1 it could be concluded that only galvanic interaction played a small role (decrease of 9%) in retarding the gold dissolution rate. This effect, however, was not very significant, which means that hematite had very little effect on the dissolution rate of gold in cyanide.

Similar results (Table 1) were obtained when Br_2(DMH) was used as lixiviant. In this case, however, the effect was so insignificant that it can be concluded that hematite has no effect on the dissolution rate of gold in bromine solutions.

The rate of dissolution did not change noticably when the two electrodes were in separate reactors in thiourea solution. This indicates that galvanic interaction was insignificant. Whereas the hematite was not readily soluble in cyanide, it seemed that hematite dissolved to such an extent in thiourea that the dissolution of gold was affected adversely when the electrodes were in the same container (Table 1).

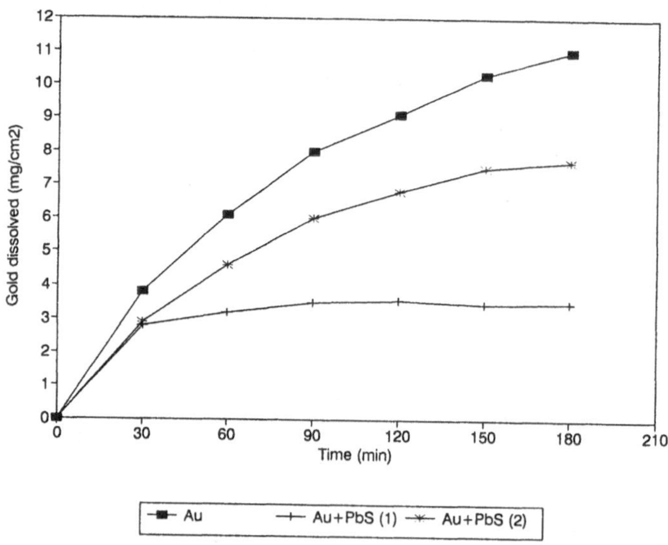

Figure 11 : Dissolution rate of gold in contact with galena (electrode area = 3.36 cm^2) in bromine solution.

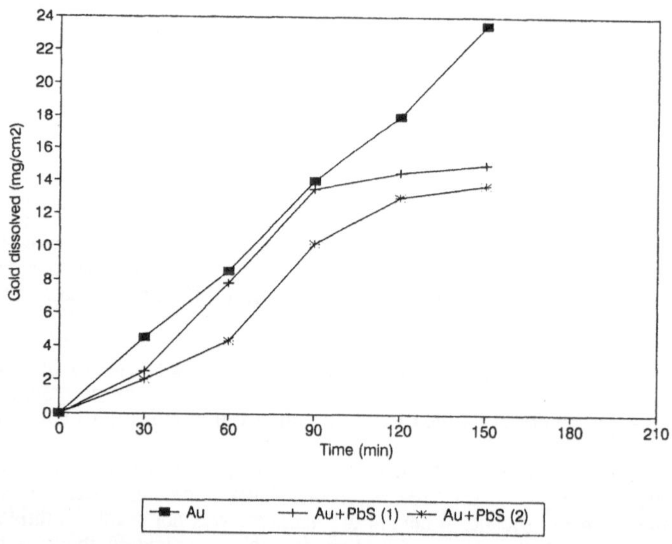

Figure 12 : Dissolution rate of gold in contact with galena (electrode area = 3.36 cm^2) in thiourea.

4.0 CONCLUSIONS

- The specific dissolution rate of gold on its own in thiourea solution was significantly higher than that in bromine solutions, which in turn was higher than that in cyanide.
- The gold electrode on its own in cyanide and bromine solutions passivated gradually, while the acidic thiourea prevented such behaviour.
- The rate of dissolution of gold in contact with copper and mild steel in cyanide and bromine solutions was inhibited significantly due to the galvanic interaction between these substances and gold. Similar results were obtained with pyrite and chalcopyrite in cyanide solutions. Mild steel showed some galvanic interaction, whereas copper showed very little interaction in thiourea.
- Additional film formation in cyanide decreased the rate even further, but not to such a large extent as galvanic interaction. Hematite had very little effect on the rate of dissolution in all three lixiviants used.
- A reasonable degree of galvanic interaction between gold and galena in separate reactors inhibited the rate of dissolution.
- Galena showed some degree of galvanic interaction, whereas the other minerals showed very little interaction in bromine and thiourea solutions. However, when the gold and galena electrodes were in the same reactor in cyanide solution, the dissolution rate was enhanced due to the action of Pb(II) ions on the surface of the gold. Galena in contact with gold in thiourea probably liberates sulphur species. This enhanced the rate of dissolution in the early stages of the reaction, but gradually formed a film on the gold, which inhibited the rate at a later stage.
- The large decrease in the rate of dissolution when steel or copper was in contact with gold in thiourea in the same reactor was caused mainly by film formation.
- The increase in the rate of dissolution when gold was in contact with pyrite or chalcopyrite in thiourea was caused probably by the ability of liberated sulphur species to regenerate thiourea.
- When gold was in contact with pyrite in bromine medium in separate reactors, galvanic interaction enhanced the leaching rate of gold, while film formation led to an additional enhancement.

From this study it is evident that the effect of galvanic interaction on the rate of gold dissolution depends mainly on the following two interactive factors:

- the mineralogy of the ore to be treated, and
- the lixiviant used.

5.0 ACKNOWLEDGEMENTS

Gratitude is expressed to the technical assistance provided by Mr P.J. Hoff and the technical staff of the Department of Metallurgical Engineering at the University of Stellenbosch.

6.0 REFERENCES

1. Cathro, K.J. and Koch, D.F.A., "The anodic dissolution of gold in cyanide solution", **J. Electrochem. Soc.**, 111 (12), (1964), 1416-1420.

2. Habashi, F., "Kinetics and mechanism of gold and silver dissolution in cyanide solutions", **Department of Metallurgy, Montana College of Mineral Science and Technology, Burea of Mines and Technology**, State of Montana, Bulletin 59, (1967), 42p

3. Groenewald, T., "Potential applications of thiourea in the processing of gold", **J.S. Afr. Inst. Min. Metall.**, 77 (11), (1977), 217.

4. Shultze, R.G., "New aspects in thiourea leaching of precious metals", **J. Metals**, 36 (6), (1984), 62.

5. Yen, W.T. and Wyslouzil, D.M., "Pressure oxidation and thiourea extraction of refractory gold ore", **Gold 100: Proceedings of an International Conference on Gold, Vol.2 - Extractive Metallurgy of Gold**, Eds. C.E. Fivaz and R.P. King, S.Afr.Inst.Min.Metall., Johannesburg, (1986), 579.

6. Van Meersbergen, M.T., Lorenzen, L. and Van Deventer, J.S.J., "The electrochemical dissolution of gold in bromine medium", **Minerals Engineering**, 6 (8-10), (1993), 1067-1079.

7. Van Deventer,J.S.J., Reuter, M.A., Lorenzen, L. and Hoff, P.J., "Galvanic interaction during the dissolution of gold in cyanide and thiourea solutions", **Minerals Engineering**, 3 (6), (1990), 589-597.

8. Lorenzen, L. and Van Deventer, J.S.J., "Electrochemical interactions between gold and its associated minerals during cyanidation", **Hydrometallurgy**, 30 , (1992), 177-194.

9. Mrkusic, P.D.P., "The recovery of gold from sulphidic and arsenic ores mainly from the Baberton area, **NIM Report No 911**, (1970 and 1975), 35p.

10. Lorenzen, L., "Galvanic interaction during the electrochemical dissolution of gold" , **M.Eng. thesis**, University of Stellenbosch, South Africa, (1984), 248p.

11. Chen, C.K. et al, "A study of the leaching of gold and silver in acidothioureation", **Hydrometallurgy**, (5), (1980).

12. Sergent, R.J. and Thanstrom, K.N., "Process for metal recovery and composition useful theirin", **U.S. patent no. 4 637 865**, (1987).

Comparative performance of porous adsorbents in presence of gold cyanide, organic foulants and solid fines

F. W. Petersen
Cape Technikon, Cape Town, South Africa
J. S. J. van Deventer
Department of Metallurgical Engineering, University of Stellenbosch, Stellenbosch, South Africa

Abstract
Activated carbon and ion exchange resins are being used extensively in the mining industry to extract metal cyanides from leached liquors. The application of ion exchange membranes and fibres as possible alternatives to carbon and resin has recently received special attention. The adsorption profile of gold cyanide in the presence of organic and inorganic substances was studied, and compared for the different adsorbents. The membrane and fibre showed much faster rates for gold uptake than the carbon or resin, and can be attributed to their large external surface areas. Results obtained showed that the kinetic parameters in a film/surface diffusion model are affected by inorganic substances and low organic solution concentrations for all the adsorbents. At higher organic concentrations both kinetic and equilibrium influences were observed. This was ascribed to permanent pore blocking of the adsorbents by organic compounds.

Furthermore, tests involving the intrusion of inert silica particles into the adsorbents indicated that the adsorption profile is affected only for activated carbon and ion exchange resins. These results were confirmed by ashing-tests performed on the adsorbents. However, the diffusion of gold cyanide to all adsorbents was inhibited owing to the physical blinding of the superficial surface by the fine particles in suspension. This result was obtained by exposing a presaturated adsorbent to suspended solids and dissolved gold cyanide. Furthermore, the effect of blinding was accounted for in the model with the introduction of an availability factor. The availability factor was determined by comparing results for a clear gold solution and pretreated adsorbents in suspended solids.
Keywords: Porous Adsorbents, Solid Particles, Metal Cyanides, Organics.

1 Introduction

Activated carbon has found increasing application in the mining industry as an adsorbent for the extraction of gold from cyanide leached liquors. The most common method for extraction gold is the carbon-in-pulp (CIP) process. Activated carbon has the advantage of being relatively selective towards base metals, but in certain applications ion exchange resins are more attractive, especially in gold plants where the pulp contains a high concentration of organic matter. Although ion exchange membranes and fibres have found increasing application in pollution control [1], it was only recently that these adsorbents were investigated as possible alternatives for gold adsorption.

Fleming and Nicol [2] and Jones and Linge [3] noted that fine suspended solids in a pulp may cause reduction in the rate of gold adsorption onto activated carbon. The same effect was observed by Jordi *et al*, [4], who concluded that adsorption by activated carbon is reduced significantly due to the physical blinding of the macropores by the fine material. Petersen and Van Deventer [5] showed a marked decrease in the adsorption rate of gold cyanide by a macroporous ion exchange resin in the presence of fine alumina and silica particles. It has generally been assumed that due to their large external surface area for adsorption, ion exchange membranes and fibres will have faster kinetics of gold uptake than ion exchange resins or activated carbon. However, fine slurry particles can externally blind the surfaces of these adsorbents, thus reducing the rate of adsorption of gold cyanide.

La Brooy *et al* [6] showed that organic substances adversely affect the adsorption of gold onto activated carbon. They attributed this effect to the blocking of pores within the carbon particle by organic compounds, thus inhibiting the diffusion of gold cyanide. Although similar results were obtained using an ion exchange resin [5], the effect of inhibition was more pronounced with activated carbon. If ion exchange membranes and fibres want to attain credibility as alternative adsorbents for gold, it will be necessary to investigate the effect of organic contaminants on their adsorptive behaviour.

A dual rate modelling approach has been used to evaluate kinetic and equilibrium parameters. These parameters have been used to compare the adsorptive performances of the different adsorbents.

2 Theoretical considerations

A dual resistance model involving both external film diffusion and intraparticle diffusion was applied to the profiles for the uptake of gold cyanide onto the different adsorbents. This model was then used to determine kinetic and equilibrium parameters so as to compare the adsorptive behaviour of the different adsorbents.

The following main assumptions have been made in the development of the model:

- The carbon and resin particles can be treated as equivalent spheres, while the fibre can be treated as an infinite cylinder for modelling purposes.
- Isothermal conditions are assumed during adsorption
- End effects are ignored for the fibre and membrane and diffusion is assumed to occur in one dimension.

- Accumulation of gold cyanide in the liquid phase within the pores of the adsorbents is negligible.
- It is assumed that the adsorption reaction on the carbon, and the ion exchange reaction on the resin, fibre or membrane occur instantaneously, so that equilibrium exists at the solid-liquid interface.
- The diffusivity of metal cyanide remains constant during a run, and is independent of the position inside the adsorbents.
- When the concentration of inert solids in the bulk liquid is significant, temporary 'blinding' of the external surface of the adsorbents may occur. An availability factor β is defined to account for this reduction in the effective surface area.

All mass balance equations developed for carbon, resin and fibre will be represented in radial co-ordinates, while those for the membrane will be recast in Cartesian co-ordinates.

2.1 Material balance inside adsorbent (A)

Activated Carbon and Resin:

$$\frac{\partial Q}{\partial t} = D\frac{\partial^2 Q}{\partial r^2} + \frac{2D}{r}\frac{\partial Q}{\partial r}$$

Membrane:

$$\frac{\partial Q}{\partial t} = D\frac{\partial^2 Q}{\partial x^2}$$

Fibre:

$$\frac{\partial Q}{\partial t} = D\frac{\partial^2 Q}{\partial r^2} + \frac{D}{r}\frac{\partial Q}{\partial r}$$

2.2 Material balance for the liquid phase in a batch reactor (B)

Activated Carbon and Resin:

$$\frac{dC}{dt} = \frac{k_f A\beta}{V}(C_s - C), \text{ where } A = \frac{6M_c}{\rho_c d_c} \text{ (carbon)}$$

$$A = \frac{6\varepsilon V_R}{d_r} \text{ (resin)}$$

Membrane:

$$\frac{dC}{dt} = \frac{4M_m k_f\beta}{\rho_m d_m V}(C_s - C)$$

Fibre:

$$\frac{dc}{dt} = \frac{4 M_f k_f \beta}{\rho_f d_f V} (C_s - C)$$

2.3 Boundary condition: No accumulation occurs at external surface of adsorbent (C)

Activated Carbon, Resin and Fibre:

$$\frac{\partial Q}{\partial r} \Big|_{r=R} = \frac{k_f \beta}{\Phi D} (C - C_s), \text{ where } \Phi = \rho_c \text{ (carbon)}$$

$$\Phi = \frac{1}{\varepsilon} \text{ (resin)}$$

$$\Phi = \rho_f \text{ (fibre)}$$

Membrane:

$$\frac{\partial Q}{\partial x} \Big|_{x=a} = \frac{k_f \beta}{\rho_m D} (C - C_s)$$

2.4 Equilibrium expressions to relate C_s and Q (D)
(See Table I and Table II for values of constants)

The Freundlich expression fitted the equilibrium for CARBON, MEMBRANE AND FIBRE:

$$Q_e = B\, C_e^n$$

The Langmuir expression fitted the equilibrium for RESIN:

$$Q_e = \frac{K_1\, C_e}{K_2 + C_e}$$

Equations [A] were transformed to ordinary differential equations by the use of average metal loadings in the pores. The value of C_s is guessed and the calculation of equation [A] repeated with a fourth order Runge-Kutta method until equation [C] is satisfied. The value of β can be determined only by comparing the results for clear gold cyanide solutions and pretreated adsorbents in a silica suspension. Hence, it is not possible to determine k_f and β simultaneously in a slurry.

3 Experimental

Adsorption-tests were carried out in a tank of internal diameter 11 cm, height 15 cm, and three evenly spaced baffles of width 1 cm. Agitation was provided by a flat blade impeller of width 6 cm and height 5 cm, driven by a Heidolph variable speed motor. At stirring speeds higher than 100 r.p.m., the fine inert particles as well as the adsorbents were well-mixed, and no sedimentation occurred.

Potassium aurocyanide dissolved in deionized distilled water was used as the adsorbate. Excess potassium cyanide was added in order to maintain a desired ionic strength and free cyanide level. Potassium ethyl xanthate (PEX) and sodium ethyl xanthate (SEX), both dissolved in water, were used as organic foulants. In experiments designed to quantify the effect of fine inert particles on the rate of gold loading, precipitated silica with a particle size smaller than 15 μm was used.

A Varian Techtron AA-1275 atomic absorption spectrophotometer was used for the analysis of gold cyanide solution. The organic foulant concentrations were determined by measuring the ultraviolet absorbance on an LKB Ultrospec II E ultraviolet spectrophotometer at 303 nm.

The pH of all solutions was monitored by using a Beckman Chem-Mate pH-meter. A muffle furnace was used for ashing tests to determine the intrusion of inert particles into the adsorbents.

The following adsorbents were used:

(a) A coconut shell activated carbon, Le Carbone G210 AS, with an average diameter of 1.4 mm and apparent density of 838 kg.m^{-3},

(b) A macroporous ion exchange resin, Duolite A161, with an average diameter of 0.8 mm and a wet-settled density of 700 kg.m^{-3},

(c) An ion exchange membrane, Ionac 3475, cut into 1 x 1 cm squares and treated by washing alternately with a 5% sodium hydroxide solution and a 5% sulphuric acid solution, and

(d) Three types of ion exchange fibres, namely:
   (i) Polypropylene-based strongbase fibre
   (ii) Polypropylene-based weakbase fibre
   (iii) Actilex B 402 weakbase fibre

Only selected results will be presented in this paper, although most of the observations will be discussed qualitatively.

4 Results and discussion

4.1 Effect of solid fines on internal gold diffusion

Adsorption tests were conducted on activated carbon, ion exchange resins, ion exchange fibres and ion exchange membranes by pretreating these adsorbents with 3 g/L of silica-sand for different time intervals, after which these adsorbents were rinsed with water and exposed to a clear 20 ppm gold cyanide solution. Results obtained showed that the adsorption profile for the uptake of gold cyanide was unaffected when ion exchange fibres and membranes were used. This indicated that the fine silica particles did not intrude into the pores of these adsorbents, and hence could not inhibit gold adsorption.

However, in the case of activated carbon and ion exchange resin, a reduction of the rate of gold adsorption was observed, as illustrated in Figures 1 and 2.

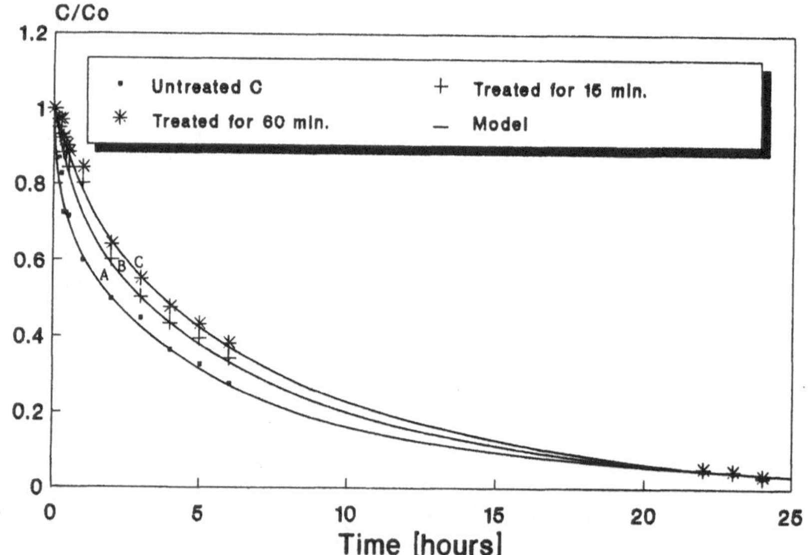

FIGURE 1: Inhibited mass transfer (kinetic parameters) of gold cyanide to activated carbon owing to fouling by silica particles. (C_o = 20 mg Au/L; V = 1,0 L; N = 250 rpm; M = 1,0 g)

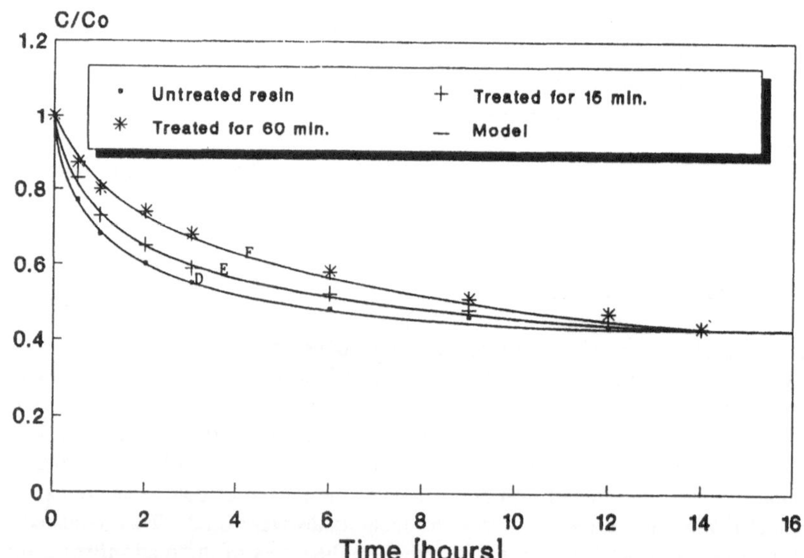

FIGURE 2: Inhibited mass transfer (kinetic parameters) of gold cyanide to resin DU A161 owing to fouling by silica particles. (C_o = 20 mg Au/L; V = 1,0 L; N = 250 rpm; V_R = 2,0 ml)

Table I. Sensitivity of **kinetic** parameters for gold adsorption onto different adsorbents

Curve	Adsorbents	Conditions	$k_f \times 10^5$ [m/s]	$D \times 10^{12}$ [m^2/s]	β*
A	Carbon:	Untreated	3,74	4,3	-
B	Carbon:	Pretreated with silica for 15 min.	3,71	3,8	
C	Carbon:	Pretreated with silica for 60 min.	3,70	3,2	
D	Resin:	Untreated	2,95	3,1	-
E	Resin:	Pretreated with silica for 15 min.	2,92	2,74	
F	Resin:	Pretreated with silica for 60 min.	2,90	2,2	
G	Carbon:	Clear Au-solution	3,6	2,9	1,00
H	Carbon:	3g silica/1L Au	3,6	2,9	0,89
I	Carbon:	5g silica/1L Au	3,6	2,9	0,85
J	Carbon:	12g silica/1L Au	3,6	2,9	0,82
K	Fibre:	Untreated	3,4	3,2	-
L	Fibre:	Pretreated with SEX for 24 hours	3,4	2,8	
M	Membrane:	Untreated	4,7	1,9	-
N	Membrane:	Pretreated with SEX for 2 hours	4,7	1,7	
O	Membrane:	Pretreated with SEX for 24 hours	4,7	1,5	

***Availability factor**

These experiments were performed with an initial gold concentration of 20 mg/L. From the model parameters given in Table I, it can be seen that the values for the film transfer coefficient k_f decrease slightly with an increase in the period of pretreatment, while large differences in the values for the surface diffusivity D were observed. It is interesting to note that the equilibrium of adsorption for both these adsorbents was not influenced. It is suggested that the inert material penetrated the porous matrices of the carbon and resin particles, partially blocked off the pores and hence hindered the diffusion of gold cyanide into the porous structure. Furthermore, the inhibition of gold cyanide was more prominent with activated carbon, as reflected in the values of the surface diffusivity D in Table I, and can be attributed to its larger pore sizes. It is also interesting to note from Table I that the k_f values for activated carbon are higher than those for resin beads. Earlier work [7] has shown that rough particles, such as activated carbon, have significantly higher values of k_f.

4.2 Effect of solid fines on external mass transfer

As mentioned earlier, the presence of inert particles reduces the rate of gold adsorption onto activated carbon and ion exchange resins. However, with the adsorbent particles suspended in the bulk of the fine material, it is also possible that shielding or blinding of the adsorbents can hinder the gold cyanide from reaching the external surface. In this study a systematic approach was taken to elucidate this phenomenon. Petersen and Van Deventer [5] showed that the loading of inert silica onto a macroporous resin and activated carbon reached an equilibrium after a period of 3 hours. Therefore, samples of resin and activated carbon were pretreated with 3 g silica/L for 5 hours to ensure complete saturation of the adsorbents. As mentioned earlier, in the case of ion exchange membranes and fibres, no intrusion has taken place, so that these adsorbents were not pretreated prior to adsorption.

All the different adsorbents were then contacted with solutions of gold cyanide containing different concentrations of silica particles in suspension. Normally, it could be argued that any decrease in the kinetics of adsorption associated with an increase in silica concentration would be caused by a reduction in mixing efficiency. In these experiments a high stirring speed of 1000 r.p.m. was used, and it was found that at stirring speeds of between 300 and 1000 r.p.m., the rate of gold adsorption by the adsorbents was independent of stirring speed. This means that the effect of mixing on the film coefficient k_f could be ignored.

It is evident from Figure 3 that an increase in the concentration of silica in suspension reduced the rate of mass transfer of gold cyanide to activated carbon significantly, but did not influence the attainment of equilibrium. Similar results were obtained for resin, ion exchange membrane and ion exchange fibre. This inhibition of mass transfer could not be ascribed to *blocking of the pores* (in view of thorough pretreatment of resin and carbon, or the fact that no intrusion has taken place in the case of membrane and fibre), or *lower mixing efficiency* (in view of the high stirring speed). Hence, the decrease in the rate of adsorption of gold cyanide is ascribed to the partial blinding of the external surface of the adsorbents. This resulted in a reduced availability of the external surface of the adsorbents, and was accounted for by the introduction of an availability factor β. The availability factor was found to decrease exponentially with an increase in concentration of silica in suspension. The decrease in the values of β was more prominent for the ion exchange fibre and membrane than for carbon and resin. This can be ascribed to the fact that fibres and membranes have a larger external surface area exposed to the silica particles, thereby increasing the degree of blinding.

Scanning electron microscopy showed severe cracks on the external surface of the membrane when exposed to a slurry of fine particles. Figures 4, 5 and 6 show an untreated membrane, a membrane pretreated for 1 hour, and a membrane pretreated for 3 hours respectively. Such attrition effects can ultimately reduce the adsorption capacity of the membrane, and hence can contribute to β indirectly.

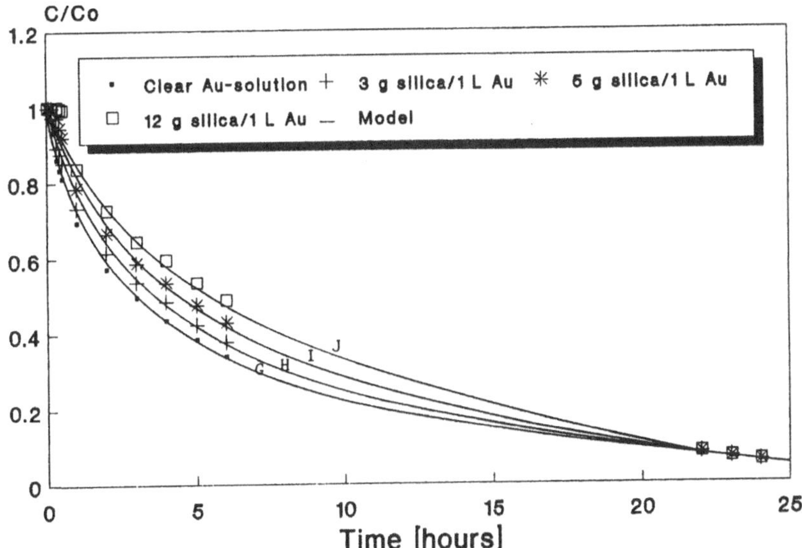

FIGURE 3: Effect of concentration of silica (kinetic parameters) in a slurry on the mass transfer of gold cyanide to carbon containing 18 mg SiO_2/1g due to pretreatment with 2 g of silica/L. (C_o = 20 mg Au/L; V = 1,0 L; N = 1 000 rpm; M = 1,0 g)

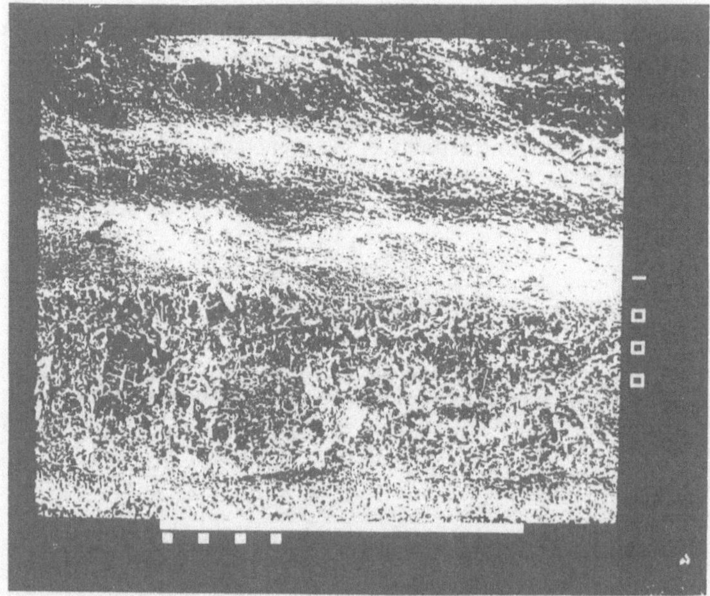

FIGURE 4: Scanning electron micrograph of an untreated membrane surface.

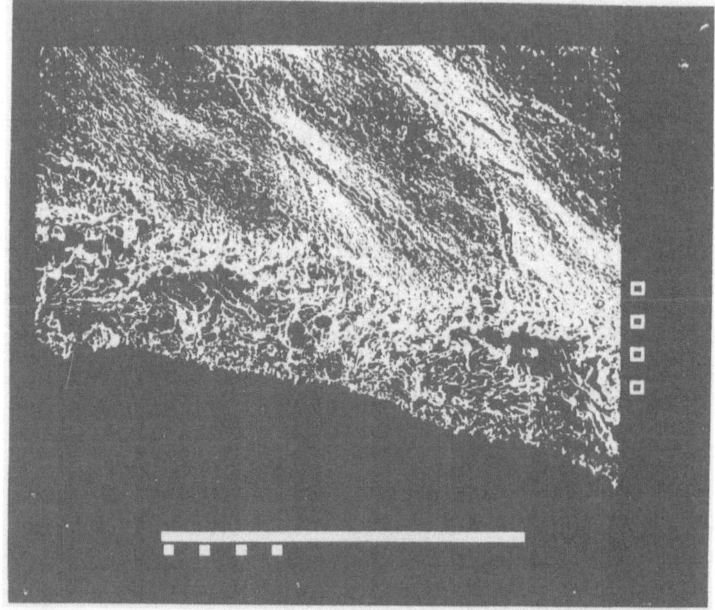

FIGURE 5 Scanning electron micrograph of membrane surface after pretreatment with a silica sand slurry for 1 hour.

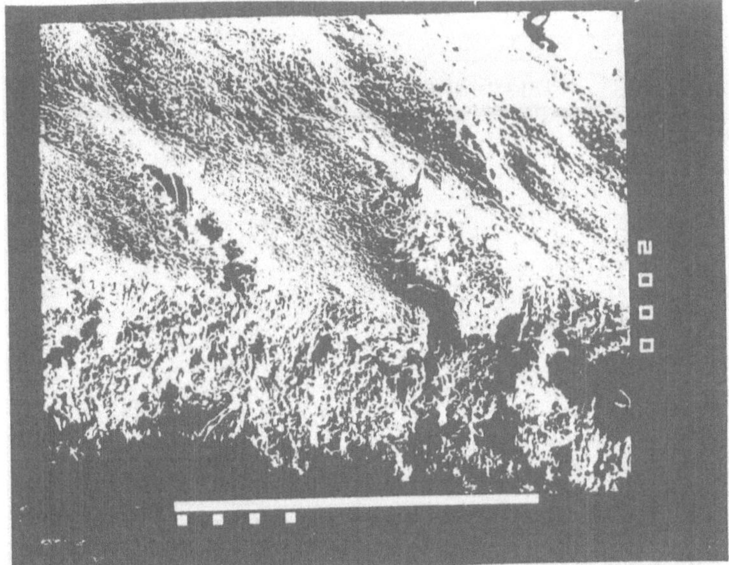

FIGURE 6 Scanning electron micrograph of membrane surface after pretreatment with a silica sand slurry for 24 hours.

4.3 Effect of organic substances on internal gold diffusion

4.3.1 Low organic loadings on adsorbents

In these tests the different adsorbents were pretreated with either PEX or SEX solution, washed with distilled water, and contacted with a clear 20 ppm gold cyanide solution. The average loading of organic compound on these adsorbents was 54 mg organic compound/g adsorbent. UV-spectrophotometry confirmed a loss of only 4% of organic loading from the adsorbents during the adsorption of gold cyanide. Hence, it is clear that organics were adsorbed fairly irreversibly.

Figure 7 illustrates clearly that the presence of organics on fibre influenced the rate of gold uptake, but had no effect on the equilibrium loading. Figure 8 shows similar results for the ion exchange membrane, but the effect was less profound. It is worth noting that the film transfer coefficient was insensitive towards the diffusion of gold cyanide, while profound differences in the surface diffusivity were observed for the untreated and pretreated adsorbents. The fact that the equilibrium of adsorption was unaffected implies that no competition for active sites existed, and a pore blocking mechanism is proposed. This mechanism acted severely in the case of activated carbon, and can be explained in terms of the differences in pore size.

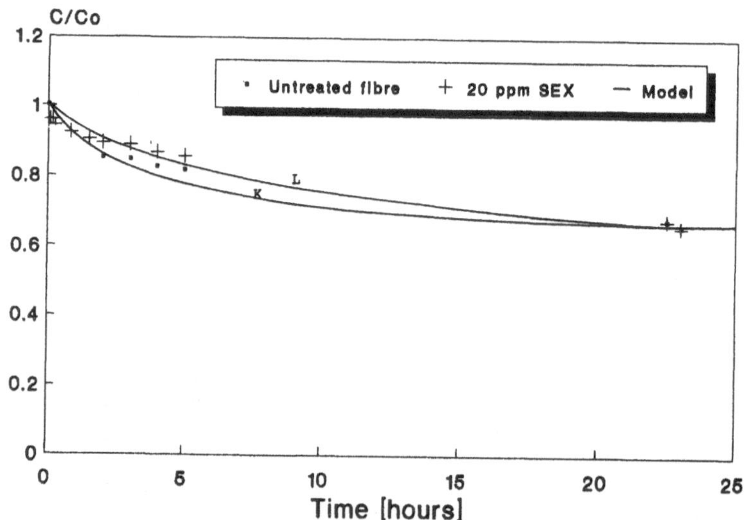

FIGURE 7: Inhibited mass transfer (kinetic parameters) of gold cyanide to ion-exchange fibre owing to fouling by sodium ethyl xanthate (SEX) to a level of 54 mg SEX/g. (C_o = 20 mg Au/L; V = 1,0 L; N = 250 rpm; M_f = 0,2 g)

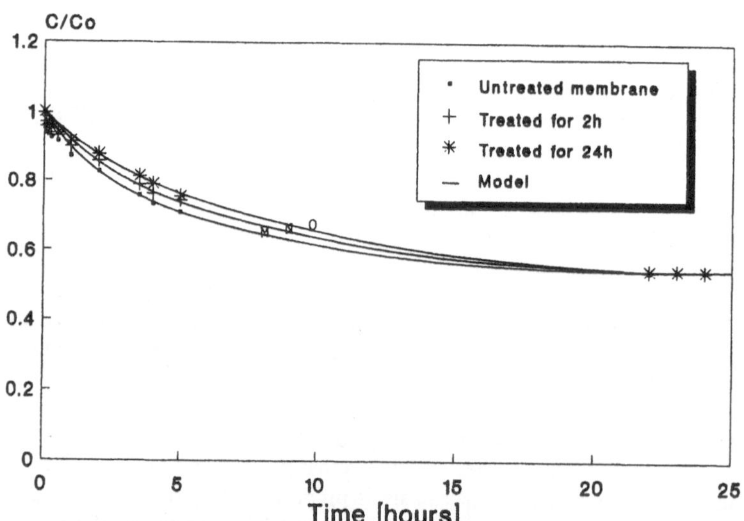

FIGURE 8: Inhibited mass transfer (kinetic parameters) of gold cyanide to ion-exchange membrane owing to fouling by sodium ethyl xanthate (SEX) to a level of 54 mg SEX/g. (C_o = 20 mg Au/L; V = 1,0 L; N = 250 rpm; M = 0,2 g)

4.3.2 High organic loadings on adsorbents

Adsorption tests were performed in a similar manner as before, but at higher loadings of organic foulants. Results presented graphically in Figure 9 show that both the rate of adsorption and the equilibrium loading were affected for the adsorption of gold cyanide onto activated carbon. Interestingly, similar results were obtained for membranes and fibres. Furthermore, model parameters in Table II show a decrease in the value of the surface diffusivity D with an increase in the period of pretreatment, and hence an increase in organic loading on the adsorbent. Moreover, the values indicate that a possible mechanism can be competitive adsorption. However, it is also possible that organic foulants merely block certain regions in the adsorbents, thus rendering them inaccessible to gold adsorption.

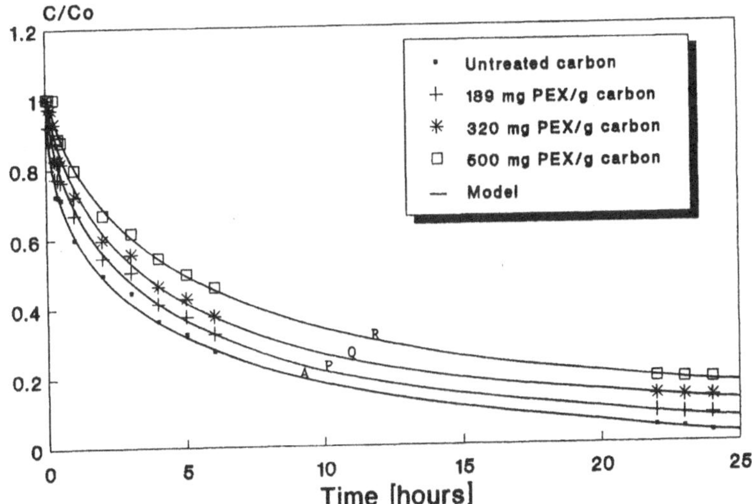

FIGURE 9: Effect of concentration of potassium ethyl xanthate (PEX) on the mass transfer (kinetic and equilibrium) parameters of gold cyanide to carbon.
(C_o = 20 mg Au/L; V = 1,0 L; N = 250 rpm; M = 1,0 g)

Table II. Sensitivity of **kinetic and equilibrium** parameters for gold adsorption onto activated carbon

Curve	Adsorbent	Conditions	$k_f \times 10^5$ [m/s]	$D \times 10^{12}$ [m^2/s]	B	n
P	Carbon:	189 mg PEX/g Carbon	3,74	3,7	26,0	0,52
Q	Carbon:	320 mg PEX/g Carbon	3,71	3,2	24,2	0,55
R	Carbon:	500 mg PEX/g Carbon	3,71	2,9	22,1	0,58

5 Conclusions

Ion exchange membrane and fibre showed much faster rates for gold uptake than the carbon or resin, which can be attributed to their large external surface areas. It was observed that the kinetic parameters in a film/surface diffusion model were affected by inert solids and low loadings of organic substances. At higher organic concentrations both kinetic and equilibrium influences were observed. This was attributed to permanent pore blocking of the adsorbents by organic compounds or competitive adsorption.

Furthermore, tests involving the intrusion of inert silica particles into the adsorbents indicated that the adsorption profile was affected only for activated carbon and ion exchange resins. This effect was more prominent for activated carbon in view of its larger pore size. However, the diffusion of gold cyanide to all adsorbents was inhibited due the physical blinding of the superficial surface by the fine particles in suspension. Membranes and fibres were more sensitive to the availability factor than resin or carbon. Severe attrition was observed when exposing membrane surface to inert particles in suspension.

6 Acknowledgement

This work was financially supported by MINTEK.

7. Nomenclature

2a	membrane thickness [m]
A	total external area of adsorbent particles in reactor [m^2]
B	constant in Freundlich expression
C	solution phase concentration [mg Au/L]
D	surface diffusion coefficient [m^2/s]
d	diameter of particle [m]
K_1, K_2	constants in Langmuir isotherm expression
k_f	external film transfer coefficient [m/s]
M	mass of adsorbent [kg]
N	stirring speed (r.p.m.)
n	exponential parameter in Freundlich isotherm
Q	metal loading on adsorbent [g Au/kg adsorbent or g Au/m^3 wet-settled resin]
r	radial variable [m]
R	radius of adsorbent particle [m]
t	time variable [s]
V	volume of liquid in the reactor [m^3]
V_R	volume of wet-settled resin [m^3]

x	distance coordinate [m]

Greek Letters

β	surface availability factor
ρ	apparent density [$kg.m^{-3}$]
ϕ	conversion factor for carbon [$kg.m^{-3}$] or resin [m^3 of resin/m^3 of wet-settled resin]
ϵ	volumetric fraction of resin particles in volume of wet-settled resin

Subscripts

c	carbon
e	equilibrium
f	fibre
m	membrane
r	resin
s	surface

8 References

1. Asolekar, S.R. Deshpande, P.K. and Kumar, R., "A model for a foam-bed slurry reactor",A.I.Ch.E Journal, vol. 34, no. 1, 1988, pp. 83-94.
2. Fleming, C.A. and Nicol, M.J., "The absorption of gold cyanide onto activated carbon. III. Factors influencing the rate of loading and equilibrium capacity", J.S. Afr. Inst. Min. Metall., vol. 84, no. 4, 1984, pp. 85-93.
3. Jones, W.G. and Linge H.G., "Effect of ore pulp on the adsorption rate of gold cyanide on activated carbon", Hydrometallurgy, vol. 22, 1989, pp. 231-238.
4. Jordi, R.G., Young, B.D. and Bryson, A.W., "Gold adsorption on activated carbon and the effect of suspended solids", Chem. Eng. Commun., vol. 102, 1991, pp. 127-147.
5. Petersen, F.W. and Van Deventer, J.S.J., "Inhibition of mass transfer to porous adsorbents by fine particles and organics", Chem. Eng. Commun., vol. 99, 1991, pp. 55-75.
6. La Brooy, S.R., Bax, A.R., Muir, D.M., Hosking, J.W., Hughes, H.C. and Parentich, A., "Fouling of activated carbon by circuit organics" Gold 100: Proc. Int. Conf. on Gold, vol. 2: Extractive Metallurgy of Gold, C.E. Fivaz and R.P King (eds), SAIMM, 1986, pp. 123-132.
7 Van Vliet, B.M. and Young, B.D., "The use of fractal dimension to quantify the effect of surface roughness on film mass transfer enhancement", Chem. Eng. Commun. vol. 69, 1988, pp. 81-94.

Reductive sorption methods for extraction of noble metals from solution

Yu. A. Tarasenko
A. A. Bagreev
G. V. Reznik
V. V. Strelko
Institute for Sorption and Problems of Endoecology, Academy of Sciences of Ukraine, Kiev, Ukraine

Abstract
Interactions of active carbons (AC) with solutions of noble metals (palladium, platinum and gold) are investigated. A electrochemical model is proposed to explain the spontaneous reduction of ions on AC' surface. The adsorption of noble metals on AC studied by methods of sorption measurements, potentiometry, flow microcalorimetry and X-ray photoelectron spectroscopy. It was shown, that ion exchange, complex formation, and electrochemical processes play an important role in adsorption of noble metals on AC. The selectivity of noble metals extraction from multicomponent solutions by AC was determined by a reductive mechanism of adsorption.
Keywords: adsorption, noble metals, active carbon, electrochemical mechanism, selectivity of extraction, microcalorimetry. XPS.

1 Introduction

Active carbons are widely used as adsorbents for noble metals extraction from mineralized solutions [1]. During the last fifteen years in Kiev Institute for Sorption and Problems of Endoecology (ISPE) the research in field of new type spherical granulated AC (SCN, SCS, etc.) synthesis, investigations and applications was carried out. It was shown [2-4], that in contrary to technical carbons, synthetic AC are characterized with high chemical stability, increased mechanical strength, developed surface and porosity that may be regulated , surface chemistry being easily changed, high electrophysical and electrochemical characteristics, compatibility with body fluids, etc. It was shown also [5] , the electrochemical activity of synthetic AC enables the reductive sorption (RS) - the spontaneous process of noble metals' ions reduction simultaneously with sorption proceeding. This property is very important in the field of hydrometallurgy applications.

2 Electrochemical mechanism of reductive sorption process

The reasons for electropositive metal ions' reduction on AC' surface were discussed in the number of communications (for example, see [6,7]). It was suggested, the metal phase formation is due to the chemical reactions proceeding.

On the contrary, in our investigations the electrochemical hypothesis for the nature of this process is proposed [5,8-11]. Within the framework of this hypothesis, the electropositive metal ions reduction is a result of the spontaneous electrochemical reaction proceeding, conjugated with reaction of AC' matrix oxidation. By means of the electroconductive carbon matrix the electron balance between the conjugated Red/Ox reactions is realized. The special feature of processes in this case is space division of electron transfer in reduction and oxidation reactions, Fig. 1.

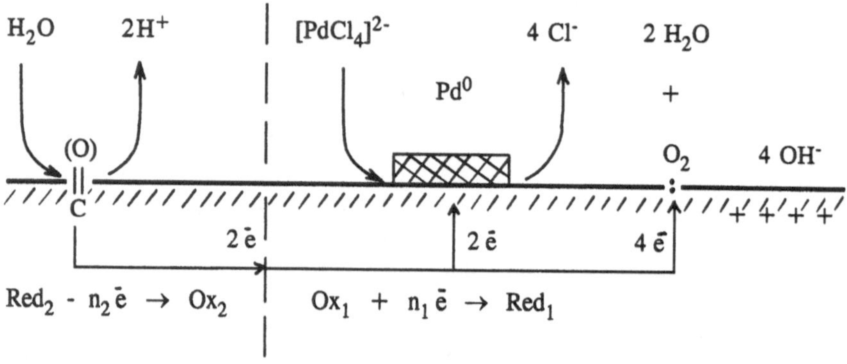

Fig. 1 The reductive sorption scheme presented in the case of palladium ion reduction.

A theoretical description is based on the irreversible thermodynamics extremal principles. Simulating of RS process was carried out by the application of necessary conditions for steady state stability in heterogeneous electrochemical systems [9,12]. Such a methodology enabled us to research the evolution of stationary state parameters for the systems under consideration.

Calculations were performed for the case when RS on AC's external surface proceeds. Formation of metal islands and division of all chemically active surface (S) on «cathodic» and «anodic» parts were taken into account. So, the closed set of equations was formed:

$$\frac{dC_1}{dt} = -\frac{1}{n_1 F V} i_1 (C_1 (t)) S_c(C_1 (t)), \qquad C_1 (t=0) = C_1 (0); \qquad (1)$$

$$i_1 = \frac{n_1 F D_1 C_1}{d} \frac{n_1 F}{R T} (E_1{}^0 - E); \qquad (2) \qquad i_2 = n_2 F D_2 C_2 / d; \qquad (3)$$

$$i_3 = i_0 n_3 F (E - E_3{}^0) / R T; \qquad (4) \qquad S_c + S_a = S_0; \qquad (5)$$

$$\partial P / \partial E = 0; \qquad (6) \qquad \partial P / \partial S_a = 0; \qquad (7)$$

$$P = \frac{1}{T} [i_1 S_c (E_1{}^0 - E) + i_2 S_0 (E_2{}^0 - E) + i_3 S_a (E - E_3{}^0)]; \qquad (8)$$

Equation (1) defines the metal ions' concentration (C_1) changes by RS proceeding in the limited solution' volume (V). Equations (2-4) describe the polarization dependences of current densities for cathodic - metal ions' (i_1) and oxygen (i_2) reduction and anodic - carbon' corrosion (i_3) reactions. Cathodic (S_c) and anodic (S_a) areas' balance is represented by (5), and the entropy production (P) in RS-systems by (8) is determined.

Calculations for Pd-ions RS were performed [9,12]. As a result, it was possible to establish:

- thermodynamical criterion for reduction or oxidation reactions proceeding on the AC's surface (forming or dissolution of metal film);
- possibility for metal ions' reduction on the carbon granules' external surface or in its volume;
- RS-systems parameters in steady states and their evolution with boundary conditions changes;
- the RS-processes control principles.

3 Experimental methods and materials

The spherical granulated AC's SCN (nitrogen containing) and SCS (sulphur containing) - type and their oxidized SCNo and SCSo modifications were used. Samples were prepared by synthetic resin carbonization followed by steam activation [3,4]. Porous structure was investigated by the mercury intrusion method (Pore-Sizer-9700 «Micromeritic») and the nitrogen adsorption (77 K). The specific surface area by BET-method was measured. Carbons characteristics represented in Table 1.

CP reagents were used.

Adsorption of metal ions on AC was measured by the «separate samples» method in static conditions at 20 C. Experiments with Au, Ag, and Ni were carried out in cyanide solutions (pH 10) and Pt, Pd, Ru, Rh and Ni - in chloride ones (pH 0.0-2.0). The metal ions content in solutions was determined by atomic absorption and spectrophotometric methods. The amount of metals absorbed (A, mg/g) was calculated from the decrease of metal concentration in solutions when the sorption was finished.

The state of metals, adsorbed on AC' surface was studied by X-ray photoelectronic spectroscopy (XPS) on IEE-15 instrument (Varian, USA). The spectra were calibrated by the AC' 1S-carbon line (284.3 eV). Error in determination of the bond energy (E_b) was ± 0.2 eV. Preparation of samples and scanning of spectra were carried out by the method [13,14].

Metal ions' adsorption heat effects on AC were measured on an LKB-2107-30 flow sorption microcalorimeter. Experimental procedure in [15,16] was described. Preliminary consecutive fixation of heat effects for interactions of carbon matrix with distilled water and the background solution (ion exchange) made it possible to separate the amount of heat, associated only with the interaction of metal ions with carbon .

Potentiometric measurings were performed with Pt-electrode in standard cell. Experiments were carried out in aerated solutions and in oxygenless medium (argon atmosphere).

Potentiometric measurings were carried out and adsorption characteristics, heat effects and the metal' surface states were investigated to determine the possible mechanism of electropositive metals adsorption on AC (electrochemical reduction, complex formation, ion exchange).

Table 1. The porous structure characteristics of active carbons

Samples of carbon	Total pore volume (cm^3/g)	Volume of micropores (cm^3/g)	Volume of mesopores (cm^3/g)	BET surface area (m^2/g)	Specific surface of mesopores (m^2/g)
SCN	1.17	0.45	0.58	960	100
SCNo	0.89	0.42	0.40	520	76
SCS-2	0.85	0.32	0.42	780	54
SCS-4	0.96	0.30	0.47	880	90
SCS-6	1.24	0.46	0.46	1000	172

4 Results and Discussion

Electrochemical reduction is thermodynamically enabled if the metal-ions' equilibrium potential is more positive then AC surface' potential. So, the value of AC surface' stationary potential and kinetics of its formation are the important factors of RS process' control. This factors are dependent on the conditions of RS proceeding: molecular oxygen or another oxidation agent presence in solution, AC' potential fixation with electronegative metals (Zn, Al) etc.. So, Fig. 2 presents the kinetic dependences for stationary AC surface' potential formation in aerated solutions for activated (1) and oxidized (2) samples of SCN-carbon, the pH-dependences of stationary (3) and initial potentials (4) and the scale of standard potentials for a different metals reduction.

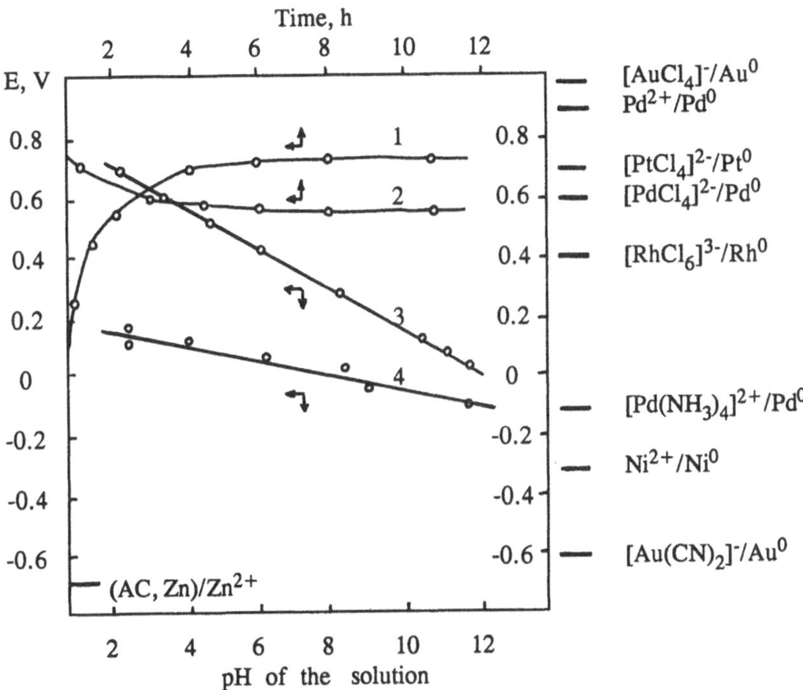

Fig. 2 Kinetics of AC' surface potential formation (1 - SCN, 2 - SCNo, pH 2) and pH-dependences of the stationary (3) and initial potential (4) for activated AC.

The comparison of Red/Ox processes' potentials for different metals shows, that electrochemical reduction is possible for $[AuCl_4]^-$, $[PtCl_4]^{2-}$, $[PdCl_4]^{2-}$ and is forbidden for $[RhCl_6]^{3-}$, $[Pd(NH_3)_4]^{2-}$, $[Au(CN)_2]^-$, Ni^{2+}.

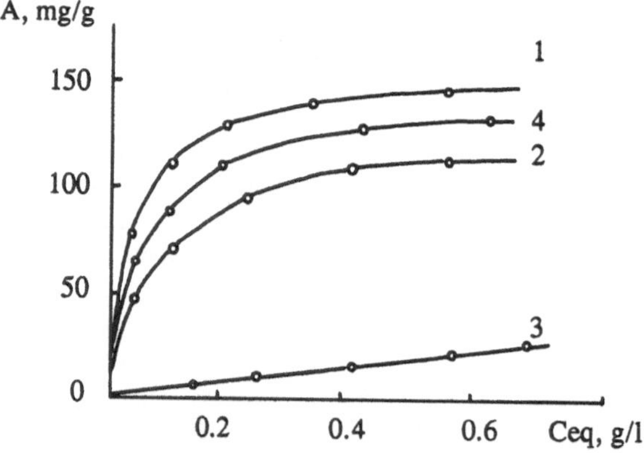

Fig. 3 Adsorption capacity of SCN carbon dependence on the concentration of palladium (1), platinum (2), nickel ions (3), palladium with a background nickel (20 g/l) (4) from hydrochloric acid solutions.

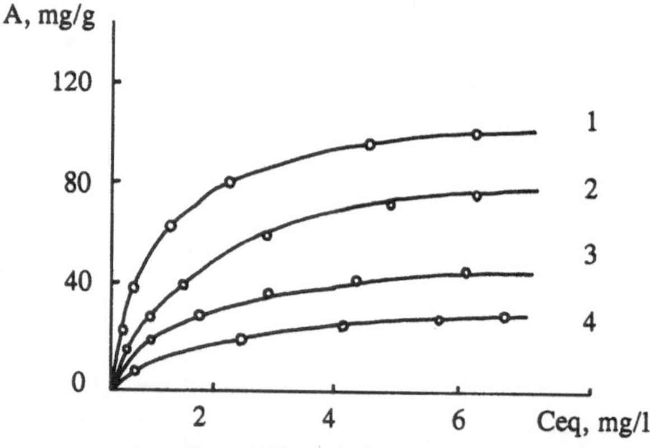

Fig. 4 Dependence of the adsorption capacity of different carbons on the gold ions concentration in cyanide solutions: 1 - SCN, 2 - SCS-4, 3 - AC without nitrogen or sulphur, 4 - SCNo.

Concentration dependences of the AC' adsorption capacity for different metals is represented on Fig. 3 and Fig. 4.

The comparison of adsorbed amounts for noble metal ions (Pd, Pt, Au) and unnoble ones indicates [16, 17] (Fig. 3), that in individual solutions the limit sorption of noble metals 10-20 times exceeds this one for unnoble ions.

The presence in system electronegative metals together with noble ones, only insignificantly lowers adsorption of noble metals by AC, for example Pd of an excess of Ni (curve 4, Fig. 3), Au and Ag in presence of unnoble impurities (Table 2).

Table 2. The selectivity of adsorption of gold and silver from multicomponent cyanide solutions

Sample of sorbent	Golg		Silver		Impurity metals (Cu,Zn,Fe,Ni,Co)		Ksel (%)
	Ceq	A	Ceq	A	Ceq	A	
Initial solutions	60.3		23.7		180.5		
SCS-2	0.3	12.0	1.3	4.5	171.2	1.9	90
SCS-4	0.5	12.0	2.1	4.3	180.4	0.1	99
SCS-6	0.1	12.0	0.9	4.6	168.9	2.5	87
AM-2B	0.9	11.9	3.3	4.1	89.6	18.1	47

Ceq - equilibrium concentration of metal in solution (mg/l); A - the capacity of adsorbent (mg/g); Ksel - coefficient of selectivity; the ratio of the adsorption capacity for gold and silver to the total adsorption of metals, %; AM-2B - ion exchange resin.

The modifying of AC' matrix by nitrogen, sulphur and oxygen atoms produce the effects:
- increasing of the Au and Ag adsorption (N about 4%, S about 1%);
- decreasing of the capacity of AC (oxygen containing groups).

A comparison of the differential heat effects for adsorption and desorption of Pd, Pt, Ni (Fig. 5) shows that although adsorption of both noble metal anions and nickel cations from acid chloride solutions on AC is accompanied by exothermic effects, the magnitude of that for Pd and Pt approximately 15-20 times greater, than for Ni. As for Au in cyanide solutions, the heat effect 20 times exceeds that for Ni and Zn [13]. Calculated entalpies of adsorption for Pt, Pd, and Ni are the next: $-\Delta H = 31.8$ kJ/mol, $-\Delta H = 30.7$ kJ/mol, $-\Delta H = 4.5$ kJ/mol.

The difference in the reversibility of the adsorption processes of the compared ions is also specific. While a heat effect of desorption is not observed for Pd, the values of the exothermal adsorption effect and the endothermal desorption effect practically coincide for nickel. This indicates the complete reversibility of nickel adsorption and virtually irreversible adsorption of palladium under the experimental conditions.

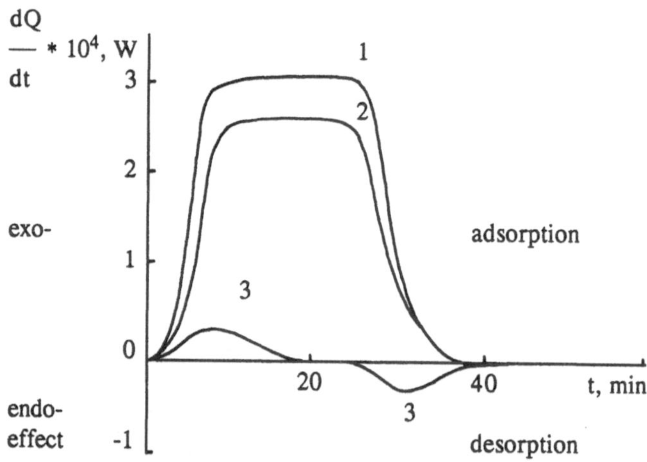

Fig. 5 Differential heat effects of adsorption and desorption of platinum (1), palladium (2), and nickel (3) ions on SCN carbon from hydrochloric acid solutions.

The chemical state studies of adsorbed by AC noble metals indicates (Fig. 6) that palladium, platinum and gold ions interaction with carbon matrix is connected with creation of surface complex with the charge transfer from AC matrix to the noble metal ion. In case of Au adsorption the surface chemical reaction proceed causing the cluster complexes creation [11]. In case of palladium and platinum ions adsorption the electrochemical reduction and surface metal-carbon complexes creation may take place [11,16]. If nickel adsorption proceed only donor-acceptor interactions are observed [13].

An analysis of the reasons for selective absorption of noble metals by active carbons was carried out with consideration of their ion exchange and reducing function and also the reactivity of the carbon matrix. As the example it was considered the palladium and nickel adsorption from the hydrochloric acid solutions [16]. In aerated solutions AC perform the oxygen gas electrode functions as follows:

$$[C_x...O] + H_2O \rightarrow [C_x{}^{2+}...2\,OH^-], \tag{9}$$

$[C_x{}^{2+}...2\,OH^-]$ is the positive charged activated carbon surface with OH^- ions in the outer coating of the double electrical layer.

In hydrochloric acid solutions of $PdCl_2$, the $[PdCl_4]^{2-}$ form is dominating, while nickel is present in the cation form. That is why at first the anion exchange will occur according to the scheme:

$$[C_x{}^{2+}...2\,OH^-] + [PdCl_4]^{2-} \rightarrow [C_x{}^{2+}...PdCl_4{}^{2-}] + 2\,OH^- \tag{10}$$

Even if there is an ion-exchange mechanism only, the adsorption of palladium anions will exceed that of nickel cations. It is the first reason for sorption selectivity.

It was shown that during the noble metals adsorption besides ion exchange the surface complexes formation took place. The metal-carbon bonds' become more stable as the metal atomic number increases, so palladium, platinum and gold complexes will be more stable than nickel ones. So, the surface complexes-forming reactions proceeding is the second reason for selective adsorption of noble metals.

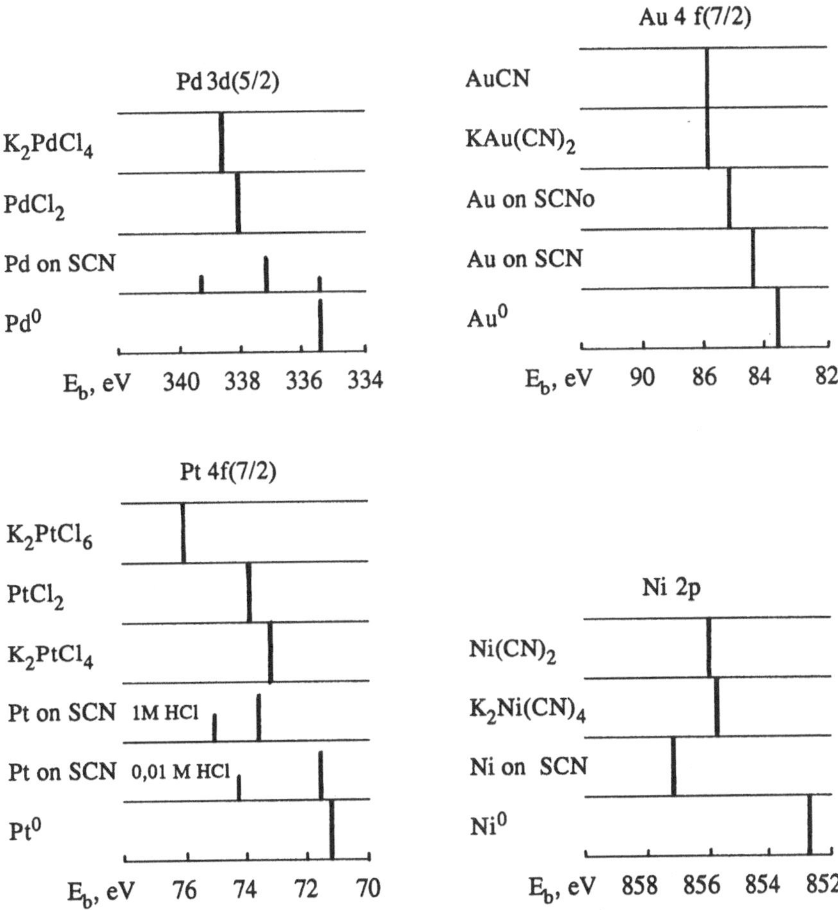

Fig. 6 Diagrams of XPS-spectra of palladium and platinum adsorbed on AC from hydrochloric acid solutions and gold and nickel from cyanide solutions.

Finally, if the metal-ions' equilibrium potential is more positive then AC surface potential, the electrochemical reduction is thermodynamically enabled. That is why palladium and platinum ions adsorption is accompanied with their reduction to the metallic state, whereas nickel ones cannot be reduced under these conditions. Thus the electrochemical reduction of palladium and platinum ions is the third factor, causing the selectivity of adsorption.

So, the selectivity of palladium and platinum adsorption from chloride solutions and of the gold adsorption from cyanide ones is based on the processes, common in their nature: the formation of surface complexes with charge transfer and subsequent formation of cluster states in the case of gold or reduction to the metal state in the case of palladium and platinum.

The reductive sorption methods of electropositive metal ions extraction have been developed to solve the problems of hydrometallurgy, protection of the environment and medicine.

5 References

1. Mc.Dougall, G.J., Hancock, R.D. (1980) **Miner. Sci. and Eng.**, 12, p.85.
2. Strelko V.V.(1990) In.: **Intern. Carbon Conf., Carbon-90**, Paris, p.16.
3. Nikolaev, V.G., Strelko, V.V. (1979) **Hemosorption on Activated Carbons**. Kiev, Naukova Dumka, 1979 (in Russian).
4. Strelko, V.V., Korovin, Yu.F., Kartel', N.T., et al., (1984) **Ukr. Khim. Zhurn. (Soviet Progress in Chemistry)**, 50, p.1157.
5. Tarasenko, Yu., Suprunenko, K., Dudarenko V., et al., (1990) In.: **Intern. Carbon Conf., Carbon-90,** Paris, p.52.
6. Garten, V.A., Weiss D.E., (1957) **Rev. Pure and Appl. Chem.**, 7, p.69.
7. Simonov, P.A., Semikolenov, V.A., Likholobov, V.A., et al., (1988) **Izv. Acad. Nauk USSR, ser. khim. (Russian Chemical Bulletin)**, N12, p.2719.
8. Tarasenko, Yu.A., Bagreev, A.A., Dudarenko, V.V., et al., (1989) **Ukr. Khim. Zhurn. (Soviet Progress in Chemistry)**, 55, p.233.
9. Tarasenko,Yu.A., Reznik, G.V., Bagreev, A.A., (1989) **Ukr. Khim. Zhurn. (Soviet Progress in Chemistry)**, 55, p.249.
10. Strelko,V.V., Tarasenko,Yu.A., Lavrinenko-Ometsinskaya,E.D., Bagreev,A.A., (1991) **Ukr. Khim. Zhurn. (Soviet Progress in Chemistry)**, 57, p.1065.11.
11. Tarasenko, Yu.A., Bagreev, A.A., Yasenko, V.V., (1993) **Zhurn. Phyz. Khimii (Russian Journal of Physical Chemistry)**, 67, p.2328.
12. Tarasenko, Yu.A., Antonov, S.P., Bagreev, A.A., Reznik, G.V. (1989) **Ukr.Khim. Zhurn. (Soviet Progress in Chemistry)**, 55, p.1179.
13. Dudarenko, V.V., Strelko,V.V., Nemoshkalenko, V.V., et al. (1985) **Ukr. Khim. Zhurn. (Soviet Progress in Chemistry)**, 51, p.708.
14. Nefedov, V.I. (1984) **X-ray Spectroscopy of Chemical Compounds**. Moscow, Khimiya, (in Russian).
15. Strelko, V.V., Dudarenko, V.V., Tarasenko, Yu.A., et al., (1986) **Ukr. Khim. Zhurn. (Soviet Progress in Chemistry)**, 52, p.1157.
16. Strelko, V.V., Tarasenko, Yu.A., Bagreev, A.A., et al., (1991) **Ukr. Khim. Zhurn. (Soviet Progress in Chemistry)** 57, p.920.
17. Tarasenko, Yu.A., Dudarenko,V.V., Mardanenko,V.V., et al., (1989) **Zhurn. Prikl. Khim. (J. Applied Chemistry)**, 61, p.1489.

Dissolution of gold in oxidized bromide solutions

R. B. E. Trindade
P. C. P. Rocha
J. P. Barbosa
CETEM—Center for Mineral Technology, Cidade Universitária, Rio de Janeiro, Brazil

Abstract
Bromide stabilizes the auric ion by forming a stable complex in acidic aqueous solution. The oxidizing agent normally used in this process is the bromine. This is, however, a corrosive liquid with high vapour pressure and this is the main reason why it has not found acceptance yet as an industrial oxidant. In the present work, ferric ion, hydrogen peroxide and sodium hypochlorite were tested as alternative oxidants, alone or in combination, to evaluate the dissolution of gold in a bromide-containing aqueous solution. The advantage is that this is a cleaner technology and thus environmental protection costs may be minimized. Using the rotating disc technique, the effects of rotating speed, leaching time, temperature, oxidant concentrations, used together or separately, in an aqueous bromide solution with different concentrations of this complexant were evaluated. Ferric ion alone is able to dissolve gold, but kinetic rates are not elevated. Ferric ion and hydrogen peroxide can dissolve gold at higher rates. Hydrogen peroxide alone will not oxidize the gold and therefore the complex cannot be formed. Sodium hypochlorite alone, however, is able to promote the dissolution of gold in a sodium bromide solution. One of the main conclusions is that the gold dissolution obtained with the bromide solution as used in this work may, in some cases, be as good as or even superior to those obtained with the cyanidation process.
Keywords: Gold, Bromide, Rotating Disc, Cyanide-free reagent.

1 Introduction

Most of the gold contained in ores nowadays is recovered by the cyanidation process, a technique known since the end of the last century, when in 1887 a British patent was granted to John Steward MacArthur (Habashi). With the implementation of the cyanidation process, the world gold output increased more than four times in the period from 1901 to 1950 compared to the period from 1851 to 1900. While this process still keeps the overwhelming dominance it has acquired over the past one hundred years, cyanide is ineffective and/or not economically adequate for use with certain "refractory" ores such as those containing carbon or some sulphide minerals like arsenopyrite, pyrite or stibnite. These ores must usually undergo a pretreatment involving flotation, oxidative roasting or pressure leaching. Moreover, cyanide is a highly toxic chemical and, therefore, must be disposed of with extreme care to avoid environmental damage (which may increase costs); also, cyanide leaching suffers from slow kinetics mainly caused by the low solubility of oxygen in water. These are the main reasons why there has recently been an increase in the search for alternative leaching reagents to treat gold and silver ores (Hiskey, Dadgar, Chen, Barbosa-Filho). Among these lixiviants, thiourea, bromine and iodine, besides thiosulphate and to a lesser extent thiocyanate, have been the most investigated (Monhemius, von Michaelis). The present work is concerned with the system bromine/bromide for the dissolution of gold.

2 Bromine/bromide as lixiviants for gold

Since the 19th century it has been known that bromine is capable of oxidizing gold. Bromide stabilizes the auric or aurous ion by forming a stable complex in acidic aqueous solution. The oxidizing agent normally applied in this process was the bromine. This is, however, a very corrosive liquid, with a high vapour pressure (yielding harmful fumes) and thus it has not found acceptance as an industrial oxidant (Dreisinger, Kirk-Othmer Enc.).

Not much work dealing specifically with the bromine/bromide system for gold dissolution are found in the literature. Fink and Putnam have patented a process in which gold dissolution in a bromine-containing solution is made possible by the addition of bromide and chloride ions. In a later patent (von Michaelis, Kalocsai) the use of a cation source (such as NH_4^+) was suggested to enhance the rate of gold dissolution in a aqueous solution containing a source of bromine in combination with a strong oxidizing agent. Pesic and Sergent have recently studied the mechanism of the dissolution of gold in the system bromine/bromide ion using a rotating disc. In their work, bromine was used as the main oxidizing agent.

In an acidic solution containing the bromide ion (which may be added as a sodium salt, e.g. NaBr) metallic gold may be oxidized to a region where either the complex $AuBr_2^-$ or $AuBr_4^-$ is stable. This situation may be visualized in the Eh vs. pH diagram of the system $Au\text{-}Br^-$ in aqueous solution (FIGURE 1).

The stability of the bromide-gold complex with respect to chloride and iodide, which also exhibit the ability to dissolve gold, is midway between the chloride and iodide gold complexes. It is clear that $AuBr_4^-$ is the predominant species, although within a narrow region (between 0.79 V and 0.86 V SHE, approximately) the complex $AuBr_2^-$ may be formed.

If bromine is used as the oxidizing agent the dissolution may be represented by the overall reaction :

$$2Au + 3Br_2 + 2Br^- = 2AuBr_4^- \qquad (1)$$

where the bromide ion is the complexing agent. It is known (Liebhafsky), though, that in water bromine undergoes hydrolysis, with the formation of hypobromous acid as follows:

$$Br_2 + H_2O = HOBr + H^+ + Br^- \qquad (2)$$

Therefore, HOBr probably plays the role of the oxidizing agent; the gold may thus be dissolved according to:

$$2Au + 3HOBr + 5Br^- + 3H^+ = 2AuBr_4^- + 3H_2O \qquad (3)$$

As seen above, this process is, however, not practically feasible because of the difficulty of handling and working with the liquid bromine (Dreisinger, Kirk-Othmer Enc.). In the present work, some alternative oxidants (ferric ion, hydrogen peroxide and sodium hypochlorite) were tested, alone or in combination, to evaluate the dissolution of gold in a bromide-containing aqueous solution. Although thermodynamically feasible, the use of this leaching system is not known to have been tested in the extractive metallurgy of gold. The immediate advantage is that, without the use of bromine, the technology is cleaner and thus environmental protection costs may be minimized.

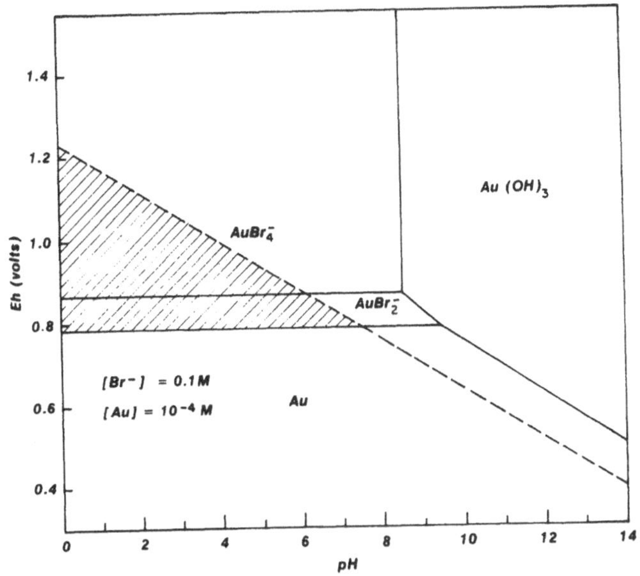

Fig. 1 - Eh vs. pH diagram for the gold-bromide system at 25°C.

3 Experimental

The experimental method chosen for the leaching tests was the rotating disc technique. A pure gold disc, with a diameter of 10.5 mm (area = $8.66 \times 10^{-5} m^2$), was mounted onto a Teflon rod, in such a way that only one surface of the metal was exposed to the leaching solution. The rod was further screwed into a stirring shaft and rotated by a precise motor (PAR Electrode Rotator, Model 636) so that the rotating speed could be easily changed. The gold disc was polished immediately before each experiment with a 6 micron Hyprez Spray diamond lapping compound placed on a silk cloth, and thoroughly rinsed afterwards with doubly distilled water. Reagent grade chemicals and distilled, deionized water were used in all tests. Experiments were carried out in a 150 ml glass reactor, with 100 ml of initial solution. The rotating disc could be lowered into the solution through a central neck in the lid; other necks were used to accommodate the pH electrode and the Eh reference electrode (Ag-AgCl). Electric contact was possible from behind the disc, so that the actual potential on the gold surface could be measured. Samples of 5 ml were collected every 15 minutes and tests normally ran for 1 hour, unless specifically stated otherwise. The amount of gold dissolved from the disc was calculated from analysis by atomic absorption spectrophotometry in all cases. Temperature was varied with the aid of a water bath.

Most of the leaching trials were carried out using the following standard conditions, except when the specific parameter was being investigated:

NaBr: 0.1 M
$FeCl_3$: 0.06 M
H_2O_2 (30%): 0.01 M (1 ml/ l of solution)
rotating speed: 600 rpm
temperature: 25°C
pH: 1.6 - 1.8 (natural)

FIGURE 2 is a schematic representation of the equipment used in the rotating disc tests.

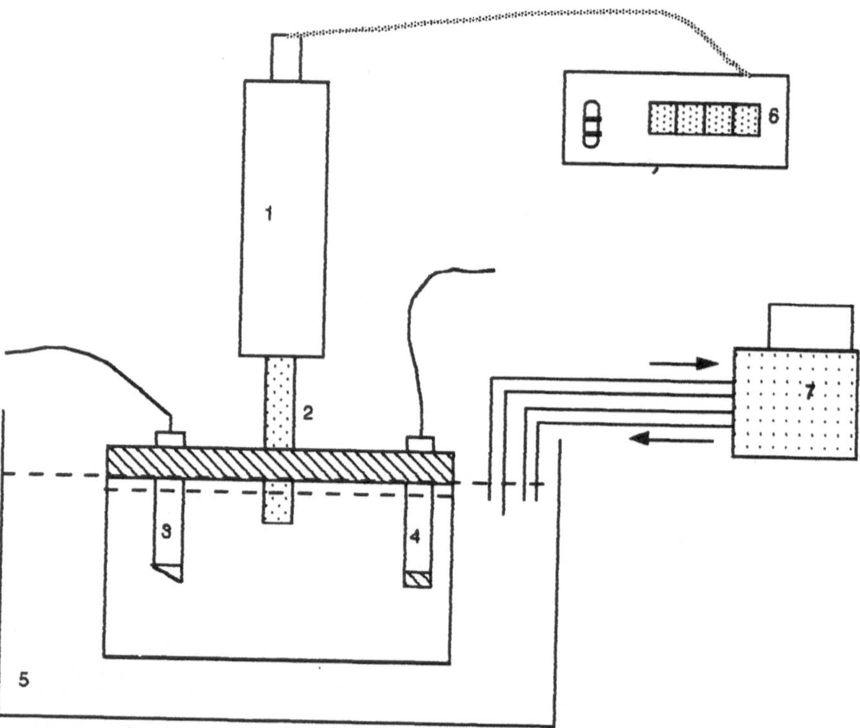

Fig.2 - Schematic representation of the reaction equipment: 1 - electrode rotator; 2 - Teflon rod with gold disc on bottom end; 3 - reference electrode (Ag-AgCl); 4 - pH electrode; 5 - water bath; 6 - rotating speed adjustor; 7 - constant temperature water circulator.

4 Results and discussion

4.1 Preliminary investigation

As seen above, one of the objectives of the present work is to introduce an alternative oxidizing agent that could replace bromine. The gold would then be stabilized in solution by the presence of the bromide ion with the formation of one of the gold-bromo complexes. The alternative oxidizing agent should, therefore, be able to increase the solution potential (in acidic pH) to at least 0.85 V (SHE), as indicated in FIGURE 1. Preliminary tests were carried out with the following oxidizing agents: ferric chloride, ferric sulphate, hydrogen peroxide and sodium hypochlorite, used together or individually. In this set of tests the rotating speed was kept at 600 rpm.

Hydrogen peroxide was initially tested alone in a solution containing 0.1M NaBr. The potential observed on the gold surface in this situation remained between 0.65 and 0.74 V (SHE), which is insufficient to allow the formation of the gold complex. A possible explanation for this observation is catalytic decomposition of the hydrogen peroxide in a solution containing Br- ions (Bray). At room temperature, in acidic pH, hydrogen peroxide oxidizes bromide to bromine as

$$H_2O_2 + 2Br^- + 2H^+ = Br_2 + 2H_2O \quad (4)$$

but also reduces the bromine to bromide (thus, bromine is indeed a strong oxidizing agent, capable of oxidizing peroxide itself) according to

$$H_2O_2 + Br_2 = O_2 + 2Br^- + 2H^+ \quad (5)$$

The sum of reactions 4 and 5 yields the decomposition of hydrogen peroxide to water and oxygen

$$2H_2O_2 = 2H_2O + O_2 \quad (6)$$

In this way, reaction (6) would prevent the peroxide from acting as an effective oxidizing agent for the process. The experimental results suggest that this mechanism might actually be occurring.

Ferric ion was tested as 0.2 M sulphate and chloride solutions, and as 0.1 M chloride only, in a solution containing 0.1 M NaBr. The potential on the gold surface remained between 0.88 and 0.90 V (SHE), which is, at least thermodynamically, sufficient to permit the formation of the gold complex. FIGURE 3 shows the amount of gold dissolved with time, under these experimental conditions. The arrow indicates the point where 5 ml/l of hydrogen peroxide (30%) was added. It is clear that this caused a dramatic increase in the rate of gold dissolution, which was to be expected since the potential on the gold surface increased to above 1.00 V (SHE). The lower $FeCl_3$ concentration (0.1 M) was responsible for less gold dissolution.

It is further noticeable that, while some gold was dissolved with Fe^{3+} as the sole oxidizing agent, the rate was low: 5.70 x 10^{-11} mol.cm^{-2}.s^{-1}. With the addition of the peroxide the rate was raised to 1.02 x 10^{-9} mol.cm^{-2}.s^{-1}, which is comparable to the values obtained in the cyanidation process, i.e. 2.86 x 10^{-9} mol.cm^{-2}.s^{-1} (Trindade). Also more gold dissolved with $FeCl_3$ than with $Fe_2(SO_4)_3$. FIGURE 4 depicts the amount of gold dissolution with 0.2 M $FeCl_3$, with 1 ml/l hydrogen peroxide 30% added from the beginning, as well as with 0.4 M $FeCl_3$ (the arrow indicates the point of addition of 0.5 ml/l hydrogen peroxide 30%).

Again the positive effect of hydrogen peroxide is observed. In the former instance the rate was 11.4 x 10^{-9} mol.cm^{-2}.s^{-1}, and in the latter (considering only the region after the peroxide addition) the rate reached 29.2 x 10^{-9} mol.cm^{-2}.s^{-1}. With peroxide from the beginning the rate is around 4 times higher than that observed in the conventional cyanidation process.

The results using sodium hypochlorite are seen in FIGURE 5. It is interesting that, unlike with the hydrogen peroxide, 5 ml/l of NaClO 3.4% alone increases the potential on the gold surface to around 950 mV (SHE), which allows the formation of the gold complex. Indeed, the rate in this case was found to be 1.19×10^{-9} mol. $cm^{-2}.s^{-1}$.

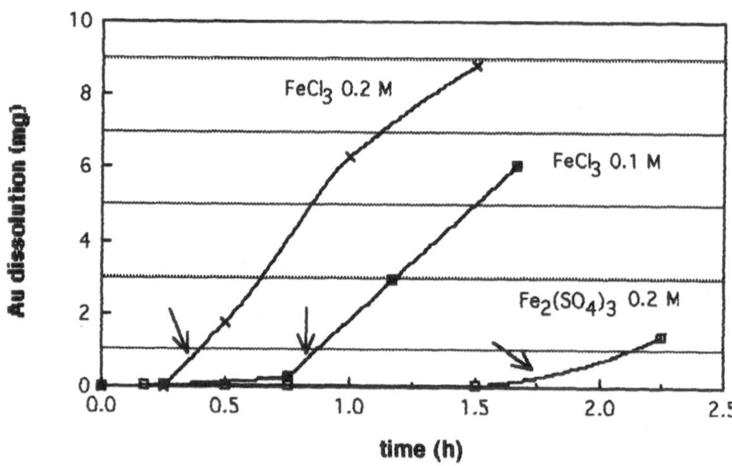

Fig.3 - Gold dissolution in 0.1 M NaBr and different concentrations of $FeCl_3$ and $Fe_2(SO_4)_3$ (the arrow indicates the point of H_2O_2 addition).

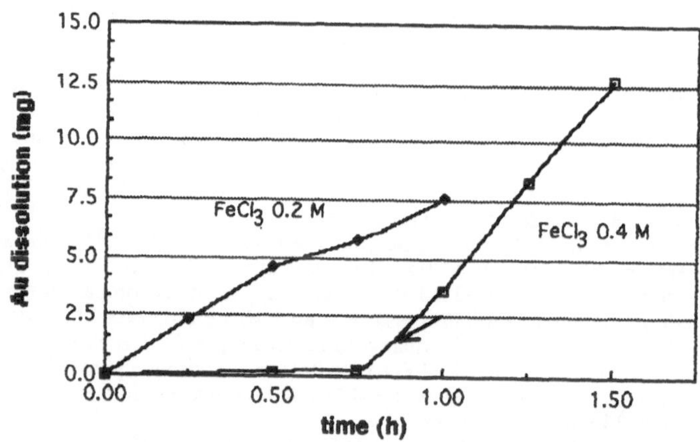

Fig. 4 - Gold dissolution in 0.1 M NaBr with $FeCl_3$ and 1ml/l of H_2O_2 30% (the arrow indicates the point of H_2O_2 addition, in one case only).

With 0.06 M $FeCl_3$ and the same NaClO concentration the amount of gold dissolved increased, although it was lower than with 1 ml/l H_2O_2 30%, as may be seen in FIGURE 5 (3.34×10^{-9} against 6.58×10^{-9} $mol.cm^{-2}.s^{-1}$).

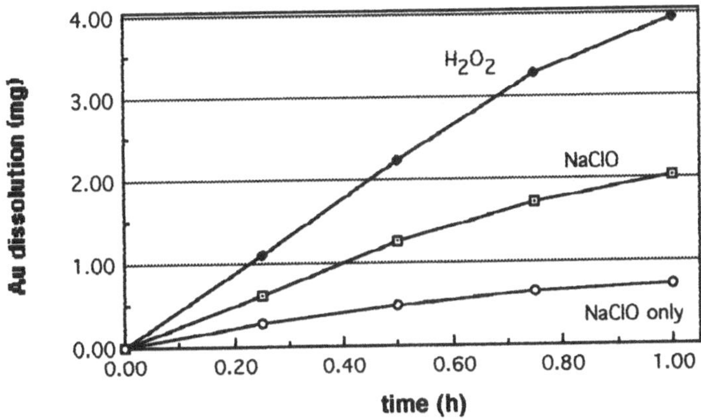

Fig.5 - Gold dissolution in 0.1 M NaBr and 0.06 M $FeCl_3$ with either 1ml/l H_2O_2 30% or 4 ml/l NaClO 3.4%, and with 0.1 M NaBr and 5 ml/l NaClO 3.4% only (no $FeCl_3$).

The results obtained in these experiments clearly indicate that a combination of ferric chloride - hydrogen peroxide may be used to dissolve gold, with possible interesting results. Therefore, further experiments were carried out in order to investigate some of the parameters involved with the system bromide + ferric chloride + hydrogen peroxide. These results are reported in the following items. Further experiments with NaClO will be published in a separate work.

4.2 Effect of ferric chloride concentration

The effect of ferric chloride concentration was investigated in solutions containing 0.1M NaBr and 1 ml/l H_2O_2 30% in the range from 0.02 M to 0.40 M $FeCl_3$ (5.4 g/l to 108.2 g/l). In all cases the potential measured on the gold surface was above 0.90 V (SHE), which indicates that $AuBr_4^-$ was the complex most probably formed (FIGURE 1).

The dissolution of gold with time as a function of $FeCl_3$ concentration is displayed in FIGURE 6. The results show that the dissolution of gold increases with increasing ferric chloride concentration.

The reported rates of gold dissolution were calculated from the slope of this type of plot. All the results could be fitted linearly, with a correlation factor equal to or greater than 0.980 in most cases. Also, there was no indication of uneven attack or pitting on the gold surface after each experiment, which further suggests that the metal surface area was constant throughout the leaching tests.

The rate of gold dissolution with 0.40 M $FeCl_3$ (2.92×10^{-8} $mol.cm^{-2}.s^{-1}$) is approximately 13 times that obtained with the lowest $FeCl_3$ concentration (0.02 M) (2.28×10^{-9} $mol.cm^{-2}.s^{-1}$). Nevertheless, it should be noticed that the rate obtained with 0.02 M $FeCl_3$ is already comparable to that obtained in the conventional cyanidation process and that this concentration, from a practical point of view, is perhaps more realistic (0.02 M = 5.4 g/l; 0.40 M = 108.2 g/l).

A reaction order plot was determined for $FeCl_3$ concentration in the range investigated. This plot is depicted in FIGURE 7 and displays good linearity with a slope of 0.82. This result indicates a near first order dependence on iron concentration.

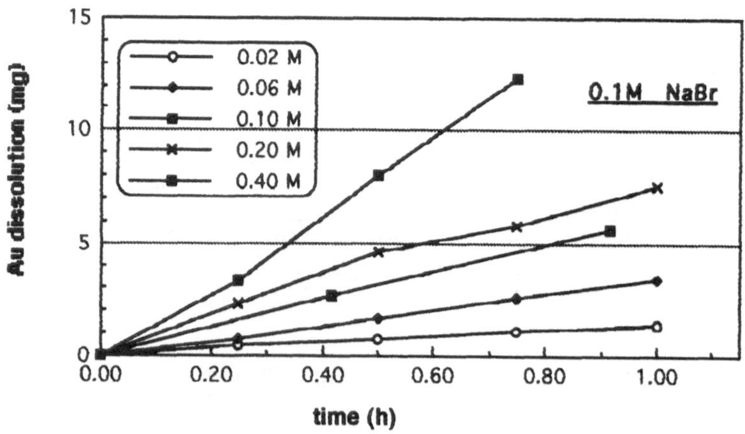

Fig. 6 - Effect of $FeCl_3$ concentration on gold dissolution in 0.1M NaBr and 1 ml/l H_2O_2 30%.

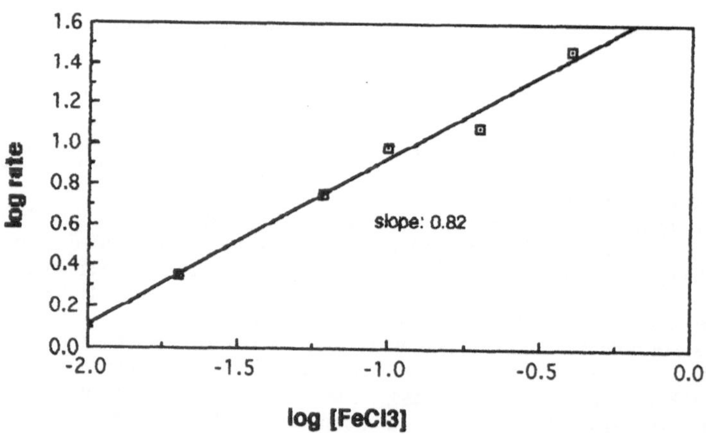

Fig. 7 - Reaction order plot for the ferric ion ($FeCl_3$).

4.3 Effect of bromide concentration

The effect of bromide concentration on gold dissolution was studied in a solution containing 0.06 M $FeCl_3$ and 1ml/l H_2O_2 30%. Bromide was added as NaBr (sodium bromide) and three different concentrations were used: 0.05 M, 0.10 M and 0.2 M.

The dissolution of gold with time as a function of NaBr is given in FIGURE 8.

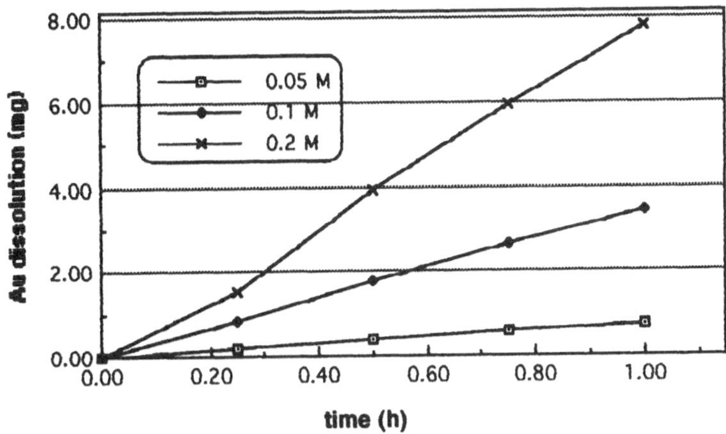

Fig. 8 - Effect of NaBr concentration on gold dissolution woth 0.06 M $FeCl_3$ and 1ml/l H_2O_2 30%.

As with $FeCl_3$, the amount of gold dissolved was found to increase with increasing NaBr concentration. The rates were calculated from slope of the linear regression of the plot "gold dissolution vs. time". The reaction order plot is depicted in FIGURE 9.

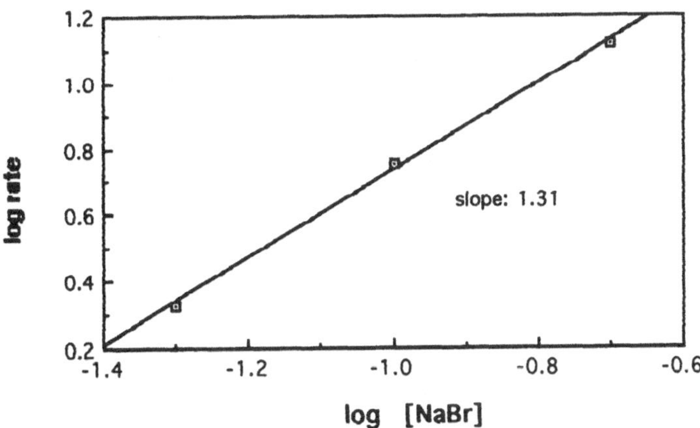

Fig. 9 - Reaction order plot for the bromide ion (NaBr).

The order of the reaction was calculated as 1.31. Whilst also indicating a near first-order dependence on the bromide concentration, this value is moderately higher than that obtained for ferric ion. This highlights the importance of the presence of bromide, without which gold dissolution would not be possible.

4.4 Effect of rotating speed

One of the characteristics of the rotating disc technique is that it has been verified by Levich (Churchill, Southampton Group, Levich) that two boundary layers near the solid (disc) surface can be established. One, not adjacent to the disc surface, was called "hydrodynamic boundary layer", in which the velocity of the fluid equals that of the bulk value; and the other, around a thousand times smaller than the first one, called "diffusion boundary layer", where there is a concentration gradient of any particular species, and mass transfer is accomplished by molecular diffusion only. It has also been demonstrated by Levich that, if the surface area of the rotating disc is considerably greater than the thickness of the disc, the magnitudes of both the hydrodynamic and diffusion boundary layers have constant values over the entire surface of the disc. Therefore, in the rotating disc system all points on the surface are uniformly accessible to the flowing liquid, rendering it quite appropriate for most mechanistic investigations.

Moreover, Levich proposed an equation to predict the mass flux of the solution to the disc surface:

$$J = 0.62 D^{2/3} \nu^{-1/6} \omega^{1/2} C_0 \tag{7}$$

where

J = mass flux onto the surface of the rotating disc ($mol.cm^{-2}.s^{-1}$)
D = diffusion coefficient of the active species ($cm^2.s^{-1}$)
ν = kinematic viscosity of the liquid (viscosity/density) ($cm^2.s^{-1}$)
ω = rotating speed ($rad.s^{-1}$)
C_0 = concentration in the bulk of the solution ($mol.cm^{-3}$)

The equation above, known as the Levich equation, is also employed as a test to confirm whether the dissolution occurring at the disc surface is entirely transport controlled. If this is the case, a plot of J versus $w^{1/2}$ will be linear passing through the origin. If, however, the process is other than purely transport controlled the plot will either not pass through the origin or will deviate from linearity.

FIGURE 10 shows that linear rates were obtained for rotating speeds from 50 to 1200 rpm. FIGURE 11 (the Levich plot) is the relationship between the rate and the square root of the rotating speed. It is interesting to note that two regions could actually be characterised, according to the rotating speed of the disc. For values up to around 240 rpm the dissolution was controlled by mass transport (i.e., the plot is linear and passes through the origin). For higher rotating speeds, however, the plot is still linear but will not pass through the origin. This indicates a transition in the mechanism of control of the reaction, from diffusional to mixed or chemically controlled. This result displays agreement with that reported recently (Pesic) for gold dissolution in a bromide-containing solution (and bromine as the oxidizing agent), which also exhibited two regions of reaction control.

4.5 Effect of temperature

The effect of temperature was investigated in the following fixed conditions: 0.06 M $FeCl_3$, 0.1 M NaBr, 1ml/l H_2O_2 30% and 600 rpm. Temperatures ranging from 15 to 45°C were investigated. The results are shown in FIGURE 12, which is a plot of gold dissolution with time.

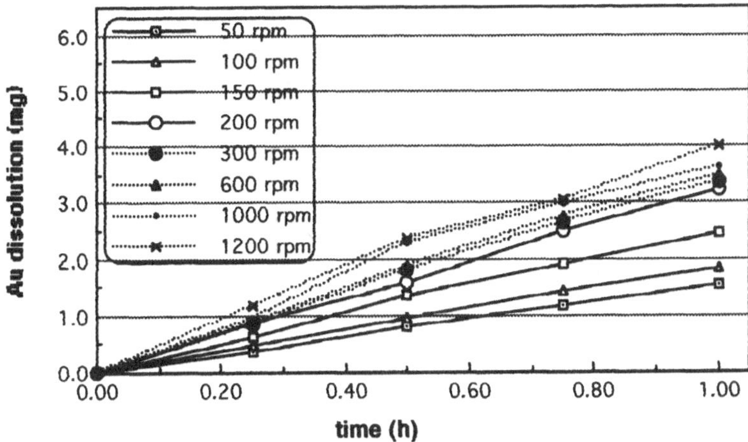

Fig.10 - Gold dissolution at various rotating speeds in 0.1 M NaBr, 0.06 M $FeCl_3$ and 1 ml/l H_2O_2 30%.

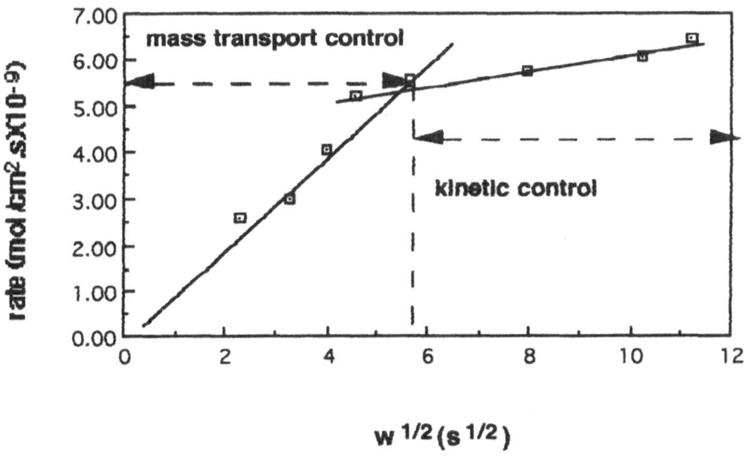

Fig.11 - Gold dissolution rate dependence on square root of rotating speed (Levich plot), showing two regions of rate of reaction control.

The rates were calculated from the slope obtained in FIGURE 12 and were plotted according to the Arrhenius plot (log rate x T^{-1}) seen in FIGURE 13. As may be perceived, good linearity was obtained. The relatively high activation energy, 28 kJ/mol, indicates that under these experimental conditions the temperature is an important parameter and that a process other than pure diffusion probably controlls the reaction rate on the gold surface. At 15°C, for instance, the rate was found to be 3.97×10^{-9}, while at 45°C it reached 10.0×10^{-9} mol.cm^{-2}.s^{-1}.

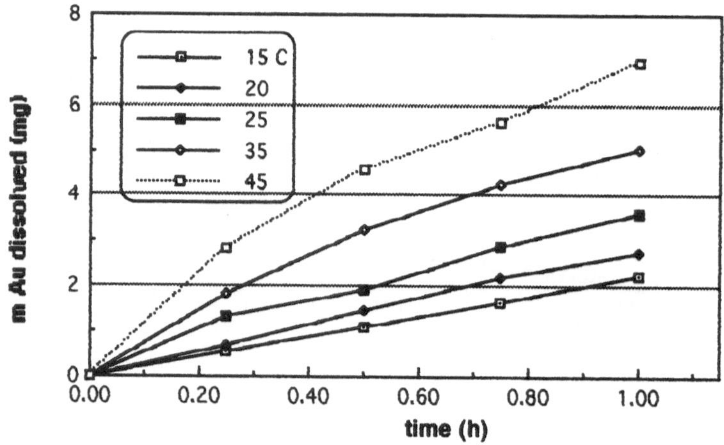

Fig. 12 - Effect of temperature on gold dissolution in 0.1 M NaBr, 0.06 M $FeCl_3$ and 1ml/l H_2O_2 30% (600 rpm).

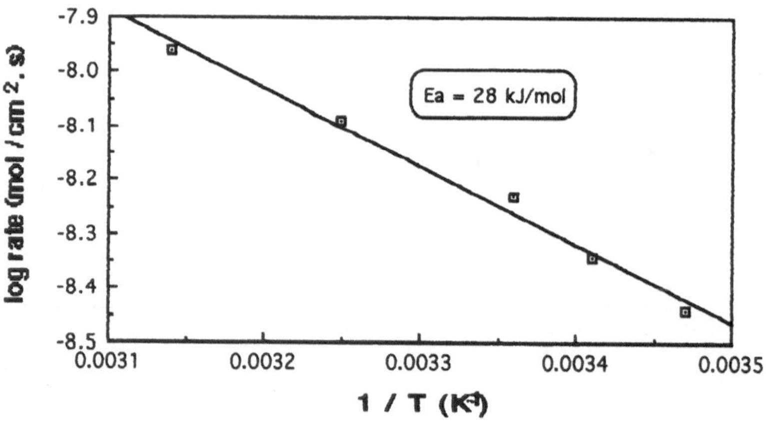

Fig. 13 - Arrhenius plot for the dissolutionh of gold in 0.1 M NaBr, 0.06 M $FeCl_3$ and 1 ml/l H_2O_2 30% (600 rpm).

5 Conclusions and final comments

The system bromine/bromide has been known to be able to dissolve gold since the last century. The use of bromine (Br_2), however, as the oxidizing agent for the process, is, in principle, not adequate in practical terms. The present work showed that the dissolution of gold is possible in a bromide-containing aqueous solution, if an oxidizing agent, other than bromine, capable of increasing the redox potential to above around 0.85 V (SHE), such as ferric chloride, is used. While thermodynamicaly feasible, the use of this leaching system is not known to have been previously tested in the hydrometallurgy of gold.

From the results obtained in the present work and from related literature, the following conclusions may also be drawn:

1 - hydrogen peroxide is not able, by itself, to elevate the redox potential of a bromide-containing solution sufficiently to allow the formation of the gold complex.
2 - the inability of the peroxide alone to provide suitable oxidizing conditions for the process is possibly explained by a mechanism of catalytic decomposition in the presence of bromide ions.
3 - the ferric ion alone is able to act as an oxidizing agent in the process, probably with formation of the complex $AuBr_2^-$.
4 - the rate of gold dissolution in the system Fe^{3+}/Br^- is low, with apparently little interest in practical terms.
5 - the combined effect of the pair Fe^{3+}/H_2O_2 as oxidizing agents for the process is efficient to dissolve gold, with results comparable to or higher than those obtained in the conventional cyanidation process.
6 - ferric ion is apparently more efficient if added as $FeCl_3$ rather than as $Fe_2(SO_4)_3$.
7 - while the reaction order is nearly first order with respect to both bromide and ferric ions, changing the bromide concentration will have a more pronounced effect on the gold dissolution than changing the ferric concentration (reaction order: NaBr = 1.31; $FeCl_3$ = 0.82).
8 - the rate of gold dissolution in 0.1 M NaBr, 0.06 M $FeCl_3$ and 1 ml/l H_2O_2 30% is around 6.0×10^{-9} mol.cm^{-2}.s^{-1}, which is around 3 times higher than the values obtained in the conventional cyanidation process.
9 - the use of NaClO to replace H_2O_2 also gives rise to gold complex formation, but the rates were found to be lower. However, it is noteworthy that, unlike peroxide, the use of hypochlorite as the sole oxidizing agent is able to oxidise gold, even if at low rates (1.19×10^{-9} mol.cm^{-2}.s^{-1}).
10 - the reaction rate follows the Levich equation up to a rotating speed of 240 rpm; from this point onwards, the dissolution rate shows a region of kinetic control (linear but not passing through the origin).
12 - temperature is an important parameter. The rate increased about 2.5 times with a temperature raise of 30°C.
13 - in a solution containing 0.1M NaBr, 0.06M $FeCl_3$ and 1 ml/l H_2O_2 30%, the activation energy was calculated as 28 kJ/mol.

Further work should be carried out involving the use of different ores, changing reagent concentrations and the main oxidizing agent. Sodium hypochlorite has already given interesting results and will be examined in tests with the rotating disc system and with ore samples. The economic aspect of the process is, of course, relevant, and will be dealt with as the research continues.

6 References

Barbosa-Filho, O. and Monhemius, A.J.(1989) Thermochemistry of thyocyanate systems for leaching gold and silver ores, in **Precious Metals'89** (eds. M.C. Jha and others), TMS- AIME, 307 .

Bray, W.C. and Livingstone, R.S.(1923) The catalytic decomposition of hydrogen peroxide in a bromine-bromide solution and a study of the steady state, **Journal of the AmericanChemical Society**, 45, 1251.

Chen. C.K. et al. (1980) A study of the leaching of gold and silver by acidothioureation **Hydrometallurgy**, 5, 207.

Churchill, M. and Laxen, P.A.(1966) The rotating-disk system and its applications in the dissolution of gold, **Research Report no.16**, National Institute for Metallurgy, Johannesburg, South Africa.

Dadgar, A.(1989) Extraction of gold from refractory concentrates, **TMS-AIME annual meeting**, Feb. 27 - Mar. 2, Las Vegas, USA .

Dreisinger, D. (1989) Enviromental issues in the aqueous processing of gold, **Journal of Metals**, 36.

Fink, C.G. and Putnam, G.L.(1942) Bromine process for gold ores, US Patent 2283198.

Habashi, F. (1987) One hundred years of cyanidation, **CIM Bulletin**, 80, 108.

Hiskey, J.B. and Atlury, V.B. (1988). Dissolution chemistry of gold and silver in different lixiviants, in **Mineral Process and Extractive Metallurgy Review**, Gordon and Breach Publ., 4, 95.

Kalocsai, G.I.Z. (1987) Dissolution of noble metals, US Patent 4684404.

Kirk-Othmer Encyclopedia of Chemical Technology (1982) 3rd. ed., 4, John Wiley & Sons, New York, USA.

Levich, V.G.(1962) **Physicochemical Hydrodynamics**, Prentice-Hall, Englewood Cliffs, New Jersey, USA.

Liebhafsky, H.A. (1934) The equilibrium constant of the bromine hydrolysis and its variation with temperature, **Journal of the American Chemical Society**, 56, 1500.

Monhemius, A.J.(1987) Recent advances in the treatment of refractory gold ores, proc. XII **International Meetting on Mineral Treatment and Hydrometallurgy**, Rio de Janeiro, 2, 281.

Pesic, B. and Sergent, R.H. (1993) Reaction mechanism of gold dissolution with bromine, **Metallurgical Transactions B**, 24B, 419.

Southampton Electrochemistry Group (1990) **Instrumental Methods in Electrochemistry**, Ellis Horwood Ltd., West Sussex, England.

Trindade, R.B.E. and Monhemius, A.J. (1993) The use of anthraquinone as a catalyst in the cyanide leaching of gold, **Minerals Engineering**, 6, 565.

von Michaelis, H.(1987) Alternative leach reagents, **Engineering and Mining Journal**, 42.

Leaching of gold from sulphide concentrates with thiosulphate/polysulphide produced by disproportionation of elemental sulphur in ammoniacal media

Zhu Guocai
Fang Zhaoheng
Chen Jiayong
Institute of Chemical Metallurgy, Chinese Academy of Sciences, Beijing, China

Abstract

A process in the leaching gold from sulfide ores with the oxidized products of the disproportionation of elemental sulfer and OH^- ions in the presence of phase transfer catalysis based on the research of gold leaching with mixed solutions of thiosulfate and polysulfide was developed. Over 90% of the gold present was extracted in 2 hours from two concentrates tested under the conditions proposed.

1 Introduction

For about 100 years, the preferred hydrometallurgical process for the extraction of gold has been alkaline cyanide leaching because of process economy and simplicity. However, this traditional gold leaching process is now facing the problem of increasing restriction on to enviromental pollution. Much more attention has recently been paid to the development of processes using non-cyanide lixiviants, such as chlorine, thiourea, thiosulfate and polsulfide. In addition to the technical problems involved, low capital and operating costs are also important for the commercial application of a new process. Elemental sulfur is cheap and rather widely available. A mixed solution of thiosulfate and polysulfide can be produced by the disproportionation of elemental sulfur with OH^- ions in ammoniacal solution through the phase transfer catalysis, as[1,2]

$$4S^0 + 6OH^- - S_2O_3^{2-} + 2S^{2-} + 3H_2O$$

$$(x-1)S^0 + S^{2-} - S_x^{2-}$$

This paper will describe and discuss the thermodynamics associated with Au-S-H_2O system and the dissolution of gold from sulfide concentrates in mixed solutions

Work supported by National Science Foundation of China.

of thiosulfate and polysulfide or a solution derived from the mixed solution obtained as the product of disproportionation of the elemental sulfur in ammoniacal media.

2 Thermodynamics

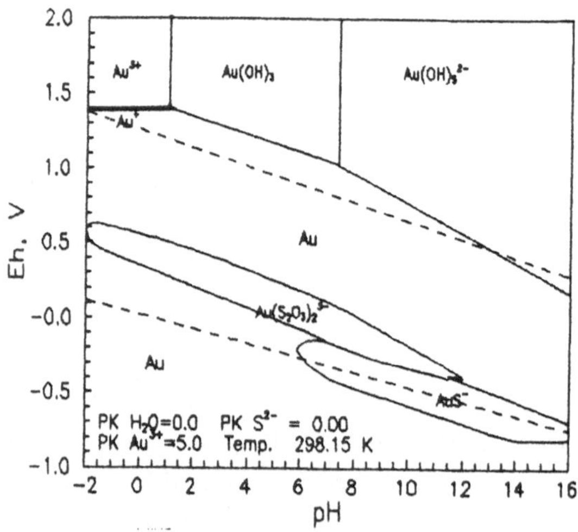

Fig. 1 Limited equilibrium Eh-pH diagram for Au-S-H_2O system[4]

The equilibrium Eh-pH diagram of S-H_2O system indicates that $S_2O_3^{2-}$ is not a thermodynamically stable species and has no dominant area[4]. At the same time, the gold complex with thiosulfate as ligand has no dominant area in the Eh-pH diagram for the system Au-S-H_2O although the existence of the species as $Au(S_2O_3)_2^{3-}$ has been proved[4]. The disagreement may come because the thermodynamic calculations are based on the assumption that all possible reactions in a defined system are in complete equilibrium. When heterogeneous reactions are involed, the process kinetics are very important. In our calculations for the Au-S-H_2O system, based on the thermodynamic data from ref[3], sulfide can be oxidized to be the valent species only. The limited equilibrium Eh-pH diagram is shown in Fig.1. It should be observed that both AuS^- and $Au(S_2O_3)_2^{3-}$ have separate dominant areas.

On the other hand, it is possible that both complexes of AuS^- and $Au(S_2O_3)_2^{3-}$ can co-exist in the system under certain conditions. The ratio of the concentration of both complexes would vary with the redox potential, pH and activities of the components in the solution. For instance, the concentration ratio of AuS^- and $Au(S_2O_3)_2^{3-}$ will be 3.5×10^4 when the concentrations of sulfide and thiosufate are equal to 1 mol/l respectively.

3 Experimemtal

Concentrates from both Jaoyuan (ore I), Shangdong province and Hetai(Ore II), Guangdong Province were used in our study. The analysis of principal elements in the two concentrates is listed in Table 1.

Table 1 Elemental analysis of the concentrates used in the experiments

Concentrate	Cu(%)	S(%)	Au(g/t)	Ag(g/t)
Ore I	0.018	31.4	92.8	36.0
Ore II	4.4	28.5	54.6	38.0

The leaching experiments were carried out in a stoppered glass reactor. Openings were provided for insertion of a thermometer, a sampling tube, an aeration tube and a glass condenser cooled with cold water respectively. An electromagnetic stirrer was used in all experiments. The reactor was placed in a thermostat to control the reaction temperature. Solution samples were drawn from the reactor and filtered immediately. All chemicals used were A.R. or C.P. grade. The polysulfide solution was prepared before use by adding elemental sulfur to a $(NH_4)_2S$ solution based on the mol ratio required.

4 Results and Discussion

4.1 Leaching with thiosulfate solution

Fig.2 shows the percent of gold leached in 3 h by solutions of 3 mol/l ammonia with different concentrations of $S_2O_3^{2-}$. It can be seen that the gold leached increases with increasing thiosulfate concentration in the low concentration range regardless of the addition of ammonium sulfate to the solution or not. In the range of high $S_2O_3^{2-}$ concentration, the dissolution of gold in the solution containing ammonium sulfate is different from that without addition. The gold leached decreases significantly with increasing the thiosulfate concentration when ammonium sulfate is present. This probably results from the decrease of ammonia concentration in the solution after addition of ammonium sulfate[4].

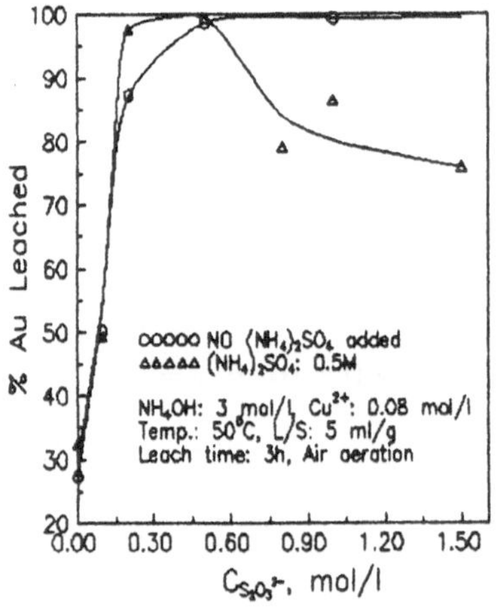

Fig. 2 Effect of thiosulfate concn. on gold leaching. Ore I.

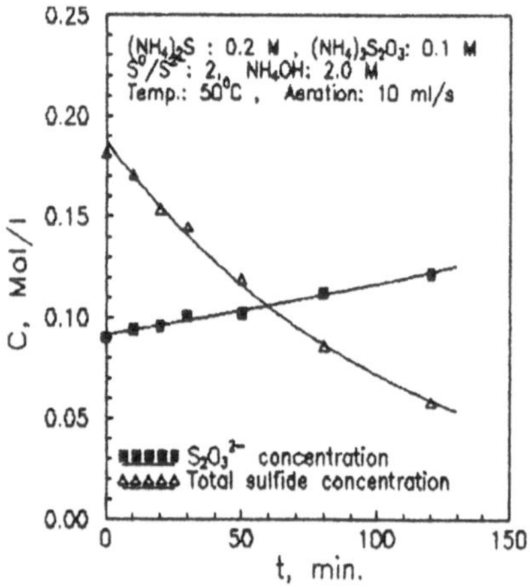

Fig. 3 Effect of aeration of the concn. of S_x^{2-} and $S_2O_3^{2-}$ in the mixed solution.

4.2 Leaching with the mixed solution of thiosulfate and polysulfide

Experimental data obtained under the condition of non-aeration are listed in Table 2. It appears that polysulfide is the effective lixiviant in the mixed solution, which agrees with the calculation by thermodynamic analysis. Therefore it is quite possible that the solution obtained from the disproportionation of sulfur should be treated to convert $S_2O_3^{2-}$ into S^{2-} or vice versa before used for gold leaching.

With aeration using compressed air into the mixed solution, the concetration of the polysulfide, or the total sulfides, decreases significantly, as shown in Fig.3. Meanwhile, the concentration of the thiosulfate increases by a small amount. Based on this fact, a process for gold leaching with the mixed solution after aeration has been proposed.

It has been pointed out that the low concentration thiosulfate solution is not satisfactory for gold leaching from the concentrate due to the low dissolution rate and low recovery of gold. Experiments show that over 90% of gold is leached with the mixed solution with aeration and with the presence of cupric ammonium complexes. The results are listed in Table 3.

Table 2 Results of gold leaching with the mixed solution of thiosulfate and polysulfide without aeration, Ore I.

No.	leaching conditions*				Au leached
	$(NH_4)S$, mol/l	$(NH_4)_2S_2O_3S$, mol/l	S^0/S^{2-}	time, h	%
1	2	1	0	3	73.7
2	2	1	0	5	86.0
3	2	1	0	7	85.8
4	2	1	1	3	90.1
5	2	0	1	3	92.7
6	2	0	1	3	92.0

*L/S=5ml/g, 50°C leaching temperature.

Table 3 Results of leaching experiments with aerated mixed solutions of sulfide and thiosulfate*

	Oxidation					Leaching		Extracted
Ore	$[(NH_4)_2S]_0$	$[(NH_4)_2S_2O_3]_0$	S^0/S^{2-}	time, h	SO_3^{2-}	time, h	Aeration	Au, %
I	0.2	0.1	0	3	0	2	Y**	72.1
	0.4	0.2	0	4	0	2	Y	96.5
	0.2	0.1	1	4	0.05	3	Y	60.8
	0.4	0.2	1	4	0.05	3	Y	96.6
II	0.2	0.1	1	4	0.05	3	Y	53.3
	0.4	0.2	0	4	0.05	3	Y	90.2

*Oxidation conditions: 2 mol/l NH_4OH, L/S=5ml/l, 50 °C leaching conditions: 50 °C, 0.08 mol/l Cu^{2+}. The flow rate of aeration with the compressed air is 10 ml/s in all experiments. ** Y means with aeration.

4.3 Leaching with the products of disproportionation of elemental sulfur in ammoniacal media

Phase transfer catalysis was used to accelerate the disproportionation of elemental sulfur[2]. The product of disproportionation followed by aeration was used to leach the gold from the ores.

The results of experiments indicate that the disproportionation of elemental sulfur in the system of CCl_4 and aqueous NaOH solution with the addition of quaternary ammonium ions $C_{16}H_{33}(CH_3)_3N^+$ forms a solution of thiosulfate and polysulfide. For instance, under the conditions of 1 mol/l NaOH, 1 mol/l sulfur in organic phase and at 65°C, an aqueous solution containing 0.20 mol/l $S_2O_3^{2-}$ and 0.302 mol/l S_x^{2-} was obtained in 4 hours. The solution produced was then oxidized by aeration for 6 hours at a flow rate of 20 ml/min per liter solution and used for gold leaching. The results show that 91.7% and 92.5% of the gold were extracted from Ore I and II respectively.

A process flowsheet for gold leaching from ores is therefore proposed, as shown in Fig.4.

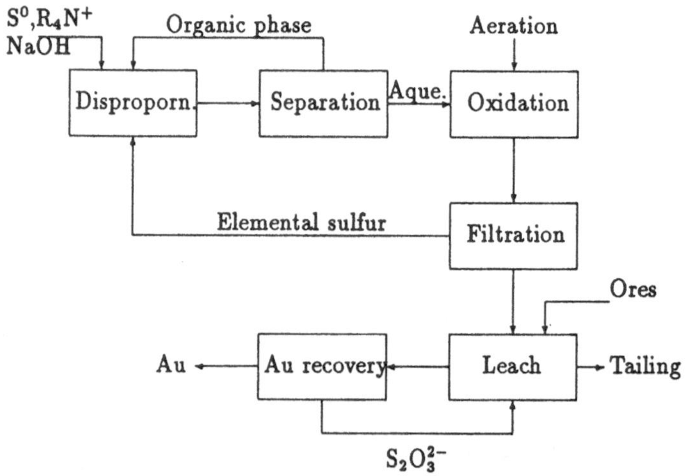

Fig. 4 The schematic flowsheet of the proposed leaching process.

5 Conclusion

It can be concluded that gold can be extracted from the sulfide ores by the mixed solution of thiosulfate and polysulfide of low concentration. A process for the leaching gold from ores using the product of disproportionation of the elemental sulfur in the presence of phase transfer catalysis is therefore proposed. The product obtained from the disproportionation of elemental sulfur in ammoniacal solution can be oxidized by air first and then used for leaching. Over 90% of the gold in the sulfide ore was extracted in 2 hours from two concentrates tested under the conditions proposed.

6 References

1. Pryor,W.A. (1962) **Mechanism of Sulfur Reactions.** McGraw-Hill Pub., New York.
2. Deng, Tong and Chen J.Y. (1988) **J. Chem. Ind. and Eng.** (China), **39**, 538.
3. Barin, I. and Knacke, O. (1973) **Thermochemical properties of Inorganic Substances.** Berlin, Springer.
4. Zhu,Guocai. (1993) **Ph. D. Dissertation.** Institute of Chemical Metallurgy, Chinese Academy of Sciences.

Adsorption of gold–thiourea complex on Greek lignite

A. I. Zouboulis
K. A. Kydros
K. A. Matis
Department of Chemistry, Aristotle University, Thessaloniki, Greece

Abstract
It was shown by laboratory bench scale experiments that it is possible to use lignite (brown coal) of Greek origin for the effective adsorption of gold/ thiourea complexes from aqueous solutions, instead of the commonly applied activated carbon. The former is cheaper and therefore, needs no regeneration. Several parameters were examined, including the dispersion pH, contact time, lignite particle size fraction, temperature and the initial concentrations of lignite, gold and thiourea. Adsorption isotherms of the Langmuir type and a typical Arrhenius plot were obtained and a comparison with powdered activated carbon of the same size fraction was attempted.
Keywords: Gold, Lignite (Brown Coal), Adsorption, Thiourea, Isotherms.

1 Introduction

For the recovery of gold from leach liquors, granular activated carbon in the size range of 1-3 mm (usually of coconut shell origin) is commonly applied (Canning and Woodcock, 1982). The adsorption of gold cyanide onto activated carbon has been studied extensively - see for example (Fleming and Nicol, 1984). The interaction between gold cyanide and high surface area activated carbon has also been examined (Dixon *et al.*, 1978). Wood charcoal and possibly, brown coal were also mentioned as alternative, cheaper adsorbents; work using them is reported to be in progress with the main aim of avoiding the recycling of carbon. Nevertheless, it is suggested that these are relatively soft solids and have rather low gold-loading characteristics. They could probably be used once, then burned and smelted for gold recovery. Recently, a Bulgarian low calorific brown coal has been proposed for gold adsorption (Gaidarjiev and Spassova, 1992).

As a potential substitute for cyanidation, due mainly to severe pollution constraints related to tailings disposal, water quality and environmental impact, a thiourea-leaching route has been developed and successfully tried on different materials and ores, some of which were refractory (Deschenes, 1986), and used in commercial practice (Chadwick, 1986). The dissolution of gold in thiourea solutions is a process involving the formation of a soluble cationic gold(I) complex ($Au[CS(NH_2)_2]_2^+$), as well as the oxidation of the lixiviant itself through several intermediate products. Thiourea, compared to cyanide, presents the advantages of lower toxicity, greater rates of gold dissolution and lower inhibition due to the presence of certain metal ions or other co-dissolved species, while the large lime additions required for cyanidation would not be needed (Deschenes, 1987). On the other hand, the main disadvantage of thiourea is that reagent consumption and hence costs are relatively higher.

The extraction of the gold/thiourea complex from aqueous liquors after adsorption onto activated carbon is also well known and compared with cyanidation (Fleming, 1987). Other possible methods which have been examined include ion exchange with resins, cementation, electrowinning, solvent extraction, pressure reduction under hydrogen, etc. (Deschenes, 1986). A relatively innovative technique for gold separation (from thiourea solutions) is ion flotation (Zouboulis *et al.*, 1993). Only a few papers exist, which examine potential recovery techniques for gold and the majority of them contain very few details.

Low rank coals, such as lignite (or brown coal) or peat and other carbonaceous material produced from waste products, are known to adsorb and remove metallic ions, particularly in effluent treatment (McLellan and Rock, 1988; Cullen and Siviour, 1978; Srivastava *et al.*, 1989). Elsewhere, lignite has been studied as an alternative adsorbent to activated carbon, in the wastewater treatment of mineral or metal processing plants in order to remove organics (Rice and Ikwue, 1990), while peat was applied to the treatment of oil-in-water emulsions (Viraraghavan and Mathavan, 1988).

Greek lignite, which is mainly used for electric power generation after combustion, is found principally in the Ptolemais area (North Western Greece), where more than 50% of total country's energy, is said to be produced (G.R.P.C, 1983). In 1985, 27.2 millions tons of lignite were mined in this area, while the utilizable deposits were estimated as 2,580 million tons (equivalent to about 350×10^6 tons of oil). In the present work, the main parameters affecting the adsorption of gold from dilute thiourea solutions onto Greek lignite are examined, i.e. the amount of lignite in dispersion, gold concentration in solution, pH of the adsorption medium, adsorption time and temperature.

In the conventional cyanide-leaching route for gold recovery from minerals, following the adsorption of gold on powdered activated carbon, flotation has been examined as a subsequent solid/liquid separation method (Aswegen, 1981). The significance of recovering the dissolved gold quantitatively and hence, to limit the gold losses in waste streams was stressed. As a further improvement towards a more efficient separation step, the flotation of powdered activated carbon previously loaded with the gold-thiourea complex could be considered (Zouboulis *et al.*, 1994). Flotation could also solve the problem of separating any coal fines produced during the process, due to the softness of lignite. Carbon may be lost due to attrition, but it is worth noting that the softer a carbon is, the higher its activity towards gold adsorption (Yannopoulos, 1991). An extensive review describing the application of different flotation techniques for the separation of fine particulate matter was recently published (Matis and Mavros, 1991).

2 Experimental

The bench scale adsorption experiments were conducted in a 250 mL stirred glass cell, immersed in a constant temperature bath (299 K), where a lignite dispersion was mixed by a mechanical stirrer (200 rpm). A contact time of 10 min was usually, considered appropriate according to preliminary experiments, unless otherwise stated. In most experiments the dispersion pH was set in the acidic region (usually, 1.5) in accordance with other literature findings (Bilston *et al.*, 1990) and was regulated by the addition of sulphuric acid or sodium hydroxide solutions and monitored continuously. The dispersion was then filtered by means of a G_5 glass Gooch crucible, having a mean pore diameter of 1-1.5 μm. The resultant solution was assayed by standard Atomic Absorption Spectrophotometric procedures, in order to determine the remaining gold concentration and hence, the percentage recovery (Re%) was calculated applying the formula:

$$Re\% = (1 - C/C_o) \tag{1}$$

where C and C_o were the final and the initial gold concentrations respectively.

The aqueous solutions tested were prepared by the addition of acidified thiourea to gold trichloride trihydrate solutions, having an initial gold concentration of 20 mg/L, unless other stated. Thiourea, $CS(NH_2)_2$ (denoted hereafter as TU), was added at 20-fold higher concentrations than that of gold, i.e. 400 mg/L (5.26 mM), as found to be appropriate for effective leaching kinetics of gold (Pyper and Hendrix, 1981). It has been reported that in acid thiourea solutions the sole gold species is always $Au(TU)_2^+$, since the reduction of Au(III) to Au(I) by thiourea is fast and quantitative (Groenewald, 1976); hence, no further oxidant such as ferric ions, was added. The formamidine disulphide, which is formed through a reversible reaction and represents the actual oxidizing agent of the solution, has a reduction potential value of 0.42 V, independent of pH value from 0 to 4.3 (Pyper and Hendrix, 1981). All reagents were purchased from Merck (analytical grade).

The lignite used in these experiments had the proximate analysis shown in Table 1; the initial moisture content was around 18% and the ash content (before desliming) 44.5%. The samples were crushed and following sieving, the particle size range of -1000 +500 μm was separated for the adsorption experiments. Various dispersion concentrations were tried. A comparison was attempted with conventional granular coconut activated carbon, supplied by Trans-Pacific Carbon Corp., having a size range from 1.410-0.350 mm and iodine number 1000. A comparison was also attempted with lignite fines, having a size range less than 45 μm. Obviously, in the finer ranges the surface area, which is available for adsorption, is increased. For a study of the adsorption isotherms three temperatures were selected, i.e. 299, 320 and 336 K.

Table 1. Chemical analysis of the lignite samples used in this study

Inorganic Constituents	Content (% w/w)
Inorganic CO_2	9.73
Carbon	32.60
Hydrogen	2.90
Nitrogen	0.90
Sulphur	0.66
Oxygen	12.45
Lowest Calorific Value	13.8 kJ/kg

3 Results and discussion

Preliminary kinetic results for the adsorption of gold onto lignite are given in Figure 1. It was found that by increasing the lignite concentration of the dispersion from 0.5 to 10 g/L, lignite loadings (mg gold/g lignite) were decreased, although gold recoveries were increased reaching values around 70%. In the case of powdered activated carbon, one-hour period was used, which is comparable with other

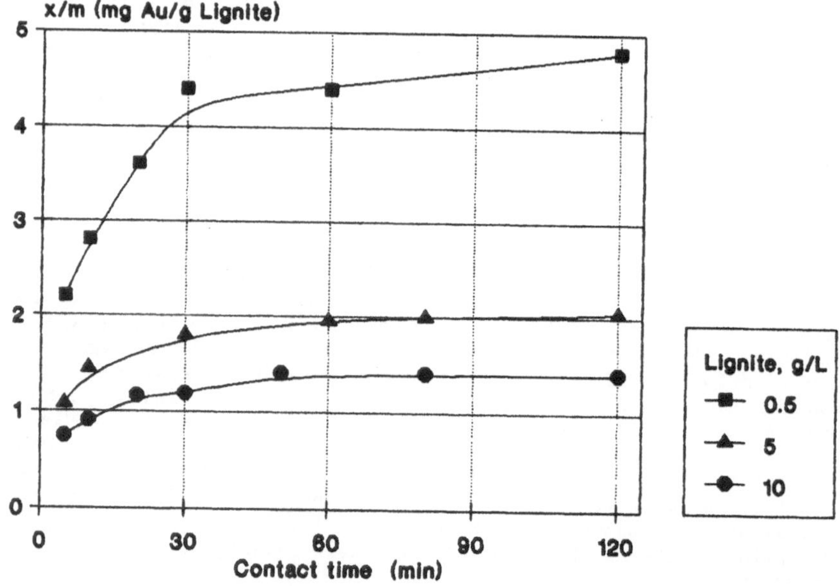

Fig.1. Adsorption studies in thiourea solution; kinetics of gold adsorption on lignite as a function of lignite concentration.

literature reports (Deschenes, 1986). Contact time seems to affect the adsorption rate higher than the equilibrium loading capacity. The slope of the curves suggests that the adsorption rate is high during the first 30 min and decreases to zero after about one hour.

Gold adsorption on lignite as a function of dispersion pH was studied over the whole pH range and was compared under the same conditions with the conventional activated carbon. In Figure 2 results are presented, only up to pH 6. Only slight difference was found at the very acidic pH value of 1 and at pH values over 6. As the solution pH is increased above pH 4, formamidine disulphide decomposes irreversibly to elemental sulphur and cyanamide (Raudsepp and Allgood, 1987). It may be expected that the decomposition products are also deposited on lignite. In the alkaline pH range a sharp increase of gold recoveries was noticed, which may be attributed to the increased adsorption capacity of the finely divided elemental sulphur formed at these pH values.

Figure 3 presents the behaviour of -45 μm lignite particles compared with the previously examined fraction (-1000 + 500 μm); the smaller size fraction is considered to be rather inconvenient from a practical point of view. As was expected higher gold recoveries were obtained using the finer fraction. Nevertheless, because of several problems, with the use of fines which were foreseen in the subsequent separation, this particle size range was avoided (Matis and Zouboulis, 1994).

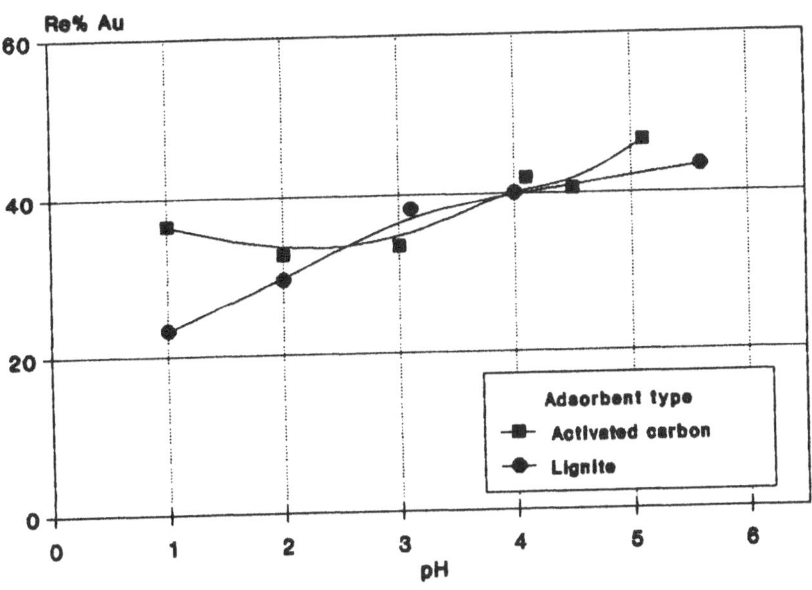

Fig.2. Effect of dispersion pH on gold recovery. Comparison of lignite with granular activated carbon of similar fraction size (-1000 +500 μm); concentration of both adsorbents 0.25 g/L.

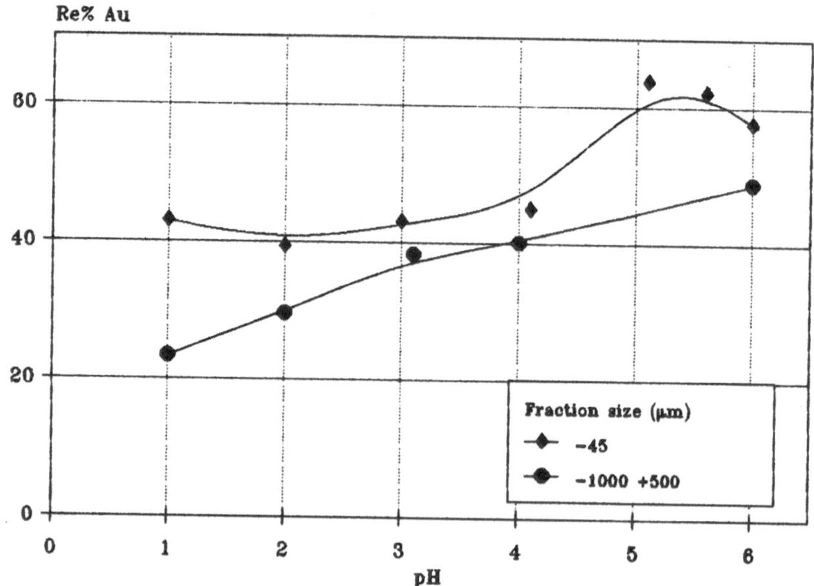

Fig.3. Effect of dispersion pH on gold recovery. Comparison of two different lignite fractions, used at concentrations of 0.25 g/L.

Further fundamental information on this method, which appeared to be effective and its performance was required. Parameters affecting the adsorption of gold-thiourea complex should be comparable with the more frequently applied case of cyanide-gold complex adsorption (Fleming, 1987). Therefore, variables such as the adsorption characteristics, gold content, pH, temperature, etc., should influence the process. In industry however, it is preferable for the gold-loading activated carbon circuits to operate at room temperature. In the case of activated carbon, elution takes place at higher temperature (Fleming, 1987).

Figure 4 shows the effect of increasing initial lignite concentration in the pulp on gold adsorption. The initial gold concentration was in the range 5-20 mg/L. The curves obtained have a similar shape, showing that when the same quantity of lignite was added, gold recovery decreases slightly as the initial gold concentration increases. As shown in this figure, to obtain over 80% recovery from an auriferous solution of 2 mg/L within 10 min, requires a charge of lignite around 20 g/L. It should be noted for comparison reasons, that 20 g/L activated carbon used in thiourea solutions, gave over 90% gold extraction (Deschenes, 1986). Also, an amount of 8 g/L GAC was applied for auriferous ores containing 10.55 g Au/ton (Schmidt *et al.*, 1993).

The effect of gold content is shown in Figure 5. In the gold concentration range investigated, the 20 g/L lignite concentration was adequate to obtain over 80% recovery.

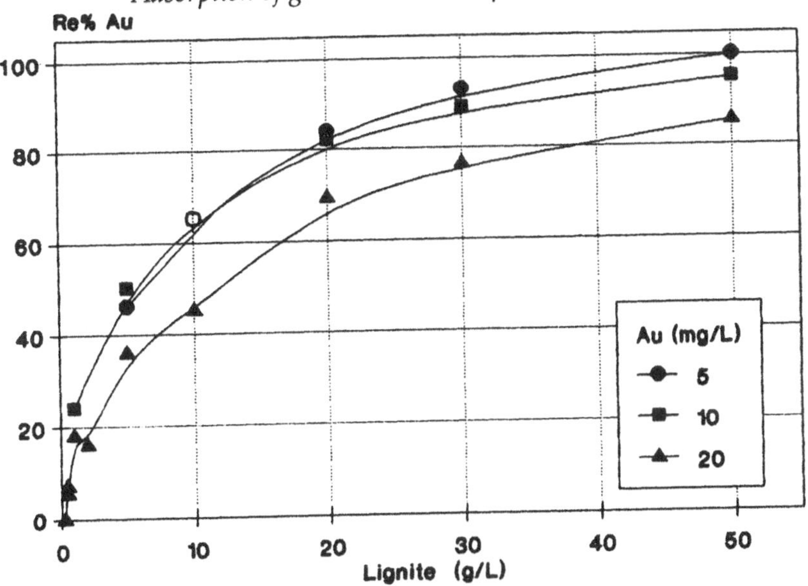

Fig.4. Effect of lignite concentration on gold recovery, at a solution pH of 1.5, for three different initial gold concentrations.

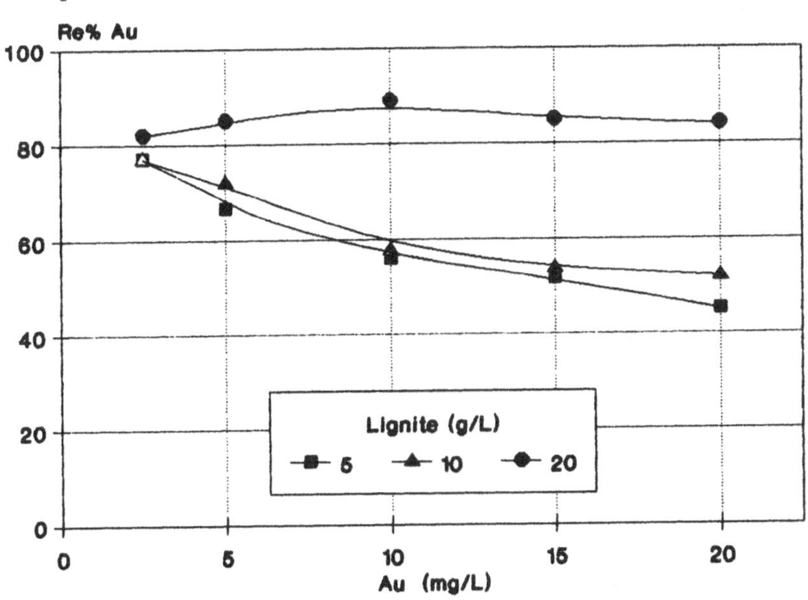

Fig.5. Effect of initial gold concentration on gold recovery, for three different initial lignite concentrations (contact time: 30 min).

As previously mentioned, the thiourea used in the experiments described was in excess (20-fold) of the respective gold concentration. The influence of the initial thiourea concentration on the gold-thiourea adsorption of complex onto lignite was further examined and it is presented in Figure 6. It is observed that the loading capacity of lignite was reduced by an excess of thiourea and the extraction of gold was also decreased. In other words, the two processes of gold leaching and adsorption were somehow antagonistic. Similar results have been reported in the literature for activated carbon (Fleming, 1987).

It has been suggested that during the process the gold-thiourea cationic complex was loaded onto the carbon surface without undergoing any chemical change, such as reduction of gold(I) to metallic gold (McDougall and Fleming, 1987). There is no reason to assume a different adsorption mechanism in the case of lignite. Apparently, this may be considered as a drawback (perhaps the major) of the process in the broad sense, whichever adsorbent is applied, because it would involve high losses of thiourea; unless, some way of regenerating the leaching agent downstream, i.e. by electrolysis in a cell incorporating a separator between the electrodes can be found (Yannopoulos, 1991).

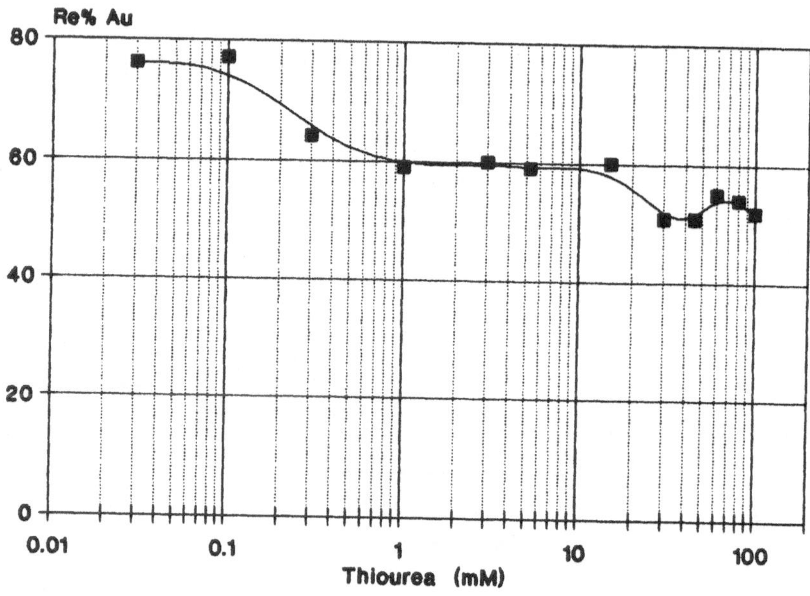

Fig.6. Effect of thiourea concentration on gold recovery; lignite concentration 10 g/L (contact time: 30 min).

Some simple calculations were also performed related with gold adsorption and presented in Figure 7. An adsorption isotherm equation of the following form (Langmuir) was adopted:

$$\frac{1}{Q} = \frac{1}{Q_m K} \cdot \frac{1}{C_{eq}} + \frac{1}{Q_m} \tag{2}$$

where Q is the quantity of gold adsorbed on lignite (mg/g or mole/g), Q_m the maximum monolayer adsorption, K the equilibrium constant for the adsorption reaction (L/mg or L/mole) and C_{eq} the equilibrium concentration in the solution (mg/L or mole/L). The isotherms could be used in order to characterize the type of coal and of adsorption. The above equation (2) was tested for three temperatures (Figure 7a) and the obtained equations are given in Table 2. The correlation coefficient was of the order of 0.99.

Similar results are typical in the literature - see for example (Dixon *et al.*, 1978). The Langmuir model, as it is known, assumes that the surface of the adsorbent consists of adsorption sites, each having a specific area, that all adsorbed species react only with a specific surface site and not with each other, and that adsorption is limited only to a monolayer; the assumptions were considered reasonable for the present case.

Table 2. Adsorption isotherms of gold-thiourea complex onto lignite

(i) **Adsorption isotherm calculations**: Equation (2)-see also Figure 7a

Temperature: 299 K	$1/Q = 0.010 + 0.41\ (1/C_{eq})$
-"- : 320 K	$1/Q = 0.011 + 0.69\ (1/C_{eq})$
-"- : 336 K	$1/Q = 0.012 + 1.42\ (1/C_{eq})$

(ii)

Temperature (K)	Qm (mol/g)	K (L/mol)
299	5.0×10^{-4}	4700
320	4.6×10^{-4}	3170
336	4.2×10^{-4}	1670

where Q_m: maximum monolayer adsorption
K : equilibrium constant for the adsorption reaction

(iii) Equation (3) (from Figure 7b)

$$\ln K = -0.745 - \frac{-2772}{R} \cdot \frac{1}{T}$$

Further, from thermodynamics (van't Hoff equation) using the temperature dependency of the equilibrium constant, previously described, gives:

$$\ln K = \ln A - \left(\frac{\Delta H}{R}\right)\left(\frac{1}{T}\right) \quad (3)$$

where A is the intercept, ΔH the heat of adsorption (kJ/mol) and R the universal gas constant: 8.314 kJ/kmol/K.

This permits (Figure 7b) fitting of the respective equation with a correlation coefficient of 0.97, from which ΔH was found equal to -23.1 kJ/mol, i.e. exothermic-see Table 2; which indicates physical adsorption, as expected. A value of 13.9 kJ/mol for the enthalpy of reaction in the case of activated carbon has been reported (Fleming, 1987).

Generally, the equilibrium loading for gold adsorption is proportional to the precious metal concentration in solution, while the temperature effect was not very important. Similar observations have been published for cyanide (Dixon *et al.*, 1978) and thiourea (Deschenes, 1986) solutions, where the equilibrium capacity of the activated carbon decreases, when the temperature was increased.

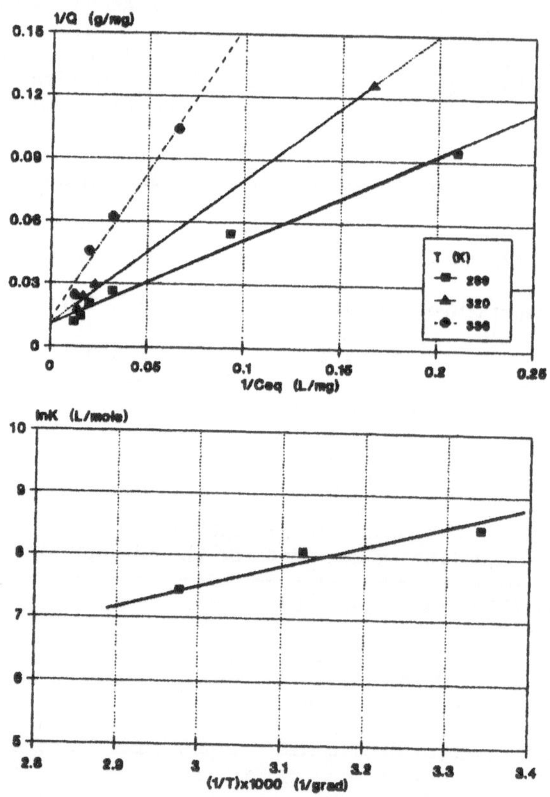

Fig.7. a) Adsorption isotherms (Langmuir type).
b) Typical Arrhenius diagram of gold adsorption on lignite.

4 Conclusions

The amount of gold adsorbed by lignite from an aqueous solution containing excess thiourea, is a function of several parameters: the amount of carbon in suspension, gold concentration in solution, temperature, time of adsorption and pH of the adsorption medium. If these parameters are kept in their optimal values, it is possible to recover almost all the gold from an aqueous solution. A summary of the optimum conditions and results is given in Table 3. Further work is warranted in order to transfer these preliminary results to a pilot-scale unit.

Table 3. Optimum conditions and results of the studied process

Parameter	Values
Gold concentration (mg/L)	20
Lignite addition (g/L)	20
Lignite fraction (μm)	-1000 +500
Thiourea concentration (mM)	up to 15
pH value of the dispersion	1.5
Contact time (min)	45
Gold recovery (%)	over 80

5 Acknowledgements

Thanks are due to the Chemistry student E. Tassiopoulos for his help in the experimental work and to the Greek Public Power Corp. for the supply of the lignite samples.

6 References

Aswegen, van P.C. (1981) Recovery of dissolved gold from rotary filter residues by the addition of fine activated carbon powder followed by flotation, in **Congress Inter. Mineralurgie**, 11th G.E.D.I.M., St. Etienne, pp. 510-519.

Bilston, D.W., Bruckard, W.J., McCallum, D.A., Sparrow, G.J. and Woodcock, J.T. (1990) Comparisons of methods of gold and silver extraction from Hellyer pyrite and lead-zinc flotation middlings, in **Sulphide Deposits - Their Origin and Processing**, (ed P. Gray), The Inst. of Min. and Metal., London, pp. 207-221.

Canning, R.G. and Woodcock, J.T. (1982) Innovations and problems in gold recovery from ores and mineral products, in **Proceedings, The Australian Inst. Min. and Metal. Conf.**, Melbourne, pp. 365-381.

Chadwick, J. (1986) Jamestown to be the largest U.S. gold milling operation. **Inter. Mining**, Sept., 24-26.

Cullen, G.V. and Siviour, N.G. (1978) Removing metals from waste solutions with low rank coals and related materials. **Water Res.**, 16, 1357-1366.

Deschenes, G. (1986) Literature survey on the recovery of gold from thiourea solutions and comparison with cyanidation. **C.I.M. Bulletin**, 79 (895), 76-83.

Deschenes, G. (1987) Investigation of the potential techniques to recover gold from thiourea solution, in **Gold Metallurgy** (eds R.S. Salter, D.M. Wyslouzil and G.W. McDonald), Pergamon Press, Toronto, pp. 359-377.

Dixon, S., Cho, E.H. and Pitt, C.H. (1978) The interaction between gold cyanide, silver cyanide, and high surface area charcoal, in **Symp. Series - Fundamental Aspects of Hydrometallurgical Processes**, A.I.Ch.E., N. York, pp. 75-83.

Fleming, C.A. (1987) The recovery of gold from thiourea leach liquors with activated carbon, in **Gold Metallurgy**, (eds R.S. Salter, D.M. Wyslouzil and G.W. McDonald), Pergamon Press, N. York, pp. 259-277.

Fleming, C.A. and Nicol, M.J. (1984) III. Factors influencing the rate of loading and the equilibrium capacity. **J. South Afr. Inst. Min. Metal.**, 84 (4), 85-93.

Gaidarjiev, S. and Spassova, S. (1992) Amide-based technologies in flotation and hydrometallurgy of precious metals, in **Innovations in Flotation Technology** (eds P. Mavros and K.A. Matis), Kluwer Academic, Dordrecht, pp. 283-291.

Groenewald, T. (1976) The dissolution of gold in acidic solutions of thiourea. **Hydrometallurgy**, 1, 277-290.

Gr. Pub. Power Corp. (1983) Production and utilization of Greek lignites, in **Proceedings - Exploitation of Low Calorific Value Solid Fuels**, Inter. Meeting, Ptolemais (Greece).

Matis, K.A. and Mavros, P. (1991) Foam/froth flotation - Part II - Removal of particulate matter. **Sep. & Purif. Methods**, 20, 163-198.

Matis, K.A. and Zouboulis, A.I. (1994) An overview of the process, in **Flotation Science and Engineering**, (ed K.A. Matis), Marcel Dekker, N. York (1994), preprint.

McDougall, G.J. and Fleming, C.A. (1987) Extraction of precious metals on activated carbon, in **Ion Exchange and Sorption Processes in Hydrometallurgy**, (eds M. Streat and D. Naden), Soc. Chem. Ind., London, pp. 56-126.

McLellan, J.K. and Rock, C.A. (1988) Pretreating landfill leachate with peat to remove metals. **Water, Air and Soil Pollution**, 37, 203-215.

Pyper, R.A. and Hendrix, J.L. (1981) Extraction of gold from finely disseminated gold ores by use of acidic thiourea solution, in **Extraction Metallurgy '81**, Inst. Min. Metal., London, pp. 57-75.

Raudsepp, R. and Allgood, R. (1987) Thiourea leaching of gold in a continuous pilot plant, in **Gold Metallurgy**, (eds R.S. Salter, D.M. Wyslouzil and G.W. McDonald), Pergamon Press, N. York, pp. 87-95.

Rice, N.M. and Ikwue, A.E. (1990) Treatment of solvent-extraction raffinates with lignite as an alternative adsorbent to active carbon, **Trans. Inst. Min. Metal.**, 99, C131-C136.

Schmidt, R., Moya, S.A., Foerster, J.E. and Varela, A.R. (1993) Carbon adsorption and leaching of gold by acid thiourea. **Intern. J. Min. Proces.**, 38, 257-265.

Srivastava, S.K., Tyagi, R. and Pant N. (1989) Adsorption of heavy metal ions on carbonaceous material developed from the waste slurry generated in local fertilizer plants. **Water Res.**, 23 (9), 1161-1165.

Viraraghavan, T. and Mathavan, G.N. (1988) Treatment of oil-in-water emulsions using peat, **Oil & Chem. Pollution**, 4, 261-280.

Yannopoulos, J.C. (1991) **The Extractive Metallurgy of Gold**. Van Nostrand Reinhold, N. York.

Zouboulis, A.I., Kydros, K.A. and Matis, K.A. (1993) Recovery of gold from thiourea solutions by flotation. **Hydrometallurgy**, 34, 79-90.

Zouboulis, A.I., Kydros, K.A. and Matis, K.A. (1994) Flotation of the gold/thiourea complex adsorbed on activated carbon. **Hydrometallurgy**, 35, in press.

Solution purification

Norzink removal of cobalt from zinc sulphate electrolytes

Kjetil Børve
Norzink AS, Odda, Norway
Terje Østvold
Institute of Inorganic Chemistry, NTH, University of Trondheim, Trondheim, Norway

Abstract
Removal of cobalt from the zinc sulphate solution prior to electrowinning is essential in the hydrometallurgical zinc production process. In order to improve the cobalt removal process, Norzink has initiated a research program.

In this paper the removal of cobalt from industrial zinc sulphate solution by zinc dust cementation is investigated with respect to some important factors.

The rate of cobalt cementation on zinc dust is studied as function of initial concentrations of copper and antimony. The effects of temperature, pH, zinc dust grain size and presence of copper cement have also been determined.

Residues both from laboratory experiments and industrial (Norzink) cobalt cementation process are investigated with respect to morphology and phase relations.
Keywords: Antimony, cadmium, cementation, cobalt, copper, electrolytic, production, zinc.

1 Introduction

In the hydrometallurgical zinc production process, the removal of impurities of the zinc sulphate solution prior to electrowinning is one of the most important and complex steps.

A failure in the purification process will give significant production loss in the cell house due to low current efficiency and bad zinc cathode quality [1]. The cost of zinc dust consumption is considerable and should be minimized.

Many investigations have been conducted to understand the cementation mechanism, but the chemistry has not been completely clarified. Tozawa et al. [2] have in a recent paper given a survey over proposed reactions for cobalt cementation with zinc metal.

MacKinnon [3] reported that the amount of cobalt cemented from solution increases with decreasing particle size of the zinc dust. In the same paper the effect of initial pH was investigated, and pH below 4 gave reduced cobalt cementation. This conclusion is confirmed by Xiong and Ritchie [4]. They suggest that this is due to the blocking action of hydrogen atoms adsorbed on the reacting surface. In another paper by Xiong and Ritchie [5] the formation of basic salts of zinc at pH > 4.5 is discussed. Colloidal zinc hydroxide is formed in the intermediate vicinity of the precipitant. This will act as a secondary inhibitor, which may reduce the number of active sites on the cathode.

While the kinetics of removal of cobalt from zinc sulphate electrolytes by zinc cementation are studied in considerable detail, less emphasis has been made on the study of the residue itself.

Fischer-Bartelk, Lange and Schwabe [6] report that additives like Cu^{2+}, Sb^{3+} and Cd^{2+} when reduced, form intermetallic phases with Co and Zn. These phases impart a more positive potential to the cementation of cobalt. Compounds like CoSb and $CoSb_2$ as suggested by Fontana and Winand [7] may also promote the cementation of Co due to their higher stability relative to pure Co. Tozawa et al. [2] observed from their SEM and X-ray analyses of deposited materials that Sb and Cu were codeposited with Co on edges of the SbCu alloy. No CoSb alloy was observed. With As present, however, CuAsCo alloy codeposition was observed, together with a very pronounced effect of Cu^{2+} additions up to 100 mg/l on the rate of Co^{2+} removal from the electrolyte. Additions of Cu^{2+} up to 5 mg/l gave a limited increase in the Co^{2+} deposition rate when Sb^{3+} was present.

Lawson and Nhan [8] showed that CoAs and $CoAs_2$ compounds were formed during reduction of Co^{2+} with Zn in a solution containing both Co^{2+} and As^{3+}. At temperatures above 92 °C $CoAs_2$ was formed. Below this temperature CoAs seemed to be the stable product.

In a very recent paper Lew, Dreisinger and Gonzalez-Dominguez [9] have studied, in a laboratory cell, the kinetics of the copper-antimony process for cobalt removal. Both reaction mechanism and morphology of the residue were studied. They found that a combination of Cu and Sb addition to the electrolyte showed a significant increase in the cobalt deposition rate.

At a given Sb level of 1.5 mg/l, additions of Cu up to 90 mg/l enhanced the reaction rate. When the Sb level was increased to 3 mg/l the initial reaction rate increased but levelled off to a lower value at reaction times above 50 min. They also observed that changes in the initial cobalt concentration did not change the reaction kinetics which they claimed to be under surface or chemical control. In agreement with earlier observations [3-5] they found high cobalt cementation rates at $pH \geqq 4$ and at high temperatures (up to 90 °C). Their SEM and WDX analyses indicated that Co and Cu were present in approximately 1:1 ratios in the precipitate.

Fontana et al. [10] together with Fischer-Bartelk et al. [6] claimed that Cd^{2+} in the electrolyte seems to promote Co cementation. An important part of the Co cementation process therefore seems to be connected with the formation of alloys which reduce the potential for Co reduction. In addition H_2 evolution, which reduces the yield in the Zn dust cementation process, may create additional problems. If the gas absorbs on the precipitated metal and blocks further reduction of Co^{2+} this may reduce the rate of cementation. If the Co alloys formed are unaffected by the H_2 (g) formed during the reduction process, these alloys may promote the cementation process.

A study of the residue formed during the Co cementation process is, for the above mentioned reasons, of vital importance for an understanding of the process.

Normally each zinc plant has its own specific operating conditions in the cementation process, even though the basic principles are the same for all plants. Removal of cobalt is the major challenge in the purification process and is probably the element which is most in focus. Practice shows that zinc powder cementation of cobalt is impossible without using additives [5]. Fugleberg et al. [11] report on the technical use of arsenic as additive while Rodier [12] reports on the use of antimony. Norzink uses antimony as additive for cobalt removal.

2 Solution purification at the Norzink plant

Today Norzink has 4 purification steps. In the first step copper is cemented from a typical level of 0.4 - 0.8 g/l down to a level of approx. 100 mg/l. The solution temperature at the beginning of the first step is 75 °C. The copper cement is not removed before the second step.

In the second step the major amount of cobalt and cadmium is removed. Antimony is added (3 mg/l) before the second step. The residue from the first and second step is separated from the solution in a thickener. Before the third step antimony (2 mg/l), copper (50 mg/l) and spent electrolyte are added to the solution (thickener overflow). Antimony is added as a solution of potassium antimony tartrate in both steps.

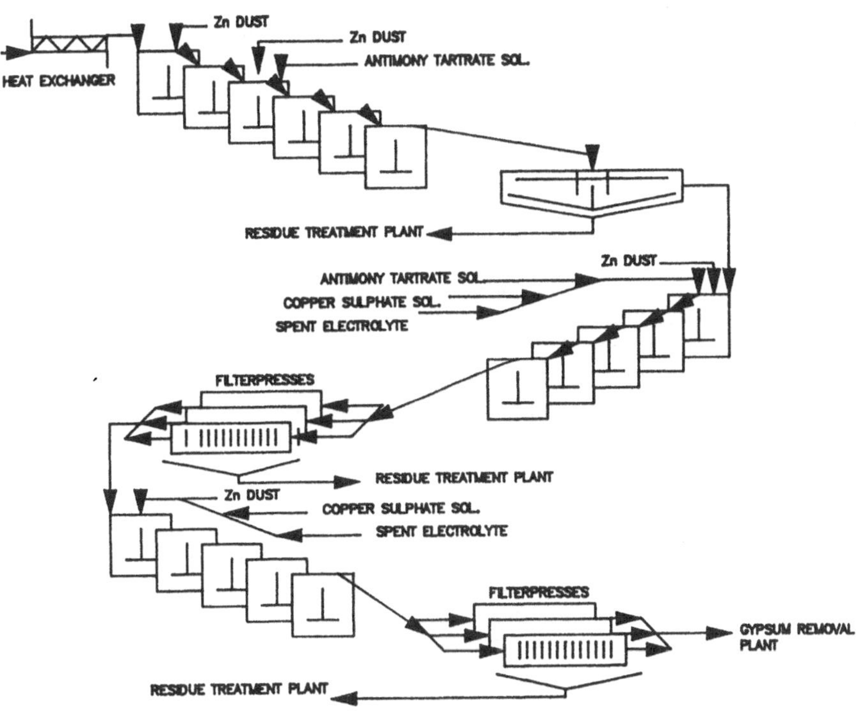

Fig. 1. Flowsheet of the Norzink purification process.

In the third step the last amount of cobalt and cadmium is removed. The residue after the third step is removed from the solution by filter presses having polypropylene plates and frames. The filter medium is polypropylene and paper.

The fourth step is a polishing and security step. The solution entering the fourth step is added spent electrolyte and some copper (25 mg/l). Residue is removed from the solution by the same type of filter presses as after the third step. Normally the solution is sufficiently pure after the third step.

The total zinc dust consumption in the purification process is about 5.5 g/l.

The principle for the Norzink purification process is shown in Fig. 1.

In order to improve the purification process Norzink has initiated a research program. Some of the results are presented in this paper.

3 Experimental

All experiments were run in batch with industrial made zinc sulphate solution. The experimental apparatus consisted of a 3 l glass beaker with a plastic stirrer. The stirrer velocity was constant and the same in all experiments (170 rpm). In each test the solution volume was 2 l. During the experiment the beaker was covered with a plastic foil. The temperature was kept constant (± 2 °C) through each experiment using a temperature controller.

The zinc dust, which was industrial air atomized made from SHG zinc, was added in one portion.

During the run, solution samples were withdrawn at different reaction times, filtered and analysed for rest concentration of cobalt in solution.

Analyses were carried out using an atomic absorption spectrophotometer for Cu, Cd, Sb, Zn and a colorimetric spectrophotometer for Co.

A larger batch of neutral leach overflow was treated with zinc dust to precipitate most of the copper in the solution from approx. 500 mg/l to approx. 5 mg/l. The copper cement was removed by filtration from this solution which then was the basis for the cementation experiments reported.

In the different experiments the start concentrations of copper, antimony, and cobalt were adjusted to desired levels by addition of $CuSO_4$, $CoSO_4$ and $KSbC_4H_4O_7$ solutions.

4 Results and discussion

It is well known that industrial removal of cobalt is complex and depends on many interactive factors. Some relevant and important parameters which influence the cobalt cementation rate are studied in this communication. The effect of

- initial Cu and Sb concentration
- removal of copper cement
- temperature

- pH
- zinc dust grain size

on the cementation process are investigated.

Residues from the Norzink purification process and from laboratory tests were also studied by SEM and EDS using a DMS 940 Zeiss instrument.

4.1 Initial antimony and copper concentrations

Totally 16 experiments were run with 4 different initial copper concentrations and 6 different initial antimony concentrations. Temperature during each run was 75 °C and pH = 4.7. 2.5 g/l zinc dust was used, grain size 40-100 µm made from SHG zinc. Initial concentrations of other elements in the solution were: Zn 155 g/l; Cd 590 mg/l and Co 15 mg/l.

In Fig. 2 the rest concentration of cobalt is plotted as a function of initial antimony concentration after 120 min. reaction time at 4 different initial copper concentrations. We may observe:

- with low initial copper concentration in the range of 5 - 50 mg/l, it is important to have an initial antimony concentration close to 3 mg/l to obtain a low rest concentration of Co.
- at higher copper concentration, 50 - 100 mg/l, the cobalt cementation is not so dependent of the initial antimony concentration as long as the initial antimony concentration is ≥ 3 mg/l.

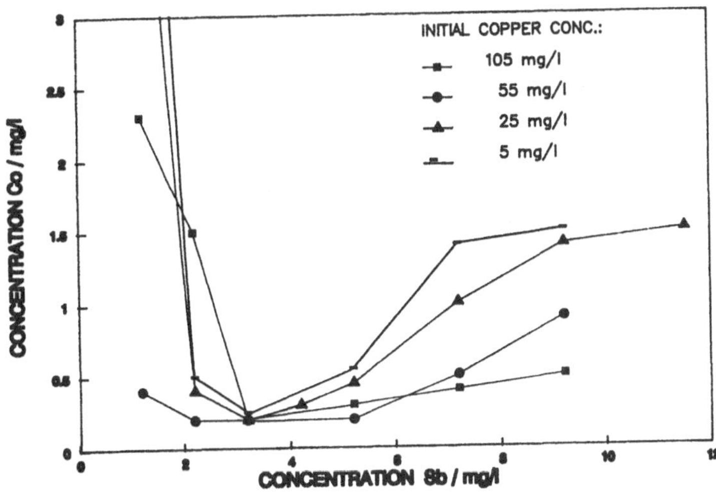

Fig. 2. The effect of initial antimony and copper concentration on the cobalt concentration after 120 min. reaction time. Zn dust 2 g/l, 40-100 µm; temp. 75 °C; Co 15 mg/l; Cd 590 mg/l and pH 4.6.

• at the given conditions it is shown that optimal cobalt cementation is given when the initial concentration of copper is in the range of 50 - 100 mg/l and initial antimony concentration is in the range of 3 - 4 mg/l.

In Fig. 3 the cobalt concentration is plotted as function of reaction time at 4 different initial copper concentrations at the same initial antimony concentration.

This plot shows that the kinetics of cobalt cementation is increasing with increasing initial copper concentration in the beginning of the cementation. However, after 120 min. reaction time the lowest cobalt concentration is obtained with an initial copper concentration of 55 mg/l.

4.2 Copper cement removal

In the Norzink purification process the copper cement is not removed from the solution prior to cobalt cementation. In Fig. 4 results of four experiments are shown. In two of the runs the copper residue was absent. In the other two runs 2 g/l copper cement was added just before the zinc dust cementation was initiated. The experimental conditions were the same as before apart from the concentration of cobalt 8 mg/l, copper 140 mg/l and antimony 4 mg/l.

The results indicate that the cementation of cobalt will improve if the copper cement is removed prior to cobalt cementation.

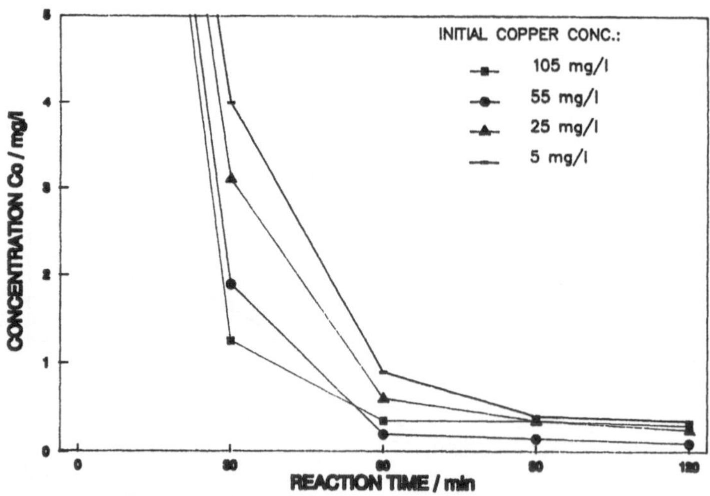

Fig. 3. The effect of initial copper concentration on the cobalt concentration as a function of reaction time. Zn dust 2 g/l; 40-100 μm; temp. 75 °C; Co 15 mg/l; Cd 590 mg/l; Sb 3 mg/l and pH 4.6.

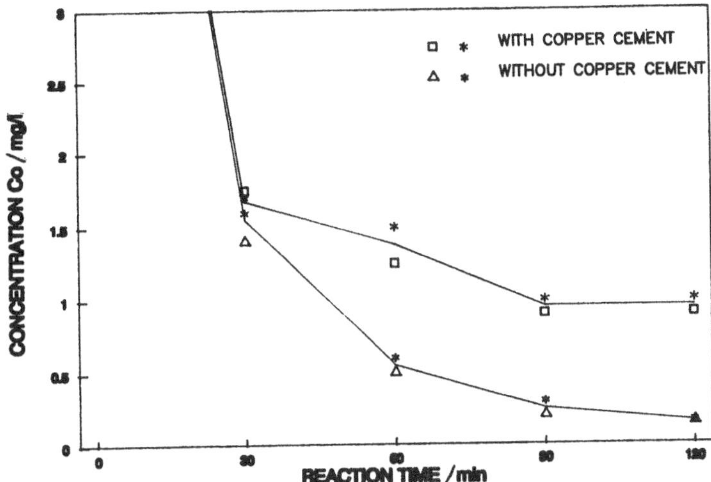

Fig. 4. The effect of copper cement removal prior to cobalt cementation. Zn dust 2.5 g/l, 40-100 μm; temp. 75 °C; Co 15 mg/; Cd 590 mg/l; Sb 3 mg/l; Cu 140 mg/l and pH 4.6.

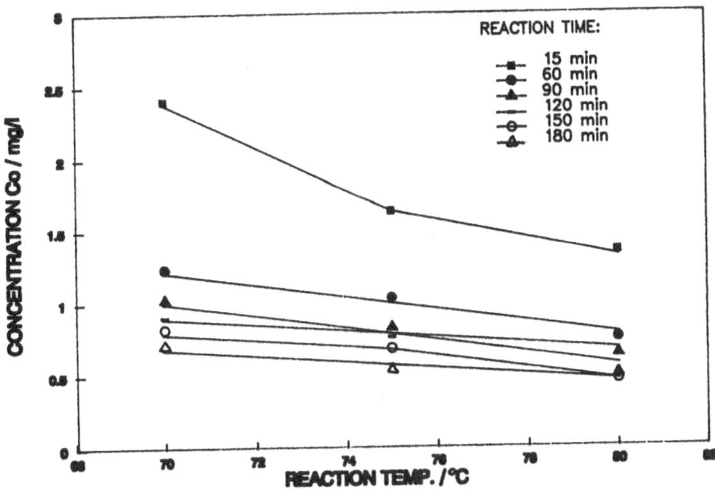

Fig. 5. The effect of cementation temperature at different reaction times on the cobalt concentration. Zn dust 2.5 g/l; 40-100 μm; temp. 75 °C; Co 15 mg/l; Cd 590 mg/l; Cu 125 mg/l; Sb 5 mg/l and pH 4.6.

4.3 Temperature

The effect of varying the cementation temperature is shown in Fig. 5. At the given conditions it is clearly seen that increased temperature increase the cementation kinetics of cobalt. After 120 min. reaction time the best result is achieved at 80 °C. The temperature, however, is less significant at higher reaction times.

4.4 pH

Four experiments were run to investigate the importance of pH in connection with cobalt cementation.

In two of the runs the initial pH was adjusted to pH = 4 by addition of sulphuric acid. In the other two runs sodium hydroxide was added, and the initial solution became slightly white due to basic zinc sulphate formation.

In Fig. 6 the concentration of cobalt is shown as function of reaction time. The plots show that in the first 60 min. the experiments with pH = 4 have somewhat lower cementation kinetics. After 60 min., however, the cementation process has reached the same cobalt concentration and is thus independent of the initial pH.

The results from these experiments indicate that hydrogen evolution which can block the sites for cobalt deposition, is more detrimental to the cementation process than the formation of zinc hydroxide which may block the zinc dust particle.

In the technical cobalt cementation process, it is therefore probably more important to prevent situations with low pH < 4 than higher pH. Formation of larger amount of basic zinc sulphate will, however, increase the load on the residue treatment plant and should be minimized.

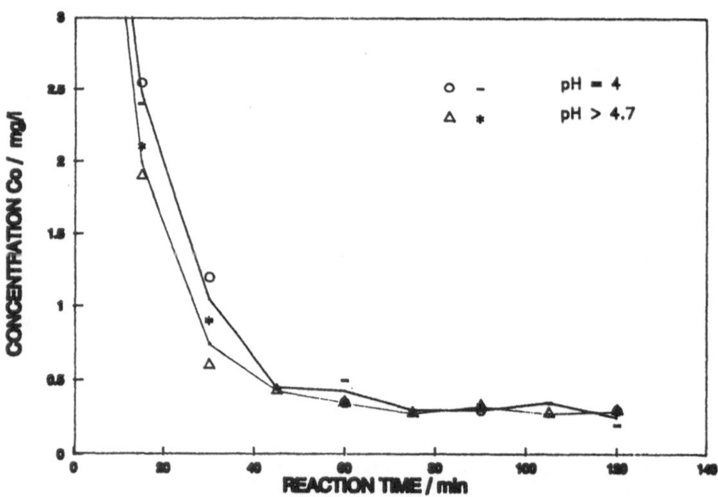

Fig. 6. Cobalt concentration as a function of reaction time and pH. Zn dust 3 g/l; 40-100 µm; temp. 75 °C; Co 15 mg/l; Cd 590 mg/l; Sb 7 mg/l; Cu 120 mg/l.

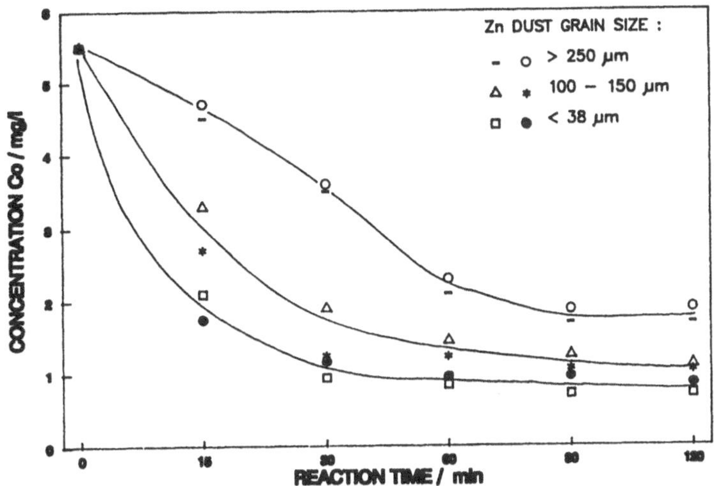

Fig. 7. Cobalt concentration as a function of reaction time and zinc dust grain size. Zn dust 1.5 g/l; temp. 75 °C; Co 15 mg/l; Cu 110 mg/l; Cd 590 mg/l; Sb 5.8 mg/l; pH 4.6.

4.5 Zinc dust grain size

Three different zinc dust grain sizes were used to investigate the effect on cobalt cementation. In Fig. 7 the cobalt cementation rate is shown for zinc dust grain size larger than 250 mm, between 100 - 150 mm and finer than 38 mm. The results show that the rate of cementation increases with decreasing zinc dust particle size. This is in agreement with data reported in the literature [3]. It is, however, interesting to notice the change in reaction rate as the reaction time increases.

4.6 A SEM investigation of residue from Zn cementation of Co

Some papers referred to earlier in this communication and the present experimental data, indicate that the concentration of Cu^{2+} in the electrolyte seems to be of major importance for a successful removal of Co^{2+}. We have therefore decided to study the influence of the Cu^{2+} concentration on the phases formed in the residue. Both residues formed in laboratory tests and in the technical process are investigated. We know from published results and from our own investigation, however, that temperature, pH, zinc dust particle size, ion concentrations and residence time may also be of importance. The effect of these variables on the morphology and phase relations of the residue will be discussed in a later communication.

4.7 Residue from the Norzink process

The Norzink residue investigated was obtained as given below:

- To the initial solution; Zn 155 g/l, Cd ≈ 0.59 g/l; Cu ≈ 0.49 g/l; Co ≈ 0.005 g/l; Sb ≈ 0.005 g/l and Ge ≈ 0.001 g/l, 0.5 g/l Zn dust was added to precipitate most of the Cu in solution. Temperature: 75 °C; pH = 4.5-5 and the reaction time ≈ 30 min!
- To this solution where mainly the Cu^{2+} concentration is reduced ($C_{Cu^{2+}}$ ≈ 100 mg/l) 2.5 g/l Zn dust was added together with 3 mg/l Sb as $KSbC_4H_4O_7$ and the cementation was continued for about 90 min in altogether 6 stirred tanks with overflow of 45 m^3 each (Fig. 1). The flowrate was 160 m^3/hr.

After this cementation process the impurity concentrations were as follows; Cu ≈ 0.01 mg/l; Cd ≈ 10 mg/l; Co ≈ 0.5 mg/l and Sb ≈ 0.1 mg/l.

A SEM (DSM 940 Zeiss) picture of this residue is shown in Fig. 8. Large platelets of $ZnSO_4$ (s) can be observed together with material from the cementation process. To obtain a proper element analysis of precipitated material from the cementation process it was decided to make X-ray mapping (EDS) measurements over a certain area. Polished samples of the precipitate, crushed and washed in boiling water, were prepared in epoxy for this purpose.

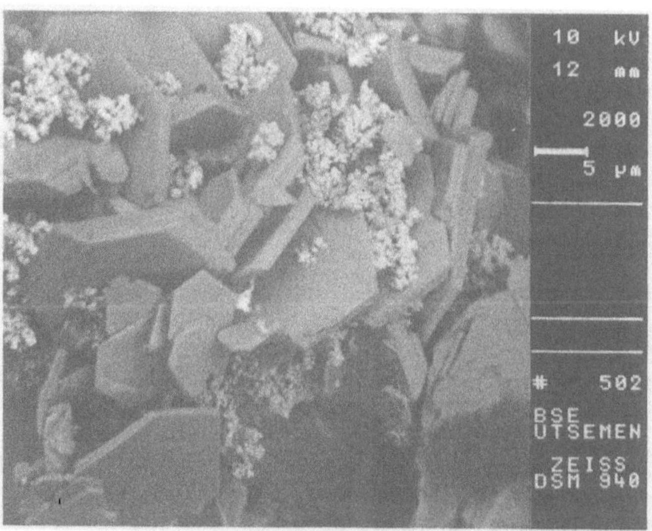

Fig. 8. A characteristic X-ray image (SEM) of residue from the first Co removal step in the Norzink cementation process. The residue was washed with cold distilled water. Magn.: 2000.

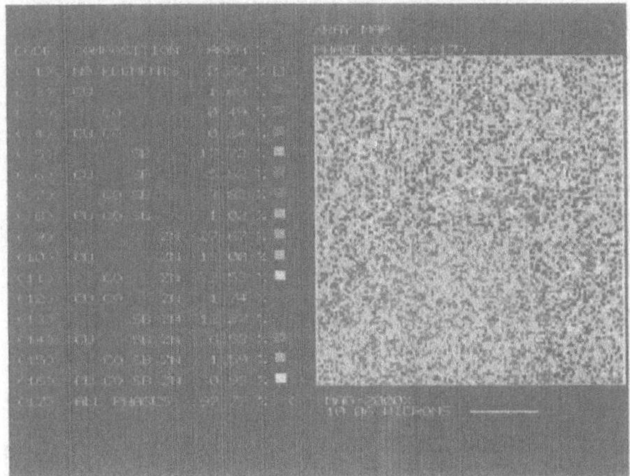

Fig. 9. X-ray mapping (EDX) over the same sample as shown in Fig. 8. The elements Zn, Cu, Co and Sb are considered. Magn.: 2000. Analysis time 5 hrs.

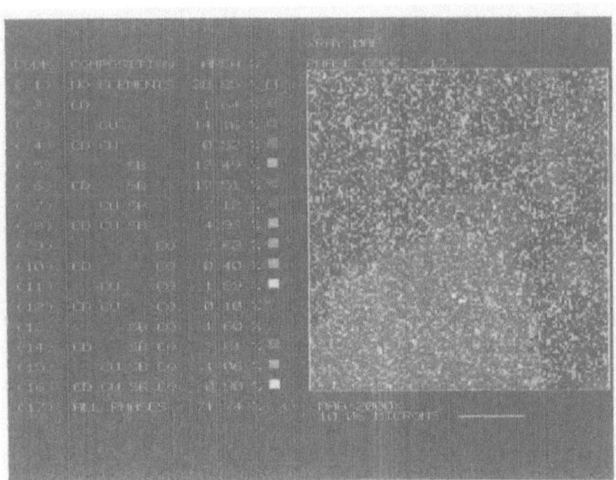

Fig. 10. The same X-ray mapping (EDX) as shown in Fig. 9. The elements Cd, Cu, Co and Sb are considered.

In Figs. 9 - 10 X-ray mappings (EDX) are shown. The analysis time was 5 hrs. In Fig. 9 a mapping over a given area is shown and the elements Zn, Co, Cu and Sb are considered. The Cu area is seen as dark blue spots and is analysed as a mixed Cu-Sb field. The blue area consists mainly of Sb, Sb with Cu and Sb with Co. Outside this area Zn is the main element together with Cu, Sb and Co on Zn.

When the elements Cd, Cu, Co and Sb are considered as shown in Fig. 10, the same area as the blue area of Fig. 9 is shown to contain Cd with Sb, Cd with Cu and Sb and Cd with Sb and Co. There also seems to be a tendency for Cu to be precipitated close to this Cd-Sb field.

The above observations are in fair agreement with data from laboratory experiments reported in the literature [1,2,3,5,9].

4.8 Residue from laboratory experiments

To investigate the effect of the concentration of Cu^{2+} during precipitation of Co with Zn dust, 5 experiments were performed. The initial cementation conditions in the first 4 were:

T = 75 °C; pH 5.2; Zn 150 g/l; Zn dust 2.5 g/l; Cd 570 mg/l; Co 7 mg/l and Sb 8 mg/l.

The Cu concentrations varied from 32 mg/l to 265 mg/l.

The cementation process took place in a well-stirred container of 2 l. The temperature was controlled to ± 2 °C. After the cementation process the aqueous phase was analysed for Cu^{2+}, Co^{2+}, Cd^{2+} and Sb^{3+} and the residue was separated by filtration.

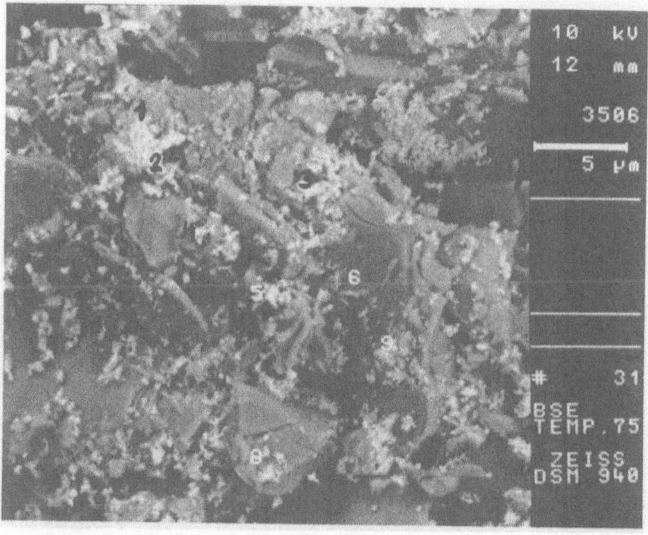

Fig. 11. A characteristic X-ray image of residue obtained in a laboratory experiment. Quantitative X-ray analyses were performed at the points marked 1-9.
Cementation conditions: temp. 75 °C; Zn 150 g/l; Zn dust 2.5 g/l; Cd 570 mg/l; Co 17.5 mg/l; Sb 1.8 mg/l and Cu 240 mg/l. Magn.: 3500.

A representative SEM picture of the residue is shown in Fig. 11. The etched Zn particle together with some platelets of $ZnSO_4$ (s) and precipitated material from the cementation process can be observed. The residue was crushed and washed in boiling water before the picture was taken. In the points numbered 1-9 quantitative X-ray analysis was performed. Results are given in Table 1.

Chi-Sad is a measure of the analytical accuracy. Low values of Chi-Sad (≈ 1-10) indicate high accuracy. The inaccuracy in the present results is due to an uneven surface. Quantitative analysis requires polished surfaces. The present data, however, give an estimate of the concentration of the given elements in each point.

Sulphur is observed as expected since it was difficult to remove all $ZnSO_4$ from the residue. Apart from this observation it is difficult to draw any firm conclusion on the basis of the data in Table 1. In points 1 and 2 Chi-Sad is relatively low indicating more reliable data. In these points S is low and Cd and Cu are high. Sb and Co seem to be associated with high values of Cd and Cu. This is in qualitative agreement with residue data reported earlier in this paper. Co removal also seems to be at a maximum at an initial Cu^{2+} concentration around 80 mg/l, and a maximum amount of Co is also found in the residue for this initial Cu^{2+} concentration.

A linear scan (EDX) is shown in Fig. 12 of the last laboratory cementation sample. This sample was obtained under somewhat different conditions than described for the 4 experiments just discussed. It can clearly be seen that Cu and Co have maximum depositions in the same areas. Such a tendency was also indicated by Lew et al. [9]. As and Sb have a higher tendency to coprecipitate together with Cd, and Sb a negative tendency to coprecipitate with Co. These observations have a definite bearing on the reaction mechanism for Co cementation. It is, however, too early to give a definite conclusion.

Table 1. Quantitative X-ray analysis of residue from Zn cementation in a laboratory experiment with initial Cu^{2+} concentration 82.2 mg/l. Analysis time 240 sec.

Point no.	Element concentration (atom %)						
	Zn	Cu	Cd	S	Co	Sb	Chi-Sad
1	45.64	14.05	33.31	5.70	0.86	0.43	16.98
2	32.36	15.02	45.39	4.81	1.05	1.37	14.65
3	56.90	9.20	22.68	10.38	0.85	0.00	32.05
4	55.08	10.46	23.78	9.65	0.53	0.50	49.26
5	52.23	11.28	26.43	8.56	0.90	0.61	34.27
6	80.31	4.17	5.63	9.57	0.32	0.00	36.29
7	68.24	4.31	8.95	17.81	0.36	0.32	111.95
8	64.15	5.77	12.55	16.75	0.37	0.41	85.67
9	66.52	8.60	14.40	8.78	1.27	0.42	36.46

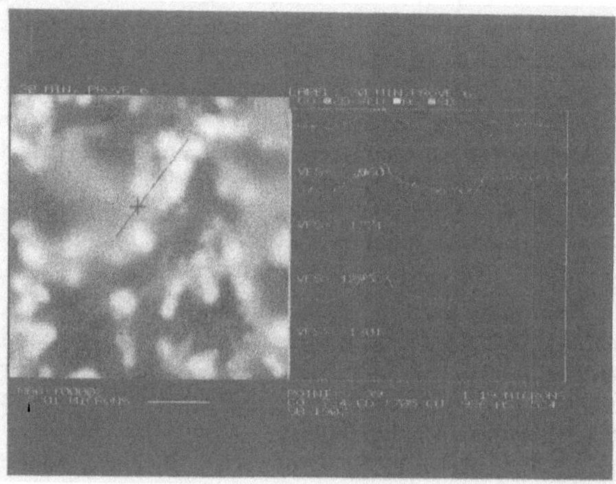

Fig. 12. Positioning and results of compositional line scan in cementation residue as obtained from a laboratory experiment.
Cementation conditions: temp. 80 °C; Zn 150 g/l; Zn dust 2.5 g/l; Cd 515 mg/l; Co 7 mg/l; Sb 8 mg/l and Cu 82.6 mg/l. Magn.: 10000.

5 Conclusions

There seems to be an optimal combination of initial copper and antimony concentrations for cobalt cementation with zinc dust. At the experimental conditions reported in this paper the optimal concentration of copper is 50 - 100 mg/l, and of antimony concentration 3-4 mg/l.

Copper cement should be removed prior to cobalt cementation.

A maximum cementation rate of cobalt was achieved at 80 °C.

In the technical cobalt cementation processes it is probably more important to prevent situations with low pH ($pH < 4$), rather than high pH.

The cobalt cementation rate increases with decreasing particle size of the zinc dust used in cementation.

The kinetics of cobalt precipitation seem to be strongly associated the Cu^{2+} concentration of the electrolyte. Cobalt is, however, found in the residue in connection with all the elements copper, cadmium and antimony.

There is some experimental evidence indicating that copper seems to precipitate close to cadmium-antimony containing grains.

6 References

1. Fukubayashi, M.N., OKeefe, T.J. and Clinton, W.C., **Bureau of Mines Report of Investigations 7966**, United States Department of the Interior, (1974).
2. Tozawa, K., Nishimura, T., Akahori, M. and Malaga, M.A., **Hydrometallurgy** 30 (1992) pp. 445-461.
3. MacKinnon, D.J., **Mines Branch Research Report R 264,** Department of Energy, Mines and Resourses, Ottawa, Canada. (1973).
4. Xiong, J. and Ritchie, I.M., **Proc. Int. Symp. Electrochem. in Mineral and Metal Processing II.** (ed. P.E: Richardson and R. Woods); The Electrocehemical Society (1988) pp. 383-.
5. Xiong, J. and Ritchie, I.M., **Proceedings 1st Int. Conf. on Hydrometallurgy Academia Sinica** (1988) pp. 632-636.
6. Fischer-Bartelk, C., Lange, A. and Schwabe, K., **Electrochim. Acta** 14 (1969) pp. 829-844.
7. Fontana, A. and Winand, R., **Metallurgie** XI (1971) pp. 162-.
8. Lawson, F. and Nhan, LeThi, **Hydrometallurgy Soc. Chem. Ind.**, London (1981) pp. 64-73.
9. Lew, R.W., Dreisinger, D.B. and Gonzales - Dominigues, J.A., Proceedings of the International Symposium on Zinc, 10-13 Oct. 1993, Hobart, Tasmania. **The Australian Institute of Mining and Metallurgy.** Publication Series No. 7/93, pp. 227-240.
10. Fontana, A., Martin, J., van Severen, J. and Winand, R., **Metallurgie** XI (1971) pp. 168-.
11. Fugleberg, S., Järvinen, A. and Sipila., V., **Lead-Zinc-Tin, TMS-AIME World Symposium on Metallurgy & Enviromental Control** (1980) pp. 157-171.
12. Rodier, D.D., **Lead-Zinc-Tin, TMS-AIME World Symposium on Metallurgy & Enviromental Control** (1980) pp. 144-.

Cementation of mercury(II) from chloroalkali effluents by use of Zn and Fe

J. M. R. de Carvalho
A. Anacleto
Department of Chemical Engineering, Instituto Superior Técnico, Technical University of Lisbon, Lisbon, Portugal

ABSTRACT

The production of chlorine and sodium hydroxide electrolytically by the mercury process using brine as raw material, creates a solid waste problem. Detoxification of this waste can be carried out by a leaching process with an oxidizing agent in a chloride medium, resulting in exhausted sludges and in a leachate rich in mercury.
In the present study cementation is applied to remove mercury from the leachate by reductive precipitation, using as sacrificial metal - zinc and iron powder.
The influence of the pH and the dosage of sacrificial metal was determinated in two different leachates: leachate A with 196.0 ppm Hg and B with 5.7 ppm Hg.
The optimum pH to operate with iron is less than 2, however with zinc the best result is obtained for pH values between 3 and 6. Since the leaching step operates at pH 8.5 zinc was selected as the sacrificial metal.
Values of sacrificial metal dosage between 1.5 and 146 mol Zn/ mol Hg were tested. The best results were obtained in the range 30-50 mol Zn/mol Hg.
Other experiments were done using a stirred reactor and a zinc rotating disk. Parameters such as temperature and rotational speed were studied. The results show a diffusion controlled process with an activation energy of 14.6 KJ/mol.

INTRODUCTION

A prime concern of the environmental engineer is the removal of small concentrations of toxic substances from industrial wastewaters. Cementation is the term used to describe the electrochemical precipitation of a metal, usually from an aqueous solution, by a more electropositive metal. The process has been used in industry for a long time, not only in hydrometallurgy [1], but also in the purification of process streams and wastewaters [2]. To fit such a method of precipitation successfully into a practical process, it is necessary to have some knowledge of reaction kinetics.
When examining the possibility of using a cementation technique after a sludge leaching process, it was found that although the use of metallic iron and zinc was reported by several authors, very little was known about these precipitation reactions in a chloride medium. The leaching process was developed to extract mercury from a solid residue produced at Uniteca, a chloroalkali plant placed on the west coast of central Portugal [3,4]. Two different leaching solutions are obtained (one with 196.0 ppm Hg and other with 5.7 ppm Hg) and this work studied the possibility of eliminating this metal from the leachates by cementation.

The advantages of cementation include simple control requirements, low energy consumption and recovery of valuable or toxic metals in a harmless and reusable form. Its main disadvantage is the consumption of the more electropositive metal when the solutions have low pH values.

Ballester et al. [5] performed the cementation of the tetrabromomercuriate ion ($HgBr_4^{2-}$) from dilute solutions, using powder metals such as iron, zinc and aluminium. The initial pH was very low (<1), and iron provides the best results with an efficiency of 99.9%. The same results were obtained over a wide range of initial concentration of mercury (500-3000 ppm) The study of temperature and stirring speed suggested that the process is diffusion controlled.

Rickard and Brookman [6] proposed cementation as a suitable technique for removal of mercury chloride from chemical process and chloralkali wastes. A pH value between 4 and 6 was established as the optimum working range. Starting with an initial concentration of 500 ppm Hg, reductions to 0.5 ppm Hg were achieved. At the end of the cementation the removal of zinc from solution was carried out by precipitation with sodium carbonate at pH 9, followed by filtration. Economical figures included indicate that the process is financially attractive.

Power and Ritchie [7] carried out kinetic studies and electrochemical measurements for $Hg(NO)_3$ on copper discs. Their experiments started with a mercury concentration of 500 ppm. Results from both techniques, gave evidence of a diffusion controlled process.

According to studies by the New Jersey Zinc Co. [8] carried out using a zinc dust bed, solutions containing 10 ppm of ionic mercury were purified to 0.02 ppm in 13 sec and to 0.005 ppm in 110 sec. The treatment was effective for streams within the pH range 5 to 10.

Ball et al. [17] claimed in their patent a method for the recovery of soluble forms of mercury in a chlorinated solution by reduction with iron powder. A starting solution containing about 80 g/l Hg and 0.8 g/l Se can be treated with an excess (38%) of fine (< 44 microns) iron powder to produce a barren solution of 2.5 ppm Hg.

Bro and Lang [9] provide a systematic description of the hydraulic characteristics of an electrochemical displacement reactor. The cementation was described as a means to remove mercuric ions from chloro-alkali plant waste brines. The entering solution contains 5% NaCl, with a pH between 5-8 and an initial mercuric ion concentration of 0.2 ppm. Using fine grade zinc powder it was possible to attain final mercury concentrations lower than 1 ppb.

Waltrich [10], described an apparatus for the treatment of effluent containing very low and even trace amounts of mercury in both elemental form and as mercuric ion. The author refered the use of zinc, magnesium, aluminium and iron as reactive metals. First the mercury solution passed through the bed then the feed was stopped and the vessel was submitted to a relatively high vacuum (0.5-2.0 Torr), while the bed itself was heated. The mercury condensed and was recovered at an extremely high degree of purity. The absence of oxygen under vacuum conditions prevents the oxidation of the reactive metal so that the bed can be used and regenerated almost indefinitely. The typical effluent containing 50 ppm Hg, leaves with 5-10 ppb Hg and consumes 0.33 lb Zn/ lb Hg.

Kimura [2] described a method for the collection and removal of mercury (II) from chemical laboratory and industrial wastewater. By this procedure the mercury ion in an extremely wide concentration range (0.1 ppm - few thousands ppm) was removed with zinc powder. The maximum removal rate was extremely good, namely 5g of Hg (II) could be collected and removed from complex solutions with 1 g of zinc powder. Ezhkov and Baiborodov [11] published work about cementation of mercury by metallic iron and by antimony, in sulphide-alkali solutions, where the potential of the metals in solution was measured.

THEORY

The basic cementation reaction for the mercury(II) - zinc system is:

$$Hg^{2+} + Zn^{o} = Hg^{o} + Zn^{2+} \qquad (1)$$

Essentially, the system is a set of shortcircuited electrolytic microcells and the cementation reaction can be considered in terms of respective half cells:

$$Hg = Hg^{2+} + 2\,e^{-} \qquad (2)$$

$$Zn = Zn^{2+} + 2\,e^{-} \qquad (3)$$

The reaction is electrochemical, in the sense that the electrons are not exchanged at the same site, but rather the half-cell reactions are separated by an arbitrary, finite distance which necessitates that the solid phase be a conductor or semi-condutor. Also cementation should be distinguished from electrolytic deposition in which the source of electrons is from a generator rather than a less noble metal.
In this way E, the difference of electrode potential is the electromotive force in the cementation process. The Nernst equation (4), relates electrode potential and activities. Thus in the reaction (1) between mercury with zinc, we have:

$$E = E^{o} - \frac{RT}{zF} \ln \frac{a_{Zn^{2+}}}{a_{Hg^{2+}}} \qquad (4)$$

where E^{o} is the standard electrode potential, which can be calculated from the standard electrode potentials of the two half cells (2 and 3).
The Gibbs free energy (5) change for a reversible reaction is equal to the net work obtained from the system at constant temperature and pressure. In an electrochemical reaction this net work is electrical work, so:

$$-\Delta G = z\,F\,E \qquad (5)$$

where z is the number of electrons transferred in the electrochemical reaction, F is the Faraday constant. Most of the common cementation reactions exhibit large negative free energy changes.

The equilibrium constant, K at temperature T is given by:

$$K = \exp\left(\frac{-\Delta G_T^o}{RT}\right) \tag{6}$$

Although cementation reactions are thermodynamically very favourable, in pratice equilibrium conditions are rarely approached and the extent of reaction is determined by reaction rate rather than by the equilibrium thermodynamics [18]. So, the initial objective of any kinetic study of a particular cementation reaction should be to determine which step (or steps) is rate controlling under the process conditions. Wadsworth [15] in a detailed analysis of cementation reactions in terms of electrochemical theory, established that the cementation rate is directly proportional to the difference of standard half-cell potencials of metals involved in the reaction. On the other hand, the back-reaction kinectics becomes important only when the difference of standard half-cell potencial is greater than 0.5 volts [21].
Levich [13] established the equation below (7) that can interpret the cementation kinectic produced with a disk of area A, rotating at an angular velocity ω in a solution of kinematic viscosity ν, when a process is controled by the difussion, D, of reactant ions of concentration C.

$$J = 0.62\, D^{2/3} n^{-1/6} w^{1/2} AC = kAC \tag{7}$$

where J (mol $s^{-1}m^{-1}$) is the flux of ions on to the disk .

EXPERIMENTAL

All the solutions referred as synthetic were prepared from reagent-grade chemicals and deionized water. Mercury solutions were prepared from $HgCl_2$ and all of them except in the chloride study contained 100 g/l NaCl. The real solutions referred as leachates were produced by a leaching process using sodium hypochlorite. There were two different leachates: leachate A with 196.0 ppm Hg(II) and leachate B with 5.7 ppm Hg(II).
The results presented in this paper can be grouped in i)Erlenmayer experiments, ii) reactor experiments with powder, iii) reactor experiments with disks.
-The Erlenmayer experiments were done in a 0.250 dm^3 erlenmayer flask, using 0.200 dm^3 solution. During these tests no variables were monitored, and it was only possible to take samples at the beginning and end of the experiment. The flasks were shaken at 100 osc/min, at ambient temperature.
-The reactor experiments with powder were carried out in a 1 dm^3 (0.011 m diameter) vessel, fitted with a cover designed to hold a thermometer, a pH probe, a potential probe and a stirring shaft. The cover also contained a N_2 inlet and a sampling port from which solution was withdrawn at certain intervals for analysis. Deareation with nitrogen was carried out for 30 minutes before the beginning of the experiment. Zinc

powder of 99.99% of purity and with a mean size of 11 μm was used. The tests were done using a 0.9 dm^3 volume of solution with samples of 5 cm^3.

-The reactor experiments with rotating disks were carried out in a small cylindrical reactor, with a capacity of 0.250 dm^3. The apparatus was the same as that described above, except that the central stirring shaft was replaced by a supported disc. In this case the volume of solution used was 0.213 dm^3 and samples were 3 cm^3. In temperature studies temperature probe was connected with the pH monitor to provide a pH correction (needed due to high temperatures). Zinc disks with (0.0271 m diameter) and a stated purity better than 99.9%, were pressure fitted into a Pespex disk holder, such that only one circular face of the disk was exposed to the solution. The tapered end of the disk was drilled and threaded to accommodate a 1/4 inch diameter shaft protected with a Teflon film. Just before the start of the experiment, the exposed surface of the metal disk was wet-polished using 800 grade silicon carbide paper, followed by 1000 grade silicon paper, followed by washing with deionized water and ethanol and rapid drying. The disc was then polished with 6 micron diamond paste followed by 10 sec of immersion in a 10% acid nitric solution.

Zinc and iron were analysed by atomic absorption spectrophotometry (Perkin Elmer 3000), but mercury was analysed by a cold-vapour technique in a Coleman MAS 50. The reaction vessel was placed in a thermostat water bath to mantain a constant temperature.

RESULTS AND DISCUSSION

Reactor experiments with powder

In this set of experiments, six different tests were done varying the initial concentration of mercury (1 , 10, 50, 100, 250, 500 ppm Hg). The results of mercury concentration against the time are represented in Figure 1. It is seen that after a few minutes of cementation, higher initial concentrations correspond to lower residual mercury levels. In the test where the initial soluble mercury was about 1 ppm, it can be observed, that no mercury was cemented so the concentration remained constant during all the test.

These results can be understood by reference to the Nersnt equation (4). At zero time there is no soluble zinc, so if we start with a higher mercury concentration, there is a higher difference of electrode potential and consequently a faster reaction [15,21].

Concerning zinc dissolution, we can say that it is proportional to mercury removal with a molar ratio between 1 and 1.5. For example in the test with an initial mercury(II) concentration of 500 ppm, at the end of 3 minutes we have 0.24 ppm of Hg and 168.8 ppm of Zn(II) in solution.

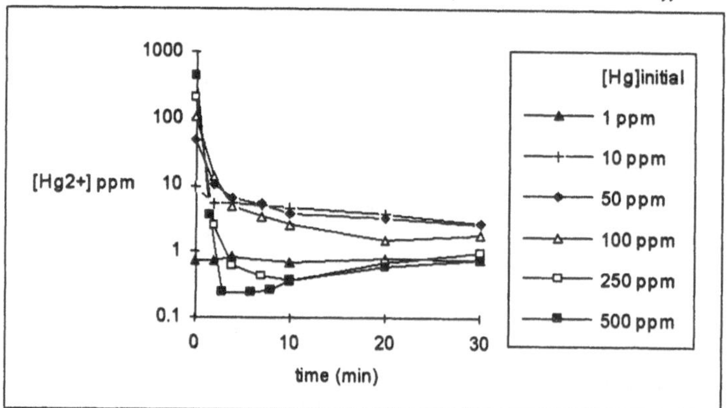

Fig. 1 - Variation of mercury concentration with time at different initial mercury concentrations. (Experimental conditions: sacrificial metal - Zn; T=20°C; [NaCl]=100 g/l; stirring speed=500 rpm; mol Zn/mol Hg=10; pH=4.0 (pH was controlled during all reaction))

Cementation on to a rotating disk

With the disk the following parameters were studied: temperature, stirring speed and atmosphere. In all disk experiments, we observed that the plot of $\ln([Hg]/[Hg]_o)$ versus time corresponds to a straight line. This result indicates that the reaction is first order in relation to Hg. Assuming this kinetic model (8), and using the integrated form: (9) it is possible to calculate *k* the kinetic constant for the reaction:

$$-V\frac{d[Hg]}{dt} = kA[Hg] \tag{8}$$

$$2.303\log\frac{[Hg]}{[Hg]_o} = -k\frac{A}{V_o}t \tag{9}$$

Where A is the superficial area of the disk in contact with the solution, V_o the initial solution volume, [Hg] ionic mercury(II) soluble at time reaction t. The influence of stirring speed and temperature on cementation kinetics can be quantified from kinetics constant values.

The change of the kinetic constant with the temperature (Figure 2) follows the Arrhenius law (10), so this set of experiments allows the calculation of the activation energy.

$$k = A\exp\left(\frac{-Ea}{RT}\right) \tag{10}$$

where A is the pre-exponential factor and R the gas constant, T the absolute temperature.

Ea = 14.6 KJ/mol.

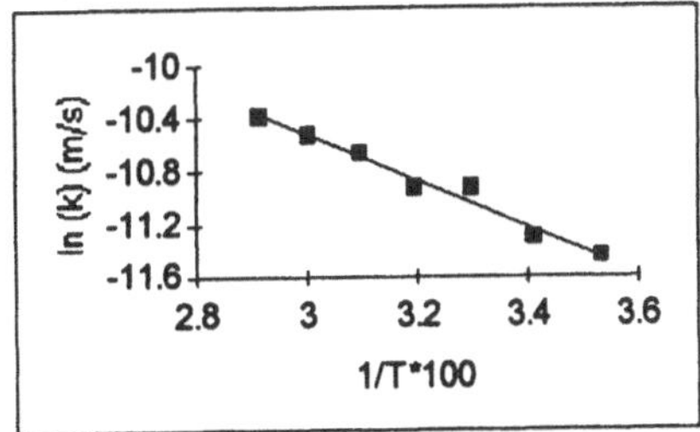

Fig. 2 - Arrhenius plot for mercury cementation on pure zinc. (Experimental conditions in each run at different temperature: initial [Hg]=100 ppm, [NaCl]=100 g/l, N_2 atmosphere, stirring speed= 1000 rpm, pH (initial)= 4.0)

This result can be compared with those obtained by other authors in Table 1. When the value is between 12 and 24 KJ/mol, the system exhibits typical values of a diffusion controlled process [12].

Concerning metals different from mercury, it is well documented in several works [12, 20,21] that although there are some cases of reaction control in cementation, the majority are diffusional controled.

System (medium)	Ea (KJ/mol)	Author
Hg(II)/ Cu (NO_3^{2-})	16	[7]
Hg(II)/Fe (Br^-)	17.7 (250 rpm)	[19]
Hg(II)/Fe (Br^-)	13.7 (500 rpm)	[19]
Hg(II)/Fe (Br^-)	7.1 (750 rpm)	[19]

Table 1 - Activation energy obtained in several works.

The plot of kinetic constant values versus square root of stirring speed is presented in Figure 3. The straight line suggests that the process follows Levich equation (7), ie. the process is diffusion controlled as already observed by the Arrhenius law.

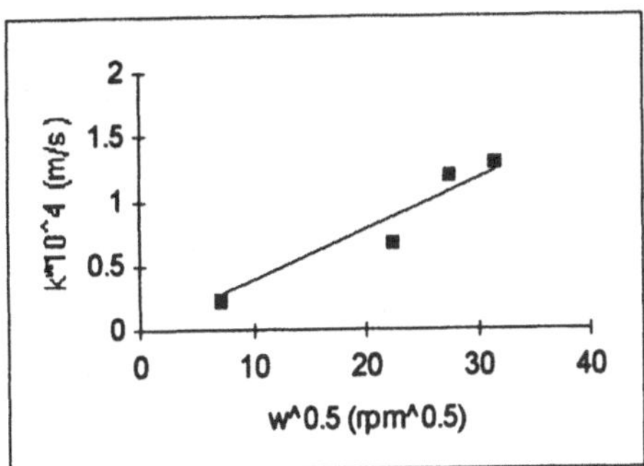

Fig. 3 - Influence of stirring speed in the cementation of mercury with zinc. (Experimental conditions in each run at different stirring speed: initial [Hg]=100 ppm, [NaCl]=100 g/l, N_2 atmosphere, T= 50 C, pH (initial)= 4.0)

A value for the diffusion of mercury(II) (D) ions into the zinc surface can therefore be obtained using equation (7) and the results shown in Fig.3, by assuming that the disk surface is equiaccessible [16]. In order to carry out this calculation, it is necessary to know the kinematic viscosity which we have assume to be equal to that of pure water [14].

$$D = 2.63 * 10^{-9}\ m^2/s$$

According to Wadsworth [15], the range of D values at typical cementation solution concentrations would be expected to fall between approximately 0.5 to 2 $*10^{-9}$ m^2/s. On the other hand Strickland [16] collected many diffusion coefficients which show a range between 0.73 to 1.95 $*10^{-9}$ m^2/s. Power [7] obtained for the system Hg(II)/Cu the value of diffusion coefficient of 0.64 $*10^{-9}$ m^2/s.

Another parameter that influences cementation kinetics is reactor atmosphere. Equations (11) and (12) show possible side-reactions when oxygen is present that are unfavourable to cementation. Equation (13) is a favourable reaction that can occur when hydrogen is present. It should be remembered that the main reaction (1) in a chloride medium has an electrochemical potential of 0.48 V.

	E (Volts)	
$2\ Zn + O_2 + 4\ H^+ = 2\ Zn^{2+} + 2H_2O$	1.99	(11)
$2\ Zn + O_2 + 2\ H_2O = 2\ Zn^{2+} + 4OH^-$	1.16	(12)
$Zn^{2+} + H_2 + 2OH^- = Zn + 2\ H_2O$	0.07	(13)

For oxygen reduction, reactions 11 and 12 have a high positive potentials in acidic solutions as indicated by the E values and occur preferentially to most metal ion reduction when oxygen is present in solution.

In Table 2, the kinectic constants of three mercury(II) cementation tests are shown, all with the same conditions except the atmosphere used. The results show that the presence of a reducing agent such as hydrogen increases the rate, on the other hand it is seen that cementation carried under atmospheric conditions produces worst results due to the reasons above. The slight difference between the nitrogen and hydrogen results does not make the use of hydrogen very attractive.

Atmosphere	k (m/s)
Hydrogen deareation	8.0 E-5
Nitrogen deareation	7.5 E-5
No deareation - Atmospheric air	6.4 E-5

Table 2 - Kinectic constants of mercury(II)/zinc cementation under different atmospheres. (Experimental conditions: initial [Hg]=100 ppm, [NaCl]=100 g/l, T= 50 C, pH (initial)= 4.0, stirring speed=1000 rpm)

Leachate cementation in Erlenmayer flasks

With the real solutions (leachates) three parameters were studied: pH, sacrificial metal dosage and type.
In order to determine the importance of these factors and to select an optimum pH for futher investigations, the leachate was divided into equal portions and the pH was adjusted by addition of 1.0N HCl or NaOH. In Figure 4 the final mercury(II) in solution is presented for different initial values of pH. In Figure 5 the corresponding values of zinc and iron dissolved are presented. Relative to leachate A (Fig.4.A) it is possible to conclude that with zinc, pH values 3 to 6 are the most favourable, wheras with iron the optimum pH is around 2.
Looking now at sacrificial metal dissolution, (Fig.4-B) as should be expected, acid conditions correspond to high metal consumption. In this way zinc cementation at pH 6 is a good compromise between efficient mercury(II) removal and low sacrificial metal dissolution.

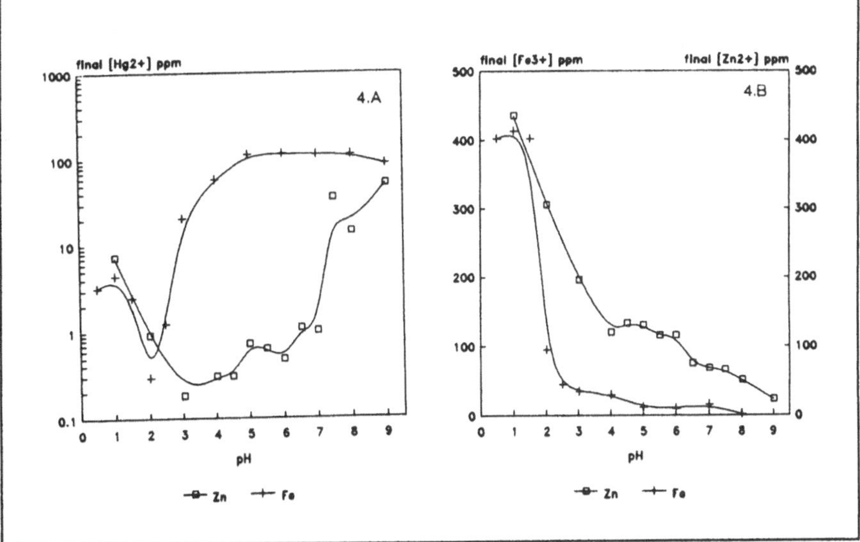

Fig.4 - Mercury(II)/zinc and mercury(II)/iron cementation of leachate A at different pH. Fig. 4.A- Final mercury(II) concentration versus initial pH. Fig 4.B - Final zinc(II) and iron(III) concentrations versus initial pH. (Experimental conditions: initial [Hg]= 196 ppm, T=19°C, stirring speed=100 osc./min, reaction time = 30 min, metal dosage=40).

The same pH study applied to leachate B, gave the results presented in Fig. 5. The conclusions are that the optimum pH range for zinc is 2.5 to 4 and for iron is pH < 1. Concerning zinc and iron dissolution the results are the same obtained with leachate A. Since the leaching process is carried out in an alkaline medium (pH 8.5) zinc was chosen as the sacrificial metal for the remainder of the work.

Rickard et al. [6] carried out a similar study applied to a synthetic solution of mercury(II). Zinc was used as sacrificial metal and their results show that the optimum pH working range was between 4 and 6.

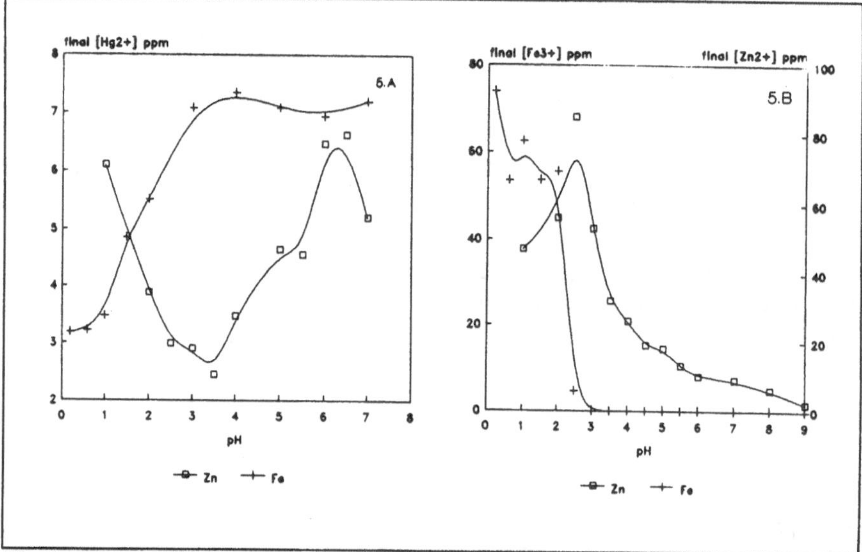

Fig.5 - Mercury(II)/zinc and mercury(II)/iron cementation of leachate B, at different pH. Fig.5.A - Final mercury(II) concentration versus initial pH. Fig 5.B - Final zinc(II) and iron(III) concentration versus initial pH. (Experimental conditions: initial [Hg]= 196 ppm, T=19°C, stirring speed=100 osc./min, time reaction = 30 min).

Another very important factor in cementation kinectics is the surface area of the sacrificial metal. This factor is closely related with two other ones: granulometry and purity. A set of tests was carried out where the same powder was utilized at different dosages (1.5, 15, 20, 30, 40. 45, 50, 58, 73, 109, 146 mol Zn / mol Hg). The results presented in Figure 6 are referred to leachate B, and show that between 30 and 50 molZn/ mol Hg, should be used.

Comparing the disk and powder experiments it is possible to make three comments: i) the metallic surface area in contact with the solution is much better defined with the disk (if it is well polished) than with powder. With powder there is always a granulometric distribution, and is necessary to know the specific area. ii) with the disk we have generally smaller areas and consequently longer reaction times. iii) If morphologic studies are intended is advisable to use the disk.

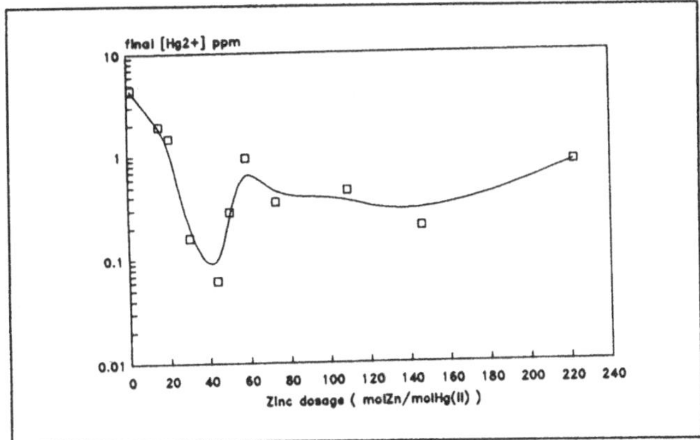

Fig.6 - Mercury(II)/zinc cementation of a leachate B with different zinc dosage. (Experimental conditions: initial [Hg]= 5.7ppm, T=19°C, pH (initial)= 3.0, stirring speed=100 osc./min, time reaction = 30 min).

CONCLUSIONS

Zinc can be used effectively for the recovery of mercury(II) ions from aqueous chloride solutions.

The process is sensitive to variations in initial concentration of mercuy, being more favourable at higher concentrations than at lower ones. The best efficiencies were obtained in the range of 10 to 500 ppm initial mercury concentration.

The cementation of mercury with zinc is a diffusion controlled reaction, with an activation energy of 14.6 KJ/mol and a diffusion coefficient of $2.63 * 10^{-9}$ m^2/s.

The presence of air in the reactor should be avoided, because off the negative effect that oxygen can have by side-reactions, mostly at acidic conditions.

Cementation of leachates was done using zinc and iron powder. Iron needs very acidic conditions to get good results, but zinc could produces good results in the pH range 3 to 6 in leachate A and 2.5 to 4 in leachate B. The best zinc dosage was established: 30 to 50 mol Zn/mol Hg.

REFERENCES

1 - Spedden, H. R.; Malouf E. E.; Prater D. J.; Cone-type precipitators for improved copper recovery; Journal of Metals, Oct 1966, 1137-1141

2 - Kimura M., Oho M., Wada Y., Fukui T.; Removal of mercury(II) ion from wastewater using zinc powder as a collector.; Nippon Kagaru Naishi 1989, vol.11, 1942-1948.

3 - Carvalho, Jorge; Pedroso, Cristina; Remoção de mercúrio de resíduos sólidos industriais, 3ª Conferência Nacional sobre a Qualidade do Ambiente, 1992 vol.2, 718-724.

4 - Carvalho, Jorge; Pedroso, Cristina; Reis, Lídia; Mercury removal from industrial sludges, International Chemical Engineering Conference, ChemPor 93, Porto Portugal, 1993, 243-250.

5 - Ballester A., Otero E., Gonzalez F.; Mercury extraction from cinnabar ores using hydrobromic acid,; Hydrometallurgy, 21 (1988) 127-143

6 - Rickard M., Brookman G.; Removal of mercury from industrial wastewaters by metal reduction,; Proceedings 26th Ind. Waste Conf. Purdue Univ. 1971, 713-721

7 - Power, G.P.; Ritchie I. M.; Metal Displacement (Cementation) reactions: the Mercury(II)/Copper system.; Electrochemical Acct, 1977 Vol.22, 365-371.

8 - Anonymous; A zinc dust bed can remove mercury from industrial waste streams.; "Chementator", Chem. Eng. 1971 Feb. 22, 63.

9 - Bro P., Lang Ko C.; Pressure drop and corrosion in zinc filters from wastes streams ; Environmental Science & Tech; 1974 8 (10) Oct 925-930

10 - Waltrich, P. F.; Removal of Mercury from effluents streams.; US 3,704,875; No 29- La; Nouette, K. H.; Heavy metals removal.; Chem. Eng. 1977, Oct 17, 73-80

11 - Ezhkov, A. B., Baiborodov P. P.; Potentiometric study of mercury (contact deposition) from sulphide-alkaline solutions; Gridomet. Tsvet. Redk. Metal. 1971 V. 139, 143 139-143.

12 - Strickland, P. H.; Lawson F.; Cementation of copper with zinc from dilute aqueous solutions; Proc. Aust. Inst. Min. ; No 236, Dec 1970, 25-34

13 - Levich V.G., Physicochemical Hydrodynamics. Prentice-Hall, New Jersey 1962

14 - Handbook of Chemistry and Physics, 52nd Edn (Edited by R. C. Weast). Chemical Rubber Co.,Cleveland (1971)

15 - Wadsworth, M.E.; Reduction of Metals in solution.; Transactions of the Metallurgical Society of AIME, 1969 Jul, 245, 1381-1394

16 - P. H. Strickland, F. Lawson ; The cementation of metals from dilute aqueous solution.; Proc. Aust. Inst. Min. Met. ; No 237, March 1971, 71-79

17 - Ball D., Boateng D.; Method for the recovery of Hg from mercury containing material.; US 5,013,358

18 - Strickland P. H.; Lawson F.; The measurement and interpretation of cementation data.; International Symposium on Hydrometallurgy Cap. 13; 293-330

19 - Calvo F. A., Ballester A., Otero E., Gonzalez F. ; Cementation de las soluciones del ion tetrabromomercurato.; Anales de Qumica, 1987, vol. 83, 390-396

20 - Vorob'ev E. E.; Apparatus for purifying solutions by cementation; USSR SU 1,404,541 (Cl. C22B3/02) 23 Jun 1988; No 54; 73/52625

21 - Miller J.; An analysis of concentration and temperature effects in cementation reactions,; Miner. Sci. Eng. 5, 3 July 1973 242-254

Removal of antimony and bismuth from copper tankhouse electrolytes

D. C. Cupertino
ZENECA Specialties, Blackley, Manchester, England
P. A. Tasker
ZENECA Specialties, Blackley, Manchester, England
M. G. King
ASARCO, Inc., Salt Lake City, Utah, U.S.A.
J. S. Jackson
ASARCO, Inc., Salt Lake City, Utah, U.S.A.

Abstract
An efficient process for the selective removal and recovery of antimony and bismuth from copper tankhouse electrolytes by means of solvent extraction has been demonstrated. This new process uses a new ligand, DS5834, developed by ZENECA Specialties and incorporates a novel stripping procedure. The same stripping procedure has also been used to regenerate chelating resins currently used at certain refineries which allows much greater efficiency in operation by improved recovery of the impurities.
Keywords: Antimony, bismuth, copper tankhouse electrolyte, impurities, purification, solvent extraction

1 Introduction

In the electrolytic refining of copper, impurities such as antimony and bismuth are introduced into the electrorefining solution from the impure copper anodes. As the copper anodes are dissolved into the electrolyte solution during the electrolysis process, the impurities are released and their concentration gradually increases causing them to be deposited on, and therefore incorporated into, the refined copper cathode. The conventional process used by the industry to control the levels of antimony and bismuth in the tankhouse, involves multistage electrolytic "depositing" cells which recover the copper, arsenic, antimony and bismuth as various mixed deposits from a bleed taken from the tankhouse. This process has many disadvantages, mainly in the limited rate at which the electrolyte can be cycled through the cells, the loss of copper back to a smelter to be recovered, the formation of toxic arsine gas, and the formation of toxic deposits and slimes which have to be disposed of.

A number of methods have been proposed for the removal of antimony and bismuth from copper electrolyte solutions. Work carried out by ASARCO [1], demonstrated that various grades of di-2-ethylhexylphosphoric acid (D2EHPA) could be used in a solvent extraction process to remove the antimony from electrolyte solution. Unfortunately, the reagents used had very poor stability and rapidly lost the ability to extract antimony. Other workers have also reported the use of organophosphorus reagents for solvent extracting antimony and bismuth, for example Norddeutsche

Affineries [2] and more recently by Dreisinger and Leong [3]. However, the reagents mentioned in these reports are also not entirely satisfactory in respect of their solubility in the organic solvents generally used for the extraction or their stability to the highly acidic conditions found in the solutions. In particular, the extractant mono-2-ethylhexylphosphoric acid (M2EHPA) has been shown to be a good extractant for antimony (Dreisinger) but has poor solubility and stability characteristics. Henkel have reported the use of hydroxamic acids to extract arsenic antimony and bismuth, but these reagents cannot easily be stripped [4]. It has also been reported that very strong acids are required in order to strip antimony from a metal-loaded organic phase containing a phosphorus reagent of the type currently being investigated. Typically 180 g/l HCl has been used, whilst other workers have reported the use of strong sulphuric acid solutions (600-900 g/l H_2SO_4) to strip antimony [2]. Less forcing conditions have been reported to strip bismuth from loaded organic solutions by e.g. using weak complexing agents such as citric and tartaric acids, but the subsequent removal of antimony required the use of stronger acids (H_2SO_4). Also, to recover antimony and bismuth products from strongly acid solutions, would require the complete neutralisation of all the excess acid before metal species could be precipitated, and this would make such a process very expensive. Stripping of metal-loaded organic solutions by direct precipitation of the metal sulphides from the organics, has also been reported [4], but this requires expensive, and hazardous, hydrogen sulphide technology.

Ion exchange processes using amino-methylene-phosphonic acid resins have also been described and are currently being used in a number of Japanese refineries [5-8]. However, a major problem associated with the use of ion exchange resin is the need to use a large excess of a strongly acidic stripping solution, typically elution with 4-6 molar hydrochloric acid has been reported, in order to liberate the metal and regenerate the resin. Consequently, large amounts of base are required to treat the excess acid used to strip the resin in order to recover the antimony and bismuth as solid products. This large consumption of acid and base makes these ion exchange processes expensive to operate.

In a joint project between ASARCO and ZENECA, a new efficient process for the selective removal and recovery of antimony and bismuth by solvent extraction has been developed. The process uses a new organophosphorus extractant DS5834, developed by ZENECA, which has been shown to be efficient at removing the impurity metal ions. The new reagent is a formulation of a type similar to M2EHPA, Fig. 1, but with much lower water solubility, and hence greater stability. The extractant is completely selective over copper, nickel, arsenic and iron(II), has fast extraction kinetics, rapid phase separation characteristics and is completely stable in contact with fresh, hot electrolyte solution.

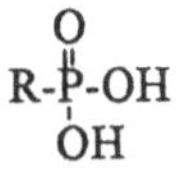

$$R\text{-}\overset{\overset{\displaystyle O}{\|}}{\underset{\underset{\displaystyle OH}{|}}{P}}\text{-}OH$$

R= 2-ethylhexyl, M2EHPA
iso-$C_{18}H_{37}$-, DS5834

Figure 1. Monophosphate ester extractant

The process also incorporates a novel stripping procedure, which allows the metal-loaded organic solution to be stripped using dilute solutions of hydrochloric acid. The same stripping procedure has also been used to regenerate chelating resins currently used at certain refineries. This new stripping procedure allows the economic recovery of the antimony and bismuth as solid products, by neutralisation of the metal loaded strip solution. The solvent extraction process has been tested in a continuous mini-rig trial by ASARCO, at their Amarillo refinery, using fresh, hot electrolyte solution. These tests showed that the antimony level was reduced from 600ppm to 134ppm over the period of the test.

2 Experimental

Initial testing was carried out using both synthetic electrolyte solutions and plant electrolyte supplied by ASARCO. The synthetic electrolytes contained approximately 170 g/l H_2SO_4, 120 g/l $MgSO_4$, 0.5 g/l Sb and 0.2 g/l Bi. The Sb and Bi were in their +3 oxidation states. A typical analysis for the plant electrolyte is shown in Table 1. and contained 80% of the Sb in the +3 oxidation state and 20% in the +5 oxidation state, the Bi was predominantly in the +3 oxidation state. The mini-rig trials were carried out at ASARCO's Amarillo refinery using fresh electrolyte solution.

Table 1. Typical analysis of plant electrolyte

Metal	Cu^{2+}	Ni^{2+}	$Fe^{2+/3+}$	$As^{3+/5+}$	$Sb^{3+/5+}$
Value g/l	50.5	18.9	0.27	7.1	0.5-0.6

In all the work described here, the extractant was used as a solution in Oroform SX-7, a kerosene diluent purchased from Philips Petroleum. McCabe-Thiele distribution isotherms were carried out at 60°C, with aqueous to organic ratios being chosen to give a reasonable distribution range. Both the organic and the aqueous phases were sampled and analysed for their metal value by inductively coupled plasma (ICP) or atomic absorption (AA) spectroscopy. Metal-loaded organic solutions used for stripping experiments were prepared by contacting the organic solution twice with fresh electrolyte with an aqueous to organic ratio of 5. Kinetics measurements were made using a baffled mixer at 60°C.

Metal-loaded ion exchange resin used for the stripping experiments was prepared by repeated contact of the resin with synthetic electrolyte solution. Samples of the loaded resin were stripped by washing with hydrochloric acid solutions containing calcium chloride.

3 Results and discussion

3.1 Solvent extraction

All the solvent extraction studies were carried out at 60°C since, in practice, it would not be desirable to change the normal operating temperature of the tankhouse electrolyte. The results for antimony and bismuth extraction are shown in Figures 2-5. These results show that the observations made by ASARCO, that certain sources of D2EHPA extracted antimony, was due to mono-2-ethylhexylphosphoric acid (M2EHPA). This extractant is present in the commercial sources of D2EHPA as an impurity. Pure D2EHPA itself does not extract antimony. This result has been confirmed by Dreisinger and Leong. However, Figure 3. shows that the ability of D2EHPA/M2EHPA mixtures to extract antimony is rapidly lost when it is contacted repeatedly with fresh electrolyte solution. ^{31}P nmr studies have shown that this loss of activity is almost certainly due to M2EHPA having some solubility in the aqueous phase where it is rapidly hydrolysed to phosphoric acid and 2-ethylhexanol in the presence of catalytic metal ions, and the strong acid in the electrolyte.

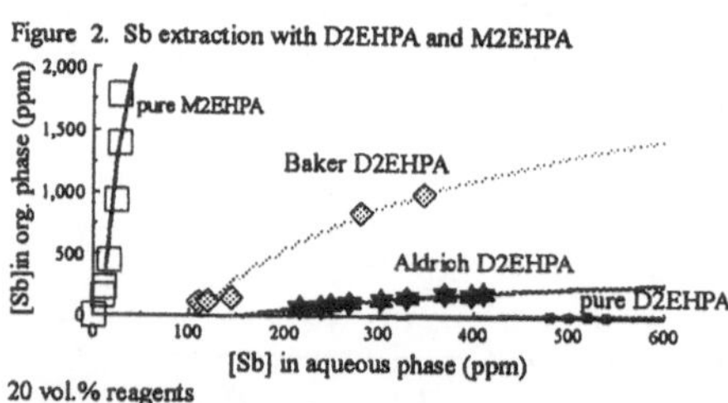

Figure 2. Sb extraction with D2EHPA and M2EHPA

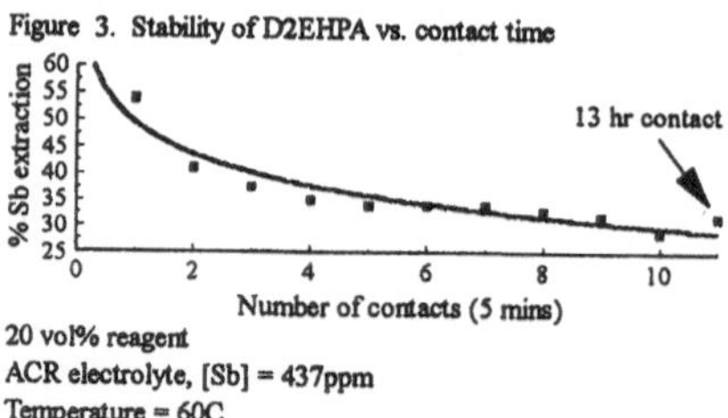

Figure 3. Stability of D2EHPA vs. contact time

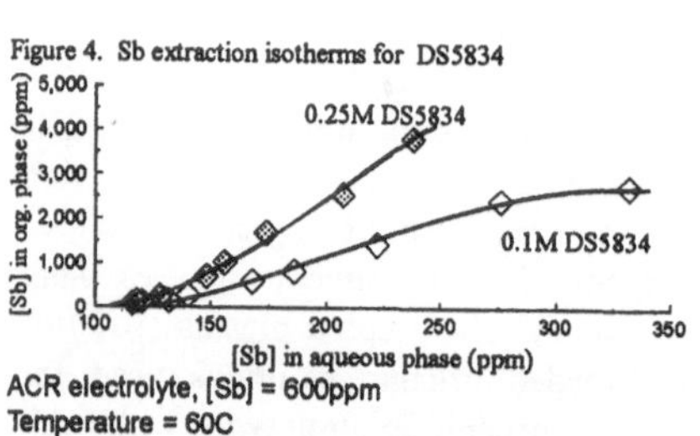

Figure 4. Sb extraction isotherms for DS5834

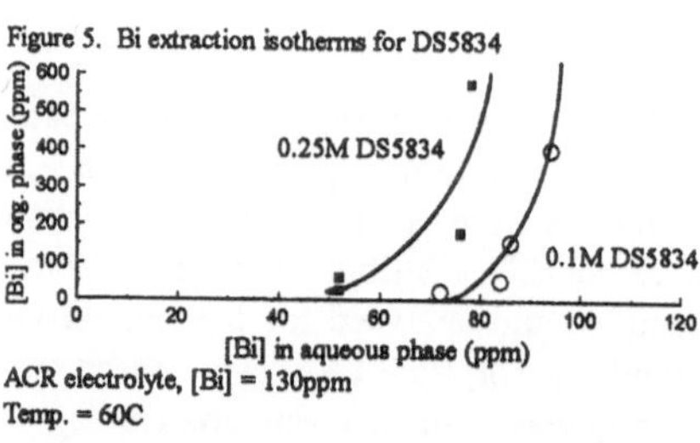

Figure 5. Bi extraction isotherms for DS5834

The extraction isotherms determined for DS5834 using synthetic electrolyte, confirm that the reagent is a very good extractant for Sb^{3+} and Bi^{3+}. The isotherms determined using plant electrolyte are consistent with approximately 20% of the Sb being present in oxidation state 5+ which is extracted only slowly. The presence of the Sb^{5+} in the plant electrolyte therefore limits the ultimate Sb level remaining in the aqueous raffinate after extraction to approximately 0.12 g/l. The reagent did not extract any copper, nickel, arsenic or sulphuric acid.

3.2 Rate processes- kinetics of extraction, stripping and phase separation

The reagent DS5834 extracts Sb^{3+} with very fast kinetics (ca. <0.5 minutes to reach equilibrium), whilst the rate of extraction of bismuth is only slightly slower, reaching equilibrium in approximately 2 minutes. Stripping was also found to be very fast for both metals. The rate of phase separation was also very fast and comparable to other commercial solvent extraction reagents.

3.3 Stability of DS5834

An indication of the stability of DS5834 has been obtained by continuously stirring a 0.1M solution in Oroform SX-7, with plant electrolyte, at 780 rpm in a baffled jar. The solutions were mixed at an aqueous to organic ratio of 1 and a temperature of 60°C. Samples were withdrawn periodically during the course of the experiment, which extended over a period of 105 days. The organic solutions were analysed for isooctadecanol, an assumed decomposition product, using gas chromatography. The results were compared with an analysis of the original DS5834 solution that had not been contacted with the electrolyte. Very small differences were observed between the original and final solutions confirming that the extractant was not decomposing at any significant rate under the test conditions. As a final test, the final organic solution was stripped with 6M HCl and reloaded by contacting with fresh electrolyte at an aqueous to organic ratio of 1. Comparison of the antimony and bismuth loading with fresh organic solution loaded under the same conditions gave identical results.

3.4 Stripping studies on DS5834

It was found that the stripping of antimony and bismuth from DS5834 can be accomplished using 6M hydrochloric acid. However, a much more efficient process, which involved contacting metal-loaded organic solutions with a strip solution containing dilute hydrochloric acid, and an alkaline or an alkaline earth metal chloride (e.g. $CaCl_2$) was identified. The results of these experiments are summarised in Figures 6-10.

The stripping of bismuth was demonstrated to be very efficient for all the conditions used. Antimony was only partly stripped when strip solutions containing less than 5 moles of total chloride ion were used. The stripping reaction appears to be independent of temperature over the range investigated.

The facile stripping of the antimony and bismuth using only dilute acid solutions containing high chloride ion concentrations, is believed to be favoured by the mass action effect of the chloride ion according to the idealised reaction scheme (1).

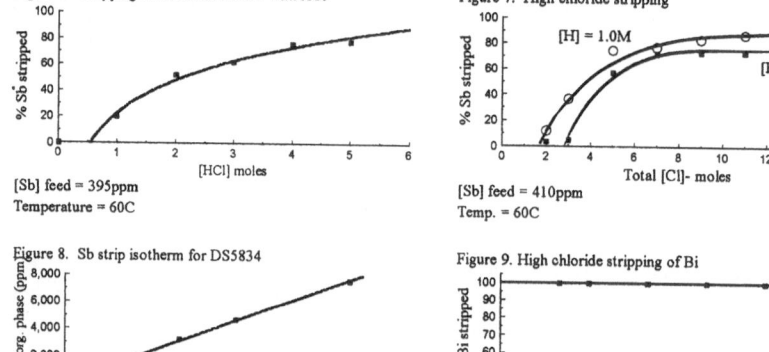

Figure 6. Stripping of Sb from DS5834 with HCl

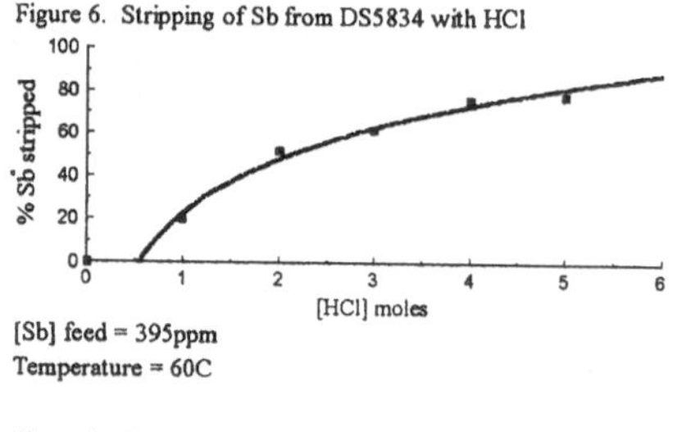

[Sb] feed = 395ppm
Temperature = 60C

Figure 7. High chloride stripping

[H] = 1.0M
[H] = 0.1M
% Sb stripped
Total [Cl]- moles

[Sb] feed = 410ppm
Temp. = 60C

Figure 8. Sb strip isotherm for DS5834

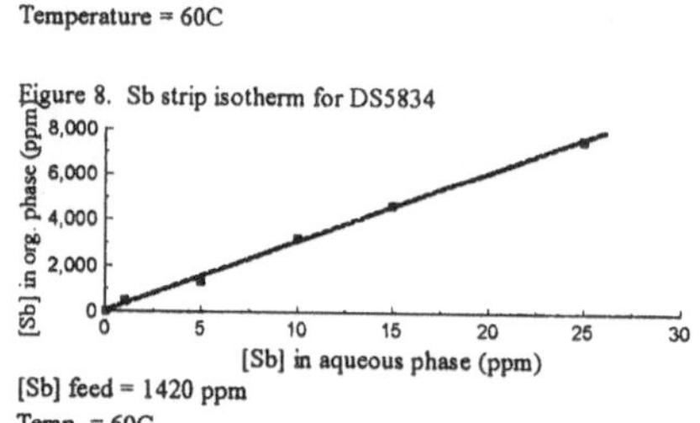

[Sb] feed = 1420 ppm
Temp. = 60C
[HCl] = 0.5M, [Cl] = 6M

Figure 9. High chloride stripping of Bi

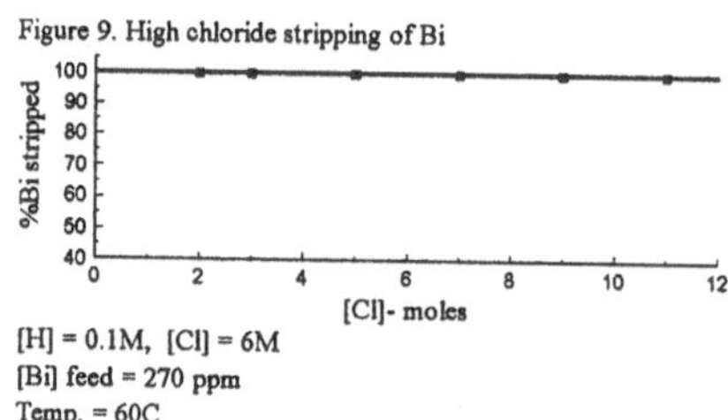

[H] = 0.1M, [Cl] = 6M
[Bi] feed = 270 ppm
Temp. = 60C

Figure 10. Bi strip isotherm for DS5834

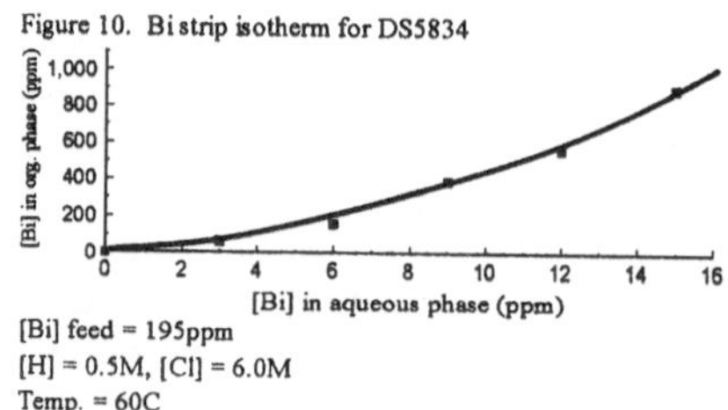

[Bi] feed = 195ppm
[H] = 0.5M, [Cl] = 6.0M
Temp. = 60C

$$L_3M_{(org)} + 3H^+_{(aq)} + 6Cl^-_{(aq)} \rightleftharpoons 3LH_{(org)} + MCl_6^{3-}{}_{(aq)} \qquad (1)$$

$LH = DS5834 \qquad M = Sb^{3+}$ or Bi^{3+}

In an industrial process, such a stripping procedure would be particularly advantageous, since the dilute acid/high chloride strip solution need only contain a slight excess of acid strength above the stoichiometric requirement for stripping the metal. Consequently, processing of the metal loaded strip solution to recover the antimony and bismuth requires only small additions of base (e.g. lime) to neutralise the excess acid present and precipitate the metals as their solid oxychlorides (MOCl, where M = Sb or Bi). The stripping solution can be regenerated for reuse by taking a small bleed to control the chloride ion concentration and reacidifying with fresh hydrochloric acid.

3.5 High chloride stripping of chelating resins

The use of dilute acid/high chloride stripping liquor to regenerate ion exchange resins has been demonstrated as part of this current study. A commercially available chelating resin, Duolite C-467, which has been reported as having good loading characteristics for Sb and Bi [3] was loaded by contacting with synthetic electrolyte solution. 1g portions of the metal loaded resin were separately contacted with 15ml portions of aqueous strip solutions containing either different concentrations of HCl (between 0.5-7M), or solutions containing 0.5M HCl and different concentrations of chloride ions (total chloride concentration was between 1 and 7M). The results are presented in Table 2. and show that almost the same level of stripping as 6M HCl solution can be achieved using 0.5M HCl containing 6.5M Cl^- ions as the stripping liquor.

Table 2. Stripping of Duolite C-467 resin

[HCl]M	% metal recovered Sb	Bi	0.5M HCl/ Total [Cl^-]	% metal recovered Sb	Bi
1.0	1.3	38.5	1.0	1.3	32.7
2.0	14.2	72.8	2.0	2.5	66.6
3.0	52.8	84.0	3.0	8.1	83.3
4.0	71.2	85.7	4.0	26.3	91.2
5.0	81.8	89.3	5.0	57.7	93.6
6.0	88.0	92.4	6.0	73.5	99.5
			7.0	79.7	100

3.6 Minirig trials

McCabe-Thiele data obtained for the extractant DS5834 have been computed to give predicted circuit parameters for a range of possible operating conditions. The results were used to design a small scale minirig plant for continuous tests using fresh plant electrolyte. The parameters were calculated solely on Sb response with the goal of reducing the Sb concentration from 0.6 g/l in the feed electrolyte to about 0.13 g/l in the raffinate. A stage efficiency of 95% has been assumed for both extraction and strip. The McCabe-Thiele construction for a circuit based on two extraction and two strip stages, and the predicted Sb mass balance diagram are shown in Figures 11 and 12.

The solution compositions fed to the solvent extraction circuit were as follows:

Extractant solution: DS5834 - 0.25M in Oroform SX-7
Aqueous Feed: Amarillo electrolyte (0.6 g/l Sb, 0.142 g/l Bi, 0.275 g/l Fe)
Aqueous Strip Solution: 0.5M HCl, 2.75M $CaCl_2$

Figure 11. McCabe-Thiele constructions for DS5834

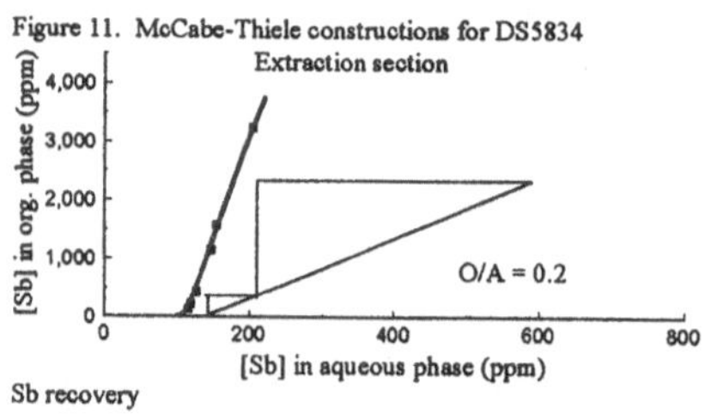

Sb recovery

McCabe -Thiele constructions for DS5834
Stripping section
[Sb] in aq. phase (ppm)
14,000
12,000
10,000
8,000
6,000
4,000
2,000
0
O/A = 5.0
0
500
1,000
1,500
2,000
2,500
3,000
3,500
[Sb] in organic phase (ppm)

Sb recovery

Figure 12. Calculated mass balance for minirig trials

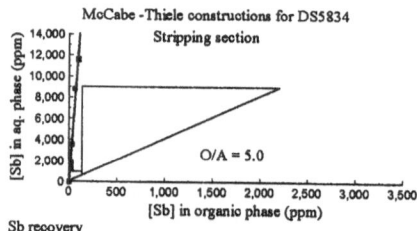

% Efficiency 95 95
O to A 0.21 0.21
electrolyte 0.600 g/l 0.130 g/l
aqueous
EXTRACTION
organic
0.202 g/l
1 Extraction stages 2
0.367 g/l
organic 2.296 g/l 0.021 g/l
0.189 g/l
organic
STRIPPING
aqueous
1 Strip stages 2
0.739 g/l
10.00 g/l 0.000 g/l
O to A 4.4 4.4
% Efficiency 95 95

Figure 13. Minirig average aqueous phase concentrations and overall percent extractions

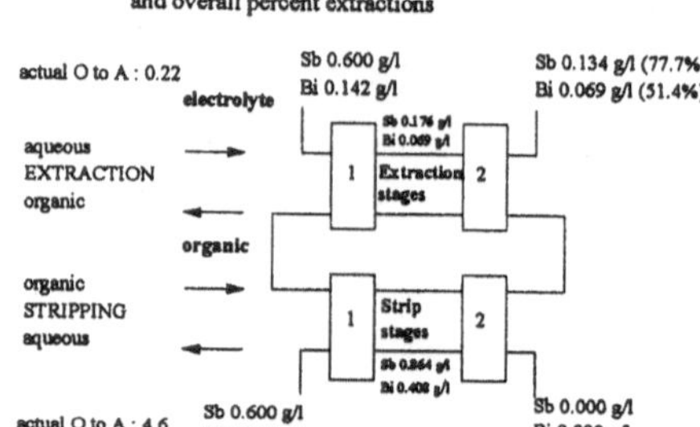

Figure 14. Minirig trials, average extraction vs. cycle time

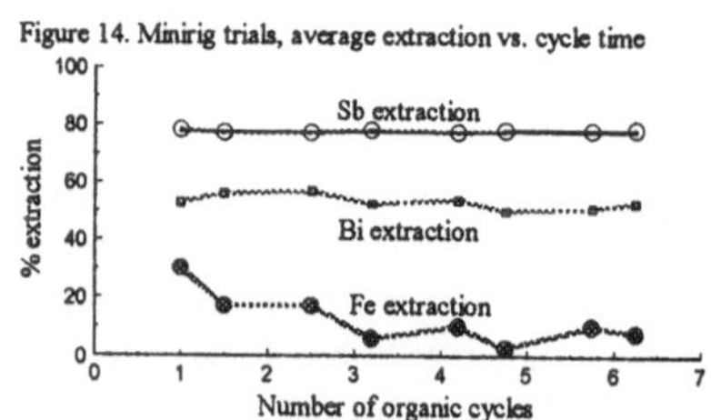

The minirig trial was run over a period of 32 hours giving over 6 complete cycles of the organic through the system. Figure 13. shows the average aqueous phase concentrations and overall percent extractions. Over the period of the test, a small amount of iron built up in the organic phase which was not stripped. Separate experiments on the loaded organic solution have shown that the iron was efficiently stripped by 5M HCl solution. A bleed from the organic phase could therefore, be used to control the iron level in the solution. The percent extraction of Sb, Bi, and Fe versus the number of organic cycles is presented in Figure 14.

3.7 Plant processes

Figure 15. summarises a possible flowsheet for a solvent extraction process to remove and recover antimony and bismuth products from a copper refining electrolyte solution. The stripping process can also be incorporated into an ion exchange circuit to regenerate the resin and recover the metals. This makes the process much more economical to run.

Figure 15. Flow sheet for Sb/Bi recovery using solvent extraction

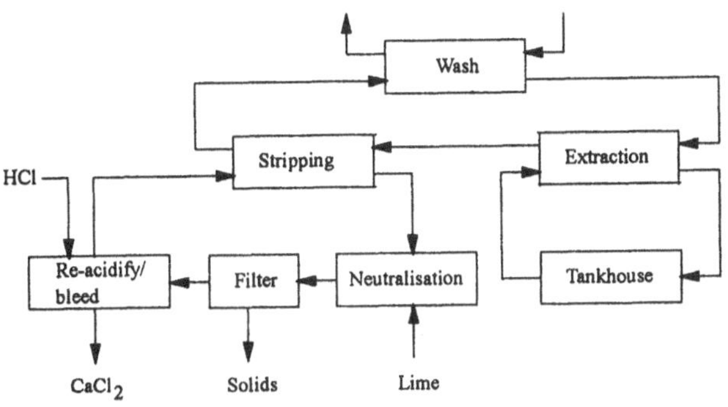

4 Conclusions

The monophosphoric acid ester M2EHPA, which is an impurity in commercial grades of D2EHPA, is a good extractant for antimony and bismuth but is rapidly degraded in use. The ZENECA developement extractant DS5834 has been shown to be efficient at selectively removing both antimony and bismuth from copper tankhouse refinery electrolyte. A stripping procedure which can be used to strip metal loaded organic solutions as well as ion exchange resins, has been developed. This procedure allows the economical recovery of antimony and bismuth containing products from metal loaded strip solutions. Continuous extraction and stripping tests have confirmed the effectiveness of the DS5834 extractant for removing antimony and bismuth from fresh plant electrolyte.

5 References

1. ASARCO Reports 4830, August 1977, 4967, September 1980 and 4989, March 1981, **Personal communication,** September 1989.
2. Schulze, R. (1975) **German Offenlengungsschrift** 25 15 862, April 11.

3. Dreisinger, D.B., Leong, B.J.Y. (1992) **CANMET Report,** No.0748, January 1992.
4. Schwab, W., Kehl, R. (1989) **U.S. Patent** 4,834,951, May 30 .
5. Sasaki,Y., Kawai, S., Takasawa, y., Furuga, S. (1991), **Proceedings of Copper'91**, Volume III, Pergamon Press, Elmsford, N.Y., 245-254.
6. Shibata, T., Hashiuchi, M., Kato, T.(1987), **TMS-AIME,** Warrendale, PA, 99-116.
7. Takashi, O., Hosaka, H., Kasai, S. (1985), **U.S. Patent,** 4,501,666, February 26.
8. Nagai, T.M., Echigo, Y. **Japanese Patent,** JP 59162108.

Novel solvent extraction reagents for recovery of zinc from sulphate leach solutions

R. F. Dalton
P. M. Quan
ZENECA Specialties, Blackley, Manchester, England

Abstract

Despite the ready availability of numerous zinc sources, the process of solvent extraction-electrowinning (SX-EW) has as yet found little favour in the zinc industry. It is suggested that this is largely due to the previous lack of availability of suitable selective reagents having the particular features required to enable them to fit in to a solvent extraction-electrowinning based process for zinc, particularly with regard to strength and selectivity for zinc over iron. This paper describes work on the development of a novel class of selective extractant for the recovery of zinc from weakly acidic sulfate leach solutions and suggests how such reagents might be integrated into new zinc processes.

Introduction

The process of solvent extraction combined with electrowinning (SX-EW) is now common place in the copper industry with more than 1.5 million tons of copper produced by this means annually. The reasons for this are quite clear:

SX-EW utilizes vast amounts of low grade ores and tailings as feedstock which can not be treated economically by other means. SX-EW is considered the most economical means of producing pure cathode copper, with some leach-SX-EW operations having overall operating costs as low as 30c/lb (US). SX-EW is also perceived to be more environmentally friendly than smelting, there being no gaseous emissions.

Most copper produced by SX-EW is recovered from dilute sulfate solutions produced by sulfuric acid leaching of oxidic ores or by biological and ferric sulfate leaching of low grade sulfide ores. Crucial to these developments have been the provision by the chemical industry - including Zeneca - of highly selective reagents for isolation of copper from these leach solutions by solvent extraction. The extraction reaction (eq.1) is reversible so that stripping of the copper by spent tankhouse electrolyte regenerates the reagent and provides a pure concentrated copper sulfate solution for recycling to electrowinning.

$$Cu^{++} + 2RH \rightleftharpoons R_2Cu + 2H^+ \qquad \text{(eq.1)}$$

By comparison, solvent extraction finds little use in the recovery and processing of zinc and so far is confined to only a few specific and comparatively small examples. Most zinc is produced by the traditional roast-leach-electrowinning process using high grade primary sources. This is despite the ready availability of alternative primary and secondary sources of zinc and zinc bearing wastes and residues.

Typical examples are:

Smelter wastes
Oxide fume
EAF dusts
Low grade ores and concentrates
Complex ore concentrates
Galvanizing residues
Tankhouse bleed solutions
Industrial and chemical wastes

It is suggested that the reason why solvent extraction has not been adopted so readily in the zinc industry is due in part to the non availability in the past of suitable selective extractants, particularly with regard to selectivity for zinc over iron. Previous publications have described our work on the development of a selective extractant for the recovery of zinc from chloride leach solutions (1,2) and one of these (2) also introduced work on a new class of reagent for the recovery of zinc from sulfate solutions. This paper describes progress of our work aimed at the development of new selective reagents for the solvent extraction of zinc from sulfate leach solutions and how these reagents might be integrated into a sulfate based solvent extraction-electrowinning process for zinc.

Reagents for the Solvent Extraction of Zinc from Sulfate Solution

Ideally, a reagent for the recovery of zinc by solvent extraction from weakly acidic sulfate feed solutions should operate on a hydrogen ion cycle in an analogous manner to the o-hydroxyaryloximes used extensively for the solvent extraction of copper, i.e.

$$Zn^{++} + 2RH \rightleftharpoons R_2Zn + 2H^+ \qquad \text{(eq.2)}$$

Thus under extraction conditions it should give a high distribution of zinc into the organic phase without the need for neutralisation or pH adjustment. Conversely, it should readily give up the zinc on contact with a strongly acidic zinc bearing spent electrolyte. In addition, it should also show high selectivity for zinc over a number of other metals, particularly iron.

The difficulties in achieving this are really quite formidable. The order of stability of complexes of the divalent metals across the first row of transition elements generally follow the well known Irving-Williams Series (3), usually presented as Mn<Fe<Co<Ni<Cu>Zn. The complexes of Fe(III) usually have a stability between those of Ni(II) and Cu(II) and are much higher than those of Zn. Thus, while high selectivity for copper over a range of other metals is in accord with the natural order, the attainment of high selectivity for zinc runs contrary to this. Changes in the normal order of complex stability can, however, be brought about by recognition of both the stereochemical preferences of metals in their complexes and of their preferences for certain ligands according to the classical Lewis acid-base, hard-soft donor atom rules (HSAB).

It is for instance well known that the reagent di-2-ethylhexylphosphoric acid (D2EHPA) forms stronger complexes with zinc than it does with copper. The complexes involve solvated as well as coordinated phosphate molecules in a tetrahedral configuration that is strongly favoured by zinc, but not by copper (4,5). Nevertheless, the zinc complexes of D2EHPA are quite weak and, though it has been used in some solvent extraction processes, recovery is poor unless D2EHPA is used either in the form of its sodium or ammonium salt, or interstage neutralisation is employed. A further disadvantage of D2EHPA is that it forms very strong complexes indeed with Fe(III), which virtually precludes its use wherever substantial concentrations of ferric iron are present in solution.

$$(C_8H_{17}O)_2P(=O)OH \qquad\qquad (C_8H_{17})_2P(=O)SH$$

D2EHPA DEHTP (CYANEX 302)

In addition D2EHPA also extracts calcium quite strongly which is a further disadvantage in the context of zinc recovery processes (6,7). Substitution of oxygen by sulfur in these extractants would seem a logical progression and the sulfur analogues di-2-ethylhexyl-dithiophosphoric acid and di-2-ethylhexylmonothiophosphoric acid are indeed much stronger extractants of zinc. The dithiocompound in particular is so strong that it is difficult to recover the zinc by stripping and the compound is also unstable. Attention has tended to focus on the more stable though somewhat weaker dialkyldithio- and dialkylmonothiophosphinic acids, commercial examples of which are Cyanex 301 and Cyanex 302. The major advantage claimed for the monothiophosphinic acid Cyanex 302 is its undoubted selectivity for zinc over calcium. It is, however, still a relatively weak extractant for zinc and lacks selectivity over ferric iron (8).

A programme of research work at ZENECA Specialties Research Centre has focused on the search for a novel extractant type having both the strength and selectivity requirements to enable its successful integration into an overall SX-EW process. These features have now been realised in a new class of extractant under investigation by ZENECA Specialties based on bisdithiophosphoramides (9).

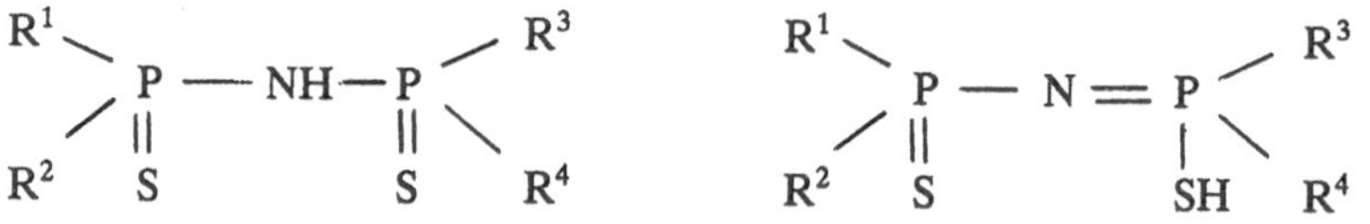

$$(R^1)(R^2)P(=S)-NH-P(=S)(R^3)(R^4) \qquad (R^1)(R^2)P(=S)-N=P(SH)(R^3)(R^4)$$

tautomeric structures of a substituted bisdithiophosphoramide

These compounds can act as chelating bidentate ligands towards zinc and the behaviour of this class of compound is exemplified herein with reference to the development formulation DS5846.

Experimental

Analysis of metals in both aqueous and organic solution was conducted by means of inductively coupled plasma spectrophotometry (ICP) using appropriate standards for calibration.
Selectivity tests were carried out by equilibrating aqueous feed solutions, of compositions as identified in the text, with solutions of the extractants in a kerosene type diluent, ESCAID 100. After contact the phases were allowed to separate, samples withdrawn and filtered through phase separation paper (organics) or Whatman filter paper (aqueous) and then analysed by ICP.

Extraction and strip isotherms were determined by equilibrating various volume ratios of organic extractant solution and aqueous feed by vigorous stirring for at least one hour in a thermostated water bath. After equilibration samples were taken of each phase, filtered and analysed.

Results and Discussion

Selectivity of DS5846

An indication of the relative strength and selectivity of DS5846 compared with D2EHPA and Cyanex 302 is given in Table I, which summarises results when a 0.2 molar solution of each reagent was contacted at an O/A ratio of 1/2 with an aqueous feed solution containing 2.95g/l zinc and 4.3g/l iron (III) at pH 2.0.

Table I Extraction of zinc and iron by different reagents.

Reagent	Metals extracted into organic phase (g/l)	
	Zn	Fe
DS 5846	4.9	0.09
Dialkylthiophosphinic acid (Cyanex 302)	0.72	0.69
Di 2-ethylhexylphosphoric acid (D2EHPA)	0.03	4.04

The strength of the iron III complex formed with D2EHPA precludes extraction of all but a few mg/l of zinc and, with Cyanex 302, zinc and iron are extracted in approximately equal amounts. In the case of DS5846, however, zinc is extracted strongly with high rejection of iron. These are in fact believed to be the highest zinc-iron selectivity ratios ever reported for an extractant.

Though highly selective for zinc over iron, DS5846 does extract some other metals as shown in Table II. These data were obtained by contacting a 0.1 molar solution of DS5846 for one hour with an equal volume of an aqueous solution containing a mixture of metal ions, each at a level of 200-300 mg/l (or saturation amount), added as the nitrate or acetate at an initial pH of 2.0, and analyzing for the amounts of each metal in the aqueous and organic phases.

Table II Extraction of metals by DS5846 from a mixed metals solution

Metal	Concentration of metal, mg/l	
	In the aqueous phase	In the organic phase
Arsenic (III)	55	<5
Calcium (II)	271	<5
Chromium (III)	260	<5
Copper	300	<5
Lead (II)	90	<5
Magnesium (II)	300	<5
Manganese (II)	295	<5
Nickel (II)	305	<5
Bismuth (III)	<5	135
Cadmium (II)	<5	295
Mercury (II)	<1	295
Silver (I)	<5	340
Zinc (II)	15	335

The metals fall into two distinct groups - those that are rejected by DS5846, and those such as bismuth, cadmium, silver and mercury that are extracted strongly and would require special consideration if present in a zinc feed. It should be noted that DS5846 complexes strongly with copper which is extracted almost quantitatively and is not readily stripped. Copper in the feed solution should therefore be reduced to low levels, e.g. by cementation, prior to the zinc solvent extraction stage. The co extraction of some heavy metals such as cadmium can in fact be viewed as a positive advantage as long as it is possible to effect a subsequent separation.

Zinc extraction equilibria with DS5846

The behaviour of DS5846 as a zinc extractant is amply illustrated by Fig. 1 which shows extraction and strip isotherms with McCabe-Thiele constructions for the recovery of zinc from an aqueous feed containing 2.95 g/l Zn, 4.3 g/l Fe (III) at pH 2.0 using a 0.2 molar solution of DS5846 in a Kerosene type diluent (Escaid 100). Stripping was with an aqueous solution containing 180 g/l sulfuric acid and 30 g/l zinc. Interpretation using the ZENECA Specialties 'MINCHEM' computer programme (10) to generate and optimise the McCabe-Thiele constructions indicates that, assuming a counter current process based on 2 extraction and 2 strip stages with 95% efficiency in each stage, a zinc in raffinate level of 0.10 g/l is predicted, which represents a 96.5% zinc recovery. Such isotherms and metal recoveries are not unlike those obtained under similar conditions for copper using o-hydroxyaryloxime extractants.

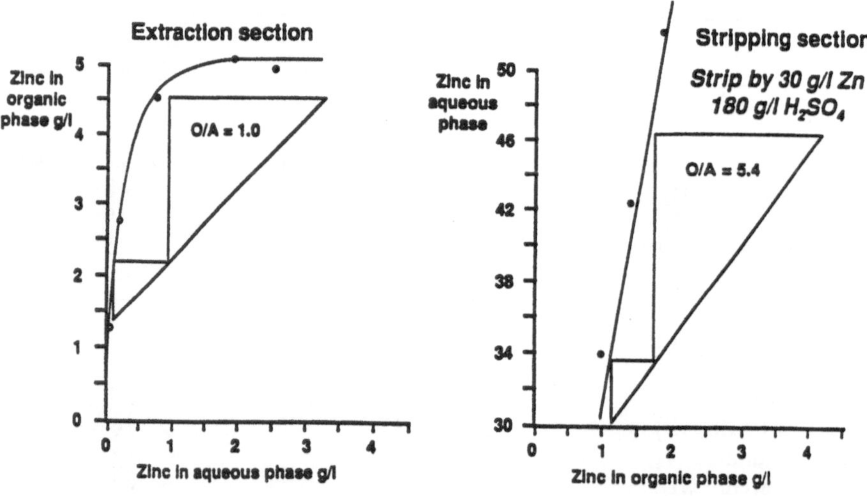

Figure 1. Recovery of Zinc from Dilute Feed Solution using DS5846

Fig 2 shows extraction and strip isotherms for the recovery of zinc under the far more exacting conditions of a feed containing 22 g/l zinc, 4 g/l ferric iron at pH 2.

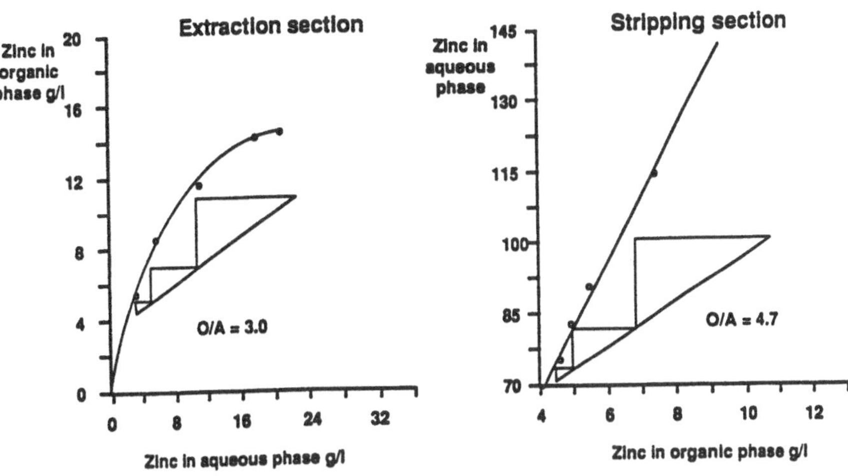

Figure 2. Recovery of Zinc from Concentrated Zinc Feed Solution using DS5846.

Extraction was carried out using a 0.5 molar solution of DS5846 and stripping was with a simulated zinc electrolyte containing 180 g/l sulfuric acid and 70 g/l zinc. The computed McCabe Thiele constructions indicate that using 3 extraction and 3 strip stages it is possible to obtain a zinc in raffinate concentration of less than 3g/l, representing an 87% recovery while attaining a fresh electrolyte composition exceeding 100g/l zinc.

Small scale pilot plant operation

A laboratory scale pilot trial has been carried out using the conditions outlined for generation of the McCabe-Thiele data shown in Fig 2 above. The circuit, feed and exit compositions from stage to stage are as shown in Fig. 3.

The Kinetics of extraction and stripping of zinc with DS5846 are slower than the familiar o-hydroxyaryloxime based copper recovery systems and the equilibria are also sensitive to temperature. It has been found to be advantageous to carry out extraction and stripping at 40°C and this condition was adopted for the small scale pilot trial. In this trial a regenerated zinc electrolyte concentration of greater than 100 g/l zinc was achieved, with an iron concentration in electrolyte of 127 mg/l.

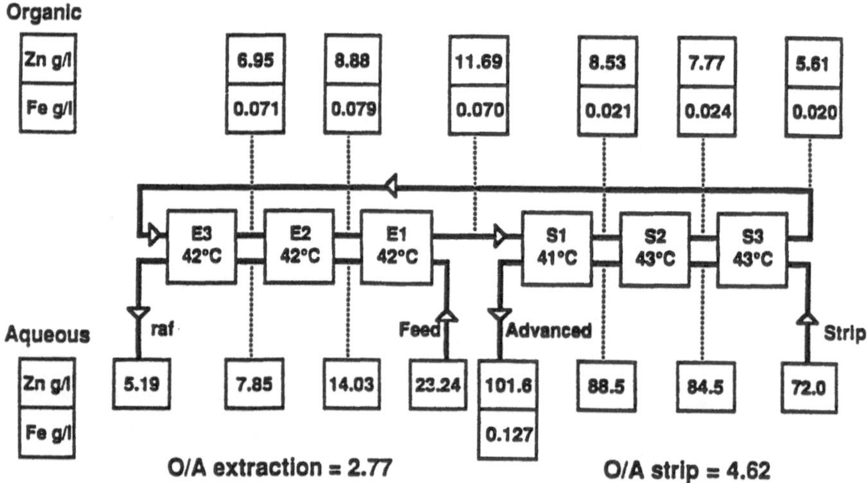

Figure 3. Laboratory Scale Pilot Trial of Zinc Recovery using DS5846.

Removal of minor metals

Iron Although it has been clearly demonstrated that the ZENECA development reagent DS5846 is capable in practice of giving very high zinc-iron ratios, it is recognised that the stringent conditions placed on electrolyte purity for the successful electrowinning of zinc mean that even this amount of iron transfer is likely to be considered excessive. Work is in hand to improve on this aspect, both by development of the reagent itself and by further process technology. A number of options for reducing iron are under evaluation including a reductive scrubbing procedure carried out on the loaded organic phase. The effectiveness of this is illustrated by the data in table III which show the rate of removal of iron from the organic phase at 25°C using the reductive scrub procedure.

Table III Rate of removal of iron from loaded organic phase by reductive scrubbing

Time (min)	Iron concentration in organic (mg/l)
0	100
4	2
15	0.3

<u>Cadmium</u> As indicated earlier (Table II), cadmium is extracted strongly by DS5846 which ensures its removal from zinc process streams.

It is not readily stripped by sulfate electrolyte solutions so cadmium transfer to the electrolyte is less than 1 mg/l. This combination of features could be considered highly advantageous. However, without some means of removing cadmium from the organic phase it would eventually build up to higher concentrations, lowering the zinc loading capacity of the extractant. One possible means of removing the cadmium from the organic phase is by stripping with 5 molar hydrochloric acid, and a strip isotherm is shown in Fig 4.

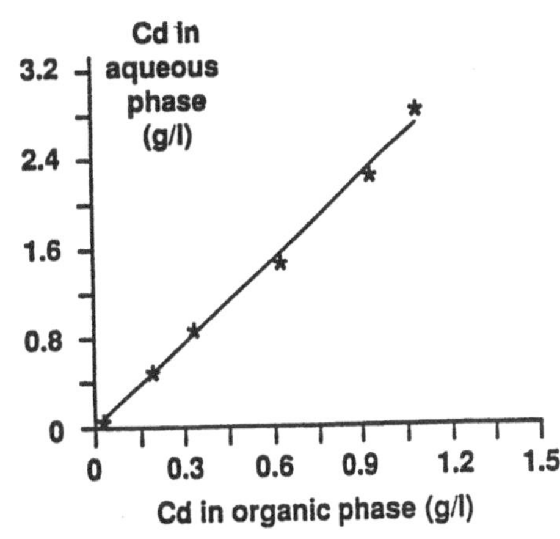

Figure 4. Cadmium strip isotherm

<u>Proposed zinc recovery process</u>

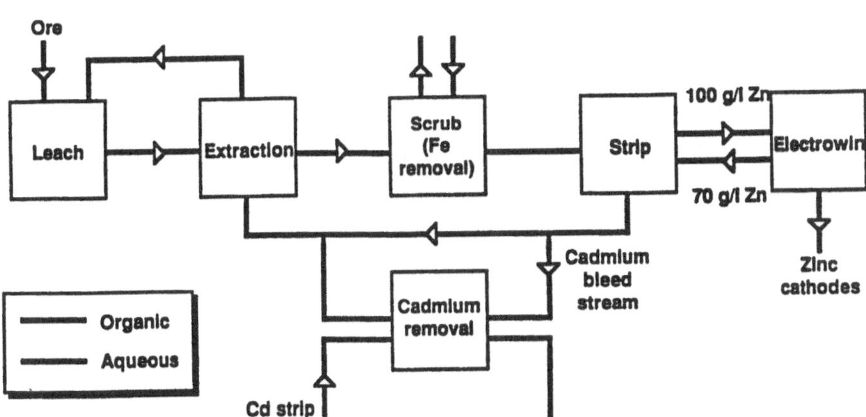

Figure 5 Proposed Zinc Recovery Process using DS5846

Fig. 5 shows a proposed flow sheet for a possible process for the recovery of zinc by solvent extraction - electrowinning, based on the properties of DS5846. There may well be alternative options and places in the circuit where iron and cadmium can be reduced to the low levels demanded for zinc electrowinning and these are being explored further. We believe, however, that the extraction and stripping properties of DS5846, and its ability to effect a major separation of zinc and iron, represents a major step forward and greatly facilitates the development of new options for zinc recovery based on solvent extraction.

Conclusions

The new development reagent DS5846 is a strong extractant of zinc from aqueous sulfate solutions showing high selectivity for zinc over iron.

The hydrometallurgical properties of DS5846 have been confirmed by continuous laboratory scale trials and ancillary processes are being developed to enable its satisfactory integration into a proposed zinc recovery process.

The properties of DS5846 are sufficiently promising to suggest that serious consideration should be given to the recovery of zinc by solvent extraction from a variety of sulfate containing process liquors and feed streams.

References

1. Dalton,R.F., Burgess, A., and Quan, P.M., ACORGA ZNX50: A New Selective Reagent for the Solvent Extraction of Zinc from Chloride Solutions. Hydrometallurgy, (1992), 30, 385-400.

2. Dalton,R.F. and Quan P.M. Novel Solvent Extraction Reagents - The Key to new Zinc Processing Technology. Conference Proceedings World Zinc'93. Ed I.G Matthew pp 347 - 355. Australasian Institute of Mining and Metallurgy.

3. Irving, H., and Williams, R.J.P., The Stability of Transition-Metal Complexes. J.Chem. Soc. (1953), 3193-3210.

4. Stoyanov Ye.S., and Mikhailov,V.A., Coordination Chemistry of Di (2-ethylhexyl) phosphoric acid. Proceedings if International Solvent Extraction Conference - ISEC'88, Vol I, 195. USSR Academy of Sciences.

5. Koncar M, Bart H.J. and Marr,R. Extraction of Zinc by Bis (2-ethylhexyl) phosphoric acid - Influence of Activity and High Loading. Proceedings of International Solvent Extraction Conference - ISEC'88, Vol II, 175. USSR Academy of Sciences

6. Cox, M., and Flett, D.S. Modern Extractants for Copper, Cobalt and Nickel Chem. and Ind (London) (1987) 188-193.

7. Bukowsky, H., et al. Hydrometallurgy, (1992). Vol 28, 323-329.

8. Rickleton, W.A. and Boyle, R.J., The Selective Recovery of Zinc with New Thiophosphinic acids, Solvent Extraction and Ion Exchange, 1991 vol 9, No. 1, pp 73-84.

9. Campbell, J., Dalton, R.F., and Quan, P.M. EP 0 573 182 A1 assigned to ZENECA Ltd. pub. 08.12.1993.

10. Morrison, J., and Townson, B., Mathematical Modelling of Metallurgical Systems using the MINCHEM II Computer Program, in Proceedings of Extraction Metallurgy ¬89, (The Institute of Mining and Metallurgy, London, 1989).

Recovery of acids and bases used in metal treatment processes by diffusion dialysis

G. P. Herz
Tokuyama Soda Company, Tokyo, Japan
C. Byszewski
Graver Water, Union, New Jersey, U.S.A.
M. Jaffari
Malek, Inc., San Diego, California, U.S.A.

Abstract
The recovery of waste acid has been successfully achieved for a number of years using various processes. DD (Diffusion Dialysis) has been shown to be an attractive process on a commercial scale, because of its simplicity, reliability and relatively low operating cost.

DD has been particularly successful in the recovery of relatively expensive acids such as HF and HNO_3. However, with increasingly strict control of waste stream quality, the recovery of even less expensive acids has been found of interest.

Recovery of acids by means of DD is in the range of 80 - 90%, treating acids in a concentration range of 1 - 7 Normal. It is possible to attain salt content as low as 10 ppm, by the use of Neosepta AHS-2.

Recently, new membranes have been developed which can be used in the DD process for the recovery of bases. Base permeable membranes are of particular interest to industries where basic surface treatment is involved, such as surface etching of aluminium in the aerospace industry. An industrial plant for the treatment of aluminium etchant has recently been built. This involves a system which maintains the purity of the etching solution and controls the metal level without giving rise to a waste stream; DD is an important component of the system.
Keywords: Acid recovery, caustic recovery, diffusion dialysis, membranes, pickling liquor, steel mill.

1 Introduction

Diffusion Dialysis (DD) is a membrane process that has been successfully used for many years for the recovery of acids. A similar method for the recovery of caustic has been assiduously sought by industry, but could not be realised due to the absence of suitable membranes. Recently, Tokuyama Soda Company developed membranes that are highly effective and so DD for caustic recovery is now also an established technology.

With the growing awareness of environmental protection, free disposal of aqueous wastes is being more and more limited. Accordingly, methods for the recovery of waste acids and bases have been developed and are being increasingly implemented.

Table 1. HF/HNO_3 recycling by diffusion dialysis process

DD Process	Operating Parameters	
Design basis	Treatment of HF/HNO_3 solution from pickling bath containing iron as a major impurity, in order to keep steady-state concentration in the bath of about:	
	Total Free Acids	3.1 N
	• Free HNO_3	2.2 N
	• Free HF	0.9 N
	Fe	12.5 g/l
	Related Fe drag-out	11.2 kg/hr
	Operating Temperature	30 - 40°C
Diffusion dialyser proposed	1 unit	TSD 50-2000
	Effective surface area of membrane	1,000 m^2
Performance (see Table 2)	Acids recovered	> 85 %
	•Free HNO_3	> 98 %
	•Free HF	> 54 %
	Metal reject	> 89 %

Table 2. Performance and benefits of a DD unit for HF and HNO_3 recovery

Performance			Benefits	
	Waste acid	Recovered acid		FF* per year (10^3)
Flow rate l/h	1000	960	Running costs	500
HNO_3 g/l	150	141		
HF g/l	20	12.5	HNO_3 recovered	850
Fe g/l	20	2.1	HF recovered	1250
			Savings on neutraliser $Ca(OH)_2$	500
			Savings on sludge treatment	15
Type of unit: TSD-50-2000			Total savings	2615
Type of membrane: AFN				

* FF - French Francs

2 Principle of Diffusion Dialysis

DD is a separation process based upon the natural diffusion of ions from a region of high concentration to a region of low concentration across a semi-permeable ion exchange membrane.

For acid recovery, anion exchange membranes are used. These are permeable to anions associated with hydrogen (e.g. H_2SO_4) but impermeable not only to cations (as are all anion exchange membranes) but also to anions associated with salts, e.g. $Fe_2(SO_4)_3$. The principle and its application to acid recovery has been described in detail by Noma [1].

In a corresponding manner, cation exchange membranes are used for caustic recovery. These membranes are permeable to cations associated with hydroxide, e.g. NaOH, but impermeable to cations associated with salts, e.g. NaCl. In other words, such membranes permit the migration of free bases while blocking salts. The principle of acid and caustic DD is shown in Figures 1 and 2 respectively [2].

3 Operation of the DD Process

Waste acid or base, containing metal ions, is passed through a cell in an up-flow direction. The cell is separated from an adjacent cell by a membrane as just described. Water simultaneously passes through the adjacent cell in a downward direction. Because of the high concentration of ions in the waste acid or caustic cell, diffusion pressure causes them to seek their way into the dilute phase in the water cell. The acid or base succeeds in passing through the membrane, while the metal salts are blocked. Because of the continuous passage of fresh water through the water cell and of waste acid or base through the acid or caustic cell, the difference in the concentration between the two cells (and, hence, the diffusion pressure) remains sufficiently great for the process to continue.

Actual equipment consists of a number of such cell pairs arranged in a "stack" of alternate dialysate chambers (carrying the feed liquor) and diffusate chambers (carrying the water and receiving the acid or base molecules during the diffusion process). The chambers are clamped between end plates somewhat as in a filter press. The membranes are separated by spacers, usually made of plastic; the cell thickness is normally in the range of 1-2 mm [3,4].

In large installations multiple stacks comprising several hundred cells are used, with an individual membrane area per cell of over a square meter, resulting in a total membrane surface for the stack of several hundred square meters [5,6].

The DD stack is the heart of the DD plant, which also includes tanks for storage of the raw waste acid or base, recovered acid or base, water and waste solution, a filtration system of some kind (briefly discussed in Section 4.2) and the pumps necessary for propelling the various solutions. Figure 3 is a schematic diagram of the general arrangement [7]. A control system of some kind is also necessary.

As mentioned above, DD requires no external force to operate. The migration through the membrane is simply due to dialysis and osmotic pressure. (Dialysis is the transfer of solute molecules across a membrane by diffusion from a concentrated to a dilute solution.

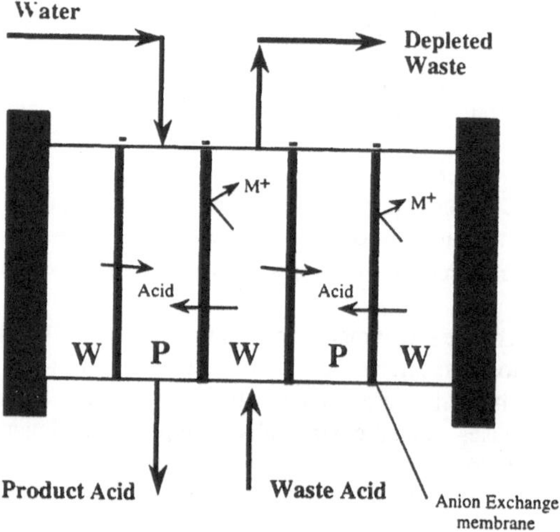

Figure 1. Acid diffusion dialysis stack

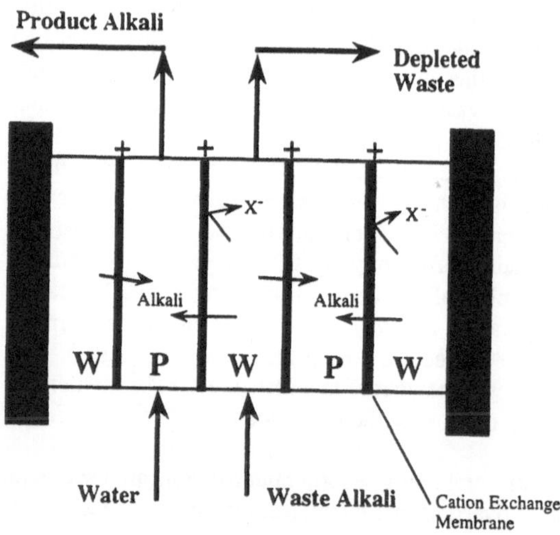

Figure 2. Alkali diffusion dialysis stack

Osmosis is the simultaneous diffusion of solvent molecules through the membrane in the opposite direction). The only energy required is the pumping force needed to move the liquids through the dialysis stack. This is low compared to the pumping pressures associated with processes that oppose osmotic pressure, such as reverse osmosis and

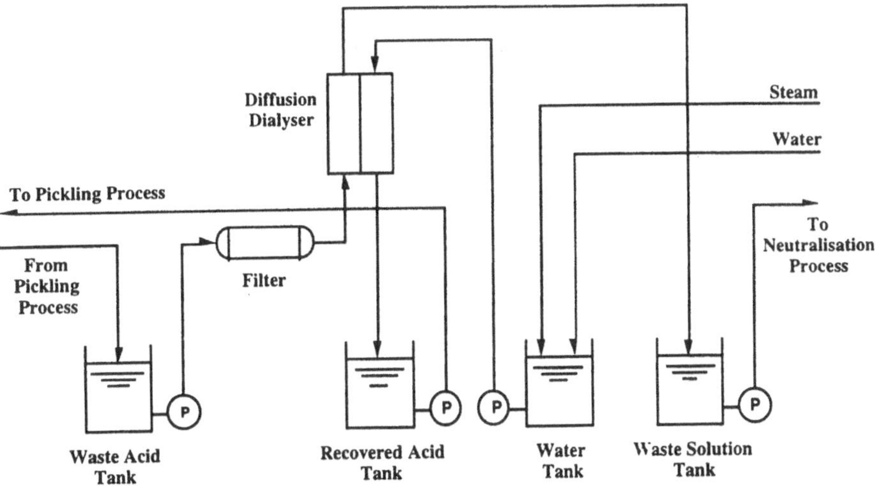

Figure 3. Flow diagram of diffusion dialysis [7]

ultrafiltration. In fact, the pumping costs are almost negligible, in view of the very low flow rates in the DD compartments (about 1 litre per hour per m^2 of membrane). The electricity costs associated with electrodialysis are also eliminated.

As with many ion exchange and membrane processes, there are certain concentration and composition ranges for which particular treatment methods are best suited. The recovery of waste acids is an application for which diffusion dialysis is often the best and most economical solution. Base recovery is relatively new and, for the moment, experience is limited to specific applications where special processes have led to very efficient systems and wider applications are probable as experience develops.

4 Acid Recovery

4.1 Operating plants

DD was originally developed in Japan. More than 50 diffusion dialysis plants for acid recovery have been constructed to date, ranging in capacity from 0.5 to 50 m^3/day of solution to be treated. Of the industrial plants in operation, the majority are being used for the treatment of pickling acid [8].

Diffusion dialysis is slowly becoming accepted in Europe, where plants for the recovery of H_2SO_4 from TiO_2 as well as for pickling acid recovery are in use or being built [9]. The recovery of sulfuric acid from titanium plants appeared to be of great interest a few years ago, but a substantial switch away from sulfuric acid leach systems has, at least temporarily, slowed down the growth of this application.

4.2 Acids suitable for treatment with DD

As a general guide-line, it can be stated that DD is suitable for the recovery of the following solutions:

• Types of acid:	Strong
• Free acid concentration:	1-7 Normal
• Temperature:	15-45°C
• Metals and/or other inorganic salts:	Almost any concentration, (as long as in solution)

The temperature range is an approximation. In fact, the higher the temperature, the better the diffusion, but the life of the membrane and other constituents of the stack which are mainly made of plastics and rubber, is reduced with increasing temperature.

Normally, filtration of the feed solution is required, to reduce the content of suspended solids reaching the DD system to below 1 ppm. Accordingly, a DD recovery system usually includes a pre-filtration arrangement which may be a simple gravity filter or may be somewhat more complex, comprising, for example, a gravity filter or filter press followed by a cartridge filter. In the case of solutions which are extremely difficult to filter (for example, TiO_2), carbon cartridge microfiltration has been successfully utilised.

4.3 Material balance

The nature of the process taking place inside the DD plant has been discussed above. To follow the process in detail, the best description is an actual material balance, which describes the quantities and composition of the streams entering and leaving the DD stack.

The material balance for a typical case of recovery of 1000 litres/hour of a mixture of nitric and hydrofluoric acid from a pickling bath is shown in Figure 4 [9].

For the purpose of this section, as well as Section 4.6, all data presented refer to the same quantity and composition of acid to be treated and to the same diffusion dialyser system.

4.4 Cost of operating a DD plant for acid recovery

Tables 1, 2 and 3 give the design basis, performance, running costs and savings for a DD plant to treat 1000 litres per hour of pickling acid as described above [9].

4.5 Space requirements for a diffusion dialysis plant

A DD stack suitable for the treatment of 1000 litres/hour of HNO_3/HF pickling acid as described in Section 4.1, has the following dimensions [8,9]: • Length: 6000mm, • Width: 730mm, • Height: 1900mm. The size of a complete plant, including filtration units, tanks, pumps, control equipment, etc. will depend upon numerous factors, such as the shape of the space available (in some cases a relatively high room of limited area will be available, in other cases another shape would be preferable). As a very rough guide-line, it can be stated that the following dimensions would be typical [8,9]:: • Length: 9000mm, • Width: 2000mm, • Height: 2000mm.

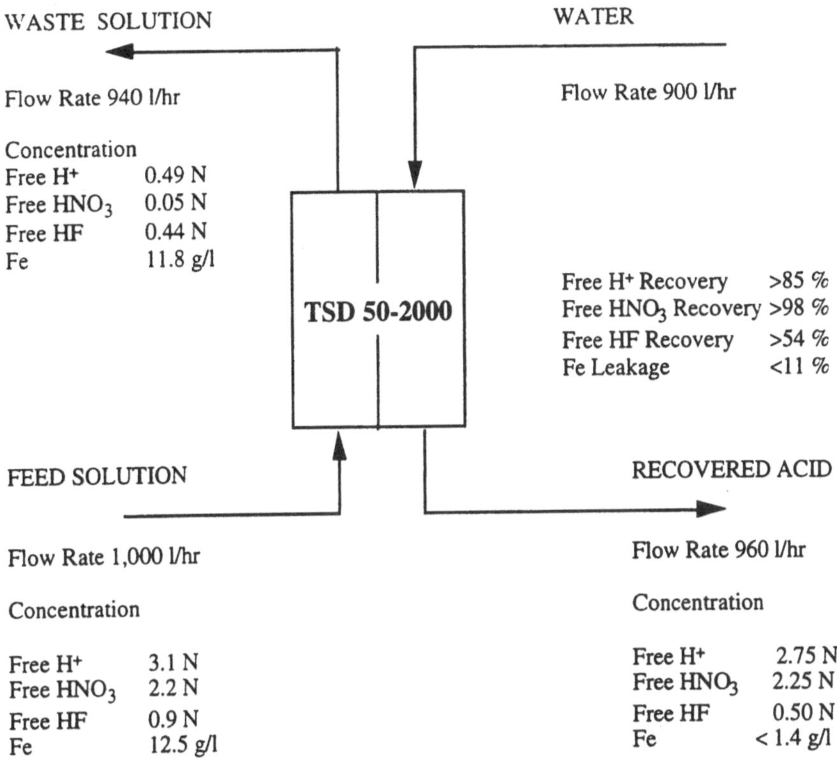

Figure 4. Mass balance for diffusion dialysis of nitric/hydrofluoric pickling acid [9]

Table 3. Breakdown of operating costs

Operation	Cost (FF*)
• Replacement of stack components (membranes, gaskets, spacers, etc.) Assuming a 5-year lifetime for such items	350,000
• Deionized water:	70,000
• Preheating of water, replacement of prefiltration cartridges, etc.:	80,000
• Total operating costs:	500,000

* FF - French Francs

4.6 Quantity and quality of acid obtained from the DD system

- Recovery of free acid: 80-90%
- Concentration of acid: >80-90% of influent concentration
- Residual metals: Usually 1-10 % for pickling acid.

However, with speciality membranes, e.g. Neosepta AHS-2 produced by Tokuyama Soda Company, maximum metals/salt concentrations below 10 ppm can be obtained. When such special membranes are used, the DD plant combines acid recovery with acid purification above the quality of fresh acid.

4.7 Waste-stream from the DD plant

While the recovery of acid by the DD plant is very good, there is a waste-stream containing the metals removed. However, it must be considered that, in the case of untreated waste acid, the entire acid volume has to be treated, contributing not only the metals, but also the acid and neutralising agent to the waste-stream. In the case of a DD recovery scheme, over 80% of the acid is recycled, leaving a solution of metals in a relatively low acid concentration requiring correspondingly low neutralising volumes.

Further concentration could be accomplished, for example, by placing an electrodialysis unit behind the diffusion dialyzer. This is, in fact, currently being considered in one of the major Japanese steel mills.

In any event, to further increase the effectiveness of the process from an environmental standpoint, suitable treatment of the waste stream from the DD plant could be added. Alternatively, a largely closed system such as described below for caustic recovery can possibly be developed for some applications.

5 Recovery of Sodium Hydroxide from Spent Aluminium Chemical Baths

5.1 Background

As mentioned above, the theory of caustic recovery by DD [10-13] is similar to that of acid recovery. The difference lies only in the membranes used which, of course, in the case of caustic recovery, are cation exchange membranes; these permit the passage of free base while blocking anions and salts.

From a practical standpoint, DD could not be used for caustic recovery in the past, due to the lack of suitable membranes. While ion exchange membranes stable in the presence of even very high concentrations of strong acids have been available for years, no membranes resistant to caustic existed. It is only recently that Tokuyama Soda Company have developed membranes with such stability.

Because of the relatively short time that such membranes have been available, the experience with caustic recovery is much more limited than that with acid recovery. Nevertheless, a process for the recovery of caustic from aluminium etching has been developed on an industrial scale. This process is very interesting, because it not only represents the first commercial scale DD waste treatment process for a caustic system, but produces no waste stream, being a closed cycle process except for the production of

$Al(OH)_3$, a marketable product.

5.2 Theory

Spent alkali etchant consists mainly of NaOH and aluminate metal ions in the form of $NaAlO_2$, which is formed by the solution of aluminium metal in NaOH. $NaAlO_2$ is an unstable, soluble salt that decomposes when pH is reduced by the removal or dilution of NaOH, according to the following equation:

$$NaAlO_2 + 2H_2O \Longleftrightarrow Al(OH)_3 + NaOH$$

Precipitation of $Al(OH)_3$ is slow; accordingly, DD can first be used to recover NaOH, with $Al(OH)_3$ being recovered downstream of the DD stack by precipitation.

5.3 DD process

The DD process operates in a similar manner described above for acid recovery. The alkaline etchant flows upwards, while water flows downwards; the two streams are separated by a cation exchange membrane permeable only by free bases and water. NaOH diffuses across the membrane from the high concentration in the etchant to the low concentration in the water. Water, on the other hand, migrates into the waste stream compartment by osmotic pressure.

The outlet product concentration can attain that of the inlet feed; in fact, it can slightly exceed this concentration, as a result of the common ion effect: a higher driving force is created by the total sodium ion concentration, including the sodium ions associated with the aluminium ions, thus making possible a base product more concentrated than the feed alkali. This resulting base product contains typically less than 1% aluminium and is suitable for recycling to the etching tank.

5.4 Crystallisation process

The characteristics of a sodium aluminate/sodium hydroxide solution are well known and are the basis of the Bayer Process. Upon dilution with water, the mixture composition can be moved into the supersaturated region of the mixture (below the curve) as depicted by the equilibrium data shown in Figure 5. Once supersaturated, a crystalline $Al(OH)_3$ product can be precipitated. To achieve a crystal that can easily precipitate, seeding, as well as a complex crystalliser are usually required.

The use of a DD stack to remove the free caustic from the etchant solution constitutes an entirely new discovery and adds a new dimension to this process. As the etchant is depleted of caustic, the equilibrium of the mixture is altered along a different trajectory, as shown in Figure 5, resulting in a more highly supersaturated solution. This gives rise to an enhancement of the nucleation of seed crystals in the mixture, enabling it to crystallise and precipitate consistently with the formation of fine pure white $Al(OH)_3$ crystals without the need for manual seeding. Mixture of almost any feed composition can be dialysed to produce a supersaturated solution that will easily crystallise.

The process operates as a continuous treatment of the etching solution to maintain a constant level of metal salt in the etching solution, removing an amount of aluminium equal to that introduced into the tank each day. On this design basis there is no need to recover 90-95% of caustic. (In acid recovery systems discussed above, maximum recovery of acid is desired). It is necessary to remove only as much caustic from the spent etchant as is required to shift the equilibrium of the mixture sufficiently to the supersaturated state to attain the desired precipitation. Once crystallisation occurs, the remaining solution is sufficiently free of aluminium to be suitable for recycling to the etch tank.

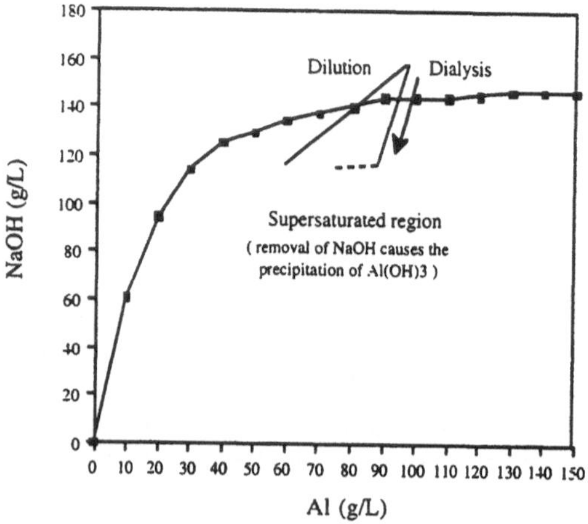

Figure 5. Saturation data of aluminium in sodium hydroxide

5.5 Overall process

Figure 6 depicts the entire process schematically. The etchant is filtered to remove smut, then sent to the DD for caustic recovery and simultaneous formation of a supersaturated etchant solution which is sent to a simple stirred tank crystalliser where $Al(OH)_3$ is crystallised and precipitated. The underflow from the stirred tank is a 35% slurry which is dewatered in a plate and frame type filter press. The recovered alkali and the supernatant solution from the crystalliser are returned to the etching tank for recycling. More than 99% of the caustic is recovered.

The precipitate from the crystalliser is pure white crystalline $Al(OH)_3$, which after washing and reduction of pH to 10 or below, is sold. (In USA, the current value of this product is around US $40/ton). The smut is essentially composed of CuS, a solid used as an additive to the milling bath to control the etching rate, and is returned to the etching tank.

The remaining product from the process is water, which can be eliminated by evaporation or which can be recycled as make-up water for the DD stack.

5.6 Field performance

A Tokuyama Soda TSD 10-300 stack, using 30m^2 of membrane, has been in use at

Caspian Chemicals in San Diego, California, for almost two years. The installation has recently been increased three-fold by the addition of a TSD 25-250 stack, using 62.5m^2 of membrane [11]. The TSD 10-300 stack has been processing 46.2 l/hr of spent etchant containing 16.5% NaOH and 10.6% aluminium with the following results:

Caustic product stream:
- 35% of free caustic removed from the solution.
- 17% NaOH produced at a rate of 15.5 l/hr.

Treated salt product stream:
- 9% NaOH and 8% aluminium at a rate of 59.4 l/hr.

Over 100 tons of $Al(OH)_3$ have been produced and sold by late 1993. Major savings as a result of reduced fresh caustic purchases and virtual elimination of the need to dispose of spent caustic have been additional benefits of the process.

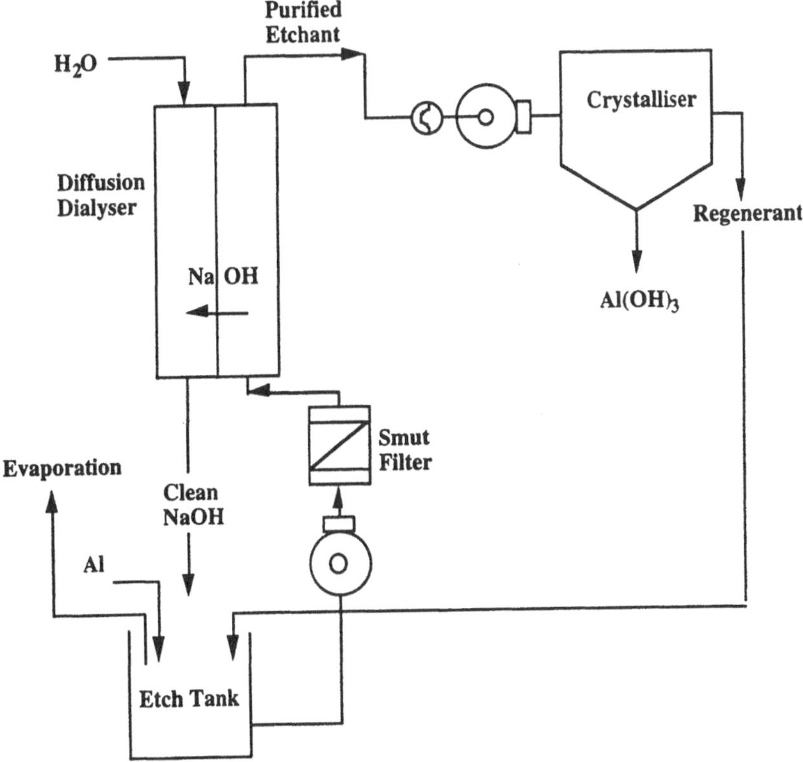

Figure 6. Diffusion dialysis etchant recovery system

5.7 Material balance

Table 4 presents a typical material balance. The crystalliser treats about 1.4 tons/week of 30% (per volume) slurry. This is sent to a filter press for dewatering and washing. Table 5 depicts the composition of the $Al(OH)_3$ cake produced.

Table 4. Base diffusion dialysis material balance

Input/ Output	Mass (%)	Volume/Mass of product
• Etchant feed		1.11 m^3/d
NaOH	16.5	
Al	10.6	
• Water		0.69 m^3/d
NaOH	1.0	
• Base product from DD		0.37 m^3/d
NaOH	17.0	
Al	3.0	
• Depleted etchant from DD		1.43 m^3/d
NaOH	9.0	
Al	8.0	
• Crystalliser supernate		1.32 m^3/d
NaOH	14.0	
Al	3.8	
• Cake $Al(OH)_3$ 100% basis		188 kg 45 wt% cake 70 wt% cake (when blown dry)
• Blended base product		1.69 m^3/d
NaOH	15.0	
Al	3.6	

Table 5. Recovered $Al(OH)_3$ by-product analysis

Component	Content (%)
Aluminium	25
Moisture	13
Sodium	0.002
Iron	0.0002
Copper	0.01
Other Metals	0.4
pH	8.5

5.8 Investment and operating costs

Table 6 depicts investment and operating costs of a complete plant capable of the treatment of 1670 m^3 of aluminium etchant per year.

Under the conditions described, the payback period for such a plant is less than two years. When coupled with smoother operation of the entire etching plant, better quality control due to consistent quality of etchant and environmental improvements, the process is obviously a very desirable one.

Table 6. Diffusion dialysis etchant recovery system cost benefit analysis

• Capital cost, $4Q93	Including smut filtration, diffusion dialysis - skid mounted, crystallisation, solids handling	$760,000
• Annual operating costs, $/year	Labour	$50,000
	Maintenance @ 8.5%TDC	$40,000
	Membrane/ Technical service	$100,000
	Power, 37KW	$15,000
	Total annual costs, $/year	$205,000
• Annual operating credits, $/year	Disposal credit (440,000 g/y)	$440,000
	Caustic recovery @$400/Ton	$160,000
	$Al(OH)_3$ cake sales $40/Ton	$12,000
	Total operating credits, $/year	$612,000
• Net credits per year		$407,000
• Simple pay back		~2 years

6 Conclusion

DD is a proven technique for acid recovery and should be considered whenever any waste acid is to be treated.

DD has now also established itself for use in caustic recovery applications. A waste stream free system for the treatment of aluminium etchants has been developed and is operating satisfactorily on an industrial scale.

Further studies concerning both acid and base recovery, including additional processing of waste streams remaining after DD treatment and/or the development of suitable waste-stream free systems where possible, will undoubtedly further promote the acceptance of DD as a standard technology.

7 References

1. Noma, Y. (1989) Recovery of Metal Ions and Acid by Ion-Exchange Membranes, *Extraction Metallurgy '89*, 1081-1105.
2. Use of Dialysis in Metal Finishing Process, Technical Bulletin, Asahi Glass Co.
3. Kobuchi, Y., Motomura, H., Noma, Y. and Hanada, F. (1984) Application of the Ion Exchange Membranes/Acid Recovery by Diffusion Dialysis, Paper presented at Europe-Japan Congr. on Membranes and Membrane Processes, Stresa, Italy.
4. Kawate, H., Tsuzura, K. and Shimizu, H. (1991) Chapter 1.10, *Ion Exchangers* ed. Dorfner, K., de Gruyter, Berlin, New York.
5. Recovery of Acid in a Process of Plating - Diffusion Dialysis Process, Technical Bulletin, Tokuyama Soda Co. Tokyo, Japan.
6. Sato, J., Onuma, M., Motomura, H., Noma, Y. (1984) Recovery of Nitric Acid and Hydrofluoric Acid from the Pickling Solution by Diffusion Dialysis, Jitsumu Hyomen Gijitsu (The Metal Finishing Society of Japan).
7. Diffusion Dialyzer for Acid Recovery, Technical Bulletin, Tokuyama Soda Co.
8. Tokuyama Soda Co., Private Communication.
9. Eurodia, Private Communication.
10. Herz, G. (1991) Treatment of Pickling Acids with Ion Exchange and Related Processes, Chapter 2.9, *Ion Exchangers*, ed. Dorfner, K., de Gruyter, Berlin, New York.
11. Byszewski, C. and Jaffari M. (1993) Diffusion Dialysis for the Recovery of Aluminium from Caustic Etchant / Milling Solutions, *Separation Science & Technology*.
12. Davis, T.A. (1991) Recovery of Sodium Hydroxide and Aluminium Hydroxide from Etching Wastes: U.S. Patent 5,049,233. The Graver Company.
13. Jaffari, M. and Byszewski, C. (1992) New Membrane-Based Caustic Recycle Process: Paper presented at the *7th Annual Aerospace Hazardous Materials Management Conference*, Sept.

Hydrometallurgy '94

Hydrometallurgy '94

Papers presented at the international
symposium
'Hydrometallurgy '94'
organized by the Institution of Mining
and Metallurgy
and the Society of Chemical Industry,
and held in Cambridge, England,
from 11 to 15 July, 1994

SCI

Published for the Institution of Mining and Metallurgy
and the Society of Chemical Industry
by Chapman & Hall

SPRINGER-SCIENCE+BUSINESS MEDIA, B.V.

First edition 1994

Originally published by Chapman & Hall in 1994
Softcover reprint of the hardcover 1st edition 1994
ISBN 978-94-010-4532-2 ISBN 978-94-011-1214-7 (eBook)
DOI 10.1007/978-94-011-1214-7

A catalogue record for this book is available from the British Library

♾ Printed on permanent acid-free text paper, manufactured in accordance with ANSI/NISO Z39.48-1992 and ANSI/NISO Z39.48-1984 (Permanence of Paper).

Contents

Organizing committee

Dr A.J. Monhemius *(Chairman)*
Dr D.S. Flett
Dr N.M. Rice
Dr T.V. Arden
Dr R.F. Dalton
Professor D.J. Fray
D. Gosden
D. Naden
Dr C.V. Phillips
K.J. Severs

List of Sponsors

Generous support of the conference by the following organizations is acknowledged with gratitude:

Bechtel Limited
BHP Minerals International Inc.
Davy International
Outokumpu Metals & Resources International
Royston Lead Plc
RTZ Limited
Solvay Interox Ltd.
Techpro Mining & Metallurgy Limited
Wenmec Systems, Inc.
ZENECA Limited

Foreword

'Hydrometallurgy '94' is the fourth in the series of international conferences on hydrometallurgy that started in 1975 in Manchester. The preceding two conferences in the series, held in 1981 and 1987, respectively, were organized by the Solvent Extraction and Ion Exchange Group of the Society of Chemical Industry (SCI): this group also initiated the organization of 'Hydrometallurgy '94'. Following preliminary discussions about the scope of the conference by the SX–IX Group committee some two and a half years ago, however, it was soon concluded that to cover adequately the current field of hydrometallurgy and its associated disciplines it would be necessary to broaden the scope of the subject matter compared with the earlier conferences. Additionally, owing to an overlap of interests in the field of hydrometallurgy, it was felt appropriate to approach the Institution of Mining and Metallurgy (IMM) with a proposal for joint organization of the conference. IMM responded swiftly and positively to the suggestion and a committee that comprised members of both bodies was assembled to begin the task of organizing a meeting that would equal and, it was hoped, exceed the high standards that had been set by previous 'Hydrometallurgy' conferences.

Since the 'Earth Summit' in Rio the concept of sustainable development has been much in vogue. The associated ideas of cleaner technology, recycling and waste minimization have particular relevance to the extraction and processing of metals and other mineral products. The scientific principles of inorganic and physical chemistry on which are based most of the techniques and processes that are used in hydrometallurgy are precisely those which have to be employed to clean up the excesses of the past, to treat the effluents of today and to design the cleaner processes of the future. Thus, the separation of ionic species in solution by selective precipitation, ion exchange or solvent extraction—techniques that are very familiar to the hydrometallurgist—can be readily adapted to the treatment of industrial effluents and other waste waters containing toxic metals and other undesirable solutes. The thermodynamic principles that are used to measure and quantify the relative stabilities and instabilities of phases, solid, liquid and gaseous, and which, for the hydrometallurgist, are most visibly embodied in the ubiquitous

Eh–pH diagrams, are just those on which judgements have to be based about the environmental compatibility and stability of process wastes destined for long-term disposal.

Thus, it seemed to the Organizing Committee to be entirely appropriate to try to reflect the broad applicability of the principles and processes of the discipline by giving 'Hydrometallurgy '94' the subtitle, 'Environmentally Sustainable Technology'. The first circular and initial publicity, in which these ideas were put forward, seem to have struck a chord with workers in the field, as they resulted in well in excess of 150 abstracts being submitted for consideration. The Organizing Committee, though very gratified by this excellent response, was then faced with a dilemma: 'Hydrometallurgy' conferences have traditionally been run in single session so that all delegates could attend the whole conference. This format, however, severely restricts the number of papers that can be accommodated. The alternative—to go to parallel or multiple sessions to increase the number of papers presented—would fragment the audience and lose the intimacy and solidarity of interest that have characterized previous conferences in the series.

Eventually, a compromise was reached: the major part of the conference would retain the single-session format, but there would be some parallel oral sessions, plus a major poster paper session, in order to increase the number of papers that could be accepted. In spite of these changes, the Organizing Committee had the unenviable task of selecting for eventual presentation at the conference no more than half of the abstracts submitted. The results of this process are contained in this book, comprising 78 papers by authors from 30 countries, which we believe presents a comprehensive picture of the current state-of-the-art and future trends in the technology of hydrometallurgy and its rapidly expanding role in the field of environmental engineering.

For their help in bringing all this to fruition I am very grateful to my colleagues on the Organizing Committee here in London for their hard work, particularly in refereeing the papers. The members of the overseas advisory board have also made an important contribution to the event by soliciting support and providing publicity for the conference in their own countries or regions.

'Hydrometallurgy '94' is the first major conference for which the IMM and the SCI have collaborated in joint sponsorship. The division of responsibilities between the two Conference offices was clearly defined by agreement at the outset, IMM taking responsibility for the editing, refereeing and production of the proceedings, whereas SCI is dealing with the organization and management of the conference itself. To date, this arrangement has worked extremely smoothly and I wish to pay tribute to the dedicated efforts of the staff of both organizations. This experience augurs well for future collaboration between SCI and IMM, which have a number of areas of common interest.

Another major difference between 'Hydrometallurgy '94' and previous meetings in the series is the change of venue to Cambridge. We hope that, in choosing Churchill College, we have provided a setting of tranquillity that will enable delegates to obtain maximum benefits from the high quality of the papers that are being presented and from the company of their colleagues from all over the world. The picturesque town of Cambridge provides a wide choice of historic settings for the social events, which we hope will help to make 'Hydrometallurgy '94' a memorable event.

Finally, I wish to record two important votes of thanks from the Organizing Committee: first, to the sponsoring companies, listed elsewhere in this volume, whose generous donations have enabled us to provide first-class social events while still keeping the registration fees to reasonable levels; and, second, to all authors for their contributions to this volume and to the conference—your work has made ours worthwhile.

Dr. A. J. Monhemius
Chairman, Organizing Committee
London, April, 1994

Contributors*

L. M. Abrantes (409), Departamento de Química, Faculdade de Ciências de Lisboa, Lisbon, Portugal
Marcela Achimovičová (209), Institute of Geotechnics of the Slovak Academy of Sciences, Košice, Slovakia
Katerina Adam (291), Laboratory of Metallurgy, National Technical University of Athens, Athens, Greece
Alain Adjemian (3), Directorate General for Science, Research and Development, European Commission, Brussels, Belgium
S. Agatzini-Leonardou (193), Department of Mining and Metallurgical Engineering, National Technical University of Athens, Athens, Greece
M. Aguilar (725), Chemical Engineering Department, Universitat Politècnica de Catalunya, Barcelona, Spain
F. J. Alguacil (939), Centro Nacional de Investigaciones Metalúrgicas, Madrid, Spain
J. Alstad (701), Department of Chemistry, University of Oslo, Oslo, Norway
A. Anacleto (579), Department of Chemical Engineering, Instituto Superior Técnico, Technical University of Lisbon, Lisbon, Portugal
M. T. Anthony (13), St. Barbara Consultancy Services, Essex, England
Keith Atkinson (1011), Camborne School of Mines, University of Exeter, Redruth, Cornwall, England
A. A. Bagreev (517), Institute for Sorption and Problems of Endoecology, Academy of Sciences of Ukraine, Kiev, Ukraine
W. Bahl (869), Ruhr-Zink, Datteln, Germany
Peter Baláž (209), Institute of Geotechnics of the Slovak Academy of Sciences, Košice, Slovakia
A. Ballester (369), Departamento de Ciencia de los Materiales e Ingeniería Metalúrgica, Facultad de Ciencias Químicas, Universidad Complutense de Madrid, Madrid, Spain
K. Barbetti (253), A. J. Parker Cooperative Research Centre for Hydrometallurgy, Chemistry Centre (WA), Department of Minerals

*Initial page number(s) of authors' contributions given in brackets.

and Energy, Bentley, Western Australia
J. P. Barbosa (527), CETEM—Centre for Mineral Technology, Cidade Universitária, Rio de Janeiro, Brazil
O. Barbosa-Filho (425), Department of Materials Science and Metallurgy, Catholic University, Rio de Janeiro, Brazil
D. W. Barr (337), Advanced Technical Development, Bundoora, Victoria, Australia
Denise Bauer (675), Laboratoire de Chimie Analytique associé au CNRS, Ecole Supérieure de Physique et de Chimie Industrielles, Paris, France
E. Ben-Yoseph (683), Israel Chemicals, Ltd., IMI Institute for Research and Development, Haifa Bay, Israel
B. C. Blakey (159), Department of Chemical Engineering and Applied Chemistry, University of Toronto, Toronto, Canada
M. L. Blázquez (369), Departamento de Ciencia de los Materiales e Ingeniería Metalúrgica, Facultad de Ciencias Químicas, Universidad Complutense de Madrid, Madrid, Spain
D. V. Boger (971), Advanced Mineral Products Centre, Department of Chemical Engineering, University of Melbourne, Victoria, Australia
C. F. Bonney (313), Mineral Industry Research Organisation, Lichfield, England
Kjetil Børve (563), Norzink AS, Odda, Norway
B. S. Boyanov (859), Department of Chemistry, University of Plovdiv, Plovdiv, Bulgaria
I. Bustero (655), INASMET, San Sebastián, Spain
C. Byszewski (613), Graver Water, Union, New Jersey, U.S.A.
S. A. Cale (949), Knight Piésold and Partners, Ashford, Kent, England
C. Caravaca (939), Centro Nacional de Investigaciones Metalúrgicas, Madrid, Spain (presently, Department of Mineral Resources Engineering, Royal School of Mines, Imperial College of Science, Technology and Medicine, London, England)
J. M. R. de Carvalho (579), Department of Chemical Engineering, Instituto Superior Técnico, Technical University of Lisbon, Lisbon, Portugal
J. F. Castle (229), RTZ Consultants, Ltd., Bristol, England
A. M. Chekmarev (219), D. Mendeleev University of Chemical Technology of Russia, Moscow, Russia
Chen Jiayong (541), Institute of Chemical Metallurgy, Chinese Academy of Sciences, Beijing, China
T. T. Chen (125), CANMET, Ottawa, Canada
Y. Cheng (655), Department of Chemistry, University of Liverpool, Liverpool, England
André Chesné (635), Laboratoire de Chimie Nucléaire et Industrielle,

Ecole Centrale Paris, Châtenay-Malabry, France
S. V. Chizhevskaya (219), D. Mendeleev University of Chemical Technology of Russia, Moscow, Russia
Y. Choi (711), Delft University of Technology, Faculty of Mining and Petroleum Engineering, Department of Raw Materials Technology, Delft, The Netherlands
R. Cierpiszewski (675), Academy of Economics, Poznan, Poland
M. J. Collins (869), Sherritt, Inc., Fort Saskatchewan, Canada
J. L. Cortina (725), Chemical Engineering Department, Universitat Politècnica de Catalunya, Barcelona, Spain
Paz Cosmen (741), CIEMAT, Madrid, Spain
M. C. Costa (409), Departamento de Química, Faculdade de Ciências de Lisboa, Lisbon, Portugal
Gérard Cote (675), Laboratoire de Chimie Analytique associé au CNRS, Ecole Supérieure de Physique et de Chimie Industrielles, Paris, France
M. Cox (219), Division of Chemical Sciences, University of Hertfordshire, Hatfield, England
D. C. Cupertino (591), ZENECA Specialties, Blackley, Manchester, England
S. A. Curran (441), Dublin Institute of Technology, Dublin, Ireland
I. Dalrymple (1075), Environmental Division, E. A. Technology, Capenhurst, Chester, England
R. F. Dalton (601), ZENECA Specialties, Blackley, Manchester, England
R. P. Das (253), Regional Research Laboratory, Bhubaneswar, India
Francisco Delmas (1075), Materials Department, Instituto Nacional de Engenharia e Tecnologia Industrial, Lumiar, Lisbon, Portugal
J. S. J. van Deventer (483, 501), Department of Metallurgical Engineering, University of Stellenbosch, Stellenbosch, South Africa
D. Dimaki (193), Department of Mining and Metallurgical Engineering, National Technical University of Athens, Athens, Greece
O. N. Dimitropoulou (463), Laboratory of Metallurgy, National Technical University of Athens, Athens, Greece
R. I. Dimitrov (859), Department of Chemistry, University of Plovdiv, Plovdiv, Bulgaria
J. Doyle (1035), CRA—Advanced Technical Development, Bundoora, Victoria, Australia
A. W. L. Dudeney (325), Department of Mineral Resources Engineering, Royal School of Mines, Imperial College of Science, Technology and Medicine, London, England
Gérard Durand (635, 655), Laboratoire de Chimie Nucléaire et Industrielle, Ecole Centrale Paris, Châtenay-Malabry, France

J. E. Dutrizac (125), CANMET, Ottawa, Canada
Saskia Duyvesteyn (887), University of California, Berkeley, California, U.S.A.
W. P. C. Duyvesteyn (887), The Minerals Laboratory, BHP Minerals, Reno, Nevada, U.S.A.
D. R. East (961), Knight Piésold and Co., Denver, Colorado, U.S.A.
S. H. Eberle (767), Kernforschungszentrum Karlsruhe, Institute for Radiochemistry, Water Technology Division, Karlsruhe, Germany
S. Elinson (837), Institute of Physical Organic Chemistry of the Belarus Academy of Sciences, Minsk, Belarus
Blanca Escobar Miguel (385), Departmento de Ingeniería Química, University of Chile, Santiago, Chile
P. K. Everett (913), Intec Pty., Ltd., Chatswood, New South Wales, Australia
Fang Zhaoheng (541), Institute of Chemical Metallurgy, Chinese Academy of Sciences, Beijing, China
L. H. Filipek (961), Knight Piésold and Co., Denver, Colorado, U.S.A.
N. P. Finkelstein (683), Israel Chemicals, Ltd., IMI Institute for Research and Development, Haifa Bay, Israel
D. S. Flett (13), St. Barbara Consultancy Services, Essex, England
B. F. Foley (441), Dublin Institute of Technology, Dublin, Ireland
S. A. Foster (795), Arthur D. Little, Inc., Cambridge, Massachusetts, U.S.A.
Matthias Franzreb (767), Kernforschungszentrum Karlsruhe, Institute for Radiochemistry, Water Technology Division, Karlsruhe, Germany
G. Friedman (1059), Israel Chemicals, Ltd., IMI Institute for Research and Development, Haifa Bay, Israel
A. R. Gee (325), Formerly, Department of Mineral Resources Engineering, Royal School of Mines, Imperial College of Science, Technology and Medicine, London, England (now BHP Minerals, Mayfield, New South Wales, Australia)
D. Gilroy (655), E. A. Technology, Capenhurst, Chester, England
Inés Godoy Ríos (385), Departmento de Ingeniería Química, University of Chile, Santiago, Chile
C. Gómez (369), Departamento de Ciencia de los Materiales e Ingeniería Metalúrgica, Facultad de Ciencias Químicas, Universidad Complutense de Madrid, Madrid, Spain
F. González (369), Departamento de Ciencia de los Materiales e Ingeniería Metalúrgica, Facultad de Ciencias Químicas, Universidad Complutense de Madrid, Madrid, Spain
J. T. Gormley (777), Knight Piésold and Co., Denver, Colorado, U.S.A.
M. D. Green (971), Advanced Mineral Products Centre, Department of

Chemical Engineering, University of Melbourne, Victoria, Australia
R. J. Grolman (1087), E. P. & P., Chicoutimi, Quebec, Canada
N. J. de Guingand (971), Advanced Mineral Products Centre, Department of Chemical Engineering, University of Melbourne, Victoria, Australia
J. J. Gusek (777), Knight Piésold and Co., Denver, Colorado, U.S.A.
C. J. Haigh (1035), Charlestown, New South Wales, Australia
A. K. Haines (27), Minerals Technology, Gencor, Ltd., Johannesburg, South Africa
R. M. Hamilton (795), National Rivers Authority, Exeter, England
L. J. F. Harris (922), Process Systems and Safety Department, Babcock King–Wilkinson, Ltd. (formerly, Babcock Contractors, Ltd.), Crawley, England
G. P. Herz (613), Tokuyama Soda Company, Tokyo, Japan
J. B. Hiskey (43), Materials Science and Engineering Department, University of Arizona, Tucson, Arizona, U.S.A.
E. M. Ho (1105), A. J. Parker Cooperative Research Centre in Hydrometallurgy, Murdoch University, Perth, Western Australia
J. E. Hoffmann (69), Jan H. Reimers and Associates, Houston, Texas, U.S.A.
W. H. Höll (767, 983), Kernforschungszentrum Karlsruhe, Institute for Radiochemistry, Water Technology Division, Karlsruhe, Germany
G. C. Holywell (1087), Alcan International, Ltd., Kingston Research and Development Centre, Kingston, Ontario, Canada
J. S. Jackson (591), ASARCO, Inc., Salt Lake City, Utah, U.S.A.
M. Jaffari (613), Malek, Inc., San Diego, California, U.S.A.
A. Jakubiak (675), Institute of Chemical Technology and Engineering, Poznan Technical University, Poznan, Poland
M. A. Jordan (337), Camborne School of Mines, Faculty of Engineering, University of Exeter, Redruth, Cornwall, England
Diego Juan (1123), Department of Chemical Engineering, University of Murcia, Cartagena, Spain
K-P. Jüngst (767), Kernforschungszentrum Karlsruhe, Institute for Technical Physics, Karlsruhe, Germany
Roland Kammel (209), Institute of Metallurgy, Technical University, Berlin, Germany
P. F. Kavanagh (441), Dublin Institute of Technology, Dublin, Ireland
Ke Jia-Jun (807), Institute of Chemical Metallurgy, Academia Sinica, Beijing, China
F. M. Kimmerle (1087), Alcan International, Ltd., Arvida Research and Development Centre, Jonquière, Quebec, Canada
M. G. King (591), ASARCO, Inc., Salt Lake City, Utah, U.S.A.
Chris Kiranoudis (463), Laboratory of Process Analysis and Design,

National Technical University of Athens, Athens, Greece
O. M. Klimenko (219) , D. Mendeleev University of Chemical Technology of Russia, Moscow, Russia
L. Kogan (1059) , Israel Chemicals, Ltd., IMI Institute for Research and Development, Haifa Bay, Israel
Constantine Komnitsas (291, 351, 361) , Laboratory of Metallurgy, National Technical University of Athens, Athens, Greece
Antonios Kontopoulos (291, 463) , Laboratory of Metallurgy, National Technical University of Athens, Athens, Greece
Pertti Koukkari (139) , Kemira Oy, Helsinki, Finland
Theodoros Kritikos (463) , Laboratory of Process Analysis and Design, National Technical University of Athens, Athens, Greece
Mária Kušnierová (209) , Institute of Geotechnics of the Slovak Academy of Sciences, Košice, Slovakia
K. A. Kydros (547) , Department of Chemistry, Aristotle University, Thessaloniki, Greece
J. Kyle (1105) , A. J. Parker Cooperative Research Centre in Hydrometallurgy, Murdoch University, Perth, Western Australia
S. Lallenec (1105) , A. J. Parker Cooperative Research Centre in Hydrometallurgy, Murdoch University, Perth, Western Australia
A. A. Latre (1025) , Instituto de Beneficio de Minerales, Facultad de Ingeniería, Salta, Argentina
P. J. Leggo (815), Environmental Minerals (U.K.), Linton, Cambridge, England
Li Deqian (627) , Changchun Institute of Applied Chemistry, Changchun, China
H. Liao (159) , Department of Chemical Engineering and Applied Chemistry, University of Toronto, Toronto, Canada
H. T. Lieuw (711) , Delft University of Technology, Faculty of Mining and Petroleum Engineering, Department of Raw Materials Technology, Delft, The Netherlands
Houyuan Liu (887) , The Minerals Laboratory, BHP Minerals, Reno, Nevada, U.S.A.
L. Lorenzen (483) , Department of Metallurgical Engineering, University of Stellenbosch, Stellenbosch, South Africa
Ma Gengxiang (627) , Changchun Institute of Applied Chemistry, Changchun, China
M. Makwana (869) , Sherritt, Inc., Fort Saskatchewan, Canada
Dimitrios Marinos-Kouris (463), Laboratory of Process Analysis and Design, National Technical University of Athens, Athens, Greece
Zacharias Maroulis (463), Laboratory of Process Analysis and Design, National Technical University of Athens, Athens, Greece
S. Martínez (939), Centro Nacional de Investigaciones Metalúrgicas,

Madrid, Spain
I. M. Masters (869), Sherritt, Inc., Fort Saskatchewan, Canada
K. A. Matis (547), Department of Chemistry, Aristotle University, Thessaloniki, Greece
M. McClaren (993), Terra Gaia Environmental Group, Vancouver, British Columbia, Canada
C. F. McDonogh (825), Solvay Interox Research and Development, Widnes, England
R. O. McElroy (993), Fluor Daniel Wright, Vancouver, British Columbia, Canada
J. McLoughlin (441), Dublin Institute of Technology, Dublin, Ireland
C. McNamee (441), Dublin Institute of Technology, Dublin, Ireland
Meng Shulan (627), Changchun Institute of Applied Chemistry, Changchun, China
M. T. van Meersbergen (483), Department of Metallurgical Engineering, University of Stellenbosch, Stellenbosch, South Africa
Isabelle Michelet (755), Laboratoire de Chimie Nucléaire et Industrielle, Ecole Centrale Paris, Châtenay-Malabry, France
J. L. Mier (369), Departamento de Ciencia de los Materiales e Ingeniería Metalúrgica, Facultad de Ciencias Químicas, Universidad Complutense de Madrid, Madrid, Spain
N. Miralles (725), Chemical Engineering Department, Universitat Politècnica de Catalunya, Barcelona, Spain
P. B. Mitchell (1011), Camborne School of Mines, University of Exeter, Redruth, Cornwall, England
A. J. Monhemius (177, 425), Department of Mineral Resources Engineering, Royal School of Mines, Imperial College of Science, Technology and Medicine, London, England
Pedro Moya (395), Departmento de Ingeniería Química, University of Chile, Santiago, Chile
J. C. Mugica (655), INASMET, San Sebastián, Spain
D. M. Muir (1105), A. J. Parker Cooperative Research Centre in Hydrometallurgy, Murdoch University, Perth, Western Australia
K. Mullins (441), Minmet Plc, Dublin, Ireland
Bosko Nikov (1153), "Zletovo" Metallurgical and Chemical Company, Titov Veles, Former Yugoslav Republic of Macedonia
Carlos Nogueira (1075), Materials Department, Instituto Nacional de Engenharia e Tecnologia Industrial, Lumiar, Lisbon, Portugal
Terje Østvold (563), Institute of Inorganic Chemistry, NTH, University of Trondheim, Trondheim, Norway
P. Owens (441), Dublin Institute of Technology, Dublin, Ireland
E. Ozberk (869), Sherritt, Inc., Fort Saskatchewan, Canada
A. P. Paiva (409), Departamento de Química, Faculdade de Ciências

de Lisboa, Lisbon, Portugal
V. G. Papangelakis (159), Department of Chemical Engineering and Applied Chemistry, University of Toronto, Toronto, Canada
N. L. Papassiopi (291, 463), Laboratory of Metallurgy, National Technical University of Athens, Athens, Greece
Dominique Pareau (635), Laboratoire de Chimie Nucléaire et Industrielle, Ecole Centrale Paris, Châtenay-Malabry, France
J. Parkes (1075), Faraday Centre, Carlow, Ireland
Ioannis Paspaliaris (463), Laboratory of Metallurgy, National Technical University of Athens, Athens, Greece
Andrés Perales (1123), Department of Chemical Engineering, University of Murcia, Cartagena, Spain
F. W. Petersen (501), Cape Technikon, Cape Town, South Africa
C. V. Phillips (337), Camborne School of Mines, Faculty of Engineering, University of Exeter, Redruth, Cornwall, England
N. L. Piret (229), Stolberg Consult GmbH, Neuss, Germany
R. E. Pocovi (1025), Instituto de Beneficio de Minerales, Facultad de Ingeniería, Salta, Argentina
F. D. Pooley (291, 351, 361), University of Wales, Cardiff, Wales
N. A. Postlethwaite (795), Marcus Hodges Environmental, Ltd., Exeter, England
M. V. Povetkina (219), D. Mendeleev University of Chemical Technology of Russia, Moscow, Russia
R. Püllenberg (869), Ruhr-Zink, Datteln, Germany
P. M. Quan (601), ZENECA Specialties, Blackley, Manchester, England
Mohamed Rakib (755), Laboratoire de Chimie Nucléaire et Industrielle, Ecole Centrale Paris, Châtenay-Malabry, France
I. Raz (1059), Israel Chemicals, Ltd., IMI Institute for Research and Development, Haifa Bay, Israel
G. V. Reznik (517), Institute for Sorption and Problems of Endoecology, Academy of Sciences of Ukraine, Kiev, Ukraine
N. M. Rice (273), Department of Mining and Mineral Engineering, University of Leeds, Leeds, England
J. J. Robinson (253), A. J. Parker Cooperative Research Centre for Hydrometallurgy, Chemistry Centre (WA), Department of Minerals and Energy, Bentley, Western Australia
P. C. P. Rocha (527), CETEM—Centre for Mineral Technology, Cidade Universitária, Rio de Janeiro, Brazil
E. G. Roche (1035), Pasminco Research Centre, Boolaroo, New South Wales, Australia
J. F. Rodríguez (939), Estaños de Zamora S.A., Villaralbo, Zamora, Spain

M. L. Ruiz (741), CIEMAT, Madrid, Spain
Mohammed Samar (635), Laboratoire de Chimie Nucléaire et Industrielle, Ecole Centrale Paris, Châtenay-Malabry, France
Angel Sanhueza (395), Departmento de Ingeniería Química, University of Chile, Santiago, Chile
K. L. Sandvik (1049), Department of Geology and Mineral Engineering, NTH, University of Trondheim, Trondheim, Norway
A. M. Sastre (725), Chemical Engineering Department, Universitat Politècnica de Catalunya, Barcelona, Spain
J. W. Scheetz (777), Brewer Gold Company, Jefferson, South Carolina, U.S.A.
D. J. Schiffrin (655), Department of Chemistry, University of Liverpool, Liverpool, England
A. A. Shunkevich (837), Institute of Physical Organic Chemistry of the Belarus Academy of Sciences, Minsk, Belarus
F. A. Silva (655), Department of Chemistry, University of Porto, Porto, Portugal
O. A. Sinegribova (219), D. Mendeleev University of Chemical Technology of Russia, Moscow, Russia
Hannu Sippola (139), GEM Systems Oy, Espoo, Finland
O. A. Skaf (1025), Instituto de Beneficio de Minerales, Facultad de Ingeniería, Salta, Argentina
V. S. Soldatov (837), Institute of Physical Organic Chemistry of the Belarus Academy of Sciences, Minsk, Belarus
Song Wenzhong (627), Changchun Institute of Applied Chemistry, Changchun, China
Tomislav Stojadinovic (1153), "Zletovo" Metallurgical and Chemical Company, Titov Veles, Former Yugoslav Republic of Macedonia
Petar Stojanov (1153), "Zletovo" Metallurgical and Chemical Company, Titov Veles, Former Yugoslav Republic of Macedonia
V. V. Strelko (517), Institute for Sorption and Problems of Endoecology, Academy of Sciences of Ukraine, Kiev, Ukraine
Anna Sundquist (139), Kemira Oy, Helsinki, Finland
Trygve Sverreson (1049), NOAH Langøya A/S, Holmestrand, Norway
P. M. Swash (177), MIRO Arsenic Research Group, Department of Mineral Resources Engineering, Royal School of Mines, Imperial College of Science, Technology and Medicine, London, England
J. Szymanowski (675), Institute of Chemical Technology and Engineering, Poznan Technical University, Poznan, Poland
S. Tamburini (655), Istituto di Chimica e Tecnologia dei Radioelementi, Padua, Italy
Yu. A. Tarasenko (517), Institute for Sorption and Problems of Endoecology, Academy of Sciences of Ukraine, Kiev, Ukraine

P. A. Tasker (591), ZENECA Specialties, Blackley, Manchester, England
N. Taylor (273), School of Chemistry, University of Leeds, Leeds, England
N. P. Tidy (291), University of Wales, Cardiff, Wales
R. B. E. Trindade (527), CETEM—Centre for Mineral Technology, Cidade Universitária, Rio de Janeiro, Brazil
S. D. Ukeles (683, 1059), Israel Chemicals, Ltd., IMI Institute for Research and Development, Haifa Bay, Israel
G. Van Weert (711), Delft University of Technology, Faculty of Mining and Petroleum Engineering, Department of Raw Materials Technology, Delft, The Netherlands
Tomás Vargas (395), Departmento de Ingeniería Química, University of Chile, Santiago, Chile
S. Vigato (655), Istituto di Chimica e Tecnologia dei Radioelementi, Padua, Italy
R. C. Villas Bôas (107), CETEM—Centre for Mineral Technology, Cidade Universitária, Rio de Janeiro, Brazil
Adriaan de Villiers (961), Knight Piésold and Co., Denver, Colorado, U.S.A.
Nikolaos Voros (463), Laboratory of Process Analysis and Design, National Technical University of Athens, Athens, Greece
C. P. Waller (1011), Camborne School of Mines, University of Exeter, Redruth, Cornwall, England
Wang Zhonghuai (627), Changchun Institute of Applied Chemistry, Changchun, China
T. E. Warner (273), Department of Mining and Mineral Engineering, University of Leeds, Leeds, England
A. Warshawsky (725), Organic Chemistry Department, Weizmann Institute of Science, Rehovot, Israel
Bradford Wesstrom (69), Phelps Dodge Refining Corporation, El Paso, Texas, U.S.A.
J. V. Wiertz (385, 395), Departmento de Ingeniería Química, University of Chile, Santiago, Chile
T. R. Wildeman (961), Colorado School of Mines, Golden, Colorado, U.S.A.
G. L. Yan (701), North China University of Technology, Beijing, China (presently, Department of Chemistry, University of Oslo, Oslo, Norway)
A. I. Zouboulis (547), Department of Chemistry, Aristotle University, Thessaloniki, Greece
Zhu Guocai (541), Institute of Chemical Metallurgy, Chinese Academy of Sciences, Beijing, China

Recommended separation processes for ion-absorbed rare earth minerals

Li Deqian
Wang Zhonghuai
Song Wenzhong
Meng Shulan
Ma Gengxiang
Changchun Institute of Applied Chemistry, Changchun, China

Abstract
The ion-absorbed rare earth minerals in China may be classified as a light rare earth ore, middle yttrium-concentrate europium ore, transition ore and heavy rare earth ore according to content of Y_2O_3, La_2O_3 and Eu_2O_3 in the oxides. The systematic studies have been carried out on the optimizations of combined separation processes and parameters of technique for the separation of individual rare earths. The rule of optimization of combined processes and parameters has been discussed. The several recommended flow-sheets of separation of individual rare earths based on the major technique of HEH/EHP solvent extraction including extraction and ion-exchange chromatography, selective reduction have been obtained.
Key word: Rare earths, Solvent extraction, Chromatography separation, HEH/EHP.

1 Introduction

The south of China has plentiful ion-absorbed rare earth resources that have characteristic of easy mining, simple extracting technique, low level of radioactivity and a good assortment of elements etc.. We can classify ion-absorbed rare earth minerals as light rare earth, middle yttrium-concentrate europium, transition and heavy rare earth ore according to content of Y_2O_3, La_2O_3 and Eu_2O_3 in the oxides. These ion-absorbed minerals constitute the middle-heavy rare earths (such as Sm, Eu, Gd, Tb, Dy, Er, Tm, Lu and Y etc.) that are rare in general rare earth minerals, and are the most important resources of developing high technique and new materials. In present work we classify ion-absorbed rare earth minerals as four types in view of this characteristic of various composition and types in ion-absorbed minerals. The optimizing combination and technical parameters of separation processes for individual rare earth have been studied. The several recommended flowsheets of rare earth separation have been obtained.

2 Results and discussion

Since di(2-ethylhexyl) phosphoric acid (DEHPA) is applied to the commercial production of europium, demonstrated (Cupta, 1992) in 1965 by the Molybdenum Corp. of America, the method that proved most attractive to these chemists of rare earth hydrometallurgy was that of solvent extraction, because separation was cleaner than by the precipitation method and operation time was shorter than that of the ion-exchange method. Nowadays solvent extraction is major method of the commercial production of individual rare earths. Late in 1970′s our laboratory has systematically investigated the extraction separation of rare earths by 2-ethylhexyl phosphonic acid mono-2-ethylhexyl ester (HEH/EHP), that has the advantage of higher mean separation factor, easier stripping, no emulsion formation, in comparison with the well known DEHPA, especially ammoniated HEH/EHP, which can improve the separation factor and extraction capacity of rare earths (Li, 1980). Since the 1980s ammoniated HEH/EHP has been widely applied to the hydrometallurgical industry of rare earths in China.

Ion-absorbed rare earth minerals can be classified as a light rare earth (La_2O_3~30%, Y_2O_3~8-10%, Eu_2O_3~0.5%), middle yttrium-concentrate europium ($Y_2O_3$20~30%, Eu_2O_3>0.8%), transition ($Y_2O_3$40~50%, $La_2O_3$10~20%), and heavy rare earth (Y_2O_3~60%, La_2O_3<3%) ore according to the content of Y_2O_3, La_2O_3 and Eu_2O_3 in the oxides (Table 1.). A optimization of separation process trend and technological parameters. the former bases on the comprehensive survey of product type and quality requirement, adopted ore type and the characteristic of separation technique of individual rare earths etc., The latter has been studied by Xu (1982).

Table 1. Examples of ion-absorbed rare earth minerals

	type	of	ore	REO(%)
	light	middle	transition	heavy
La_2O_3	29.84	28.01	10.21	2.18
CeO_2	7.18	2.02	0.55	<1.0
Pr_6O_{11}	7.14	7.03	2.64	1.08
Nd_2O_3	30.18	24.01	8.91	3.47
Sm_2O_3	6.32	4.52	3.43	2.34
Eu_2O_3	0.51	0.91	<0.5	<0.1
Gd_2O_3	4.21	4.02	5.52	5.69
Tb_4O_7	0.46	0.58	1.15	1.13
Dy_2O_3	1.77	3.48	7.65	7.48
Ho_2O_3	0.27	0.40	1.75	1.60
Er_2O_3	0.80	1.27	4.17	4.26
Tm_2O_3	0.13	0.25	0.70	0.60
Yb_2O_3	0.62	1.35	3.41	3.34
Lu_2O_3	0.13	0.30	0.68	0.47
Y_2O_3	10.07	21.82	48.58	64.10

The optimum standard of separation process is simple and easy technique, high purity and yield of product, low cost and the flexibility of process trend etc..

2.1 Separation Process of Light and Middle Yttrium Concentrate Europium Ore

If a scheme of product for two rare earth ore is given in Table 2., the recommended optimum process of separating rare earths is shown in Fig. 1.. This recommended flowsheet bases on the major technique of HEH/EHP solvent extraction(Li, 1985,1986) including extraction and ion-exchange chromatography (pen, 1985, Ling, 1985), selective reduction.

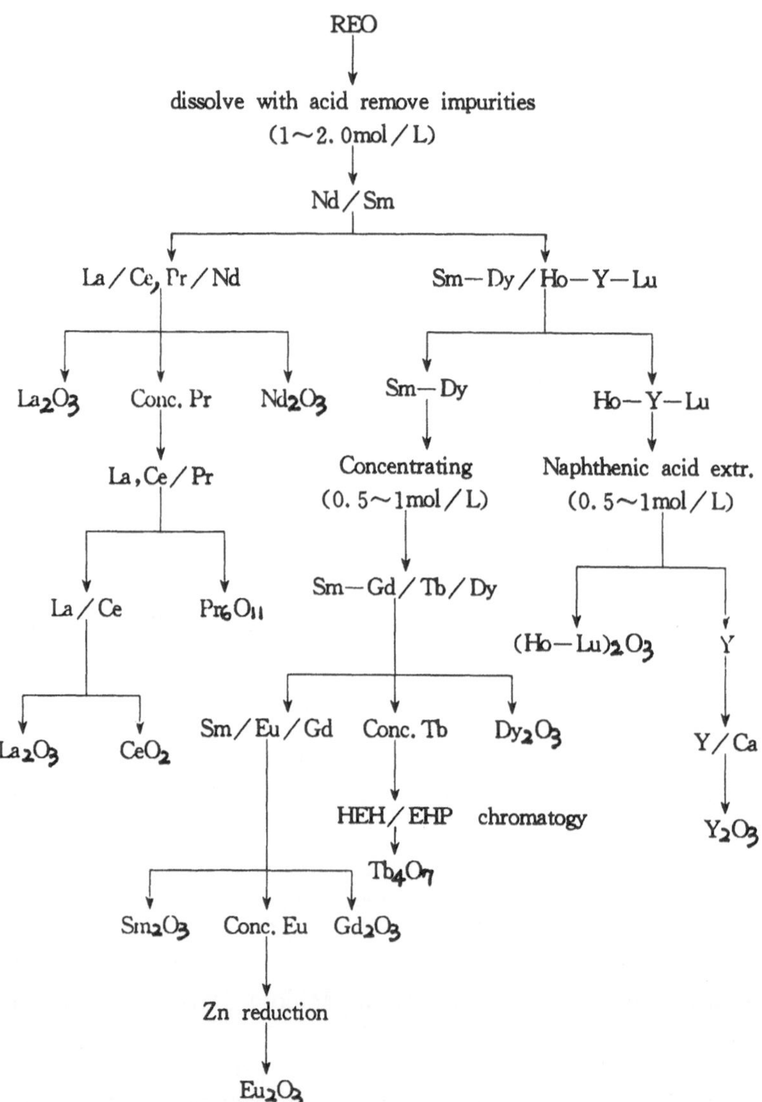

Fig. 1. Flowsheet of recommended process for light or middle yttrium-conc. europium ore

Table 2. Scheme of product

oxide	Purity(%)	oxide	purity(%)
Y_2O_3	99.99(fluorescent)	Eu_2O_3	99.99(fluorescent)
La_2O_3	99.9	Gd_2O_3	>99
Pr_6O_{11}	>99	Tb_4O_7	>99.95 (fluorescent)
Nd_2O_3	>99	Dy_2O_3	>99
Sm_2O_3	99-99.9	concentrated heavy REO	

2.2 Separation Process of Transition and Heavy Rare Earth Ore

Separation technique of high purity Y_2O_3 by naphthenic acid solvent extraction has been applied to the commercial production (Dai, 1985), but the separation of high purity Y_2O_3 from the transition ore ($Y_2O_3<50\%$, $La_2O_3>10\%$) with naphthenic acid extraction has not been reported previously.

Table 3. Separation factor between La and Y(β^{La}_{Y})in kinetic test

stage		distribution La(I)	coefficient Y(I)	β^{La}_{Y}	$\bar{\beta}^{La}_{Y}$
10	(extraction)	1.67	0.976	1.71	
12		1.41	0.980	1.44	
14		1.46	0.970	1.51	
36		1.14	0.716	1.59	1.66
37		1.21	0.660	1.83	
38		1.17	0.630	1.86	
39	(stripping)	0.967	0.727	1.33	
40		0.950	0.762	1.25	
50		0.875	0.685	1.28	1.24
52		0.935	0.695	1.35	
54		0.964	0.697	1.38	
56		1.01	1.22	0.83	

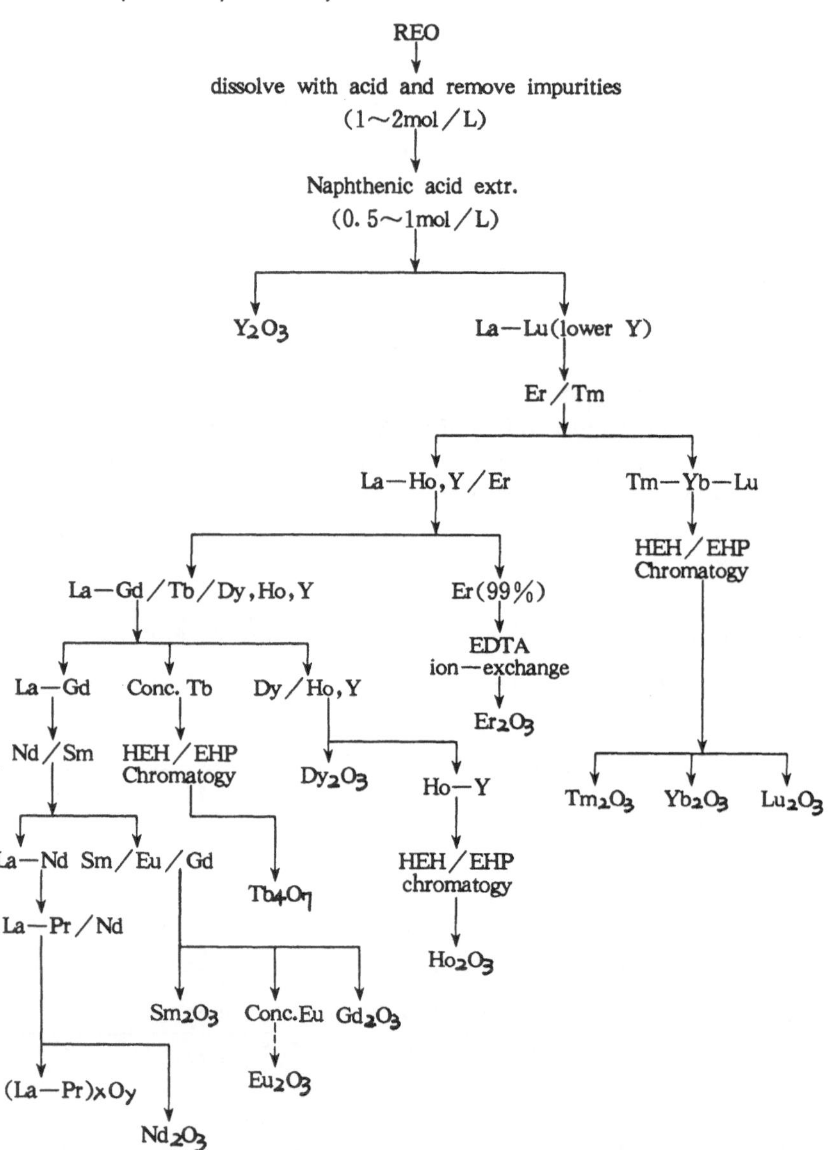

Fig. 2. Flowsheet of recommended process for trans. or heavy RE ore

The major problem is a small separation factor between Y and La. We have completed long-time kinetic test of separating Y_2O_3 with naphthenic acid in mixer-settler stages, and observed the distribution behaviour and accumulation rule of La in every stage(Table 3.). A optimum parameter of separation technique for high purity Y_2O_3 with naphthenic acid has been obtained, Table 4.and Fig. 2. show the scheme of product and recommended optimum process for transition or heavy rare earth ore.

Table 4. Scheme of product for trans. or heavy RE ore

Oxide	purity(%)	oxide	purity(%)
Y_2O_3	99. 99(fluorescent)	Ho_2O_3	>99. 95
Nd_2O_3	>99	Er_2O_3	>99. 95
Sm_2O_3	>99	Tm_2O_3	99—99. 99
Gd_2O_3	99—99. 99	Yb_2O_3	>99. 95
Tb_4O_7	99. 99(fluorescent)	Lu_2O_3	99—99. 95
Dy_2O_3	99—99. 95		

3 Acknowledgement

The project was supported by the National Science Foundation of China.

4 References

Cupta, C. K. and Krishnamurthy, N. (1992) Extractive metallurgy of rare earths. International Materials Reviews, 5, 214.

Dai,Z. R. Wang, C. X. Wang, Z. Y. (1985) Solvent extraction separation of high purity yttrium. CN pat. 85 1 02220B.

Li, D. Q. , Wan, X. , Lin, D. Z. , Xie, Y. F. Lin, S. X. Wang, X. H. Li, H. and Ji, E. R. (1980) Extraction separation of rare earth elements, scandium and thorium with HEH/EHP. Proc. ISEC' 80, Liege, Belgium, pp. 80—202.

Li D. Q. , Xu, W. and wang, Z. H. (1985), Liquid—liquid extraction separation of rare earth with phosphonates. CN pat. 85 1

02244B.

Li D. Q. , wang, Z. H. and Xu, W. (1986) Liquid—liquid extraction separation of dysprosium. CN pat. 86 1 08135A.

Ling, S. X. and Yuan, F. (1985) Chromatography separation of high purity Ho_2O_3 and Dy_2O_3. CN pat. 85 1 01611B.

Peng, C. L. , Niu, W. and Piao, Z. X. (1985) Chromatography separation method of high purity Tm_2O_3 and Lu_2O_3. CN pat. 85 1 01874B.

Xu, G. X. Li, B. G. and Ni, Y. M. (1982) Solvent extraction separation of rare earths. Rare Metals, 1(1), 10(ch.)

Purification of waste waters containing heavy metals by surfactant liquid membrane extraction

Mohammed Samar
Dominique Pareau
Gérard Durand
André Chesné
Laboratoire de Chimie Nucléaire et Industrielle, Ecole Centrale Paris, Châtenay-Malabry, France

The aim of this work is to find a process for the purification of waste waters containing cadmium, lead and mercury.

A complete study of the metal extraction by the liquid membrane process using HDEHP as the extractant and SPAN 80 as the surfactant is performed in terms of extraction and stripping efficiencies, emulsion breaking and swelling. A process flowsheet with a stage of mixer- settler as the contactor is proposed leading to high concentration factors if the solution has a significant ionic strength (for instance gas washing solution). In contrast when the solution contains only traces of ionic species the important swelling leads to smaller concentration factors; this can be avoided by using another surfactant (ECA 4360).

Another system with Cyanex 301 and Aliquat 336 as the extractants and ECA 4360 as the surfactant is tested: the stripping of mercury is impossible, but this system is however a promising one when the solution contains only lead and cadmium. Its major advantages indeed are: a reduced swelling giving high concentration factors and the unnecessary pH control. Further studies are now in progress to improve the proposed processes.

Keywords: heavy metals, surfactant liquid membrane, water purification

Introduction

The aim of this work is to find a process for the purification of waste waters containing cadmium, lead and mercury. The pH of these solutions depends on their origin (1 to 4 generally).

A bibliographic survey lead us to select acidic reagents as the extractant (1, 2, 3). A comparison was made between a phosphoric acid (di- 2- ethylhexyl phosphoric acid or HDEHP) and a dithiophosphinic acid (CYANEX 301), which are commercial compounds. They are both cationic exchangers and extract metallic cations present in an aqueous solution. These cations are exchanged for protons which are released in the aqueous phase leading to its acidification.

The surfactant liquid membrane process which was introduced by Li (4) was used because of two major advantages: possibility of treating diluted solutions and high concentration factors. Its principles are described in detail in many papers (5, 6, 7).

Di- 2- ethylhexylphosphoric acid (HDEHP)

Several preliminary experiments showed that the extraction of the three metallic cations begins to be significant at pH 3; for lower values of pH, the cationic exchange is difficult due to the rather high concentration of protons.

The stripping solutions were nitric acid for cadmium and lead and EDTA (diammonium salt of ethylenediaminetetracetic acid) for mercury or mixtures of the three metals.

The model solutions contained Cd, Pb or Hg at different pH. They were prepared by dissolving the metal nitrates in water. The pH was adjusted by addition of nitric acid.

The organic phase was a mixture of HDEHP, SPAN 80 (surfactant) and dodecane. The stripping phase was nitric acid or EDTA.

The emulsion was prepared by dispersing a given volume of the stripping phase in the organic solution by means of a rotating turbine. The emulsion is then dispersed in the aqueous phase containing the pollutants in an agitated vessel.

Cadmium was studied more precisely than the other metals in order to find out the optimal conditions for a good transfer. These conditions were afterwards tested on the other metals.

1 Case of cadmium

1.1 Cadmium extraction

The experiments were generally performed in the following conditions:

. volume ratio of the emulsion ORG/AQ: 1
. emulsion volume: 20 mL
. volume of the aqueous phase to be purified: 200 mL
. stripping phase: nitric acid 0.5 mol/L
. initial cadmium concentration: 100 mg/L
. concentration of HDEHP: 7% (in weight)
. concentration of SPAN 80: 5%
. rotating speed during the transfer step: 200 rpm

During the transfer of Cd, the pH of the aqueous external phase was kept to the desired value by addition of ammonium acetate.

Several chemical and operating parameters were studied.

1.1.1 Influence of the external phase pH

The experiments were performed for different values of pH (figure 1); in the following figures (Cd)ext is the concentration of cadmium in the aqueous external phase at time t.

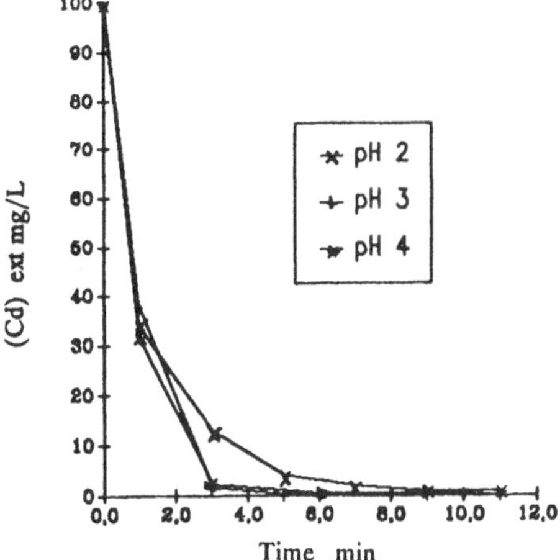

Figure 1 Influence of the external pH on the extraction of cadmium

The extraction kinetics increases when the pH increases. The extraction is indeed slower for pH 2, the cationic exchange being more difficult when the acidity increases. In contrast the extraction is very satisfactory between pH 2.5 and 4: the extraction yield is greater than 99.8%. At pH 6 the emulsion is no more stable.

The optimal range of pH is then 3 to 4. All the following experiments were performed at pH 3.

1.1.2 Influence of the extractant concentration

The results are presented in figure 2. The Cd concentration in the external phase decreases with time and increasing HDEHP concentration (up to 7%). Past this concentration the emulsion is not stable. For concentrations smaller than 3% the extraction is poor. The optimal concentration is 5 to 7%.

1.1.3 Influence of the surfactant concentration

Several concentrations of SPAN 80 were used and the kinetics of cadmium extraction was measured (figure 3); HDEHP concentration is 7%.

For lower SPAN concentrations (< 2%) the cadmium concentration in the external phase increases after an initial fall; this is due to the breaking of the emulsion, leading to a partial mixing of the internal phase with the external. By increasing the surfactant concentration it is possible to limitate the breaking of the emulsion: the range 4 to 5% of SPAN seems to be convenient to obtain a stable emulsion. Using greater concentrations is not favourable due to the

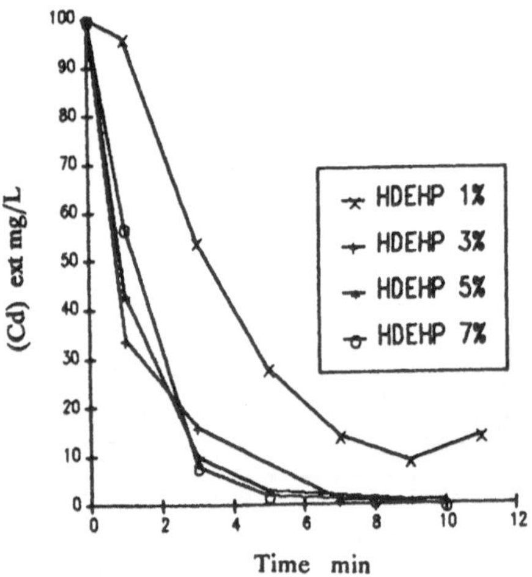

Figure 2 Influence of HDEHP concentration on the extraction of cadmium

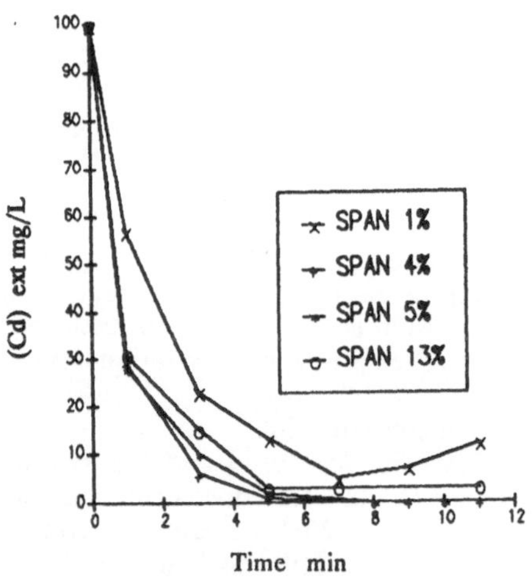

Figure 3 Influence of SPAN concentration on the extraction of cadmium

increasing viscosity of the organic solution resulting in a slower transfer and higher residual concentrations.

1.1.4 Influence of the emulsification rotating speed

By increasing the rotating speed of the emulsificator the droplet diameter in the emulsion decreases resulting in a greater internal interfacial area. Two different emulsions were tested to extract cadmium, with no significant differences (diameter between 5 and 14 μm).

1.1.5 Influence of the rotating speed in the tranfer step

In these experiments HDEHP concentration was 7% and SPAN concentration 5%.

As expected the greater the rotating speed is, the better the extraction (figure 4). The greater agitation indeed results in a bigger external interfacial area and then a quicker transfer. But it is not possible to increase the rotating speed over 400 rpm without a significant breaking of the emulsion (see paragraph 1.3.5). So an intermediate value of the rotating speed must be adopted (200 to 300 rpm).

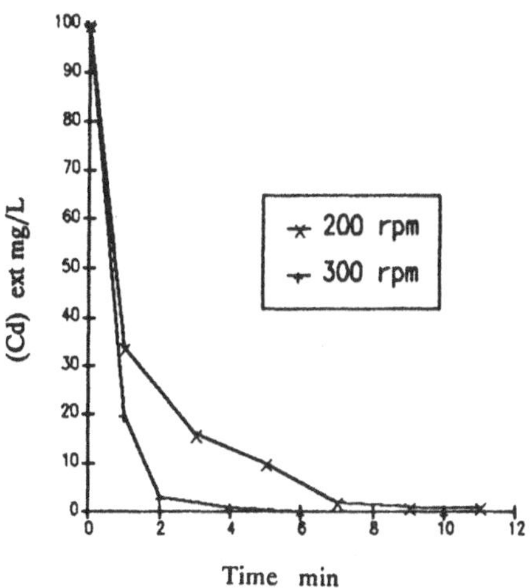

Figure 4 Influence of the rotating speed on the extraction of cadmium

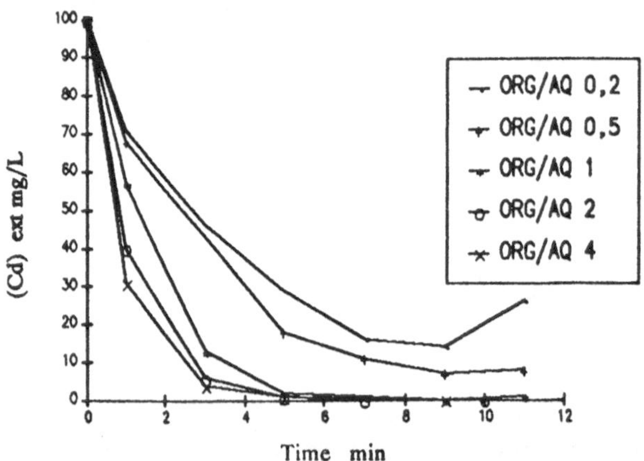

Figure 5 Influence of the ratio ORG/AQ on the extraction of cadmium

1.1.6 Influence of the volume ratio of the emulsion
This ratio ORG/AQ (ratio of the volumes of the organic and the internal phases) was varied from 0.2 to 4 and the kinetics of cadmium extraction was measured (figure 5). In these experiments HDEHP and SPAN were equal respectively to 5 and 3%.

For ratios lower than 0.5 the extraction of cadmium is not quantitative and the stability of the emulsion is rather bad. The emulsion viscosity in this case is much higher than for greater ratios; this is due to the higher proportion of dispersed phase in the emulsion and these conditions are therefore not acceptable.

For ratios greater than 0.5, the cadmium extraction and the emulsion stability become better, leading to a good transfer efficiency. From 5 in contrast the extraction is much slower due to the greater thickness of the organic membrane.

The optimal range is then 0.5 to 4.

1.1.7 Influence of the volume ratio of the aqueous external and internal phases
This ratio B was varied from 20 to 40 with HDEHP and SPAN concentrations of 5 and 2% respectively and a volume ratio of the emulsion of 1. As expected the cadmium extraction slightly increases when the studied ratio decreases due to the greater volume of the emulsion, as can be seen on the following table where (Cd) is the concentration of metal remaining in the external phase after 4 minutes of contact; but the differences are very limited in this range of B values.

Ratio B	20	30	40
(Cd) mg/L	0.6	1.1	1.9

1.2 Cadmium stripping

Two parameters were studied: the volume ratio of the emulsion (ORG/AQ) and the concentration of nitric acid. The other conditions were: HDEHP 5%, SPAN 3%, ratio B 20, rotating speed in the transfer step 300 rpm, contacting time 4 min, pH of the external phase 3.

1.2.1 Influence of the volume ratio of the emulsion

The nitric acid concentration in the stripping phase was 0,5 mol/L.

The stripping efficiency was estimated by measuring the final cadmium quantity in the internal phase for different volume ratios.

ORG/AQ	1	2	3	4
% Stripping	98	92	88	85.

The stripping efficiency decreases as the volume ratio increases due to the smaller quantity of stripping phase and to the more limited internal interfacial area.

A ratio of 1 seems to be convenient.

1.2.2 Influence of the nitric acid concentation

The ratio ORG/AQ was in this case 1. For acid concentrations between 0.5 and 2 mol/L, the stripping of cadmium is quantitative (> 98%). But when the concentration of nitric acid is greater than 2 mol/L, the emulsion swells due to the osmosis and the internal phase is diluted resulting in a less effective stripping; in addition the emulsion stability is poorer (paragraph 1.3.4).

1.3 Breaking of the emulsion

This phenomenon was measured by contacting the emulsion with an external phase composed of pure water. The variation of the pH of this phase with the time is an indication of the emulsion breaking.

The influence of several parameters was studied: SPAN concentration, HDEHP concentration, ratio ORG/AQ, nitric acid concentration, rotating speed in the tranfer step.. The breaking rate BR is the volume proportion of the internal phase which is ejected in the external solution.

1.3.1 Influence of SPAN concentration

Experiments were performed with HDEHP 5%, a ratio ORG/AQ of 1, a rotating speed of 300 rpm, 10 minutes of contact, nitric acid 0,5 mol/L and different concentrations of SPAN; as expected the stability of the emulsion increases with SPAN concentration, but the differences are limited showing that the two concentrations of SPAN are convenient:

SPAN %	3	5
BR %	2.0	1.5

1.3.2 Influence of HDEHP concentration

The conditions were the same as previously; the SPAN concentration was 5%.

HDEHP %	3	5	7	13
BR %	1.0	1.5	2.0	7.0

HDEHP has an unfavourable influence on the stability of the emulsion. It is probably due to the interfacial properties of HDEHP which favours O/W emulsions and is therefore opposed to the SPAN action.

1.3.3 Influence of the ratio ORG/AQ

The stability slightly increases with increasing ORG/AQ as can be seen on the following table. This result is expected because the ejection of the internal phase is easier when its proportion in the emulsion is higher.

The contacting time was in these experiments 10 minutes.

ORG/AQ	1	2	3	4
BR %	1.8	1.5	1.3	1

1.3.4 Influence of nitric acid concentration

When the nitric acid concentration increases, the stability of the emulsion decreases, maybe due to a reaction of the acid with SPAN, resulting in a partial loss of its surfactant properties (see following table for 10 minutes of contact).

HNO_3 mol/L	0,5	1	2
BR %	1.8	2.1	7.5

From the stability point of view a concentration of nitric acid greater than 1 mol/L is prohibited.

1.3.5 Influence of the rotating speed in the transfer step

As expected the stability greatly depends on the rotating speed; when the external interfacial area is increased by an increased rotating speed, the ejection of the internal phase is facilitated. This can be seen on the following results with a contacting time of 10 minutes:

Rotating speed rpm	300	400	500
BR %	1.8	6.3	8.5

In terms of stability the rotating speed must not exceed 300 rpm.

1.4 Optimal conditions

They are the following:

. SPAN 80: 5%

. HDEHP: 5%

. ORG/AQ: 1

. Rotating speed of the transfer step: 300 rpm
. Nitric acid: 0.5 mol/L
These conditions were then tested on the other metals.

2 Case of lead

In the conditions previously determined, a study of the extraction of lead versus the pH of the external phase was performed. At pH from 1 to 3 the extraction of lead is very satisfactory; the extraction yield for pH 3 is 99.5%. The stripping efficiency is very good too (99%).

3 Case of mercury

The extraction of mercury was studied for different pH of the external phase, all the conditions being the same than for lead. The extraction is slower than for the other metals but it is probably due to the bad stripping. The stripping efficiency is indeed only 16% for nitric acid 0.5 mol/L. Even for acid concentrations greater than 6 mol/L, the stripping is not quantitative (< 85%). In the same time the extraction decreases when the nitric acid concentration increases because of a significant emulsion breaking.

Other mineral acids (hydrochloric, perchloric) were tested without success. In contrast a solution of diammonium salt of ethylenediaminetetracetic acid (EDTA) is convenient as can be seen on the following table where the concentration of EDTA is 0.014 mol/L.

Internal phase pH	4	5	5.5
Stripping efficiency %	78	89	92

For pH greater than 6 the emulsion is not stable.

The influence of the EDTA concentration was then studied at pH 5.5; the results are presented on the following table:

EDTA mol/L	0.014	0.030	0.060	0.090
Stripping Efficiency %	92	93	96	96

A concentration of 0.06 mol/L EDTA seems to be convenient; tests on the stripping of the other metals gave good results (> 97% for Cd and Pb).

The solution of EDTA 0.06 mol/L at pH 5.5 is then selected as the internal phase for the further experiments with the three metals.

4 Kinetic study of the purification of a solution containing the three metals

The initial external aqueous solutions contained 100 mg/L of Cd, Pb and Hg at pH 3. The organic phase was composed of 5% HDEHP, 5% SPAN 80 and dodecane. The internal solution contained EDTA 0.06 mol/L at pH 5.5. The volumes of the internal, the organic and the external phases were respectively 10, 10 and 200 mL.

The results are represented on figure 6. The extraction is good and the stripping quantitative (> 93%) for the three elements.

The transfer equation can be written as follows:

$- V\,dC = E\,(C - C^{*})\,a\,dt$ with:

. V the external phase volume

. t the time

. C the concentration of the metal in this phase at t

. C^{*} the concentration in the aqueous phase if this phase would be in equilibrium with the emulsion; in the particular case of the emulsion liquid membrane process $C^{*} = 0$

. a the external interfacial area

. E the global transfer coefficient.

By integrating one can obtain the following relationship:

$\ln C/C_0 = - E\,a\,t / V$

C_0 is the initial concentration of the metal.

$\ln C/C_0$ is plotted versus t in figure 7; a straight line is obtained for t > 3 min and for the three metals. The assumption on the form of the transfer equation is therefore verified except in the first minutes of the contact. The initial great decrease in concentration is probably due to an instantaneous extraction in the organic phase followed by slower steps (diffusion across the organic membrane, stripping).

For t > 3 minutes, K (= E a / V) and E were calculated for the three metals, by estimating a with a mean emulsion drop diameter 10^{-3} m: a = 0.12 m^2

Metal	Cd	Pb	Hg
K s^{-1}	0.0042	0.0022	0.0030
E m/s	7 10^{-6}	3.7 10^{-6}	5 10^{-6}

By calculations developped elsewhere (8), the limiting step of the process was identified; this is the diffusion of the complex metal/HDEHP through the organic membrane.

5. Continuous experiments

These studies were performed in a laboratory scale pulsed column: 1 m height, 2.5 cm diameter with a packing of 40 disks and crowns.

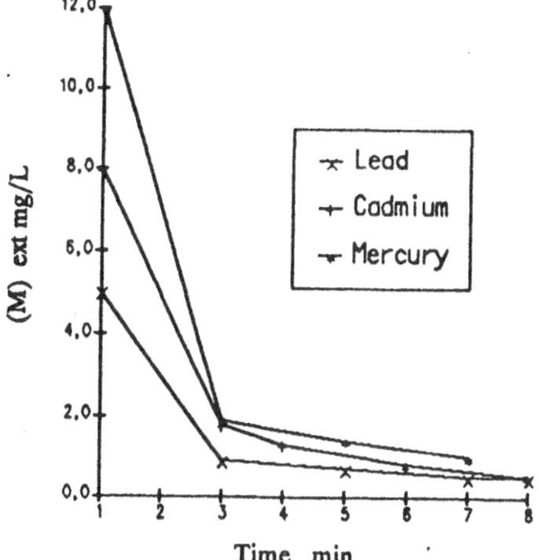

Figure 6 Extraction of the three metals

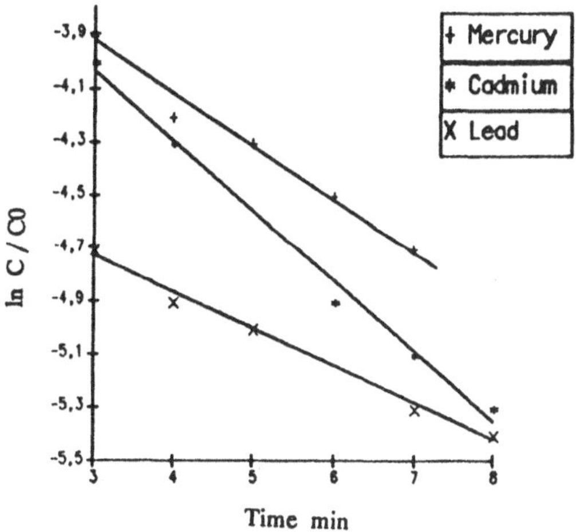

Figure 7 Kinetics of the extraction of the three metals

5.1 Preliminary hydrodynamic study

A preliminary hydrodynamic study was conducted in order to determine hydrodynamic acceptable conditions. Two total specific flowrates were used: 1.2 and 2.2 L h^{-1} cm^{-2}, with a flowrate ratio external phase / emulsion of 10. The amplitude a and the frequency f of the pulsation were varied; a convenient countercurrent flow was obtained for af varying from 100 to 140 cm/min.

5.2 Transfer of the metals

The optimal conditions found in the batch experiments lead to flooding. So complementary experiments were conducted to find suitable operating conditions for the column.

The studied parameters were the following:

. volume ratio of the emulsion ORG/AQ: 1 to 7
. flowrate ratio B aqueous external phase / emulsion: 8 to 10
. EDTA concentration in the internal phase: 0.05 to 0.15 mol/L
. total specific flowrate: 0.5 to 2 L h^{-1} cm^{-2}
. pulsation intensity af: 100 to 140 cm/min

The other chemical parameters were the same as previously.

The following optimal conditions were found:

. ORG/AQ = 7
. flowrate ratio B = 8
. total specific flowrate: 0.8 L h^{-1} cm^{-2}
. pulsation intensity 110 cm/min
. EDTA 0.15 mol/L

The flowrate ratio B is decreased in comparison with the batch experiments to give a good extraction efficiency (8 here to 10 in the batch study). But if the volume ratio of the emulsion is kept to 1 as in the batch study, the difference between the densities of the two phases (external and emulsion) is too small to allow a good counter current operation. So the proportion of stripping phase in the emulsion is decreased; but in the same time the stripping efficiency falls, which can be compensated by an increase of EDTA concentration.

The total specific flowrate is smaller than in solvent extraction processes. This is probably due to the presence of the surfactant which lowers the interfacial tension at the external interface; the emulsion drops for a given flowrate and a given pulsation intensity are smaller than in solvent extraction resulting in a more frequent flooding.

An experiment was performed in the optimal conditions with a solution containing 100 mg/L of the three metals at pH 3.

The residual concentrations in the raffinate were:

Cd 9.5 mg/L **Pb 8.4 mg/L** **Hg 10.3 mg/L**
pH 2.

The extraction is not quantitative because of the acidification of the raffinate which prevents a further transfer.

This raffinate was then neutralized to pH 3.8 and another experiment was performed on it in the same conditions.

The final solution contained only **0.13 mg/L Cd, 0.09 mg/L Pb, 0.20 mg/L Hg (pH 3.7).**

These concentrations can be reduced at will by increasing the height of the column.

6. Process

6.1 Pulsed column

A process flowsheet is now proposed with two columns, the second treating the raffinate of the first after a partial neutralization (figure 8). In order to improve the elimination of the three metals the pH of the raffinate after neutralization must be as high as possible.

The loaded emulsion coming from the first column is split by electrocoalescence giving the organic solution which is recycled to the emulsification step and the internal solution containing the concentrated metals. It is possible to recycle EDTA by acidifying this solution; the acid form of EDTA is precipitated and separated by filtration. The solid is then dissolved in ammonia to prepare a new stripping phase for the emulsification step.

The acid concentrate of the metals is recovered in the filtration step. The concentration factor of the metals is about 10 to 20, less than expected (56) due to the swelling of the emulsion. This phenomenon is a drawback which cannot be avoided because of the presence of the surfactant.

In order to reduce the swelling it is necessary to minimize the contacting time. From this point of view it seems more favourable to use a stage of mixer settler as the contactor.

6.2 Studies in a mixer settler

Different contacting times were used for the purification of the initial solution at pH 3. The pH was fixed during the transfer by addition of ammonium acetate. With the given volume ratios (ORG/AQ = 7.6; external solution / emulsion = 5), the expected concentration factor is 38 (without swelling). F is the ratio of the measured concentration factor and the theoretical one, the difference between F and 1 being an estimation of the emulsion swelling. The swelling rate is equal to ($Vi - Vi_0)/ Vi_0$ where Vi_0 is the initial volume of the internal phase and Vi its final volume.

Contacting time (min)	2	3
F	0.55	0.40
% Swelling	78	143

Swelling is very important and occurs in the first minutes of the contact. It cannot therefore be avoided by minimizing the contacting time without decreasing the extracting efficiency.

Swelling is mainly due to osmosis (water transport) from the external phase to the internal, due to the difference of their ionic strengthes; the extraction from a solution of higher ionic strength was then tested in order to verify this hypothesis. In the model solution was added ammonium nitrate. The results are presented in the following table; the theoric concentration factor is 33.

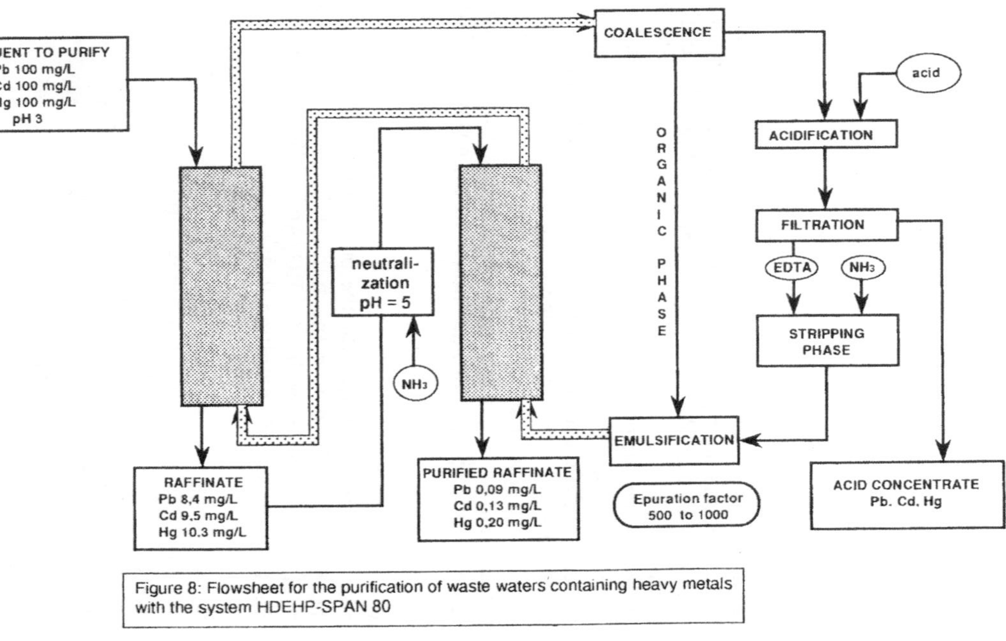

Figure 8: Flowsheet for the purification of waste waters containing heavy metals with the system HDEHP-SPAN 80

NH_4NO_3 mol/L	0.15	0.25	0.25
Time min	3	3	2
F	0.85	0.85	1.0
% Swelling	15	15	0

The presence of ammonium nitrate slowers the water transport; swelling is less pronounced. The results are satisfactory with 2 minutes of contact where no swelling was observed.

7. Conclusion

HDEHP is a convenient extractant for a liquid membrane process if the solution containing the pollutants has a significant ionic strength (> 0.1). In this case by using a single stage of mixer- settler with 2 min of contact a concentrated solution of the pollutants (concentration factor > 50) is obtained.

The use of a pulsed column presents two major drawbacks:

. no possibility of fixing the external phase pH, so a slower extraction and the need of 2 columns with an intermediate neutralization

. an important swelling leading to rather small concentration factors, due to the higher residence time.

A mixer settler must therefore be preferred. This system is adapted to the treatment of waste waters containing salts, for example gas washing solutions. The general flowsheet is similar to that proposed in figure 8, the columns being replaced by a stage of mixer- settler.

In contrast when the ionic strength of the solution approaches zero, the swelling is important and the concentration factor in the column or in the mixer settler is limited (< 20). Another system must be preferred: studies are now in progress with other surfactants having less affinity for water (ECA 4360 for instance). The first experiments indicate the possibility of reducing drastically the swelling and thus obtaining more concentrated solutions.

CYANEX 301

Cyanex 301 is able to extract very effectively cations in acidic media (pH 1); it is a major advantage in comparison with HDEHP. The extraction of the pollutant traces at pH 1 or 1.5 will lead to a negligible variation of pH; so there is no need for a neutralization of the exchanged protons.

1. Solvent extraction experiments

Preliminary experiments were conducted in solvent extraction to find suitable stripping solutions.

1.1 Cyanex 301 alone

Cyanex 301 was diluted at 5% (weight) in dodecane (about 0.16 mol/L). This organic solution was contacted with a solution containing 100 mg/L of Cd, Pb and Hg at pH 1. The extraction of the three metals was quantitative. Several stripping solutions were tested; the results are presented in the following table:

Stripping Solution	% Str Cd	% Str Pb	% Str Hg
EDTA 0.1 mol/L pH 7	0	0	0
idem pH 9	0	0	0
idem pH 10	0	0	0
HCl 3 mol/L	7.7	0.6	0
HCl 6 mol/L	100	100	0

The stability of the metal complexes in the organic phase is very important, especially in the case of mercury, for which no stripping solution was found.

This extractant is therefore not suitable for the extraction of the three metals by a liquid membrane process; the lack of stripping of Hg would lead after a while to an accumulation of this metal in the membrane and the interruption of the extraction.

1.2 Mixture of Cyanex 301 and Aliquat 336

Cyanex 301 reacts with quaternary ammonium chlorides to give Cyanex salts which are good extractants as Cyanex alone, but give facilitated stripping reactions.

The following notations are adopted:

HA: Cyanex 301

R_4NCl: Aliquat 336 (methyltricaprylammonium chloride)

The formation reaction of the Cyanex salt is the following:

$$HA + R_4NCl \rightleftharpoons R_4NA + HCl$$

The extraction reaction can then be written:

$$2\,R_4NA + M^{2+} + 2\,Cl^- \rightleftharpoons MA_2 + 2\,R_4NCl$$

and the stripping reaction is the reverse one, facilitated by the formation of R_4NA.

Equimolar mixtures of HA and R_4NCl (0.08 mol/L) were prepared in dodecane. The extraction of the three metals was quantitative for pH > 1. As previously several stripping solutions were tested as presented in the following table:

Strip solution	% Str Cd	% Str Pb	% Str Hg
EDTA 0.1 mol/L pH 5	0.4	0.6	0
idem pH 7	1.2	15.2	0
idem pH 9	15.3	74.2	0
idem + NaOH 0.05 mol/L	100	100	0
idem + NaOH 0.1 mol/L	100	100	0
TEA 0.1 + NaOH 0.1 mol/L	29.7	100	0
idem + NaOH 0.5 mol/L	40.0	100	0
idem + NaOH 1 mol/L	47.2	100	0
idem + NaOH 3.5 mol/L	94.4	100	0

TEA: triethanolamine

As previously the stripping of mercury is impossible, showing the remarkable stability of the metallic organosulfide. This has to be related to the well known strong affinity of mercury for sulfur. Cyanex 301 is then prohibited as an extractant for mercury and cannot be used for industrial purposes when the solution to be purified contains this metal. Despite this disappointing result Cyanex 301 is an excellent extractant for cadmium and lead; the stripping is rather easy when Cyanex is used in mixture with a quaternary ammonium salt.

Liquid membrane experiments were then performed in the case of cadmium and lead in order to find conditions for good extraction and stripping.

2. Liquid membrane experiments

Batch experiments were conducted as for HDEHP.

The internal phase was a mixture of EDTA and NaOH whose stripping efficiency was prooved in the solvent extraction experiments. The composition of this solution was EDTA 0.1 mol/L and NaOH 0.05 mol/L. The organic solution contained equivalent concentrations of Cyanex 301 and Aliquat 336 (0.06 mol/L) in dodecane. Various concentrations of surfactant were added.

2.1 Choice of the surfactant

The concentration of SPAN 80 was varied from 3 to 10% with a ratio ORG/AQ of the emulsion equal to 1. In all cases the emulsion was not stable. This can be attributed to a chemical reaction of SPAN with NaOH (beginning of the ester saponification) whose product does not exhibit the same surfactant properties as SPAN itself.

ECA 4360 was then choosen as the surfactant; concentrations from 1 to 3% gave stable emulsions which could afterwards be used to extract the two metals. The breaking rate is always very limited (< 2%).

2.2 Extraction of the metals

Several parameters were studied: the external phase pH, the volume ratio ORG/AQ of the emulsion, the volume ratio of the external phase and the emulsion in order to find suitable conditions for the complete extraction of the two metals.

The general conditions were:

. initial concentrations of lead and cadmium: 100 mg/L
. concentration of ECA 2%
. volume of the external phase 150 mL
. volume of the emulsion 25 mL
. ratio ORG/AQ of the emulsion 2
. theoretical concentration factor 18
. contacting time 3 min

2.2.1 Influence of the external phase pH

For pH in the range 1 to 3 the extraction of the two metals is quantitative: the concentrations of metal remaining in the external phase are presented on the following table:

pH	1	2	3
(Pb) mg/L	0,12	0,10	0,11
(Cd) mg/L	0,15	0,14	0,13

2.2.2 Influence of the ratio ORG/AQ

This ratio was varied from 2 to 5; in all cases the extraction is quantitative (remaining concentrations about 0.1 mg/L).

2.2.3 Influence of the ratio external phase / emulsion

Different volumes of external phase (from 100 to 200 mL) were used with the same volume of emulsion (25 mL; ORG/AQ = 2); the extraction was always quantitative.

2.3 Stripping of the metals

In the previous experiments the emulsion was broken and the two resulting phases were analysed. The concentrations of the metals in the internal phase and in the organic one (after stripping) were measured.

In all cases the organic phase contained less than 2% of the metals, showing a very good stripping efficiency. The concentrations of the metals in the internal phase confirmed this result and by calculating the concentration factor one could see that swelling was limited compared with the previous case. This is due to the less important affinity of ECA for water in comparison with SPAN.

As an example are presented the results obtained for the influence of the ratio ORG/AQ on the stripping efficiency (volume of external phase 150 mL, volume of emulsion 25 mL). The metal concentrations in the internal phase, the concentration factors (theoretical T and measured M), their ratio F are given along with the swelling rate.

ORG/AQ	2	5
(Cd)int mg/L	1310	2990
(Pb)int mg/L	1320	2980
T	18	36
M	13	30
F	0.72	0.83
% Swelling	36	19

In comparison with HDEHP the emulsion swelling is very reduced: F is here about 0.8 to 0.4 in similar conditions for HDEHP.

As a conclusion these few experiments show that the mixture of Cyanex 301 and Aliquat 336 is convenient for an effective purification of a solution containing traces of cadmium and lead. Further investigations are now in progress to design a suitable process flowsheet, which will certainly be similar to that proposed for HDEHP. Two contactors are studied and compared: the pulsed column and a stage of mixer- settler.

General conclusion

At this step of the work a complete process can be proposed for the purification of solutions containing cadmium, lead and mercury. For chemical systems containing HDEHP and SPAN 80, a stage of mixer- settler is preferred for the possibility of pH control during the operation and the reduction of the contacting time leading to the minimization of the emulsion swelling. It however seems better to use ECA 4360 for limiting the swelling and obtaining a higher concentration factor, especially when the solution has a reduced ionic strength.

Attempts to replace HDEHP by mixtures of Cyanex 301 and Aliquat 336 were deceiving; the stripping of mercury was impossible, prohibiting the use of this system when the solution contains Hg. But in the absence of it this system is convenient and does not need a control of pH which is a major advantage. In addition the swelling is limited allowing high concentration factors.

This work then proposes acceptable industrial solutions to purify waste waters containing heavy metallic cations; further studies are now in progress to improve them.

Bibliography

1. M. SHOEMACKER Ann. New York Acad. Sci., 65, 504 (1957)
2. R. G. HOLDICH, G. J. HANSON Hydrometallurgy 14, 387 (1985)
3. J. M. SING, S. K. GOGIA, S. N. TANDON Hydrometallurgy 9, 97 (1982)

4. N. N. LI, US Patent 3, 410, 794 (1968)
5. J. DRAXLER, R. MARR, Chem. End. Process, 20, 319- 29 (1986)
6. J. DRAXLER et al, Proc. Int. Solvent Extr. Conf., 553- 60, Munich (1986)
7. D. HARTMANN Thèse de l'Ecole Centrale Paris (1989)
8. M. SAMAR Thèse de Doctorat Paris (1992)

Electroassisted separation of metals by solvent extraction and supported-liquid membranes

D. J. Schiffrin
Department of Chemistry, University of Liverpool, Liverpool, England
Y. Cheng
Department of Chemistry, University of Liverpool, Liverpool, England
F. A. Silva
Department of Chemistry, University of Porto, Porto, Portugal
S. Vigato
Istituto di Chimica e Tecnologia dei Radioelementi, Padua, Italy
S. Tamburini
Istituto di Chimica e Tecnologia dei Radioelementi, Padua, Italy
D. Gilroy
E. A. Technology, Capenhurst, Chester, England
I. Bustero
INASMET, San Sebastián, Spain
J. C. Mugica
INASMET, San Sebastián, Spain

Electroassisted separation methods based on the use of electrically conducting supported liquid membranes and of liquid/liquid extraction under control of the interfacial Galvani potential are described. The control of interfacial potentials and the use of metal ion specific ligands have been employed to achieve specificity in metal ion separations and for the stripping and concentration of metal ion from dilute streams. It is shown that a supported liquid membrane can be used as an ion selective membrane for the electrolysis of specific metal ions. The interfacial potential for electroassisted solvent extraction is determined by the two-phase Nernst equation, which relates the distribution of ionic species to the Galvani potential and to the standard Gibbs energies of ionic transfer between an aqueous and an organic phase immiscible with water. It is shown that specificity in separations can be achieved in solvent extraction by controlling the Galvani potential.
Key Words: Electroassisted separations, solvent extraction, supported liquid membranes, ion transfers, liquid/liquid interface, electrochemistry.

1. Introduction

The use of phase transfer catalysts under interfacial potential control to carry two phase redox reactions has been recently demonstrated [1]. Transfer of ionic species across a liquid-liquid interface can also be controlled by applying interfacial potential differences or by the use of ionic partition equilibria. The first approach is equivalent to electrodialysis using membranes which are specific to a particular ion by virtue of the use of appropriate ligands in the organic phase. Solvent extraction assisted by an applied potential difference across the liquid/-

liquid interface is a separation technique which can be of significance in both metal ion separations and effluent treatment.

Supported Liquid Membranes (SLM) containing a carrier have recently received considerable attention for the extraction and refining of metals because of their potentials in industrial-scale separations and concentration processes [2-9] from dilute solutions. The technique uses a porous polymer membrane soaked with an organic solvent that separates the feed and the stripping solutions. A mobile extractant is incorporated in the membrane liquid phase and the solute of interest is extracted from the feed solution by the mobile carrier, generally using a proton gradient as the thermodynamic driving force. The metal ion-carrier complex formed at the feed side permeates through the membrane phase and is then stripped by the receiver solution. Advantages over traditional solvent extraction separation methods is the simultaneous extraction and stripping operation in the same unit [10-11], the uphill transport of solutes [12], high selectivity, the possibility of using expensive ligands due to the small volumes of organic phase required and low operating costs.

In the present work, the fundamental and applied aspects of electroassisted separation techniques have been investigated. The specificity and efficiency of metal separations are largely dependent on the properties of the ionophores used. For this reason a new family of ligands based on Schiff bases have been designed, prepared and tested. The electrochemical properties of their metal complexes at the liquid/liquid interface as well as ligand selectivities have been investigated.

2. Principles of the methods used

2.1 Solvent extraction under interfacial potential control

There is a fundamental difference between classical chemical solvent extraction and electroassisted separations. In the former case, the separation of metal ion M^{2+} occurs through the reaction:

$$M^{2+}(w) + LH_2(o) \rightarrow 2H^+(w) + ML(o) \qquad \text{feed} \qquad (1)$$

followed by

$$ML(o) + 2H^+(w) \rightarrow LH_2(o) + M^{2+}(w) \qquad \text{strip} \qquad (2)$$

where LH_2 represents the acid form of the ligand; *w* and *o* refer to the aqueous and organic phases, respectively. The sequence involves the transfer of the H^+ ion from the feed to the strip solution as the thermodynamic driving force.

In the electroassisted case, the transfers are:

$$M^{2+}(w) + L(o) \rightarrow ML^{2+}(o) \qquad \text{feed} \qquad (3)$$

followed by

$$ML^{2+}(o) \rightarrow L(o) + M^{2+} \qquad \text{strip} \tag{4}$$

L represents the ligand. In this case, the driving force is the applied electrical potential across the interface and the ionically conducting S.L.M. is a membrane separating the two compartments of an electrochemical cell. Thus, it is possible to achieve control over the separation efficiency not only by the choice of ligand, but also by the applied potential difference. The potential differences of interest are those appearing across the aqueous-organic solution interfaces. The driving force for ionic extraction into the membrane is the Galvani potential difference between the feed solution and the membrane phase, $\Delta_o^w\phi_f$, whereas stripping is electroassisted by the corresponding potential difference at the organic phase-receiving solution interface. Conductivity inside the membrane is achieved by the incorporation of hydrophobic electrolytes into it.

The Galvani potential difference between two immiscible electrolyte solutions for electroassisted extraction can be achieved not only by imposing externally an electrical potential, but also by the partition equilibrium of ionic species between the organic and aqueous phases. The partitioning ion is called the potential determining ion and the Galvani potential difference is given by the two phase Nernst equation [13]:

$$\Delta_o^w\phi = \Delta_o^w\phi_i^o + \frac{RT}{z_iF}\ln\left(\frac{a_i^{(o)}}{a_i^{(w)}}\right) \tag{5}$$

where $\Delta_o^w\phi$ is the inner potential difference ϕ(w)-ϕ(o), ϕ = Galvani potential and a_i is the activity of ion *i* and z_i is the charge of the ion. The standard Galvani potential, $\Delta_o^w\phi_i^o$, is defined by [13]:

$$\Delta_o^w\phi_i^o = -(1/z_iF)\,\Delta_o^w G_{t,i}^{0,o\rightarrow w} \tag{6}$$

where $\Delta_o^w G_{t,i}^{o\,0,o\rightarrow w}$ is the standard Gibbs energy of transfer of ion *i* between the aqueous and the organic phase.

2.2 Electrochemistry of assisted ion transfers

When the applied potential is large enough to overcome the unfavourable Gibbs energy of transfer of a particular ion, its transfer from the aqueous to the organic phase can take place. In consequence, ions can be specifically extracted by applying appropriate potential differences between an aqueous and an organic phase. However, if a suitable ionophore *L* is present in the organic phase, the formation of a complex between metal ion M^{2+} and *L* results in a shift of the transfer potential of M^{2+} given by [6]:

$$\Delta(\Delta\phi) = -\frac{RT}{2F}\ln[K_{ML}a_{M^{2+}}(w)] \tag{7}$$

$a_{M^{2+}}(w)$ denotes the activity of M^{2+} in water and K_{ML} is the stability constant of the ionophore-metal complex. Thus, the Gibbs energy of transfer is modified by that of complex formation; ions having a transfer potential outside the potential window of the base electrolytes can be transferred within the accessible polarisation range if suitable ligands are chosen. The importance of this is the high degree of control on extraction that can be achieved by both the applied potential and by the ligand chemistry.

2.3 Electrodialysis

Supported Liquid Membranes (SLM) are attractive alternatives to solvent extraction for the selective electroassisted removal and concentration of transition metal ions. In this technique, an ionically conducting SLM separates the feed and strip compartments in which the anode and cathode electrodes are placed. The membrane contains a ligand and a potential difference is imposed across it. Figure 1 represents schematically the transfer process when the chelating agent is specific

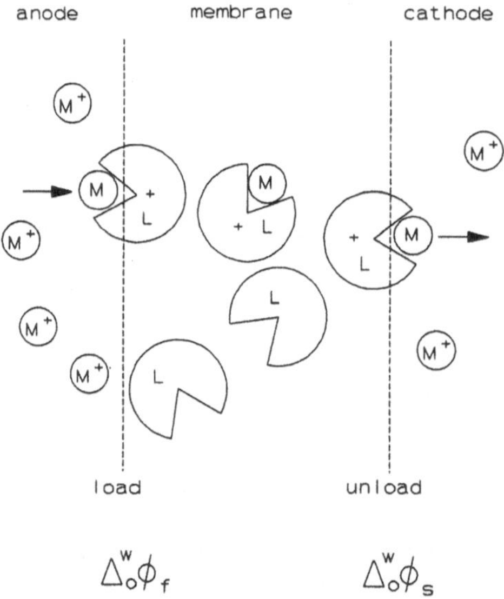

Figure 1 Schematic representation of electroassisted ion transfer across a supported liquid membrane. L = chelating ligand, M^+ = metal ion, f and s refer to the feed and strip solutions, respectively.

to metal ion M^+. When an appropriate ligand is present, the membrane becomes conducting with respect to one specific ion and the current observed between the feed and strip solutions is transported by the ion in question. This allows the electroassisted separation of a target ion. Obviously, the ion must be transported across the two membrane/solution interfaces and the bulk membrane phase and therefore, the transfer phenomena are different from those in solvent extraction due to the existence of the externally applied electric field. A further requirement of the method is the need to have ligands that form a charged metal ion complex, so that the extraction process can proceed under potential control.

3. Experimental

3.1 Preparative techniques for the new ligands

2,6-Diformyl-4Z-phenyl (Z = Cl, CH_3, $C(CH_3)_3$) was prepared according to the literature [14]. The ligands synthesised were characterised by ir, ^1H, ^{13}C, ^{31}P nmr spectroscopy and mass spectroscopy.

The Schiff bases have been prepared by the condensation reaction of stoichiometric amounts of the formyl- and polyamine- precursors in methanol solution. The solid compounds were purified by trituration with pentane, hexane or diethylether; where possible a chromatographic purification technique was used. The phosphorus containing ligand, P_2N_2 was synthesised according to the literature [15] by demetallation in benzene with NaCN of the deprotonated nickel(II) complex. This was obtained from the metal template catalysed condensation of *bis*(2-phenylphosphinophenylamido)nickel(II)[16]with1,3-*bis*(toluene-*p*-sulph-onyloxy)propane [17] in the presence of anhydrous potassium carbonate. The crude P_2N_2 product was purified by chromatography (alumina column and benzene as eluent) and recrystallised from ethanol/benzene. A summary of the ligands synthesized is shown in the Appendix.

3.2 Electrochemical methodology

A four electrode potentiostat was used for the cyclic voltammetric study of ionic transfers across the liquid/liquid interface [18]. The applied potentials refer to the cell:

CELL (A):

Ag | AgCl′ | TBACl(w) | TBATPB(o) | σ | LiCl(w) | AgCl | Ag
or SCE SCE

The organic base electrolytes were TBATPB and TPAsTPBCl; LiCl, Li_2SO_4 and $MgSO_4$ were used in the aqueous phase. The transfer Galvani potentials are referred to the potential of CELL(A) by:

$$E_A = E_{AgCl/Ag} - E'_{AgCl/Ag} + \Delta_o^w\phi - \Delta_o^w\phi_{TBA^+} \tag{8}$$

where $\Delta_o^w\phi_{TBA^+}$, the potential of the ion selective liquid/liquid junction, is calculated from equation (5); $\Delta_o^w\phi$ is the Galvani potential difference of the interface investigated. From the half wave potential of the ion transfer reaction, the standard Gibbs energy of transfer is obtained from equation (6). Both commercial and synthetic ligands and several transition metal ions were investigated. Experimental details are described elsewhere [1,13,18].

3.3 Solvent extraction

For developing the technique, the system studied was composed of an organic solution containing the organic base electrolyte and the ligand, placed in contact with an aqueous solution of the metal ion. One of the salts used in the organic and the aqueous phases had a common ion in order to fix the interfacial potential. Different ratios of the common ion were used to vary the potential difference established by the partition equilibrium across the interface (see equation (5)). The ionic partition systems chosen to establish the interfacial Galvani potential were TEATPB/TEACl and TMATPB/TMACl (Fluka, parum). The potential range that could be readily achieved for $\Delta_o^w\phi$ was -25 to 95 mV for TEA^+ and 50 to 170 mV for TMA^+ as partitioning ions.

The organic electrolytes used were TMATPB, TEATPB, and TPATPB. TMACl, TEACl and TPACl (all Fluka, purum) were employed in the aqueous phase.

DPT, Terpy and TPTZ were the commercial ligands (Aldrich) employed for electroassisted ion transfer studies. The metal ions studied were Cu^{2+}, Cd^{2+} and Ni^{2+}, all of which were added to the aqueous phase as their sulphate salts; the pH was adjusted to 2.5. The cell used in the electroassisted solvent extraction experiments is represented by:

CELL (B):

SCE | y M XCl(w) | y M XB + Ligand(o) | σ | z M XCl + Metal(w) | SCE

where X^+ is the potential determining ion and σ is the interface where ion partition takes place. The extraction experiments were carried out by placing 10 ml of aqueous phase containing 10 ppm of Cu^{2+} and different concentrations of TEACl in contact with 50 ml of the organic phase containing 2 mM TEATPB and the ligand, DPT. Several ligand/metal ratios were used. The interfacial potential was monitored during the course of the extraction experiments. The values of $\Delta_o^w\phi$ measured were in agreement with those predicted from equation (5). Activity corrections were carried out as described in reference [1]. The solutions were stirred for 1 hour, and the composition of the phases was analysed by atomic absorption spectroscopy.

3.4 Electrodialysis

Figure 2 shows the cell employed for electroassisted separations using a single hydrophobic membrane impregnated with the organic phase. Two platinum electrodes were used as current feeders and the membrane containing the solvent,

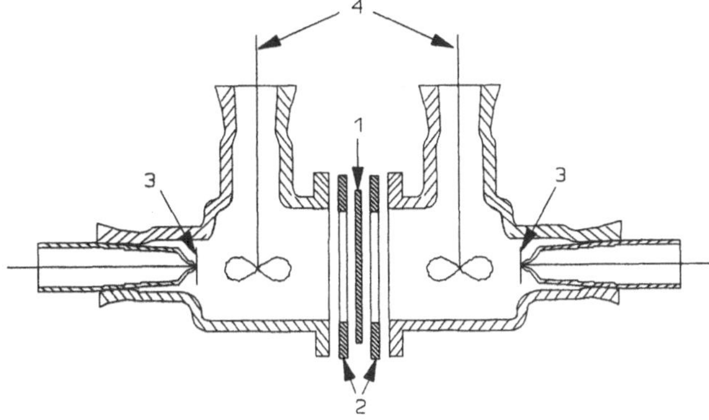

Figure 2 Cell used for the study of the electroassisted ion extraction across a supported liquid membrane. 1) supported liquid membrane; 2) silicon rubber seals; 3) Pt electrodes and 4) stirrers.

ligand and electrolyte was placed between the two compartments. The porous PTFE membrane was impregnated with a solution of TPTZ in nitroanisole containing TPAsTPBCl as conducting salt.

In order to investigate the possibility of scaling up the process a two membrane system was also designed in which the organic phase was contained by two hydrophobic porous membranes. Each aqueous compartment, of 130 cm^3 volume, was constructed with a circular glass flange to support one side of the organic membrane phase compartment. Platinised titanium electrodes were inserted into the aqueous solutions, which were usually 0.5 M Na_2SO_4. The anodic compartment also contained copper cations at a starting concentration of 500 - 1000 ppm. The organic phase was restricted between the two membranes by a polypropylene spacer, through which the organic solution was circulated. Two solvents, 1,2-DCE and benzonitrile, were used as the membrane liquid phase which contained the ligand DPT and TBATPB. Several commercially available microporous membranes were investigated as the liquid phase support.

Experiments were conducted by passing a constant current between the two electrodes and samples. The membrane was soaked overnight in the organic solution before being assembled between the two halves of the cell.

4. Results and Discussion

4.1 Ligands for electroassisted transfer

Many ligands have been designed and synthesised not only to mimic the functions of natural carriers for the transport of metal ions but also for the development of new methodologies in separation science.

Macrocyclic and macroacyclic ligands have been synthesised in the present work to verify the influence on metal ion binding and selectivity of the nature of the donor atoms, their relative position in the cavity, the number and size of chelate rings, the flexibility of the ring, the shape of the coordinating moiety and their planar or tridimensional architecture [19,20]. For macrocyclic systems, the cavity size, which may greatly influence the ability to discriminate between metal ions, has been investigated as an additional parameter. In the present work, a synthetic approach for the synthesis of macrocyclic or macroacyclic Schiff bases, their reduced analogues and related compounds has been developed. These ligands were originally designed for d- or f- metal ions coordination. Polydentate, macrocyclic, "end-off", side-off or acyclic ligands were prepared by condensation of the formyl precursors with the appropriate polyamines [26]. However, for some related phosphino-containing ligands, a different synthetic approach was utilized. In these compounds the donor atoms sets, their planar or tridimensional arrangement and the dimension of coordination cavities was modulated in order to synthesise a series of ligands with specific coordination ability and enhanced selectivity towards particular metal ions [26].

The ligands synthesised belonged to the following classes:

1. Macrocyclic or macroacyclic Schiff bases containing different donor atoms and therefore, different shapes and cavity size.
2. Phosphine containing ligands which are especially useful for transition metal ions in low oxidation states, for instance, P_2N_2 (see the Appendix for abbreviations).
3. Schiff bases containing crown-ethers as pendant groups.
4. Switchable ligands. These compounds have been synthesised by inserting groups which undergo redox-process with the consequent modification of their coordination properties and/or of the charge of the complexes formed. Dihydroxyphenyl-, ferrocene- and disulphide- groups have been chosen as redox centres for redox switchable ligands.

4.2 Facilitated ion transfer investigations at the liquid/liquid interface

The understanding of the ionic transfer process at the aqueous/non-aqueous interface is very important for solvent extraction and electrodialysis using supported liquid membranes. This section presents results on fundamental information on transport processes at the liquid/liquid interface required for the design of electroassisted ionic separations. Some commercially available and the synthetic ligands described above have been used to study the facilitated transfer of transition metal ions. In what follows, the ligand P_2N_2 will be used as an example to described the

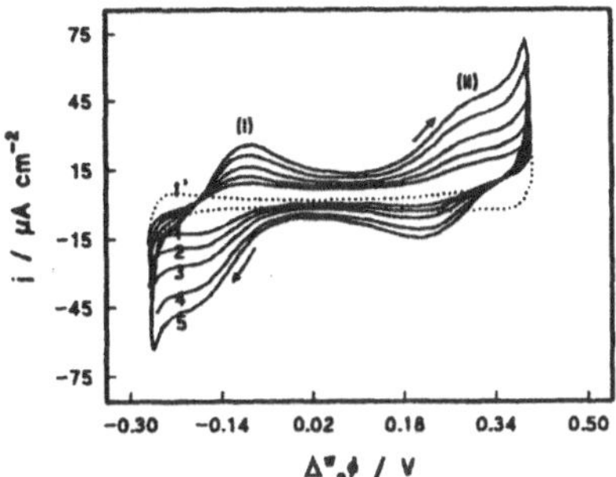

Figure 3 Cyclic voltammograms of $1.0\text{x}10^{-4}$ M copper(II) transfer facilitated by 1.0×10^{-3} M P_2N_2 in 1,2-DCE at different sweep rates: (1) 12; (2) 25; (3) 50; (4) 100 and (5) 150 mV s^{-1}. Dotted line (1'): CV obtained in the absence of copper in the aqueous phase, υ = 50 mV s^{-1}. The solution compositions were: 0.01 M TPAsTPBCl in the organic phase and 0.05 M LiCl in the aqueous phase.

type of experiments required to characterise a ligand. Results for other ligands will be given only in tabular form.Figure 3 shows the cyclic voltammograms for ion transfer of Cu^{2+} between water and 1,2-DCE. When the organic phase does not contain the P_2N_2, only capacitive currents can be observed (Fig. 3(1')). The increase in current observed at both potential limits correspond to the transfer of the base electrolyte ions. Therefore, copper ion transfer is not possible in the

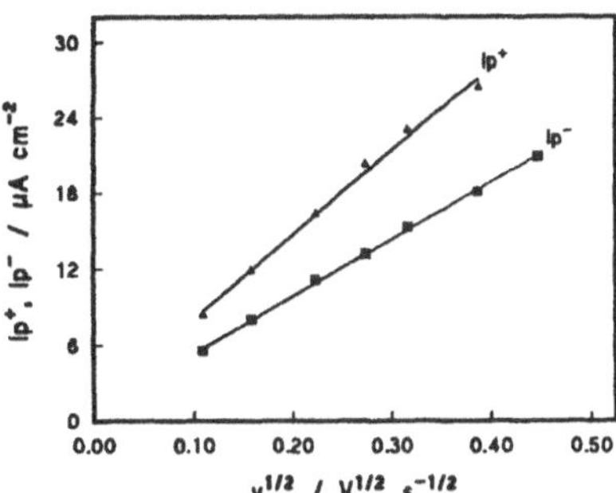

Figure 4 Dependence of the positive (I_p^+) and negative (I_p^+) peak current on the square root of the sweep rate (υ) for the results in Figure 3.

potential window available when the ligand is absent. In the presence of both $CuSO_4$ and P_2N_2 one well defined peak with a half wave potential of $\Delta_o^w\phi$ = -0.150 ± 0.005 V was observed. This peak was detected when either P_2N_2 or $CuSO_4$ was present in excess. At a constant P_2N_2 concentration, the peak current is proportional to the square root of the sweep rate (Figure 4) and also to $CuSO_4$ concentration.

The peak separation was always close to 60 mV for sweep rates between 12 to 100 mV s^{-1}, corresponding to the theoretical value for the transfer of monovalent ions [25] indicating fast, reversible ionic transfer. This means that the transferred species must be singly charged, and therefore, that coupled anion transfer must occur since the complexation of Cu^{2+} by the neutral ligand P_2N_2 should give a doubly charged complex. In order to investigate this unusual effect, experiments were carried out using Li_2SO_4 or $MgSO_4$, instead of LiCl as the aqueous supporting electrolyte. In these cases, no complex transfer current could be detected. Addition of excess LiCl to the sulphate electrolytes resulted in the appearance of a transfer peak located at the same half wave potential as that observed for the LiCl solution (Figure 5). This clearly illustrates that Cl^- is involved in the complexation and transfer of Cu^{2+}.

Figure 6(a) shows the linear dependence of the half wave potential on -log($C_{P_2N_2}$). The slope was found to be 62 ± 5 mV. A linear relationship between the half wave potential and -log(C_{Cl^-}) with a slope of 64 ± 5 mV was also observed (Figure 6(b)). These results indicate that the reaction order (P_i) for assisted Cu^{2+} transfer are $P_{Cl^-} = +1$ and $P_{P_2N_2} = -1$. However, the ***positive*** shift in the half wave potential with increase in chloride concentration is opposite to that observed in the complexation between metal ions and neutral ligands [21-23]. The effect of chloride can be two-fold: a) Ion pairing of hydrated Cu^{2+} with chloride; the further reaction of $CuCl^+$ with the ligand in the organic phase is energetically more favourable than that of Cu^{2+} due to the large desolvation Gibbs energies involved in the transfer and further coordination of the naked ion; and b) Ion pairing between protonated $P_2N_2H^+$ and Cl^-, giving a neutral form of the ligand, P_2N_2HCl in the organic phase. The following reaction can best explain the experimental results:

$$CuCl^+(w) + P_2N_2HCl(o) \rightarrow H^+(w) + Cl^-(w) + Cu(P_2N_2)(Cl)^+(o) \qquad (9)$$

This considers that the species transferred is the $CuCl^+$ complex, which on coordination to P_2N_2HCl results in proton (and hence, Cl^-) displacement. When the ligand is in excess in the organic phase, the process is controlled by diffusion of aqueous $CuCl^+$. For the other extreme situation, i.e. $C_{Cu^{2+}} >> C_{P_2N_2}^{(O)}$ and $C_{Cl^-} >> C_{P_2N_2}^{(O)}$, the half wave potential was independent of the metal ion concentration when C_{Cl^-} and $C_{P_2N_2}^{(O)}$ were fixed. This kind of reaction at a liquid/liquid interface was observed by Matsuda *et al.* [24] for the transfer of Ba^{2+} facilitated by 18C6.

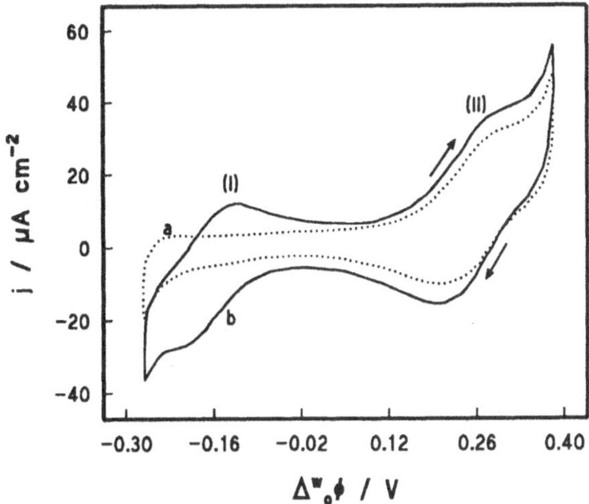

Figure 5 Cyclic voltammograms obtained in the absence (a), and presence (b) of 2.0 x 10^{-3} M LiCl in the aqueous phase. Base electrolyte in the aqueous phase was 0.05 M Li_2SO_4. The concentrations of Cu^{2+} and P_2N_2 were 1.0 x 10^{-4} and 2.0 x 10^{-3} M, respectively; υ = 50 mV s^{-1}. Peak I corresponds to assisted Cu^{2+} transfer; peak II is related to proton transfer.

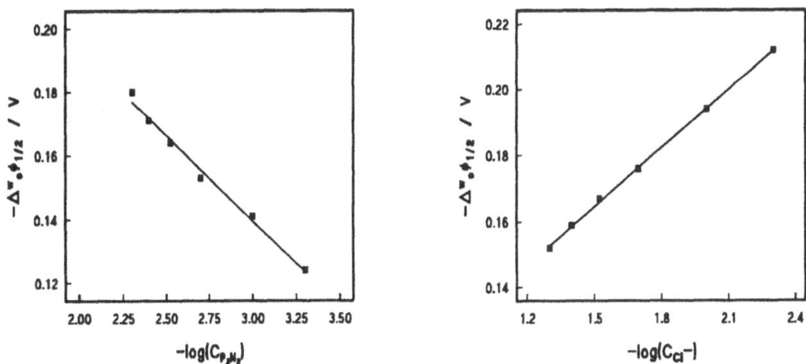

Figure 6 (a) Variation of the half wave potential with the concentration of P_2N_2 in the organic phase at υ = 50 mV s^{-1}. Organic phase: 0.01 M TPAsTPBCl; aqueous phase: x M LiCl + y M $MgSO_4$ + 1.0 x 10^{-4} M $CuSO_4$. The ionic strength in the aqueous phase was kept constant at 0.05 M. (b) Dependence of the half wave potential on the concentration of Cl^- in the aqueous phase. The conditions are same as in (a), but with $1.0x10^{-3}$ M P_2N_2 in the organic phase.

The chelating ability of P_2N_2 for the transfer of other metal ions has also been investigated for Cd^{2+}, Co^{2+}, Mn^{2+}, Ni^{2+} and Zn^{2+} ions. When 0.01 M TPAsTPBCl + 1.0 x 10^{-3} M P_2N_2 and 0.05 M LiCl were used as supporting electrolytes, the presence of Cd^{2+}, Co^{2+}, Mn^{2+}, Ni^{2+} or Zn^{2+} in the aqueous solution resulted in cyclic voltammograms which were indistinguishable from those of the base electrolytes. However, addition of copper(II) to the above mentioned metal ions mixture yields the copper(II) ion transfer current and no interference with copper transfer current could be seen due to the presence of the other metal ions. This indicates that P_2N_2 can be used for specific electroassisted separations of copper(II).

These results were confirmed by solvent extraction by fixing $\Delta_o^w\phi$ at 64 mV with TMA^+ as the partition ion. Results are given in Table I, which show that cadmium, nickel and zinc are not extracted from the aqueous to the organic phase, while copper is. These results are in good agreement with those obtained from cyclic voltammetry showing the usefulness of electrochemical methods for the study of metal extraction.

Table I. Extraction of metal ions with 1.0 x 10^{-3} M P_2N_2. The interfacial potential was fixed at 64 mV with tetraethylammonium as the partition ion; the aqueous phase contained 0.05 M LiCl and 5 x 10^{-4} M of the respective metal ion. The results are presented as the amount of metal found in the aqueous phase after contacting with the organic phase for the indicated period of time.

Metal Ion	1 hour (%)	2 hour (%)
Cu^{2+}	80	65
Cd^{2+}	97	98
Ni^{2+}	102	103
Zn^{2+}	100	99

Table II summarises the results obtained with other synthetic ligands. It can be seen that most ligands synthesised (Table II) show the required coordinating ability to one or two metal ions, indicating their possible use in electroassisted metal ion separations.

Table II Electrochemical behaviour of the ligands studied, showing the transfer of metal ions across the H_2O/1,2-DCE interface.

Ligand	**aqueous solution**			**metal**					
	HCl	**LiOH**	**LiCl**	**Pb^{2+}**	**Cu^{2+}**	**Ni^{2+}**	**Co^{2+}**	**Co^{2+}**	**Zn^{2+}**
K243	T	T	T*	T	T	T*	T	T*	T
K246	T	NT	NT	NT	T	NT	NT	NT	NT
K247	T	NT	NT	T	T	T*	T*	T*	NT
K248	T	NT	NT	T*	T	NT	NT	NT	NT
K250	T	NT	NT	T	T	T*	T*	T*	T*
K251	T	NT	NT	T	T*	T	T*	T*	T
K252	T	NT	NT	T	T	T	T*	T*	T
K253	T	NT	NT	T	T	T	T	T*	T
K254	T	T	T	--	--	--	--	--	--
K255	T	NT	T**	T	T	NT	NT	NT	NT
K264	T	T	T**	T	T*	NT	NT	NT	T*
K265	T	NT	T**	T	T	NT	T	T	T*
K266	NT	NT	NT	T	T*	NT	NT	NT	NT
K267	T	T	NT	T	T	NT	NT	NT	NT

T a transfer peak can be observed under that experimental condition
NT no transfer peak is observed under that experimental condition
* voltammograms with small and broad current peaks, with peaks at the end of potentialwindow or peaks that gradually disappear
** voltammograms with insignificant current peaks

4.3 Electroassisted solvent extraction

4.3.1 Electroassisted transfer of Cu^{2+} and Cd^{2+} by DPT

Electroassisted Cu^{2+} solvent extraction experiments with DPT as ligand were carried out at a Galvani potential $\Delta_o^w\phi$ controlled at 93 mV. The results are given in Table III. Previous cyclic voltammetric experiments showed that extraction should be feasible at this potential. Extraction was largely dependent on the ligand/metal ion ratio, stirring time and the controlled interfacial potential. It can be seen that 95% of Cu^{2+} is extracted with 1 hour stirring time and a ligand/metal ratio of 3:1. These results show that the considerations resulting from the two-phase Nernst equation (equations (5) and (7)) can be simply applied to the prediction of specific metal ion separations.

Table IV shows a summary of Cd^{2+} extraction results obtained with different values of $\Delta_o^w\phi$. It was found that the use of TMA^+ as partition ion gives an appropriate potential range under which Cd^{2+} ion transfer assisted by DPT was observed. However, in this case it was necessary to use a large concentration of ligand to obtain an extraction efficiency of 95% of Cd^{2+}.

Table III Percentages (%) of extraction of Cu^{2+} by DPT for different ligand-metal ion ratios. The applied Galvani potential was 93 mV.

stirring time	DPT :Cu			
(Hours)	1:1	2:1	3:1	10:1
1	55	85	95	100
2	60	90		
3	60			

Table IV Percentages (%) of extraction of Cd^{2+} by DPT; the applied Galvani potential was 167 mV.

Stirring Time	DPT :Cd			
(Hours)	2:1	10:1	30:1	50:1
5	0	15	70	95
15	0	25	71	
25	0	26		

4.3.1 Electroassisted transfer of Ni^{2+} by Terpy.

The results obtained are presented in Table V. As for the Cd^{2+}-DPT system, a high ligand to metal ion ratio is required for achieving high separation rates in a single operation.

4.3.2 Separation of Cu^{2+}, Cd^{2+} and Ni^{2+}

Based on the results presented above, the separation of Cu^{2+}, Cd^{2+} and Ni^{2+} is possible using suitable ligands and applying the appropriate interfacial potential by ionic partition. The separation process proposed is the following:

1. Transfer of 100% of Cu^{2+} to the organic phase using DPT as ligand and TEATPB/TEACl as ionic partition system.
2. Transfer of 100% of Cd^{2+} to the organic phase using a large concentra-

Table V Percentages (%) of extraction of Ni^{2+} by Terpy. 2 mM TEA^{+} in the organic phase and 10 mM in the aqueous phase were used to fix the interfacial potential.

Stirring Time		**Terpy**	**:Ni**		
(Hours)	**1:1**	**2:1**	**3:1**	**10:1**	**20:1**
1	21	30	43	50	91
4	29	54	75	89	99
8	38	85	94	98	
16	50	82	98		
24	51	86	98		

tion of DPT (50:1) and TMATPB/TMACl as the ionic partition system.

3. Ni^{2+} stays mainly in the aqueous phase although a small quantity of Ni^{2+} (15%) is extracted to the organic phase.

4.4 Electrodialysis

This section describes results obtained with the cell shown in Figure 2. The ligands chosen were previously investigated by cyclic voltammetry as described above. The membrane was impregnated with ligand dissolved in nitroanisole.

4.4.1 Nickel ion transfer

Electroassisted nickel ion transfer using TPTZ has been studied. The changes in concentration of metal ions in the catholyte as a function of electrodialysis time are presented in Figure 7. It can be seen that TPTZ selectively removes Ni^{2+} in preference to Cr^{3+} and Fe^{2+} from a mixture at a rate of approximately 7.9 mg cm^{-2} hr^{-1}. Co^{2+} is also transferred although at a lower rate, of 2.6 mg cm^{-2} hr^{-1}. The current efficiency for Ni^{2+} transfer by TPTZ was more than 50% and only 18% for Fe^{2+} and Cr^{3+}, respectively.

4.4.2 Copper ion transfer

Using 0.01 M DPT as an extractant and 0.1 M TBATPB as the organic base electrolyte, copper transfer from the anodic to the cathodic compartment was observed with 30% current efficiency. Presumably, proton transfer is responsible for the low current efficiency observed. Attempts were made to increase the operating current density, but this resulted in removal of the membrane organic phase. Unfortunately, preliminary experiments with the double membrane system showed too much retention of Cu^{2+} ion in the organic phase, which could not be easily

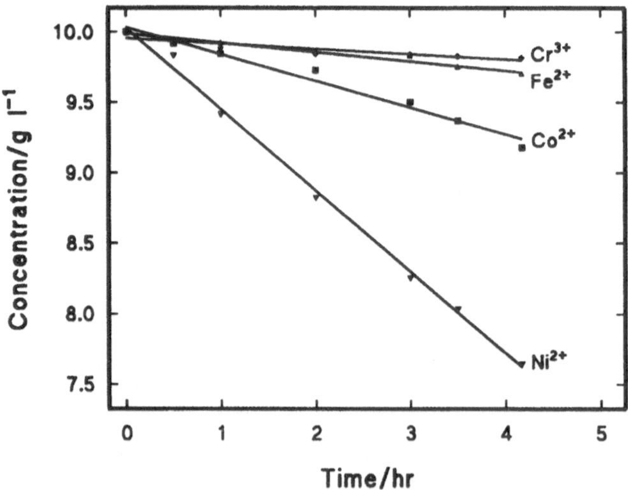

Figure 7 Electroassisted separation of Ni^{2+}, Co^{2+}, Fe^{2+} and Cr^{3+} ions through a supported liquid membrane containing 20 mM TPAsTPBCl and 10 mM TPTZ (extractant) nitroanisole solution. Applied current density = 4.5 mA cm^{-2}. The concentration is that in the anodic compartment. The initial concentration of all salts was 10 g dm^{-3}.

stripped to the cathodic compartment. Experiments with a thinner organic layer and a more appropriate organic solution composition are now being performed in order to overcome the above problem.

5. Conclusions

Most of the ligands derived from Schiff bases showing that synthesis in the present work exhibited coordination abilities towards transition metal ions with good selectivity. The ligands are hydrophobic enough to reduce ligand losses when they are incorporated in the organic phase of a liquid membrane. Ion transfers across the water/1,2-DCE interface facilitated by ionophores have been observed and investigated for several systems. The results obtained provide a good guidance for separations of metal ions by means of solvent extraction and supported liquid membranes. Selective removal of metal ions from mixtures in solution has been achieved successfully using both methods described above. The work described above clearly shows the usefulness of electroassisted separation strategies and current work is aimed at solving the problems that have been identified for practical applications:

1. Recycling of potential determining ions.
2. Practical solvents.
3. Current efficiency.

4. Current density.

Acknowledgement
This work has been carried out under the BRITE-EURAM Recycling of Non-Ferrous Metal programme, DG XII, Contract Number MA2R-CT91-0012-C.

Appendix

P_2N_2: *N,N'-bis*[2-(diphenylphosphino)phenyl]-propane-1,3-diamine
K243: 4-chloro-2,6-*bis*[2-phenylsulphide)ethyliminomethyl] phenol
K246: 4-hydroxyl-[2(2-phenylsulphide)ethyliminomethyl] phenol
K247: 4-methoxyl-[2(2-phenylsulphide)ethyliminomethyl] phenol phenol
K248: 4-chloro-2,6-*bis*[2(2-pyridyl)methyliminomethyl] phenol
K250: 6-hydroxyl-[2(2-pyridyl)methyliminomethyl] phenol
K251: 4-hydroxyl-[2(2-pyridyl)methyliminomethyl] phenol
K252: 6-methoxyl-[2(2-pyridyl)methyliminomethyl] phenol
K253: 6-ethoxyl-[2(2-pyridyl)methyliminomethyl] phenol
K255: 4-chloro-2,6-*bis*[2(2-pyridyl)methyliminomethyl] phenol
K264: 4-methyl-2,6-*bis*[4(benzo-15 crown 5)iminomethyl] prydine
K265: 2,6-*bis*[4(benzo-15 crown 5)iminomethyl] prydine
K266: 6-hydroxyl-2[4(benzo-15 crown 5)iminomethyl] phenol
K267: 3-methoxyl-2[4(benzo-15 crown 5)iminomethyl] phenol
DPT: 3-(2-Pyridyl)-5,6-diphenyl-1,2,4,triazine
Terpy: 2,2':6,2''-terpyridine
TPTZ: 2,4,6-tri(2-pyridyl)-s-triazine
18C6: 18-crown-6
TBATPB: tetrabutyl ammonium tetraphenyl borate
TPAsTPBCl: tetraphenyl arsonium tetrakis(4-chlorophenyl)borate
TEA: tetramethyl ammonium
TMA: tetramethyl ammonium
TPATPB: tetrapropylammonium tetraphenylborate
1,2-DCE: 1,2-dichloroethane

6. References

1. Cunnane, V.J. Schiffrin, D.J. Beltran, C.A. Gablewicz, G. and Solomon, T. (1988) The role of phase transfer catalysts in two phase redox reactions. **J. Electroanal. Chem., 247**, 203.
2. Huang, T.C. and Tsai, T.H. (1991) Separation of cobalt and nickel ions in lithium nitrate solutions by solvent extraction and liquid membrane with HEHEHP kerosene solution. **Acta Chem. Scand., 45**, 383.
3. Honaker, C.B. and Freiser, H. (1962), Kinetics of extraction of Zn dithizone. **J. Phys. Chem., 66**, 127.
4. McClellan, B.E. and Freiser, H. (1964) Kinetics and mechanism of extraction of zinc, nickel, cobalt and cadmium with diphenylthiocarbazone, di-o-tolyl-

thiocarbazone and di-α-naphthylthiocarbazone. **Anal. Chem., 36**, 2262.

5. Dietz, M.L. and Freiser, H. (1991) Role of the interface in the kinetics and mechanism of nickel extraction with certain halogen- and alkyl-substituted 8-Quinolinols. **Langmuir, 7**, 284.
6. Koryta, J. (1979) Electrochemical polarization phenomena at the interface of two immiscible electrolyte solutions. **Electrochim. Acta, 24**, 293; (1984) Electrochemical polarization phenomena at the interface of two immiscible electrolyte solutions-progress since 1978. **29**, 445; (1988) Electrochemical polarization phenomena at the interface of two immiscible electrolyte solutions-progress since 1983. **33**, 189.
7. Vanysek, P. and Buck, R.P.(1984) New development in liquid/liquid interface transport. **J. Electroanal. Chem., 163**, 1.
8. Koryta, J. and Skalicky, M. (1988) An electrochemical approach to salt extraction kinetics. **J. Colloid Interface Sci., 124**, 44.
9. Koryta, J. and Skalicky, M. (1987) Kinetics of salt extraction into the high-permittivity oil phase, Part III. **J. Electroanal. Chem., 229**, 265.
10. Flett, D.S. (1977) Chemical kinetics and mechanisms in solvent extraction of copper chelates. **Acc. Chem. Res., 10**, 99.
11. Albery, J.W. Choundhery, R.A. and Fisk, P.P. (1984) Kinetics and mechanisms in the solvent extraction of copper. **Faraday Discuss. Chem. Soc., 77**, 53.
12. Kihara, S. Suzuki, M. Maeda, K. Ogura K. and Matsui, M. (1986) The transfer of anions at the aqueous/organic solvents interface studied by current-scan polarography with the electrolyte dropping electrode. **J. Electroanal. Chem., 210**, 147.
13. Girault, H.H. and Schiffrin, D.J. (1989) electrochemistry at liquid/liquid interfaces, in **Electroanalytical Chemistry**, Vol. **15**. (ed. A. Bard), Marcel Dekker Inc., New York, 1989.
14. Lehn, J.M. (1988) Supramolecular chemistry-scope and perspectives molecules, supermolecules and molecular devices (Nobel lecture). **Angew. Chem. Int. Ed. Engl., 27** 89.
15. Cooper, M.K. Duckworth, P.A. Hambley, T.W. Organ, G.J. Henrick, K. McPartlin, M. and Parekh, A. (1989) Synthesis of a new P_2N_2 ligand N,N'-bis[2-diphenylphosphino)-phenyl]propane-1,3-diamine. **J. Chem. Soc. Dalton Trans.**, 1067.
16. Ansell, C.W.G. McPartlin, M. and Tasker, P.A. Cooper, M.K. and Duckworth, P.A. (1983) X-ray structure of *cis*-bis[(o-amidophenylphosphine] nickel(II) acetone solvate. **Inorg. Chim. Acta, 76**, L135.
17. Ansell, C.W.G. Cooper, M.K. Dancey, K.P. Duckworth, P.A. Hendrick, K. McPartlin, M. and Tasker, P.A. (1985) template synthesis of a new P_2N_2 macrocyclic ligand *via* direct alkylation of co-ordinated amido nitrogen atoms; X-Ray structure analysis of the free ligand and its neutral Ni(II) complex. **J. Chem. Soc., Chem. Commun.**, 439.
18. Cheng, Y. and Schiffrin, D.J. (1991) Electron transfer between bis-pyridine-(*meso*-tetraphenylporphyrinato) iron(III) and ruthenium(II) and the hexa-

cyanoferrate couple at the 1,2-dichloroethane/water interface. **J. Electroanal. Chem., 314**, 153.

19. Henrick, K. Tasker P.A. and Lindoy, L.L. (1985) Specification of bonding cavities in macrocyclic ligands. **Progr. Inorg. Chem., 33** 1. (ed. S.J. Lippard), J. Wiley Publisher, 1985.
20. Lindoy, L.F. (1989) The development of mixed donor macrocyclic systems for metal-ion discrimination. **Pure and Appl. Chem., 61** 1575.
21. Wang, E. and Liu, Y. (1986) Electrochemistry of cadmium ion at the water-nitrobenzene interface. **J. Electroanal. Chem., 214**, 465.
22. Fan, R. and Wang, X. (1989) Semi-differential cyclic voltammetry of cadmium ion transfer across the water/nitrobenzene interface facilitated by 2,2′-bipyridine. **J. Electroanal. Chem., 261**, 77.
23. Yoshida, Z. and Freiser, H. (1984) Ascending water electrode studies of metal extraction. Role of kinetics in the faradaic ion transfer of metal-phenanthroline complex ions across an aqueous-organic solvent interface. **Inorg. Chem., 23**, 3931.
24. Matsuda, H. Yamada, Y. Kanamori, K. Kudo Y. and Takeda, Y. (1991) Facilitated effect of neutral macrocyclic ligands on the ion transfer across the interface between aqueous and organic solutions. I. Theoretical equation of ion-transfer-polarographic current-potential curves and its experimental verification. **Bull. Chem. Soc. Jpn., 64**, 1497.
25. Bard, A.J. and Faulkner, L.R. (1980) **Electrochemical Methods, Fundamentals and Applications**, Wiley, New York.
26. (a) Vigato, P.A. Tamburini S. and Fenton, D.E. (1990) The activation of small molecules by dinuclear complexes of copper and other metals. **Coord. Chem. Rev., 106**, 25. (b) Fenton D.E. and Vigato, P.A. (1988) Macrocyclic Schiff base complexes of lanthanides and actinides. **Chem. Soc. Rev., 17**, 69.

Equilibrium and kinetic studies of copper extraction from chloride solutions with pyridine carboxylates

J. Szymanowski
Institute of Chemical Technology and Engineering, Poznan Technical University, Poznan, Poland
A. Jakubiak
Institute of Chemical Technology and Engineering, Poznan Technical University, Poznan, Poland
Gérard Cote
Laboratoire de Chimie Analytique associé au CNRS, Ecole Supérieure de Physique et de Chimie Industrielles, Paris, France
Denise Bauer
Laboratoire de Chimie Analytique associé au CNRS, Ecole Supérieure de Physique et de Chimie Industrielles, Paris, France
R. Cierpiszewski
Academy of Economics, Poznan, Poland

Equilibrium studies carried out at constant and non-constant water activity in chloride media demonstrate the superiority of the commercially available and well-tailored extractant ACORGA CLX-50 over studied pyridine carboxylate models, especially in respect of the selectivity of copper(II) extraction versus iron(III), zinc(II), lead(II) and hydrochloric acid. The ability of pyridine monocarboxylates for extracting copper(II) increases as follows : decyl picolinate < decyl nicotinate < decyl isonicotinate, whereas that of pyridine dicarboxylates ranges in the following order : dioctyl pyridine-3,5-dicarboxylate < dioctyl pyridine-2,4-dicarboxylate = dioctyl pyridine-2,5-dicarboxylate. Model pyridine monocarboxylates are much less selective for copper(II) than ACORGA CLX-50, especially with respect to iron(III). Among studied pyridine monocarboxylates, only decyl nicotinate shows a strong tendency to extract zinc(II). Pyridine monocarboxylates strongly extract HCl and can be ranged in the following order as far as their affinity for HCl is concerned : decyl picolinate < decyl nicotinate < decyl isonicotinate. On the other hand, ACORGA CLX-50 can extract only very small amounts of HCl. The rate of copper(II) extraction depends upon the ionic strength of the aqueous phase, concentration of reagents, extractant hydrophobicity and position of the esterified carboxylic group(s). For instance, decyl nicotinate extracts copper(II) much more rapidly than decyl isonicotinate. The flux of copper(II) decreases as the length of the alkyl group is increased in the homologous series of alkyl nicotinates, i.e. when the hydrophobicity of the extractant is increased.
Keywords : ACORGA CLX-50, chloride solutions, copper extraction, kinetics, pyridine carboxylates, thermodynamics, water activity.

1 Introduction

Hydroxyoximes are well established extractants [1] which are used in several plants to recover copper from dilute acidic sulphate solutions. By using them, copper can be recovered from oxide ores and wastes economically. However, such reagents are not of great interest for the processing of sulphidic ores. Indeed, the latter can be successfully leached with ferric chloride, but the resulting concentrated copper solutions cannot be treated with hydroxyoximes as the extraction of copper(II) by hydroxyoximes is basically a cation exchange and thus encounters the major problem of acid release which imposes a neutralizing step in the process. Solvating type extractants must therefore be used.

Pyridine carboxylates which are solvating type extractants were proposed by ICI Specialties, now ZENECA Specialties, for copper extraction from chloride solutions and the CUPREX Metal Extraction Process was successfully used at the pilot plant scale for the processing of sulphidic copper ore concentrates [2-4]. ACORGA CLX-50 which is commercially available from ZENECA Mining Chemicals was used as an extractant. It contains hydrophobic pyridine carboxylates, probably esters of pyridine-3,5-dicarboxylic acid of unknown structure and composition.

Equilibrium of copper extraction with model pyridine carboxylates and ACORGA CLX-50 was described recently [5,6]. It was found that the efficiency of copper extraction is strongly dependent on the water activity (a_w) and total concentration (σ) of ionic or molecular species dissolved in the aqueous phase. Extraction of copper with ACORGA CLX-50 is described by a more complex model than the extraction with individual pyridine carboxylate models.

The aim of this work was to obtain additional information about the extraction of copper(II) with individual model pyridine monocarboxylates and ACORGA CLX-50, including transfer of hydrochloric acid, selectivity for copper(II) versus iron(III), zinc(II) and lead(II) and kinetics of copper(II) extraction.

2 Experimental

Individual pyridine monocarboxylates (decyl picolinate, decyl isonicotinate and decyl, dodecyl and tetradecyl nicotinates) were synthesized by us as reported previously [5]. ACORGA CLX-50 was used as delivered by ZENECA Mining Chemicals. Some extraction data concerning pyridine dicarboxylates (dioctyl pyridine-2,4-dicarboxylate, dioctyl pyridine-2,5-dicarboxylate, dioctyl pyridine-3,5-dicarboxylate) are also presented. The formula of the various pyridine carboxylates considered in the present work have been given elsewhere [5,6].

Extraction of copper from aqueous chloride solutions with pyridine carboxylates was studied at non-constant and constant water activity as described in previous papers [5,6]. In the experiments carried out at non-constant water activity, chloride concentration was adjusted by addition of LiCl, whereas in the experiments at constant water activity (and constant σ), LiCl, NaCl, $LiNO_3$, $NaNO_3$ and $Mg(NO_3)$ were used to adjust chloride concentration, water activity and the total concentration of ionic or molecular species dissolved in the aqueous phase [6].

The selectivity of copper extraction was investigated by using aqueous solutions which initially contained 25 g L^{-1} Cu(II), 65 g L^{-1} Fe(III), 30 g L^{-1} Zn(II), 0.164 g L^{-1} Pb(II) and 0.1 mol L^{-1} HCl. Chloride ion concentration, adjusted by addition of LiCl, was equal to 2, 5 or 8 mol L^{-1}. The organic solutions contained 0.6 mol L^{-1} ACORGA CLX-50 in kerosene or 0.2 mol L^{-1} model pyridine monocarboxylates in toluene.

The ascending drop method was used for kinetics studies. Two columns having the same inner diameter (25 mm) but different heights, namely 8.5 and 160 cm, were used. The former was used as a reference column to eliminate ending effects, i.e., the contribution to extraction during drop formation. The aqueous phase contained 0.01

to 0.1 mol L^{-1} $CuCl_2$ and an appropriate amount of LiCl to adjust the ionic strength. The organic phase contained 5 x 10^{-3} mol L^{-1} of the extractant dissolved in toluene.

All the experiments were carried out at room temperature. Metal concentrations were determined in the aqueous phase by Inductively Coupled Plasma Atomic Emission Spectrometry with an ICP 1500 Plasma Therm Inc. equipment.

3 Results and discussion

Extraction of copper from chloride solutions with pyridine carboxylates can be described by the two following equations [6] :

$Cu^{2+}_{aq} + 2\,Cl^{-}_{aq} + 2\,EXT_o = CuCl_2(EXT)_{2\,o}$ (Extraction constant K_{ex})

with $K_{ex} = [CuCl_2(EXT)_2]_o\,[Cu^{2+}]_{aq}^{-1}\,[Cl^-]_{aq}^{-2}\,[EXT]_o^{-2}$

$Cu^{2+}_{aq} + i\,Cl^{-}_{aq} = CuCl_i{}^{(2-i)}{}_{aq}$ (Formation constant β)

with $\beta = [CuCl_i{}^{(2-i)}]_{aq}\,[Cu^{2+}]_{aq}^{-1}\,[Cl^-]_{aq}^{-1}$

where subscripts "aq" and "o" denotes aqueous and organic phases, respectively. Such a chemical model gives a good agreement with the experimental equilibrium data [6].

The ability of pyridine carboxylates to extract copper can be characterized both by the chloride ion and/or extractant concentrations needed to extract a given amount of copper, e.g., $[Cl^-]_{50}$ and $[EXT]_{50}$ needed to obtain 50% extraction with a phase volume ratio equal to 1. A series of typical extraction curves is given in Fig. 1.

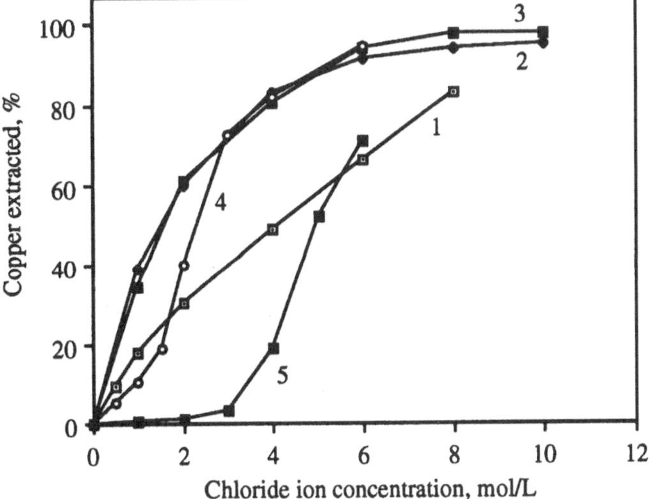

Fig. 1 Extraction of copper(II) (initially 10^{-2} mol L^{-1}) by various 0.2 mol L^{-1} pyridine carboxylates in toluene or ACORGA CLX-50 in kerosene, at non-constant water activity. (1) dioctyl pyridine-3,5-dicarboxylate; (2) dioctyl pyridine-2,5-dicarboxylate; (3) dioctyl pyridine-2,4-dicarboxylate; (4) decyl nicotinate; (5) ACORGA CLX-50.

Examination of $[Cl^-]_{50}$ values presented in Table 1 (column 1) for the system in which copper and extractant concentrations are kept constant and chloride concentration is adjusted by addition of appropriate amounts of LiCl shows that the extraction ability of ACORGA CLX-50 differs significantly from that of decyl nicotinate, dioctyl pyridine-2,4-dicarboxylate and dioctyl pyridine-2,5-dicarboxylate. It is however of interest that $[Cl^-]_{50}$ value for ACORGA CLX-50 is close to that corresponding to dioctyl pyridine-3,5-dicarboxylate, which tends to confirm that the active substance of ACORGA CLX-50 is a dialkyl pyridine-3,5-dicarboxylate.

The extraction properties of pyridine carboxylates depend significantly upon water activity. The latter decreases when the electrolyte concentration is increased. For instance, decyl pyridine monocarboxylates efficiently extract copper(II) even at high water activity, i.e., at moderately low electrolyte content ($[Cl^-]_{50}$ = 2 mol L^{-1}), whereas a medium of low water activity is needed to obtain a significant extraction of copper(II) with ACORGA CLX-50 ($[Cl^-]_{50}$ = 5 mol L^{-1}). Indeed, the extraction ability of ACORGA CLX-50 increases sharply when the water activity falls from 0.8 to about 0.6. Examination of the results given in columns 2 and 3 (Table 1) once more demonstrates that for obtaining similar yields of copper extraction, the water activity of the aqueous phase should be much lower in the case of ACORGA CLX-50 than for the model pyridine monocarboxylates.

Among pyridine monocarboxylates, the extraction ability increases in the following order : decyl picolinate < decyl nicotinate < decyl isonicotinate (column 3 in Table 1). Obviously, the difficulty of stripping copper(II) from pyridine monocarboxylate solutions increases in the same order. The extraction ability of pyridine dicarboxylates increases as follows : dioctyl pyridine-3,5-dicarboxylate < dioctyl pyridine-2,4-dicarboxylate = dioctyl pyridine-2,5-dicarboxylate (column 1 in Table 1).

Due to the high $[Cl^-]_{50}$ value obtained for ACORGA CLX-50 when operating at non-constant water activity, the stripping of copper(II) from ACORGA CLX-50

Table 1. Characteristic data for extraction of copper(II) (initially 10^{-2} mol L^{-1}) by various pyridine carboxylates in toluene or ACORGA CLX-50 in kerosene.

Extractant	$[Cl^-]_{50}$ (mol L^{-1}) at non-constant water activity for [EXT] = 0.2 mol L^{-1}	$[Cl^-]_{50}$ (mol L^{-1}) at constant water activity for [EXT] = 0.2 mol L^{-1}	$[EXT]_{50}$ (mol L^{-1}) at constant water activity (a) $[Cl^-]$ = 2 mol L^{-1} (b) $[Cl^-]$ = 6 mol L^{-1}
Decyl isonicotinate	-	0.4 (a_w = 0.83)	0.14 (a_w = 0.83) [a]
Decyl nicotinate	2.5	1.2 (a_w = 0.83)	0.22 (a_w = 0.83) [a]
Decyl picolinate	-	-	0.53 (a_w = 0.83) [a]
Dioctyl pyridine-2,4-dicarboxylate	1.7	-	-
Dioctyl pyridine-2,5-dicarboxylate	1.7	-	-
Dioctyl pyridine-3,5-dicarboxylate	4.4	-	-
ACORGA CLX-50	5.1	0.83 (a_w = 0.62)	0.15 (a_w = 0.62) [b]

solutions can be achieved with pure water, which is not possible with the other studied extractants. This clearly shows that ACORGA CLX-50 is a well-tailored reagent permitting both the extraction of copper from the aqueous solutions of low water activity obtained by leaching of sulphidic ores with ferric chloride and its cheap stripping with pure water.

Selectivity of copper extraction versus Fe(III), Zn(II) and Pb(II) is demonstrated in Table 2. Satisfactory results were obtained only for commercial reagent ACORGA CLX-50 although even in this case small amounts of iron(III), zinc(II) and lead(II) were also transferred to the organic phase. The effect of chloride concentration upon the selectivity of copper extraction with ACORGA CLX-50 is negligible under studied extraction conditions. It can be noticed that the three pyridine monocarboxylates extract equivalent amounts of iron(III). Conversely to the other pyridine monocarboxylates, decyl nicotinate has a strong tendency to extract zinc(II).

Undesired extraction of hydrochloric acid from aqueous solutions of various HCl concentrations is shown in Figs. 2 and 3. Pyridine monocarboxylates extract HCl strongly. The extraction ability increases as the basicity of nitrogen atom present in the aromatic ring increases, i.e., according to the following order : decyl picolinate < decyl nicotinate < decyl isonicotinate. One extractant molecule can extract several molecules of HCl as in the case of aliphatic amines [7]. The maximum amount of HCl transferred to the organic phase by decyl isonicotinate corresponds to a HCl / extractant molar ratio of 3.65. The precipitation of pyridine monocarboxylate hydrochlorides can be observed, especially when kerosene is used as a diluent instead of toluene.

ACORGA CLX-50 has only a weak tendency to extract HCl. This tendency is further depressed by the presence of copper(II) which is preferentially extracted. ACORGA CLX-50 can be dissolved both in aromatic and aliphatic diluents and the precipitation of hydrochlorides was not observed even at high aqueous HCl concentrations.

Table 2. Extraction of copper(II) by various 0.2 mol L^{-1} pyridine carboxylates in toluene or 0.6 mol L^{-1} ACORGA CLX-50 in kerosene in the presence of iron(III), zinc(II) and lead(II) : concentrations of metals in the organic phase after extraction. (initial aqueous solution : 25 g L^{-1} Cu(II), 65 g L^{-1} Fe(III), 30 g L^{-1} Zn(II), 164 ppm Pb(II) and 2, 5 or 8 mol L^{-1} Cl^-)

Extractant	Chloride concentration mol L^{-1}	Cu(II) g L^{-1}	Fe(III) g L^{-1}	Zn(II) g L^{-1}	Pb(II) ppm
ACORGA CLX-50	5	17.9	0.4	0.4	4
	8	17.0	0.9	0.8	9
Decyl picolinate	2	4.3	3.3	0.2	14
Decyl nicotinate	2	2.5	1.7	2.9	14
Decyl isonicotinate	2	3.2	3.0	0.1	14

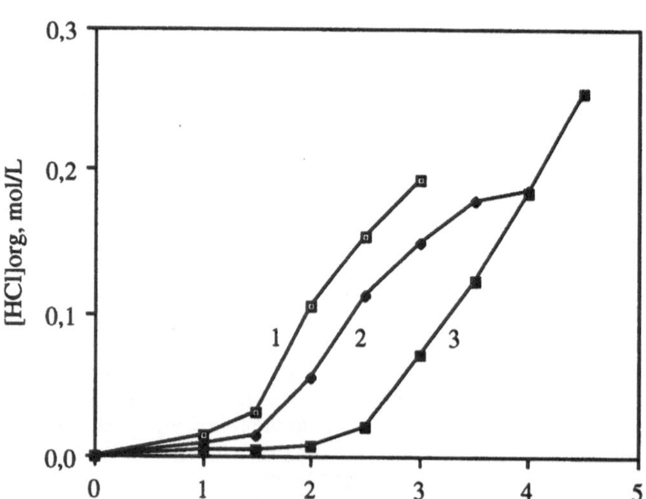

Fig. 2. Extraction of HCl by 0.2 mol L^{-1} pyridine monocarboxylates in toluene, at non-constant water activity. (1) decyl isonicotinate; (2) decyl nicotinate; (3) decyl picolinate.

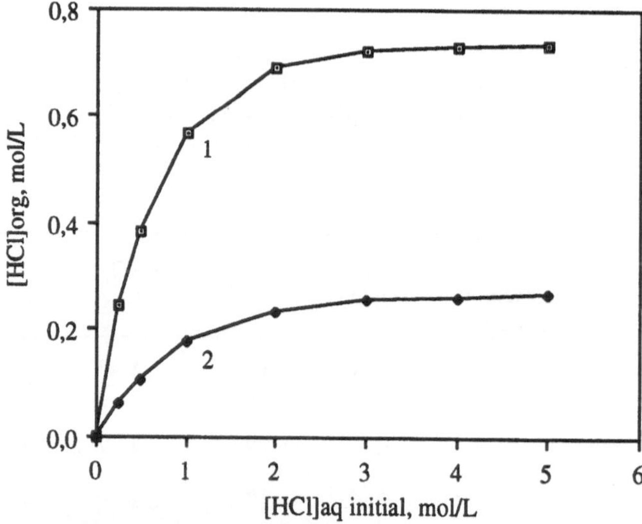

Fig. 3. Extraction of HCl by 0.2 mol L^{-1} (1) decyl isonicotinate in toluene or (2) ACORGA CLX-50 in kerosene, at constant water activity ($a_W = 0.61$).

Table 3. Rate of copper(II) extraction with 5.0×10^{-3} mol L^{-1} decyl isonicotinate in toluene, at room temperature

Copper(II) concentration mol L^{-1}	Ionic strength mol L^{-1}	Flux µg m^{-2} s^{-1}
0.10	4	172
0.05	4	67
0.01	4	22
0.10	2	81
0.10	0.5	22

Table 4. Rate of copper extraction with various 5.0×10^{-3} mol L^{-1} alkyl nicotinates in toluene, at room temperature, ($[CuCl_2] = 0.1$ mol L^{-1} and I = 4 mol L^{-1})

Alkyl group	Flux µg $m^{-2}s^{-1}$
$C_{10}H_{21}$	547
$C_{12}H_{25}$	470
$C_{14}H_{29}$	416

The rate of copper(II) extraction depends not only upon the reagent concentration, but also upon the ionic strength (or more generally of a_w and σ) (Table 3). For decyl isonicotinate, the two following relationships were obtained $J = -2.93 + 1690\,[Cu^{2+}]$ and $J = -1.92 + 43.2\,I$ where J denotes the flux of copper(II) through the interface in µg m^{-2} s^{-1} and I is the ionic strength in mol L^{-1}. The coefficients R^2 were equal to 0.974 and 0.999, respectively.

The structure of the extractant, i.e., position of the carboxylic group and the length of the alkyl chain, also affects the extraction of copper(II) (Table 4). For instance, decyl nicotinate extracts copper(II) significantly more rapidly (J = 547 µg m^{-2} s^{-1}) than decyl isonicotinate (J = 172 µg m^{-2} s^{-1}) (Tables 3 and 4). The flux J also decreases when the length of the alkyl group is increased in the alkyl nicotinate series, i.e., when the hydrophobicity of the extractant is increased. This can be expressed by the following relationship : $J = 870.6 - 32.75\, n_C$ with a coefficient R^2 equal to 0.991 and where n_C denotes the number of carbon atoms in the alkyl group of studied nicotinates.

Taking into account that the interfacial activity of pyridine monocarboxylates is weak and almost negligible in the concentration region presently considered [8], it is likely that the flux J of copper(II) from the aqueous phase to the organic phase mainly depends upon the aqueous concentration of the extractant which itself increases when the length of the alkyl chain is decreased.

4 Acknowledgement

The authors thank ZENECA Mining Chemicals for samples of ACORGA CLX-50 and for permission to publish the results obtained with this reagent. The work was partly supported by a Polish KBN grant.

5 References

1. Szymanowski, J. (1993) **Hydroxyoximes and Copper Hydrometallurgy**. CRC Press, Boca Raton, USA.
2. Dalton, R.F., Price, R., Quan, P.M. and Stewart, D. (1982) Extraction of metal values. **Eur. Pat. Appl.** EP 57,797.
3. Dalton, R.F., Price, R., Hermana, E. and Hoffman, B. (1987) The CUPREX process - a new chloride-based hydrometallurgical process for the recovery of copper from sulphidic ores, in **Separation Processes in Hydrometallurgy**, ed. G.A. Davies, Ellis Horwood Limited, Chichester, 1987, pp. 466-76.
4. Dalton, R.F., Diaz, G., Price, R. and Zunkel, A.D. (1991) The CUPREX metal extraction process : recovering copper from sulfide ores. **J.O.M.**, 43, 51-6.
5. Szymanowski, J., Jakubiak, A., Cote, G., Bauer, D. and Beger, J. (1993) Synthesis and extraction properties of pyridinecarboxylic acid esters, in **Solvent Extraction in the Process Industries (Proc. ISEC'93)**, eds. D.H. Logsdail and M.J. Slater, SCI - Elsevier Applied Science, London, 1993, pp. 1311-8.
6. Cote, G., Jakubiak, A., Bauer, D., Szymanowski, J., Mokili, B. and Poitrenaud, C. (1994) Modelling of extraction equilibrium for copper(II) extraction by pyridine carboxylic acid esters from concentrated chloride solutions at constant water activity and constant total concentration of ionic or molecular species dissolved in the aqueous phase. **Solvent Extraction and Ion Exchange**, in press.
7. Eyal, A.M. and Baniel, A.M. (1991) Recovery and concentration of strong mineral acids from dilute solutions through LLX. I. Review of parameters for adjusting extractant properties and analysis of process options. **Solvent Extraction and Ion Exchange**, 9, 195-210 and references therein.
8. Szymanowski, J., Cote, G., Sobczynska, A., Firgolski, K. and Jakubiak, A. (1994) Interfacial activity of decyl pyridinemonocarboxylates. **Solvent Extraction and Ion Exchange**, in press.

Cadmium removal from phosphoric acid—Israeli experience

S. D. Ukeles
E. Ben-Yoseph
N. P. Finkelstein
Israel Chemicals, Ltd., IMI Institute for Research and Development, Haifa Bay, Israel

Over the last few years, a number of European countries have imposed limits on the amount of cadmium that may be present in fertilisers that are consumed in these countries. This paper presents details of a process developed by IMI to remove cadmium and other heavy metals from wet process phosphoric acid (WPA). The process is based on the precipitation of cadmium from WPA through simultaneous addition of iron powder and a dithiophosphinate reagent to concentrated WPA exiting from the concentration stage. The cadmium-containing precipitate can be filtered or settled. This process has been successfully demonstrated in the laboratory, in pilot operation (60 liters concentrated WPA per hour), and on a plant scale (60 tons P_2O_5 per day).
Keywords: Arsenic, Cadmium, Chemical reagents, Copper, Filtration, Reduction, Settling, Wet Process phosphoric acid

1 Introduction

During the last decade, there has been a growing awareness thought Western Europe concerning the impact that fertilizers can have on the environment. One of these concerns relates to the element cadmium, one of the minor components in phosphate rock. Cadmium is highly toxic and causes death in some forms of water life at concentrations of 0.1 ppm or less. It also tends to accumulate in organisms over an extended period because of an inability to be easily eliminated. At present, the World Health Organization has recommended that the weekly human intake of cadmium should not exceed 400-500 micrograms. Over the last few years, a number of European countries have imposed limits on the amount of cadmium that may be present in fertilizers produced or used in those particular countries.

The quantity of cadmium carried over into phosphate fertilizers depends especially upon the origin of the phosphate. Characteristic cadmium contents in various rocks are presented in Table 1 [1]. As is apparent from the table, the phosphate rocks of volcanic origin (Russia, South Africa) contain very small amounts of cadmium, while sedimentary rocks have a wide range of cadmium concentrations.

Table 1. Average cadmium contents of phosphate rocks according to country of origin

Country	Phosphorous	mg Cd / kg rock
Israel	14.2 %	10 - 40
Jordan	14.6 %	5 - 10
Morocco (Khourigba)	14.2 %	18
Morocco (Youssofia)	14.6 %	40
Syria (Khneifiss)	13.9 %	6
Togo	15.7 %	55
Tunisia (Gafsa)	13.2 %	50
USA (Florida)	14.4 %	8 - 10
USA (Texas Gulf)	14.4 %	45
Russia (Kola)	17.5 %	0.15
South Africa (Palfos)	17.5 %	0.15

While it is known that fertilizers are only one of the sources of cadmium in soil, many companies and universities have been involved in the development of cadmium removal processes from phosphate rock or phosphoric acid. Some of the techniques suggested in the past for cadmium removal from phosphoric acid are summarized below:

1. Cadmium precipitation from phosphoric acid using hydrogen sulphide under pressure. Because pressure is needed, such a process would be technically difficult and expensive to operate [2].
2. Solvent extraction using amino-salts e.g. tridodecylamine hydrochloride that give extraction efficiencies that increase with the wet process acid concentration [3]. The high cost of the extractant would be one of the main drawbacks to this process.
3. Wet process acid is treated with dithiodiphosphoric acid ester [4], followed by the adsorption of the resulting cadmium-containing precipitate on a porous support such as active carbon, perlite, kieselguhr, carbon black (obtained by the pyrolysis of the acetylene or gasification under oxygen pressure of heavy oils), silicates or amino-silicates, in particular zeolites. The initial patent had the following disadvantages: (a) the process required a large amount of reagent, (b) the expensive adsorbent had to be regenerated by stripping, for example, with hydrochloric acid. A later process improvement [5] included a reduction step, but as pointed out later on, this does not significantly improve the process.
4. An ion flotation process which consists of the following steps: pretreatment (cooling, clarification and reduction), complexing of the cadmium ion with a dithiophosphate salt, flotation and removal of precipitate [6].

5. Bierman et al. [7] describe a solvent extraction process for the selective recovery of Cd, Mo, Zn and Ni from WPA, where the metal-containing carrier solution is contacted with a thio-organophosphinate extractant to remove metal values as a solid precipitate. The concentration of the extracted metal species is determined by the extractant concentration in the solvent, which can be selectively adjusted to recover a series of metal values in a sequence of stages. Some of the disadvantages of this process are that: (a) it would be very expensive for WPA containing copper and other metal ions, and (b) the phase separation would be very difficult in the presence of precipitates.

2 Process principle of cadmium removal for phosphoric acid

The process [8] described in this paper refers to the removal of cadmium and copper ions from either dilute (25-29% P_2O_5) or concentrated (52-54% P_2O_5) phosphoric acid by the precipitation of cadmium as a cadmium-dithiophosphinate complex, that can be subsequently filtered or settled. The treatment is carried out either in one or two reactors where a reducing agent and a thio-organophosphine reagent are simultaneously introduced into the wet process phosphoric acid, in the temperature range of 50-75°C. The cadmium-containing solids (after solid/liquid separation) give a relatively small amount of sludge which is washed and ultimately disposed of at a waste disposal site.

A reduction step is needed due to the following considerations: Cd^{+2} forms a compound with dithiophosphate ion corresponding to $Cd(DTP)_2$, i.e., a 1:2 stoichiometry. However, both Cu^{+2} and Cd^{+2} ions are present in Israeli phosphoric acid at about the same concentration. It is known that both copper and cadmium ions possess a strong affinity for dithiophosphates (DTP) and dithiophosphinates (DTPN). For dithiophosphates (DTP), it has been observed [6], using polarographic measurements made on synthetic phosphoric acid solutions, that copper reacts first quantitatively with dithiophosphates, i.e., cadmium begins to precipitate only after total copper consumption. Thus, the presence of copper ions in industrial phosphoric acid would cause excessive reagent consumption, through a combination of complexation and oxidation-reduction [6]. Another consumer of the dithiophosphate reagent, as pointed out in the same reference, is the ferric (Fe^{+3}) ion which when present in sufficient concentration, reacts according to reaction 1.

$$2Fe^{+3} + 2DTP^{-} \rightarrow 2Fe^{+2} + (DTP)_2 \downarrow \qquad (1)$$

Thus, an iron powder (or other reducing agents such as hydrazine) reduction step is necessary for the following reasons:

1. To achieve cementation of the Cu^{+2} species to copper metal and thus prevent Cu^{+2} from reacting with the dithiophosphate/dithiophosphinate reagents.
2. To reduce Fe^{+3} to Fe^{+2} and thus prevent reaction 1 from occurring. Aside from this reduction taking place (see reaction 2), there is also a competing reaction of the iron metal with the acidic protons to release hydrogen (reaction 3).

$$Fe^0 + 2Fe^{+3} \rightarrow 3\ Fe^{+2} \qquad (2)$$
$$Fe^0 + 2\ H^+ \rightarrow Fe^{+2} + H_2 \downarrow \qquad (3)$$

As will be shown later in this paper, similar behaviour has been observed with the dithiophosphinate reagent used in our work.

3 Experimental section

3.1 Batch tests

A mixture of industrial phosphoric acid (52-54% P_2O_5) or synthetic phosphoric acid (in which each sample contained added Cd and Cu ions) and 2% process gypsum (as an adsorbent) were heated to the required temperature in a round-bottom flask connected to a cooled condenser to avoid concentration of the acid. (The choice of gypsum was made due to its inherent presence in wet process acid production.) In several experiments, only the solids from post precipitation (about 0.7% solids) were present. This slurry was stirred by a double bladed pitched turbine to achieve complete mixing. To this mixture was added iron powder and the sodium diisobutyl dithiophosphinate reagent at 55°C, and these were stirred for a given period of time. In some experiments, the treatment step was divided into 2 parts: first the iron powder was added and this was followed by the addition of the dithiophosphinate reagent. Solid/liquid separation was achieved either by a filtration step or by settling of the treated slurry.

3.2 Continuous tests

While one and two reactor systems were studied, the final preferred system was the one reactor system. The laboratory continuous experiments were carried out in a reactor 2.2 litres in volume, having an oil-heated outer jacket to allow for operation up to 75°C. A metering pump fed 52-54% P_2O_5 acid to the heated reactor, while the dithiophosphinate reagent was delivered by a syringe pump. Pre-weighed iron powder samples were added to the reactor periodically, so that proper dosing was assured. Thorough mixing was achieved by high stirring speeds and the use of a single stage pitched blade turbine mixer. Samples were removed at specific intervals and either filtered, or allowed to settle at 55°C. When 2 reactors were used, the wet process acid and iron powder were added to the first reactor, and the dithiophosphinate reagent was added to the second reactor.

3.3 Pilot description

After obtaining encouraging laboratory scale results, a pilot unit was constructed for use in different plant locations. It consisted of a single reactor (constructed from PVC) having an active volume of 50 litres to which was fed concentrated acid (54% P_2O_5) at a flow of about 60 litres per hour. In this pilot, the following parameters were examined: dosages of the dithiophosphinate reagent and iron powder, reactor temperature, residence time, type and extent of mixing, and source of iron powder. A schematic diagram of the pilot is shown in Figure 1.

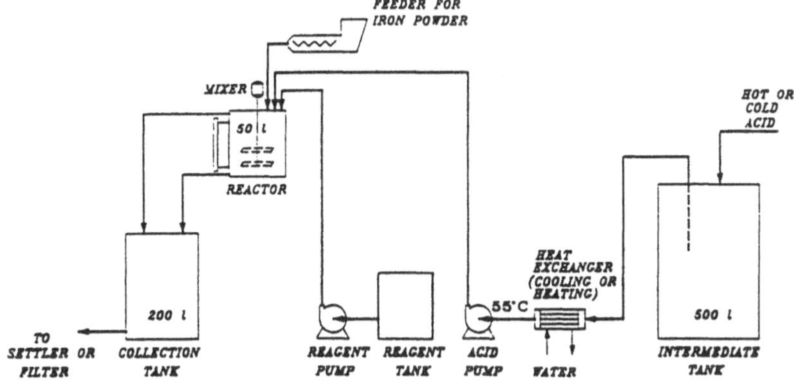

Fig. 1 Schematic description of pilot plant unit for cadmium precipitation from wet process acid

Previous pilot tests had been carried out on diluted WPA (27% P_2O_5), but the test using concentrated acid was chosen in order to give flexibility in the treatment of the acid by allowing for partial treatment of the overall acid production.

3.4 Washing tests

Tests were carried out to determine whether the cadmium and copper adsorbed on the solids present in phosphoric acid are washed out during a simulated washing of the solids. Solids isolated by vacuum filtration from the 27% P_2O_5 acid after treatment with the dithiophosphate (DTP) or dithiophosphinate (DTPN) and iron powder, were successively washed with solutions containing 11% and 3% P_2O_5, and then with water, using a 1:1 dry solids/wash liquor ratio. The 3 wash solutions were subsequently analysed for Cd and Cu.

4 Results and discussion

4.1 Batch results

4.1.1 Reagent screening tests

The results of reagent screening tests, a partial listing being given in Table 2, showed that the most promising commercial reagents for cadmium removal were DTP (Aero 3477) and DTPN (Aerophine 3418A) reagents. A brief description of the above reagents is given in Appendix 1.

4.1.2 Reagent dosages (dithiophosphinate and iron powder)

Table 2, shows that a large dosage of the dithiophosphinate reagent (0.3% or 10 kg reagent per ton P_2O_5), even in the absence of the iron powder reducing agent, was capable of removing virtually all of the Cd and Cu ions present in solution. This is shown for concentrated WPA (52% P_2O_5) in Figure 2, where the cadmium and copper concentrations in the acid decreased with increasing dithiophosphinate reagent, in the absence of the iron powder.

Table 2. Reagent screening tests

Reagent test results for 30% P_2O_5 acid

Reaction conditions: 60°C — Solids: 30% gypsum

Original acid ppm	Reagent	%	% Fe	Cd ppm	Cu ppm
Cd 36 Cu 40	Mercaptobenzthiazole**	0.5	0	27	1
Cd 19 Cu 27	Aero 3894**	0.075	0.34	19	1
Cd 22 Cu 28	Cyanex 471X**	0.15	0.34	22	1
Cd 36 Cu 40	Aerophine 3418A*	0.3	0	1	<1
Cd 20 Cu 28	Aerophine 3418A*	0.15	0.34	<1	1
Cd 24 Cu 24	Aero 3477*	0.28	0	<1	<1
Cd 22 Cu 28	Senkol (65)**	0.3	0.34	21	1
Cd 22 Cu 29	Aero 4037**	0.3	0	23	30

* Composition given in Appendix 1 ** Composition given in Appendix 2

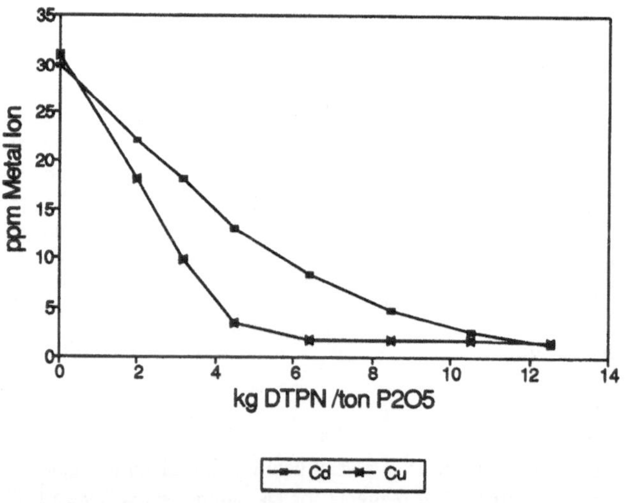

Fig.2 Cadmium and copper ion concentrations present in treated concentrated phosphoric acid (52% P_2O_5) as a function of dithiophosphinate dosage. No iron powder added. Temperature: 55°C

The addition of an excess amount of iron powder (Fe) allowed the use of only 2.5 kg dithioposphinate reagent per ton P_2O_5 to remove virtually all of the Cd and Cu from the filter acid (30% P_2O_5). Data are presented in Table 3 to show the influence of increasing reagent dosage at a fixed dosage of reducing agent. In the range of 1.2-2.5 kg dithiophosphinate reagent per ton P_2O_5, a significant decrease in the extent to which cadmium and copper were removed was observed. In this table, results show that: (a) cadmium and copper ions are removed also when 1% solids are used; and (b) virtually the same results are obtained whether the iron powder is added first followed by a later addition of the dithiophosphinate reagent or both reagents are added simultaneously.

Table 3. Effect of DTPN dosage and percent solids

Reagent test results for 30% P_2O_5 acid

Reaction conditions: 60°C Iron powder: 11 kg/ton P_2O_5

Original acid ppm	% Solids	DTPN Reagent %	kg/t P_2O_5	Cd ppm	Cu ppm
Cd 22 Cu 29	30	0.038	1.2*	12	1
Cd 20 Cu 28	30	0.075	2.5*	1	1
Cd 20 Cu 28	30	0.15	5.0*	<1	1
Cd 18 Cu 30	30	0.075	2.5*	2	0.7
Cd 22 Cu 29	1	0.075	2.5**	0.4	0.7

Notes: * DTPN reagent and iron powder added together
** Fe powder added 3/4 hour before DTPN

When iron powder is used as the reducing agent, the liberation of hydrogen gas occurs because of the metal's reaction with acid. The amount of hydrogen is very small, perhaps due to hydrogen gas reacting immediately as a reducing agent, as well as the iron powder's competing reactions with Cu^{+2} and Fe^{+3}. In later pilot and plant tests (where good air circulation was present), we were unable to detect any release of hydrogen.

4.1.3 Effect of temperature

The effect of reaction temperature was studied extensively for the concentrated acid (52% P_2O_5). Three temperatures were studied: 25°C, 55°C and 80°C. The results are presented in Table 4 for dosages of (a) 4 kg iron powder/ton P_2O_5 and 3 kg dithiophosphinate/ton P_2O_5 and (b) 8 kg iron powder/ton P_2O_5 and 2 kg dithiophosphinate/ton P_2O_5. The cadmium removal was most significant at 55°C for both of these sets of experiments. At temperatures well below 55°C, the efficiency of cementation of the copper ion will decrease while at temperatures of 75°C-80°C, the onset of dithiophosphinate decomposition begins to be an important consideration.

Table 4. Effect of temperature on Cd and Cu removal

Test results for 52% P_2O_5 acid with gypsum solids
Reagent: Dithiophosphinate   Mixing time: 2 hours

Original acid ppm	Temp. °C	% Solids	Reagent kg/t P_2O_5	Fe kg/t P_2O_5	Cd ppm	Cu ppm
Cd 25 Cu 33	25	0.7	3	4	15	3.5
Cd 28 Cu 31	55	2	3	4	4	1.5
Cd 25 Cu 33	80	2	3	4	16	1.2
Cd 25 Cu 33	25	0.7	2	8	19	4.0
Cd 28 Cu 31	55	2	2	8	7	1.1
Cd 25 Cu 33	80	2	2	8	18	1.2

4.1.4 Washing tests

The aim of the washing is to recover the P_2O_5 from the solids. The gypsum after the cadmium removal treatment and filtration contains (a) cadmium still present in non-dissolved phosphate rock (b) cadmium precipitated during reaction with the dithiophosphinate reagent and (c) cadmium from the treated WPA entrained in the solids.

Two comparative tests were made to determine whether the Cd-DTP or the Cd-DTPN are washed from the gypsum cake after they are trapped on it during the treatment process. 332 g wet process acid (30% P_2O_5) containing about 36 ppm Cd and 40 ppm Cu were mixed at 60°C with 143 g of $CaSO_4 \bullet 2H_2O$. Results are summarized in Table 5. In experiment A-1, 1.12 g iron powder and 2.5 g Aero 3477 (DTP) were added to the acid slurry, while in experiment A-2, 1.12 g iron powder and 2.5 g Aerophine 3418A (DTPN) were added to the acid slurry. After mixing for 1 hour followed by filtration, each of the filter cakes was washed successively with 11% P_2O_5, 3% P_2O_5, and water. The wash waters were analyzed for cadmium and copper. It was calculated that at least 50% of the total cadmium present in the gypsum solids was washed when the dithiophosphate (DTP) reagent was used while only about 4% of the cadmium was liberated when the dithiophosphinate (DTPN) was the reagent of choice. The same advantage has been observed for the DTPN reagent when 52% P_2O_5 acid was used.

Table 5. Summary of washing tests

Cadmium removal from 30% P_2O_5 acid
Experiment A-1 Dithiophosphate (DTP)
Wash 1: 13 ppm Cd
Wash 2: 1 ppm Cd
Wash 3: 1 ppm Cd
Experiment A-2 Dithiophosphinate (DTPN)
Wash 1: <1 ppm Cd
Wash 2: <1 ppm Cd
Wash 3: <1 ppm Cd

4.2 Continuous laboratory bench scale testing

An important consideration in the process development is the treatment cost per unit ton P_2O_5. Once the treatment had been shown to be feasible in batch tests with both dilute (25% P_2O_5) and concentrated (54% P_2O_5) acid, the research programme focused on verifying optimum operating conditions i.e. highest reagent efficiency at lowest cost.

4.2.1 Residence time

The effect of residence time under continuous operation was studied in a single reactor at three different residence times: 5, 30, and 60 minutes. The reagent dosages were 4 kg iron powder/ton P_2O_5 and 2 kg dithiophosphinate/ton P_2O_5 in the three continuous runs. A summary of the results is shown graphically in Figure 3. When the residence time was 60 minutes, a distinct improvement in the final Cd level was observed. At this residence time, the level of the copper ions decreased in the acid by 94% of its original concentration (from 32 ppm to 1 ppm), while that of the cadmium ions was lowered by about 82% of its initial level (from 28 ppm to 5 ppm). It was also noted that after 60 minutes, the total iron concentration in the acid was higher than after shorter periods, which could mean that more extensive copper cementation and Fe^{+3} reduction to Fe^{+2} had occurred. P_2O_5 concentrations in the treated acid remained virtually unchanged from their initial concentration.

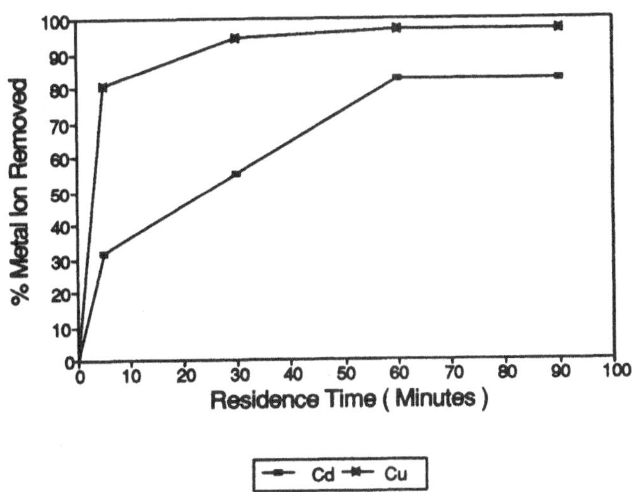

Fig. 3 Percent metal ion removed versus residence time (minutes). Continuous run carried out at 55°C. Dosages: 4 kg iron powder/ton P_2O_5 and 2 kg dithiophosphinate/ton P_2O_5

A residence time of 2 hours, (results not presented in this paper), did not succeed in improving the extent of copper cementation. Thus, the 60 minute residence time was chosen for later pilot operation.

4.2.2 Iron powder and dithiophosphinate reagent dosages

Comparisons were made at varying iron powder and dithiophosphinate dosages to determine the effect of dosage on the extent of cadmium removal in the treated acid. Continuous runs were made using dosages of 3 and 4 kg iron powder per ton P_2O_5 at a constant dosage of 2 kg dithiphosphinate per ton P_2O_5. The graphical comparison between the 2 runs is shown in Figure 4, where the extent of cadmium removal is shown as a function of running time. Both runs were made at 60 minutes residence time and at a temperature of 55°C. At a dosage of 4 kg iron powder per ton P_2O_5, slightly more than 80% of the cadmium ion was removed after about 3 hours of running time, while almost 70% removal was achieved when the dosage was 3 kg iron powder per ton P_2O_5. (It is interesting to note that a stoichiometric calculation shows that only 2 kg iron powder per ton P_2O_5 is needed to reduce Fe^{+3} to Fe^{+2} and Cu^{+2} to Cu^0).

Similarly, the sensitivity of the cadmium removal process to dithiophosphinate concentration was compared by decreasing the dithiophosphinate dosage in a continuous run from 2 kg /ton P_2O_5 to 1 kg /ton P_2O_5, at a constant iron powder dosage of 4 kg/ton P_2O_5. The comparison is shown in Figure 5, where the extent of cadmium removal is shown as a function of running time. For 1 kg dithiophosphinate per ton P_2O_5, a maximum of 15% of the cadmium ion was removed.

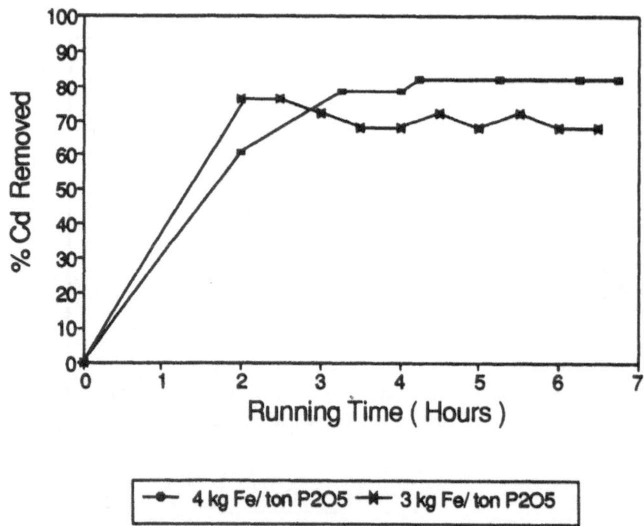

Fig. 4 Percent cadmium ion removed versus running time (hours), at varying iron powder dosage. Conditions for the continuous run: residence time 1 hour, and dithiophosphinate dosage of 2 kg/ton P_2O_5

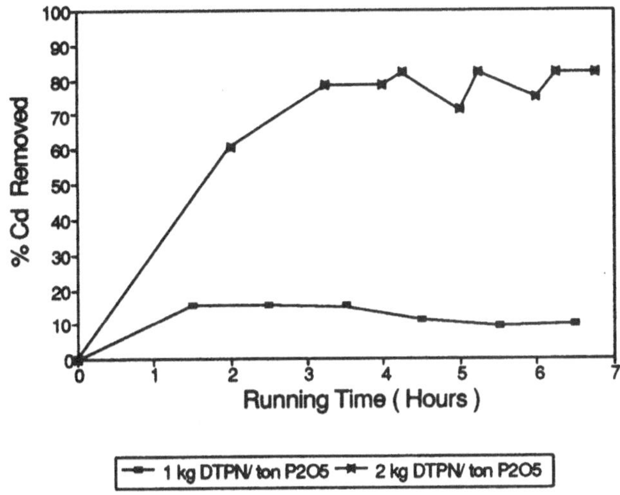

Fig. 5 Percent cadmium ion removed versus running time (hours), at varying DTPN dosage. Conditions for the continuous run: residence time of 1 hour and iron powder dosage of 4 kg/ton P_2O_5

4.2.3 Temperature

As previously noted for batch tests, the use of a treatment temperature below 40-45°C will decrease the efficiency of the cadmium removal process, probably due to a decrease in the kinetics of the cementation reaction. A comparison of the extent of Cd removal at 35°C and 55°C is shown in Figure 6, where each test was made using a dosage of 2 kg dithiophosphinate per ton P_2O_5 and 4 kg iron powder per ton P_2O_5. (The choice of 35°C as a test temperature was made as this can be the temperature of the product acid after it remains in the storage tank for a day or two.) It was seen that about 40% of the cadmium ion could be removed at 35°C, while about 80% cadmium removal was achieved at 55°C. In subsequent pilot and plant tests, the temperature range of 55-60°C has been proven to be the most successful range in which to operate. Operation of the process above 60°C should probably be avoided for the following reasons: (a) A high temperature would most likely accelerate the decomposition of the dithiophosphinate reagent, (b) excessive amounts of dissolved iron would be present in the product acid, thus causing post-precipitation problems, and (c) the oxidation of the dithiophosphinate reagent by Fe^{+3} would accelerate at higher temperatures, thus decreasing the amount of reagent available for reaction with Cd.

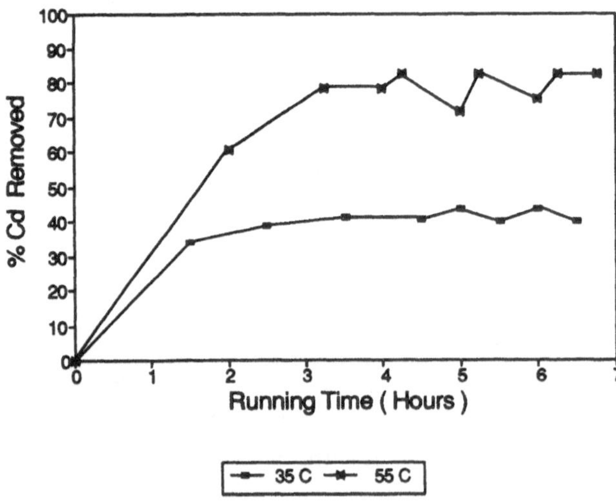

Fig. 6 Percent cadmium ion removed versus running time (hours) at 2 temperatures. Conditions for the continuous run: residence time of 1 hour, and iron powder dosage of 4 kg/ton P_2O_5 and dithiophosphinate dosage of 2 kg/ton P_2O_5

5 Pilot testing with concentrated phosphoric acid

The above process was tested extensively in continuous pilot operations, both for dilute WPA (27% P_2O_5) and concentrated acid (52-54% P_2O_5), the latter pilot previously described in Section 3.3.

The following conclusions were drawn from the pilot tests:

(a) Extremely efficient mixing of the acid is required to homogenize the added iron powder and to prevent it from floating on the surface of the acid.

(b) The same temperature and residence time requirements (55°C and 60 minutes, respectively) as previously determined, were found to be valid for this scale of operation.

(c) The use of 3-3.5 kg of iron powder and 2-2.5 kg dithiophosphinate reagent, each dosage per ton P_2O_5, lowered the Cd and Cu concentrations from 27 ppm to 2-4 ppm, and from 37 ppm to 2 ppm, respectively.

(d) There was no indication of any build-up of hydrogen gas that could be formed via the reaction of iron powder with the acid. Thus, there is no danger of explosion during the treatment.

(e) The crystals of the Cd-DTPN precipitate are very small. It is possible to filter them (along with other solids present in the wet process acid) either by vacuum or pressure filtration. When the solid/liquid separation is by sedimentation, long settling times are needed due to the small crystal size.

The treatment presented has already been in operation in Israel on a plant scale for several years. To the best of our knowledge, there has been no cadmium removal process for WPA implemented until now.

5.1 Untreated versus treated wet process phosphoric acid

Presented in Table 6 is a detailed comparison of the composition of the concentrated acid prior to and after the cadmium removal treatment. The total iron level rose from 0.18% to just below 0.3%, while there was a small increase in the overall organic matter composition due to the addition of the organic reagent. It should also be noted that the arsenic concentration in the treated acid was also significantly lowered, apparently through complexation with the dithiophosphinate reagent. From experiments carried out, but not reported here, it appears that this reagent is more selective to the arsenic in the acid than it is to the cadmium ion.

Table 6. Phosphoric acid before and after heavy metal ion removal using the IMI process

Species	Before treatment	After treatment
$\%P_2O_5$	52.9	52.4
%Fe	0.18	0.28
$\%Fe^{+2}$	0.046	0.185
%Mg	0.36	0.35
%Al	0.17	0.18
%Na	0.070	0.053
%F	0.19	0.16
%C	0.072	0.094
ppm Cd	23	4
ppm Cu	33	2
ppm As	17	2.1
ppm Ca	100	70
ppm K	370	420
ppm Cl	180	190
ppm Mo	18	17
ppm V	240	230
ppm Ni	43	42
ppm Zn	760	720
ppm Pb	<1	<1
ppm Cr	140	140
ppm Si	240	180

5.2 The use of sodium diisobutyl dithiophosphinate

The sodium diisobutyl dithiophosphinate reagent significantly differs in its structure from the dithiophosphate reagent, as it contains different organic functional groups. In addition, there is a basic difference in the behaviour of these two types of reagents toward the adsorbent used. Whereas the metal ions (principally cadmium, copper and arsenic) removed by the dithiophosphinate reagent are firmly retained on the adsorbent, such is not the case with the dithiophosphate reagents. This property of the dithiophosphinate reagent is important when considering the manufacturing process for wet process acid, where calcium sulphate dihydrate (in the case of a dihydrate process) is produced. After its filtration, the gypsum retains a substantial amount of phosphoric acid, which must be recovered prior to its disposal. This washing is executed in several stages with dilute wet process acid, followed by washing with water. The efficiency of these washings will actually determine the total yield of WPA recovered from the original phosphate rock. Thus, when using the dithiophosphinate reagent, in which the cadmium complex and copper metal are firmly retained on the adsorbent (e.g. calcium sulphate dihydrate), a substantially complete washing of the phosphoric acid can be achieved without removing the heavy metal impurities adsorbed on the solid.

6 Conclusion

The process developed by IMI and described in this report can be used for the removal of various heavy metals from phosphoric acid. The process is relatively simple and flexible for use in WPA plants. It can also be used at a client site for acid purchasers who require quantities of low cadmium phosphoric acid. The small amount of cadmium-containing solids resulting from the process can be disposed of at permitted sites.

7 References

1. Singh B.R., Unwanted Components of Commercial Fertilisers and their Agricultural Effects. Proceedings No. 312. The Fertiliser Society.
2. Baechle H.T. and Wolstein F., Cadmium Compounds in Mineral Fertilisers. The Fertiliser Society, London, 4 October, 1984.
3. Frankenfeld K., Ruschke P., Eich G. (Buddenheim), Removing Cadmium from Acidic Phosphorous Pentoxide containing Solutions. Ger. Offen. DE 3327394, 14 February 1985.
4. Gradl R., Schimmel G., Krause W., Heymer G. (Hoechst), Removal of Heavy Metal Ions from Wet Process Phosphoric Acid, Ger. Offen DE3202658, 4 August 1983.
5. Schimmel G., Gradl R. (Hoechst), Process for Purifying Phosphoric Acid, Ger. Offen DE 3434611, 21 December 1984.
6. Jdid E., Blazy P., Bessiere J. and Durand R., Removal of Cadmium contained in Industrial Phosphoric Acid using the Ionic Flotation Technique in Trace Metal Removal from Aqueous Solution. Edited by R. Thompson, Royal Society of Chemistry.
7. Bierman L.W., Polinsky S.M., Hempel D.A., Humberger R.B. (Simplot), Process for Recovery of Cadmium and Other Metals from Solution, U.S. Patent No. 4,511,541, 16 April 1985.

8. Ukeles S.D., Ben-Yoseph E. and Finkelstein N.P., Method for the Removal of Cadmium, Copper and Other Heavy Metal Ions from Phosphoric Acid, Israeli Patent No. 85751, Eur. Pat. Appl. EP 333,489, 20 September 1989.

Acknowledgement

The authors wish to thank the Managements of Rotem Amfert Negev Ltd. and Israel Chemicals Limited, IMI Institute for Research and Development for permission to publish this report.

Appendix 1

Aerophine 3418A

Aerophine 3418A promoter is a 50% aqueous solution of sodium diisobutyldithiophosphinate having the following formula:

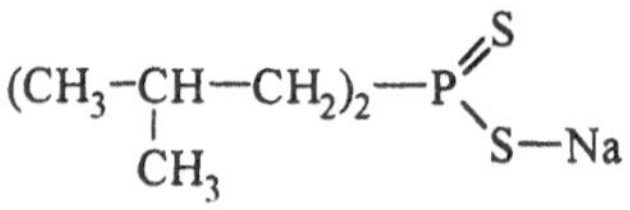

$$(CH_3-\underset{\substack{|\\CH_3}}{C}H-CH_2)_2-P\begin{matrix}\nearrow\!\!\!\!\nearrow S\\ \searrow S-Na\end{matrix}$$

The reagent is a colourless to green or yellow mobile liquid, slightly alkaline, completely soluble in water, has a specific gravity of 1.1 and a boiling point of 106°C at 760 mm Hg. This reagent is advertised by Cytec (Cyanamid) as being a flotation agent for metallic sulphides (Cu, Zn, Pb and Ag) which provides the strong collecting power of xanthates coupled with the selectivity against pyrite of the dithiophosphates. A patent was applied for in 1985 by J.R. Simplot [7] for a process to selectively remove cadmium from wet process phosphoric acid.

Aero 3477

The Aero 3477 promoter (sodium diisobutyl dithiophosphate) has the following formula:

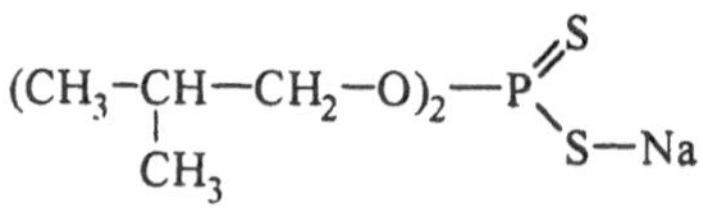

$$(CH_3-\underset{\substack{|\\CH_3}}{C}H-CH_2-O)_2-P\begin{matrix}\nearrow\!\!\!\!\nearrow S\\ \searrow S-Na\end{matrix}$$

This is a strong fast-acting promoter for precious metals and copper, silver and zinc sulphide ores. The aqueous solution is colourless to yellow and has the following physical properties: pH 10-13, specific gravity (30°C) 1.12, and boiling point of 103°C.

Appendix 2

Formulas of reagents tested

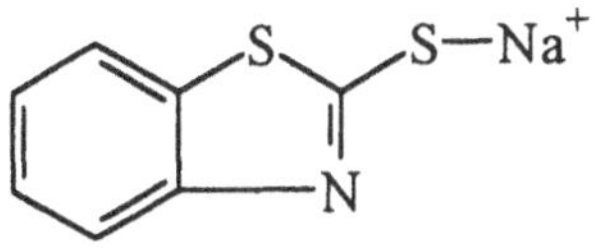

$C_6H_4(S)(N)C{-}S^{-}Na^{+}$

A. Mercaptobenzthiazole, sodium salt

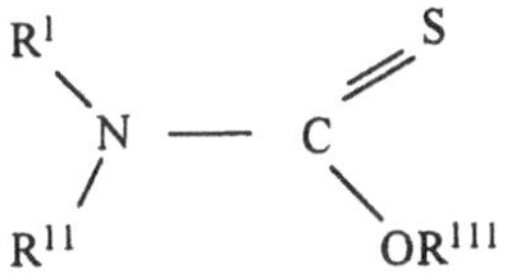

$(R^{I})(R^{II})N{-}C({=}S)OR^{III}$

B. Aero 3894 (Alkyl - alkyl thionocarbamate)

$R_3P{=}S$

R = Alkyl group
C. Cyanex 471X

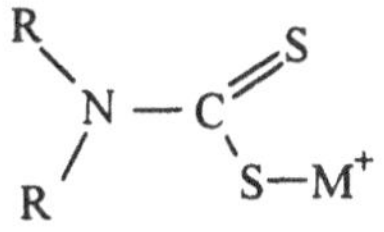

$R_2N{-}C({=}S){-}S^{-}M^{+}$

R = butyl group
D. Senkol 65 (available from Karbochem)

E. Aero 4037
A dithiophosphate - thionocarbamate formulation, only partially water soluble.

Solvent extraction of rhodium with Kelex 100

G. L. Yan
North China University of Technology, Beijing, China (presently, Department of Chemistry, University of Oslo, Oslo, Norway)
J. Alstad
Department of Chemistry, University of Oslo, Oslo, Norway

Abstract
Solvent extraction of rhodium, ruthenium and iridium with Kelex100 was investigated in this paper. Under the conditions ([Cl^-]=0.20M, [Kelex100]=20% (Wt), pH=6.05), Rh is extracted more than 80%, Ir and Ru less than 40% at phase ratio 1:1.
Keywords: Kelex100, rhodium, ruthenium, iridium.

Introduction

It is well known that the aqueous chemistry of platinum group metals, especially rhodium, is extremely complex and leads to difficult separation problems. Solvent extraction of rhodium is still one of the most difficult tasks because of its tendency to form strongly hydrated ions in aqueous solution. In chloride media rhodium is usually present as mixed aquo-chloro complexes and the composition of its complexes depends on the concentration of chloride and hydrochloric acid in the solution[1]. There are some reports on the extraction of rhodium as compounds of its hydrated cation[2][3] even from high chloride concentrations[4], but the hydrated cation tends to convert with time to aquo-chloro complexes which affect the extraction of rhodium.

Benguerel and Demopoulos have recently reported on extraction of the rhodium-tin(II) chloride complex with Kelex100 [5] the same extracting agent as we have applied in the present study on the separation of rhodium from ruthenium and iridium.

Experimental

Reagents

Kelex100 supplied by Ashland Chemical Company, Ohio, U. S. A., was diluted to 20% in Isopar-M from ESSO with 10% 1-decanol(Merck). $RhCl_3$ was from Fluka, Swtzerland, $IrCl_3$ from Alfa Inorganics Ventron, Mass., USA, and Ru powder(200 mesh) from Aldrich Chem. Co Wi, USA. Tracers of ^{99}Rh, ^{101}Rh, ^{102}Rh and ^{103}Ru were produced by irradiating Ru powder with 15 MeV protons and by irradiating Mo with 18 MeV α particles respectively in the cyclotron at the Institute of Physics, University of Oslo. The tracers of ^{192}Ir and ^{194}Ir were produced by neutron activation of 2-5 mg of the metal chloride in the JEEP II reactor at Institutt for Energiteknikk, Kjeller, Norway.

Analysis Procedure

The concentrations of metals in both aqueous and organic phases at a preselected phase contact time were determined by measuring γ-intensity in samples of equal volumes of the two phases on a Ge detector with a multi-channel analyzer. The half-lifes and γ-energies of the active radio-tracers of Rh, Ru and Ir are given in Table 1. The standard deviation of the measurements of samples with γ-spectrometry was <5%.

Table 1. Properties of radionuclides utilized as tracers

nuclide	half-life	energy of γ-radiation,keV
^{99}Rh	16d,4.7h	528,353,90
^{101}Rh	4.4d,3.3a	307,127,198
^{102}Rh	2.9a,206d	475,628
^{103}Ru	39.35d	497
^{192}Ir	74d,241a	317,468
^{194}Ir	171d,19.15h	483,328

Extraction Procedure

A 5 mL portion of the aqueous phase containing stable metal ions labelled with tracers, chlorides and acid in appropriate concentrations was contacted with an equal volume of a solution of Kelex100 in a separation funnel, which was shaken by hand for a predetermined time. The concentrations of Rh, Ru and Ir in feed were about 100 mg/l. The ionic strength of the aqueous solution was adjusted with $NaClO_4$ and all the experiments were carried out at room temperature.

Results and dicussion

Distribution ratio as a function of aqueous pH

The aim was to investigate the variation in extraction of Rh, Ru and Ir on the acidity. The results, given in Figure 1, show that Rh^{3+} can be extracted from Ru^{3+} and Ir^{3+} at pH=6.05 and $[Cl^-]$=0.20 M in the aqueous phase and [Kelex100]=20%(Wt), 10% 1-decanol in Isopar-M as the organic phase. The shaking time was 10 min. The extraction yields of Rh^{3+}, Ru^{3+} and Ir^{3+} are 81.3%, 13.0% and 27.5% respectively at pH 6.05 at which point ruthenium began to precipitate and the distribution ratio of ruthenium decreased. The extraction of rhodium, however, was not affected. The loaded organic phase was completely stripped with 0.5 M HCl.

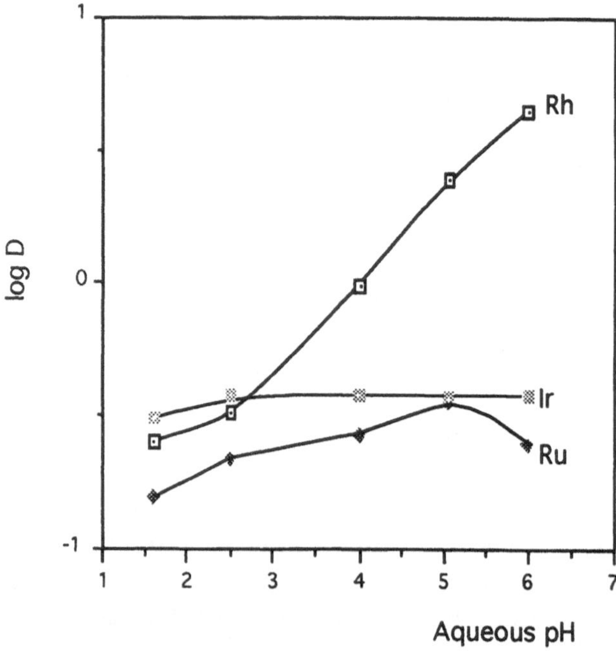

Figure 1. Distribution ratios of Rh, Ru and Ir as a function of aqueous pH; phase contact time 10 min.; aqueous chloride 0.20 M; organic phase 20% Kelex100 and 10% decanol as modifier in Isopar-M.

Distribution ratio as a function of shaking time

The experiments were carried out at [Kelex100]=20%(Wt) and [decanol]=10% in Isopar-M as organic phase, and $[Cl^-]$= 0.20 M in pH=5.50 aqueous phase. The results, given in figure 2, show that the extraction equilibrium for rhodium was reached in about 10 min, ruthenium and iridium in about 3 min.

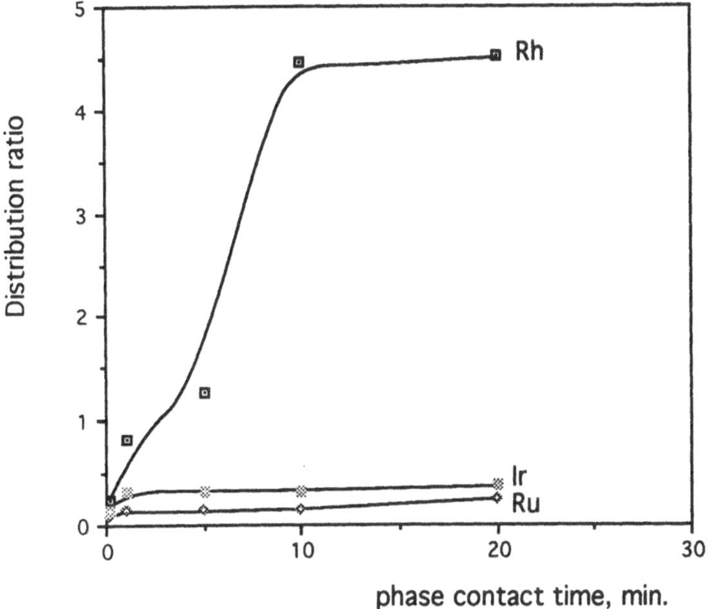

Figure 2. Distribution ratios of Rh, Ru and Ir as a function of phase contact time; aqueous pH 5.50; aqueous chloride 0.20 M; organic phase 20% Kelex100 and 10% decanol as modifier in Isopar-M.

Distribution ratio as a function of chloride concentration

As it is well known that most platinum group metals in chloride media tend to form chloro-complexes whose compositions depend strongly on chloride concentration and pH value in the solution. Rhodium, however, forms hydrated ions in solution and is much more inert in its extraction behaviour than other precious metals. It is necessary to determine the extraction dependence of rhodium, ruthenium, and iridium on the concentration of chloride. The experiment was carried out at [Kelex100]= 20% (Wt) and [decanol]=10% in Isopar-M as organic phase, pH=4.50, shaking time 10 min and the ionic strength was kept at 1.0 with $NaClO_4$. The results, given in Figure 3, show that the distribution ratios of rhodium, ruthenium and especially iridium decrease as the concentration of chloride in the solution increases.

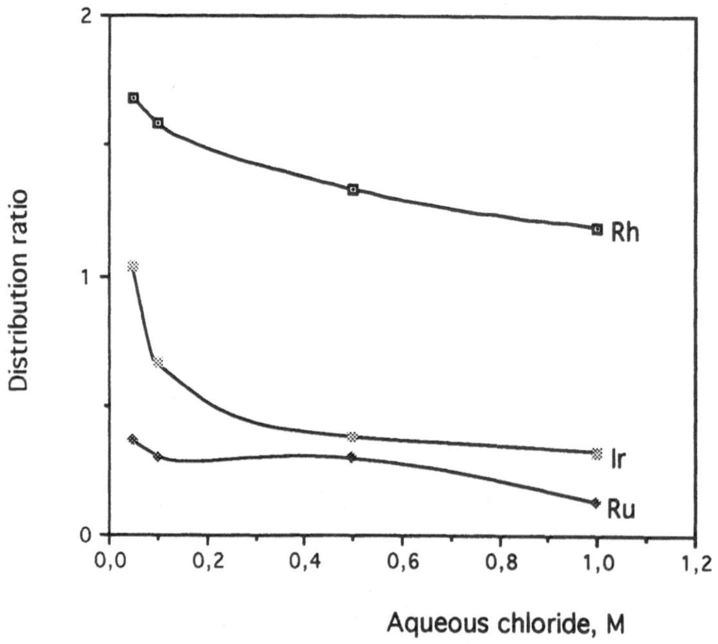

Figure 3. Distribution ratios of Rh, Ru and Ir as a function of chloride concentration in aqueous phase; organic phase 20% Kelex100 and 10% decanol as modifier in Isopar-M; phase contact time 10 min.; aqueous pH 4.50; ionic strength **I**=1.0 ($NaClO_4$).

Conclusion

From the experiments above it is seen that Rh^{3+} can be extracted from Ru^{3+} and Ir^{3+} with Kelex100 and the procedure is simple. Although ruthenium begins to precipitate at a pH value of about 6, it does not affect the extraction of rhodium.

Acknowledgements

The authors are very grateful for the financial support from *Stefi og Lars Fylkesakers Vitenskapelige Stiftelse* and the help from staff and students in Nuclear Section of Department of Chemistry, University of Oslo.

References

1. Edwards, R. I. and te Riele, W. A. M.(1983) Commercial Process for Precious metals, **Handbook of Solvent Extraction,** Wiley Interscience, New York.
2. Ali Khan, M. and Morris, D. F.(1967) **Journal of Less-Common Metals,** 13, 53.
3. Jiang, M. and Wang, X. Y.(1986) Solvent extraction of rhodium and iridium with HDEHP, **Conference of Precious Metals,** Kunming, China.
4. Yan, G. and Alstad, J. (1993) Separation of rhodium by fast solvent extraction, **International Solvent Extraction Conference,** York, England.
5. Benguerel, E. and Demopoulos, G. P. (1993) A novel solvent extraction system for rhodium, **International Solvent Extraction Conference proceedings,** York, England.

Effluent treatment

Removal of free cyanide from solution with silver-impregnated activated carbon

Y. Choi
H. T. Lieuw
G. Van Weert
Delft University of Technology, Faculty of Mining and Petroleum Engineering, Department of Raw Materials Technology, Delft, The Netherlands

Abstract

An earlier investigation by Van Weert and de Jong [1] showed that it was possible to obtain a free cyanide concentration below 10 ppb when silver impregnated activated carbon (SIAC) was contacted with a 150 ppb CN^- solution. The formed argentocyanide complex was adsorbed on the activated carbon. No silver losses occurred. Based on these promising results, investigation of the cyanide/SIAC system was extended to industrially more relevant 20 ppm CN^- starting solutions. To study the kinetics, batch tests were conducted varying the silver content of the activated carbon, stirring speed, carbon/CN^- ratio, and the gas atmosphere used in the SIAC process.

The results of these batch experiments suggest that cyanidation of the silver in or on the activated carbon is the rate limiting step. The variation of the stirring speed has no influence on the rate of cyanide adsorption. The rate of cyanide adsorption is increased by increasing the SIAC/CN^- ratio, but not by increasing the Ag/CN^- ratio. At a SIAC/CN^- weight ratio of 1600, 20 ppm of free CN^- can be reduced to less than 10 ppb within 30 minutes. Maximum loading occurs at a C/CN^- weight ratio ≅ 142, with approximate 2.7 - 3.0 wt% Ag in the carbon. The rate of cyanide adsorption is not effected by the gas atmosphere during the impregnation thermal treatment process at 350°C.
Keywords: Activated carbon, cyanide, silver cyanide, silver.

Introduction

Environmental awareness has induced authorities to tighten up plant discharge regulations. In several countries such regulations impose extremely low cyanide discharge levels. The gold and silver industry, which uses cyanide to recover these precious metals, has adopted several systems for the reduction of cyanide in their mill discharges. These systems can be classified as SO_2 assisted oxidation, natural degradation, acidification-volatilization-reneutralization, oxidation, and biological treatment [2]. However, cyanide reduction by the first three processes does not appear to meet the strict regulatory requirements. The usefulness of the biological treatment system is limited to regions without frigid winter conditions. The only system in commercial use which claims reduction of the cyanide concentration to 0.1 ppm or less is the oxidation with hydrogen peroxide. However, this process requires an expensive reagent which cannot be reused. In view of the shortcomings of these many alternatives, a cyanide treatment system using silver impregnated activated carbon (SIAC) was explored. This system appears to have the following advantages: It removes cyanide from solution, leaves no oxidation products, is selective, has low cost and is compatible with gold and silver milling practice.

The earlier study by Van Weert and de Jong [1] proved that it was possible to reduce the cyanide concentration from 150 ppb to below 10 ppb by using silver impregnated activated carbon. Batch experiments have now been carried out to investigate some of the factors [3] which might affect the rate of cyanide adsorption, such as the silver content of the activated carbon, the stirring speed, the activated carbon/CN^- ratio, and the gas atmosphere used during the silver impregnation process.

Background

Cyanide removal will occur when activated carbon is contacted with an aqueous solution containing free cyanide. This is caused by either the adsorption of cyanide or its catalytic oxidation. Although the reaction of free cyanide with molecular oxygen is slow, Adams [4] and Weber and Corapcioglu [5] have shown that the rate of cyanide removal will increase when activated carbon is present in cyanide solution due to the carbon-catalyzed oxidation reaction. When SIAC is present in a cyanide solution, the removal of cyanide from solution is thought to be principally due to the formation and adsorption of the argentocyanide complex. There have been a number of theories put forward to explain the mechanism of dissolution of silver in a cyanide solution [6]. Recently, Li and Wadsworth [7] have investigated the rate processes of silver cyanidation. The charge transfer steps plus the diffusion of the reactants from the bulk solution to the silver surfaces were identified as rate-limiting. Also, several mechanisms have been proposed to explain the adsorption of the argentocyanide complex at the surface of the activated carbon [6, 8]. Frumkin et al. [9] proposed the adsorption of $Ag(CN)_2^-$ at the carbon surface due to the electrostatic interaction with the carbon surface. Scott and Gross [10] proposed the adsorption of $Ag(CN)_2^-$ as ion pair or as a neutral molecule. Cho and Pitt [11, 12] proposed multi- layer adsorption of the argentocyanide complex at the activated carbon surface. It is thought that the mechanism of CN^- removal with SIAC occurs in three steps such as the dissolution of the silver in the cyanide solution, the formation of the argentocyanide complex, and the adsorption of the argentocyanide complex on the activated carbon. The emphasis of this work has not been on mechanisms, however, but on process development.

Experimental

In batch experiments, the removal of free cyanide in solution in contact with SIAC was measured. A stirred flask reactor of 3 liter capacity containing 2 liter solution was used. The solution in the reactor was stirred by a glass turbine impeller with a variable speed motor. Samples of the solution were withdrawn periodically for chemical analysis.

Two analytical procedures were used for the determination of free cyanide. For analyzing cyanide concentration at the ppb level, the silver indicator technique was used [1, 3, 13, 14]. Clyster et al. [14] reported relative standard deviations of 1.1 and 9.5% at cyanide levels of 100 and 10 ppb, respectively. At the ppm CN^- level, titration of cyanide with $AgNO_3$ was used. The analysis of dissolved silver was conducted with an Atomic Absorption Spectrometer. The supporting analyses consisted of electromicroprobe, X-ray diffraction, and X-ray fluorescence to reveal the silver morphology and distribution on the impregnated activated carbon.

The activated carbon used in this study was a NORIT product named ROX 0.8. The manufacturer's data state that ROX 0.8 is a micro and meso porous carbon, derived from extruded peat, with a surface area of 900 m^2/g as determined by the BET method. This activated carbon is NORIT's standard grade for the gold milling industry. Table 1 lists the specifications of this product.

Table 1 Specifications of ROX 0.8

Specifications	ROX 0.8
bulk density (g/l)	410
moisture (%)	2
total specific surface area by BET (m^2/g)	900
ash content (%)	3
total pore volume (ml/g)	1
acid extractable matter	0.8
calcium (%)	0.02
iron %)	0.02
hardness (%)	92 ball pan
iodine number (mg/g)	1050
pH	4 - 7

The silver impregnation procedure consisted of four steps. First, the ROX 0.8 was cleaned. This was done by boiling the carbon in distilled water for one hour. In this way the air and fines were removed from the activated carbon particles. After the cleaning stage, the activated carbon was transferred into an $Ag(CN)_2^-$ solution (700 ppm as Ag) for four hours at room temperature. The third step was the filtering and drying of the impregnated activated carbon for one night at 100°C in an oven. The final step of the impregnation process was the decomposition of the silver cyanide present on the activated carbon. To achieve this, the impregnated carbon was placed for one hour in a tube furnace at 350°C with a hydrogen atmosphere after which the carbon was cooled in argon for a further hour. In separate tests, hydrogen was replaced by nitrogen and air to investigate the influence of these gas atmospheres in the decomposition step on subsequent cyanide adsorption.

The batch experiments were carried out after de-aerating the SIAC by boiling it in distilled water for fifteen minutes. Two liter of solution were prepared, one batch with 20 ppm CN^-, the other batch with 20 ppm of Ag^+ (as $Ag(CN)_2^-$), in the three liter flask. Therefore, the initial concentration of CN in the second batch of solution was less than 20 ppm. The pH of the solution was adjusted with 1 M NaOH to 11.5. While contacting the cyanide or silver cyanide solution with a known amount of impregnated activated carbon, or with as received activated carbon, the decrease of the cyanide concentration versus time was measured as follows. The solution sample was diluted and analyzed with the silver indicator technique. This procedure was checked by comparing cyanide concentration measured with the selective electrode to the cyanide concentration measured by titration. The deviation was below 5%.

Experimental Results and Discussion

1. The influence of silver on the removal of cyanide with as received activated carbon.

The adsorption rates of silver cyanide and free cyanide on activated carbon were investigated. The experimental conditions were initial concentration of 20 ppm of Ag^+ (in the form of $Ag(CN)_2^-$) or 20 ppm CN^-, activated carbon of 0.5 gram/liter, stirring speed of 570 rpm, pH of the solution of 11.5, and room temperature. Figure 1 shows a great difference in affinity of free cyanide and $Ag(CN)_2^-$ for as received ROX 0.8 activated carbon. These results confirm that at these high cyanide levels (20 ppm versus 150 ppb used in [1]) $Ag(CN)_2^-$ also adsorbs much better and faster on the activated carbon than free cyanide. It may be interesting to note that, if the activated carbon obtained after 4 hours in this experiment had been furnaced in H_2, the silver content would have been 2.9 wt%.

A series of experiments were carried out to investigate the effect of cyanide adsorption on the adsorption rate of silver cyanide. The experimental conditions were the same as in the previous experiment.

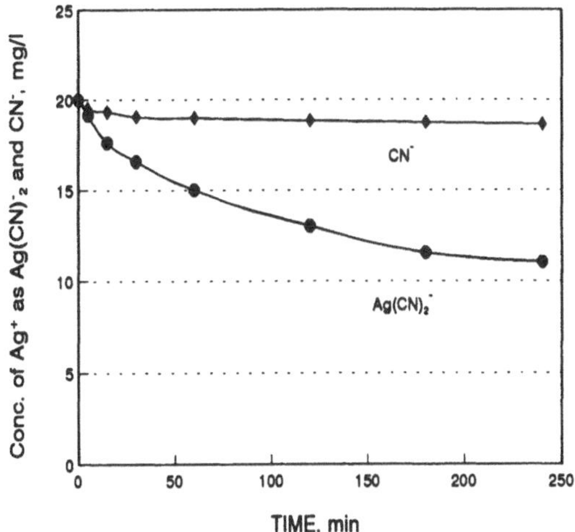

Figure 1. Comparison of adsorption rates of silver cyanide and free cyanide (initial concentration of Ag^+ (as $Ag(CN)_2^-$) and CN^-: 20 ppm, ROX 0.8 activated carbon: 0.5 gram/liter, pH: 11.5, stirring speed: 570 rpm, room temperature).

The as received activated carbon was impregnated in a 20 ppm CN^- solution for 24 hours after which it was transferred into a 20 ppm Ag^+ ($Ag(CN)_2^-$) solution and the decrease of the Ag^+ monitored. The results shown in Figure 2 prove that the adsorption of $Ag(CN)_2^-$ is not hindered by prior adsorption of free cyanide, if any. There are three possible explanations for this behavior. One explanation could be that the argentocyanide complex drives out the free cyanide. Since no free cyanide was detected during the experiment, this explanation is not persuasive. The other explanation is that activated carbon provides enough active sites for both $Ag(CN)_2^-$ and CN^-. Thirdly, free CN^- is not adsorbed but oxidized.

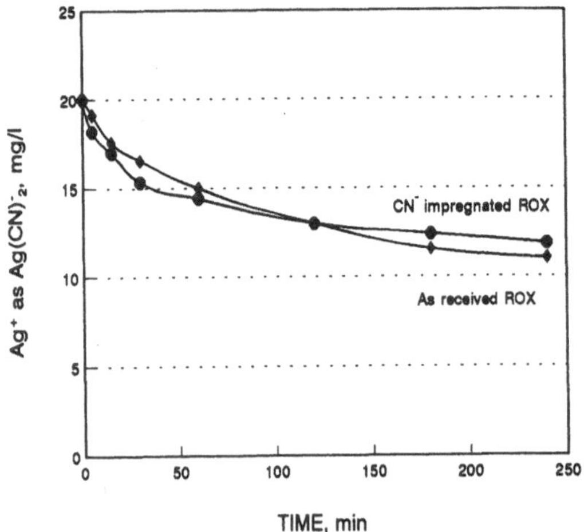

Figure 2 Effect of prior adsorption of free cyanide on the adsorption rate of silver cyanide (initial concentration of Ag^+ (as $Ag(CN)_2^-$): 20 ppm, ROX 0.8 activated carbon: 0.5 gram/liter, pH: 11.5, stirring speed: 570 rpm, room temperature).

2. Cyanide removal with Ag impregnated activated carbon.

Figure 3 shows the experimental results of the rates of adsorption of 20 ppm cyanide on as received activated carbon, furnaced activated carbon in hydrogen atmosphere at 350°C without silver impregnation, and silver impregnated activated carbon. The experimental conditions were the same as previous experiment. The amount of silver impregnated on the activated carbon was thus varied from 0 to 2.7 wt%. It is obvious that the removal of cyanide from the solution is increased when silver is present on the activated carbon. But the increase in the rate is not very significant. The effect of heat treatment on activated carbon is not discernable. This was expected; the activated carbon structure is considered stable at 350°C. Also, the variation of the silver content on the activated carbon has no significant effect on the adsorption rate of cyanide. This is further confirmed by the results shown in Figure 3, where the removal rates of cyanide are compared with 2 g instead of 0.5 g SIAC with 0.7 and 2.7 wt% Ag used per liter solution.

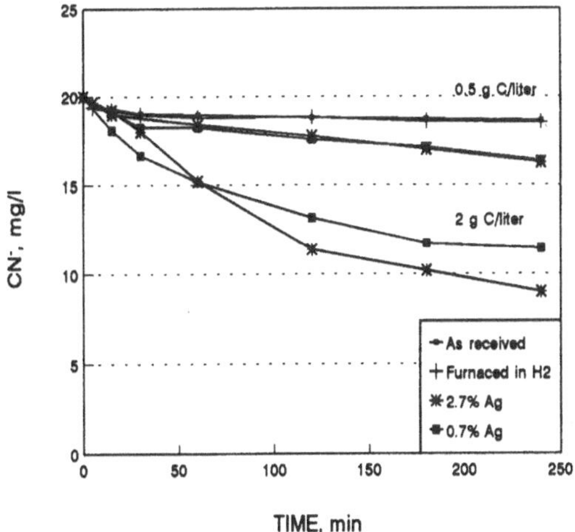

Figure 3. Effect of silver content of impregnated activated carbon on the adsorption rate of free cyanide (initial concentration of CN^-: 20 ppm, ROX 0.8 activated carbon: 0.5 and 2 gram/liter, amount of silver impregnated: 0.7 and 2.7 wt%, pH: 11.5, stirring speed: 570 rpm, room temperature).

It was found that under conditions with excess free CN^- in solution some silver from the impregnated activated carbon was solubilized (1 mg Ag^+/liter after 4 hours in the 0.5 g C/liter test in Fig. 3). The effect of this on column operation will be discussed in a future publication [15].

Next, the effect of oxygen concentration in the solution on the adsorption rate of free cyanide was studied by conducting experiments where air or pure O_2 was bubbled through the solution at room temperature. It was found that the effect of oxygen concentration in solution is not significant. At 15°C and air at atmospheric pressure 8.2 mg of O_2 are dissolved in 1 liter of water. This concentration is enough for over 100 mg of CN^- to be removed via following dissolution-complexation reaction: $2Ag^o + 4CN^- + H_2O + ½O_2 \rightarrow 2Ag(CN)_2^- + 2OH^-$. Therefore, it can be expected that the rate of silver dissolution is not limited by oxygen transfer from the gas to the aqueous phase, since the initial concentration of CN^- is 20 ppm in this study.

Since the carbon to CN^- ratio appeared significant, a series of experiments were carried out to investigate the effect of the amount of impregnated activated carbon on cyanide adsorption rates. The experimental conditions remained the same. The amount of SIAC was varied from 1 to 64 g in the 2 liter solution. The amount of silver in the impregnated activated carbon was 3.0 wt%. These results are shown in Figure 4.

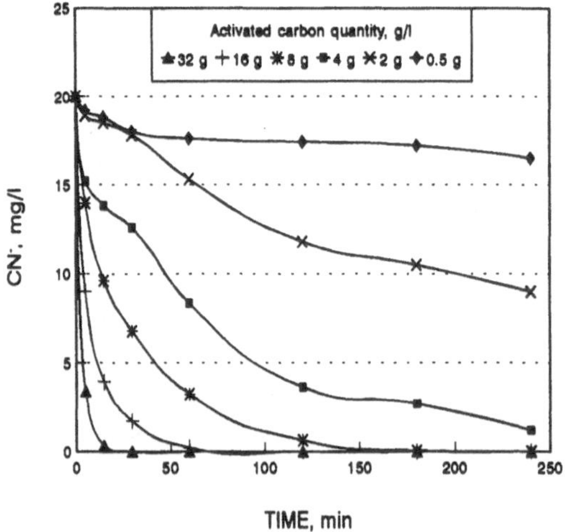

Figure 4. Effect of impregnated activated carbon quantity on the adsorption rate of free cyanide (initial concentration of CN^-: 20 ppm, amount of silver impregnated: 3.01 wt%, pH: 11.5 stirring speed: 570 rpm, room temperature)

An increase in the amount of impregnated activated carbon caused a very significant increase in the rate of removal of CN^-. Also, the cyanide concentration of the solution decreased to below 10 ppb with the 64 g of SIAC after 15 minutes reaction time. This result suggests that the process investigation of cyanide removal with SIAC should focus on columns [15].

It has been concluded that the maximum loading of cyanide is about 270 μM per gram of SIAC with 4 hours adsorption time (test 0.5 g of carbon in Fig. 4). Therefore, it can be calculated that 1 g CN^- would require about 142 g SIAC. It can be seen from Figure 5 that an increase in the carbon/cyanide ratio results in a near proportional increase in the initial rate of cyanide adsorption. Since an increase in the silver content of the impregnated activated carbon does not lead to a higher rate of cyanide removal, it is likely to say that the greater available surface area is responsible for the rate increase.

3. The film resistance.

To investigate whether film transport is a rate limiting factor in removal of cyanide, the effect of stirring speed on the cyanide adsorption rate was investigated. The experimental conditions remained the same. The amount of silver in the impregnated

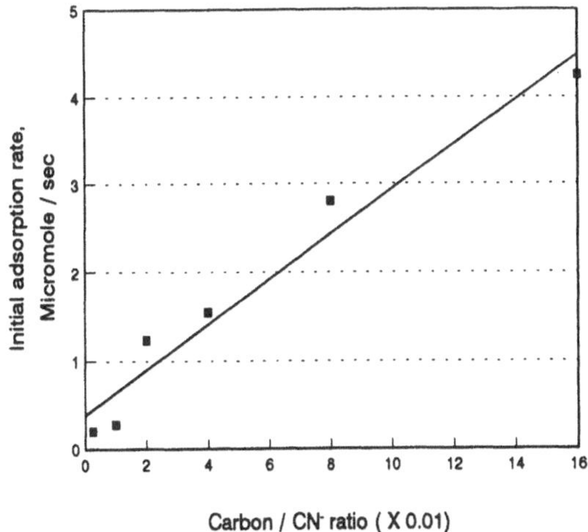

Figure 5. Initial adsorption rate as a function of carbon/cyanide weight ratio.

carbon was 2.7 wt%. Stirring speeds of 265 and 570 rpm were used. No difference in adsorption rate was found. Under both these stirring speed conditions, the activated carbon particles are well mixed and fully suspended and external boundary layer transport is not a rate limiting factor in the CN^- uptake process. However, the process is quite likely to have internal mass transport control, which can only be elucidated by simulation with computer models.

4. The influence of the gas atmosphere during the reduction step on the rate of CN^- removal.

The effect of various gas atmospheres during the furnacing step of the silver impregnation process on the subsequent adsorption rates of cyanide and silvercyanide was also studied. Figure 6 shows the experimental results for cyanide. Effects of air, nitrogen, and hydrogen were investigated. The cyanide adsorption rate remained unchanged when impregnated activated carbon was treated with these gases. This means that activated carbon surface is not affected during the decomposition stage. This was confirmed when the adsorption rates of cyanide on as received activated carbon and the same activated carbon furnaced in hydrogen at 350°C without impregnation were compared. When the same experiments were conducted with a 20 ppm Ag^+ as $Ag(CN)_2^-$ solution, a small effect was found in the adsorption rate of the

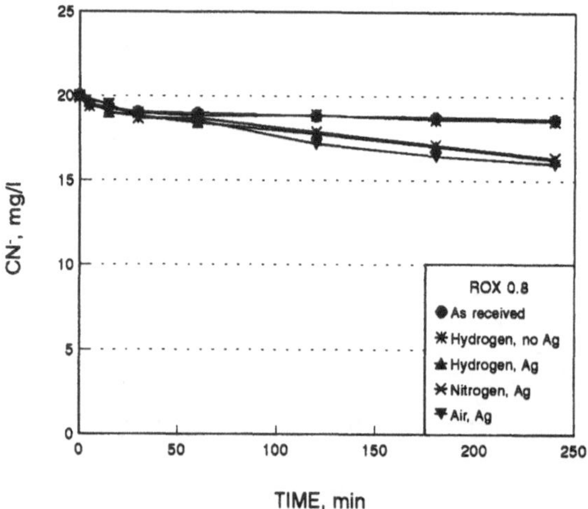

Figure 6. Effect of gas during impregnation process on the adsorption rate of free cyanide (initial concentration of CN^-: 20 ppm, ROX 0.8 activated carbon: 0.5 gram/liter, amount of silver impregnated: 2.7 wt%, pH: 11.5, room temperature).

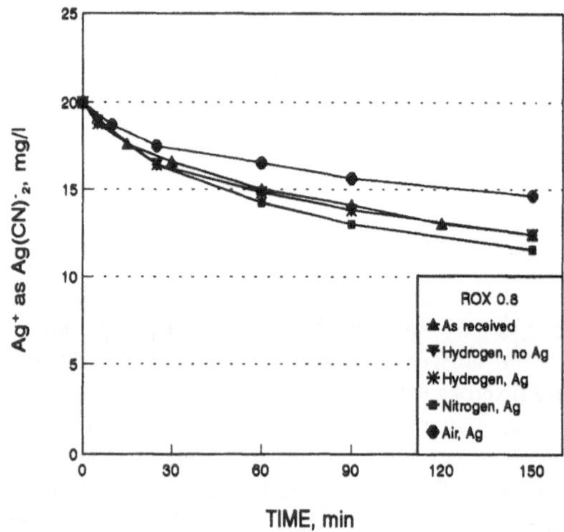

Figure 7. Effect of gas atmosphere during impregnation process on the adsorption rate of silver cyanide (initial concentration of Ag^+ (as $Ag(CN)_2^-$): 20 ppm, ROX 0.8 activated carbon: 0.5 gram/liter, amount of silver impregnated: 2.7 wt%, pH: 11.5, room temperature).

argentocyanide complex, as can be seen from Figure 7. The lowest adsorption rate of $Ag(CN)_2^-$ was found with air treated impregnated activated carbon, followed by the impregnated activated carbon treated with hydrogen and nitrogen. The reason for this variance is not clear. The adsorption rates of $Ag(CN)_2^-$ on as received activated carbon, activated carbon furnaced in H_2 without impregnation, and the impregnated activated carbon furnaced in H_2 were almost identical.

5. Morphology.

The insensitivity of the rate of cyanide adsorption to variations of the silver content of SIAC can be explained as follows. A higher silver content does not mean that a larger silver area is obtained, which would be easily accessible by the cyanide. Silver deposition could take two possible forms: a mono-layer and silver crystallites. The latter will have a smaller surface area. X-ray diffraction of SIAC showed strong silver reflections. These reflections prove that at least a fraction of the silver on the SIAC is present as crystallites. Scanning in the axial direction of a particle resulted in silver detection. However, when spot analyses were conducted on the radial cross section of a ROX extrudate which had been broken in two, no silver response was obtained. Regrettably since silver distribution profiles of only a few particles were made, no firm conclusion can be given. The presence of silver as a mono-layer cannot be excluded since the XRD or EMP is not able to detect silver when it is present as a mono-layer. Another very interesting aspect is that X-ray diffraction analyses of batches of silver impregnated activated carbon which were furnaced in an inert nitrogen environment and in an oxidizing air atmosphere showed very strong silver reflections. Whether the CN^- or the activated carbon plays a role in silver crystallite production also remains as a subject for study.

Conclusions

Experiments have been carried out to study free and silver cyanide adsorption behavior from solution on silver impregnated activated carbon. The results reveal the following information over the cyanide/silver impregnated activated carbon system:

1. Free cyanide can be reduced from 20 ppm to below 10 ppb with silver impregnated activated carbon. (SIAC).
2. The rate of cyanide adsorption is not affected by the stirring speed. The solution bulk and boundary transport are not rate limiting factors under the chosen experimental conditions.
3. Variation of the silver content of the impregnated activated carbon has no significant influence on the cyanide adsorption rate. The cyanidation of silver appears the rate limiting factor because mechanisms other than the formation of the argentocyanide complex can be neglected for the removal of cyanide in the described experiments with a run time of 4 hours and less.

4. The rate of free cyanide adsorption remains unchanged when SIAC is furnaced in different gas atmospheres during its preparation at 350°C.
5. An increase in the carbon/cyanide ratio results in a nearly proportional increase in the rate of free cyanide adsorption.
6. Results from these batch experiments suggest that to remove free cyanide effectly from solution with SIAC, a packed column should be used.

Acknowledgements

This work has been made possible by the Research Fellowship Fund of the Delft University of Technology supporting one of the contributors (Dr. Y. Choi). Appreciation is expressed for the support, and the provision of activated carbon ROX 0.8, by NORIT N.V.

References

1. Van Weert, G. and de Jong, I. (1992) Trace Cyanide Removal by Means of Silver Impregnated Activated Carbon, Proceedings of the Symposium on Emerging Process Technologies for a Cleaner Environment, S. Chander (Editor), SME, Feb. 1992, pp. 161-165.
2. Scott, J.S. (1984) An Overview of Cyanide Treatment Methods for Gold Mill Effluents, Proc. Cyanide and the Environment Conf., D. van Zyl (Editor), Tucson, Arizona, Vol. 2, pp. 307-327.
3. Lieuw, H.T. (1992) The Removal of Cyanide from an Aqueous Solution in a Fixed Bed of Silver Impregnated Activated Carbon, Master Thesis, Delft University of Technology.
4. Adams, M.D. (1990), The Chemical Behavior of Cyanide in the Extraction of Gold - 1. Kinetics of Cyanide Loss in the Presence and Absence of Activated Carbon, J. S. Afr. Inst. Min. Metall., Vol.90, No. 2, Feb., pp.37-44.
5. Weber, W.J. Jr. and Corapcioglu O. (1982) Catalytic Oxidation of Cyanides, Proc. 36th Ind. Waste Conf., Butterworth Publisher, pp. 500-508.
6. Habashi, F. (1970) Principles of Extractive Metallurgy, Vol. 2, Gordon and Breach Science Publishers, pp. 24-34, 183-193.
7. Li, J. and Wadsworth, M.E., (1992) Rate Processes of Silver Cyanidation, Proceedings of EPD Congress, J.P. Hager, (Editor), TMS, pp. 257-271.
8. Smit, R.W. (1987) Die Adsorption von Gold und Silber an Aktivkohle, Dissertation, Technischen Universität Clausthal.
9. Frumkin, A., Burstein, R. and Lewin, P. (1931) Uber Aktivierte Kohle, Z. Physik. Chem., A157, pp. 442-446.
10. Scott, J.W. and Gross, J. (1927) Precipitation of Gold and Silver from Cyanide Solution on Charcoal, U.S. Bureau of Mines, Washington DC, Tech. Paper No. 378.

11. Cho, E.H. and Pitt, C.H. (1979) Kinetics and Thermodynamics of Silvercyanide Adsorption on Activated Charcoal, Metall. Trans., June, Vol. 10B, pp. 159-164.
12. Dixon, S., Cho, E.H. and Pitt, C.H. (1978) The Interaction between Gold Cyanide, Silver Cyanide, and High Surface Area Charcoal, Fundamental Aspects of Hydrometallurgical Processes, AIChE Symposium Series, Vol. 74, pp. 75-83.
13. Frant, M.S., Ross, J.W. Jr. and Risemann, J.H., (1972) Electrode Indicator Technique for Measuring Low Levels of Cyanide, Anal. Chem., Vol. 44, No. 13, pp. 2227-2230.
14. Clyster, H., Adams, F. and Verbeek, F. (1976) Potentiometric Determinations with the Silver Sulphide Membrane Electrode, Part I. Determination of Cyanide, Anal. Chim., Acta, Vol. 83, pp. 27-38.
15. Choi, Y., Lieuw, H.T. and Van Weert, G. In preparation.

Removal of heavy metal ions from liquid effluents by solvent-impregnated resins

J. L. Cortina
Chemical Engineering Department, Universitat Politècnica de Catalunya, Barcelona, Spain
A. Warshawsky
Organic Chemistry Department, Weizmann Institute of Science, Rehovot, Israel
N. Miralles
Chemical Engineering Department, Universitat Politècnica de Catalunya, Barcelona, Spain
M. Aguilar
Chemical Engineering Department, Universitat Politècnica de Catalunya, Barcelona, Spain
A. M. Sastre
Chemical Engineering Department, Universitat Politècnica de Catalunya, Barcelona, Spain

The paper describes a study of metal extraction with Impregnated Resins prepared by adsorption of a bifunctional extractant, O-methyl-dihexyl-phosphine-oxide O'-hexyl-2-ethyl phosphoric acid (BL) into polymeric macroporous supports of Amberlite XAD2-type (XAD2-BL). The extraction of these metal ions involves the formation of mixed species in the resin phase with a general composition $ML_2(HL)_{q,r}$ and $ML_{(2-t)}(NO_3)_t(HL)_{q,r}$. The application of these resins in column process allow the extraction efficienly of Zn(II), Cu(II) and Cd(II) and elution of them with HCl solutions. For the three metal ions, enrichment factors of 10 could be obtained in the elution step.
The kinetic studies on the extraction process at relatively high concentrations show parabolic kinetic curves and suggest a particle diffusion control for the extraction reaction. On the other hand, at low metal concentration the linear dependence of the resin loads on contact time suggests a regime of film diffusion control for the three metal ions.

1 Introduction

The disposal and dispersion of waste metals to the environment is undesirable for both environmental and strategic reasons. Furthermore, waste metal recovery for reuse could be economical and socially desirable because it will contribute directly to metal resource conservation and environmental protection. This increasing concern towards optimization of industrial process dealing with metals, impose the need for the development of advanced separation techniques and, in particular, for liquid wastes (Paterson, 1990).

In this context, the preparation of selective adsorption systems by physical immobilization of metal extractants, on macraporous resins to give impregnated resins (SIR) has been presented as a technological alternative to solvent extraction and ion exchange technologies (Warshawsky, 1981; Tavlarides, 1987). The idea behind the development of impregnated resins is the ability to combine the selectivity and specificity of conventional liquid extractants with the advantages of a discrete polymer support material, thus tailor-making adsorbents for a specific separation process, usually in the field of hydrometallurgy (Muscatello, 1988; Wakui, 1989) and analytical chemistry (Abollino, 1990; Horwitz, 1987).

Since the pioneering work of Warshawsky (Warshawsky, 1971, 1974) and Kroebel (Kroebel, 1971) on impregnated resins, the development and application of these systems in metal extraction processes has been intensively investigated for hydrometallurgical separation and recovery applications (Muscatello, 1988; Gonzalez-Luque, 1983; Yoshizuka, 1990). Impregnated resins containing organophosphorous compounds have also been used as stationary phases in the extraction of actinides for nuclear reprocessing (Shoen, 1982; Apostolidis, 1991; Louis, 1984). In the past several years our corresponding research groups have been working in the development of Solvent Impregnated Resins (SIR) for recovery and separation of metal ions from dilute solutions. The impregnated resins have been characterized physically and chemically and their behaviour in the extraction of base metals has been studied (Cortina, 1992, 1993 and 1994 (a); Warshawsky, 1986, Arad-Yellin, 1990). However, the application of these systems in industrial scale equipment using fixed column or fluidized bed technology requires a knowledge of the operating hydraulic behaviour, the equilibrium data and kinetics of metal extraction processes.

The paper describes a study of metal extraction with Solvent Impregnated Resins prepared by direct adsorption of a bifunctional extractant, incorporating a phosphine oxide and phosphoric acid di-ester funcionalities, O-methyl-dihexyl-phosphine-oxide O'-hexyl-2-ethyl phosphoric acid (BL) into high surface polymeric macroporous supports of Amberlite XAD2-type (XAD2-BL).

Experimental measuraments of the distribution of Zn(II), Cu(II) and Cd(II) in batch experiments and the sorption and desorption isotherms from synthetic solutions of these metal ions in nitrate media with XAD2-BL resins were performed. Another research objective was to study the extraction kinetics of Zn(II), Cu(II) and Cd(II) from nitrate solutions. For this purpose, solutions of these metal ions were prepared and contacted with the XAD2-BL resin. The resin loading as a function of the contact time was monitored.

2 Experimental

2.1 Reagents

O-methyl-dihexyl-phosphine-oxide O'-hexyl-2-ethyl phosphoric acid (BL) was synthesized as was described previously (Warshawsky, 1986, Arad-Yellin, 1990).

$$(BL):(C_6H_{13})_2\text{-}(PO)\text{-}CH_2O\text{-}(PO)(OH)\text{-}OC_8H_{17}$$

Stock solutions of Zn(II), Cu(II) and Cd(II) (1g.dm^{-3}) were prepared by dissolving the corresponding salts (Merck, A.R. grade) in water. Sodium nitrate, sodium hydroxide and nitric acid (Merck, AR grade), were used for the preparation of the different solutions.

Amberlite XAD-2 Resin supplied by Rohm and Haas, size 0.3-0.9 mm, was used.

2.2 Impregnation Process

The impregnated resins were prepared according to a modified version of the dry impregnation method (Cortina, 1992). The amount of extractant impregnated, $[BL]_r$ (mol.kg^{-1} dry impregnated resin), was evaluated after washing a known amount of resin with ethanol, which completely elutes the ligand, and subsequent titration with NaOH.

2.3 Batch Extraction Procedure.

The extraction of Zn(II), Cu(II) and Cd(II) was carried out with batch experiments at 25°C. Samples of 0.2 g of XAD2-BL resins ($[BL]_r$= 0.17 and 0.64 mol.kg^{-1} dry SIR), were mixed mechanically in special glass stoppered tubes with an aqueous solution (20 mL) having composition 0.1 M (M^{2+}, H^+, Na^+)NO_3^- until equilibrium was achieved. After phase separation with a high-speed centrifuge the equilibrium pH was measured using a Metrhom AG 9100 combined electrode. Metal content in both phases was determined by Atomic Absorption Spectrophotometry. A Perkin-Elmer 2380 AAS with air-acetylene flame was used.

2.4 Column extraction procedure.

The column was packed with known amounts of swollen resin. A XAD2-BL resin ([BL]=0.64 mol.Kg^{-1}) was slurry-packed in an Omnifit borosilicate glass column (10 mm i.d., 150 mm legth) fitted with porous 25 micron polyethylene frits and teflon end pieces. A peristaltic pump at the column entrance delivered solution at constant flow rate of 1 ml/min. Metal ion was determined by following the change in concentration with throughput of the samples collected to follow the extraction histories.

After each metal extraction experiment, the flow of the metal solution is stopped and the resin washed succesively with water, 0.2 M hydrochloric acid solution at flow rate of 1 ml/min through the resin bed in the column. Metal ion concentration was determined in all the samples.

2.5 Metal Extraction Kinetic measurents

The testwork was performed using the shallow bed technique on a micro scale. According to this technique (Helfferich, 1962), an aqueous metal solution is passed at high flow rate through a thin layer of resin beads in a column. The objective of this procedure is to avoid the formation of a concentration gradient along the resin bed. Thus, the composition of the external solution remains practically constant throughout the experiment. The flow of the solution is periodically stopped and the resin is washed an analyzed to provide data on resin composition as a function of time. A typical resin bead contained 40 beads (2-3 mm depth) and the flow rate employed was 200 ml/h.

The resin composition was determined by passing volumes of 25 or 50 ml of 0.2 M hydrochloric acid solution at low flow rate 50 ml/h through the thin layer of resin beads in the column. After appropiate dilution, the metals were analyzed by atomic absorption spectrophotometry. The resuls are expresed as mmol of metal per gram of dry impregnated resin XAD2-BL.

The impregated XAD2-BL resin was used in all experiments. Before use the resins were hydrated and sieved. The fraction of largest beads, having a radius between 0.5 to 0.7 mm, was selected for the test work.

3 Results and discussion.

3.1 Metal Extraction Reactions.

The distribution of Zn(II), Cu(II) and Cd(II) between the resin phase containing BL (XAD2-BL) and the aqueous phase can be obtained directly as:

$$D=\frac{[M(II)]_r}{[M(II)]}=\frac{([M(II)]_t-[M(II)])x(V/m_r)}{[M(II)]} \quad (1)$$

where $[M(II)]_r$ denotes the total concentration of M^{2+} in the resin phase in mol.kg-1 and [M(II)] its total concentration in the aqueous phase, $[M(II)]_t$ is the initial total concentration of metal in the aqueous phase, V denotes the volume of aqueous phase, and m_r the mass of dry impregnated resin..

Metal distribution data with XAD2-BL resins are plotted as log D versus pH in 0.1 M $NaNO_3$ and given in Figure 1. These figures show that the distribution functions are straigh lines of slope between 2 and 1 depending on the metal and extractant concentration in the resin phase.

Accordingly, the extraction of these metal ions with XAD2-BL resins can be described with the following general reaction:

$$M^{2+}+(2+q-t)\,HL_r+tNO_3^- \rightleftharpoons M(NO_3)_tL_{2-t}(HL)_{q,r}+(2-t)\,H^+ \quad (2)$$

Following the approach developped by Marcus (Marcus, 1966) in the study of metal extraction reactions in ion exchangers the corresponding stoichiometric equilibrium constant of the extraction process (β^*_{2qt}) is defined as:

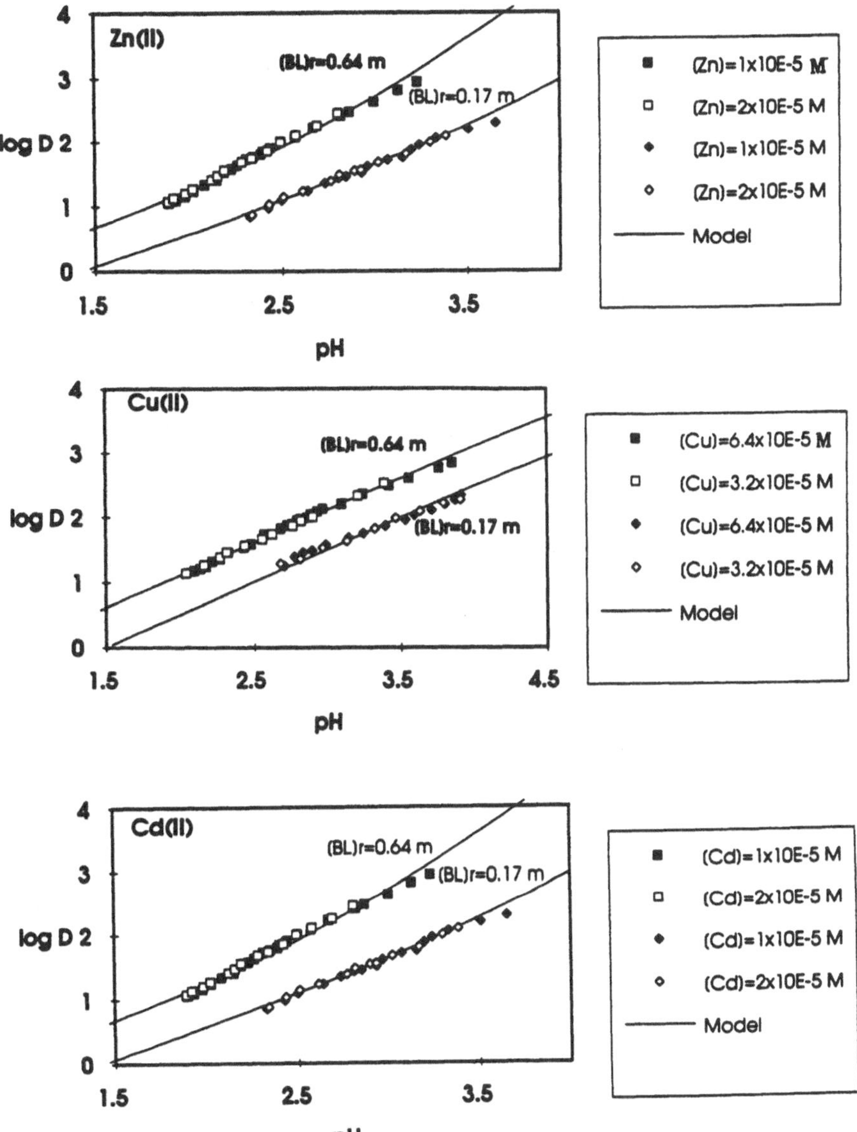

Figure 1. Variation of the distribution coefficient of Zn(II), Cu(II) and Cd(II) as a function of pH at two total concentrations of BL in the resin phase (marked within the diagram) and total metal concentrations (marked at the side). The full-drawn lines have calculated using the constants by given Cortina et al (Cortina, 1994a).

$$\beta^{*}_{2qt}=\frac{[ML_{(2-t)}(NO_3)_t(HL)_q]_r[H^+]^{(2-t)}}{[M^{2+}][NO_3^-]^t[HL]_r^{(2+q-t)}} \tag{3}$$

In order to relate the metal composition in the resin phase to the metal distribution data (D values) the formation of a simplest species may be assumed. So, if only one species of the type $ML_{(2-t)}(NO_3)_t(HL)_q$ is formed the distribution coefficient for (M^{2+}) becomes:

$$D=\Sigma_q\Sigma_t\beta^{*}_{2qt}[HL]_r^{(2+q-t)}[NO_3^-]^t[H^+]^{-(2-t)} \tag{4}$$

Analysis of the experimental data obtained in the distribution equilibrium studies were performed using the computer program LETAGROP-DISTR (Liem, 1971) to determine the composition of the extracted species in the resin phase. A detaliled description of the speciation of the systems will be published later (Cortina, 1994b).

Accordinly to these results the extraction process of Zn(II) Cu(II) and Cd(II) could be explained assuming the following reactions (eq5-10) that involve the formation of the species $ZmL_2(HL)$ and $ZnL(NO_3)$ for Zn(II), $CuL(NO_3)$ and $CuL(NO_3)(HL)$ for Cu(II) and $CdL_2(HL)_2$, $CdL(NO_3)(HL)$ for Cd(II).
For Zn(II):

$$Zn^{2+}+3HL_r \rightleftharpoons ZnL_2(HL)_r+2H^+ \tag{5}$$

$$Zn^{2+}+HL_r+NO_3^- \rightleftharpoons Zn(NO_3)L_r+H^+ \tag{6}$$

For Cu(II)

$$Cu^{2+}+2HL_r+NO_3^- \rightleftharpoons Cu(NO_3)L(HL)_r+H^+ \tag{7}$$

$$Cu^{2+}+HL_r+NO_3^- \rightleftharpoons Cu(NO_3)L_r+H^+ \tag{8}$$

For Cd(II):

$$Cd^{2+}+4HL_r \rightleftharpoons CdL_2(HL)_{2,r}+2H^+ \tag{9}$$

$$Cd^{2+}+HL_r+NO_3^- \rightleftharpoons Cd(NO_3)L_r+H^+ \tag{10}$$

3.2 Column extraction procedure.

Adsorption isotherms. Breakthrough curves ($[M]/[M]_o$] versus volume (V)) for Zn(II), Cu(II) and Cd(II) with XAD2-BL resins are shown in Figure 2. For each experiment a 0.1 M $NaNO_3$ solution containing $1x10^{-4}$M of Zn(II), Cu(II) and Cd(II) was passed downwards at a flow rate of 1ml/mim at 20 °C. The effective capacity (corresponding to $[M]/[M]_o$= 0.5 ($C_{0.5}$)) and dynamic capacity ($[M]/[M]_o$= 0.05 ($C_{0.05}$)) were calculated from sorption isotherms for the tree metal ions and are collected in Table 1.

Table 1. Efective (C_{05}) and dinamic (C_{50}) capacities of XAD2-BL columns in the extraction of Zn(II), Cu(II) and Cd(II) from 0.1 M $NaNO_3$.

Metal	$C_{0.05}$(mg)	$C_{0.5}$(mg)
Zn(II)	2.12	3.931
Cu(II)	3.48	6.45
Cd(II)	2.77	4.43

Elution isotherms. Desorption of the bed was studied with 0.2 M HCl. Samples of metal ions of the eluate were collected at the exit of the column to obtain the elution curve. Figure 3 show the extent of elution (based on the total metal ion in the resin) and the metal ion content (in mg of M(II)) in the eluate as a function of eluate volume. In table 2, the preconcentration factors defined as the ratio between the volume of metal ion passed through the column and the volume of eluant needed to elute the 95% of metal ion retained on the column are collected.

Table 2. Loading of metal, in mg, enrichment factors and recovery percentages of Zn(II), Cu(II) and Cd(II) in column experiments with XAD2-BL resins.

Metal	mg M(II)	%Recovery	Preconcentration Factor
Zn(II)	3.74	95.5	10
Cu(II)	4.43	99.8	14
Cd(II)	6.34	98.2	11

For the three metal ions preconcentration factors of 10 have been obtained. This means that the metal extraction step is followed by a regeneration step of the column in which the chloride solution obtained is 10 times concentrated.

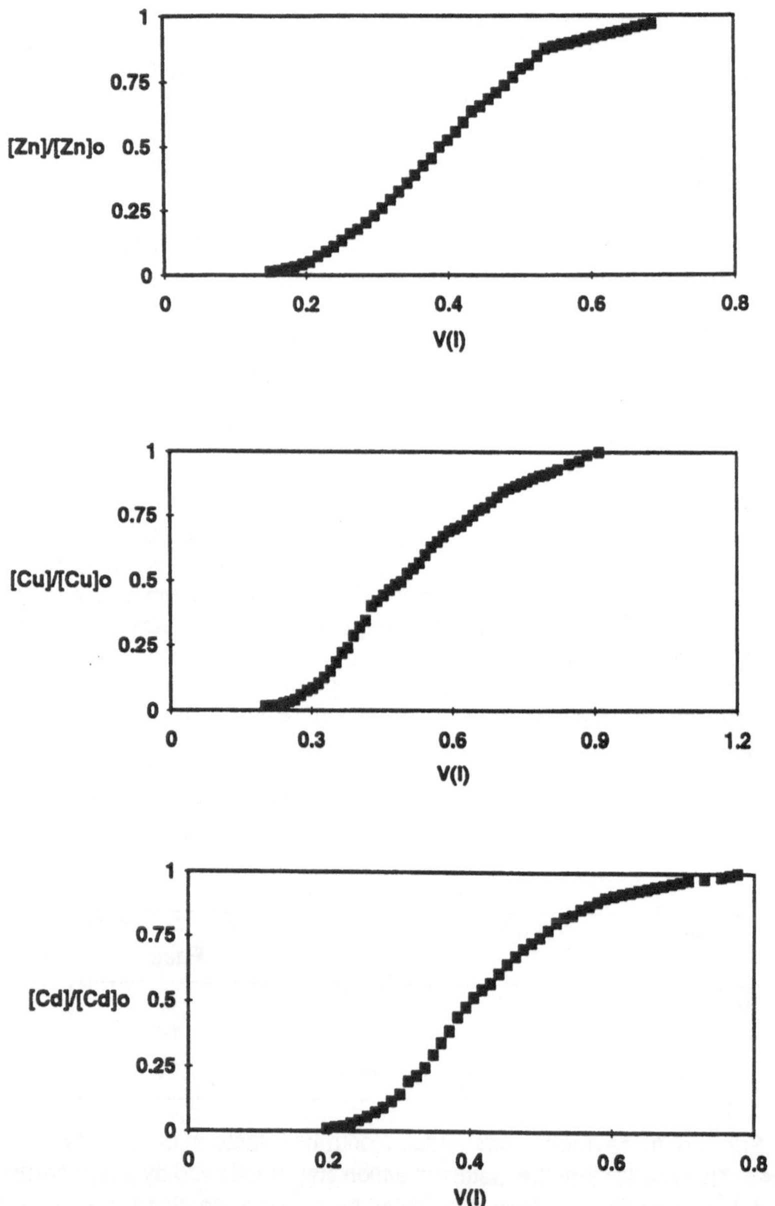

Figure 2. A breakthrough curve for the extraction of Zn(II), Cu(II) and Cd(II) with resin XAD2-BL from nitrate media.

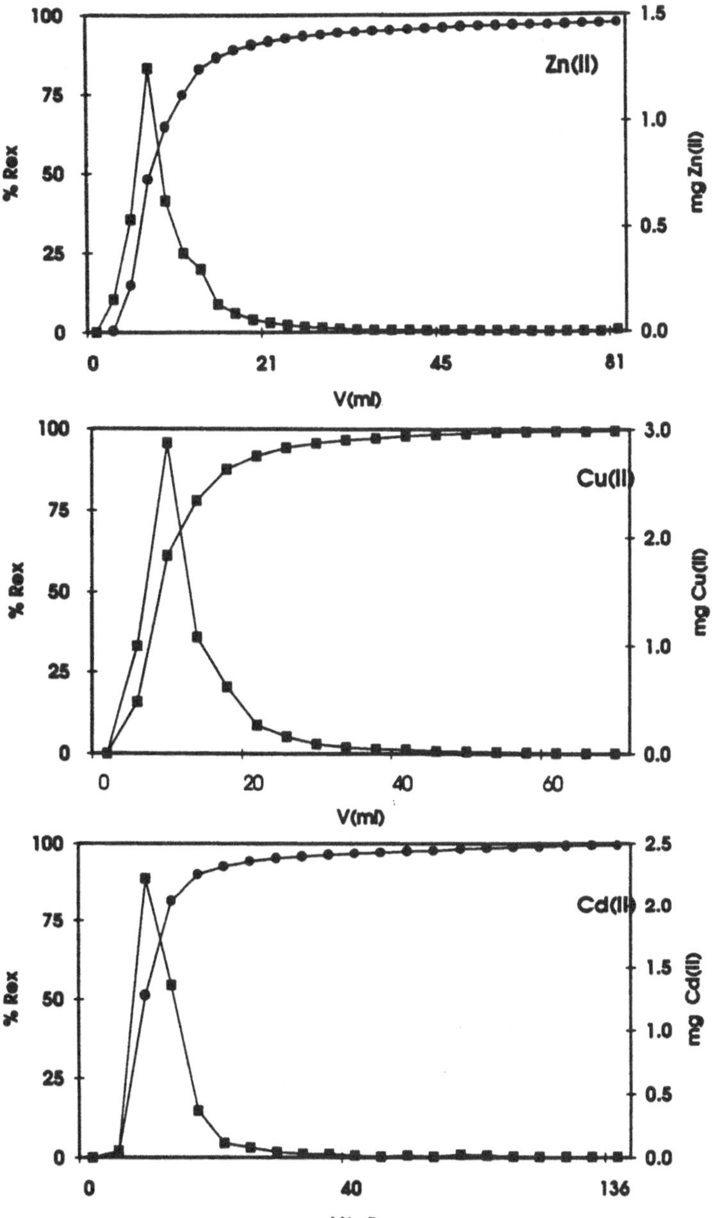

Figure 3. Elution history of Zn(II), Cu(II) and Cd(II) with 0.2 M HCl for metal loaded resins M(II)-XAD2-BL.

3.3 Metal Extraction Kinetic studies.

In all cases the rate of extraction was measured under conditions simulating those in a countercurrent extraction process, i.e., conditions in which the concentration of metal in solution remains approximately constant as the metal loading on the resin increases and approaches its equilibrium value. These conditions were obtained using the shallow technique described above.

Figure 4 shows the extraction kinetics of Zn(II), Cu(II) and Cd(II) from single element solutions in the form of resin loading as a function of contact time. This series of experiments was run at a relatively high concentration ($1x10^{-2}mol.l^{-1}$) with the purpose of promoting particle diffusion control. Only the the first hour of the process was studied in this experiment.

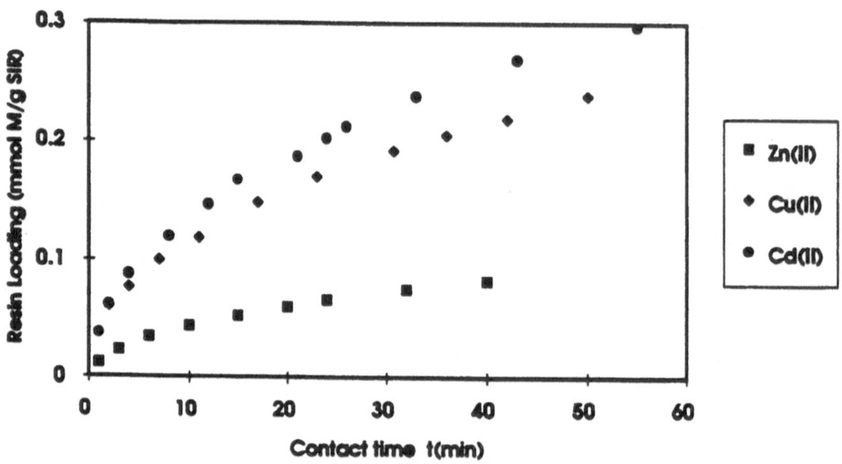

Figure 4. Metal loading rates from high aqueous metal concentration ($1x10^{-2}M$) for Zn(II), Cu(II) and Cd(II) with XAD2-BL resins.

The fractional approach to equilibrium as a function of time is shown in Figure 5. Cadmium and copper do, in fact, approach equilibrium at a faster rate than zinc. The parabolic kinetics observed in all cases strongly suggest a particle diffusion control of the extraction kinetics since the metal concentration in the external solution remained constant. From these observations it appears reasonable to conclude that both cadmium (II) and copper (II) exhibit higher diffusivities in the resin matrix than Zinc (II) ions. Further confirmation would to be found in kinetic parameters obtained from an adequate modelling of the extraction kinetics (Cortina, 1994c).

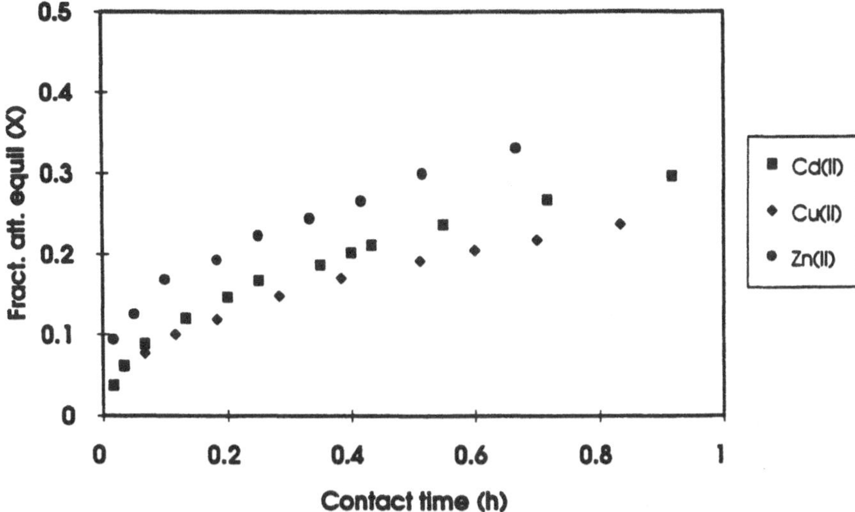

Figure 5. Rate approach to equilibrium at high aqueous metal concentration ($1x10^{-2}$M) for Zn(II), Cu(II) and Cd(II) with XAD2-BL resins.

Figure 6 shows the extraction of Zn(II), Cu(II) and Cd(II) from single-element solutions at a concentration equivalent to that used for the column extraction experiments, i.e., $1x10^{-4}$ mol.l^{-1}. Only the the first hour of the process was studied in this experiment.

The relationship between resin loading and contact time is clearly linear. From the slopes of lines, the individual extraction rates (mmol/g/h) were obtained. They were for zinc, 0.1579; for copper, 0.1189; and for cadmium, 0.0222 mmol/g/h. The relative order of extraction rates under these conditions is:

$$Zn(II) = Cu(II) > Cd(II)$$

Figure 7 shows the fractional approach to equilibrium as a function of contact time for the last series of experiments. The linearity of the graphs suggests a regime of film diffusion control since the Nerst diffusion layer and the extrenal solution concentration arte constant.

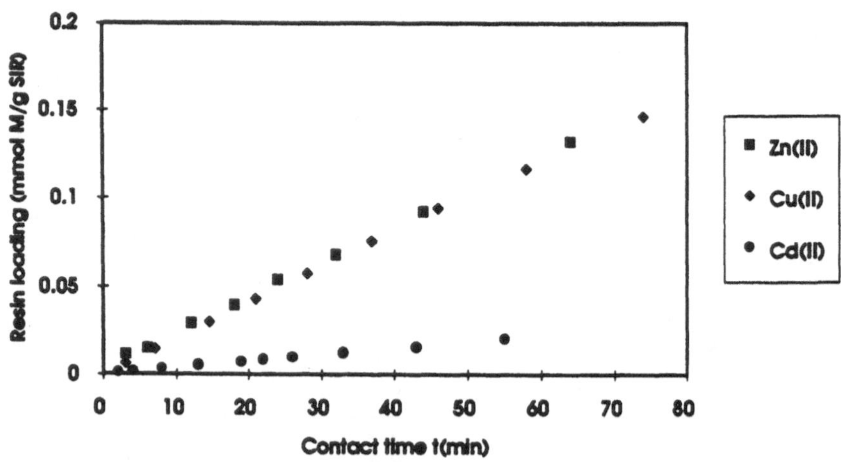

Figure 6. Loading rate of metal at low aqueous metal concentration ($1x10^{-4}$M) for Zn(II), Cu(II) and Cd(II) with XAD2-BL resins.

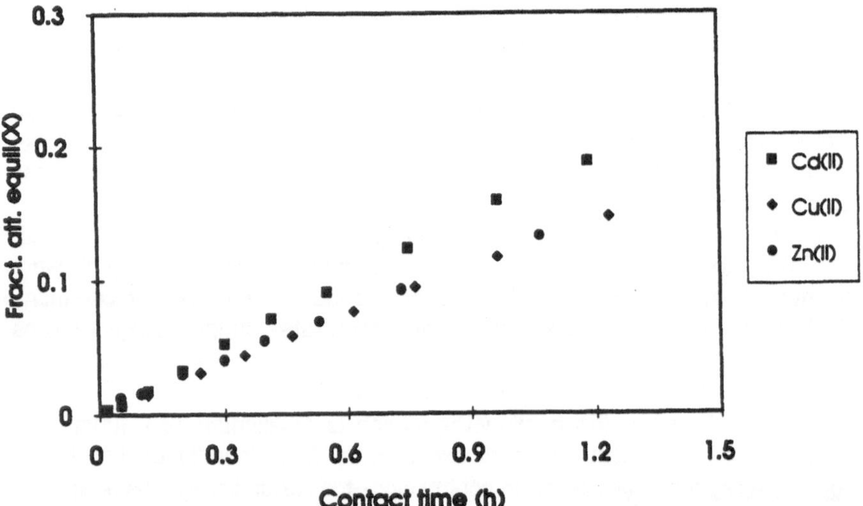

Figure 7. Fraction attainment to equilibrium at low aqueous metal concentration ($1x10^{-4}$M) for Zn(II), Cu(II) and Cd(II) with XAD2-BL resins.

4 Conclusions

The use of O-methyl-dihexyl-phosphine-oxide O'-hexyl-2-ethyl phosphoric acid physically immobilized into Amberlite XAD2 supports show a high afficiency in the extraction of Zn(II), Cu(II) and Cd(II) increasing with the increase of pH and extractant concentration in the resin phase. The extraction of these metal ions involves the formation of mixed species in the resin phase with a general composition $ML_2(HL)_{q,r}$ and $ML_{(2-t)}(NO_3)_t(HL)_{q,r}$ where q,t values take different values depending on the metal.

The application of these resins in column process allow the extraction efficienly of Zn(II), Cu(II) and Cd(II) and elution of then with hydrochloric acid solutions. For the three metal ions, enrichment factors of 10 could be obtained in the elution step.

Finally, the kinetic studies on the extraction process at relatively high at relatively high concentrations show parabolic kinetic curves and suggest a particle diffusion control for the extraction reaction. The differences in the loading dependece on time allow to concluded that the resin diffusivity of the metal ions folow the order Cd(II) > Cu(II) > Zn(II).

On the other hand, at low metal concentration the linear dependence of the resin loads on contact time suggests a regime of flim diffusion control for the three metal ions. The relative order of extraction rates under these conditions are: Zn(II) > Cu(II) > Cd(II).

5 Acknowledgement

This work was supported by CICYT Project MAT 93-6212 (Ministerio de Educación y Ciencia de España). The authors are indebted Mr. Abraham Deshe, Weizmann Institute of Science, for synthesis of BL. Finally, one of the authors (J.L. Cortina) acknowledges with thanks, his stay in the Department of Organic Chemistry The Weizman Institute of Science with Prof. Abraham Warshawsky.

6 References

Abollino, O., Mentasti, E., Porta, V., Sazarnani C. (1990) Immobilized 8-oxine units on different solid sorbents for the uptake of trace metals, **Anal. Chem.** 62, 21-26.

Apostolidis, C., Bekelund H. and Glatz, J.P., (1991) Actinide separation by high pressure cation exchange. The neptunium case, in **New separation Chemistry Techniques for Radioactive Waste and other Specific**

Applications (eds L. Cecille, M. Casarci and L. Pietrelli), Elsevier, London (UK), pp. 213-218.

Arad-Yellin, R., Zangen, M., Gottlieb H. and Warshawsky, A. (1990) Bifunctional phosphoric acid-phosphine oxide extractants: Synthesis and complexes with uranium-(IV) and (VI) and iron (III). **J. Chem. Soc. Dalton Trans**, 2081-2088.

Cortina, J.L., Sastre, A., Miralles, N., Aguilar, M., Profumo, A. and Pesavento,M. (1992b) Solvent Impregnated resins containing Cyanex 272. Preparation and application to the extraction and separation of divalent metals. **React. Polym.**, 18, 67-75.

Cortina, J.L., Sastre, A., Miralles, N., Aguilar, M., Profumo, A. and Pesavento,M. (1993) Solvent Impregnated resins containing di-(2,4,4-trimethylpentyl)phosphinic acid. II. Study of the distribution equilibria of Zn(II), Cu(II) and Cd(II). **React. Polym.**, 21, 67-75.

Cortina, J.L., Miralles, N., Aguilar, M., Sastre, A. (1994a) Solvent Impregnated resins containing di-(2-ethylhexyl)phosphoric acid. II. Study of the distribution equilibria of Zn(II), Cu(II) and Cd(II). **Solv. Extr. Ion Exch.** (in press).

Cortina, J.L., Miralles, N., Aguilar, M., Warshawsky, A. (1994b) Solid-liquid distribution studies of divalent metals from nitrate media using impregnated resins containing a bifunctional organophosphorous extractant (O-methyldi-hexyl-phosphine oxide O'Hexyl-2-ethyl phosphoric acid, (submitted for publication to **Reactive Polymers**).

Cortina, J.L., Miralles, N., Aguilar, M., Warshawsky, A. (1994c) Kinetic studies ofthe extraction of Zn(II), Cu(II) and Cd(II) with impregnated resins containing a bifunctional extractant O-methyl-dihexyl-phosphine-oxide O'-hexyl-2-ethyl phosphoric acid (BL) immobilized into Amberlite XAD2 (XAD2-BL). (manuscript in preparation).

Gonzalez-Luque, S. and Streat, M. (1983) Uranium sorption from phosphoric acid solutions using selective ion exchange resins. **Hydrometallurgy**, 11, 207-225.

Helfferich, F. (1962), in **Ion Exchange**, Mc Grawn-Hill, New York, USA.

Horwitz, E., Dietz, M., Nelson, D., La Rosa, J. and Fairman, W. (1990). Concentration and separation of actinides from urine using a supported bifunctional organophosphorous extractant, **Analytica Chimica Acta**, 238, 263-271.

Kroebel. R. and Meyer, A. (1971) **West German Patent**, 2, 162, 951.

Liem, D.H. (1971) High speed computers as a supplement to graphical methods. 12. Application of LETAGROP to data for liquid-liquid distribution equilibria. **Acta Chem. Scand**, 25, 1521-1543.

Louis, R. and Duyckaerts, G. (1984) Some parameters affecting the extraction chromatographic performance of TBP-impregnated macroporous XAD4 columns for americium(III)-europium (III) separations. **J. Radional. Nucl. Chem.**, 81, 305-315.

Marcus, Y. (1966) Ion-Exchange studies of complex formation .**Ion Exchange and Solvent Extraction**, A series of Advances Series, Vol 1, Chapter 3 (eds J.A. Marinsky), Marcel Dekker.

Muscatello, A.C., Navratil, J.D. (1988) Plutonium removal from nitric acid waste streams. **J. Radional. Nucl. Chem. Letters**, 128, 463-477.

Paterson, J.W. (1990) Metal Separation and Recovery, **Metal Speciation Separation and Recovery**, Lewis Publishers Eds, INC, pp. 27-39.

Shoen, J., Ochsenfeld, W. (1982) Separation processes using polymeric resins in reprocessing of nuclear fuels. **Angew. Makromol. Chem**, 109/110, 215-222.

Tavlarides, L.L., Bae, J.H. and Lee, C.K. (1987) Solvent extraction, membranes and ion exchange in hydrometallurgical dilute metals separation. **Sep. Sci. Technol.**, 22, 581-597.

Yoshizuka, K., Sakamoto, Y., Baba, Y. and Ionue, K., (1990) Distribution equilibria in the adsorption of coabalt (II) and nickel (II) on Levextrel resin containing Cyanex 272, **Hydrometallurgy**, 23, 309-315.

Wakui, Y., Matsunaga H. and Suzuki, T.M. (1989) Selective recovery of trace scandium from acid aqueous solution with 2-ethylhexyl hydrogen-2-ethylhexyl phosphonate-impregnated resin. **Anal. Sci**, 5, 189-193.

Warshawsky, A. (1971) **S. Afr. Patent Appl.**, 71/5637 (23/7/71).

Warshawsky, A. (1974) Solvent impregnated resins in hydrometallurgical applications. **Trans. Inst. Min. Metall.**, 83, C101.

Warshawsky, A. (1981) Extraction with solvent impregnated resins. **Ion Exchange and Solvent Extraction**, vol 8, J.A. Marinsky and Y. Marcus, Eds., Marcel Dekker, New York, pp. 229-310.

Warshawsky, A., and Arad-Yellin, R. (1986) **Israeli Patent Application**, 79-999 and 80-000, YEDA Company, Israel.

Use of a fluidized-bed ion-exchange system for heavy metals removal from waste streams

Paz Cosmen
M. L. Ruiz
CIEMAT, Madrid, Spain

Abstract
Ion exchange, using solid resins, is one of the best methods to reduce heavy metals concentration in waste streams down to levels fulfilling the every day stricter regulations on the disposal of waste waters.

Both, capital investment and resin inventory are lower in a fluidized-bed ion exchange system as compared with conventional fixed bed process. It has other additional advantages as, for instance, to be practically a continuous process able to handle dirty effluents with moderate quantities of suspended solids.

A great experience in fluidized-bed ion exchange has been gathered by CIEMAT researches in the last few years. On the basis of the acquired experience on this process, its application to heavy metals removal from waste waters, has been studied. In this paper, a fluidized-bed ion exchange process applied to heavy metals removal from waste streams is described.

There is a significant number of electroplating, mining and other industries that produce large volumes of liquid wastes containing heavy metals. In some instances, these wastes are discharged to river streams or to the ocean with no control. Frequently, these waste streams are diluted to fulfil regulations, which implies a substantial increase in water consumption. On the contrary, these liquid wastes may be treated by ion exchange in order to remove heavy metals allowing treated water to be recycled.

In the case of heavy metals removal from waste waters, pH is a critical parameter. In order to obtain an adequate pH to transfer metals from waste to resin, it is necessary to transform the resin to its sodium form before it enters into the loading column (by treating it with a caustic soda solution).

In this paper, a new system, in which the pH is controlled during resin loading is proposed. The first results obtained in Ni/Cu separation in one stage column with simultaneous pH control are presented. Thus, the excellent possibilities offered by fluidized bed ion exchange applied to the removal, recovery and separation of heavy metals from waste streams may be realized.
Keywords: Fluidized-bed, heavy metal, ion exchange, waste streams.

1 Introduction

Ion exchange using solid resins, is one of the best methods to reduce heavy metals concentration in waste streams. For many years lime has been used for heavy metals removal from waste waters. However, sufficiently low metal values are not reached for final disposal.

Although liquid-liquid solvent extraction can be applied, it uses to be more expensive than ion exchange and an additional pollution problem might be caused by organic entrainment. Reverse osmosis and electrodialysis may also be applied although these technologies have not been proved at industrial scale yet.

Both, capital investment and resin inventory are lower in a fluidized-bed ion exchange system as compared with conventional fixed bed process and it is a practically continuous process able to handle dirty effluents with moderate quantities of suspended solids.

Experience in fluidized-bed ion exchange has been gathered by CIEMAT researches in the last few years. This process was used for uranium recovery, and a 5-10 m3/h Demonstration Plant was built in 1987, which has been in normal operation since then.

On the basis of the acquired experience on this process, application to heavy metals removal from waste waters has been studied. In this case, pH is a critical parameter. To obtain an adequate pH to transfer metals from waste to resin, it is necessary to transform the resin to its sodium form before it enters into the loading column (by treating it with a caustic soda solution). To avoid this treatment, a new system of pH control in the loading column is proposed.

In this paper, a fluidized-bed ion exchange process applied to heavy metals separation and/or removal from waste streams is depicted.

2 Objectives

The objective of the work presented in this paper is to remove heavy metals from syntetic solutions by using a fluidized-bed ion exchange system, so that treated wastes fulfil regulations on waste disposal.

3 System description

The process is divided in two steps as follows: One loading step carried-out in a fluidized-bed column provided with six stages, divided by perforated plates, and one elution step performed in a fixed-bed column. A diagram of the system is represented in figure 1.

3.1 Loading step.

The feed containing heavy metals enters at the column bottom and flows up making the resin to fluidize due to the high feed flow rate. Mixing of resin and solution is produced in this way and heavy metals transfer from the solution to the resin.

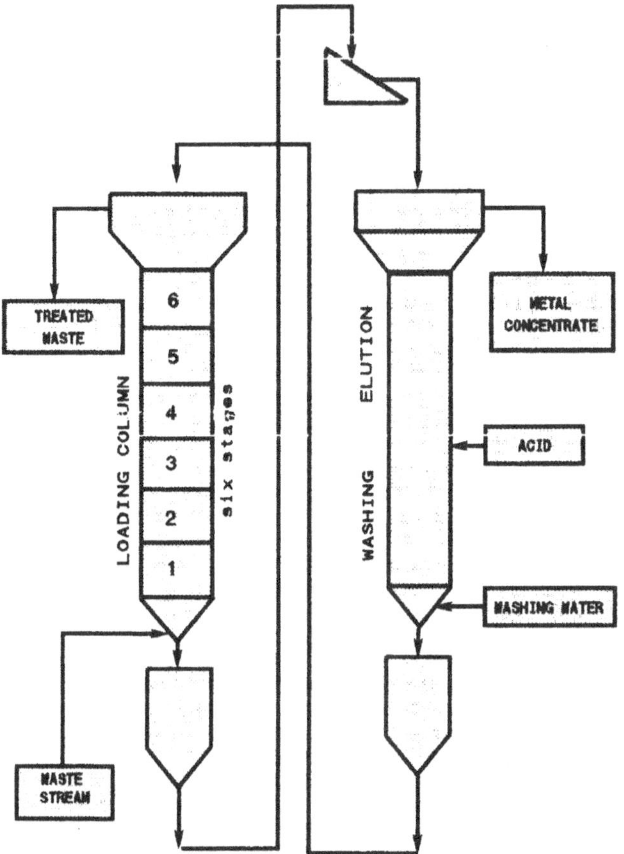

Figure 1. Continuous ion exchange in fluidized bed.

The purified waste (raffinate) leaves the column at the top. At this point the solution has a very low concentration of heavy metals and it is adequate to be discharged to the river, in some cases, after a suitable pH adjustment.

Periodically, once a cycle ends, the waste water flow is stopped which makes the resin to settle on each interstage perforated plate separator, whose special design prevents resin to leak down to the lower stage, when there is no flow.

Once the resin has settled down, a discharge valve at the bottom of the loading column is opened. A downwards flow is established in such a way that resin passes from each stage to the lower one through the holes of the perforated plate separator. Likewise, the loaded resin corresponding to the first stage is transferred to an outer tank located under the loading column. This discharge operation takes from 10 to 30 seconds and is conveniently controlled by level probes in the collecting tank.

When the discharge operation finishes, the waste stream feed flow is restored for a new loading cycle to start again. At the same time, the discharged volume of loaded resin is replaced by an equivalent volume of unloaded resin, which is air-lifted from a surge tank located under the elution column. The loaded resin containing the heavy metals is sent by air-lift to a static screen located above the elution column, where accompaning liquor is separated from the loaded resin and sent back to the feed tank, and resin flowing over the screen is transferred by gravity to the top of the elution column.

3.2 Washing and elution steps

Both steps, elution and washing, are carried-out in the same column, which is divided into two zones: the top one, where the elution of the resin is accomplished, and the bottom one, where the washing of the resin is carried-out, removing the remainder acidity.

Acid is introduced at the lowest part of the elution zone, where it is mixed with the washing liquor. Water is introduced at the lowest part of the washing zone.

The loaded resin, containing the heavy metals, is introduced at the column top, where it comes into contact with the acid (aqueous solution of hydrochloric acid or sulphuric acid), so that the transfer of metals from resin to acid is produced.

The unloaded resin flows down through the column reaching the washing zone, where the remainder acid is countercurrently removed by the washing water. The acid solution containing the heavy metals (eluate) leaves the elution column at the top.

When the elution cycle finishes, the unloaded resin is discharged to the unloaded resin surge tank located under the elution column. Elution acid and washing water feeds to the column are not stopped while discharging. Afterwards, unloaded resin is transferred to the loading column, just when a new cycle has to begin.

In the elution column, fluidization of the resin is not achieved since fluxes of elution acid and washing water are much lower than in the loading column.

The difference between the waste stream flow and the elution acid plus washing water flows causes a significant concentration of heavy metals. This feature constitutes the key advantage of the fluidized-bed ion exchange system for treatment of liquid wastes as compared with competing technologies.

As the efficiency in loading and elution operations is 100 %, the concentration factor is equal to the ratio of waste stream flow-rate to acid plus water flow-rates.

3.3 Process automation

This plant has been designed to be automatically controlled since the various cycles of the process are short and decisive. The most critical one is the column discharge, with periods of aproximately 30 seconds.

The plant is connected to a PLC with connection to a personal computer. A program is in charge of controlling and managing the plant. Plant information is presented in the computer screen by means of menus selected by the operator. Besides, it is possible to manage the plant from the computer in a manual mode, too.

4 Application to heavy metals removal

Nowdays, there are resins which are adecuate to remove all the heavy metals. Within such a broad scope, the study was started undertaking those metals causing problems in the mining and electroplating industries as for example: copper, zinc, cadmium, manganese, cobalt and nickel. The proposed process applied to heavy metals removal is represented in figure 2.

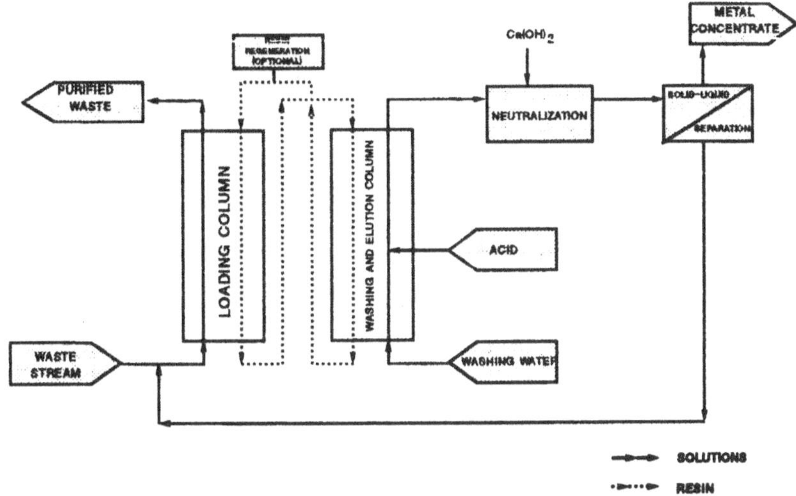

Figure 2. Flow-sheet applied to heavy metals removal

In general, these metals are present in solution as cations. Consequently, the use of cationic resins to retain them is the obvious first choice. The process will be represented by the following reactions:

Loading: $M^{+n} + n\ R\text{-}H \rightarrow MR_n + n\ H^+$ (1)

Elution: $MR_n + n\ X\text{-}H \rightarrow n\ R\text{-}H + MX_n$ (2)

$X = Cl^-$ or $SO4^=$; R-H = resin in H^+ form.

It can be observed that in the loading reaction protons are released, making waste acidity to increase. This is not convenient because according with reaction 1, a pH decrease makes the reaction to go in the opposite direction, decreasing metal extraction efficiency. One way of overcoming this problem is to convert the resin into its sodium form after leaving the elution column (regeneration), what is easy to achieve by mixing the resin with a caustic soda solution before it enters the loading column. The reactions of the process are as follows:

Loading: $M^{+n} + n\ R\text{-}Na \rightarrow MR_n + n\ Na^+$

Elution: $MR_n + n\ X\text{-}H \rightarrow n\ R\text{-}H + MX_n$

Regeneration: $R\text{-}H + NaOH \rightarrow R\text{-}Na + H_2O$

In this way, during loading, heavy metal ions are exchanged by sodium ions instead of protons, and no increase in acidity is produced.

It can be noticed in figure 2 that lime precipitation of heavy metals is applied to the metals concentrate (eluate), whose volume is in the range of 100 to 1000 times lower than the liquid waste (depending on the metal concentration in the waste). Consequently, this operation is performed at a much lower cost.

Needless to say that direct lime precipitation does not achieve metals concentration in the treated waste low enough to fulfil waste disposal regulations, and that pH is increased requiring further pH adjustment, with the corresponding extra cost.

Once the lime precipitation has been carried-out, the resulting water is recycled to the loading column after a suitable solid-liquid separation.

In this process a sludge which contain the metals is obtained. In general this is the most economic solution when there are different metals present in the efluent. But there can be cases where the metals are valuable and their recovery might be interesting. In fact, separation of the various metals may be achieved by controling the pH in the loading and/or elution column.

5 pH influence on metal extraction

The pH influence on metal extraction efficiency was studied. The feed used in these tests had a concentration of 1 mEq/l of each metal (Cu, Cd, Ni, Co, Mn and Zn) in sulphate form. Resin used was Duolite C-467 from Rhom-Haas. This is a resin which contains aminophosphonic group.

The procedure for securing equilibrium data was as follows: 400 ml of the aqueous solutions were contacted with 2 ml of resin. The pH of the aqueous solutions were adjusted to the desired pH by adding NaOH or H_2SO_4. Mixing of both phases was performed by a mechanical stirrer for 1 hour after the desired value of pH was obtained. Then, samples of the aqueous phase were directly analyzed.

The final aqueous solutions are represented in table 1 and in figure 3 the pH influence on metal extraction efficiency is represented, where a substantial influence of pH on metal extraction can be observed. On the other hand, it is worth to notice the ease with which metal separation can be achieved by controlling the pH during the loading stage.

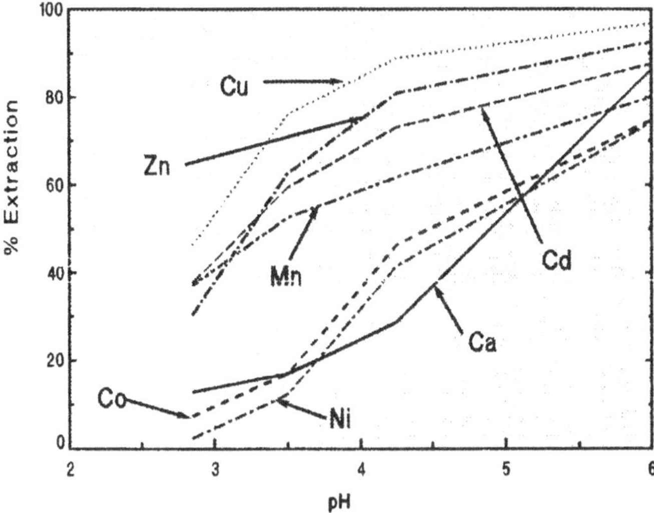

Figure 3. pH influence against metal extraction

Table 1. Metal concentrations against pH, in final aqueous solutions.

Ca	Cd	Co	Cu	Mn	Ni	Zn	M^{2+}, (mg/l)
9.4	37	20.5	20.5	17.5	20.5	23.0	Feed
8.2	23.0	19.0	11.0	11.0	20.0	16.0	pH=2.85
7.8	15.0	17.0	4.9	8.3	18.0	8.6	pH=3.50
6.7	10.0	11.0	2.3	6.7	12.0	4.4	pH=4.25
1.3	4.6	5.2	0.65	3.5	5.3	1.7	pH=6.00

6 Pilot plant description

The pilot plant used for continuous testing, that has a throughput of 200 l/h of waste water, is shown in figure 4. This plant consists of two columns (loading column and washing plus elution column). The fluidized-bed loading column (C-1) is made up of six 1 meter high and 120 mm diameter stages.

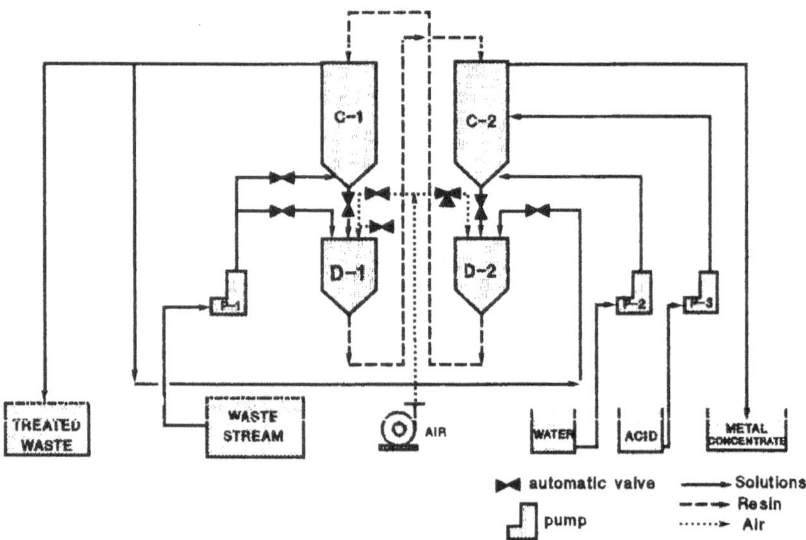

Figure 4. Pilot Plant flow-sheet

The washing and elution column (C-2) has two different parts. The elution part is the top of the column and the washing part is the bottom of the column. This column is 5 m high and 70 mm diameter. Both columns are made of transparent PVC and operate cyclically. The transfer bins are 600 mm high and 150 mm diameter glass cylinders. They are provided with level probes to control the resin volumes to be discharged and transferred.

Resin motion is obtained by means of compressed air. An adequate liquid volume must go with the resin during this motion. All pilot plant operations may be manually or automatically controlled.

7 Continuous pilot plant tests

Tests were done to remove or separate different heavy metals, with satisfactory results. Each test was run for 60 hours with cycles between resin discharges of 90 minutes for test 1 and 135 minutes for test 2. H_2SO_4 2N was used to elute the resin. The acid was mixed with the washing water at the flow rate ratio of 1:1. Consequently, the actual elution acidity was H_2SO_4 1N.

In the first test, resin was regenerated (mixed with a NaOH solution) before entering the loading column, so that there was no pH decrease. Due to the high pH, total extraction of heavy metals was accomplished.

In the second test, resin enters into the loading column as it leaves the elution column (in protonated form). Due to the low pH produced by protons release, the separation of Co, Ni and Ca from the rest of the metals is obtained.

The input and output average results, once the steady state was reached, are shown in figure 5 (test 1) and in figure 6 (test 2). Samples of resin and solution from each stage were taken at the end of the last cycle of the two tests. The results are shown in table 2 (test 1) and in table 3 (test 2).

Referring to test 1, in figure 6 it is worth noticing that the treated waste has no heavy metals at all. The volume of the metal concentrate is aproximately 100 times lower than the volume of waste water and it contains 100 % of the heavy metals initially present in the waste stream. It can be observed that two stages are enough to remove all the heavy metals under the conditions used in this testing.

Referring to test 2, the treated waste contains the total of Co, Ni and Ca. On the other hand, the metal concentrate contains the rest of the metals (Cd, Cu, Zn and Mn) in a volume 100 times lower than the volume of waste water. It can be observed that six stages are necessary to separate the heavy metal in two groups, under the conditions of test 2.

On the other hand, these conditions were carefully selected to obtain a good metal separation, but many times it would be not possible to reach it, without pH control.

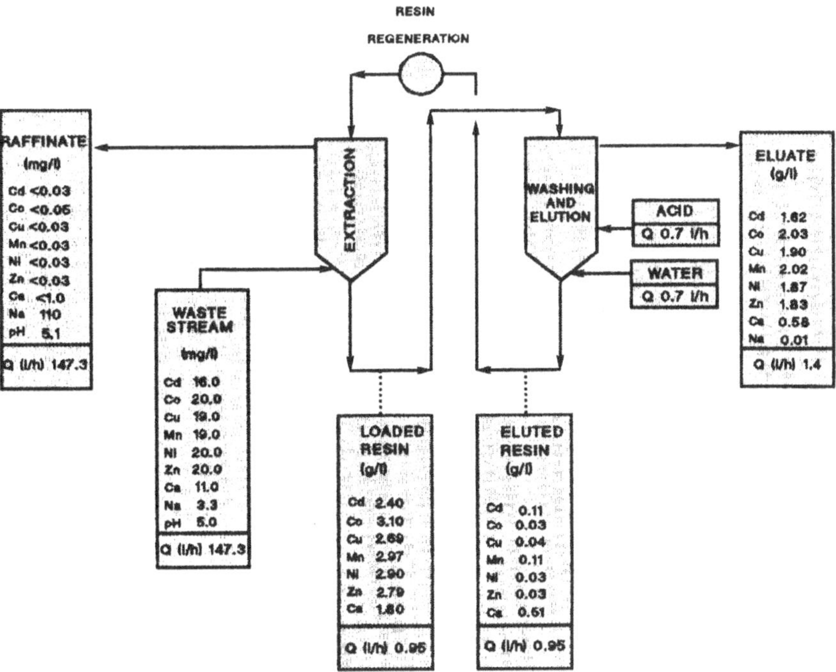

Figure 5. Input and output average results in test 1

Table 2. Metal concentration in resin and solution in each stage in test 1

Stage	Ca	Cd	Co	Cu	Mn	Ni	Zn	pH
				Resins, g/l				
1	2.63	2.20	3.14	2.67	2.62	3.40	2.56	-
2	1.61	0.28	1.80	0.066	0.75	2.40	0.132	-
3	0.49	0.15	0.059	0.051	0.158	0.084	0.035	-
4	0.48	0.15	0.029	0.053	0.158	0.035	0.035	-
5	0.50	0.14	0.026	0.050	0.151	0.028	0.033	-
6	0.49	0.13	0.029	0.048	0.142	0.033	0.039	-
				Aqueous solutions, mg/l				
1	7	0.90	13.0	0.10	3.50	18.00	0.60	5.2
2	<1	0.04	0.20	<0.03	0.11	0.25	<0.03	6.4
3	<1	<0.03	<0.05	<0.03	<0.03	<0.03	<0.03	7.4
4	<1	<0.03	<0.05	<0.03	<0.03	<0.03	<0.03	7.1
5	<1	<0.03	<0.05	<0.03	<0.03	<0.03	<0.03	6.3
6	<1	<0.03	<0.05	<0.03	<0.03	<0.03	<0.03	5.1

Table 3. Metal concentration in resin and solution in each stage in test 2

Stage	Ca	Cd	Co	Cu	Mn	Ni	Zn	pH
				Resins, g/l				
1	0.99	6.64	0.45	3.565	2.85	0.28	3.41	-
2	1.33	4.80	0.55	0.989	2.85	0.30	1.90	-
3	1.91	3.10	0.75	0.277	2.38	0.37	1.13	-
4	2.74	1.65	0.91	0.080	1.54	0.41	0.61	-
5	2.76	0.78	1.25	0.029	0.82	0.53	0.31	-
6	2.80	0.25	1.35	0.017	0.30	0.72	0.11	-
				Aqueous solutions, mg/l				
1	10.5	16.5	13.0	2.25	11.0	13.0	5.75	3.3
2	12.5	11.5	12.5	1.40	9.1	13.0	3.95	3.1
3	14.0	7.7	14.0	0.31	7.1	13.0	2.60	3.1
4	16.0	3.2	15.5	0.17	3.6	13.5	1.14	2.9
5	17.0	1.5	16.5	0.03	1.8	14.0	0.57	2.9
6	8.2	0.3	9.7	<0.03	0.4	10.8	0.11	2.8

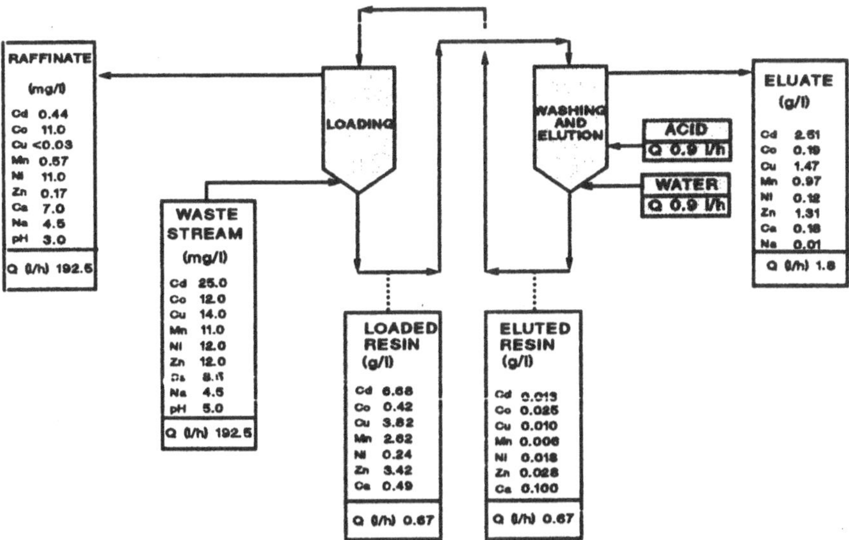

Figure 6. Input and output average results in test 2

8 Ni and Cu separation in pilot plant

The above-mentioned results with no pH control may be only obtained in very special instances. In order to have metal separation, a different system with pH control is proposed, which is schematically represented in figure 7. This plant consists of a fluidized- bed loading column, made up of 1 meter high and 120 mm diameter stage, where pH can be automatically controled during the loading cycle, by means of NaOH adition. Column is made of transparent PVC and it opetares continuously.

This process can be interesting due to the fact that the latest tendencies are focused toward Centralized Treatment Plants, where many effluents with broad spectrum of heavy metals may arrive. If a technology, able to treat a blend of them all is available, one can take advantage of the economy of scale, making possible valuable metal recovery at lower treatment costs.

Tests to separate Cu and Ni from a solution that contains both metals at the same concentration (referred to equivalents) have been carried-out. Solution containing Ni and Cu in sulphate form, past through the loading column, and pH is automatically controlled by NaOH adition. Samples were taken during the test at different times. The feed flow rate was 100 l/h. Their results are represented in table 4.

These first results are very promising (referred to only one stage), and we are working in this subject to improve the results obtained.

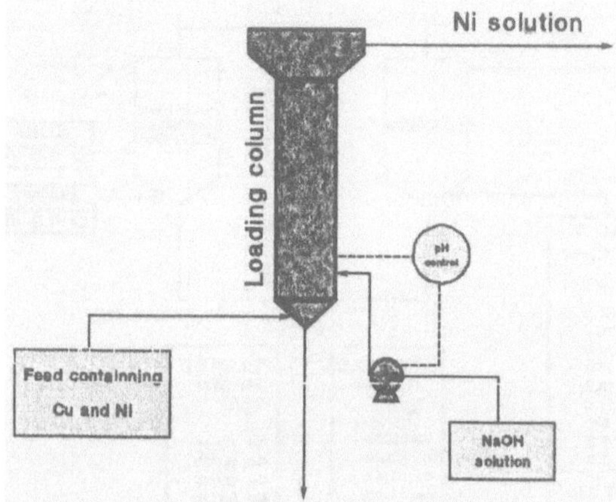

Figure 7. Pilot plant diagram for Ni/Cu separation

Table 4. Evolution of the efluent.

Time, minutes	Ni, mg/l	Cu, mg/l	pH
0 (Feed)	68	76	5.00
25	5	3	2.73
45	9	2	2.80
55	17	3	2.85
80	41	6	2.94
110	62	11	3.00
140	68	14	3.05
185	71	21	3.05
215	71	24	3.10
275	68	32	3.10
335	68	37	3.10
395	68	44	3.10

9 Conclusions

In this paper, the excellent possibilities offered by fluidized-bed ion exchange technology as applied to the removal / recovery / separation of heavy metals from waste streams, may be realized.

With resin DUOLITE C-476 (from Rohm & Haas), copper, zinc, manganese, cadmium, nickel and cobalt can be removed. The treated wastes are totally metal free and they fullfil the strictest regulations on the disposal of these effluents.

On the other hand, heavy metal separation can be achieved by pH control in the loading column. First results of Ni/Cu separation obtained in one stage loading column are promising. A improvement of the process may be obtain, by using a multistage column with a pH control in each stage.

Both pilot plants are ready now to carry-out tests with actual waste waters proceeding from different industrial activities. These pilot plants can also be used for removing either anions (using anionic resins) or organic matter (using adsorbent resins or active carbon).

10 Acknowledgements

We wish to thank Carlos Capdevilla and Alberto Quejido for their collaboration in the project, giving all the analytical support.

11 References

1. M.J.Slater; B.H. Lucas. CIM Bulletin. August 1978. pp 117-123.

2. Botella,T; Gasós,P. Hydrometallurgy,17. 1986. pp 91-112.

3. Himsley,A. Hydrometallurgy'81. Proc. Soc. Chem. Ind. Symp. 1981. E3/1-E3/14.

4. Kadlec,V and Hubner,P. "Ion Exchange for industry". Ed Michael Streat. 1988. pp 148-155.

5. Jakubec,K; Haman,J. Hydrometallurgy,17. 1986. pp 27-38.

Application of ultrafiltration assisted by complexation to treatment of industrial waste waters

Gérard Durand
Mohamed Rakib
Isabelle Michelet
Laboratoire de Chimie Nucléaire et Industrielle, Ecole Centrale Paris, Châtenay-Malabry, France

Abstract
A process for purifying waste waters containing heavy and common metals has been studied. The physicochemical conditions of complexation of metallic cations by a polyethyleneimine macroligand have been first determined. Separation has been achieved by ultrafiltration with inorganic membranes. The influence of physicochemical and hydrodynamic parameters have been studied in order to propose a whole flow-sheet applicable to treatment of 1 m^3/h of waste waters. This work shows the applicability of ultrafiltration assisted by complexation to lower greatly the level of heavy metals in waters.
Keywords: complexation, macroligand, metallic cations, process, ultrafiltration, waste waters.

1 Introduction

A major environmental problem is the presence of heavy metals in industrial waste waters, essentially because water is often recycled not only for industrial but also for domestic uses. The aim of this work is to study a new process intended to lower heavy metals concentrations in waters.

The process studied is founded upon membrane separations, more precisely ultrafiltration assisted by complexation (UFAC). Ultrafiltration is commonly used to separate species of high molecular weight present in solution, for example in food industry [1], but this process is not applicable to separations of ionic species such as metallic cations like heavy metals, because their ionic radii are too small. In UFAC, metallic cations are firstly complexed by a macroligand in order to increase their molecular weight [2]. These compounds which are then bigger than pores of the selected membrane can be retained and flow out in the retentate whereas water of the permeate is then purified from heavy metals [3].

We have studied the applicability of this process to treatment of waters issued from scrubbing of industrial wastes incineration fumes. In the classical process of treatment these waters are purified from their metals by precipitation of the corresponding hydroxides with milk of lime. After a solid-liquid separation, effluents are rejected in the environment, because they are accordance with regulations. They contain minor amounts of metallic chlorides such as Fe, Cu, Pb, Ni, Cd, Zn with concentrations included between 10^{-5}-10^{-7} mol/L (Pb, Zn, Cd) and 10^{-8}-10^{-9} mol/L (Fe, Cu, Ni). According to the expected lowering of regulation concerning heavy metals wastes, it is necessary to perfect new selective processes to improve purification of waste waters. So this work has been done on liquid effluents containing minor amounts of heavy metals, in order to propose a process allowing to lower their concentration in waste waters.

2 Experimental

Metal chlorides (Ca, Cu, Ni, Zn, Pb, Cd) and hydrochloric acid used were pure grade reagents.

The macroligand used was a polyethyleneimine supplied by BASF. Its molar mass is comprised between 30 000 and 40 000 Daltons (atomic mass unit). Its capacity is about 7 eq/kg. The range of concentration used was 1 to 10 g/L.

The ultrafiltration membranes were inorganic asymetric composite ones (Carbosep M2) manufactured by Tech-Sep. Their molecular cut-off is about 15 000 Daltons. Two types of tubular modules were used: a micromodule (area 40 cm^2) and a pilot one (area 0,32 m^2). The separation was carried out by tangential filtration. The velocity of fluid along the membrane was from 1 to 5 m/s. The transmembrane pressure varied from 2 to 6 bars.

An ICP emission spectrometer (Jobin-Yvon JY 38) and/or an atomic absorption spectrometer (Varian Spectraa 300P) were used for analysis of metallic cations.

3 Results and Discussion

3.1 Rejection rate

The first step of an UFAC process is to find the best physicochemical conditions of complexation by macroligand of species to be captured by membrane [4].

The formation of a metallic complex ML_x according to

$$M + xL \Longleftrightarrow ML_x \tag{1}$$

is characterized for each cation M by the stability constant

$$K_x = \frac{[ML]}{[M][L]^x} \tag{2}$$

which is pH dependent, because M acts as a weak acid, macroligand L acts as a weak base, and because ML_x is sometimes amphoteric. In UFAC the complexation can be measured by rejection rate R_M

$$R_M = 1 - \frac{C_p}{C_i} \cong 1 - \frac{C_p}{C_r} \tag{3}$$

where C_p, C_i and C_r are the concentrations of C_M (= [M] + [ML_x]) respectively in the permeate, along the membrane and in the retentate.

In the present case where acid effluents (hydrochloric acid media) are first neutralized by milk of lime, we must also study the influence of chloride concentration on the stability of complexes ML_X (through the influence of Cl^- on M). Thus it is possible to calculate the rejection rate vs physicochemical parameters

$$R_M = \frac{\Sigma K_X \left[\frac{C_L}{1 + K_a [H^+]}\right]^X}{1 + \Sigma \beta_i^{OH} [OH^-]^i + \Sigma \beta_j^{Cl} [Cl^-]^j + \Sigma K_X \left[\frac{C_L}{1 + K_a [H^+]}\right]^X} \tag{4}$$

Where K_X is the stability constant of ML_X
K_a is the protonation constant of L
β_i^{OH} are the cumulative constants of complexes MOH
β_j^{Cl} are the cumulative constants of complexes MCl
C_L is the total macroligand concentration

This expression shows that rejection rate grows from low pH to higher one and can reach the maximum value 1 if the first three terms of denominator are negligible. The presence of chloride ions is always unfavorable because the corresponding term $\Sigma \beta_j^{Cl} [Cl^-]^j$ is independant of pH and its value may be high. Table 1 shows the experimental values of R_M vs chloride concentrations.

Table 1. Influence of chloride concentration on rejection rate

Rejection rate	pH / $[Cl^-]$ (mol/L)	1,5	3	5	6,5
	0	*65*	99	98,3	98
R_{Cu}	0,2	*33*	96	95	94,8
	0,4	*31*	94,5	93,5	93,5
	0	*61,5*	*48,5*	96	95,5
R_{Ni}	0,2	*18*	*14*	93	92,5
	0,4	*12,5*	*12*	91	90,5
	0	*65*	*46*	96,5	98
R_{Zn}	0,2	*15,5*	*12,5*	94	95,5
	0,4	*9*	*7*	92	94
	0	*68*	*35*	*51*	95
R_{Pb}	0,2	*0*	*0*		91
	0,4	*0*	*0*	*44*	89
	0	*63*	*47,5*	94	97
R_{Cd}	0,2	*13*	*14*		94
	0,4	*1*	*4*		91,5

It appears that the influence of chloride is great at low pH, when the rejection rate without chloride is not maximum. On the contrary when R_M is near maximum, the influence of chloride is weak. Nevertheless, it can be seen that at pH 6.5 which is the operating value, there is a R_M decrease of 4 to 6% with chloride concentrations 0.2-0.4 mol/L. Experimental results obtained are shown on figure 1

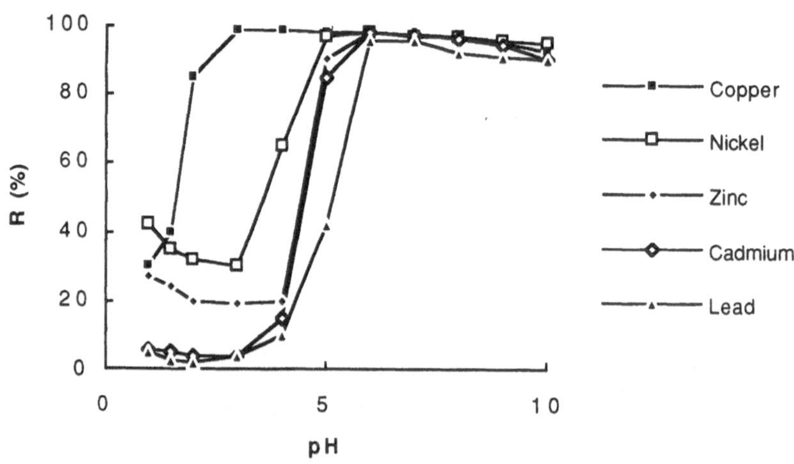

Fig 1. Experimental values of rejection rate vs pH
([PEI] = 5g/L, [Cl^-] = 0.2 mol/L, 20 ppm of each metal)

This figure shows that stability constants K_X are not the same for all metals because they are not complexed in the same pH range; but it also appears that there is a pH range where R_M is maximum for all, allowing to complex all cations (6< pH <7).

In these conditions feed solution is purified from its metallic cations which flow out the ultrafiltration module with retentate. The only species which cross the membrane and flow out in permeate are the non-complexed cations and the free part of complexed one according to the stability constant K_X and macroligand concentration in solution.

3.2 Operating parameters

Performances of an ultrafiltration module depends not only of physicochemical conditions, but also of operating parameters, themselves under the dependance of experiment [5-6]. The apparatus used is shown on figure 2.

We have operated with a recirculation loop, mainly to prevent membrane fouling. In these conditions, the functioning of ultrafiltration module is characterized by 4 parameters:

- the extraction factor α which depends on the flow ratio of permeate and feed and is controlled by the valve V and by the pressure pump P_1,

$$\alpha = \frac{D_p}{D_a} \tag{5}$$

- the rejection rate R_M which depends on the pair macroligand-membrane (i.e. molar weight of macroligand vs membrane cut-off) and physicochemical conditions. With fixed conditions, rejection rate is bound to permeate and retentate concentrations (3).

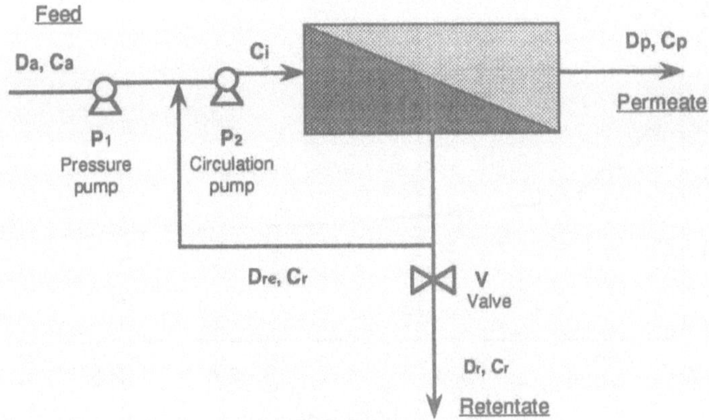

Fig. 2 Experimental device and operating parameters

- the metal concentration in the permeate

$$C_p = C_a \frac{(1-R_M)}{(1-\alpha R_M)} \tag{6}$$

- the metal concentration in the retentate

$$C_r = C_a \frac{1}{(1-\alpha R_M)} \tag{7}$$

which characterize the efficiency of ultrafiltration separation. These two last expressions show that with fixed physicochemical conditions (R_M), it is possible to modulate permeate and retentate concentrations by modifying the operating parameter α.

3.3 Permeate flow

This is an important practical characteristic of an industrial process assigned to purify waste waters. So we have studied the influence of some parameters on the permeate flow.

3.3.1 Permeate flow vs macroligand concentration

We have shown that a concentration of 1 to 2 g/L of macroligand is adequate to complex all cations of waste waters considered (ratio $\frac{[L]}{\Sigma[M]} > 40$) But with the apparatus used (fig. 2), the concentration of macroligand increases in the recirculation loop, and then the macroligand concentration along the membrane will be higher. The curve representing the permeate flow vs macroligand concentration is given figure 3. It shows that the permeate flow is not much influenced by variations of ligand concentrations above 5 g/L. In our experimental conditions the concentration of the macroligand in the recirculation loop is in the range 5 to 20 g/L.

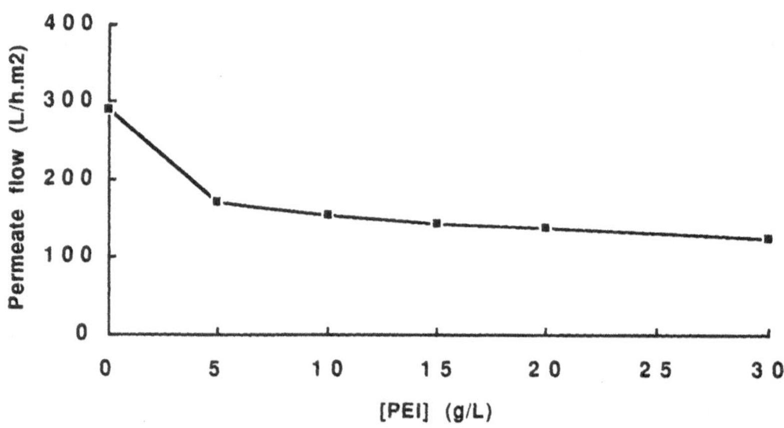

Fig 3. Permeate flow vs macroligand concentration ($\Delta P = 4$ bars, v = 4 m/s, T = 25 °C)

3.3.2 Permeate flow vs transmembrane pressure
We have measured the experimental variation of D_p vs transmembrane pressure for pure water and for different concentrations of PEI. In all cases a growth of transmembrane pressure lead to an almost linearly increase of the permeate flow (fig. 4). The comparison

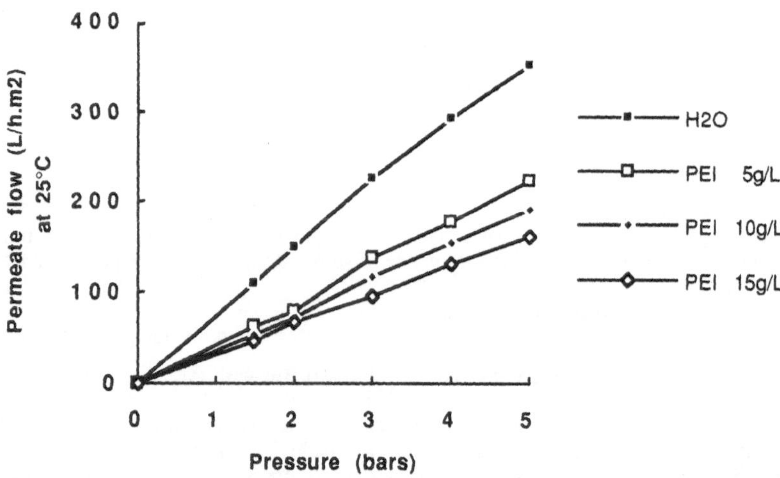

Fig 4. Permeate flow vs transmembrane pressure

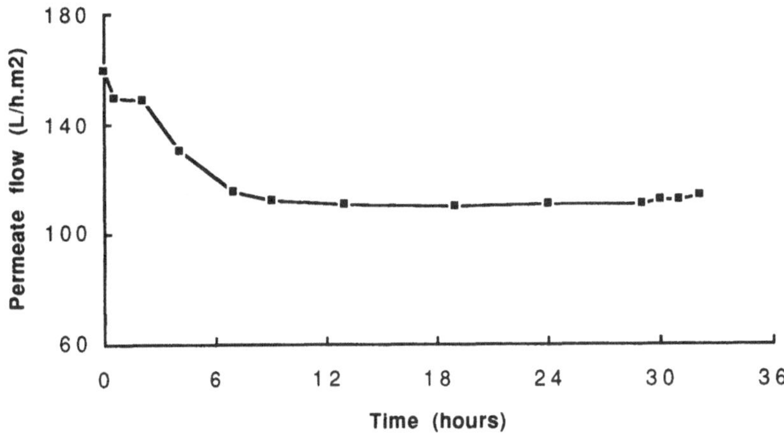

Fig 5. Permeate flow vs time (ΔP = 4 bars, v = 4 m/s, T = 25 °C)

of these curves shows that for a given pressure, permeate flow is strongly lowered from pure water to water with macroligand. But when water contains increasing PEI concentrations the decrease of D_p for a given pressure is not so important: 40 L/h/m^2 for 4 bars.

3.3.3 Permeate flow vs time

An other important factor influencing the permeate flow is time, mainly in consequence of membrane fouling. In the experimental conditions chosen we have mesured the variation of D_p during 32h (fig. 5). It appears that after an important decrease at the beginning of experiment, the permeate flow remains almost constant from 9 to 32h. We conclude that with these experimental conditions, after 9h of functioning there is no increase of membrane fouling.

3.4 Permeate concentration

Concentration of cations in the permeate is a characteristic of the efficiency of process, so we must operate under the best conditions to lower the level of metallic cations in effluents. The expression (6) shows that this concentration depends on both operating factor α, and physicochemical conditions by R_M. In the process, after complexation by macroligand, R_M is constant. But it is possible to vary C_p by acting on the value of α. Figure 6 represent the variation of permeate concentration vs extraction factor for different values of rejection rate.

The shape of these curves are the same for different values of rejection rate. Nevertheless, we see that the range of variation of permeate concentration vs extraction factor decreases when rejection rate increases. In all cases it is necessary to operate with low values of α to deliver low heavy metals concentrations.

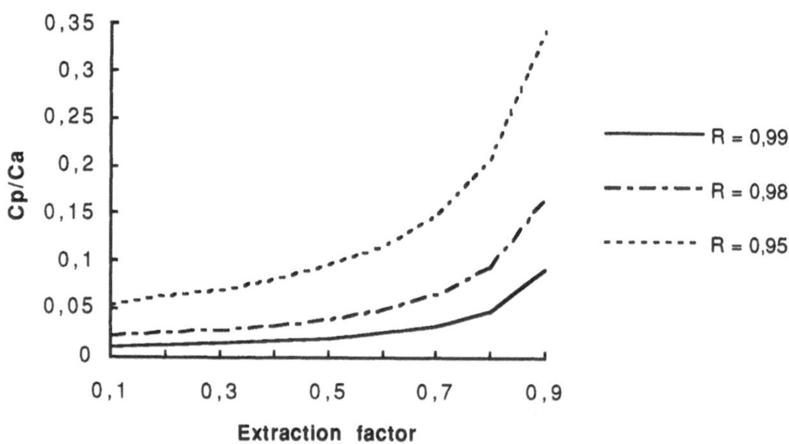

Fig 6. Permeate concentration vs extraction factor for different rejection rate

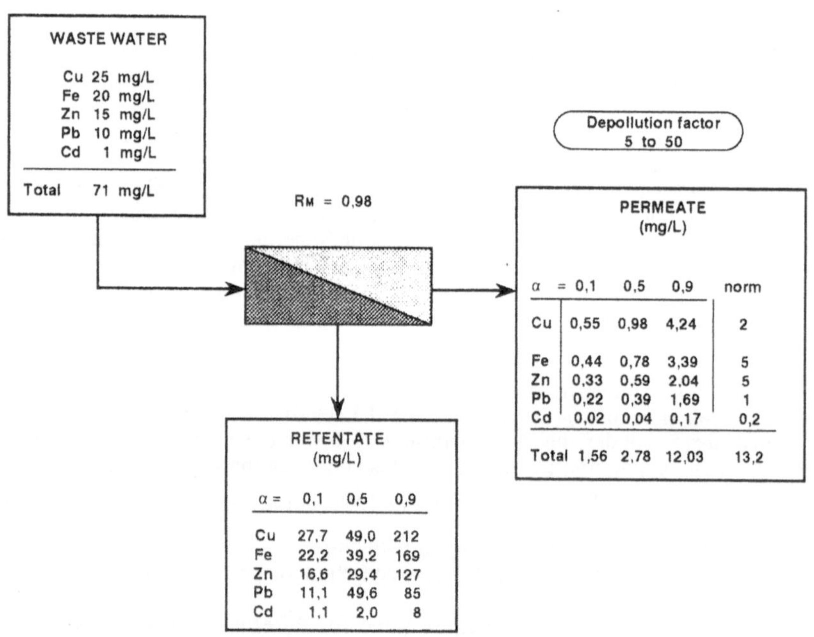

PERMEATE (mg/L)

α =	0,1	0,5	0,9	norm
Cu	0,55	0,98	4,24	2
Fe	0,44	0,78	3,39	5
Zn	0,33	0,59	2,04	5
Pb	0,22	0,39	1,69	1
Cd	0,02	0,04	0,17	0,2
Total	1,56	2,78	12,03	13,2

RETENTATE (mg/L)

α =	0,1	0,5	0,9
Cu	27,7	49,0	212
Fe	22,2	39,2	169
Zn	16,6	29,4	127
Pb	11,1	49,6	85
Cd	1,1	2,0	8

Fig. 7 Performances of an ultrafiltration module

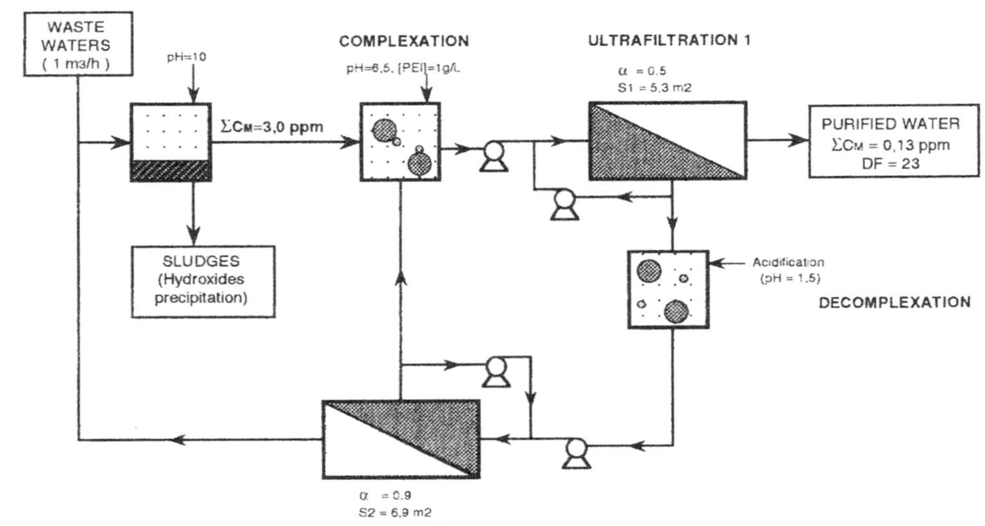

Fig. 8 Flow diagram for treatment of waste waters containing heavy metals

In order to examine the depolluting possibilities of an ultrafiltration module, we have calculated its performances under these conditions. A synthetic composition of waste waters has been chosen and each metal concentration calculated in permeate and retentate for a retention rate of 98% (fig. 7). Considering only concentrations in permeate, it appears that whatever the value of α, waste waters are purified: depollution factor varies from 5 to 50. The best conditions of depollution are when α is low. But a practical problem may appear in this case because a value of $\alpha = 0.1$ means that 90% of feed goes into the recirculation loop. Then in this case the concentration of heavy metals in permeate is low, but permeate flow is also small. In an industrial plant, optimization is necessary between these two parameters because the aim will be to purify a maximum volume of waste waters.

Considering now retentate, it appears that metal concentrations are either slightly higher or very higher than that of feed. From a practical point of view it is interesting to concentrate the retentate for further treatment.

3.5 Flow-sheet

According to the previous results, a whole flow-sheet for treatment of 1 m^3/h of waste waters containing metallic cations in hydrochloric acid media has been calculated (fig. 8). The first operation is the precipitation of metal hydroxides by milk of lime leading to sludges and to an effluent containing about 3 ppm of metallic cations (solubility of metal hydroxides at pH 10). This one feeds a complexation unit of metals by macroligand (PEI 1 g/L), at pH 6.5 (cf fig. 1). Effluents issued from the complexation unit enters then an ultrafiltration module which membrane area is 5.3 m^2. The separation occurs with an operating extraction factor of 0.5. Thus the total metal concentration in permeate is 0.13 ppm, which corresponds to a depollution factor of 23. The retentate is then acidified up to pH 1.5 to separate metals from ligand. This solution enters then a second ultrafiltration unit which membrane area is 6.9 m^2 and functioning with a value of $\alpha = 0.9$. This value allows a high concentration factor at one and the same time for retentate, returned to the complexation unit, and permeate joined to the feed entering the precipitation unit.

4 Conclusions

This work shows that ultrafiltration assisted by complexation is a promising separative technique applied to purification of effluents containing heavy and common metals. Coupled with a classical precipitation step, it allows to greatly lower the level of pollution in waste waters.

This work is still in progress, principly to appreciate the loss of macroligand at the first ultrafiltration step, to calculate more precisely the flow diagram and to calculate the industrial cost of such an industrial plant.

5 References

[1] KREULA M., KIVINIEMI L., VUDRINEN E., HEIKONEN M., Milchwissenschaft, (1974) **29**, 129-137.

[2] ENNASSEF K., PERSIN M., DURAND G., Analusis (1989) **10**, 565-575

[3] RAKIB M., STAMBOULI M., DESMARES S., DURAND G., Fourth World Congress of Chemical Engineering, (1991) **3**, 6-4

[4] KORTLY S., SUCHA L., Handbook of Chemical Equilibria in Analytical Chemistry, Hellis Horwood (1985)
[5] CHERYAN M., Ultrafiltration Handbook, Technomic Pub. Co. (1986)
[6] RAUTENBACH R., ALBRECHT R., Membranes Processes, John Wiley (1989)

Acknowledgements

The authors are grateful to EMC Services which partly supported this work.

Application of strong magnetic fields for heavy metal removal

Matthias Franzreb
Kernforschungszentrum Karlsruhe, Institute for Radiochemistry, Water Technology Division, Karlsruhe, Germany
W. H. Höll
Kernforschungszentrum Karlsruhe, Institute for Radiochemistry, Water Technology Division, Karlsruhe, Germany
S. H. Eberle
Kernforschungszentrum Karlsruhe, Institute for Radiochemistry, Water Technology Division, Karlsruhe, Germany
K-P. Jüngst
Kernforschungszentrum Karlsruhe, Institute for Technical Physics, Karlsruhe, Germany

Abstract
The applicability of high gradient magnetic separation (HGMS) for the removal of heavy metals from mixed solutions of para- and diamagnetic ions (nickel and zinc) was studied. The heavy metals were precipitated as phosphates by the addition of Na_3PO_4. The susceptibilities of the precipitations showed the expected linear dependence with respect to the mass fraction of nickel. However, the separation experiments with a 1.8 Tesla magnetic separator demonstrated that even small fractions of nickel result in separation efficiencies comparable to those of pure nickel phosphate. The experimental results were compared with theoretical filter calculations based on deep filter theories and single wire models.
Keywords: Heavy metals, high gradient magnetic separation, modelling, phosphates, susceptibility measurements.

1 Introduction

Magnetic separation techniques for the removal of ferromagnetic particles are the subject of numerous scientific investigations and industrial applications, for example water polishing in nuclear power plants and phosphate removal [1,2]. This has particularly been the case since the development of high gradient magnetic separation in the late seventies [3,4]. A high gradient magnetic separation device uses filamentary ferromagnetic material which is placed in a strong magnetic field. Examples for packing materials are stainless steel wool or expanded metal meshes. Besides the effect of producing high magnetic field gradients the matrix offers a large effective surface area. The main advantages of the HGMS process are the selectivity for ferro- and paramagnetic materials, the possibility to remove particles in the micrometer range and the feasibility of high filter velocities. In addition, HGMS equipment is robust and insensitive to high temperatures [5].

Several investigations exist for the applicability of HGMS techniques to diamagnetic particles in which fine-grain ferromagnetic materials (e.g. magnetite) are added to the solution [6,7]. In combination with flocculants such as $Fe_2(SO_4)_3$ the diamagnetic components bind to the seeding material and can be removed. Afterwards the seeding material can be recycled, a process, however, which requires additional chemicals and technical equipment. On the other hand, by not recycling the seeding material there would be an high increase of the heavy metal sludge, something which is unsuitable with increasing disposal costs. The present investigation uses another approach: the combined removal of para- and diamagnetic heavy metals without any seeding. The experiments carried out here examined the magnetic separation of Ni-Zn mixed phosphates. The magnetic properties of the compounds produced are characterised by mass susceptibility measurements. Experiments with laboratory scale HGMS equipment yield the corresponding data for the separation behaviour of the particles. At the end of the paper the experimental results are compared with the theoretical predictions.

2 Theory

For the description of the macroscopic behaviour of magnetic filters the equations for deep bed filtration are used. Assuming one-dimensional piston flow the following mass balance equation for a differential filter segment is obtained [8,9].

$$\frac{\partial \sigma}{\partial t} + \frac{\partial(\varepsilon c)}{\partial t} + v_f \frac{\partial c}{\partial z} - D_d \frac{\partial^2 c}{\partial z^2} = 0 \tag{1}$$

Introducing the filter time τ and ignoring the axial dispersion, eq.(1) simplifies to:

$$\frac{\partial \sigma}{\partial \tau} + v_f \frac{\partial c}{\partial z} = 0 \tag{2}$$

The kinetic approach results by assuming that the increase of the local filter loading is proportional to the particle concentration penetrating the corresponding filter element per time unit.

$$\frac{\partial \sigma}{\partial \tau} = \lambda \cdot v_f \cdot c \tag{3}$$

The probability that a particle is captured in the filter element decreases with increasing loading of the filter matrix. The simplest way to describe this relationship is given by the following empirical expression:

$$\lambda = \lambda_0 \left(1 - \frac{\sigma}{\sigma_s}\right) \tag{4}$$

Here, λ_0 represents the filter coefficient of the clean matrix and σ_s the saturation loading. Eq.(4) assumes that the filter coefficient is independent of the particle concentration. Herzig et al [9] showed that with this assumption the spatial derivation of the loading can be expressed as follows:

$$\frac{\partial \sigma}{\partial z} = -\lambda \cdot \sigma \tag{5}$$

The combination of eqs.(2) - (5) finally yields the particle concentration in the filter as a function of time and axial location.

$$\frac{c}{c_0} = \frac{\exp\left\{\sigma_s^{-1} c_0 v_f \lambda_0 \tau\right\}}{\exp\left\{\sigma_s^{-1} c_0 v_f \lambda_0 \tau\right\} + \exp\left\{\lambda_0 z\right\} - 1} \tag{6}$$

However, before this equation can be used for predicting the separation behaviour of magnetic filters, an expression must be found in which the filter coefficient λ_0 is linked with the magnetic properties of the filter matrix and the particles. Comparing the expression of eq.(6) for a clean matrix with a corresponding expression which Watson [10,11] derived regarding a single wire of the separation matrix, the following relationship is obtained:

$$\lambda_0 = \frac{4}{3} \frac{f_M R_C}{\pi a^2} \tag{7}$$

Here f_M is the fraction of the filter volume which is filled by transversally arranged wires and 'a' stands for the diameter of the matrix wires. Assuming that the magnetic field of the HGMS equipment is strong enough to achieve the saturation magnetisation of the matrix, the dimensionless capture radius R_C can be approximated by [3]:

$$R_C = \frac{3}{4}\sqrt{3}\left|\frac{v_m}{v_l}\right|^{1/3} \left\{1 - \frac{2}{3}\left(\frac{v_m}{v_l}\right)^{-\frac{2}{3}}\right\} \tag{8}$$

The capture radius describes the range in which the centre of a particle must be for its trajectory to end at the wire surface. v_m is the particle velocity which adjusts if the magnetic force is in equilibrium with the streaming resistance.

$$v_m = d_p^2 \left(\chi_p - \chi_l\right) \cdot M_S \cdot H_0 / (18 a \eta) \tag{9}$$

The saturation loading strongly depends on the water content and the morphology of the particles produced. Therefore, no general equation for its approximation exists and the saturation loading must be determined experimentally.

3 Experimental

The experimental procedure separates into three stages: production of mixed heavy metal phosphates; measurement of the susceptibilities; and separation experiments using laboratory scale HGMS equipment. To achieve constant precipitation conditions an artificial model-water containing 2 mmol/l $CaCl_2$, 1 mmol/l Na_2SO_4 and 3 mmol/l $NaHCO_3$ was used. The water was slightly supersaturated with $CaCO_3$, however, no turbidity appeared in the pure model water even more than a week after preparation. A stoichiometric amount of Na_3PO_4 and 1 mg/l of the coagulant aid Separan were added to the model-water for the ensuing precipitation of the heavy metals as phosphates. The pH-value of the solution was about 9. Immediately the heavy metal mixtures which were dissolved as chlorides in a small amount of distilled water, were added under vigorous stirring conditions. After 30 s the rate of stirring was reduced and continued at a moderate speed for 1h to support the flocculation. The generated particles were filtered out, dried, sieved into narrow size-fractions and re-suspended into the liquid, because of the strong dependence of the separation efficiency with particle size. The precipitation experiments were supported by numerical calculations with AWASA, a computer program for the prediction of the equilibrium composition of aquatic multi-species systems [12]. The calculations show that for the conditions stated here the precipitates formed are phosphates and not hydroxides. The theoretical predictions were confirmed by susceptibility measurements. For example, the value of the mass susceptibility of pure nickel phosphate is given in the literature as $28 \cdot 10^{-6}$ cm^3/g, whereas the experimental measurements of the precipitates which resulted from the addition of only nickel gave values of $26.5 \cdot 10^{-6}$ cm^3/g.

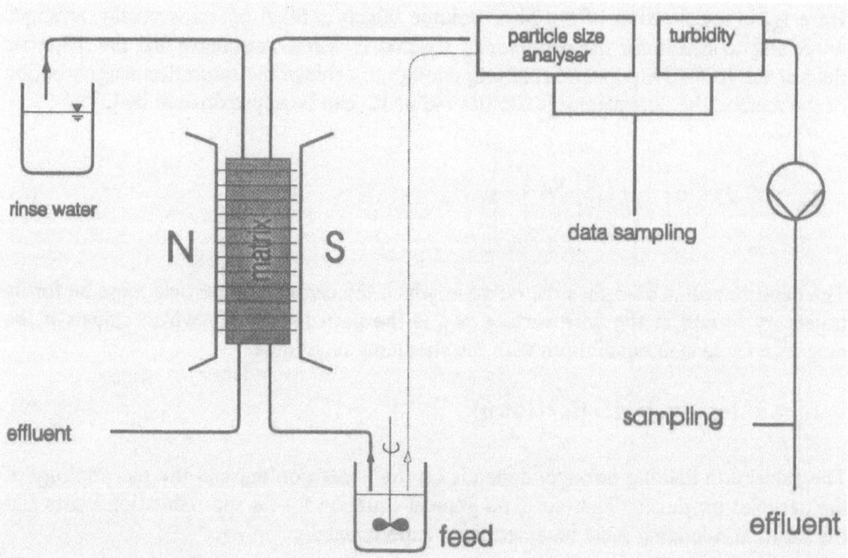

Fig. 1 Scheme of the HGMS-equipment

For the susceptibility measurements the magnetic susceptibility balance Mark-II of Johnson Matthey was used. The balance is capable of measuring volume susceptibilities in the range of 10^{-7} to 10^{-3}. In addition the balance offers the option to calculate the mass susceptibility if the height and the weight of the sample is entered. The relationship between the volume susceptibility and the mass susceptibility is given by:

$$\chi_g = \frac{\chi_V}{\rho} \tag{10}$$

The separation experiments were conducted with laboratory scale HGMS equipment which is shown in Figure 1. The main piece of the equipment is a FRANTZ electromagnet with a maximum field of 1.8 Tesla. The gap width between the poles of the magnet is 1 cm, the length of the poles is 23 cm and their depth is 2 cm. In this gap an acrylic glass cuvette is placed which itself contains a ferromagnetic matrix. The matrix consists of three pieces of gold plated steel nets. The volume fraction of the cuvette which is filled with transversally arranged wires is 3.6 %. Transversal wires are directed perpendicular to the flow direction and also to the magnetic field and they are responsible for the main part of the magnetic separation ability of the matrix. However, in addition to the transversal wires the void volume of the cuvette is partly filled with axial wires and a frame which keeps the steels nets together. The flow velocity inside the magnetic separation unit was 5 cm/s (180 m/h) which is rather high, even for magnetic filters. The separation efficiency was controlled by turbidity measurements and sampling at short time intervals. The duration of the separation cycles was about 2 to 7 minutes depending on the particle concentration and their susceptibility. During this time the turbidity showed a slightly increasing plateau value, which was followed by a steep breakthrough. After the breakthrough the experiment was stopped, the magnet was switched off and the matrix was cleaned by counter current backwashing.

4 Discussion

The mass susceptibilities of different Ni-Zn mixed phosphates are shown in Figure 2. The susceptibilities show the expected linear dependence on the mass fraction of nickel (m_{Ni} related to the sum of m_{Ni} and m_{Zn}) beginning with a slightly negative value for pure zinc phosphate up to a value of $26.5 \cdot 10^{-6}$ cm^3/g for pure nickel phosphate. However, this simple relationship is not necessarily true. If two heavy metals form a new compound in molecular dimensions, e.g. a spinel structure, the susceptibility of the 'mixed' particles can strongly deviate from this simple superposition rule. During the experiments no separation between nickel and zinc phosphates was observed. That means the degree of zinc removal was in the same range as the removal of nickel and the precipitates could be treated as one species with the measured average susceptibilities.

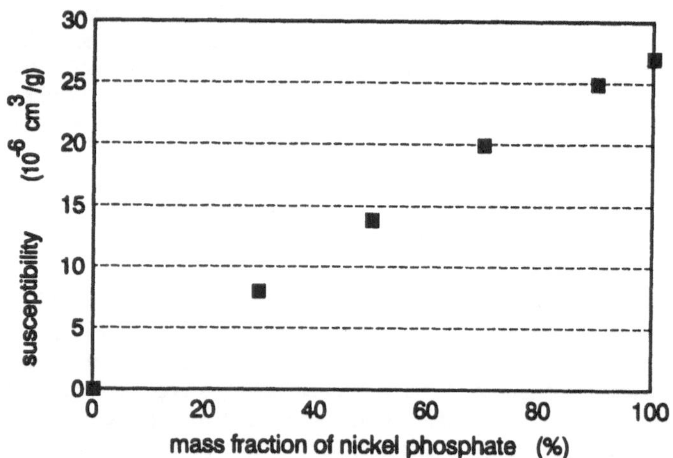

Fig. 2 mass susceptibilities for different Ni-Zn phosphates

Figure 3 shows the separation efficiencies of nickel-zinc mixed phosphates with different nickel fractions. Beginning at 0 % for pure zinc phosphate the separation factor strongly increases with an increasing nickel fraction. Even with a nickel fraction of only 20 % the separation factor reaches nearly that of pure nickel phosphate.

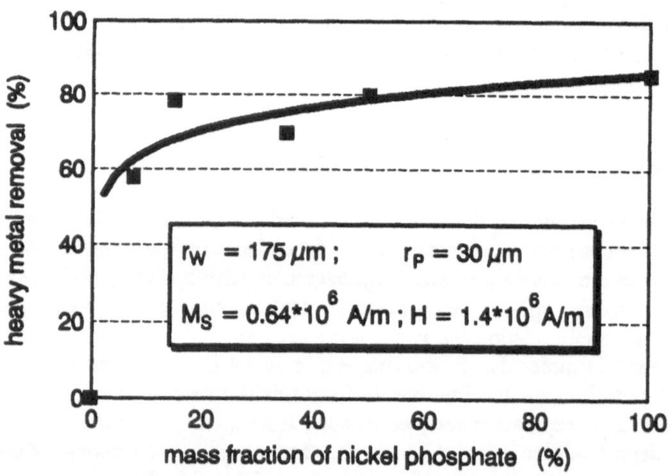

Fig. 3 Experimental separation efficiencies for different Ni-Zn-phosphates

However in contrast to the theoretical calculations the heavy metal removal does not exceed 85 %, even for pure nickel phosphate. The results of the theoretical calculations are plotted in Figure 4. The Figure shows that with nickel mass fractions greater than 40 % the heavy metal removal should be better than 90 %. Despite the numerous simplifications which are used in the calculations such as ignoring the detachment of the already separated particles, the mutual influence of the wires or assuming a constant void volume, the discrepancy results from the non-idealities of the separation matrix used. The width of the separation zone is only about 5 mm, so a disturbing influence of unimpeded flow in the fringe areas is probable. This assumption is supported by the fact that in the experiments an increase of the mass fraction of nickel from 20 % to 100 % has nearly no impact on the extent of heavy metal removal.

5 Summary

The investigations show that the approach to directly remove heavy metals from mixtures of para- and diamagnetic ions by precipitation as phosphates is an interesting alternative to the use of ferromagnetic seeding materials. The method offers the advantage that no recycling of the seeding material is needed and that the amount of sludge is not increased. The observed mass susceptibilities of the particles produced are in good agreement with the theoretical calculations, whereas the separation efficiencies for particles with higher susceptibilities are worse than the predictions. The main reason for this discrepancy is probably due to the unfavourable geometry of the matrix used. Therefore the next stage for this investigation should be experiments using an HGMS equipment with a separation matrix of larger cross-sectional area.

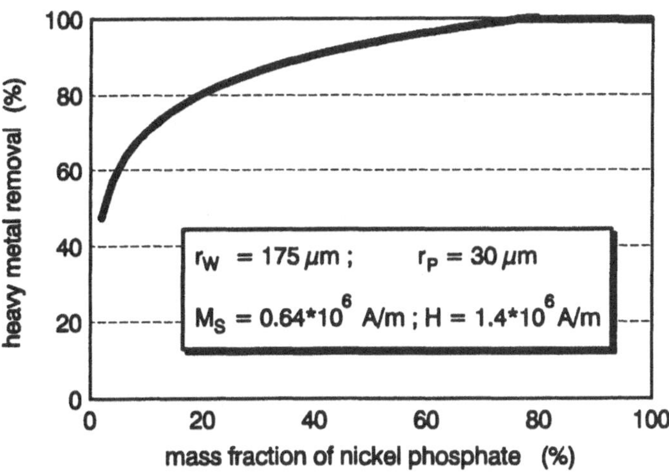

Fig. 4 Theoretical separation efficiencies for different Ni-Zn-phosphates

6 Notation

a	m	diameter of the matrix wires
c	g/m^3	particle concentration
c_0	g/m^3	particle concentration in the feed
D_p	m^2/s	dispersion coefficient
d_p	m	particle diameter
f_M	---	filter volume fraction filled by transversally arranged wires
H	A/m	magnetic intensity
M_s	A/m	saturation magnetisation
R_c	---	capture radius
r_w	m	radius of the separation matrix wires
r_p	m	average particle radius
v_f	m/s	superficial filter velocity
v_l	m/s	filter velocity
v_m	m/s	magnetic velocity
V_p	m^3	particle volume
z	m	axial filter coordinate

Greek Letters

ε	---	void fraction of the filter volume
η	Ns/m^2	viscosity
λ	m^{-1}	filter coefficient
λ_0	m^{-1}	initial filter coefficient
σ	g/m^3	matrix loading per filter volume
σ_s	g/m^3	saturation matrix loading
τ	s	filter time $(\tau = t - z/v_F)$
χ_g	cm^3/g	mass susceptibility
χ_v	---	volume susceptibility

7 References

1. Emory, B.B. (1982) High Gradient Magnetic Filtration for Impurity Removel in Nuclear Power Reactor Coolant Circuits. IEEE Transaction on Magnetics, 18, 1686-1688.
2. Smit Nymegen (1989) Magnetic Separation; Professional Journal for the Chemical and Petrochemical, Pharmaceutical and Food Industries, International Issue, 8-10.
3. Gerber, R. and Briss, R.R. (1983) High Gradient Magnetic Separation, John Wiley&Sons Ltd, Chichester.
4. Oberteuffer, J.A. (1974) Magnetic Separation: A Review of Principles, Devices and Applications, IEEE Transactions on Magnetics, 10, 223-238.
5. Appelton, A.D. and Dobbing, P.P. (1978) A Discussion of Some Aspects of High Gradient Magnetic Separation, IEEE Conference Publication No. 142, Second conference on advances in magnetic materials and their applications, 65-68.
6. Anderson, N.J., Blesing, N.V., Bolto, B.A. and Jackson, M.B. (1987) The Role of Polyelectrolytes in a Magnetic Process for Water Clarification. Reactive Polymers, 7, 47-55.
7. deLatour, C. and Kolm, H. (1975) Magnetic Separation in Water Pollution Control, IEEE Transactions on Magnetics, 11, 1570-1572.

8. Iwasaki, T. (1937) Some Notes on Sand Filtration. Journal of the American Water Works Association, 29, 1591-1602.
9. Herzig, J.P., LeClerc, D.M. and LeGoff, P. (1970) Flow of Suspensions Through Porous Media -Application to Deep Filtration. Industrial Engineering Chemistry, 62, 8-35.
10. Watson, J.H.P. (1973) Magnetic Filtration. Journal of Applied Physics, 44, 4209-4213.
11. Watson, J.H.P. (1975) Theory of Capture of Particles in Magnetic High-Intensity Filters, IEEE Transactions on Magnetics, 11, 1597-1599.
12. Hennes, E. (1983) Entwicklung und experimentelle Überprüfung eines komplexchemischen Gleichgewichtsmodells für Gewässer am beispiel des Rheinwassers, Thesis, University of Karlsruhe, Germany.

Design and construction aspects of pilot-scale passive treatment systems for acid rock drainage at metal mines

J. J. Gusek
Knight Piésold and Co., Denver, Colorado, U.S.A.
J. T. Gormley
Knight Piésold and Co., Denver, Colorado, U.S.A.
J. W. Scheetz
Brewer Gold Company, Jefferson, South Carolina, U.S.A.

Abstract
A pilot scale system is a logical and practical step in evaluating the applicability of passive treatment technologies (constructed wetlands) to remediation of acid rock drainage (ARD) at metal mining operations. This paper presents two case studies in the US of the phased development of pilot scale passive treatment facilities at:

1. An abandoned copper/sulfur mine in California
2. A heap leach and open pit at an eastern US gold mine that is engaged in closure activities

Design of pilot scale systems for the two sites were based on "Proof-of-Principle" test results conducted with actual ARD samples and candidate passive treatment system construction materials. The diversity of the two case histories with respect to ARD quality and final system configuration show that passive treatment systems are more complicated than simply routing ARD through open plant-filled basins (aerobic cells), limestone-filled galleries (anoxic limestone drains) and bogs filled with organic material (anaerobic cells). The chemistry of the ARD guides the overall design layout and the cell configurations are controlled by geography and established engineering principles.
Keywords: Passive treatment, constructed wetlands, acid rock drainage, metals, case studies, metal mining.

1 Introduction

In the past several years, both coal and hard rock mining companies and governmental agencies in the United States and elsewhere have been examining and/or utilizing passive treatment systems to remediate water quality problems associated with acid rock drainage (ARD). These systems are typically comprised of man-made wetlands or bogs that capitalize on naturally-occurring geochemical and biological processes to immobilize metals within the passive treatment system substrate and raise the pH of the water and thus remediate ARD situations.

The formation of ARD is also a natural process. In the presence of water and bacteria, oxidizing sulfide minerals such as pyrite produce sulfuric acid; iron and other metals are released into the water. The problem can be associated with both coal and hard rock operations where previously-buried sulfide minerals are exposed to oxygen and water. Below a pH of 4.5, some of these natural oxidizing reactions appear to be catalyzed by the action of natural bacteria, Thiobacillus ferrooxidans, which accelerates the pyrite oxidation process and lowers the pH even further. Other metal sulfides present will also be oxidized as a result of "incongruent reactions", releasing additional metals such as copper, lead, zinc and manganese into solution [1].

1.1 Passive Treatment Systems - Phased Design Approach

Since about 1990, Knight Piésold and Co. has been following a phased approach to passive treatment design for ARD remediation with three increasingly-larger stages of investigation:

- Proof of Principle Laboratory Testing
- Bench Scale Testing
- Pilot Scale Systems

This paper focuses on the design and construction aspects of pilot scale systems, with special emphasis on the case histories of passive treatment systems installed at two sites in America:

- The Leviathan Mine, California - an abandoned copper/sulfur mine consisting of about 202 hectares (500 acres) of disturbed land at an elevation of 2,100 meters (7,000 feet) on the eastern slopes of the Sierra Nevada Mountains. The site has steep slopes and is located at the headwaters of a river that is a regional drinking water supply source. One ARD source (of many on site) is collected in a rock gallery/pipe drain beneath a concrete lined diversion channel. The passive treatment construction site is on a narrow bench adjacent to a perennial stream.
- The Brewer Gold Mine, South Carolina - a heap leach and open pit mine engaged in closure activities in the Eastern US. The 10 hectare (25 acre) site includes several pits which may be backfilled, a waste rock pile and six heap leach pads, located along the crest of a flat hilltop at an elevation of about 180 meters (600 feet). Two construction sites were used for the passive treatment systems: one adjacent to the pit edge, another adjacent to a heap leach pad.

Water chemistry, design layout and construction constraints at each site posed challenges that significantly influenced design and construction methods. Table 1 shows the chemical characteristics of the ARD between the two sites. The design challenges and their related solutions will be discussed in a series of case histories that follow.

Table 1
ARD Chemical Characteristics

Water Quality Parameter	Leviathan Mine	Brewer Mine Pit ARD	Brewer Mine Pad 5 ARD
pH (s.u.)	4.5	2.3	2.5
Iron mg/liter	340	600	300
Copper mg/liter	-	100	22
Aluminum mg/liter	34	90	50
Arsenic mg/liter	0.1	-	-
Nickel mg/liter	0.2	-	-
Sulfate mg/liter	1,809	2,531	3,375
Manganese mg/liter	-	2.2	2.5
Zinc mg/liter	-	0.3	0.2

1.2 Passive Treatment Methodologies

Two fundamental ARD passive treatment methodologies are available. They are surface-flow *aerobic* wetlands and subsurface-flow *anaerobic* wetlands. An aerobic wetland system may include an Anoxic Limestone Drain (ALD) to add alkalinity and buffer iron hydrolysis reactions. Sizing in both cases is a function of the flux of metals to the wetland. The wetland type or combination of wetland types used to treat the drainage is a function of the chemistry of the drainage and its flow characteristics. Figure 1 shows the inter-relationships of the two basic methodologies as interpreted by the US Bureau of Mines for application at coal mining sites. Metal mine sites, as discussed in this paper, typically include "compost" or anaerobic wetlands for heavy metal removal.

1.2.1 Aerobic Wetlands

This type of wetland treatment works best to remove iron, manganese, arsenic and selenium from solution as long as the acidity of the ARD is not high. The dominant process at work on ARD flowing into a surface flow aerobic wetland is oxidation of iron and precipitation of iron hydroxides. Arsenic and selenium typically adsorb on to the iron hydroxide particles and settle out. Manganese forms as an oxide after iron has been removed.

Plant-assisted reactions appear to aid the metal-removal performance of the system, perhaps by increasing oxygen and hydroxide concentrations in the surrounding water through photosynthesis-related reactions in the plant root zone. If necessary, an ALD may be installed upstream of an aerobic wetland to add alkalinity and raise pH if dissolved oxygen, and iron (III) are absent in the ARD. Excessive amounts of dissolved aluminum can limit the application of ALD's.

Aerobic cells are sized based on "area loading factors". For iron-bearing ARD with favorable alkalinity, an aerobic wetland can remove from 6 to 11 grams per day per square meter (gdm) of surface; some pilot systems have removed manganese at a rate of about 4 gdm, assuming algal mats are employed. Aerobic cells in series can remove about 50 mg/liter of dissolved iron per cell due to dissolved oxygen depletion limitations. In other words, if an incoming iron concentration is 150 mg/liter, the total required area should be subdivided into three cells in series to provide the necessary aeration for complete iron hydrolysis. Locally-available wetland vegetation is typically employed, but it is planted on wide centers and allowed to naturally fill in gaps.

Anoxic Limestone Drains are sized and configured based on retention time, water chemistry, limestone chemistry, and hydraulics. Retention time varies based on the amount of desired alkalinity. Retention times can vary from about 30 minutes to two days. Dissolved oxygen, Fe(III) and significant levels of aluminum in the influent water have been found to be fatal to an ALD's longevity because iron hydroxide and aluminum hydroxide precipitates foul the void spaces in the limestone, restricting ARD flow. Limestone with a high percentage (90 percent or more) of calcium carbonate is desirable. Particle gradation will affect reactive surface area; a nominal particle size of about 2 to 4 cm is recommended for full scale ALD's. ALD's are typically buried to exclude invasion of atmospheric oxygen.

1.2.2 Anaerobic Wetlands

Anaerobic wetlands have been demonstrated to be most effective for treating very acidic waters (pH less than 5.5) and typical high dissolved heavy metal concentrations associated with metal mines. Bacterial reduction of sulfate and precipitation of metal sulfides are the important chemical processes in anaerobic passive treatment. The bacterial reactions involve the reduction of sulfate to hydrogen sulfide gas. The gas, bubbling up through the ARD-saturated and organic-rich substrate, precipitates metals such as iron, copper, lead, zinc and cadmium as sulfides, essentially reversing the reactions that occurred to produce ARD. Bicarbonate is a by-product of sulfate reduction by the bacteria which appear to function best above pH 5.5. Accordingly, the bacteria form their own buffering system at a pH of about 7 by consuming H^+ at lower pH's and producing H^+ at pH's above 7.

Anaerobic Cells are sized based on an area loading factor (which is suspected to be pH-dependent) and a volumetric loading factor, 0.3 moles of dissolved metals/cubic meter of organic-rich substrate/day. As the ARD flows through the substrate in an anaerobic cell, hydraulic considerations (Darcy's Law) must be satisfied to prevent ponding of water on the cell surface. Surface vegetation is not required, but it may be added for aesthetics.

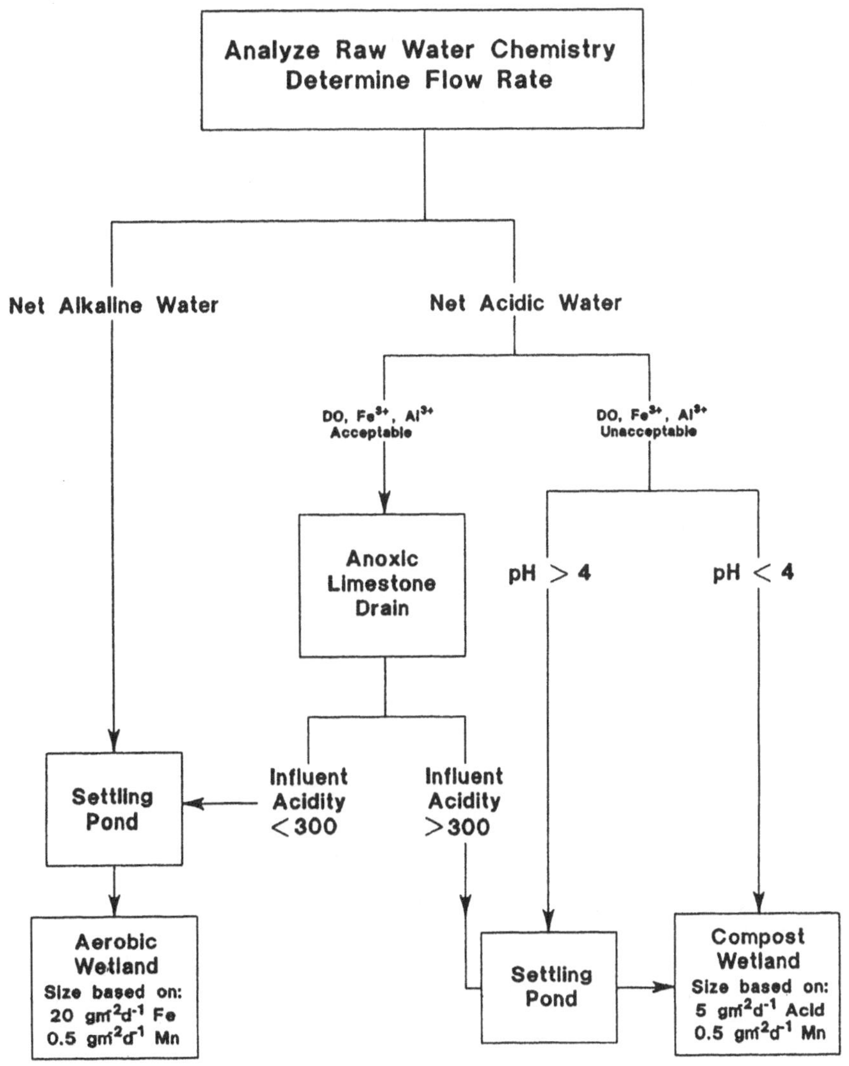

Reference: US Bureau of Mines

FIGURE 1
FLOWCHART FOR DESIGNING & SIZING
PASSIVE "ARD" TREATMENT SYSTEMS

2 Proof of Principle Laboratory Tests

The first step in Knight Piésold's established approach to passive treatment design includes characterizing the drainage chemistries, locating candidate substrate materials, and performing laboratory "proof-of-principle" tests to determine the optimum biogeochemical treatment processes. With this information, bench or pilot-scale systems are designed to put into practice the theories developed in the laboratory. Typical proof of principle tests are conducted in 250 ml clear plastic screw-cap culture bottles for anaerobic evaluation and 250 ml open erlenmeyer flasks for aerobic tests. Various amounts of ARD, substrate and bacterial inoculum are added to the bottles and the evolving chemistry of the water in the bottles/flasks is monitored for up to six weeks.

Soil- or plant-based material, hereafter called "substrate" comprises the heart of passive treatment systems. Substrates for aerobic and anaerobic cells, which include bacterial inoculum, typically have different properties. Anaerobic substrates should contain at least 30 percent organic matter, have some acid neutralizing capacity, be moderately permeable ($K >= 1 \times 10^{-4}$ cm/sec) and have a consortium of sulfate-reducing and other bacteria to reduce the sulfate to sulfide and iron (III) to iron (II). Animal manure, especially cow manure, has been found to be an excellent source of sulfate reducing bacterial inoculum.

Properties for aerobic substrates are typically less restrictive than anaerobic substrates. Typical aerobic substrates include native soils that provide a medium for plant growth and may include algal growths that can function as inoculum.

For the Leviathan site, a graduate student at the University of Nevada at Reno evaluated samples of horse manure, dewatered sewage sludge and locally grown alfalfa for evidence of sulfate reducing bacteria. Samples were analyzed for pH, Eh, color change and odor (H_2S) and sulfate at the end of six weeks incubation. Test results indicated that horse manure, inoculated with local cow manure, would be the best available anaerobic substrate. While the Leviathan system would include aerobic cells, no aerobic substrate testing was conducted.

Proof of principle testing at the Brewer Mine was conducted by Brewer personnel under the guidance of Knight Piésold. Two source waters from the site (from the pit and from a heap leach pad) were used in 25 tests using ten different candidate materials including composted and fresh turkey litter, composted hardwood and softwood sawdusts, a phosphate rock reject product, fresh cow manure, bog sediment and several inert soils. Turkey litter, composted hard wood and the phosphate rock were selected for the substrate recipe based on the test results and local availability (in quantity). Cow manure was selected as an inoculum. Substrate material substitutions during construction are discussed in the Brewer Mine case history.

3 Case Histories

The option to use bench scale testing was considered in both projects. However, current understanding of passive treatment system performance and the proof of principle test results provided sufficient basis for pilot-scale design development. The following general designs for each respective site were adopted as shown schematically on Figure 2:

Leviathan Mine: Anoxic Limestone Drain followed by three aerobic cells feeding into an anaerobic cell

Brewer Mine: Two anaerobic cells; one for pit water, one for heap leach Pad 5 water

Combined, both projects utilized the standard passive treatment components. The design challenges and their associated solutions are discussed in individual case histories that follow.

3.1 Case History No. 1, Leviathan Mine, California

As shown on Figures 2 and 3, the Leviathan passive treatment system consists of a 380 liter tank plumbed to direct flow to either an ALD, the first aerobic cell or a by-pass pipe. The ALD contained about two tonnes of minus 1 mm limestone. The ALD was lined with 0.040 inch (40 mil) thick PVC geomembrane and covered with about 0.6 meters of clayey soil. The ALD discharged through an air trap to a 3-compartment aerobic cell (overall dimensions 3 meters by 8.8 meters. The three compartments were partially filled with local soils and plants. The last aerobic cell discharged to an anaerobic cell (115 sq. meters, 1.1 m deep) filled with horse manure. Cow manure was used as a bacterial inoculum. The system was designed to treat about 2 liters of ARD per minute.

The site offered several design challenges. Those challenges and the responding actions follow.

Remote Site, Harsh Climate Electrical power was not available for pumping. The remoteness of the site, about 25 kilometers from the nearest paved road, necessitated low site maintenance requirements. During the winter, deep snow makes access to the site very difficult. Components needed to be capable of winter operation without the application of external heat sources.

The remoteness of the site mandated a totally passive system which consisted of gravity flow to a constant head tank to the ALD, thence to the aerobic cell, thence to the anaerobic cell. Site winterization included insulation by burying pipes and the tank in soil and fiberglass insulating of above-ground portions of the head tank. Data from the previous winter suggested that the significant snow cover at the site provides some insulation. The entire system has yet to weather a full winter, but early indications show that flow can be maintained in the diversion pipeline during winter conditions. It is unknown how the aerobic system will work in severe winter conditions. The anaerobic system performance may decrease if substrate temperature drops below 10 degrees C.

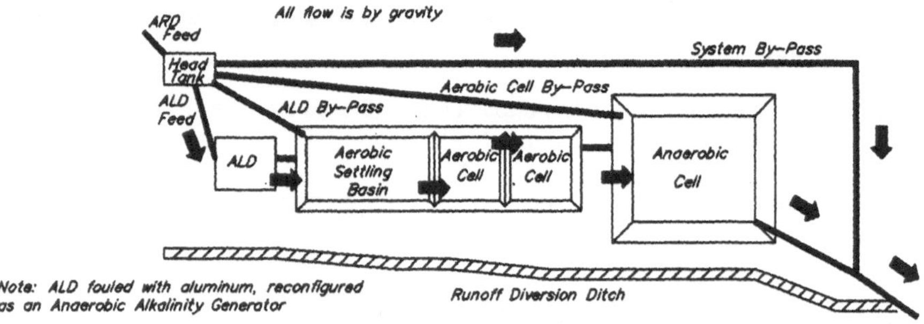

Leviathan Mine, California

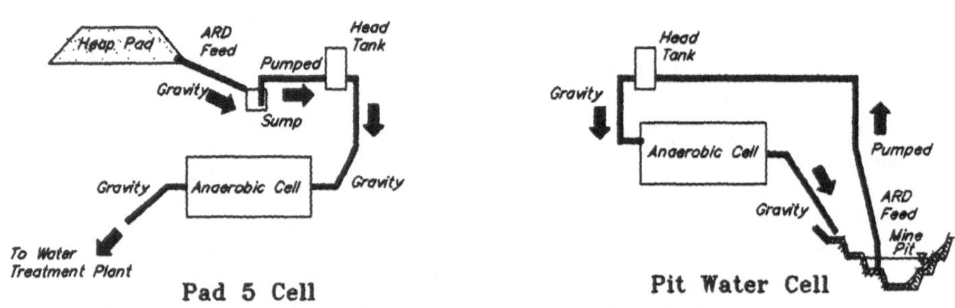

Brewer Mine, South Carolina

FIGURE 2
PASSIVE TREATMENT SYSTEM SCHEMATIC LAYOUTS

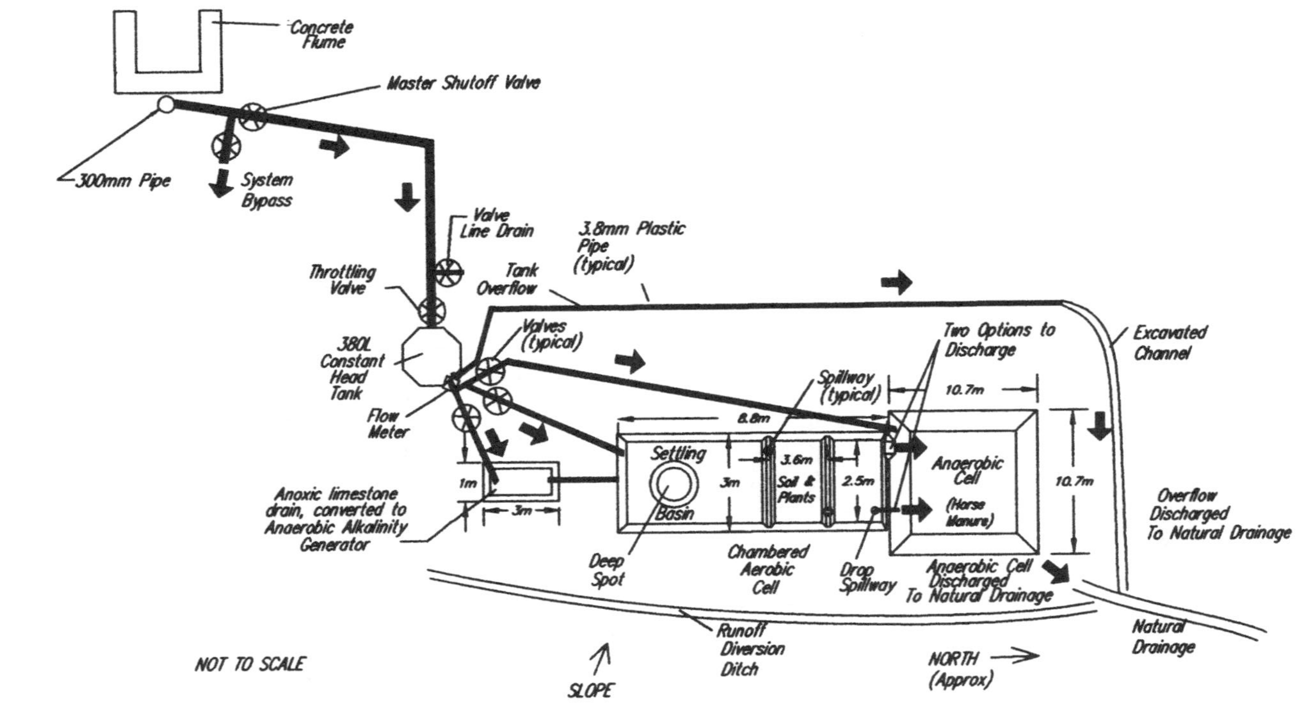

FIGURE 3
LEVIATHAN MINE
PILOT SCALE
PASSIVE TREATMENT SYSTEM

Water Diversion The ARD source discharged from a drain pipe, free falling to Leviathan Creek. Diversion should ideally occur prior to air contact to preserve Fe(II) and Al in solution. Furthermore, the diversion point was nearly 6 meters higher than the head tank elevation; the excess hydraulic head required dissipation. Throttling valves should not offer the opportunity to introduce air into the water.

A field fit diversion assembly was inserted about a meter into the drain pipe and secured in place. The assembly included a flared flexible rubber drill casing "basket" which conformed to the inside of the drain pipe which had some iron hydroxide accumulation. A poor fit on the upper portion of the flared rubber allowed excess water to continue discharging out the drain pipe. The backwater from the assembly minimized air entry, preserving low dissolved oxygen levels and Fe (II) in solution. The assembly is connected to a 38 mm diameter plastic pipe which at design flows of 60 liters per minute, provides a significant friction loss on the available 6 meters of hydraulic head. A throttle valve, at the bottom of the 75 meter long pipe run, is located just upstream of the head tank.

The head tank was covered and sealed to prevent air contact with the water. All pipes exiting the head tank to the atmosphere pass through air traps.

System Flexibility Because the system was also to be available for graduate student thesis work, it needed to be flexible to allow uncomplicated modification to test different cell configurations.

Flexibility is provided first by multiple gate valves and separate feed pipes from the head tank to ALD, aerobic cell and anaerobic cell as shown on Figure 3. Discharges from the three aerobic cells can in the future be diverted to a runoff diversion ditch, constructed parallel to the long axis of the site. Normally, the aerobic cells are connected in series and discharge to the anaerobic cell. Flow in anaerobic cell is controlled with a variable height discharge. The anaerobic cell discharges to a natural drainage channel which provides more aerobic/polishing treatment for iron removal downstream of the by-pass confluence. This feature was unintentional, but beneficial.

The system by-passes nearly 90 percent of the diverted flow, so additional cells could be constructed at a later date if additional space were available.

Budget Restrictions Preliminary cost estimates (about $3,500 US for materials only) suggested that the system size would be limited to treating about 4 liters per minute of flow. Available surface area restrictions discussed below conflicted with this value, however. Budget limitations also precluded the inclusion of elaborate flow rate measuring or sampling appliances. Flow control and measuring needed to be simple and durable.

Student construction labor, free horse manure from the university's agricultural school helped to keep costs down. System instrumentation included a floating indicator type flow meter and plumbing configurations that allowed implementation of bucket and stopwatch methods. At each open channel spillway in the aerobic cell compartments, "flaps" of geomembrane liner material were added to facilitate bucket and stopwatch flow measurements.

Maximum Use of Available Surface Area The overall area available was about 12 meters by 51 meters. A treatment flow rate of four liters per minute was found to be too high for the site. Area loading values showed that the removal of all 300 mg/l of dissolved iron aerobically was not feasible. A balance would need to be achieved between aerobic and anaerobic components with respect to iron removal.

With a fixed size ALD (two tonnes), the surface areas of the aerobic and anaerobic cells were proportioned by trial and error to fill the available space. Ultimately, the aerobic cells were sized for the removal of 100 mg/l of iron if the water passed through the ALD first (11 gdm's). The balance of the original 300 mg/l of iron would be removed as iron sulfide in the anaerobic cell. The precipitation of the some of the iron in the aerobic cells would hopefully be sufficient for the co-precipitation of arsenic. Aluminum would also settle out in the aerobic cells if it had not fouled the ALD. The system was ultimately sized to treat about 2 liters per minute.

3.1.1 Leviathan System Results

The system was commissioned in early May, 1993. Typical preliminary results show 99% removal of iron, 90% removal of arsenic, 99% removal of nickel and 100% removal of the aluminum. The pH of the effluent is 6.7 to 7.2. The anaerobic cell effluent has about 800 to 1000 mg/liter of excess alkalinity, capable of neutralizing more ARD if mixed with other flows. This actually occurs when the anaerobic discharge combines with the 58 liters per minute of bypassed flow.

The ALD plugged with aluminum precipitates after several months of operation and several unsuccessful rebuilding efforts. Ultimately, the ALD was modified into an "anaerobic alkalinity generator" in an attempt to mimic the performance of the anaerobic cell's neutralization of the by-pass flow. Results of the ALD modification are pending as the paper goes to press. A photo of the as-constructed system is shown on Figure 4.

3.2 Case History No. 2, Brewer Mine Pit and Heap Pad Waters

As shown on Figure 2, the site has two anaerobic cells. Both are lined with geosynthetics that were ordered as prefabricated sheets; no field welding was required. Both cells are filled with mixture of composted turkey litter, sawdust, phosphate rock reject and cow manure inoculum. The cells treat 3.9 to 9.5 liters of ARD per minute (pit and pad 5 flows, respectively). ARD is pumped to constant head tanks (about 19,000 liters capacity each) and thence flows by gravity into the cells. The respective cells discharge by gravity back to the mine pit and the heap pad sump. Both cell bottoms cover about 370 square meters. Substrate depths vary from 610 mm for the pit cell to 732 mm for the pad cell. A detailed drawing of the pad cell is shown on Figure 5. The pit cell was virtually identical in configuration. Photos of the pad cell are shown on Figure 6.

FIGURE 4
LEVIATHAN MINE, CALIFORNIA

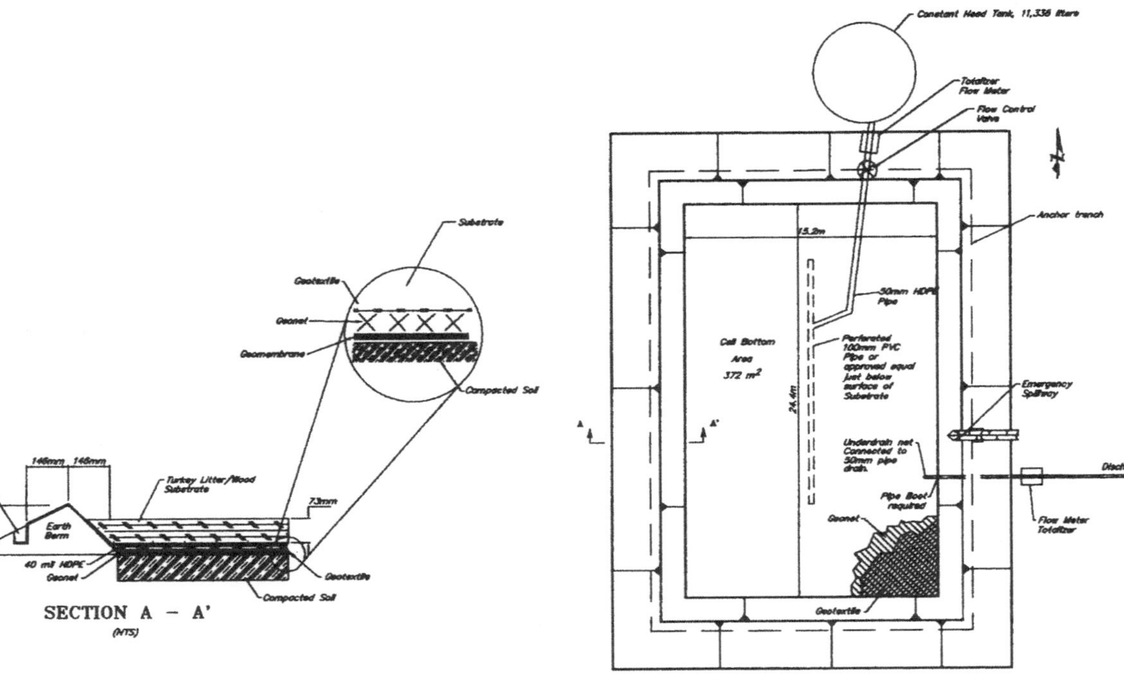

NOTES:

1. Geonet covers bottom of cell, does not extend up the sides.
2. Geotextile covers geonet and is folded beneath the geonet bottom for minimum of 150mm clearance.

FIGURE 5
BREWER GOLD MINE
PIT WATER
ANAEROBIC PILOT CELL

HEAP LEACH ANAEROBIC CELL LAYOUT

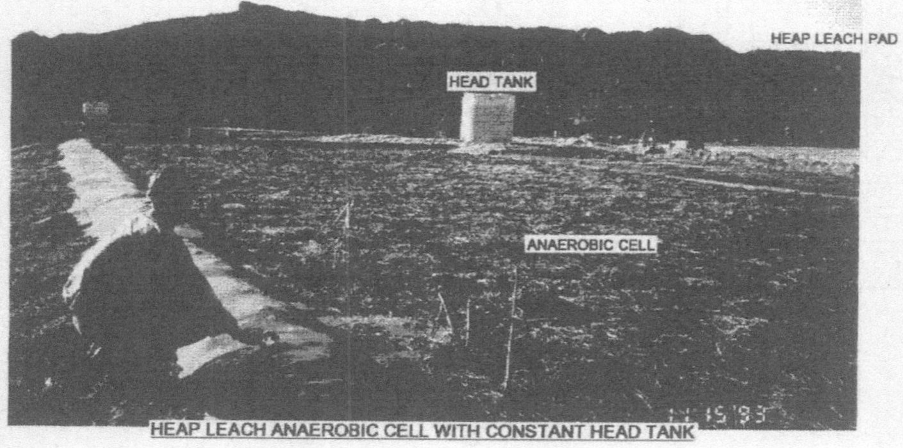

HEAP LEACH ANAEROBIC CELL WITH CONSTANT HEAD TANK

FIGURE 6
BREWER GOLD MINE
SOUTH CAROLINA

This project differed from the previous two case histories in that space was never a constraint. Furthermore, pumps were either in place or easily installed to keep the constant head tanks full. The problems encountered prior to and during construction follow.

Substrate Procurement Proof-of-principle test results indicated that composted and fresh turkey litter, composted wood chips, phosphate rock and cow manure would provide the best substrate mix. Procurement of composted wood chips became an issue when local sources appeared to be primarily from soft wood trees. The testing had suggested that hard wood chips would be better due to the a higher pH of the material and the lower likelihood of fermentation with hard wood. The hard wood chips also were a little more coarse-grained which yielded better hydraulic conductivities than the soft wood. The lesson learned was to identify abundant sources of all potential substrate materials early in the testing/design process. The best material may be in short supply.

Substrate Characteristics: Laboratory vs Field Several difficulties were encountered in scaling up from the proof of principle tests to the pilot field tests. Substrate materials with ambient moisture content were used in lab tests. In the field, substrate materials stored outside were subject to weather variations (mostly rain) that caused the field moisture to vary. Field proportioning by volume (based on lab proportioning by weight) was shown to be very sensitive to the moisture content of the substrate components. It is recommended that laboratory studies be conducted on a dry-weight basis and the data carried through for estimating field quantities.

On a bulk scale, mixing coarse-grained materials such as animal manures, wood chips and phosphate rock typically results in volumetric shrinkage as the smaller grained particles of one component fill in the voids of another coarser material. Passive treatment designs usually require a certain volume of substrate. It is recommended that from 10 to 25 percent additional volume of materials be ordered to account for the shrinkage effect. At the Brewer mine, a four cubic meter front loading mine shovel or a two cubic meter front end loader were used to proportion and mix the substrate materials.

Substrate Installation Overall cell dimensions were selected so that substrate materials could be placed within the reach of a tracked backhoe. This approach worked well for the pad 5 cell. During its construction, substrate was transported about 100 meters from a stockpile and dumped into the cell with a two cubic meter front end loader and subsequently graded with the backhoe. Thus, earthmoving equipment did not have to work atop the substrate and compaction was avoided.

Site topography at the pit cell allowed limited trackhoe placement. Here, substrate was hauled in 50 tonne mine trucks and end-dumped into one end of the cell. After a ramp of substrate was constructed to protect the geomembrane liner, a light tracked dozer was used to distribute the substrate. Ultimately, the ramp had to be removed with hand labor to minimize damage to the liner. The dozer was allowed to work on top of the substrate because laboratory testing had showed that compaction at 90 percent Standard Proctor density still provided adequate hydraulic conductivity in the substrate.

In the pit cell, final grading of the substrate was accomplished with a crude plow-blade constructed of wooden beams. The blade was advanced via a rope attached to the tramming backhoe. The wooden blade was repositioned for each pass by hand laborers. This innovation worked reasonably well but would have been more effective with perhaps a dedicated winch and return cable and perhaps a fabricated metal plow.

Cost Control To minimize costs, Brewer Gold personnel procured all materials directly, constructed all the earthworks, and installed all the geomembrane and plumbing. In addition, Brewer personnel conducted all proof-of-principle testing on site with nominal guidance from Knight Piésold.

Variable ARD Quality Water quality test results used in the design were based on proof-of-principle data which was less than year old. At startup, pit water and pad 5 water qualities were found to be significantly poorer than those assumed for the design. The cells were originally designed to treat 19 liters of ARD per minute. To account for the increased metal loading observed at startup, flows were trimmed to 3.9 and 9.7 liters per minute for the pit and pad cells, respectively.

3.2.1 Brewer Mine Results

The passive treatment cells were commissioned in early September, 1993. Typical results 2 months after startup for both cells showed 99% removal of copper, 85% removal of iron, 99% removal of aluminum, pH of effluents 6.5 to 7.0. Both cell effluents have about 300 to 600 mg/liter of excess alkalinity, capable of neutralizing more ARD if mixed with other flows. During the first months of winter weather coupled with cold weather, the pad cell is not functioning as well as the pit cell (whose water quality is the worse of the two). Data are being studied to ascertain the cause of the performance difference between the two cells.

4 Summary

The diversity of the two case histories with respect to water quality and final system configuration show that passive treatment systems are more complicated than routing water through open basins (aerobic cells), limestone-filled galleries (anoxic limestone drains) and organic-filled bogs (anaerobic cells). The chemistry of the ARD drives the overall design layout and the cell configurations are driven by geography and established engineering principles.

The systems constructed were pilot scale; however, the step from pilot to full scale is facilitated if one considers these pilot systems as "modules" operated in parallel to treat the entire volume of ARD. ARD flows are typically cyclic; during periodic low flow periods, modules could be taken off line and serviced/repaired. Servicing could conceivably include metals recovery and substrate refurbishment.

The longevity of passive treatment systems remains as a topic of further study. The two systems described may provide the tools for further understanding of passive treatment system long term performance.

5 Acknowledgements

The authors gratefully acknowledge the following persons or companies for contributing information or permission to publish this paper.

- Dr. Thomas R. Wildeman, Colorado School of Mines
- Dr. Glenn Miller, University of Nevada at Reno
- Eric Taxer, California Regional Water Quality Control Board
- Brewer Gold Company

6 References

1. Wildeman, T.R., G.A. Brodie, and J.J. Gusek. (1993) **Wetland Design for Mining Operations.** Bitech Publishing Co., Vancouver, BC, Canada.

Wheal Jane—can wetlands technology cope?

R. M. Hamilton
National Rivers Authority, Exeter, England
N. A. Postlethwaite
Marcus Hodges Environmental, Ltd., Exeter, England
S. A. Foster
Arthur D. Little, Inc., Cambridge, Massachusetts, U.S.A.

Abstract
Wheal Jane is an abandoned tin mine in South West England. Closure and subsequent flooding led to the generation of a large volume of acidic, metal-rich drainage water. Most of this is pumped and treated by conventional means at extremely high cost. There is an urgent need for a technically- and cost-effective long term solution. Wholly active methods have been rejected. Passive methods used in North America have been evaluated for their suitability for transfer. Acidity and metals concentrations can be reduced but no single system capable of dealing with the conditions at Wheal Jane exists. A strategy has been adopted to develop an integrated, passive system on a pilot scale. Methods to be tested include anoxic limestone drains - to increase alkalinity, oxidation ditches - for initial metal removal, a combination of aerobic and anaerobic wetlands - to further increase alkalinity and remove metals, and rock filters - to remove manganese. Successful methods will be included in recommendations for a full size system. However, it is recognized that a wholly passive system may be insufficient and that some conventional treatment will have to be incorporated into the final design.
Keywords: Acid mine drainage, treatment methods, wetlands, Wheal Jane.

1 Background

The closure of Wheal Jane, a tin mine in South West England, occurred in March 1991. De-watering pumps were removed and ground water level recovered within the year. Initial discharge in November 1991 was low but subsequently increased in response to rainfall. The rock at Wheal Jane is sulphide rich and the normal oxidation processes resulted in the mine water being extremely acidic and with high concentrations of dissolved metals.

The scale of the problem is enormous. During dry periods, the discharge from the mine can be less than 9 Megalitres (2 million gallons) a day. The ground water catchment draining directly to Wheal Jane is about 4.4 km^2 (1.7 square miles), most of which has been mined for heavy metals. There is an immediate response to rainfall from surface drainage as well as a delayed response, starting 20 to 21 days later, through the ground water. After prolonged, heavy rainfall, discharge has been as high as 42.3 Ml/d (9.3 mgd). This is not necessarily the maximum flow, only that which has been observed since closure. Theoretical calculations of rainfall events with specific return periods indicate the total volume falling on the catchment could be as high as 462,000 m^3 (102 million gallons) for a 1 in 100 years, 24-hour event. Attenuation patterns are unknown but if they are short, it is clear that peak discharge could easily exceed 10 mgd.

The mine water is acidic, the pH being about 3. It has not markedly improved since closure. Initial concentrations of metals were extremely high. During recovery, samples from the middle and lower parts of the water column in the main shaft were usually considerably worse than surface water, but quality changed rapidly with time. Iron and zinc concentrations frequently exceeded 2 g/l each. During January 1992 the discharge occurred accidentally through an adit draining an area immediately adjacent to Wheal Jane. Its quality is indicated in Table 1. The quality of water in the shaft has improved with time, as shown in Figure 1 for iron [1]. Other metals follow a similar trend but still present a substantial total concentration.

Table 1. Chemical quality of the discharge, 14 - 30 January 1992

pH	2.6 - 3.1
Fe	1720 - 1900
Zn	1260 - 1700
Al	170 - 197
As	26 - 29
Mn	11 - 25
Cu	14 - 18
Ni	4.2 - 5.1
Cd	1.4 - 1.9
Pb	0.2 - 0.3
Cr	less than limit of detection

Except for pH, all units in mg/l dissolved.
From: Hamilton et al, 1994.

To limit the impact of this discharge on the environment, a temporary control system has been installed. This involves pumping surface water from the mine, treating it with hydrated lime and a flocculant aid, and settling the resultant metal hydroxides in a large tailings dam. This system has a number of constraints. Pumping capacity has been restricted to 15 Ml/d (3.3 mgd) because a large volume of low density sludge is produced and this is utilizing the capacity of the tailings dam relatively quickly. The cost of conventional pumping, treatment and sludge disposal is high, almost £2 million in the first year. Available resources are limited and it is clear that a technically- and

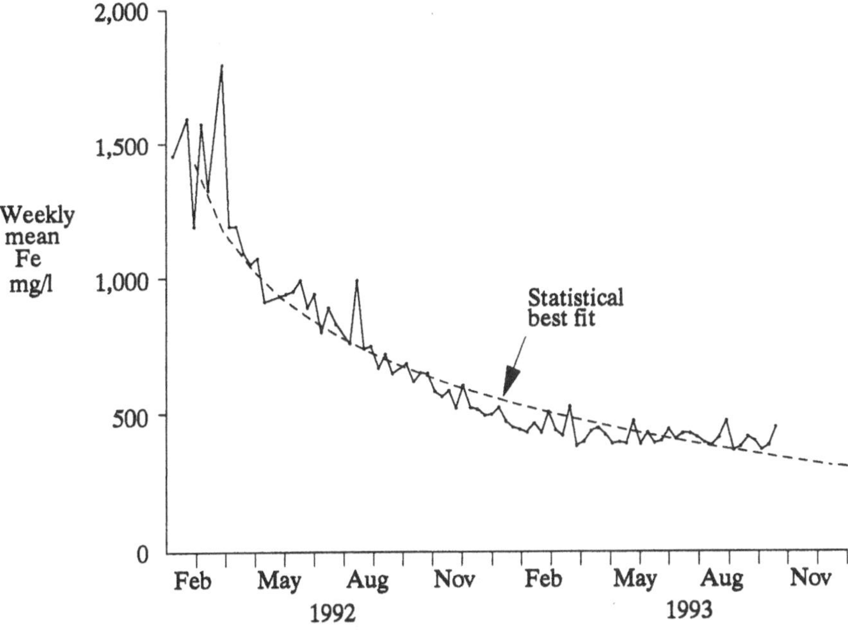

Fig.1 Improving raw water quality at Wheal Jane. Mean weekly iron concentrations, February 1992 to October 1993. From Hamilton et al (1994).

cost-effective long term solution is needed. Moreover, public opinion was understandably outraged when the accidental discharge polluted a substantial area of the Carrick Roads and adjacent Falmouth Bay. It is unlikely that cessation of treatment would be politically or socially acceptable.

2 Criteria for long term solution

The discharge will continue for ever, although there are possibilities for limiting the flow, and it is likely that poor water quality will continue for the foreseeable future. Consequently, any long term solution will have to operate for an extremely long time. Hence the operational and maintenance costs will have to be low. Since the legislation specifically exempts from control the discharges from abandoned mines, the cost of providing long term treatment ultimately rests with the Government. It is unlikely that any Government will support an expensive project of this nature for all time. For the same reason, the capital cost of a long term solution must be affordable. It must be implemented within a few years, before the storage capacity of the tailings dam is exceeded, and it would be sensible for the transition from temporary to permanent treatment system to be phased so that the inevitable operational problems of commissioning can be overcome without jeopardising environmental water quality. The long term system must deliver a discharge which enables water quality standards and environmental objectives to be met, and finally, if sludge is generated, there must be a viable route for its disposal.

A range of methods is available to treat acidic mine water in the Wheal Jane situation. Preliminary evaluations of the most likely active and passive methods were carried out and initial conclusions were reached on how to progress.

3 Active methods

Active methods are defined as those requiring considerable and continuing input of resources, whether in terms of man's time or consumables. Those considered in the first instance were pH neutralization, electro-chemical and reverse osmosis methods, and the use of limestone aggregate permeable plugs.

3.1 Neutralization

pH control is followed by coagulation, flocculation and sedimentation. The choice of neutralizing agent is governed by cost, availability, basicity, reaction time and sludge characteristics. Lime is cheap and readily available. Much of the iron is precipitated following aeration, and other metals such as arsenic and cadmium are co-precipitated with the ferric hydroxide. However, there are problems with settling the low density ferric hydroxide floc, requiring the use of expensive flocculant aid and large settlement and sludge drying areas. Limestone produces sludge with a higher solids content but requires a longer reaction time and cannot achieve the high pH necessary to precipitate most of the metals. Sodium based alkalis are expensive. Overall, such a system would have high capital, operation and maintenance costs. A large amount of metalliferous sludge would be produced, the disposal of which would be difficult and costly, as a consequence of its designation as a hazardous waste. The problem of solid-liquid separation may be addressed by the use of hydrocyclones, but at Wheal Jane the method was ineffective. Sludge recirculation in a high density sludge facility was also considered but this increased the costs and labour requirement with relatively little benefit. With conjunctive use of the tailings dam, this method is the best short term solution, but the high costs and demand for labour make it unsuitable as the only long term solution.

3.2 Electrochemical processes

There are several proprietary systems available, all of which have high capital, operating and maintenance costs. Payback is dependent on metal recovery efficiency, particularly where there are many metals, and on their market prices. Currently, these are low. Overall, this was not considered to be a viable option, principally on grounds of high cost.

3.3 Reverse osmosis

This method has high capital, operating and maintenance costs. The technology is unproven for dealing with flows of the magnitude experienced at Wheal Jane. The end product would be a concentrated waste stream which would need further treatment. Overall, this was not considered as a viable option.

3.4 Limestone aggregate permeable plugs

In the early 1970s, the U.S. Environmental Protection Agency sponsored studies on the use of limestone aggregate as permeable plugs for mine openings [2]. The seal is initially porous but, as mine drainage water passes through, metal oxide and possibly calcium sulphate are precipitated within the pore spaces, reducing the permeability of the seal. During this process, other metals would also be precipitated. Where metals are in high concentration and the water is well oxygenated, as at Wheal Jane, metal oxide armouring of the limestone would occur rapidly. This would reduce the surface area available for subsequent reaction and also the rates of neutralization and precipitation. Furthermore, the area around Wheal Jane has numerous connected shafts and adits, and the ground is relatively weak as a consequence of the early extensive and shallow mining activity. Plugging the most obvious drainage routes would simply divert the problem elsewhere. This option was not considered to be viable.

It is recognized that this is not a comprehensive list of potential active treatment methods. During the course of the past three years, many proposals have been made and the opportunity is being taken to test some of those likely to meet the majority of the criteria for the long term solution.

4 Passive methods

Passive treatment methods are defined as those using physical, chemical and biological processes requiring little or no intervention from man, once commissioning has been completed. During this initial evaluation, attention was focused on North America, where many experimental, pilot and full scale treatment programmes have been implemented to deal with acid drainage from coal and metal mines.

A comprehensive literature search and review was undertaken with the objectives of providing an understanding of the processes involved and an indication of where successful programmes were being operated. It was immediately apparent that there were two approaches. In the Appalachian region, a combination of methods including constructed aerobic and anaerobic marshes are being used at over 300 sites where coal mine discharges occur. In the Western U.S. and Canada, the use of sulphate-reducing bacteria in anaerobic "bioreactors" has been successful in dealing with metal mine effluent.

Initially, eleven projects were evaluated in terms of the potential for method transfer to Wheal Jane. Subsequently, eight were assessed in detail. Issues addressed were site history and system design, the geochemistry of the source, water quality, the system ecological profile, the vegetation profile, microbe management, hydrologic setting and hydraulic engineering, scaling up and sizing implications, and functional effectiveness. During the course of this exercise, information from other projects became available. Finally, visits were made for on-site discussions with operators at eight sites, including some where large volumes of industrial and municipal waste water were being treated by constructed wetlands. The sites investigated are shown in Table 2.

Table 2. Sites and programmes evaluated

Coal:	Simco, Ohio	P		
	Daniel Boone National Forest, Kentucky	P		
	Jones Branch, Kentucky		D	
	Friendship Hill, Pennsylvania	P	D	
	Pennfield, Pennsylvania	P		
	Bark Camp Run, Pennsylvania		D	
	Howe Bridge, Pennsylvania		D	V
	REM, Pennsylvania			V
	Morrison, Pennsylvania			V
Metal:	Big Five Tunnel, Colorado	P	D	V
	Eagle, Colorado	P	D	
	Quartz Hill Tunnel, Colorado	P		
	Dunka, Minnesota	P	D	
	Grey Eagle, California	P	D	
	Bell Copper, British Columbia, Canada	P		
	U.S. Bureau of Mines "Biofix" Programme	P		
Flow:	Orlando, Florida			V
	Lakeland, Florida			V
	American Crystal Sugar, Hillsboro, North Dakota			V
	Minot, North Dakota			V

P = Preliminary evaluation
D = Detailed evaluation
V = Visit

Coal mine drainage in the Appalachians is typically dominated by iron, with aluminium, manganese and zinc being less important. Other metals may be present in low concentrations and pH may vary from less than 2 to over 7. Shallow, aerobic marshes planted with *Typha latifolia* are widely used and are highly efficient at removing iron. Although extreme conditions can be treated, such systems are more effective when the inflow pH is greater than 5.5 and the dissolved iron concentration is less than 50 mg/l. Various rules of thumb have been calculated. For example, to remove 100 grams of iron a day at pH 3.0 to 3.5, an aerobic marsh of 25 m^2 is required, but at pH greater than 6, only 5 m^2 is needed. Oxidation and hydrolysis also generates acidity. If the pH drops below 3, or when the iron concentrations are high, aerobic wetland efficiency is much reduced. Excess alkalinity provides valuable buffering capacity and this can be obtained by pre-treatment using anoxic limestone drains.

Providing the mine water is de-oxygenated, passing it through an air-tight chamber packed with high quality $CaCO_3$ limestone will raise the pH by adding up to 200 mg/l alkalinity whilst retaining the metals in solution. On emergence, oxidation precipitates metal (usually iron) hydroxides and generates some acidity but this is buffered by the alkalinity added during passage through the anoxic limestone drain. For example, at Howe Bridge, pH was raised from 3.07 to 4.95 through the drain, losing less than one unit after oxidation[3]. Details are shown in Table 3.

Table 3. Wetland performance, Howe Bridge, Pennsylvania

	Inflow	ALD	Oxidation Pond	Marsh
pH	3.07	4.95	4.02	6.75
Alk	40.0	190.0	-----	86.7
Fe	261.0	225.0	180.0	25.0
Al	0.4	0.2	<0.2	<0.2
Mn	35.6	34.6	32.2	12.7

ALD = Anoxic Limestone Drain

Except for inflow, all samples taken downstream of ALD, oxidation pond or marsh.
Except for pH, all units in mg/l.

The success of anoxic limestone drains is dependent on several factors including limestone quality and size, the elimination of oxygen infiltration, flow rate and the lack of aluminium in the inflow. Initial tests at Wheal Jane indicated that limestone suspended at depth in one of the main shafts would not become armoured with a coating of metal oxides. It was concluded that anoxic limestone drains and oxidation ponds would be suitable for testing, although the requirement for a low rate of flow means that a considerable area of land might be needed at Wheal Jane. If de-oxygenated water cannot be guaranteed, some form of pre-treatment will be necessary. An organic slurry pond may offer an inexpensive approach, although there are no examples of these being used in acid mine water remediation systems.

Many metals are precipitated during oxidation and, providing residence time is sufficient, will settle in the oxidation ditch or pond. Details for some metal settlement at Howe Bridge are also shown in Table 3. During subsequent marsh treatment, passing the water through organically rich anaerobic substrates has been found to be effective, as at Howe Bridge. This process increases the alkalinity, giving additional buffering capacity and, if zinc is present, encouraging its removal. In anaerobic substrates, sulphate-reducing bacteria promote alkaline conditions suitable for their own activity, a beneficial bi-product of which is the removal of metals as insoluble sulphides. The removal of manganese is achievable only at high pH. Dense algal blooms in open water or encrustations on rock filters are capable of providing high pH, but are obviously dependent on the availability of light and a suitable temperature regime. All these methods are likely to provide some benefit at Wheal Jane.

In the western states of America, a few constructed wetlands have been used at sites generating relatively small volumes of acid, metal-rich water from abandoned hard rock mines. Here, the emphasis has been on anaerobic systems employing sulphate-reducing bacteria. Much of the work has been on isolating bacterial processes, designing substrates which promote sustained growth of bacteria and developing systems which maximise water-substrate contact without the risk of blocking flow. A range of substrates has been investigated and the best are spent mushroom compost and a mixture of aged cow manure and straw. These provide a high natural buffering capacity and a pH between 8 and 9. They have large amounts of organic carbon available for bacteria and which act as adsorption sites for metals. A good supply of sulphate is essential; the concentration must exceed 30 mg/l. Horizontal and downflow systems have been shown to be inefficient, and pre-soaking the substrate appears to minimise the creation of preferred flow pathways. Two small stages are better than one large unit.

Sulphides can be destroyed by acidic or aerobic conditions and it is important to maintain several centimetres of water over organic substrates to prevent or minimise oxidation. Flow control is critical. An anaerobic bioreactor at Grey Eagle, an abandoned copper and gold mine in California, has performed well [4]. Flow is small, 330 gpd, and bottom fed through two metres of substrate within a tank of 3m x 3m x 11m. Some results are shown in Table 4. Anaerobic bioreactors are successful in low flow situations but the problems of significant scaling-up have not so far been addressed. The concept is attractive but the practice may be difficult to use at Wheal Jane.

Several methods were considered as unsuitable for direct transfer to Wheal Jane. These included aerobic wetlands of the type designed for treating organic waste. Although capable of dealing with large volumes, they have not been used to treat metal loads and acidity levels such as those found at Wheal Jane. However, if acidity and metal loadings could be reduced, aerobic wetlands may have a part to play, particularly when anaerobic areas are included. Abiotic sulphide reaction chambers were rejected on the basis of high costs, health and safety risks, and the generation of a waste stream which would need further treatment. Similarly, immobilized biological reactors, ion exchange methods and floating reed mats were rejected on the grounds of high costs, the generation of hazardous waste with little or no commercial value, or ineffectiveness.

None of the methods considered suitable for transfer has been operated at a scale appropriate to Wheal Jane. The overall conclusion is that there is no single passive method which can be adopted. A system will have to be developed.

Table 4. Bioreactor performance, Grey Eagle, California.

	Inflow	Bioreactor
pH	3.4	6 -7
Cd	0.12	0.002
Cu	130	2.5
Ni	1	0.2
Fe	320	4.5
Zn	42	0.13
SO_4	2900	1100

Except for pH, all units in mg/l.

5 Strategy

The strategy adopted is to test, at a pilot scale, those methods considered suitable, within an integrated system of acidity reduction (anoxic limestone drains), metal removal (oxidation ditches, aerobic wetlands) and final polishing (anaerobic wetlands, rock filters), and to consider the successful ones for inclusion in a recommendation to Government for a long term solution. Consideration was given to a small scale research approach and to the option of building a full scale solution immediately but neither was favoured owing to timescale difficulties and high risks of failure. The step of pilot to full scale was considered to be an acceptable risk. Options for flow control, to limit the amount of water needing treatment, and for active methods not previously evaluated will also need to be addressed.

Even so, the preferred strategy has a number of risks attached to it. Some or all of the potentially suitable methods may be ineffective in the Wheal Jane situation. If raw water quality does not improve naturally, some of the methods may be near the limit of their capability. A further constraint to success is the amount of land available, only 43 hectares (107 acres), all of it in the flood plain and much of it already contaminated. Calculations based on published removal rates for American systems and for theoretical flows and metal concentrations at Wheal Jane indicate that a substantial area of constructed wetland may be needed (Table 5).

Table 5. Calculated vegetated wetland areas in acres, based on published removal rates and estimated loading at Wheal Jane.

Total metals mg/l	Flow mgd	Total metals load g/d	Removal g/d/m²: Influent pH: Source reference:	Fe 4.0 3.0 [5] [6]	Fe 150 3.0 [7]	Fe 10 4.0 [6]	Fe 13.6 6.5 [8]	Ni 0.066 7.0 [9]
100	0.1	50 000		3.1	0.08	1.3	0.9	189
	1.0	500 000		31.3	0.80	12.5	9.2	1884
	4.0	2 000 000		125.0	3.20	50.0	36.8	7576
200	0.1	100 000		6.3	0.16	2.5	1.8	379
	1.0	1 000 000		62.5	1.60	25.0	18.0	3788
	4.0	4 000 000		250.0	6.40	100.0	72.0	15152
300	0.1	150 000		9.4	0.24	3.8	2.8	588
	1.0	1 500 000		93.8	2.40	37.5	28.0	5882
	4.0	6 000 000		375.0	9.60	150.0	112.0	22728
	Assumes 1 gall. = 5l	Assumes Fe = 80%		Many wetlands reviewed	System over-loaded	Somerset (US) wetland	Simco. Mean flow 144 000 gpd	Dunka. Natural marsh

6 Conclusions

Passive methods offer the opportunity for dealing with acidity and metal contamination, but the scale of the Wheal Jane problem in terms of flow and metal loading, and the constraint of land availability indicates that an ideal long term solution will be difficult to achieve. Much depends on the outcome of testing the behaviour of the chosen methods under various flow and loading conditions. During predicted low flows, it is likely that all the discharge could be treated by a full scale passive system in the area available. However, at predicted high flows it seems probable that insufficient passive treatment capacity would be available. The options then to be considered would be full treatment of a proportion with the remainder being untreated, allowing hydraulic overload to reduce overall treatment efficiency, and the incorporation of an active method within the system. The first two involve a compromise on meeting environmental quality standards, whilst the third would increase costs.

It may be that if raw water quality continues to improve naturally with time, a full scale passive system will be adequate. Thus the compromise on environmental quality standards or on cost may be for a limited period, the length of which cannot presently be predicted.

7 References

1. Hamilton, R. M., Waite, R. R. J., Postlethwaite, N. A. & Cambridge, M. (1994). Wheal Jane, its abandonment and treatment of the resultant discharge. Presentation at the International Land Reclamation and Mine Drainage Conference and the Third International Conference on Abatement of Acidic Drainage, April 1994, Pittsburgh.

2. Pearson, F. H. & McDonnell, A. J. (1975). Limestone barriers to neutralize acidic streams. Proc. Am. Soc. Civil Engineers 101, 425-441.

3. Kepler, D. & McCleary, E. (1992) pers. comm.

4. Wildeman, T. (1992) pers. comm.

5. Hammer, D. A. (1990). Constructed wetlands for acid water treatment - an overview of emerging technology. Presentation at the Annual Meeting of the Geological Association of Canada and the Mineralogical Association of Canada, May 1990, Vancouver .

6. Hedin, R. S. & Nairn, R. W. (1990). Sizing and performance of constructed wetlands: case studies, in Proceedings of the Mining and Reclamation Conference, ed Skousen, J. G., Sencindiver, J. C. & Samuel, D. E., 1990, West Virginia University.

7. Chalfant, G., Halverson, H.G., Demeritt, M.E. & Wade, G. L. (1991). A constructed wetland to remove metals from acid mine drainage. USDA Forest Service, Daniel Boone National Forest, Winchester, Kentucky.

8. Stark, L. R., Stevens, S. E., Webster, H. J. & Wenerick, W. R. (1990). Iron loading efficiency and sizing in a constructed wetland receiving mine drainage, in Proceeding of the Mining and Reclamation Conference, ed Skousen, J. G., Sencindiver, J. C. & Samuel, D. E., 1990, West Virginia University.

9. Eger, P., Melchert, G., Antonson, D. & Wagner, J. (1991). The use of wetland treatment to remove trace metals from mine drainage at LTV's Dunka Mine. Minnesota Department of Natural Resources, Division of Minerals.

Retention of cadmium in clay minerals by a hydrothermal method

Ke Jia-Jun
Institute of Chemical Metallurgy, Academia Sinica, Beijing, China

Abstract
Cadmium is one of the most hazardous heavy metals. The retention of heavy metal Cd by clay mineral has been studied in this work. The experiments were carried out by transforming cadmium into sulphide by adding sulphur under hydrothermal conditions of temperature ranges from 150° to 240°C. It has been shown that the retention of cadmium increased with hydrothermal temperature and with the amount of sulphur added. Only traces of cadmium are leached out at pH 3.0 after hydrothermal treatment. It appears that an approach of hydrothermal sulphidizing in neutral or slightly acid suspension is more suitable for fixation of heavy metal Cd in contaminated waste water, soil and sludge.
Keywords: Retention of Heavy Metal, Cadmium, Clay Mineral, Hydrothermal Method.

1 Introduction

The disposal without attendant pollution of industrial effuents containing heavy metals is becoming a matter of increaing concern, because many heavy metals form stable complexes with biomolecules and their presence in even small amounts can be detrimental to plants and animals. The phenomenon of some soils containing clay mineral which can be adsorbed dissolving metals to a considerable extent has been thoroughly studied by Weaver and Pollard (1973), Wada (1985) and Ziper et al.(1988). The experiments for the adsorption of heavy metals on bentonite, a clay with a high content of montmorillonite, and then treatment of this slurry under hydrothermal conditions to retain heavy metals have been carried out by Ke and Sørenson (1992). It was shown that the fixation increased with temperature and with the amount of alkali added,and the order of retention was Ni > Cu > Zn > Cd, which reflects the adaptation of metals ion to the silicate lattice of bentonite clay mineral. Cadmium as one of the most hazardous pollutants is not satisfactorily fixed by the silicate lattices of clay mineral. In such a case, a more promising approach is transforming cadmium into sulphide, which can be done by adding a small amount of sulphur under hydrothermal conditions as shown in the following reactions. According to the disproportionation of elemental sulphur under hydrothermal conditions, it follows:

$$4S^{o} + 4H_2O \rightarrow 3H_2S + SO_4^{2-} + 2H^{+} \tag{1}$$

Meanwhile, the reaction of Cd^{2+} with S^{2-} produces a precipitate of CdS,

$$Cd^{2+} + S^{2-} \rightarrow CdS \tag{2}$$

CdS formed resists acid attack at pH>1. The objective of the present work is to investigate,from a practical stantpoint, the influence of some parameters on the transformation of Cd into sulphide under hydrothermal conditions.

2 Experimental

The clay mineral sample used in experiments is bentonite for industrial use, which is obtained from Nei Monggol, China. This is identified by X-ray diffractometer as montmorillonite mineralogically. The physical properties of bentonite powder sample are as follows: white colour, particle size -100 mesh > 97%, bulk density 0.8 g/cm^3. The chemical composition of bentonite sample is shown in Table 1.

Table 1. Chemical composition of bentonite sample

Composition	%
SiO_2	56.24
Al_2O_3	14.37
Fe_2O_3	1.45
FeO	0.28
MgO	2.18
CaO	3.68
Na_2O	0.45
K_2O	0.69
H_2O^+	5.13
H_2O^-	15.25
Total	99.72

Static adsorption was carried out by mixing bentonite with solution containing Cd^{2+} to prepare the samples for hydrothermal sulphidizing tests. 100 g of bentonite powder was suspended in 1 litre solution of 0.02 M Cd^{2+} in the form of sulphate at ambient temperature and with agitation at 500 min^{-1} for 2 h. After filtration, bentonite sample with adsorbed Cd^{2+} was used for subsequent tests of hydrothermal treatment. The aim of these tests is to investigate the transformation of cadmium into sulphide under hydrothemal conditions.

The autoclave is made of stainless steel, which has a ϕ 40 mm × 60 mm cylindrical chamber with a Teflon liner and is closed with a screw cap. The hydrothermal sulphidizing experiments were carried out using 3 g Cd-containing bentonite sample and adding apposite amount of sulphur powder in 25 ml H_2O. The autoclave was placed in an electric oven at desired temperature. When the prescribed time had elapsed, the autoclave was cooled in a water bath. After cooling and filtration, the wet filter cake was used for subsequent leaching tests to evaluate the results of fixing Cd.

The solution samples were analysed for Cd by atomic absorption spectroscopy with a Perkin-Elmer Model 4000 AAS. The clay minerals were identified by X-ray diffraction.

3 Results and discussion

The clay mineral montmorillonite is made up of negatively charged silicate layers and has a typical composition $(Al,Mg)_{2-3}(Si,Al)_4O_{10}(OH)_2 \cdot nH_2O$ with interlayer cations compensating the positive charge deficiency of silicate layers. The chemical composition of this clay is variable due to considerable atomic substitution possible in this clay, but the major constituents are always Si, Al, Mg and water along with considerable amounts of exchangeable cations. These exchangeable cations are present along with water molecules inside the interlayer spacings of the clay framework as shown in Fig. 1. The interlayer space in montmorillonite clay is substantial, permitting it to act as a host compound and, therefore, it is possible to adsorb some exchangeable cations present in the interlayer spacings of the clay mineral.

The static adsorption results of varying the initial concentrations of Cd in the solution are shown in Fig. 2. The graph is seen to have the shape of a usual adsorption isotherm concluding that available sites become gradually more scarce. The wet filter cakes of clay mineral with adsorbed Cd were used for autoclave treatment tests. Experiments were carried out using the sample of bentonite containing Cd 11.8 mg/g. The pH of this slirry was 7.0.

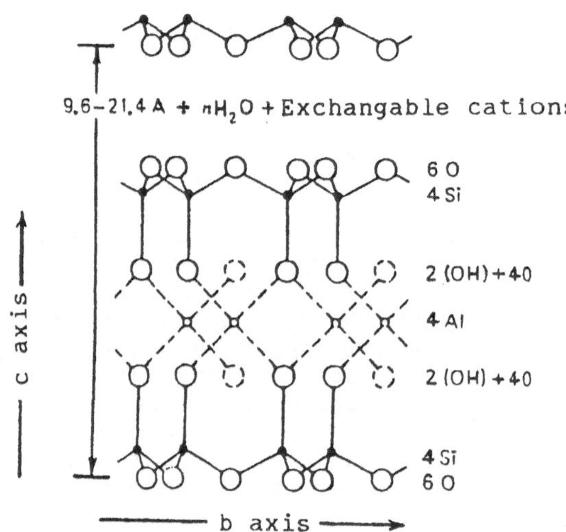

Fig.1. Structure of montmorillonite.

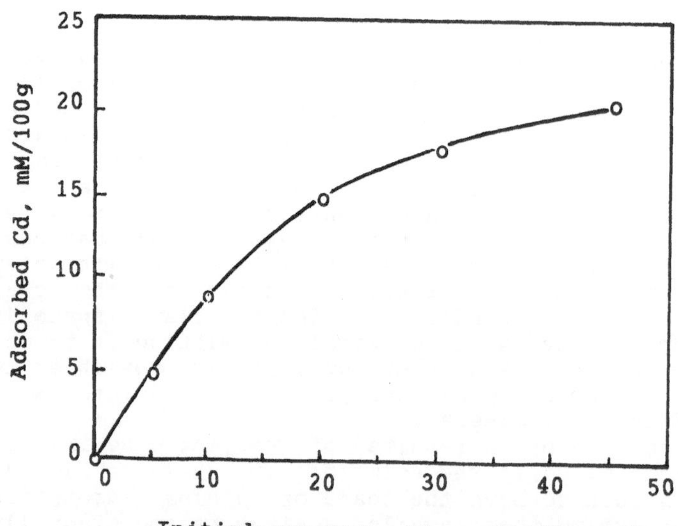

Fig.2. Adsorption of Cd on bentonite versus initial concentration in solution.

Four portions, each corresponding to 3 g dry Cd-bentonite were placed in the autoclaves with the addition of 0.2 g sulphur and 25 ml H_2O after adjusting the pH stepwise down to 4.0 with H_2SO_4. The autoclaves were placed in the oven for 2 h at 240°C whereupon the content was filtered. The resulting pH and Cd content in the filtrates are shown in Table 2. It can be seen that with the addition of sulphur, the pH of the slurry formed after autoclave treatment is $\leqslant 3.0$, because H^+ is produced by reaction (1). The analyses of the filtrates indicate <0.1 ppm Cd, i.e., cadmium is rendered insoluble.

Table 2. Sulphidization of Cd at 240°C

pH of slurry		ppm Cd in filtrates
Initial	After treatment	
7.0	3.0	<0.1
6.0	2.7	<0.1
5.0	2.5	<0.1
4.0	2.3	<0.1

Conditions: 3 g Cd-bentonite and adding 0.2 g sulphur in 25 ml H_2O at 250°C for 2 h.

After hydrothermal treatment,the filtered solid samples (each of 0.5 g dry substance) were suspended in water in the flasks placed in a laboratory shaker for the leaching tests,respectively. The leaching solutions were acidified with H_2SO_4 until a stable pH of 3.0 was reached for simulating an acid rain condition. The volumes of solution were adjusted to 100 ml, after which samples were drawn at selected time intervals and analyzed for Cd. The results of leaching for two days are shown in Table 3. It is shown that only traces of Cd are leached out since 0.1 ppm Cd in solution corresponds roughly to 0.2% of the total Cd present. This agrees well with the supposed formation of CdS, which resists acid attack at pH>1.

Table 3. Leaching of Cd after sulphidizing at 240°C

Initial pH of slurry	ppm Cd in leaching solution
7.0	0.12
6.0	0.11
5.0	0.10
4.0	0.10

Condition: 3 g dry Cd-bentonite in 25 ml H_2O adding 0.2 g sulphur with various initial pH of slurry at 240°C for 2 h and then 0.5 g dry substance leached in 100 ml, pH 3.0 solution for 2 days (25°C).

Hydrothermal sulphidizing experiments with varying the amounts of sulphur were carried out at a lower temperature of 150°C. The results of leaching for 2 days are shown in Table 4. It is seen that ≤0.5 ppm Cd are leached out at pH 3.0 when the amounts of sulphur added are more than 0.1 g of sulphur in 3 g of dry Cd-bentonite.

Table 4. Leaching of Cd after sulphidizing at 150°C

Adding amounts of sulphur,g	ppm Cd in leaching solution
0.05	2.43
0.1	0.50
0.2	0.40

Conditions: 3 g dry Cd-bentonite in 25 ml H_2O adding sulphur at 150°C for 2 h and then 0.5 g dry substance in 100 ml, pH 3.0 solution leached for 2 days (25°C).

The influence of hydrothermal temperature on the transformation of Cd into CdS was carried out within the temperature range of 150° to 240°C under the conditions of 3 g Cd-containing bentonite and adding 0.2 g sulphur in 25 ml H_2O for 2 h. The results of leaching for 2 days are shown in Table 5. It is seen that the Cd contents in leaching solution decrease with the increase of temperature, i.e., the increase of temperature advances the transformation of cadmium into sulphide under hydrothermal conditions.

Table 5. Leaching of Cd at 25°C after sulphidizing

Hydrothermal temperature,°C	ppm Cd in leaching solution
150	0.40
180	0.30
210	0.24
240	0.12

Conditions: 3 g dry Cd-bentonite in 25 ml H_2O adding 0.2 g sulphur at various temperature for 2 h and then 0.5 g dry substance in 100 ml pH 3.0 solution leached for 2 days.

4 Conclusion

The experimental results have been shown in this work that the hydrothermal sulphidizing in neutral or slightly acid suspension seems more suitable for insolubilizing Cd. The retention of Cd increase with the amounts of sulphur added and hydrothermal temperature (150° to 240°C). Only traces of Cd are leached out at pH 3.0 after hydrothermal treatment. It appears that this method may be adapted for fixation of Cd in contaminated waste water, soil and sludge.

5 References

Ke, Jia-Jun and Sørensen, E. (1992) Fixation of heavy metals in clay and sludge under elevated temperatures and pressures. Mineral Processing and Extractive Metallurgy Review, 9, 107-124.

Wada, K. (1985) Adsorption phenomena on soil: Exchange and adsorption of ions by clay minerals. Japanese Society of Soil and Fertillizer (in Japanese), pp. 5-57.

Weaver, C. E. and Pollard, L. D. (1973) The chemistry of clay mineral. Elsevier, Amsterdam, pp. 107-118.

Ziper, C., Komarneni, S. and Baker, D. E. (1988) Specific cadmium sorption in relation to the crystal chemistry of clay minerals. Soil Sci. Soc. Am. J., 52, 49-53.

Use of zeolitic tuff for control and recycling of effluent

P. J. Leggo
Environmental Minerals (U.K.), Linton, Cambridge, England

Abstract

The discovery of abundant deposits of sedimentary zeolites has, in the last 20-30 years, provided industry with a new technology to control and re-cycle waste pollutants.

Zeolites are naturally occurring alumino-silicates which have unique sorbtive properties that can be used extensively for adsorption and ion-exchange reactions. As silicates these minerals are very stable to changes in pH and can be cyclically regenerated either by elution or in the case of gaseous exchange, by a pressure/temperature inversion.

Although the technological applications of these minerals became apparent in the 1950's UK industry has done little to adopt and develop their use. Undoubtedly the absence of easily mined deposits in the UK has been a factor in the lack of interest and zeolite technology has mainly been pursued in overseas countries that have their own abundant supplies.

Considerable progress has been made , over the last twenty years, in the treatment of aqueous effluent. Plant has been developed that cyclically recovers ammonia-nitrogen and phosphate complexes from municipal sewerage converting it to a form that can be used as agricultural fertilizer . A similar treatment for gaseous effluent such as SO_2 and NO_x is discussed and progress in the use of zeolitic tuff to control heavy metal pollutants and chlorinated hydrocarbons is also reviewed.

The list of uses for natural zeolite has become longer each year and it is now time for UK industry to show more of an interest in these *environmentally friendly silicates.*

Key words: Uses of zeolites, agricultural uses, crystal chemistry, gas absorption, ion exchange,sedimentary zeolites, water purification, zeolite technology.

1 Introduction

In this review attention is focused on the natural zeolite minerals , their mode of occurrence and industrial applications. This family of nearly forty species of hydrated alumino-silicate has unique sorptive and diffusive properties which makes it particularly useful to industry as a means of removing and recycling waste products. Britain and in general Western Europe have fallen behind countries such as Japan,U.S.A, and some of Eastern Europe in the use of zeolite technology, although British mineralogists and chemists have made valuable contributions in its understanding. This paradoxical situation has developed as a result of the lack of interest shown by the British minerals industry probably fostered by the fact that zeolite deposits suitable for economic extraction are not found in the UK and at the moment are only known to occur in few countries throughout Western Europe.

Although zeolites have been known for over two hundred years their value as chemical absorbents was only fully realized in the 1950's. In the last forty years zeolite technology has been developed on an industrial scale in the USA and Japan and plants recycling waste gases produce by-products such as oxygen, carbon dioxide,argon, ammonia and sulphuric acid .

Considerable progress has also been made in the treatment of industrial aqueous effluent. As early as December 1971 the Japanese soap and detergent industry employed ion-exchange methods using zeolitic tuff to remove ammonia-nitrogen complexes from waste water which results in a greatly enhanced purification over conventional sand filtration. This same technology is used in the U.S.A for the extraction of ammoniacal nitrogen from sewage and agricultural effluent with the profitable recycling of the by-products.

Having mentioned a few of the broader aspects it is important to state some of the more specialized uses of the mineral . Radioactive caesium and strontium from the low-level waste streams of nuclear power generating plants can be removed effectively and heavy metal ions can likewise be extracted from polluted waste water. Zeolites have been employed as additives to soil to control moisture content and malodour from the waste products of farm animals. Direct addition to chemical fertilizers improves the nitrogen retention of soils by promoting a slower release of ammonium ions and phosphate up-take can be controlled in the same way. Similarly, heavy metals such as copper, lead, zinc and cadmium can be ion-exchanged from soils by natural zeolite additives limiting the extent to which these elements are transferred to the food chain.

All these environmentally advantageous processes can be made readily available provided an economic source of natural zeolite is available. Zeolite material can be relatively easily synthesized but to produce suitable material in large bulk would be prohibitively expensive. It is therefore necessary to locate cheap and accessible natural sources of the material to make zeolite cost effective as a control agent for chemical pollution .

2 Mineralogical details

The naturally occurring zeolite minerals cover a wide range of chemical and structural types. The name, derived from the ancient Greek for "boiling stone", refers to the fact that when a zeolite is heated to about 200°C water is expelled from its surface. This phenomena is due to the presence of loosely bound water molecules within the crystal lattice.

Zeolites are classified as framework silicates having a fundamental structural unit of silica, aluminium and oxygen atoms arranged in tetrahedral fashion with each silica or aluminium atom surrounded by four oxygen atoms. The structure of the framework is then formed by the mutual sharing of all the oxygen atoms at the vertices of each tetrahedra. The tetrahedra can then be thought of as being configured into secondary building units to produce different types of framework structure. Using this approach the natural zeolites can be classified into five groups having either ring or chain structures, [2] .

In the case of a zeolite the alumino-silicate framework has additional features in the form of open channels or pores filled with water molecules and exchangeable cations, and it is these features that give the mineral its unique properties, Fig.1. In chemical studies the term "molecular" sieve has been introduced to describe this structural property.

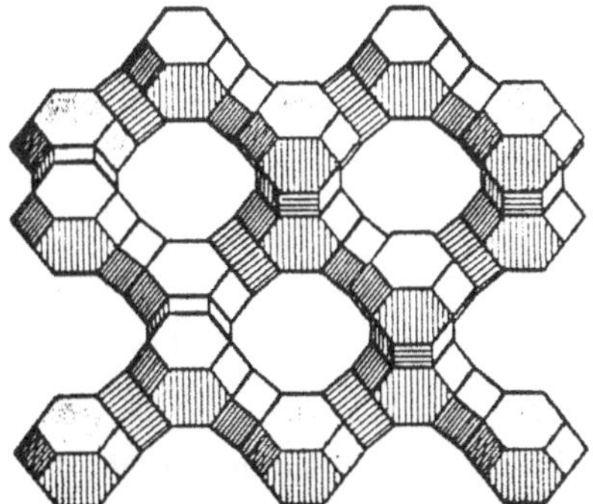

FIG.1 Alumino-silicate framework of Faujasite viewed along [110] to show the wide channels in the structure. The corners of each polyhedrón represent the centres of $(Si,Al)O_4$ tetrahedra but oxygen atoms are not represented (after Bergerhoff et al.,1958).

The term "exchangeable" refers to the exchange of cations, of alkali or alkaline earth elements (Na, K or Mg. Ca, Sr, Ba) that can take place at low temperatures, ie below 100°C, between the mineral and fluids passing through the pore channels.

While it is not the aim of this paper to cover the structural complexity and chemical diversity of the natural zeoloite minerals it is important to realize that they are a complicated mineral family and that their applications are specific to their crystal chemistry. This fact cannot be over-emphasized as much of the misunderstanding and consequent misuse of the mineral has been due to poor identification resulting in the wrong mineral being used for a particular application [3].

The early work of mineral classification was done on well formed crystals obtained from "vuggy" cavities within lava flows. Such material analyzed optically by X-ray techniques and chemical methods was used to identify thirty three different zeolite species which make up five structural groups [2]. Deposits that are economic to mine, however, are volcano- sedimentary in origin and contain zeolite crystals that are of only micron size. To identify and classify such material X-ray diffractometry and scanning electron microscopy are two fundamental tools although many other forms of analysis are used to measure and investigate their chemical and physical properties.

Table.1, gives the names together with the physical and chemical features of the natural sedimentary zeolites most commonly found in volcano-genic sediments .

Table 1. Natural sedimentary zeolites. (After Gottardi 1978)

Group (SBU)	Species name	Typical cell content	Most abundant cation
S4R	Phillipsite	$(CaO_{0.5},Na,K)_6[Al_6Si_{10}O_{32}].12H_2O$	Ca or Na or K
	Analcime	$Na_{16}[Al_{16}Si_{32}O_{96}].16H_2O$	Na
	Laumonite	$Ca_4[Al_8Si_{16}O_{49}].16H_2O$	Ca
S6R &D6R	Chabazite	$Ca_2[Al_4Si_8O_{24}].13H_2O$	Ca or Na
	Erionite	$(K_2,Ca,Mg,Na_2)_{4.5}[Al_9Si_{27}O_{72}].27H_2O$	Ca or Na or K
	Mordenite	$Na_8[Al_8Si_{40}O_{96}].24H_2O$	Ca or Na
	Ferrierite(rare)	$Na_{1.5}Mg_2[Al_{5.5}Si_{30.5}O_{72}].18H2O$	Mg
	Heulandite	$Ca_4[Al_8Si_{28}O_{72}].24H_2O$	Ca
	Clinoptilolite	$Na_6[Al_6Si_{30}O_{72}].24H_2O$	Na

Additional notes:
In Table.1 the column headed Group(SBU) refers to the classification of the alumino-silicate tetrahedral structural building units as follows:-
S4R....................Single four member ring
S6R...................Single six member ring
D6R...................Double six member ring
4-1.....................Natrolite unit structure
5-1.....................Mordenite unit structure
4-4-1.................Stilbite unit structure
For brevity details of the space group classification has been omitted. The chemical content of the typical unit cell represent ideal compositions but in nature there is considerable replacement of cations in the external positions which result in controversy over mineral names in some cases. These points are mentioned to avoid confusion in reading Table.1 .

3 Geological occurrence

Zeolite minerals are found in volcanic rocks and sediments derived from such rocks. The species and quantity of zeolite present in such rocks depend to a large extent on the chemical characteristics of the host rocks and the physio-chemical conditions of the depositional environment. Features such as the pH of the circulating groundwater ,texture and temperature of the host rock environment and age of the deposit are all of fundamental importance.

Although beautifully crystallized, macro-size, zeolites minerals are commonly found in lava flows derived from the interaction of circulating groundwater with the cooling rock. All economic zeolite deposits are found in volcanogenic sediments.

Fine grained volcanic ash and glass shards are the basic source materials of such deposits and in some circumstances thick accumulations, measured in tens of meters, of near mono-mineralic zeolite have been discovered. Recent geological studies have shown that the most likely process of formation involves the deposition of hot volcanic glass fragments in hydrologically restricted or closed systems. This condition is satisfied when volcanic ash is deposited in saline water either in land-locked basins or a shallow marine environment, [4] .

Research studies have shown that apart from the chemical characteristics of the depositional environment secondary effects due to changes in temperature and pressure during burial result in alteration and zonation of zeolite species as thermodynamic equilibrium is approached. This secondary behaviour is time dependent but as a general rule the more valuable economic deposits are Quaternary in age, ie. found in the most recent geological formations as disequilibrium low temperature assemblages. In the countries of Western Europe, Italy stands alone in having important zeolite deposits due to extensive "modern" volcanicity and the presence of saline depositional conditions. Similar geological environments also exist to a lesser extent in Germany and France and possibly new examples will be discovered in other Western European countries.

In comparison Eastern European countries such as Hungary, Slovakia, Bulgaria, Yugoslavia, Greece and Turkey have many economic deposits some of which are currently being mined by open-cast methods and produce excellent material for industrial use.

Nevertheless much geological and mineralogical work is still needed before the best sources can be identified for future use in British industry.

4 Industrial applications

The industrial uses of natural zeolites, over the last twenty years, can be grouped into three broad categories:-

1. Gas absorption.
2. Water purification.
3. Agricultural uses.

4.1 Gas absorption

Natural sedimentary zeolites, ie. clinoptilolite, mordenite and ferrierite have a higher resistance to acid conditions, being stable at pH=2, than synthetic zeolite. This property makes them very valuable in the absorption of stack gases such as SO_2, CO_2, H_2S and NO_X. Table 2. lists specific zeolites and their uses as gas absorbers.

Experimental work carried out at the Kasaka Smelting Plant, Akita Prefecture, Japan using clinoptilolite to absorb SO_2 from waste gases has shown that high recovery can be obtained by "the temperature-swing" process, [5] . In this process the input gas carrying some 1700 ppm SO_2 is passed over dry clinoptilolite which results in nearly 100% adsorption of the SO_2. The charged clinoptilolite then falls down through a pre-heating zone after which all the SO_2 is removed from the zeolite by raising the temperature to near 290°C. After desorption the re-activated mineral is removed to a stock vessel by an air lift and then recycled. This and similar processes are now being developed and used in other countries for the recycling of the products shown in Table 2, [6] .

Table 2. Gas Purification (after Tsitsishvili, 1988) [5]

Zeolite	Use
Chabazite	Removal of CO_2, SO_2 and H_2S from waste gases.
Clinoptilolite	Removal of H_2S from waste gases.
Clinoptilolite	Removal of H_2O from SO_2.
Mordenite	-ditto-
Mordenite	Removal of NH_3 from crude gas (product of coal gasification).
Chabazite	-ditto-
Phillipsite	-ditto-
Ferrierite	-ditto-
Clinoptilolite	-ditto-
Clinoptilolite	Removal of NOX from waste gases.
Mordenite	Removal of Nitrosyl from Chlorine.

4.2 Water purification

The ion-exchange properties of natural zeolites have long been used to soften hard water by the removal of calcium ions but in recent years more attention has been paid to the selective absorption of other ionic species. Successful experiments in the US and other countries have shown that ammonium and phosphate ions can be removed and re-cycled from wastewater. In Hungary pilot plants have been developed that can remove 90% ammonium ion (NH_4^+) from drinking water, with a capacity of 50 meters^3/day, using columns packed with clinoptilolite. Cyclic processes have also been developed in which ammonia is air-stripped from a charged zeolite (Chabazite or Phillipsite) in which condition the mineral is then available to absorb phosphate ions. The cycle is completed by eluting the phosphate with a sodium chloride solution and after further washing with fresh water the ion-exchange unit is reused in the next cycle. In this case the cost of water purification is estimated at 1 p / meter^3 and the recovered material can be recombined as magnesium ammonium phosphate and sold as a high quality chemical fertilizer, [7] .

Italian research has shown that volcanic tuff rich in Phillipsite, ie."Neapolitan yellow tuff", has a remarkable selectivity for the (NH_4^+) ion even in the presence of varying concentrations of interfering cations (Na,K,Mg and Ca), [8]. In this case the origin of the pollutant arises from the use of ammonium salts in leather tanning and is removed by passing the wastewater through a fixed bed of zeolite which was found preferable to the direct addition of zeolite to the effluent.

The use of natural zeolite in the treatment of municipal wastewater to remove NH_4^+ is developed on a large scale in the USA, [9]. Installations operating at Tahoe-Truckee, California, Upper Occoquan, Virginia and Denver, Colorado are well known examples. On these sites clinoptilolite is employed to remove NH_4^+ which is recovered as a 40% solution of ammonium sulphate which is of agronomic value.

Similar industrial process in Japan, Hungary and Italy have been developed to exchange NH_4^+ and phosphate ions (PO_4^{3-}) from sewage effluent which substantially reduces its COD (chemical oxygen demand). Considerable work has also been conducted on the extraction of heavy metal ions from aqueous solution. Results have shown that clinoptilolite is an excellent sorbent for such purposes and Bulgarian published work on simultaneous sorption kinetics in multi - component systems, ie. aqueous solutions containing lead, cadmium, zinc & copper (Pb,Cd,Zn & Cu) ions, established the affinity sequence on Na-exchanged clinoptilolite as $Pb^{++} > Cd^{++} > Zn^{++} > Cu^{++}$. Although still at an early stage of development it is possible, once having the equilibrium and kinetic information, to design sorption apparatus to work under a range of experimental conditions and to extrapolate the results for practical purposes.

The high ion-exchange selectivity and resistance to radiation damage have made zeolites such as clinoptilolite, mordenite, erionite and chabazite very useful for the separation and purification of ^{137}Caesium(Cs), from radioactive wastewater. Other long-lived fission products such as ^{90}Strontium(Sr), ^{106}Ruthenium(Ru), ^{129}Iodine(I) and ^{144}Cerium(Ce) have also been studied and it has been found that traces of ^{137}Cs & ^{90}Sr can be 99% removed by passing contaminated wastewater through zeolite filter beds. This technology has now been in use for over twenty years in the nuclear industry for the treatment of low- and intermediate-level radioactive waste and some 100,000 tons has been used for such purposes at Chernobyll.

A novel use for natural zeolite has been developed in Azerbaijan,SSR,USSR in which chemically modified clinoptilolite and mordenite are used to remove organic chlorine compounds such as :- trichlor ethylene, dichlor ethane,dichlor ethylene, chloroform and others from industrial wastewaters. Zeolites modified by methyl amine hydrochloride are found to have increased sorptive properties for organic chlorine compounds of some 35 to 40%. Natural zeolites of the type used appear to have a valuable role to play in controlling the release of such toxic compounds into the environment.

4.3 Agricultural uses

Natural zeolites have been found to have a variety of uses in such areas as:- soil conditioning, animal husbandry, poultry and fish-breeding.

Again clinoptilolite is a mineral which has due to its high ion-exchange and retention ability and its absorptive affinity for water been successfully applied to arable farming. Current agrochemical techniques involving addition of highly soluble chemical fertilizers invariable result in contamination of surface water. This effect has occurred on a regional scale in areas such as East Anglia and other areas of intense arable farming throughout Western Europe to the detriment of the environment. Top soil dressing with zeolitic tuff containing clinoptilolite and mordenite has been found to prevent rapid loss of fertilizer with significant increase in crop yields, [10]. On the introduction of zeolite soil moisture contents are stabilised throughout periods of low rainfall and the use of chemical fertilizers can be kept to a minimum. This approach has proved to be very economic in countries which experience annual dry conditions and is now being used in Australia. The slow release of water, nitrogen, and phosphorus from zeolite treated soil will offer advantages to Third World countries which have dry climatic conditions and will benefit the Western World in limiting the use of unnecessary chemical fertilizers.

5 Market Factors

At present, despite the versatility and huge market potential natural zeolite has yet to find a market in the U.K. Synthetic zeolite, on the other hand, is a well established product being used for a range of specific purposes, eg. as catalysts in the hydrocarbon industry and as a builder in phosphate-free laundry detergents, and is continuing to find new applications.

In the last eight years the consumption of zeolites in detergents, due to the ban and limitation of use of sodium tripolyphosphate (STPP) in many countries, has increased some 500% and is destined to continue this trend into the future.

In terms of capacity (tonnes per year) European synthetic zeolite producers supply 760,000 tonnes in comparison to 316,000 tonnes produced in the Western Pacific Region which reflects the European anti-phosphate legislation. However, the demand is bound to grow as environmental attitudes become more widespread.

Although no reliable production statistics exist for natural zeolites the world production in 1986 was estimated to be in the order of 300,000 tonnes valued at some US$33 million.

It is unlikely to be very much higher today but research now being done in mineral beneficiation and treatment is likely to produce a product equal and better in some cases to synthetic zeolite.

The small market in the natural material is due, to a large extent, to the lack of control on the quantity and quality of the material and are both factors which have undermined confidence in its use. With the extension and development of zeolite technology cost will become more of a factor and attention will become focus on natural materials.

6 Conclusions

As natural silicates the zeolites are unique in their ion-exchange and sorptive properties and clearly their use in controlling environmentally damaging pollutants should not be ignored. The problem in their ready use results from the mixed nature of the available deposits as the source rocks are volcanic tuffaceous sediments which can contain more than one type of zeolite together with rock fragments, glass shards and other alumino-silicate minerals. The zeolite minerals in such deposits are commonly cryptocrystalline and need mineralogical identification and examination of their crystal chemistry . But having made this comment the problem is no different to that which faces the clay industry and can be solved with the same laboratory facilities as those routinely employed by this industry.

The philosophy of re-cycling chemical pollutants, where possible, could offer an attractive incentive to water companies who have the problem of water treatment and it would seem worthwhile to study the feasibility of such operations for use in the U.K.

References

1. Bergerhoff,G.,Baur,W.H. & Nowacki,W. (1958) **Uber die Kristallstruktur des Faujasites.** Neues Jahrbuch fur Minerallogie.Monatshefte, p.193.
2. Gottardi,G. (1978) Mineralogy and Crystal Chemistry of Zeolites. **Natural Zeolites,Occurrence,Properties,Use.** Pergamon Press,Oxford, pp.31-44.
3. Mumpton,F.A. (1988) Development of Uses for Natural Zeolites; A critical commentary. **Occurrence Properties and Utilization of Natural Zeolites.** Akademiai Kiado,Budapest,1988. pp.333-366.
4. Sheppard,R.A.,and Grude,A.J.(1968) Distribution and Genesis of Authigenic Silicate Minerals in Tuffs of Pleistocene Lake Tecopa , Inyo County,California. **US Geological Survey Professional Paper 597**,1968.
5. Minato, H.(1988) Occurrence and application of natural zeolites in Japan. **op.cit.** Akademiai Kiado,Budapest.1988. pp.395-418.
6. Tsitsishivili,G.V.(1988) Perspectives of Natural Zeolite Applications. **op.cit** Akademiai Kiado,Budapest,1988.pp.367-393.
7. Ciambelli.P.,Corbo.P.,Liberti,L.,and Lopez,A.(1988) Ammonium recovery from urban sewerage by natural zeolites. **op.cit.** Akademiai Kiado,Budapest.1988. pp.501-509.
8. Nastro,A and Colella,C.(1983) Column ion exchange data for ammonium removal from water by phillipsite tuff. **Ing.Chim.Ital.19,** 1983. pp.41-45.
9. Mercer,B.W., Ames,L.L.Jr., Touhill,C.J., Van Slyke,W.J., and Dean,R.B.(1970) Ammonia removal from secondary effluent by selective ion exchange. **Journal of Water Pollution ,Control Fed.42,** R95-R107.
10. Leonard,D.W.(1979) The role of natural zeolites in industry. **Association of Mechanical Engineers ,Fall Meeting,** Tucson , Arizona,1979.

NO_x suppression with hydrogen peroxide in the metals industry

C. F. McDonogh
Solvay Interox Research and Development, Widnes, England

Abstract

Metal dissolution processes involving nitric acid are major sources of NOx. Technologies using hydrogen peroxide have been developed and proven on an industrial scale to suppress the emission of NOx from these processes. No changes are made to the dissolution process. Case histories are presented to illustrate the effectiveness of hydrogen peroxide with particular reference to the pickling of stainless steels.
Keywords: Hydrogen Peroxide, Nitric Acid, NOx Suppression

1 Introduction

There has been a considerable increase in recent years in the legislation affecting the discharge of effluents from industrial processes. The Environmental Protection Act of 1990 in the U.K. for example, increased the responsibility of Her Majesty's Inspectorate of Pollution (H.M.I.P) to the regulation of 200 categories of industry, 5000 major industrial plants and 8000 premises.

Concerning the use of nitric acid, guidelines issued by H.M.I.P. [1] in August 1993 indicated a maximum NOx emission limit of 300 mg/m^3 (≡ 160 ppm), irrespective of whether the emissions originated from metal dissolution processes, stainless steel pickling, or nitric acid manufacture.

2 NOx generation

When nitric acid reacts with a metal it is reduced to nitrous acid (HNO_2), which is, in turn, in equilibrium with a mixture of nitrogen oxides. The principal components of the mixture are NO, nitrogen monoxide, and NO_2, nitrogen dioxide, the mixture itself being referred to as NOx.

$$Fe + 4H^+ + NO_3^- \rightarrow Fe^{3+} + NO + H_2O \quad 1$$
$$Fe + 6H^+ + 3NO_3^- \rightarrow Fe^{3+} + 3NO_2 + H_2O \quad 2$$

Nitrogen dioxide is a brownish-red toxic gas with a pungent odour. It is in equilibrium with its dimer, dinitrogen tetroxide, N_2O_4, the equilibrium position being temperature dependent and shifting almost all the way to nitrogen dioxide at 100°C. At temperatures above 150°C it dissociates to nitrogen monoxide and oxygen. It has a toxic limit value of 5 ppm and a lethal dose of 200 ppm.[2]

Nitrogen monoxide is colourless and odourless with no irritating effects. It reacts, however, with haemoglobin resulting in cyanosis. It has a TLV of 25 ppm.

NOx itself is involved in the production of 'photochemical smog' and is a major air pollutant. It has been held partially responsible, along with its sulphur analogue SOx, for the production of acid rain.

3 NOx emission reduction techniques:

Four basic approaches are currently in use to reduce NOx emissions to the atmosphere. These are:

1. Reduction [3][4]
2. Scrubbing [5][6][7][8]
3. Adsorption [9][10]
4. Suppression [11][12]

3.1 Reduction:

Reduction can be non-selective or selective and is normally restricted to major sources of potential NOx emissions such as nitric acid production units. The principle is the addition of a fuel to the NOx gases, which are then passed over the surface of a catalyst to form nitrogen and water vapour. The method is very efficient, but the capital costs usually prohibit its use in metal dissolution and pickling processes. There are two types, selective and non-selective. Non-selective catalytic reduction involves the addition of hydrogen, or a hydrocarbon such as natural gas, to the NOx, heating the mixture to >200°C (>480°C in the case of natural gas) and passing the mixture over a catalyst, usually platinum deposited on a ceramic sponge. NOx reduction efficiencies of 70-90% are achieved. A problem with the technique is that the process is non-selective, ie. all oxides, and oxygen is reduced in the process, leading to high fuel consumptions. The selective catalytic reduction process uses ammonia to reduce NOx to elemental nitrogen at temperatures in excess of 250°C.

3.2 Scrubbing

Scrubbing the emitted gases is probably the most common form of NOx removal practiced in metal dissolution and metal pickling processes. The gases are passed, counter-current, down a tower through which absorbing liquor is cascaded. The liquor composition can vary from simply water, through sodium hydroxide solution, to exotic mixtures of sodium thiosulphate and potassium permanganate. The scrubbing method is generally quite effective at removing NOx from gas streams, but is generally ineffective at reducing the emission levels to below the current consent limit. A further problem is the formation of solutions of nitrous acid or nitrites.

$$NO + NO_2 \rightarrow N_2O_3 \quad (3)$$

$$N_2O_3 + NaOH \rightarrow 2NaNO_2 + H_2O \quad (4)$$

$$2NO_2 + 2NaOH \rightarrow NaNO_2 + NaNO_3 + H_2O \quad (5)$$

These generally require some form of treatment prior to discharge. One method of scrubbing NOx emissions without generating an effluent is with hydrogen peroxide. This oxidises NOx to nitric acid, which can be returned to the process.

3.3 Absorption

Adsorption processes involve the use of molecular sieves to adsorb NOx. Although the process can be used commercially it has found little favour in industry and is rarely encountered.

3.4 Suppression

Suppression of NOx emissions from taking place is obviously the most efficient method of controlling pollution, and is preferable to any bolt-on, end-of-pipe treatment to an existing process.

There is a direct relationship between the nitrite concentration of a solution and the concentration of NOx emitted from that solution, as shown in figure 1. The addition of a chemical which will reduce the nitrite concentration in the liquor will therefore reduce the amount of NOx emitted by that liquor. There are two principal methods of carrying out this reaction, one is by the addition of urea to the bath, and the other the addition of hydrogen peroxide.

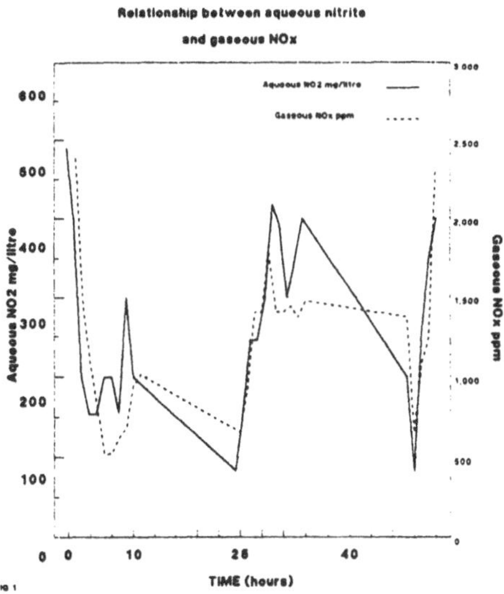

FIG 1

NO_2

$$2NO_2\,(g) \rightleftharpoons N_2O_4\,(g) \rightleftharpoons N_2O_4\,(aq)$$

$$N_2O_4\,(aq) + H_2O \rightarrow HNO_2 + H^+ + NO_3^-$$

$$HNO_2 + H_2O_2 \rightarrow HNO_3 + H_2O$$

$$2NO_2 + H_2O_2 \rightarrow 2HNO_3$$

NO/NO_2

$$NO + NO_2 \rightleftharpoons N_2O_3\,(g) \rightarrow N_2O_3\,(aq)$$

$$N_2O_3\,(aq) + H_2O \rightarrow 2HNO_2$$

$$2HNO_2 + 2H_2O_2 \rightarrow 2HNO_3 + 2H_2O$$

$$NO + NO_2 + 2H_2O_2 \rightarrow 2HNO_3 + H_2O$$

NO

$$NO\,(g) \rightarrow NO\,(aq)$$

$$NO\,(aq) + H_2O_2 \rightarrow NO_2\,(aq) + H_2O$$

$$3NO_2 + H_2O \rightarrow 2HNO_3 + 2H_2O$$

$$2NO + 3H_2O_2 \rightarrow 2HNO_3 + 2H_2O$$

3.4.1 Reaction with Urea

Urea reacts with nitrous acid to form carbon dioxide and nitrogen.

$$CO(NH_2)_2 + 2HNO_2 \rightarrow CO_2 + 2N_2 + 3H_2O \qquad 6$$

There is a further reaction, however, which urea undergoes in hot aqueous solution. It is hydrolysed to ammonia, which then reacts with nitric acid to form ammonium nitrate.

$$CO(NH_2) + H_2O \rightarrow CO_2 + 2NH_3 \xrightarrow{2HNO_3} 2NH_3\,(NO_3) \qquad 7$$

The overall reaction of adding urea to a nitric acid solution is therefore a reduction in nitrite concentration (and hence NOx emissions) and an increase in nitric acid consumption.

3.4.2 Reaction with Hydrogen Peroxide

Hydrogen peroxide reacts with nitrous acid to form nitric acid. It also reacts with other nitrogen oxides as shown in Figure 2.

$$HNO_2 + H_2O_2 \rightarrow HNO_3 + H_2O \qquad 8$$

As calculated by Halfpenny & Robinson [13] in acidic solution the reaction rate

$$\begin{aligned} R &= -d[H_2O_2] \\ &= K[H_2O_2][HNO_2][H^+] \\ &= 8.4\text{x}10^3 \text{ mol/l/min at } 15°C \end{aligned}$$

NOx removal techniques

Treatment	Catalytic reduction by NH_3	Gas Scrub. by NaOH soln.	Gas Scrub. by H_2O_2	Addn. to bath H_2O_2	Addn. to bath $CO(NH_2)_2$
Capital Cost	very high	high	high	low	low
Variable Cost	low	low	high	high	low
HNO_3 Usage	no influence	no influence	lower	lower	higher
NO_x Redn.	very high	high	very high	very high	low
By-prods	no problem	difficult to discharge	rec	rec	no prob
Surface Quality	no influence	no influence	no influence	better	worse

fig 3

Hydrogen peroxide undergoes catalytic decomposition in the presence of transition metals, its stability is also reduced at elevated temperatures.

$$2H_2O_2 \rightarrow 2H_2O + O_2 \qquad 9$$

Thus, as there are no nitric acid consuming side reactions, the use of hydrogen peroxide should prove to be beneficial provided the decomposition reaction can be minimised.

A summary of the NOx removal methods is shown in Figure 3.

The use of hydrogen peroxide in NOx suppression from metal dissolution processes can be illustrated by the following four examples. The first three use stainless steel pickling processes as examples, whilst the fourth uses a chemical milling process.

4 Stainless Steel Pickling

A brief description of a typical stainless steel pickling process would explain the techniques adopted. When stainless steel is processed, ie. whenever work carried out on it, stresses are built up in the metal. These stresses are created whether the metal is being rolled into sheet, drawn into wire, or bent into tube and the normal method of relieving these stresses is by annealing, or heating in a furnace. Heating the metal in an annealing furnace not only relieves the stresses, but also oxidises the metal surface. An oxide layer is built up consisting principally of three types of iron oxide, these being Wustite, Magnetite and Haematite (FeO, Fe_3O_4 and Fe_2O_3). A second reaction also occurs which results in the migration of chromium from the surface of the steel into the oxide film. This results in a chromium depleted layer beneath the oxide layer. This oxide film and chromium depleted layer are removed by dissolving them in acid, and this dissolution stage is termed pickling. A variety of pre-pickling treatments may be applied to the steel, such as electrolysis in acid sulphate and sodium dichromate, scale crushing, grit blasting or shot blasting, but the final acid pickling bath is almost universally used. The bath consists typically of 10-15% nitric acid and 2-4% hydrofluoric acid. In the case of tubes, pickling can be carried out at ambient temperature, as tubes are conventionally pickled by batch processes in open tanks, where a lower temperature and longer times reduce the rate of emission of NOx fumes, whereas coil or sheet pickling is carried out continuously at 45-70°C in closed tanks. A typical processing route is shown in figure 4.

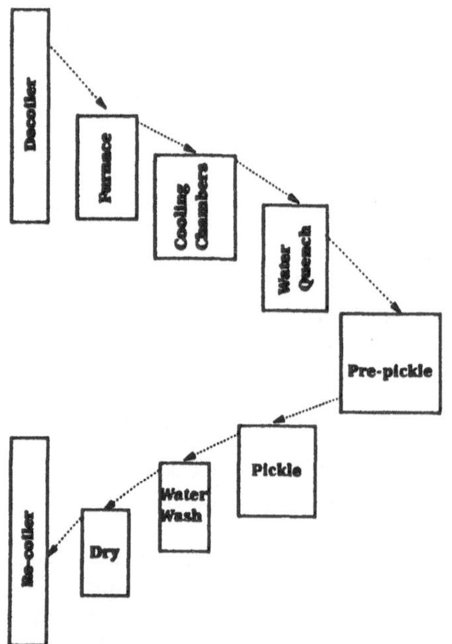

figure 4 Typical anneal and pickle route for stainless steel

Austenitic steels are generally endothermic in pickling character, whilst ferritic steels are exothermic due to the differences in chemical composition. It is usual, therefore to include some form of temperature control in the bath. The normal method is to incorporate a recirculation loop, whereby the bath liquor is recirculated through either a heater or a cooler, the recirculation speed being up to ten bath changes per hour. It has been stated that the injection of hydrogen peroxide into this recirculation loop would provide maximum mixing efficiency[14] and hence minimum decomposition, and trials in three steel mills in Europe have shown this to be the case.

4.1 Peroxide Injection into a Recirculation Loop

During the trials NOx emissions were measured at the entry point to the scrubber using a conventional Du Pont or Signal NOx analyser. Care was taken to ensure complete removal of hydrogen fluoride prior to the sample entering the analyser to minimise fogging of the optics. A 35% solution of hydrogen peroxide was then injected into the recirculation line after the recirculation pump. The injection point was positioned close to the actual tank and a non-return valve installed to prevent back-mixing of pickling liquor in the hydrogen peroxide line. The initial injection rate was 1.6 litres/minute into a 15% HNO_3/3% HF solution at 60°C contained in a 25 m^3 bath with a liquor recirculation speed of approximately 35 m^3/hour. NOx measurements were taken at 20-30 minute intervals and the hydrogen peroxide injection rate slowly reduced. It was noted that, in this particular trial, there was an apparent time lag of 5 hours between making any change to H_2O_2 injection rate and observing a change in NOx emissions. This was probably due to the speed of pickle liquor recirculation as well as the normal time lag which would be expected from adding a relatively small volume of one reactant to a very large volume of a second reactant. The issue was also clouded by the variations in steel types being processed, these being 302, 304, 310, 316 and presented in random order to the annealing line, the type of steel having a considerable effect on the amount of NOx liberated. The results are shown in Figure 5.

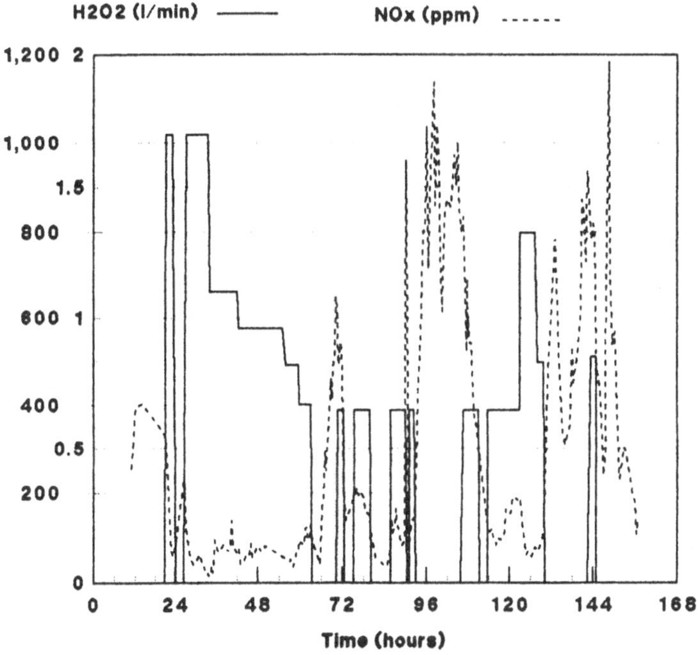

fig 5 Effect of H2O2 addition on NOx emission

For this bath configuration it was found that the optimum hydrogen peroxide injection rate was 0.7 litres hydrogen peroxide (35%) per minute, this achieving 80% NOx suppression efficiency with a final NOx concentration of 150-200 ppm being emitted to the scrubber. A crucial part of the apparatus design proved to be the position of the non-return valve. (NRV) When the NRV was positioned one metre away from the recirculation line pickle liquor reacted with the hydrogen peroxide to cause decomposition, as seen from the photograph in Figure 6. Simply moving the NRV up to the recirculation line resulted in increasing the NOx suppression efficiency from 50% to 80% without increase in peroxygen consumption.

Fig 6 Decomposition caused by poor NRV positioning.

4.2 Injection via Sparge Pipe

The second example is of a stainless steel pickling line which was not fitted with a recirculation loop. The steel being treated was ferritic, mainly 410 and 430 grade. Pickling of this type of steel is normally carried out at 45-50°C due to its reactive nature, the reaction being exothermic and NOx emissions of 2500-3000 ppm being not uncommon. A simple sparge pipe, made from 1" diameter polypropylene tubing with 1/8" holes drilled at 6" intervals was inserted in the bath. Due to the large amount of insoluble scale experienced on this line the tube was arranged with the holes pointing at 45° downwards to the horizontal. The sparge pipe was positioned at the steel entry end of the bath, just low enough beneath the steel sheet to prevent accidental collision with the sheet. Again, hydrogen peroxide was injected initially at 1.6 l/minute and reduced incrementally until the desired NOx emissions were found. Results are shown in Figure 7. Both of these trials resulted in commercialisation of the technology.

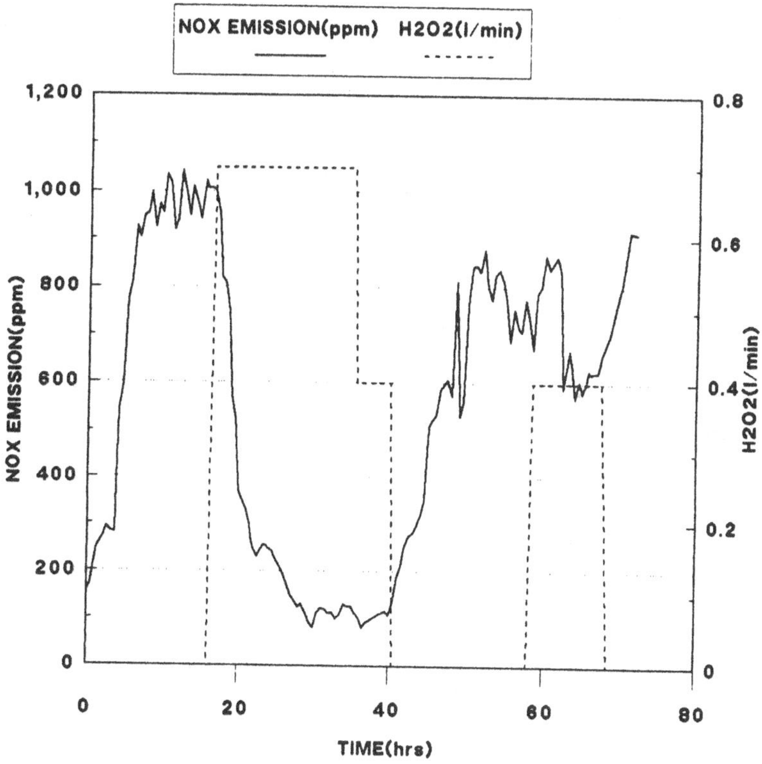

Fig 7 Line without a Recirculation Loop

NOx Emission vs H2O2

4.3 Injection into a Spray Pickling Facility

The third example is that of NOx suppression from a stainless steel plate pickling line. In the case of plate pickling the steel plate is passed through a spray chamber, the pickling liquor being sprayed onto both sides of the plate. The pickle liquor, which consists of 15% nitric acid and 3% hydrofluoric acid, is returned from the spray facility to a sump, from where it is recycled to the spray chamber. A pilot trial was carried out at this facility by injecting hydrogen peroxide into the liquor recirculation line immediately before it split into two branches. It was found that NOx emissions were reduced within 10 minutes of peroxide being introduced, indicating that the NOx suppression reaction took place at the site of NOx generation.

5 Chemical Milling of Superalloys

Chemical milling is the term applied to the dissolution of discrete parts of a metal casting in order to produce a complex-shaped item. The technology is used in the aerospace industry, for example, for the production of components from superalloys. A typical chemical milling bath might consist of a solution containing 30-40 g/l HCl, 100-120 g/l HNO_3, 10-30 g/l HF and contain transition metal nitrates and chlorides and is operated at >60°C.

The injection of hydrogen peroxide into this system is equally effective at suppressing NOx as it is in stainless steel pickling. However more consideration of the system was required prior to the trial taking place. It is well known that acidic solutions of chloride and nitrates have the potential to form nitrogen trichloride under certain conditions. The addition of hydrogen peroxide increases the potential and hence considerable laboratory work was necessary in order to investigate the customer's actual operating conditions and quantify the risk of NCl_3 formation. Although in this particular case it was found that the risk was minimal, this does serve to illustrate the point that hydrogen peroxide can be an oxidising agent, a reducing agent, or a complexing agent depending on reaction conditions. These potential side reactions must, therefore, be considered when assessing the overall process of NOx suppressions.

6 Nitric Acid Replacement

Nitric acid contributes an acid and an oxidant to the metal dissolution process. Theoretically, therefore, the replacement of the acid contribution by another acid, for example sulphuric acid, and the oxidant contribution with hydrogen peroxide, will result in a nitric acid-free metal solubilising solution. This has been successfully achieved in both stainless steel pickling and in other metal dissolution processes.

Examples of nitric-free stainless steel pickling liquors are to be found in the UG3P process invented by Ugine Guegnon of France,[15] and in the S333 process of Solvay Interox.[16] These processes rely on the oxidising nature of the ferric ion, and the ferric ion content of the pickling liquor is maintained at a minimum concentration of 15 g/l. This is achieved by the introduction of an oxidising agent to oxidise the ferrous iron formed during the pickling reaction to ferric. The oxidising agent chosen is normally hydrogen peroxide, as this does not introduce any foreign ions into the bath, and the method of addition is essentially the same as that used in NOx suppression techniques. The actions of HF/H_2O_2 mixtures on stainless steel is considerably less aggressive than that of HF/HNO_3, and hence additives are also added to the bath, such as nonionic surfactants, stabilisers to prolong the life of the hydrogen peroxide, etc. Although considerable progress has been made in the development of nitric-free pickling techniques for stainless steel since the mid '80's considerable scope for development still remains, and the majority of stainless steel is still pickled by the conventional HNO_3/HF process.

Other metal dissolution processes which have been developed by Solvay Interox include lead dissolution and platinum group metals dissolution, previously carried out in either nitric acid or aqua regia but now possible in hydrogen peroxide based solutions.

7 Conclusions

It has been shown, on full scale trials, that hydrogen peroxide is remarkably effective in suppressing NOx emissions from metal dissolution processes involving nitric acid. It has also been demonstrated that, in some instances, hydrogen peroxide will replace nitric acid as an oxidant resulting in completely NOx and nitrate-free reactions.

1. Chief Inspector's Guidance to Inspectors Guidance Note IPR 4/11 "Processes for the manufacture or recovery of nitric acid and processes involving the manufacture or release of acid forming oxides of nitrogen H.M.S.O. 1993.

2. Lewis R.J; Sax's Dangerous Properties of Industrial Materials 1993 Van Nostrand Reinhold.

3. Publicity material issued by Steuler, Germany, on selective catalytic reduction technology.

4. Publicity material issued by Degussa AG on Desonox Technology.

5. Rosenburg H.S. et al "Control of NOx emissions by Gas Stack Treatment". EPRI FP-925 (1978).

6. "Urea as a pollution controlling Agent" Nitrogen 95 (1975) 32 published by British Sulphur Corporation.

7. Adrian J C et al "A process for reduction of NOx Content in Flue Gas" Paper presented at 2nd International Conference on the Control of Gaseous Sulphur and Nitrogen Compound Emission, University of Salford, April 1976.

8. "Pollution Control" Nitrogen 95 (1975) 44 published by British Sulphur Corporation.

9. Buck B J et al (1976) Proc. Environ.Symp. 157-168.

10. German patent Application DE3226840 A1.

11. Karlsson H.T et al "Control of NOx in Steel Pickling". Environmental Progress 3 (1) 40-43.

12. European Patent 0267 166 B1.

13. J.Chem.Soc. 1952 (2) 928-938.

14. Swedish Patent 8305648-1.

15. French Utility Certificate 2 551 465

16. International Patent Application WO 93/08317.

Application of fibrous ion exchangers in air purification from acidic impurities

V. S. Soldatov
I. S. Elinson
A. A. Shunkevich
Institute of Physical Organic Chemistry of the Belarus Academy of Sciences, Minsk, Belarus

Abstract
Sorption of HCl, HF, SO_2 and H_2S from their mixtures with air have been studied in dynamic conditions on two types of novel fibrous ion exchangers: a strong base one of type 1 on polypropylene-styrene matrix and a weak base exchanger with amido-amino groups. Influence of the air flow rate, sorbate concentration and air flow humidity have been studied. Conditions for the ion exchangers regeneration with water or soda solutions are described. It has been found that side effects such as sorbate molecules association, water sorption and oxidation of the substances sorbed strongly influence their sorption - desorption behaviour. Characteristics of industrial air filtering plants with fibrous ion exchangers are presented.
Keywords: fibrous ion exchangers, sorption, regeneration, air purification.

Introduction

Many hydrometallurgical processes are accompanied by air pollution with volatile acids, anhydrides and aerosols. Fibrous anion exchangers can be one of the powerful means to solve this problem.

The possibility of use of ion exchangers for air purification from this type of pollutants directly follows from their properties. Sorption of SO_2, CO_2, NH_3, HF, HCl by ion exchange resins have been described in a number of publications [1-7]. Their high sorption capacity and selectivity in combination with positive effect of air humidity is a good premise for development of new air purification technologies. Nevertheless, ion exchange resins did not find wide application in this field.

Air purification technologies normally deal with large volumes of purified air. Even processes of the smallest scale require treatment of thousands of cubic meters of air per hour. It means the flow rates in these processes must be high which require very high sorption rates.

Consideration of requirements for application of ion exchange technologies for air purification shows that they may be practical only if the following parameters of the filtering plants can be achieved:

- air flow rate — >2 m/min;
- pressure drop — <800 Pa;
- protection time — >4 hours;
- regeneration time — <<protection time;
- operation volume of a middle size apparatus (10 000 m^3/h) — <1 m^3;
- filtering layer thickness — <15 mm;
- filtration area in 1m^3 — >20 m^2;
- number of working cycles — >1000;
- total exploitation time — >2 years.

Being good sorbents from a chemical standpoint, ion exchange resins have poor technological properties in relation to gaseous processes. They are usually produced in granular form with a particle size of 0.25-1 mm. This size is too large to provide sufficient sorption rates, as well as mechanical and osmotic stability. The latter is particularly important for cyclic gaseous processes because the sorption stage occurs in the gaseous phase while the following regeneration stage takes place in aqueous solutions. Frequent drying and moistening causes severe osmotic shocks disintegrating the particles. Slow sorption makes it necessary to enlarge the filtering layer thickness, which, in turn, results in increasing air flow resistance of the system. Application of macroporous resins does not save the situation. The rate of the overall mass transfer process remains too low for the gaseous processes as well as unsatisfactory osmotic and mechanical properties. There are serious technical difficulties in meeting the requirement of a large filtering area in a small volume with the possibility of periodic regeneration.

Fibrous ion exchange materials in the form of cloth or non-woven felt with a filament thickness below 50 microns can satisfy all the above requirements.

It has been shown [8] that the rate of sorption in this case is one or two orders of magnitude higher than that for industrially produced resins of similar chemical structure. These materials are elastic to some degree and have outstanding osmotic stability [9]. In a number of studies on the application of fibrous ion exchangers in gaseous ptocesses their advantages have been proved [7,10,11]. In our previous publications preparation, properties and application of the Fiban type ion exchangers have been reported [12,13].

New data on application of strong- and weak base Fiban ion exchangers for purification of air from several common acidic pollutants (HCl, HF, SO_2, H_2S) will be described in the present paper.

Experimental

Materials

The ion exchangers used were produced by Ecofil-Deco Ltd*. In the laboratory experiments they were used as staple cotton.

Strong base fibre Fiban A-1 is a polypropylene based monofunctional ion exchanger containing trimethyl benzyl ammonium groups. Exchange capacity in Cl^- form was 2.80 m-eq/g, water uptake was 0.71 g H_2O/g. The filament thickness was 40 ±5 μ.

Weak base Fiban AK-22-G is a polyacrylic fibre based ion exchanger containing the amido-amino group R-CONH($CH_2CH_2NH)_2H$ and carboxylic groups. It was computed from potentiometric titration curves of its free-base form that the contents of the first group was 3.50 m-mole/g and that of the second one was 1.5 m-eq/g. Since the anion exchanging group contained two protonised nitrogen atoms, the full anion exchange capacity was 7.00 m-eq/g.

Water uptake was 0.69 g H_2O/g in the free base form. The filament diameter was 25±5 μ.

Apparatus and procedure

A schematic diagram of the system for study of the sorption processes is given in Fig. 1.

It consists of three blocks: (a) gas flow feed; (b) gas flow conditioning; (c) sorption measurements.

The system provided constant gas flows (within ±3%) with rates up to 350 l/h (the flow rate is denoted as V m/min), sorbate concentration ranged between 3 and 1000 mg/m^3 and relative humidity P/Po was in the range of 0-0.98.

A gas cylinder *1* (in case of H_2S or SO_2) or chemical gas generator (HCl, HF) *2* were joined to the mixer *4* through a pressure stabilizer *3*. The pressure was held constant automatically and monitored with a precision manometer *5*. The final adjustments were done with a precision valve *6*.

The gas flow containing the sorbate was mixed with air flow in mixer *4* with air coming from compressor *7* through filters *8* and *9* which removed particles and oil micro-drops. A pressure drop stabiliser *10* and regulator *11* provided constant air flow. In order to prepare a gas mixture with desired humidity the gas flow was separated into two parts one of which was dried

---

* Address: P.O. Box 47, Minsk 220050, Republic of Belarus. Fax: (0172) 352543

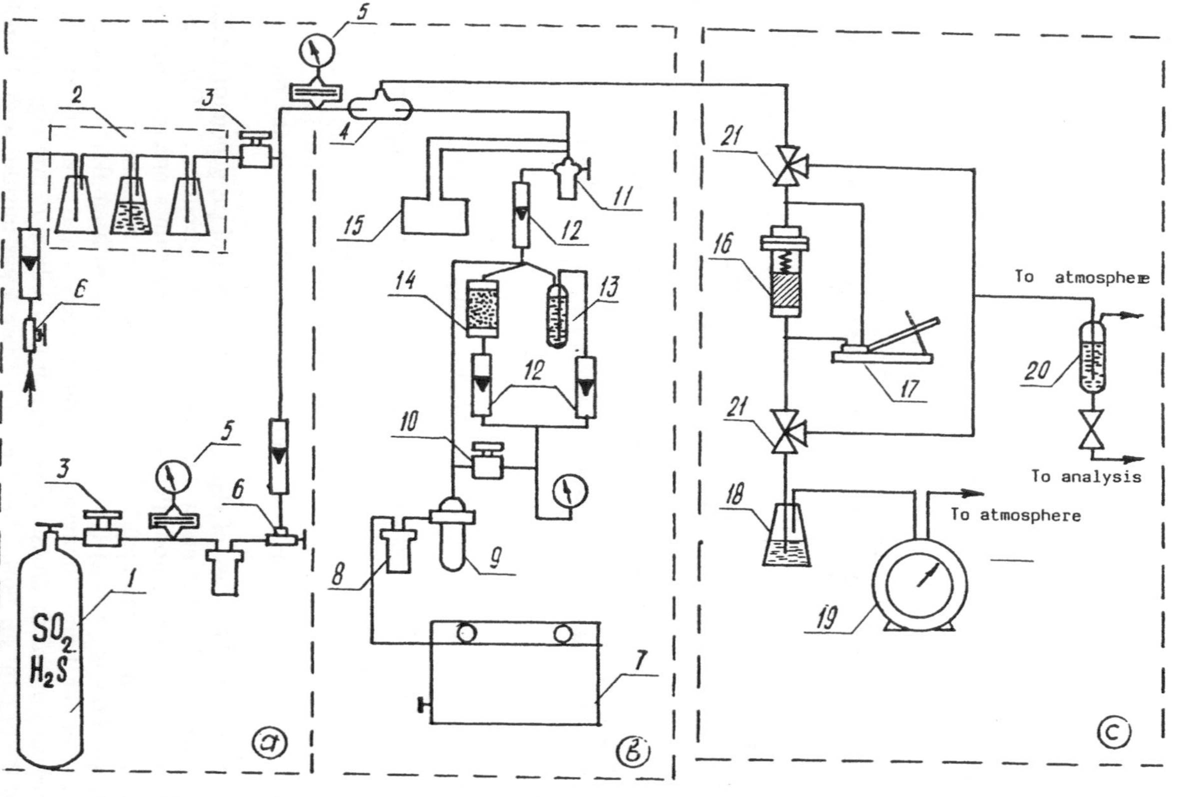

Fig. 1. Scheme of the system for study dynamics of sorption

over silica gel in dryer *14*, and the other was saturated with water vapours in bubbler *13*. The flow rate was measured with rotameter *12*. To measure the gas flow humidity humidity meter *15* was used. The dry and humid flows were mixed in desired proportions and forwarded to mixture *4* where preparation of the conditioned gas mixture was completed. The air flow from mixer *4* was forwarded to the sorption block, cell *16*, containing the fibrous ion exchanger. The aerodynamic resistance was controlled by differential manometer *17*. Coming from the sorption cell *16*, the gas passed through neutraliser *18* before release to atmosphere. Total volume of air passed was measured by counter *19*.

Concentration of the sorbates was determined in the cell outlet by analysis of the solutions obtained after passing the air through the NaOH solutions (HCl, HF), H_2O_2 in the case of SO_2, or $CdCl_2$ in the case of H_2S, in control vessel *20*.

Selective electrodes were used for determination of Cl^- and F^- in the solutions. The SO_2 concentration was determined by automatic titration of H_2SO_4 formed by the reaction $H_2O_2+SO_2=H_2SO_4$. H_2S was determined by automatic titration of HCl formed by the reaction $H_2S+CdCl_2=CdS\downarrow+2HCl$.

The filtering cell used in all experiments was a cylinder with diameter 24 mm.

The thickness of the filtration layer h was 5-15 mm, the density of the fibrous ion exchanger package was 0.1 g/cm^3. The fibrous layer in the cell was kept compressed by a spring with a porous piston. The temperature of the experiments was 18±2 C^o. In all cases before sorption experiments the ion exchangers were converted into hydroxylic or carbonate form followed by a water wash. When necessary, the ion exchangers were equilibrated with water vapour by keeping them in a desiccator over saturated salts or sulphuric acid solutions with a desired water vapour pressure. The results of the experiments were expressed as breakthrough curves with the co-ordinates: relative concentration of a sorbate C/Co-time (Fig. 2).

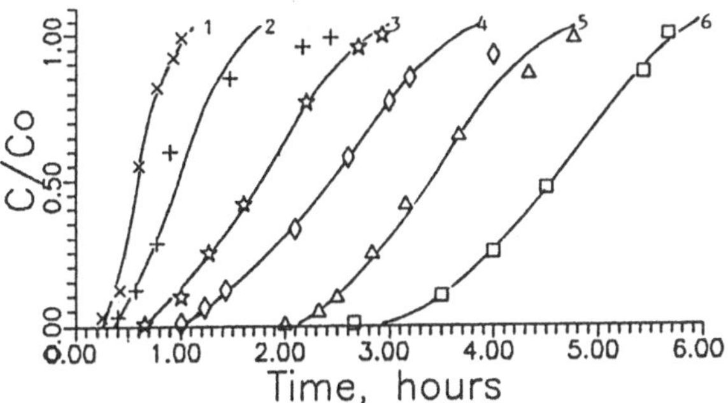

Fig. 2 Breakthrough curves for SO_2 with Fiban A-1. Conditions: V=7.0 m/min, P/Po=0.85, h=7 mm. Curves are related to different SO concentrations: 1 - 600, 2 - 400, 3 - 200, 4 - 150, 5 - 100, 6 - 50 mg/m^3.

The main parameters calculated from the breakthrough curves were dynamic activity a_d and equilibrium dynamic activity a_e. The value of a_d was computed as the content of substance absorbed per gram of dry ion exchanger before the breakthrough point which was defined as C/Co=0.05. The value a_e is the content of substance absorbed per gram of dry ion exchanger before its concentration reaches that of the initial gas flow. It was computed by integration of the breakthrough curve and was checked by the following desorption.

Results and discussion

The main parameters necessary for development of the technology of air purification and construction of the filtering plants are: (1) air flow rate, (2) concentration of the impurity and (3) air flow humidity; also, conditions of filter regeneration (4), are critical. In Table 1 the experiments done on different ion exchangers and with different substances are summarised with respect to the four parameters given below.

Table 1. Survey of experiments on sorption

	HCl	HF	SO_2	H_2S
Fiban A-1			1,2,3,4	1,2,3,4
Fiban AK-22	1,2,3,4	1,2,3,4	1,2,3,4	

It was found that sorption-desorption behaviour of each of these substances was very individual and requires separate consideration.

HCl Sorption by the weak base ion exchanger

Dependence of dynamic activities at different air flow rates is presented in Fig 3. Large differences in the dynamic and equilibrium sorption are characteritic for this system. No dependence of these parameters on the flow rate was found within range of 8-17 m/min.

The equilibrium activity corresponds to sorption of 5.5 m-eq HCl per gram of the ion exchanger.

The dynamic activity corresponds to sorption of only 1 m-eq HCl/g indicating relatively slow sorption of HCl thereafter by the ion exchanger.

Values of a_e as a function of concentration of HCl in the gas phase are given in Fig. 4. It is seen that no tendency to saturation of the ion exchanger is observed. A value 7.8 m-eq/g occurs for 700 mg HCl/m^3, which exceeds the exchange capacity of the exchanger. The super-equivalent sorption is characteristic for the acids and is possible due to association of the

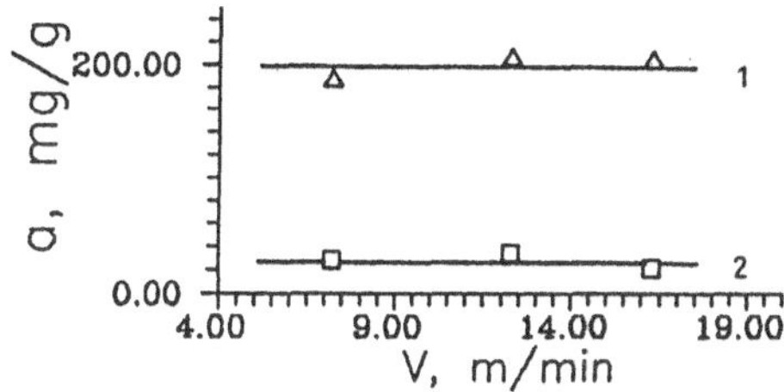

Fig. 3. Dynamic activities as a function of flow rate.
Conditions: C=51.3 mg/m^3, P/Po=0.59, h=7.5 mm. 1 - equilibrium, 2 - breakthrough. Sorption of HCl by Fiban AK-22-G.

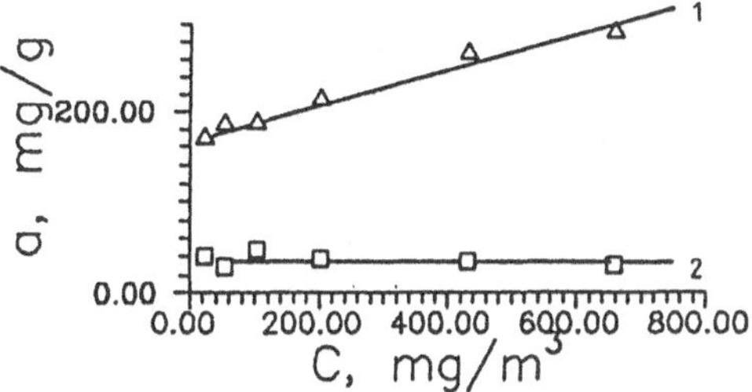

Fig. 4. Dynamic activities as a function of concentration.
Conditions: V=7.1 m/min, P/Po=0.62, h=8 mm. 1 - equilibrium, 2 - breakthrough. Sorption of HCl by Fiban AK-22-G.

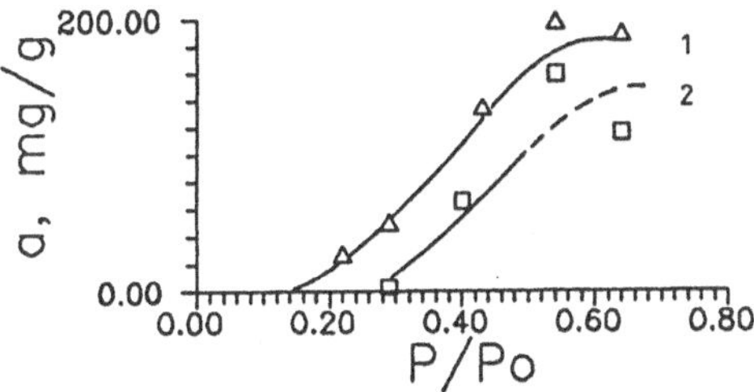

Fig. 5. Dynamic activities as a function of air humidity.
Conditions: V=7.1 m/min, C=105 mg/m^3, h=9 mm. 1 - equilibrium, 2 - breakthrough. Sorption of HCl by Fiban AK-22-G.

acid ion pairs caused by strong hydrogen bonding. In view of that, we tried to present a total amounts of acids sorbed as a sum of "strongly" and "weakly" sorbed. The weakly sorbed part can be washed out by a relatively small volume of water. In the present work we conditionally considered the acid washed out from 1 g of the ion exchanger by 150 ml of water in dynamic conditions as being "weakly" bound.

The "weakly" bound HCl is represented by easily hydrolysed Cl^--form of ion exchanger and associated HCl molecules by eqn's of the type:

$$RNH_2+nHCl \rightarrow RNH_3^+Cl^- \cdot (n-1)HCl+aq \rightarrow RNH_3^+Cl^-+HCl \cdot aq$$

It appeared that the quantity of the strongly bound HCl is independent of its concentration in the equilibrium gas phase. This opens the possibility of use of water as a regenerant for HCl if the Cl^- - form (with strongly bound HCl) of the ion exchanger is used initially in the sorption-desorption cyclic process.

This was utilised in a cyclic experiment in which HCl was washed out with water and in each following experiment desorption was done with the solution obtained after regeneration in the previous cycle. The conditions for the experiment were as follows: mass of the ion exchanger was 0.7060 g dry: H=8 mm; C_{HCl}=100 mg/m^3; P/Po=0.6, V=7 m/min; wash water volume was 150 ml. The ion exchanger in the first cycle was used in the carbonate form. The results of this experiment are given in Table 2.

As follows from the Table, water can be used as an efficient desorbent in the recycle processes. Sorption efficiency appears to be insensitive to the concentration of HCl in the regeneration solution in the range studied.

Table 2. Regeneration characteristics of Fiban AK-22-G in cyclic sorption HCl - desorption process

No cycle	a_e, mg/g	Total, mg/g desorb.	% in desorb.	mg/g desorb. in cycle	a_d, mg/g	mg/g desorb. at break-through
1	123.5	23.2	0.01	23.2	49.3	
2	122.1	95.3	0.04	72.1	73.6	23.6
3	125.9	179.9	0.08	84.6	67.2	25.9
4	122.2	267.9	0.12	88.0	77.2	32.2

In order to establish the highest possible concentration of HCl which can be obtained after regeneration, a similar experiment was done with HCl solutions of higher concentrations. The results are presented in Table 3.

Table 3. Regeneration characteristics of the HCl sorption process with desorption by HCl solutions

Cycle No	% in desorpt. sol	a_e , mg/g	a_d, mg/g
1	0.12	122.2	77.2
2	2	127.1	86.9
3	4	122.2	72.9
4	6	117.9	56.3
5	8	44,0	5.3
6	0.12	95.5	67.9

As seen from the Table, a significant drop in dynamic activity was observed only when regeneration HCl solution concentration reaches 8%. This means that it is realistic to obtain 6-7% solution of the recovered HCl by starting the process with water as regenerant.

It is necessary to note, that the lower a_d and a_e values obtained in cycle 6 compared to cycle 1 are caused by the fact that some HCl was retained by the layer of the ion exchanger after regeneration in cycle 5.

More complete regeneration of the ion exchanger can be achieved with soda or alkali solutions. The additional quantity of sorbed HCl in this case will be 167 mg/g (4.72 m-eq/g) independent of the HCl concentration in vapour phase, as seen from Fig. 4.

Sorption of HCl by ion exchange fibres is strongly affected by humidity of the gas phase. An example of this dependence is given in Fig. 5. It is seen that sorption starts from a relative humidity of 0.15 and reaches maximum value at P/Po=0.5.

HF sorption by weak base ion exchanger

Data on HF sorption are presented in Fig's 6-8 and Table 4.

Table 4. HF sorption characteristics

HF, mg/m^3	P/Po	a_e, mg/g	Desorbed H_2O, mg/g	Desorbed soda, mg/g
16.9	50.9	232	136	96
42.4	38	219	132	87
102	44.9	257	161	95
253	45.8	303	200	103
449	51.6	267	174	93
919	56.2	425	327	98

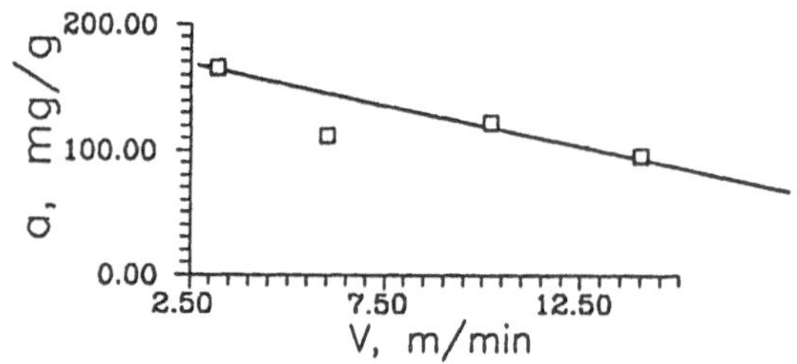

Fig. 6. Dynamic activity (a_d) sorption as a function of flow rate.
Conditions: C=52.8 mg/m^3, P/Po=0.65, h=9 mm.
Sorption of HF by Fiban AK-22-G.

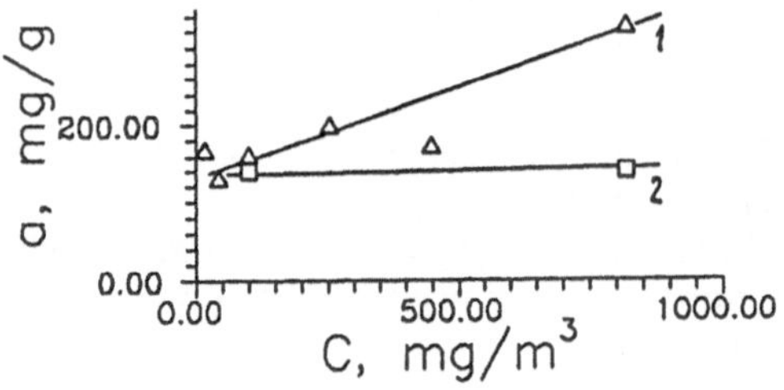

Fig. 7. Dynamic activities as a function of concentration.
Conditions: V=8 m/min, P/Po=0.50, h=8 mm.
1 - equilibrium, 2 - breakthrough. Sorption of HF by Fiban AK-22-G.

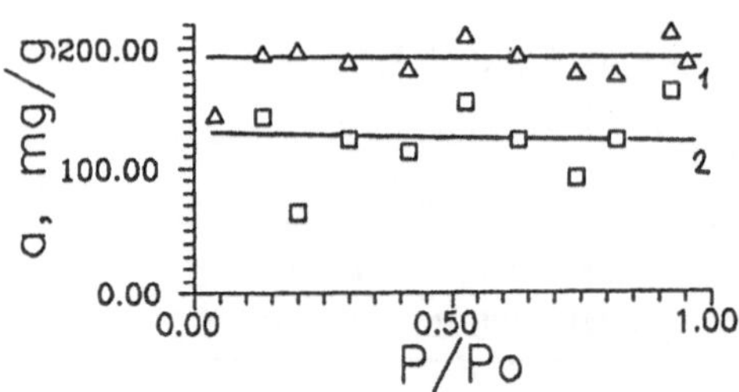

Fig. 8. Dynamic activities as a function of air humidity.
Conditions: V=9.8 m/min, C=52 mg/m^3, h=8 mm.
Sorption of HF by Fiban AK-22-G.

The sorption behaviour of HF has much in common with that for HCl described above. The quantities of strongly sorbed HCl and HF (removed by soda) appeared to be independent of concentration of the acids in the gas phase and equal in both cases to 4.45 m-mole/g. Consequently, the quantity of weakly bound HF is higher than for HCl. This value for HCl is 1.1-3.5 m-mole/g in the concentration range studied, while for HF this value is ranging between 6.7 and 15.9 m-mole/g. This is probably due to stronger association of HF molecules and is a good pre-condition for water-regeneration technology.

The other peculiarity of the HF sorption is that sorption is practically independent of the gas phase humidity in the range P/Po=0.1-1. As it is known, HF is a strong solvation agent and can behave similarly to water molecules in solution. This could be the reason for the weak dependence of HF sorption on air humidity.

SO_2 sorption by weak base ion exchanger

One of the peculiarities of the SO_2 sorption-regeneration process is that water is an active chemical participant in the sorption since SO_2 is sorbed as HSO_3^- and/or SO_3^{2-} ions, e.g.:

$$SO_2 + H_2O \rightleftarrows H_2SO_3 + 2RNH_2 \rightleftarrows$$

$$(RNH_3^+)_2\ SO_3^{2-} + H_2SO_3 \rightleftarrows 2RNH_3^+\ HSO_3^-$$

The other peculiarity is that SO_2 can be spontaneously oxidised in air-aqueous systems to form SO_3. This process is slow in air and requires special catalysts. It is much faster in aqueous media, where the easiest oxidising species is the HSO_3^--ion. Therefore it can be expected that this process would occur in the ion exchanger phase as well as in the regeneration solution.

$$RSO_3^{2-}(HSO_3^-) + 1/2\ O_2 = RSO_4^{2-}(HSO_4^-)$$

The set of data on the sorption-desorption process is presented in Fig's 9-11. The following peculiarities have been observed.

The critical humidity value is substantially higher than in the case of HCl and HF and corresponds to P/Po=0.5. Saturation is achieved in the case of equilibrium sorption at P/Po=0.75. It means that for SO_2 removal from air, additional humidification may be required in some cases.

High humidity critical value can be caused by the fact that one water molecule is required to form a hydrosulphite ion from an SO_2 molecule and some additional amount of water forms the diffusion medium in the ion exchanger phase needed to provide accessibility of the functional groups for

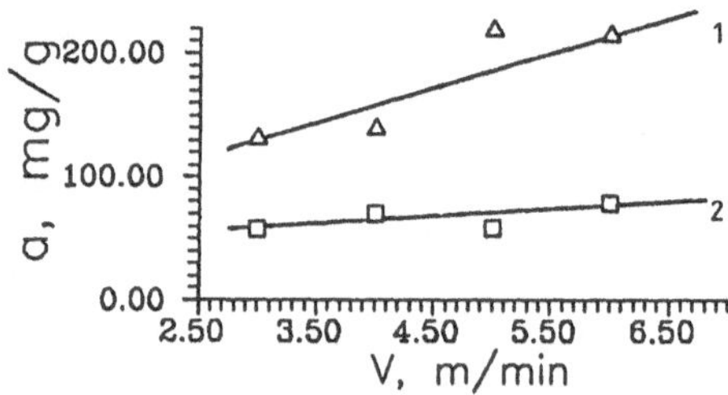

Fig. 9. Dynamic activities as a function of flow rate. Conditions: C=220 mg/m^3, P/Po=0.98, h=5 mm. 1 - equilibrium, 2 - breakthrough. Sorption of SO_2 by Fiban AK-22-G.

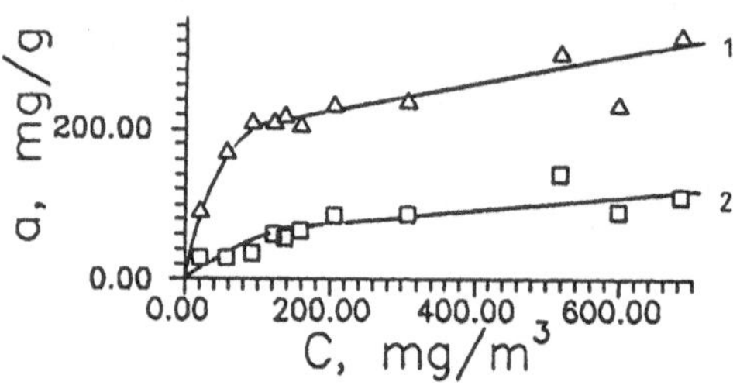

Fig. 10. Dynamic activities as a function of concentration. Conditions: V=7 m/min, P/Po=0.95, h=5 mm. 1 - equilibrium, 2 - breakthrough. Sorption of SO_2 by Fiban AK-22-G.

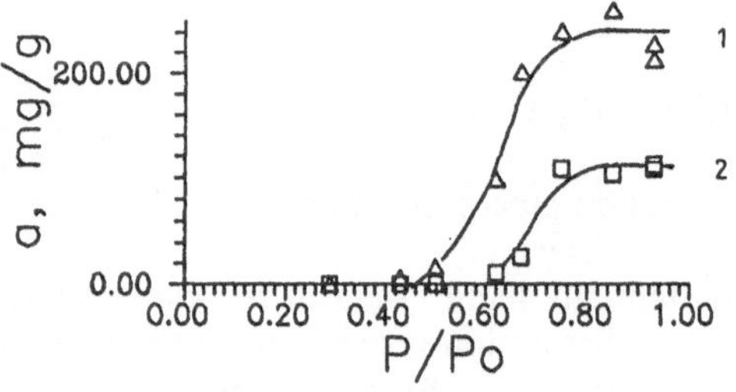

Fig. 11. Dynamic activities as a function of air humidity. Conditions: V=7.0 m/min, C=210 mg/m^3, h=5 mm. 1 - equilibrium, 2 - breakthrough. Sorption of SO_2 by Fiban AK-22-G.

the SO_2 molecules or $HSO_3^-(SO_3^{2-})$-ions. As isopiestic studies have shown, critical air humidity in this case corresponds to 2 water molecules per one functional group of the ion exchanger. In the cases where the reaction with water does not occur, critical humidity usually corresponds to one water molecule per reacting functional group (e.g. HCl), which is sufficient to form the diffusion medium.

Regeneration was done by passing 3% soda solutions through the cell after the sorption cycle at a rate of $2.8 \cdot 10^{-2}$ m/min. Sulphite and sulphate ion concentrations were determined in the effluent. In all cases the average degree of conversion of sulphite into sulphate ions was 55-72%. Since oxidation could take place only during the sorption process (not more than 8 hours), this was evidence of a marked catalytic activity of the ion exchanger in the oxidation process. For comparison in other work [14] it has been found that it needed 120-140 hours to oxidise ammonium sulphite into sulphate up to the equilibrium state in presence of electron exchanger EO-7.

SO_2 sorption by strong base ion exchanger

The processes occuring in this case can be formulated as follows:

$$SO_2 + H_2O \rightleftarrows H_2SO_3 + (RN(CH_3)_3^+)_2CO_3^{2-} \rightleftarrows$$

$$(RN(CH_3^+)_3)_2SO_3^{2-} + H_2CO_3 \rightleftarrows (RN(CH_3)_3^+)_2SO_3 + H_2O + CO_2$$

$$(RN(CH_3)_3^+)_2SO_3^{2-} + H_2SO_3 \rightleftarrows 2\ RN(CH_3)_3^+HSO_3^-$$

The set of experimental data on the influence of different factors on the dynamics of SO_2 sorption by Fiban A-1 ion exchanger is presented in Fig's 12-14.

Compared to the weak base ion exchanger, the sorption in this case is characterised by much steeper breakthrough curves. The sorption isotherms are very steep at low concentrations of SO_2 in the gas phase and already at 30 mg/m^3 SO_2 the ion exchanger becomes completely converted into the sulphite form and further is oxidised into the sulphate form.

An important feature of this process is extremely high oxidation rate in the presence of strong base ion exchanger. It was established that the degree of oxidation reaches practically 100% during the sorption stage. It was found to be controlled by the external mass transfer and can be completed in several minutes if the air flow is sufficient. Even at static conditions the time of complete conversion is 1.5 hour. Similar to the sorption process, air humidity strongly influences the rate of oxidation. The oxidation process is accelerated with increasing humidity up to P/Po=0.5 when it reaches limi-

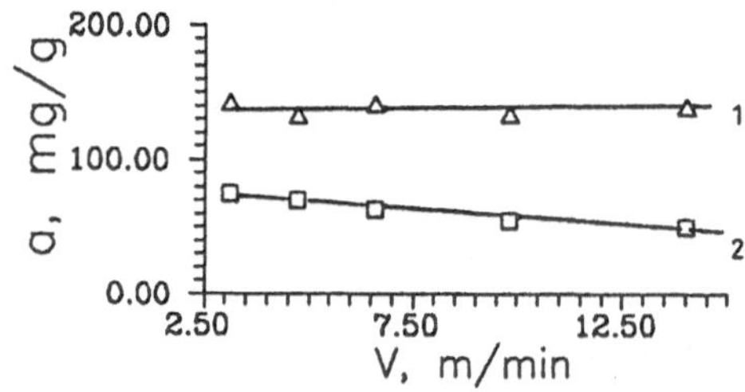

Fig. 12. Dynamic activities as a function of flow rate. Conditions: C=200 mg/m^3, P/Po=0.85, h=7 mm. 1 - equilibrium, 2 - breakthrough. Sorption of SO_2 by Fiban A-1.

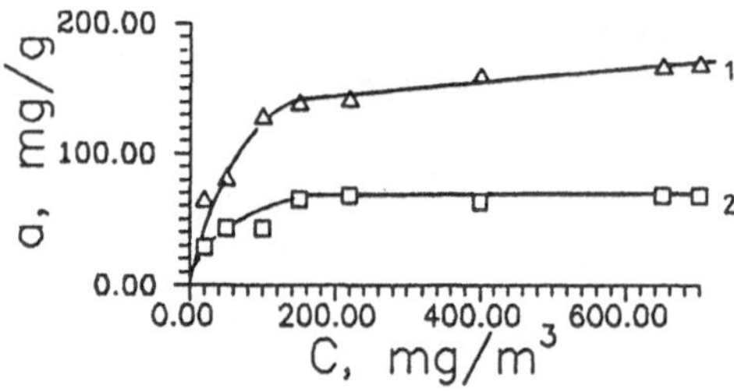

Fig. 13. Dynamic activities as a function of concentration. Conditions: V=6.6 m/min, P/Po=0.85, h=7 mm. 1 - equilibrium, 2 - breakthrough. Sorption of SO_2 by Fiban A-1.

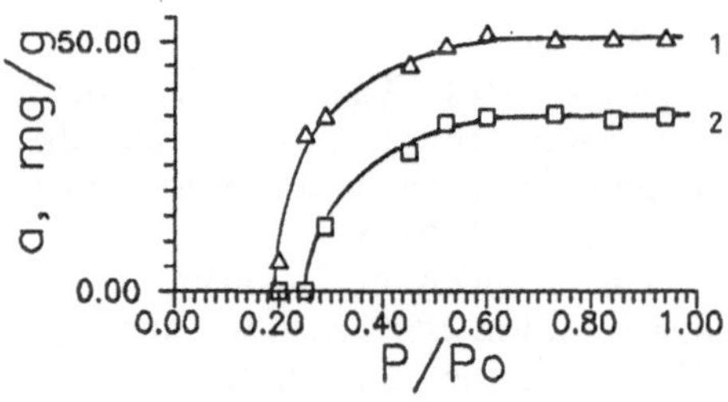

Fig. 14. Dynamic activities as a function of air humidity. Conditions: V=6.6 m/min, C=200 mg/m^3, h=8 mm. 1 - equilibrium, 2 - breakthrough. Sorption of SO_2 by Fiban A-1.

ting values. More detailed this process will be described in another paper.

Oxidation of sulphite into sulphate is a process of great practical importance because sulphite toxicity is much higher than that of sulphate and additional oxidation is usually required to satisfy sanitary standards.

H_2S sorption by strong base ion exchanger

Sorption of H_2S by OH-form of Fiban A-1 fibre was studied. The main process is:

$$2\ RN(CH_3)_3^+OH^- + H_2S \rightleftarrows (RN(CH_3)_3^+)_2\ S^{2-} + 2H_2O$$

$$(RH(CH_3)_3^+)_2 S^{2-} + H_2S \rightleftarrows 2\ RN(CH_3)_3^+\ HS^-$$

In Fig's 15-16 data on H_2S sorption are presented. Preliminary experiments have shown that only strong base ion exchangers can be used in a practical purification process. Sorption of H_2S by Fiban AK-22 does not exceed 10 mg/g. A very important factor complicating this process is the presence of CO_2 in the atmospheric air. Being a stronger acid than H_2S, H_2CO_3 generates bicarbonate ion which strongly affects H_2S sorption. Therefore all experiments have been done either in the absence of CO_2 or by strictly controlled its concentration in the gas phase.

It is seen from Fig. 15 that the dynamic activity of Fiban A-1 in the H_2S sorption is high enough for its practical application in air purification even in the presence of CO_2 concentration of 915 mg/m^3 which is slightly higher than in natural air. A marked improvement was observed in the absence of CO_2.

This process is less sensitive to air humidity as seen from Fig. 16. There are two reasons for that. First of all, the OH-form of the strong base ion exchanger at very low air humidity (P/Po=0.1) already contains 4 water molecules per equivalent, which is sufficient for the H_2S sorption. Further, as it follows from the reaction equation the reaction between H_2S and the OH-form of ion exchanger causes release water which improves the kinetics of this process.

Desorption of H_2S from the ion exchanger was done by 4% NaOH solution.

It was established, that an oxidation process accompanies ion exchange substitution of sulphide (bisulphate) ions in the ion exchanger. Formation of several sulphur containing species was observed. Thus 25% of total amount of sulphur in the eluate was made up of $SO_3^{2-}(HSO_3^-)$, $S_2O_3^{2-}$, HS^- and other reduced forms and 13% was SO_4^{2-}-ion. The rest of the sulphur could not be eluted from the ion exchanger. Probably it presents in the ion exchanger as elementary sulphur. Its total amount was determined according to Shöeniger method. Total amount of the sorbed H_2S was in good agreement with the sum of the species determined separately.

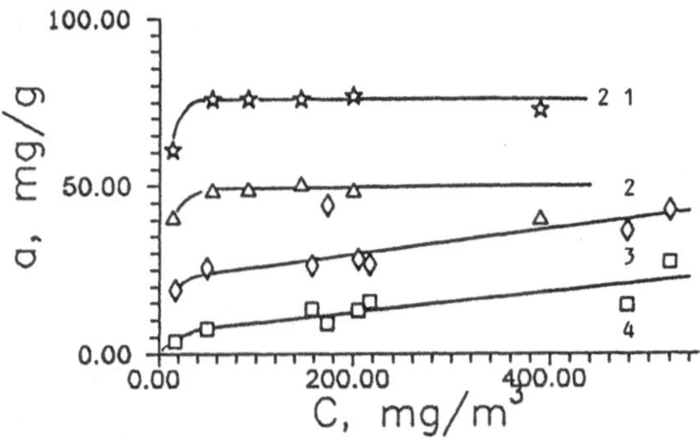

Fig. 15. Dynamic activities as a function of concentration. Conditions: V=3.5 m/min, P/Po=0.90, h=15 mm. 1,3 - equilibrium, 2,4 - breakthrough. Sorption of H_2S by Fiban A-1. 1,2 - in absence of CO_2, 3,4 - 915 mg CO_2/m^3.

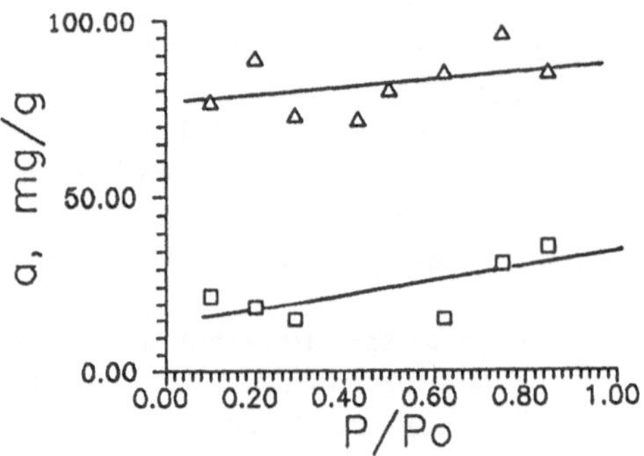

Fig. 16. Dynamic activities as a function of air humidity. Conditions: V=3.5 m/min, C=200 mg/m^3, h=12 mm. 1 - equilibrium, 2 - breakthrough. Sorption of H_2S by Fiban A-1.

Industrial application of Fiban ion exchangers in air purification

Technologies of air purification with different Fiban ion exchangers have been worked out on a pilot plant with capacity 300 m^3/h, and with a filtration area of non-woven ion exchange materials of 1 m^2. Sorption and regeneration stages have been optimised and technical documentation necessary for construction of industrial filtering plants have been obtained.

Several types of filtering plants, named FIF (Fiban Ion exchange frame Filter) have been developed and installed in about 20 different factories in Belarus, Russia, the Ukraine and some other states formed after the breakup of the Soviet Union. The total capacity of the equipment installed is about 0.5 million m^3/hour.

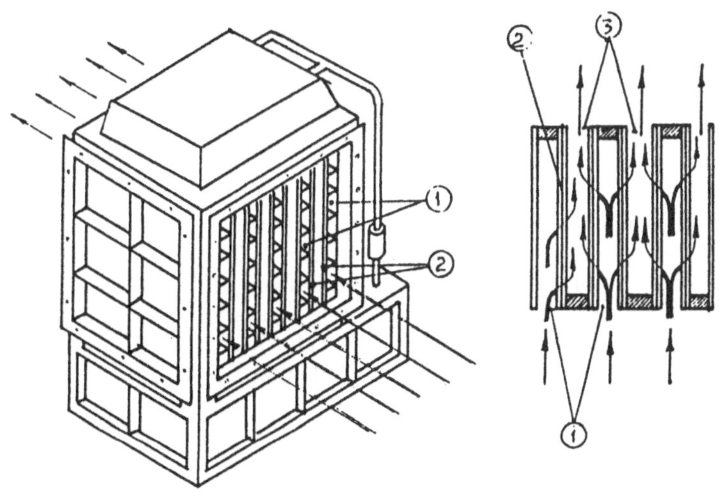

Fig. 17. FIF filtering plant scheme. View of the filtration block and its cut from the top. Direction of the air flow is shown by the arrows. The purifying air enters the filtering plant through inlet slit *1* and penetrates through the layers of the fibrous ion exchanger *2* into the outlet slit *3*.

The filtering plant scheme and their characteristics are presented in Fig. 17 and Table 5. As seen from the Table, industrial filtering plants FIF type, equipped with Fiban ion exchangers satisfy all requirements formulated above.

Fiban Ion exchange frame Filters (FIF) specification

Specification	Brand			
	FIF-0.5	FIF-5	FIF-10	FIF-20
Capacity, m^3/h	500	5000	10000	20000
Sorbent mass, kg	3	35	80	200
Filtering area, m^2	2.3	15	30	80
Aerodynamic resistance, max., Pa	500	500	600	800
Consumption of regenerating solution, m^3/h	0.05	0.2	0.6	1.2
Regeneration time, h	2	2	2	2
Overall dimensions, m	0.5x0.6x0.7	1.4x1.0x1.7	1.7x1x2.3	2 x1.7x 2.1
Mass, kg	90	418	745	1206
Linear velocity of filtering, m/s	0.06	0.09	0.09	0.07

Conclusions

Fibrous ion exchange materials Fiban have properties allowing their successful application in air purification from acidic impurities of different nature. In spite of the fact that sorption - desorption processes basically are of an ion exchange nature, regularities of each particular substance are not entirely predetermined by their acidity. The behaviour of each of the substances is rather individual and is strongly affected by several side effects such as water sorption, sorbate molecules association and oxidation.

References

1. Nikandrov, G.I. (1975) Purification of gases from sulphur dioxide by ion exchange sorbents, **Industrial and Sanitary Purification of Gases** (USSR), 4, pp. 14-8.
2. Nizovtseva , O.P. and Shunkevich A.A. (1978) Ammonia sorption by fibrous cation exchangers in the H-form. **Izvestiya Akademii Nauk BSSR**, seria Khimicheskikh Nauk, 5, pp. 62-5.
3. Türhölmez, S. (1965) Beseitigung der Geruchbelästigungen durch Austausch-Adsorption mittels Kunstharz-Jonenaustauschern. **"Wasser, Luft und Betrieb"**, 11, 737-743; 12, pp. 812-6.
4. Hashida, J. (1974) Studies of ion exchange resins adsorbing gases. Bull. Osaka Munic. Res. Techn. Inst., 1.

5. Gostomczyk, M.A., Kuropka, J. (1977) Investigation on sorption of acid gases on anion exchangers, **"Environment Protection Engineering"**, No 1-2, 135-144.
6. Vulikh, A.I., Zagorskaya, M.K., Bogatyrev, V.L., Ksenzenko, V.I., Luk'yanova, G.V. Application of ion exchange resins to sorption and purification of gases, in **"Ion Exchange Materials and their Application"**, Alma-Ata, 1968, 228-231.
7. Vulikh, A.I., Olovyanikov, A.A. and Nikandrov, G.A. Gas purification as a new scope of ion exchangers, (ed. M.M. Senjavin), **Ion Exchange**, Nauka, Moscow, 1981, 214-229.
8. Petruzzelli, D., Tiravanti, G., Liberti, L., Sergeev, G. and Soldatov, V.S. Chloride/Sulphate Exchange Kinetics on Fibrous Resins. A Comparative Study with Spherical Exchangers, in **Ion Exchange Processes: Advances and Applications**, The Royal Society of Chemistry, Cambridge, 1993, 167-179.
9. Soldatov, V.S., Tsigankova, A.V., Elinson, I.S., Tsigankov, V.I. and Shunkevich, A.A. (1988) Chemical, osmotic and frost resistance of strong base fibrous anion exchangers. **Journal "Applied Chemistry"**, USSR, 11, 2465-2472.
10. Zverev, M.P. **Chemosorption Fibres**, Khimiya, Moscow, 1981.
11. Soldatov, V.S., Elinson, I.S. and Shunkevich, A.A. Purification of air form acid gases (SO_2) by non-woven strong base filtering materials (ed. L. Pawlowski), **Chemistry for protection of the environment**, Elsevier, Amsterdam-Oxford-New York-Tokyo, 1986, 369-386.
12. Soldatov, V.S. (1984) New fibrous ion exchangers for purification of liquids and gases (ed. L. Pawlowski), **Chemistry for protection of the environment**, Elsevier, Amsterdam-Oxford-New York-Tokyo, 1984, 353-364.
13. Soldatov, V.S., Shunkevich, A.A. and Sergeev, G.I (1988) Synthesis, structure and properties of new fibrous ion exchangers, **Reactive Polymers**, 7, 159-172.
14. Miagkoi, O.N., Meleshko, V.P., Serdyukova, M.N. a.o. **Theory and Practice of Ion Exchange Processes**, Voronehz, 1972, 138-143.

Processes

Hydrometallurgical processing of complex copper-zinc concentrates

B. S. Boyanov
R. I. Dimitrov
Department of Chemistry, University of Plovdiv, Plovdiv, Bulgaria

Abstract
A flow-sheet of copper-zinc concentrates processing is suggested on the basis of industrial and laboratory tests. It includes roasting in fluid bed at 780-800 ^oC, standard or high acidity and high temperature leaching, sulphutizing or treatment of the cake in Waelz furnace. The achieved total copper and zinc recovery is 92-96%.
Keywords: Copper-zinc concentrates, hydrmetallurgical processing, flow-sheet

1 Introduction

In the process of enrichment by flotation of copper-zinc ores which are complex from the point of view of their chemical and mineralogical analyses, copper concentrates are produced with high zinc contents (8-12%) and zinc concentrates with high copper contents (3-5%). Often, a bulk flotation is performed, as well, where the copper-zinc concentrates produced contain 10-25% zinc and 8-15% copper [1]. The complex copper-zinc concentrates are treated by either pyrometallurgical, hydrometallurgical or combined methods [2-5], depending on their quantity and content, as follows:

1. Blending the copper-zinc concentrates with the zinc concentrates in a zinc plant.
2. Independent roasting and processing the calcine in Waelz furnaces.
3. Independent roasting and leaching the calcine by the classical flow-sheet.
4. High temperature and high acid leaching of the zinc calcine.

Investigations of the treatment of Bulgarian copper-zinc concentrates have been carried out on the basis of literature references and our previous results on the oxidation of metal sulphides [6-8], roasting of copper and zinc concentrates [9,10], and solid-phase interactions, which take place in these processes [11-13].

When the quantity of a raw material is not large, processing methods that provide incorporation in the operating flow-sheet of existing plants, are the more rational ones. Such an approach has been applied in the search for an appropriate technological flow-sheet for the processing of copper-zinc concentrates, aiming at their comprehensive treatment under the normal operating conditions of the running zinc plants. After some preliminary laboratory investigations, industrial trials have been performed for the processing of copper-zinc concentrates.

2 Results and discussion

2.1 Roasting of copper-zinc concentrates in fluid-bed rouster

It has been established that due to the high Cu and Fe contents and the increased contents of SiO_2 and Pb, it is impossible for the process of roasting to be carried out at a temperature of 920-950 °C, which is typical for roasting zinc concentrates [14,15]. The behaviour of the concentrate, established in previous roasting tests, shows that it resembles that of copper concentrates. At 650-700 °C the rate of oxidation is low and the productivity of the furnace is low as well. With increase in temperature above 800 °C there is increase in the coarse fraction of the calcine, the hydrodynamics of the bed are upset, and the quantity of the non-stoichiometric copper-zinc ferrites increases. A temperature of 780-800 °C for roasting the copper-zinc concentrates was found to be the optimum from this point of view.

Industrial scale studies were carried out in a fluid-bed reactor of height 11.6 m, hearth area 22 m^2, height of the weir sill 1.1 m, effective cross-section 0.98 m^2 and surface of the forechamber 0.96 m^2.

The regime of operation was the following: temperature of the forechamber 780-800 °C; temperature in the bed 770-800 °C; consumption of air in the forechamber 1450 nm^3/h; consumption of air in the central part 7500 nm^3; consumption of air in the air-cooler 650 nm^3/h; pressure drop across the fluid-bed 14.5-15.5 kPa. The quantity of dust, caught in the system for coarse dust collection, represents 40-50% of the input charge; dust content of the gas after the cyclones - 3.5 g/nm^3; SO_2

content in the roaster gases 3.5-4%; productivity of the furnace - 2.7-3.0 Mg/m^2 calcine over 24 hours.

During the industrial scale experiments of roasting the copper-zinc concentrates, the furnace was running initially with zinc concentrates at 960 ^{o}C. After 2 hours the temperature was reduced to 800 ^{o}C and feeding of the copper-zinc concentrates begun. After 48 hours of operation, the composition of the calcine from the air-cooler corresponded to the copper-zinc concentrate feed (Table 1). After treatment of the requisite quantity of copper-zinc concentrate, the roasting of zinc concentrate was renewed with gradual increase of the temperature back to 960 ^{o}C.

The sieve analysis of the calcine showed that no coarsening of the calcine was observed. More than 50% of the calcine had a particle size of < 0.102 mm. By means of chemical and X-ray diffraction analysis and Mösbauer spectroscopy the following phases were identified in the calcine: ZnO, CuO, $(Zn,Cu)Fe_2O_4$, Fe_3O_4, Fe_2O_3, $PbSO_4$, SiO_2 (α-quartz), CaO, MgO, Al_2O_3.

Roasting of copper-zinc concentrates is also possible in a tunnel kiln [9].

Table 1. Average analysis of the input copper-zinc concentrate and the generated calcine and dusts (%)

Component	Concentrate	Calcine from air-cooler	Dust from cyclone
Zn	23.62	24.85	29.67
$Zn_{a.s.}$	-	19.12	23.64
$Zn_{w.s.}$	-	8.45	10.43
Cu	7.12	7.97	7.01
Fe	12.46	13.38	15.60
S_S	29.05	0.52	0.39
S_{SO4}	-	9.35	8.40
Pb	2.75	3.16	2.75
SiO_2	3.02	3.21	2.97

$Zn_{a.s.}$ - zinc, soluble in 7 % solution of H_2SO_4;
$Zn_{w.s.}$ - zinc, soluble in water.

2.2 Hydrometallurgycal processing of copper-zinc calcines

Solutions of sulphuric acid and spent zinc electrolyte were used in the leaching of the copper-zinc calcine. The tests, carried out in conditions close to the industrial classical hydrometallurgical ones (t=70 ^{o}C, τ=30 min.), confirmed the considerable influence of the final acidity on the recovery of the metals. At the final content of sulphuric acid of 2-3 g/dm^3 (pH = 1.2-1.4) good dissolution of the copper-zinc calcine was achieved (Table 2).

The recovery of the metals increased with increases of the temperature and the initial concentration of the acid. Depending on the ratio of the copper-zinc calcine to the solution, the copper and zinc concentrations in the leach solution varied in the range 12-27 g/dm^3 and 63-108 g/dm^3, respectively.

Table 2. Results from the recovery of the copper-zinc calcine

N^o	Test conditions				Metal recovery (%)	
	t(^oC)	(min)	H_2SO_4 (g/dm^3) at inlet	H_2SO_4 (g/dm^3) at outlet	η_{Zn}	η_{Cu}
1	70	30	80	pH = 3.90	70.1	38.1
2	70	30	80	4.5	81.8	67.6
3	70	30	100	4	82.1	67.7
4	95	60	140	25	87.6	80.0
5	95	420	140	18	91.4	86.9
6	70	30	131*	pH = 4.10	75.0	41.7
7	70	30	131*	pH = 2.00	84.8	79.3
8	102	332	182	15	94.9	96.2

131* - spent zinc electrolyte

The contents of Fe and SiO_2 in the solutions increased with increases in the initial and final acidities. At a final acidity of under 0.5 g/dm^3, iron concentrations are 1-1.5 g/dm^3 and SiO_2 - up to 0.2 g/dm^3. At high final acidity, the iron concentration is in the range 6-11 g/dm^3 and SiO_2 - 1.5-2 g/dm^3. The iron in solution is mainly in the trivalent form (91-96% of the total quantity), which mainly precipitates in the form of jarosite.

Depending on the type of raw material and the leaching conditions, the filter cake contains (in mass %): Zn - 9.5-11; $Zn_{a.s.}$- 2-3; $Zn_{w.s.}$- 1.6-2.5; Cu - 3.55-7.5; Fe - 23.8-30.8; Pb - 5.7-10.1; SiO_2 - 4.45-7.0; S - 4.7-7.75; S_S - 0.3 and should be further treated for better recoveries of the components.

The possibility of removing iron through high temperature (90 ^oC) hydrolysis [16,17] was investigated. The neutralization of the solution (Zn - 77.4 g/dm^3; Cu - 12.86 g/dm^3; Fe - 15.29 g/dm^3) was carried out with fumed oxides which had a high content of acid-soluble zinc and a low content of iron (in mass %): Zn - 60.15; Pb - 10.85; Fe - 0.32; S - 8.05; C - 0.42; F - 0.32; Cl - 0.19; Cd - 0.06. The final pH of the solution should be 1.5-1.9.

Jarosite cations R^+ (K^+, Na^+, $NH_4{}^+$) were added in the form of sulphates in stoichiometric quantities with respect to the iron in the solution for the formation of jarosite having the general formula: $RFe_3(SO_4)_2(OH)_6$ (Table 3).

The highest degree (85.6%) of hydrolysis was achieved by adding K_2SO_4 whereby 2.2 g/dm^3 of iron remained in solution, which met the requirements for its further treatment.

Table 3. Results of the high temperature hydrolysis

Cation	pH	Precipitation (%)			Electrolyte content after precipitation (g/dm^3)		
		Zn	Cu	Fe	Zn	Cu	Fe
K^+	1.80	12.0	53.0	85.6	110.3	6.05	2.20
Na^+	1.71	11.0	56.4	79.3	111.6	5.35	3.12
NH_4^+	1.65	9.2	50.2	78.4	114.3	6.38	3.31
-	1.90	11.6	49.8	59.6	110.8	6.46	6.34

The copper content of the solution was rather high (5-6.5 g/dm^3) and copper can be removed by cementation with zinc powder. The filter cakes, produced as a result of the hydrolysis are also of high copper percentage (7.5-8%) and may be further processed in a lead plant or in Waelz kilns.

2.3 Processing of copper-zinc cakes by sulphatization

The possibility of sulphatizing the copper-zinc cakes by sulphuric acid, containing (in mass %): Zn - 32.00, Cu - 5.60, Fe - 11.44, SiO_2 - 4.42, S - 8.20, Pb - 3.30 was investigated.

The cake was granulated and samples were calcined at 450-650 ^oC for different times. Dissolution of the calcined granules was carried out at room temperature with water, slightly acidified with sulphuric acid (5 g/dm^3).

The dissolution results show considerable influence of the temperature and the duration of the heat-treatment on the recoveries of copper and zinc. The best results were obtained at 600 ^oC and 120 min.: η_{Cu} - 82.6%, η_{Zn} - 74.4%. The solution obtained had the following composition (in g/dm^3): Cu - 6.44, Zn - 15.80, Fe - 2.60 and the content of the obtained cake was (in mass %): Cu - 0.90, Zn - 4.50, Fe - 30.48.

2.4 Flow-sheet for processing of copper-zinc concentrates

Roasting of the copper-zinc concentrates is carried out at the temperature of 800 ^oC in a fluid-bed furnace (fig.1). The calcine obtained is recovered by spent electrolyte and the solution of 1.5 g/dm^3 acidity is thickened. The overflow contains 12.92 g/dm^3 Cu and

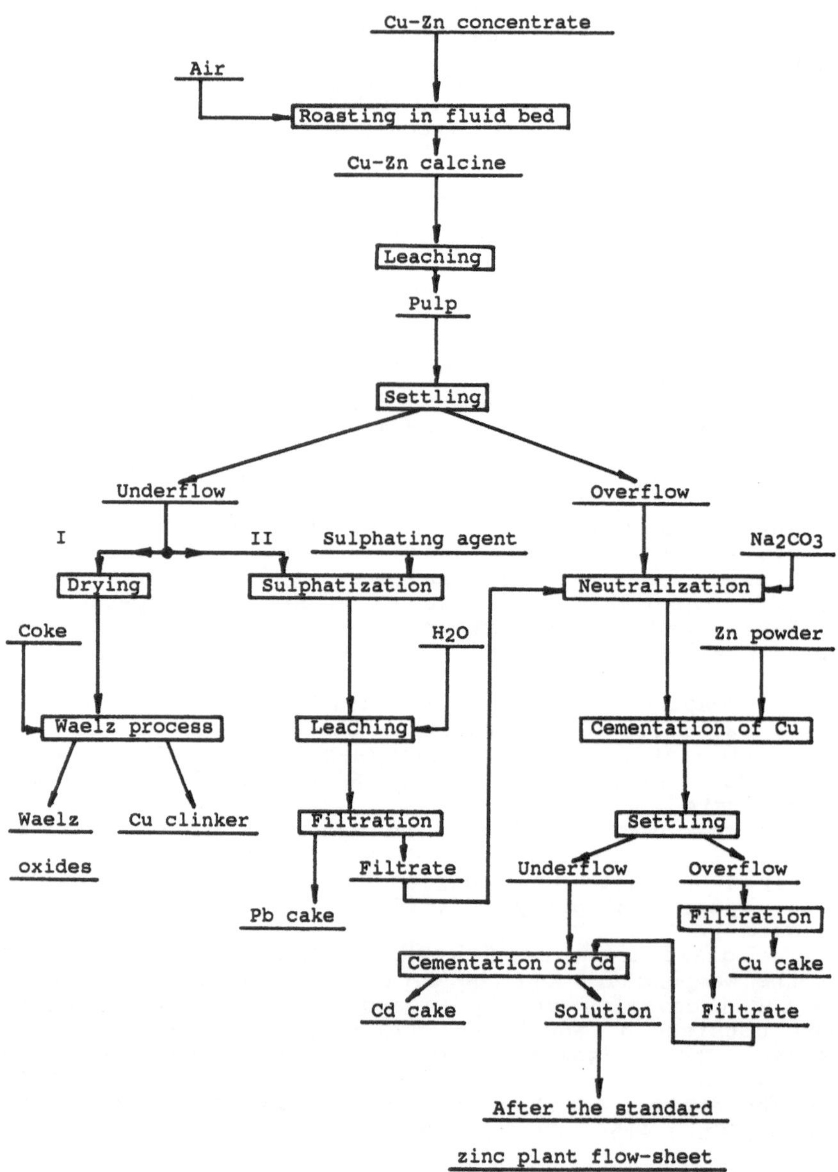

Fig.1. Flow-sheet of the copper-zinc concentrate processing: I,II - first and second variants

146 g/dm^3 Zn. It is neutralized with Na_2CO_3 to pH=4.8-5.0, aiming at avoiding the formation of AsH_3 during cementation with zinc powder. Then Cu^{2+} is removed from the neutral solution by cementation. The cement copper contains (in mass %): Cu - 21.06; Zn - 11.58; Cd - 0.42; Fe - 4.12. The cadmium is then precipitated by cementation and the cadmium precipitate is processed by the standard flow-sheet for cadmium production. After the settling of the suspension the underflow is filtered and the filter cake, containing (in mass %): Zn - 15.95; $Zn_{a.s.}$- 9.12; $Zn_{w.s.}$- 2.94; Cu - 3.20; Fe - 16.50; Cd - 0.11; Pb - 5.46 is treated by the Waelz process while the filtrate is processed by the suggested flow-sheet.

3 Conclusions

3.1 When the quantity of copper-zinc concentrates is not large it is possible to use the same furnace, passing from roasting of zinc to roasting of copper-zinc concentrates and vice versa. An optimum technological regime has been developed for this purpose.

3.2 A temperature of roasting of 780-800 °C and a degree of desulphurization of 75% and S_S < 0.5% are recommended for concentrates containing 7-10% Cu and 25-30% Zn. Under these conditions the furnace operated steadily without coarsening of the calcine and disturbing the behaviour of the fluid-bed.

3.3 The following degree of dissolution (in %) was achieved with the copper-zinc calcine: Zn - 86.7; Cu - 72.9; Cd - 88.0. The gold, silver and the rest of the copper, contained in the calcine, are recovered in the copper plant (from the clinker).

3.4 It is possible to carry out high acidity, high temperature leaching of the copper-zinc calcine, whereby iron is removed from the solution by high temperature hydrolysis in the form of jarosite $RFe_3(SO_4)_2(OH)_6$ (R^+ = K^+, Na^+, NH_4^+) at 90 °C, for 4.5 h and final pH = 1.6-1.8.

3.5 The recovery of 80% Zn and 85% Cu into solution and simultaneous precipitation of iron was possible by sulphatizing of the filter cake produced after dissolution under the established optimum conditions. The total copper and zinc recovery increased to 92-96% when the operations of sulphatization or treatment of the cake in Waelz furnace, are included in the flow-sheet.

4 References

1. Naboichenko, S.S. and Smirnov, V.I. (1974) **Hydrometallurgy of Copper**, Metallurgizdat, Moscow, pp.234-241
2. Monhemius, J. (1981) Hydrometallurgical processing of complex materials, **Chemistry and Industry**, 20, 410-420
3. Myzenkov, F. A., Mechev, V. V., and Kalnin, E. I. (1990) New method of copper-zinc and zinc ore and waste materials processing, **Tsvetnye Metally**, 11, 38-41
4. Sergeev, G. I., Khudyakov, I. F., Sergeeva, L .V., Plushnikova,Z.A., and Bakin, I.V. (1989) Comparable analysis of CuZn and Zn concentrates granulometric characteristics in the process of roasting and crushing kinetics in fluid bed, **Tsvetnaya Metallurgiya**, 6, 28-32
5. Eliseev, I. S., Babadzhan, A. A., Lebed, B. V., Naboichenko, S.S., and Khudyakov, I.F. (1974) On the choice of a method for copper-zinc flotation products processing method, **Tsvetnye Metally**, 3,3-8
6. Dimitrov, R. and Vanyukov,A. (1970) Investigation of kinetics and mechanism of ZnS oxidation, **Tsvetnye Metally**, 3, 7-11
7. Dimitrov, R. and Bonev, I. (1986) Mechanism of zinc sulphide oxidation, **Thermochimica Acta**, 106, 9-25
8. Dimitrov, R. and Boyanov, B. (1983) Investigation of the oxidation of metal sulphides and sulphide concentrates, **Thermochimica Acta**, 64, 27-37
9. Dimitrov,R. (1986) Behaviour of the particles during zinc concentrates roasting in fluid bed, **Tsvetnye Metally**, 12, 39-41
10. Dimitrov, R., Khekimova, A., and Draganov, N. (1975) Investigation of fluid bed roasting of copper sulphide, **Zhurnal Prikladnoi Khimii**, 1, 108-112
11. Boyanov, B. and Dimitrov, R. (1991) Ferrite behavior in the processing of complex copper-zinc sulphide concentrates, **Proceedings of the First Europian Metals Conference**, Brussels, September, 1991, 119-126
12. Boyanov, B. and Dimitrov, R. (1990) Kinetic study of the process of ferrite formation involving zinc, copper and cadmium oxides, **Proceedings of the 1st International Symposium Interprogress-Metallurgy**, Kosice-Bratislava, June, 1990, 182-185
13. Dimitrov, R. and Boyanov, B. (1984) Investigation of solid State Interactions in the Systems ZnO-Fe_2O_3, $ZnFe_2O_4$-CuO and $ZnFe_2O_4$-CaO, **Mining and Metallurgy Quarterly**, 31, 67-80

14. Zak, M.S., Khodov, N. V., Serebrennikova, E. Ya., Doverman, A.I., and Roshanovskii, V.M. (1989) Oxidation roasting of zinc concentrates in fluid bed, **Tsvetnye Metally**, 7, 62-65
15. Lebed, B.V., Guzairov, R.S., Tulenkov, I.P., Sergeev G.I., and Abramich I.L. (1980) Industrial tests of copper-zinc semi-finished products roasting in fluid bed furnaces, **Tsvetnye Metally**, 9, 30-33
16. Steintveit, G. (1970) Die Eisenfällung als Jarosit und ihre Anwendung in der Nassmetallurgie des Zinks, **Erzmetall**, 23, 532-537
17. Zyuzikov,V.E.(1992) Potassium jarosite precipitation in continious process, **Tsvetnye Metally**, 3, 15-17

Integration of Sherritt Zinc Pressure Leach Process at Ruhr-Zink refinery, Germany

M. J. Collins
Sherritt, Inc., Fort Saskatchewan, Canada
E. Ozberk
Sherritt, Inc., Fort Saskatchewan, Canada
M. Makwana
Sherritt, Inc., Fort Saskatchewan, Canada
I. M. Masters
Sherritt, Inc., Fort Saskatchewan, Canada
R. Püllenberg
Ruhr-Zink, Datteln, Germany
W. Bahl
Ruhr-Zink, Datteln, Germany

Abstract
The third commercial application of the Sherritt Zinc Pressure Leach Process commenced operation at the Ruhr-Zink refinery in March 1991. Integration of pressure leaching with the existing roast-leach-electrowinning facility has increased zinc production capacity by at least 50 000 tonnes per year slab zinc. The Sherritt Zinc Pressure Leach Process was selected for the plant expansion, to achieve high zinc recovery from a wide range of zinc concentrates while meeting the stringent environmental regulations governing the Ruhr-Zink refinery. The project was initiated in mid 1988 with a test program and engineering study to establish its technical and financial viability. The detailed design engineering and construction activities were started in the first quarter of 1989, and continued through 1990. Following commissioning activities in January and February, the pressure leach plant started producing zinc in March 1991. This paper describes highlights of the flowsheet development program and the design engineering phase, and results for the first three years of commercial operation.
Keywords: commercialisation, commissioning, engineering design, process development, Ruhr-Zink, Sherritt, sulphur flotation, zinc pressure leach.

1 Introduction

Ruhr-Zink GmbH operates an electrolytic zinc plant in Datteln, located in the northern part of the Ruhr industrial area in Germany [1,2]. The plant, which was the first in Germany to produce special high grade zinc using the roast-leach-electrowin process, was commissioned in 1968.

The Sherritt Zinc Pressure Leach Process, in which zinc concentrate is treated directly with zinc electrowinning spent electrolyte under oxygen pressure to oxidize sulphide sulphur to elemental sulphur and extract zinc into the solution [3 to 8], was first commercialised at the Trail, British Columbia zinc refinery of Cominco Limited in early 1981 [9,10]. Kidd Creek Mines (now Falconbridge Limited, Kidd Creek Division) commissioned the second commercial zinc pressure leach plant at Timmins, Ontario in 1983 [11,12]. Both cases involved expansion of zinc production capacity at existing roast-leach-electrowin refineries.

Ruhr-Zink started investigating the feasibility of expanding zinc production at Datteln from 135 000 to 200 000 tonnes per year slab zinc in 1986. The investigation was completed in 1987, and the Sherritt Zinc Pressure Leach Process was selected over the installation of additional roasters and expansion of the sulphuric acid plant because of lower cost, higher zinc recovery and the superior ability to meet environmental and work place hygiene requirements. Ruhr-Zink engaged Sherritt in July 1988 to conduct a test program with feeds available to Ruhr-Zink, to demonstrate the viability of the Sherritt Zinc Pressure Leach Process in meeting the objectives of Ruhr Zink, and to generate process criteria for preparing the engineering design of the pressure leach plant. The test program was completed in October 1988 and was followed by the process engineering design, which was completed in January 1989.

In March 1989, Ruhr-Zink management decided to implement the project and the detailed design engineering started immediately. The detailed design engineering and construction continued through 1989 and 1990. The operating manuals for the zinc pressure leach and sulphur recovery circuits were prepared by Sherritt during the second half of 1990, and the plant was commissioned in January and February 1991. The zinc pressure leach plant started producing zinc sulphate solution in March 1991, and has been in operation since then. Greater than one third of the Ruhr-Zink zinc input was extracted in the pressure leach plant in 1993.

2 Test program

The major process objectives were high extraction of zinc to a solution suitable for treatment in the existing zinc refinery, production of a high grade lead-silver concentrate containing less than 2% elemental sulphur for sale to a lead smelter, and recovery of elemental sulphur in a saleable form. The test program, which was carried out at the Sherritt Research Centre in Fort Saskatchewan, consisted of three major parts. In the first part, batch leach tests were carried out to determine the amenability of the potential feed materials to pressure leaching, and to define operating conditions for continuous pressure leaching testwork. The second part of the program consisted of a ten day continuous miniplant test, including pressure leaching and sulphur flotation steps, to demonstrate the applicability of the process in meeting Ruhr-Zink's objectives. Liquid-solid separation tests were conducted by an equipment vendor during the ten day continuous test run. The feasibility of recovering a saleable elemental sulphur product from the sulphur flotation concentrate was evaluated in the third part of the program.

2.1 Test materials

The feed materials for the testwork were three zinc concentrates, a bulk lead-zinc concentrate, and reduction residue from the Datteln plant. The compositions of these feed materials are summarized in Table 1. Twelve different blends of feed materials were examined during the continuous pressure leaching test run. The blends contained 8 to 12% Fe, 2 to 5.5% Pb, 30 to 36% S, 39 to 48% Zn, 1100 to 2500 g/t As, 3 to 20 g/t Hg, 1 to 9 g/t Se and 2 to 5 g/t Te. Reduction residue, containing 38% elemental sulphur, accounted for 20% of the feed in two of the blends.

Table 1. Chemical analyses of feed materials used for the test program

Feed	Fe (%)	Pb (%)	Ag (%)	S (%)	Zn (%)
Zinc Concentrates	9 to 12	2 to 3	0.01 to 0.02	30 to 33	47 to 49
Bulk Concentrate	13	17	0.035	31	30
Reduction Residue	7	2	0.030	50	12

2.2 Batch pressure leach tests

Six batch pressure leach tests were conducted on individual concentrate feeds and a further eight tests were conducted on blends of concentrates prior to the continuous test run. The effects of acid to zinc mole ratio, retention time, additive concentration and concentrate particle size were investigated. Zinc extractions in excess of 99% were achieved in one hour of pressure leaching for all feeds and feed blends tested, provided that the feeds were reground to at least 80% passing 45 mm. Iron precipitation in the autoclave was minimized, allowing for the recovery of a high grade lead-silver concentrate from the pressure leach residue, with an acid to zinc mole ratio in the autoclave charge of at least 1.7:1, to give a minimum of 60 g/L free sulphuric acid in the discharge solution.

2.3 Continuous pressure leach test

Continuous pressure leaching was conducted in a multi-compartment horizontal autoclave with an operating volume of 30 L. Feed slurry, at about 70% solids, was pumped to the first compartment of the autoclave, and spent electrolyte lixiviant was split between the first two compartments to maintain the target temperature profile. Leached slurry was discharged to a pressurized flash tank, and from the flash tank to an atmospheric conditioning tank. Slurry from the conditioning tank was directed to either a thickener or a continuous flotation circuit.

A total of 900 kg of feed solids was treated with 5 100 L of spent electrolyte in 240 hours of operation of the continuous pressure leaching circuit. The concentrate feed rate averaged 3.75 kg/h over the test period, corresponding to using a scale up factor of 4 000 for the commercial plant design. A summary of the operating parameters and results for the nineteen different operating periods during the continuous test run is provided in Table 2. Other variables included the rates of addition of calcium lignosulphonate and quebracho surfactants, and the pressure. The temperature in each case was 145 to 150°C in the first autoclave compartment, and 150°C in subsequent compartments.

Table 2. Summary of continuous pressure leach test results

Period	Feed Blend	Acid to Zn Mole Ratio	Retention Time (min)	Discharge H_2SO_4 (g/L)	Extraction Iron (%)	Extraction Zinc (%)
1	A1	1.71:1	60	64	66.3	96.8
2	A1	1.71:1	90	63	73.4	98.0
3	A2	1.57:1	43	56	60.4	96.0
4	A2	1.54:1	43	50	60.8	96.0
5	A2	1.48:1	87	53	63.1	98.1
6	A2	1.46:1	87	50	61.1	98.5
7	A2	1.72:1	87	62	66.9	98.2
8	B	1.79:1	89	55	73.9	99.4
9	D	1.68:1	90	63	81.0	98.8
10	E	1.68:1	94	63	75.9	92.4
11	F	1.72:1	94	66	68.3	91.0
12	I	1.65:1	91	61	74.5	97.5
13	I	1.61:1	92	61	78.0	97.9
14	G	1.78:1	89	59	84.1	99.0
15	H	1.78:1	58	58	89.3	99.4
16	H	1.79:1	57	59	89.7	99.2
17	L	1.74:1	91	61	79.7	98.9
18	K	1.64:1	94	61	73.6	98.7
19	M	1.69:1	95	60	74.2	98.9

A retention time of 90 minutes in the autoclave was typically required to achieve 98 to 99% zinc extraction. The anomalously low zinc extraction values obtained for feeds E and F, which each consisted of 100% of the same zinc concentrate, but with different particle size distributions, were explained by the presence of residual defoamer in the drum used to ship the concentrate sample. In subsequent batch pressure leach tests with an uncontaminated sample, 98% zinc extraction was achieved, and it would appear that the defoamer interfered with the action of the calcium lignosulphonate and quebracho surfactants in the continuous tests.

Net iron extractions in the autoclave ranged from 60 to 90%, and depended primarily on the composition of the feed material and to a lesser extent on the acid to zinc mole ratio, which was maintained at 1.46:1 or greater. As expected, iron extraction was highest for the low pyrite feeds, and lowest for the blends containing the largest fraction of bulk lead-zinc concentrate, which was high in pyrite. As was observed in the batch tests, iron hydrolysis and precipitation was largely prevented by maintaining a terminal acidity of 60 g/L H_2SO_4 or greater.

Conversion of sulphide sulphur in the feeds to elemental sulphur averaged 85%, with 8% oxidation of sulphide to sulphate and 7% unreacted (primarily as pyrite).

2.4 Sulphur flotation tests

Pressure leach discharge slurry from the continuous leaching operation was treated by froth flotation to recover separate sulphur and lead-silver fractions. Conditioning tank discharge slurry, containing 5 to 7% solids, was floated directly in the continuous flotation circuit, while both conditioning tank discharge slurry and thickened conditioning tank discharge solids were treated in batch flotation tests. The elemental sulphur floated well in all cases, and no frothers, collectors or other flotation reagents were required.

Six minutes of collection was required in the batch tests with conditioning tank discharge slurry to produce flotation tailings containing 2% elemental sulphur or less. Recovery of elemental sulphur to the rougher-scavenger concentrate was generally greater than 99%, while recovery of lead to the lead-silver concentrate (flotation tailings) ranged from 75 to 85% and recovery of silver to the lead-silver concentrate was about 70%.

Although elemental sulphur in the lead-silver concentrate represents a loss of this value, and an undesirable contaminant, lead and silver values in the sulphur concentrate are recoverable by recycle to the process. Elemental sulphur is separated from the sulphur concentrate by heating to melt the elemental sulphur, followed by pressure filtration. The filter cake, containing the unleached sulphides as well as oxidic minerals containing lead and silver that are entrained in the sulphur concentrate, is recycled to the roasters in the commercial design.

The continuous flotation test circuit contained one rougher, one cleaner and two scavenger flotation cells. The continuous flotation tests largely confirmed the results of the batch tests. Sulphur floated readily to leave a lead-silver concentrate typically containing between 20 and 40% lead and between 0.6 and 2.0% elemental sulphur. Recovery of lead to the lead-silver concentrate ranged from 78 to 95%, while recovery of elemental sulphur to the sulphur concentrate ranged from 99.0 to 99.8%.

The grade of the lead-silver concentrate depended primarily upon the quantity of gangue in the feed to the autoclave. The reduction residue and one of the zinc concentrates were high in silica, which reported largely to the lead-silver concentrate.

The cleaner concentrate recovered in the continuous flotation tests contained between 85 and 91% elemental sulphur, with the grade of this concentrate depending primarily upon the pyrite content of the feed to pressure leaching. Unleached sulphides are in intimate contact with elemental sulphur in the pressure leach residue, and float along with elemental sulphur. Since oxidation of pyrite in the autoclave is less complete than other sulphide minerals, feed materials that are high in pyrite result in residues with relatively high quantities of unleached sulphides, and produce lower grade sulphur flotation concentrates.

2.5 Liquid solid separation

Liquid-solid separation tests confirmed that both the conditioning tank discharge slurry and the lead-silver concentrate slurry were amenable to thickening, and the sulphur cleaner concentrate slurry was amenable to filtration and washing. In order to reduce the moisture content of the washed sulphur concentrate cake, to prevent frothing in the molten sulphur pit, steam purging of the cake on the belt filter was required in the commercial design.

2.6 Sulphur recovery tests

Sulphur cleaner flotation concentrates produced in the continuous flotation tests were melted at 150°C and filtered in a laboratory pressure filter of cross sectional area 0.005 m^2. The pressure filtration cakes contained between 27 and 45% elemental sulphur, representing a recovery of about 90% elemental sulphur to the filtrate. The overall recovery of sulphide sulphur in the feed to elemental sulphur in the product, considering 85% conversion of sulphide sulphur to elemental sulphur in the pressure leach, 99% recovery in flotation and 90% recovery in melting and filtration, was about 75%.

Arsenic, mercury, selenium and tellurium have an affinity for elemental sulphur and are common contaminants in sulphur recovered from metallurgical processes. The elemental sulphur filtrates produced in the tests contained less than 1 g/t As, 5 to 17 g/t Hg, 1 to 15 g/t Se and less than 1 g/t Te, corresponding to less than 0.1% of the arsenic, 15 to 30% of the mercury, 40 to 50% of the selenium and less than 5% of the tellurium in the feed to the autoclave.

2.7 Conclusions from the test program

The test program demonstrated that zinc extraction in excess of 98% could be obtained from all of the available feed materials in 90 minutes of pressure oxidation at 150°C. An acid to zinc mole ratio of 1.7:1 was sufficient to leave 60 g/L H_2SO_4 in the discharge solution and to minimize the formation of iron precipitates in the autoclave.

It was further demonstrated that, after conditioning, the autoclave discharge slurry was amenable to the separation of elemental sulphur and lead-silver concentrates by flotation, and without the addition of flotation reagents. Recovery of elemental sulphur to the sulphur concentrate was 99% or greater, and a lead-silver concentrate containing 2% elemental sulphur or less could readily be obtained. The grade of the lead-silver concentrate depended upon the relative quantities of lead and gangue minerals in the feed to the autoclave.

The sulphur flotation concentrate was amenable to melting and filtration to separate an elemental sulphur byproduct.

Sufficient data was collected in the test program to enable design of the commercial pressure leach plant.

3 Zinc pressure leach process description and plant integration

The zinc recovery process in operation at Datteln prior to the plant expansion [2] included roasting of zinc concentrates, neutral leaching of calcine, hot acid and super hot acid leaching of neutral leach residue, zinc dust purification of neutral leach liquor, and electrowinning of zinc from the purified solution (see Figure 1). The residue from super hot acid leaching was sold to a lead smelter as a high grade lead-silver concentrate. Zinc concentrate was added to the hot acid leach liquor in an atmospheric reduction step, to reduce ferric iron in the solution to the ferrous state prior to neutralization and precipitation of iron as hematite. The reduction residue, containing residual zinc sulphide and a significant quantity of elemental sulphur, was treated in the roasters.

The test program demonstrated that an increase in zinc production was feasible at Datteln by direct pressure leaching of zinc concentrate, without an increase in sulphuric acid production or sulphur dioxide emissions, and that the product streams resulting from pressure leaching were suitable for integration with the existing process. The integrated process, as designed, is shown in Figure 2, and a schematic equipment flow diagram for the zinc pressure leach plant is shown in Figure 3.

The concentrate feed to the pressure leach circuit is reground in a ball mill and thickened to a slurry containing about 70% solids. The thickened slurry is combined with reduction residue and leach additives for feed to the autoclave, where the slurry is contacted with a portion of the spent electrolyte from electrowinning, under oxygen pressure. Treatment of the reduction residue in the autoclave rather than in the roasters allows for recovery of the sulphur content of this residue as elemental sulphur rather than as sulphuric acid, and allows for an increase in the quantity of zinc concentrate that may be treated in the roasters. The leach additives are required to disperse molten elemental sulphur in the autoclave, to prevent coating of unreacted zinc sulphide particles with elemental sulphur, and to control agglomeration of solids upon discharge of the pressure leach slurry.

The pressure leach slurry is discharged to a flash tank, where its temperature is decreased from about 150°C to about 120°C by controlling the operating pressure. The flashed steam is recovered for heating process solutions elsewhere in the plant. The flash tank discharge slurry is directed to an atmospheric conditioning tank, where the temperature is further decreased to about 80°C, and elemental sulphur in the slurry freezes as small pellets, which are amenable to separation from the slurry by flotation.

In the flotation circuit, the conditioned pressure leach slurry is treated directly in primary flotation cells. The primary flotation tailings slurry is thickened, and the thickener underflow slurry is the feed to conventional rougher, scavenger and cleaner cells in the secondary flotation circuit. The cleaner concentrate is combined with the primary flotation concentrate and the combined slurry is filtered and washed on a vacuum belt filter. This sulphur concentrate is melted and filtered under pressure to yield an elemental sulphur byproduct. The sulphide filter cake is recycled to the roasters for recovery of the zinc, lead, silver and sulphur (as sulphuric acid) values. The scavenger flotation tailings slurry, containing a substantial portion of the lead and silver values in the pressure leach residue, is combined with the super hot acid leach slurry for feed to the old lead-silver concentrate thickener. The primary flotation tailings thickener overflow solution, the product solution from the pressure leach circuit containing zinc and iron extracted in the autoclave, is split between the leaching steps in the old part of the plant.

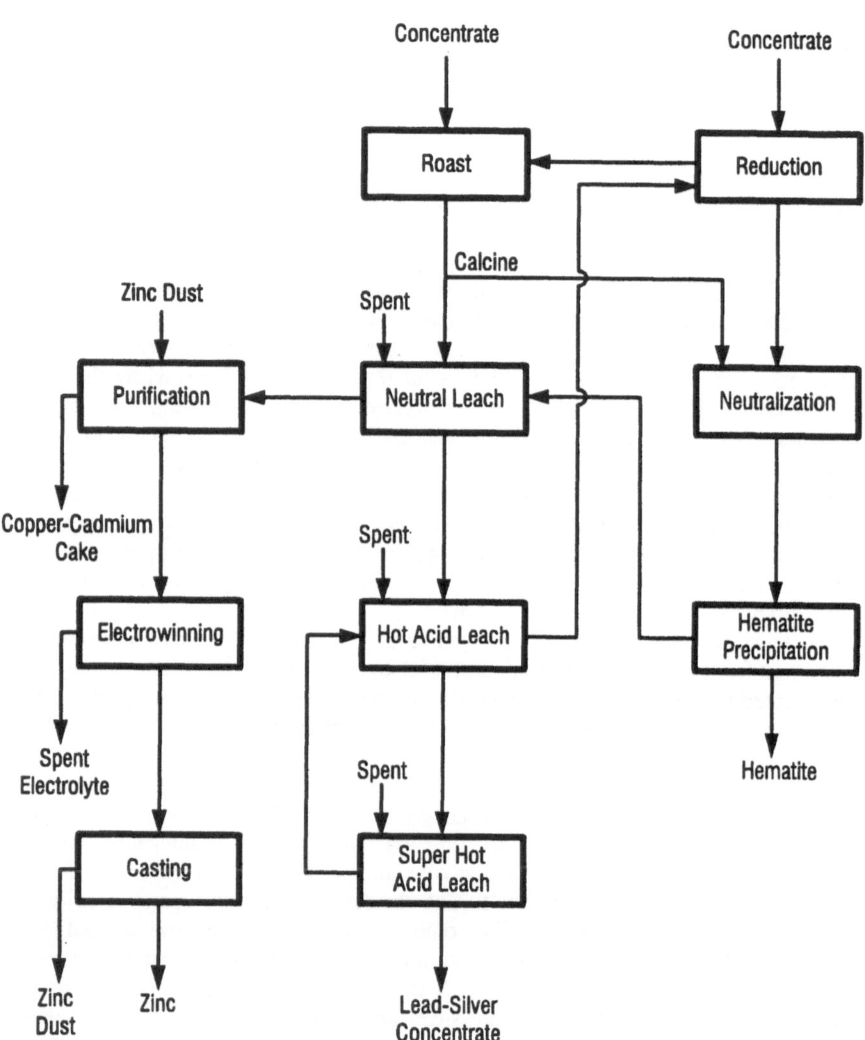

Figure 1. Ruhr-Zink flowsheet prior to pressure leaching

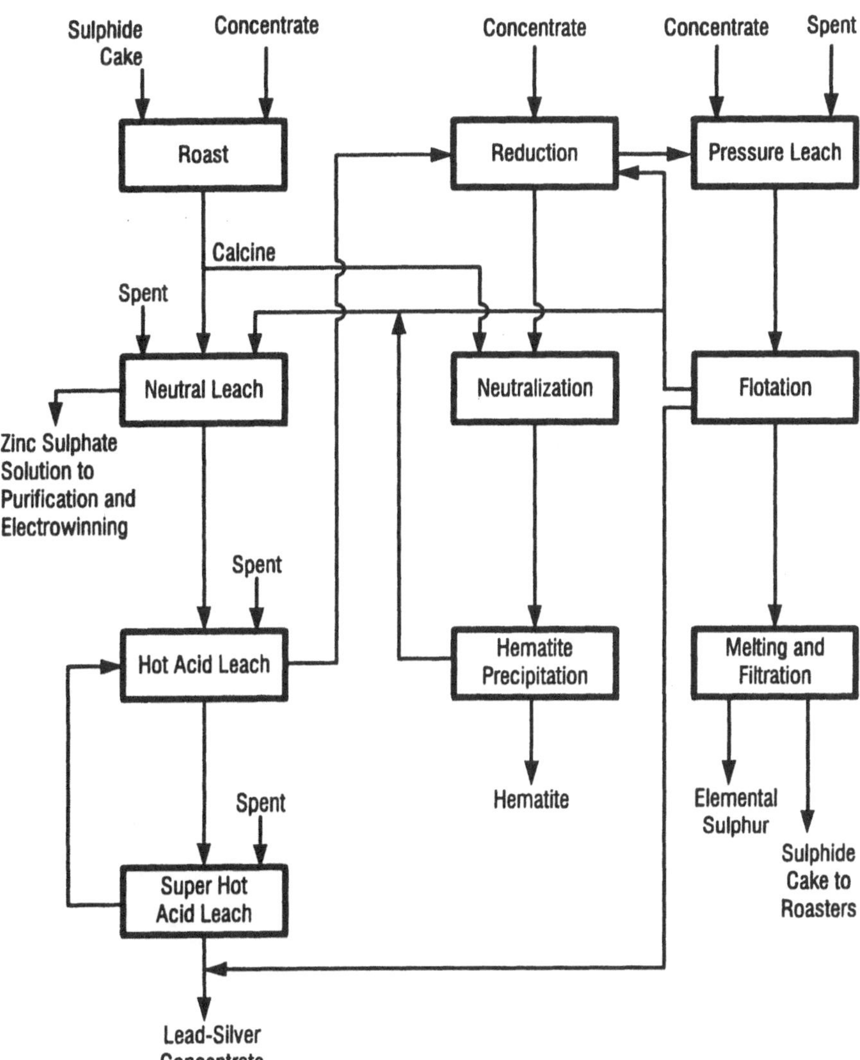

Figure 2. Integrated Ruhr-Zink flowsheet including pressure leaching

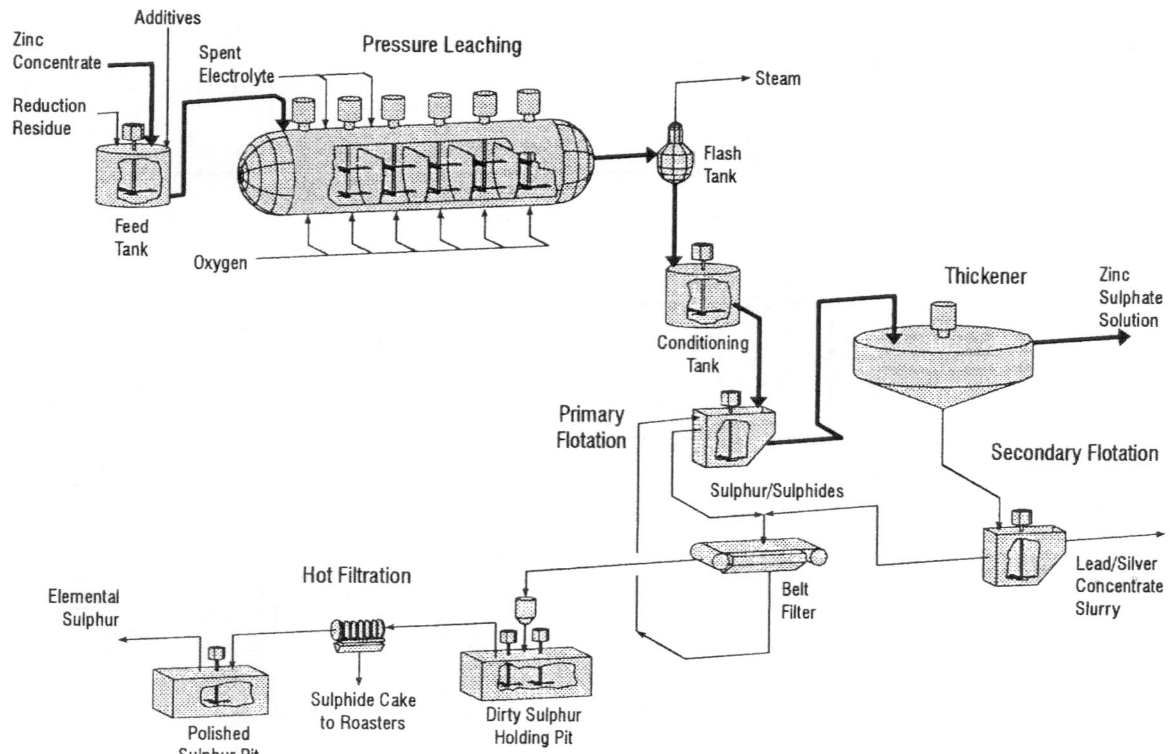

Figure 3. Equipment flow diagram for the zinc pressure leach plant

4 Engineering design

The design of the Ruhr-Zink commercial pressure leach plant was based on the criteria outlined below.

1. Confirmed process design criteria established during the batch and continuous test programs.
2. Proven technology and engineering design as practised at two commercial zinc pressure leach plants and Sherritt's experience in autoclave technology.
3. Meeting the requirements of the objectives established by Ruhr-Zink.
4. Compliance with the strict environmental abatement regulations applicable in Germany.

4.1 Environmental compliance

The procurement of the required licenses to satisfy Germany's environmental regulations proceeded in conjunction with the engineering design of the pressure leach plant. A report outlining the proposed process technology and plant design prepared by Sherritt was submitted by Ruhr-Zink to the environmental authority of the region to obtain the license to proceed with the plant installation. During the detailed design phase, a detailed environmental audit of the plant design was prepared and submitted to the local environmental authority. A group of personnel from the regional and local environmental authorities and Ruhr-Zink, accompanied by Sherritt personnel, visited Cominco's Trail operation to gain better understanding of the environmental impact of a typical zinc pressure leach plant. Subsequently, Ruhr-Zink was granted all necessary permits to construct and operate a pressure leach plant.

4.2 Process engineering design

A process engineering design report was prepared by Sherritt for the Ruhr-Zink pressure leach plant. The report contained sufficient engineering data to enable Ruhr-Zink to prepare overall detailed cost estimates for economic evaluations. The report also formed the basis for the detailed design of the commercial pressure leach plant.

The process design criteria established during the laboratory test program and the overall design basis provided by Ruhr-Zink were used for the plant design. The pressure leach plant was designed to extract 48 000 t/y of zinc into solution. The feed materials for the pressure leach plant consisted of a blend of three zinc concentrates, a bulk concentrate and the reduction residue produced in the existing reduction step.

Process flowsheets, detailed material and energy balances, and piping and instrumentation diagrams, to convey the control philosophy of the plant, were included in the process engineering design report. Preliminary equipment specifications, plant layout drawings, operating requirements and the capital cost estimate of the pressure leach plant battery limits were also provided to permit Ruhr-Zink to confirm the economic viability of the project.

4.3 Detailed engineering design and construction

In early 1989, Ruhr-Zink made the decision to proceed with the construction of the pressure leach plant. The responsibilities for the execution of the project were divided between Ruhr-Zink, Sherritt and Lurgi. Sherritt was responsible for the detailed design of the plant, which included feed preparation, pressure leaching, sulphur concentrate flotation and elemental sulphur melting and filtration. Ruhr-Zink was responsible for the detailed design of the feed material grinding, elemental sulphur solidification and utilities, and reagent supply circuits. The procurement of the equipment was carried out by Lurgi. Ruhr-Zink was also responsible for the plant construction and the overall management of the project, which included the modification and expansion of the solution purification, electrowinning, and melting and casting facilities.

Prior to the commencement of the detailed design of the plant, Ruhr-Zink confirmed the feed blend to the plant, which consisted of two zinc concentrates, a bulk concentrate and the reduction residue. The design capacity of the plant was finalized to extract 50 000 t/y of zinc into solution and process design criteria were revised accordingly for the plant design. The process flowsheet selected for the commercial plant design is illustrated by the equipment flow diagram shown in Figure 3. The commercial plant design included a second stage of flotation of pressure leach discharge solids and an elemental sulphur solidification circuit, both of which were added during the detailed design engineering phase.

The detailed engineering design commenced with the revisions to the process flowsheet and the material and energy balances to reflect changes in the feed blend, the plant capacity and the circuit modifications. Detailed piping and instrumentation diagrams were prepared, incorporating instrumentation and control philosophies practised at the existing Ruhr-Zink refinery. Because the available space at the proposed location of the pressure leach plant was limited, finalization of the layout of the plant required several revisions.

Sherritt prepared specifications and drawings of all special and standard equipment in sufficient detail for obtaining vendor quotations. The selection of the equipment was based on Sherritt's past experience and meeting Ruhr-Zink's requirements. The materials of construction selected for the plant design reflected the operating temperature and acid content of the process fluids.

The pressure leach autoclave is fabricated from a carbon steel shell, lined with lead and acid resistant bricks. It has six agitators, with two agitators in the first compartment. The autoclave is designed to provide sufficient retention time to achieve greater than 99% zinc extraction. The autoclave discharge slurry is partially depressurized in an agitated flash tank and the flash tank discharge slurry is cooled and conditioned in a brick lined carbon steel tank fitted with cooling coils and an agitator.

The complete sulphur concentrate flotation circuit and the vacuum belt filtration circuit consists of vendors' standard design equipment which satisfied the specified process duties. The flotation circuit consists of standard box type cells equipped with rotor mechanisms through which supercharged air is introduced. The vacuum belt filter is designed to provide two stages of washing and a final steam drying stage to minimize the cake moisture. The flotation cells and the feed end of the belt filter are covered and vented to a wet-scrubber. The air passing through the belt filter is also scrubbed prior to discharging through the vacuum pump and to the atmosphere. These provisions ensure a good work place environment and satisfy the local environmental regulations.

The sulphur concentrate cake is melted in two melting cyclones each designed for half of the plant capacity. A major portion of the heat required to melt the elemental sulphur is provided by shell and tube heat exchangers. The molten dirty sulphur is collected in a below ground pit which is equipped with steam coils and agitators. The molten dirty sulphur is filtered in a pressure leaf type filter. The filter consists of a horizontally retractable cylindrical tank and a filter pack which holds the stainless steel filter leaves. The molten sulphur filtrate is collected in a clean sulphur pit which is also fitted with steam coils and an agitator. The filter cake, consisting of unleached zinc sulphide, pyrite and elemental sulphur, is passed through a pug mill prior to being recycled to the roasters. The clean molten sulphur is solidified into pastilles, which are stored in a silo and shipped out by trucks. The melting cyclones, the two sulphur pits and the molten sulphur filters are appropriately covered and vented to a wet-scrubber.

5 Commissioning of the plant

A comprehensive set of operating manuals for all circuits within Sherritt's design responsibilities was prepared by Sherritt, and was available for the commissioning phase. The actual commissioning of the pressure leach plant started in January 1991. Ruhr-Zink adopted a phased approach to the commissioning of the pressure leach plant, with the pressure leach and the sulphur flotation circuits started first. The sulphur concentrate cake produced from the belt filter was transferred directly to the roasters while construction of the sulphur concentrate melting and filtration circuit and the sulphur solidification circuit was being completed.

Both process and equipment challenges were faced during commissioning of the new plant. A modification to the atmospheric leach circuit in the existing plant immediately prior to commissioning of the pressure leach plant resulted in the production of excess reduction residue, and this material was treated in the autoclave during the early stages of commissioning, to upgrade it for the recovery of the lead and silver values. The feed blend to the autoclave initially contained up to 60% reduction residue, which in turn contained 4 to 6% zinc, compared with the process design basis of 17% reduction residue containing 20 to 25% zinc. The pressure leach residue produced from this feed blend did not respond well to the separation of sulphur and lead-silver concentrates by flotation. Improved flotation results were obtained after increasing the amount of zinc concentrate in the feed blend.

Equipment challenges included the spent electrolyte heat exchanger, which was not delivered on time. In order to avoid delay in the commissioning of the plant, the heat required for raising the spent electrolyte temperature was provided by direct steam injection into the autoclave. In the meantime, the piping for the supply of the quench solution was not connected to the last three compartments of the autoclave due to late delivery of piping materials. As a result, the control of temperature in the autoclave was not as steady as desired. Installation of the spent electrolyte heat exchanger and the quench solution piping allowed for temperature control according to the design.

Additional equipment difficulties included the autoclave agitator seals, which initially leaked due to a manufacturing defect and had to be replaced, and the autoclave dip pipes, which were damaged two months after startup. The dip pipes were repaired according to a modified design and replaced. In the flotation area, braces had to be installed to eliminate vibration of the primary flotation cells, and the original carbon steel rubber lined rotors were replaced with Alloy 904L.

These equipment deficiencies were corrected during the first three months of operation and by mid 1991 steady production was achieved. At the same time, the sulphur concentrate melting and filtration circuit and the sulphur solidification circuit were commissioned.

6 Operating results

The operation of the zinc pressure leach plant since the completion of the commissioning phase can be divided into two periods. The first period ended on May 3, 1993, when zinc production was halted to implement changes to the entire Ruhr-Zink plant production flowsheet. In particular, the hematite plant, which had been in operation since 1979 [2], was shut down due to the high cost of operation. The reduction step was no longer required, and treatment of reduction residue in the zinc pressure leach plant was discontinued when the pressure leach plant restarted on June 28, 1993. The flowsheet for the present process is shown in Figure 4. Zinc-iron concentrate from the neutral leach and neutralization steps is sold for further processing and metals recovery.

The feed composition and the operating results for the pressure leach plant are compared for the two periods in Table 3. Up to May 3, 1993, the autoclave availability was 95%, as was the availability of the zinc pressure leach plant as a whole. The feed blend contained 50 to 60% zinc concentrate, with the balance being reduction residue, and the zinc content in the feed varied between 30 and 40%. Zinc extraction in the autoclave was consistently greater than 97%, and often exceeded 99%. Since production of reduction residue was discontinued, the feed blend to the autoclave has contained 45 to 50% zinc, and zinc extraction has increased to 98% on average. The availability of the autoclave has been 95% since June 28, 1993.

Recovery of elemental sulphur in the pressure leach discharge solids to sulphur product has been in the range of 85 to 90% since 1991, and discontinuing the treatment of reduction residue has not affected this value. The lead concentrate contained 30 to 35% Pb, 5 to 6% Fe, 8 to 10% SiO_2, less than 2% elemental sulphur and 1 to 2% Zn when treating reduction residue, and 18 to 22% Pb, 7 to 9% Fe, 10 to 12% SiO_2, less than 5% elemental sulphur and 1 to 2% Zn without reduction residue in the autoclave feed.

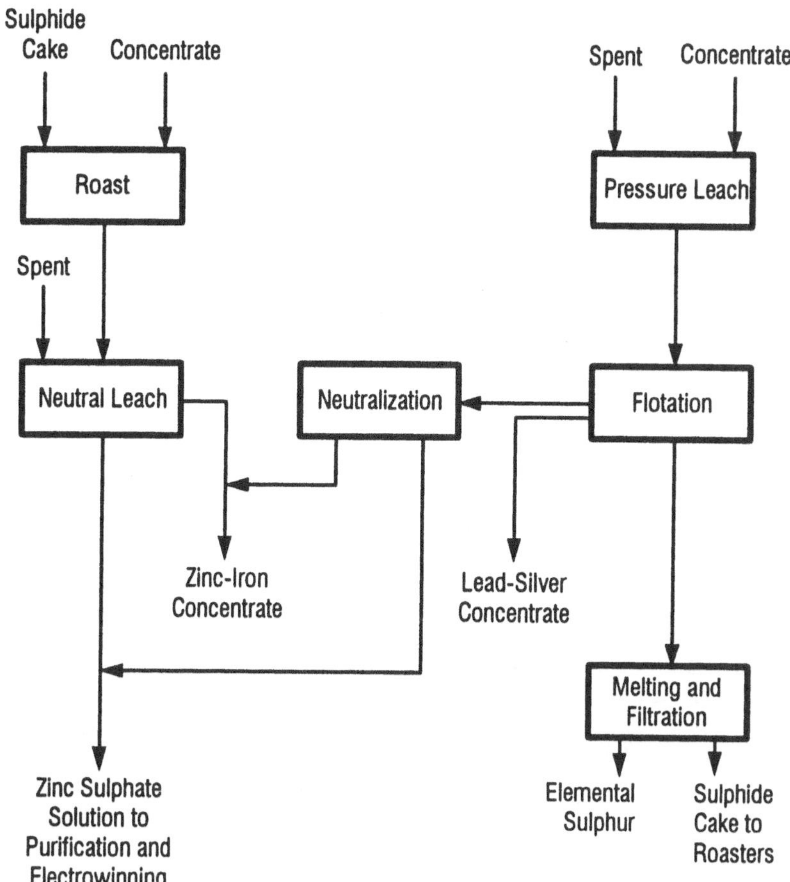

Figure 4. Present flowsheet at Datteln

Since completing the overall plant modifications in June 1993, production in the zinc pressure leach plant has been 10 to 15% higher than the design capacity of 50 000 t/y zinc.

Table 3. Operating results

Period	1991 to May 1993	June 1993 to Present
Feed Blend, %		
Concentrate	50 to 60	100
Reduction residue	40 to 50	-
Zinc Content of Feed Blend, %	30 to 40	45 to 50
Autoclave Availability, %	95	95
Zinc Extraction, %	>97	>97
Sulphur Recovery, %	85 to 90	85 to 90
Lead Concentrate Analysis, %		
Lead	30 to 35	20
Iron	5 to 6	8
Silica, SiO_2	8 to 10	10 to 12
Elemental Sulphur	<2	<5
Zinc	1 to 2	1 to 2

7 Conclusions

Integration of the Sherritt Zinc Pressure Leach Process at the Ruhr-Zink refinery was initiated in mid 1988 with the decision to proceed with a test program. Within three years the process development studies, basic engineering, detailed design engineering, construction and commissioning activities were completed and steady operation at the pressure leach plant was achieved. Since mid 1991, the zinc pressure leach plant has become an essential part of the refinery, and now accounts for greater than one third of the total zinc production at Ruhr-Zink.

8 Acknowledgments

The authors wish to thank the management of Sherritt Inc. and Ruhr-Zink GmbH for permission to publish this paper.

9 References

1. von Röpenack, A. (1990) Future changes in the physico-chemistry of zinc electrowinning, in **Lead-Zinc '90**, ed. T.S. Mackey and R.D. Prengaman, TMS, Warrendale, PA, pp. 641-652.

2. von Röpenack, A. (1986) Hematite - the solution to a disposal problem - an example from the zinc industry, in **Iron Control in Hydrometallurgy**, ed. J.E. Dutrizac and A.J. Monhemius, Ellis Horwood Limited, Chichester, pp. 730-741.

3. Collins, M.J., Chalkley, M.E., Masters, I.M. and Ozberk, E. (1993) The Sherritt two-stage zinc pressure leach process with hematite precipitation, in **World Zinc '93**, ed. I. Matthew, AIMM, Parkville, Victoria, pp. 315-323.

4. Chalkley, M.E., Collins, M.J. and Ozberk, E. (1993) The behaviour of sulphur in the Sherritt zinc pressure leach process, in **World Zinc '93**, ed. I. Matthew, AIMM, Parkville, Victoria, pp. 325-331.

5. Chalkley, M.E., Collins, M.J., Masters, I.M. and Ozberk, E. (1993) The integration of the Sherritt zinc pressure leach process with commercial iron precipitation processes, in **Hydrometallurgy: Fundamentals, Technology and Innovations**, ed. J.B. Hiskey and G.W. Warren, SME, Littleton, CO, pp. 1273-1296.

6. Berezowsky, R.M.G.S., Collins, M.J., Kerfoot, D.G.E. and Torres, N. (1991) The commercial status of pressure leaching technology. **JOM**, 43(2), pp. 9-15.

7. Makwana, M. and Collins, M.J. (1991) Advantages of the Sherritt zinc pressure leach process, in **Lead and Zinc in the 1990s: World and Latin America**, Chameleon Press, London, pp. 89-112.

8. Collins, M.J., Doyle, B.N., Ozberk, E. and Masters, I.M. (1990) The zinc pressure leaching process: applications, in **Lead-Zinc '90**, ed. T.S. Mackey and R.D. Prengaman, TMS, Warrendale, PA, pp. 293-311.

9. Ashman, D.W. and Jankola, W.A. (1990) Recent experience with zinc pressure leaching at Cominco, in **Lead-Zinc '90**, ed. T.S. Mackey and R.D. Prengaman, TMS, Warrendale, PA, pp. 253-275.

10. Parker, E.G. and Romanchuk, S. (1980) Pilot plant demonstration of zinc sulphide pressure leaching, in **Lead-Zinc-Tin '80**, ed. J.M. Cigan, T.S. Mackey and T.J. O'Keefe, AIME, New York, pp. 407-425.

11. Mollison, A.C. and Moore, G.W. (1990) Sulphide pressure leaching at Kidd Creek, in **Lead-Zinc '90**, ed. T.S. Mackey and R.D. Prengaman, TMS, Warrendale, PA, pp. 277-291.

12. Johnston, B.H. and Doyle, B.N. (1986) Start up and operation of the Kidd Creek zinc sulphide pressure leaching plant. **Minerals and Metall. Processing**, February 1986, pp. 1-7.

Recovery of platinum group metals from oxide ores—TML Process

Saskia Duyvesteyn
University of California, Berkeley, California, U.S.A.
Houyuan Liu
The Minerals Laboratory, BHP Minerals, Reno, Nevada, U.S.A.
W. P. C. Duyvesteyn
The Minerals Laboratory, BHP Minerals, Reno, Nevada, U.S.A.

Abstract
The Great Dyke, Zimbabwe, contains over 800 million ounces of platinum and approximately 20 million ounces of rhodium. However, about 10% of the ore is oxidized and cannot be processed with current recovery methods. A conventional flotation-smelting process has been proposed to recover the platinum group metal (PGM) values from the sulfide ores. With the development of a low temperature roasting, acidic bromine leaching process by The Minerals Laboratory of BHP Minerals, substantial PGM recoveries from oxide and mixed oxide/sulfide ores are now possible. This patented TML Process for platiniferous oxide ores will yield recoveries of 95% for gold, 85% for platinum and over 65% for rhodium values.
Keywords: Bromide Leaching, Platinum, Gold, Rhodium, PGM's, Oxide Ores.

1 Introduction

The Great Dyke, Zimbabwe, contains in the range of 800 million ounces of platinum and 20 million ounces of rhodium, Prendergast and Wilson (1989). A joint venture of BHP Minerals and Delta Gold recently completed a feasibility study for the development of the sulfide reserves from the Hartley Complex of the Dyke. However, about 10% of the ore is oxidized and cannot be processed with current recovery methods for sulfide ores. Hence this oxide ore is either not mined or discarded as waste. However, with the development of a novel leaching process by The Minerals Laboratory(TML) of BHP Minerals, substantial platinum group metal (PGM) recoveries from oxide and mixed oxide/sulfide ore are now possible.

PGM dissolution is generally thought to be only possible with aggressive chemicals. Recent studies by Bonucci and Parker (1984), Letowski and Distin (1985) and Demopoulos (1989), have shown that combinations of 12*M* chloride with strong oxidants can solubilize PGM's. The well-known *aqua regia* method, Papademetriou and Grasso (1970), combines three essential conditions that are required for PGM dissolution: (a) a high acidity, (b) a high oxidation potential and (c) a high concentration of complexing ions. The TML innovative leaching technique is based on the use of sulfuric acid for acidity control, bromide ions for the formation of PGM complexes, and bromine for redox potential control. From the work by Dadgar (1989) and Dadgar et al. (1990) on the application of bromine leaching for gold extraction, it was determined that **low** bromide concentrations (0.1*M* Br^- rather than 12*M* Cl^-) would complex gold. The Stillwater process which was recently developed by the USBM by Baglin et al. (1985), showed that 12*M* Cl^- was needed for PGM dissolution. The low bromide levels encouraged the development of a milder, and hence, more feasible method of recovery. As a result **The TML Process** was developed for platiniferous oxide ores that can yield recoveries of 95% for gold, 85% for platinum and 65% for rhodium. The process has been patented, Duyvesteyn et al. (1992a) and (1992b).

2 Background and Theory

2.1 Ore Characterization

Mineralogical analysis of the Hartley ore showed the following components: PGM's, silicate gangue minerals, oxides and base metal sulfides. The ore minerals, which contain the PGM's, make up less than 0.01 *vol*%, while the silicate gangue minerals constitute about 95 *vol*% and the remainder of ore is made up of oxides and base metal sulfides. The gangue minerals consist primarily of hydrated silicates, talc being the main component. The mineral oxides found were iron(II) oxide, ilmenite, chromite and wolframite, while the sulfides consisted of pyrrhotite, chalcopyrite and pentlandite.

2.2 Differential Thermal Analysis and X-Ray Diffraction Analysis

X-ray diffraction analysis (XRD) was used to analyze raw Hartley ore and its calcine from roasting at 300°C, 850°C and 1050°C. The patterns, given in Figure 1, showed major peaks identifying chlorite and talc, which disappeared upon calcination. The characteristic peaks for chlorite started to disappear at 300°C and the lack of formation of any new peaks indicated that the decomposition products were amorphous. The talc peaks began to change at 850°C, but did not completely disappear until 1050°C.

As indicated by DTA in Figure 2, major endothermic reactions occurred at 425°C, 525°C, 650°C and 875°C. The negative deflection starting at 425°C represents the decomposition of goethite and the formation of hematite, whereas the decomposition of chlorite starts at 525°C and is completed at about 650°C. At this temperature the recrystallization of chlorite begins.

Endothermic steps:

300 to 425°C: Dehydration of chemically bound and hydroxyl water.
Decomposition of goethite and formation of hematite.
450 to 525°C: Transformation of crystalline chlorite to an amorphous phase.
600 to 650°C: Decomposition of chlorite.
800 to 850°C: Decomposition of talc.

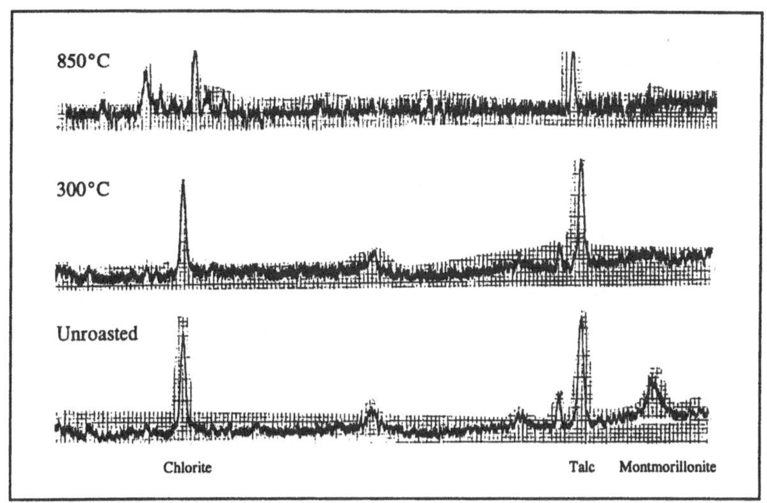

Fig.1. XRD analyses of Hartley oxide ore.

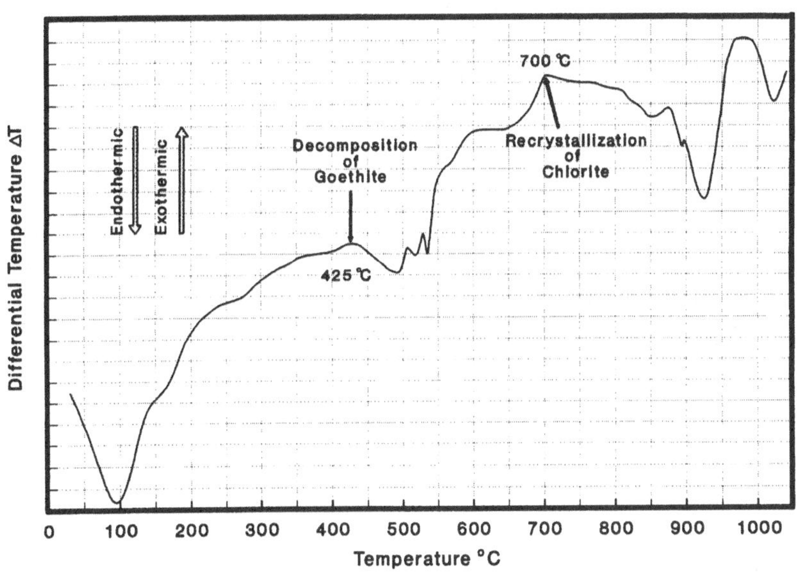

Fig.2. Differential thermal analysis (DTA) of Hartley oxide ore.

Exothermic steps:

540 to 600°C: Oxidation of ferrous to ferric iron.
650 to 700°C: Recrystallization of chlorite.
930 to 970°C: Recrystallization of talc.

2.3 Roasting Mechanism

Thermodynamic calculations showed that it was not practical to directly leach PGM minerals with halogen and oxygen. Remy (1956) reported that platinum sulfide (PtS) was even insoluble in aqua regia. It has also been shown that the platinum oxides and rhodium trioxide were insoluble in dilute acid.

To recover PGM's from Hartley oxide ore, two changes have to take place: first, the PGM compounds must be converted to their elemental state before they can be brought into solution (minimize the chemical lock-up) and secondly, the lixiviant must be able to access the PGM's through the surrounding minerals (minimize the physical lock-up). This means that the silicate matrix must be broken down. During roasting of the oxide ore five major transformations occur which play a role in PGM extraction: (a) decomposition of PGM minerals (at higher temperatures only), (b) decomposition of hydrated minerals, (c) decomposition of goethite, (d) decomposition of silicates and (e) recrystallization of various decomposition products. Thus roasting achieves the following:

(a) Decomposition of the PGM minerals, such as sulfides, arsenides, tellurides, etc., to obtain elemental platinum group metal and to avoid over-oxidation.
(b) Oxidation of ferrous and other lower valence elements decreasing the consumption of oxidant during leaching.
(c) Break down of the silicate matrix which encloses the PGM minerals providing access for the lixiviant.

The two primary minerals that undergo decomposition are sulfides and arsenides:

$$PtS_{(s)} + O_{2\,(g)} \longrightarrow Pt_{(s)} + SO_{2\,(g)} \quad (1)$$

$$PtAs_{2\,(s)} + {}^{3}/_{2}\,O_{2\,(g)} \longrightarrow Pt_{(s)} + As_2O_{3\,(g)} \quad (2)$$

Referring to Figure 2, the dehydration of the silicate minerals helps to break down the PGM-enclosing matrix, making the PGM's more accessible to the leachant. The XRD showed that chlorite starts decomposing at 300°C and converts into an amorphous phase, but the decomposition of talc does not occur until 850°C. Decomposition of chlorite and talc takes place according to:

Chlorite: $$(Mg,Al,Fe)_{12}(Si,Al)_8O_{20}(OH)_{16} \longrightarrow (Mg,Al,Fe)_{12}(Si,Al)_8O_{28} + 8H_2O \quad (3)$$

Talc: $$Mg_3SiO_{10}(OH)_2 \longrightarrow Mg_3SiO_{11} + H_2O \quad (4)$$

Goethite is an iron weathering product and elements present in the weathering solutions, including PGM's, adsorb onto its hydrated oxide surface. Roasting the ore above 425°C results in the decomposition of the goethite into hematite, which upon recrystallization, locks up most of the elements, including the PGM's, that were adsorbed onto its surface.

$$2FeOOH \longrightarrow Fe_2O_3 + H_2O \tag{5}$$

Although chlorite decomposition starts at a temperature of 300°C, it does not recrystallize until a temperature of 750°C is attained. Upon recrystallization, it locks up PGM's, reducing the leachant's access. This recrystallization process also results in agglomeration, which was confirmed by the increase in particle size of the calcine roasted at higher temperatures. Over-oxidation of the ore must also be avoided because of the formation of the very inert platinum and rhodium oxides, PtO and Rh_2O_3, which cannot be leached in the mild lixiviant of the TML Process.

2.4 Leaching Mechanism

Although bromine's ability to leach PGM's has been known since the 1880's, its industrial application ended in the early 1900's. With the new innovation in stable, inexpensive bromine compounds, the application of bromine leaching has become feasible for a number of applications in the precious metals industries.

2.4.1 The Chemistry of Geobrom 3400

Geobrom 3400® is a sodium bromide solution in which liquid *bromine* is dissolved. It is a proprietary reagent manufactured by Great Lakes Chemical Company. Its exact composition is not known. However, its behavior in different environments is known. Geobrom 3400 hydrolyses in water to from hypobromous acid.

$$\text{Geobrom 3400} + H_2O \longrightarrow HOBr + Br^- \tag{6}$$

HOBr, which is a powerful oxidant, the strength of which increases as the pH decreases, has reduction products either being bromide, Br^-, or bromine, Br_2. HOBr decomposes rapidly in solution to form bromide and bromate.

$$3HOBr \longrightarrow 2Br^- + BrO_3^- + 3H^+ \tag{7}$$

Bromine can be produced from the reaction products:

$$HOBr + Br^- + H^+ \longrightarrow Br_2 + H_2O \tag{8}$$

Bromic acid, $HBrO_3$, is a strong acid with a dissociation constant of $2x10^{-1}$. Hence bromate, as produced in reaction (7), will be the predominate species at pH > 0.70. Bromate is also a powerful oxidizing agent in acidic aqueous solution, the strength of which increases as the pH decreases. Although BrO_3^- is a powerful oxidizing agent, the speed of its reactions are reported to be moderately slow.

2.4.2 Bromine Leaching of Platinum

It is known that platinum and other PGM's will form divalent $[PtCl_4]^{2-}$ or tetravalent metal complexes $[PtCl_6]^{2-}$ with halide ions, having a higher affinity for the heavier halide ions according to the following order: $I^- > Br^- > Cl^- > F^-$. However, heat and light readily decompose the tetravalent iodine complex in aqueous solution, making it unsuitable for leaching purposes. Due to platinum's higher affinity for bromine, bromine leaching can be carried out in a milder environment. Since platinum bromide complex acid, H_2PtBr_6, is deliquescent and stable while other complex salts are less soluble, it is logical to select the bromide complex acid as platinum carrier for a halide leaching system.

The leaching is performed by adding sodium bromide and Geobrom 3400 to a sulfuric acid solution. The acid reacts with the metal oxides and silicates, which are generally represented as, MeO and $Me_w^{+a}Si_xO_y(OH)_z$, to unlock the PGM's.

$$H_2SO_4 + MeO \longrightarrow Me_2SO_4 + H_2O \tag{9}$$

$$\tfrac{wa}{2} H_2SO_4 + Me_w^{+a} Si_xO_y(OH)_z \longrightarrow \tfrac{wa}{2} Me_{a/2}SO_4 + xSiO_2 + \tfrac{(wa+z)}{2} H_2O \tag{10}$$

The reaction of sulfuric acid with sodium bromide provides the hydrobromic acid needed for platinum bromide complex formation. The free bromine from the Geobrom 3400 and the hydrobromic acid react with the platinum metal to form the stable complex, $[PtBr_6]^{2-}$, that can be recovered from the leach solution.

$$H_2SO_4 + 2NaBr \longrightarrow Na_2SO_4 + 2HBr \tag{11}$$

$$Pt^o + 2Br_2 + 2HBr \longrightarrow H_2PtBr_6 \tag{12}$$

The stability of the platinum bromide complex is aided by the presence of hydrobromic acid, implying an excess of sulfuric acid and sodium bromide may be required. The E_h-pH diagram for bromine in aqueous solution, as given by A.J. Bard in "Encyclopedia of the Electrochemistry of The Elements" shows, that very low pH's are required with redox potentials above 750 mV to maintain $HBrO_3$ in solution.

3 Experimental

3.1 Ore Preparation

Since mineralogical investigations had found that the particle sizes of most PGM mineral particles were less than 10 microns, fine grinding was carried out to liberate the minerals of interest. Both wet and dry grinding were tested and the results are given in Table 1.

Table 1: Particle Size Distribution of Oxide Ore

Sample	Particle Size <38 μm	38-75 μm	75-150 μm	>150 μm	Total
Dry Milling (% w/w)	91.7%	5.2%	1.3%	1.8%	100%
Wet Milling (% w/w)	86.9%	12.7%	0.4%	---	100%

The different particle size fractions were sent for analysis to determine their PGM content. The results are shown in Table 2.

Since the particles smaller than 38 μm contained the largest percentage of PGM's, a grinding program was designed so almost all the particles were of this size. This program is equivalent to the dry mill grinding, described above, which yielded 92% of the particles with a size <38 μm. All samples were prepared following this method. Before roasting, the ground ore was screened using 70 mesh, so that all particles had a size less than 212 μm.

Table 2: PGM content for size distribution of particles

Element	Particle Size <38 μm	38-75 μm	75-150 μm
Gold	0.55 ppm	0.42 ppm	0.25 ppm
Platinum	3.90 ppm	1.74 ppm	0.17 ppm
Rhodium	0.3 ppm	0.1 ppm	0.0 ppm
Gold Distribution	89.8%	10.0%	0.2%
Platinum Distribution	93.9%	6.1%	0.0%
Rhodium Distribution	95.4%	4.6%	0.0%

3.2 Equipment and Operation

The ground Hartley Oxide ore was roasted, initially in a muffle furnace and later in a rotary furnace. Since the muffle furnace gave erratic results, a change-over was made to the rotary furnace where temperature control was much more consistent and the gas environment could be controlled and adjusted.

The following sections outline the ranges of the major parameter for this investigation:

Roasting:	temperature	- 150°C to 1050°C
	time	- 15 min to 2 hours
	atmosphere	- air, oxygen, nitrogen
	oven type	- muffle or rotary kiln
	flow rate	- 0.5 L/min to 6 L/min

Leaching:	temperature	- 25°C, 40°C, 70°C
	time	- 2 hours to 24 hours
	sulfuric acid	- 50 g/L, 100 g/L
	sodium bromide	- 10 g/L, 20 g/L, 40 g/L
	ORP or Redox	- uncontrolled or >800 mV
	solid concentration	- 240 g/L or 400 g/L
Preleaching:	temperature	- 25°C, 40°C, 70°C
	time	- 2 hours or 4 hours
	sulfuric acid	- 50 g/L, 100 g/L, 150 g/L
Recovery:	carbon	- synthetic and process solutions
		- with solutions, CIP and CIL
	resin	- synthetic and process solutions
		- in beaker and IX column

The bromine leaching was performed in an enclosed glass flask using a magnetic stirrer and a heating plate. Geobrom 3400, containing 34% free bromine, was used to control the redox potential. Samples were taken from the flask by syringe. Immediate high pressure filtering stopped the liquid-solid reaction. The recovery of gold, platinum and rhodium was calculated using the assay data of the leachate, which was analyzed by ICP, while the feed calcine and the residue were analyzed by a proprietary fire assay technique.

After initial testing, a standard bromine leaching procedure was established as follows:

Leaching:	temperature	- 70°C
	time	- 2 hours
	sulfuric acid	- 100 g/L
	sodium bromide	- 10 g/L
	ORP	- >800 mV
	solid concentration	- 240 g/L
Preleaching:		- none

The Stillwater Process required a preleaching stage to achieve high rates of extraction. Therefore, investigations were performed to determine the effectiveness of such a step for the Hartley oxide ore. Using the Stillwater Process preleaching conditions as a starting point, a Hartley Oxide sample was preleached with sulfuric acid followed by a standard bromine leach. It was found that, in contrast to the Stillwater leach, the preleaching step has a negative effect on the initial platinum dissolution rate. It was interesting that the Pt extraction with the preleaching stage was directly proportional to the time elapsed, while without preleaching the maximum Pt extraction was obtained almost instantaneously.

4 Results and Discussion

4.1 Overview

Initial leaching test work showed promise as 50% of the platinum in ore, which was not pretreated, was dissolved in a sodium bromide/sulfuric acid solution where the E_h was controlled at about 800 mV vs SCE. However, the one major draw back was the excessive Geobrom (bromine) consumption. The origin of the high consumption was traced to the presence of ferrous iron in the oxide mineral matrix. This problem was solved by the introduction of a simple roasting procedure with higher temperatures breaking down the platinum arsenide minerals and destroying the hydrated silicate matrix and where air could be used to oxidize the ferrous iron. This procedure was tested and resulted in a reduction in the Geobrom consumption, down to about 10% and platinum recoveries increased to around 70%. However, overall PGM recoveries were, though acceptable, not yet very high. Since processing of Great Dyke sulfide ore gave an 80% PGM recovery, it was decided that similar recoveries should be obtained from oxide ores. Furthermore, rhodium dissolution never exceeded the ten percent mark. Higher roasting temperatures were investigated to further minimize Geobrom consumption and to better recoveries, but platinum extractions did not improve.

Regrinding of leach residue was, to our surprise, somewhat successful in enhancing the platinum extractions, indicating that agglomeration during the roasting procedure might have locked up some platinum that could have been liberated to begin with. Fortunately, a rather large number of roast-leaching experiments had been conducted. Low temperature roasting (300°C and lower) had initially been rejected because of high Geobrom consumption, but an improvement in the dissolution of rhodium led to an approach whereby rather than maximizing PGM recovery, the dollar recovery was used as a guideline. This resulted in a conclusion that lower temperatures, rather than higher ones, were called for. It was found that a significant benefit was gained by roasting at a temperature lower than 425-450°C: platinum extraction improved from 65-70% to 85-90%, whereas rhodium recovery was raised from 10-20% to 60-70%. The postulation that rhodium is associated with an hydrated iron fraction was confirmed when efforts were made to physically concentrate PGM's by means of flotation.

4.2 Roasting

4.2.1 Roasting Temperature and Time

Since DTA and XRD analyses showed that the mineral phases were temperature-sensitive, investigations were made into the effect of roasting temperature on the PGM dissolution. Figure 3 shows the optimum roasting temperatures for gold, platinum and rhodium.

Although gold dissolution is slightly higher with a roasting temperature in the range of 675°C, it remains around the 90-95% level. However, a rapid decline in platinum extraction is noted around 700°C, which corresponds to the recrystallization temperature of chlorite. From the graph it can also be established that a

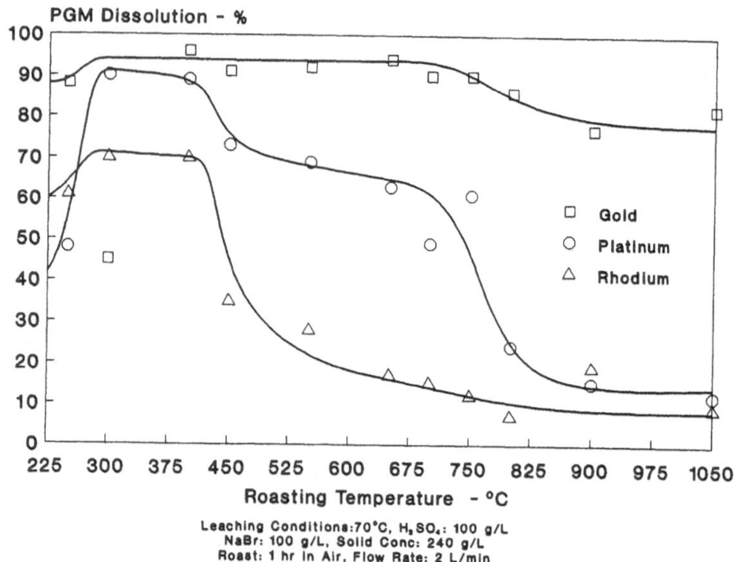

Fig.3. Effect of roasting temperature on PGM dissolution.

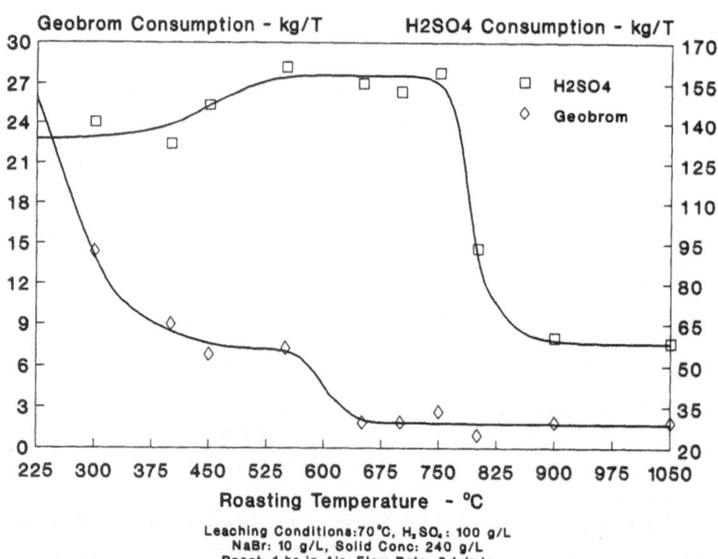

Fig.4. Effect of roasting temperature on Geobrom and acid consumption.

high extraction window, in particular for rhodium, exists in the 300-450°C temperature range. It is important to note that this range is below the decomposition temperature of goethite. It can be postulated that most of the rhodium and some of the platinum values in the weathered oxide exist in species that have been adsorbed onto the hydrated iron oxide surface of goethite, FeOOH.

It is important to consider the influence of the roasting temperature on reagent consumption, which sets the extent of oxidation. At roasting temperatures less than 450°C only partial oxidation of lower valence metal ions, sulfides and arsenides occurs and as shown in Figure 4 results in a substantial consumption of Geobrom 3400 during the leaching process.

Specifically, the conversion of ferrous oxide to ferric oxide and the removal of the hydroxyl ions from the hydrated silicates must occur to decrease both the Geobrom 3400 and acid consumption. From Figure 4, it can be established that a roasting temperature greater than 550°C is desired to minimize the Geobrom 3400 consumption, but this does not correspond to the highest platinum and rhodium recovery. The increased Geobrom 3400 consumption due to a decrease in roasting temperature from 425°C to 300°C, is compensated for by the sharp increase in recovery of the platinum, and more importantly, the rhodium. It has been estimated that with Geobrom costing $0.70 per pound and platinum and rhodium selling for $400 and $4000 per ounce, respectively, one percent platinum or rhodium recovery represents the same value as one pound of Geobrom for platinum and three quarter pounds Geobrom for rhodium. This assumption does not include the potential benefits of bromide/bromine regeneration and recycle.

Figure 4 also shows the sulfuric acid consumption as a function of roasting temperature. Although a roasting temperature above 750°C is optimum for minimizing sulfuric acid consumption, the corresponding platinum and rhodium dissolution is very low. Below 750°C, the acid consumption is relatively constant. However, there is a slight dip in the acid consumption in the temperature range 300-425°C.

A simplified cost analysis of the PGM extraction as a function of temperature is shown in Figure 5. Taking about 150 data points, the platinum and rhodium recoveries and the Geobrom 3400 and sulfuric acid consumptions were averaged for each roasting temperature and plotted to indicate overall trends. Surprisingly, the rhodium recovery curve and the Geobrom cost curve follow each other, remaining constant at temperatures greater than 700°C and increasing identically below 700°C. The cost of sulfuric acid remains relatively constant until it begins dropping above 750°C. A significant drop occurs in the platinum recovery curve at 700-900°C, but at temperatures less than 700°C, it steadily increases. At lower temperatures, the increase in rhodium recovery almost compensates for the increase in Geobrom consumption. As the sulfuric acid consumption is relatively constant, the increase in platinum recovery at lower temperatures suggests the roasting temperature should remain in that range. These cost data do not include the potential benefits of bromide/bromine regeneration and recycle, as indicated.

In conclusion, it can be said that the hypotheses outlined earlier should be relatively accurate, resulting in a recommendation of roasting temperatures ranging

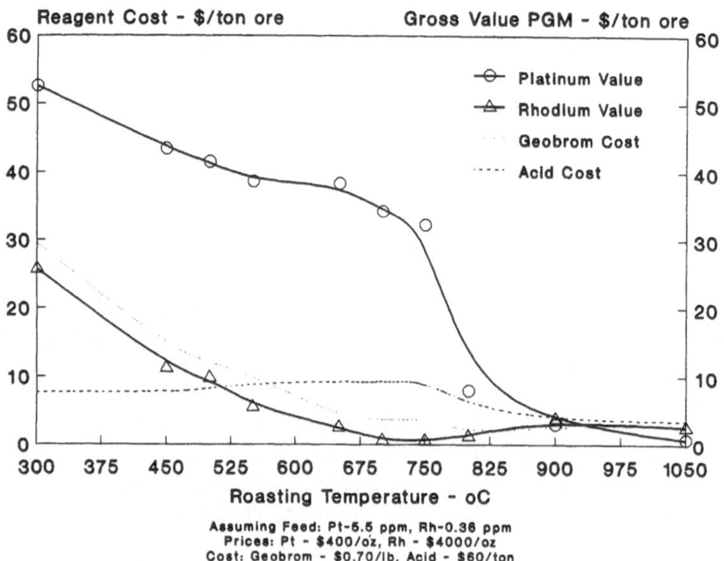

Fig.5. Simplified cost analysis utilizing metal values and reagent costs.

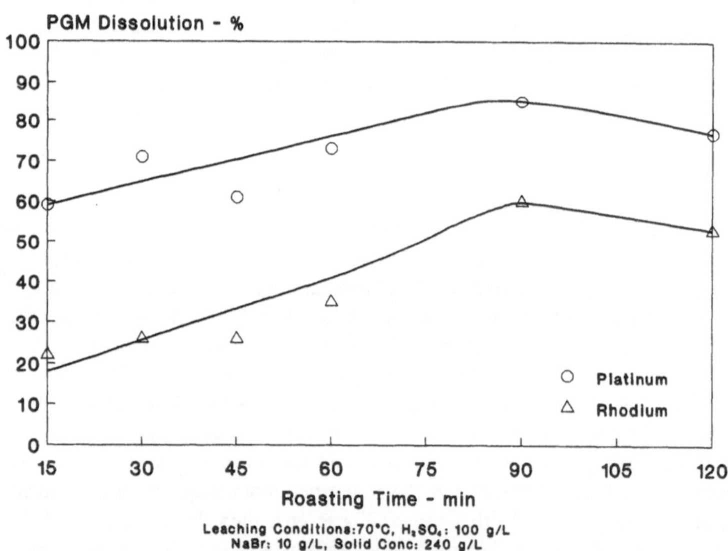

Fig.6. Effect of roasting time on PGM dissolution

from 300°C to 425°C. As expected, lower platinum recoveries were noted at temperatures above 700°C and increases in platinum and rhodium recoveries below a roasting temperature of 425°C.

As Figure 6 indicates, platinum recovery appears to reach a maximum at 90 minutes roasting for temperature of 450°C and 550°C. Further work will be necessary to determine the most beneficial roasting time for the recommended temperature range of 300-425°C.

4.2.2 Gas Atmosphere and Flow Rate

Two types of roasting ovens were used in the investigations, a muffle furnace (fixed bed, natural ventilation) and a rotary kiln (moving bed, forced ventilation). As depicted in Figure 7, use of the muffle furnace resulted in higher dissolutions of gold, platinum and rhodium than with the rotary kiln. There is no clear explanation for this difference, but it is felt that the actual temperature in the muffle furnace might have been lower than was indicated by the thermocouple. It was also felt that the composition of the gas atmosphere might have had an influence. Therefore, most of the roasting work was carried out in the rotary kiln as it allows for variation of the atmospheric parameters, such as gas composition and flow rate, necessary for an in-depth investigation. It is suggested that further tests be performed with the muffle furnace for the temperature range 300-425°C to confirm our postulations.

The effect of the atmosphere gas was investigated by roasting the ore in air, in oxygen at different partial pressures and in nitrogen environments. Table 3 shows the PGM recoveries and reagent consumptions under three typical atmospheres.

This table shows that both atmospheres of no oxygen and excessive amounts of oxygen is detrimental for PGM extraction. Oxygen reduces the acid consumption, whereas nitrogen results in excessive bromide consumption. Figure 8 shows the effect of oxygen partial pressure on the extraction of gold, platinum and rhodium, when roasting at 500°C. In the range of 0% to 15-20%, there is a noticeable increase in PGM recovery. Above a concentration of 20%, the decrease in extraction is negligible for gold and platinum and minimal for rhodium. It is suspected that the 100% oxygen environment does not increase recovery because it encourages over-oxidation of the ore, converting platinum metal to a very stable platinum oxide, which is not leachable in mildly acidic conditions.

Table 3: Effect of atmosphere on PGM recovery and reagent consumption

Gas Atmosphere	Gold Recovery (%)	Platinum Recovery (%)	Rhodium Recovery (%)	Acid (kg/ton)	Geobrom (kg/ton)
Air	94%	80%	43%	161	9.0
Oxygen	96%	73%	30%	109	9.2
Nitrogen	49%	62%	44%	144	82.9

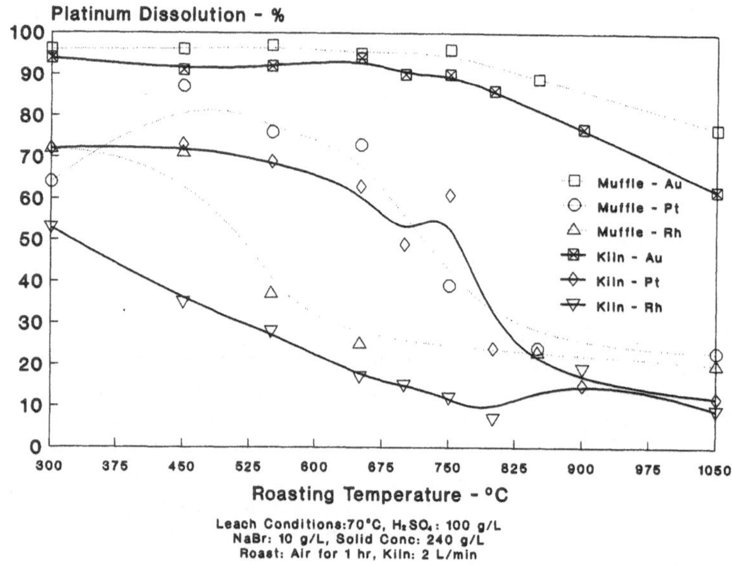

Fig.7. Effect of oven type on PGM dissolution.

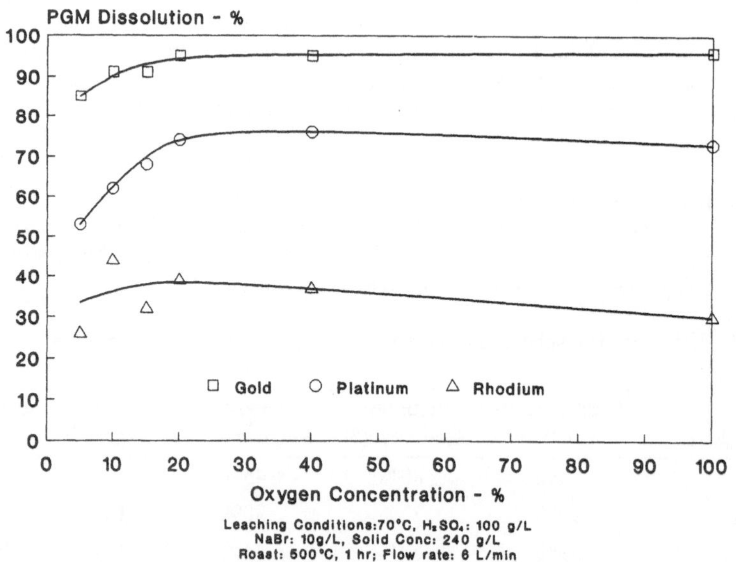

Fig.8. Effect of oxygen concentration on PGM dissolution.

The consumption of Geobrom 3400 is related to the roasting environment as given in Figure 9. With no oxygen present in the environment, approximately 10 times more Geobrom 3400 is required than if some oxygen is available.

Investigation into the effect of the flow rate on the PGM dissolution proved inconclusive. It was determined that, under the range investigated, the flow rate did not have any noticeable influence on PGM recovery, as shown in Figure 10.

4.3 Leaching

4.3.1 Leaching Temperature and Time

Leaching the ore at three temperatures, 25°C, 40°C and 70°C, proved, not unexpectedly, that a higher leaching temperature leads to a higher PGM dissolution and a higher reagent consumption, as shown in Figure 11. Although changing the leach temperature from 40°C to 70°C caused a 25% increase in the H_2SO_4 consumption, platinum dissolution rose 10% and rhodium dissolution rose 4½ times. Leaching at 70°C is recommended because the increased H_2SO_4 cost is compensated for by the increased platinum and rhodium recovery. Even higher temperatures might be more beneficial, but as the use of rubber lined equipment is anticipated for a commercial plant, no such tests were carried out.

As can be expected, PGM dissolution increases with time, as in Figure 12. Since the increase is not very significant, a time of two hours was determined to be optimal taking industrial and economic factors into account.

4.3.2 Concentration of Leaching Agents

Sulfuric acid concentration - According to Equations (10), (11) and (12) the function of sulfuric acid in bromine leaching is to unlock PGM's by reaction with base metal oxides and hydroxyl radicals in the silicate matrix, and to react with sodium bromide to produce hydrobromic acid, which stabilizes the platinum bromide complex, $[PtBr_6]^{2-}$. Thus higher sulfuric concentration should be favorable for platinum recovery. This relationship is shown in Figure 13. To maximize PGM extraction and to minimize acid consumption, the optimal starting sulfuric concentration was determined to be 100 g/L.

Sodium Bromide Concentration - Figure 14 suggests higher NaBr concentrations increase the extraction of gold, platinum and rhodium. Further investigation should be completed to determine the most beneficial concentration level. Although the sodium bromide is not consumed during the leaching process, a low sodium bromide concentration is called for since losses are anticipated with tailings disposal as well as with the waste water treatment circuit.

Geobrom 3400 Consumption - The solubility of bromine in water is about 3.43 grams per 100 mL at 25°C and about 3.58 grams per 100 mL at 54°C. Excess use of Geobrom will increase its consumption. Initially, a standard amount of Geobrom 3400 was added during leaching. Close analysis showed that it was only necessary to add Geobrom 3400 drop wise to keep the ORP of the leach solution above the critical level of 800 mV. As the consumption of the Geobrom 3400 varied significantly for different leaching conditions, it became one of the

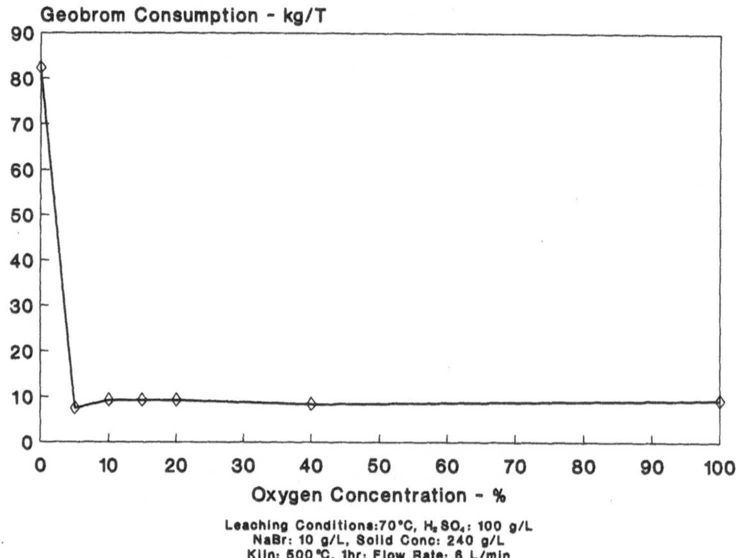

Fig.9. Effect of oxygen concentration on Geobrom consumption.

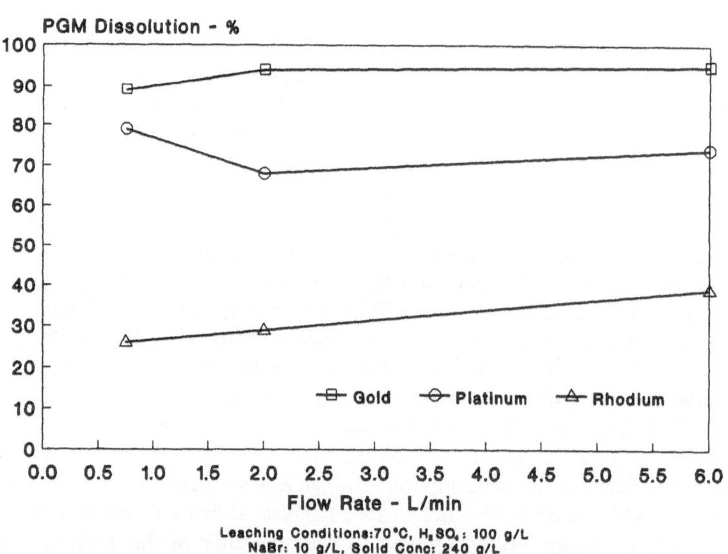

Fig.10. Effect of flow rate on PGM dissolution.

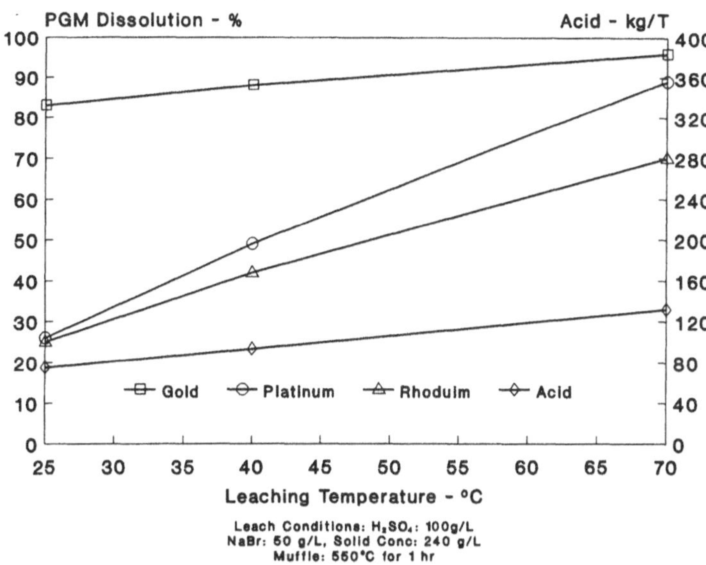

Fig.11. Effect of leaching temperature on PGM dissolution and acid consumption.

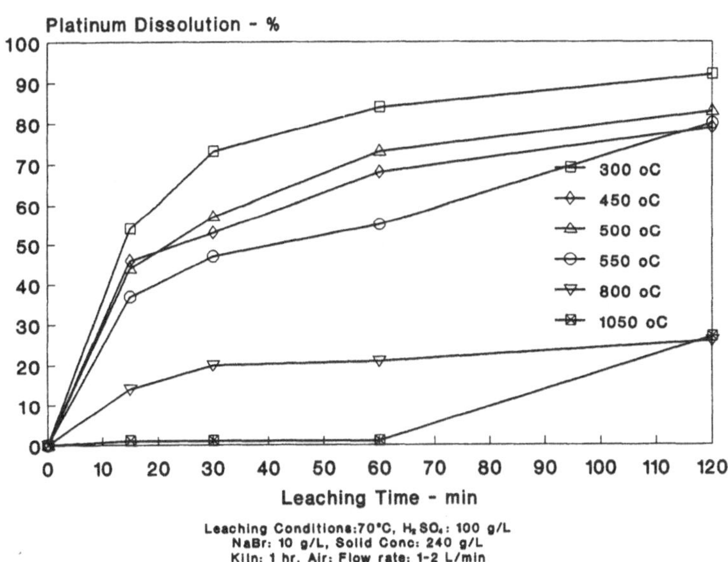

Fig.12. Effect of leaching time on platinum dissolution.

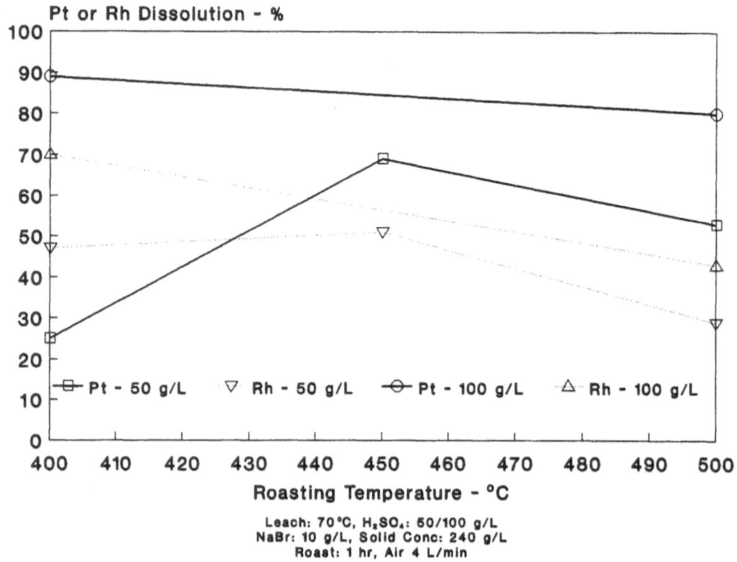

Fig.13. Effect of acid concentration on PGM dissolution.

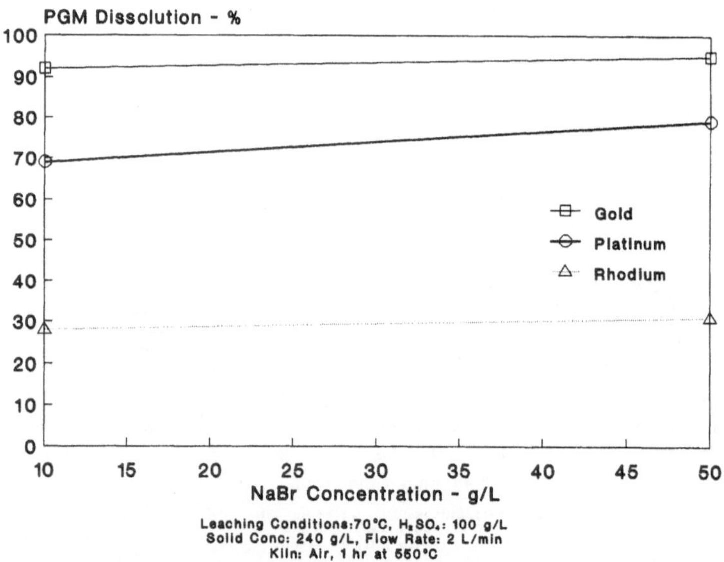

Fig.14. Effect of sodium bromide concentration on PGM dissolution.

significant factors in calculating the economics for each condition.

An interesting trend was noticed when plotting average platinum dissolution versus average Geobrom 3400 consumption, as shown in Figure 15. It is clear that an increase in the amount of Geobrom 3400 required corresponds to an increase in platinum dissolution.

4.3.3 Leach Solution Composition

Table 4 shows the leachate composition of different calcines under standard bromine leaching conditions. It was found that the leaching reactivity of major metallic elements, with exception of gold, decreased with the increasing temperature. This verified the general influence of the phase transfer of silicates matrix on the leaching reactivity.

Table 4: Leachate composition of calcines roasted at different temperatures

Metal	Temperature				
	300°C	450°C	550°C	800°C	1050°C
Al (g/L)	3.35	1.98	2.19	0.45	0.49
Au (mg/L)	0.10	0.08	0.10	0.15	0.12
Cr (g/L)	0.15	0.11	0.11	0.03	0.01
Cu (g/L)	0.26	0.11	0.11	0.01	0.02
Fe (g/L)	8.55	6.09	6.16	0.61	0.59
Mg (g/L)	3.55	3.26	3.60	0.60	0.35
Ni (g/L)	0.46	0.28	0.37	0.06	0.02
Pt (mg/L)	1.41	1.16	1.17	0.31	0.31
Rh (mg/L)	<0.01	<0.01	<0.01	<0.01	<0.01

4.3.4 Solids Concentration

Most tests were performed with 240 g/L solids. A few tests were done with 400 g/L but no noticeable difference in PGM dissolutions were observed. From an acid consumption point of view, the highest possible solids concentrations should be recommended.

4.4 Recovery of Metal

4.4.1 Adsorption onto Activated Carbon

Based on similar work done on the loading of gold from bromide solutions by Taylor and Sergent (1990), the use of activated carbon for PGM recovery was investigated.

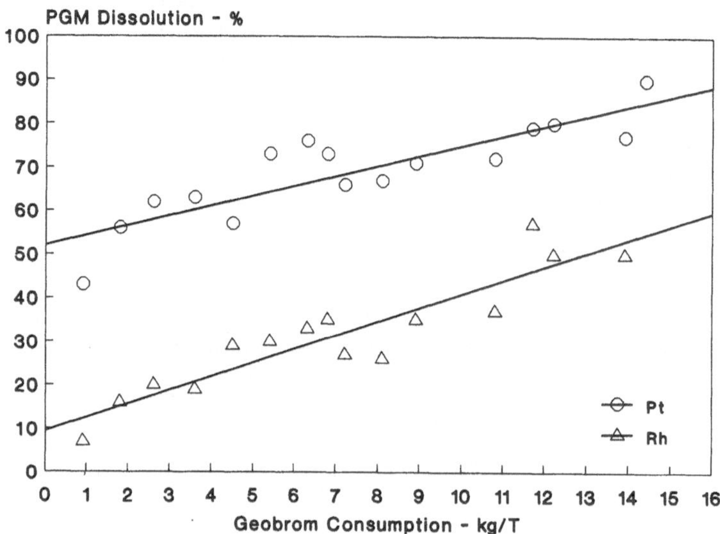

Fig.15. PGM dissolution as a function of Geobrom consumption.

Carbon loading was tested with a synthetic solution that was made by dissolving platinum sponge. The platinum adsorption was over 95% after the first 30 minutes and 99% after one hour. This indicated that the loading kinetics are substantially faster when compared to gold loading from cyanide leach solutions where, generally 24 hours are required for a 99% loading efficiency.

An investigation was performed on the carbon-in-pulp method for platinum recovery, by adding carbon to a standard large scale leaching test. It was noted that the addition of the carbon caused an immediate drop in the ORP. The leaching recoveries for platinum, gold and rhodium were 87%, 91% and 51%, and the carbon loading efficiencies were 90%, 99% and 1%, respectively. The inability of the carbon to load rhodium needs further investigation. On one hand this may give an opportunity to separate platinum from rhodium or alternatively this may preclude the use of carbon for adsorption of all PGM's. These results indicate that further work is needed to determine the effectiveness of CIP. High carbon loading was possible indicating the opportunity to ash carbon for platinum recovery.

4.4.2 Ion Exchange Resin

Previous work (Pt sponge dissolved in H_2SO_4, NaBr and Geobrom) had shown the effectiveness of adsorption of gold bromide complexes onto resin and it was felt that the TML Process solution might respond in a similar fashion. The resin used for the adsorption test was DOWEX-21K, a common resin produced by the Dow Chemical Corporation. It is a strong-base resin containing tri-methyl-amine functional groups with a copolymer of styrene and divinyl benzene matrix.

Initial tests showed that the resin was able to remove most of the platinum from leachate. Elution of the resin was not complete. After a second cycle adsorption had dropped to about half the level of the first cycle and further work on this method is needed. Rhodium may be recovered by ion exchange after carbon recovery of gold and platinum.

4.4.3 Bromide/Bromine Recovery

At 50 cents a pound for a 45% solution of sodium bromide, the leach solutions represent a significant value that needs to be recirculated. The counter current decantation washing of the leach residue is an effective way to minimize chemical losses. However, waste water processed in the water treatment plant should also be processed for bromide recovery. Ion exchange has been successfully tested by others and was considered to have potential for Hartley. The production of a rather concentrated bromide stream also provides the added advantage that bromine can be produced or that Geobrom can be regenerated. Both steps would result in significant savings in chemical costs. The recovery of bromine/bromide involves two distinct steps: first the recycling of the bromide ions and then the subsequent electrochemical regeneration of bromine.

4.4.4 Bromide Recovery

Amberlite IRA-68 (Rohm and Haas), a weak base anion exchange resin, was used for bromide recycling. After adsorption of platinum, leachate with a bromide concentration around 8 g/L was pumped through a column filled with IRA-68 resin. The loading of bromide on the resin was considered finished when the ORP and pH of outlet solution approaching the ORP and the pH of the feed solution. Then the column was washed with de-ionized water till a neutral solution was obtained. In the stripping procedure 1*M* sodium hydroxide solution was passed through the column till the pH of the outlet solution increased sharply. The bromide assay showed that IRA-68 resin had a good capacity and selectivity for bromide. More work still needed to be done.

Table 5: Bromide concentrations with IRA-68 resin ion exchange resin

Solution	Br^- in g/L
Barren Solution	0.20
Wash Solution	<0.01
Eluate	9.19

4.4.5 Bromine Regeneration

Although no test work was carried out relative to platinum leaching, it has been successful and economic in pilot-scale operation from a gold ore leach/recovery circuit, as shown by Howarth et al. (1991).

The Chloropac system from Electrocatalytic Inc. was tested for bromine regeneration. It consists of an outer sub-assembly of two titanium pipes separated by an insulating spacer. The inner titanium cylinder was positioned concentrically. The inside of the outer pipes and the outside of inner pipe formed the electrode surfaces. The anodic portion of the electrode had a platinum coating designed to catalyze the electro-oxidation of halide ions, and direct current was passed across the anode and cathode connectors. The current efficiencies were 80-90% and unaffected by bromide tenor within 2.5 to 5% range provided the electrolysis was run to a low conversion (<15%). This electrolytic regeneration of bromine in a gold leach/recovery circuit dramatically improved the process economics over cyanidation.

4.5 Flowsheet and Cost Estimates

A conceptual process flowsheet is presented in Figure 16. A preliminary engineering study indicated capital and operating costs similar to plants that process refractory gold ores.

4.6 Final Process Conditions

The following parameters have been determined to yield a gold recovery of 90%, a platinum recovery of 85% and a rhodium recovery of 70%:

Roasting:	temperature	- 300 to 700°C
	time	- 60 to 90 minutes
	atmosphere	- air
	oven type	- rotary kiln
	flow rate - 2 L/min	
Leaching:	temperature	- 70°C
	time	- 2 hours
	sulfuric acid	- 100 g/L
	sodium bromide	- 10 g/L
	ORP (Geobrom 3400)	- >800 mV
	solid concentration	- 240 g/L or 400 g/L
Preleaching:		- none

5 Conclusion

The work resulted in the following findings:

(a) Mineralogical investigations HAVE indicated that the dominant platinum phases in Hartley oxide ore were sperrylite ($PtAs_2$) and braggite (PtPdNiS). The gangue minerals were found to primarily consist of hydrated silicates (talc and chlorite etc.), iron(II) oxide, goethite, and minor amounts of ilmenite and chromite.

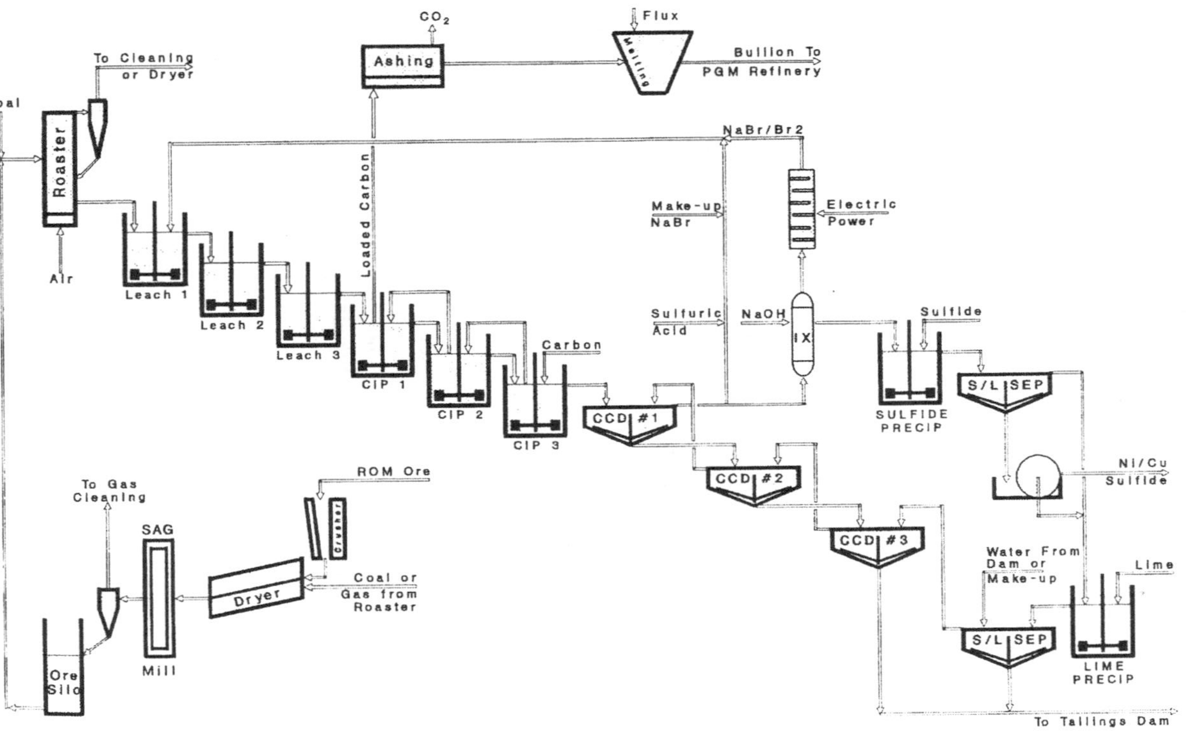

Fig.16. The TML Process for Hartley oxide ore.

(b) XRD, DTA and leaching investigations indicated that the phase transfer of gangue minerals were sensitive to temperature, and their influence on leaching was conclusively established. Low temperature roasting, in range of 300 to 425°C, caused a phase transfer of chlorite from crystalline to amorphous state. This increased the reactivities of PGM and base metals in the leaching procedure. Roasting at temperatures higher than 425°C resulted in the partial lock up of platinum, but especially rhodium, in the hematite phase that resulted from the decomposition of goethite. Roasting at even higher temperature (>700 °C) leads the recrystallization of the silicates, which resulted in a sharply reduced leaching reactivities of PGM and other metals. The optimum roasting temperature window for maximum PGM extraction was determined to be in the range of 300 to 425 °C in an atmosphere of air. The higher the roasting temperature, the lower the bromine consumption.

(c) The acidic bromine/bromide lixiviant system effectively leached PGM and gold from Hartley oxide ore due to the formation of the stable, deliquescent bromide complex acid. Ninety-five percent of gold, 85-90% of platinum and 70% of rhodium were solubilized at 70°C within three hours by a lixiviant consisting of 10 gpl sodium bromide plus 100 gpl sulfuric acid and with the redox potential over 800 mV maintained with Geobrom 3400. These leaching conditions were much milder than the aqua regia and other hydrochloride system.

(d) The aqueous platinum and gold bromide complexes were readily loaded onto activated carbon. The loading capacity was dependent on the concentration of platinum complex in the leach solution. Carbon stripping test work was not carried out as high loading capacities were obtained, which provided an opportunity to ash the carbon rather than to strip it. Dowex21K, an anion ion exchange resin, could also load the platinum bromide complex, but stripping and regeneration was not possible as the resin degraded quickly due to the oxidation by the residual free bromine in solution.

(e) Amberlite IRA-68, a weak base anion exchange resin, showed the ability and selectivity to recover the residual bromide in the leachate. This further increased the economic advantage of bromine/bromide lixiviant system.

6 Acknowledgement

The authors would like to acknowledge the contributions by Doug Ellsworth and his staff in the development of the various analytical techniques for PGM determinations, by Puru Shrestha in assisting with the mineralogical evaluation and interpretation and by Sue Martineau in carrying out the extensive bench scale test program.

7 References

Baglin, E.G. Gomes, J.M. Carnahan, T.G. and Snider, J.M. (1985), **Recovery of Platinum, Palladium, and Gold from Stillwater Complex Flotation Con centrate by a Roast-Leaching Procedure**, BuMines RI 8970.

Bonucci, J.E. and Parker, P.D. (1984), Recovery of PGM from Automobile Catalytic Converters, **Precious Metals, Mining, Extraction, and Processing,** TMS, Warrendale, PA.

Dadgar, A. (1989), Refractory Concentrate Gold Leaching: Cyanide vs. Bromine, **JOM**, 12, 37-41.

Dadgar, A. Sanders, B.M. McKeown, J.A. Sergent, R.H. and Jacobson, R.H. (1990) Leaching and Recovery of Gold from Black Sand Concentrate and Electrochemi cal Regeneration of Bromine in **Proceedings of the Annual Meeting of the AIME**, Reno, NV.

Demopoulos, G.P. (1989) Refining of Platinum Group Metals, **CIM Bulletin**, 82, 165-170.

Duyvesteyn, W.P.C. Liu, H. and Duyvesteyn, S. (1992a) **Dissolution of Platinum Group Metals from Platinum Metals Containing Materials**, US Patent Application 845,068, June 1992.

Duyvesteyn, W.P.C. Liu, H. and Duyvesteyn, S. (1992b) **Recovery of Platinum Group Metals from Oxide Ore**, US Patent Application 896,675, June 1992.

Howarth, J. Dadgar A. and Sergent H. (1991) Electrochemical Regeneration of Bromine in a Gold Ore Leach/ Recovery Circuit in **Proceedings of EPD Congress**, New Orleans, LA (ed. D.R. Gaskell), TMS, 709-718.

Letowski F.K. and Distin, P.A. (1985) Platinum and Palladium Recovery from Spent Catalyst by Aluminum Chloride Leaching, **Recycling and Secondary Recovery of Metals**, TMS, Warrendale, PA.

Papademetriou, T. and Grasso, J.R. (1970) Recovery of Precious Metals from South African Matte", **Engelhard Ind. Techn. Bulletin**, 3, 121-129.

Prendergast, M.D. and Wilson, A.H. (1989) The Great Dyke of Zimbabwe II: Mineralization and Mineral Deposits, **Transactions IMM**, 185, C21-42.

Remy, H. (1956) **Treatise on Inorganic Chemistry**, Translated by J.S. Anderson, (ed. by J. Kleinberg), Vol. II, Elsevier, New York, NY.

Taylor, P. and Sergent, H. (1990) The Adsorption of Gold onto Activated Carbon from Bromide Solutions, **Proceedings of the Annual Meeting of the AIME,** Los Angeles, CA.

Development of Intec Copper Process by an international consortium

P. K. Everett
Intec Pty., Ltd., Chatswood, New South Wales, Australia

Abstract

The Intec Copper Process is a low cost, chloride based, hydrometallurgical method of converting copper sulphides to very high purity copper (+99.99%) and elemental sulphur. The process has a number of novel steps including the purification, without solvent extraction, and electrowinning at high current densities, to result in the low cost structure determined in a very extensive study by a major international mining company.

The technology has been developed at a time when sulphur dioxide pollution regulations are becoming much more stringent. The fact that sulphuric acid can now have a large negative value in the smelter economics, means that there is a large incentive for development of the Intec Copper Process.

To facilitate the development of the process, a consortium of eleven international mineral companies has been formed.

The process utilises a multi-stage countercurrent leach at 80-85°C with the addition of air at atmospheric pressure to precipitate iron as a goethite type compound, followed by a two stage purification process separating impurities in metallic and other forms. Purified cuprous chloride/sodium halide solution is fed to an electrolytic diaphragm cell where high purity copper granules are produced at current densities of 1,000-1,500 A/m^2 without the need for conventional electrode stripping. Power consumption is approximately half that of the cupric route. The product is suitable for the direct powder metallurgical conversion of the granules to products such as wire or strip. All the steps in the "furnace free" conversion of copper mineral to product have been demonstrated. It is recognised that other marketable forms are required and work has been carried out to convert the copper granules to rhondelles, briquettes and sheet form similar to cathode.

The process can treat low grade and contaminated concentrates without penalty, thereby offering greater returns to the miner with much greater flotation recoveries and decreased grinding costs. Silver may be recovered as metal, with the high oxidation potential of the electrolyte allowing the efficient extraction of gold. The gold is selectively precipitated on activated carbon with negligible copper and silver contamination.

This paper describes the process and the plans of the consortium to take it to commercialisation.

Keywords: Elemental sulphur, hydrometallurgical, international consortium, low cost copper.

1 Introduction

There are several new factors affecting the economic and environmental performance of copper smelters. These include the increasingly severe regulations governing sulphur dioxide emissions, and also the difficulty of marketing sulphuric acid which may have a severe impact on smelter viability.

There has probably never been as much incentive for the development of a process which:

- Has low capital and operating costs
- Is environmentally friendly
- Forms elemental sulphur rather than sulphur dioxide
- Operates at atmospheric pressure and moderate temperature
- Recovers precious metals
- Will accept low grade and contaminated concentrates
- Can recover associated base metals in metallic form
- And most importantly, recover copper in very high purity form.

Copper smelters normally require a high grade feed (+20% copper) and are unable to handle many metallic and other contaminants. The net regulatory-cost burden being carried by U.S. smelters in 1987 was estimated by Rothfeld and Towle (U.S. Bureau of Mines - 1989) [1] to be as high as U.S. 7.5 cents per pound of copper, which is over 45 per cent of their total estimated operating cost. In Europe, it has been estimated that a negative U.S.$100/t value for sulphuric acid will cost copper producers U.S. 4.5 cents/lb of copper (up to 30% of their operating costs).

Many attempts have been made to overcome the problems of the smelting processes, resulting in the expenditure of hundreds of millions of dollars in research to find alternatives, usually with processes which recover the sulphur in elemental form. These alternative processes have usually been uneconomic due to the requirement for high pressures and/or temperatures, expensive reagents, expensive materials of construction and the production of an impure product. The most successful of these processes, Duval's "Clear" process (Schweitzer and Livingston, 1982) [2] used pure oxygen in titanium autoclaves at up to three and a half times atmospheric pressure and temperatures up to 140°C. The product was impure and required electro-refining.

Another development, the Dextec copper process [3], used a slurry electrolysis technique at 80-85°C with air at atmospheric pressure to reject iron, thereby allowing the use of inexpensive materials of construction such as fibreglass and polypropylene. The product, while produced at low power costs, did not meet top market specifications.

It was thought that the reason for being able to operate the Dextec process under such mild conditions, was due to the role of the anodes in the slurry. This view was supported by a university study and a U.S. patent (Kruesi, 1972) [4] showing the importance of current density on the anodes in accomplishing the leaching of the copper mineral in the slurry. The Dextec process was successfully tested in a nine cell pilot plant by Anglo American Corporation of South Africa and a derivative of the Dextec Copper Process was also piloted by Boliden Metal A.B. in Sweden.

Extensive research by Intec into reaction mechanisms in 1989 showed that the presence of the anodes was not necessary to accomplish the conversion of chalcopyrite to ionic copper under such mild conditions. This insight into the reaction mechanisms allowed Intec to separate the leaching and electrowinning operations and, following development of new purification technology, resulted in the development of the Intec Copper Process. The process produces L.M.E. "A" grade copper, with the advantage of mild operating conditions, and substantial other economic advantages achieved by the ability to optimise each of the unit operations of the process separately. The process recovers copper from the monovalent state in high intensity electrolytic diaphragm cells with the generation, in the anode compartment, of oxidised species at an oxidation potential high enough to leach gold. These oxidised species (halogen complexes referred to as "halex" and described later in this paper), allow the storage of anodic energy in soluble form, with considerable advantages over conventional methods, such as ferric and cupric ions, and chlorine gas.

In 1990, over a period of approximately nine months, BHP Minerals and BHP Engineering carried out a very extensive study of the economics and chemistry of the Intec Copper Process, at scales of 15,000, 50,000 and 200,000 t.p.a. of copper production. Figure 1 shows the economic comparisons. Smelting was not compared at the 15,000 t.p.a. scale and oxidative leaching/SX/electrowinning was not compared at the 200,000 t.p.a. scale. The results showed that the Intec Process was greatly superior to the best available alternative hydro-metallurgical processes and also the best available smelting processes selected by BHP. Since the BHP study, substantial improvements have been made to further reduce the capital and operating costs of the Intec Copper Process.

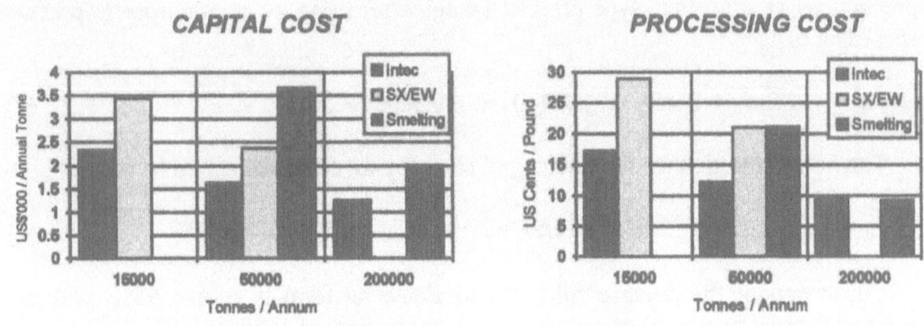

Fig. 1 **BHP Minerals economic comparison**

The reasons for the low capital and operating costs of an Intec copper plant can be understood when the relevant sections of the plant are examined in detail.

The high cost of sulphur dioxide containment is eliminated, with the sulphur reporting to the leach residue in elemental form. No account was taken, during the BHP revenue calculations, of the value of this saleable by-product.

Copper is electrowon from the monovalent state requiring approximately half the power compared to electrowinning from the divalent state. Coupled with this two times productivity increase, the electrowinning cell uses high current densities of four to six times conventional for even greater productivity. The combined effect of these two factors is an electrowinning cell with approximately eight to twelve times the productivity per unit of electrode area, compared to the conventional sulphate system.

Conventional electrode stripping is eliminated with the particulate copper recovered in slurry form. In addition, because of the absence of gas evolution, considerable problems with the formation of acid mist (in the sulphate system) which are currently the subject of severe E.P.A. regulations, are eliminated.

Operation is at temperatures below 100°C and at atmospheric pressure, allowing the use of common materials of construction such as polypropylene and fibreglass.

Solvent extraction is not required.

2 Process description

During most electrowinning operations, the reaction at the anode (positive electrode of the cell) is wasted, with the oxygen produced being vented to the atmosphere. The Intec technology stores the anodic energy for conversion of the sulphides to soluble form with the production of elemental sulphur. The process uses a strong sodium chloride solution with anodically generated oxidants to leach mineral and produce high purity copper with separate silver and gold products. A generalised flow sheet is shown in Figure 2. The primary oxidants are cupric ion and halogen complexes ("halex"), referred to later in the paper, generated at the anode of the electrowinning cell.

The three stage process uses a counter current leach at 80-85°C where iron is precipitated as a goethite type product (akaganeite) using air at atmospheric pressure according to:

$$Fe^{2+} + 2Cu^{+} + 3/4O_2 + 1/2H_2O \rightarrow FeOOH + 2Cu^{2+} \qquad (1)$$

The overall reaction for the leaching of chalcopyrite can be shown to be according to:

$$CuFeS_2 + Cu^{2+} + 3/4O_2 + 1/2H_2O \rightarrow 2Cu^{+} + FeOOH + 2S^0 \qquad (2)$$

Conversion of the reactive sulphides to elemental form is around 95%, with any sulphate formed finally reporting to the leach or purification residues.

Gold may be leached and precipitated as metal, with negligible contamination from copper or silver, on activated carbon at very high loadings. A purification stage produces a separate metallic silver product and various mixtures of contaminants. The electrowinning stage produces high purity copper at the cathode and cupric ion at the anode. In the event that lead and zinc concentrations are high in the feed, the circuit may be modified to produce lead and zinc in metallic form. Because the process can treat low grade and contaminated concentrates (e.g. 10% Cu with As, Sb, Bi, Hg etc.), a large number of savings can be made for the miner. These include:

1. Reduced mining costs where selective mining is practised.
2. Reduced grinding costs.
3. Smaller flotation plant.
4. Increased flotation recoveries for production of low grade concentrate.
5. Elimination of smelter penalties.

The chemistry of the leaching process has been described in more detail in a paper by Everett and Moyes [5]. The chemistry of the purification process remains confidential at this stage, pending final patent approvals.

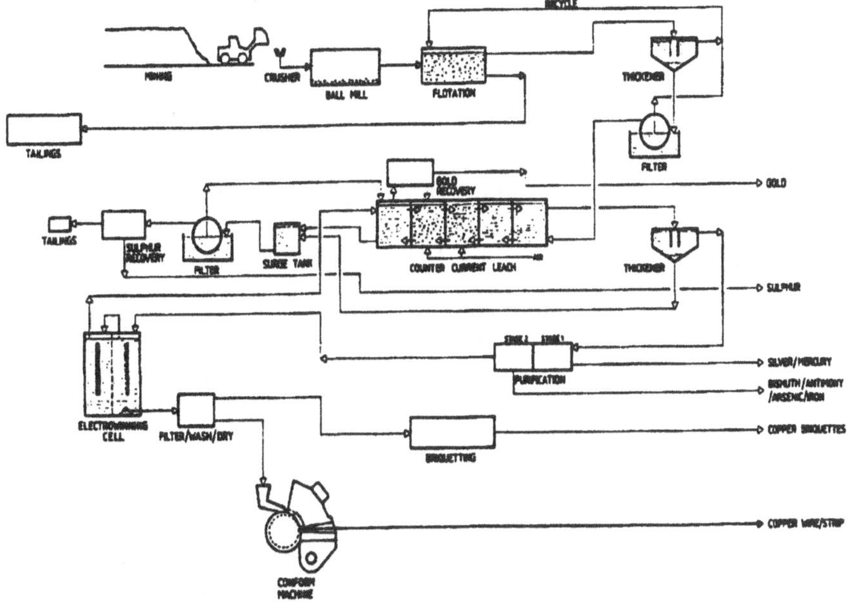

Fig. 2 Intec copper production process

3 Electrowinning

A number of novel features of the electrowinning operation have helped to reduce costs.

3.1 Cathodes

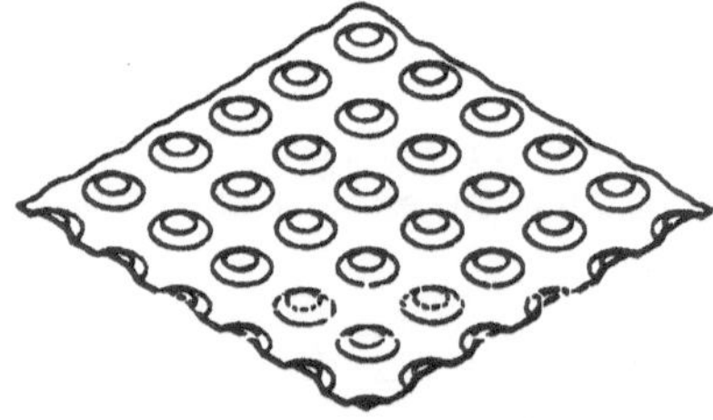

Fig. 3 Dimpled copper sheet

Fig. 4 Finished cathode section

The cathodes are formed from copper sheet which has been dimpled on a 10mm square grid as shown in Figure 3. The final cathode unit is assembled from two of these dimpled sheets, measuring 1.5m x 1.25m, placed "back to back" in a mould where rubber is vulcanised between the sheets and around the dimples to form a very rigid (bend resistant) sandwich-type structure as shown in Figure 4.

This design leaves growth sites approximately 2mm diameter at 10mm spacing to allow copper particles to grow in a fashion similar to that shown in Figure 5, which is a photograph from above the catholyte after 60 minutes plating. Normally the particles would not be allowed to grow as large as this. A wiper moves across the face of the cathode approximately once every 20 minutes. The particles, because of the high stress concentration at the surface of the cathode, detach very easily. Vibrators have been used successfully with previous designs. It was felt however, that wipers ensured a much higher degree of confidence that particles could not grow across to the membrane.

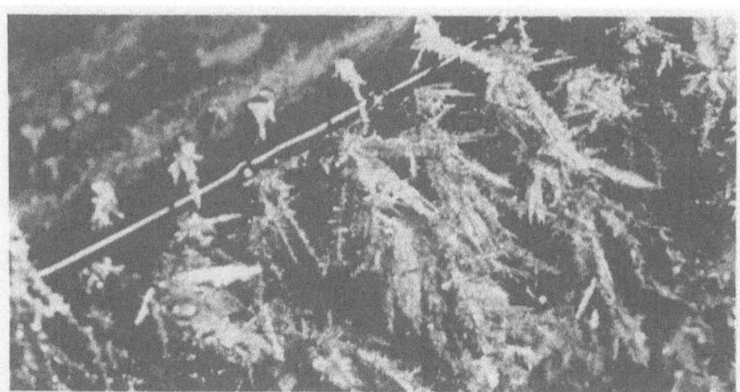

Fig. 5 Copper product growing at the cathode surface

3.2 Membrane

The membranes initially used in this cell were made from ground sulphonated polystyrene resin beads bonded with an ion permeable material to a polyester cloth substrate. These membranes were not suitable for work with highly oxidising solutions and subsequent work was with "Nafion" membranes, made by DuPont in the US. These membranes are P.T.F.E. based with carboxylic and sulphonic acid groups in the structure. They are expensive (approximately US\$700/m^2) and because of their non-stick nature, make chemical bonding to a fibreglass mount extremely difficult. A clamping system was developed using teflon gaskets and titanium bolts. Intec has recently developed its own membrane which is approximately $^1/_{10}$ the cost of Nafion and uses a glass cloth substrate. The glass cloth is bonded to a fibreglass frame using vinylester resin. Intec's proprietary coating is applied to the cloth as a final step. The membrane, apart from being much cheaper, does not swell in water and is much easier to assemble.

3.3 Anodes

The anodes are of the "zero gap" type as illustrated in Figure 6. A layer of titanium mesh, coated with Ruthenium/Iridium oxides, is applied to each side of vertical copper cored titanium conductors. Since the conductivity of titanium is only a fraction of that of copper, the copper cored conductors allow a much smaller cross section than that required using only titanium.

3.4 The cell

The general arrangement of the cell is shown in Figure 7 with 18 cathodes and 17 anodes in membrane bags. A round fibreglass tank, which is much easier to build than a square tank, has a conical bottom which allows easy collection of the copper product. The copper product is pumped from the cell as a slurry to a washing and drying system. It is particularly important to completely wash the product free of chloride before drying in a steam (oxygen free) atmosphere to ensure that the product meets top market specifications.

Fig. 6 Plan view of anode structure

3.5 The product

Although the granules are suitable for the direct conversion to products by powder metallurgical techniques such as 'Conform', it is anticipated that this use of the product will be small at first, but likely to grow rapidly due to the large cost reductions resulting from this method of making copper products.

Samples of the granules have been converted to 'rhondelles' and trials are to be carried out with briquetting machines. One major Australian copper consumer is quite happy to accept 'rhondelles' or 'briquettes' and samples are to be sent to the Intec Copper shareholders and other copper consumers around the world. Samples of granules have been pressed into other shapes and work will be carried out shortly to directly convert the copper granules to sheet form.

3.6 Cathode and anode reactions

The feed to the catholyte is purified cuprous chloride/sodium halide solution, produced from Intec's proprietary purification process. Because of the lack of contaminating impurities, it is possible to electrowin at high current densities, without inclusion of impurities in the product. Normally a current density of 1000 amps/m^2 of superficial cathode area is used, however operation has been carried out at 1500 amps/m^2. The choice of current density is governed by the extra cost of power at the higher current densities. The copper content of the catholyte feed is 80gpl and the cell catholyte concentration is 30gpl. Spent catholyte is fed to the anode compartments where the remaining cuprous oxidises to cupric and halogen complexes are produced at the anode in amounts equivalent to 20gpl of cupric, so that the cell reactions are balanced.

Storage of the anodic energy in this fashion, rather than allowing the evolution of oxygen, has an additional benefit in that it avoids tank house acid mist which is becoming a very serious hygiene problem.

Figure 8 shows the oxidation potential of "halex" electrolytes with increasing amounts of stored electrical energy.

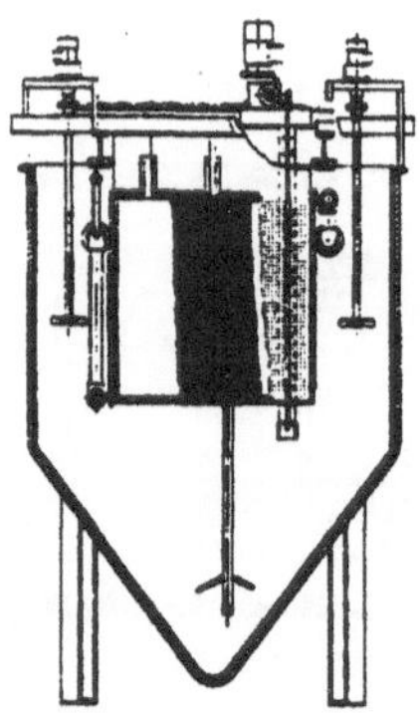

Fig. 7 Cell general arrangement

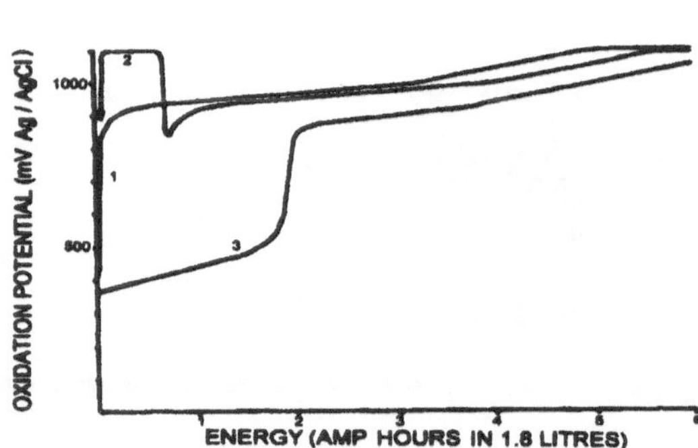

Fig. 8 Storage of electrical energy in "halex" electrolyte

Curve 1 shows the oxidation potential of an electrolyte containing 280gpl NaCl and 28gpl NaBr (equivalent to 21.74gpl Br^-). The first part of the curve shows the oxidation of Br^- and Cl^- to $BrCl_2^-$ according to:

$$Br^- + 2Cl^- \rightarrow BrCl_2^- + 2e^- \qquad (3)$$

The second part of the curve shows the decreasing current efficiency as the free Br^- content decreases and the evolution of chlorine gas becomes a competing reaction, raising the oxidation potential.

Curve 2 shows the oxidation potential of 280gpl NaCl solution without Br^- with immediate chlorine gas evolution, followed by addition of NaBr which stopped the gas evolution and reduced the oxidation potential to that of the "halex" reaction.

Curve 3 shows the oxidation potential of 280gpl NaCl plus 28gpl NaBr plus 12gpl Cu^+.

The energy stored by the electrolytes, above 600mV (Ag/AgCl) which is sufficient to leach gold and attack many oxidation resistant minerals such as pyrite and arsenopyrite, can be shown to be equivalent to the formation of $BrCl_2^-$.

4 The consortium

The difficulties in commercialising new technology are well known to most researchers. Following discussions between Intec Pty. Ltd. and Resource Finance Corporation Ltd., a leading Australian merchant bank in the resources sector, it was felt that the commercialisation of the process could best be achieved by the involvement of a number of major copper producers. A new company was formed, Intec Copper Pty. Ltd., and a number of major mining companies were invited to take equity.

The consortium presently consists of the following companies.

- BHP Minerals *(Australia)*
- Bangkok Cable Co. Ltd. *(Thailand)*
- Dowa Mining Co., Ltd. *(Japan)*
- Magma Copper Company *(U.S.A.)*
- Marubeni Corporation *(Japan)*
- Mitsui Mining & Smelting Co., Ltd. *(Japan)*
- Nippon Mining & Metals Co., Ltd. *(Japan)*
- Rio Algom Limited *(Canada)*
- RTZ Limited *(U.K.)*
- Sumitomo Metal Mining Co., Ltd. *(Japan)*
- Western Mining Corporation Ltd. *(Australia)*

Following the first Technical Review Committee meeting, it is felt that the formation of the consortium has been very beneficial, allowing cross fertilisation of ideas from many companies from a number of countries.

5 Commercialisation program

A pilot plant has been constructed at the Intec laboratory in Sydney, Australia and runs on a continuous basis producing 55kg per day of better than L.M.E. "A" Grade copper, using a 1,000 amp electrowinning cell. The plant will enable the study of specific applications of the Intec Copper Process to investors' ores and concentrates. Sufficient quantities of product for test marketing purposes are being generated during pilot campaigns. In addition, sufficient quantities of residue are being generated for environmental evaluation and sulphur recovery tests.

Intec has a commercial sized cell at its laboratory, which uses 25,000 amps of current. The cell is to be re-equipped with the latest Intec design and technology for membranes, cathodes, wipers etc., and run for extended periods to confirm the reliability of the system, prior to construction of the demonstration plant. The budget for this program is A$2.5 million.

A second stage of the project entails the engineering and construction of a 3,500 t.p.a. plant located at Port Kembla, NSW about 100km south of Sydney on the premises of MM Metals Ltd. This demonstration plant at a port side facility will allow the importation of trial shipments from overseas for test purposes before building plants in other countries. The cost of this plant is estimated at approximately A$8 million and will have a cash operating surplus.

6 References

1. Rothfeld, S.W. Towle, 1989 ,U.S. Bureau of Mines, Engineering and Mining Journal, October.
2. Schweitzer & R. Livingston, 1982, "Duval's" CLEAR Hydrometallurgical Process, presented at the 1982 AIME Annual Meeting, Dallas, TX (Feb 14-18)
3. Everett, 1981, "The Dextec Copper Process" Extraction Metallurg. '81 (London, U.K: IMM, p149)
4. Kruesi, U.S. Patent N° 3,673,061, Cyprus Metallurgical Processes Corporation
5. Everett & Moyes, "The Intec Copper Process", AusIMM Extractive Metallurgy of Gold and Base Metals Conference, Kalgoorlie, Western Australia, 26-28 October 1992.

Introduction to spray roasting process for hydrochloric acid regeneration and its application to mineral processing

L. J. F. Harris
Process Systems and Safety Department, Babcock King-Wilkinson, Ltd. (formerly, Babcock Contractors, Ltd.), Crawley, England

Abstract
The development, current state and economics of the spray roaster process for the regeneration of spent hydrochloric acid are described. Primarily used in conjunction with steel pickling, the system has applications in mineral processing where hydrochloric acid can be used either for the leaching of gangue oxides or for the extraction of valuable materials from suitable feedstocks. The use of hydrochloric acid regeneration ensures virtually complete recovery of the acid values thus permitting the leaching of minerals to be carried out in an environmentally acceptable way by obviating the need to discharge chloride solutions as waste products. Examples are included from the Company's experience to illustrate how leaching and filtration need to be integrated with regeneration in order to develop an economic approach.
Keywords: Hydrochloric Acid Regeneration, Spray Roasting.

1 Introduction

Hydrochloric acid can be used to leach a variety of metal oxides from minerals thereby forming an acidic aqueous metal chloride solution. Such an operation would be carried out in order either to upgrade an ore by leaching out impurities or to extract a valuable component that could be recovered from the leachate or spray roasted to form an oxide.

For the majority of the commonly occurring metal salts, most or all of the acid value in the spent leachate can be recovered for re-use by means of hydrochloric acid regeneration using the spray roasting process.

The commercial use of this technique for the regeneration of hydrochloric acid from aqueous solutions has been established now for over thirty years. The process is based on the fact that many metal chlorides will hydrolyse in the presence of air and water vapour at high temperatures to produce a metal oxide and hydrogen chloride gas. The oxide forms a by-product which may or may not be of value depending on its composition; the hydrogen chloride gas is absorbed in water to form an 18-20% hydrochloric acid stream for re-use in the leaching process or for use elsewhere.

Currently, about 200 such spray roasting plants have been installed since 1959, the vast majority processing acidic ferrous chloride solution produced from the HCl pickling of strip steel. There have been, in addition, a small number of commercial plants built in connection with direct mineral processing although these do include the world's largest installation at Mobile, Alabama, designed by the Company and commissioned in 1976.

2 Early development of the process

The process was initially developed during the period 1955-57, with the construction and operation of a pilot plant. The development was carried out by the Company in response to a need for a process to form refractory-grade magnesium oxide from either of two solutions containing magnesium chloride resulting from the processing of Dead Sea brine [1].

There are a number of ways in which magnesium chloride solution can be hydrolysed to form magnesium oxide.
eg Reactions with alkali:-

$$MgCl_{2\,(aq)} + Ca(OH)_{2\,(aq)} \longrightarrow Mg(OH)_{2\,(c)} + CaCl_{2\,(aq)}$$

$$MgCl_{2\,(aq)} + 2NaOH_{\,(aq)} \longrightarrow Mg(OH)_{2\,(c)} + 2NaCl_{\,(aq)}$$

In the above examples, both reactions occur at ambient temperature in the aqueous phase and hydrated magnesia is precipitated. The chloride solution is washed out of the magnesia and returned to the sea or elsewhere - so the chloride in the filtrate will be of nil value without extra evaporation and crystallization.

The first reaction, with lime, forms part of the well-known sea water process for magnesia production. Clearly a convenient outlet is needed for the waste chloride solution as well as a source of lime.

Alternatively, it was known that crystalline magnesium chloride, eg $MgCl_2.6H_2O$, could be roasted in air and water vapour to form MgO and hydrogen chloride gas:-

$$MgCl_{2\,(c)} + H_2O_{\,(g)} \xrightarrow{\text{Heat}} MgO_{\,(c)} + 2HCl_{\,(g)}$$

In this example two useful products are formed, viz magnesium oxide and hydrogen chloride gas which can be absorbed in water to form aqueous acid. To use this procedure, the intention will have to be to market the by-product acid elsewhere on a non-recycle basis. This reaction uses only heat and was first demonstrated industrially in 1861 but found over many years to be very difficult to effect practically when heating the crystals in different types of roasting furnaces because of the formation of intermediate molten or sticky phases and localised corrosion.

It was proposed that by spraying magnesium chloride as a solution into a hot countercurrent gas stream, it would be possible to effect all these difficult physical and chemical changes whilst the magnesium chloride particle was suspended in a gas stream. The two products would be gaseous HCl and inert magnesia, both of which could be collected. In this way, the intermediate semi-reacted or sticky products would not come in contact with the surfaces of a reactor and temperatures could be maintained locally in excess of the dew point of hydrochloric acid thereby obviating corrosion risks.

Following some promising initial tests, a batch operated spray roasting pilot plant was devised and erected. Tests were initially carried out on molten bischofite brine and, later, on other metal chloride solutions. The latter included tests on ferrous chloride solutions and a mixed chloride from clay leaching. As a result of this work, a pilot plant was erected at the Fullers Earth Union works at Redhill, Surrey, and this was followed by a commercial scale plant in 1961. A second, continuously-operated, pilot plant was also erected at this time by ourselves to replace the first batch pilot plant. A great deal of development work took place over the next fifteen years or so covering the behaviour of a large number of different metal chloride solutions and various related process improvements.

3 Process description

Process developments that have been implemented include the provision of direct heat recovery from off-gases, tail gas scrubbing for HCl emission reduction and various continuous oxide handling techniques. The process design parameters are well established and the Company has installed plants covering the capacity range 0.2 m^3/hr to 51 m^3/hr, the latter comprising three streams. Installation takes between typically twelve and eighteen months depending on capacity and circumstances. A typical modern installation as shown in Fig 1 may be described as follows:-

3.1 Preconcentration and spray roasting

Chloride solution is delivered to the sump of the preconcentrator vessel where it mixes with the circulating liquor. The preconcentrator acts as a loop heat exchanger/ reactor in which the circulating solution is partially evaporated by direct heat exchange with the spray roaster off-gas.

A bleed of the preconcentrated liquor is then sprayed by means of one or more inlet spray pipes into the upper zone of the spray roaster vessel where flash evaporation takes place, the spray being in contact with the rising hot gas from below. Water evaporates into the gas stream such that the exit gas mixture is cooled to about 350°C, some 72-77% of the sensible heat in the inlet products of combustion being exchanged countercurrently with the chloride solution and product oxide.

The key to the process is the spray roaster itself (Fig 2). This comprises a large vertical cylindrical steel vessel with the wall and conical base being suitably refractory lined for protection against the high temperature in the firing zone. In addition, the vessel and its off-gas ducting are externally insulated to maintain the steel temperature above the dew point of aqueous hydrochloric acid.

The reactor is directly fired with a suitable fuel-air mixture corresponding typically to a 20-50% excess air mixture and a combustion temperature in the range 1250-1500°C. The hot gas enters the cylindrical base of the roaster tangentially at a desired velocity, the number of burners and inlet ports being selected to suit the roaster size and heat demand. Commonly heavy fuel oil or natural gas will be used depending on costs and location.

As the evaporating droplets fall by gravity into the high temperature zone of the roaster further evaporation occurs followed by the formation of an outer thin shell of oxide produced by a combination of crystallization and hydrolysis. In the case of iron chlorides, typical overall hydrolysis reactions that occur under oxidizing conditions in the hottest zone are as follows:-

$$2FeCl_2 + 2H_2O_{(g)} + \tfrac{1}{2}O_2 \longrightarrow Fe_2O_{3(c)} + 4HCl_{(g)}$$

$$2FeCl_3 + 3H_2O_{(g)} \longrightarrow Fe_2O_{3(c)} + 6HCl_{(g)}.$$

Product oxide passes through the outlet cone into the oxide discharge system.

As noted previously, other metal chlorides react similarly. High conversions of chloride to oxide are found, providing the reactor sprays are well maintained. For example, the conversion of the following common metal cations in chloride solutions is complete or nearly complete under normal spray roasting conditions: Fe^{++}, Fe^{+++}, Mn, Al, Mg, Ni, Co, Sn. Zinc needs special consideration and calcium

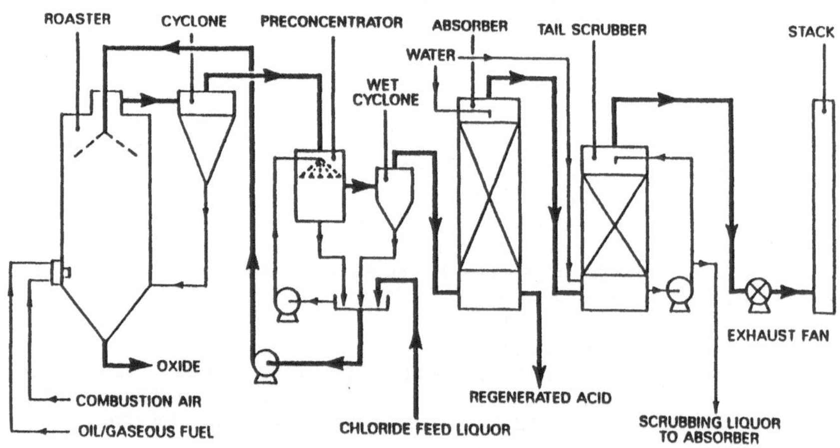

Fig. 1 TYPICAL FLOWSHEET - BABCOCK W-D SPRAY ROASTER PROCESS

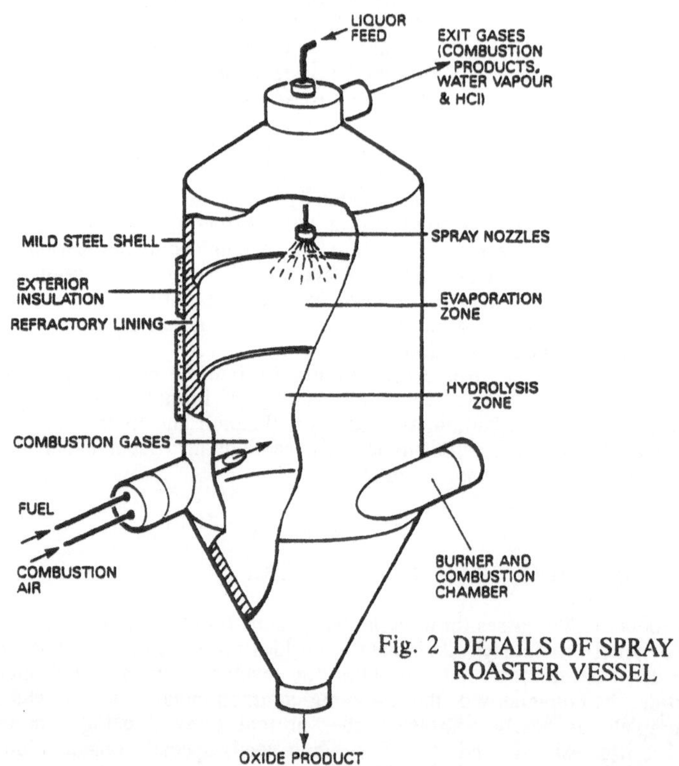

Fig. 2 DETAILS OF SPRAY ROASTER VESSEL

reacts only partially if at all. The alkali metals, Na and K do not react to any significant degree.

These reactions take place in the lower zone of the reactor at temperatures generally in the range 650-850°C.

In the case of mixed chloride feed solutions, mixed oxides and ferrite compounds will form on hydrolysis, eg nickel ferrite, $NiO.Fe_2O_3$, will be formed from a solution of nickel and ferrous chlorides. Unconverted Ca, Na and K will remain as soluble chlorides in the product oxide but these contaminants can be washed out of the oxide.

In the case of solutions containing significant levels of calcium ions, special steps, eg reaction with sulphuric acid, can be undertaken to minimize the loss of potential HCl which would otherwise report as calcium chloride in the oxide.

Depending primarily on solution concentrations, the roaster exit gas will first pass to a dry cyclone for separation and recycle of entrained dust. The partially cleaned gas then passes through the preconcentrator in which the sensible heat drop in the gas between about 350°C and 95°C is used to evaporate approx 25% of the water content in the feed. The limitation on the preconcentration of feed solutions is the solubility of the salts present. This must not be exceeded in the preconcentrator otherwise crystallization will occur causing spray blockage problems.

Most of the residual dust present in the entering gas is also scrubbed and dissolved by reaction with free hydrochloric acid present in the circulating solution. This solution approaches thermal and chemical equilibrium with the preconcentrator off-gas, as a result of which its free hydrochloric acid concentration will usually be somewhat higher than that in the feed solution.

3.2 HCl recovery

The hydrogen chloride present in the off-gases from the preconcentrator is then absorbed adiabatically in a packed tower using a countercurrent flow of water of a suitable quality, thereby producing a sub-azeotropic 18-20% w/w acid solution. The absorbing water is commonly wash water from the leaching process containing a proportion of metal chloride salts. Such salts will be present because of the inevitable inefficiency of the liquid-solid separation step immediately downstream of the main leaching plant.

In the conventional adiabatic absorbers used in spray roasting applications, the heat of solution of the HCl gas is dissipated by using it to evaporate surplus water from the absorber feed water and to heat the product acid. The maximum acid strength produced is limited to about 20-21% w/w because of the presence of the $HCl\text{-}H_2O$ azeotrope. Attempts to make stronger acid merely result in HCl gas passing through the absorber. For sub-azeotropic conditions, the regenerated acid produced will be relatively stronger than the total equivalent acid content of the spent liquor. Thus the regeneration process can remove, to an extent, surplus water from the spent liquor.

The gas leaving the HCl absorber, stripped of the bulk of its HCl content, then passes to a high circulation rate tail scrubber for removal of the final traces of HCl. The cleaned off-gas is then discharged to atmosphere through an exhaust fan which maintains the roaster gas train under a slight negative pressure for operational safety reasons. The weakly acidic tail scrubber water and stack condensate are both recycled to the HCl absorber.

Plant efficiency in terms of HCl recovery will commonly be in excess of 98%. It is generally calculated by measuring two sources of losses. These can be summarized as:-

> Hydrochloric acid present in the emitted stack gas. The limit will be governed by the local statutory regulation.

Equivalent hydrochloric acid lost as unconverted chloride in the oxide. Defective sprays allow large particles to form which are difficult to convert fully.

Housekeeping losses can also occur due to spillages outside the spray roaster limits, stray fume emissions, etc.

3.3 Oxide Handling

There are numerous methods of oxide handling that can be adopted depending on the application. For some mineral processing applications when a waste mixed oxide is produced, a wet slurry system has been used. The oxide is discharged from the base of the spray roaster into a slurrying vessel in which it is wetted by a copious flow of circulating water to form a 10-20% w/w slurry. This is pumped to a lagoon for oxide settling and recirculation of the slurry water. Other methods, commonly used for high purity iron oxide handling, include pneumatic conveying and buffer storage of the oxide powder.

It is also feasible to pelletize the oxide with water to form reasonably hard 10-15 mm pellets. High purity ferric oxide, produced from steel pickle liquor is sold for ferrite or pigment use but iron-based mixed oxides, typically containing 50-90% Fe_2O_3, have virtually always been regarded as a waste although attempts to find markets have been made.

Oxide that is to be dumped should be mixed with sub-soil first because of its poor load-bearing properties. Nowadays, it will probably be necessary to carry out leaching tests but this was not the case in the past. Some constituents, eg unreacted $CaCl_2$, can clearly be leached from the oxide but the majority of metallic components present will remain as fixed oxides.

4 Leaching procedures

4.1 General principles

The regenerated acid is produced hot at about 85-90°C. For ore leaching the acid will usually be required hot so it will be advantageous to conserve its temperature. Only occasionally with extremely exothermic leaching reactions is it necessary to cool the acid in order to slow down the initial reaction. Leaching with HCl is generally restricted to the extraction of metals or metal oxides. Sulphide, phosphate and carbonate minerals, the latter generally calcined, have also been processed by this means.

An example of HCl leaching would be the upgrading of 54% TiO_2 ilmenite by the extraction of iron and other oxides to form a 92-96% TiO_2 synthetic rutile on calcination, the TiO_2 content of the mineral being largely unattacked.

Another example would be the leaching of tin present in a reduced cassiterite lode ore. Leaching produces a mixed chloride solution from which tin can be extracted by cementation with scrap aluminium. The spent chloride solution is then spray roasted to regenerate the acid, the oxide being a waste product.

As a general rule, titania, silica, zirconia and chromite are largely unattacked by hydrochloric acid. Many metallic oxides are substantially leached eg iron, manganese, nickel, tin, magnesia, calcium, etc. Some oxides, eg alumina, exhibit variability in the proportions leached.

There are a number of methods by which leaching can be effected depending on the application and the acid to ore ratio needed. This ratio will depend on the proportion of material to be leached from the ore and is usually calculated theoretically and then demonstrated by experiment.

One feature that all such leaching processes will have in common is that of HCl fume release from various sources - such fumes must be extracted from the equipment and scrubbed with water. The weakly acidic liquor thus produced is mixed with other plant filtrates and wash water for use as absorber water.

All the methods used for leaching with hydrochloric acid are comparatively simple. Complicated vessels with internal metallic appendages are not used, internal surfaces being generally non-metallic and suitable for the corrosive conditions. Materials such as tantalum, ceramics or graphite can be used for special internals.

The most important single factor relating leaching to regeneration economics concerns the degree of utilisation of the acid. In countercurrent leaching, which is very difficult to achieve in practice, it will generally be possible to reach a final HCl strength of, perhaps, 2-4% w/w compared with 18-20% in the feed acid. Thus, the degree of acid utilisation will be 80-90%. Maximising this will minimise the volume of spent liquor to be processed - this will reduce significantly the operating costs for regeneration as described later.

With single stage co-current (ie 'batch') leaching, on the other hand, a very low final free HCl is more difficult to achieve. This is because the final upgraded solid, containing only a small proportion of semi-inaccessible residue to be extracted, becomes difficult to leach and the reaction rate falls substantially as the free HCl level falls. If the reaction rate subsides to a non-economic level before completion it might be necessary to increase the acid:ore ratio.

Another factor of great importance is the oxidation state of the iron present. Where ores are being upgraded, iron tends to be the major impurity present and should be in an easily leachable form particularly when a substantial amount is to be removed in relation to the leached ore mass. This can be done, for example, by ensuring (a) that any spinel bond present is broken and (b) any ferric oxide, Fe_2O_3, present is reduced to the easily leached ferrous oxide, FeO. In general, ferric oxide will only leach very slowly once the free acid level has fallen to below, say, 10%.

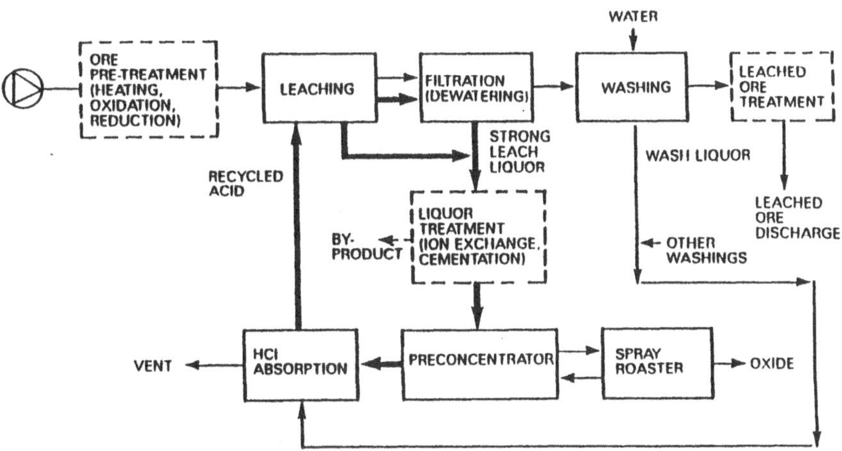

Fig. 3 LEACHING OF MINERALS USING 20% w/w HYDROCHLORIC ACID TYPICAL FLOW DIAGRAM

Fig 3 shows a generalized schematic of a leaching system integrated with an HCl regeneration plant. This shows how two liquor streams are formed. The strong leachate stream containing a high concentration of chlorides with little or no additional dilution water passes directly to the preconcentrator of a spray roaster for regeneration or to a metal separation stage (eg ion-exchange, solvent extraction, cementation) prior to spray roasting. The residual chloride in the leached solids left after initial dewatering approximates to around 10% of the total metal chloride production. A suitable washing procedure, possibly using countercurrent stages, enables a low chloride content filter cake to be formed (typically containing < 0.25% Cl). The resulting weak filtrate is then used for acid absorption mixed with other acidic washings as necessary.

This approach ensures that all the metal chlorides produced are regenerated. If washings are not recycled then there would be a substantial HCl make-up requirement and an effluent disposal problem would be created.

4.2 Practical examples

Included below are brief mentions of some of the leaching techniques that have been used in connection with HCl regeneration schemes with which the Company has had contractual or developmental involvement.

In many other cases the Company has carried out feasibility studies based on clients' experimental work in order to develop the economic integration of leaching with regeneration.

4.2.1 Two-stage fluidised bed leaching of pre-reduced ilmenite with countercurrent transfer of acid and ilmenite

In this example, freshly reduced ilmenite containing 98% of the iron present as FeO is leached under atmospheric pressure at 100°C in two one-hour batch stages. The feed ilmenite is first leached for one hour using a batch of partially spent acid from Stage 2. Following this, the slurry is allowed to settle and the now fully spent acid decanted and sent for regeneration. The partially leached ilmenite is then re-fluidised using a fresh batch of 18% acid from the regeneration plant and the leaching is completed.

This technique results in the residue of undissolved gangue oxide remaining at the end of the second leach being in contact with reasonably strong acid; this ensures that a higher proportion of gangue oxides are leached. In Table 1 are reported some typical data for this technique showing the overall proportions leached of various oxides present in the reduced ore. This shows that ferrous, manganese and magnesium oxides leach substantially whereas the TiO_2 present is largely unattacked.

The leached ilmenite particle contains some 40-50% voids initially filled with partially spent acid. After decantation and initial dewatering, the residual acid present in the pores and around the particles is washed out preferably countercurrently till the desired residual chloride content in the leached ilmenite has been reached.

4.2.2 Two-stage rotating bed leaching of partially reduced ilmenite with co-current split flow of the acid (Benilite Process)

In this commercial example, the acid requirement is split into two similar sized fractions and successive leaches are carried out, the final acid in each half of the operation being nearly spent. There is a lower liquid:solid volumetric ratio used per stage with this method than in the former method and contact between the phases is effected by rotation of the acid brick lined spherical digesters.

Leaching is carried out under a pressure of up to 3.4 bar g and a temperature of 146°C. Each stage lasts up to four hours. The pre-reduction of the ilmenite upstream of leaching was less effective with this process than with the fluidized bed process, hence the need for longer leaching times.

Table 1. Ilmenite Upgrading

Reduction with H_2 of oxidized ilmenite, followed by two one hour countercurrent fluidized bed leach stages using 18-20% w/w regenerated hydrochloric acid.
Overall Proportions leached (range of results found with different ilmenites)

	%		%
TiO_2	0.3 to 4	As_2O_3	46.5
Fe	96.9 to 99.6	Nb_2O_5	Trace
(present largely as FeO in reduced ore)		ZrO_2	Trace
Cr_2O_3	20.0 to 42.1	SiO_2	Trace
V_2O_5	48.1 to 73	Al_2O_3	20.9 to 61.5
P_2O_5	14.0 to 55.6	CaO	50.0 to 95.4
MnO	96.1 to 99.9	MgO	81.4 to 100.0

NB The proportions leached of the minor components vary with the ilmenite.

After decantation of the bulk of the strong liquor after the second leach stage, dilution and cooling water is added to the digester and the contents slurried and pumped to the belt filtration stage.

In both these first two examples, the washing and fume scrubbing circuits were arranged to ensure that there was no dilution of the spent liquor and the wash water production did not exceed the demand of the regeneration plant. Both methods are practicable for the leaching of dense minerals which are otherwise difficult to agitate.

4.2.3 Rotary kiln leaching

This technique was used for the co-current leaching of silica sand where the required acid:ore ratio was so low that the acid was substantially retained by the particles during their passage through the vessel. Hydrochloric acid with regeneration has been used for this application but its use would not normally be economic. Only a very low acid:mineral volumetric ratio was required because of the small proportion of leachable oxides present in the silica. Oxides leached included iron (substantially), calcium, aluminium, potassium and sodium.

The small volume of liquor associated with the solid meant that it was impracticable to produce a substantially strong liquor stream separate from the washings. The spray roaster feed solution was, as a result, too dilute for economic use.

4.2.4 Stirred tank leaching

This technique has been used for minerals of low specific gravity which are easily suspended by agitation. Batch leaching is usual but this could be converted to a continuous co-current system by the use of a series of overflowing reactors.

A commercial example would be the use of regenerated hydrochloric acid for the production of activated earths. Bentonite is leached batchwise in a stirred tank at 100°C using live steam. Substantial proportions of the iron, alumina, calcium and magnesium present are dissolved, the internal surface area of the mineral being increased considerably by this means. The product is very porous and is used as an adsorbent.

After leaching, the acidic slurry is pumped to a filter and washed with water to remove the chloride solution. Traditionally, there was no separation of the initial strong filtrate from the weaker filter washings. As a result, the mixed feed to the regeneration plant was weak and the solution had to be pre-evaporated before regeneration.

4.2.5 Direct gas contacting

In at least two instances, it has been found practicable to absorb the HCl content of the spray roaster off-gas in a reacting slurry of a mineral. The mineral is thus directly leached at about 95°C by the freshly dissolved acid. Examples of this include phosphate rock and magnesite dissolution.

In both cases the leaching rate is still sufficiently fast at the comparatively low free acid concentrations prevailing in the strong chloride solution formed in the vessel. The solution is bled from the system, suitably purified and, in the case of magnesia, then fed to the spray roaster for oxide formation and acid regeneration.

5 Purification of the spent leach liquor

5.1 General

The leach liquor may contain one or more components that need to be removed because of their intrinsic value, eg tin, or because of their nuisance value in spray roasting operations eg insoluble matter or excessive concentrations of zinc or sodium. The techniques used for this in connection with regeneration schemes include:-

- Filtration: Removal of suspended solids.
- Cementation: Copper and tin can be removed from solution by cementation with scrap iron or scrap aluminium. The post cementation solution is then spray roasted to produce an impure waste oxide.
- Ion Exchange or Solvent Extraction: Separation of zinc.
- Crystallization: Separation of excessive concentrations of NaCl.

As can be seen, these are the same methods as used for metal recovery from the leachate produced by ferric chloride leaching of sulphide ores.

5.2 Zinc recovery

As an example of the necessity for one of the above techniques, it may be noted that excessive concentrations of zinc chloride cannot be handled in spray roasters because a certain proportion of the zinc entering the roaster vessel evaporates as zinc chloride faster than the material can hydrolyse to form the oxide. The zinc chloride that does react forms zinc ferrite, $ZnO.Fe_2O_3$. If the feed concentration of zinc chloride is too high, then liquid zinc chloride can condense on cooling in the ducts and cyclone causing blockages.

In the case of a plant designed by the Company for its own use to regenerate acid from a mixed ferrous/zinc spent pickle liquor arising from sources including galvanizers, it was anticipated from our feasibility studies that the zinc level in the feed would be excessive. As a result, a moving bed ion-exchange plant was installed and operated for zinc separation as shown in Fig 4 [2]. The zinc was recovered as zinc sulphate but the value of the amount of zinc recovered from the solution did not, at the small production scale, cover the cost of the operation. These costs included that caused by dilution of the pickle liquor with recycled resin wash liquor. This extra water has to be evaporated in the regeneration plant although the effect can be minimized with good design.

Eventually the volatilization problem was circumvented by careful blending of zinc-bearing and zinc-free solutions to minimize zinc concentration in the mixed feed to the spray roaster.

More recently, there has been development of solvent extraction processes for zinc removal from chloride media, including ferrous chloride solution. Extractants used include TBP and ACORGA ZNX50. The subject has been frequently revisited over the years but, unless there is a high concentration of zinc in the feed as could occur, for example, in the leaching of a zinc-iron material, recovery is unlikely to be economically attractive.

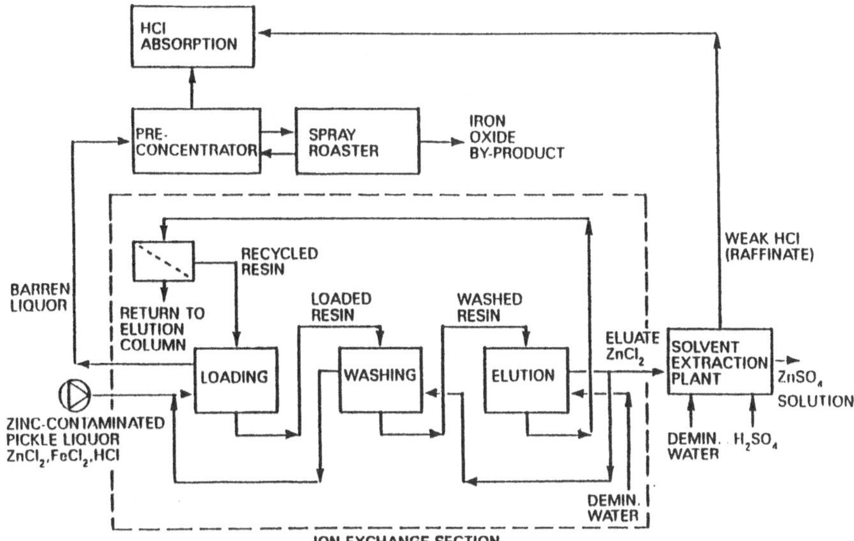

Fig. 4 OUTLINE FLOW DIAGRAM OF THE 'METSEP' PROCESS FOR REMOVING ZINC FROM FERROUS CHLORIDE PICKLE LIQUOR

6 Economic factors

6.1 General

The following principal factors govern the economic aspects of HCl regeneration in a given situation.

- Ease of supply of make-up HCl.
- Cost of make-up HCl.
- Cost and ease of disposal of spent liquor as an alternative to regeneration.
- Plant capacity.
- Acid:ore ratio.
- Spent liquor strength.
- Fuel cost.
- Ease and costs of oxide disposal.
- Operation and maintenance.

6.2 Comments on the economic factors

6.2.1 Ease of supply and cost of make-up HCl

This is usually supplied as 28-36% w/w HCl.

1. If HCl regeneration is **not** installed, then there will clearly be a 100% make-up requirement. For large plants, this is not a practicable option and the decision might be made to use sulphuric acid. With sulphuric, the corresponding effluent can be neutralized with lime to precipitate calcium sulphate. Hence there would be a low TDS in the aqueous effluent compared with that produced from a chloride-based solution.
   Supply adequacy is not usually a problem in most industrialized areas but in remote locations, HCl is not readily available.
2. For many years, in the UK and the USA the cost of hydrochloric acid plus transport has been very low compared with that found in certain non-industrialised areas. The cost of chlorine, to which the price of HCl is related, has been recently gradually increasing so there could be a slow upward movement.

6.2.2 Cost of disposal of spent leach liquor (after metal extraction or when the liquor contains only waste constituents)

The spent liquor will contain some free acid and varying concentrations of base and heavy metals. In the past, as an alternative to regeneration, many users of HCl were able to discharge their spent effluent to drainage without significant treatment. This situation no longer applies and it will be necessary to neutralise the free HCl with lime and raise the pH of the solution to precipitate the metals as hydroxides. The clear liquor, after settling and filtration of the sludge, will contain substantially calcium chloride and possibly also $MgCl_2$ or NaCl. This solution is quite suitable for seawater disposal but, in some locations, eg Germany and arid areas of the USA, disposal via drainage or groundwater is no longer being allowed and costly effluent schemes are required. In the UK the solution would have to be neutralised by the user or off-plot by an effluent treatment contractor.

6.2.3 Plant capacity

Because of the generally low cost of hydrochloric acid, it will not normally be economic to operate a low capacity spray roaster, the capital related and operating charges total outweighing the value of the acid and the neutralization costs. In such a case pollution restrictions will be the main drive behind the use of HCl regeneration and this has occurred in the USA for small steel stock-holding/ pickling operations. The regeneration plant capacity ought to exceed about 3 m^3/h for strong liquors to ensure that the operation is economic.

6.2.4 Acid:ore ratio

Where the acid requirement per unit weight of ore is low, it will be difficult to effect a good liquid-solid separation. Hence, it will be necessary to wash the leached ore extensively and this may produce a weak liquor requiring a larger roaster for any given acid demand. In such circumstances it is usually better to use sulphuric acid, followed by neutralisation. For large capacity installations with higher acid:ore ratios, eg ilmenite upgrading plants, a good case for regenerating HCl can usually be made.

6.2.5 Liquor strength

It is advantageous to maintain the leach liquor at as high a strength as possible to minimise the volume. Most of the heat duty of the spray roaster is in the evaporation of water and unnecessary dilution should be avoided. Examples of the typical dilution that occurs are (i) live steam heating of leaching vessels (condensation), (ii) dilution caused by scrubbing in ion-exchange or solvent extraction circuits and (iii), addition of washings to the strong liquor. Pre-evaporation can be carried out indirectly with waste heat being used to heat the leach liquor. Water is then evaporated by means of an air-blown packed column or by other means.

6.2.6 Fuel
On large capacity plants, the main operating cost item is fuel. Low cost gas usually exists only on an integrated steel works in the form of coke oven gas or on a very large complex as a synthetic fuel gas although, in some parts of the world, natural gas is also inexpensive. One economic possibility is to use waste chlorinated hydrocarbons to supply all or part of the fuel demand. The chlorine content would form HCl on combustion hence supplying the HCl make-up requirement and probably leaving some for export.

6.2.7 Oxide disposal
If the oxide is a waste product, then an allowance must be included for disposal.

6.2.8 Operation and maintenance
The plants are basically easy to operate requiring no sophisticated control equipment, only conscientious operation. Small plants will require only one operator with occasional assistance from another, larger plants will require two or, rarely, three operators and one of these will usually be involved in tank farm or other duties.

Maintenance is a difficult item to quantify since costs are dependent on good housekeeping. Hydrochloric acid is very corrosive and spillages not cleaned up can prove expensive mistakes. The plant's internal surfaces at the 'wet end' are all built in corrosion-resistant materials eg rubber-lined steel, glass-reinforced plastic, PVC, ABS, titanium, etc. Operating experience on the best run plants indicate that on-stream times in excess of 90% can be achieved and that, for a plant several years old, the overall maintenance cost of labour plus materials approximates to 3% to 6% on average of the plant's installed value.

7 Economic summary

An idea of the economics involved is shown below considering a conceptual case of a single large gas fired spray roaster on a UK site and cost basis processing some 17 m^3/h of hot, concentrated spent leach liquor. It is also assumed that the facility is attached to an existing works and tankage is available. The capital cost is for a complete finished plant including structures, erection on prepared foundations, design fees, etc. In sub-section 2, the regeneration cost is compared with the costs for the alternative case of neutralization of spent liquor and the purchase of fresh acid demonstrating the economic advantage of regeneration in this example.

Outline of Regeneration Plant Economics
(based on US$1.5 = £1 sterling)

1. Basis

Capital Cost	$\$9 \times 10^6$
Operating Year	7500 hr (on line)
Spent Liquor Flow Rate	17 m^3/hr

2. Regeneration Costs - Comparison with Disposal of Spent Liquor

	$/annum
· Regeneration (total of fixed and variable costs)	5,339,000
· Disposal of Spent Liquor (neutralization and bought-in acid)	12,173,000
Net annual saving in Regeneration Costs compared with Disposal	6,834,000

The breakdown of these cost elements is provided in Sub-sections 3, 4 and 5.

3. Regeneration - Fixed Annual Costs

   Typically, these include:-

   - Depreciation
     - Rates and Insurance
     - Interest Charges
   - Operating Labour (2 men/shift)
   - Maintenance
   - Overheads

   Estimated Sub-total $2,455,000

4. Regeneration - Principal Variable Costs

		Unit Cost	Annual Quantity	$/annum
(i)	Fuel: Natural Gas	$4.25/GJ (gross)	0.361 x 10^6 GJ	1,534,000
(ii)	Power	8.4 c/kWh	5.1 x 10^6 kWh	428,000
(iii)	Process Water	Normally available in form of wash water		
(iv)	Make-up HCl (32%)	$75/t	1,920 tonnes	144,000
(v)	Oxide Disposal	$30/tonne	25,950 tonnes	778,000
	Sub-total			2,884,000
	Total of Fixed and Variable			5,339,000

5. Alternative - Disposal of Spent Liquor

		Unit Cost	Annual Quantity	$/annum
(i)	Equivalent 32% HCl Recovered (to be purchased)	$75/t	96,012 t	7,201,000
(ii)	Disposal of spent liquor (in-house neutralization and unrestricted free discharge of neutral chloride solution)	$30/t**	165,750 t*	4,972,000
*	Excludes any wash liquor disposal costs			
**	Including capital charges			
Sub-total				12,173,000

Based on the above principles the economics of HCl regeneration for various plant capacities has been compared with the alternative of neutralization and disposal of the spent liquors together with the purchase of fresh acid for leaching purposes. This comparison is shown in graphical form in Fig 5 which shows, for a given volumetric feed rate of spent liquor, the cost to produce regenerated acid. For ease of comparison with the cost of bought-in acid, the regenerated acid cost is calculated on a 32% w/w basis, the normal commercial concentration.

The data shows that, for larger scale operations at least, the regeneration of HCl from spent leach liquors makes economic sense as well as being environmentally more acceptable than the alternative of neutralisation and disposal.

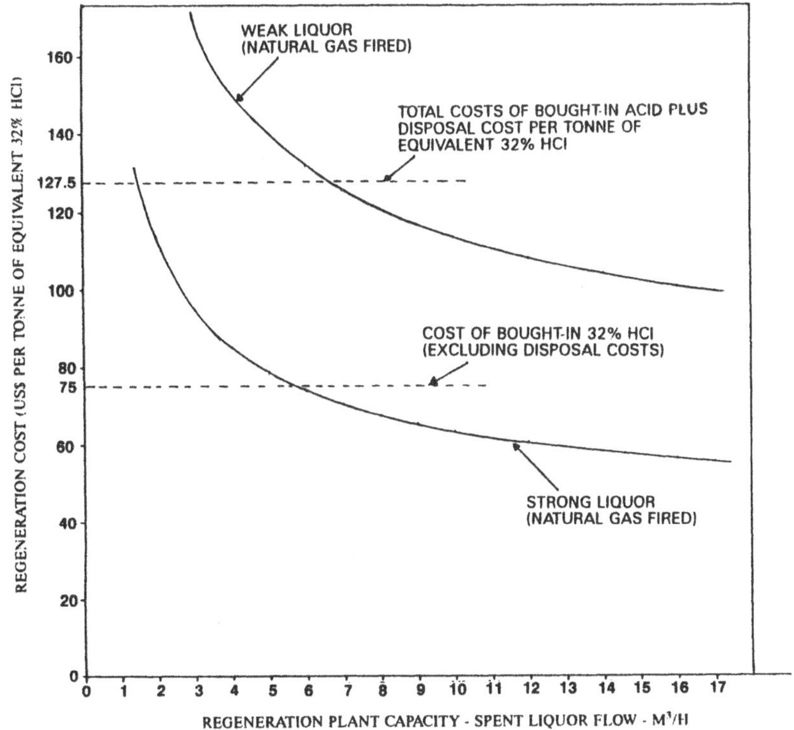

Fig. 5 GRAPH TO SHOW TOTAL REGENERATION COST AGAINST PLANT CAPACITY (TYPICAL)

8. References

1. Utilization of the Dead Sea Minerals (A Review), J.A. Epstein, **Hydrometallurgy, 2** (1976), 1-10.
2. The Metsep Process for the Separation and Recovery of Zinc, Iron and Hydrochloric Acid from Spent Pickle Liquors, T.H. Tunley and T.D. Sampson, **Jnl of the South African Inst Mining & Metallurgy**, 76, 10, May '76, 423-427 inc.

Production of high-grade tin in a new electrolytic plant

J. F. Rodríguez
Estaños de Zamora S.A., Villaralbo, Zamora, Spain
F. J. Alguacil
Centro Nacional de Investigaciones Metalúrgicas, Madrid, Spain
S. Martínez
Centro Nacional de Investigaciones Metalúrgicas, Madrid, Spain
C. Caravaca
Centro Nacional de Investigaciones Metalúrgicas, Madrid, Spain (presently, Department of Mineral Resources Engineering, Royal School of Mines, Imperial College of Science, Technology and Medicine, London, England)

At the end of 1992 the installation of a plant to obtain high grade tin by means of electrorefining was concluded in Spain. The plant is expected to produce about 1800 Ton/year of metallic tin. After a year of operation, the present work describes the main features of this new plant, presently unique in Europe for this metal, together with the more interesting results obtained in terms of operability and final tin purity.
Keywords: Electrolytic production, high grade tin.

1 Introduction

Tin became a valuable and useful metal for mankind centuries ago, and its importance has been greatly enhanced in recent times by the prohibition, first in USA and later in the EEC, of the use of lead in the food industry especially where beverages use corks and thus overcaps. In these circumstances tin became the alternative in lead substitution.

Once the metal has been chosen, the next problem was to find a technique which would yield tin of the required grade. Very often tin is obtained from raw materials by means of pyrometallurgical techniques which do not give the necessary purity. Electrolytic refining has a major advantage over other methods in that nearly all the impurities commonly found in pyrometallurgically produced tin can be reduced to the low levels necessary to meet, for example grade A specifications; and, also very important this can be achieved in a single operation.

The major disadvantage of electrorefining is the large quantity of metal locked up in the process, a figure normally referred to as the stock ratio, or otherwise defined as the weight of tin in electrodes/weight of tin produced per day. Sometimes the relative inflexibility of the process is also regarded as disadvantageous.

2 Background of Estaños de Zamora S.A.

Estaños de Zamora S.A., located near the city of Zamora in the Northwest of Spain, had been producing tin by smelting of cassiterite raw materials. The operation is carried out, after the relevant conditioning operations, by means of five sumerged electric arc furnaces. The crude tin is then refined to eliminate some impurities such as copper, lead and iron. A very pure tin is produced but not adequate to meet the high grade requirements as defined today for use as described above. Thus investment was made in a new electrorefining plant, which after a number of laboratory studies and other investigations under a wide variety of operating conditions is today operating. It should also be noted that laboratory studies are continuing to obtain some more information about the electrolytic behaviour of tin.

3 The electrolytic plant

Basically speaking the plant could be divided into six sections: anode production, electrolytic cells, electrolyte circulation, cathode production, melting and ingot casting and lastly anodic slimes treatment.

Crude tin as obtained in the pyrometallurgical plant is cast into 100 kg anodes which are inserted into the cells by crane. The average composition of the anodes is given in Table 1. The immersed anode dimensions are 0.7m·0.7m·0.025m.

Table 1. Anode composition

Element	%	Element	%
Sn	99.377	As	0.052
Pb	0.3	Zn	0.001
Cu	0.02	Bi	0.007
Fe	0.08	Cd	N.A.
Sb	0.1	In	N.A.

Pure tin metal is plated on the cathodes, for which the immersed dimensions are 0.7m·0.7m·0.001m, and cast in a convenient crucible. Impurities are collected as anode slimes.

The electrolytic plant, consists of 56 polypropylene cells in 4 rows of 14 cells. These cells are arranged with working platforms between rows and walkways along the end of the cells rows. Also there is a clear height at the bottom of the cells for access to cells and electrolyte supply lines.

Electric connections between cells are made by the Walker series system, in which the anodes of one cell are connected with the cathodes of the next cell in the series. Connections of cathodes and anodes are made by the use of copper bars.

The electrolyte is elevated by pumping from the main storage tank to the working electrolyte tank which is situated higher than the cells. In the main tank the electrolyte is heated to 25°C by electric resistance, although coil or plate exchangers can be used. The heated electrolyte is fed into the bottom of each cell at a 15 liter per minute flow rate, overflowing from there and returning by gravity to the main storage tank, thus closing the circuit.

Although the current electrolyte composition is protected by Patent, it can be said that the electrolyte was built up from sulphuric acid and stannous oxide until the required composition was reached. An alternative method to obtain the electrolyte - only experimented with at a very small scale- is by means of an electrolytic cell in which pure sulphuric acid is put into contact with metallic tin at a high current density. Table 2 gives overall values of the electrolyte composition.

Table 2. Electrolyte composition

Tin as Sn^{2+}	20-24 gpl
Sulphuric acid	100-120 gpl
Organic additives	1 gpl
-Xanthates	
-Animal glue	

Organic additives are used to control the growth of dendrites and are of the main parameters to obtain good plant performance. Additions, if necessary, of the corresponding electrolyte component are carried out in the main storage tank.

The anodes are removed by crane from the cells every five days for brushing in the anode brush machine. The slimes, also refered to as black mud, accumulate on the anodes and contain the impurities of the crude metal. These slimes are removed to a slime recovery tank, washed, dried and stored for eventual further treatment if considered necessary or economically feasible.

After 10 days of continous operation, the anodes have lost about 75% of their initial weight and are removed from the cells, brushed free of mud, washed with water and recycled to the corresponding anode smelting crucible for casting into new anodes. Sometimes and depending of the operational conditions some half-spent or semi-agoted anodes are also reused.

Cathodes are removed after 10 days, and placed in the cathode smelting crucible to obtain high grade tin. Part of the metal is used to form starter sheets, whereas the remainder is cast into high grade tin ingots to be sold to the market.

New cathode starter sheets are made by pouring molten pure tin onto an inclined steel plate. Cathodes are placed in the cells allowing 120mm space between two consecutive cathodes and putting the anodes exactly in the middle between two cathodes. Each cell contains 16 cathodes and 15 anodes. The production of the plant can be quantified as 0.6 kg Sn/Kwh.

There are three factors which are critical to obtain good working efficiency in the plant. According to the data available the results are summarized in Table 3, in terms of the control of the cathodic deposit. The three main parameters considered are: (1)

current density, (2) stannous ion concentration and (3) organic reagents concentration to avoid crystalline growth.

Table 3. Control of cathodic deposit

(3) \ (1) / (2)	Low / High	Medium / Medium	High / Low
Low	Oriented deposit and homogeneus	Oriented deposit semidendritic	Dendritic
Medium	Oriented deposit and homogeneus	Oriented deposit and compact	Oriented deposit and compact
High	Oriented deposit and homogeneus	Oriented deposit and compact	Non-oriented deposit and compact

According to operational data, the levels of each one of these parameters described in Table 3 are:

Current density:
Low: 0.2-0.8 $A \cdot dm^{-2}$
Medium: 0.8-1 $A \cdot dm^{-2}$
High: 1-2 $A \cdot dm^{-2}$
Stannous ion concentration:
Low: 15-20 gpl
Medium: 20-24 gpl
High: >24 gpl
Organic reagents:
Low: 0.3-0.8 gpl
Medium: 0.8-1 gpl
High: 1-1.5 gpl

Another parameter that can affect the current efficiency is the anode slime layer, although the removal of the anodes every 5 days for cleaning makes this parameter not as critical as the others mentioned above.

The main features of the new plant are summarized below:

Electrolyte:
Sn^{2+} 20-24 gpl
H_2SO_4 100-120 gpl
Organic additives 1 gpl
Volume 81.9 m^3
Temperature 25°C
Electrolyte flow rate per cell 15 liters per minute
Anode:
Weight 100 kg
Sn 99.377%
Life 10 days
Brushing 5 days

Current density:
Effective 0.8 A·dm^{-2}
Cell voltage 0.3 V
Stock ratio:
12.9
Tin production:
1800 Ton/year considering 250 days effective work

Table 4 shows the average cathode composition obtained since the start up of the plant; it can be seen that high grade tin can be obtained.

Table 4. Cathode composition

Element	%
Sn	99.982
Pb	0.002
Cu	0.0008
Fe	0.0016
Sb	0.005
As	0.003
Zn	0.0007
Bi	0.005
Cd	Traces
In	Traces

Figure 1 shows a general schematic flowsheet of the electrolytic refinery, Figures 2 and 3 are photographic views.

It should be noted that at the present time no final decision has been taken on the treatment of the anodic slimes. Accordingly, with the expected tin production, it is calculated that approximately 14 Ton/year of slimes will be produced. The composition of the slimes is not yet very well defined but major constituents are tin, lead, and antimony and to a lesser extent iron, arsenic, copper and bismuth. The mud can be either reduced to alloy, drossed or blended and sold as a solder or a bearing metal alloy. Another possibility for is it to be sold as it is to bearing metal manufacturers. The adoption of one or another option, or simply the safe disposal of the slime, will depend of the prevailing market prices.

4 Conclusions

During the period of time in which the plant has been working, the results obtained showed that the technique of electrolysis is reliable for obtaining high grade metallic

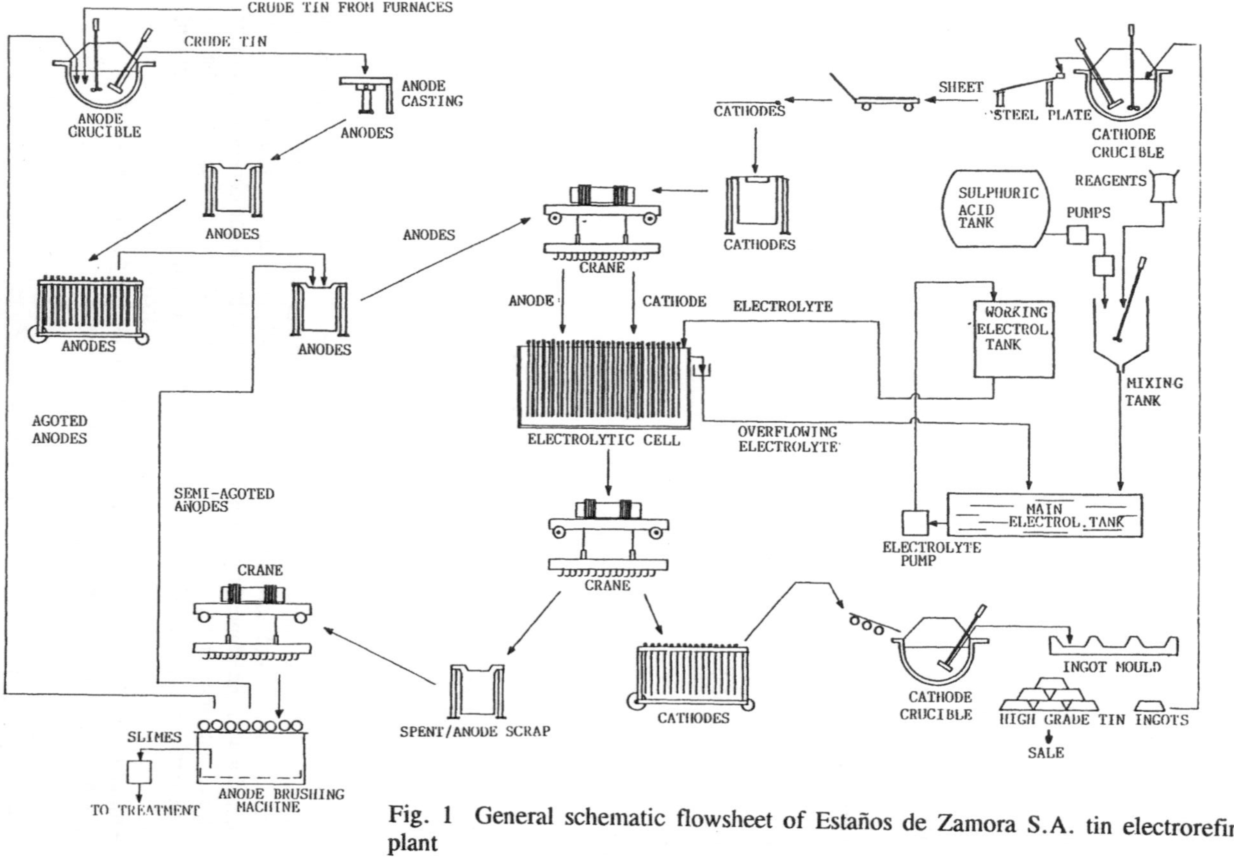

Fig. 1 General schematic flowsheet of Estaños de Zamora S.A. tin electrorefining plant

Fig. 2 Tin electrolytic cells with the working electrolyte tank in the upper background

Fig. 3 Overall view of some electrolytic cells during the phase of charge of anodes and cathodes

tin for its use under several purity restrictions. No very great problems have been encountered in the operation and management of the different parts of the plant. Electrolytic tin as obtained in the plant presents a higher degree of purity than that obtained by other refining techniques.

Note from authors

The only recent reference to similar work is that presented by: Wright, P.A. (1982) **Extractive Metallurgy of Tin**, especially chapter 11 and references therein. Elsevier, Amsterdam. Partial financial support by the CDTI (Spain) is acknowledged.

Waste disposal

Towards zero-discharge mining: minimization of water outflow

S. A. Cale
Knight Piésold and Partners, Ashford, Kent, England

Abstract
Increasing environmental pressures have resulted in the call for a zero discharge policy to be adopted by many mining operations located in sensitive areas. The achievement of this policy is feasible in many climatic regions, without excessive cost, especially if account is taken in the initial design. Retrofitting is usually more difficult.

The paper considers the major pollutant carrier, water, and give examples of protection measures which can be adopted. Particular attention is paid to different climatic conditions, arid, temperate, tropical, etc., to illustrate the prominent role that climate plays in a zero discharge facility.

The paper explains the use of a water balance program which has been developed to enable alternative scenarios to be examined quickly for inclusion in the environmental assessment, which is now inevitably required before any new mining development can proceed.

Experience drawn from operating on all continents is used to demonstrate the problems which can arise and how they may be overcome.
Keywords: mining development, environmental pressures, water, water balance, zero discharge, climatic conditions

1 Introduction

Sustained and increasing environmental awareness has resulted in calls for mining operations to adopt a "zero discharge policy" in certain circumstances. This is particularly relevant to water, which is often seen as the major pollution carrier.

Significant volumes of water are moved as a result of mining operations and it is the control of this activity which will be examined and discussed. It is readily accepted that other potential sources of environmental concern are also present, e.g. gaseous discharges, solid waste, etc, but these are outside the scope of this paper.

It is necessary at this point to define what might be considered zero discharge. Certainly, total containment of many substances is feasible, especially solid waste discharges, as demonstrated by the nuclear industry. However, the cost of secure storage, or full treatment, cannot normally be supported by precious or base metal mining ventures. Water presents a number of special problems so in practice zero discharge is often taken to mean minimum discharge of potentially polluting effluent to water courses and groundwater.

Underseepage into the ground is always likely to occur, and is usually unseen, and as a result is frequently ignored, even though it is highly mobile. Disposal management systems to control seepage are available and employ impermeable liners, sub-aerial deposition and drained repositories, as necessary, but again their use can be restricted by financial constraints. However, simple precautions and careful operational techniques can often reduce seepage to manageable and acceptable proportions.

2 Water Polluting Sources

For any new development, and frequently for an existing site, an Environmental Assessment will have been carried out to identify potential problem areas and suggest mitigation measures. Even if an EA has not been undertaken the major sources of potentially polluted water in a conventional mining operation are readily recognised and are normally:-

Open Pit/Mine — Seepage and rainfall run-off will normally be pumped out, and may possibly contain heavy metals, oils, etc. If the water is uncontaminated it may be discharged directly into watercourses providing simple measures are taken to settle suspended solids. It can be a useful source of make-up water in marginal cases.

Water Storage — A water storage dam is a useful and often essential facility for balancing seasonal supply variations against a constant demand, but is frequently excluded from developments on the grounds of cost. It can be used as a domestic water source providing it is located in an area which is kept reasonably free of pollutants. Unless fully lined, some seepage into the ground will occur.

Plant Area — Drainage and spillage from plant areas and roads is very likely to contain polluting substances which must be dealt with properly. Major flows in the drainage system of the plant area, haul roads and housing areas usually occur intermittently following heavy rainfall. Drainage inevitably contains a high concentration of particulate solids, and is also likely to contain pollution derived from accidental chemical, fuel and oil spillage.

The direct discharge of drainage flows to water courses is becoming less acceptable and routing to a holding pond where solids can be settled is relatively straightforward. The treatment of potentially polluted water is less straightforward and is site specific depending on the chemicals known or expected to be present.

Waste Dumps Seepage and run-off from waste and overburden dumps is often affected by acid drainage derived from oxidisation of pyritic minerals in the waste material. Drainage from the dumps can exhibit an extremely low pH, to the extent that it will leach metals from the waste or underlying rock and carry a cocktail of pollutants into nearby water courses. Severe deterioration of water quality can occur which may be treated by passive systems if other methods are unsuitable.

Satisfactory construction of such dumps is essential during their working lives, especially with regard to location and profiling. Suitable abandonment procedures may also be necessary to maximise rainfall run-off, such as capping, and thereby minimising water and air infiltration. These measures must also be combined with adequate provision for the collection and proper disposal of drainage emanating from the dumps. Bund walls have been shown to be very effective in many instances as traps for controlling silt transportation.

Tailings Dam As the ore usually contains only a small percentage of commercially valuable minerals, large quantities of tailings waste are produced. The tailings can contain a wide range of ancillary minerals which may undergo chemical decomposition in the same way as waste dumps. Residual reagents are also likely to be present.

Seepage will occur into the ground unless the facility is specifically designed for its minimisation. Decant and spillway flows are potentially controllable providing the hydrology of the depository is understood.

Tailings disposal facilities are frequently designed not only to receive and accommodate solid waste but also to act as a sump or water storage dam for the collection of drainage and run-off. In such cases they provide a mixing/retention and settlement pond for the water recycling system. This function conflicts with other design requirements which are intended to maximise deposited densities by restricting the volume of water in a tailings dam.

The tailings depository frequently fulfils a key role in the water balance equation and is regularly the starting point for modelling the overall water system for a mine and determining the potential outflows to the environment.

Fig 1 shows a typical mining operation and indicates where the major movements of water occur. Clearly a number of additional or alternative water routes and facilities, for example backfill operations, are possible which can complicate water balance calculations. A simplified version of this typical operation, in which the tailings storage dam and water retention reservoir have been combined, has been used in the numerical examples discussed later.

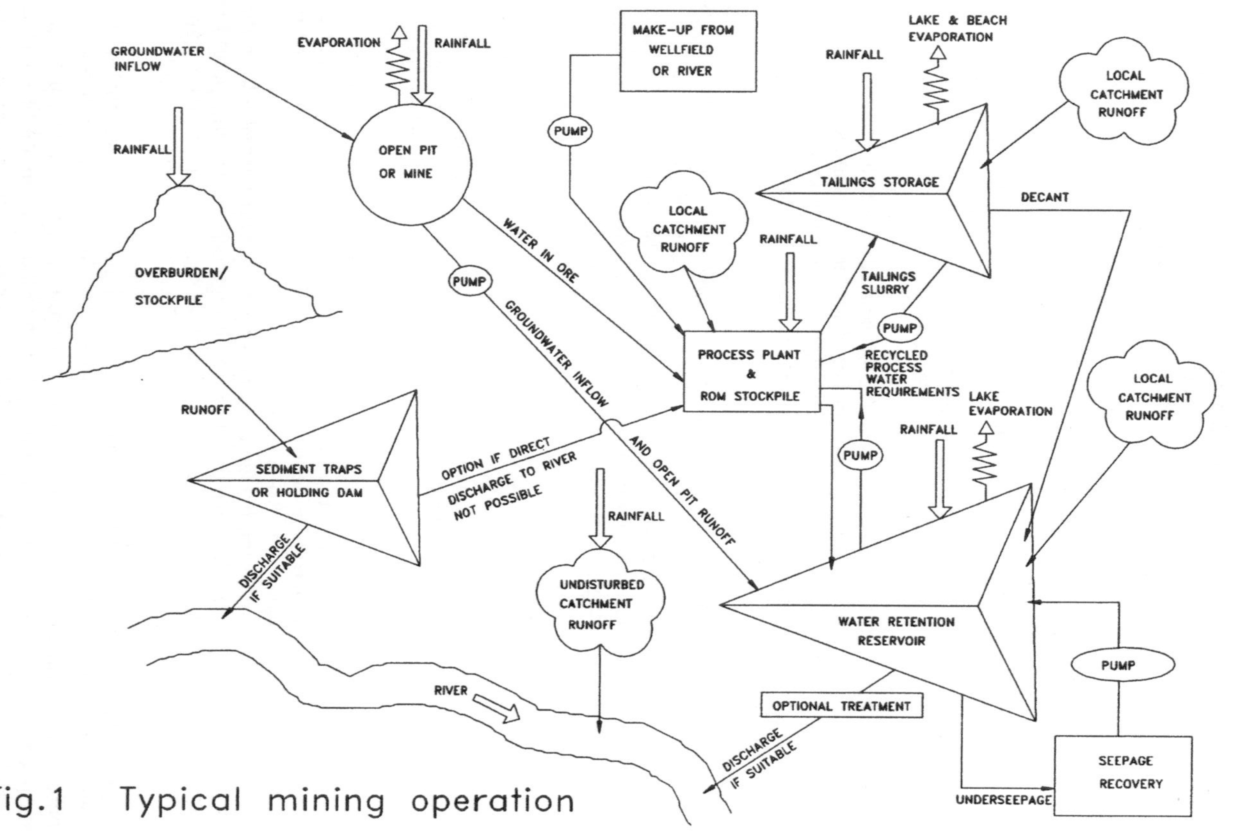

Fig.1 Typical mining operation

The first and overriding principle which should be adopted if it is wished to minimise discharges is clearly to recycle as much water as possible. This policy brings with it the need to consider the increasing concentration of substances which will not degrade during recirculation and their possible effects on the process.

The second principle should be to contaminate as little water as possible by diverting clean water away from or around sources of pollution.

If both these policies are not adhered to it will be impossible to approach minimum discharge in adverse climatic conditions and large volumes of water will either have to be evaporated or treated before discharge to the environment. Once-through systems are commonplace but are becoming more problematical as beneficiation processes employ increasingly complex metallurgical techniques and reagents which present disposal problems.

When siting facilities at a new mine flexibility in the layout of the plant and appurtenant works often exists and it is possible to arrange for the gravitation of most liquid pollutants to a downstream sump. In such cases, and notwithstanding other considerations, locating the tailings disposal area or water storage dam downstream of all other facilities - and in particular waste rock dumps - can be beneficial. It will also be advantageous to locate facilities which could cause pollution in as few sub-catchment areas as possible. This might be seen as concentrating the potential problems but has the advantage of limiting the number of separate mitigation measures which must be devised.

The most important factor in the water balance equation is the climatic conditions applicable to the area in which the mine is located. In particular the relationship between rainfall and evaporation is critical and determines whether a site can be considered as a candidate for a zero discharge policy without extensive capital works. It is imperative, therefore, to have a thorough understanding of the overall water balance for any operation at the outset. It should be noted that in this paper account will be taken only of average hydrological conditions in the examples. In reality a full range of conditions would be examined, including drought and heavy rainfall sequences, and monthly data would be used to determine maximum storage, discharge or make-up requirements.

3 Mine Water Balances

The key items which must be taken into account in the overall mine water balance are as follows:-

3.1 Inflows

- Importation of water for domestic use or mineral processing purposes
- Rainfall run-off over urban housing and industrial processing plant areas and disturbed ground surrounding the mine, waste rock dumps, tailings dams, natural ground and so forth
- Drainage water derived by pumping from underground or open pit workings
- Water arriving at the plant in the ore.

3.2 Outflows

- Decant discharge, spillage and seepage from tailings dams, other waste repositories or water storage dams
- Water retained within the tailings deposit
- Evaporation from water and tailings dams
- Domestic sewage effluent
- Water leaving the plant in concentrates or lost via the smelting process
- Plant drainage

Computer models have been developed to undertake the calculations necessary to analyse any situation on a monthly basis and also to carry out sensitivity analyses to examine the situation under a wide variety of operating and climatic conditions, which can vary considerably at any single location. The models are usually tailored to meet the specific circumstances prevailing at a mine. However, the results will only be as accurate as the input data, which must be determined as accurately as possible.

The major unknown is likely to be seepage, however, if data for an existing mine can be obtained it is possible to adjust the parameters in the model to match the historical evidence. Modern computing power permits ever increasing sophistication to be incorporated in water balance models, including allowance for some of the small inflows/outflows mentioned above. However, the main weakness in any model remains the accuracy of the input data. Separate routines are included which enable internal water transfers to be calculated e.g. recycled flows.

Liquid flows are often highly polluted with heavy metals or processing reagents, solids from mineral washing, or, in the case of sewage effluent, faecal matter, and require processing, treatment or dilution before they can be discharged to local water courses or into local groundwater systems if they are not to cause serious damage to the environment. It must therefore be recognised that water quality is an important factor in considering water balances, particularly with regard to recycling.

4 Legislation Affecting Discharges

There is a wide variety of legislation in most countries by which control is exerted over the discharge of water by industry. There is also a wide variety of standards of enforcement. Escalating pressure on world water resources is already resulting in an increase in legislation. The trend is expected to continue. Mining companies frequently employ their "home country" legislation when developing or operating in countries where legislation is less stringent. This tends to avoid criticism of environmental damage.

In the United Kingdom, for example, there is a substantial body of legislation governing water supply, drainage and pollution. This includes the Water Resources Act, 1963, together with the Water Acts of 1945, 1948, 1973 and 1989, Part II of the Control of Pollution Act 1974, the Land Drainage Acts of 1974 and 1976 and the Salmon and Fresh Water Fisheries Act. The EC has Directives which similarly controls pollution caused by mining and quarrying activities. The

United States has particularly stringent EPA legislation governing pollution, which often sets the trends which other countries follow.

5 Importance of Climate on Water Balance

The extent and type of operations at a mine play a large part in the magnitude of water surpluses or deficits at a particular location, but of more relevance are the magnitude of rainfall and evaporation, and their seasonal variation, although this latter aspect will not be considered in detail in this paper.

Monthly and annual rainfall and evaporation rates for some typical mining sites located in widely varying climates throughout the world are given in Table 1.

Notwithstanding considerations of pan factors and estimates of evaporation from tailings beaches, which can be considerable, the following general comments can be made.

In arid and semi-arid zones, represented typically by Oman and Central Australia respectively, where annual evaporation greatly exceeds rainfall, water for processing and other uses is at a premium and zero discharge can generally be achieved by completely evaporating excess pollution bearing water thereby effectively solidifying heavy and other metals and noxious materials. The main hydrological problem when mining in arid zones is often the provision of adequate water supplies for processing. The use of sea water as a supplementary supply is a solution which has been adopted at a number of mines.

In tropical zones, typified by Zambia, Brazil and Ghana, annual evaporation is of the order of 1300 to 2000 mm per annum and is fairly evenly distributed during the year. However, rainfall varies significantly, depending on location, from as little as 500 mm to as much as 11 000 mm per annum and can be either reasonably uniform throughout the year or concentrated in a rainy season of a few months duration. This requires much more careful consideration of overall water balances. Zero discharge may be achieved in many circumstances providing sufficient storage exists to cope with wet season inflows.

Within wet and cold temperate zones, characterised by the UK in Europe and Montana in the NW of USA, respectively, there is a marked seasonal variation in evaporation. The annual total varies considerably at locations in relative proximity to each other. In the more northern regions mid-winter evaporation rates fall to zero so that seasonal variations in the water balance assume greater importance. In addition, the metabolic rates of many biological and bacteriological processes are strongly temperature and solar energy dependent. Rates of purification by the latest techniques in wetland and reed bed water treatment therefore undergo distinct seasonal fluctuations. Accordingly seasonal storage of effluent waters to give longer residence times may be required to balance discharge rates with potential natural purification rates to achieve outflows commensurate with consent limits.

Table 1. Rainfall and evaporation at some typical mining sites

Country > Climate >	Oman Desert-Arid		Australia Semi-Arid		Zambia Tropical		Brazil Tropical		Ghana Tropical		United Kingdom Wet Temperate		USA (Montana) Cold Temperate	
Month	Rainfall (mm)	Evap. (mm)	Rainfall (mm)	Evap. (mm)	Rainfall (mm)	Evap. (mm)	Rainfall (mm)	Evap. (mm)	Rainfall (mm)	Evap. (mm)	Rainfall (mm)	Evap. (mm)	Rainfall (mm)	Evap. (mm)
January	10	104	230	195	304	124	245	99	32	185	142	18	5	10
February	71	101	235	145	244	117	208	89	75	148	107	17	0	10
March	24	166	165	165	180	157	160	100	151	138	103	36	20	30
April	19	229	40	165	39	165	81	82	175	110	68	51	164	51
May	1	384	20	135	5	168	34	95	234	104	70	72	523	84
June	0	380	15	130	1	147	6	111	265	111	69	81	511	112
July	1	313	5	140	0	168	2	136	167	115	66	74	171	132
August	1	271	5	165	1	211	10	177	88	141	81	65	61	91
September	0	243	5	205	2	239	16	178	175	142	92	64	74	58
October	1	221	20	260	23	246	84	152	236	121	119	47	30	38
November	2	164	55	255	133	193	216	120	142	120	138	27	3	10
December	28	119	125	230	261	145	347	95	59	147	147	21	8	3
Totals	158	2695	920	2190	1193	2080	1409	1434	1799	1582	1202	573	1570	630

The ratio P/Eo is a simple and useful indicator of the likely magnitude of mean annual run-off, and hence humidity, at any given locality. Very approximately the run-off at a particular location is given by the following relationship:-

$$RO = P - Eo \times \tanh(P/Eo)$$

where RO is the mean annual run-off (mm)
P is mean annual precipitation (rainfall) (mm)
Eo is the mean annual potential evaporation (mm)

The estimated run-off values given by this formula in the locations under consideration, are given in Table 2 below. The run-off is an important input to any water balance calculations and although the above formula has been found to be useful there is no substitute for actual data from stream gaugings and rainfall measurements.

The P/Eo ratio itself is valuable as a quick indicator of a site's hydrological/run-off potential.

Table 2. Approximate run-off under various climatic conditions

Location	Rainfall P (mm)	Potential Evaporation Eo (mm)	P/Eo	Run-off RO (mm)	Run-off (%)
Oman	158	2695	0.06	0.2	0.1
Australia	920	2190	0.42	50.6	5.5
Zambia	1193	2080	0.57	115.6	9.7
Brazil	1409	1434	0.98	327.5	23.2
Ghana	1799	1582	1.14	512.1	28.5
UK	1202	573	2.10	646.0	53.7
USA	1570	630	2.49	948.4	60.4

A number of typical examples of water balances in each of the areas mentioned have been examined and the results are given in the following section.

6 Numerical Examples

The marked difference in mine water balances which can occur in the various climatic zones can be illustrated by taking a typical, but simple, small mining project and applying the climatic parameters outlined in Tables 1 and 2. It is necessary to make a number of simplifying assumptions, but the overall conclusions remain valid and consistent with the results of actual fuller analyses carried out for specific projects at each location.

Table 3 shows a summary of the outcome of the analyses and indicates the likely magnitude of the surpluses and deficits, as a percentage of process requirements, which occur at various P/Eo ratios for the standard case.

Table 3. Variations in water surplus/deficit at varying P/Eo ratios

Country	Surplus/(Deficit) (%)	P/Eo Ratio
Oman	(1175)	0.06
Australia	(182)	0.42
Zambia	(84)	0.57
Brazil	43	0.98
Ghana	52	1.14
UK	75	2.10
USA	80	2.49

The data presented in Table 3 is plotted on Fig 2 as the standard case. Also illustrated is the notional best case. This is a situation which can be achieved under favourable circumstances.

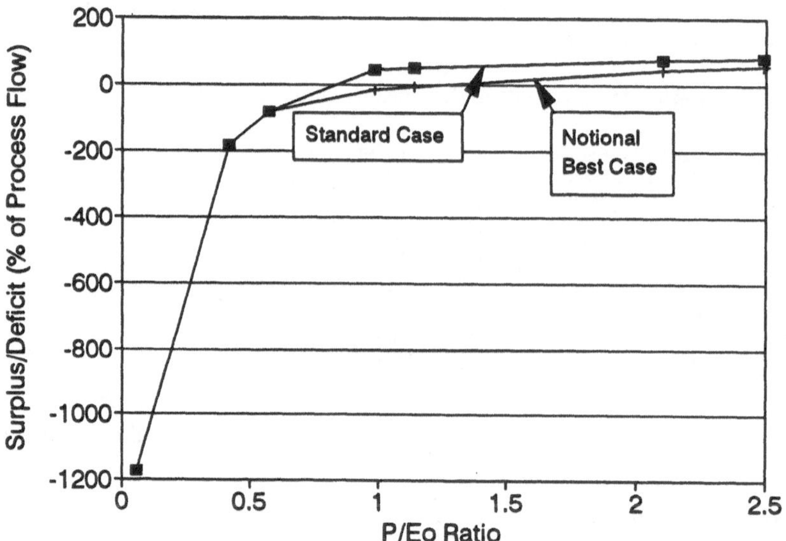

Fig. 2 Surplus/Deficit vs P/Eo Ratio

The above figure demonstrates that at low P/Eo ratios very significant deficits occur making it possible to achieve a zero discharge policy. As the P/Eo ratio approaches unity surpluses occur and become larger as the ratio increases but the percentage is small compared with the deficit examples. In a situation where the P/Eo ratio is close to unity a borderline case can be transformed from a marginally surplus to a marginally deficit case by adopting the recommendations in this paper.

It is also of interest to note that the water which can be recycled from the tailings dam for process requirements in the examples shown varies from about 30% at low P/Eo ratios to well over 100% at high ratios.

7 Discussion of Typical Cases

7.1 Arid and Semi-arid Zones

In these zones the P/Eo ratio is often between 0 and 0.5, run-off is very low in relation to precipitation and accordingly appropriate design of the tailings disposal facilities, together with the provision of ancillary evaporation pans with sufficient area, can provide a system with zero discharge. Greater attention will need to be paid to the provision of make-up water and also to supplying sufficient water at start-up. Examples of this approach are the uranium mines in Queensland and copper mining in Oman, where suitable fresh water for ore processing is not readily available.

7.2 Tropical Zones

As the P/Eo ratio increases from 0.5 to about 1 in these areas, much more careful consideration has to be given to the recycling of water, but an effective zero discharge system may still be achieved by encouraging evaporation during dry seasons by maximising water surface areas. The importation of water will be necessary in dry and average years, but a positive water balance is likely in wet years. In such instances attention should be paid to diverting clean run-off water around the pollution producing areas. Typical examples are the copper mines in Zambia and the gold mines in central Brazil.

Interest in more effective recycling methods, and the reduction of outflow from the system, often arises as a result of enforced water shortages originating from the demands of a higher ore throughput or a period of drought, rather than considerations of pollution reduction *per se*.

When P/Eo exceeds 1, zero discharge becomes more difficult to achieve, and extensive diversion of run-off will be required. The design of the tailings dam in terms of area/volume ratio assumes critical importance and selection of a site, if the choice exists, with the lowest area/volume ratio and minimum catchment area will assist. Only in particularly dry years will zero discharge be possible and provision for discharge during wet periods must be made. The gold mines in Ghana provide ideal examples of this situation.

7.3 Temperate Zones

In temperate zones, and especially in areas of high rainfall, the P/Eo ratio is generally above 1.5, often exceeds 2 and rises over 2.5-3. Run-off is high and there is generally a relatively large water surplus requiring disposal. Even with recycling, evaporation rates are insufficiently high to enable surplus water to be evaporated. In a dry year a surplus is still likely. Examples are the tin mines in the UK and copper mines in high mountain regions of the USA.

8 Conclusions

To achieve zero discharge attention has firstly to be paid to the climate at the mine site. As a first benchmark the P/Eo ratio gives a useful indication as to whether advantageous circumstances exist. Even if unfavourable much can be done to enable outflows to the environment to be minimised, these are :-

- Collect all potentially polluted drainage from rock dumps and other areas and dispose of it in the tailings dam, this is easier if it is located downstream.
- Reduce the requirement for the importation of water for processing purposes by recycling as much water as possible from the tailings dam or water storage dam.
- Where the P/Eo ratio exceeds unity, i.e. rainfall exceeds evaporation, maximise the area of tailings beach and pond to enhance evaporation and adopt frequent rotational deposition techniques.
- If appropriate, collect and return all tailings dam underseepage to the tailings pond. This can be achieved by pumping from sumps and/or from interception wells sited at convenient low points in the groundwater phreatic surface.
- In cases where zero discharge is unlikely to be achieved, as indicated by the P/Eo ratio being significantly in excess of unity, divert as much clean water as economically possible around pollution causing sites.
- If surplus water is inevitable and it cannot be discharged directly, make provision to treat the water by chemical or biological processes before discharging to the environment.

Design of a tailing facility to mitigate potential acid rock drainage

D. R. East
Knight Piésold and Co., Denver, Colorado, U.S.A.
L. H. Filipek
Knight Piésold and Co., Denver, Colorado, U.S.A.
Adriaan de Villiers
Knight Piésold and Co., Denver, Colorado, U.S.A.
T. R. Wildeman
Colorado School of Mines, Golden, Colorado, U.S.A.

Abstract
This paper presents the design of a reclamation cap for a tailings disposal facility as a means to eliminate oxidation and thereby mitigate potential acid production by sulphide minerals. Available technology and current research findings are cited in support of the design of such a facility.
Keywords: Acid rock drainage, anaerobic conditions, reclamation, subaqueous, sulphidic ore, tailing disposal facility.

1 Introduction

The long-term storage of potentially acid-producing mine and mill waste is becoming an increasingly important aspect of mining reclamation. One such project, in a mountainous region of the western USA where an excess of net precipitation occurs, is being designed for total submergence of the mill tailings both during operation and at reclamation. This paper describes the engineering and geochemical considerations of the reclamation cap design and the long-term mitigation of the area for use as a wetland.

The tailings will be stored in a side valley impoundment with diversion of the existing creek designed to remain the low point of the valley reclamation. The tailing impoundment basin is being designed with a low-permeability soil/geomembrane composite liner system and a managed peripheral spigot tailing discharge system which will form the final shape of the surface in order to provide permanent surface drainage consistent with the concept of totally submerged tailings.

The reclamation cap will comprise a 1.2-meter-thick layer of non-acid-generating crushed rock placed onto a geofabric filter layer which will be the separation layer between the tailings and the caprock. The lower 300 to 600 mm of the caprock will be a selected low-permeability zone with the upper portion consisting of an open-graded high-permeability rock designed to take all precipitation and flow-on from the surrounding hillside. Phreatic levels within the reclamation cap will be controlled by a constant elevation spillway from the synthetically-lined impoundment. The caprock will be covered by about 600 mm of growth medium consisting of highly-organic surface soil removed from the site prior to excavation of the tailing impoundment basin. This material should consume oxygen in the surface water and "condition" the drainage entering the soil, rock, and low-permeability layers to keep the tailing material isolated from oxidizing reactions.

2 Geologic and Geomicrobiological Review

Closure of a mining facility necessitates leaving solid materials in an environment in which they are stable on a long-term basis. If the minerals in the ore deposit are sulfides, then that environment should be anoxic (devoid of oxygen).[1] [2] [3] From a geologic perspective, the occurrences of metals in anaerobic sedimentary deposits such as coals or lignites, should determine the more thermodynamically stable forms of metals deposited under these conditions.[1,2,3,4] In these sediments, the stable iron minerals are pyrite or siderite, the stable manganese material is usually rhodochrosite, and base metals such as Co, Ni, Cu, Zn, As, Cd, Ag, Se, Hg and Pb exist as sulfides.[1] [2] [4] [5] Also, the same mineral association of sulfides and carbonates is seen in anoxic sedimentary ore deposits such as black shales.[1]

In investigations of recent depositional environments, marine sediments deposited in deep stagnant waters such as fjords or enclosed seas also exhibit the same mineral association as in coals and sedimentary ore deposits.[6] [7] [8] [9] In fresh water environments, anoxic conditions can be established; however the mix of minerals can be quite different because the availability of nutrients, particularly sulfate and nitrate, is often limited.[10] [11]

Application of these principles to the closure of a sulfide ore deposit leads to two conclusions:

1. If a sulfide deposit is to be closed, it best be left in a state where anoxic conditions can be maintained. This usually means storage in a stagnant water environment.
2. If a system is to be maintained undisturbed in an anoxic state, the reactants to produce sulfides and carbonates need to be readily available.

The above questions of stability do not include the issue of reaction rates, whether the constituents in the water and sediments will react rapidly enough to consume all the oxygen. This question of kinetics is controlled by the microbes that inhabit the aquatic environment. Generally, these microorganisms survive in nature by catalyzing chemical reactions that release energy to the organism.[12] [13] [14] In an aquatic environment, the progression to thermodynamic stability is almost always speeded up by microbial activity.[12] [14] [15]

In most instances, reactions include organic material, and the microbes are designated as being heterotrophic. An important example of this process is the reaction upon which the sulfate-reducing microbial consortia survive:[13]

$$2\ H^+ + SO_4^{2-} + 2\ \text{"}CH_2O\text{"} \longrightarrow H_2S + 2\ H_2CO_3$$

Here, "CH_2O" represents organic matter such as cellulose and other carbohydrates, and the sulfate is dissolved in the water. In this reaction, there is no oxygen; instead sulfate serves as the electron acceptor. This type of environment is termed anaerobic, an anoxic system where microbes are making use of other electron acceptors. Note that the reaction provides two products important to precipitating and stabilizing metals in an anaerobic environment; H_2S and H_2CO_3. Also note the reaction consumes hydrogen ions.

To promote sulfate-reducing microbial activity, it is necessary to provide the reactants in the above reaction. Mine waters are an excellent source of dissolved sulfate.[16] In most mineralized areas, an adequate concentration of sulfate will be available.[17] Organic-rich sediments are necessary for providing the organic nutrients. In the review of sedimentary situations where sulfides are formed, there is always an excess of organic debris.[1] [4] [6] [8] Consequently, another criterion for establishing the proper geomicrobiological environment is an adequate supply of organic material.

3 Geotechnical Considerations

In the closure of a tailings deposit, dry covers more than wet covers have been used to isolate the sulfidic materials from the atmosphere.[18] Nevertheless, the objectives for dry cover use are to isolate the tailings. Many of the situations where dry covers are used are in arid climates where water can be successfully diverted from the isolated tailings.

In most of the geological situations where a depositional site is isolated from the atmosphere, it is water that is the isolating agent. Principals of aquatic chemistry [2] [14] [15] [19] show that water has two properties that help this isolation:

1. The diffusion of oxygen through water is many times slower than it is through air.
2. Constituents in the water can react with the oxygen to remove it.

For these two properties to be most effective, burial by water should be continuous and the sedimentary material should not be exposed to the atmosphere for even intermittent periods of time.

In underwater disposal of tailings, most of the cases where this has occurred have been by accidental or unregulated disposal into natural bodies of water.[20] These have included fresh and salt water lakes and marine coastal bays and fjords. Also, most of these disposal situations have been in temperate zones where net water accumulation occurs and there may be considerable snow cover.[20] [21] In the past five years, these disposals have been intensively studied, especially through projects sponsored by the Mine Environmental Neutral Drainage Program (MEND) in Canada. Included in these studies have been chemical characterization of the waters and sediments in the lakes. The four lakes that have been extensively studied all show the same characteristics. The tailings are all contained under anaerobic conditions and there has been minimal release and sometimes accumulation of metal contaminants in the tailings. Natural production of organic material in the lakes has helped to seal the tailings and generate an anaerobic ecosystem. All the principles reviewed above have been verified in the analysis of the waters and sediment cores from the four lakes extensively studied in the MEND projects.[22] It is important to note that these were situations where tailings were disposed of in natural settings. Little was done to design and engineer the disposal.

There are potential problems with the deposition of tailings into bodies of water. Although molecular diffusion of oxygen may be minimal, convective diffusion caused by the lake overturning in fall can deliver oxygen to the water-sediment boundary.[14] [15] [19] This would certainly mix oxic and anoxic waters. What is needed is consistently stagnant water that will maintain its anoxic character. In addition, wind action on a large water body may also cause convective diffusion of oxygen to the water-sediment boundary. Also, in shallow lakes, wind action may cause agitation of the sediments so that sealing the surface and maintaining isolation of the tailings cannot occur.[15] [19]

What is necessary for underwater disposal of tailings is to maximize the positive aspects of the concept and minimize the potential problems by design. Maximizing the pros and minimizing the cons has been successfully applied to the design and construction of wetlands.[23]

In recent years, the principles of aquatic chemistry and geochemistry have been used in mine closure for the treatment of acid mine drainage (AMD) by use of wetlands. In the case of AMD treatment, wetlands have been constructed so that either aerobic or anaerobic conditions are created that maximize the appropriate treatment reactions.[10] [23] [24] [25] [26] In addition, it is important to control the hydraulics of the water to be treated to ensure that maximum treatment is effected. Use of constructed wetlands is an excellent example of combining geochemical and engineering principles in the design and construction of an effective treatment system.[23] Designs are made so that in one wetland cell a certain process (such as alkalinity generation) is brought to completion before the water enters the next cell where another process (such as aerobic removal of iron) is effected.[23] Natural wetlands are not used because they provide a balance of all processes rather than maximizing one. In anaerobic wetland cells, the reactions controlled by sulfate-reducing bacteria are emphasized.[24] [25] [26] Bench- and pilot-scale systems built to optimize the flow of AMD through a submerged substrate, which has been designed to promote the activity of the sulfate reducers and provide some neutralizing capacity, have successfully treated severely contaminated waters.[23] [24] [26] It has been verified that the metals are removed as sulfides and carbonates. In many of these situations, the flow of the AMD has been down through the substrate and the water covering the substrate has only been a few centimeters in depth.[10] [23] [26]

If success is possible in a natural system where geomicrobiological as well as physical processes are competing against one another, then it should be possible to achieve the same success by designing an underwater tailing disposal site. In this case, design and construction principles would try to maximize anaerobic conditions. To this end the following guidelines would be followed:

1. The tailings would be from sulfidic ores so that the closure objective is to minimize diagenetic reactions that might release contaminants, and maximize the stability of minerals that were formed under anoxic conditions.
2. The setting would be in an area where there was a net accumulation of moisture so that a water cover would be assured even in times of drought.
3. The amount of exposed water would be minimized so that release of contaminants to the environment by that route would be small. Also, this would eliminate problems with overturn, wind and hydraulic head associated with large areas and depths of free water. This would imply that the anaerobic environment would be created in the substrate of a wetland rather than in the sediment below a lake.
4. Any flow of air and water through the tailings would be minimal so that the amount of oxygen that possibly enters the system is minimal.
5. Provisions would be made to capture any water flowing through the tailings. This suggests that the flow through the buried tailings would be downward to ensure that any contaminants in the water flow away from the surface ecosystem.
6. A continuously submerged cover would be constructed in such a way that dissolved oxygen and other oxidizing agents in the water such as Fe(III) and nitrate would be completely consumed before the water reaches the tailings.
7. To effect criteria 4. and 5., the cover soil would have a low hydraulic conductivity, but be slightly higher than the hydraulic conductivity of the tailings. To effect criterion 6., the cover would have sufficient organic material in the soil to insure continuous activity of anaerobic bacteria.

4 Design

Subaqueous tailing deposition in an engineered impoundment at a site where an excess of net precipitation occurs offers a logical opportunity for long-term, "walk away" mitigation of potential acid-producing tailings. To illustrate such a design, the following case is examined:

- A precious metal mine with pyrite in the ore body.
- A net annual surplus in precipitation at the site.
- Precious metals recovery by pyrite flotation.
- A side-valley impoundment in the valley adjacent to the process facilities is selected as the most appropriate tailing disposal facility.

The potential for long-term acidic drainage from the tailings is an obvious concern to the mining company, state and federal agencies, and the general public. In response to these concerns, a fully-lined impoundment is designed analogous to a liquid containment structure.

The mineral extraction process requires that the process water be maintained at a pH above 10. The make-up water requirements, allowance for design storm containment and operational fluctuations in the supernatant water level, result in an average operational design depth for the supernatant water of no less than 1 meter (3 feet), even during a 100-year drought. Therefore, during the operating life of the mine, an alkaline water cover can be maintained over the tailings in the impoundment, thus preventing oxygen diffusion and providing neutralization for any acid formation.

Settlement tests on tailing samples demonstrated that the tailings settle and consolidate rapidly even under submerged conditions. Consolidated densities were demonstrated on the order of 16 kN/m^3 within one to two years. Therefore, conditions suitable for the support of reclamation activities on the tailings are anticipated.

The objectives of the reclamation cap design are fulfilled by the following components:

- A layer of organic material overlying the tailings
- A filter fabric overlying the organic layer (340 grams/m^2 geotextile)
- A low permeability caprock layer overlying the filter fabric
- A relatively coarser-grained caprock layer overlying the lower caprock layer
- A filter fabric overlying the caprock
- Organic-rich soils overlying the upper filter fabric
- A vegetative cover established in the organic-rich soils
- A surface water diversion channel terminating at a spillway

The purpose of each of the components of the reclamation cap as shown in Figure 1 are further explained below.

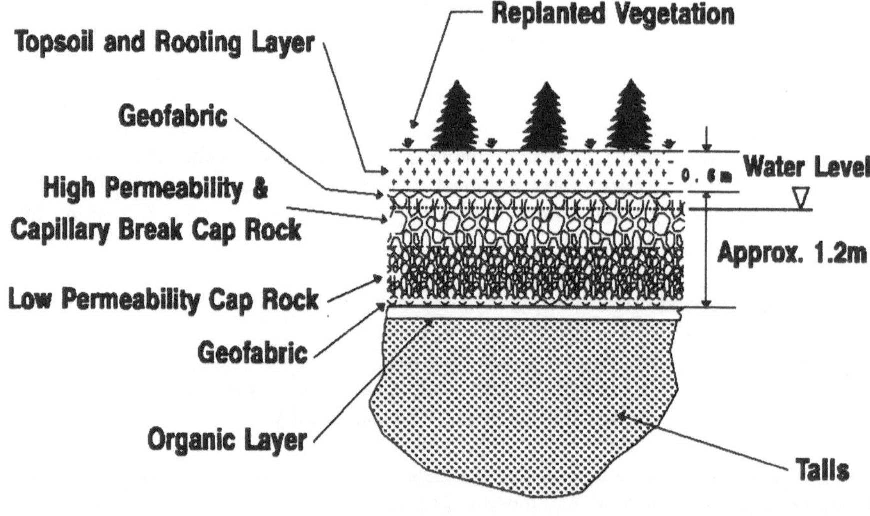

FIGURE 1

From the top down: surface water run-on and direct precipitation onto the impoundment will replenish the water cover which will be maintained within the caprock region. Such water will reach the caprock by filtering through the revegetated soil cover or more directly, at the surface diversion channel with its invert within the upper caprock region. The vegetation will have a dual effect, i.e. providing nutrients for the promotion of anaerobic bacterial activity as well as the provision of erosion resistance. Aesthetic appeal is also achieved.

The spillway at the end of the diversion channel is designed to provide level control of the water within the caprock and within the impoundment. The filter fabric is designed to prevent migration of fine material from the soil cover to the caprock. The upper caprock region (coarse material) is provided to act as a capillary break to water within the lower (finer) caprock, and to maximize circulation of direct precipitation and run-on to the facility from adjacent areas.

The lower caprock region has as its main function the maintenance of stagnant water over the tailings. This is achieved by selecting material for the lower caprock to provide hydraulic conductivity of one or two orders lower than the overlying, coarser caprock.

The filter fabric underlying the caprock confines the organic material layer and also serves as a construction base for the caprock. Light construction equipment will be able to spread materials over the filter fabric, with limited penetration of the caprock into the tailings.

Finally, the organic layer is provided as an oxygen consumer to assure anoxic conditions at the tailings/cap interface. It may be noted that even without the organic layer, research has indicated that anoxic conditions may be expected at the top of the tailings. [20] [21] [22]

The tailings themselves should also be considered as integral to the reclamation cap design. As a silty, consolidated material, the hydraulic conductivity is expected to be very low.

4 Conclusion

In conclusion, appropriate wetlands technology and research under the MEND program prove the effectiveness of a water cover to mitigate acid production from sulfidic tailings. A man-made subaqueous disposal facility offers the opportunity to maximize and control those positive aspects for acid generation mitigation which researchers have found in tailings disposed of in natural aqueous environments. A mine site where an excess of net precipitation occurs may be an ideal environment for application of subaqueous tailing disposal technology.

5 References

1. Maynard, J.B. (1983) **Geochemistry of Sedimentary Ore Deposits.** Springer—Verlag, New York, USA, 305 pp.
2. Stumm, W., and Morgan, J.J. (1981) **Aquatic Chemistry, 2nd Ed.** John Wiley & Sons, New York, USA, 780 pp.
3. Lindsay, W.L. (1979) **Chemical Equilibria In Soils**. John Wiley & Sons, New York, USA, 449 pp.
4. Boushka, V. (1981) **Geochemistry of Coal**. Elsevier Publishing Co., New York, USA, 284 pp.

5. Valkovic, V. (1983) **Trace Elements In Coal, Vol. 1.** CRC Press Inc., Boca Ration, Florida, USA, 210 pp.
6. Berner, R.A. (1964) Iron sulfides formed from aqueous solution at low temperatures and atmospheric pressure. **J. Geol., 72**, pp. 293-306.
7. Berner, R.A. (1967) Thermodynamic stability of sedimentary iron sulfides. **American Journal of Science, 265**, pp. 773-785.
8. Jorgensen, B.B. (1983) **Microbial Geochemistry In The Microbial Sulfur Cycle,** W.E. Krumbein, ed., Oxford: Blackwell Sci. Publishers, pp. 91-124,
9. Filipek, L.H. (1986) Influence of iron and manganese on the chemical partioning of copper, zinc and chromium during early diagenesis in outer continental shelf sediments from the Gulf of Mexico. In **Studies in Diagenesis**, U.S. Geological Survey Bull. 1578, pp. 31-50.
10. Machemer, S.D., Reynolds, J.S., Laudon, L.S., and Wildeman, T.R. (1983) Balance of sulfur in a constructed wetland built to treat acid mine drainage in Idaho Springs, Colorado, **Applied Geochemistry,** vol. 8, pp. 587 - 603.
11. Herlihy, A.T., Mills, A.L. and Herman, J.S. (1988) Distribution of reduced inorganic sulfur compounds in lake sediments receiving acid mine drainage. **Applied Geochememistry,** vol. 3, pp. 333-344.
12. Ehrlich, H.L. (1981) **Geomicrobiology**. Marcel Dekker, Inc., New York, USA, 393 pp.
13. Postgate, J.R. (1979) **The Sulphate—Reducing Bacteria**. Cambridge University Press, New York, USA, 220 pp.
14. Libes, S.M. (1992) **An Introduction to Marine Biogeochemistry.** John Wiley & Sons, New York, USA, 734 pp.
15. Wetzel, R.G. (1983) **Limnology, 2nd Ed.** Saunders Publishing, Philadelphia, Pennsylvania, USA, 767 pp.
16. Wildeman, T.R. (1991) Drainage from coal mines: Chemistry and environmental problems. In **Geology in Coal Resource Utilization**, D.C. Peters, ed., Tech Books, Fairfax, Virginia, USA, pp. 499-511.
17. Runnells, D.D., Shepherd, T.A., and Angino, E.E. (1992) Metals in water. In **Environmental Science & Technology**, 26, pp. 2316-2323.
18. Hutchison, I.P.G., and Ellison, R.D. (1992) **Mine Waste Management**. Lewis Publishers, Boca Raton, Florida, USA, 654 pp.
19. Drever, J.I. (1988) **The Geochemistry of Natural Waters, 2nd, ed.** Prentice Hall, Englewood Cliffs, New Jersey, USA, 437 pp.
20. Robertson, J.D. (1989) **Subaqueous Disposal of Reactive Mine Wastes: An Overview.** CANMET, Ottawa, Ontario, Canada, MEND Program Report 2.11.1a.
21. Robertson, J.D. (1991) Subaqueous disposal of reactive mine waste: An overview of the practice with case studies. CANMET, Ottawa, Ontario, Canada in **Proceedings of the 2nd International Conference on the Abatement of Acidic Drainage**, vol. 3. pp. 185-200.
22. **A Critical Review of MEND Studies Conducted to 1991 on Subaqueous Disposal of Tailings.** (1992) CANMET, Ottawa, Ontario, Canada, MEND Program Report 2.11.1d.
23. Wildeman, T.R., Brodie, G.A. and Gusek, J.J. (1993) **Wetland Design for Mining Operations.** Bitech Publishing, Vancouver, B.C., Canada, pp. 300.

24. Wildeman, T.R., Updegraff, D.M., Reynolds, J.S. and Bolis, J.L. (1994) **Passive Bioremediation of Metals from Water using Reactors or Constructed Wetlands. In: Emerging Technology for Bioremediation of Metals,** J.L. Means and R.E. Hinchee, eds. Lewis Publishers, Boca Raton, Florida, USA, pp. 13-25.
25. Filas, B.A., and Wildeman, T.R. (1992) The use of wetlands for improving water quality to meet established standards, in **Proceedings from Successful Mine Reclamation Conference,** Nevada Mining Association, Reno, Nevada, USA, pp. 157-176.
26. Emerging Technology Summary (1993). **Handbook for Constructed Wetlands Receiving Acid Mine Drainage,** U.S. EPA Report EPA/540/Sr-93/523.

Exploitation of shear and compression rheology in disposal of bauxite residue

M. D. Green
N. J. de Guingand
D. V. Boger
Advanced Mineral Products Centre, Department of Chemical Engineering, University of Melbourne, Victoria, Australia

Abstract
The paper presents an overview of the research conducted in the Department of Chemical Engineering at the University of Melbourne in conjunction with Alcoa of Australia Limited. The first part of the presentation reviews the shear rheology of bauxite residue suspensions and emphasises how an understanding of the fundamental flow properties of concentrated bauxite residue suspensions can be exploited in developing a semi-dry waste disposal scheme, which has been developed and proven by Alcoa in Western Australia and is now under consideration by a number of international alumina producing companies. The paper compares the rheology of highly concentrated bauxite residue suspensions in shear to results obtained in compression using techniques for compressive yield stress measurement developed at the University of Melbourne. It is shown how the compressive yield stress properties are particularly useful in the design and specification of thickeners for dewatering red mud suspensions.

The techniques presented for both shear and compression rheology measurement are applicable not only in the disposal of bauxite residue but are also useful in examining dewatering processes in other mineral system applications and for other waste disposal problems, for example, in the disposal of coal tailings.
Keywords: Shear rheology, compression rheology, disposal techniques, bauxite residue, red mud.

1 Introduction

Bauxite residue or red mud is produced in huge quantities during the production of alumina in the Bayer process. Alcoa's Western Australian refineries currently produce about 5.5 million tonnes of alumina annually [1] which accounts for a significant percentage of the world alumina production. The ore used in these refineries is considered to be of low grade by world standards with the result that for every tonne of alumina produced, approximately 2 tonnes (dry weight) of residues are produced. This equates to an immense volume of red mud which must be disposed of each year. This bauxite residue is primarily composed of iron oxide, silica and undigested alumina, the remainder being other mineral oxides. The traditional method of disposal of this red mud was to pump a low concentration slurry of between 20 and 30 wt% solids to huge dyked settling ponds. In the last decade, however, the most economic and environmentally sound procedure for the alumina industry has been shown to involve highly concentrating the red mud before it is disposed of to drying beds. Dewatering of a typical mud slurry to a concentration of 50-65 wt% solids is the usual objective. The many advantages of such a 'semi-dry' disposal scheme over the traditional 'wet' disposal methods are discussed by Nguyen and Boger [2-3].

Two interdependent studies must be completed before a semi-dry disposal scheme for a concentrated suspension system is able to be considered for implementation. Firstly, it must be proven that such a suspension can be successfully handled at such a high concentration. This means that it must be able to be pumped from the plant to the prepared disposal area with the least expenditure of energy, be able to be started up from a standstill in a pipeline and also be able to be spread over the available disposal area without blockage of the pipeline outlet. This study entails a full characterisation of the shear rheology of the system under the full range of expected processing conditions. If necessary, appropriate additives may be required to be used in order to meet these conditions. The shear rheology of concentrated red mud suspensions has been extensively investigated [3-5] and the pumpability and spreadability of such suspensions has been established. From these studies, the viability of such disposal schemes has been demonstrated and in fact proven by Alcoa in Western Australia [6]. The techniques used for the determination of the shear rheology of red mud and the main results from these studies will be briefly presented in this paper.

The second study that must be made is that the feasibility for the dewatering of such a suspension to the required concentration at the required rate must be determined. Numerous methods for the dewatering of suspension systems are available but for plants with a high throughput, usually thickeners are the only practical option. Additives, if any, that are used to achieve the required shear rheology must also be compatible with the achievement of a highly concentrated product, thus the two studies must be undertaken simultaneously. The production of these highly concentrated suspensions in a thickener dictates the successful exploitation of the compression zone. In order to design such a thickener, a good understanding of the compression rheology of the suspension system is required. For red mud, the compression rheology has been studied by De Guingand [7] and Green et.al. [8]. Such a system has complex chemistry and thus methods to increase the solids concentration must be determined empirically. Knowledge of these dewatering techniques have been directly exploited by Alcoa and these techniques and results will be briefly discussed here.

The techniques used for the characterisation of these red mud suspensions for a highly concentrated semi-dry disposal scheme which are presented in this paper are equally applicable to almost any concentrated slurry or suspension system. At the present moment, these techniques are yet to be implemented in any other mineral processing industry so it is the objective of this paper to demonstrate how the waste disposal problems of Alcoa are being solved using these techniques and how they may be easily applied to other systems. Hopefully, this should prompt further exploitation of these techniques across a diverse range of applications in the mineral/ hydrometallurgical industry.

2 Shear rheology

2.1 Measurement techniques

The characterisation of the shear rheology of concentrated mineral suspension systems requires the use of specialised measuring equipment. Red mud exhibits highly time dependent, non-Newtonian behaviour which is best measured using a capillary rheometer [4, 9-10]. The capillary rheometer has been shown to be broadly applicable to a wide range of concentrated suspension systems such as pigments, mineral slurries and pastes, mine tailings and brown coals [2, 11].

The shear yield stress of highly concentrated suspensions is best measured using the vane technique [12-13]. This technique uses a slowly rotating vane, which when started from a standstill, the maximum torque generated may be directly related to the shear yield stress. The vane technique has been demonstrated to be generally applicable to materials which exhibit a shear yield stress.

Using these techniques, the shear rheology of concentrated mineral suspensions may be fully characterised. The shear rheology of red mud and the effect of various physical and chemical changes on the suspension system will now be briefly discussed in relation to their industrial application.

2.2 Shear results

The shear rheology of mineral suspensions depends strongly upon the solids concentration. The flow properties of the system in question should thus be closely examined around the proposed final processing concentration after dewatering. This will determine whether or not it is able to be handled at that concentration. For red mud, the objective is to reach a final solids concentration after dewatering of 50-65 wt%. At these concentrations, the red mud suspensions are shear thinning, which means that the viscosity decreases with increasing shear rate, they exhibit a shear yield stress and they are also strongly thixotropic. Thixotropy implies that the flow properties of the suspension are also dependent upon the time of shear and upon the previous shear history that the suspension has experienced. The shear yield stress is the minimum shear stress that must be exceeded to initiate flow [3].

The shear rheology of highly concentrated red mud is thus fairly complicated and hence must be carefully measured and controlled. Firstly, the effect of shear history on the flow properties must be evaluated. This is done by shearing the suspension by mixing, then measuring the shear stress-shear rate using the capillary rheometer at increasing times of shear. It was found that a 64 wt% filter cake became more fluid with increasing duration of shear which reached an equilibrium state after sufficient time had elapsed (> 100 hours) [3]. Similar behaviour of the shear yield stress, measured by the vane technique, was also observed. These results clearly demonstrate that a pumpable consistency is able to be achieved given sufficient mechanical agitation.

Once the shearing is removed, the rate of recovery to the original solid-like state should also be recorded by measuring the increase in the shear yield stress over time. It was found that the rate of increase is fairly rapid at first but decreases with longer resting times. This recovery curve is directly compared with the breakdown curve for a 68 wt% red mud suspension in Figure 1. The time scale for the breakdown of the

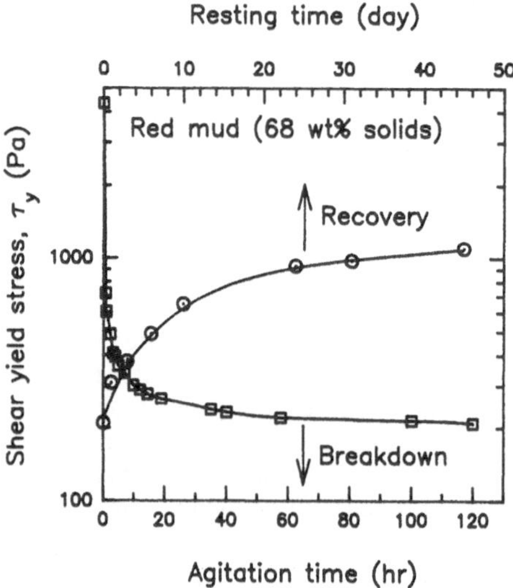

Fig. 1 Direct comparison between reduction of yield stress with mixing time (breakdown) and increase with resting time (recovery). (From Nguyen and Boger, 1984).

structure of a concentrated red mud suspension is thus very rapid compared with the rate of structural recovery. This property of red mud is directly exploited when pumping the suspension from the final dewatering operation to the disposal area. The suspension is quickly broken down by a high shear centrifugal pump and then pumped by high pressure pumps in a low viscosity, low shear yield stress equilibrium state to the disposal area. Once there, the suspension is able to spread over the available area before structural recovery of the suspension begins. A stable drying bed is therefore established which is able to dry to 65-70 wt% solids, thus enabling future reclamation of the land.

Having evaluated the thixotropic behaviour of red mud, the equilibrium state flow properties must now be determined. This is done by shearing the suspension until the viscosity is minimised, then measuring the shear stress-shear rate data. Concentrated red mud suspensions have been found to be shear thinning from these data. This shear stress-shear rate information can be directly used for pipeline design and scale up. The pumping energy calculations for red mud over a range of solids concentrations have been made from these data and an optimum solids concentration at which the pumping energy is minimum has been identified. This concentration increases with throughput and decreasing pipe size. These trends, shown in Figure 2, are taken from Nguyen and Boger [3].

The equilibrium shear yield stress of the suspension which is pumped to the disposal area is an important parameter in the design of the pumping and distribution system.

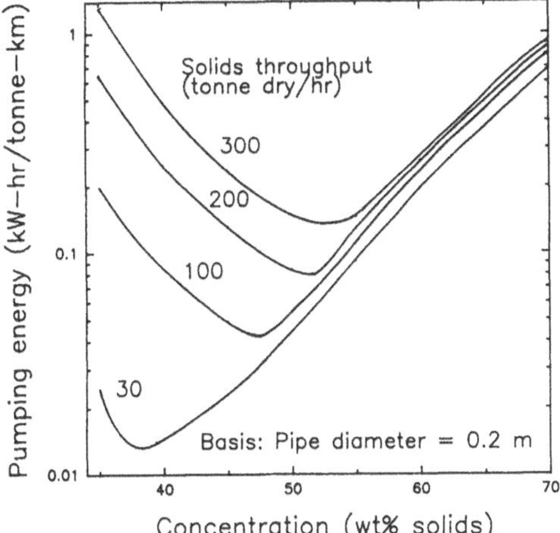

Fig. 2 Calculated specific pumping energy as a function of solids throughput and solids concentration at a fixed pipe size for red mud suspensions.

The yield stress should be low enough to enable the pipeline to be started up from a standstill but high enough to operate in the laminar flow regime and thus prevent solids deposition problems and pipeline erosion problems from occurring. The ability for the slurry to spread over the available disposal area is also enhanced by a lower yield stress. An optimum shear yield stress should thus be determined for each particular disposal scheme.

It may be necessary to adjust the chemistry of the suspension in order to achieve the required high solids loading in conjunction with the required optimum (relatively low) shear yield stress for the system. It is the very fine particles is the suspension which have been found to give most of the suspension's rheological characteristics. Additives may thus be used to control the surface chemistry of the these particles in order to reduce the yield stress. Several options may be examined to accomplish this; for example, the dissolved salt concentration, the pH or the amount and type of specific additives may be adjusted. Due to the nature of the alumina process, the chemistry of red mud is extremely complex and little has been done to directly exploit the surface chemistry of fine red mud particles. This, however, may not be the case for other mineral systems.

The shear history of the suspension also has another significant influence on the shear rheology. This is illustrated in Figure 3 which shows shear stress-shear rate data for red mud of two different histories [4]. The upper curve is for red mud that was just completely mixed and the lower curve is for the same red mud that was centrifuged, then reslurried and completely mixed. A dramatic drop in the shear yield stress for the centrifuged red mud is evident. The suspension structure has thus been completely changed by the centrifugation resulting in a more fluid suspension.

It has thus been demonstrated that careful control of the shear rheology of highly concentrated mineral suspensions is of vital importance in the pumping and disposal

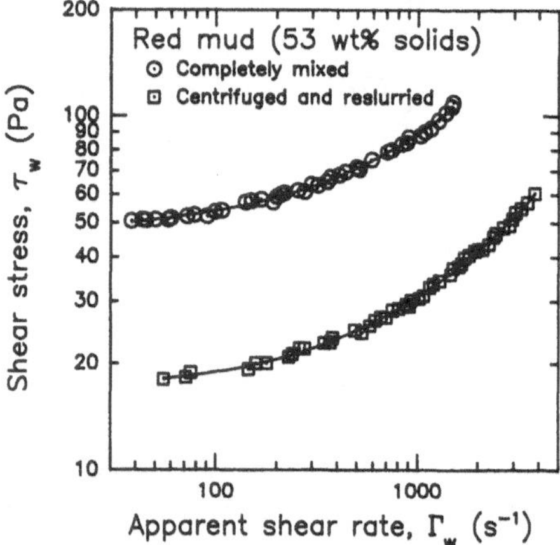

Fig 3. Reduction in red mud consistency as a result of centrifugation.

of such suspensions. A wide range of rheological states may be attained for any particular mineral system by simple modification of the processing conditions. These modifications include changes to the processing equipment and processing times used, which is especially pertinent to mixing and pumping equipment, and the use of additives in the suspension. The effects of these modifications to the suspension in question may be evaluated by the complete characterisation of the shear rheology in the laboratory. These results may then be directly applied in the design of the pipeline and pumping system.

3 Compression rheology

3.1 Compressive yield stress

Utilisation of the compression zone in a continuous gravity thickener can have a dramatic effect on the achievable product solids concentration. The compression zone of a flocculated suspension in a thickener begins when the concentration reaches the gel point, ϕ_g, which is the point at which flocs leave the hindered settling zone and first come in mutual contact. A network structure is then developed in this zone and compression of the structure is due to the transmitted network pressure of the overlying continuously linked structure. Flocs at the bottom of the thickener are thus subjected to the weight of the entire mass of flocs in the compression zone, thus resulting in a maximum concentration for the thickener.

The compressibility of a particular suspension in the compression zone may be quantified by the compressive yield stress, $P_y(\phi)$, which is a direct function of concentration, ϕ. The compressive yield stress is a measure of the strength of the sediment structure and may be defined as the value of the network pressure at which the suspension, at a volume fraction, ϕ, will begin to yield and collapse. This means that if the stress applied to a suspension at a particular concentration exceeds the compressive yield stress for that concentration, then the suspension will consolidate, given enough time, to a new concentration at which the compressive yield stress for this concentration will now equal the new applied stress.

The simplest constitutive equation to describe this process is:

$$\frac{D\phi}{Dt} = 0, \qquad P < P_y(\phi) \qquad (1)$$

$$\frac{D\phi}{Dt} = \kappa(\phi)\ [P - P_y(\phi)], \qquad P \geq P_y(\phi) \qquad (2)$$

where D/Dt is the material derivative and $\kappa(\phi)$ is the dynamic compressibility, from Buscall and White [14]. From this premise, the compressive yield stress may be calculated using the equilibrium sediment height technique which is also described in the Buscall and White paper. Essentially, this technique uses a low speed centrifuge to measure the equilibrium height of a suspension placed in a centrifuge tube at progressively increasing speeds. The raw data generated is thus a curve of the equilibrium height of the sediment versus the gravitational acceleration at the base of the tube. From this curve, the compressive yield stress as a function of concentration (up to the maximum produced in the centrifuge) may be determined from the solution of the appropriate compression equations. The solution of these equations is an involved numerical exercise which is discussed in references [7-8, 14]. A revised algorithm for this solution is currently under development.

The compressive yield stress at a particular concentration may be directly related to the height of the compression zone, H_s, which would be required to yield that concentration which is the underflow from a gravity thickener. This height would be underestimated for a real thickener since it models ideal settling behaviour and assumes that the thickener operates at equilibrium and that there is no liquid drag. However, this height may be used as a basis for preliminary design of the thickening operation since it gives an indication of the size of the unit that would be required to produce the desired product concentration.

Compressive yield stress results may be equally applied to any thickening operation which utilises the compression zone. In a pressure or vacuum filtration system, the compressive yield stress equates to the applied stress required to reach that particular concentration. Likewise, in a centrifugal filtration device, the gravitational acceleration required may be determined.

3.2 Compression results

The exploitation of the compression zone is dramatically illustrated for red mud in Figure 4. For a typical unflocculated red mud in a conventional thickener with no compression zone, the bottom concentration is 44 wt% [7]. A compression zone of 1 m would result in a bottom concentration of 56 wt% and a height of 5 m would yield

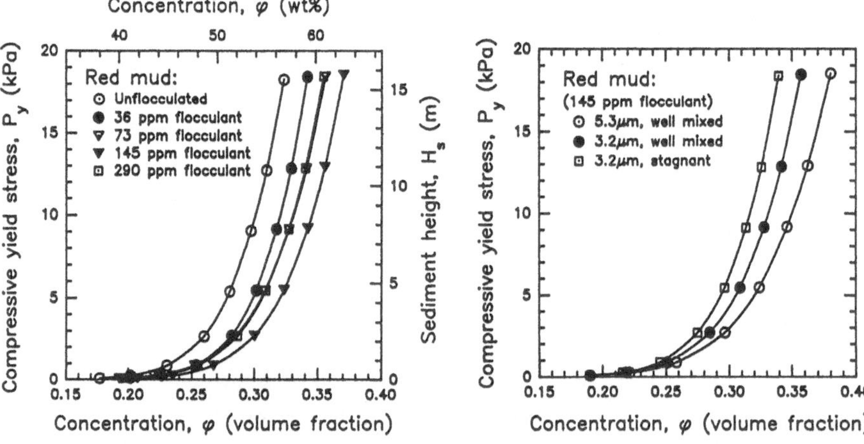

Fig 4. Compressive yield stress versus concentration for a range of flocculant concentrations.

Fig 5. Effect of particle size and shear history on the compressive yield stress for flocculated red mud suspensions.

66 wt%. This is a significant improvement on conventional thickening technology which does not adequately utilise the compression zone.

The addition of flocculants has a major effect on the compression dewatering characteristics of a suspension. For the same red mud used above in Figure 4, with the addition of 145 ppm of flocculant, for a sediment height of 5 m, an increase in bottom concentration from 66 wt% to 74 wt% is possible. This flocculant concentration of 145 ppm was found to be the optimum concentration to yield the minimum P_y for any given ϕ. This would produce the maximum bottom concentration in a continuous thickener. The simple exploitation of the compression zone used in conjunction with the correct concentration of flocculant has thus increased the thickener bottom concentration of red mud from 44 wt% to 74 wt%, a massive improvement.

Control of the mean particle size also has a large impact on the compressibility of the suspension. It was found [7] that an increase in the d_{50} mean particle diameter of red mud from 3.2 μm to 5.3 μm dramatically increases the sediment compressibility as is demonstrated in Figure 5. The effectiveness of flocculation on the compressibility, however, is diminished by the increased particle diameter. Optimisation of the mean particle size and the degree of flocculation would thus maximise the product bottom concentration in a thickener.

The shear history of the suspension is another important factor affecting the compression rheology. Samples of red mud agitated before compression were found to produce a more compressible sediment than samples left stagnant before testing [7]. This effect is shown in Figure 5 for the 3.2 μm red mud samples. The increased resistance to compression in the stagnant samples was probably because inter-floc bonds were allowed to re-form in the sediment during the stagnation period thus producing a stronger sediment. This effect may be directly exploited in industry since mixing the suspension before feeding it to a compression thickener would immediately yield a higher bottom concentration. Alternatively, it would enable a higher throughput to be maintained from the thickener. Multiple compression thickeners placed in series with intermediate mixing between the stages could also be used as a way to increase the ultimate product concentration from the thickening operation.

From the above study, it is clear that detailed knowledge of the compression rheology of the suspension in question is essential to the optimisation of the dewatering process. Understanding process factors such as the depth of the compression zone as determined by the throughput rate, the flocculant concentration, the mean particle size, and the shear history of the suspension on the ultimate product concentration, is crucial. The correct manipulation of these process conditions would result in a more efficient thickening operation with obvious economic advantages.

4 Comparison between compression and shear rheology

It has been demonstrated that there is an inter-relationship between the shear and compression rheology of concentrated suspensions [7-8]. A qualitative comparison may be seen in Figure 6 for two unflocculated red muds of similar physico-chemical characteristics where the compressive and shear yield stresses are shown on the single plot. The compressive yield stress was roughly two orders of magnitude greater than the shear yield stress in this case. The shape of the two curves, however, were similar which indicates that the two yield stresses respond in a similar manner to concentration changes. At high concentrations, almost vertical curves were apparent indicating solid-like behaviour.

This inter-relation between the shear and compressive yield stresses has yet to be fully investigated. It is hoped that a general correlation between the two yield stresses could be derived which would allow an approximate value for the compressive yield

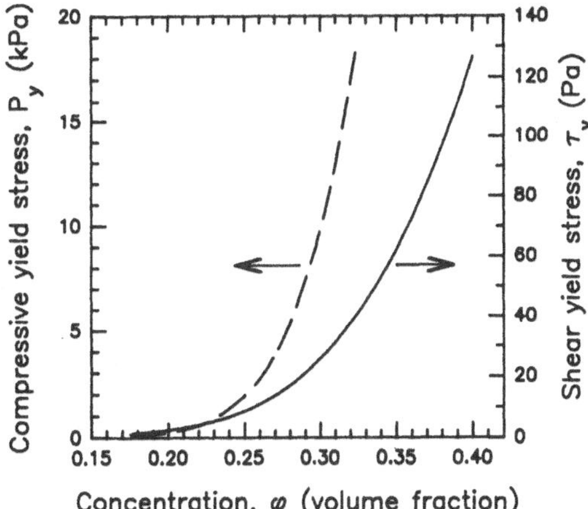

Fig 6. Compressive yield stress and shear yield stress as a function of concentration for unflocculated red mud.

stress to be calculated from the shear yield stress. The shear yield stress, which is a much quicker and simpler measurement to make, may then be used to quickly locate the concentration region which may be feasibly achieved using compression. Detailed compression measurements around this concentration region of interest may then be made in order to determine the compressive yield stress curve and the corresponding compression zone height required to reach this concentration.

5 Other applications

The techniques outlined in this paper for the characterisation of the shear and compression rheology of concentrated red mud suspensions have also been successfully applied to other mineral systems. Similar studies for the waste disposal of coal tailings have been made at the University of Melbourne [15]. The coal tailings consist primarily of a slurry of fine coal particles and montmorillonite clay suspended in a liquor. These suspensions, like red mud, also exhibit shear thinning behaviour with a shear yield stress. Pumping energy calculations for this system again revealed that an optimum concentration for the pumping of the slurry exists which corresponds with the transition from the turbulent to the laminar flow regime. Studies are currently in progress in which the surface chemistry of the coal tailings are being manipulated in order to improve the compression rheology.

Other unpublished work has also been completed on lead and zinc concentrates and tailings for pipeline transportation and waste disposal.

6 Current work

A comparative study of the three high density super thickeners in use by Alcoa in Western Australia is currently under way with the intention of determining an optimum semi-dry disposal scheme for possible future applications. The largest of these operations uses a 90 m diameter, 10 m deep compression thickener which has a design capacity of 450 tonne/hr, uses 60 g of flocculant per tonne and produces an underflow concentration of about 50 wt% which is disposed to a 70 hectare drying area.

7 Conclusions

Exploitation of the shear and compression rheology of bauxite residue or red mud has been successfully used in a semi-dry disposal scheme which has been implemented by Alcoa in Western Australia. Measurements of the shear rheology of this mineral suspension using the capillary rheometer and the vane technique, enable the pumping and disposal system of highly concentrated red mud to be designed. Measurements of the compression rheology using a low speed centrifuge and the equilibrium sediment height technique enable the appropriate dewatering operation to be designed. The optimisation of this entire dewatering and disposal scheme may be achieved by careful controlled modification of the shear and compression rheology with the assistance of various additives and processing techniques. These techniques are equally applicable to a wide variety of concentrated mineral suspension systems.

Acknowledgments

This research was supported by Alcoa (Australia) during the time period of 1978-1982, 1985-1986 and is now supporting the current work in this area. The work is part of the activities of the Advanced Mineral Products Centre at the University of Melbourne, a Special Research Centre of the Australian Research Council.

References

1. Cooling, D.J. and Glenister, D.J. (1991) Practical aspects of dry residue disposal, in **Light Metals 1992**, (ed. E.R. Cutshall), The Minerals, Metals & Materials Society.
2. Nguyen, Q.D. and Boger, D.V. (1984) Exploiting the rheology of highly concentrated suspensions. **Proceedings of the Ninth International Congress on Rheology**, Mexico City, Mexico, 153-171.
3. Nguyen, Q.D. and Boger, D.V. (1986) The rheology of concentrated bauxite residue suspensions - A complete story. **Proceedings of the International Conference on Bauxite Tailings**, Kingston, Jamaica, 53-65.
4. Nguyen, Q.D. (1983) Rheology of concentrated bauxite residue suspensions. Thesis (Ph.D), Monash University, Victoria, Australia.
5. Nguyen, Q.D. and Boger, D.V. (1985) Thixotropic behaviour of concentrated bauxite residue suspensions. **Rheologica Acta**, 24, 427-437.
6. Marunczyn, M. and Laros, T.J. (1992) Bauxite residue disposal at Alcoa of Australia using the EIMCO Hi-Density thickener. **Proceedings of the International Bauxite Tailings Workshop**, Perth, Western Australia, Australia, 31-42.
7. De Guingand, N.J. (1986) The behaviour of flocculated suspensions in compression. Thesis (M.Eng.), University of Melbourne, Victoria, Australia.
8. Green, M.D., De Guingand, N.J., Nguyen, Q.D. and Boger, D.V. (1992) The shear and compression rheology of bauxite residue - An overview. **Proceedings of the International Bauxite Tailings Workshop**, Perth, Western Australia, Australia, 116-125.
9. Nguyen, Q.D. and Boger, D.V. (1980) Energy requirement for the transportation of highly concentrated red mud suspensions to waste disposal. **Proceedings of the Symposium on Rheology in Conversion and Conservation of Energy**, British Society of Rheology (Australian Branch), Melbourne, Australia, 55.
10. Tiu, C., Nguyen, Q.D., Uhlherr, P.H.T. and Boger, D.V. (1981) Rheology of highly concentrated red mud. **Proceedings of the Second A.P.C. Chemical Engineering Congress**, Manilla, Philippines, 57-62.
11. Leong, Y.K., Creasy, D.E., Boger, D.V. and Nguyen, Q.D. (1987) Rheology of brown coal-water suspensions. **Rheologica Acta**, 26, 291.

12. Nguyen, Q.D. and Boger, D.V. (1985) Direct yield stress measurement with the vane method. **Journal of Rheology**, 29, 335-347.

13. Nguyen, Q.D. and Boger, D.V. (1983) Yield stress measurement for concentrated suspensions. **Journal of Rheology**, 27, 321-349.

14. Buscall, R. and White, L.R. (1987) On the consolidation of concentrated suspensions I: The theory of sedimentation. **Journal of the Chemical Society, Faraday Transactions 1**, 83, 873-891.

15. De Kretser, R.G. and Boger, D.V. (1992) Rheological properties and dewatering of slurried coal mine tailings. **Proceedings of the Sixth National Conference on Rheology**, Clayton, Victoria, Australia, 23-26.

Leaching of heavy metals from contaminated concrete rubbish material

W. H. Höll
Kernforschungszentrum Karlsruhe, Institute for Radiochemistry, Water Technology Division, Karlsruhe, Germany

Abstract

Heavy metal-contaminated concrete rubbish material was eluted in order to remove the heavy metals and to allow the reuse of the solid material. The treatment concept consisted of one or two elution cycles with water and/or organic complexing agents. Copper, nickel, and zinc were eluted almost completely with only negligible remaining amounts whereas the efficiency for lead was smaller. The residual contamination of the solid material is very small and should allow its reuse. Recovery of the heavy metals without selective uptake is most effective if a chelating ion exchange resin with iminodiacetate functional groups is applied. Application of acrylic anion exchangers seems to allow a very selective elimination of copper-bearing species from tartrate complex systems.
Keywords: Heavy metals, concrete rubbish material, ion exchange, recovery.

1. Introduction

In a former metallurgical factory in South-West Germany copper was recovered from pieces of electrical cables. In the recovery process the coating of the wire was burnt off in the first processing step. The pure metal wires were then dissolved in sulfuric acid and finally recovered by means of electrolysis. The last step took place in large concrete tanks. During the years of operation the coating of the concrete became damaged followed by substancial invasion of sulfuric acid and copper sulfate into fissures and holes leading to precipitates of solid copper sulfate in the solid material. Furthermore, corners of the tanks were covered with green or blue precipitates up to several centimeters in thickness.

When the factory was closed in the eighties considerable environmental problems arose from the remediation of the entire site. Apart from serious dioxine problems within the factory and in the close environment caused by burning off the PVC coatings of the wires large parts of the buildings and the soil in the area were contaminated with heavy metals, mainly copper. About 17,000 tons of hazardous waste had to be brought to a special underground deposit for hazardous waste to avoid the leaching of both the organic and inorganic contaminants.

2. Concept of treatment

The problems with heavy metal-contaminated solid waste are mainly due to the fact that such material has to be discharged to special landfill sites where the available space is rather limited. As a consequence, the disposal of such solids becomes very expensive with costs amounting approximately DM 1000 per ton.

Problems and costs can be avoided if the heavy metal contamination can be eliminated effectively thus allowing the solids to be reused. However, only few concepts exist so far. One proposal applies concentrated hydrochloric acid for dissolution of the heavy metal salts. Lime is added stepwise to the eluant to achieve the separate precipitation of different heavy metal hydroxides. Although tested in the pilot scale there were too many drawbacks mainly related to the use of a concentrated acid [5]. In a second proposal the solids are eluted by means of sodium nitrilo-triacetate [6]. In this case, however, a complexing agent with a poor biodegradability is applied. The concept has not yet been tested on a larger scale.

Unlike these existing proposals, the general strategy of treatment comprises of (a) the selective elimination of the heavy metal contaminations, (b) no dissolution or destruction of the solid matrix thus allowing its reuse (in the case of rubbish material e.g. for road construction purposes), (c) the application of non-polluting or environment-friendly eluants, and (d) the recycling of the the heavy metals which requires their separation.

As a consequence, the following treatment concept was to be studied:

- Elution of the broken material by means of water to remove easily soluble sulphate compounds which form the main part of the contamination.

- Treatment of the pre-eluted material by means of tartaric or citric acid for dissolution of heavy metals which are sorbed or bonded to the concrete matrix.

In both steps the liquid phase was to be recycled after passing through an ion exchange column in which heavy metal or heavy metal complex species were to be eliminated.

Tartaric or citric acids were selected because of their biodegradability which does not require the complete rinsing-out of remaining quantities in the solid. Furthermore, it had been found that tartrate solutions allow a very selective separation between copper and other heavy metal complexes by means of anion exchangers [6].

3. Experimental

3.1 Experimental setup

The treatment concept has been studied at the laboratory scale in a plant which is schematically shown in Figure 1. The plant consists of two columns in which the contaminated solids and the ion exchange resin are stored. In the smaller installation both columns had an inner diameter of 5 cm. The height of the column for the contaminated material was 20 cm, that of the ion exchange column amounted to 50 cm. The liquid phase had a volume of 10 litres which was recirculated for 3.5 to 4 hours by means of a membrane pump, the throughput being 9 to 10 l/h.

3.2 Solid material

During the experiments solid material samples from the factory were eluted. Since the samples were from different tanks and/or from different places from within the tanks they differed greatly in their composition. As a consequence, expermental results are not reproducible, each experiment leads to unique results. Furthermore, when the solid material was collected only strongly contaminated pieces were selected which are not representative for the entire rubbish material. Because of the high heavy metal content of the samples of up to 50 g/kg only small quantities of 300 g were treated in one experiment.

3.3 Eluant solutions

In the first elution step of the treatment, tap water, softened tap water, as well as distilled water were used. KNa-tartrate and Na-citrate solutions were prepared from softened water. The concentrations amounted to 1 % in the first experiments (equalling approximately 30 mmol/l) and 10 mmol/l in the later experiments.

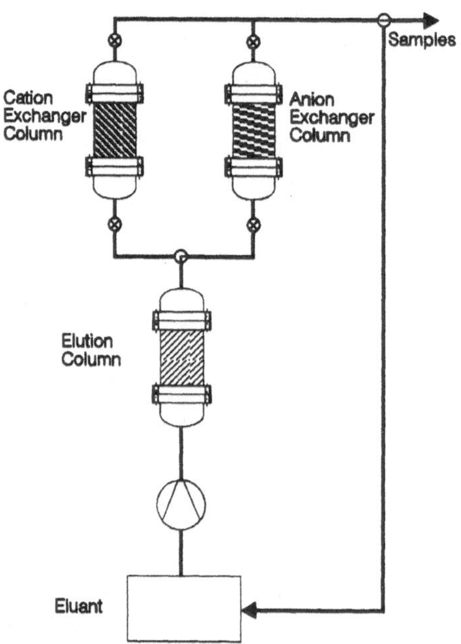

Figure 1: Scheme of laboratory treatment plant

3.4 Ion exchange resins

During the elimination of copper and other heavy metals by means of water in the first step, the heavy metals appear as cationic species in the effluent of the treatment column. Thus they were eliminated by chelating resins. Two different exchange resins were applied.

In most of the experiments the imino-diaceate resin LEWATIT TP 207 was applied. According to the manufacturers recommendations the resin was used in the mono-sodium form. This form is achieved by contacting the free-acid form of the resin with an amount of sodium hydroxide which corresponds to half of the capacity [1]. In further experiments the chelating resin DUOLITE ES 346 with amidoxime functional groups was tested. This resin was applied in the protonated form after treatment with hydrochloric acid.

After each experiment the resin was regenerated for both the measurement of the amount of heavy metals eluted and for the following experiment. LEWATIT TP 207 was regenerated with hydrochloric acid (1 mol/l) and then treated with sodium hydroxide. For regeneration of DUOLITE ES 346 only hydrochloric acid (1 mol/l) was applied.

In principle both exchange resins can also be used for the elution by means of tartrate or citrate solutions. In the case of tartrate solutions, however, it has been found that copper-rich tartrate complexes are very selectively adsorbed by acrylic anion exchangers with only negligible amounts of nickel, lead, or zinc species adsorbed at the same time. Therefore, the macroporous strong-base resin AMBERLITE IRA 958 in the chloride form was applied in some experiments during the elution by means of tartrate solutions. For regeneration concentrated sodium chlorde solutions were used.

In the first experiments the resin volumes were 2.4 l of LEWATIT TP 207 in the first step and 0.63 l of AMBERLITE IRA 958 in the second step. In the further experiments only one resin column was applied which contained a resin volume of 0.75 l each.

3.5 Residual load

After elution in the column as described above the solid material was immersed into hydrochloric acid in order to estimate the remaining amounts of heavy metals which were not eluted. Samples were taken from the eluting liquid, from the regenerant of the resins and from the concentrated acid solution. In all samples copper, nickel, lead and zinc were determined by means of atomic absorption.

4. Results

The total contents of the solid samples differed strongly depending on the individual origin. The range of contents of copper, nickel, lead, and zinc in ten experiments is listed in Table 1. The results demonstrate that copper and nickel are the main contaminants with up to 10 % of the weight of individual samples. The contents of zinc and lead are much smaller, but were still larger than tolerated for road construction.

Table 1: Range of heavy metal contents in the samples.

Metal	Total amounts in 300 g sample, mg
Copper	370 - 29,400
Nickel	2,000 - 19,000
Lead	10 - 170
Zinc	130 - 370

Results of the two-step elution of one solid sample are summarized in table 2. Only 37% of the copper was eluted by means of softened water and immobilized by the cation exchanger. The remainder was eliminated during the second treatment step and sorbed onto the anion exchanger in the form of tartrate complex species. It must therefore be concluded that in this sample part of the contamination did not consist of copper sulphate. Nickel is well eliminated by the cation exchanger, however, the nickel load of the anion exchanger is almost negligible. Thus, the selective elimination of copper tartrate species and the effective separation from other heavy metals becomes obvious. The residual load of the solid material is very small (table 2).

Table 2: Results of the two-step elution, initial mass of sample: 300 g. Eluants: water (step 1), KNa-tartrate (step 2). Resins: LEWATIT TP 207 (step 1) and AMBERLITE IRA 958 (step 2).

	Copper, mg	Nickel, mg
First step, amount on resin	8,690	1,310
First step, amount in liquid	23.7	n.m.
Second step, amount on resin	14,000	19.3
Second step, amount in liquid	522	79
Residual load	5.9	1.7

Table 3: Initial and residual amounts of copper, nickel, lead, and zinc for different eluants.

Metal	Water	KNa-tartrate	Na_3-citrate
	Initial load, mg / residual load, mg		
Copper	18,650 / 144	10,240 / 44	397 / 15
Nickel	5,530 / 38	16,780 / 116	1,980 / 8
Lead	126 / 89	15 / 4.7	11.5 / 3.5
Zinc	90 / 2	316 / 3	278 / 1

The further experiments were carried out with only one single elution step in order to test the efficiency of different eluants and different ion exchange resins. The first of these tests demonstrated that most of the contamination could be eliminated by means of normal tap water. The amount of heavy metals eluted in the second step with a complexing agent was more or less negligible. This is obviously due to the predominant presence of easily soluble sulphates in these samples. KNa-tartrate or Na-citrate also allowed a very efficient elimination if applied in the first step. Table 3 shows that for copper, nickel, and zinc the residual load normally amounted to less than 1 % of the original load (sometimes even much less) Lead is much less mobile and shows up to 70 % residual load.

According to existing regulations rubbish material for road construction purposes has to be tested by eluting two samples of 60 g and 140 g with 2 litres of distilled water each [4]. Considering the remaining loads with heavy metals as found after immersing the solids into concentrated acid it can be concluded that the remediated material is suitable for reuse.

Table 4: Results of the elution with water, KNa-tartrate, and Na-citrate. Resin: LEWATIT TP 207.

Metal	**Amount in eluant, mg**	**Amount in regenerant, mg**	**Residual load, mg**	**Total, mg**
		Eluant: Water		
Copper	355	18,150	144	18,650
Nickel	1,125	4,368	38.3	5,530
Lead	9.4	28.8	88.7	126
Zinc	30	58.8	2	90
		Eluant: KNa-tartrate		
Copper	10,160	38	44	10,242
Nickel	16,550	111.5	115.5	16,777
Lead	7.8	2.4	4.7	15.9
Zinc	310	2.7	3	315.7
		Eluant: Na-citrate		
Copper	68.9	312.5	14.9	396.3
Nickel	970	1002	7.6	1979.6
Lead	4.0	4.0	3.4	11.4
Zinc	26.7	250	0.7	277.4

The optimum recovery of heavy metal species was achieved when tap water was applied in the elution and the iminodiacetate resin LEWATIT TP 207

TP 207 was applied for separation. In these experiments more than 97 % of copper was recovered, for nickel, lead, and zinc the respective amounts were smaller. The relatively poor elimination of the latter metals might also be due to the high total heavy metal content of the samples and to the relatively small resin volume of 750 ml. If KNa-tartrate or Na-citrate are applied for elution the heavy metals undergo the formation of more or less stable complexes with the organic molecules. Due to the formation of relatively stable complexes the elimination with tartrate and citrate solutions was poorer and amounted to only 75 % even for copper. With Na-citrate only one single experiment was carried out in which the sample contained an unusually small total amount of heavy metals. Some results are summarized in table 4.

Table 5: Results of the elution with water. Resin: DUOLITE ES 346.

Metal	Amount in eluant, mg	Amount in regenerant, mg	Residual load, mg	Total, mg
Copper	2985	2340	20.0	5345.0
Nickel	9780	80.4	11.7	9872.1
Lead	3.0	4.0	1.95	8.95
Zinc	190	3.5	0.36	193.9

Table 6: Results of the elution with KNa-tartrate. Resin: AMBERLITE IRA 958.

Metal	Amount in eluant, mg	Amount in regenerant, mg	Residual load, mg	Total, mg
Copper	10160	37.7	44.1	10241.8
Nickel	16550	111.5	115.5	16777
Lead	7.8	2.4	4.7	14.9
Zinc	310	2.7	3.0	315.7

With the chelating resin DUOLITE ES 346 the elimination was much less effective. For copper and lead the percentage was in the range of 40 % of the total amount, whereas the elimination of nickel and zinc was below 10 % (see table 5).

From earlier experiments with pure copper tartrate systems it could be expected that a considerable uptake of copper-rich complex anions on an acrylic anion exchange resin would occur [2, 3]. Thus, a very selective separation of copper would become possible. A corresponding result was found in the very first experiments where the KNa-tartrate concentration was 1 %. Unfortunately, this selective uptake was not found in the experiments with concentrations of 10 mmol/l. In the latter experiments the uptake of copper containing species was more or less negligible (table 6). The difference between both results might be caused by unfavourable copper/tartrate ratios in the experiments. This has to be clarified by further investigations.

5. Conclusions

The main objective of the treatment of contaminated solids has been the selective elimination of heavy metals to allow the treated solids to be reused. Secondary pollution due to the eluants was to be avoided. The process has been demonstrated using strongly contaminated material in which the heavy metal salts were easily accessible and well soluble.

For the solid material used in the experiments the mere elution of the main contaminants copper and nickel was almost total and independent of the kind of eluant. To a certain extent, not quantified in the investigations, there was also a dissolution of the matrix of the solid material. With respect to the separation of the heavy metals from the eluant, the results demonstrate that chelating ion exchange resins with iminodiacetic acid functional sites allow a satisfactory elimination which, however, is non-selective. Such a selective sorption of copper-bearing species seems to be possible from tartrate solutions by means of acrylic strong-base resins. However, the optimum conditions still have to be found. The treated solids meet the standards for rubbish material to be reused for road construction purposes.

As a consequence of the experiments on a laboratory scale it can be seen that this process allows a successful treatment of this particular solid material. Based on these results the process will now be scaled up to demonstrate its technical feasibility.

6. References

[1] BAYER AG, LEWATIT Manual, 1977

[2] HÖLL, W. H. (1991) Spaltung von Schwermetallkomplexen an Anionenaustauschern, **Vom Wasser**, 77, 35 - 45.

[3] HÖLL, W. H. (1993) Unusual column behaviour of heavy metal complexes on anion exchange columns, in: **ION EXCHANGE PROCESSES: Advances and Applications** (ed. A. Dyer, M. J. Hudson, P. A. Williams), Royal Society of Chemistry, 1993, pp. 181 - 190.

[4] VERKEHRSMINISTERIUM BADEN-WÜRTTEMBERG (1991) Verwaltungsvorschrift des Verkehrsministeriums und des Umweltministeriums über vorläufige Lieferbedingungen für aufbereiteten Straßenaufbruch und Bauschutt zur Verwendung im Straßenbau in Baden-Württemberg, **Gemeinsames Amtsblatt, des Landes Baden-Württemberg** 39, 1183 - 1187.

[5] Müller, G., Riethmayer, S. (1982) Chemische Entgiftung: das alternative Konzept zur problemlosen und endgültigen Entsorgung Schwermetall-belasteter Baggerschlämme, **Chemiker-Zeitung**, 106, 289 - 292.

[6] Dehnad, F. (1993) Verfahren zur Entfernung von Schwermetallen aus Baggergut und anderen Korngütern unter Kreislaufführung der Prozeßwässer, **Wasser Abwasser Praxis**, 1, 48 - 50.

Processing of electric arc furnace dust via chloride hydrometallurgy

R. O. McElroy
Fluor Daniel Wright, Vancouver, British Columbia, Canada
M. McClaren
Terra Gaia Environmental Group, Vancouver, British Columbia, Canada

ABSTRACT

Production of steel in the electric arc furnace (EAF) generates a dust byproduct containing non-ferrous metals. In particular, zinc, lead and cadmium are volatilized at steelmaking temperatures and report as oxides or ferrites in the dust collected from EAF off-gas. Direct recycle of dust is impractical, while - in North America at least - disposal in secure landfills is becoming prohibitively expensive.

The paper presents the results of a phased process development program for the treatment of this dust, including:

- preliminary scoping/process definition study
- laboratory scale testing of key unit operations
- preliminary feasibility study

The first stage of the process sequence is atmospheric leaching of EAF dust with a ferric chloride solution:

$$2\ FeCl_3 + 3\ MO + H_2O \rightarrow 2\ FeO{\bullet}OH + 3\ MCl_2$$

(M = Zn, Pb, Cd, etc.)

The slurry product is then treated in an autoclave under conditions that result in formation of a free-settling, filterable residue consisting mainly of hematite and silica. Results of hot CCD and filtration/washing tests indicate that the residue meets USEPA and Canadian criteria for disposal in sanitary landfills. Possibilities for alternate uses of residue are also discussed.

Process unit operations described include:

- leaching and filtration
- primary lead recovery by precipitation of $PbCl_2$
- zinc recovery by solvating solvent extraction
- bleed stream treatment by sulphide precipitation of cadmium and other heavy non-ferrous metals

Components of the feasibility study to be presented include materials of construction, flowsheet development, overall (estimated) capital/operating costs and revenue sources.

INTRODUCTION

Flue dust is produced in varying quantities and compositions during manufacture of steel using electric arc furnace (EAF), basic oxygen or open hearth technologies. Testwork and process development described in this paper have been directed to processing of EAF which, typically, has relatively high contents of non-ferrous metals.

Electric arc furnace technology can be used to manufacture steel from raw material including any proportion of scrap. Consequently it is the process of choice for "mini-mills" producing steel in regions where there is no primary steel industry.

Dust is generated in EAF steelmaking by a variety of mechanisms including droplet ejection from the turbulent bath and vaporization. The latter mechanism is particularly important for non-ferrous metals such as zinc, lead and cadmium which are largely vaporized at steelmaking temperatures in the 1600°C range.

Table 1 presents the major element composition of two samples of EAF dust used in the testwork described herein, and ranges for carbon steelmaking dusts reported in a recent review of flue dust processing (1). Zinc and lead contents of the test samples are at the high end of their ranges, as expected for a regional mill (Western Canada Steel, Vancouver, B.C.) which processed secondary materials almost exclusively.

Table 1: EAF Compositions, wt %

Component	Test Samples		Reported Ranges (1) for EAF Carbon Steel
	A	B	
Zinc	26.5	24.4	11 - 27
Lead	6.9	3.1	1.1 - 3.8
Cadmium	0.08	0.05	0.03 - 0.15
Iron	21.8	19.9	25 - 47
Manganese	-	2.5	2.5 - 4.6
Calcium (as CaO)	-	3.7	1.9 - 10
Magnesium (as MgO)	-	2.0	0.8 - 2.9
Silica (as SiO_2)	-	14.8	2.7 - 5

The data in Table 1 also indicate some of the problems with utilization or disposal of EAF dust:

- Recycle to steelmaking is precluded by a combined non-ferrous metal oxide and slag materials content > 50%.

- Zinc content is low compared to conventional zinc industry raw materials and a significant proportion of zinc is present as ferrite ($ZnO \bullet Fe_2O_3$).

- Levels of "toxic" heavy metals, particularly lead, are far too high to permit disposal in sanitary landfills.

Many North American secondary steel mills have opted to store their production of EAF dust, typically 10-20 kg of dust per tonne of steel produced, pending identification of more economic (or at least less costly) processing methods than are currently available.

Terra Gaia Environmental Group Ltd. became involved in EAF dust processing in relation to rehabilitation of an industrial site formerly occupied by a secondary steel mill. After review of available technologies and options, Terra Gaia contracted with Fluor Daniel Wright Ltd. to undertake development of technology which would economically address both the specific and general problems relating to economic treatment/disposal of EAF dust.

The initial study objectives were to produce non-ferrous metals in saleable form(s) while generating a residue meeting current (and probable future) criteria for disposal in sanitary landfills.

To date, this venture has proceeded through two campaigns of laboratory testwork, generation of preliminary process flowsheets, equipment lists and capital/operating cost estimates.

SUMMARY OF TEST PROGRAMS

Hydrometallurgical treatment was selected for developmental testwork on the basis that pyrometallurgical processes are reasonably well developed but prohibitively expensive at small scale. Also, in North America at least, the primary waste generator (i.e., steel mill) retains legal liability for metallurgical wastes even after contracted secondary treatment. Thus the "permanent" secure landfill/disposal of slag generates a permanent liability for the steel producer.

Initial atmospheric leaching tests using sulphuric acid and sulphuric-hydrochloric acid mixtures served mainly to confirm results of previous investigations, specifically:

- incomplete dissolution of zinc ferrite
- poor settling/filtration behaviour of leach residues
- extensive co-dissolution of iron with excess (vs. non-ferrous metals) hydrochloric acid

Consideration of the high temperature (i.e., pressure autoclave process) behaviour of acid chloride brines suggested a possible approach to both the chemical (i.e. incomplete zinc recovery) and engineering (poor settling/filtration) problems encountered in atmospheric pressure leaching. Initial tests of this approach were based on use of (excess) ferric chloride as a source of hydrochloric acid at atmospheric pressure:

$$2\ FeCl_3 + 3\ MO + H_2O \rightarrow 2\ FeO{\bullet}OH + 3\ MCl_2$$

(M = Zn, Pb, Cd, etc.)

followed by pressure treatment of the atmospheric leach pulp at 175°C and corresponding steam pressure to complete leaching of zinc ferrite while converting the products of ferric chloride hydrolysis (represented as goethite for convenience) to more settleable and filterable hematite:

$$2\ FeO{\bullet}OH \xrightarrow[\text{90 minute retention}]{175^{\circ}C} Fe_2O_3 + H_2O$$

Testing under these conditions resulted in:

- significantly improved settling and (hot) filtration properties of insoluble residue
- zinc and lead extractions > 99% with acceptable (3 displacement) washing of filter cake

It should be noted that both filtration behaviour and lead extraction are contingent on conducting the filtration hot ($T \geq 80°C$) to prevent crystallization of lead chloride which will result in blinding of the filter membrane. This presents problems with residue filtration which have been addressed in preliminary design.

Autoclave leaching of steel industry blast furnace dusts in hydrochloric acid has been previously reported (4). However, the maximum temperature (140°) and retention time (60 minutes) appear to be sub-optimal, since consistent high (> 99%) zinc and lead extraction appear not to have been achieved.

Results of further scoping testwork on the acid-ferric chloride leaching of several flue dust samples indicated:

- no useful effect of extended atmospheric pre-leaching prior to autoclave processing

- time-temperature dependence of zinc ferrite leaching in the range of 140-175°C (see Table 2)

- primary recovery of lead as $PbCl_2$ by cooling/crystallization of hot filtered leach solution

- acceptable (i.e. > 99%) zinc recovery to leach solution with a residual solution $FeCl_3$ content of about 5 g/L of ferric iron

- acceptable zinc-lead recoveries from leach tests in an acid chloride brine containing inert chlorides ($CaCl_2$, $MgCl_2$, NaCl, KCl) in quantities calculated to represent recycle of about 80% of process solutions)

- batch solvent extraction/stripping tests with a commercial reagent (Acorga ZNX-50 (5)) to confirm selectivity of this reagent for zinc extraction from complex EAF dust leach solutions

TABLE 2: Lead-Zinc Extractions vs. Autoclave Leach Parameters

Test Code	Temperature, °C	Time, min.	% Extraction	
			Zn	Pb
L-12	175	90	99.3	99.5
L-13	175	45	80.3	95.7
L-15	140	180	98.2	98.5

In terms of the initial testwork objective, results presented in Table 3 indicate that process residues can pass the USEPA Toxic Characteristic Leaching Procedure which is a key requirement for waste acceptability in a sanitary landfill.

TABLE 3: Results of Residue TCLP* Test

	Concentration (milligrams/litre)	
	Test Leachate	USEPA Criterion
Arsenic	<0.02	5.0
Cadmium	0.005	1.0
Chromium	<0.005	5.0
Lead	0.45	5.0
Mercury	<0.02	0.1

Test Code PRA-15, 300 g/L $FeCl_3 \cdot 6H_2O$, 150 g/L dust, 90 minutes at 175°C
* USEPA procedure - buffered acetic acid leach

At this stage, study effort was shifted to preliminary engineering as a basis for estimates of capital and operating cost. With such cost data and independent revenue estimates, Terra Gaia will have the basic data required to assess site specific opportunities for economic treatment of EAF dust.

PRELIMINARY DESIGN

Preliminary design is discussed with reference to simplified flowsheets presented as Figures 1 and 2. These flowsheets outline two concepts for recovery of zinc after solvent extraction (SX):

- concentration to commercial standard (60 or 70 wt % $ZnCl_2$ solutions) for direct sale, and
- concentration followed by conversion to zinc sulphate using spent electrolyte from an (assumed) adjacent conventional electrolytic zinc plant.

Direct electrolysis of SX strip solutions (as proposed by the reagent developers (5)) is a technically feasible alternative, but preliminary cost estimates indicated that this approach is economically questionable at the 20,000 and 100,000 tonnes/year scales of operation (dry dust basis) used for estimating.

Production/sale of zinc chloride is considered only for the 20,000 tonnes/year case, since markets for this material are limited. The 100,000 tonnes/year case is based on recycle of hydrochloric acid from zinc chloride → sulphate conversion as shown in Figure 2.

Leaching and Primary Filtration

The simplified flowsheet (Figure 1) and equipment lists assume dry bulk transport of EAF dust to the processing facility and pneumatic handling of dust from the transport unit through storage silos to a plant feed silo. From the feed silo, dust is metered by a screw feeder into a high intensity mix tank where it is slurried with recycle acid brine solutions.

FIGURE 1: SIMPLIFIED LEACHING AND FILTRATION FLOWSHEET

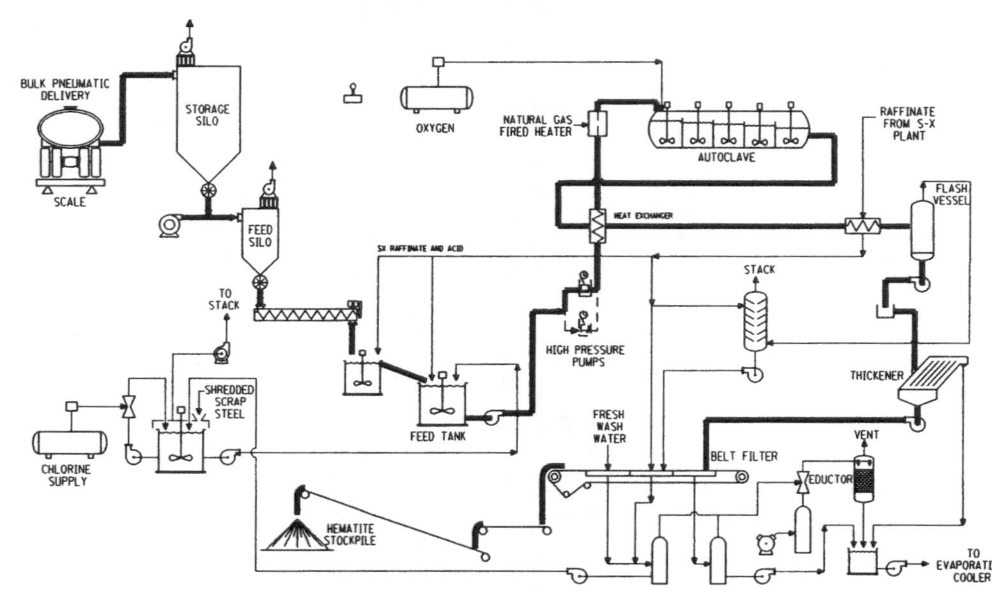

FIGURE 2: SIMPLIFIED ZINC CHLORIDE CONCENTRATION/CONVERSION FLOWSHEET

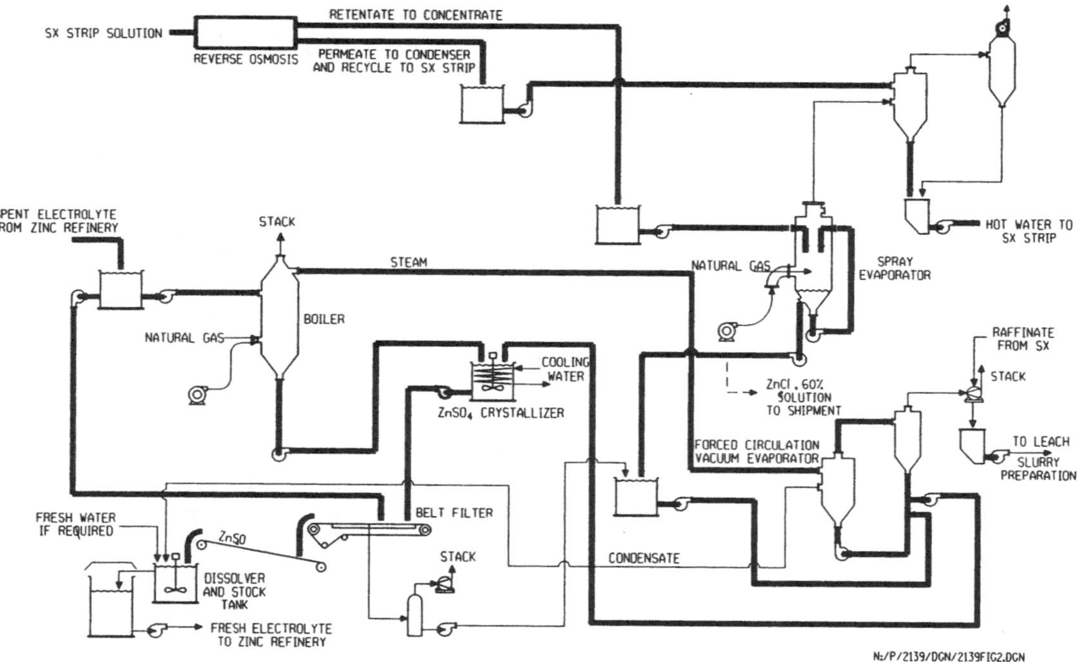

In the 20,000 tonnes/year case, with sale of zinc chloride solution, chlorine supply is by purchased liquid chlorine. An obviously attractive alternative would be spent chloride pickle liquor which may itself represent a disposal problem.

For the 100,000 tonnes/year case, hydrochloric acid from zinc chloride → sulphate conversion (Figure 2) is a major internal recycle stream.

From the mix tank, pulp passes by gravity to a leach feed stock tank where pulp density and chemistry are adjusted with solution additions.

From the stock tank, leach pulp is pressurized by a centrifugal pump and high pressure diaphragm slurry pump. Pressurized slurry is indirectly heated to reaction temperature by heat exchange with autoclave discharge and a gas-fired heater. Slurry then passes through a five compartment, 90 minute design retention time autoclave operated at 175°C and an oxygen partial pressure sufficient to ensure complete oxidation of any ferrous chloride in the leach feed.

Autoclave discharge is heat exchanged with autoclave feed and recycle SX raffinate before depressurization via a ceramic choke to a flash tank. Flashed slurry passes to an insulated lamella thickener. Thickener underflow is filtered and washed on a continuous belt filter; vacuum is provided by an air eductor to avoid flashing problems with a liquid ring pump which would be typically used for this service.

Lead Crystallization/Iron-Copper Reduction

Combined filtrate and thickener overflows are pumped to an evaporative cooler/crystallizer; $PbCl_2$ slurry product passes to a settler. Underflow is filtered and water washed to a clean crystalline $PbCl_2$ product for sale or cementation with scrap iron to a crude lead metal. Settler overflow passes to an agitated vented tank where shredded steel scrap is added to reduce residual ferric chloride:

$$2\ FeCl_3 + Fe^\circ \rightarrow 3\ FeCl_2$$

as required to avoid co-extraction of iron in zinc SX. Reduced solution passes via a clarifying filter to zinc SX.

Zinc Solvent Extraction

Zinc SX (Figure 3) uses Acorga ZNX-50 reagent in a 3 x 2 (extraction-strip) circuit with hot water stripping to recover a commercially pure $ZnCl_2$ solution containing 60 g/L zinc. Raffinate (after carbon column sorption of entrained SX reagent) is split into recycle (≈80%) and bleed (≈20%) streams.

Bleed Stream Treatment

Bleed solution sufficient to maintain a constant inert ((Ca, Mg) Cl_2, (Na, K)Cl) chloride content passes to a treatment circuit comprising:

- lime/air treatment
- iron/manganese precipitate settling and batch filtration
- sulphide (NaHS) precipitation of residual "toxic" heavy metals (Zn, Cd, Pb, Cu) with high (300%) sludge recycle to improve filterability
- batch filtration/washing of sulphide residue
- containerized shipment of mixed heavy ("toxic") metal sulphides to a smelter
- recycle or disposal of batch filtered iron-manganese precipitate
- disposal of inert chloride brine to sanitary sewer

Design basis testing of these unit operations is incomplete; cost estimating is based on FDW design files and published data.

Treatment of Strip Solution

Hot, acidified (pH $\leq$ 2) water stripping of loaded ZNX50 reagent generates commercially pure $ZnCl_2$ solution containing up to 60 g Zn/L. For both recovery options, this solution is stage concentrated by:

- reverse osmosis
- gas-fired spray evaporation

to 60-70% $ZnCl_2$. Reverse osmosis of such low pH solutions is at the limit of vendors anticipated capabilities for developmental systems. Multiple effect evaporation is a higher operating cost alternative.

FIGURE 3: SIMPLIFIED SOLVENT EXTRACTION FLOWSHEET

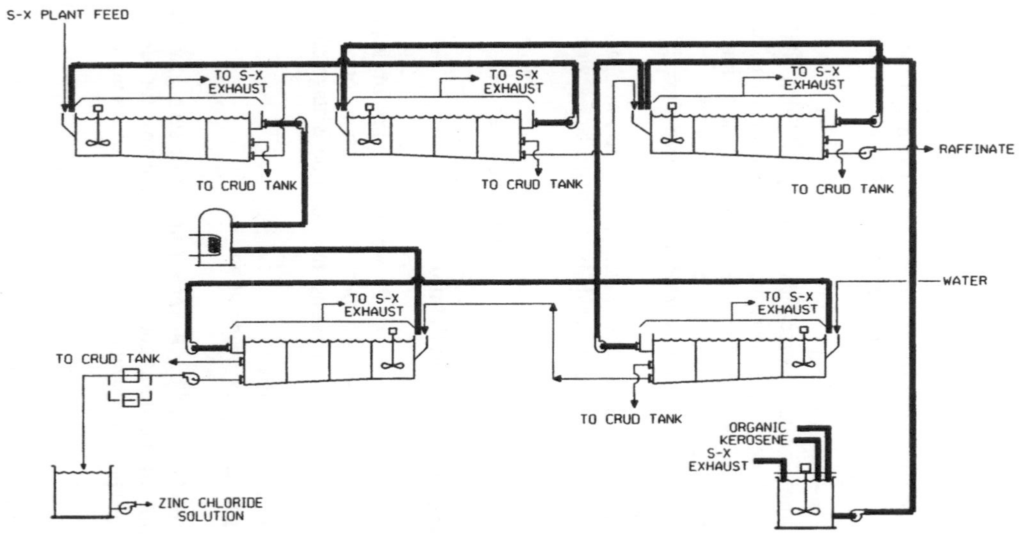

S-X PLANT FEED
TO S-X EXHAUST
TO S-X EXHAUST
TO S-X EXHAUST
TO CRUD TANK
TO CRUD TANK
TO CRUD TANK
RAFFINATE
WATER
TO S-X EXHAUST
TO S-X EXHAUST
TO CRUD TANK
TO CRUD TANK
ORGANIC
KEROSENE
S-X EXHAUST
ZINC CHLORIDE SOLUTION

For the 20,000 tonnes/year case, commercial 60% or 70% $ZnCl_2$ solution is the final zinc product.

For the 100,000 tonnes/year case (see Figure 2) additional unit operations include:

- gas-fired preconcentration of spent $ZnSO_4$-H_2SO_4 electrolyte with steam generation
- crystallization of zinc sulphate hydrate from a mixed $ZnCl_2$-$ZnSO_4$-H_2SO_4 solution
- filtration/washing and resolution of $ZnSO_4$
- evaporation of constant boiling hydrochloric acid from crystallizer filtrate
- condensation and recycle of hydrochloric acid with zinc SX raffinate

Design basis testing of these unit operations is also incomplete. Conservative allowances have been made for energy and cooling water requirements, but test data are required to confirm acceptably low chloride levels in zinc sulphate solution for return to the final electrolyte purification stages of a conventional electrolytic zinc plant.

MATERIALS OF CONSTRUCTION

Processing EAF dust in acid chloride brines requires careful attention to materials of construction, especially wetted components of process equipment. Estimates (see following section) are based on:

- butyl rubber lining (BRL) of all atmospheric pressure vessels
- polymer, fibre reinforced polymer (FRP) or BRL piping for atmospheric pressure components
- titanium for wetted metal in atmospheric, heat exchange evaporation and pressure autoclave components
- selective use of niobium-titanium (Nb-Ti) alloys in autoclave oxygen sparge tubes
- appropriate structural use of FRP for atmospheric pressure solution exposure
- epoxy coated concrete drainage floors and pillars with a working layer of quartz aggregate
- <u>no</u> exposure of ferrous alloys to process solutions

PRELIMINARY COST ESTIMATES

Capital Costs

Capital costs for the 20,000 and 100,000 tonnes/year cases have been estimated on the following basis:

- process equipment and first fill reagents per vendors budget quotes for major items
- piping (31% of equipment; file based factor)
- electrical (13% of equipment; file based factor)
- instrumentation (15% of equipment; file based factor)
- buildings per preliminary layout with allowance for corrosion resistant materials
- non-process and miscellaneous by allowance
- factors for spares and commissioning
- level site with good foundation conditions
- installation costs at West Coast North American labour rates (unionized, U.S. $45/hr, all in) and productivity
- engineering at 8% of direct cost
- construction management at 6% of direct cost
- contingency at 20% of total estimate

Excluded were owner's costs, site acquisition, working capital, escalation and interest during construction.

Utilities (electric power, fresh water, sanitary sewer, pipeline natural gas) were assumed to be available at battery limits.

On this basis, capital costs for processing to concentrated (60%) $ZnCl_2$ solution were estimated as:

	EAF Feed → Zinc Chloride Product Plant Capacity - EAF Dust, dry basis	
	100,000 tonnes/year	**20,000 tonnes/year**
Project total capital cost (U.S. $, 4th quarter 1993)	U.S. $30,000,000	U.S. $10,500,000

For conversion of concentrated zinc chloride to zinc sulphate, the additional estimated cost for the 100,000 tonnes/year (dust) capacity plant was U.S. $13,200,000.

Total capital costs for the selected cases are thus:

	Total Capital Cost (U.S. $, 4th quarter 1993)
100,000 tonnes/year dust capacity zinc sulphate product	U.S. $43,200,000
20,000 tonnes/year (dust) capacity concentrated zinc chloride product	U.S. $10,500,000

Operating Costs

Operating cost estimates include:

- operating/maintenance/supervisory and administrative labour with 35% burden
- operating suppliers/utilities (U.S. $)
    - utilities (electric power @ $0.03/kWh, natural gas @ $2.25/$10^6$ Btu)
    - chlorine @ $206/tonne
    - oxygen @ $98/tonne
    - Acorga ZNX 50, diluent, lime, sodium hydrosulphide and miscellaneous reagents at vendor quote or published prices with transport allowance
- mobile equipment allowance

- maintenance supplies @ 4% of total capital cost
- allowance for miscellaneous changes including communication/computing/office/government-regulatory affairs and environmental monitoring
- 10% contingency

Excluded are property taxes, insurance and costs (transport, tipping fees) associated with disposal of washed leach residue, and sewer charges.

On this basis, total estimated operating costs are:

Case	Total Operating Cost (U.S. $, 4th quarter 1993)
100,000 tonnes/year (dust) capacity $ZnSO_4$ product	U.S. $92/tonne (dust)
20,000 tonnes/year (dust) capacity concentrated (60%) $ZnCl_2$ product	U.S. $194/tonne (dust)

REVENUE POTENTIAL

Potential revenue sources from processing EAF dust include:

- disposal fees chargeable to EAF dust generators
- concentrated $ZnCl_2$ (60% or 70%) solution (20,000 tonnes/year plant) (or)
- zinc in concentrated sulphate solution (100,000 tonnes/year plant)
- mixed ((Zn, Pb, Cu, Cd) S) sulphide concentrate from bleed stream treatment
- washed primary leach residue as concrete/tile colouring agent or pigment

Development of markets for primary leach residue will have a strong effect on project economics since any sale of this material would result in corresponding reduction in tipping fee disposal cost.

Revenue potential is highly site specific; prospects are being investigated by Terra Gaia on the basis of test data and evaluation of product by potential consumers.

FUTURE DIRECTIONS

At the manuscript deadline (March 1, 1994), Terra Gaia anticipates the following sequence of project development:

- upgraded engineering and capital/operating cost estimates
- continued lab/pilot testing of critical operations
- site specific project economic analysis
- joint venture financing of a demonstration project

Some critical components of the processes described are subject to U.S.A. and multi-national patent application by Terra Gaia.

ACKNOWLEDGEMENTS

Permission of Terra Gaia and Fluor Daniel Wright to publish this article is gratefully acknowledged. Significant technical contributions to the project (also gratefully acknowledged) were provided by Mr. R.J. Johnston (formerly of Fluor Daniel Wright) and staff of the testing laboratories: Process Research Associates, Vancouver, B.C. and Lakefield Research, Lakefield, Ontario.

REFERENCES

1. Nyireuda, R.L., The Processing of Steelmaking Flue-Dust: A Review, Minerals Engineering 4, No. 7-11, p. 1003-1025, 1991.

2. Duyvesteyn, W.P.C., and M.C. Jha, Mixed Lixiviant for Separate Recovery of Zinc and Lead from Iron Containing Waste Materials. U.S. Patent 4,614,543, 1986.

3. Duyvesteyn, W.P.C., and M.C. Jha, Zinc Recovery from Steel Plant Dusts and Other Zinciferous Materials. U.S. Patent 4,572,771, 1986.

4. van Weest, G., A van Sandwijk and S. Honiugh, "The Treatment of Iron Making Blast Furnace Dust by Chloride Hydrometallurgy", Hydrometallurgy: Fundamentals, Technology and Innovations: Proceedings of the International Symposium of Hydrometallurgy,

4th, 1993, Salt Lake City, Utah, Society for Mining, Metallurgy and Exploration, 1993, pp. 931-946.

5. Dalton, R.F., A Burgess and P.M. Quan ACORGA ZNX50 - A New Selective Reagent for the Extraction of Zinc from Chloride Solution, Hydrometallurgy Theory and Practice, W.B. Cooper and D.B. Dreisinger ed., Elsevier, 1992, p. 385-400.

Prediction of water contamination arising from disposal of solid wastes

P. B. Mitchell
C. P. Waller
Keith Atkinson
Camborne School of Mines, University of Exeter, Redruth, Cornwall, England

Abstract

The dispersion of metal-bearing wastes into the natural environment has caused widespread contamination and pollution of soils. Poor regulated disposal has also put ground water quality at risk. In a world of diminishing soil and water resources, the identification and remediation of the most degraded sites has assumed a greater importance than ever before. This paper describes the extent and nature of soil contamination, its impact on ground water quality and the significance of metal contaminant partitioning between the solid and aqueous phases, and between inorganic and organic soil components.

Four methods that may prove useful in the prediction of the water contamination potential of metal-bearing solid wastes are reviewed: soil solution analysis, determination of "plant-available" metals, analysis of metal partitioning between soil components by sequential extraction, and leaching tests. Of these, sequential extraction tests in parallel with leaching studies appear to present the most flexible option for predicting the long-term release of metals from contaminated wastes and soils, under steady-state or evolving chemical and physical conditions.

Keywords: Aqueous extractants, leach tests, metal partitioning, plant-available metals, soils, solid wastes, water contamination.

Introduction

Metal-bearing wastes are produced by almost every industry, from those exploiting ore reserves to those manufacturing consumer goods. Solid, metal-bearing wastes are often disposed of directly in landfills or at other suitable (and unsuitable) land-based sites. Liquid effluents may also transfer metals to soils and other solids that they contact, while aerial deposition of fine, metal-bearing particulate matter is a significant contributor to soil contamination and pollution. Due to the relatively strong interaction between the extraneous metals and the soil, input often exceeds output (output occurring mainly through leaching), resulting in a build-up of metals in the soil. Depending on their concentration, these additional metals may be classified as contaminants or pollutants. A *contaminant* has been defined as "a substance found in a given medium at a concentration higher than that which one would expect from other considerations and where the source of the additional concentration of the substance appears to be human activity" [1]. A contaminant is also a *pollutant* if the observed contamination is high enough to cause harm to some organism. Proving that pollution, rather than contamination, has occurred is often a difficult task, particularly retrospectively, and only the term contamination is used here.

Despite the current trend towards increasingly stringent legislation, the legacy from decades (or centuries) of poorly regulated disposal still requires careful study. It has been noted that, based on current estimates, there are 75,000-100,000 potentially contaminated sites in the United Kingdom, 150,000-200,000 in Germany and 110,000 in the Netherlands [2]. Globally, the number of sites is likely to be far more forbidding. Even with strict legislation, the sheer volume of wastes requiring disposal continues to degrade the natural environment. For example, approximately 17,800 million tonnes of mine waste and tailings are produced annually, nearly equalling the annual tonnage of sediments naturally eroded into the oceans [3]. The priority list developed by the US Environmental Protection Agency [4] contained approximately 1,000 sites posing significant environmental or health risks. Approximately 40 % of these sites reported problems with metals, and of these, 40 % were landfill sites. Undoubtedly, land and water resources everywhere are under extreme pressure from mankind. A US Library of Congress Report [5] listed 1,360 well closures over a 30 year period, of which 619 arose from metal contamination. In the long-term, these pressures must be removed or diminished, to avoid potential catastrophe. Long-term solutions rest mainly with the politicians, but also with society in general, and are outside the scope of this paper. The most appropriate short-term solution must be the protection of remaining resources (ultimately the responsibility of the legislators), and the rehabilitation of those areas that have not yet been irreversibly degraded.

In a world of limited financial resources, some means of ranking contaminated sites is required, to allow those having the highest actual, or predicted, impact to be targeted for immediate remedial action. A site's water contamination potential is a useful index for the purposes of ranking, as it reflects (a) the capacity of the contaminated waste or soil to degrade ground and surface water quality (and thereby limit the possible end-uses for the water) and (b) the potential for harm to soil-plant ecosystems (where only the more soluble metal fractions have any biological significance [1]).

Before the possible methods of determining water contamination potential are examined, the nature of soil contamination must first be explored in terms of metal-soil component interactions, which ultimately decide the aqueous mobility of metal contaminants.

Soil components and the fate of contaminants

Soil components

Inorganic soil components result from weathering of the parent rock. Principal components include iron and aluminium oxides and oxy-hydroxides (eg: gibbsite, goethite and limonite), quartz, aluminosilicates such as micas (eg: biotite and muscovite) and clay minerals (eg: illite, montmorillonite and vermiculite). Depending on the nature of the pedogenesis, other mineral species may also be present (eg: zeolites, calcite, gypsum, pyrite, phosphates, halides and nitrates).

Of the inorganic components, clay minerals have the greatest significance in determining the ion exchange behaviour of soils [6]. Their cation exchange capacity (CEC) is determined by the extent of isomorphous substitution and exposed surface area. CEC ranges from 2-6 meq/100g for kaolinite (low surface area, minimal isomorphous substitution) up to 200 meq/100g for vermiculite (high surface area, significant isomorphous substitution).

If the distribution of ionic species between liquid and solid phases in natural permeable media is primarily governed by ion exchange [7] [8] [9], clay minerals must play a significant role in the aqueous mobilisation of contaminants. If, however, in reality, systems are so complex that the partitioning of contaminants between liquid and solid phases can only be described in terms of ion exchange, adsorption, precipitation, dissolution, and chemical and biochemical transformations [6], the clay minerals may have a reduced part in determining the aqueous mobilisation.

Organic soil components are a mixture of non-decayed, partially decayed and fully decayed substances derived mainly from vegetative biomass. Their CEC (averaging several hundred meq/100g) derives from the dissociation of acid functional groups, which is a function of solution pH, the pKa of the functional group, and interactions with nearby functional groups [6] (cf: inorganic soil components, where the permanent negative charges arising from isomorphous substitution are independent of pH).

Fate of contaminants

Although soils exposed to metal contaminants are often considered irreversibly contaminated, the metals in fact have a number of fates. Although it is more usual for the surface horizons (which may also have the greatest concentration of organic matter) to be the most contaminated the metals may infiltrate deeper into the soil, through the unsaturated zone to eventually reach the ground water. The rate and extent of infiltration is related to the physical and chemical properties of the metal contaminant, the flow of water, and (time-dependent) interactions with the soil components. As reactions between the metal contaminants and soil components occur over variable timescales (from microseconds for ion exchange to years for the incorporation of contaminants in mineral structures) [10], there is a corresponding variation in times between waste disposal or soil contamination, and ground water contamination, which may not occur at all, or take months, years, or even centuries to manifest itself. It is this lag time that has lead many sites to be described as "time bombs". Metals may also move into surface waters, carried by eroded particulate matter. Even in the absence of human activity there is a natural and continuous transfer of metals between soil, air, water and biota. Therefore, it is not that metal contaminants are being released from anthropogenic wastes that is important, but rather the **rate** of release. Human activity tends to increase the rate beyond that normally seen in "natural" processes. When the rate exceeds the capacity of the receiving body to disperse, dilute or otherwise attenuate the contaminants, contamination and pollution are likely to occur.

Changes in environmental conditions affecting contaminant mobilisation

Wastes and soils that are apparently stable and that exhibit minimal leaching of metal contaminants, may become unstable with time, in response to two types of stimuli: external (eg: seasonal variations in climate) or internal (eg: evolving waste/soil chemistry). The partitioning of metals between soil components, and between the solid and aqueous phase must therefore be considered to be in a state of flux. The challenge is to incorporate an assessment of the likely response of contaminated wastes or soils to a wide range of environmental conditions in the predictive methodology.

The effects of changing environmental conditions on aqueous metal mobilisation can be simplistically demonstrated in two well known examples: (a) the exposure of sulphidic metalliferous mining wastes to a weathering environment, and (b) the chemical (and microbial) evolution of landfills.

(a) Sulphidic metalliferous mining wastes are stable under reducing conditions. However, when exposed to a weathering (oxidising) environment, the sulphides oxidise to produce soluble sulphates and acidity. Under acid conditions, the mobility of metals such as Cd, Cu, Hg, Pb and Zn is increased [3].

(b) In municipal solid waste landfills, an aerobic phase (pH 7 to 8), is followed by an acidic anaerobic phase (pH as low as 5) with increasingly reducing conditions. The landfill evolves from this acetic phase to a methane generating (methanogenic) phase. Typically, the concentration of metals in the leachate is greater during the acid decomposition phase than during the methanogenic phase. However, landfills are dynamic systems, and beyond 30 years of evolution there is a lack of experience with regard to the landfill chemistry. When one considers that work by some researchers [11] has indicated that the transformation of organic materials will continue over a geological time scale (10^3 to 10^7 years), this lack of predictive knowledge may pose considerable problems for future generations.

Unlike these two examples, the chemistry of many waste types has not yet been investigated, and even for those that have been studied in depth, a number of important questions remain about their long-term environmental impact. Therefore, one possible predictive method is the reduction of the complex characteristics of contaminated wastes and soils into more easily understood component parts. Theoretically at least, the response of the whole system may then be estimated from integrating the response of the component parts, possibly by comparison with a range of leaching procedures.

The prediction of water contamination potential

What is required is a flexible and relatively inexpensive diagnostic technique the prediction of water contamination potential. Unfortunately, the development of such a technique has posed considerable problems. The determination of metal concentrations in interstitial pore water has been unsuccessful. Instead, a number of aqueous extractants have been employed to calculate the concentration of plant-available (easily solubilised) metals and the division of metal contaminants between soil components. Leaching procedures have also been developed. The relative benefits and drawbacks of these techniques is examined below.

Soil solution analysis

Soil solution can be recovered using advanced filtration methods, or by displacement with a neutral salt solution that neither affects the soil solution pH nor exchanges metals held by the solid phase.

However, weak extractant solutions such as dilute acetic acid, ammonium acetate and EDTA are more commonly used. These attempt to simulate the extractive capacity of plant roots, which have access to greater concentrations of metals than those present only in soil solution. Although often termed "soil solutions", the resulting leachates overestimate actual soil solution concentrations.

The metal concentration of soil solutions (either interstitial or measured by weak extractants) varies according to field conditions, and is sensitive to the mode of recovery. As such the analysis of soil solutions represents only a short-term "snapshot" of what is actually a dynamic process, and is unsuitable for use as a predictive technique to assess the long-term movement of metals between the solid and aqueous phases.

Determination of "plant-available" metals

"Plant available" metals are generally easily solubilised, and therefore there are some parallels between them and those metals likely to cause water contamination in the short-term (and possibly longer). "Plant-available" metals include those held in the soil solution and at cation exchange sites, and those that are readily released into the aqueous phase from other inorganic and organic soil components. "Plant-available" determinations do not address the long-term movement of metals between soil components, and can not be used to predict the movement of metal contaminants from stronger to weaker binding sites (or vice versa), which may occur as environmental conditions develop or change.

The most common extractant used for "plant-available" metals is 0.5 M acetic acid [12]. However, the degree of correlation between acetic acid-extractable metals and plant response has been variable, as the application of a universal extractant fails to recognise the importance of soil pH, CEC, texture and so on. Therefore, perhaps it would be better to apply it only as a predictor of short-term release from contaminated wastes and soils.

Analysis of metal partitioning between soil components by sequential extraction
Sequential extractions provide information on the partitioning of metals between soil components. Generally between three and eight extractants are used, beginning with the least powerful-most specific, and ending with the most powerful-least specific [13]. Once metal partitioning has been determined, the origin, mode of occurrence, biological and physico-chemical availability, mobilisation and transport of the metals can be better understood [14].

Some authors have assumed that extractants are specific to sharply defined soil components. Other workers have been less confident [13], and probably rightly so. Many variables affect the capacity of extractants to liberate metals from soil components. There are few, if any, extractants for which there is not some contradictory information available. Some of these discrepancies undoubtedly arise from the poor reporting of procedure. Sometimes even the most basic experimental information, such as the pH of the extractant solution, is absent. Sequential extractions developed for one material may not be applicable to others and sometimes too little thought is given to the limitations of the technique or the veracity of the data. Other contradictions arise from the limited specificity of some extractants, or from problems with the fresh immobilisation of metals released by an extractant, by other, unaffected, soil components.

These reservations aside, as long as the limitations of the extractants are recognised, and experimental conditions are carefully controlled (and reported), sequential extraction procedures can produce valid estimates of metal partitioning.

Table 1 indicate some of the extractants that have been, and are being, used by various researchers, either sequentially or in isolation. Extractants are in an approximate order of weakest first, (or in order of increasing binding power of the soil component). Components in square brackets, denote those where there is some debate as to the extractability by that particular extractant. The table is summarised from Beckett's excellent 1989 review of extractants used for trace metal studies in soils, sewage sludges and sludge-treated soils [13], and a number of other sources [15] [16] [17] [18] [19] [20] [21]. Note that Table 1 represents just a small selection of a very large field of research, and the list is not exhaustive.

Table 1: Summarised data for soil components and their reported extractants

Extractant	**Soil component**
Deionised or distilled water	Water soluble
Potassium chloride, potassium nitrate, calcium chloride, calcium nitrate and magnesium chloride	Exchangeable, neutral salt exchangeable, water soluble or non-specifically adsorbed
DTPA	Exchangeable and organically bound metals, "precipitates"
Ammonium acetate and ammonium chloride	Exchangeable metals
Potassium and sodium fluorides	"Adsorbed" metals, [organic matter]
Lead nitrate and copper acetate	"Specifically bound" metals
Acetic acid	Carbonates, some iron and manganese oxides, "acid-soluble" metals, "specifically bound" metals and decomposed organic matter
Sodium and potassium pyrophosphate	Organically bound metals [some aluminium, iron and manganese oxides, some metal carbonates and copper sulphides]
Sodium hypochlorite	Organic matter
Hydrogen peroxide	Organic matter, sulphides, carbonates and oxides
Sodium hydroxide	Organic matter, some metal carbonates
Sodium citrate	Iron oxides
Oxalic acid	Free oxides and primary minerals
EDTA	Carbonates, "inorganic precipitates", amorphous and crystalline oxides and hydroxides of iron, organically bound metals
Acid ammonium oxalate	Iron and aluminium oxides, hydroxides and oxyhydroxides, aluminium and iron from organic complexes, zinc and copper carbonates, [sulphides]
Hydroxylamine hydrochloride	Manganese oxides
Sodium dithionate	Iron oxides, organic-iron complexes
Strong, mineral acids	Residual minerals

Sequential extraction procedures, based on Viets' notion of "pools" of elements in soils [22], have been developed. Viets' soil "pools" (or components) are:

A: Soil solution (water soluble)
B: Exchangeable
C: Sorbed and organically bound
D: Bound, occluded in oxides and secondary clay minerals
E: Residual (primary mineral lattice)

The strength of fixation and resistance to release under changing environmental conditions increases from A to E. Wastes containing metals in pools A, B or C will release higher concentrations of metals into the aqueous phase than those in pools D and E, and therefore will pose a greater threat to underlying strata and to ground waters. Each "pool" contains related soil components. An analysis of the theory of metal partitioning between soil components, and its affect on the choice of extractants, their sequence of use, and the difficulties that arise in the use of certain extractants is outlined below.

Pool A: Soil solution (water soluble) This apparently simple phase, is in fact relatively complex. It can be recovered in a number of ways, for example by the addition of distilled water to the soil and filtering of the resulting sludge; the addition of 0.1 M magnesium chloride solution, or by direct centrifugation of larger samples. There is considerable variation in the metal concentrations of solutions recovered, even by these relatively similar procedures.

Pool B: Exchangeable This fraction is easily removed from soil components by ion exchange. The metals in this pool tend to be those adsorbed onto clay minerals, iron and manganese oxides and hydroxides, organic matter and other colloids. They are held in place by weak (Van der Waals) ion-dipole or dipole-dipole attraction and are very susceptible to changes in pH. If conditions become more acid, the metals tend to desorb and go into solution. Soluble salts are commonly used as extractants, with the salt cation replacing the adsorbed metal. Examples are ammonium acetate, sodium acetate or magnesium chloride. However, some research has indicated that ammonium acetate may also attack carbonates, releasing associated metals.

Pool C: Sorbed and organically bound This phase is often further subdivided into adsorbed metals and organically bound metals. Stronger covalent bonds or chemical associations bind the metals to the surface of silica, iron, manganese and aluminium compounds, and carbonates. Most authors have adopted the use of acid salt solutions for the extraction of metals in this phase. By far the most common extractant is sodium acetate in acetic acid (at pH 5). This will not, however, extract metals in the organic phase. Finding an extractant to specifically remove organically bound metals is far more taxing. The forces binding metals and organic substances vary from physical adsorption to strong chelation.

It is important to remove the organic fraction prior to attacking the hydrous oxide fraction, as some organic substances may solubilise metals and influence subsequent stages of extraction [23].

The two routes to determining metals in the organic fraction are the oxidative decomposition of organic matter or the chelation of the metals. A common oxidant is hydrogen peroxide, but some authors have stated that it does not achieve complete decomposition, that it will "damage" amorphous constituents and clay minerals and solubilise manganese oxides and oxidise sulphides. It can, therefore, be considered non-specific. An alternative oxidant is sodium hypochlorite, which is a more effective decomposer, but also decomposes sulphides, and thus must also be ruled out.

Chelating agents (such as EDTA (ethylenediaminetetra-acetic acid) and DTPA (diethylenetriaminepenta-acetic acid)) do not destroy the organic matter, but preferentially chelate the metals. They do, however, also remove some metals associated with oxides [24]. Pyrophosphate salts also chelate organically bound metals with a reduced effect on the oxides.

The persistence of the organic matter may affect subsequent extractions, as the partial release of organic compounds may solubilise metals. Therefore, following pyrophosphate extraction, the organic matter must be decomposed, while minimising the attack of other components. If sodium hypochlorite is used, only sulphides are also decomposed. The metals previously bound as sulphides precipitate out as hydroxides at the high pH generated by the hypochlorite. The hydroxides can then be recovered with a weak acid leach [23] (other soluble, exchangeable and adsorbed metals already having been removed), and the sulphide phase determined by back calculation.

Pool D: Bound, occluded in oxides and secondary clay minerals This fraction contains metals held at sites which are usually more resistant to weathering than pools A, B and C. Metals bound by manganese oxides and iron oxides can be differentiated by utilising the lower stability of the former in hydroxylamine hydrochloride. Chao and Theobald [25] found that 0.1 M hydroxylamine hydrochloride in 0.01 M nitric acid (pH 2) dissolved 85 % of the manganese oxide, but only about 5% of the iron oxide.

A number of extractants have been used to remove the amorphous iron oxide fraction. Chao and Theobald [25] used 0.25 M hydroxylamine hydrochloride in 0.25 M hydrochloric acid to extract amorphous iron oxides. A more common extractant for amorphous iron is "Tamm's reagent" (ammonium oxalate at pH 3.5 [23] [26]), but this extractant is light sensitive, and lighting conditions must be carefully controlled, a point often ignored in the reporting of data.

Pool E: Residual and primary minerals Extraction methods for this phase are the same as those used for the wet chemical analysis of total metals and generally strong mineral acids such as perchloric, nitric and hydrofluoric are utilised.

There appears to be a tendency for metals to migrate, with time, sequentially from pool A to pool D, indicating that with time the binding between the contaminants and soil components becomes stronger. This could be interpreted as proof that older wastes may represent less of a hazard for ground waters than more recent wastes, although it would be unwise to assume this without detailed site investigations. As an example, Tuin and Tels [27] used a sequential extraction procedure to investigate the partitioning of metals in contaminated clay soils before and after remedial cleaning. They noted differences between soils polluted artificially in the laboratory, and soils from two genuine waste sites. In the artificially contaminated soils, the majority of the added metals were held by the exchangeable, carbonate and oxide fractions. Metals in the waste site soils were found mainly in the organic and residual fractions. This difference may have arisen from the preferential leaching of metals held at exchangeable, carbonate and oxide fractions, or from the immobilisation of metals via the aging of amorphous oxy-hydroxides and oxides, the formation of insoluble oxides, or diffusion to interior lattice sites. The cleaning procedures, using hydrochloric acid or EDTA, effectively removed metal contaminants with the exception of those in the residual fraction. Tuin and Tels [27] questioned whether further remediation to remove these final resistant metals should be implemented if they cannot be removed by successive leaches with HCl or EDTA. In practice, would environmental conditions ever change in such a way as to liberate metals from the residual fraction? This is an important question. Returning soils to a pristine condition can be prohibitively expensive, and may not even be technically feasible without destroying the soil. If the residual fraction can be shown to be stable under a wide range of environmental conditions this may be all that is required.

Leaching tests

Leaching tests differ from sequential extractants in that they assess the response of the whole system, rather than its component parts. However, they are more time consuming, and predicting the release of metals under varying environmental conditions generally requires some simulation of those conditions under a laboratory regime.

The available waste leaching tests have been summarised as (1) agitated extraction tests, (2) non-agitated extraction tests, (3) sequential chemical extraction tests and (4) concentration build-up tests [28]. Dynamic tests involve the periodic renewal of the leaching agent to drive contaminant mobilisation, and provide information on mobilisation kinetics. Dynamic tests are the preferred approach for the investigation of complex wastes.

In the Netherlands, the Standard Leaching Test is used to estimate the amount of heavy metals leaching from solids [29]. The test simulates leaching by percolating rainwater and consists of three sub-tests: (a) column tests, (b) dynamic cascade tests and (c) maximum leachability tests.

(a) the column test is conducted using upwardly flowing demineralised water acidified to pH 4 with nitric acid. The ratio of water flow to sample weight is calculated from the anticipated rainfall and thickness of the contaminated layer in the field. Dissolved metals are measured in the resulting leachate.

(b) in the dynamic cascade test a known weight of sample is shaken with 20 times that weight of demineralised water (acidified with nitric acid to pH 4) for 23 hours, after which the leachate is recovered, and analysed for dissolved metals, and the process repeated a further four time with fresh acidified water (ie: in total the sample is leached with one hundred times its weight of acidified water).

(c) maximum leachability is determined by stirring a known weight of sample with one hundred times that weight of demineralised water. During the four hour leaching period, the solution phase is held at pH 4. The leachate is then analysed for dissolved metals.

In the United States, the Environmental Agency extraction procedure (EP) [30] has been superseded by the toxicity characteristic leaching procedure (TCLP). The TCLP was designed to simulate the leaching that an industrial waste undergoes when co-disposed with refuse in a sanitary landfill. It allows the determination of leachable inorganic and organic constituents in solid, liquid and mixed wastes, using a dilute acetic acid solution. If the concentrations of certain prescribed elements in the leachate exceed the acceptable regulatory concentrations then the waste is classified as hazardous. The test was developed from a model of a landfill containing 5 % solid industrial wastes and 95 % municipal wastes. The choice of acetic acid as the leachant is based upon its presence during the aerobic phase of landfill chemistry development.

Sullivan and Yelton [31] attempted to correlate metal immobilisation and release using three alternative methods: (a) sequential extraction to determine metal partitioning, (b) TCLP and (c) humidity cell weathering to simulate the maximum metal release [32]. With the sequential extraction it proved possible to predict metal release based on the metal fractionation. The TCLP results did not realistically assess the potential hazards. The humidity method required months to obtain data, and was a rather intensive procedure. When the authors compared the results of the sequential extraction with the TCLP and humidity cell data, it was shown that leachate concentrations could be predicted from the sequential extraction data, and the latter may therefore represent a rapid way of assessing metal release, if some form of validation is also available. The TCLP may not be applicable to wastes disposed of at non-landfill sites, but it does serves as a regulatory point of reference [31] in a field that is notoriously short of realistic assessments of the hazards associated with solid wastes, which are still often based on total metal concentrations.

Conclusions

Despite the extensive literature on trace and major elements in soils and solid wastes, and the existance of legislation in many countries concerning the permissible concentrations of contaminants, there does not appear to be a clear definition of procedures necessary to assess the water contaminating potential of such materials. Frequently quoted metal contents are regarded as "total metals" and not necessarily those that might be easily leached into ground and surface waters. When there is a clearly defined, universally recognised, definition of the basis for "permissible concentrations", then procedures can be established for deriving values for comparison to these concentrations. It appears that what is required is a flexible leaching procedure, linked to a detailed analysis of the partitioning of the metal contaminant within the soil or waste material so that both the likely affects of changing environmental conditions on metal partitioning and the behaviour of the overall system under a range of environmental conditions can be predicted.

However, more work is required to identify truly selective extractants and to assess the rate of metal movement between soil and waste components, and between the solid and aqueous phases.

Soils and wastes must be better characterised in terms of their reactive components, and their physicochemical interactions with metals in the aqueous phase must be described in greater detail. The complexity of soils and anthropogenic wastes continues to inhibit a comprehensive understanding of metal-solid interactions, and until this hurdle is overcome, predicting the long-term water contaminating potential of solid wastes and contaminated soils will remain a difficult and uncertain process.

References

1. Davies, B.E. (1992) Trace metals in the environment: retrospect and prospect, in **Biogeochemistry of Trace Metals**, (ed. D.C. Adriano), pp. 1-17.

2. Hasselman, J.F. (1992) Site investigations in Europe, in **Proceedings of Site Investigations for Contaminated Sites**, 21-22 September, London, IBC Technical Services Ltd, p 2.

3. Förstner, U. (1993) Dispersion of contaminants from landfill operations, in **Migration and Fate of Pollutants in Soils and Subsoils**, (ed. D. Petruzzelli and F.G. Helfferich), Springer-Verlag, Berlin, Heidelberg, pp. 435-454.

4. Anonymous. (1986) **National Priorities List Fact Book**. HW 7.3, US Environmental Protection Agency, Washington DC.

5. Anonymous. (1980) Groundwater strategies. **Environmental Science and Technology**, 14, 1030-1035.

6. Petruzzelli, D. and Lopez, A. (1993) Solid-phase characteristics and ion exchange phenomena in natural permeable media, in **Migration and Fate of Pollutants in Soils and Subsoils**, (ed. D. Petruzzelli and F.G. Helfferich), Springer-Verlag, Berlin, Heidelberg, pp. 75-92.

7. Bolt, G.H. and Bruggenwert, M.G.M. (1978) **Soil Chemistry Vol. 5A: Basic elements**. Elsevier Science Publishing Company, Amsterdam.

8. Bolt, G.H. (1982) **Soil Chemistry Vol. 5B: Physico-chemical models**. Elsevier Science Publishing Company, Amsterdam.

9. Sposito, G. (1981). **The Thermodynamics of Soil Solutions**. Oxford University Press, Oxford.

10. Amacher, M.C. (1991) Methods of obtaining and analyzing kinetic data, in **Rates of Soil Chemical Processes**, (ed. D.L. Sparks and D.L. Suarez), Soil Science Society of America, Madison, WI, pp. 19-60.

11. Lichtensteiger, T., Brunner, P.H. and Langmeier, M. (1988) **Klärschlamm in Deponien**. EAWAG Project No.30-681. EC-COST.

12. Williams, J.H. (1979) Metal concentrations in soil and crop phytotoxicity, in **Reclamation of Contaminated Land Proceedings**, Society of the Chemical Industry Conference, Eastbourne, Sussex, pp. C6/1-12.

13. Beckett, P.H.T. (1989) The use of extractants in studies on trace metals in soils, sewage sludges, and sludge-treated soils, in **Advances in Soil Science, Volume 9**, Springer-Verlag, New York, pp. 143-176.

14. Tessier, A., Campbell, P.G.C. and Bisson, M. (1979) Sequential extraction procedure for the speciation of particulate trace metals. **Analytical Chemistry**, 51, 844-851.

15. Matthews, M.R. (1990) **Distribution and dynamics of trace metals in soils and vegetation of a small contaminated catchment in mid-Wales**. PhD Thesis, Department of Environmental Science, University of Bradford.

16. Park, J.S. (1990) **The development of analytical methods to elucidate the form of Cu, Zn, Cd and Pb in soil contaminated by sewage sludge and other effluents**. PhD Thesis, Robert Gordons Institute of Technology, Aberdeen, CNAA.

17. Miller, W.P., McFee, W.W. and Kelly, J.M. (1983) Mobility and retention of heavy metals in sandy soils. **Journal of Environmental Quality**, 12, 579.

18. Miller, W.P. and McFee, W.W. (1983) Distribution of cadmium, zinc, copper and lead in soils of industrial northwestern Indiana. **Journal of Environmental Quality**, 12, 29.

19. Johnston, S.E. and Barnard, W.M. (1979) Comparative effectiveness of fourteen solutions for extracting arsenic from four western New York soils. **Soil Science Society of America Journal**, 43, 304-307.

20. Bradford, G.R., Bair, F.L. and Hunsaker, V. (1971) Trace and major element contents of soil saturation extracts. **Soil Science**, 112, 225-230.

21. Halstead, R.L., Finn, B.J. and MacLean, A.J. (1969) Extractability of nickel added to soils and its concentration in plants. **Canadian Journal of Soil Science**, 49, 335-342.

22. Viets, F.G. (1962) Chemistry and availability of micronutrients. **Journal of Agriculture and Food Chemistry**, 10, 174.

23. Hoffman, S.J. and Fletcher, W.K. (1978) Selective sequential extraction of Cu, Zn, Fe, Mn, and Mo from soils and sediments, in **Proceedings of the 7th International Geochemical Exploration Symposium**, (ed. J.R. Watterson and P.K. Theobald), Association of Exploration Geochemists, Rexdale, Ontario, Canada, pp. 289-299.

24. McLaren, R.G. and Crawford, D.V. (1973) Studies on soil copper. I. The fractionation of copper in soils. **Journal of Soil Science**, 24, 172-181.

25. Chao, T.T. and Theobald, P.K. (1976) The significance of secondary iron and manganese oxides in geochemical exploration. **Economic Geology**, 71, 1560-1569.

26. Kersten, M. and Förstner, U. (1986) Chemical fractionation of heavy metals in anoxic estuarine and coastal sediments. **Water Science and Technology**, 18, 121-130.

27. Tuin, B.J.W. and Tels, M. (1990) Distribution of six heavy metals in contaminated clay soils before and after extractive cleaning. **Environmental Technology**, 11, 935-948.

28. Stegemann, J. (1991) Compendium of waste leaching tests, **Wastewater Technology Centre News Letter No. 18**, 6-7.

29. Versluijs, C.W., Aalbers, Th.G., Adema, D.M.M., Assink, J.W., van Gestel, C.A.M. and Anthonissen, I.H. (1988) Comparison of leaching behaviour and bioavailability of heavy metals in contaminated soils cleaned up with several extractive and thermal methods, in **Contaminated Soil '88**, (ed. K. Wolf, W.J. van den Brink and F.J. Colon), Kluwer Academic Publishers, pp. 11-21.

30. Federal Register. (1980) **40 CFR Part 261, Appendix II**, v. 45, no. 98, May 19, p 33122, 33127-33129.

31. Sullivan, P.J. and Yelton, J.L. (1988) An evaluation of trace element release associated with acid mine drainage. **Environmental Geology and Water Science**, 12, 181-186.

32. Caruccio, F.T. (1968) An evaluation of factors affecting acid mine drainage production and the ground water interactions in selected areas of western Pennsylvania. **Proceedings of the Second Symposium of Coal Mine Drainage Research**, Bituminous Coal Research, Inc., Monroeville, PA.

Improved process for concentration of ulexite and boric acid production

R. E. Pocovi
A. A. Latre
O. A. Skaf
Instituto de Beneficio de Minerales, Facultad de Ingeniería, Salta, Argentina

Abstract
South America reserves of boron minerals are concentrated in the Central Andes. Ulexite is one of the most important borates found in these deposits. The mineral is associated with impurities such as sand, clays, chlorides and sulfates. Boric acid is produced by leaching the ulexite with sulfuric acid and it is recovered through a process of liquid-solid separation and then, by a cooling crystallization process.

The residual slurry is rejected and it is necessary to reduce its adverse environmental effects. This paper reports an improved process for ulexite concentration and boric acid production. It has been developed to minimize the pollution problems found in the present industrial process used in Argentina.
Keywords: Ulexite, Borates, Boron Minerals, Boric Acid, Sodium Sulfate.

1 Introduction

Boron ore reserves in South America are the fourth in the world and they are of about 30 million ton. of B_2O_3. The main deposits in Argentina are located in the provinces of Salta, Catamarca and Jujuy in the Northwest of Argentina near its border with Chile and Bolivia. They are situated in a high plateau called Argentina Puna at about 4,000 meters above sea level. The main borate deposits are shown in Figure 1.

The most abundant borate in this region is ulexite, $Na_2O.2CaO.5B_2O_3.16H_2O$, with a B_2O_3 content of 43%. It is found superficially in salt deposits (evaporite salt deposits), in the form of nodules or plates with a fibroid structure. The fibers are very soft and the nodules are of easy disintegration in the individual fibers. The mineral is found as an ore, in which the waste gangue consist of impurities, such as clays, sand, sodium chloride and sodium sulfate. It is extracted by manual or mechanical methods and after a manual removing of the main gangue materials, it is sun dried at places near the salt deposit. It is so because of the advantage of the dry, sunny and windy weather at the Puna region. The final product has a B_2O_3 content of about 22% - 28%.

Ulexite is used for the boric acid production, which is useful as raw material in the manufacture of various kinds of glass, textile fiberglass and frits. Other nonglass uses correspond to the manufacture of soaps, detergents, fluxing material in welding, etc.

Total installed boric acid production capacity in Argentina is about 30,000 tpa, divided in eight plants. Seven of them are situated in the Lerma Valley, where the city of Salta is located, which is an agricultural area with farming of vegetables, fruits and tobacco. Borates and boric acid have relatively low toxicity for humans and animals. Because small quantities of boron are essential for all plant life, it is added in trace levels to fertilizers; however quantities well below concentrations affecting humans and animals are so extremely toxic to vegetation that boron compounds are also used commercially as herbicides. Because of the soil pollution produced by the boric acid plants, numerous public protests have been recently raised against the presence of the boric acid plants in the Lerma Valley.

Fig.1. Main borate deposits in the Central Andes

Table 1. Typical composition ranges of ulexite ore

Components	Composition ranges (% w/w)
B_2O_3	22 - 28
Cl^-	4 - 15
$SO_4^=$	1 - 6
Insoluble in acid	9 - 34

2 Boric acid present production process

Boric acid is a white crystalline powder produced by leaching the ulexite, or other borate, with a sulfuric acid solution of pH=3. The reaction is:

$$Na_2O.2CaO.5B_2O_3.16H_2O + 3\ H_2SO_4 \dashrightarrow 10\ H_3BO_3 + Na_2SO_4 + + 2\ [CaSO_4.2H_2O] \quad (1)$$

Typical composition ranges of the ulexite ore used as raw material are given in Table 1. The reaction is carried out in agitated tanks, usually cylindrical in form, where paddle agitators are normally used and swirling is prevented with baffles. The system temperature is kept in a level between 60°C and 80°C with steam coils to prevent the boric acid crystallization in the leaching tanks. This is necessary because its solubility vs. temperature curve has a very large positive slope.

The sodium sulfate and the soluble ore impurities such as sodium chloride remain in solution, but calcium sulfate, in the form of gypsum, remains with clays and sand as insoluble residual solids.

Boric acid is recovered using a sedimentation process (a liquid-solid separation process) with an adequate flocculant addition, to separate the clear hot boric acid solution from the residual acid slurry of clays, sand and gypsum. Of course, the sedimentation tanks must be heated with steam coils to avoid the boric acid precipitation. The hot solution is sent to agitated batch crystallizers where it is cooled with water, using cooling coils, to get the boric acid crystallization. The boric acid crystals are separated and washed by centrifugation from the mother liquor. Finally, the boric acid is dried with hot air in a direct rotary dryer. Generally, parallel flow of air and solids is used in the dryer to avoid the H_3BO_3 dehydration to HBO_2 (metaboric acid), which starts to occur if the crystals are heated to more than 95°C.

A flow sheet of the present production process is shown in Figure 2.

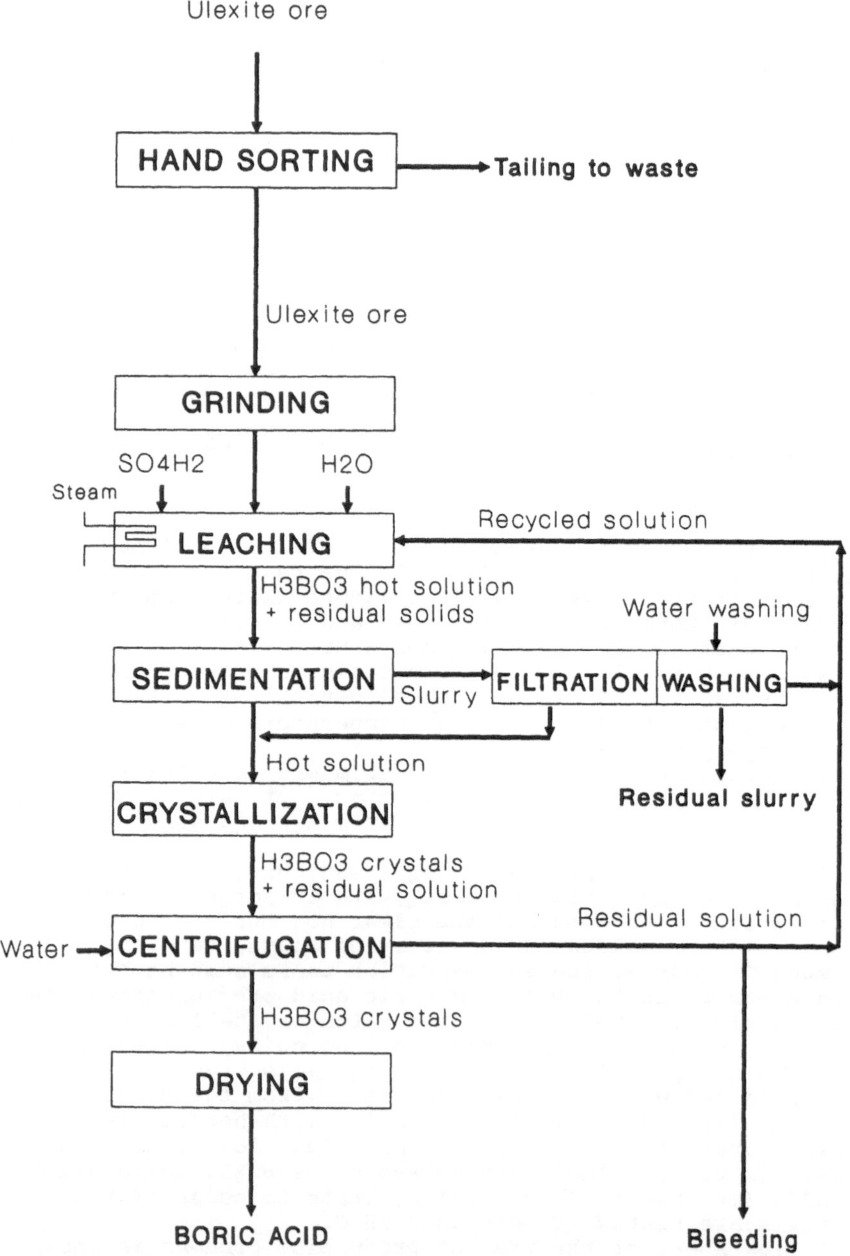

Fig.2. Boric acid present production process

The chemical composition of the residual solution sent back to the leaching tank is controlled through periodical bleedings. This step is necessary to prevent the boric acid crystals impurification due to the co-precipitation of Na_2SO_4 and NaCl: the drained flow is the required amount to keep the concentration of these salts below the saturation condition. The waste residual solution brings an important decrease in the B_2O_3 recovery and, according to the explanation given in the introduction of this paper, it produces pollution in the ground water, rivers and soil in the surroundings of the industrial plant. Another important pollution problem is due to the residual acid slurry.

3 Boric acid alternative production process

In order to solve, or to minimize the economical and pollution problems found in the present industrial process used in Argentina, an alternative boric acid production process has been studied. The differences of this process with the present one are the following:

Concentrated ulexite is used as raw material instead of ore, with a grade between 36% to 40% of B_2O_3.
Anhydrous sodium sulfate is obtained as a by-product using a fractional crystallization process: vacuum crystallization for H_3BO_3 and evaporating crystallization for Na_2SO_4.

Experimental work in laboratory and pilot plant scales has been done to get the required technical information to develop the proposed production process.

3.1 Ulexite solubility

The solubility of ulexite in water has been studied experimentally and it was found that the mineral is soluble in a small degree, only 4.0 % at 20°C. Its solubility increases with the temperature. This fact is very important because it indicates that the losses of ulexite in a wet concentration process, due to its solubility, would be very low.

3.2 Ulexite concentration

The concentration experimental work has been done using ulexite samples with chemical compositions that correspond to the feed composition ranges given in Table 2. A series of concentration tests was carried out using gravity concentration equipments to find out information about the B_2O_3 grade in the concentrate product and the recovery of B_2O_3. The gravity concentration process was chosen to take advantage of the large difference between the sedimentation velocities of ulexite particles and impurities particles present in the ore used as process feed. This is because ulexite is a light mineral, with a specific gravity of 1.7.

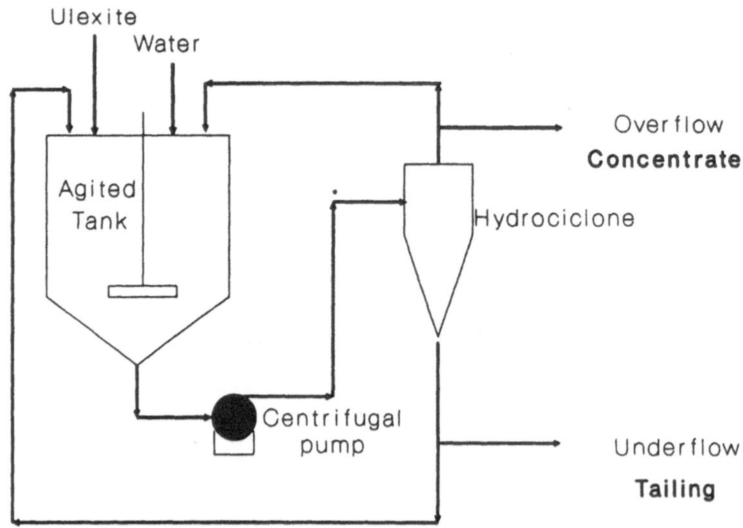

Fig.3. Ulexite concentration. Experimental equipment

Wet concentration equipments were selected to perform the separation process because they have two advantages regarding the dry concentration equipments: a) The soluble ore impurities are dissolved in the process water, and then they are separated from the ulexite, b) A wet concentration process is more efficient than a dry process.

The best results were obtained using a hydrocyclone as concentration device. The experimental equipment used is represented in Figure 3. A pulp with 10% of solids concentration is prepared in an agitated tank and then it is pumped with a centrifugal pump to the hydrocyclone. The ulexite concentrate is the overflow, and the underflow is the residual product. Samples were taken from both streams to determine the B_2O_3 grade in the concentrate, tailing and the recovery of B_2O_3. Some typical results are shown in Table 2.

Table 2. Summary of results

Product	Weight (%w/w)	Chemical Analysis (%w/w) B_2O_3	Cl^-	$SO_4^=$	Insol	Recovery (%B_2O_3)
Feed	100	22-28	4-15	1-6	9-34	100
Concentrate	47-58	36-40	.2-.9	.4-.7	4-16	78-87
Tailing	15-40	7-22	1.4-2	1.9-7	38-65	8-17
Losses	9-30	-	-	-	-	5-13

It can be seen that the concentrates obtained have B_2O_3 grades of 36/40%, the B_2O_3 recovery ranges from 78% to 87% and that the impurities contents have been largely reduced. It is a very important fact that the elimination of Cl^- as soluble impurity, is about 95%.

Consequently, the studied concentration process is convenient to prepare the feed for the boric acid manufacture, because it minimizes slurry production and avoids the coprecipitation of NaCl with the boric acid crystals.

3.3 Fractional crystallization of sodium sulfate

It is well known that when two or more solutes are present in a solution it is often possible to crystallize one of them and leave the others in solution. The solubility equilibrium diagram of H_3BO_3-Na_2SO_4-H_2O system at different temperatures is shown in Figure 4. For example at 40°C, solutions represented by any point of line AB are saturated with respect to boric acid but not with respect to Na_2SO_4. The points of line BC represents solutions saturated with respect to Na_2SO_4 but not with respect to H_3BO_3. Point A corresponds to the solubility of H_3BO_3 in water and C to Na_2SO_4 solubility in water. Point B corresponds to a saturated solution of both solutes at 40°C. It can be seen that there is a higher influence of temperature on the H_3BO_3 solubility than on the Na_2SO_4 solubility, and it is observed a little influence of one of the solutes in the solubility of the other one. The reverse solubility of Na_2SO_4 over this temperature range (40/100°C) is important for the fractional crystallization process.

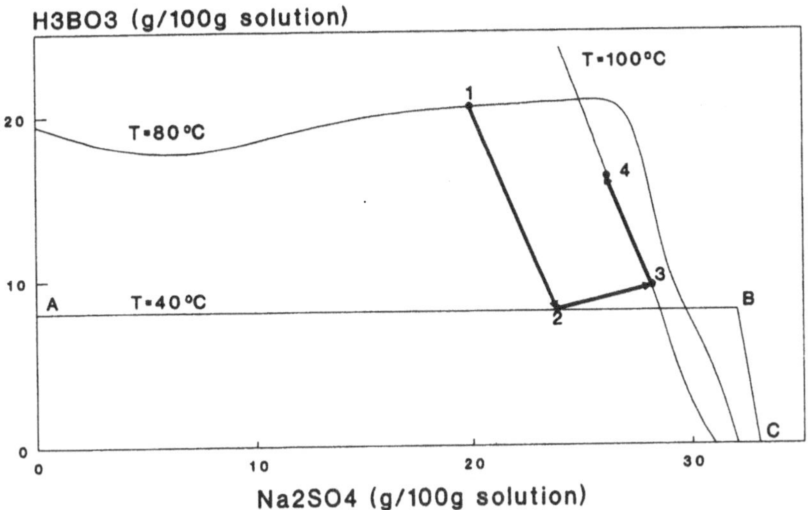

Fig.4. Equilibrium diagram H_3BO_3-Na_2SO_4-H_2O

The composition changes of a solution in different experimental processing stages are shown in thick lines. Line 1-2 represents the vacuum crystallization of boric acid; point 1 represents the original composition of the H_3BO_3 saturated solution at 80°C and point 2 the final composition of the residual solution at 40°C. Here, the solution cools adiabatically to the boiling temperature that corresponds to the vacuum in the vessel (40°C is the boiling temperature under a pressure of 0.075 bar). The cooling results in crystallization, not only because of the cooling of the solution, but also because of evaporation. Line 2-3 represents the evaporation of the residual solution at 100°C in the next process stage. The concentration of both solutes increases in the same ratio until the composition 3 is reached. Here the solution is saturated with respect to sodium sulfate. Further evaporation precipitates sodium sulfate and the composition of the solution moves from 3 to 4. Point 4 represents the composition of the final residual solution of sodium sulfate crystallization.

3.4 Alternative production process

According to the results of these studies, it is possible to develop an alternative process to produce boric acid using ulexite concentrate instead of ore and fractional crystallization of H_3BO_3 and Na_2SO_4. This permits the partial recovery of the latter and avoids the pollution problems caused by the periodical bleeding of residual solutions. This process is represented in Figure 5.

It can be seen that the whole process is divided in two separated stages: a) ulexite concentration; and b) boric acid production. Preliminary economical evaluations have been carried out and they indicate that the following is the best alternative:

> The ulexite concentration is carried out near the ore deposit (salt deposit). Then the tailing and the process water are recycled to the salt deposit and pollution problems are avoided. Besides this, the transportation cost is decreased (in \$/ton. B_2O_3) because most of the gangue is not transported uselessly through several hundred kilometers of mountain roads.
>
> The boric acid production plant is built in the Lerma Valley, near Salta city, where sulfuric acid, fuel gas, electric energy, etc. are easily available.

The ulexite concentration plant is very simple. It consists of grinding by a hammer mill, dispersion in an agitated tank, concentration in hydrocyclones, filtration and drying. The water bleeding is necessary to avoid the saturation of the process water with sodium chloride, because in such a case the ore sodium chloride would not be dissolved.

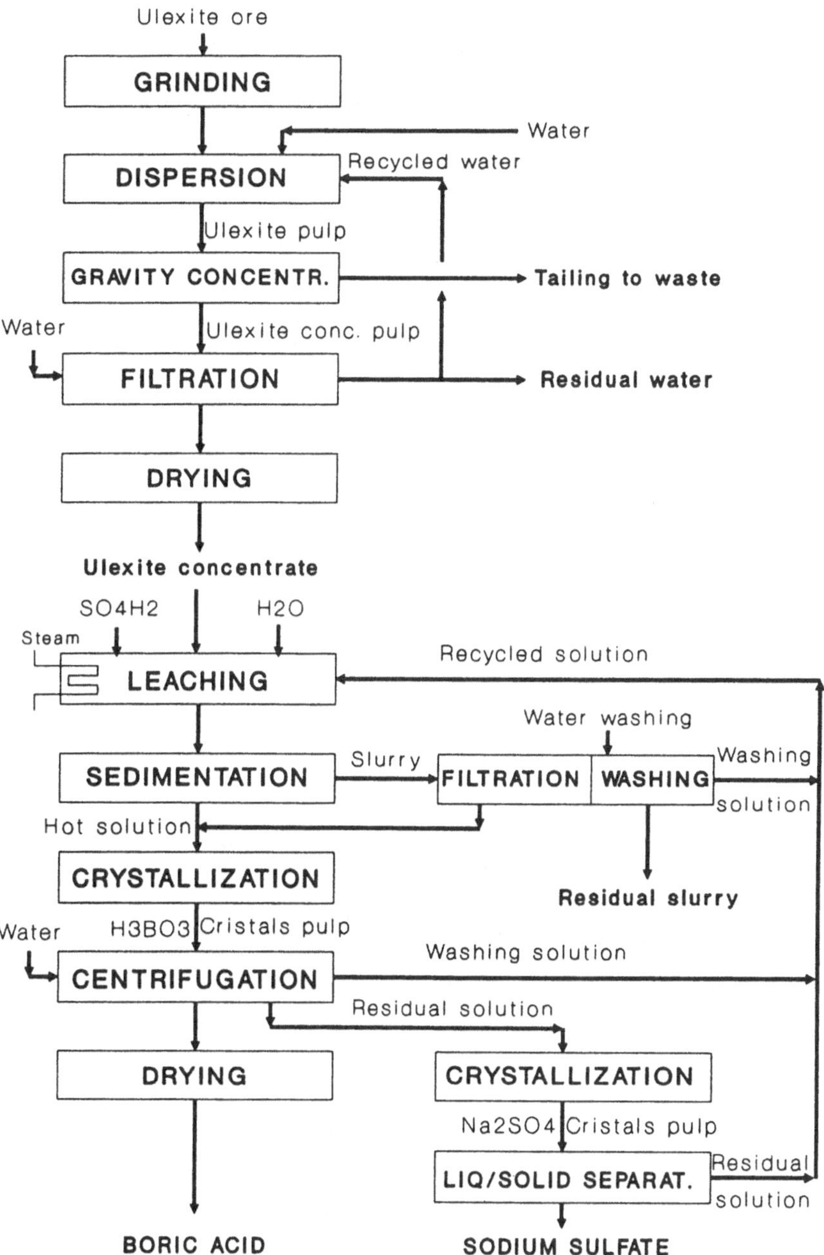

Fig.5. Alternative production process

4 Conclusion

Mass balances of the present boric acid production process and the proposed alternative were performed with a computer. A summary of the most important figures is shown in Table 3. In the alternative process the ulexite has been concentrated up to a grade of 38% B_2O_3, and there is a sodium sulfate production of about 0.4 ton.Na_2SO_4/ton. H_3BO_3 produced. These changes diminished the process residual solids from about 1.4 to 0.7 ton. clay/ton. H_3BO_3 prod., and B_2O_3 in the effluent from 0.20 to 0.02 ton. B_2O_3/ton. H_3BO_3 prod. It is important to point out that the bleeding of residual solution has been eliminated and the only residual stream is the washed clay slurry, whose deposits are totally controlled by state organisms.

The results obtained should be considered as preliminary. At present, experiments are carried out at greater scale to get more information which is necessary to complete the technical and economical project evaluation.

Table 3. Most important figures of both process

	Production process	
Important parameters	Present	Alternat.
Feed to the plant	ore	concentr.
H_3BO_3 recovery(in boric acid plant)	74%	96%
H_3BO_3 recovery(in the whole process)	56%	80%
Boric acid losses in the slurry	10%	4%
Boric acid losses in the bleeding	16%	-
Ton. feed/ton. H_3BO_3 prod.	2.8	1.6
Ton. H_2SO_4/ton. H_3BO_3 prod.	0.63	0.54
Ton. solids(in slurry)/ton. H_3BO_3 prod.	1.4	0.71
Ton. effluents/ton. H_3BO_3 prod.	5.4	1.1
Ton. NaCl in effluent/ton. H_3BO_3 prod.	0.32	0.002
Ton. Na_2SO_4 in effluent/ton. H_3BO_3 prod.	0.30	0.06
Ton. B_2O_3 in effluent/ton. H_3BO_3 prod.	0.20	0.02
Ton. Na_2SO_4 produced/ton. H_3BO_3 prod.	-	0.41

5 References

Adams, R.M. ed. (1964) **Boron, Metallo-Boron Compounds and Boranes.** Interscience Publishiers, New York.

Lyday, P.A. (1985) Boron, in **Minerals Facts and Problems.** Bureau of Mines, United States Department of the Interior, Washington, pp. 91-103.

Seidell, A. Linke, W. (1958) **Solubilities of Inorganic and Metal-Organic Compounds, Vol.I, IV Ed..** American Chemical Society, Washington, D.C., pp. 272-281.

Decontamination of site of a secondary zinc smelter in Torrance, California

E. G. Roche
Pasminco Research Centre, Boolaroo, New South Wales, Australia
J. Doyle
CRA—Advanced Technical Development, Bundoora, Victoria, Australia
C. J. Haigh
Charlestown, New South Wales, Australia

Abstract

Upon closure of the Pasminco Incorporated facility in Torrance, California, it was found that the soil was contaminated with a number of metals, including zinc, lead, copper and nickel. Under local environmental regulations the soil is classified as hazardous and therefore remediation of the site was indicated.

As the site is within the county of Los Angeles a number of constraints have to be met by a soil decontamination process. In particular the supply of water is severely restricted, and limits on the discharge or disposal of wastes are tight.

Pasminco Research Centre has developed a flowsheet, based on leaching the soil with acidified brine, which is designed to operate within these limitations. In addition to removal of the metals to an acceptable level, the process is cost effective and meets all United States Federal environmental requirements.

INTRODUCTION

In the 1970's Pasminco Incorporated acquired the Pacific Smelting Company. One of the assets of this company was a secondary zinc facility located in the city of Torrance, Los Angeles County, California. It was determined in the late 1980's that the operation was not viable and in early 1989 the facility was demolished and the site levelled prior to its intended sale. This could not be completed, however, as it was found that a significant part of the soil on the site was contaminated.

A survey of the site was carried out in early 1990 by International Technology Corporation (IT) to determine the extent of contamination[1,2]. IT identified three types of contamination which would have to be addressed during a clean up operation. These were:

- Contamination of the soil by diesel and gasoline residues near to the location of several old underground storage tanks.
- Presence of very small quantities of gasoline, trichloroethylene (TCE) and related species in the ground water.
- Widespread contamination of the surface soil by zinc, lead and other base metals.

IT found that these three forms of contamination affected separate parts of the site, and it would be possible to consider each as a single isolated clean up problem. In late 1989 and early 1990 Pasminco Incorporated began to examine a range of technical proposals for each of the three clean up tasks.

SELECTION OF A STRATEGY FOR REHABILITATION OF THE SITE

The contamination of the soil by zinc, lead and other metals represents the largest and most difficult task of rehabilitation. This is due in part to the strength of environmental regulation in the U.S.A. and particularly California, coupled with the problem of carrying out sensitive work within the bounds of the city of Torrance.

The location of the site in such an area imposed some unusual difficulties which had to be addressed during the formulation of a flowsheet. The tightness of the local environmental regulations coupled with the length of time required for arranging permits can be an obstacle, as is establishment of acceptable target clean up levels. In addition, major water supply problems have affected the county of Los Angeles for the last few years, and the large amounts of water usually required for a soil cleaning process were thought to be unobtainable.

Under local legislation the soil is classed as a hazardous waste and hazardous waste treatment plants require special permitting, which in some cases can take several years to obtain. An additional problem is presented by the fact that the site is located in a mixed industrial and residential setting, and therefore there is an added public relations challenge to be met when operating a "hazardous waste" treatment plant in such an area. Although both of these issues present a major obstacle to the operation of a plant it was still necessary to consider operating a soil treatment facility on the site itself.

THE NATURE OF THE PROBLEM

The survey of the distribution of the metals in the soil carried out by IT in early 1990 identified six elements of concern which exceeded either the U.S.E.P.A. Total Threshold Limit Concentration (TTLC) or ten times the Soluble Threshold Limit Concentration (10 x STLC) at one or more locations across the site. A total of 55,000 tons of soil was classified in this way. The six metals, together with their approximate average and range of concentrations in this tonnage of soil are given in Table 1 below.

Table 1: Concentration of Metals in the Torrance Site Soil

Element	Analysis (ppm)			TTLC limit (ppm)	10 x STLC limit (ppm)
	Average[1]	Low	High		
Zn	49,000	300	299,000	5,000	2,500
Pb	4,700	32	16,500	1,000	50
Cu	3,100	19	39,600	2,500	250
Ni	765	8	12,600	2,000	200
Sb	<5	<3	1,500	500	150
Cd	20	<0.3	210	100	10

Note 1: Composite sample assembled by Decommissioning Corporation from 11 site survey samples.

PROPOSING A PROCESS

Early in 1990 Pasminco Incorporated retained Decommissioning Corporation of Sparks, Nevada, to investigate in small scale laboratory experiments a range of solvents to dissolve the contaminant metals from the soil[3]. The solvents tested included sulfuric acid, hydrochloric acid, acetic acid, citric acid, ethylenediaminetetraacetic acid (EDTA), ammonium chloride and hydrochloric acid/sodium chloride/calcium chloride mixtures. Decommissioning Corporation found that sulfuric acid, hydrochloric acid and acetic acid each could remove the zinc from the soil to a low level. As all of the metals in the soil, particularly the lead, were soluble in acetic acid to a useful extent, Decommissioning Corporation proposed a soil treatment flowsheet based on this reagent.

In parallel with this initial work, Pasminco Incorporated commissioned Mosmans Mineraaltechniek B.V. in Holland to examine the prospect of reducing the tonnage of material which would have to be leached, by preconcentrating the metals using grinding, flotation and/or gravity separation[4]. Unfortunately this line of development proved impractical at an early stage and work ceased in early 1991.

Decommissioning Corporation's acetate flowsheet consisted of three main stages, including leaching of the soil with acetic acid to extract zinc and other metals, purification to remove the minor metals, then electrowinning of the zinc from the purified zinc acetate solution with simultaneous regeneration of the original acetic acid.

At this point Pasminco Research Centre (PRC) was asked to review the proposed flowsheet to determine whether it would be a practical answer to the soil treatment problem. The conclusion was that it would not fit the requirements of the task for the following reasons:

- Electrowinning zinc from zinc acetate solution was undeveloped technology and due to the poor conductivity of the solution was unlikely to be successful.
- Acetate ions are not likely to be acceptable either in the treated soil or in an effluent from the process. There also is no easy method for destroying acetate ions in either case.
- Acetic acid is an expensive reagent with a pungent smell. The latter point alone would make its use difficult to justify in a residential area.

In response to the acetic acid flowsheet PRC proposed an alternate flowsheet which appeared to overcome the above limitations. The alternative flowsheet, which was based on the use of acidified brine as the leaching agent, has proven to be the most acceptable and technically feasible of those examined.

THE BASIC FLOWSHEET

It is important to consider how the whole of a process will fit together, especially in an environmentally sensitive setting such as is the case here. If a proposed process fails to meet even one environmental requirement then it would have to be rejected. In this case there were a number of difficult constraints which had to be met by the flowsheet. In detail these included the following points:

- No toxic or hazardous effluents or wastes may generated in any quantity. Any byproducts of the process must be benign.
- The metals recovered from the soil must either be converted directly into saleable products or into a form which is compatible with further processing, for example at an existing zinc smelter.
- Because of the water supply problems in the Los Angeles area the use of towns water must be kept to an absolute minimum.
- As effluent disposal costs are high in California the amount of effluent to be disposed down either the storm water or the sewerage systems must also be minimised.
- Due to its location in the middle of a city the impact of the plant upon the

local area must be kept low. In practical terms this means that the plant must be odourless, non-toxic, quiet and inconspicuous.

- While fulfilling these points the cost of the plant must still be kept as low as possible.

With these in mind the generalised flowsheet which PRC considered to be the best answer to the rehabilitation problem is given in Figure 1 below:

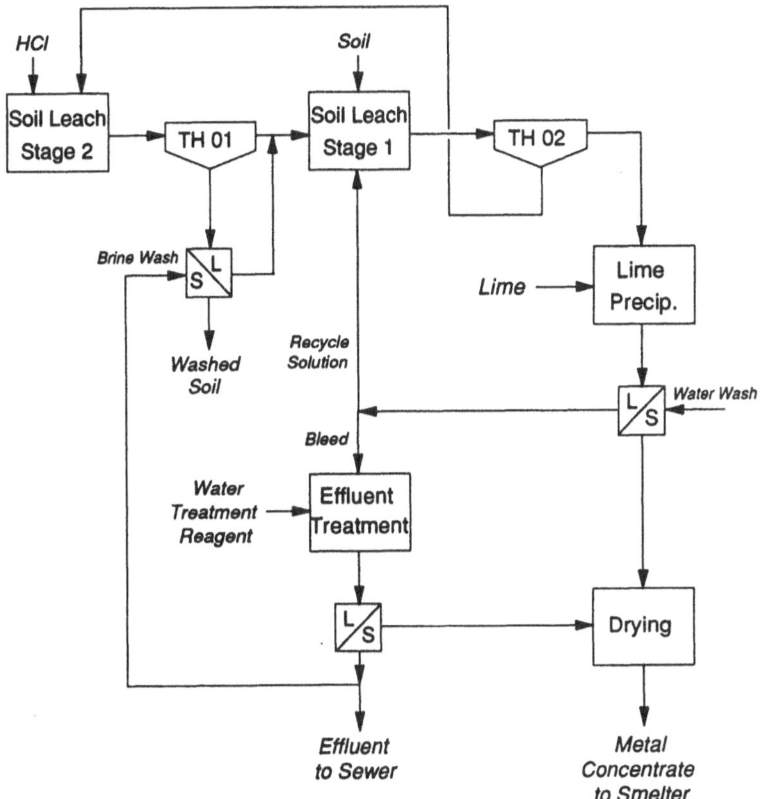

Figure 1 - Schematic Flowsheet

The key features of the process are:

- The soil is leached with a solution hydrochloric acid and calcium chloride in a two stage counter current leaching arrangement.
- The leached soil is filtered and is exhaustively washed with clean brine to displace all of the soluble metals.
- The main portion of the metals in solution are precipitated using lime to give a precipitate containing zinc, lead, copper and other metal hydroxides.
- The metals precipitate is filtered and is washed with chloride free process water. The washed filter cake is dried before transport off site.
- As much as possible of the filtrate from this stage is recirculated to the leaching stages.
- Some filtrate is diverted to an effluent treatment plant to produce an effluent capable of disposal to the sewer system and to generate clean brine for washing the filtered soil.
- Finally the metals precipitate is transported to a zinc smelter for conversion into saleable metal.

The selection of these features in the final version of the flowsheet was driven by a variety of restrictions which became apparent during the development of the process. Development was carried out by integrating bench scale experimental work with computer modelling on the Metsim™ metallurgical modelling package. The findings from this procedure are described below.

1. The Leaching Stages

Three leaching systems were tested during the experimental work: HCl/NaCl, $HCl/CaCl_2$ and $HCl/MgCl_2$. Selection of one system in preference to the others depended upon the outcome of the metal precipitation testwork, since for example if sodium sulfide were to be used as the precipitation reagent the result would be a build up of NaCl in the circulating solution, and therefore the HCl/NaCl would be the leaching system used. Similar arguments also hold for the $HCl/CaCl_2$ and the $HCl/MgCl_2$ systems.

During the experimental work no difference was observed between each of the above mixtures for leaching, although in theory the $HCl/CaCl_2$ should have an advantage due to the higher solubility of calcium chloride. In practice the concentration of calcium chloride which can be used is limited as above a certain concentration there is a detrimental effect upon thickening and filtration related to the higher viscosity of the solution.

The chemistry of leaching is similar for each of the three systems:

$$Zn(OH)_2 + 2HCl \dashrightarrow ZnCl_2 + H_2O \qquad \textbf{E1}$$

Analogous reactions can be written for the other metal species contained in the soil.

In practice this equation is simplistic, since some species such as zinc, copper and lead form a variety of anionic chloro complexes. This is especially important in the case of lead, since the formation of lead chloro complexes overcomes the low solubility of lead chloride and prevents precipitation of the lead from solution by the small quantities of sulfate which usually occur in soil.

Experimental work on the leaching step was carried out by PRC, Decommissioning Corporation and Minproc Technology Incorporated during the course of development of the process. A representative set of results from this work is given in Tables 2a and 2b.

Table 2a: Representative Results From Two Stage Soil Leaching Testwork

Testwork By		Zn (%)	Pb (%)	Cu (%)	Fe (%)	Ni (%)	Sb (%)	Cd (%)
Decommissioning Corp. Feb. 1991	Start	4.9	0.47	0.37	2.95			
	Stage 1	0.74	0.16	0.14	2.76			
	Stage 2	0.11	0.02	0.02	2.00			
Pasminco Research Centre Feb. 1992	Start[1]	7.49	0.75	0.52	2.25	0.16	0.06	0.004
	Stage 1	2.66	0.14	0.10	3.58	0.08	0.21	<0.001
	Stage 2	2.24	0.06	0.03	1.71	0.03	0.04	<0.001
Minproc Technology Dec. 1992	Start	7.05	0.61	0.74				
	Stage 1	5.64	0.35	0.63				
	Stage 2	0.54	0.06	0.07				

Note 1: Average of three analyses of a single sample of soil. Inconsistencies in the analysis reflect the inhomogeneity of the metal distribution in the soil and the particle size distribution (all samples were screened to -4mm but were not ground.

Table 2b: Experimental Conditions

Testwork By		HCl (g/L)	Brine Loading		Solids Loading (% w/w)	Temp. (°C)	Time (h)
			Salt	g/L			
Decommissioning Corp. Feb. 1991	Stage 1	50	NaCl	165	50	25	24
	Stage 2	50	NaCl	165	50	25	24
Pasminco Research Centre Feb. 1992	Stage 1	20	NaCl	250	20	25	3
	Stage 2	107	NaCl	250	30	25	3
Minproc Technology Dec. 1992	Stage 1	15	$CaCl_2$	200	40	60	0.33
	Stage 2	100	$CaCl_2$	100	45	60	3

The original work by Decommissioning Corporation was intended to be a preliminary assessment of the process, and therefore it was only necessary to use roughly appropriate conditions. The work by PRC and Minproc, however, was carried out using conditions intended to be as close as possible to those expected to be employed in the final plant. Computer modelling of the flowsheet was used to estimate these conditions initially, and as better data became available the model was adjusted to give results which closely simulated the experimental data.

The experimental programme carried out at PRC showed at an early date that consistent removal of metals from the soil could only be guaranteed if a two stage counter current leaching process was used. The selection of this approach was also supported by the need for the process to be robust, to be able to handle wide variations of soil composition (e.g. 0.03-30% Zn), and to optimise the consumption of the hydrochloric acid, which is the largest single operating cost for the plant.

Counter current leaching is relatively difficult to control, however, which poses a problem for the actual operation of the plant. On the other hand there is a real advantage in having the strongest HCl concentration available to leach the soil down to the lowest practical metal concentration. The counter current approach is the only realistic way of achieving this.

As is apparent from the above Tables it was generally found that the $HCl/CaCl_2$ or HCl/NaCl leach could remove zinc to well below any limit which would be reasonable to set. The single exception to this occurred when a refractory zinc aluminate known as "gahnite" ($ZnAl_2O_4$) appeared in the soil. This mineral, which is responsible for the poor zinc result obtained by PRC in Table 2b, could not be leached even using extremely harsh conditions such as 100 g/L HCl at 95°C for several hours.

The presence of gahnite in the soil is not considered to be a problem, however, as it has appeared in very few of the soil samples which had been collected. It is also clear that its refractory nature means that it does not represent an environmental problem, since the zinc is not bioavailable or mobile.

2. Filtration and Washing of the Leached Soil

It was found during the experimental work that an important link existed between good washing of the soil and the results obtained in the Toxicity Characteristic Leaching Procedure (TCLP) test[5]. Without good washing the TCLP targets cannot be achieved, since enough metal is contained in the liquor entrained in the filter cake to exceed the limit, even if there is no metal remaining in the soil itself. In practical terms this requires a wash efficiency of the order of 99%, or at least 4 wash displacements each of 70% wash efficiency.

The pressure and vacuum filtration rates determined for the leached soil, and the results of the TCLP test after washing are summarised below in Tables 3a and 3b.

Table 3a: Settling and Filtration Rates for the Leached Soil

Testwork By	Leach Soil From:	Settling Test Results[1]		Vacuum Filtration Rate (kg/m²/h)	Pressure Filtration Rate (kg/m²/h)
		Rate (m/h)	Final S.L. (% w/w)		
Pasminco Research Centre Feb. 1992	Stage 1	3.0	70		
	Stage 2	5.0	65	100	500
Minproc Technology Dec. 1992	Stage 1	6.0	46		
	Stage 2	0.3	47	190	Note 2.

Note 1: PRC results are an average from several tests. PRC settling tests were conducted with 6-30 ppm of Magnafloc 333 as flocculant. Minproc settling tests were conducted using 7-22 ppm Percol 351 as flocculant. "Final S.L." denotes final thickener underflow solids loading.

Note 2: For preliminary design purposes, assumed by Minproc to be approximately twice the vacuum filtration rate.

Table 3b: TCLP Test Results (in ppm)

Conditions of Washing on the Filter	Zn	Pb	Cu	Fe	Ni	Sb	Cd
5 displacement wash using 250 g/L NaCl solution.	36	6.3	6.4	2.2	3.7	0.3	<0.1
5 displacement wash with 250 g/L NaCl solution followed by repulping in 250 g/L NaCl solution to 50% w/w and refiltering.	2.5	1.8	0.25	0.1	0.4	0.2	<0.1
TCLP Test Limit[1]:	-	5	25	-	20	15	1.0

Note 1: No standard limit is available for zinc or iron, although in certain cases a limit of 50 ppm for each has been applied.

3. Precipitation of the Metals

This area of the flowsheet represented one of the most important challenges, since it has a direct impact upon the water balance and the need for towns water, the effluent quality and disposal approach, and also upon the overall process economics.

A number of proposals were examined before the final strategy was selected. These included using the following reagents and mixtures:

- H_2S gas
- Na_2S, NaHS or CaS
- Lime (CaO or $Ca(OH)_2$)
- MgO
- "UltraClear" (a reagent containing sodium dimethyldithiocarbamate)

Each of these reagents are capable of precipitating the heavy metals which are leached from the soil. The relevant chemical equations for zinc precipitation are given below. The equations for the other important metals are similar.

$$ZnCl_2 + H_2S \dashrightarrow ZnS + 2HCl \quad \textbf{E2}$$

$$ZnCl_2 + Na_2S \dashrightarrow ZnS + 2NaCl \quad \textbf{E3}$$

$$ZnCl_2 + Ca(OH)_2 \dashrightarrow Zn(OH)_2 + CaCl_2 \quad \textbf{E4}$$

$$ZnCl_2 + MgO + H_2O \dashrightarrow Zn(OH)_2\ MgCl_2 \quad \textbf{E5}$$

$$2ZnCl_2 + (CH_3)_2NCS_2Na + 2H_2O \dashrightarrow 2ZnS + (CH_3)_2NCO_2Na + 4HCl \quad \textbf{E6}$$

These chemical equations highlight the reasons for and against the selection of each of these reagents. For example, although H_2S is a toxic, smelly gas it has two important advantages: when it reacts it regenerates hydrochloric acid, thereby greatly reducing operating costs, and also it does not produce a byproduct which has to be rejected from the process. In the case of Na_2S the process produces salt (i.e. NaCl) which has to be disposed of to maintain the mass balance. Similar arguments also affect the use of MgO, lime and CaS. UltraClear (see **E6**), which was proposed by Minproc Technology for the effluent treatment duty, is essentially a non-toxic, easy to handle alternative to H_2S gas with similar advantages. However it is quite expensive, and is therefore used primarily for final effluent polishing stages.

Magnesium oxide initially looked extremely promising, since a byproduct "kiln dust" from calcining natural magnesite was sourced by Minproc Technology at only US$25 per short ton. By contrast lime costs approximately US$150 per short ton in southern California. Although the kiln dust was proven to be quite capable of precipitating the metals from solution to low levels, unfortunately it was also found that the precipitate could not be produced with greater than 22% Zn (dry basis). For treatment in a zinc smelter, and to reduce transport costs Pasminco Incorporated required at least 30% Zn in the metals precipitate, and therefore MgO had to be abandoned.

The final version of the process uses slaked lime ($Ca(OH)_2$) for the main precipitation duty. As can be seen in the flowsheet this results in a solution which is mostly recycled to the leaching stages, and therefore does not need to be especially "clean". On the other hand it is essential that both the final effluent from the process and the "clean brine" wash solution contain very low concentrations of heavy metals. This is achieved by treating a bleed stream with UltraClear as shown.

As mentioned the effect of using a precipitation reagent such as lime is that a salt is produced, in this case calcium chloride. It is apparent from equation **E4** that for every two moles of HCl added to the process for leaching purposes it is necessary to use at least one mole of $Ca(OH)_2$, which in turn results in the production of one mole of $CaCl_2$. It is necessary to reject this calcium chloride from the circuit to prevent it from building up to very high concentrations.

Although the conventional approach is to discard the excess calcium chloride in the form of an effluent, this is not entirely practical in this case, since for every tonne of water contained in the discarded effluent it is necessary to add one tonne of clean water to the process. As there is a tight limit on the amount of water which can be obtained in Los Angeles a compromise approach was adopted, whereby the leached soil would be washed with cleaned brine rather than fresh water. This meant that part of the calcium chloride would be rejected by entrainment in the soil, and the amount of water needed by the process could be minimised. Although this approach is not ideal it is considered to be the best available, particularly as calcium, magnesium and sodium chloride brines have been used on the site for many years to suppress dusting. A high chloride content in the treated soil is considered acceptable for this reason.

4. Filtration, Washing and Drying of the Metals Precipitate

Metal hydroxide precipitates are often difficult to handle. In this case it was found that the filtration rate was adequate (although slow) at 320 $kg/m^2/h$, but that the filter cake contained as much as 60% moisture. The high moisture content leads to two handling problems: firstly there are significant amounts of chloride present in the residual solution entrained in the filter cake and this can make treatment at a zinc smelter difficult and expensive. Secondly, transport of the metals filter cake at only 40% solids content is costly.

To overcome these two problems it is necessary to both wash the filter cake with fresh water and to dry it. Unfortunately it is not practical to allow the washed cake to dry in an open stockpile due to the need to avoid dusting, and therefore the cost of a drying kiln has to be borne.

5. Effluent Treatment and Soil Washing

Some solution does have to be discarded from the circuit to maintain the calcium chloride mass balance in addition to the material entrained in the filtered soil, and because of the tight environmental limits imposed in California it is necessary to further treat the filtrate from the lime precipitation stage before any of it can be disposed. In the high chloride solution it proved impossible to get acceptable removal of all metals using lime by itself, so UltraClear was adopted for the effluent treatment stage, despite the extra cost. To minimise the cost only a bleed stream is treated rather than the whole of the filtrate, although this does require some additional capital equipment.

6. Transport and Processing of the Metals Precipitate

One alternative considered for further processing of the metals concentrate is shipment of the material to a smelter such as the Pasminco ISF zinc smelter at Newcastle, NSW. The main problem with this approach is clearly the cost of shipping the material from Los Angeles to Australia. However the ability to control the rehabilitation process at all points is considered worth the additional cost.

TARGETS FOR CLEAN UP AND DEVELOPMENT OF THE SITE

Although at present the limits for removing the heavy metals from the Torrance Site soil have not been finalised, Pasminco is satisfied that the treated soil would pass the TCLP test. As Los Angeles is located in an area with low average rainfall it is reasonable to assume that natural environmental leaching will be quite restricted or will not occur at all, and therefore the TCLP test conditions can be regarded in this case as being very pessimistic.

The degree of clean up of a contaminated site is usually related to its environmental impact. For example sites of special environmental sensitivity would require stringent limits to protect wildlife or flora under threat. In this case the contaminated site is located within the bounds of the county of Los Angeles and is reserved for commercial or heavy industrial use and therefore environmental sensitivity is less of an issue.

TECHNICAL DEVELOPMENT OF THE FLOWSHEET

A number of steps were completed from the formulation of the flowsheet through to the development of a practical plant design. In summary these were:

- Outline the basic flowsheet, and the options to be examined (such as different metals precipitation reagents).
- Model the flowsheet variants using the Metsim™ metallurgical modelling package.
- Use the results of the modelling work to design and carry out experiments to determine the metallurgical performance of each unit operation.
- Develop a final flowsheet based on the experimental results.
- Complete any experimental work required to test the final flowsheet.
- Compile a plant design based on this flowsheet, and estimate capital and operating costs to a ±25% order of precision.

Some discussion of the stages in development of the plant is worthwhile. The modelling work was carried out at a very early stage to determine the best way of minimising water usage and to produce sets of experimental conditions to be tested in the laboratory. When the experimental results were obtained these were used to refine the model until the final version of the flowsheet fitted all of the requirements.

Additional experimental work was also carried out by Industrial Compliance Technologies (ICT) of Denver, Colorado under the supervision of Minproc Technology Incorporated to collect the experimental data required to size equipment and estimate the capital and operating cost[6]. The final detailed flowsheet used in the design of the plant is included below (Figure 2).

Lastly it is worth mentioning that second hand tanks and equipment would be used wherever possible. As the life time for the plant is only about 1 year there is no need to select expensive materials. For example, rubber lined tanks would be used rather than fibre-reinforced plastic (FRP) or stainless steel, even though these are the usual materials for a plant operating in a high chloride environment.

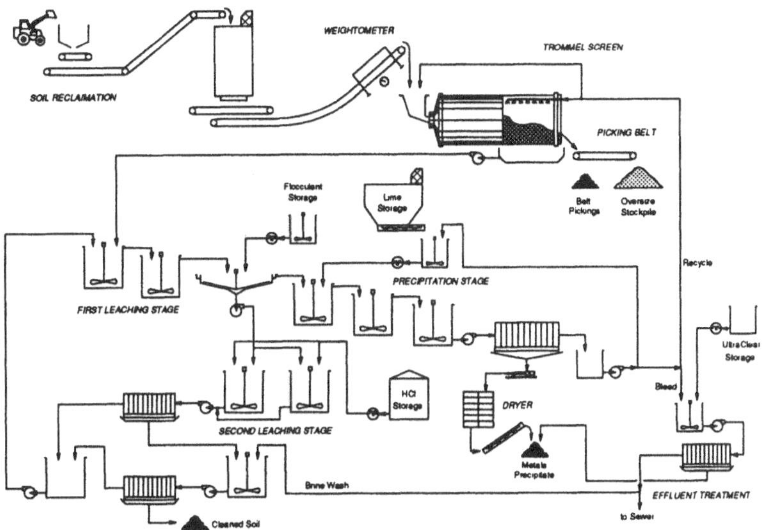

Figure 2 - Process Engineering Flowsheet

CONCLUSIONS

Environmental rehabilitation projects of the type described here are becoming more common as time goes by. The goal of this paper is to give an example of the type of development procedure which may be required to evolve a practical and appropriate answer to such a problem. We trust that we have shown that much care needs to be taken when proposing a metallurgical answer to an environmental problem. Along this line of thought we have considered it more important to explain the reasoning behind the selection of a single viable process from a large number of alternatives, rather than to present a large amount of experimental detail.

We believe it is significant that the process has been shown to pass the TCLP test and would be an effective process in treating soils contaminated with heavy metals. The authors hope that this example will be of use to metallurgists and engineers facing similar tasks in the future.

ACKNOWLEDGMENTS

Pasminco Research Centre gratefully acknowledges the contribution of Minproc Technology Incorporated, International Technology Incorporated, Decommissioning Corporation Incorporated, and Kappes Cassiday and Associates for their contributions to the project described in this paper.

REFERENCES

1. Anon., October, 1992, Feasibility Study Pasminco Incorporated Property Torrance, California, (International Technology Corporation, Irvine, California).

2. Anon., December, 1993, Health Risk Assessment for Closure and Development Pasminco Incorporated Property Torrance, California, (International Technology Corporation, Irvine, California).

3. Anon., 21 February, 1991, Pasminco Torrance Report of Investigations Chloride Leaching Testwork, (Decommissioning Corporation, Sparks, Nevada).

4. Mosmans, C., 27 March, 1991, Preliminary Tests to Purify the Soil of the Torrance California Site, Report to Pasminco Incorporated from Mosmans Mineraaltechniek B.V.

5. U.S. Federal Register, 55 (61), 29 Mar 1990, Rules and Regulations, pp 11863-11875.

6. Anon., January, 1993, Pre-Feasibility Study Torrance Site Clean-Up, (Minproc Engineers Incorporated, Englewood, Colorado).

Disposal of inorganic wastes from sulphate titanium dioxide process at Langöya, Norway

K. L. Sandvik
Department of Geology and Mineral Engineering, NTH, University of Trondheim, Trondheim, Norway
Trygve Sverreson
NOAH Langøya A/S, Holmestrand, Norway

ABSTRACT
Langöya is an idyllic island in the Oslofjord, Norway. The island 3 km long 500 m wide, consists of cambro-silurian limestone, which was mined for cement production, most of the island is therefore today lower than the sea surface. It is a hope one day to restore the island to purely recreational use. The quarry should for this purpose preferably be filled in. It is thereby ideally suited for disposal of inorganic wastes during the next 40 -50 years. The most demanding disposal undertaking so far is the neutralization of the waste acid from Kronos Titan which started in 1989. The operations, as well as some of the pilot testwork operated to design the plant is described here. The major neutralizing agent is limestone, mined and ground at the island while CaO is used for precipitation of ferrous iron hydroxide at pH 8.

Because of the large amounts of gypsum precipitated, Langöya offers the additional possibility for disposal of other metal hydroxides, acid sludge etc. which effectively can be sealed off from the outside after neutralization.
Keywords: Waste Acid, Metal Precipitation, Acid Disposal, Neutralization.

BACKGROUND
It is easy to forget, but the elder of us should still remember that in the not too distant past most chemical plants discharged their effluent into the nearest stream. The most lucky ones situated at the seaside could continue such disposal longer than others. Kronos Titan in Fredrikstad belonged to the last group. The company was and still is producing titanium dioxide pigments from ilmenite ($FeTiO_3$). The ilmenite concentrate comes from the mine Titania A/S. In addition to titanium it contains some extra iron oxide in solid solution as well as magnesium in the ilmenite mineral. Some unwanted minerals are in addition not completely separated from the ilmenite so there will be minor amounts of heavy metals. In the process ilmenite is dissolved by sulphuric acid. Much of the iron

sulphate is precipitated by evaporation and sold for water purification. Titanium hydroxide is removed from the solution by hydrolysis at high temperature. Some of the sulphuric acid is recirculated after concentration by evaporation. The remaining ferrous iron containing waste acid which contains the impurities of other elements has to be disposed of. The quantity now is about 130 000 cubic metres per year. An analysis is given in table 1.

Table 1: Analysis of waste acid from Kronos Titan.

Solution density	1350 kg/m^3	$CaSO_4$	0.072%
H_2SO_4	310 g/l 23.0%	$Cr_2(SO)_3$	0.065%
$FeSO_4$	146 g/l 11.0%	$NiSO_4$	0.0058%
$MgSO_4$	2.4%	$ZnSO_4$	0.0039%
$TiOSO_4$	0.86%	$As_2(SO_4)$	0.0002%
$MnSO_4$	0.20%	$CuSO_4$	0.0001%
$VOSO_4$	0.25%	$PbSO_4$	0.00003%

At the time when the waste disposal permit had to be renewed, Aker Norcem cement company abandoned their limestone quarry on the island of Langöya in the Oslofjord not far from Fredrikstad. The floor of the quarry is lower than the sea level due to the mining operations and should preferably have been filled in. The idea of combining the two events was therefore not far fetched. A good plan for disposal required a good process for neutralization and some guarantee that the disposed material would stay in place for the foreseeable future. In Norway we tend to consider the foreseeable future in this respect to be to the next ice age.

The suitability of the island from a geotechnical point of view had also to be established.

PROCESS IDEAS AND TESTWORK.

The waste acid had to be neutralized and the dissolved metals to be precipitated.

Ritcey has reviewed many such processes. According to him precipitation with limestone reduces the volume of precipitate compared to precipitation by burnt lime and limestone also increases settling rate. Final raising of the pH by lime does not change this. Possible methods for neutralization could be;

1.First neutralization by $CaCO_3$ and then pH increase by CaO to 8 which will precipitate sufficient divalent iron hydroxide. (The ferrous route.)

2. Neutralization by $CaCO_3$ and oxidation of iron from the ferrous to the ferric form by air in a flotation cell. The ferric iron would be precipitated by a pH obtainable by

limestone. (The limestone route.)
3. Combinations of 1 and 2. (The ferric route.) Ferric hydroxide may have a better long term stability than ferrous hydroxide.

Common for all processes were that gypsum would be the main and stabilizing precipitate. An excess of neutralizing agent had to be used to ensure longtime chemical stability.

Pilot tests reported by Rein were carried out at the Mineral Dressing Laboratory on all alternatives in continuous operation after NORCEM had done the exploratory testwork. Requirements of the precipitation process were that 80 % of the iron had to be precipitated in the reaction vessels and that pH at the end of the process should be above 8.5. A flowsheet used in the laboratory is shown in figure 1.

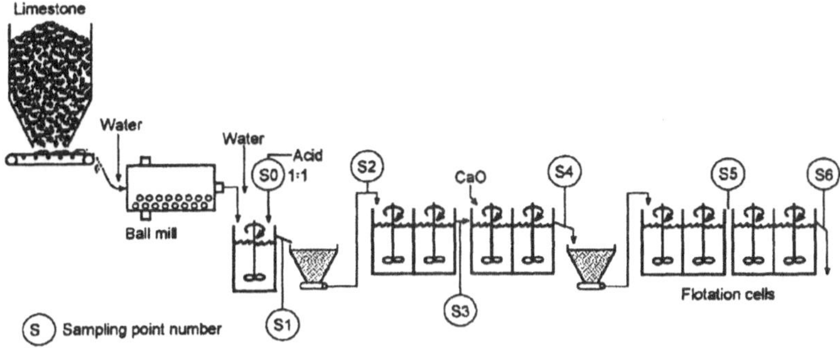

Figure 1. Pilot plant flowsheet for acid ferrous sulphate neutralization.

The flow through the plant was 12 l/h waste acid, but dilution by limestone slurry and water brought the flow at the point of primary neutralization to 48 l/h. Particle size of the limestone after grinding was 90 % finer than 70 micron. The particle size was unimportant for reaction rates provided 90 % weight is passing a sieve not exceeding 150 micron. For neutralization an excess of limestone of about 10 % counted as $CaCO_3$ was used. Addition of lime brought the excess of neutralizing agent to about 40 %. Except for the first reaction vessel Aker flotation cells were used for mixing and aeration. Sampling for Fe^{2+} and pH took place at each transfer point, numbered from 1 to 6, that is after the mixing vessel, after the pump sump and after the 2.nd, 4.th, 6.th and 8.th flotation cells. The amount of ferrous iron in solution was determined by titration of the supernatant after centrifugation. Figure 2 shows the development of pH and ferrous iron with time.

The assumed starting concentration of ferrous ion 14.1 g/l is based upon 1:1 dilution. Dilution was required to keep the received acid well enough dissolved for constant feeding. The drop in ferrous iron ion concentration in the mixing tank is due to further dilution and the spread in values at sampling point 1, reflect the variation in acid strength with time.

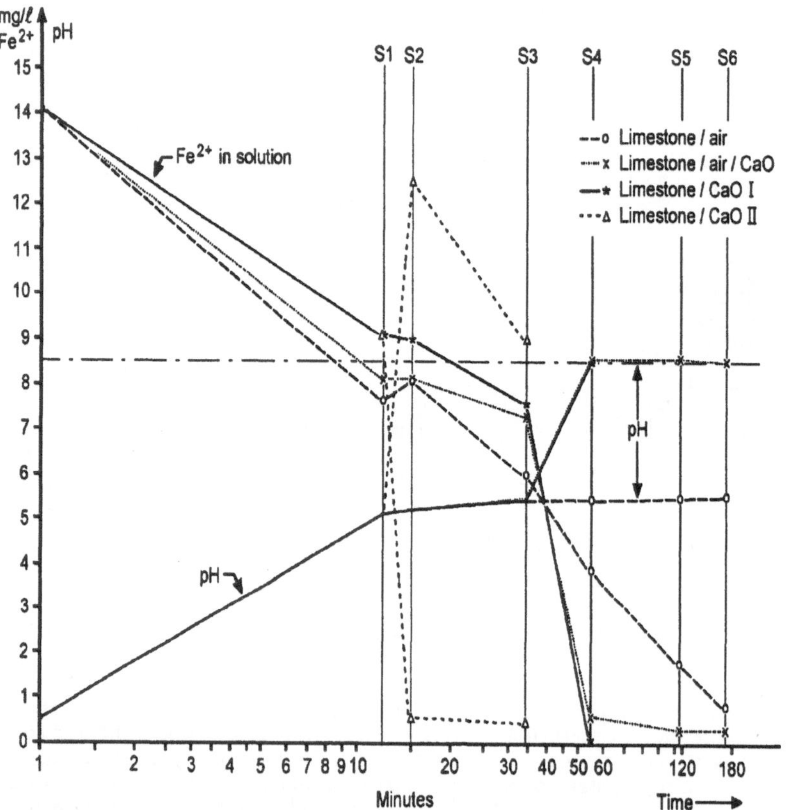

Figure 2. pH and ferrous iron concentration in solution as a function of residence time in the pilot circuit.

It can be seen that it is quite difficult to bring the pH up to 6 due to saturation by CO_2, even when the slurry is aerated, when limestone only is used. Earlier batch tests had indicated that it was easier to reach pH 6 with increased limestone addition (50 % surplus).

Oxidation of the last grams of iron is quite slow in

this case. If CaO is added, ferrous iron in solution disappears more rapidly by precipitation or oxidation. The amount of iron left in solution corresponds rather well to what may be calculated by using a solubility product for ferrous hydroxide of $3.2*10^{-14}$. It can be deduced from the curves in figure 2, that complete oxidation without burnt lime requires a long retention time (several hours). This is also true for oxidation followed by lime addition to remove the last of ferrous ions from solution.

The efficiency of ferrous/ferric oxidation is an interesting point. For oxidation 100-120 l of air was used per l of acid. As the iron content of undiluted acid was about 30 g/l it meant that only 10 % of the oxygen present reacted with iron according to:

$$2Fe^{2+}+2OH^{-} +1/2\ O_2 + H_2O =2Fe(OH)_3$$

The testwork indicated the retention times necessary for the alternative processes, and showed the fastest way to precipitation of iron. It was clear that the ferrous route gave the lower retention times and thereby large savings in process equipment as well as lower operation costs because aeration not was necessary.

GEOTECHNICAL STABILITY OF THE DISPOSAL AREA AND THE PRECIPITATE.

A cross section of the island is shown in figure 2. The wall of the quarry was quite impregnable by water because it was a consolidated limestone and because of the products formed between limestone and ions in sea water. Plugging of fissures by gypsum and ferric hydroxides would seal the area even better.

The stability of the settled sludge was tested on site by preparing test dams filled with ferric and ferrous based slime. This was done on an early stage, before tests to design the process were finished. The experience is that the test material conforms well to the precipitate produced today. The ferrous sludge appeared to form the mechanically most stable residue after settling and consolidation, it had more and deeper evaporation cracks, possibly formed by escaping CO_2. The gypsum texture consisted of unorientated needles giving it a relatively high strength (like clay), regardless of 130 % water content. The oxidised sample had a structure consisting of platelets, less permeable to water. At 20 cm depth the shear strength was 24 kPa for the ferrous and 17 kPa for the ferric residue as reported by Strømme and later Bryhni. This was another reason for choosing the ferrous route.

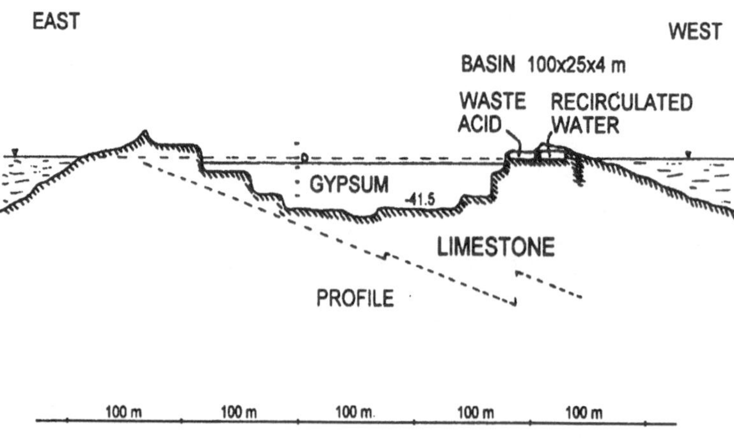

Figure 3. Cross section of the Langöya disposal site.

OPERATIONS

Lime for neutralization of the acid is mined at the site and crushed and wet ground in a ball mill. The waste acid about 150-170 000 tons a year (max. 130 000 m^3) is received in a basin blasted in the limestone structure. Some dilution water may be added to this basin to prevent precipitation of sulphates. The solution is fed to the neutralization plant. This plant operates much the same way as the pilot plant when the ferrous route was tested. Total water dilution including the water from the grinding mill is reduced to a factor of 2.5 to 3 from 4 used in the pilot circuit. Dilution is necessary to keep the slurry rheology manageable after precipitation of gypsum which takes place in the first reactor after addition of limestone in a slight excess compared to the stoichiometric quantity. The pH in the solution saturated with CO_2 is about 4. CO_2 gas has to be vented off at this stage. In the following vessel the reaction with limestone is completed. Burnt lime is then added to bring the pH up to 8 which was shown sufficient to precipitate the ferrous iron and other metal hydroxides. The total amount of neutralizing agent is about 115 % of what is necessary for full neutralization. This ensures longtime stability of the precipitate. In figure 4 the flowsheet is given. Reaction vessel no 1 has a volume of 80 m^3 and the following 40 m^3. The flows shown in the figure were the original, chosen after the pilot tests. Optimization during operation has shown that the plant can handle thicker slurries and that extra retention times are beneficial to ensure full precipitation and limit the use of lime and limestone. Today the total flow is in the range 65 m^3/h, which means that the retention time in the first step is about 1.5 hours and in the second about 1 hour.

PROJECT KRONOS TITAN - ACIDTREATMENT PLANT - BURNT LIME PROCESS

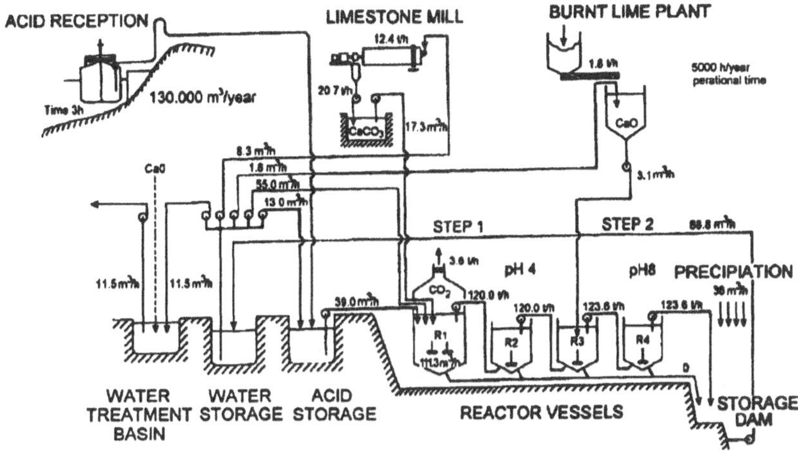

Figure 4. Flowsheet, acid neutralization process.

The effluent from the fourth reaction vessel is discharged to the disposal area.

The water in the system is the water received with the waste, the natural precipitation and the circulating water. Clarified water is pumped to a holding basin from where water to the operation is taken. pH is here 6.5-7 and the iron content 0.2-2 mg/l. Excess water is pumped to a cleaning basin where alkali may be added to precipitate the last traces of metals. The metal content of the decanted discharge is given in table 2, together with the limits for the plating and metal producing industries.

Table 2: Analysis of water discharge from the disposal area to the sea.

	limits mg/l	measurement fall 1992 mg/l
Fe	5	0.2
Cu	1	0.085
Cr	1	< 0.02
Cr^{6+}	0.1	< 0.02
Zn	3	0.088
V	1	0.54
Pb	0.1	< 0.03
Cd	1	< 0.005
Mn	10	3.5
pH		6.5-7

The sludge settles in the disposal area which presently occupies about 250 000 square metres. The total volume of the quarry 9.3 mill m^3 could therefore last the next 40 to 50 years before it is filled up at the present disposal rate. The island may then be restored to something resembling the original before the mining operations began.

It also affects the operations of the plant that part of the island was in 1988 declared a nature reserve because of the interesting botany and fossils in areas that still are untouched. Much of the shore is also used for recreation by boating people during the summer time.

Langöya was until March 1993 owned by the Aker-company NORCEM A/S. At this time the company NOAH took over the property and the processes located at Langöya to develop a good alternative for the disposing of inorganic hazardous wastes and other waste from the Norwegian industry.

OTHER POSSIBILITIES

Because of the large amount of gypsum that effectively seals off the other precipitates, but also because of the reducing environment and a stable pH situation, the possibility exists for depositing smaller amounts of inorganic wastes originating from other places.

Permission is given to add 10 % by weight of other hydroxides.

The divalent iron in the acid will reduce Cr^{6+} ions which are present in small amounts, to Cr_2O_3 which is insoluble as well as from other sources. The excess of divalent iron also presents interesting possibilities, such as precipitation of arsenopyrite. This requires a low concentration of the sulphide ion as referred by Ritcey (1).

Stabilization of Hg containing waste and recovery of some metals are other fields where research are underway. The reducing conditions existing should give living conditions for bacteria reducing sulphate to sulphide. Such reactions are observed elsewhere. They may be a factor contributing to the pH increase in the Løkken minewater from 2.5 to 4.9. Copper content dropped from 48 to 0.42 mg/l at the 430 m level during the three years after the mine was abandoned. Arnesen et al., Norton describes wetland treatment of mine effluent where sulphur reducing bacteria account for the precipitation and fixing of heavy metals. This is another good argument for precipitating iron in the ferrous state. The possibility of turning the interior of the island into a wetland after it is filled up may be an alternative to the "original state". Coastal wetland has to a large degree been filled in Norway so migratory birds have problems to find places to rest and so on.

CHEMICAL STABILITY

There are many methods suggested to evaluate longtime chemical stability of disposed material. However it is not so easy to find the method suitable for a given problem, as it should give a realistic picture of what may take place over a long time under the local conditions. It has also to satisfy the control authorities. This means the test should preferably already be used by the control authorities in some other country. A number of methods were evaluated and the two that appeared most relevant were tested out by Veie. One was the DIN 38414 s4 test where the sample was agitated for a given time in a rotating container. In this case distilled water or sea water was used and the leaching was run for 24 hours in two steps. The other was the TVA Eluatest taken from Switzerland where the sample suspended in a column in water is bubbled through with CO_2 for 2*24 hours. This leaching should resemble the effect of rain water percolating through for a long period of time. A resume of results are given in table 3.

Table 3: Stability of some metals deposited together with gypsum from the plant in standard leaching tests. Concentration in the leach solution is given in mg/l.

	DIN 38414 s4				TVA Eluatest	
	In dist. water		In sea water		CO_2 saturation	
	1.st leach	2.nd leach	1.st leach	2.nd leach	1.st leach	2.nd leach
Fe	0.1	< 0.05	0.2	0.2	646	113
Cr	< 0.05	< 0.05	0.1	0.1	0.1	0.1
Zn	< 0.01	< 0.01	< 0.01	< 0.01	2.6	1.0
Mn	0.9	0.14	3.0	1.10	10.0	7.8
Cu	0.07	0.05	0.08	0.08	0.08	0.05
pH	6.7	6.9	7.3	7.3	6.1	6.6

According to the DIN norms leach water originating from the deposit should contain acceptable levels of heavy metals. The leaching with CO_2 is a tougher requirement. This test leached out very much iron and also a large quantity of manganese. The other metals were even with this test within permissible limits, possibly because the original concentration was low. Oxidation of iron to the ferric form would undoubtly have given more resistance to leaching by CO_2.

CONCLUSION

The final effluent from many chemical processes are containing acids and dissolved metals. Their safe disposal is a major problem that has to be solved according to well established chemical knowledge and local conditions. The disposal plant at Langöya is one demonstrated acceptable solution for such a problem. Long range stability should be is considered good both from a geotechnical and chemical point of view. In addition the availability of ferrous iron offers the possibility of safe disposal of other metal ions, such as 6 valent chromium and arsenates.

There are investigations going on to study the application of waste alkaline materials and the possibilities of recycling materials which are present in the process.

REFERENCES

Arnesen, R.T, Iversen E.R, Källqvist S.T, Laake M, Lien T. and Christensen B.: Monitoring water quality during filling of the Lökken Mine: A possible role of sulphate-reducing bacteria in metals removal. Second International Conference on the Abatement of Acid Mine Drainage. Montreal Sept. 1991.

Bryhni O.: Langöya gypsum disposal, geotechnical investi gations. Norwegian Geothechnical Institute. Report 87766-1. Nov. 9. 1987. (In Norwegian)

Norton P.J: Engineered wetland and AMD. . Min. Env. Management 1. 3. (sept.93). pp.12-13.

Rein A: Neutralization of spent acid. Mineral Processing Laboratory Report Trondheim 23.08.88. (In Norwegian)

Ritcey G.M.: Tailings Management. Elsevier, Amsterdam 1989.

Sverreson T: Langöya - disposal of inorganic waste from industry and other sources. Paper NIF course Trondheim Jan. 1993, Storing and disposal of waste from produc tion and special waste. (In Norwegian)

Strömme: Geotechnical consideration of the gypsum disposal at Langøyà. Report 11.11.1987 (in Norwegian)

Veie,A.: Chemical stability of disposed material. Student project work. Spring 93. University of Trondheim. (in Norwegian).

Effluent treatment in IMI Phosphoric Acid Process

S. D. Ukeles
I. Raz
G. Friedman
L. Kogan
Israel Chemicals, Ltd., IMI Institute for Research and Development, Haifa Bay, Israel

We present in this paper, in general terms, a description of the effluent treatment section of the IMI Phosphoric Acid Process that was developed by IMI. The IMI process, for production of high-grade technical phosphoric acid, involves the attack of phosphate rock by hydrochloric acid. It has been commercialized in Israel and in India. Process improvements and the need to reduce the quantity of solid waste produced by the Wet Phosphoric Acid Process to meet environmental regulations, have recently made the IMI process also attractive for fertilizer acid production. Five tons of phosphogypsum are formed for every ton of phosphoric acid produced by the Wet Phosphoric Acid Process, while the quantity of solid wastes in the IMI process is 10-15 times less. The paper includes: (a) a discussion of the effluents and the solids formed during the neutralization stage; (b) a presentation of effluent treatment results using different neutralizing agent combination; and (c) characteristic flow-sheets for the effluent treatment.
Keywords: Brines, Calcium chloride, Calcium phosphates, Effluents, Neutralizing, Phosphoric acid, Settling, Waste treatment.

1 Introduction

A brief description of the IMI Phosphoric Acid Process, which is used for the production of technical phosphoric acid or fertilizer grade acid [1], is given prior to presenting various aspects of the effluent treatment for this process. A schematic diagram of the process for fertilizer grade acid production is shown in Figure 1.

In addition to the most frequently used sulphuric acid, both hydrochloric and nitric acids can also be employed to digest phosphate rock to produce phosphoric acid. When the latter acids are used, it becomes necessary to separate the product acid from the soluble calcium salts solutions.

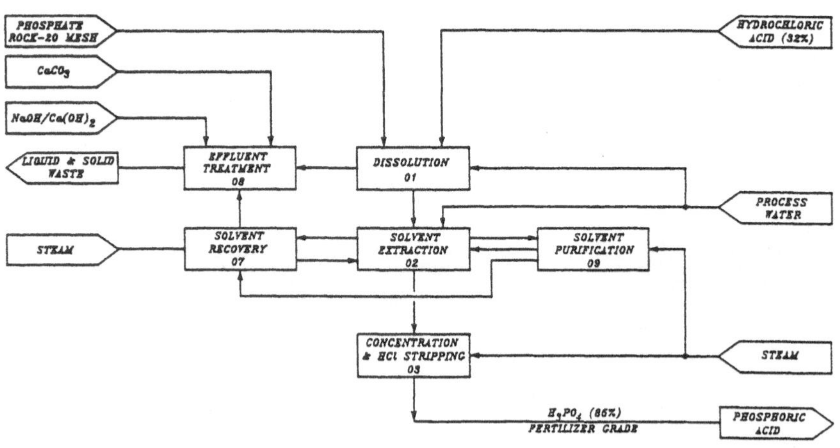

Fig. 1 The IMI process for production of fertilizer grade phosphoric acid from phosphate rock and hydrochloric acid

A block flow diagram of the rock dissolution section of the IMI process is shown in Figure 2.

The dissolution can be represented generally by the decomposition of fluorapatite (1):

$$Ca_{10}F_2(PO_4)_6 + 20HCl \rightarrow 2HF + 6H_3PO_4 + 10CaCl_2 \quad (1)$$

The insoluble residue is washed by counter-current decantation (CCD). Depending upon the water balance, a variety of techniques can be utilized to separate the insoluble materials from the reaction mixture, including sedimentation, filtration or centrifugation. The recovered liquor is returned to the rock dissolution section and the sludge is sent for effluent treatment.

The clarified liquor from the dissolution step contains phosphoric acid, hydrochloric acid, calcium chloride and a large variety of metal chlorides (usually at ppm levels) originating from the phosphate rock. The acids are separated from the salts by solvent extraction with iso-amyl alcohol, giving a dilute purified product and leaving a spent brine (SB) which contains all of the metal salts and small amounts of acids. Treatment of the stripped spent brine is referred to later in this article. The solvent content of the spent brine is recovered by stripping with live steam under vacuum in a distillation column. The recovered solvent is returned to the solvent circuit. The solvent extraction section of the IMI process is shown in Figure 3.

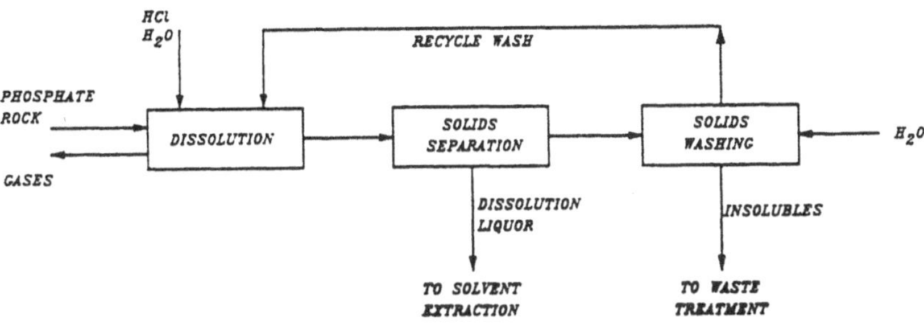

Fig. 2. Block flow diagram of the rock dissolution section of the IMI process

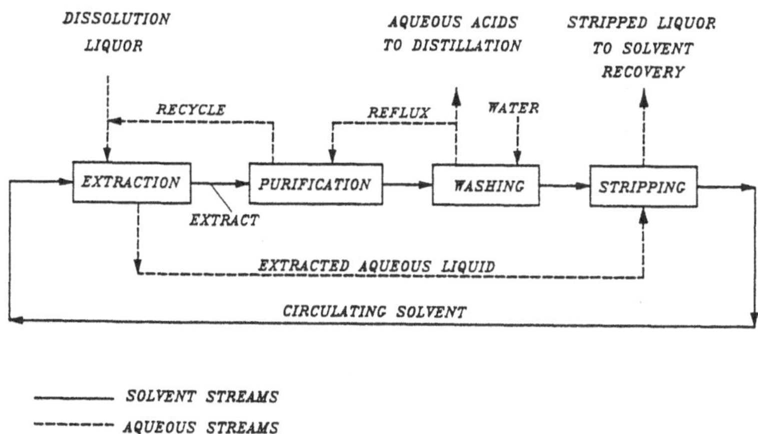

Fig. 3. Block flow diagram of the solvent extraction section of the IMI process

HCl is distilled from the dilute purified product and the phosphoric acid obtained is finally purified by another solvent extraction cycle, leaving an aqueous effluent stream referred to here as the metal removal wash (MRW). The last step of the process is further concentration of the phosphoric acid.

2 Description and treatment of the effluents

2.1 Gaseous effluents

The gases produced in the acid attack reaction consist mainly of carbon dioxide and contain small amounts of HCl and volatile fluorides (HF). These gases are scrubbed with water and sodium hydroxide prior to release into the atmosphere. The solution obtained is treated together with the other liquid effluents. The gases that are evolved in the solvent extraction section and in the concentration system are water-scrubbed to eliminate the solvent vapours.

2.2 Liquid effluents

The liquid effluents consist primarily of: (a) the stripped spent brine (SB) from the solvent recovery section and (b) wash of the metal removal (MRW) section containing mainly ferric, cupric, and zinc chlorides, and hydrochloric acid. There are also minor quantities of scrubber liquors from various sections, which contain HF, HCl, sodium hydroxide, sodium chloride and sodium fluoride.

The stripped spent brine is the largest of all the streams that have to be treated. The phosphate rock composition has a determining influence on the spent brine composition. In Table 1, a range of species compositions is presented for the stripped spent brine. This stream contains mostly calcium chloride, and small quantities of magnesium and sodium chlorides, fluosilicic acid, hydrochloric acid, phosphoric acid, and metal ion impurities.

Table 1. Typical spent brine analysis

Species	Concentration (g/l)	Species	Concentration (g/l)
P_2O_5	1-4	SiO_2	0.5-3.2
Ca	94-100	Mg^{+2}	2-3
Cl^-	175-191	Fe	0.2-0.4
H^+(1)	0.2-0.3	Al	0.52-0.65
F-	0.5-5.8	SO_4	35 mg/l

Note: g/l - grams per litre

All the liquid streams are mixed with solid $CaCO_3$ to attain a pH 2-3, after which the treatment is completed with either (a) a calcium hydroxide slurry or (b) a sodium hydroxide solution to achieve a pH of 7. Equations describing the effluent treatment are given in Table 2.

Table 2. Relevant equations in IMI phosphoric acid effluent treatment

$2HCl + CaCO_3 \rightarrow CaCl_2 + H_2O + CO_2$
$2HCl + Ca(OH)_2 \rightarrow CaCl_2 + 2H_2O$
$6H_3PO_4 + 10Ca(OH)_2 \rightarrow 3Ca_3(PO_4)_2 \bullet Ca(OH)_2 \downarrow + 18H_2O$
$Ca^{+2} + 2F^- \rightarrow CaF_2 \downarrow$
$MCl_2 + Ca(OH)_2 \rightarrow M(OH)_2 \downarrow + CaCl_2$

During neutralization to pH 7, P_2O_5 is precipitated as hydroxyapatite, fluoride is precipitated as calcium fluoride, and the metal ions are precipitated as metal hydroxides. Thus, the solution after neutralization contains mainly calcium chloride, a small amount of magnesium chloride, sodium chloride, and a few parts per million of metal ions. After addition of flocculant (3-10 ppm), the slurry is generally sent to a thickener in which the solids are concentrated by a factor of 2-2.5, after several hours of settling. The underflow is centrifuged, the solids being sent to a waste pond (see section 2.3). The combined centrifugate and the thickener overflow, which is a clear solution of high quality calcium chloride, are dealt with by either: (a) sea or river disposal or (b) calcium chloride production. The treated effluents can be directed to a river or to the sea, depending upon ecological regulations, since the toxicity of calcium chloride is similar to that of common table salt. The second alternative allows calcium chloride to be recovered as calcium chloride flakes by concentration, flaking and drying. Possible uses for calcium chloride are: de-icer for pavements, dust control and roadway base stabilization, industrial processing, and in oil and gas well fluids.

2.3 Solid effluents

The solid effluents are composed of insoluble matter that is not dissolved by the hydrochloric acid during phosphate rock attack, which is 7-10% of the initial phosphate rock weight. As mentioned above, these solids are separated by decantation and washing. The underflow from the second countercurrent decanter (CCD) contains about 20-27% solids, whose composition is greatly influenced by the composition of the phosphate rock used in the particular process. The solid effluents contain mainly calcium phosphates, silica and calcium fluoride; magnesium, aluminium, iron and other metal chlorides, as well as insoluble organic matter are also present. The liquid adhering to these solids contains calcium chloride, hydrochloric acid, phosphoric acid, soluble fluoride and metal ion impurities.

The underflow from the second countercurrent decanter can be filtered. The filtrate, consisting of diluted dissolution liquor is returned to the rock dissolution section while the filtered wet solids are treated with a calcium hydroxide slurry. The solids are separated by filtration, then combined with solid wastes from the spent brine and metal wash treatments and finally sent together to an earth built pond, from which can be separated all of the neutralized liquid effluents. This pond should be large enough to require dredging only once in several years.

The filtrate from this section is combined with the other treated liquid effluents (section 2.2) and sent together to the sea or to calcium chloride flakes production, as mentioned above.

3 Experimental methods for treatment of liquid effluents

Laboratory studies were conducted using batch and continuous operation. For batch tests, 0.5-1.0 litre of the preweighed effluent sample was mechanically stirred at room temperature (~22-24°C) in a round bottom flask. The effluent pH was initially determined and subsequently monitored during the neutralization-precipitation.

Solid calcium carbonate was gradually added until a pH in the 2.0-2.7 interval was achieved. The neutralization to pH 7 was attained by the addition of calcium hydroxide slurry (10-30% solids) or by the addition of a 40% sodium hydroxide solution.

Continuous tests were carried out in a 3-reactor system for spent brine neutralization. Each reactor had an active volume of 1 litre and was constructed of PVC. The residence time in each reactor was 1 hour. Calcium carbonate was added to the first reactor by a solids feeder. Calcium hydroxide slurry was added to the third reactor by a metering pump. The pH was monitored by a glass electrode in each reactor.

Settling tests were carried out in 100 ml graduate cylinders at room temperature for both the spent brine and metal removal wash effluents after neutralization. The slurry height as a function of time was monitored, following the addition of the appropriate flocculant.

4 Results

4.1 Batch tests - spent brine treatment

Table 3 presents experimental results of reagent consumptions from the batch neutralization of the spent brine effluent using the calcium carbonate/calcium hydroxide scheme. The initial spent brine had a composition that is within the ranges given in Table 1. A comparison of treated spent brine effluents is given in Table 4 for two neutralizing procedures: (a) calcium carbonate to pH 2 followed by calcium hydroxide slurry addition to pH 7; (b) calcium carbonate addition to pH 2 followed by NaOH (40% solution) to pH 4 or to pH 7.

Table 3. Reagent consumptions - treatment of spent brine with calcium carbonate and calcium hydroxide

Expt. No.	$CaCO_3$ g/l SB	pH	$Ca(OH)_2$ g/l SB *	pH	Comments
SBN-14B	20.0	2.0	11.8	7.0	Solid calcium hydroxide
SBN-16	20.8	2.0	6.6	7.0	30% calcium hydroxide
SBN-19	17.0	2.0	6.0	7.0	10% calcium hydroxide
SBN-20	18.5	2.0	5.3	7.0	10% calcium hydroxide

* On a 100% basis

Table 4. Comparison of treated spent brine compositions using calcium carbonate and $Ca(OH)_2$ or NaOH

Species	$Ca(OH)_2$ Neutralization	NaOH Neutralization	
	SBN-14B (to pH 7)	SBN-13 (pH 4)	SBN-4 (pH 7)
Ca^{+2}	107 g/l	101.0 g/l	101.9 g/l
Cl^{-}	189 g/l	188.8 g/l	196.3 g/l
$H(1)^{+}$	Not detected	Not analyzed	1 mg/l
P_2O_5	10.9 mg/l	8.4 mg/l	4.4 mg/l
Al^{+3}	<20 mg/l	<10 mg/l	<10 mg/l
Fe^{+3}	Not detected	65 mg/l	Not detected
Mg^{+2}	Not analyzed	Not analyzed	2.24 g/l
SO_4^{-2}	Not analyzed	Not analyzed	23 mg/l
Na^{+}	Not analyzed	Not analyzed	6.0 g/l

4.2 Continuous tests - spent brine treatment

The continuous laboratory scale tests were run: (a) to verify reagent consumptions determined in batch tests and (b) to determine the effect of temperature on settling rate. Two comparison runs were made: (a) Run SBN-23 at 22°C; and (b) Run SBN-24 at 40-50°C in the first reactor, with the successive unheated second and third reactors being at 38°C and 32°C, respectively. The pH in the first reactor was 1.2-1.3, in the second reactor the pH was 2.3 and in the third reactor the pH was 7.0-7.1.

A summary of the treated spent brine composition for Run SBN-23 is given in Table 5. It is observed that at pH 7, no P_2O_5 is left in the calcium chloride effluent, while only a limited amount of Al^{+3} ions remain. No differences were noted between batch and continuous tests regarding either reagent consumption or settling rate. In addition, no benefits were obtained from raising the temperature in the first reactor to 40-50°C.

Table 5. Composition of treated spent brines - continuous test

Species	Initial	SBN-23	
	g/l	to pH 2	to pH 7
Ca^{+2}	94	Not analyzed	96.7 g/l
Cl^{-}	177	Not analyzed	175.5 g/l
$H(1)^{+}$	0.31	Not analyzed	< 1 mg/l
P_2O_5	4.2	0.34 g/l	Undetected
Al^{+3}	0.65	0.09 g/l	<10 mg/l
Fe^{+2}	0.21	0.11 g/l	Undetected
Mg^{+2}	2.7	Not analyzed	2.17 g/l
SiO_2	2.46	1.18 g/l	0.07 g/l

4.3 Effect of initial calcium and fluoride concentrations on sedimentation of the neutralized slurries

One of the interesting aspects of dealing with the neutralized slurries is the influence of the initial concentrations of Ca^{+2} (mainly $CaCl_2$) and F^{-} (mainly SiF_6^{-2}) on the settling rate and on the compaction ratio of the treated spent brine effluent. The compaction ratio is the ratio of the overflow (O/F) clarified liquid to the underflow (U/F) solids after a given period of gravity settling, usually 24 hours. A series of batch experiments were run at varying initial Ca^{+2} and F^{-} ion concentrations (Table 6) to observe these influences.

Table 6. Neutralization of spent brines with varying Ca^{+2} and F- compositions using 20% calcium hydroxide

Expt. No.	Initial Ca^{+2} g/l	F^{-} g/l	SiO_2 g/l	Base Consumption g/l SB	% Solids	Compaction Ratio 24 Hours
SB-7	57.3	0.35	0.36	7.54	0.72	Blank 3.54:1 Flocc.3.87:1
SB-8	95.0	0.37	0.19	6.94	-	Blank 4.10:1 Flocc.3.65:1
SB-9	95.0	5.12	1.87	16.4	1.73	Blank 0.10:1 Flocc.0.11:1
SB-10	57.5	5.04	1.92	16.0	1.80	Blank 1.11:1 Flocc.1.10:1

Also presented is a series of settling curves that summarize the influence of the initial calcium and fluoride concentrations on the sedimentation behaviour.

Figure 4 shows the settling behaviour from 3 settling experiments carried out on the treated spent brine slurry from Batch SB-7. The settling rate was found to be the most rapid when anionic flocculant AF-1 was used, followed by anionic flocculant AF-2, with the slowest settling rate being observed with the blank (no flocculant added). Figure 5 presents a comparison between the settling rates for SB-7 and SB-8, and shows that in the presence of an increased concentration of Ca^{+2} (i. e. by the addition of $CaCl_2{\bullet}2H_2O$ in SB-8), a lower settling rate is observed when compared to the settling rate for SB-7. The compaction after 24 hours was the same for SB-7 and SB-8 when flocculant was used.

The settling curves for SB-9 and SB-10 are shown in Figure 6. The time scale in this figure corresponds to about 3 days in comparison with almost 6 hours in Figures 4 and 5. Thus at high fluoride concentrations, the flocculant was highly ineffective.

These settling results indicate the following trends:

1. Increasing the calcium ion concentration from 57 g/l to 95 g/l will decrease the settling rate, apparently caused by small changes in the specific gravity and in the kinematic viscosity of the solution. The compaction ratio after 24 hours was apparently unaffected by the increase in the calcium chloride concentration.

2. Increasing the fluoride (and concurrently silica) concentrations, had a highly negative influence on settling rates and compaction ratios after 24 hours, as the amount of solids was doubled.

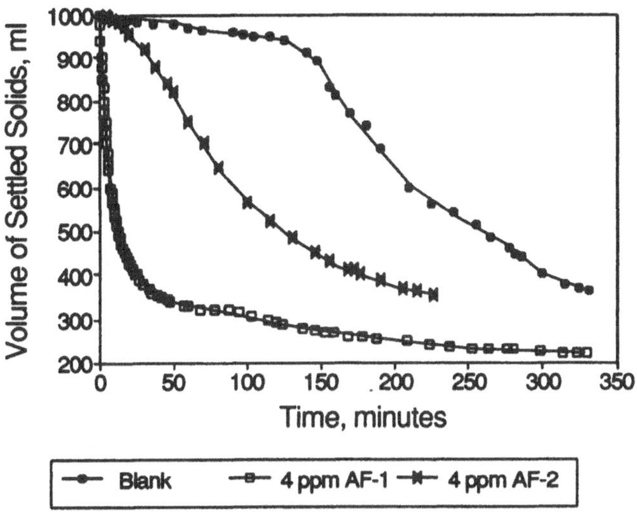

Fig.4. Volume of settled solids versus time for SB-7. Temperature: 25°C

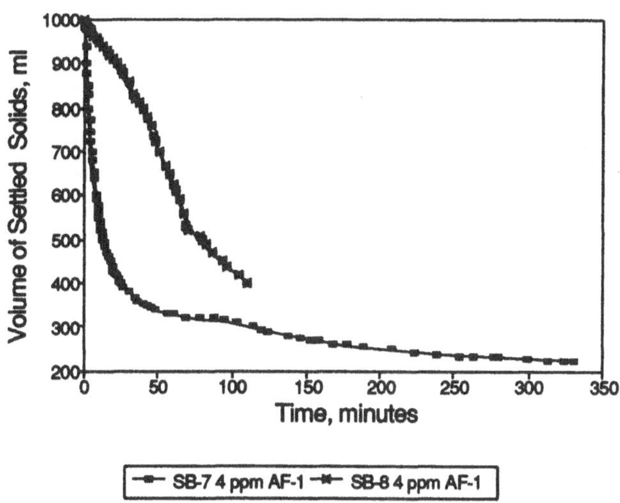

Fig. 5. Volume of settled solids versus time for SB-7 and SB-8. Temperature: 25°C

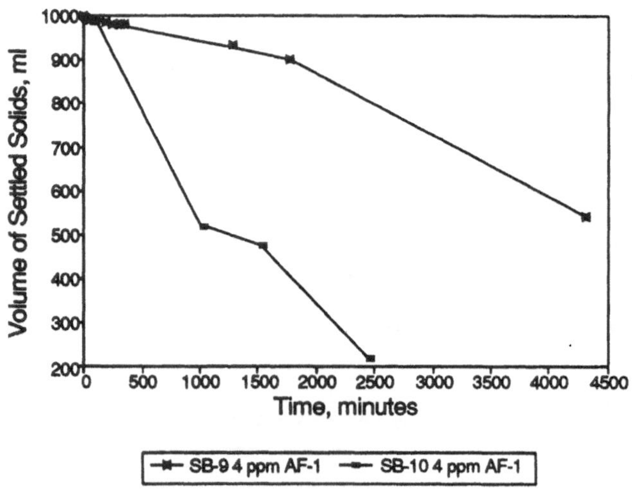

Fig. 6. Volume of settled solids versus time for SB-9 and SB-10. Temperature:25°C

4.4 Metal-removal wash treatment

In addition to the spent brine stream described, in many cases there is also a metal-removal treatment section (using solvent extraction) from which emanates a wash stream requiring effluent treatment. This wash stream contains primarily Fe^{+3} in the chloride form. A series of batch tests using different combinations of neutralizing agents was carried out on a metal-removal wash (MRW) effluent containing about 5 g/l Fe^{+3}, the high iron concentration being due to the use of an iron-rich phosphate. The results, presented in Table 7, show that several types of treatment are viable for this type of stream. A characteristic analysis of the metal removal wash is given in Table 8, using calcium carbonate and sodium hydroxide to treat this effluent.

Table 7. Treatment of metal-removal wash with $CaCO_3$ - base combinations

Exp	$CaCO_3$ g/l MRW	pH	base (100%) g/l MRW	pH	Comments
MRW-1	23.1	2.6	6.2 g NaOH	7.5	NaOH solution
MRW-2	21.5	2.6	6.9 g NaOH	7.3	NaOH solution
MRW-9	16.6	2.0	9.3 g $Ca(OH)_2$	6.5	30% $Ca(OH)_2$
MRW-10	26.5	2.7	3.2 g $Ca(OH)_2$	7.0	10% $Ca(OH)_2$
MRW-11	27.9	2.6	3.3 g $Ca(OH)_2$	7.0	10% $Ca(OH)_2$

Table 8. Analysis of the treated metal-removal wash

Component	Metal-removal wash
$CaCl_2$ % (w)	1-2
Ca^{+2} g/l	8
Mg^{+2} g/l	20 mg/l
Cl^- g/l	20
H^+ mg/l	1 (max)
Fe^{+3} mg/l	0.5
Al^{+3} mg/l	<5
Zn^{+2} mg/l	0.1
Cu^{+2} mg/l	0.2

5 Discussion

Detailed results on the treatment of two of the acidic effluent streams formed during the IMI Phosphoric Acid Process have been presented. The largest stream for treatment is the spent brine stream containing mainly calcium chloride and small quantities of HCl, P_2O_5, fluoride and heavy metal impurities. While the treatment scheme presented for this stream is a combination of solid calcium carbonate and a calcium hydroxide slurry (or concentrated sodium hydroxide), the underlying causes for using a particular type of treatment, are cost and subsequent disposal of the liquid and solid streams. It is also possible to use calcium hydroxide slurry for the entire treatment. The pros and cons for calcium hydroxide vs sodium hydroxide treatment have been discussed previously [2]. Some of the advantages of calcium hydroxide in effluent treatment are its low cost, more settleable and filterable precipitates, and a de-watered cake that is dryer. Sodium hydroxide is often favoured because of its high reactivity, ease of handling and capital costs for feeding and handling being lower than for calcium hydroxide.

The removal of fluoride from effluent solutions can often present challenges because of the presence of various interfering and complexing species which will not allow the fluoride to readily react with added calcium ion [3]. Under the above mentioned circumstances, it will be difficult to reach the theoretical solubility of fluoride ion (as calculated from the solubility product of calcium fluoride) or below, as is usually required by environmental regulations. In such a case, there are other options available such as: (a) using excess of alkali to attain a pH of 11-12 or (b) adding a phosphate source to precipitate a less soluble fluoride compound such as $Ca_{10}(PO_4)_6F_2$; or (c) adding aluminium sulphate to fulfill the role of precipitant and coagulant.

Calcium chloride flakes can be produced from the treated spent brine stream by evaporation of water, hence the importance of avoiding dilution of this stream during effluent treatment. It was found that the use of calcium carbonate above pH 2 is to be avoided because of slow and incomplete reaction. If the production of calcium chloride flakes is a desired goal, then the use of calcium hydroxide is preferred to sodium because it will give more $CaCl_2$ and no sodium impurity will be introduced. In the spent brine treatment, the neutralized solids have to be washed in order to avoid

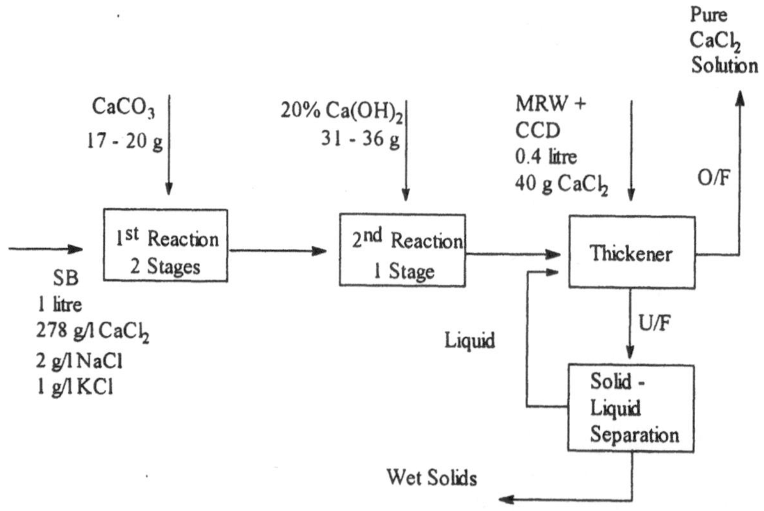

Approximate quantities after treatment	
1.28 Litre treated Spent Brine	Wet Solids
227 g/l $CaCl_2$	25-28 g solids
~ 0.5-2 g/l NaCl	
~ 0.1-1 g/l KCl	

Fig. 7. Spent brine treatment

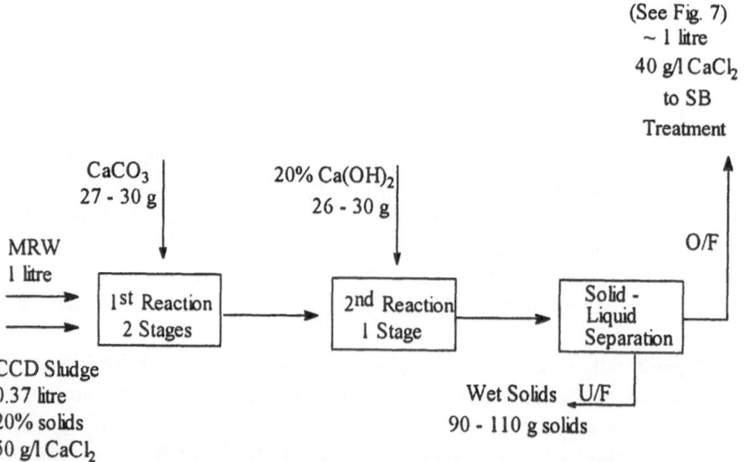

Fig. 8. MRW and CCD treatment

calcium chloride losses and also to reduce the quantities of solids to be disposed of by landfill burial. If these are not washed, calcium chloride from the solution wetting the precipitated solids will also conceivably precipitate as calcium chloride hexahydrate during evaporation in the ponds. The washing can be performed with liquids from the treatment of other effluent streams not containing calcium chloride. A proposed flowsheet is given in Figure 7.

The other two effluent streams, MRW (metal-removal wash) and CCD (countercurrent decanter), can be treated together. The CCD stream contains unreacted phosphate rock particles after the countercurrent decantation of the adhering diluted dissolution liquor. The two combined streams are also treated with calcium carbonate up to a pH of 2.6-2.7 and then with calcium or sodium hydroxide up to a pH of 7. The proposed flowsheet for this part of the effluent treatment is given in Figure 8.

6 Conclusions

We have presented a brief description of the IMI Phosphoric Acid Process along with a survey of some of the effluents (solids, liquids and gases) that are formed during the process. This process is especially advantageous because the quantity of solid wastes produced in the IMI process is 10-15 times less than in the traditional process using sulphuric acid.

Possible schemes for the treatment of the stripped spent brine and metal removal wash effluent, using batch and continuous data, show that a number of alternatives are possible. The option for producing calcium chloride flakes from the treated spent brine has also been stressed. Possible factors influencing the sedimentation behaviour of the treated streams demonstrate the importance of initial effluent composition. This was demonstrated through a comparison of a series of settling curves.

References

1. **Phosphoric Acid Manufacture Using Hydrochloric Acid.** An examination of the IMI Process. Phosphorous and Potassium No. 125 (May-June 1983), pp. 29-33.
2. Mace G.R. and Casaburi D., Lime vs Caustic for Neutralizing Power Plant Effluents, **Chemical Engineering Progress** (1977), 8, pp. 86-90.
3. Paulsen E.G., Reuducing Fluoride in Industrial Wastewater, **Chemical Engineering/Deskbook Issue**, Oct. 17, 1977, pp. 89-94.

Acknowledgement

The authors wish to thank the Managements of Rotem Amfer Negev Ltd. and Israel Chemicals Limited, IMI Institute for Research and Development for permission to publish this report.

Recycling

Metals recovery from hydrodesulphurization catalysts

Francisco Delmas
Materials Department, Instituto Nacional de Engenharia e Tecnologia Industrial, Lumiar, Lisbon, Portugal
Carlos Nogueira
Materials Department, Instituto Nacional de Engenharia e Tecnologia Industrial, Lumiar, Lisbon, Portugal
I. Dalrymple
Environmental Division, E. A. Technology, Capenhurst, Chester, England
J. Parkes
Faraday Centre, Carlow, Ireland

Abstract
Hydrodesulphurization spent catalysts used in sulphur removal from light petrol fractions contain in general cobalt or nickel and molybdenum, as sulphides, in an alumina matrix, together with some hydrocarbons, coke and very small quantities of *tramp* metals, namely lead, vanadium, arsenic and iron.

In this paper, work carried out under a European Community recycling project, envisaging the recovery of the contained metals leaving behind the inert substrate, is described. An overall flowsheet for the treatment of Ni-Mo or Co-Mo spent catalysts is proposed. The process includes pre-treatment by heating, acid leaching, molybdenum recovery by solvent extraction, aluminium removal with lime, precipitation of cobalt or nickel as hydroxides, and finally cobalt or nickel electrowinning.

Influence of aluminium and particularly molybdenum on cobalt/nickel cathodes quality is discussed.
Keywords: Recycling, spent catalysts, leaching of catalysts, molybdenum solvent extraction, cobalt electrowinning, nickel electrowinning.

1 Introduction

Large amounts of cobalt and nickel catalysts are used in petrochemical processes, namely in desulphurization, hydrogenation, steam reforming and methanation. Most of these catalysts are heterogeneous comprising cobalt or nickel and molybdenum, supported on an alumina inert substrate.

These catalysts lose activity or selectivity after use for a period of time. Some possible reasons for that are overheating, deposition of carbon, sulphur and nitrogen compounds and contamination with heavy *tramp* metals such as lead, arsenic, iron and vanadium coming from the feed stock [1,2]. Although a periodic regeneration process reactivates the catalysts, they definitely lose activity after a period of time. Then they must be rejected and substituted by fresh catalysts.

The catalysts used in hydrodesulphurization (HDS) of petrol fractions are small spheres or extrudates essentially composed of nickel or cobalt (2 to 3%) and molybdenum (8-10%) as sulphides, supported in an alumina matrix.

The spent catalysts are usually dumped. The disposal of such residues must take into account their toxicity due to the presence of heavy metals, sulphur, nitrogen and organic materials. So, the disposal of the spent catalysts can be expensive and environmentally dangerous. Furthermore, their metal content can be valuable.

The work reported in this paper, carried out in INETI (Portugal), E.A. Technology(U.K.) and Faraday Centre (Ireland), was aimed at the development of hydro-electrometallurgical processes for the recovery of valuable metals (Co, Ni, Mo) from spent HDS catalysts, leaving behind an inert residue suitable for disposal. The advantages are obvious: (1) the recovery of valuable metals turns an undesirable residue into a source of profit, (2) the recovery of metals from a secondary material decreases the consumption of raw materials and (3) the reduction or elimination of environmental impact.

Some routes for treatment of cobalt/nickel catalysts have already been investigated. Usually the first step of the processes is the calcination for the removal of organics and the transformation of insoluble sulphides into soluble species. Calcination under air access [3,4], chloride roasting [5] and carbonate roasting [6] were investigated.

Some studies were also performed on leaching of catalysts with or without pre-treatment. Acid leaching [7,8] with sulphuric and hydrochloric acids and alkaline leaching [9,10] with sodium hydroxide and ammonia were studied. An alternative to leaching is the pyrometallurgical treatment of the catalysts by chlorination with chlorine or hydrogen chloride [11,12].

The purification of the solutions containing the dissolved metals can be performed using selective precipitation [8] or solvent extraction [13,14].

The process described is this paper consists of pre-treatment of spent catalysts by calcination followed by sulphuric acid leaching to dissolve metals (Co, Ni, Mo). The purification of the solution and separation of the valuable metals was carried out by solvent extraction (in the case of molybdenum recovery) and precipitation (in the case of the removal of the aluminium dissolved from the substrate). Molybdenum was recovered by precipitation from the strip liquors as calcium molybdate or ammonium molybdate, and cobalt or nickel were recovered as metals by electrowinning.

2 Experimental work

The catalysts considered in this work, a cobalt-molybdenum (Co-Mo) and a nickel-molybdenum (Ni-Mo) alumina supported catalysts, are being used in hydrodesulphurization of light petroleum fractions in Portuguese refineries (HDS catalysts).

2.1 Characterisation of spent catalysts

Physical characterisation of HDS Co-Mo and Ni-Mo catalysts was carried out by X-ray diffraction (XRD), both on as received and calcined samples. Scanning electron microscope analyse were also performed to obtain information about the morphology of the catalysts.

Chemical characterisation was carried out by atomic emission (ICP-AES) and atomic absorption (AAS) spectrometries on dissolved samples of catalysts. In all the experimental work, concentrations of processing solutions and residues were determined by the above analytical methods.

2.2 Pre-treatment by calcination

Samples of catalysts were calcined in a muffle furnace, at different temperatures (300°C to 1100°C) and residence times (15 min to 1 h). XRD analyse were conducted to interpret the chemical and physical transformations occurring during calcination. Leaching tests on calcined samples were performed to select the best calcining conditions before hydrometallurgical treatment. The best calcining conditions were tested in a continuous rotary kiln.

2.3 Leaching of catalysts

Leaching of as received and calcined samples were performed firstly in orbitally shaken flasks at several temperatures, and using several leaching agents (sulphuric and hydrochloric acids, sodium hydroxide, hydrogen peroxide and ferric solutions). The best routes selected (acid and oxidising leaching) were tested in stirred batch reactors, at constant temperature and controlled pH and redox potential. The influence of leachant concentration and pulp density were studied. Leaching yields were determined from the analysis of solutions and leach residues. Sulphuric acid leaching on calcined catalysts was tested in a bench-scale continuous leaching plant, provided with automatic control of pH and temperature, at flow-rates of about 2.5 l/h.

2.4 Purification of leach solutions

Experimental work on molybdenum separation by solvent extraction was performed in several steps. Firstly, organic/aqueous contacts with several acidic extractants (D2EHPA, EHEHPA and CYANEX 272) were performed at O/A ratios of 1/1, in separatory funnels, to screen the best extractants in terms of capacity for Mo extraction and in terms of selectivity for Mo vs. Co. Secondly, extraction and stripping isotherms were calculated for the most promising extractants using synthetic and real leach solutions. Ammonia, ammonium carbonate and sodium hydroxide solutions were tested as stripping agents. Finally, continuous countercurrent extraction of molybdenum from real leach solutions was carried out in a bench-scale battery of PVC mixer-settlers, at flow-rates of 2 to 5 l/h.

The battery comprised an extraction section (3 stages), a stripping section (1 stage) and a equilibration section (1 stage). The equilibration section was used to convert the ammonium salt of the extractant to its acidic form, before recycling it back to extraction. The aqueous feed solution was the leachate obtained in the continuous leaching testwork. The stripping and conditioning steps were carried out with appropriate ammonia and sulphuric acid solutions, respectively.

Tests on aluminium removal from leach solutions were performed with several precipitating agents, namely calcium carbonate and calcium hydroxide. Some precipitation curves (concentration vs. pH) were determined. The precipitant selected was used for aluminium removal from Mo free leach solutions.

Further quantitative precipitation with lime of cobalt or nickel hydroxides from purified solutions was carried out to produce a feed material to be used in continuous electrowinning experiments. The pH for precipitation was optimized.

2.5 Cobalt and nickel electrowinning

Cobalt and nickel electrowinning was studied in several types of experimental cells (tank cells, divided cells and "Chemelec" cells). The influence of concentration, pH, temperature and impurities (molybdenum and aluminium) on current efficiency was evaluated. Electrochemical production of valuable salts instead of metals was considered.

The recovery of cobalt and nickel metals from solutions coming from the processing of HDS catalysts was carried out in a continuous electrolysis / dissolution integrated unit (bench-scale) composed of two stirred reactors and one electrowinning cell, at controlled flow-rates, current, potential and pH.

3 Results and discussion

3.1 Characterisation

The spent catalysts tested were black spheres or extrudates containing the metals (Co or Ni and Mo) in sulphide form, in a matrix of alumina. Some organic materials and coke (less than 5%) were present too. The chemical analysis and some characteristics of the catalysts are presented in table 1.

Table 1. Composition of the as received Co-Mo and Ni-Mo HDS catalysts.

Catalyst Ref.	Catalyst type	Composition (%)				Appearance	Dimension (mm)
		Mo	Co	Ni	Al		
UOP-S9	Co - Mo	6.8	1.8	-	42	black spheres	~ 1.5
UOP-S16	Ni - Mo	8.5	-	2.5	42	black extrudates	~ 1.5

Some minor elements including lead, arsenic, vanadium, iron and copper were detected at very low concentrations. XRD analysis of the catalysts showed that the as received samples were practically amorphous, the matrix being mainly of hydrated alumina.

3.2 Pre-treatment

Samples were calcined at different temperatures, from 300°C to 1100°C. The transformation of starting hydrated alumina of the matrix to γ and α alumina was observed. As is well known, α alumina is very refractory to chemical attack, so minimizing substrate dissolution.

Concerning cobalt and nickel, the calcination at low temperatures led to the formation of oxides and sulphates. At higher temperatures (between 500 and 800°C) the Ni and Co oxides react with alumina to yield the corresponding aluminates ($CoO.Al_2O_3$, $NiO.Al_2O_3$ and probably $CoO.Co_2O_3$). These species are spinels, which are known to be compounds very refractory to chemical attack.

At 500°C practically all molybdenum was in its trioxide form, which is readily soluble in acid media. Nevertheless, volatilisation of important quantities of MoO_3 begins above 800°C, being this a limitation for the use of high temperatures in calcination.

Calcination at medium temperatures, i.e. about 500°C, was considered the most adequate to achieve acceptable recovery yields and to obtain stable residues free of sulphur, coke and organics.

After calcination, the catalysts were ground to minus 250 μm using a laboratory disk mill, to increase the particle specific area for the following leaching studies.

3.3 Leaching

Several routes were considered for the leaching of hydrodesulphurization catalysts, namely acid leaching and alkaline leaching of calcined catalysts and oxidative leaching of as received catalysts. In acid leaching testwork, HCl and H_2SO_4 were used as leachants at several concentrations (from 2g/l to 200g/l) [15]. Figure 1 shows the main results of the tests carried out with sulphuric acid, on samples of catalysts calcined for 1 hour at 500°C.

As can be seen, yields over 90% for the valuable metals were only achieved with a substantial dissolution of the alumina matrix. However, it is possible to leach about 65-70% of Co, Ni and Mo with only 10% of aluminium co-solubilization, using low acid concentration (about 10 g/l) which seems to be a good approach.

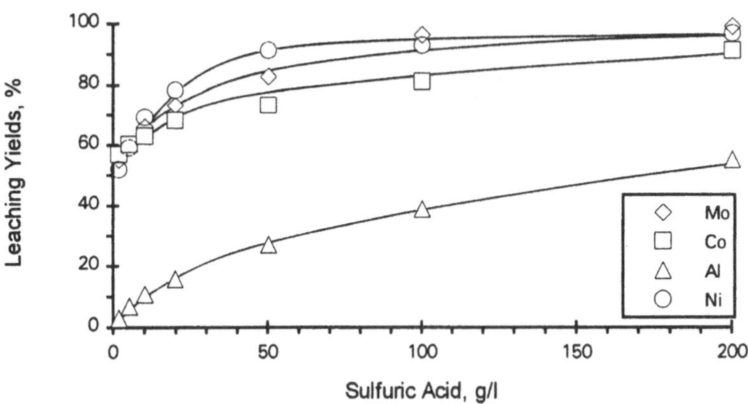

Fig. 1 Leaching of calcined HDS Co-Mo and Ni-Mo catalysts with H_2SO_4, at 80°C for 2 h; pulp density: 20 g solids/litre solution.

Alkaline leaching with concentrated sodium carbonate solutions led to selective solubilization of Mo (80%). Further acid leaching of the residue led to a recovery rate of 55% for cobalt. Nevertheless final solutions were contaminated with residual molybdenum, which must be removed. So, the use of a preliminary alkaline leaching does not eliminate the further necessity of a Mo separation step.

Oxidative leaching of as received catalysts with ferric chloride and sulphate were tested. The results were not significantly different from those obtained on acid leaching of the calcined catalysts. Additionally, the presence of iron in solution would become harmful in further operations, namely in aluminium removal and in molybdenum solvent extraction. Moreover the final residue contained the bulk of original organics, being not suitable for disposal.

So, a weak acid leaching route was selected as the best approach for the treatment of the HDS catalysts. Reaction temperature and time as well as pulp density were optimised for further continuous testwork. A bench scale continuous leaching plant was constructed and tested (figure 2). The average concentrations obtained in test runs as well as leaching yields are presented in the figure.

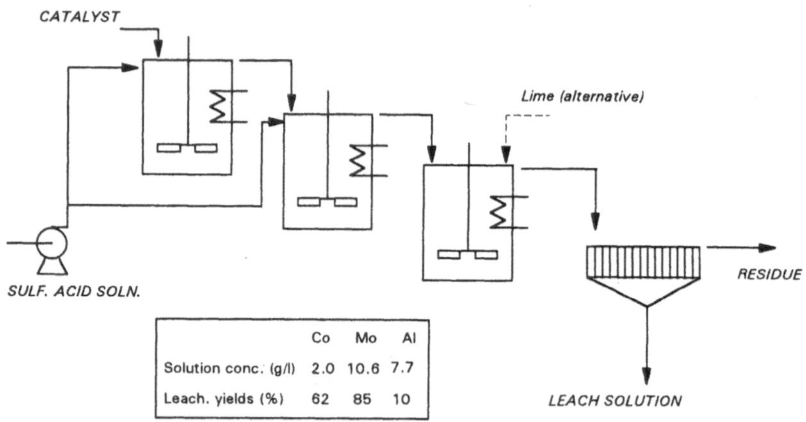

	Co	Mo	Al
Solution conc. (g/l)	2.0	10.6	7.7
Leach. yields (%)	62	85	10

Fig. 2 Continuous leaching of Co-Mo calcined catalyst, at 80°C, pH 1 and using a pulp density of about 200 g catalyst/litre solution.

3.4 Molybdenum solvent extraction

From preliminary tests on screening of extractants, it was concluded that the use of alkylphosphoric acids alone is not suitable for Mo recovery due to the persistent formation of a 3rd. phase during stripping with alkaline solutions. The mixtures D2EHPA/TBP and D2EHPA/n-decanol gave the better results. Figure 3 shows the comparative extraction behaviour of the two mixtures for molybdenum and cobalt extraction.

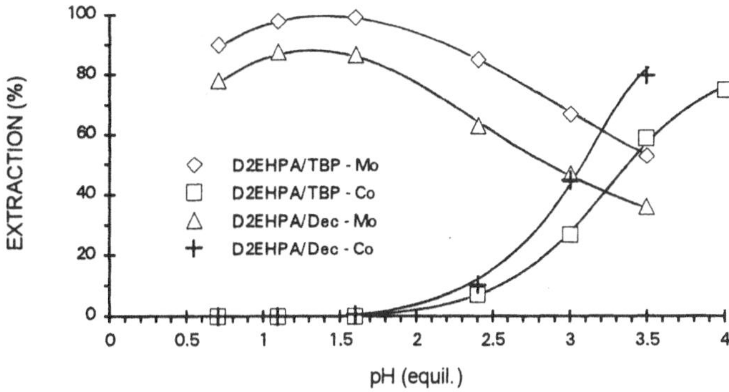

Fig. 3 Extraction of molybdenum and cobalt from leach solutions with D2EHPA 0.5M using two modifiers, TBP and n-decanol, as a function of pH. Org./aq. ratio = 1.

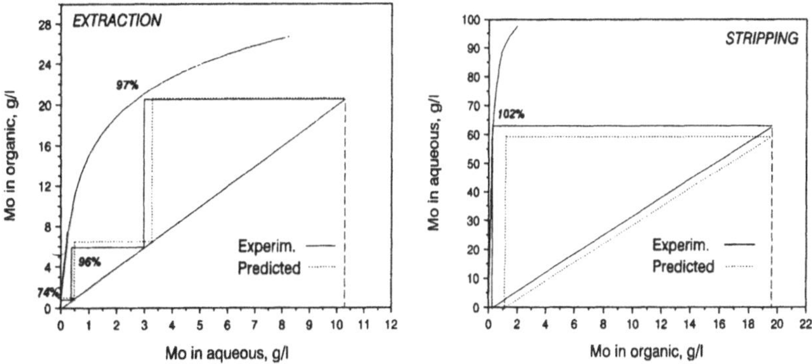

Fig. 4 Experimental and predicted stage compositions and calculated stage efficiencies (%) in the continuous solvent extraction bench-scale plant.

The mixture D2EHPA 0.5M with TBP was used in continuous countercurrent testwork given its better performance and selectivity for Mo against Co. In figure 4 are shown the McCabe-Thiele constructions, including the extraction and stripping equilibrium isotherms, the operating lines, the stage by stage concentrations and the stage efficiencies.

Using the appropriate flow ratios, practically overall extraction of molybdenum (>99%) was achieved without contamination with cobalt or aluminium. The final strip solution was a concentrated ammonium molybdate solution containing 64 g/l of Mo. This solution was crystallised to produce pure ammonium molybdate salt, which was calcined to obtain molybdenum trioxide. Alternatively, calcium molybdate can be obtained by precipitation.

3.5 Aluminium removal and cobalt / nickel precipitation

The Mo free raffinate coming from solvent extraction containing cobalt or nickel and small quantities of aluminium, was purified by selective precipitation, at pH 4 to 5. Lime and limestone were considered as precipitating agents. The results obtained showed that the use of limestone decreases losses of cobalt/nickel by coprecipitation with aluminium. Practically 90% of Al was removed from the solutions, with less than 5% of Co or Ni lost by co-precipitation.

The purified solutions obtained were quantitatively precipitated to produce a precipitate containing cobalt or nickel hydroxides and gypsum, to be used as feed in the following electrowinning bench-scale plant loop.

3.6 Cobalt and nickel electrowinning

Exhaustive batch testwork on electrowinning of cobalt and nickel was carried out envisaging the selection of the best conditions to produce pure cobalt and nickel cathodes at high current efficiencies and without interferences of impurities, namely aluminium and molybdenum [16, 17].

A continuous bench-scale plant was constructed to carry out the electrowinning of cobalt or nickel (separately). According to figure 5, a closed circuit including two basic steps - electrowinning and dissolution - was the approach used. In the electrolysis, a quantity of metal corresponding to a drop of 3.5 to 4.5 g/l in solution was plated. The acidity generated in electrowinning was then used to dissolve the equivalent Co or Ni from respective hydroxides in the dissolution step.

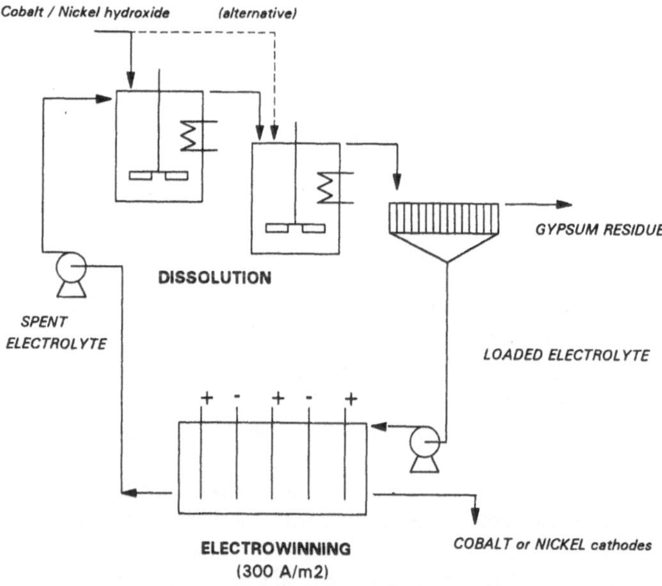

Fig. 5 Continuous integrated electrowinning / dissolution bench-scale plant.

The performance of this plant was optimized in terms of pH and flow ratios (solids/solution) to achieve a closed mass balance. Current efficiencies of 70% (in the case of nickel) and 80% (in the case of cobalt) were obtained.

Concerning cathode quality, high purity (>99%) cobalt and nickel cathodes were produced. The build-up of minor elements in the circuit was followed. It was found that the presence of molybdenum in the circuit is harmful, even at small concentrations. At about 20 mg/l, a decrease in current efficiency to about 60% was found. So, complete removal of Mo in the solvent extraction unit must be achieved. Other impurities present in the spent catalysts such as copper, iron and lead were not detected in the circuit. Aluminium and calcium, although present in electrowinning solutions at equilibrium levels of about 200 mg/l don't interfere in electrodeposition. The cathodes obtained were shiny, peelable and rarely pitted.

4 Conclusions

INETI, E.A. Technology and the Faraday Centre have developed and tested on a continuous basis a hydro-electrometallurgical process for the treatment of HDS spent catalysts, producing good quality cobalt and nickel cathodes, pure molybdenum oxide or calcium molybdate and final inert residues and effluents suitable for disposal or discharge with no environmental risk. Preliminary determinations of leachability of leach residue by water according to the method STD DIN 38414 were made, and the values obtained showed that the residue can be considered inert, providing that an efficient washing operation be made. Concerning the liquid effluents, only a waste water from precipitation of nickel and cobalt hydroxide at pH 8.5 is produced, which can be considered appropriate to discharge.

The process developed led to overall recoveries of 60 or 70% for cobalt or nickel and 85% for molybdenum, without significant dissolution of the alumina substrate, and comprises the following steps:

- Calcination at 500°C
- Acid leaching
- Molybdenum solvent extraction using a D2EHPA/TBP mixture
- Aluminium removal
- Cobalt or Nickel precipitation
- Cobalt or Nickel electrowinning

A simplified diagram of the process is shown in figure 6. The economic feasibility of the process depends strongly from the price of the metals and from the environmental regulations for disposal of industrial wastes.

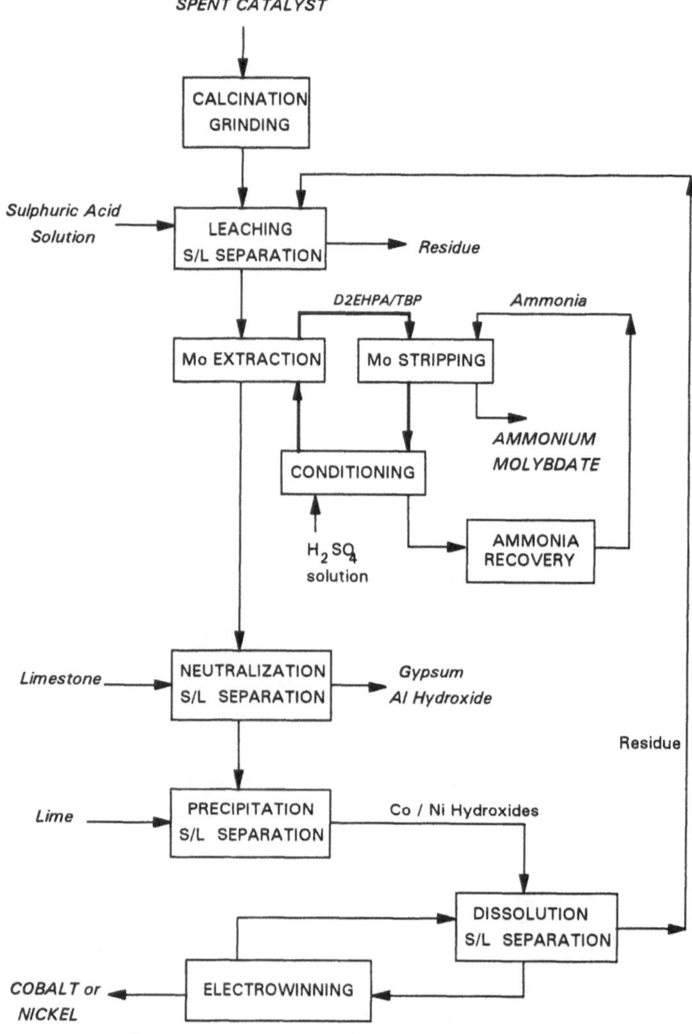

Fig. 6 Overall flowsheet of the hydro-electrometallurgical process developed for the treatment of HDS spent catalysts.

5 Acknowledgements

The authors wish to thank to the Commission of the European Communities, DG XII (Raw Materials and Recycling Programme) for the financial support, and to Petrogal E.P. for the supply of samples of spent hydrodesulphurization catalysts.

6 References

1. Surfleet, B. (1991) Recovery of Metal Values from Spent Catalysts, **Raw-Materials and Recycling E.C. Contract MA2R-CT90-0004,** 1st EAT Report, September
2. Marcantonio, P. J., (1985) Treatment of Alumina-base Catalysts, **U.S. Patent no. 4 537 751**, 27 August
3. Gutulkov, G. (1971) Method of Recovering Metals from Hydrorefining Catalysts, **U.S. Patent no. 3,567,433**, 2 March
4. Raisoni, P. R., Dixit, S. G. (1988) Leaching of Cobalt and Molybdenum from a Co-Mo /γ-Al2O3 Hydrodesulphurization Catalyst Waste with Aqueous Solutions of Sulphur Dioxide, **Minerals Engineering**, 1 (3), 225-234
5. Biswas, R. K., Wakihara, M., Taniguchi, M. (1985) Recovery of Vanadium and Molybdenum from Heavy Oil Desulphurization Waste Catalyst, **Hydrometallurgy**, 14, 210-230
6. Sebenik, R. F., Ference, R. A. (1982) Recovery of Metal Values from Spent CoMo/Al2O3 Petroleum Hydrodesulphurization and Coal Liquefaction Catalysts, **Symposium on Recovery of Spent Catalysts**, September, Kansas City, USA
7. Tilley, G. L., (1982) Method for Recovering Valuable Metals from Deactivated Catalysts, **U.S. Patent no. 4,343,774**, 10 August
8. Parkinson, G., Ushio, S., Hibbs, S. M., Hunter, D. (1987) Recyclers Try New Ways to Process Spent Catalysts, **Chemical Engineering**, 16 Feb., 25-31
9. Toida, S., Ohuo, A.,. Higuchi, K (1979) Process for Recovering Molybdenum, Vanadium, Cobalt and Nickel from Roasted Products of Used Catalysts from Hydrotreatment Desulphurization of Petroleum, **U.S. Patent no. 4,145,397**, 20 March
10. Hubred, G., Van Leirsburg D. (1984) Procedée d'Épuisement par une Solution Aqueuse d'une Solution Organic Contenant des Metaux et un Composé d'Alkylammonium Quartenaire, **Brevet Français n° 2 535 979**, 18 Mai
11. Soc. I. G. Farbenindustrie Aktiengesellschaft (1932) Procedé pour Tirer le Molybdène, le Tungstène et le Vanadium des Minerais et des Matières Similaires qui en Renferment, **Brevet Français no. 724,905**, 4 Mai
12. Jay Welsh, Y., Piquet, P. (1980) Procedé de Récuperation de Metaux à Partir de Catalysers d`Hydrodésulfuration d'Hydrocarbures, **Brevet Européen n° 0 017 285**, 15 Octobre
13. Ritcey, G. M., Ashbrook, A. W. (1979) **Solvent Extraction - Principles and Applications to Process Metallurgy**, Part II, Elsevier Scient. Publ. Comp., Amsterdam, The Netherlands
14. Coca, J., Diez, F. V., Morís, M. A. (1990) Solvent Extraction of Molybdenum and Tungsten by Alamine 336 and D2EHPA, **Hydrometallurgy**, 25, 125-135
15. Delmas, F., Nogueira, C., Coelho, M.C. (1991) The Development of Electrochemical Processes and Integrated Systems for the Recovery of Metal Value from Spent Catalysts, **Raw-Materials and Recycling E.C. Contract MA2R-CT90-0004**, 2nd. INETI Report, December

16. Dalrymple, I., Mitchell, R. (1992) The Development of Electrochemical Processes and Integrated Systems for the Recovery of Metal Value from Spent Catalysts, **Raw-Materials and Recycling E.C. Contract MA2R-CT90-0004,** 4th. EAT Report, July
17. Parkes, J., Breen, W., Dolan, L., Deegan, O. (1992) ibid., 4th. Faraday Centre Report, July

Environmentally sound hydrometallurgical recovery of chemicals from aluminium industry spent potlining

R. J. Grolman
E. P. & P., Chicoutimi, Quebec, Canada
F. M. Kimmerle
Alcan International, Ltd., Arvida Research and Development Centre, Jonquière, Quebec, Canada
G. C. Holywell
Alcan International, Ltd., Kingston Research and Development Centre, Kingston, Ontario, Canada

Abstract
Spent Potlining (SPL) is a waste material generated by all types of electrolytic cells used in the smelting of alumina oxide to make aluminium metal. This material has been traditionally stored on site in many locations. Numerous processes have been investigated to deal with this problem. Some of the solutions involve the treatment of the waste, with chemicals and high temperature incineration, primarily aimed at eliminating cyanides and reducing leach rates of soluble fluorides prior to landfilling. Alcan has taken a different approach. SPL contains recoverable chemical elements (C, F, Na and Al) which are lost in stabilisation or immobilisation processes. Using hydro-metallurgical technology, we have developed a process which recycles the chemicals contained in the SPL. The process consists of grinding and classifying the SPL before digesting it in hot dilute caustic. The resulting slurry is filtered and the residue washed and used as fuel. The filtrate is autoclaved to decompose the cyanides then evaporated to precipitate the sodium fluoride. The sodium fluoride crystals are then filtered from the caustic liquor. This liquor is recycled to the alumina refinery. The sodium fluoride can be sold as is or be redissolved and converted to calcium or aluminium fluoride. This process is not only competitive with existing SPL treatments but is also environmentally superior. The process has been piloted.
Keywords: Environment, Aluminium Industry, Spent Potlining, Chemical Recovery, Hydro-Metallurgical.

1 Introduction

Aluminium is made by the Hall-Héroult process in which aluminium oxide is dissolved in a cryolite bath then acidified with aluminium fluoride. The process is electrochemical and takes place in electrolytic cells consisting of a steel shell lined with refractory bricks, carbon block linings and carbon anodes (see Figure 1). During the life of the Hall-Héroult cells, molten fluoride salts and sodium penetrate into the carbon cathode blocks, ramming paste, the monolithic carbon structure and eventually into the alumina refractory lining or firebrick below. Pot failures occur generally after three to eight years service due to the generated stresses within the pot which allow attack of the iron collector bars and refractory lining by bath electrolyte or liquid aluminium metal. The failed electrolysis cells, or pots, are withdrawn from service in order to replace the cathode lining. As much as possible of the loose alumina, excess bath and liquid metal is removed. Once cooled, the remaining lining is broken up and scooped out of its steel shell. Iron and large aluminium pieces are manually removed and some smelters attempt segregating the carbonaceous material (first cut) from the refractory lining (second cut) but both should be considered to constitute spent potlining (SPL). SPL is now generally recognised as a hazardous substance because it contains "significant concentrations of toxic constituents that are both labile and persistent" (i.e. cyanides and fluorides). In contact with moisture, SPL also has the potential of generating ammonia, hydrogen and other explosive gases. Figure 2 shows the typical chemical composition of SPL. The exploded parts of the chart represent the water soluble salts

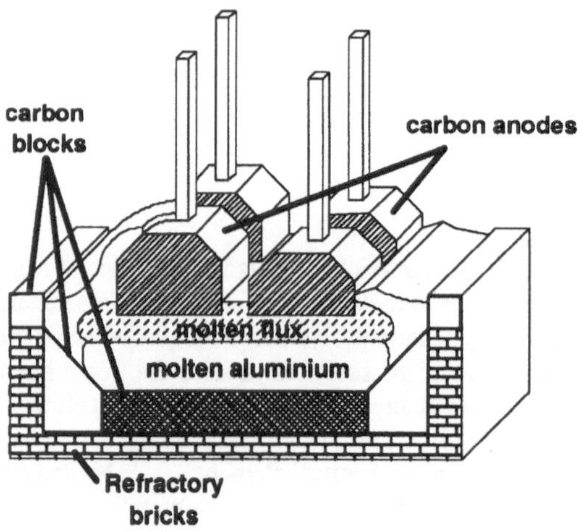

Fig. 1 Electrolytic cell construction

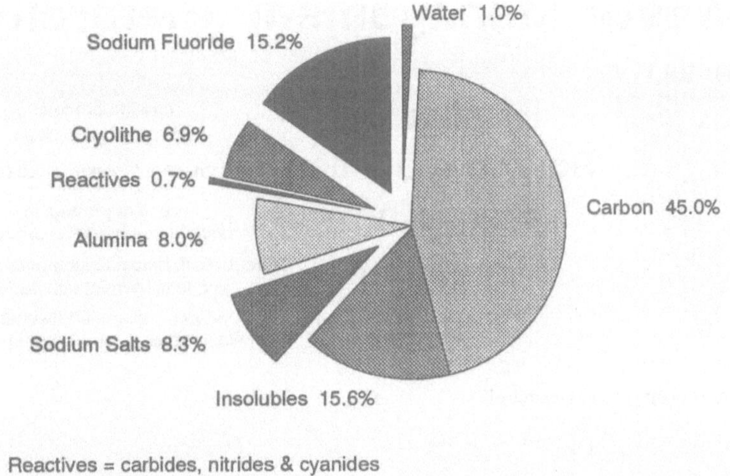

Fig. 2 Chemical composition of SPL

The production of aluminium currently generates about 35 kg of SPL per ton of aluminium or about half a million tons of SPL annually worldwide [1]. Unprocessed SPL was often landfilled as a method of eliminating the problem. Other solutions included pyro-metallurgical routes such as burning in cement kilns or use as a fluxing agent in steel furnaces. These pyro-metallurgical methods had difficulties with fluorine emissions and refractory degradation caused by the sodium present in the SPL. As the environmental liabilities of these practices became apparent, they were replaced by storage in buildings. In one location, Alcan is recycling hydrated recovered carbon (H.R.C.) into monolithic paste [2] (this a partial solution in that not all of the waste can be handled). Early methods used to recover some of the chemicals from SPL did not address all of the issues. These processes included water leaching to recover the caustic which was recycled to the alumina refining process and a caustic leach to recover caustic, aluminium and fluorine in the form of cryolite which was then recycled to the aluminium smelters. As environmental issues were addressed in the aluminium industry, these processes became non-economical and were abandoned.

In 1991, as part of its environmental policy, Alcan created a task force with the mandate of identifying the potential technologies to treat SPL. The criteria used to select the process were:

- Technical feasibility
- Capital and operating costs
- Generation of secondary waste material
- Tolerance to variations in SPL composition
- Recoverable residues and zero discharge

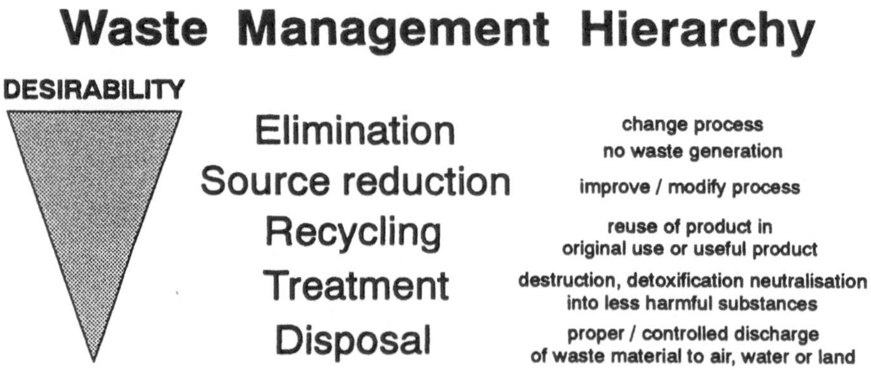

Waste Management Hierarchy

DESIRABILITY

Elimination — change process, no waste generation

Source reduction — improve / modify process

Recycling — reuse of product in original use or useful product

Treatment — destruction, detoxification neutralisation into less harmful substances

Disposal — proper / controlled discharge of waste material to air, water or land

Fig. 3 Waste management hierarchy [3]

The waste management hierarchy as promoted by the E.P.A. (USA) was used as a guideline by the task force (see Figure 3).

A summary of presently proposed options for treating SPL can be found in Figure 4 below. (Not all are operational processes.)

From Figure 4 it becomes clear that apart from landfill and stabilisation, all the other processes are pyro-metallurgical. These pyro-metallurgical routes will be discussed in more detail in the following section.

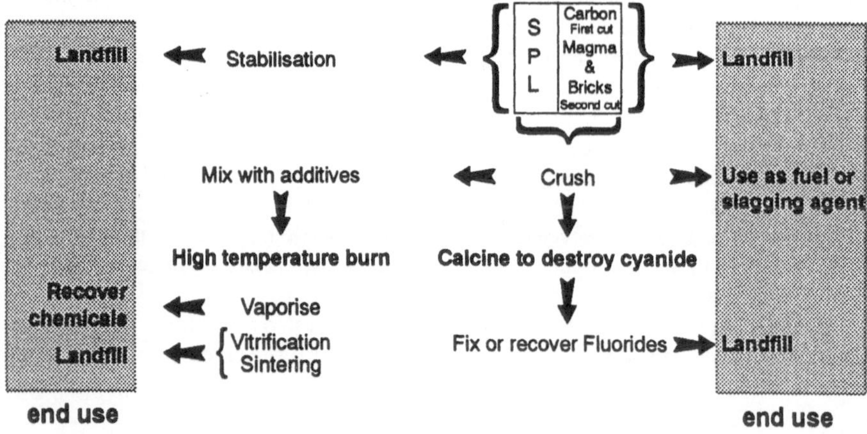

Fig. 4 Summary of SPL treatment options

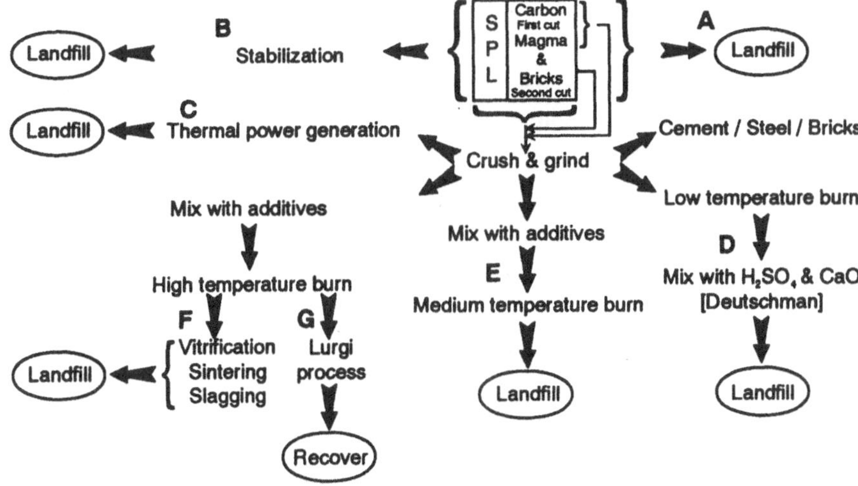

Fig. 5 Non-recovery routes for SPL treatment

2 Landfill and Pyro-Metallurgical Routes

The major non-recovery options for SPL disposal are summarised in Figure 5.

The objectives of these processes with the exception of process A & B were to destroy the cyanides and reactive compounds and reduce the solubility of the fluoride salts enough to meet leaching standards.

A. Landfilling of untreated and thus leachable SPL is not an acceptable solution. Landfilling of segregated cyanide-poor refractory bottom lining might be legally possible today in some jurisdictions, however, this material will be subject to fluoride leaching. These two options should therefore be eliminated.

B. Stabilisation through incorporation in a cement-like aggregate or an asphalt mixture or, simply mixing with gypsum, will retard ion mobility, but laboratory tests indicated that the aggregates would be unlikely to meet delisting criteria. This leaves the various pyro-metallurgical options that will be discussed in more detail.

C. Burning crushed but untreated segregated SPL for power generation, in the cement kilns, steel production, or to provide a reducing atmosphere in other industrial processes is technically feasible. But, as SPL is a hazardous waste, in some jurisdictions industrial processes used for its treatment will become hazardous waste treatment processes, with the incumbent change in regulations. Unfortunately, experience over the past few years, has shown that the effort required for environmental impact studies, to use these processes for SPL destruction, outweigh any advantages.

D. Various calcination, pyrolysis and combustion schemes have been combined with chemical stabilisation, fixation and vitrification. In the presence of water vapour, cyanides are readily destroyed at below 500°C and Deutschman [4] showed that the fluoride leach rates were much reduced when the surface of the pyrolized SPL was given a subsequent dilute sulphuric acid treatment.

E. The Reynold's thermal treatment of SPL in a rotary kiln [5] at intermediate temperatures (<900°C) with the addition of lime and other additives is the only commercial process presently in operation but involves a 2.4 fold increase in the weight to be landfilled. Of the two ultra-rapid sintering techniques, the Comalco's COMTOR process [6] leaves a leachable ash (SPLASH) for further treatment while Pechiney's SPLIT process [7] fixes fluorides through a reaction with $CaSO_4$. Reactions in fluidized bed or circulating fluidized bed combustors are still being actively pursued with the patent literature emphasising means to overcome operating problems due to agglomeration [8]. Actual combustion of the graphitized carbon requires careful temperature control to avoid coating the SPL particles with a liquid salt layer together with a judicious choice of additives to raise the melting temperature and/or separate the SPL particles.

F. High temperature incineration (>1050°C), must cope with dust emission and increased volatility of fluorides, particularly HF and SiF_4, through massive additions of lime producing slags or glassy residues whose chemistry somewhat resembles that of the slags formed during iron melting. Different equipment including plasma torches, cupola or glass making furnaces have been considered. Incineration or medium temperature treatments do not automatically eliminate subsequent leaching of fluorides.

   Pyrosulpholysis [9] has not been actively pursued by its proponents and was not reviewed in detail.

G. The elegant but technically challenging pyrohydrolysis in a rotating fluidized bed reactor proposed by LURGI [10] would not only destroy cyanides, carbides and nitrides and burn the carbon, but also transform the fluorides into HF.

   Most of these high temperature processes generally increase volumes to be landfilled and because of the high temperature these processes are expensive to build and operate.

H. One notable exception to this rule is Alcan Brasil's use of SPL as an additive to the clay of ceramic bricks that not only reduces energy consumption during the firing process but also improves the quality of the bricks [11].

2.1 Fluoride Leach Rates

The applicable analytical procedures vary significantly with the legislative jurisdiction involved. The most commonly used procedures follow the Toxicity Characteristic Leaching Procedure (TCLP) as outlined in US EPA 40 CFK 261.2. It can be summarised as extracting 100 g of neutral or slightly alkaline treated SPL with 1000 mL of a 0.1 N acetic acid buffer solution (pH 4.93 ±0.05) or 100 g of untreated or alkaline SPL with 1000 mL of 0.1 N acetic acid solution (pH = 2.88 ±0.05). Other jurisdictions, such as France's AFNOR T95J which uses three successive water extractions simulate more closely leaching by rainwater.

As illustrated in Figure 6, a direct correlation is typically found between the logarithmic leach rate from untreated SPL and the final pH of the leachate. This can be explained

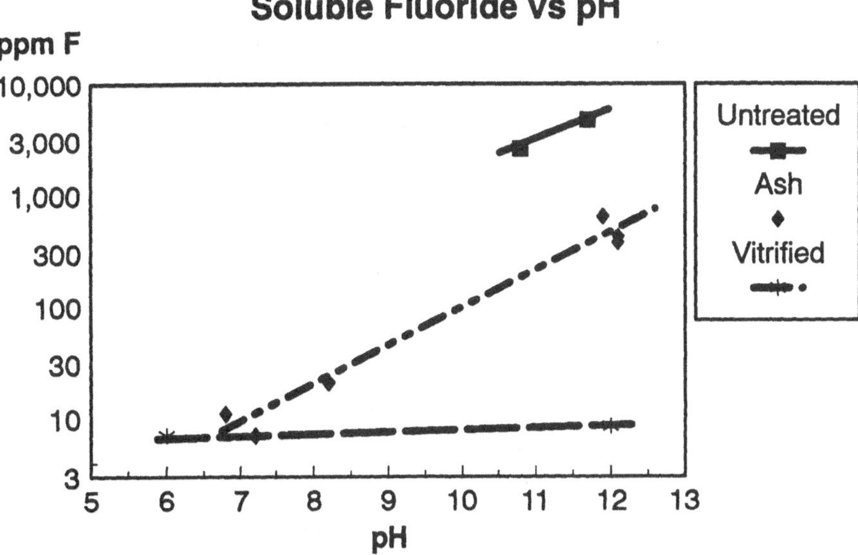

Fig. 6 Leachability of treated and untreated SPL versus pH

by the pH dependency of the solubility of the bath constituents. Treatments which fix the fluorides as crystalline CaF_2 or vitreous slags yield leach rates which are relatively independent of the final extractant.

Figure 6 also indicates a third case, where some of the fluorides in a residue produced by calcining a SPL, lime and sand mixture, were converted into the alkali-soluble calcium fluorosilicate ($Ca_4SiO_7F_2$ or cuspidine) but where no NaF or cryolite could be detected by X-ray diffraction analysis. Although the treated material would readily meet the EPA TCLP limits, if tested using the AFNOR or Australian leach procedures, it would show the similar pH dependence as untreated SPL. We conclude that some thermal treatments may still lead to residues which could leach fluorides and caustic if exposed to rain or ground waters and as such should not be landfilled as simple industrial waste.

3 Hydro-metallurgical routes

Alcan's efforts since World War II to recover the chemical values in SPL were reviewed earlier by McGeer [12] and emphasise various hydro-metallurgical approaches. Figure 7 summarises the major hydro-metallurgical routes examined in this study. The hydro-metallurgical routes described below aim to produce useful recoverable (recyclable) products. A secondary objective, the worst case scenario, was to be able to meet all leach

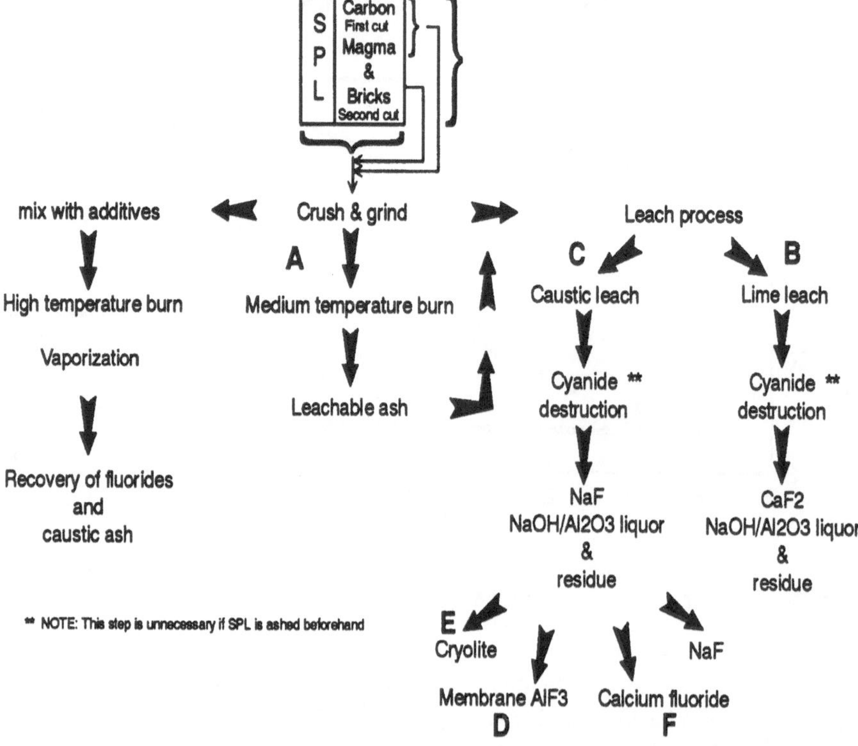

Fig. 7 Hydro-metallurgical routes for SPL treatment

tests for fluorides and cyanides so that it would be possible to landfill the products. This would also allow delisting and improved acceptability of the products.

Although segregation of the first and second cuts might simplify subsequent chemical treatment this involves additional operating and capital costs and is never perfect, if only because of the variable thickness of the magna or densified aluminium silica refractories found at the interface of the carbon blocks and insulating refractories. A number of processes have been proposed which combine a calcination step with subsequent chemical conversion. The COMTOR process constitutes an example of a medium temperature treatment designed to destroy cyanides, carbides and nitrides and leave a leachable residue.

Some of the hydro-metallurgical processes are schematised in Figure 7. All chemical extraction schemes would either have to incorporate wet cyanide destruction or be preceded by a calcination step. A simple water leach followed by crystallisation of NaF [13] could leave a residue having a lower soda content, more suitable to the cement and other industries than the original SPL, but without cyanide destruction, would hardly allow delisting. Acid attacks [14], or modern versions thereof [15], would need to cope with massive SiO_2 dissolution, disposal of highly soluble Na_2SO_4 and potential HCN emissions; they were not extensively evaluated.

A. A low or medium temperature pyrolysis to destroy cyanides could be followed by digestion with lime
B. (Mini-L process [16]) to yield a caustic solution and a mixture of carbon, refractories, CaF_2 and lime.
C. Digestion of the untreated, unsegregated SPL with caustic, followed by autoclave destruction of the cyanides [17] would extract the chemicals from the carbon and refractory inerts. There exist a number of processes to convert the chemicals into value added compounds. Following crystallisation of NaF,
D. a bipolar membrane [18] could be used to produce HF and eventually AlF_3,
E. pH adjustment and alumina addition would lead to cryolite recovery, while lime addition
F. would lead to relatively pure CaF_2, recovery of the caustic as Bayer liquor and leave a nontoxic carbon/refractory residue.

3.1 Process Choice

The decision to explore recovery routes in our development programs stems in part from the unique nature of the Jonquière complex which is located within 300 miles of the smelters producing almost 10% of the world's primary aluminium and also regroups a Bayer and an aluminium fluoride plant. It was also influenced by the fact that available or near commercial solutions (pyro-metallurgical) to treat SPL are expensive and create secondary sources of waste materials in larger quantities than the original. Table 1 illustrates the weight ratio of the residue to original SPL of the potential treatment routes examined.

Table 1. Comparison of alternate SPL treatment proposals considered

	Material In			Material Out			Landfill/ SPL Ratio
LCL & L	1 t SPL	0.2 t CaO	1.4 t H2O	0.7 t Carbon	1.7 t Bayer liquor	0.2 t CaF2	~0
LCL & M	1 t SPL	0.1 t Al(OH)3	1.3 t H2O	0.7 t Carbon	1.5 t Bayer liquor	0.2 t ALF3	~0
LCL & NaF	1 t SPL	0.3 t NaOH	0.5 t H2O	0.7 t Carbon	0.9 t Bayer liquor	0.2 t NaF	0
Mini L	1 t SPL	0.3 t CaO	1 t H2O	1.3 t Carbon	1 t Bayer liquor		0 =>1.3
Elkem	1 t SPL	1.4 t Fe2O3	0.3 t CaO	1.1 t Landfill	0.9 t Fe		1.1
Pechiney	1 t SPL	1 t CaSO4	0.2 t H2O	2.2 t Landfill			2.2
Reynolds	1 t SPL	1.4 t filler		2.4 t Landfill			2.4

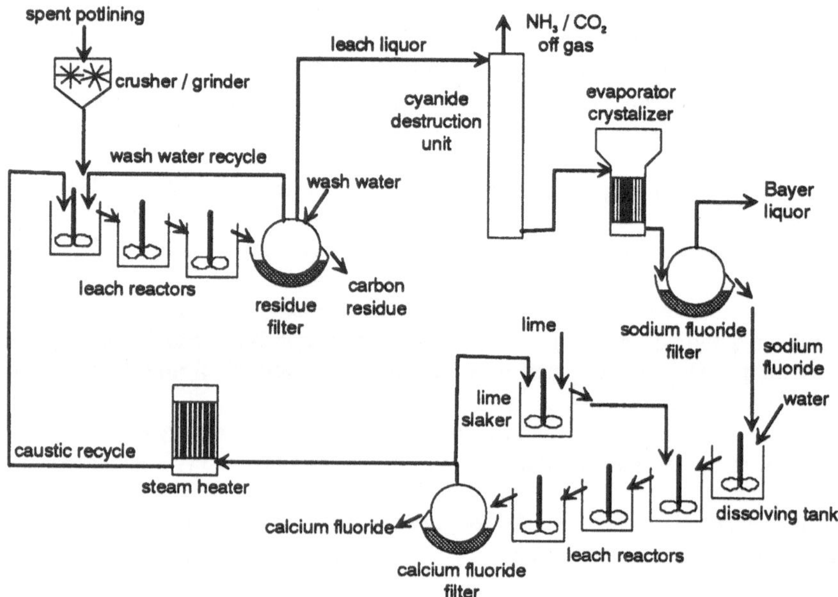

Fig. 8 Low caustic leach and lime process for SPL treatment

Moreover, according to preliminary cost analyses and bench-scale tests, the hydrometallurgical routes seemed particularly attractive. The most promising version, involving **L**ow **C**austic **L**each and **L**iming (LCLL), was piloted at 1/400 scale in the testing facilities of the Mineral Research Centre of Quebec (Ministère de l'Énergie et des Ressources) and validated the commercial feasibility of the technology, established some of the engineering specifications and confirmed the cost estimates.

3.2 The LCLL Process: General Description

The LCLL flowsheet shown in Figure 8 refers to four different operating blocks :

- Crushing and grinding
- Extraction and leaching of SPL
- Cyanide destruction and crystallisation of NaF from leachate
- Causticization of the sodium fluoride

3.2.1 Crushing and grinding

The chemical and physical composition of SPL varies greatly and it is composed of carbon blocks, bricks, steel and tramp aluminium as well as various chemical compounds. The size of SPL also varies greatly from powder to lumps several feet in diameter. A natural lubricating agent, graphitized carbon, present in the SPL renders standard crushing techniques such as jaw and gyratory crushers inefficient. Hammer mills, ball mills and autogenous grinding mills will however reduce the aggregate size to an average of 0.5 mm. This fine size is required to efficiently extract the fluorides from this waste material in the leach reactors.

3.2.2 Extraction and Leaching

The heart of the LCLL process is the extraction, leaching and washing of the SPL. This group of unit operations takes its roots from Alcan's cryolite recovery process. The finely ground SPL is digested in hot, dilute caustic solutions. Multiple, agitated cascade reactors digest the slurry extracting fluorides, sodium, alumina, free and complexed cyanides and some silica into the leach liquors. The cryolite present reacts with caustic and decomposes to soluble sodium fluoride and aluminate:

$$Na_3AlF_6 + 4\ NaOH \Rightarrow 6\ NaF + NaAlO_2 + 2\ H_2O \qquad \{1\}$$

Any intercalated sodium and any remaining aluminium metal will dissolve with the evolution of hydrogen:

$$2\ Na + 2\ H_2O \Rightarrow 2\ NaOH + H_2 \qquad \{2\}$$

and

$$2\ Al + 6\ NaOH \Rightarrow 2\ NaAlO_2 + 3\ H_2 \qquad \{3\}$$

while any aluminium nitrides and carbides will yield flammable ammonia and methane respectively:

$$AlN + NaOH + 2\ H_2O \Rightarrow NaAlO_2 + NH_4OH \qquad \{4\}$$

$$Al_4C_3 + 4\ NaOH + 4\ H_2O \Rightarrow 4\ NaAlO_2 + 3\ CH_4 \qquad \{5\}$$

Because of their nature, the densified refractories are more readily attacked than the original bricks yielding fluorides, aluminates and silicates. The fluorides present in the original bath electrolyte as CaF_2 remain insoluble and are not extracted into alkaline solutions.

The solubility of sodium fluoride and aluminate depend on the caustic concentration, as shown in Figure 9, as well as on the temperature and silica content. Therefore, the caustic concentration of the leachate will dictate the SPL/leachate ratio. A lower caustic

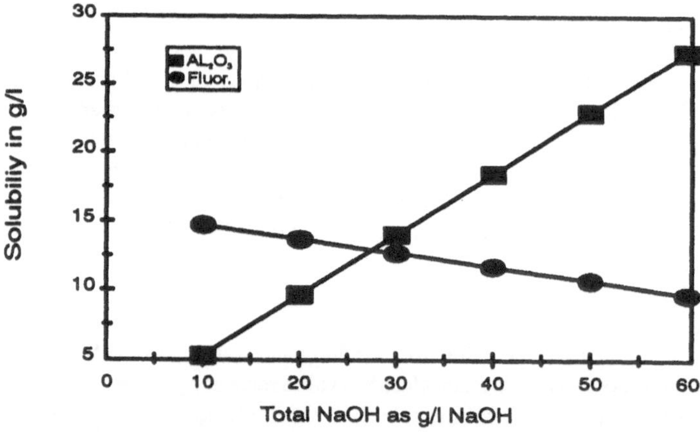

Fig. 9 Solubility of fluorides and alumina in low caustic liquors

concentration will limit the driving force to decompose cryolite while a higher concentration will limit the fluoride solubility and require a more dilute slurry.

As indicated in the fishbone diagram, Figure 10, a number of parameters can influence the fluoride extraction efficiency which is defined as:

$$\text{Extraction efficiency} = \frac{\text{Amount of fluoride solubilised}}{\text{Fluoride in SPL other than } CaF_2} \qquad \{6\}$$

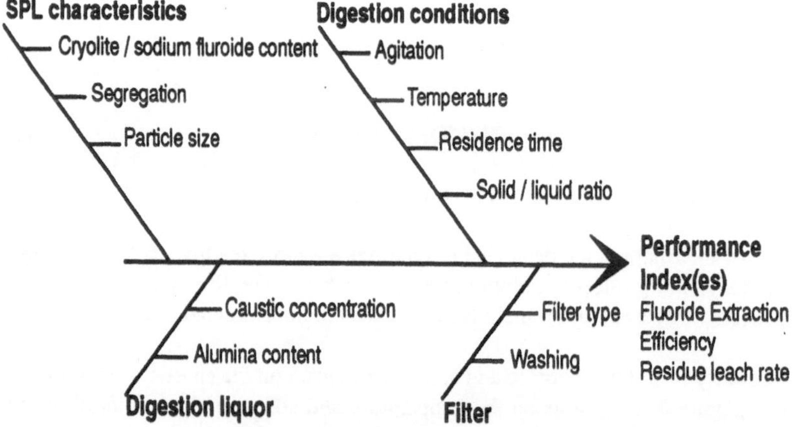

Fig. 10 Ishikawa diagram: Parameters influencing fluoride extraction efficiency

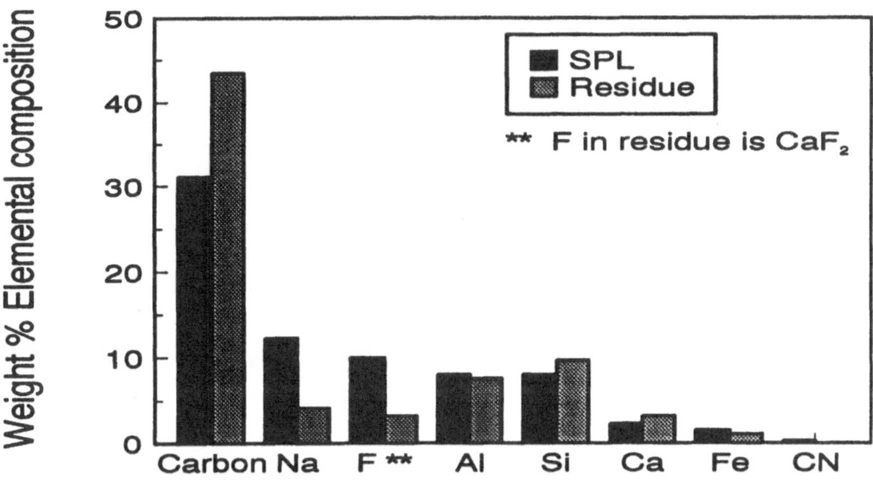

Fig. 11: Composition of original SPL and digestion residue

Extraction efficiencies obtained by varying the slurry composition, NaOH concentration, average particle size, filter type and residence time in two cascaded, well stirred reaction vessels were all in the 90% to 100% range.

The solid residue is separated by filtration and washed with suitable amounts of water. Thus segregated and decontaminated, the solids are considered a high ash industrial fuel. A comparison of the composition of the original SPL and the residue is given in Figure 11, illustrating the mineralogy. The original NaF and Na_3AlF_6 peaks (XRD spectra) have completely disappeared.

Table 2. Leachability of SPL residue as determined by MENVIQ test procedures

Contaminant	Verified concentration	Criteria to be met
Fluoride leachable[1]	<65 ppm	<150 ppm
Cyanide leachable[1]	<14 ppm total	<20 ppm
Cyanide reactive[2]	<45 mg/kg total	<250 mg/kg

1 The leachable fluorides are determined by the method MENVIQ, 1003-85, 25 May 1985. Evaluation of the characteristics of solid wastes.
2 The reactive cyanide was determined by the method MENVIQ 87.09/108-Reac. 1.1 Determined of the reactivity of cyanides and sulphides from dangerous materials.

Table 2 gives typical values for the leachable fluorides, cyanides and reactive cyanides as determined by the Quebec Ministry of Environment test procedures. Since it meets local and EPA TLCP criteria, this material, initially classified as a hazardous material can now be declassified. This declassified material has a much lowered sodium content than the initial SPL, and is expected to meet ready acceptance in a number of industrial processes as a high ash fuel.

3.2.3 Cyanide Destruction and Sodium Crystallisation from Leachate

Alkaline hydrolysis of the cyanides in a pressure vessel is inspired from the destruction in SPL leachates [19] in commercial use since 1989. In this stage, the caustic leachate from the digestion unit containing around 600 ppm total cyanides, enters a stainless steel plug flow reactor. The filtered leach liquor is slightly enriched with caustic to reach the 60 g/L NaOH level, shown to give the highest decomposition rates. In the absence of oxygen, the initial decomplexation of any ferrocyanides present:

$$2\,[Fe(CN)_6]^{4-} + 4\,OH^- \Rightarrow 2\,FeO + 12\,CN^- + 2\,H_2O \qquad \{7\}$$

is followed by the slower hydrolysis of the cyanide ion itself:

$$CN^- + 3\,H_2O \Rightarrow NH_4OH + HCOO^- \qquad \{8\}$$

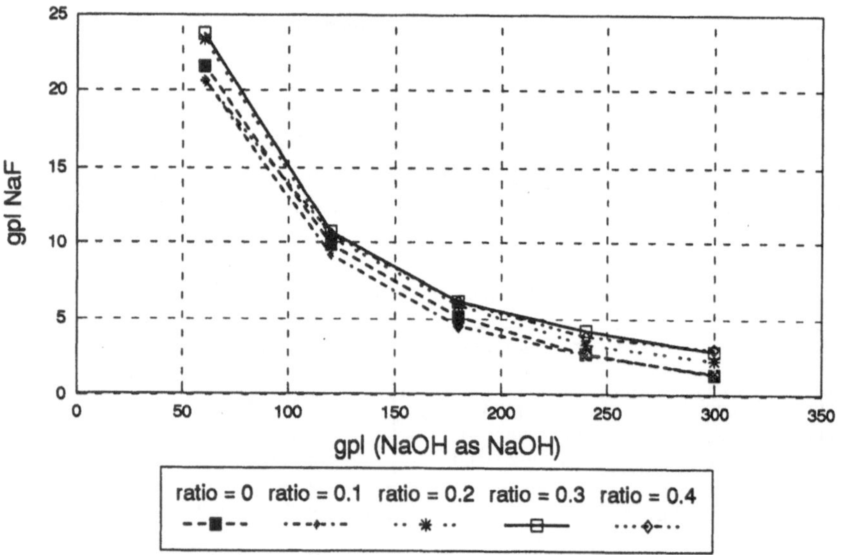

Fig. 12 Solubility of fluorides in strong caustic liquors

The liquor is flashed down and any iron oxides formed from the decomposition of complexed cyanides are filtered out.

The clean caustic leachate is fed to an evaporator/crystallizer which raises the caustic concentration to the point where NaF crystallises out. As illustrated in Figure 12, most of the fluoride can be recovered by evaporating the caustic concentration to around 225 g/L NaOH.

Although evaporating the leachate to a higher caustic concentration can further reduce the fluoride left in the liquor, the quality decreases as the alumina solubility limit is approached. The extent of evaporation is a compromise between the grade of NaF precipitated, the amount recovered and the purity of the Bayer liquor produced.

The NaF slurry from the pilot test was continuously filtered and washed to produce two value-added products: the concentrated, alumina rich caustic liquor, so-called Bayer liquor, and NaF crystals 0.5 mm in diameter of 95 % purity. A single reprecipitation of the crystals would improved their purity substantially as illustrated in Table 3.

Table 3. Composition of impure and recrystallised NaF (weight %)

	Impure		Recrystallised
	Soluble	Insoluble	
Weight %	99.2	1.0	
Element			
F	42.8	2.63	43.7
Na	52.8	16.32	54.9
Al	0.11	11.2	not detected
Si	1.0	14.6	0.53
Fe	<0.2	2.2	<0.2
CO_3^-	<0.1	<0.0	
Li	0.1	<.005	
CN^-	<0.0001		not detected
H_2O			0.27
Total	97.2	47	99.45

3.2.4 Causticization of Sodium Fluoride

Causticization is a treatment similar to that still used to neutralise wet scrubber liquors from aluminium smelters. In the LCLL process, this operation transforms the highly soluble sodium fluoride into insoluble calcium fluoride. The NaF precipitated during the evaporation step is redissolved and the co-precipitated insoluble impurities can be filtered out depending on the required purity. The sodium fluoride is then neutralised with substoichiometric additions of milk of lime in two precipitators in series. The CaF_2 generated may be sent to Alcan's fluoride plant as a feed stock to produce aluminium fluoride while the caustic liquor produced is recycled to the extraction step of the LCLL process.

4 Conclusion

In the LCLL process the cyanides are destroyed, the carbides and nitrides solubilised as sodium aluminate and all of the products extracted can be recycled. The decontaminated carbon and inert material is a potential high ash fuel for a number of industries. The Bayer liquor, containing recovered soda and aluminate is recycled to the alumina refinery. The sodium fluoride is of a commercial grade and could be marketed as is or transformed into calcium fluoride.

In a continuous operation, the pilot plant fully substantiated the expected benefits of the LCLL process. All the technical goals were achieved and in some aspects, exceeded. The robustness and the flexibility of this hydro-metallurgical process and the stability of operation was shown pilot run. Treatment costs, on a per ton basis are estimated to be considerably less than currently available or envisaged alternate pyro-metallurgical potlining treatments. The LCLL process will provide an elegant and dependable solution to one of the industry's most urgent environmental problem. In addition to being cost-effective and environmentally acceptable, the process has been proven in pilot-scale testing and is ready for a full scale unit.

A summary statement for hydro-metallurgical processes would be **why do between 800°C and 1800°C what can be done better hydro-metallurgically, at less than 180°C?**

5 Acknowledgements

The authors gratefully acknowledge the input of all of the members of Alcan's SPL task force, Mr. Jean Luc Bernier of the Arvida Research and Development Centre for his work on the pilot project, the technical support and assistance during piloting given by Dr. Arthur Plumpton, Messrs. Jacques Pérusse and Jaques Turgeon of the Centre de Recherches Minérales, Québec, and the scientific back-up from CRDA's Drs. R. Breault and V. Kasireddy.

6 References

1. Stuart J. Spiegel and Thomas K. Pelis, **Regulations and Practices for the Disposal of Spent Potliner by the Aluminium Industry** Journal of Metals, Nov. 1990, pp 70-73.
2. A. Mikkelsen, **Production of Carbon Lining for Reduction cells,** US Pat. 3 932 244
3. EPA waste minimisation opportunity assessment manual. EPA/625/7-88/003 Hazardous waste engineering research laboratory, Cincinnati, Ohio, USA 45268.
   Waste minimisation guide institute of chemical engineers, Davis Building, 165171 Railway Terrace, Rugby, Warkshire CB213HQ
4. John E. Deutschman, **Treatment of scrap lining from aluminium reduction cells,** European Pat. Appl. EP 117761
5. Dennis G. Brooks, Euel R. Cutshall, Donald B. Banker and Denis Strahan, **Thermal Treatment of Spent Potliner**, Light Metals, 1992, pp 283- 287.

6. G. A. Wellwood, Il.L. Kidd, C. G. Goodes and R. Nivens, **The Comtor Process for Spent Potlining Detoxification**, Light Metals 1992, pp 277-282.
7. J. C. Bontron, D. Laronze and P. Personnet, **Spent Potlining Insolubilisation Technology (SPLIT)**, Proceedings Light Metal Processings and Applications, 32nd Annual Conf. CIM, pp. 179-188.
8. William S. Rickman, James L. Kase and Bernard W. Gamson, **Method for the Combustion of Spent Potlining from the Manufacture of Aluminum**, US PAT 4,763,585. Ronald Stanley Tabery and Ky Dangtran, **Fluidized Bed Combustion of Aluminum Smelting Waste**, Int Pat WO 90/13774
9. C. G. Goodes, H. W. Hayden and D. J. Williams, **The treatment of spent cathode waste by pyrosulpholysis**, Light Metals 1987 paper A87-14
10. J. N. Anderson and N. Bell, **Pyrohydrolysis Process for Spent Aluminum Reduction Cell Linings**, US Pat. 4160809
11. A. C. Filho Brant, A. R. Silva, L. C. B. Martins and M. R. Paula, **Use of Spent Potlining in the red brick ceramic industry**, Light Metals 1988, pp. 731-734
12. J. Peter McGeer, **Alcan's Treatment of Potlings over the last 40 years**, Journal of Metals, 1984 p 30-34
13. G. Lever, **Treatment of Aluminum Production Fluoride containing Waste**, Norwegian Patent N08400264A
14. B. Gnyra, **Acid Attack as a Means of Treating Spent Potlining**, Light Metals, 1980 pp 683-703
15. H. Kaaber, M. Mollgaard, **A Process for recovering Aluminium and Fluorine from Fluorine containing Waste Materials**, Int. Pat. Appl. WO 92/13801
16. B. Gynra, R. R. Sood and J. D. Zwicker, **Treatment of Wastes Containing Water-leachable Fluorides**, UK Pat. Appl. GB2,056,425
17. F. M. Kimmerle, P. W. Girard, R. Roussel and J. G. Tellier, **Cyanide Destruction in Spent Potlining Leachate**, Light Metals 1989, pp. 387-394
18. G. Lever, **Preparation of Aluminum Fluoride from Scrap Aluminum Cell Potlinings**, US Pat. 4816122
19. D D. H. Bell, **Apparatus and method for Hydrolysis of cyanide containing liquids**, US Pat. 5 160 637

Recovery of vanadium from spent catalysts and alumina residues

E. M. Ho
J. Kyle
S. Lallenec
D. M. Muir
A. J. Parker Cooperative Research Centre in Hydrometallurgy, Murdoch University, Perth, Western Australia

Abstract
The various process options available to recover vanadium from spent dehydrosulphurisation catalysts, sulphuric acid catalysts, and alumina sludge residues from the Bayer Process are reviewed, and the fate of other metal impurities such as Mo, Ni, Co, Al and Fe are considered. Most processes give an impure V_2O_5 product, but selective solvent extraction of vanadium from the impure leach solution allows high purity product to be obtained. A comparison is made between the D2EHPA and Amine extractants with regard to the vanadium species and other metals extracted. The results of studies with tertiary and quaternary amines in acidic media are reported, and the relative performance of quaternary amines in extracting a range of vanadium(V) anionic species between pH 6-13 is presented. It is shown that quaternary amines offer the greatest flexibility for treating acidic, neutral, or alkaline liquors depending on the process of choice.

1 Introduction

With increasing environmental concerns and legislation over the disposal or shipping of hazardous materials, companies and countries are being forced to process their own waste products and residues. The choice of options must be suited to local conditions, markets, and existing technology. Spent catalysts represent both a waste from local petrochemical and chemical industries, and a resource to any local mineral industry. A study by the U.S.Bureau of Mines [1] found that in 1981 alone, the U.S.A. discharged spent catalysts containing 12 million kgs. of recyclable metals including Ni, Co, Mo, Cr, V, Cu and Zn supported on alumina, silica or zeolite. Spent automobile catalysts represent another specific category, with over 500,000 ozs. of platinum group metals discarded per annum, dispersed over a honeycombed alumina support.

Catalysts containing Ni, Co, Cu, Zn and platinum group metals are often processed at local copper or nickel processing plants along with primary feed. However, those containing Mo, Cr and V are processed at specific centralised facilities due to their value and scarcity of primary processing plants.

In Australia, the recovery of vanadium from spent catalysts discharged by the sulphuric acid and petroleum industries is of particular interest because of the difficulty of shipping this waste to overseas facilities. Catalysts used to promote the oxidation of SO_2 in the manufacture of sulphuric acid usually contain between 2-6% V_2O_5 on a porous silica (crystobalite) substrate. Hydrodesulphurisation catalysts used to remove sulphur from crude oils also pick up vanadium impurities in the oil and typically contain 7-15% V_2O_5 as V_3S_4, as well as 4-8% Mo, 2-3% Ni, and 1-2% Co as sulphides and oxides on an alumina substrate [2,3]. Furthermore, there is the potential to recover vanadium from local bauxitic ores.

In India, such ores are processed by the Bayer Process, and a vanadium sludge containing 6-20% V_2O_5 mixed with alumina is obtained as a residue for subsequent processing [4].

Thus all three materials are similar in composition and offer the potential for a common processing facility. This paper aims to review the processing options available, and to consider the best method of recovering pure V_2O_5 by either selective precipitation or by selective solvent extraction of vanadium(IV) or (V).

2 Vanadium Process Chemistry

Vanadium exhibits a complex variety of speciation and oxidation states in aqueous solution. The stability regions of its predominant ions and solid precipitation compounds are shown in the Eh-pH diagram (Figure 1)

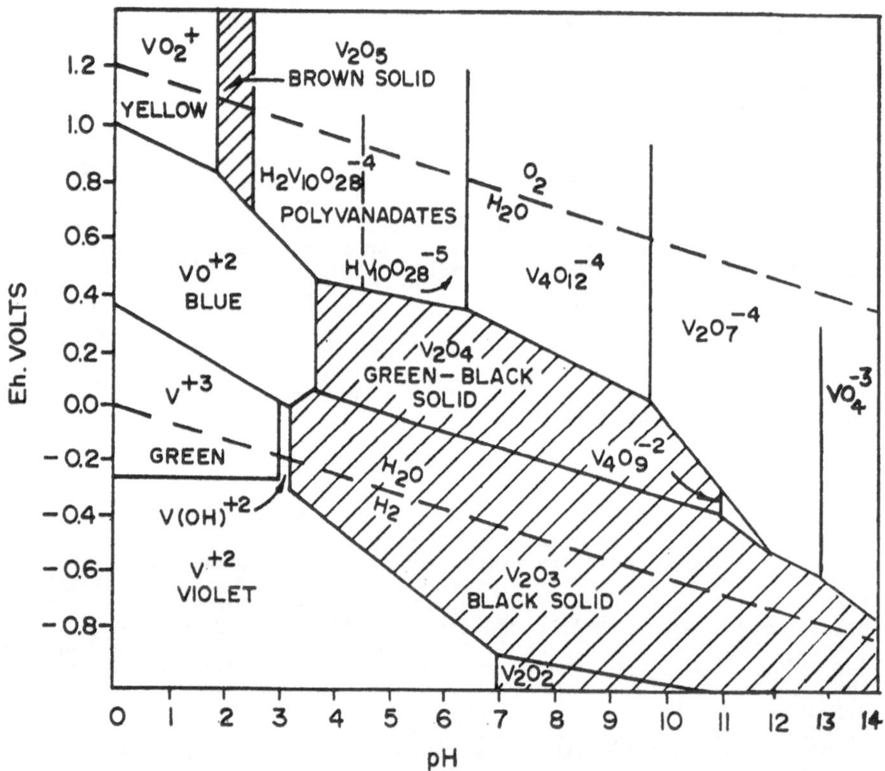

Figure 1. Stability Regions of Vanadium Ions and Solid Oxides (At 25°C. and 10^{-2}M.[V]. Adapted from Ref.[4].)

In leach liquors from the processing of catalysts, ores and residues, vanadium(IV) and (V) predominate in acid media and vanadium(V) in alkaline media. Vanadium(V) is a strong oxidant in dilute acid when it is present as $[VO_2]^+$,(E° = 1.0 V.), but it is a mild oxidant in dilute caustic as the $[V_2O_7]^{4-}$ and $[VO_4]^{3-}$ ions (E° < 0 V.). Between pH 2-12, vanadium(V) exists in a series of polymerised polyanions such as decavanadate $[V_{10}O_{28}]^{6-}$ or metavanadate $[V_4O_{12}]^{4-}$ which are partially protonated according to pH, and may be slow to transform [5,6].

Vanadium is usually precipitated from solution either as brown vanadic acid - $V_2O_5.nH_2O$ at around pH 2 (often known as "red cake"), or as ammonium metavanadate - NH_4VO_3 - at around pH 9 which is sparingly soluble in cold water. "Red cake" however, usually contains significant amounts of sodium and approximates in composition to $Na_2H_2V_6O_{17}$ [7]. Vanadate ion also forms complexes with ions of other acids to form heteropolyacids which may co-precipitate at higher pH; the more common being phosphate, molybdate, silicate and arsenate. These salts have varying molar ratios gand compositions e.g. $11K_2O.2P_2O_5.24V_2O_5$ [5]. Phosphate, in particular, is an unacceptable impurity to the ferrovanadium industry.

Both vanadic acid and ammonium metavanadate are calcined to produce fused V_2O_5 for market. Although vanadium(IV) may be precipitated as V_2O_4 at pH >3.5, it does not fuse on heating and gives a less acceptable friable powder [7].

3 Processing Options for Catalysts and Residues

3.1 Hydrodesulphurisation Catalysts

An extensive review of the literature [8] reveals a wide variety of approaches for treating these catalysts which are usually contaminated with 10-20% oil residues. The processes, mostly described in Patents, vary in their selectivity for metals and complexity of operation, but adopt one of the following approaches:

I Acid leaching with either H_2SO_4 [1,2,9-12], HCl [3,13,14], or $(COOH)_2$ [15] - often after roasting [13-20].

II Caustic leaching with NaOH [21-24] - sometimes after roasting [2,25,26].

III Salt roasting with Na_2CO_3 or NaCl or NaOH [27-33] followed by leaching with water [34] or Na_2CO_3 [35].

IV Smelting either directly [36-39] or after calcination [40]

V Anhydrous chlorination [21,41-44].

Roasting at 550-800°C. is usefully employed to burn off oil residues and to convert metal sulphides to oxides. However, high temperature roasting causes some molybdenum to volatilise as MoO_3, and produces phase changes in the alumina substrate to make it intractable to leaching.

Direct acid or caustic leaching is therefore employed, sometimes under pressure, where there is a need to recover an alumina product or to achieve the highest recovery of metals. Acid leaching enables nickel, cobalt, molybdenum and vanadium to be brought into solution readily, but subsequent separation and recovery of the metals is more complex. Alkaline leaching is more specific for molybdenum and vanadium but leaves nickel and cobalt behind with the alumina for subsequent processing in a smelter.

Salt roasting with either NaOH, Na_2CO_3, or NaCl and leaching with water or dilute Na_2CO_3, provides for a more selective leach of molybdenum and vanadium over alumina and silica under milder conditions and is similar to that adopted to process primary ores of vanadium. Following roasting with Na_2CO_3 at 750°C., Sebenik and Ference [34] carried out a caustic leach at 250°C and 34 atmos.to dissolve the alumina residue and recover pure Al_2O_3

Smelting is carried out with coke or natural gas together with added scrap iron at temperatures around 1700°C. [40], or else the catalyst is converted to a sulphide matte. It is an expensive option alone and gives a product which still needs processing further. Molybdenum volatilises off as MoO_3 above 650°C., but any vanadium would report to the slag and be difficult to recover.

Chlorination at 200-500°C. results in molybdenum, vanadium and some aluminium being volatilised, but nickel and cobalt chlorides remain in the alumina residue and are leached with water. This commercially unproven approach, which provides a rapid and clean separation of Mo and V from Ni and Co at relatively low roasting temperatures, was favoured by a U.S.Bureau of Mines investigation into methods of processing a Ni-Mo-Al catalyst [21].

3.1.1 Commercial Approaches

The two most comprehensive commercial approaches to treating spent catalysts are the Metrex Process [14,45], established in Holland in 1991, and the CRI-MET Process [24,46] which has been operating for some time in Louisiana, U.S.A. Other commercial operations at the Gulf Chemical Corp., Texas [47], and at Sumitomo Co.Ltd., Japan [48], focus on vanadium and molybdenum recovery by salt roasting with either Na_2CO_3 or NaCl in a steam + nitrogen atmosphere and leaching with water. Both these companies recover vanadium by precipitation of ammonium metavanadate. However, Gulf Chemical precipitates molybdenum as H_2MoO_4, whilst Sumitomo prefers to solvent extract molybdenum by tertiary amines and recover as MoO_3.

In the Metrex Process (Figure 2), the catalyst is roasted at 800-850°C. in air and natural gas and part of the molybdenum is volatilised. The remaining metal oxides are then leached with dilute H_2SO_4 and separated, purified and recovered by a series of undisclosed steps involving solvent extraction. Any dissolved alumina is finally removed as aluminium sulphate. The company literature does not include an option for vanadium recovery, though no doubt this could be incorporated if necessary.

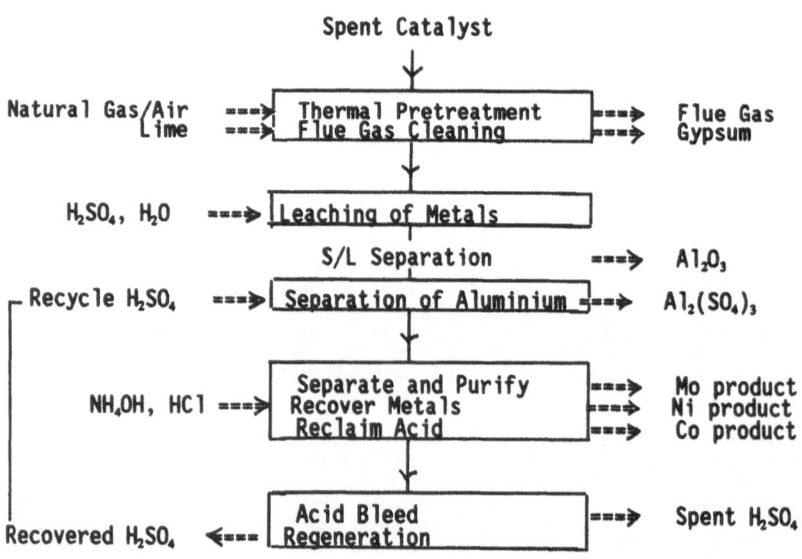

Figure 2. Schematic Flowsheet of the METREX Process (Ref [14,45]).

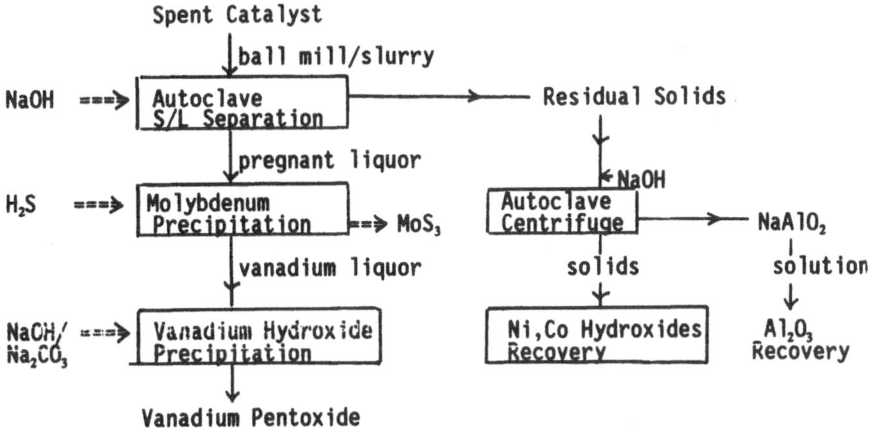

Figure 3. Schematic Flowsheet of the CRI-MET Process (Ref [24,46]).

The CRI-MET Process (Figure 3) is an alkaline pressure leach process which also treats fly-ash and other wastes. Ground catalyst is leached with NaOH/$NaAlO_2$ and air at 150-250°C to oxidise sulphides and organics, whilst the sodium aluminate inhibits the dissolution of alumina. Molybdenum is recovered from solution as MoS_3 by acidification and sparging with H_2S, and vanadium is precipitated as its hydroxide with NaOH/Na_2CO_3 addition. The first stage leach residue is then dissolved by strong caustic to leave a nickel and cobalt-rich residue for smelting, and alumina is recovered from the liquor. This is a comprehensive approach which would appear to be particularly suitable for Australia because of its well established Bayer Process and nickel smelting facilities.

3.2 Vanadium/Alumina Sludge

In the Bayer Process for recovering alumina from bauxite, about 30% of any vanadium in the bauxite is leached and is concentrated in the recycled aluminate liquors. Once it reaches a certain limit, it is precipitated as a vanadium sludge by either slowly cooling or aerating the liquor. This sludge typically contains 6-20% V_2O_5 together with alumina and phosphates [49,50], and is a useful source of vanadium which can be processed like spent catalyst.

Pattnaik et al [49], however, describe a simpler approach involving dissolution of vanadium from the sludge with hot water and precipitation of either $2FeO.V_2O_5$ or $2CaO.V_2O_5$ from the dilute filtrate containing 15 g/l V_2O_5 by addition of ferrous sulphate or excess calcium chloride whilst maintaining the pH around 1-2. The iron containing precipitate could be used directly for ferrovanadium production, whilst the calcium containing precipitate requires redissolution and recovery of vanadium as ammonium metavanadate. This approach is clearly more suited to freshly precipitated sludges, devoid of other metals, where the vanadium is readily dissolved and the alumina residue can be recycled. But the final purity of V_2O_5 is not high.

A newer approach [51], involves slurrying the same dilute filtrate with activated carbon (100 g/l) at 85°C and pH 2-3 for 4 hours, which absorbs 90% of the vanadium from solution as $[H_2V_{10}O_{28}]^{4-}$. This is subsequently stripped either as $[VO_2]^+$ with 0.9 M.H_2SO_4 at 85°C., or as $[V_2O_7]^{4-}$ with ammonium hydroxide. The acid eluate requires oxidation with chlorine prior to pH adjustment and precipitation of V_2O_5, whilst the ammoniacal eluate is oxidised by air, and precipitates V_2O_5 directly upon acidification and heating. The highest purity V_2O_5 was obtained from the ammoniacal strip, since the acid strip gave V_2O_5 contaminated with Al, Si, Ca, Mg and Fe.

3.3 Sulphuric Acid Catalysts

Spent sulphuric acid catalysts consist of a relatively pure mixture of water soluble V(IV) and V(V) salts, - probably $VO(SO_4)$ and $(VO_2)_2.SO_4$ - as well as V_2O_5, dispersed on a silica and alum substrate which is soaked with sulphuric acid. The main contaminants are potassium and iron. Alkaline leaching is selective for vanadium over iron but dissolves some silica and is obviously more costly with reagents [52,53]. Sulphuric acid leaching does not dissolve silica, and can be modified to produce a leach solution containing either V(IV) or V(V) for further processing.

Dilute H_2SO_4 has been used at 60-90°C. on its own [54], and in the presence of hydrazine [55,56], oxalic acid [57], or sulphur dioxide [58], as reductants for V(V). This prevents the precipitation of vanadic acid ($V_2O_5.nH_2O$ or $H_6V_{10}O_{28}$) which occurs in dilute acid, and maximises the extraction of vanadium. But the liquor must be re-oxidised by peroxide or sodium chlorate before V_2O_5 can be precipitated from solution.

More concentrated H_2SO_4 leaching, or roasting the catalyst in its own acid at 200°C.and water leaching, is equally effective and results in an acidic V(IV)/V(V) leach liquor for oxidation and neutralisation. The main problem is that iron(III) hydroxide co-precipitates with vanadic acid around pH 1.5-2.5 and contaminates the final product.
Some of the iron can be precipitated selectively from vanadium by firstly reducing the leach liquor to Fe(II) and V(IV) and then controlling the oxidation with peroxide to a potential around Eh 800-900 mV. At this potential, Fe(II) is oxidised (E°= 770 mV.) but not V(IV) (E°= 1000 mV.), and iron(III) hydroxide precipitates at pH 2-3. The oxidation is completed to recover V_2O_5.

However, solvent extraction of vanadium offers the greatest selectivity and highest purity V_2O_5 product.

4 Solvent Extraction of Vanadium

The solvent extraction of vanadium has been developed to treat neutral, acidic and basic solutions containing either vanadium(IV) or vanadium(V) species and is particularly useful in separating molybdenum or iron. Commercial plants have focussed on either the extraction of V(IV) as $[VO]^{2+}$ from acid solutions using Di(2-ethylhexyl)phosphoric acid (D2EHPA), or the extraction of V(V) as one of its many anionic species (Figure 1) using tertiary or quaternary amines. Tertiary amines extract vanadium more effectively at pH 2-3, but quaternary amines function over a wide pH range of 1-13.

4.1 Extraction of Vanadium(IV) with D2EHPA

Kerosene solutions containing 5-20% v/v D2EHPA extract $[VO]^{2+}$ with fast kinetics in preference to $[VO_2]^+$ at pH 1.5-2.5 [59], and are readily stripped with dilute H_2SO_4. Mechanistic studies by Sato et al [60], indicate that the extracted species is polymeric involving several D2EHPA molecules e.g. $(VO).S_4.H_2$. where SH = D2EHPA. The rate is proportional to the pH and the square root of D2EHPA concentration [61]. Hence the more concentrated solvent exhibits a much higher distribution coefficient; thus lowering the pH for extraction, and offering a greater selectivity for V(V) over Fe(III) [62]. Giavarini [62] also noted a small decrease in distribution coefficient using D2EHPA-TBP mixtures, but a much larger decrease as the concentration of vanadium increased above 2 g/l.

It is unfortunate that D2EHPA co-extracts Fe(III), as well as Al(III) and Mo(VI) [63]. Although the kinetics of Fe(III) and Al(III) extraction appear to be slow [64-65], the build-up of Fe(III) in particular is a problem because it cannot be stripped directly. By contrast, vanadium can be selectively stripped from molybdenum by pH adjustment [63].

To minimise the iron problem, and to maximise vanadium extraction with D2EHPA, leach liquors are kept fully reduced by treatment with iron powder [66], or sulphur dioxide [67] to maintain Fe(II) and V(IV) in solution. Nevertheless some Fe(II) is inevitably oxidised by air, particularly with mixer-settlers, and reports to the solvent. Column contactors have been recommended to minimise air contact [68], but do not appear to have been used commercially in this context.

Various innovative approaches have been tried to strip the Fe(III), including reductive stripping with sugars [69], caustic stripping to precipitate $Fe(OH)_3$, and a recirculating hot acid strip system using sulphur dioxide and activated carbon catalyst to reduce Fe(III) prior to recycling [70]. Just recently, Chen et al [71] reported that Fe(III) could be stripped in 20 minutes from 15% D2EHPA solutions containing 20% tri-alkyl-phosphine oxide (TRPO) using only 1 M.H_2SO_4 (Figure 4).

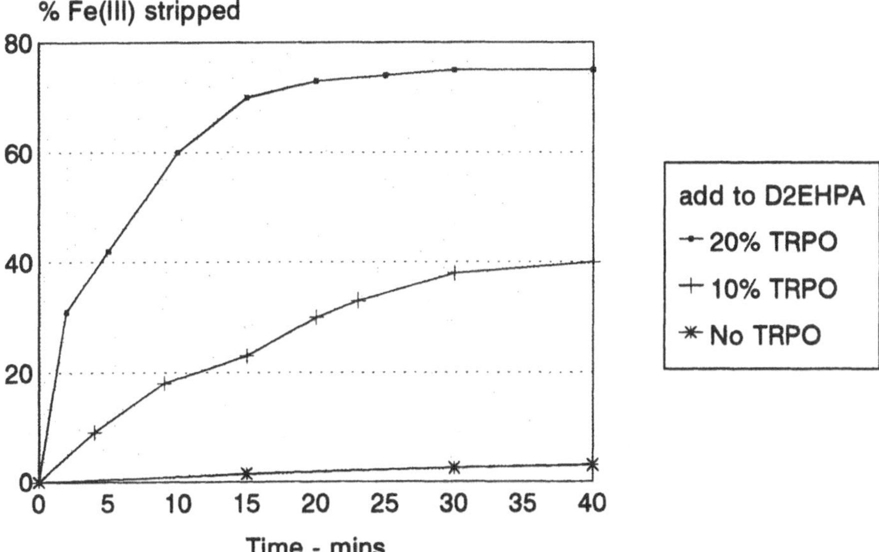

Figure 4. Effect of Tri-Alkyl-Phosphine-Oxide (TRPO) Addition on the Rate of Stripping of Fe(III) from 15% v/v D2EHPA using 1 M.H_2SO_4. (25°C., O/A = 1, Ref [71])

Similar results were reported by Ottertun and Strandell [64], using 10-15% TBP as a modifier and a 1-2 M.H_2SO_4 strip solution at 50-60°C. They found that TBP improved phase separation characteristics and had only a small negative effect on the extraction of vanadium. Extraction was best carried out at room temperature to minimise the slower co-extraction of Fe(III) and Al(III). Iron and aluminium stripped more slowly than vanadium and needed a higher concentration of acid. Thus by taking into account the differential kinetics of extraction and stripping of V(IV), Fe(III) and Al(III) it was possible to obtain an acceptably pure vanadium strip solution and V_2O_5 product (after oxidation and neutralisation), as well as scrub the D2EHPA for recycling.

4.2 Extraction of Vanadium(V) with Amines

The vanadate cation $[VO_2]^+$ readily hydrolyses even in dilute acid solutions above pH 1.5 to form oxy-anions (equation 1) which polymerise and protonate to a series of anions of general formula $[H_zV_{2x}O_{5x+z}]^{z-}$, [7].

$$2\ [VO_2]^+ \ + \ 3\ H_2O \ = \ [H_3V_2O_7]^- \ + \ 3\ H^+ \qquad (1)$$

Thus, except under strongly acidic conditions, vanadium(V) is present in solution as various anionic complexes which can be extracted by protonated tertiary amines in acid media, or by quaternary amines over a wide pH range. Stripping at neutral pH brings the highly soluble and stable decavanadate ion ($[V_{10}O_{28}]^{6-}$) back into the aqueous phase. Usually, however, stripping is carried out with ammoniacal salts at pH 9-10 and 60°C. to directly convert the decavanadate ion to metavanadate and precipitate insoluble ammonium metavanadate (AMV). This conversion is slow at ambient temperature.

Therefore amines are more flexible in their application than D2EHPA, and offer the advantage of not extracting Fe(III) and Al(III); thus providing a purer V_2O_5 product. It has been reported, however, that amines suffer from slow disengagement times, (particularly tertiary amines at concentrations > 10% v/v), and readily form cruds with any suspended solids [62]. Isodecanol is often employed as a phase modifier to improve disengagement.

Comparative studies with both types of amines [72,73], show that tertiary amines extract vanadium efficiently only between pH 1.5 - 4, whilst quaternary amines extract between pH 1.5 - 12, and are optimum between pH 6 - 9 [63,66,74]. Tertiary amines are most commonly used to extract vanadium from salt-roasted leach liquors, which are close to neutral pH, and then acidified to pH 3. The decavanadate ion is partially protonated at this pH, but is stable and does not depolymerise to precipitate vanadic acid.

With strongly acidic leach liquors from spent catalyst treatment, the pH can only be raised to about pH 1.5 before vanadic acid precipitates from solution, along with any $Fe(OH)_3$. Clearly, because the precipitation involves an equilibrium hydrolysis reaction, the exact pH will depend on the vanadium concentration and temperature. Hence it may be necessary to raise the pH during multi-stage extraction, both to avoid precipitation and to improve extraction. Around pH 1.5, however, anionic vanadium species begin to form, and thus the extraction efficiency is very dependent upon the pH of these liquors. Wilkomirsky [74] reported that <20% V was extracted from a solution containing 1.2 g/l V(V) at pH 1 using Alamine 336, but >80% V was extracted at pH 2. Agers [75] recovered 99% V in 4 stages of extraction using 5% tertiary amine from a liquor containing 4 g/l V(V) at pH 1.7 - 1.85.

Unfortunately, vanadic acid at pH <2 is a strong oxidant, comparable to chromic acid, and this slowly oxidises the amine, isodecanol, and aromatic kerosene diluents, as evidenced by the appearance of blue V(IV) ions in the aqueous phase. It is reported that the Union Carbide plant in the USA employed centrifugal contactors rather than mixer-settlers to minimise contact time and solvent degradation [66].

Quaternary amines are generally used to extract vanadium from slightly alkaline salt-roasted leach liquors around pH 9, and are stripped with ammoniacal ammonium chloride over several hours to produce AMV. However they can also be used to extract molybdenum. Bal et al [76] found that the extraction of V(V) with Aliquat 336 was optimal between pH 2 - 9, whilst Mo(VI) was extracted between pH 1 - 6. Thus it is possible to selectively extract V(V) from Mo(VI) at pH 9, and to subsequently extract Mo(VI) at pH 5. It was also noted that any phosphate in the liquor was partially co-extracted at pH <7, presumably as phospho-molybdates or phospho-vanado-molybdates, and that at pH 1 - 2 oxidation of the solvent by V(V) was evident over several days at room temperature - or over several hours at 50°C.

Stripping also showed differences. Both V(V) and Mo(VI) were stripped efficiently using 1 M. NaOH solution, but whereas molybdenum stripped immediately, vanadium stripped slowly over several hours unless the solution was heated. Hirai and Komasawa [77,78], reported that vanadium could not be stripped from quaternary amines by dilute acid at pH 2 unless the vanadium is reduced to V(IV) in the organic phase by either addition of ascorbic acid or by using an electrochemical cell.

Very little work has been carried out on the direct extraction of V(V) from either acid leach liquors at pH 1, or alkaline leach liquors at pH >10 using quaternary amines. However, Ritcey and Lucas [73] found that whilst extracting chromium from alkaline-roasted magnetite leach liquors at pH 13.2, it was also possible to extract the 1 g/l V(V) also present in the liquor with 10% Adogen 464 using 6 stages of extraction. The distribution coefficient for vanadium extraction was very low, and it was not clear whether this was due to the fundamental equilibrium property of the $[VO_4]^{3-}$ species at this pH, or due to competition from other ions.

4.3 Comparison of Tertiary and Quaternary Amines for the Extraction of Vanadium(V) from Acid Leach Liquors

In this work, the extraction behaviour of tertiary amine Adogen 364 and quaternary amine Adogen 464 (supplied by Sherex Pty.Ltd.) were compared on a synthetic acid leach liquor containing 7.9 g/l V(V) which was partially neutralised to pH 1.2. This was the highest pH which could be obtained before vanadic acid precipitated from this solution; although more dilute solutions could be neutralised further.

4.3.1 Phase Disengagement Tests

With 20% v/v Adogen 364 it was found that shake-out tests in separating funnels usually produced stable emulsions, but when mixed by an impellor in an "aqueous continuous" mode by adding organic slowly to the aqueous, phase disengagement was quite fast (35-55 secs) and unaffected by isodecanol. Slightly faster disengagement was observed when using aliphatic diluents, like Shellsol T (Shell), rather than aromatic diluents like Solvesso 150 (Exxon); and when using the chloride rather than sulphate media. With 20% Adogen 464 in Shellsol 2046 (Shell) as diluent, 10% isodecanol was required to prevent third phase formation. The disengagement time was about 1 minute in the aqueous continuous mode and 3 minutes in the organic continuous mode.

4.3.2 Extraction Tests

The effect of concentration of the two amines on the single stage extraction of vanadium at pH 1.1 - 1.2 is compared in Figure 5. This shows that the quaternary amine is the better extractant, with around 50% V extracted with 10% v/v Adogen 464. In other tests with 20% v/v Adogen 364, the percentage V(V) extracted from liquors of pH 0.9 decreased to about 30%. In all cases, the extraction was unaffected by changing the diluent, or by raising the temperature to 50°C.

The appearance of blue V(IV) ions associated with solvent oxidation was noticed at high organic loadings. It was more significant with tertiary than quaternary amines, and at 50°C. rather than at ambient temperature. Generally there was negligable V(IV) after several hours contact with the quaternary amine at 20°C.

An extraction isotherm obtained at 20°C with 20% Adogen 464 was observed to be S-shaped, as shown in Figure 6. Although quite high loadings of vanadium could be obtained in the organic phase by successive shake-outs, it was difficult to extract below 1 g/l V(V) at pH 1.05 because the proportion of anionic species decreases with concentration according to equilibrium reaction (1). Nevertheless >90% extraction was obtained in 3 stages from liquors containing 7.9 g/l V(V) without further neutralisation, and stripping was readily achieved with 1.5 $M.NH_3/NH_4Cl$ at 60°C.to recover AMV. At high organic loadings, organic degradation and crud formation became evident.

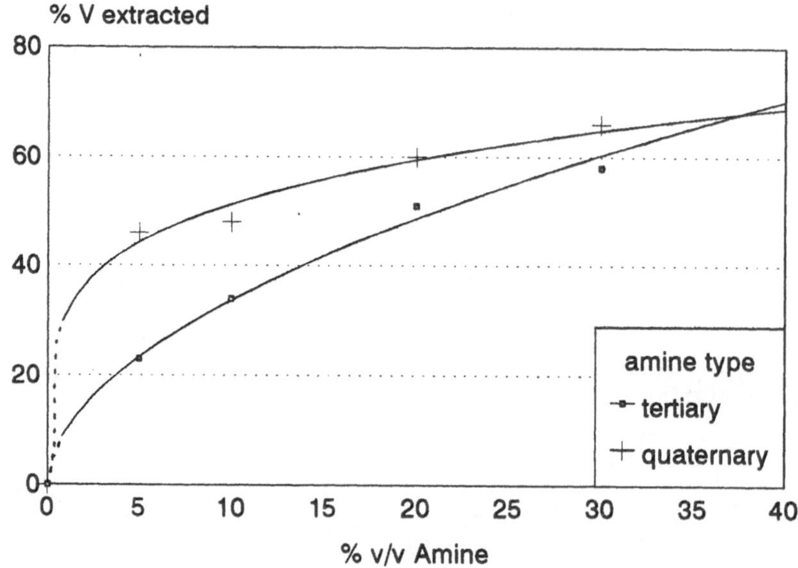

Figure 5. Effect of Amine Concentration and Type on the Extraction of Vanadium(V) from Acidic Solutions.
(Amine = Adogen 364 or 464 in Shellsol 2046; pH =1.1-1.2; [V] = 8 g/l).

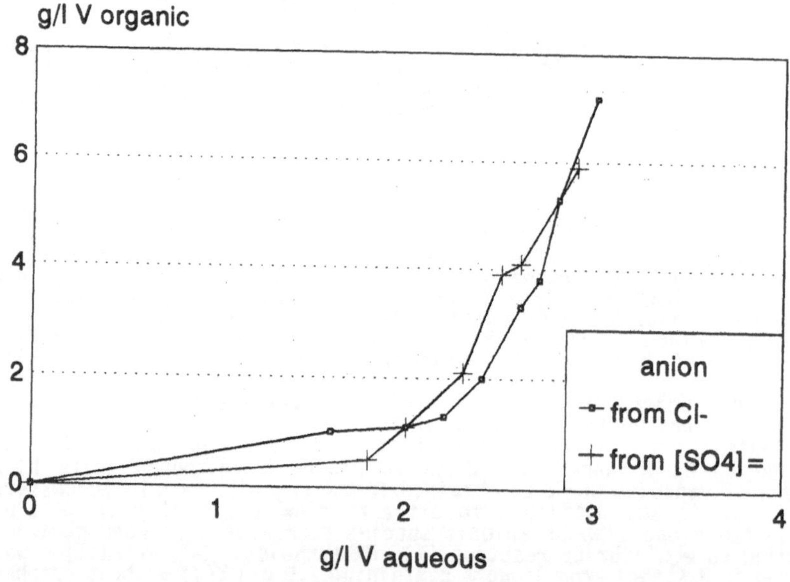

Figure 6. Extraction Isotherm for Vanadium(V) at pH 1.1 and 20°C. using 20% v/v Adogen 464 ((x) as Cl-, (o) as $[HSO_4]^-$ form).

4.4 Extraction of Vanadium(V) at pH >3 with Quaternary Amines

Because of the number of anionic vanadium(V) species which are stable between pH 3-13 (Figure 1), the extraction performance of quaternary amines will vary according to the predominant species present. The maximum saturated loading of the extractant also varies according to the charge of the ion and the number of vanadium atoms in each species. Table 1 summarises the predominant vanadium species according to their pH range and the theoretical maximum V(V) loading for 10% v/v Aliquat 336 (MW = 431) assuming a 1:1 ratio of charge to moles of amine. It demonstrates that the highest loadings are achieved with the decavanadate species, and that even with perfect extraction, loadings of the orthovanadate ion at pH 13 will be 7.5 times less.

Table 1. Calculated Vanadium Loadings on 10% v/v (0.22 M.) Aliquat 336 According to Speciation, Charge and pH.

pH	Species		V : charge	sat. loading
2.6-3.9	$[H_2V_{10}O_{28}]^{4-}$	decavanadate	2.50	29.6 g/l V
3.9-6.0	$[H.V_{10}O_{28}]^{5-}$	"	2.00	23.7
6.0-6.2	$[V_{10}O_{28}]^{6-}$	"	1.67	19.7
6.2-9.0	$[V_4O_{12}]^{4-}$	metavanadate	1.00	11.8
9.0-13.0	$[V_2O_7]^{4-}$ or $[VO_3(OH)]^{2-}$	"	0.50	5.9
>13.0	$[VO_4]^{3-}$	orthovanadate	0.33	3.95

Extraction isotherms obtained from contacting 10% v/v Aliquat 336 with 1.03 g/l V(V) at several pH values between pH 6 and pH 13.4 are shown in Figure 7. The isotherms confirm that the maximum loading is close to that calculated. However, higher loadings than predicted were obtained at pH 9-11, indicating that the decavanadate rather than the metavanadate ion is being extracted; whilst lower loadings were obtained at pH 13.4 (10 g/l NaOH) indicating some competition between hydroxide ion and orthovanadate ion.

An unusual feature is the S-shaped curve obtained at pH >9, which is not apparent when loading decavanadate ions. Such S-shaped curves have been noted previously with amine extractants [66] and have been attributed to polymerisation of the extractant. Recent fundamental studies by Belaustegi et al [79], confirm that strong aggregation occurs with tri-n-dodecyl-ammonium salts with the formation of tetramers and pentamers in toluene. This work indicates that these polymers occupy the interface and inhibit the extraction of the smaller vanadate ions until sufficient vanadium is present in the organic phase to disrupt and prevent aggregate formation. However more studies are required to resolve this phenomenon. In practical terms, more stages would be needed to completely extract vanadium if required.

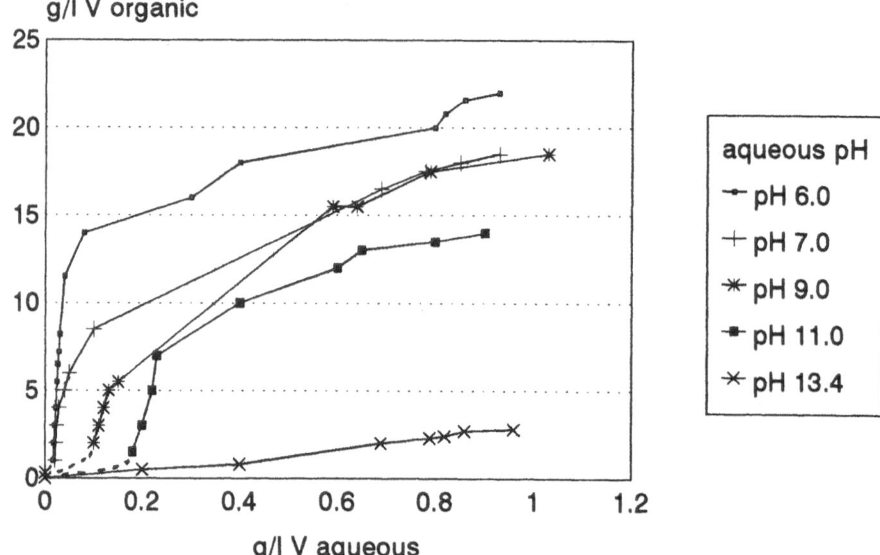

Figure 7. Effect of pH on the Extraction Isotherms of V(V) with 10% v/v Adogen 464 in Shellsol 2046. ([V] = 1 g/l.; 25°C.)

5 Conclusions

Overall, it appears that the quaternary amine extractant is versatile to acidic, neutral and alkaline solutions and is the reagent of choice to extract vanadium(V) from a variety of process options and liquors which may be obtained from treating spent catalysts and V_2O_5 sludges. Attention to temperature and contact time is required in acidic media to inhibit degradation by vanadic acid, whilst attention to hydroxide ion concentration is required in strongly alkaline media to inhibit competition with orthovanadate ion. In the presence of molybdenum(VI), selective extraction of vanadium is achieved at pH 9, whilst selective stripping is possible if co-extraction occurs at lower pH.

A mixed D2EHPA-TBP system is suitable for extracting vanadium(IV) from reduced acidic solutions, but particular attention must be given to minimise the presence Fe(III) and Al(III) in the leach solution, and their co-extraction with V(IV). This can be achieved by Eh control, excluding air and shortening the extraction time.

Intrinsically, a purer V_2O_5 product should be obtained using quaternary amines, but each catalyst and residue must be evaluated after considering the relative concentrations and value of other metals present.

6 Acknowledgements

The authors would like to thank Gordon Ritcey and Doug Flett for useful comments and advice on the solvent extraction of vanadium; and Graham Lloyd (AMMTEC Pty Ltd.), Ian Corrans (Normet Pty Ltd.) and Ramsay El Gammel (Clough Resources Pty Ltd.) for their interest and support in this work. E.Ho gratefully acknowledges the grant provided by the Australian Department of Industry, Trade and Commerce (DITAC) through AMMTEC to support her Honours project. S Lallenec gratefully acknowledges a grant by Clough Resources Pty Ltd. to support his Honours project.

7 References

1. Siemens R.E.,Jong B.W. and Russel J.H.,(1986) "Potential of Spent Catalysts as a Source of Critical Metals", **Conservation and Recycling**, 9(2), pp 189-196.

2. Biswas R.K.,Wakihara M. and Taniguchi M.,(1986) "Characterisation and Leaching of the Heavy Oil Desulphurisation Waste Catalyst", **Bangladesh J.Sci.Ind.Res.**,21, pp 228-237.

3. Sefton V.B.,Fox R. and Lorenz W.P.,(1989) "Recovery and Separation of Metals from Spent Petroleum Refining Catalyst." **US Patent** 4,861,565., 29 Aug.

4. Mukherjee T.,Chakraborty S.,Bidaye A. and Gupta C.,(1990) "Recovery of Pure Vanadium Oxide from Bayer Sludge." **Minerals Engineering** 3(3/4), pp 345-353.

5. Cotton F.A.and Wilkinson G.,(1972) **"Advanced Inorganic Chemistry. A Comprehensive Text."** 3rd.Edition, Interscience New York.

6. Druskovich D.M. and Kepert D.L.,(1975) "Base Decomposition of Decavanadate." **J.Chem.Soc.,Dalton**, pp 947-951.

7. Gupta C.K. and Krishnamurthy N.,(1992) **"Extractive Metallurgy of Vanadium"** (Process Metallurgy Series 8), Elsevier Amsterdam.

8. Ho E.M.,(1992) **"Recovery of Metals from Spent Catalysts"** Honours Literature Review, Murdoch University, Perth.

9. Rastas J.K.,Karpale K.J. and Titinen H.,(1983) "Recovery of Metal Values from Spent Catalysts used in Extracting Sulphur from Crude Petroleum." **Belgian Patent** 894,678, 31 Jan.

10. Hyatt D.E.,(1987) " Value Recovery from Spent Alumina-base Catalyst." **US Patent** 4,657,745., 14 April.

11. Fereniec L.,Adamski Z. and Sokalska G.,(1984) "Recovery of Molybdenum from Spent Molybdenum-Cobalt Catalysts." **Polish Patent** 123,436.,31 Oct

12. Tilley G.L.,(1982) "Recovering Valuable Metals from Deactivated Catalysts." **US Patent** 4,343,774., 10 Aug.

13. **"Kirk-Othmer Encyclopedia of Chemical Technology"** (3rd.Edition), Volume 5, John Wiley & Sons, New York, 1979.

14. Ward V.C.,(1989) "Meeting Environmental Standards when Recovering Metals from Spent Catalyst." **J.of Metals**, 41(1), pp 54-55.

15. Lee Fu-Ming.,Knudsen R.D.and Kidd D.R.,(1992) "Reforming Catalyst Made from the Metals Recovered from Spent Atmospheric Resid. Desulphurisation Catalyst." **Ind.Eng.Chem.Res.**, 31, pp 487-490.

16. Anon.,(1992) **"Recovery of Metals from Spent Catalysts."** Metrex Ltd., General Information Leaflet, March.

17. Anon.,(1982) "Recovery of Valuable Metals from Spent Petroleum Hydro-desulfurisation Catalysts." **Japanese Patent** 57,022,119., 5 Feb.

18. Takamatsu K.,and Kobayashi N.,(1979) "Separation of Molybdenum in Aqueous Molybdate and Vanadate." **Japanese Patent** 54,155,111., 6 Dec.

19. Anon.,(1984) "Recovery of Valuable Metals in Spent Heavy Oil Hydrodesulfurisation Catalyst." **Japanese Patent** 59,056,535., 2 April.

20. Anon.,(1982) "Recovery of Molybdenum, Vanadium, Cobalt and Nickel in Spent Petroleum Desulfurisation Catalysts.", **Japanese Patent** 57,082,122., 22 May.

21. Jong B.W. and Siemens R.E.,(1985) "Proposed Methods for Recovering Critical Metals from Spent Catalysts." in **Recycle and Secondary Recovery of Metals.**, Proc.Int.Symp., Ft.Lauderdale, Florida, A.I.M.E. Warrendale, pp 477-488.

22. Anon.,(1992) "Treatment of Desulfurisation Catalyst." **Japanese Patent** 57,003,716., 9 Jan.

23. Grzechowiak J.,Grysiewicz W.,Ostrowski A.,Randomyski B. and Walendziewski J.,(1987) "Recovery of Aluminium, Cobalt,Iron, and Nickel from Used Catalysts Containing Metals of the Ferrous Group and Possibly Molybdenum on Alumina Support." **Polish Patent** 136,713., Feb 28.

24. Wiewiorowski E.I.,Tinnin L.R. and Crnojevich R.,(1987) "Cyclic Process for Recovering Metal Values and Alumina from Spent Catalysts.", **US Patent** 4,670,229., 2 June.

25. Millsap W.A. and Reisler N.,(1978) "Cotter's New Plant Diets on Spent Catalysts...and Recovers Molybdenum, Nickel, Tungsten and Vanadium Products.", **Eng.and Min.J.**, 179(5), pp 105-107.

26. Ikeyama T.,(1987) "Recovery of High Purity Vanadium and Molybdenum Components from Oxidation-Calcined Spent Catalysts.", **Japanese Patent** 62,102,834., 13 May.

27. Biswas R.K.,Wakihara M. and Tanigushi M.,(1985) "Recovery of Vanadium and Molybdenum from Heavy Oil Desulfurisation Waste Catalyst.", **Hydrometallurgy**, 14(2), pp 219-230.

28. Fox J.S. and Litz J.E.,(1973) "Recovery of Metals from Spent Hydrodesulfurisation Catalysts." **German Patent** 2,316,837., 25 Oct.

29. Toda S.,(1989) "Alkaline Roasting of Spent Catalyst Containing Molybdenum and Vanadium.", **J.Min.and Materials Processing Inst.of Japan.**, 105(3), pp 261-264.

30. Anon.,(1983) "Apparatus for Roasting of Spent Catalysts." **Japanese Patent** 58,020,716., 7 Feb.

31. Anon.,(1980) "Metal Recovery from Spent Petroleum Refining Catalysts.", **Japanese Patent** 55,015,411., 23 April.

32. Toda S.,Ishikawa H. and Yoshida Y.,(1979) "Oxidation, Calcination of Waste Catalyst for Petroleum Refining." **Japanese Patent** 54,107,495., 23 Aug.

33. Naka I.,(1972) "Recovery of Vanadium and Molybdenum from Vanadium Containing Molybdenum Spent Catalysts." **Japanese Patent** 47,031,892., 13 Nov.

34. Sebenik R.F. and Ference R.A.,(1982) "Recovery of Metal Values from Spent Cobalt-Molybdenum/Alumina Petroleum Hydrodesulfurisation and Coal Liquifaction Catalysts: Laboratory Scale Process and Preliminary Economics." **Preprint Amer.Chem.Soc.Div.Pet.Chem.**, 27(3), pp 674 - 678.

35. Sebenik R.F.,LaValle P.P.,Laferty J.M. and May W.A.,(1985) "Recovery of Metal Values from Spent Hydrodesulfurisation Catalysts.", **US Patent** 4,495,157., 22 Jan.

36. Mueller H.R.,Krismer B. and Scottman W.,(1980) "Recovery of Valuable Metals from Catalysts." **W.German Patent** 2,908,570., 18 Sept.

37. Oshiumi T.,(1979) "Recovery of Molybdenum and Cobalt from a Spent Molybdenum-Cobalt-Aluminium Oxide Catalyst.", **Japanese Patent** 54,010,215., 25 Jan.

38. Ogui N.K.,Travkin N.S.,Lastovitskaya K.S.,Zaiko V.P. and Ryss M.A., (1971) **USSR Patent** 319,337. 2 Nov.

39. Krismer B., Mueller H.R. and Nadler H.G.,(1979 "Recycling of the Valuable Components Cobalt and Molybdenum from Hydrodesulfurisation Catalysts in Consideration of Methods Conserving the Environment.", **Erzmetall**., 32(12), pp 514 - 518.

40. Howard R.A. and Barnes W.R.,(1991) "Smelting Process for Recovery of Valuable Metals from Spent Catalysts on an Oxide Support.", **US Patent** 5,013,533., 7 May.

41. Welsh J.Y.,Picquet P. and Schyns P.,(1980) "Recovery of Metals from Catalysts Used in the Hydrodesulfurisation of Hydrocarbons.", **European Patent** 17,285., 15 Oct.

42. Gravey G.,Le Goff J. and Ganin C.,(1978) "Aluminium, Molybdenum, Vanadium, Cobalt and/or Nickel Chlorides.", **German Patent** 2,758,498., 6 July.

43. Timonova R.I. and Ivashentsev Y.I.,(1979) "Extraction of Vanadium by Hydrogen Chloride from Spent Catalyst Used in Sulfuric Acid Production **J.Appl.Chem.USSR (Eng.Transl)**, 52(1), pp 20-23.

44. Inooka M.,Shimizu T. and Nakamura M.,(1979) "Recovery of Metals from Spent Catalyst." **Japanese Patent** 54,099,704., 6 Aug.

45. O'Sullivan D.,(1992) "Facility Recovers Metals from Spent Catalyst.", **Chem.& Engng.News**, 70(43), pp 20 - 21.

46. Crnojevich R.,Wiewiorowski E.,Tinnin L.R. and Case A.,(1990) "Recycling Chromium-Aluminium Wastes from Aluminium Finishing Operations.", **J.of Metals**, 42(10), pp 42 - 45.

47. Parkinson G.,Ishio S.,Hibbs M. and Hunter D.,(1987) "Recyclers Try New Ways to Process Spent Catalysts.", **Chemical Engng.**, 94(2), pp 25- 31.

48. Swanson R.R.,Dunning H.N. and House J.E.,(1961) "Commercial Recovery of Vanadium by a Liquid Ion-Exchange Process.", **Eng.Min.J.**, 162(10), pp 110 - 115.

49. Pattnaik S.P.,Mukherjee T.K. and Gupta C.K.,(1983) "Ferrovanadium from a Secondary Source of Vanadium.", **Metall.Trans.B.**,14B.,pp 133-135

50. Thakur R.S.and Sant B.R.,(1975) "Chemicals from an Alumina Industry Waste and By-products.", **Metals & Minerals Review**, 15, pp 1- 3.

51. Mukherjee T.K.,Chakraborty S.P.,Bidaye A.C. and Gupta C.K.,(1990) "Recovery of pure V205 from Bayer Sludge.", **Minerals Engng.**, 3(3/4), pp 345 - 353.

52.Uhlemann E.,Ludwig E.,Rath L.,Stockmann V.,Kain C.,Fuertig H. and Haase R.,(1990) "Recovery of Vanadium from Spent Catalysts in Sulfuric Acid Manufacture.", **E.German Patent** 276,672., 7 Mar.

53.Budiu T.,Vatulescu R and Pal I.,(1982) "Recovery of Vanadium from a Spent Catalyst.", **Romanian Patent** 78,284., 30 Jan.

54.Hong X.,(1988) "Vanadium Recovery from Used Vanadium Catalysts.", **Xiandai Huagong**, 8(5), pp 35 - 41.

55.Reznitskii I.G.,Lieberman I.I.and Molodkina E.M., **USSR Patent**, SU 1,381,069.

56.Budavari S.,O'Neill M.J. and Smith A.(Eds),(1989) **The Merck Index.**,11th.Edition, Merck Co., USA.

57.Lasiewicz K.,(1989) "Method of Utilising Spent Vanadium-Potassium Catalysts.", **Polish Patent** 142,584., 31 Aug.

58.Badoiu R.,Segarceznu T.,Cornea A.,Dragomirescu T. and Busila G.,(1968) "Extraction of Vanadium from Spent Catalysts.", **Romanian Patent** 51,390., 8 Nov.

59.Coleman C.F.,Brown K.B.,Moore J.G. and Crouse D.J.,(1958) "Solvent Extraction with Alkyl Amines.", **Ind.Eng.Chem.**, 50, pp 1756 - 1762.

60.Sato T.,Nakamura T. and Kawamura M.,(1978) "The Extraction of Vanadium(IV) from Hydrochloric Acid Solutions by Di-(2-ethylhexyl)-phosphoric Acid.", **J.Inorg.Nucl.Chem.** 40, pp 853 - 856.

61.Ipinmoroti K.O. and Hughes M.A.,(1990) "The Mechanism of Vanadium(IV) Extraction in a Chemical Kinetic Controlled Regime.", **Hydrometallurgy**, 24, pp 255 - 262.

62.Giavarini C.,(1982) "Recovery of Vanadium from Ash-Leaching Solutions by Solvent Extraction.", **Fuel**, 61(June), pp 549 - 552.

63.Litz J.E., (1980) "Solvent Extraction of Tungsten, Molybdenum and Vanadium: Similarities and Contrasts." in **Extractive Metallurgy of Refractory Metals**, (H.Y.Sohn, O.N.Carlson and J.T.Smith Eds.), Met.Soc. AIME., Warrendale, pp 69 - 81.

64.Ottertun H. and Strandell E.,(1977) "Solvent Extraction of Vanadium (IV) with Di-(2-ethylhexyl)-phosphoric Acid and Tri-n-butylphosphate." in **Proc.Int.Solv.Extr.Conf.'77**, Toronto, C.I.M., pp 501 - 508

65.Flett D.(1992), Personal Communication.

66.Ritcey G.M. and Ashbrook A.W.,(1979) **"Solvent Extraction. Principles and Applications to Process Metallurgy. Part 2."**, Elsevier Amsterdam,

67.Biswas R.K.,Wakihara M. and Taniguchi M.,(1985) "Recovery of Vanadium and Molybdenum from Heavy Oil Desulfurisation Waste Catalyst.", **Hydrometallurgy**, 14, pp 219 - 230.

68.Ritcey G.M.(1992), Personal Communication.

69.Demopoulos G.P. and Gefvert D.L.,(1984) "Iron(III) Removal from Base-Metal Electrolyte Solutions by Solvent Extraction.", **Hydrometallurgy**, 12, pp 299 - 315.

70.Henrickson A.V. and Hazen W.C.,(1967) "Solvent Extraction Process for the Recovery of Vanadium Values.", **US Patent** 3,348,906.

71. Chen J.,Yu S.,Liu J.,Meng X. and Wu Z.,(1992) "New Mixed Solvent Systems for the Extraction and Separation of Ferric Iron in Sulfate Solutions." **Hydrometallurgy**, 30, pp 401-416.

72. Brooks P.T. and Potter G.M.,**(1974) "Recovering Vanadium from Dolomitic Nevada Shale."**, US Bureau of Mines, RI 7932.

73. Ritcey G.M. and Lucas B.H.,(1977) "Recovery of Chromium and Vanadium from Alkaline Solutions Produced by an Alkaline Roast-Leach of Titaniferous Magnetite.", in **Proc.Int.Solv.Extr.Conf.'77**, Toronto, C.I.M., pp 520 - 531.

74. Wilkomirsky I.A.E.,Luraschi A. and Reghezza A.,(1985) "Vanadium Extraction Process from Basic Steel Refining Slags.", in **Extraction Metallurgy '85** Inst.Min.& Metall., London, pp 531 - 549.

75. Agers D.W.,Drobnick J.L. and Lewis C.J.,(1962) **"The Recovery of Vanadium from Acidic Solutions by Liquid Ion Exchange."**, General Mills Technical Publication,- presented at AIME Annual Meeting, New York.

76. Bal Y.,Cote G. and Bauer D.,(1992) "The Peculiarities of Molybdenum(VI) and Vanadium(V) Extraction by Lipophilic Quaternary Ammonium Salts." in **Solvent Extraction 1990**, (T Sekine, Editor), Elsevier Amsterdam, pp 919 - 924.

77. Hirai T. and Komasawa I.,(1992) "Separation and Purification of Vanadium and Molybdenum by Solvent Extraction and Subsequent Reductive Stripping." in **Solvent Extraction 1990**, (T Sekine, Editor), Elsevier Amsterdam, pp 1015 - 1020.

78. Hirai T and Komasawa I.,(1993) "Electro-Reductive Stripping of Vanadium in a Solvent Extraction Process for the Separation of Vanadium and Molybdenum Using Tri-Octyl-Methyl-Ammonium Chloride.", **Hydrometallurgy**, 23(1/2), pp 73 - 82.

79. Belaustegi Y.,Olazabal M.A.,Fernandez L.A. and Madariaga J.M., (1993) "Interactions of Metal Extractant Reagents.V. The Aggregation of Tri-n-Dodecyl-Ammonium Hydrogenchromate and Hydrogensulphate Dissolved in Toluene.", **J.Solution Chem.**, 22(7), pp 641 - 646.

Ecological treatment of waste products from zinc hydrometallurgy

Diego Juan
Andrés Perales
Department of Chemical Engineering, University of Murcia, Cartagena, Spain

Abstract

The waste products of zinc hydrometallurgy fall into two categories:

- Ferric waste
- Toxic waste

The first of these, produced by leaching the roasted zinc, is based on jarosite, goethite or haematite depending on the process followed.

The second kind, toxic waste, arises from the purification of neutral leaching solutions, and consists of zinc powder and various cemented metals (copper, cadmium, cobalt, nickel, among others).

In this paper, two laboratory-developed methods based on the technique of solvent extraction are presented and discussed.

The ferric waste is treated by a conventional hot acid leaching process; the acid ferric leach is treated with an amine solution, and subsequently the extracted iron is precipitated directly in the organic phase. The ferric hydroxide is calcinated, and a high-quality iron oxide with 70% iron is obtained.

On the other hand, in the case of the residue from purification, it undergoes mixed leaching, in the presence of an organic phase and another aqueous one in an oxidising atmosphere at room temperature.

The zinc is recovered through stripping of the organic phase, and the metals remaining in the aqueous phase are recovered through selective extraction or through precipitation. Lead and silver are left in the leaching residue.
Keywords: Zinc hydrometallurgy, Ecological residue treatment, Solvent extraction, Environmental protection.

1 Introduction

Metal zinc is obtained through hydrometallurgy via electrolysis of sulphate solutions between insoluble electrodes. To prepare these solutions the procedure involves leaching of zinc oxide or carbonate, be it natural or artificial (obtained by oxidative roasting of the sulphur concentrates with dilute sulphuric acid arising from the electrolytic reaction).

The leaching of these oxides can be carried out in various ways but in all of these a zinc sulphate solution is obtained (with soluble metallic sulphates at pH 4.5-5 plus a sludge containing the insoluble sulphates, lead, calcium etc, along with the insolubles arising from the concentrate and iron contained in the latter in a form which depends on the process followed for its elimination (oxide+hydroxide, basic sulphates, jarosites, ferric oxides, goethites, etc) impregnated with the zinc sulphate solution [1] [2] [3] [4] [5] [6].

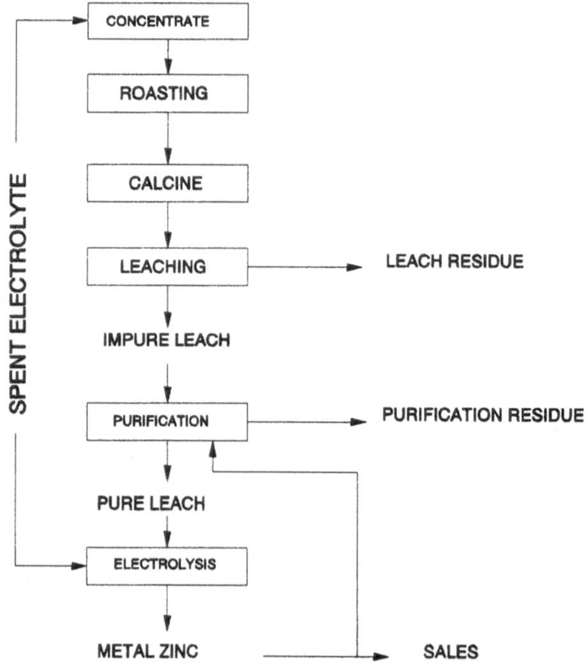

FIGURE 1 Simplified Diagram of Hydrometallurgical Method for Obtaining Electrolytic Zinc

The neutral primary solution (overflow from neutral thickeners) is not suitable for electrolysis due to the cations which accompany Zn^{2+} (Cu^{2+}, Cd^{2+}, Co^{2+}, Ni^{2+} ,etc) and these impurities have to be removed.

This purification is carried out through precipitation with zinc powder in the presence or not of overpotential reducers such as As_2O_3 or Sb_2O_3 and at temperatures of between 50°C and 90°C [7] [8] [9].

As a result of this process a purified solution is obtained, plus a residue containing minority metals in their metallic state, but in such small particles that rapid transition to solution is favoured.

Neither the leaching residue, nor that of purification is capable of passing the EPA [10] toxicity test, so the disposal of both even at controlled dumping sites is restricted and may seriously jeopardize the survival of the metallurgical industry in the near future.

As regards the purification residue, which, is the more toxic of the two, it was considered as a raw material for obtaining cadmium due to its content of this metal; other metals can also be recovered from it, such as copper, either as a metal or a preconcentrate. Unfortunately, given the low value of these metals, this solution is nor financially viable at present so the treatment of this residue has only ecological justification.

2 Global view of the problem

Individual hydrometallurgical treatment of each of the residues of this process has not provided a solution to the problem as we are always left with a residue which fails the EPA toxicity test or with a process which is not financially viable given current metal prices.

In the case of the leaching residue, thermal processes are being resorted to, through Waelz Furnaces or through Slag Fuming, with the objective of recovering zinc oxide and a slag which can be used as a filler once it has passed the EPA test [11].

The capacity of solvent extraction methods for separating metal ions in solution offers a range of possibilities which should not be underestimated when considering the options for treatment of these waste products.

One method, now being exploited in industry is the Excinres Process [12], which allows for the treatment of leach residue of the jarosite type, recovering soluble metal salts by washing and solvent extraction, and with a totally inert waste product [13].

This satisfactory introduction of the technique of solvent extraction, and its financially successful integration into basic metallurgy led us to the conclusion that one way of solving the environmental problems generated by zinc hydrometallurgy waste was to attempt to arrive at methods which would enable organic solvents to be used in metallurgical processes. The processes to be studied had to fulfil certain conditions if they were to be valid:

- No waste products or if waste was produced, this should be treatable in some other metallurgical process.
- Maximum number of metals recovered
- Maximum recovery of each metal
- Financial viability of the process.

3 Leaching residue

As can be seen in Figure 2 there are many options for neutral residue treatment. Currently a large part is treated in hydrometallurgical processes, generating an acid sludge which as indicated previously can be detoxified by the Excinres Process.

However, in this last-mentioned process waste is still produced, and, although it is free of soluble heavy metals in sulphuric medium at pH 2, may still pose problems of disposal and does not fulfil all the conditions stated previously as desirable.

The integration of a process for neutral waste treatment into zinc hydrometallurgy could be effected starting from what has already been achieved.

In fact, in all the methods of acid treatment of these residues an intermediate point is reached where two products are separated:

- A Pb-Ag Residue
- A leachate with all the soluble cations dissolved in it, and, basically Zn and Fe.

The first of these products is usable in lead metallurgy, and, in the second Fe must be removed from the solution for it to be included in zinc metallurgy.

In the classic processes (LAC [3], CONVERSION [4], JAROSITE [1], HAEMATITE [2], GOETHITE [5]), the removal of iron from the leach is achieved hydrothermally, as basic iron compounds (sulphates, jarosites, oxides) in such a way that the final product is a residue which even after thorough washing to eliminate all solubles is not accepted by the iron and steel industry.

Obviously then if Zn/Fe separation could be achieved by a method other than the

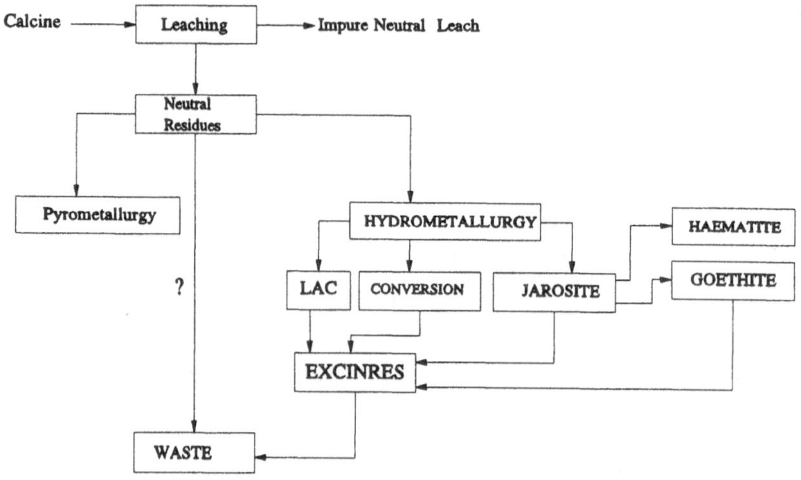

FIGURE 2 Various Alternatives for the Treatment of Neutral Leaching Residue

hydrothermal one which offered the steel industry an acceptable ferrous residue, the problem would be solved, as such a process could be easily included in zinc hydrometallurgy (Figure 3) and the waste products used in other industries.

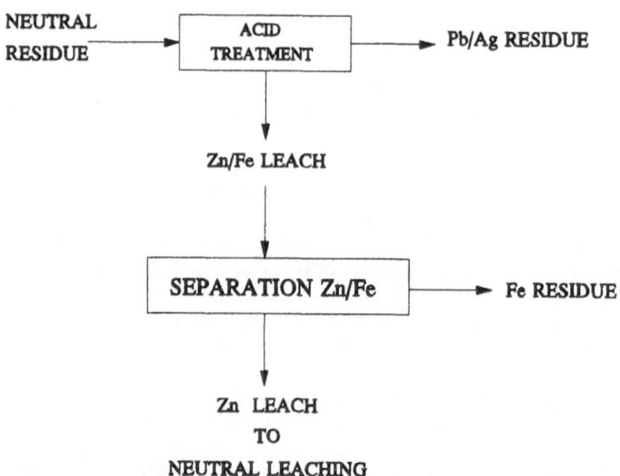

FIGURE 3 Diagram of Inclusion of the Process in Classic Zinc Hydrometallurgy

The Zn/Fe separation mentioned can be attained through organic solvent extraction. If an attempt is made to use acid reagents in this separation process then we encounter problems because of its non-specific nature although these have been solved in some cases, such as in the Versatic Acid Process presented by Colliers et al [14].

As a result of other work done by us [15] we considered that if the zinc, in a sulphate medium was not extracted with Primere 81R amine, then the Zn/Fe separation problem would be solved. The commercialisation of the iron residue would also be assured as not only would it then be accepted by the iron and steel industry but it could also be used as a pigment.

3.1 Equipment and experimental methods

For the extraction experiments the equipment, techniques, and analytical methods described in [15] were used. The zinc was analyzed via polarography in an ammonia/ammonium chloride medium [16].

For the leaching experiments a 2-litre reactor with thermostat was used, equipped with a multi-speed teflon stirrer of the filled-in anchor type.

The moist residue was wet pulped with the spent electrolyte (at the test temperature) and was transferred to the reactor.The load was topped up and when the temperature had dropped by 5°C the zero sample was taken.

The leaching loads were monitored for iron levels and were stopped when these reached maximum value.

3.1.1 Pre-experiment

The first thing to establish was if the Zn/Fe separation was possible with the extracting agent chosen (Primene 81R Amine* at 10% / Isodecanol at 10% / 80% Ibermetal D**).For this purpose a trial was carried out with solutions which contained Zn (II) and Fe (III) cations in the form of sulphates, in similar quantities to those which the real solutions would hold.

The two phases were mixed together by a mechanical stirrer (three times, for five minutes each time) until equilibrium was reached and were then analyzed.The results are shown in Table 1.

TABLE 1 Separation of Zn(II)/Fe(III) as sulphates with the amine Primene 81R

O/A	Initial g/l				Final g/l				Final g/l Zn	% Zn	
	ml	Zn	Fe	SO_4H_2	ml	Zn	Fe	pH	(corrected)	extraction	
1/1	100	85.8	36.8	29.9	99	86.5	32.5	-0.1	85.6	0.2	12.6
2.5/1	100	84.6	35.9	20.0	95	88.9	13.9	1.7	84.5	0.1	63.2
2.5/1	100	88.5	36.5	10.0	97.5	90.5	9.6	2.4	88.3	0.2	74.3
3.5/1	100	85.5	36.8	20.9	95	90.4	< 0.1	2.9	85.9	-	100

3.1.2 Technical development of a method for treating neutral residues.

In the light of the possibility existing of separating Zn (II) and Fe (III) in a sulphate medium through extraction with Primere 81R, the next step was to design a method

---

* Primary Amine, manufactured by Rohm & Haas.

** Parafinic Solvent, manufactured by Repsol S.A.

of treatment which would fulfil the conditions laid down.

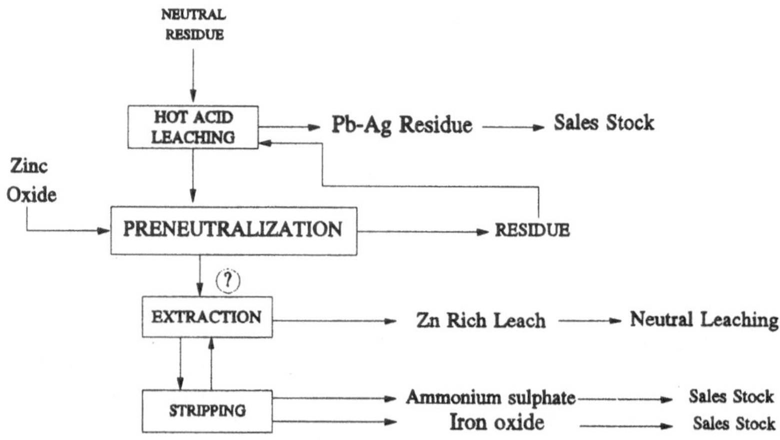

FIGURE 4 Theoretical Process to follow for the Treatment of Neutral Leaching Residue

As can be seen in Figure 4 the unkown quantity factor in this method lies in the connection between the two sub-processes, Acid-Attack and Extraction, since the conditions in which they occur are, by their very nature, opposite to each other.

In fact, the aim here is to obtain a lead residue of commercial value, and an acid Zn-Fe leach with the lowest possible acidity and maximum iron content. The ideal conditions for this are shown in Table 2, and, as stated, they appear to be contradictory.

TABLE 2 Conditions for Attacking the Residue and Fe(III) Extraction from the Leach

	Pb-Ag RESIDUE IMPROVES:	LEACH QUALITY IMPROVES:
TEMPERATURE	MAXIMUM (Ref.[10])	MINIMUM (Ref.[11])
FINAL ACIDITY	MAXIMUM (Ref.[10] [11])	MINIMUM (Ref.[9])
FINAL IRON	MINIMUM (Ref.[12])	MAXIMUM (Ref.[9])

To avoid this problem it was decided to effect the acid attack as outlined in Figure 5.

Following this plan, the work would be done in conditions of minimum acidity and maximum iron content (danger of low attack level and precipitation of jarosite and basic sulphates) at a low temperature, below that of precipitation of basic iron compounds, although there is a risk of low performance. The operating time would be short, also to prevent jarosite precipitation. It could more appropiately be described as a neutralization than an attack process.

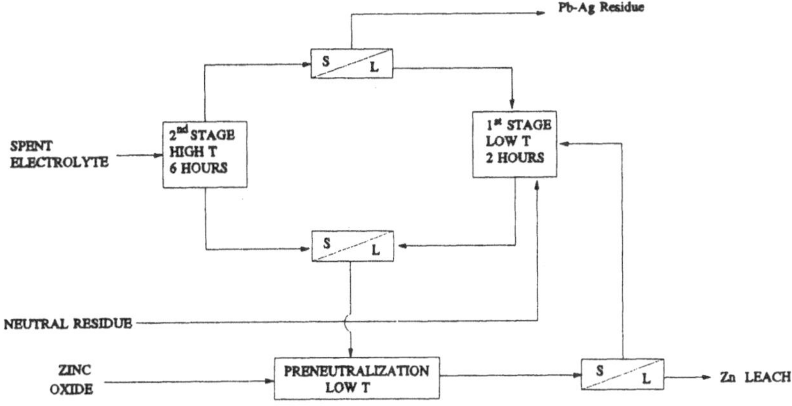

FIGURE 5 Theoretical Conditions for Leaching Residue Attack

In the second stage the attack is carried out at high temperature and acidity levels and with low iron content, so that the combination of acidity and temperature discourages the precipitation of basic iron compounds and a lead residue is obtained with the highest possible lead content, the iron compounds having been totally dissolved.

In the second stage the quality of the residue is the main concern, and in the first that of the leach.

Subsequently, the leach is pre-neutralized by the addition of roasted blende or of low quality zinc oxide, reducing the acidity to the maximum possible compatible with zero precipitation of iron compounds, attaining in this way the stated goals:

- Commercial quality Pb-Ag residue
- Low free acidity in the Zn-Fe leach
- High Fe (III) content in the Zn-Fe leach

From this moment on the union with the extraction process designed in [15] is simple, since we have two clearly defined processes:

- Residue attacking process, which causes no appreciable variation in a process of the classic jarosite, goethite or haematite type.
- Iron recovery process.

These two processes are connected in the extraction stage, as seen in Figure 6.

3.1.3 Experimental development.

3.1.3.1 Attack calculation.
To calculate the sulphuric acid needs, met by spent electrolyte, the following conditions were selected:

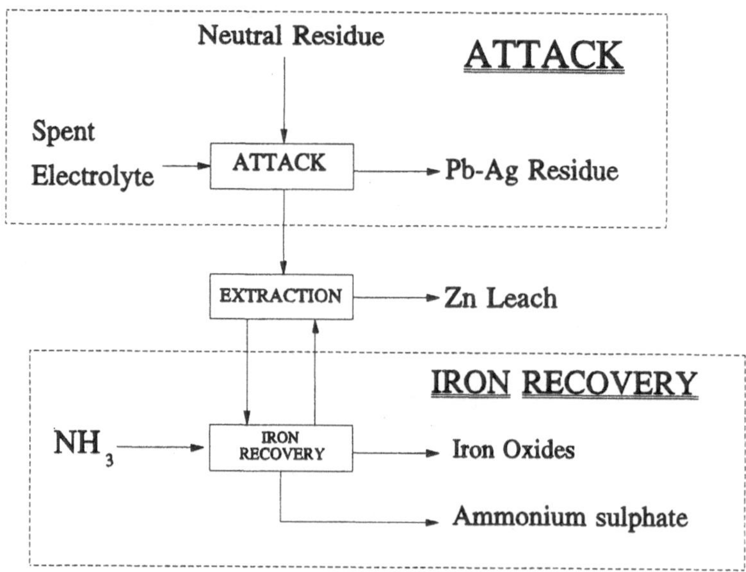

FIGURE 6 Processes used in the Method for Neutral Residue Treatment

- All matter retained in the residue in the form of soluble sulphate can be dissolved without acid consumption.
- All metals in free oxide form can be dissolved with an equivalent consumption of sulphuric acid.
- The sulphated compounds are dissolved through a REDOX process, by Fe (III) reduction, with no acid consumption.
- Zinc and iron, as ferrites, are 80% dissolved with equivalent sulphuric acid consumption.
- Free acid contained in the leach arising from the second attack, must range from 30 to 40 g/l.

3.1.3.2 Experimental method.
The operation was carried out in a thermostat controlled, two-litre, closed glass reactor.

The neutral residue, previously analyzed, was wet-pulped by means of the spent electrolyte (both residue and electrolyte were supplied by an electrolytic zinc plant), at experiment temperature, and put into the reactor, where the remaining spent electrolyte was added and the attack took place.

After this, 1% of 0.1% diluted Magnafloc 351 was added, allowed to settle for 1 hour, and the phases were separated.

In the case of a second stage attack, settled pulp was filtered and washed, whereas when it was first stage, it was prepared for the second.

Second-stage leaches were mixed together and kept, whereas first-stage ones were used at second stage, as per Figure 7.

The leaches obtained (from A-1 through to H-1) were pre-neutralized by a Slag-Fuming Process, until pH 1.1-1.2 was reached, at a temperature of 70°C.

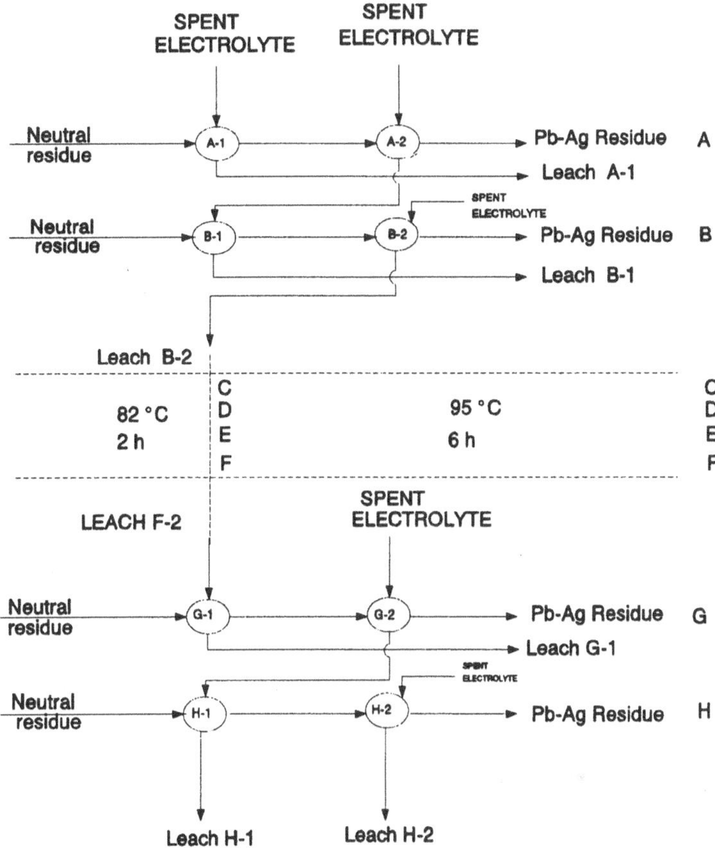

FIGURE 7 Diagram of Neutral Residue Attack

The pre-neutralized leach was summited to extraction by a solution consisting of 10% Primene 81R, 10% Isodecanol, and 80% Ibermetal D. The work was done in a 5-litre beaker, in a 2/1 organic/aqueous phase ratio and organic phase was added until the total Fe amount within the aqueous phase was less than 1g/l. The phases were separated, and the organic phase was water-washed, the washing water then being added to the aqueous extract.

Iron was precipitated into the organic phase, as shown in [15], with an ammonium sulphate/ammonia solution at a ratio of O/A = 1/1.

The phases were segregated, the aqueous phase was filtered and the iron residue was calcined, as shown in [15].

3.1.3.3 Results obtained and discussion

The tests were carried out according to the above method and as shown in Figure 8.

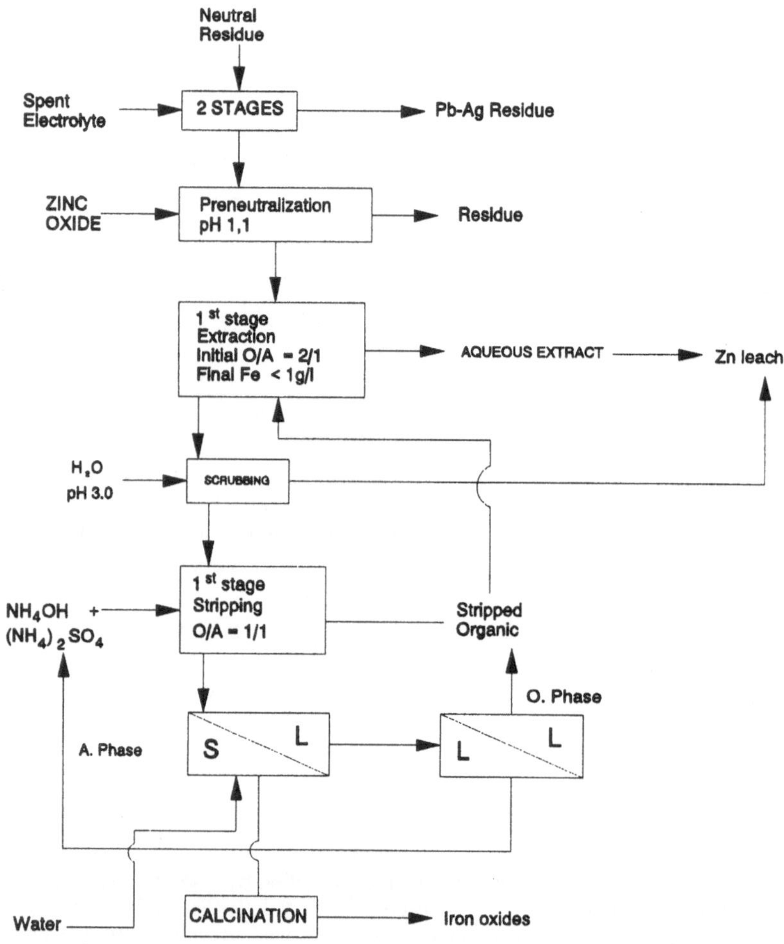

FIGURE 8 Diagram of Method of Treatment Required to Recover Iron from the Attacking Leach of the Neutral Residue

By this method, the following results (expressed in Tables 3, 4 and 5) were obtained.

TABLE 3 Results Obtained in Neutral Residue Attack

		g/l or % Content					% Recovery			
		Zn	Fe	Pb	Ag	SO_4H_2	Zn	Fe	Pb	Ag
Neutral Residue	(%)	22.0	24.0	4.8	0.0145	-	-	-	-	-
Spent Electrolyte	(g/l)	70	-	-	-	182	-	-	-	-
Pb/Ag Residue	(%)	6.8	14.7	17.1	0.052	-	8.7	14.8	100	100
Zn/Fe Rich Leach	(g/l)	102	40.1	-	-	32	91.3	85.2	-	-
Zinc Oxide	(g/l)	70	-	-	-	-	-	-	-	-
Preneutralization Residue	(%)	7.1	-	-	-	-	1.5	-	-	-
Zn/Fe Rich Preneutralization Leach	(g/l)	132	40.1	-	-	pH 1.1	-	-	-	-

TABLE 4 Results Obtained in Extraction Experiments

	ml Volume	g/l Content			pH	% Yield	
		Zn	Total Fe	Ferrous Fe		Zn	Fe
Pregnant Liquor	3000	132	40.1	0.75	1.11	-	-
Organic Stripped	7200	-	20.1	-	-	-	-
Raffinate	2780	142	0.8	0.7	3.2	-	-
Impure Loaded Organic	7213	0.17	16.4	-	-	-	-
Scrubbing Water	220	5.6	-	-	3.0	-	-
Zn Leach	3000	131.9	0.7	0.7	3.2	99.9	1.7
Pure Loaded Organic	7213	0.04	16.4	-	-	0.1	98.3

TABLE 5 Results Obtained in Stripping/Precipitation/Calcination Experiments

	Volume or Weight	g/l Contents			
		Zn	Fe	$(NH_4)_2SO_4$	NH_3
Loaded Organic	7000 ml	0.04	16.4	-	-
Aqueous Liquor	7000 ml	-	-	300	19.8
Stripped Organic	7000 ml	-	-	-	-
Product	5775 ml	0.04	20.1	360	4.3
Wet Iron Hidroxide	2190 g	-	precipitate	-	-
Washing Water	300 ml	-	-	-	-
Iron Oxide Calcinate	165 g	0.12	70.03	-	-

Because of its possible commercial value, the iron oxide obtained was analyzed, and the results were as show in table 6.

TABLE 6 Analysis of Calcined Iron Oxide

Fe: 70.03%	Cu: 80 g/t	Co < 5 g/t
Zn: 0.12%	Cd: 10 g/t	Ni < 5 g/t
S : 0.012%	In: 280 g/t	Ag < 5 g/t
	Te: 30 g/t	Sb < 5 g/t
		Mg < 5 g/t

As seen here, Indium accompanies Fe (III) during the process, and this would be a very interesting way to try to recover the former.

The calcined mixture still contains a relatively high amount of zinc, which should not be a problem, as the remedy would be to intensify the washing. Provided the Zn/Fe segregation process has been carried out, in the laboratory, and iron has been recovered in a marketable form, we would propose a global treatment process, where ammonium sulphate would be present as a by-product, arising from the SO_4^{2-} ion, extracted (from free acid within the zinc-and-iron rich leach and from that accompanying Fe (III) when this is extracted, with Primene 81R) [15].

A mass balance, working from the results obtained in the laboratory on the lines of the Figure 9 scheme, gives us the following:

TABLE 7 Results of Proposed Treatment

Consumption in m³/t or t/t of Dry Treated Residue		
Dry Residue	(t/t)	1000
Spent Electrolyte	(m³/t)	50130
Zinc Oxides	(t/t)	0.100
Sulphuric Acid 98 %	(t/t)	0.635
Ammonia 99 %	(t/t)	0.222
Water	(m³/t)	2.553
Productions in m³/t or t/t of Dry Treated Residue		
Zn Leach	(m³/t)	5.721
Pb-Ag Residue	(t/t)	0.314
Iron Oxides	(t/t)	0.316
Ammonium Sulphate	(t/t)	0.860

A scheme like this could take the form of the one in Figure 9.

As can be deduced from the above, very few variations need be introduced into the zinc plant's main production line, and no agent foreign to industrial plants is used, save the extractor in a side line of the process.On the other hand all products are reusable, so we feel that the proposed method complies with all the requirements.

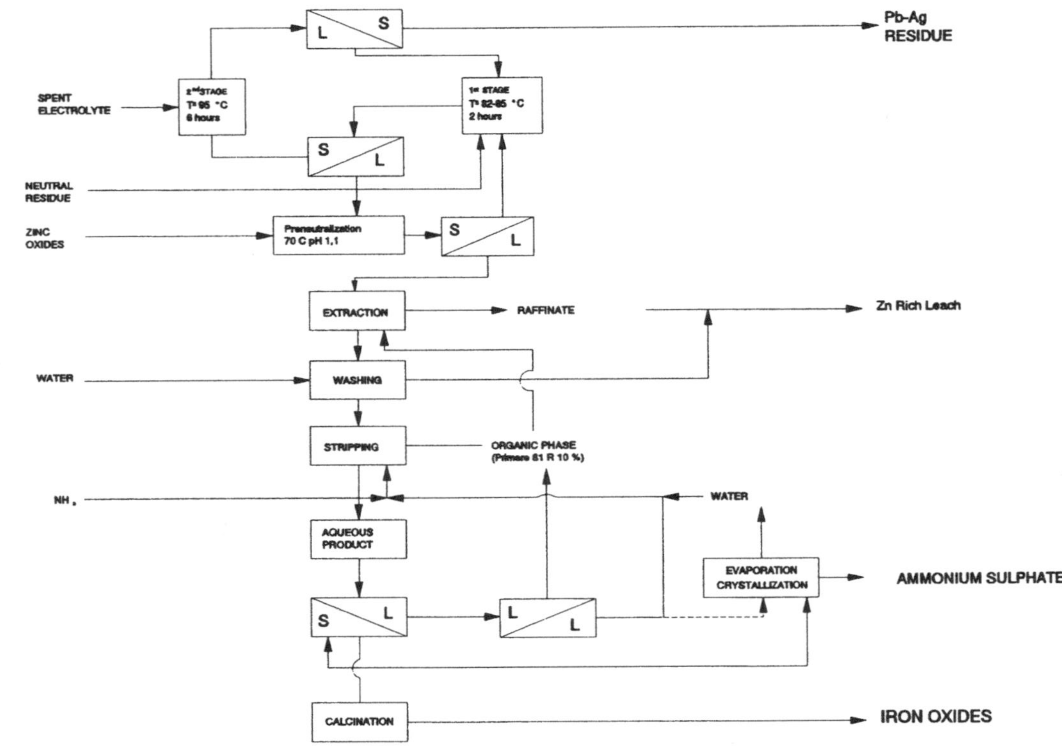

FIGURE 9 Diagram of Proposed Process

3.1.4 Financial comparison

From laboratory data on balances and plant diagrams, we carried out a financial comparison between this procedure and the Jarosite process.

To make this comparison, we have used the following base:

- Production: 100,000 tons./year.
- Efficiency: 94 % .
- Raw Blende: Zn: 55%, Fe: 8%
- Pb-Ag residue isn't given a value in either of the processes.
- The iron acids are not given a value.
- Disposal costs are considered to be, in the case of jarosite, around 50-62.5 DM/Zn tons. (60 DM/ton.)
- 3% additional losses are assumed, using the jarosite method

195000 x 0.03= 5850 t/year

(5850/100000) x 300$/t x 1.75 DM/$ = 30.71 DM/Zn ton.

Using these values and assumptions we achieved the results shown in Table VIII.

We can see here that costs in both methods are very similar at around DM 175 Zn/t, but if we consider the income from the sale of ammonium sulphate, at a very possible price of 75 DM/t, they are clearly lower in the SX process because this process then means a 30% saving.

TABLE 8 Comparatives cost of Neutral Residue Treatment via SX or Jarosite

	SX	JAROSITE
Investment	16.02 MM DM	15 MM DM
Capital Investment Charges	29.60 DM/t Zn	27.60 DM/t Zn
Direct Operating Cost	140.80 DM/t Zn	51.30 DM/t Zn
Labor Cost	10 DM/t Zn	5 DM/t Zn
Disposal Cost		60 DM/t Zn
Zn Losse Cost		30.72 DM/t Zn
TOTAL COSTS DM/t	180.40 DM/t Zn	175.06 DM/t Zn
Sales of $(NH_4)_2SO_4$ 700 Kg/t @ 75 DM/t	52.50 DM/t Zn	-
TOTAL COST DM/t Zn	127.90 DM/t Zn	175.06 DM/t Zn

Thus we think that the procedure we have developed, EXFERES (EXtracción de FE de los RESiduos-Iron's extraction from residue) fulfils the initial requirements, is on the whole environment-friendly and should show better economic results than the jarosite method.

4 Purification residue

Problems found with this waste treatment are totally different to those of the leaching residue; one is an economic problem, and the another is an environmental one.

At current metal prices the treatment of this residue is not profitable [17].

Moreover, it should not be dumped, because of its highly pollutant nature.

Further more, if we disregard economic results, and treat it, it always produces a residue which does not pass an EPA leaching test.

Therefore it is of interest to find a treatment method allowing:

- Process profitability
- Separation of cadmium treatment plant from zinc production plants
- Non-production of residues or in the case of any being produced, that it be treatable in other processes.

To fulfil the first of these conditions we must:

- Reduce if not remove zinc powder consumption
- Make the maximum number of metals involved profitable
- Recover most of the metal content, under conditions of maximum added value.

To fulfill the second requirement the process will have to allow the presence of an "filter" between the two installations preventing direct communication between them.

It is very clear to the authors that classical metallurgy (pyrometallurgy as well as hydrometallurgy] has exhausted its resources in this residue treatment.

Firstly it was thought that if a total attack were carried out against the residue we could achieve a solution of all soluble elements and a residue of Pb-Ag (Fig. 10). Subsequently the leach would be treated through solvent extraction, separating all and every one of the dissolved metals.

This way, zinc powder would not be needed in order to separate Zn/Cd thus giving value to Zn, Cd and Cu (already highly valued nowadays) as well as to Pb, Ag, Co and Ni.

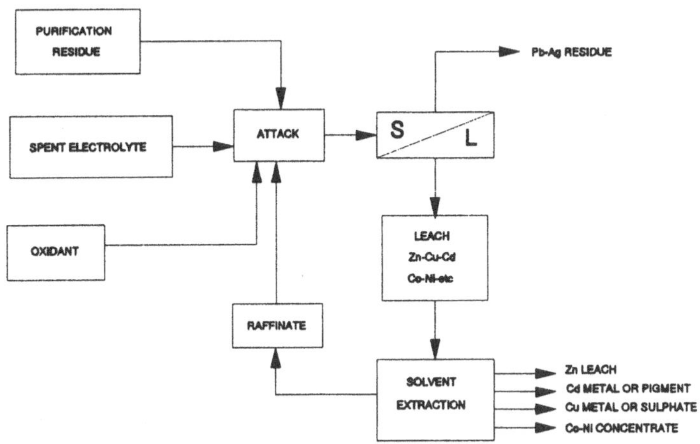

FIGURE 10 Basic Diagram of Purification Residue Treatment

Besides this, the metals, once separated, could be recovered either as crystallizated salts, as precipitates or in metallic state. Zn would be returned to the production plant after the organic phase had acted on it as a filter.

The method, thus described, has the economic handicap of needing a neutralizing agent to carry out the Zn extraction.

The method developed by Thorsen [18] would lead us to try and use the residue itself as neutralising agent, thus removing the financial inconvenience derived from the use of another neutralising agent, as seen in Figure 11, and the leach would undergo a separation treatment by organic solvents as above-stated.

To carry out Zn extraction, we thought, given our experience, of using DEHPA (2 Diethylhexyl Phosphoric Acid).

To extract copper, a bibliographic review was carried out, after consulting manufacturers, we finally opted for ACORGA M5640*.

For cadmium we used DEHPA, mainly to avoid introducing more reactives to the process.

4.1 Experimental method and equipment

To carry out the leaching experiments,we used the equipment shown in Figure 12, consisting of a 2l-or-5l-glass reservoir, an adjustable speed blade agitator and an air injection system.The extraction equipment was similar to that in the case of the leaching residue and consisted of a four glass mixer-set, working as separating tanks,equipped with teflon cocks, the dimensions being as shown in Figure 13. The agitators were of stainless steel.

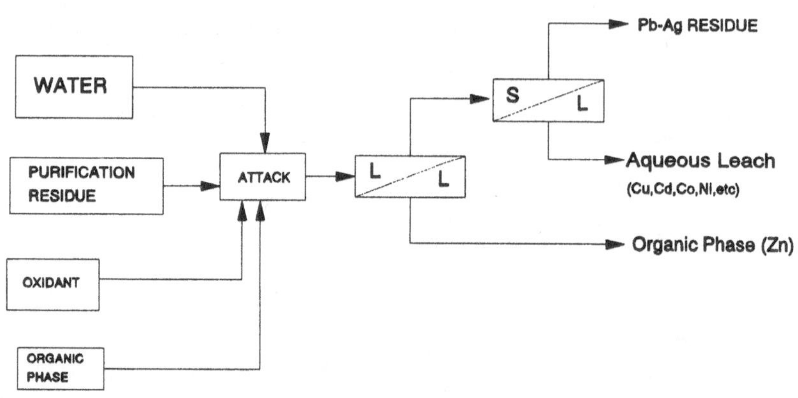

FIGURE 11 Diagram of Attack on the Residue with a Mixed Aqueous/Organic Phase

---

* Hydroxyoxime, manufactured by ICI

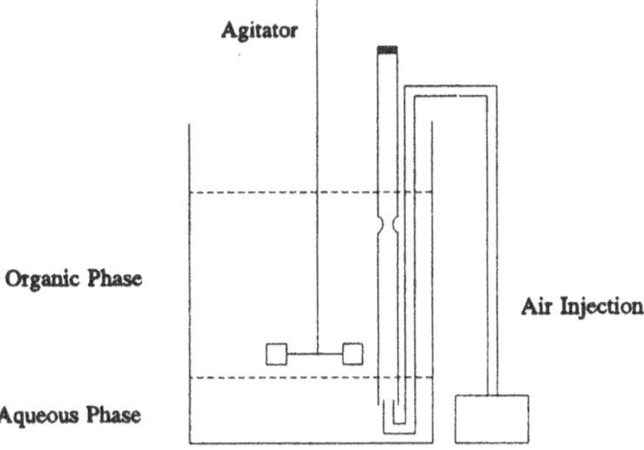

FIGURE 12 Apparatus used in Purification Residue Leaching Experiments

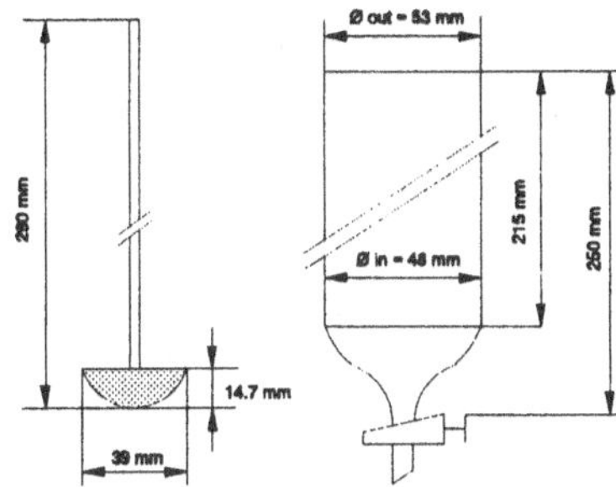

FIGURE 13 Reservoir and Agitators used in Extraction Experiments

For carrying out the leaching tests, the wet residue was suspended within the aqueous phase at pH 2 and was transferred to the reservoir, where the organic phase (DEPHA at 12.5% within Ibermetal D) was already undergoing agitation. Air injection was commenced and the operation was carried out for a pre-set time. At the end, the aqueous phase pH was adjusted to approximately 2, by the addition of sulphuric acid or ammonia.

The phases were separated and the aqueous one was filtered.

The extractions were made at free or controlled pH. In the first case, predetermined volumes of each phase were mixed, by mechanical agitation, for the time needed to reach equilibrium.After finishing the operation, the phases were separated.

In the second case, the operation was the same at controlled pH,but once the phases were balanced, we proceeded to pH control and adjustment (by addition of ammonia). The operation was repeated until the pH was constant between two subsequent stages without the addition of neutralising agent.

The analytical methods used were:

- Polarography within a ammonium chloride/ammonia medium for Zn, Cu and Cd.

- Polarography within an acetate medium for Pb (after separation as a sulphate).
- Permanganimetry as per Zimmerman for the Fe (where the Fe/Cu ratio in the sample was less than 3, Cd was removed through Fe (III) precipitation within Ammonium chloride/ammonia) [19].
- The remaining elements were analysed using techniques shown in [20] and ICP.
- The organic extracts were analysed by transformation in an aqueous solution, through extraction by CLH 1+1, and subsequent application of some of the above techniques.

4.1.1 Method

To reduce the amount of zinc and to regulate the copper and cadmium, purification residue underwent pH 3.5 leaching [9] obtaining a zinc-rich leach and a residue of copper and cadmium, which is the one we have worked on, as seen in fig.14.

This copper-and-cadmium residue, whose analysis is shown in Table IX, which was supplied by an electrolytic zinc plant, underwent total attack with controlled pH, with a mixture of an aqueous phase (washing waters and pH 2-sulphuric acid water) and an organic one (DEHPA at 12.5% within Ibermetal D) in the presence of air (as an oxidant).

$$ZnSO_4(s) + H_2O \text{------>} (Zn^{2+})_A + (SO_4^{2-})_A \quad (1)$$
$$(Zn^{2+})_A + [2RH]_O \text{----->} (Zn\ R_2)_O + [2H^+]_A \quad (2)$$
$$ZnO(s) + [2H^+]_A \text{------>} [Zn^{2+}]_A + H_2O \quad (3)$$
$$Zn(s) + \tfrac{1}{2}\ O_2\ (g) + [2H^+]_A \text{------>} [Zn^{2+}]_A + H_2O \quad (4)$$
$$MeSO_4(s) + H_2O \text{----->} [Me^{2+}]_A + [SO_4^{2-}]_A \quad (5)$$
$$MeO\ (s) + [2H^+]_A \text{------>} [Me^{2+}] + H_2O \quad (6)$$
$$Me\ (s) + \tfrac{1}{2}\ O_2(g) + [2H^+]_A \text{----->} [Me^{2+}] + H_2O \quad (7)$$

(Me = Cu or Cd)

TABLE 9 Analysis (%) of Copper-Cadmium Residue Samples used in the Experiments

	Sample 1	Sample 2
H_2O	21.5	33.3
Zn sol. H_2O	14.9	15.6
Zn sol. at pH 2	26.0	24.5
Total Zn	26.0	29.3
Cu sol. H_2O	0.2	0.2
Cu sol. pH 2	8.6	0.3
Total Cu	9.6	11.3
Cd sol. H_2O	4.8	2.3
Cd sol. pH 2	6.6	6.5
Total Cd	6.7	7.9
Fe	2.0	2.4
Pb	1.2	0.9
Insoluble	6.9	2.1
Ag	0.0170	0.0143
Co	0.0600	0.0700
Ni	0.0400	0.0600

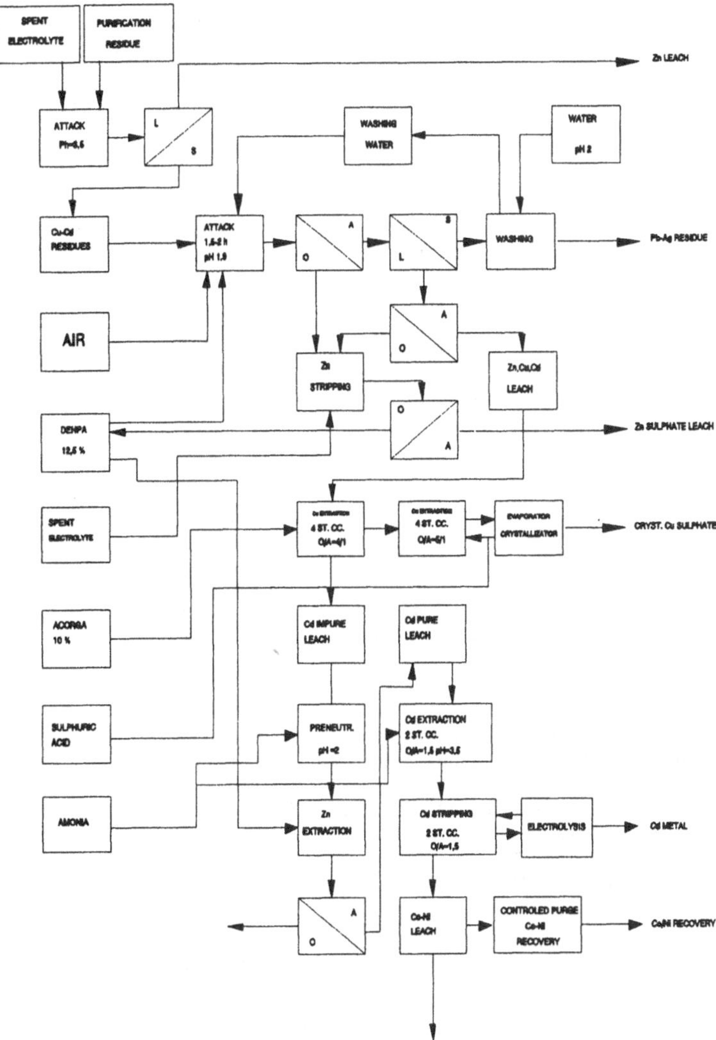

FIGURE 14 Method of Treatment Required

It is difficult to achieve the result that the protons needed for reactions (3) to (7) be equal to those produced in reaction (2), therefore, we have to control the pH, adding during the attack phase, sulphuric acid, if pH is more than 2.

After the attack and to achieve a minimum amount of copper and cadmium in the organic solution, the aqueous phase must be comprised between 1 and 2 g/l.

Once the attack is over, the phases are separated by settling, and the aqueous phase is filtered.

The filtration residue is pH 2 water-washed, mixing washing waters and attack leach. This washed residue is the lead-silver residue which is stored for sale.

Zinc is recovered within the organic phase, by spent electrolytic stripping, returning the organic phase to the attack stage.

The aqueous leach is sent for copper recovery, which is done through extraction by Acorga M5640 at 10% in Ibermetal D, as shown in Figure 11. Copper is recovered as metal or as a crystallised copper sulphide. In the strip the copper recovery stage aqueous extract contains a certain amount of zinc, which would cause problems when recovering cadmium by DEHPA extraction. To prevent this, the pH is adjusted to 2 and extraction by DEHPA at 12.5% in Ibermetal D is carried out (phase ratio O/A=0.5/1).

The organic extract is mixed with the organic phase that is included in the copper-cadmium attack residue.

The aqueous solution undergoes extraction at controlled 3.5 pH under the conditions shown in Figure 14, by DEHPA diluted at 12.5%, to recover the cadmium. Once stripped, this may be obtained either as a metal (electrolytically) or as a pigment (sulphur precipitation).

The aqueous leach arising from the cadmium-extracting stage could be:

- Recirculated to become Co and Ni rich and to carry out the recovery of these metals in a controlled purge.
- Recovered directly from the leach as a whole.

4.1.2 Results obtained and discussion

Operating as described above and taking into account that:

- The organic phase/aqueous phase ratio (O/A) at the attack rate should be from 5/1 to 10/1 to minimize crud formation.
- The amount of organic phase during the attack must be such that it becomes saturated, at the end, by the zinc arising from the residue (5-6 g/l).

A continuous treatment simulation was carried out in the laboratory, as seen in the Figures (from 15 to 20), the following results being obtained (Table 10 and 11):

TABLE 10 Analytical Results of the Proposed Process

		% (in solid) or g/l (in liquid)								g/t (sol.) or mg/l (liq.)		
		H_2O	Zn	Cu	Cd	Fe	Pb	Ins	SO_4H_2	Ag	Co	Ni
Purification Residue	(sol.)	32.04	29.20	11.30	7.90	2.40	0.90	2.10		143	700	600
Final Residue	(sol.)		7.5	3.00	0.10	12.30	17.20	40.80		2800	---	---
Zn DEHPA	(liq.)		5.9	0.10	0.04	0.19					1.5	1.5
Zn-Cu-Cd Leach	(liq.)		1.5	21.50	15.70						123	110
Cu Strip	(liq.)		0.02	26.50	--				118		0.5	3.5
Zn-Cd Leach	(liq.)		1.5	0.02	15.70						125	106
DEHPA Depur.Cd-Zn Leach	(liq.)		2.30	0.008	0.40							
Cd Leach	(liq.)		0.38	0.02	15.50						115	98
Cd Strip	(liq.)		0.30	0.008	15.30				37		7.4	1.4
Co-Ni Leach	(liq.)		0.06	0.02	0.20						108	97
Organic Loss		6 - 7 l/t dry residue										

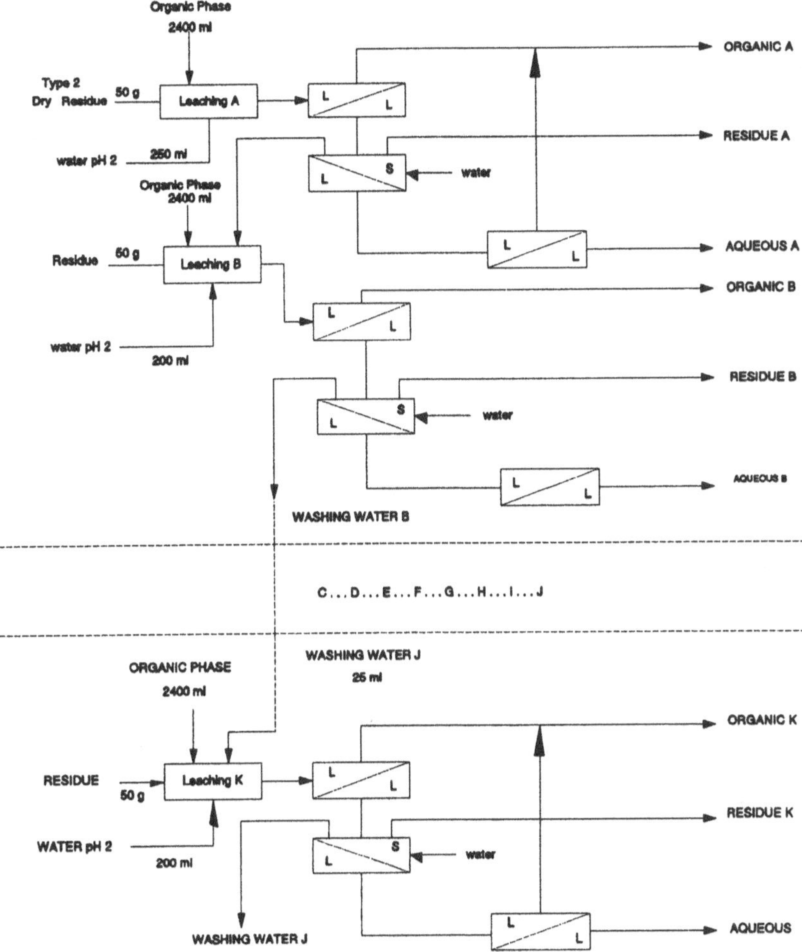

FIGURE 15 Diagram Residue Attack Simulation

Initial Residue = B + C + D + E + F + G + H + I + J + K = 500 g
Final Residue = B + C + D + E + F + G + H + I + J + K = 24 g
Loaded Organic = B + C + D + E + F + G + H + I + J + K = 24000 ml
Cu-Cd Aqueous Rich = B + C + D + E + F + G + H + I + J + K = 2500 ml

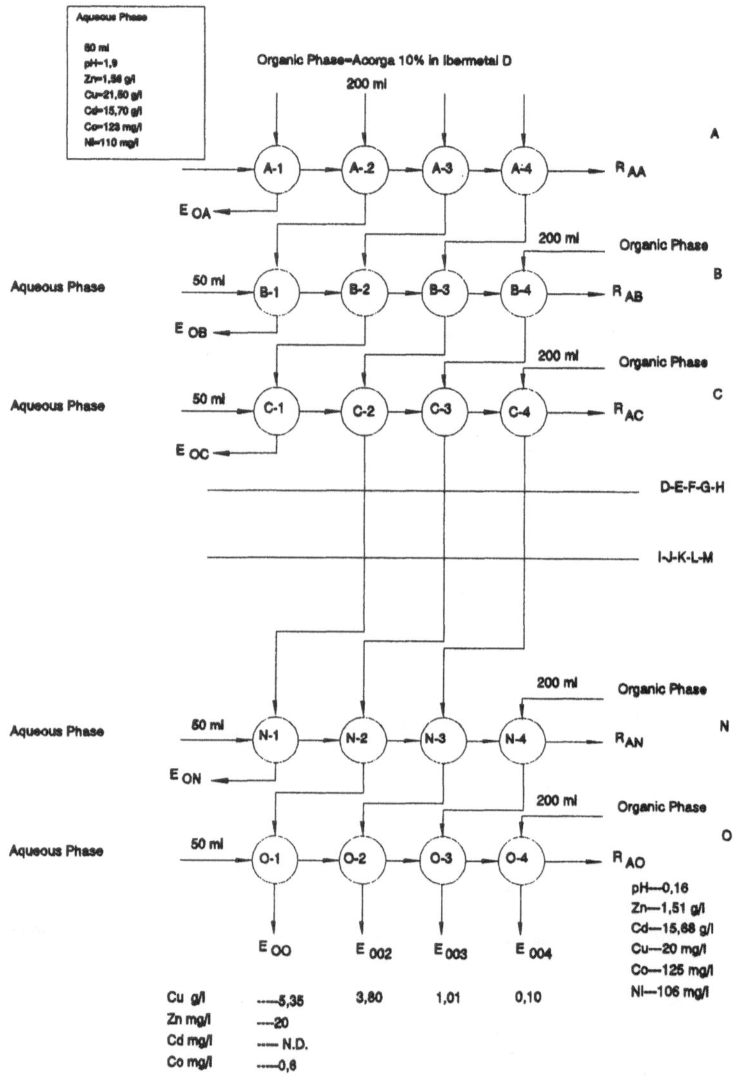

FIGURE 16.- Diagram and Results of Copper Extraction Process with Acorga M5640

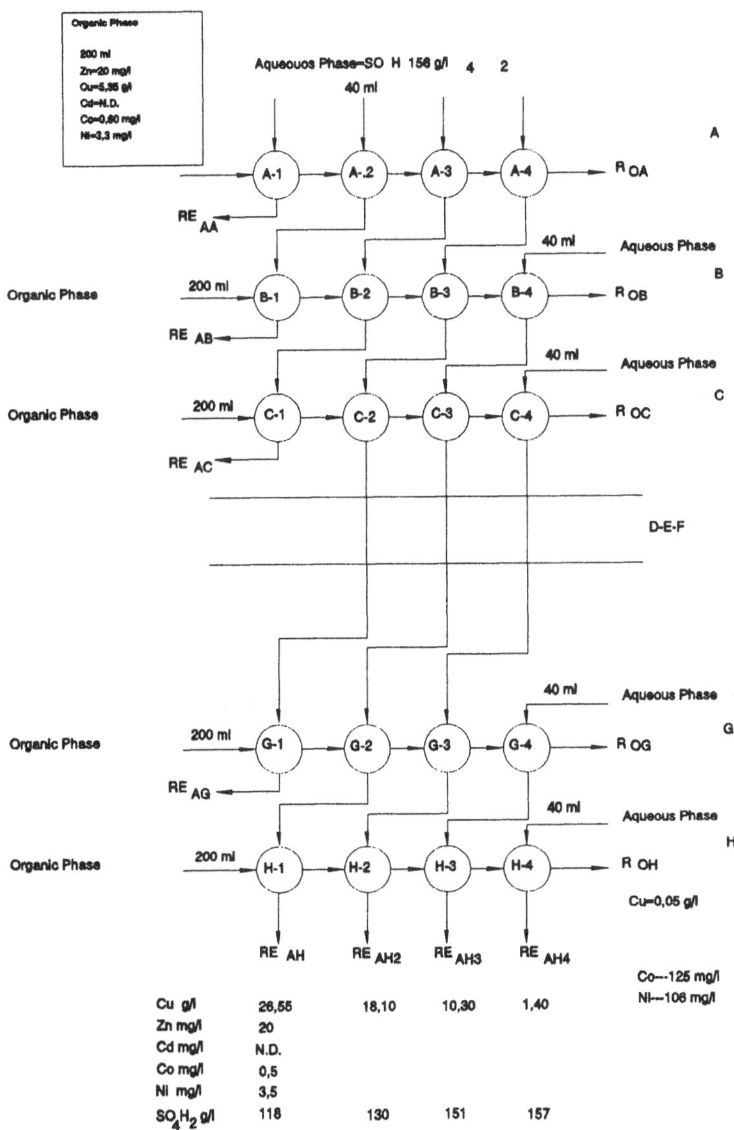

	RE AH	RE AH2	RE AH3	RE AH4
Cu g/l	26,55	18,10	10,30	1,40
Zn mg/l	20			
Cd mg/l	N.D.			
Co mg/l	0,5			
Ni mg/l	3,5			
SO_4H_2 g/l	118	130	151	157

FIGURE 17.- Diagram and Results of Copper Stripping Process with Sulphuric Acid

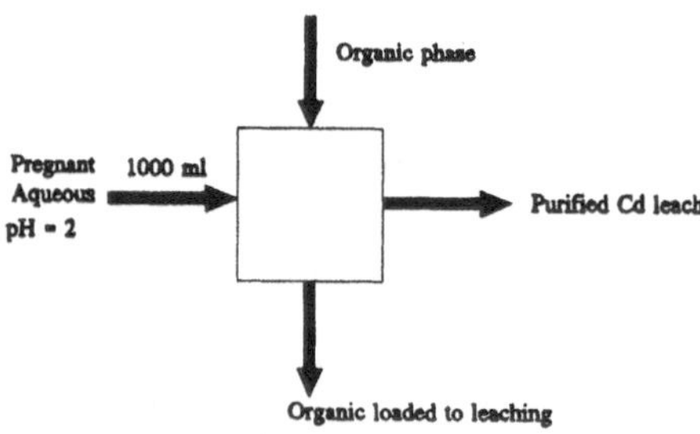

FIGURE 18 Cadmium Leach Purification

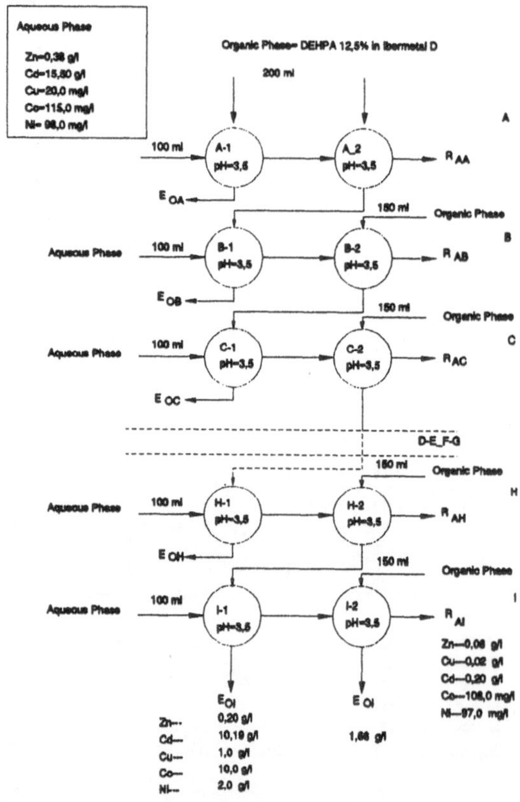

FIGURE 19 Diagram and Results of Cadmium Extraction Process with DEHPA

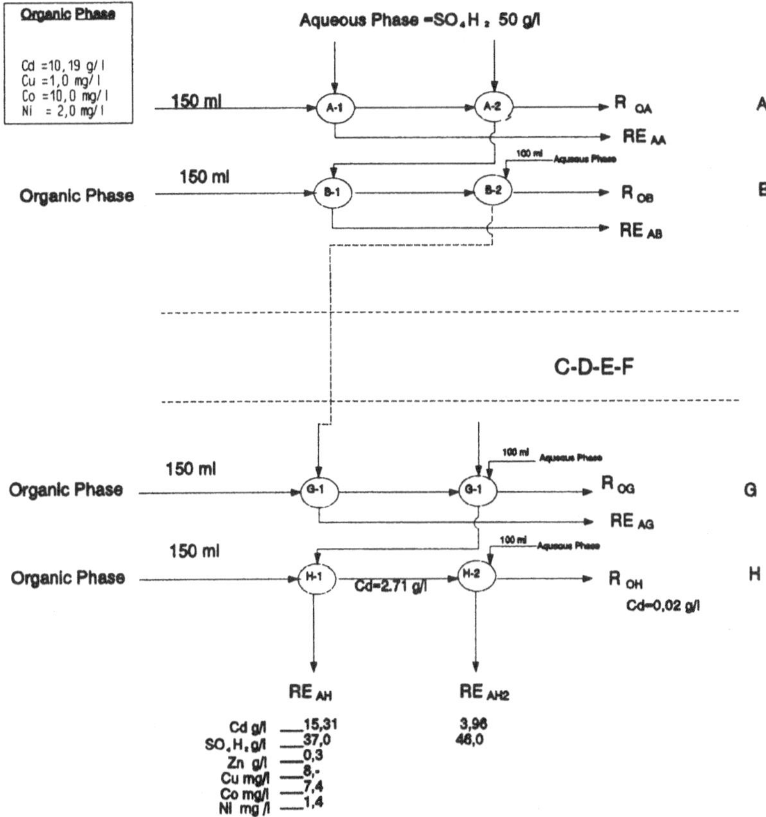

FIGURE 20 Diagram and Results of Cadmium Strippng Process with Sulphuric Acid

TABLE 11 Leaching Yield (%)

	Zn	Cu	Cd	Pb	Ag
Pb-Ag Residue	1.3	1.4	0.1	100	100
Organic Leach	97.1	4.2	2.5	-	-
Aqueous Leach	1.6	94.4	97.4	-	-
TOTAL recovery	98.7	98.6	99.9	-	-

From these results a mass balance was made of the process, calculating the recoveries, returns and losses in an industrial process and the results shown in Table 12 were obtained.

TABLE 12 Expected Results of Purification Residue Proposed Treatement

	% Metal in Pb-Ag Residue	% Return Metal to Zinc Plant	% Metal Recovery
Zn	1.3	98	98.0
Cu	1.3	3.9	93.5
Cd	0.1	2.2	97.7
Co	---	9.6	77.1
Ni	---	11.2	80.8
Pb	100	----	100.0
Ag	100	----	100.0

Technically, this method has advantages over other process as follows:

- Minimum impurity recirculation in the main circuit (zinc plant)
- Recovery and /or value increase of copper, cobalt , nickel, lead, and silver
- No zinc powder consumption
- Possibility of obtaining copper and cadmium as products with a high added value
- Full elimination of toxic and dangerous residue as required by EPA.

The only inconvenience would be the need to use diverse organic agents during the process.

4.2 Economic comparision

We compared this process with another two methods, one a one-stage low-temperature purification (antimony trioxide) and the other a two-stage high-temperature purification (arsenic troxide).

To do this an annual output of 80000 tons was assumed and an overflow of neutral thickeners was used as a base composition as follows:
Zn: 180 g/l, Cu: 0.8 g/l, Cd: 0.45 g/l, Co and Ni: function of the return

And the follows economic conditions also was used:

- Inflation rate was assumed to be 6 % and the plant service life, 10 years.
- Metal prices are those in force in 1992 (April-May).

From these results and considering that plants recover 50 % of their net actualised value and filling cost, capital reserves and circulating capital depicted in the Table 14.

TABLE 13 Summary of Annual Starting-Up Costs in M DM

	COCADEX	LOW TEMPERATURE	HIGH TEMPERATURE
INVESTMENT	7.62	5.27	4.48
CAPITAL INVESTMENT CHARGE	1.03	0.72	0.61
Direct Operating + Labor Cost	2.91	5.09	6.71
SALES	7.60	6.36	7.99
GROSS PROFITS	3.66	0.55	0.66
TAXES 35 %	1.32	0.20	0.24
NET PROFITS	2.34	0.35	0.43
NET CASH FLOW	3.38	1.06	1.04

TABLE 14 Estimated Capital Reserves in each Process Compared in M DM

	COCADEX	LOW TEMPERATURE	HIGH TEMPERATURE
Filled Cost	0.34	0.18	0.22
Capital reserves	1.90	1.59	2.00
Circulating Capital	0.76	0.53	0.45

Working with these values, we obtained the values for INTERNAL RATE OF RETURN and ECONOMIC LIFE TIME shown in the Table 15.

TABLE 15 Economic Results of the Processes Compared (TRI= Internal Rate of Return. VSE= Economic Life Time)

	COCADEX	LOW TEMPERATURE	HIGH TEMPERATURE
TRI %	34.66	11.7	12.9
VSE years	3.89	> 10	9.9

As can be observed the designed project improves on the currently used methods from an economic point of view .

In conclusion we feel that such a process as proposed in Figure 21 complies with all the conditions initially proposed and achieves all the proposed objectives.

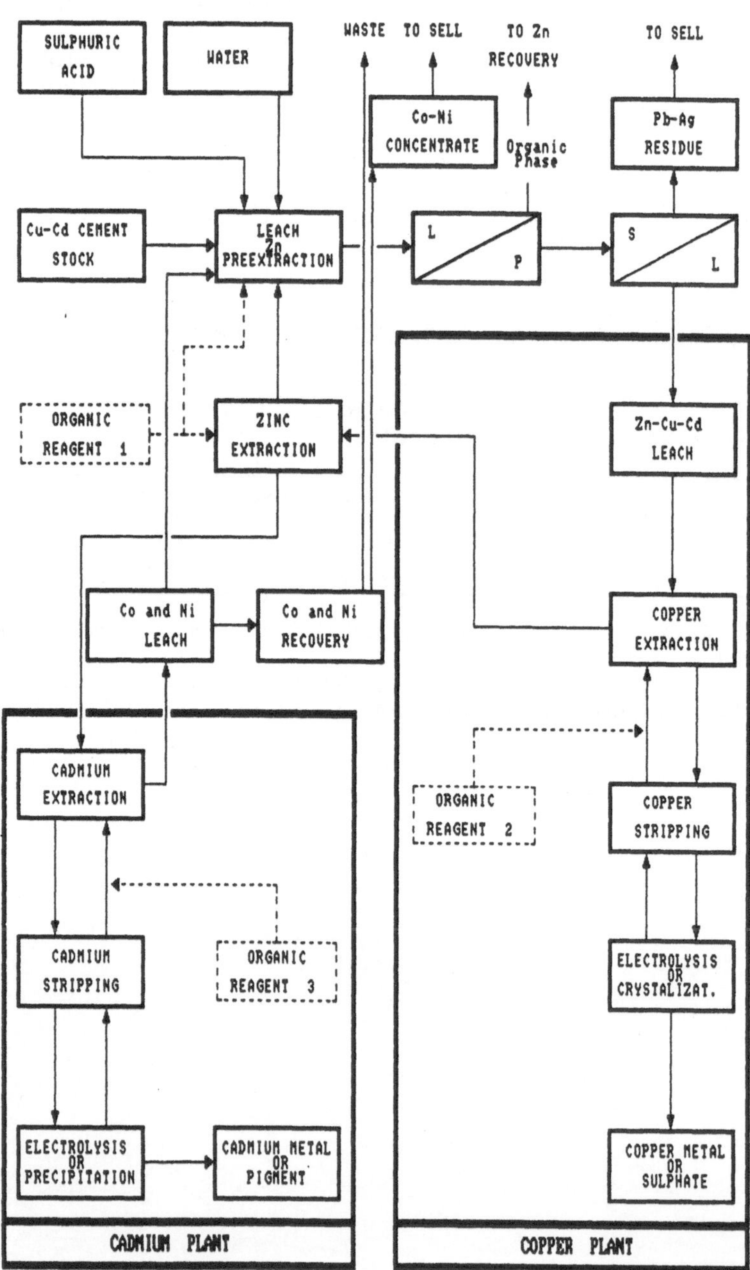

Figure 21 Diagram of COCADEX Process

5 References

1. Sitges F.and Arregui V. (1964) **Patente Española** n° 304.601.
2. André J.A. and Dalvaux R.J. (1970) Production of Electrolytic Zinc at the Balen Plant of S.A. Vieille Montagne. **World Symposium on Mining & Metallurgy of Lead & Zinc**. AIME. (St.Louis. Missouri. USA).
3. De Juan D. (1973) **Patente Española** n°412.670.
4. Rastas J.R., Huggare T.L.J. and Fugleberg S.P. (1973) **Finland Patent** n°A10/73.
5. Möller D.J.and Wiegand V. (1979) **Utilization of Electrolytic Zinc Plant Residues.** Lurgi Information .1979
6. Arregui V., Gordon A.R. and Steintveit G. (1980) The Jarosita Process: Past, Present and Future. **Symposium on Metallurgy & Environmental Control**. AIME.
7. Steintveit G. (1970) Electrolytic Zinc Plant and Residue Recovering. Det Norske Zinkkompani A/S. **World Symposium on Mining & Metallurgy of Lead & Zinc**. AIME. (St.Louis. Missouri. USA).
8. Schmidt. (1970) **Cadmium Production at the Rühr-Zinc GmbH at Datteln.** Erzmetall (Germany).
9. De Juan D. and Del Valle J.L. (1980) Un nuevo método para la recuperación del cadmio en las plantas de cinc electrolítico. **IV Congreso Internacional de Minería y Metalurgia**. (Huelva-España).
10. U.S. EPA. **Standars applicable to owners and operators of hazardous waste treatment, storage and disposal facilities**. Code of Federal Regulations, Tittle 40, Parts 260-267.
11. Piret N.L. and Castle J.F. (1990) Review of Secondary Zinc Treatment Process Option. **Symposium on Recycling of Metalliferous Materials**. IMM.
12. Selke A. and De Juan D. (1989) Zinc Recovery by Solvent Extraction. **International Symposium on Production and Technology in the Metallurgy Industries**. (Kohln. Germany).
13. De Juan D. and Del Valle J.L. (1988) El proyecto Excinres:de la ilusión a la realidad. **VIII Congreso Internacional de Minería y Metalurgia**. (Oviedo-España).
14. Collier D., Heng R., Lehman R. and Pietsch H.B. (1986) The Versatic Acid Process Solution to the Iron Problem in the Zinc Industry?. **Symposium on Iron Control in Hydrometallurgy**. IMM. (Toronto-Canada).
15. De Juan D., De Juan J. and Lozano L.J. (1992) Empleo del par Primene 81R/Fe (III) para la obtención de óxidos de hierro. **Revista de Metalurgia**. (Madrid-España).28(2), 1992, pp. 98-111.
16. De Juan D. (1966) Determinación polarográfica arbitral de cobre y cadmio. **Revista de Metalurgia**. (Madrid-España) .1966, pp. 283-285.
17. Perales A. (1993) **Estudio técnico-económico de un método de tratamiento no contaminante de los residuos de purificación en la hidrometalurgia del cinc**. Tesis Doctoral. Universidad de Murcia.
18. Thorsen G. **U.S.Patent** n° 4.008.134. **Britain Patent** n° 1.474.944.
19. Comision de quimicos de la industria minera y metalurgica alemana. (1956) **Métodos de análisis de control industrial.** Ed. Aguilar (Madrid-España).
20. De Juan D. (1971) Análisis de constituyentes en productos metalúrgicos no férreos por espectrofotometría de absorción atómica. **Información de Química Analítica**. (Madrid-España).1971, pp. 55-62.

Production of indium from Imperial Smelting Process residues

Bosko Nikov
Petar Stojanov
Tomislav Stojadinovic
"Zletovo" Metallurgical and Chemical Company, Titov Veles, Former Yugoslav Republic of Macedonia

Abstract
This paper represents a brief review of the activities and the results obtained so far within the Smelters research center aiming to recover indium from indium bearing residues including copper dross and run off zinc.

The experimental work was carried out in three main fields: leaching, solvent extraction and metallic indium precipitation.

A solution of di-2-ethylhexyl phosphoric acid (D2EHPA) in kerosene was found to be most convenient for purification and concentration of indium bearing solutions. Further purification of the strip liquor was carried out by an additional solvent extraction using TBP. Loaded TBP was stripped with water.

After the more electropositive impurities have been cemented on an indium plate, indium is cemented by means of an Al-Zn alloy. Thus, a relatively high purity indium (99.985%) was produced after the sponge was subjected to a caustic smelting.
Keywords: Residues, indium, leaching, solvent extraction, cementation, D2EHPA, TBP.

1 Introduction

Indium is one of the most dispersed elements in the earth's crust, but it is wide spread throughout the intermediate products of some lead and zinc production processes.

The distribution of indium differs from one lead and zinc producing plant to another depending on the process route applied. Significant differences are evident even among plants operating the same technology [1]. Consequently, different methods have to be applied for indium recovery depending on the nature of the raw materials to be treated.

2 Raw materials

Three types of intermediate products can be a source for indium in the ISP based lead and zinc production. These are: copper dross, liquation lead and run-off zinc.

Copper dross is an intermediate product which arises in the first stage of lead bullion refining - decopperising of lead. The dross is an eutectic mixture consisting mainly of lead and copper, but its composition is rather complex and minor quantities of metals associated with the bullion such as: arsenic, antimony, tin, bismuth, silver, indium etc. are also reported [2], [3].

The chemical composition of copper dross depends primarily on the raw materials processed in the smelter. However, the composition shown in table 1 may be considered a typical one for a manually collected dross.

Particle size distribution of the dross depends very much on the procedure applied during the decopperising process and the method used for collecting the dross [4]. The pneumatically separated dross is far more homogeneous and much finer than the dross collected manually, the particle size distribution of which is not uniform due to the substantial amount of mechanically entrained metallic lead. The particle size distribution of manually collected copper dross at the "Zletovo" ISP smelter in Titov Veles is shown in table II, together with the indium and copper content in each fraction.

Contrary to expectations in view of the physical characteristics of the pure metal, the greater part of the indium charged in the Imperial Smelting furnace goes forward with the zinc vapor and reports in the slab zinc. During the refining of the zinc, however, indium always remains in the reflux and is subsequently distributed between liquation lead and the run-off zinc. After multiple reboiling of the latter the concentration of indium increases up to 0. 5%. Further concentration in a New Jersey type column is not possible due to a poor fluidity of the reflux. However, zinc powder, zinc oxide or even zinc metal can be produced from this material in a conventional retort and the eventual residue may contain up to 5% In.

Table 1. Chemical composition of manually collected copper dross at Zletovo smelter

Element	Concentration[%]	Element	Concentration[%]
Pb	60-65	Ag	0. 05-0. 07
Cu	20-30	Cd	0. 002-0. 01
Fe	0. 5-1	Zn	1. 0-2. 5
Sn	0. 2-0. 4	Bi	0. 05-0. 15
As	0. 5-0. 8	In	0. 03-0. 12
Sb	1. 5-2. 5	S	1. 0-2. 0

Table 2. Indium and copper contents in separate copper dross fractions

Fraction [mm]	Share [%]	In [%]	Cu [%]
>2	7. 9	0. 036	14. 70
>1, <2	5. 85	0. 045	15. 30
>. 315, <1	13. 45	0. 036	13. 30
>. 210, <. 315	6. 05	0. 048	19. 35
<. 21	66. 75	0. 132	35. 40

3 Treatment of the zinc powder plant residue

GOB zinc is used for zinc powder production in the "Zletovo" smelter. A residue consisting mainly of lead, zinc and tin is removed from the retort at the end of each campaign. Indium is concentrated in this residue reaching a figure of 2%. The chemical composition shown in table 3 could be considered typical for this type of material.

Different methods for extracting indium from such an alloy were studied. In order to save energy the metal was granulated and leached with sulphuric acid at almost 100°C. Leaching efficiencies, however were very low.

Interesting results were obtained by applying anodic leaching with fluorosilicic acid: lead and tin plated on the cathode, copper on the anode and indium split between the anode and the electrolyte. It might be possible to improve this process, but severe environmental problems were experienced and the fluorosilicic program of the experimental was canceled.

Table 3. Average chemical composition of the residual metal from the zinc powder plant

Element	Composition [%]	Element	Composition [%]
Lead	72	Zinc	10
Tin	11	Arsenic	1
Antimony	1. 5	Indium	2
Iron	0. 15	Copper	2. 5

Aiming to facilitate the leaching of indium, the alloy is oxidized in an open hearth furnace until no indium is detected in the metal bath. Indium, tin, copper, zinc, arsenic and most of the antimony are concentrated in the resulting dross which usually has the composition shown in table 4.

This dross is leached for three hours with 3M H_2SO_4 at a temperature of about 100°C. Satisfactory leaching efficiencies are obtained and a liquor containing 3 g/l In, 2g/l Fe, 0. 002g/l Pb, 0. 002 g/l Cd, 0. 085 g/l As, 2 g/l Sn and 250 g/l H_2SO_4 results from the leaching. Lead, cadmium, antimony, zinc and copper are almost completely eliminated from the solution and trivalent iron is reduced to divalent by means of Na_2S. H_2S could also be used, but it is less effective compared with Na_2S.

Table 4. Chemical composition of the dross formed in an open hearth furnace

Element	Concentration [%]	Element	Composition [%]
Lead	40	Zinc	21
Tin	10	Indium	4
Arsenic	3	Antimony	1
Iron	3	Copper	6

4 Solvent Extraction

Indium is readily extracted with ethylhexyl phosphoric acid - a mixture of mono and diester diluted in kerosene (fraction close to Escaid 100) - from sulphate solutions even as acidic as our leachate (fig 1). No modifier is required for third phase prevention providing that the leachate has been previously treated with Na_2S and thoroughly filtered. Three mixer settlers were specified in the pilot plant design to ensure proper phase continuity for minimum entrainment, but only two remained in use after a short period of experimental work.

Trivalent iron is almost completely transferred to the organic phase. Tin, lead, zinc and antimony are extracted to a certain extent, but copper, cadmium and arsenic go through with the raffinate. Lead sulphate and tin oxide readily form crud if conditions are not controlled.

The strip stage is conventional from the point of view of the equipment. Due to the difference in the behavior of indium and iron in the HEHPA - HCl system (fig. 2), three

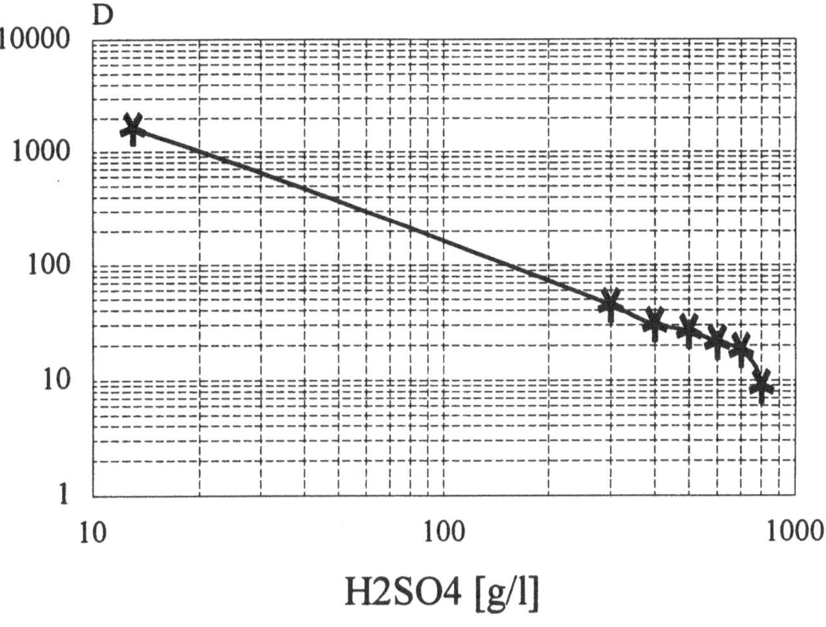

Fig. 1. Distribution coeffitient of indium as a function of sulfuric acid concentration

stages of 4 N HCl readily extract indium from the organic phase, but most of the iron remains in it. Therefore, an additional iron stripping stage is required before the extractant is returned to the extraction stage. The concentration of indium in the re-extract reaches a figure of 60 g/l, whilst the contaminants have the following concentrations: Fe - 0. 4 g/l, Cu - 2 mg/l, Pb - 40 mg/l, Cd 2 mg/l, As - 2 mg/l, Sb - 130 mg/l, Sn - 200 mg/l.

The reextract is still not pure enough either for electrolysis or for cementation. 50% (vol.) of TBP in kerosene is used for further purification. The extraction is carried out in three stages. The first one is performed at an O:A ratio of 1:50, the extractant being close to crowd loading with indium, so that further extraction of iron from the inorganic phase does not reduce indium concentration considerably. In the next two stages indium is transferred to the organic phase leaving the impurities in the aqueous solution.

The stripping stage in this section consists of a two stage mixer-settler unit and is performed by means of water in a counter current flow. The resulting aqueous strip solution contains about 57 g/l In, 4 mg/l Sb, 20 mg/l Fe, 8 mg/l Sn and only traces of other metals.

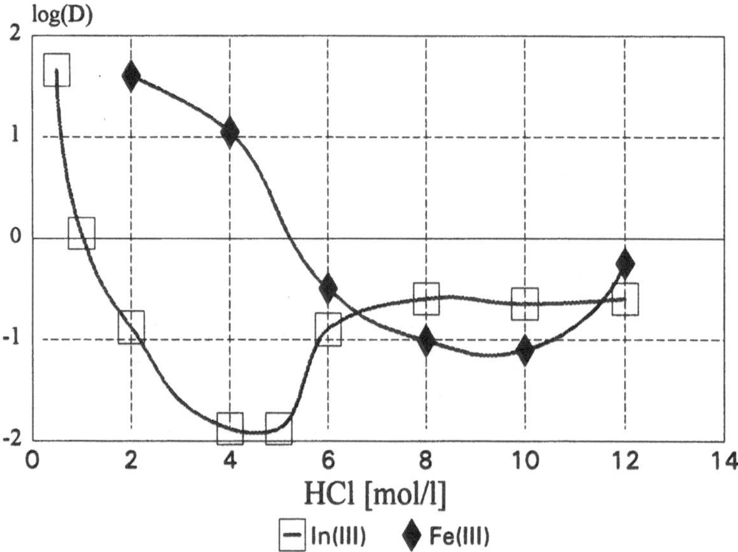

Fig. 2. Indium and copper distribution in HCl-D2EHPA systems (plot of logD vs HCl strenght

5 Cementation

Most of the tin and antimony is deposited by immersing an indium plate in the aqueous strip liquor for three hours at a temperature of 90°C. After this final operation, the liquor is ready for cementation. Zinc and aluminium are most frequently used for precipitation, however the surface of special high grade (SHG) zinc plate is pasivated at the very beginning of the process and on the other hand commercial grade aluminium is not pure enough to produce 99. 99% In. Therefore, an alloy consisting of 95% Zn and 5% Al is successfully used for this purpose.

Evident cementation occurs at 60 °C and as soon as this point has been reached, the temperature rises quickly up to 95°C. In two hours the precipitation of indium is almost completed, but in four hours there is no indium left in the solution (fig 3). Any impurities contained in the indium pregnant liquor and those from the Al-Zn alloy go through to the precipitate. The resulting indium sponge is washed, briquetted in 200 g compact metallic briquettes to minimize the surface of the sponge exposed to the oxidizing surroundings and melted in a caustic bath.

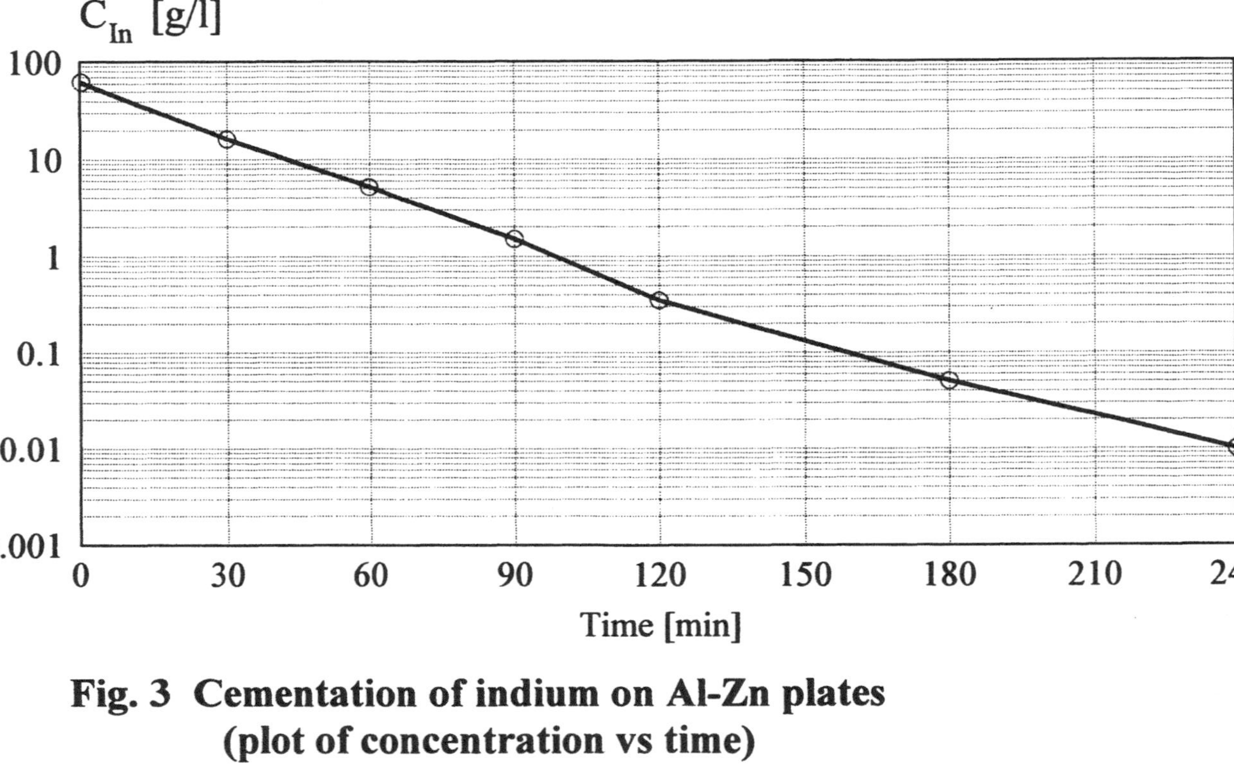

Fig. 3 Cementation of indium on Al-Zn plates (plot of concentration vs time)

The concentration of iron in the cast metal is reduced to one tenth of that in the sponge. The presence of other impurities is mainly due to the poor quality of the alloy used for cementation.

Figure 4 shows the overall process flow sheet. Further improvements are under consideration in both the metal purity and the environmental conditions, by respectively applying electrolysis rather than cementation and by eliminating arsenic from the lead residue in an early stage of dross formation.

6 Electrolysis

Electrolytic deposition of indium could be considered a better solution compared with cementation, but no attempt was made to apply it within the smelter's laboratories as suitable electrical equipment was not available. The results from experimental work undertaken at the University of Skopje confirm the above assumption. In fact it was proved that at least five 9's indium could be produced by multiple amalgamless electrolysis starting from indium anodes produced in the smelter's pilot plant.

7 Environmental

Aiming to prevent formation of arsine, tests were undertaken to eliminate arsenic from the indium rich dross. NaOH was fed into the molten indium bearing lead residue in a small (5 t) lead refining kettle. The molten bath was stirred for some time and a viscous slag was collected from the surface. Zinc and arsenic were almost completely transferred to the slag together with about one fifth of the indium content. The metal remained in the kettle was oxidized by injecting oxygen until indium was not detected in the molten lead. The resulting dross was very fine sand like dry material with a composition shown in table 5.

Table 5.

Element	Concentration [%]	Element	Concentration [%]
Copper	2	Zinc	0. 35
Tin	12	Arsenic	0. 4
Indium	6. 5	Iron	0. 03
Antimony	2		

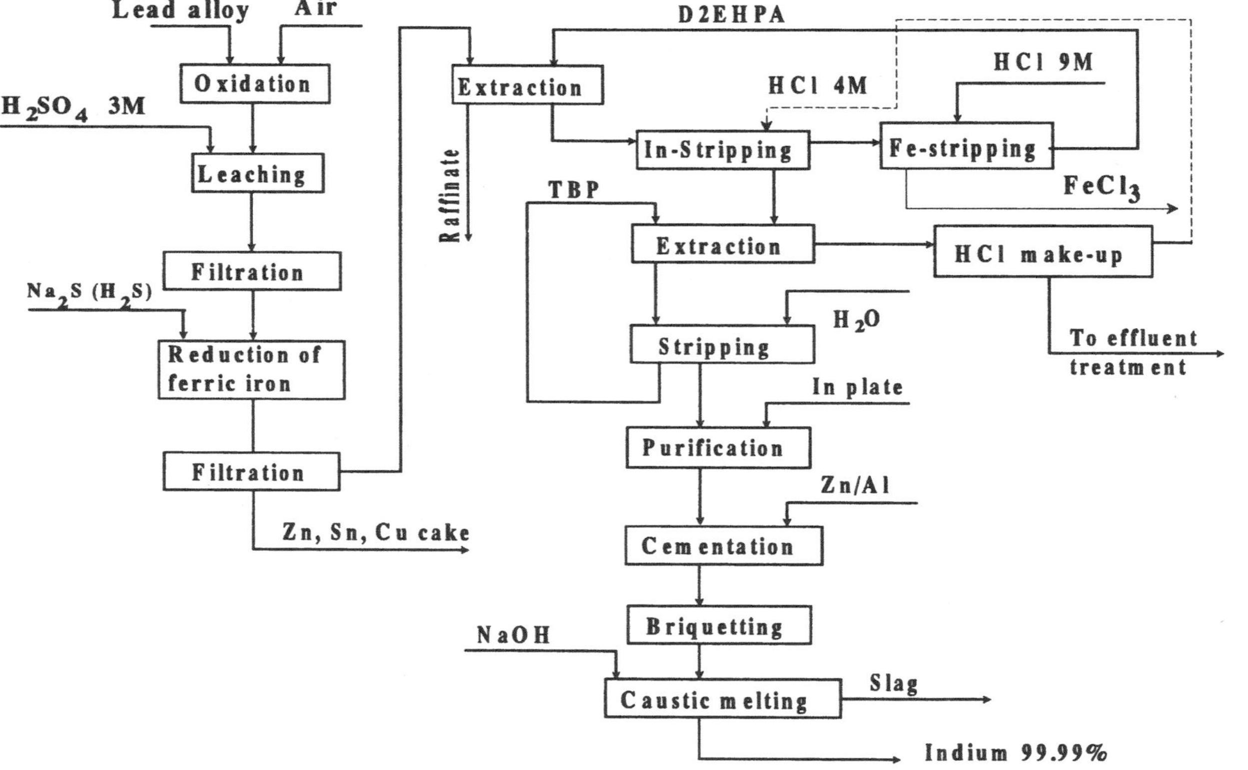

Fig. 4 Outline flowsheet for indium recovery from zinc powder plant residues

Experimental work on treating this dross is still being performed and the results are not completed yet.

8 Treatment of Other Indium Bearing Intermediaries

In the Berzelius Imperial Smelting plant in Duisburg, West Germany, copper is produced by leaching copper dross with aqueous solutions of sulfuric acid (20%) at approximately 100°C and oxygen addition [2]. As the copper dross produced in the "Zletovo" Smelter contains considerable amount of indium a number of series of experiments using oxidizing leaching with sulfuric acid were carried out. The results of this series of experiments, some of them shown in fig. 5, indicate that indium does not

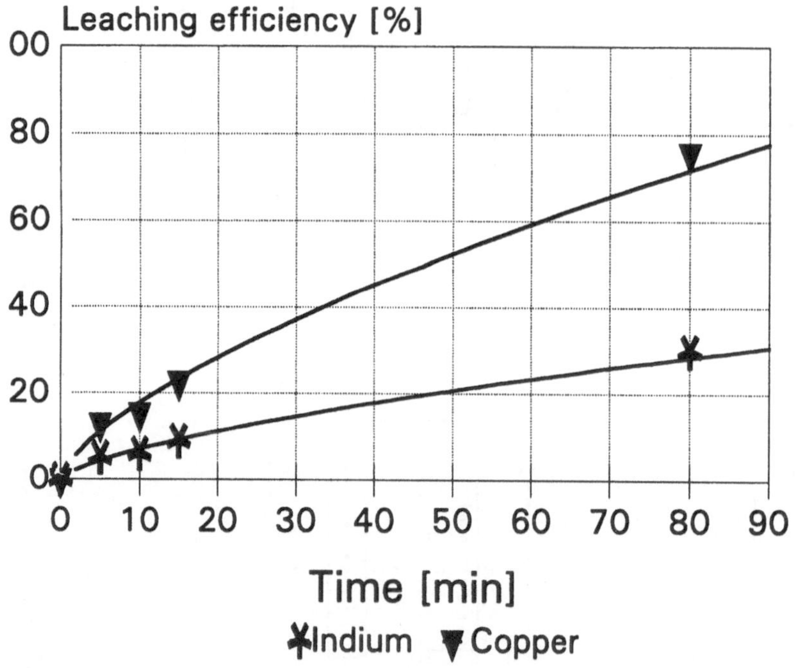

Fig. 5. Leaching of indium and copper from copper dross with 3 M/l H_2SO_4

concentrate significantly neither in the leachate nor in the sludge thus making the method inappropriate.

Satisfactory results from the point of view of the leaching efficiency have been obtained by applying sulphation roasting of the copper dross followed by water leaching. In addition, relatively low concentration of sulfuric acid in the resulting leach solution is obtained, thus reducing the problem of the effluent treatment. However, one major problem which is limiting factor for this procedure is the environment protection aspect. Most of the sulfuric acid decomposes during the roasting and would require substantial measures to be taken to protect the environment. On the other hand, the sludge is mainly lead sulfate that is normally fed back onto the sinter machine thus reducing its production capacity.

The ammoniacal leach, solvent extraction process for the copper dross was developed by the Imperial Smelting Processes Ltd. in Avonmouth, UK [1], [4], [5]. Selectivity in both the leaching and the solvent extraction steps is the major advantage of the process. Some zinc is leached together with copper, but its concentration does not exceed 10 - 15 g/l. Nickel, lead, arsenic, antimony and traces of other metal ions are found in the leech liquor, almost none of them in the pregnant electrolyte. Lead carbonate is the major constituent of the sludge. All the other components are in carbonate form as well. Such a slurry is readily treated in a short rotary furnace for lead production.

In the search for an appropriate process route for indium, the ammoniacal leaching, solvent extraction process was found interesting due to the carbonate character of the slurry. The idea was to recover indium from the residue by applying sulphuric acid leaching.

Portions of 100 g copper dross were leached with 1 l of an ammoniacal solution containing 20 g/l NH_3 at 22 ^oC. Oxygen and carbon dioxide were continuously admitted through sintered glass distributors at a flow rate of 20 l/h each. After three hours of vigorous agitation the suspension was filtered, the residue washed and then leached with 1l 2M H_2SO_4 at 75^oC for another three hours. More than 90% of the indium was leached together with almost the entire iron and the arsenic remaining from the ammoniacal leaching.

Having eliminated copper from the solution, the reduction of trivalent iron in order to prevent its coextraction with indium is possible.

Despite the satisfactory leaching efficiency and relatively simple operation, this method still has the disadvantage of converting nearly all the lead content in the dross to the sulphate form.

At present experimental work is being carried out in order to recover indium from the slag arising from smelting the carbonate residue in a short rotary furnace. Two possibilities have been investigated:

1. leaching of a mixture of oxides from the zinc powder plant and slag from the short rotary furnace and
2. separate leaching of the slag with subsequent transfer of indium to sulphate solution to be joined to the indium rich solution from the leaching of oxides .

Acknowledgments

Many persons have been involved in the implementation of this project. Particular mention must be made of A. J. Monhemius and M. J. Meixner for their advice in solvent extraction and leaching respectively. Thanks are also due to the Faculty of Technology and Metallurgy of the University of Skopje for the cooperation and their work on improving the quality of the produced indium.

References

1. Robson A. W. (1980) **Distribution and recovery of other metals in the ISF,** ISP Conference, Cracow, Poland.
2. Hopkin W. (1975) **Process for treatment of zinc - lead blast furnace dross,** Hydrometallurgy Symposium, Institution of Chemical Engineers, Manchester, U. K.
3. Tack K. (1975) **The recovery of copper from lead bullion at "Berzelius",** ISP Conference, Bristol U. K.
4. Hopkin W. (1976) **Process for treatment of zinc - lead and lead blast furnace copper dross,** 105ht AIME Annual meeting, Las Vegas, USA.
5. Hopkin W. (1975) **Recovery of copper from ISP copper dross,** ISP Conference, Bristol, UK.

Subject index